5G NR

5G NR

Architecture, Technology, Implementation, and Operation of 3GPP New Radio Standards

Sassan Ahmadi

ELSEVIER

ACADEMIC PRESS
An imprint of Elsevier

Academic Press is an imprint of Elsevier
125 London Wall, London EC2Y 5AS, United Kingdom
525 B Street, Suite 1650, San Diego, CA 92101, United States
50 Hampshire Street, 5th Floor, Cambridge, MA 02139, United States
The Boulevard, Langford Lane, Kidlington, Oxford OX5 1GB, United Kingdom

British Library Cataloguing-in-Publication Data
A catalogue record for this book is available from the British Library

Library of Congress Cataloging-in-Publication Data
A catalog record for this book is available from the Library of Congress

ISBN: 978-0-08-102267-2

For Information on all Academic Press publications
visit our website at https://www.elsevier.com/books-and-journals

Publisher: Mara Conner
Acquisition Editor: Tim Pitts
Editorial Project Manager: Joshua Mearns
Production Project Manager: Kamesh Ramajogi
Cover Designer: Greg Harris

Typeset by MPS Limited, Chennai, India

Contents

Acknowledgments

The author would like to acknowledge and thank his colleagues in 3GPP and ITU-R for their encouragement, consultation, and assistance in selection and refinement of the content as well as editing and improving the quality of the chapters.

The author would like to sincerely appreciate the Academic Press (Elsevier) production, publishing and editorial staff for providing the author with the opportunity to publish this work and for their support, cooperation, patience, and understanding throughout this long project.

Finally, the author is immeasurably grateful to his family for their unwavering encouragement, support, patience, and understanding while the author was engaged in this project.

List of Abbreviations

3GPP	third-generation partnership project
5G-AN	5G access network
5GC	5G core
5G-EIR	5G-equipment identity register
5G-GUTI	5G globally unique temporary identifier
5GS	5G system
5G-S-TMSI	5G-S-temporary mobile subscription identifier
5QI	5G QoS identifier
AA	antenna array
AAA	authorization, authentication, and accounting
AAS	active antenna system
ABS	almost blank subframe
AC	access class
ACIR	adjacent channel interference ratio
ACK	acknowledgement
ACLR	adjacent channel leakage ratio
ACS	adjacent channel selectivity
A-CSI	aperiodic-CSI
ADC	analog to digital converter
AF	application function
AKA	authentication and key agreement
AM	acknowledged mode
AMBR	aggregate maximum bit rate
AMC	adaptive modulation and coding
AMF	access and mobility management function
ANR	automatic neighbor relation
AoA	angle of arrival
AoD	angle of departure
AP	application protocol
APN	access point name
ARFCN	absolute ratio frequency channel number
ARP	allocation and retention priority
ARQ	automatic repeat request
AS	access stratum/angular spread
AUSF	authentication server function
AWGN	additive white Gaussian noise
BA	bandwidth adaptation
BC	broadcast channel (MU-MIMO)
BCCH	broadcast control channel
BCH	broadcast channel

BER	bit error rate
BPSK	binary phase shift keying
BS	base station
BSF	binding support function
BSR	buffer status report
BW	bandwidth
BWP	bandwidth part
C/I	carrier-to-interference ratio
CA	carrier aggregation
CACLR	cumulative ACLR
CAZAC	constant amplitude zero auto-correlation
CBC	cell broadcast center
CBE	cell broadcast entity
CBG	code block group
CBGTI	code block group transmission information
CBRA	contention based random access
CC	component carrier
CCE	control channel element
CDF	cumulative distribution function
CDM	code division multiplexing
CD-SSB	cell defining SSB
CFO	carrier frequency offset
CFRA	contention free random access
CGI	cell global identifier
CIF	carrier indicator field
CM	cubic metric
CMAS	commercial mobile alert service
CMC	connection mobility control
CoMP	coordinated multi-point
CORESET	control resource set
CP	cyclic prefix
CPE	common phase error
CP-OFDM	cyclic prefix-OFDM
CQI	channel quality indicator
CRB	common resource block
CRC	cyclic redundancy check
CRE	cell range extension
CRI	CSI-RS resource indicator
C-RNTI	cell RNTI
CRS	cell-specific reference signal
CSA	common subframe allocation
CSI	channel state information
CSI-IM	CSI interference measurement
CSI-RS	CSI reference signal
CSI-RS	CSI reference signal
CSI-RSRP	CSI reference signal received power
CSI-RSRQ	CSI reference signal received quality
CSI-SINR	CSI signal-to-noise and interference ratio
CW	codeword
CW	continuous wave

DAC	digital to analog converter
DAI	downlink assignment index
DCCH	dedicated control channel
DCI	downlink control information
DFT	discrete Fourier transform
DFT-S-OFDM	DFT-spread-OFDM
DL	downlink
DL TFT	downlink traffic flow template
DL-SCH	downlink shared channel
DM-RS	demodulation reference signal
DN	data network
DNN	data network name
DoA	direction of arrival
DRB	data radio bearer
DRX	discontinuous reception
DTCH	dedicated traffic channel
DTX	discontinuous transmission
DwPTS	downlink pilot time slot
EAB	extended access barring
ECGI	E-UTRAN cell global identifier
E-CID	enhanced cell-ID (positioning method)
ECM	EPS connection management
ECN	explicit congestion notification
eICIC	enhanced inter-cell interference coordination
EIRP	effective isotropic radiated power
EIS	equivalent isotropic sensitivity
EMM	EPS mobility management
eNB	E-UTRAN Node B/evolved node B
EPC	evolved packet core
ePHR	extended power headroom report
EPRE	energy per resource element
EPS	evolved packet system
E-RAB	E-UTRAN radio access bearer
E-SMLC	evolved serving mobile location centre
ETWS	earthquake and tsunami warning system
E-UTRA	evolved UTRA
E-UTRAN	evolved UTRAN
EVM	error vector magnitude
FBW	fractional bandwidth
FDD	frequency division duplex
FDM	frequency division multiplexing
FFT	fast Fourier transform
FQDN	fully qualified domain name
FR	frequency range
FR1	frequency range 1
FR2	frequency range 2
FRC	fixed reference channel
FTP	file transfer protocol
GBR	guaranteed bit rate
GFBR	guaranteed flow bit rate

GNSS	global navigation satellite system
GP	guard period
GPS	global positioning system
GPSI	generic public subscription identifier
GSCN	global synchronization channel number
GT	guard time
GTP	GPRS tunneling protocol
GUAMI	globally unique AMF identifier
GUMMEI	globally unique MME identifier
GUTI	globally unique temporary identifier
GW	gateway
HARQ	hybrid ARQ
HARQ-ACK	hybrid automatic repeat request acknowledgement
HetNet	heterogeneous network
HFN	hyper frame number
HO	handover
HSS	home subscriber server
HTTP	hypertext markup language
ICIC	inter-cell interference coordination
ICS	in-channel selectivity
ID	identity
IDC	in-device coexistence
IE	information element
IEEE	Institute of Electrical and Electronics Engineers
IETF	Internet Engineering Task Force
IFFT	inverse fast Fourier transform
IMD	inter-modulation distortion
IMS	IP multimedia sub-system
INT-RNTI	interruption RNTI
IoT	internet of things
IP	internet protocol
IQ	in-phase/quadrature
IRC	interference rejection combining
I-RNTI	inactive RNTI
ISI	inter-symbol interference
ISM	industrial, scientific, and medical
ITU-R	radiocommunication sector of the international telecommunication union
L1-RSRP	layer 1 reference signal received power
LA	local area
LCG	logical channel group
LCID	logical channel identifier
LDPC	low density parity check
LI	layer indicator
LLR	log likelihood ratio
LNA	low-noise amplifier
LTE	long term evolution
MAC-I	message authentication code for integrity
MAP	maximum a posteriori probability
MBMS	multimedia broadcast/multicast service
MBR	maximum bit rate

MBSFN	multimedia broadcast multicast service single frequency network
MCE	multi-cell/multicast coordination entity
MCS	modulation and coding scheme
MDBV	maximum data burst volume
MDT	minimization of drive tests
MFBR	maximum flow bit rate
MIB	master information block
MIMO	multiple-input multiple-output
MLD	maximum likelihood detection
MME	mobility management entity
MMSE	minimum mean-square error
MNO	mobile network operator
MRC	maximal ratio combining
MTC	machine-type communications
MU-MIMO	multi-user MIMO
N3IWF	non-3GPP interworking function
NACK	negative acknowledgement
NAI	network access identifier
NAS	non-access stratum
NCC	next hop chaining counter
NCGI	NR cell global identifier
NCR	neighbor cell relation
NCRT	neighbor cell relation table
NDS	network domain security
NEF	network exposure function
NF	network function
NGAP	NG application protocol
NGAP	next generation application protocol
NH	next hop key
NNSF	NAS node selection function
NCR	neighbor cell relation
NR	new radio
NR-ARFCN	NR absolute radio frequency channel number
NRF	network repository function
NRT	neighbor relation table
NSI ID	network slice instance identifier
NSSAI	network slice selection assistance information
NSSF	network slice selection function
NSSP	network slice selection policy
NWDAF	network data analytics function
OBUE	operating band unwanted emissions
OCC	orthogonal cover code
OFDM	orthogonal frequency division multiplexing
OFDMA	orthogonal frequency division multiple access
OMC-ID	operation and maintenance centre identity
OOB	out-of-band (emission)
OSDD	OTA sensitivity directions declaration
OTA	over-the-air
PA	power amplifier
PAPR	peak-to-average power ratio

PBCH	physical broadcast channel
PBR	prioritized bit rate
PCC	primary component carrier
PCCH	paging control channel
PCell	primary cell
PCF	policy control function
PCFICH	physical control format indicator channel
PCH	paging channel
PCI	physical cell identifier
PCRF	policy and charging rules function
PDCCH	physical downlink control channel
PDCP	packet data convergence protocol
PDN	packet data network
PDR	packet detection rule
PDSCH	physical downlink shared channel
PDU	protocol data unit
PEI	permanent equipment identifier
PER	packet error rate
PFD	packet flow description
PGW	PDN gateway
PHICH	physical hybrid ARQ indicator channel
PHR	power headroom report
PHY	physical layer
PLMN	public land mobile network
PMI	precoding matrix indicator
PMIP	proxy mobile IP
PO	paging occasion
PPD	paging policy differentiation
PPF	paging proceed flag
PPI	paging policy indicator
PRACH	physical random-access channel
PRB	physical resource block
PRG	precoding resource block group
P-RNTI	paging RNTI
PSA	PDU session anchor
PSAP	public safety answering point
PSC	packet scheduling
PSS	primary synchronization signal
pTAG	primary timing advance group
PT-RS	phase-tracking reference signal
PUCCH	physical uplink control channel
PUSCH	physical uplink shared channel
PWS	public warning system
QAM	quadrature amplitude modulation
QCI	QoS class identifier
QCL	quasi co-location
QFI	QoS flow identifier
QoE	quality of experience
QoS	quality of service
QPSK	quadrature phase shift keying

RAC	radio admission control
RACH	random access channel
RAN	radio access network
RANAC	RAN-based notification area code
RAR	random access response
RA-RNTI	random access RNTI
RAT	radio access technology
RB	radio bearer
RB	resourcc block
RBC	radio bearer control
RBG	resource block group
RDN	radio distribution network
RE	resource element
REFSENS	reference sensitivity
REG	resource element group
RF	radio frequency
RI	rank indicator
RIB	radiated interface boundary
RIM	RAN information management
RIV	resource indicator value
RLC	radio link control
RLF	radio link failure
RM	rate matching/Reed-Muller (code)
RMS	root mean square (value)
RMSI	remaining minimum system information
RNA	RAN-based notification area
RNAU	RAN-based notification area update
RNL	radio network layer
RNTI	radio network temporary identifier
RoAoA	range of angles of arrival
ROHC	robust header compression
RQA	reflective QoS attribute
RQI	reflective QoS indication
RQoS	reflective quality of service
RRC	radio resource control
RRM	radio resource management
RS	reference signal
RSRP	reference signal received power
RSRQ	reference signal received quality
RSSI	received signal strength indicator
RTP	real-time transport protocol
RTT	round trip time
RU	resource unit/radio unit
RV	redundancy version
RX	Receiver
S1-MME	S1 for the control plane
S1-U	S1 for the user plane
SA NR	standalone new radio
SAE	system architecture evolution
SAP	service access point

SBA	service based architecture
SBI	service based interface
SCC	secondary component carrier
SCell	secondary cell
SC-FDMA	single carrier-frequency division multiple access
SCH	synchronization channel
SCM	spatial channel model
SCS	sub-carrier spacing
SD	slice differentiator
SDAP	service data adaptation protocol
SDF	service data flow
SDL	supplementary downlink
SDMA	spatial division multiple access
SDU	service data unit
SEAF	security anchor functionality
SEPP	security edge protection proxy
SFBC	space-frequency block code
SFI-RNTI	slot format indication RNTI
SFN	system frame number
SGW	serving gateway
SI	system information
SIB	system information block
SIC	successive interference cancellation
SIMO	Single-input multiple-output
SINR	signal to interference plus noise ratio
SI-RNTI	system information RNTI
SISO	single-input single-output/soft-input soft-output
SLA	service level agreement
SLIV	start and length indicator value
SMC	security mode command
SMF	session management function
SMS	short message service
SMSF	short message service function
SN	sequence number
SNR	signal to noise ratio
S-NSSAI	single network slice selection assistance information
SON	self-organizing network
SPID	subscriber profile ID for RAT/frequency priority
SPS	semi-persistent scheduling
SPS C-RNTI	semi-persistent scheduling C-RNTI
SR	scheduling request
SRB	signaling radio bearer
SRS	sounding reference signal
SS	synchronization signal
SSB	SS/PBCH block
SS-RSRP	SS reference signal received power
SS-RSRQ	SS reference signal received quality
SSS	secondary synchronization signal
SS-SINR	SS signal-to-noise and interference ratio
SST	slice/service type

sTAG	secondary timing advance group
S-TMSI	S-temporary mobile subscriber identity
SU	scheduling unit
SUCI	subscription concealed identifier
SUL	supplementary uplink
SU-MIMO	single-user MIMO
SUPI	subscription permanent identifier
TA	tracking area
TA	timing advance
TAB	transceiver array boundary
TAC	tracking area code
TAD	traffic aggregate description
TAE	time alignment error
TAG	timing advance group
TAI	tracking area identity
TAU	tracking area update
TB	transport block
TCI	transmission configuration indicator
TCP	transmission control protocol
TDD	time division duplex
TDM	time division multiplexing
TEID	tunnel endpoint identifier
TFT	traffic flow template
TI	transaction identifier
TIN	temporary identity used in next update
TM	transparent mode
TNL	transport network layer
TNLA	transport network layer association
TPC	transmit power control
TPMI	transmitted precoding matrix indicator
TrCH	transport channel
TRP	total radiated power
TSP	traffic steering policy
TTI	transmission time interval
TX	transmitter
UCI	uplink control information
UDM	unified data management
UDP	user datagram protocol
UDR	unified data repository
UDSF	unstructured data storage function
UE	user equipment
UEM	unwanted emissions mask
UL	uplink
UL TFT	uplink traffic flow template
ULR-Flags	update location request flags
UL-SCH	uplink shared channel
UM	unacknowledged mode
UPF	user plane function
UpPTS	uplink pilot time slot
URLLC	ultra-reliable and low latency communications

URRP-AMF	UE reachability request parameter for AMF
URSP	UE route selection policy
V2X	vehicle-to-everything
VLAN	virtual local area network
VoIP	voice over internet protocol
VRB	virtual resource block
X2-C	X2-control plane
X2-U	X2-user plane
XnAP	Xn application protocol
Xn-C	Xn-control plane
Xn-U	Xn-user plane
ZC	Zadoff-Chu
ZCZ	zero correlation zone
ZF	zero forcing (receiver)
ZP CSI-RS	zero power CSI-RS

Introduction and Background

1 Introduction to 5G

The tremendous growth in the number and variety of connected devices and the substantial increase in user/network traffic volume and types as well as the performance limitations of 4G technologies have motivated industry efforts and investments toward defining, developing, and deploying systems for the fifth generation (5G) of mobile networks. The 5th generation of mobile broadband wireless networks have been designed to meet the challenging system and service requirements of the existing and the emerging applications in 2020 and beyond. 5G is a fast, secure, and reliable connected ecosystem comprising humans and machines, which enables seamless mobility, efficient connectivity, increased connection density, increased industrial productivity, automation, and sustainability. The future connected societies are characterized by the significant growth in connectivity and traffic density, network densification, and the broad range of new use cases and applications. As a result, there is a continuous need to push the performance envelope of the wireless systems to the new limits in order to satisfy the demands for greater network capacity, higher user throughput, more efficient spectrum utilization, wider bandwidths, lower latency, lower power consumption, more reliability, increased connection density, and higher mobility through virtualized and software-defined network (SDN) architectures. While extending the performance envelope of mobile networks, 5G includes intrinsic flexibility and configurability to optimize the network usage and services, accommodating a wide range of use cases, and business models. The 5G network architecture encompasses modular network functions that can be deployed, configured, and scaled on demand, in order to accommodate various use cases in a smart and cost-efficient manner.

The scope of 5G is best understood by scrutinizing the main usage models targeted by this megatrend namely enhanced mobile broadband (eMBB), ultra-reliable low-latency communications (URLLC), and massive machine type communications (mMTC). The high-throughput mobile broadband use case is primarily based on the evolution and enhancement of LTE technology. Evolutionary nature of mobile broadband use case implies coexistence with the existing radio access networks, but this consideration is only limited to sub-6 GHz RF spectrum.

Any mobile broadband deployments above 6 GHz are expected to use new baseband and radio technologies that improve spectral efficiency, throughput, latency, and other key performance metrics. The mission-critical machine type communication and mMTC use cases are revolutionary in nature. Technology and design choices for the latter use cases are unaffected by the burden of legacy support or backward compatibility (see Fig. 1). The Internet of Things (IoT), automotive, smart grid, traffic safety, emergency response services are some of the examples of low-latency machine type communications. Tactile Internet is an interesting aspect of this use case that is expected to create new services.

As shown in references [11,14], peak and average data rates as well as transport delay are the leading performance indicators of the high-throughput mobile broadband scenario. Interactive gaming, augmented/virtual reality, and immersive entertainment are notable consumer mobile broadband services that are in nascent state and hold significant promise of new business opportunities and user experience. The eMBB delivers over 20 Gbps downlink peak data rates with expected user experienced data rates of approximately 100 Mbps anywhere anytime. This would require substantial increase in network capacity in 2020 and beyond. Network operators are gearing up toward meeting such a dramatic increase in capacity by a combination of new spectrum utilization, spectral efficiency improvements, and ultra-dense network deployments.

The key challenge in the design and deployment of 5G systems is not only about the development of a new radio interface(s) but also about the coordinated operation in a highly heterogeneous environment characterized by the existence of multi-RAT systems, multi-layer

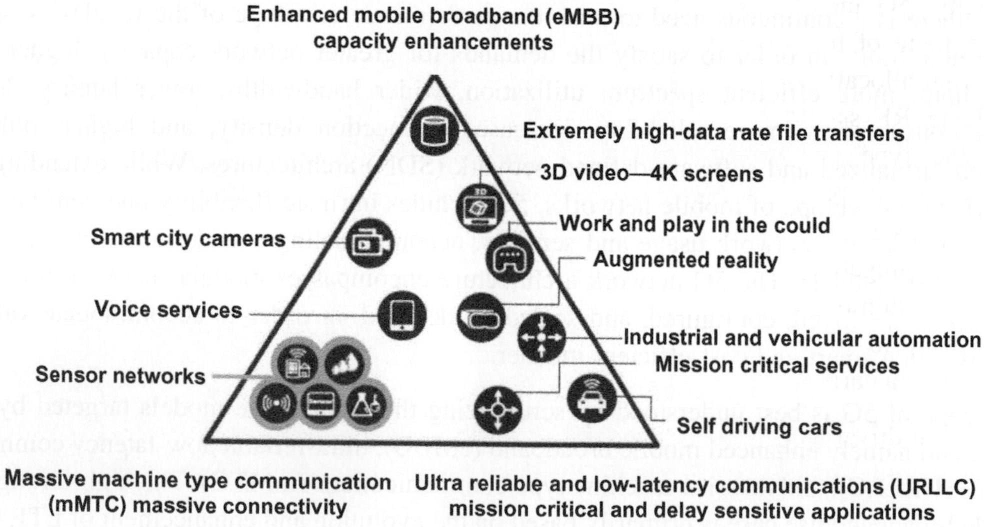

Figure 1
5G use case categories [1].

networks, multi-mode devices, and diverse user interactions. Under these conditions, there is a fundamental need for 5G to achieve seamless mobility and consistent user experience across time and space. New business models and economic incentives with fundamental shift in cost, energy, and operational efficiency are expected to make 5G feasible and sustainable. 5G also is going to offer value proposition to vertical businesses and consumers through the definition and exposure of capabilities that enhance today's overall service delivery.

In addition to supporting the evolution of the established mobile broadband usage models, 5G is going to support numerous emerging applications with widely different requirements and service attributes; that is, from delay-tolerant applications to ultra-low-latency applications, from high-speed scenarios in trains/airplanes to nomadic/stationary scenarios in home or office, and from best effort applications to ultra-reliable applications such as in healthcare and public safety. Furthermore, content and information will be delivered across a wide range of devices (e.g., smartphones, wearables) and through heterogeneous environments. Fulfilling such stringent requirements will require larger number of base stations/access nodes in the form of a heterogeneous network architecture, which can be realized as small cells and high-capacity cloud-RAN or virtual RAN architectures serving a large number of remote radio heads. Content caching and processing at the cloud edge in 5G networks will overcome bottlenecks in the network transport for delay-sensitive applications.

5G use cases and business models require allocation of additional spectrum for mobile broadband and flexible spectrum management capabilities. The new networks are employing a set of new frequency bands (licensed, unlicensed, and shared) to supplement the existing wireless frequencies, enabling larger bandwidth operation and substantially improved capacity. 5G inevitably requires some of the sub-6 GHz spectrum to be repurposed for deployment of the new technologies. Existing cellular bands can be augmented with new spectrum allocations above 6 GHz in order to create extremely wider operating bandwidths. Utilizing RF spectrum above 6 GHz for mobile broadband, low-latency mission-critical machine type communication, and low-energy massive machine type communication requires development of new radio interfaces or access technologies as well as substantial innovations in analog RF-integrated circuits and semiconductor technologies. In addition, carrier aggregation techniques are used to combine segments of spectrum that are not co-located within the same band to further improve peak data rates. 5G systems are required to support up to 400 MHz of contiguous bandwidth and to support a total bandwidth of up to 3.2 GHz via carrier aggregation beyond 6 GHz.

Advanced driver assistance systems (ADAS) and autonomous vehicles are emerging trends in the automotive space. Together, they bring a number of benefits, including improved safety, reducing collision risks and road congestions, improved fuel economy, and higher productivity for the drivers. 5G wireless technologies supporting high-speed, low-latency vehicle-to-vehicle and vehicle-to-infrastructure communications are key enablers of ADAS and autonomous vehicles. In addition, automotive industry is demanding richer infotainment options, which are adding more traffic bottlenecks to wireless networks. eCall is an initiative by the transportation safety regulatory bodies to provide immediate assistance to people involved in accidents. It is a subsystem of telematics systems being installed in vehicles. Implementation of eCall requires significant investment in infrastructure. In the case of a crash, it contacts the closest emergency center and forwards the GPS location of the crash site to the emergency center operators. eCall can be activated manually as well as automatically. It can be activated automatically, if the vehicle is involved in an accident and the airbags are inflated.

As in the previous generations, improvements in radio access technologies, in the form of either evolutionary or revolutionary, have continued to be the core focus of the research and development in 5G. However, it should not be forgotten that the proper design of network architecture and functionality along with ease of programmability/configurability have been equally important to 5G to ultimately achieve the goal of an integrated communication platform serving multiple vertical markets and diverse consumer applications. Network function virtualization (NFV) has been an underlying method to enable network operators to create network slices per end-user application or service requirements with guaranteed performance and quality corresponding to service level agreements. Both cloud and edge computing components will be needed in these network slices to address varying performance and latency requirements. Each network slice will rely on enhanced LTE and/or NR access technologies to provide reliable wireless link between the user terminals and the network.

The comprehensive research and studies conducted on 5G in the past few years and the ongoing standards development efforts are revealing a number of key technologies that have helped achieve the ambitious system and service requirements as follows (see Fig. 2):

- New spectrum: Use of large blocks of spectrum in higher frequency bands and heterogeneous carrier aggregation (particularly above 6 GHz and up to 100 GHz) in addition to unlicensed spectrum have made wider system bandwidths (up to 3.2 GHz) feasible, leading to higher peak data rates and network capacity.
- New waveforms and multiple access schemes: OFDM-based LTE air-interface may not be suitable for some use cases, and therefore a number of new waveform candidates and multiple access schemes have been studied. However, the complexity and practical limitations of some of those new candidates have convinced the 3GPP community to adopt OFDM with adaptive numerology which would allow configurable frame structures and radio resource allocations depending on use case, available spectrum, and bandwidth without affecting backward compatibility with legacy LTE systems.
- Massive MIMO: As the extension of multi-user MIMO concept to large number of antennas at the base station, it has offered a promising solution to significantly increase the user throughput and network capacity by allowing beamformed data and control transmission and interference management. The effect of considerably increased path

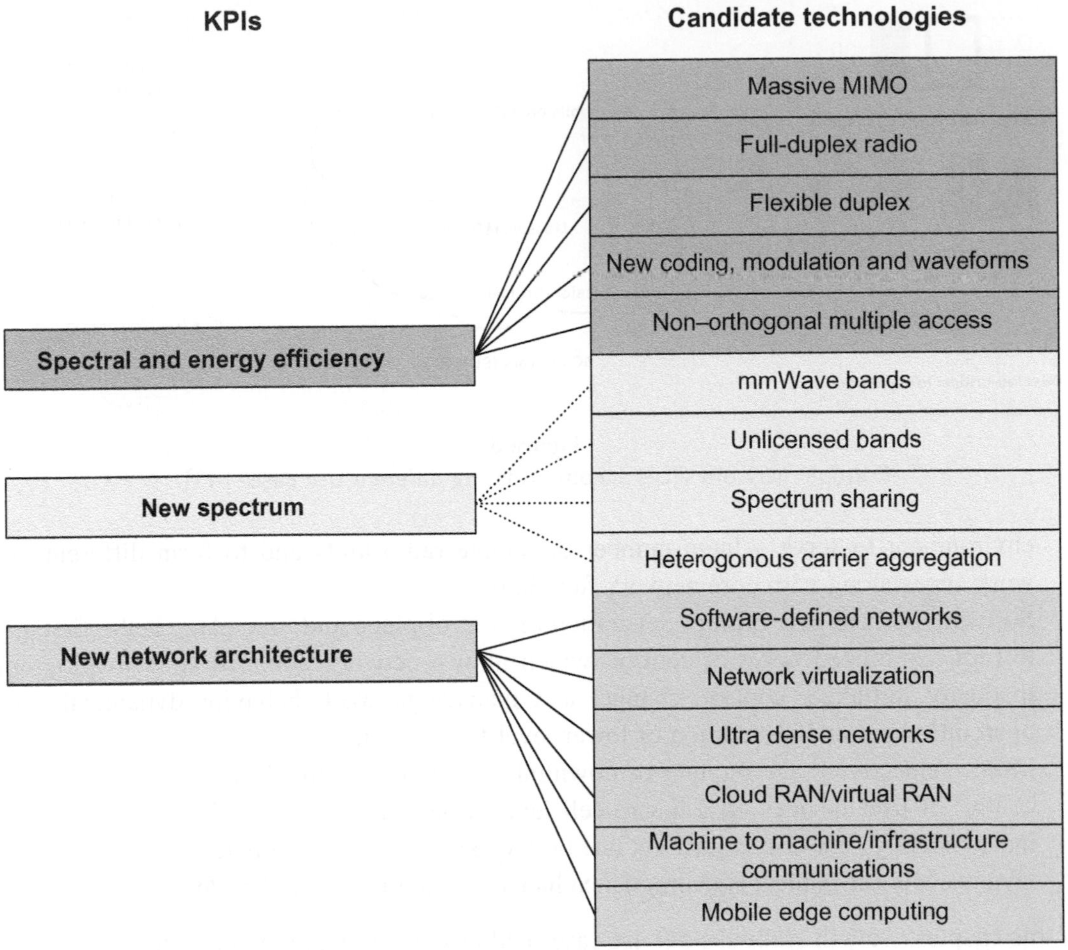

Figure 2
Mapping KPIs to candidate 5G technologies [2].

loss in very high frequencies is compensated by higher antenna gains (directivity), which is made possible by increasing number of antennas at the base station using adaptive beamforming and active antenna schemes.

- New network architecture: NFV and cloud-based radio access and core network architecture have been promising approaches to attain flexibility and versatility in 5G network design, where the objective is to run today's network functions, which are typically implemented on dedicated hardware, as virtualized software functions on general-purpose hardware comprising servers, storage, and hardware accelerators. This is further extended to the radio network by splitting base stations into remote radio units and baseband processing units (connected via optical fiber links, high-speed Ethernet cables, or wireless fronthaul) and pooling baseband functions in the virtualized

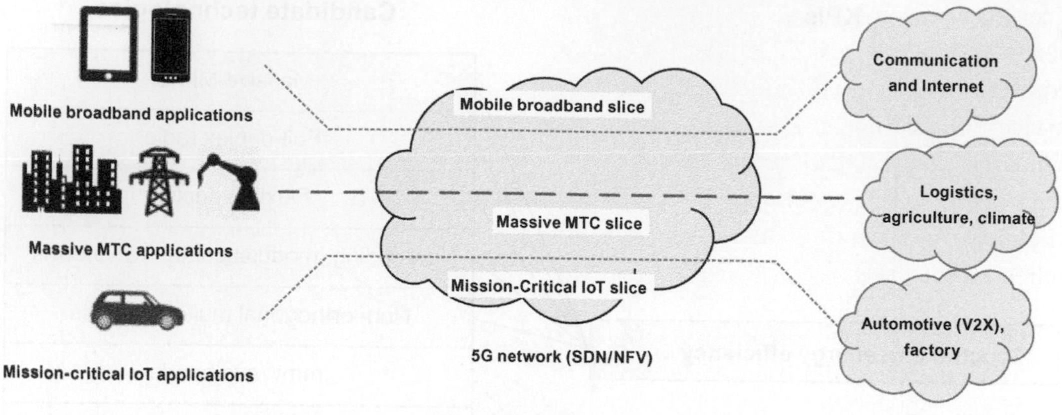

Figure 3
Example network slices accommodating different use cases [43].

environment to serve a large number of remote radio units and to form different network slices along with core network functions.

- Software-defined networking: Separation of control-plane and user-plane is the first step toward centralized network control, which allows network administrators to programmatically initialize, control, change, and manage network behavior dynamically via open interfaces and abstraction of lower level functionality.
- Multi-connectivity: Decoupling of downlink and uplink control/data paths would allow control of user devices on a macro-cell level, where user terminal data packets can be independently sent or received via one or several small cells in a heterogeneous network layout while the control signaling is exchanged with a macro-cell node.

In this chapter, we will study the 5G use cases and prominent deployment scenarios. We will further define the performance metrics and review architectural, system, and service requirements for 5G systems. The standards development activities and timelines within 3GPP and ITU-R will be discussed, and later the key aspects of spectrum allocations and regulatory issues concerning 5G systems will be described. The chapter will be concluded with a discussion about the future of wireless systems and services beyond 5G era (see Fig. 3).

2 Use Cases and Deployment Scenarios

The breakthroughs in 5G network design are expected to sustain the extensive growth and enhancement of mobile Internet and the IoT services. The use of 5G technologies in the IoT and vertical industries are providing new business opportunities for network operators. In addition, expanded and enhanced mobile Internet services have been further improving consumer experience and satisfaction. To adequately support the enhancement and expansion of mobile Internet and IoT, 5G networks have to become the primary means of network access for person-to-person, person-to-machine, and machine-to-machine connectivity. This means that 5G will need to be widely deployed and adapted to the diversity of service requirements and their unique characteristics. The fulfillment of the requirements for a service-oriented network in order to provide better user experience in a flexible and efficient manner is nontrivial. Wireless networks will need to match the performance offered by the advances in fixed networking in terms of delivering a high level of quality of service (QoS), reliability, and security. Based on the requirements, two major challenges are needed to be addressed for the design of the 5G systems, as follows:

- The system should be capable of flexible and efficient use of all available spectrum types from low-frequency bands to high-frequency bands, licensed bands, unlicensed bands, and shared bands.
- The system should be adaptable in order to provide efficient support for the diverse set of service characteristics, massive connectivity, and very high capacity. Flexible network design is required to improve spectral efficiency, to increase connection density, and to reduce latency. These requirements and challenges have also impacts on the design of 5G air interface and the network architecture.

The foundation of the 5G radio access networks has been laid on two major concepts: (1) software-defined networking and (2) network function virtualization. An elastic (adaptable) air interface would allow optimizing the network to support various applications and deployment scenarios. The 5G networks further encompass self-organization and inter-cell coordination algorithms that utilize the features, protocols, and interfaces to overcome the limitations of the cell-based network topology (realizing virtual cell concept). The 5G network architectures are typically heterogeneous in order to enable the cooperation between low-frequency wide-area and high-frequency small-area network topologies. There is a consensus in the industry that higher frequency bands are the complementary bands to deploy 5G, whereas low-frequency bands (i.e., sub-6 GHz) are still the primary bands of the 5G spectrum. High frequency also enables unified access and backhaul since the same radio resources are shared. 5G is expected to use a unified air interface and hierarchical scheduling for both radio access and backhaul which enables flexible backhauling and low-cost ultra-dense network (UDN) architectures.[1] Future radio access may also employ bands with different levels of access regulation including exclusively licensed, non-exclusively licensed

[1] UDN is the most promising solution to address high traffic rates in wireless networks. It increases network capacity, improves link efficiency, and spectrum utilization while reducing power consumption. In fact, densification of infrastructure has already been used in current cellular networks, where the minimum distance between base stations has been drastically reduced. Nonetheless, UDN faces many technical challenges, including difficulties in mobility management and backhauling (including self-backhaul). Advanced interference and mobility management schemes need to be developed at the physical and network layers to support UDN. UDNs further require comprehensive characterization in terms of cost, power consumption, and spectrum utilization efficiency [4].

(shared), and unlicensed bands. The 5G system supports both licensed and unlicensed spectrum with a flexible and unified air interface framework [35].

The ever increasing user traffic in mobile networks necessitates increasing the amount of spectrum that may be utilized by the 5G systems. High-frequency bands in the centimeter wave and millimeter wave range are being used due to their potential for supporting wider channel bandwidths and eventually the capability to deliver ultra-high data rates. Some new spectrum below 6 GHz was allocated for mobile communication at the World Radio Conference (WRC)[2] 2015, and the bands above 6 GHz are expected to be allocated at WRC 2019 [40].

The 5G systems are expected to support diverse use cases and applications that will expand beyond the current systems. The wide range of capabilities would be tightly coupled with different use cases and applications for International Mobile Telecommunications (IMT)-2020 initiative. The main use cases of IMT for 2020 and beyond systems include the following [1]:

- *Enhanced Mobile Broadband:* This use case addresses the user-centric applications for wireless access to multimedia content, services, and data. The demand for mobile broadband services has continuously increased in the past decade, leading the industry to significantly enhance the mobile broadband capabilities. The eMBB use case encompasses new application areas (with their rigorous requirements) in addition to the existing applications that strive for improved performance and increasingly seamless user experience. This usage scenario covers a range of cases including wide-area coverage and hotspots, which have different requirements. For the hotspot case, that is, for an area with high user density, very high traffic capacity is needed, while the requirement for mobility is relaxed and user data rate is higher than that of wide-area coverage. For the wide-area coverage case, seamless coverage and medium to high mobility are desired with improved user data rates compared to the existing data rates. However, the data rate requirement may be less stringent compared to the hotspot scenario.
- *Ultra-reliable and Low-Latency Communications:* This use case has stringent requirements for capabilities such as throughput, latency, and availability. Some examples include wireless control of industrial manufacturing or production processes, remote medical surgery, distribution automation in a smart grid, and transportation safety.
- *Massive Machine Type Communications:* This use case is characterized by very large number of connected devices (e.g., a sensor network) typically transmitting relatively small payloads containing non-delay-sensitive data. These devices are required to be low cost and consume very low power.

[2] WRC, http://www.itu.int/en/ITU-R/conferences/wrc.

A large number of deployment scenarios in each category of eMBB, mMTC, and URLLC are foreseen to emerge in the next few years, which require more flexibility, programmability and configurability in the design, development and deployment of 5G access and core networks. These design features help curtail increasing operators' CAPEX and OPEX for future network upgrades. In the following sections we study various use cases and the deployment scenarios supported by 5G systems.

2.1 Use Cases

In an information society, users will be provided with wide range of applications and services, ranging from infotainment services, safe and efficient transportation, as well as new industrial and professional applications. These services and applications include very high data rates and very dense distribution of devices, with stringent requirements on the end-to-end performance and user experience. The new type of challenges that arise from the emerging application areas include very low latency, very low power, very low cost, and massive number of devices. In many cases, one of the key challenges is related to support of seamless mobility (i.e., transparent to application layer) over a wide range of speeds. Mobile broadband is the most prominent use case today, and it is expected to continue to be one of the key use cases driving the requirements for the 5G systems. The challenge goes far beyond providing the users with basic mobile Internet access and covers support of rich interactive work, media, and entertainment applications in the cloud or reality augmentations (both centralized and distributed frameworks).

5G has strived to enable new usage models for mobile networks in various vertical markets, ranging from wearable electronic devices to autonomous driving. However, the approach for spectrum usage for vertical markets seems to be still open to different solutions and the most important drivers appear to be market demand, time to market, and efficient spectrum utilization. The requirements for 5G are derived from the use cases that consider how the 5G system will provide services to the end user and what services are provided. The use cases are grouped into broad categories in order to contrast the differences in their requirements. A number of use cases are related to scenarios where the user selects from a number of network options, depending on service, performance, cost, or user preferences. When the service transfers from one network configuration to another, one of the important considerations is the degree of network interoperability.

One of the 5G promises was to provide a unified experience for users across multiple devices. From an operator's point of view, 5G also aims to unify multiple access types into a single core network as well as requirements for interoperability for one device operating on different access technologies; this also introduces requirements for service continuity between devices. 5G systems provide the user with the ability to transfer existing service sessions from one device to another device with a minimal perceptible service interruption to the user.

The modern cities of this era greatly depend on a number of critical infrastructure to function properly: electricity, water, sewer, gas, etc. Critical infrastructure monitoring is a cumbersome undertaking, often requiring service levels achievable only by dedicated wireline connectivity. For instance, in order to detect a fault in high-voltage transmission lines and be able to take corrective actions to prevent cascading failures, the required communication latency is beyond what current wireless networks can achieve. Similarly, structural monitoring requires the provisioning of a large number of low-data-rate battery-powered wireless sensors, and today's wireless networks are not optimized to support this deployment model, both in terms of battery life and cost efficiency. Also, with the massive migration of the world's population toward urban environments, there is an increasing demand for cities to modernize their infrastructure and services. From water and power management to buildings and transportation, city planners will rely on new scalable, interconnected services that do not require cities to overhaul private and public infrastructure and are built for the future. 5G systems are enabling smart cities around the world to build long-term connectivity strategies that help improve livability and sustainability.

The massive number of 5G connected devices (e.g., sensors and actuators) collects and processes data to monitor critical parameters and optimize performance based on environmental conditions. Sensors also enable service providers to detect when hidden pipes and cables need repair or when an unauthorized access occurs. 5G systems are designed to support reliable low-latency communications among densely deployed devices that are subject to power constraints with wide range of data rate requirements.

Video and audio streaming, video calls, social networking, and multimedia messaging are just some of the popular communications and entertainment applications used on today's wireless networks. In addition, new applications are emerging such as real-time multi-user gaming, virtual/augmented reality (VR/AR), 3D multi-site telepresence, ultra-high-resolution video streaming (e.g., 4K and 8K video), and photo and video sharing. These applications will require significant increase in data rate, capacity, and very low transport latency that is not supported by today's wireless networks. 5G systems provide solutions that enable continued evolution of communications and entertainment applications. There has been a great interest and early work around in the industry to enable smart, connected devices and realization of the IoT. This use case is about enabling growth through reducing the cost of connectivity of smart devices to support massive deployments, enabling carrier networks to provide connectivity for both cellular and non-cellular IoT devices, and enabling the efficient use of spectrum for device-to-device (D2D) communications.

A wide range of new and diverse use cases will need to be supported by the 3GPP ecosystem. This needs to be done at the same time that the industry continues to support the traditional mobile broadband use cases. The new use cases are expected to come with a wide range of requirements on the network operation. For example, there will be different requirements on functionalities such as charging, policy control, security, mobility, speed, and availability. Some use cases such as mobile broadband may require application-specific charging and policy control while other use cases can efficiently be managed with simpler charging or policies. Some use cases also have significant differences in performance requirements, for example, power consumption and complexity, meaning that it will not be acceptable to simply provide applications with the superset of all requirements.

In order to manage the multitude of service segments and verticals in a sustainable manner, there is a need to isolate the often intrinsically different segments from each other. For example, a scenario where a large number of consumer premises' electricity meters are malfunctioning in the network should not negatively impact the services provided to eMBB users or the healthcare and safety applications. In addition, with new verticals supported by the 3GPP community, there will also be a need for independent management and orchestration (MANO) of segments, as well as providing analytics and service exposure functionality that is customized to each vertical's or segment's need. The functionality should not be restricted to providing isolation between different segments but also to allow an operator to deploy multiple instances of the same network partition. Fig. 3 provides a high-level illustration of the concept. A slice is composed of a collection of logical network functions that support the communication service requirements of particular use case(s). Some slices will be very rich in functionality, while other slices will be very minimalist, but the network slices are not arranged in any form of hierarchy. A given network slice will simply contain the functions required for a given application or class of service(s). Devices can be directed to the appropriate network slice in a way that fulfills operator and user needs, for example, based on subscription or device type. The network slicing primarily targets a partition of the core network, but it is not excluded that the RAN may need specific functionality to support multiple slices or even partitioning of resources for different network slices.

Migration toward network slicing for 5G services necessitates careful consideration of the security requirements, isolation between slices, and the specific individual needs of each operator, whether it is the hosting platform provider or network slice tenant, in ensuring the integrity of the data processing and data stored in the network slices [45].

The presence of a very large number of IoT devices on the cellular network may require radically new technology and solutions that the current 4G networks may not be able to provide. A fundamental feature needed in order to support massive IoT deployments is scalability on the device and the infrastructure sides. Current LTE networks are designed in a manner that makes such scalability technically and financially prohibitive. 5G systems are designed to provide connectivity to an extremely large number of low-cost, low-power, low-complexity devices under mMTC usage scenario. In general, IoT devices require to operate in low-power mode during communication while maintaining a high level of reliability and coverage.

Mobility-on-demand provides a selection of options, which may be dynamically assigned to a device or application according to the device and application context, or statically configured for specialized devices and applications.

Mobility-on-demand consists of two components: (1) managing mobility of active devices and (2) tracking and reaching out to devices that support a power-saving idle mode. At the same time, the requirements on mobility support may also vary based on the applications and services. While some services require the network mobility events to be transparent to the application layer in order to avoid interruption in service delivery, other applications may have specific means to ensure service continuity. The act of concealing mobility events may include some aspects such as minimizing interruption time and packet loss or maintaining the same IP address during intra-RAT or inter-RAT cell changes.

An important part of improving user experience is to collect and analyze some information from individual users in order to understand their service priorities and perception of the network service quality. In order to allow such data collection, an end-user opt-in and control element should be provided to users to specify which aspects of service performance are the most important to them and to set their preferences for data collection in order to maintain a balance between user privacy and service quality optimization. The ecosystem around user experience monitoring should include third-party entities such as data brokers and content providers, which is further considered as part of the user control. Content providers also need to give consent to collection of data on the usage of their applications and content.

The typical usage model for the development of previous generations of cellular technologies has been centered on the telephony and wireless multimedia services with human interfaces. This type of service will continue to be important in the 5G timeframe, but cellular technology is now required to efficiently support a range of other emerging applications. The IoT is widely mentioned as an umbrella for a range of applications that include remote healthcare, smart grid, smart city, intelligent transport, and smart cars. 5G systems are required to support a wide range of IoT applications including geographically and/or power constrained, low-data devices and/or sophisticated devices with requirements for large amounts of real-time data. The IoT is an area where direct communication between devices, rather than communication via the network, may be desirable.

Augmented and virtual reality services are rapidly growing markets driven by advances in device capabilities, consumer excitement about new user experiences, and a range of practical applications. These services are expected to be demanding in their requirements for high bandwidth (particularly for virtual reality) and low latency (particularly for augmented reality). Meeting users demand for these services is expected to be an important driver for 5G.

Another emerging application area for cellular systems is in public safety and mission-critical communications for first responders and other users. 5G technology should maintain and extend the mission-critical communication technology being developed as part of LTE. This area will also have specific requirements for IoT devices and wearable technology. Users' growing reliance on mobile connectivity continues to present technical and economic challenges for the provision of adequate coverage and capacity for these critical applications.

2.2 Deployment Scenarios

The study of 5G deployment scenarios is the first step toward identifying service continuity and performance requirements. In this section, we discuss the prominent scenarios considered for 5G systems deployment. While these scenarios are similar to the use cases, they are primarily focused on the deployment issues rather than user experience. As shown in Fig. 4, the initial deployments of 5G systems will likely be predominantly in the urban areas with high user density, low mobility, and extremely high-capacity demand. Over time, network services will be expanded to include suburban and rural areas with lower user density but higher mobility and wider coverage requirements. Note that as illustrated in the figure, the urban coverage consists of macrocells and small cells where macrocells may operate at sub-6 GHz frequencies and small cells may operate at above 6 GHz frequencies (e.g., mmWave bands). As the coverage is expanded to suburban and rural areas, macrocells will likely use higher transmit powers at sub-6 GHz bands in order to provide larger coverage and extended range. As a result, multi-connectivity will play an important role in early deployments of 5G systems particularly in urban and suburban areas.

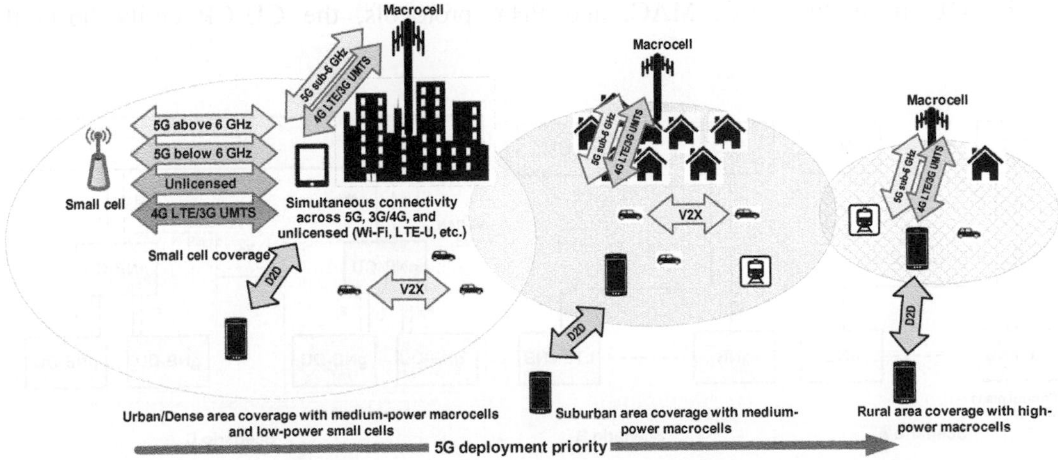

Figure 4
Multi-connectivity across frequency bands and access technologies [45].

In order to understand 5G deployment scenarios, we start our analysis with the study of high-level RAN architectures and network topologies supported by 5G. The design principles of 5G RAN architecture are based on disaggregation of the base stations and separation of the control plane (CP) and user plane (UP) entities. The architecture further supports connectivity between 3GPP LTE and 3GPP NR access and core network entities. The RAN architecture scenarios shown in Fig. 5 support co-located deployment of LTE eNBs and NR gNBs connected to either LTE core network (EPC) or 5G core (5GC) network in a non-standalone (NSA) mode as well as standalone (SA) deployment of NR disaggregated gNBs connected to 5GC. In the early deployments of 5G systems and before 5GC equipment is available, the NSA scenarios are going to be predominant and later SA deployments of 5G systems will prevail [16].

In the deployment scenarios shown in Fig. 5, gNB-CU and gNB-DU denote the central unit (CU) and distributed unit (DU) of a logical gNB, respectively, which may or may not be co-located. The logical interfaces F1 and Xn are new network reference points that comprise control plane and user plane paths between the CU and the DU(s) that will be discussed in more detail in Chapter 1. There are two gNB deployment options when considering C-RAN architectures, as shown in Fig. 6 [16,17]:

- *Collapsed gNB deployment:* In this scenario, all RAN protocols and functions are co-located within the same site. This deployment option corresponds to the current deployments of LTE systems and as such it ensures backward compatibility with the existing LTE deployments.
- *Disaggregated gNB deployment:* In this scenario, RAN protocols and functions are distributed across different sites; that is, in a C-RAN architecture, the DU and the CU may be physically apart. The CU may be further divided into CP and UP entities. The DU hosts the RLC, MAC, and PHY protocols, the CU-CP entity hosts the

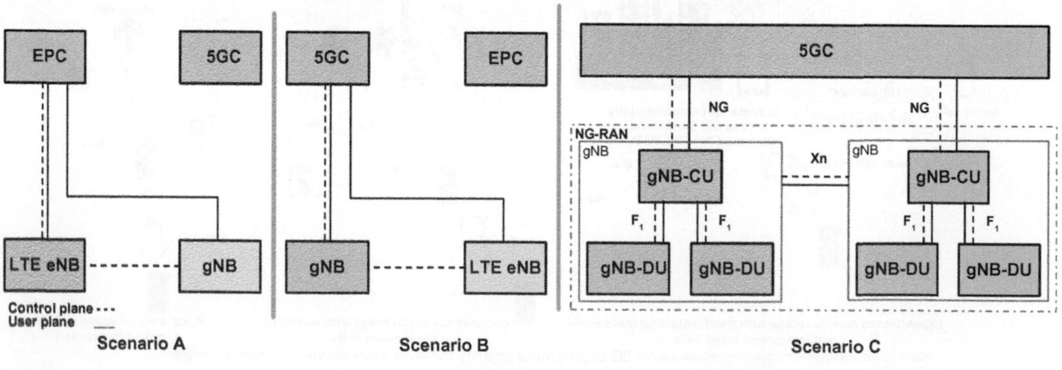

Figure 5
Different deployment scenarios of NR and LTE [16].

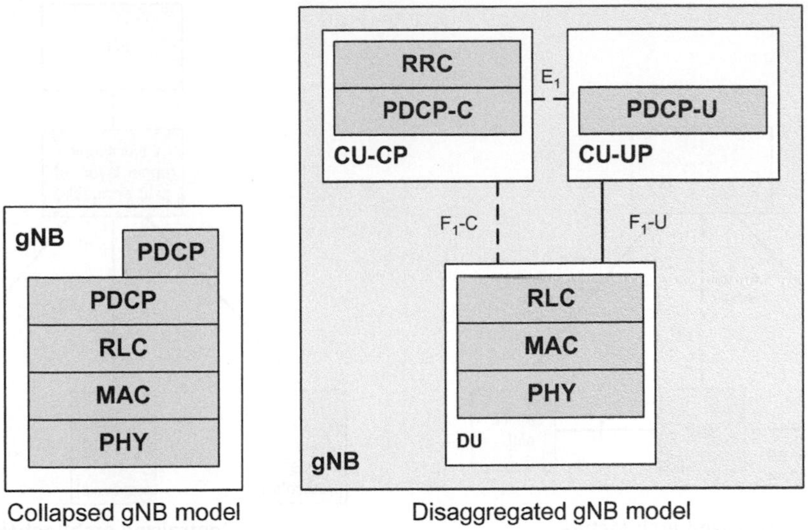

Figure 6
Different deployment options for gNB (NR base station) [16].

PDCP-C and radio resource control (RRC) protocols whereas the CU-UP entity hosts the PDCP-U (and SDAP) protocols. In the disaggregated gNB deployment, the separation of control-plane CU-CP and user-plane CU-UP entities offers the possibility of optimizing the location of different RAN functions based on the desired topology and performance. For example, the CU-CP can be placed in a location close to the DU entities. It could also be co-located with the DU, thus providing shorter latency for the critical CP procedures such as connection (re)establishment, handover, and state transition. On the other hand, the CU-UP can be centralized in a regional or national data center, thus favoring centralized implementations and providing a central termination point for the UP traffic in dual connectivity and tight interworking scenarios. An additional CU-UP can also be placed closer (or co-located) with the DU to provide a local termination point for the UP traffic for applications that require very low latency (e.g., URLLC traffic).

In the disaggregated gNB architecture shown in Fig. 6, it is necessary to coordinate the CU-CP and CU-UP entities. Some of the functions that may require a control-plane interface between CU-CP and CU-UP include CU-CP set-up, modification, and configuration of the data radio bearers (DRBs) in the CU-UP and when CU-CP entity configures the security keys in the CU-UP entity for RAN-level security activation and configuration. A new open interface between CU-CP and CU-UP has been defined to enable these functions, which is denoted as E_1. The interface E_1 is a control-plane interface, and it does not require a UP part because CU-CP and CU-UP do not exchange UP traffic.

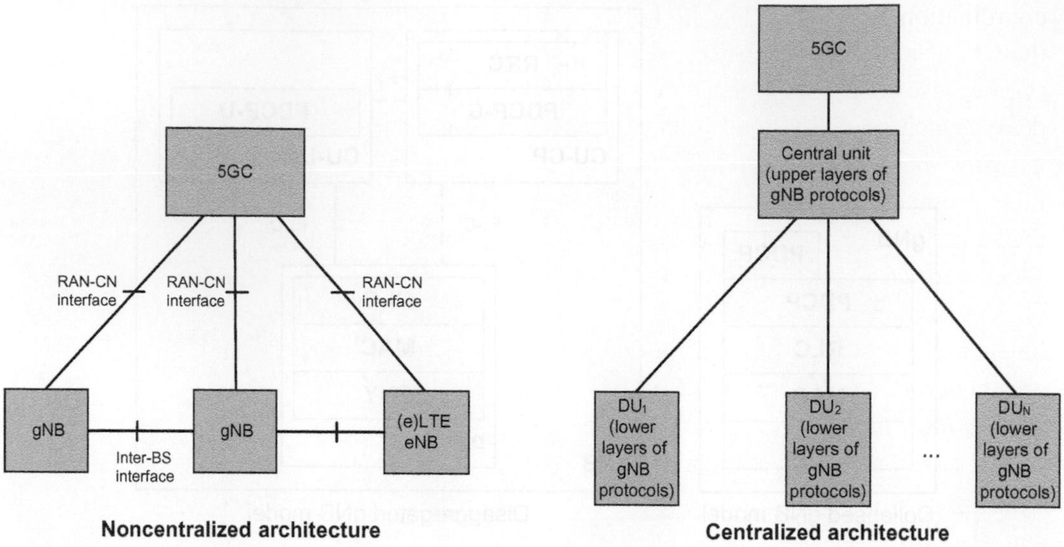

Figure 7
Illustration of the non-centralized and centralized deployment scenarios [16,17].

The new 5G RAN architecture, as shown in Fig. 7, supports the following topologies in order to enable diverse use cases and deployment scenarios [16]:

- *Non-centralized RAN architecture:* In this scenario, full user-plane and control-plane radio protocol stacks are implemented at the gNB, for example, in a macrocell deployment or an indoor (public or enterprise) hotspot environment. The gNBs can be connected to each other using any transport mechanism (e.g., Ethernet, optical fiber). However, it is assumed that the gNB is able to connect to other gNBs or eLTE eNBs[3] through standard interfaces defined by 3GPP.
- *Centralized RAN architecture:* In this scenario, the upper layers of the radio protocol stack (PDCP and RRC layers) are implemented in the CU which is typically located in the edge cloud. Different functional splits between the CU and the DU(s) are also possible, depending on the transport layer configuration. High-performance transport mechanisms, for example, optical transport networks (OTN),[4] between the CU and the lower layers of the stack located at the DUs would enable advanced coordinated multipoint (CoMP) transmission/reception schemes and inter-cell scheduling optimization, which could be useful in high-capacity scenarios, or scenarios where inter-cell

3 The eLTE eNB is the evolution of LTE (i.e., LTE Rel-15 and later) eNB that supports connectivity to EPC and 5GC.

4 OTN is a set of optical network elements connected via optical fiber links that are able to provide transport, multiplexing, switching, and management functions as well as supervision and survivability of optical channels carrying users' signals.

coordination is desired. The higher layers of NR radio protocol stack can be moved to the CU, if a low performance transport mechanism is used between the CU and the DU (s), because in this case the requirements on the transport layer in terms of bandwidth, delay, synchronization, and jitter are relatively relaxed.

- *Co-sited deployment with LTE:* In this scenario, the NR functions are co-located with LTE counterparts either as part of the same base station or as multiple base stations at the same site. Co-located deployment can be applicable in all NR deployment scenarios such as urban macro. As shown in Fig. 8, in this scenario it is desirable to fully utilize all spectrum resources assigned to both radio access technologies (RATs) using load balancing or connectivity via multiple RATs (e.g., utilizing lower frequencies as wide-area coverage layer for the cell-edge users).
- *Shared RAN deployment:* The NR supports shared RAN deployments, implying multiple hosted core operators. The shared RAN topology can cover large geographical areas as in the case of national or regional network deployments. The shared RAN architecture can also be heterogeneous; that is, limited to a few smaller areas, as in the case of shared in-building RANs. A shared RAN should be able to efficiently interoperate with a non-shared RAN. Each core operator may have its own non-shared RAN serving areas adjacent to the shared RAN. The mobility between the non-shared RAN and the shared RAN is supported in a manner similar to that of LTE. The shared RAN may operate either in the shared spectrum or in the spectrum of each host operator (see Fig. 8).

The cell layouts corresponding to the prominent deployment scenarios for NR are shown in Fig. 9. In the homogeneous deployment scenario, all cells (macro or small cell only) provide the same coverage, whereas in the heterogeneous deployment scenarios, cells of different sizes have overlapped coverage, that is, a combination of macro and small cells where typically macrocells provide wide-area coverage and are used as anchor points and small-

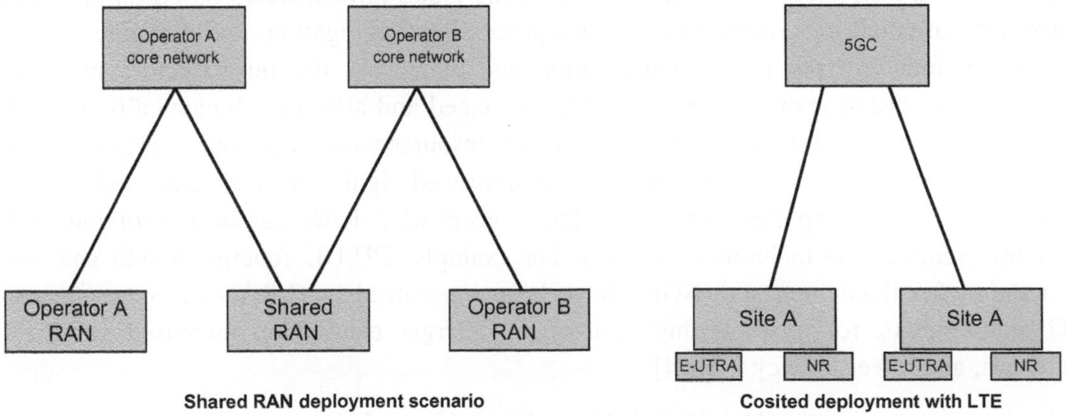

Figure 8
Cosited and shared RAN deployment scenarios [16].

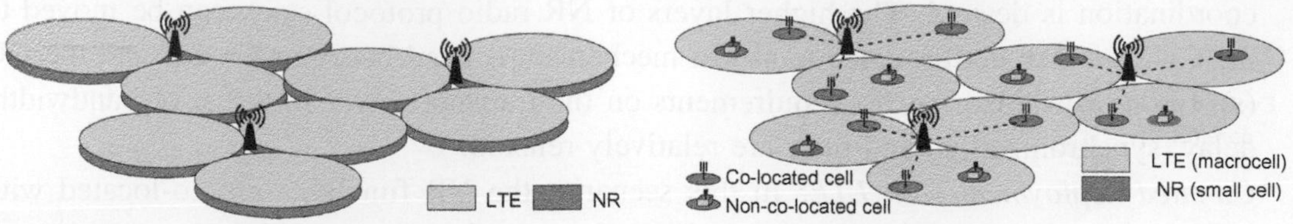

Figure 9
Co-deployment scenarios of NR and LTE [15,16].

cell access points provide high-throughput small coverage areas within macrocell coverage. In these deployment scenarios, depending on the cell layout and the location of the access nodes, both NR and LTE coverage may coexist in the same geographical area. One of the scenarios in Fig. 9 illustrates both LTE and NR cells that are overlaid and co-located, both providing similar coverage area. In this case, both LTE and NR cells are either macro or small cells. Fig. 9 further depicts a scenario where LTE and NR cells are overlaid, but not necessarily co-located, where each providing different coverage area. In this case, LTE serves macrocells and NR serves small cells. The opposite scenario is also possible which may not be practically useful for initial deployments of 5G networks. According to 3GPP definition, a co-located cell refers to a small cell together with a macro cell for which the eNBs are installed at the same location, whereas a non-co-located cell refers to a small cell alongside a macro cell for which the respective eNBs are installed at different locations.

The distributed base station model leads way to the centralized and cooperative C-RAN architectural concept where all or part of baseband functions are performed in a CU. Centralized signal processing greatly reduces the number of site equipment needed to cover the same areas served by a distributed network of base stations. Cooperative radio transmission combined with distributed antenna scheme provided by RRHs provides higher spectral efficiency and interference coordination. Real-time cloud infrastructure based on open platform and base station virtualization enables processing, aggregation, and dynamic resource allocation, reducing the power consumption and increasing the infrastructure utilization rate. The processing resources on CU can be managed and allocated dynamically, resulting in more efficient utilization of radio and network resources and improved energy efficiency (EE). Moreover, network performance is also improved significantly because collaborative techniques can be supported in C-RAN. The concept of C-RAN has been evolving in the past few years as the technology matures. For example, CU/DU functional split and next-generation fronthaul interface (NGFI) have been introduced in C-RAN to better meet the 5G requirements, for example, higher frequency, larger bandwidth, increased number of antennas, and lower latency. [23,24].

It is observed that in a typical 5G C-RAN architecture, the baseband unit often comprises a CU and one or more DUs. The principle of CU/DU functional split lies in real-time processing requirements of different functions. As shown in Fig. 10, the CU functions mainly

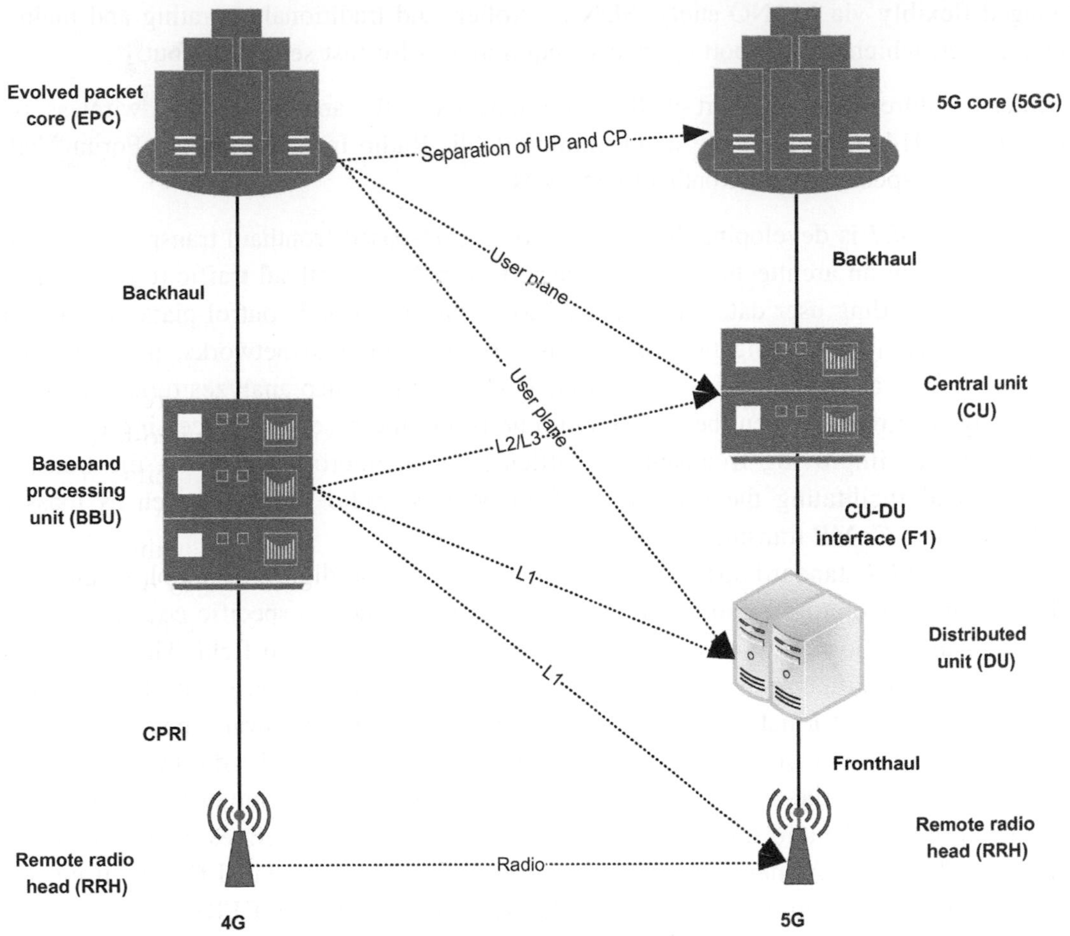

Figure 10
Evolution of single-node 4G BBU to separated CU/DU architecture in 5G [23].

include non-real-time higher layer RAN protocol processing as well as core network functions that have been moved to the edge of the network to enable mobile edge computing (MEC) services and tactile Internet. Accordingly, a DU is mainly responsible for physical layer processing and real-time processing of some or all layer 2 protocols. In order to minimize the transport capacity and latency requirements between RRU and DU(s), part of physical layer processing can be moved from DU to RRU(s). From hardware implementation point of view, the CU equipment can be developed based on general purpose processors or server platforms. The DU hardware can be implemented using customized SoCs and/or hybrid platforms, which enable real-time processing of computationally intensive functions. With NFV infrastructure, system resources (including CU and DU) can be orchestrated and managed flexibly via MANO entity, SDN controller, and traditional operating and maintenance center, which can support operators' requirements for fast service rollout.

In order to address the transport challenges among CU, DU, and RRU in 5G wireless systems, IEEE 1914 working group[5] and Common Public Radio Interface (CPRI) Forum[6] have independently specified new fronthaul interfaces.

- *IEEE P1914.1* is developing the standard for packet-based fronthaul transport networks. It is defining an architecture for the transport of mobile fronthaul traffic (e.g., Ethernet-based), including user data traffic as well as management and control-plane traffic. The group is further defining the requirements for the fronthaul networks, including data rates, timing and synchronization, and QoS. The standard also analyzes functional partitioning schemes between the remote radio units and the baseband processing units with the goal of improving fronthaul link efficiency and interoperability on the transport level, and facilitating the realization of cooperative radio functions such as massive MIMO, and CoMP transmission and reception.
- *IEEE P1914.3* standard specifies the encapsulation of the digitized complex-valued in-phase and quadrature radio signal components, and (vendor-) specific control information channels/flows into an encapsulated Ethernet frame payload field. The header format for both structure-aware and structure-agnostic encapsulation of existing digitized radio transport formats have been specified. The structure-aware encapsulation is assumed to have detailed knowledge of the encapsulated digitized radio transport format content, whereas the structure-agnostic encapsulation is only a container for the encapsulated digitized radio transport frames. The standard further defines a structure-aware mapper for CPRI frames and payloads to/from Ethernet encapsulated frames. It must be noted that the structure-agnostic encapsulation is not restricted to CPRI.
- *CPRI Forum* has developed a new specification (eCPRI) for fronthaul interface which includes increased efficiency in order to meet the foreseeable requirements of 5G mobile networks. Note that the widely used CPRI (radio over fiber) specifications were also developed by this special interest group in the past. The eCPRI specification is based on new functional partitioning of the cellular base station functions within the physical layer. The new split points enable significant reduction of the required fronthaul link capacity. The required bandwidth can scale flexibly according to the user-plane traffic transport over the Ethernet. The use of Ethernet opens the possibility to carry eCPRI traffic and other traffic simultaneously in the same switched network. The new interface is a real-time traffic interface enabling use of sophisticated coordination algorithms and guaranteeing the best possible radio performance. The eCPRI interface is meant to be future proof, allowing new feature introduction by software updates

[5] IEEE 1914 Working Group, Next Generation Fronthaul Interface, http://sites.ieee.org/sagroups-1914/.

[6] CPRI, http://www.cpri.info/.

in the radio network. In addition to the new eCPRI specification, the special interest group continues to further develop the existing CPRI specifications in order to maintain a competitive option for all deployments with dedicated fiber connections in the fronthaul including 5G wireless systems.

As shown in Fig. 11, an NGFI switch network connects CU and one or more DU(s) entities. The use of NGFI interface allows flexible configuration and deployment of CU and DU entities in different deployment scenarios. In case of ideal fronthaul, the deployment of DU can also be centralized, which can support physical layer cooperation across various transmission/reception nodes. In case of non-ideal fronthaul, the DU(s) can be deployed in a distributed manner. Therefore, a C-RAN architecture based on NGFI interface supports not only the centralized DU deployments, but also enables distributed DU deployments.

Mobile networks have traditionally been optimized for voice and data services. However, in 5G era, they have to serve a variety of devices with different characteristics and service/performance requirements. As mentioned earlier, some of the most common use cases for 5G include mobile broadband, massive IoT, and mission-critical IoT, and they all require

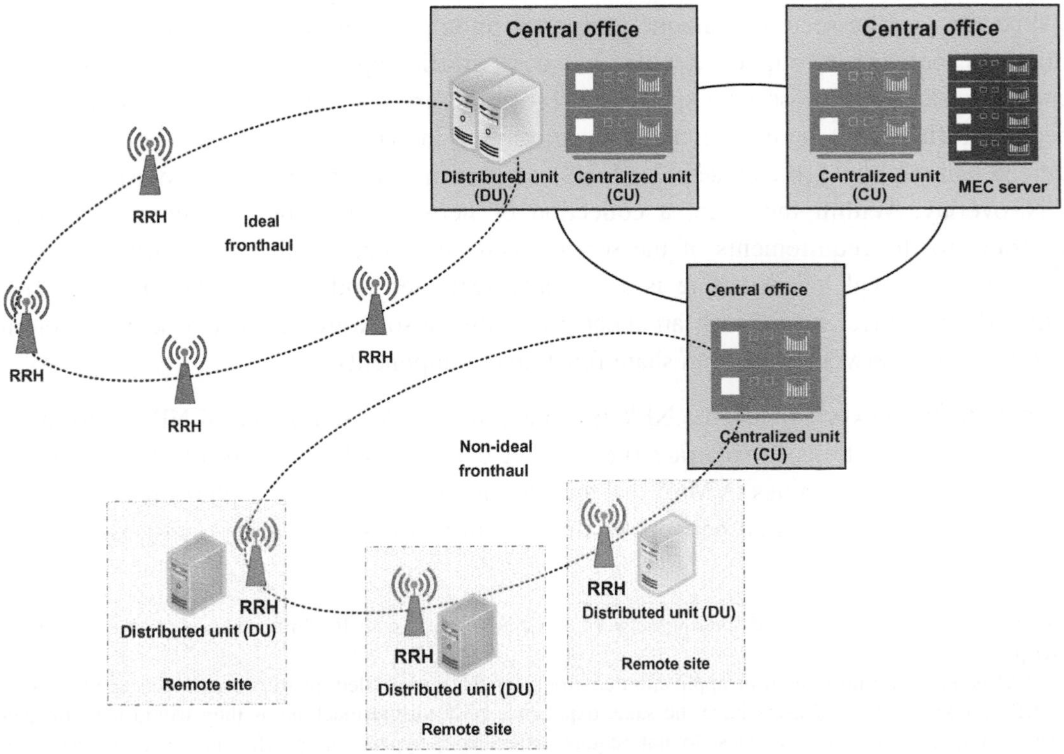

Figure 11
CU/DU-based C-RAN architecture [23–26].

different types of features and networks in terms of mobility, charging, security, policy control, latency, reliability, etc. For instance, a massive IoT service that connects stationary sensors measuring temperature, humidity, precipitation, etc. to mobile networks does not require features such as handover or location update, which have been critical in serving voice services. A mission-critical IoT service such as autonomous driving or remote-controlled robots requires a substantially low end-to-end latency in the order of a fraction of millisecond, whereas some massive MTC services are delay-tolerant.

The objective of network slicing is to transform the network from using a rather static one-network-fits-all approach to a dynamic model where multiple logical networks can be created on top of the same physical infrastructure to fulfill the diverse needs of different use cases. Network slicing is a concept that enables the operators to create service-customized (logical) networks in order to provide optimized solutions for different services which have diverse requirements in terms of functionality, performance, and isolation. The fundamental assumption in this solution is that slicing of an operator's network should be transparent to the UEs at the radio interface level. A network slice is a logical network that includes a set of network functions, which may be virtual network functions (VNFs) or physical network functions, and corresponding resources, such as compute, storage, and networking resources. A slice can also be seen as a unique profile for an application, defined as a set of services within the network to support a given use case, traffic type, or a customer. A slice may serve a particular purpose or a specific category of services (e.g., a specific use case or a specific traffic type) or even individual customers, in which case it may be created on-demand using a network-as-a-service (NaaS)[7] approach. Each network slice exists within an SDN overlay. Within the slice, a collection of network functions are chained together according to the requirements of the service using the slice. Each slice sharing the same underlying physical infrastructure is a logically separated and isolated system that can be designed with different network architecture, hardware/software and network provisioning. In some cases, network slices can share functional components.

To implement network slices, the NFV is a prerequisite. The main idea of NFV is to implement network functions in software (i.e., packet core and selective radio access functions) mapped to virtual machines (VMs)[8] that are run on commercial server platforms as opposed to processing over dedicated network equipment. In this case, the access network works as

[7] NaaS is a business model for delivering network services virtually over the Internet on a pay-per-use or subscription basis.

[8] A VM is an operating system or application environment that is installed on software, which emulates dedicated hardware. The end users have the same experience on a virtual machine as they would have on dedicated hardware. Virtual machines do not require specialized, hypervisor-specific hardware. Virtualization does, however, require more bandwidth, storage, and processing capacity than a traditional server or desktop if the physical hardware is going to host multiple running virtual machines. VMs can easily be moved, copied, and reassigned between host servers to optimize hardware resource utilization.

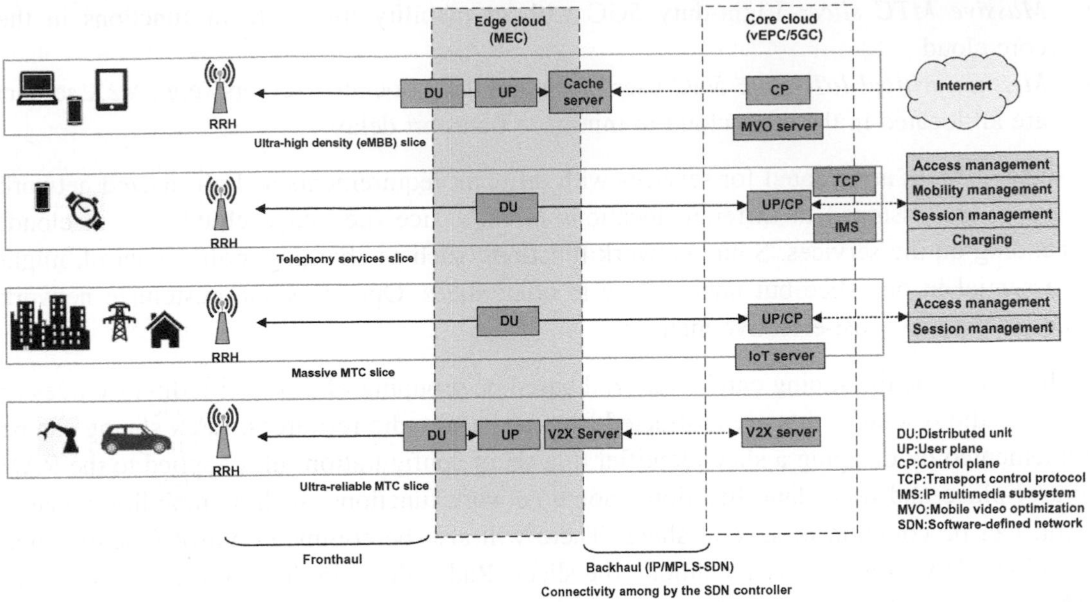

Figure 12
Illustration of network slicing concept [43].

edge cloud, whereas the core network operates as core cloud. Connectivity among VMs located on the edge and core clouds are provisioned using SDN. Network slices are automatically configured and created for each service (i.e., voice service slice, massive IoT slice, mission-critical IoT slice). Fig. 12 shows how applications dedicated for each service can be hypothetically virtualized and implemented in each slice. In the example shown in the figure slices that have been configured for the specific services contain the following components [43]:

- *Mobile broadband slice:* All virtualized DU, 5GC user-plane, and cache server in the edge cloud, and virtualized 5GC control-plane as well as mobile video optimization (MVO)[9] server in the core cloud.
- *Voice services slice:* 5GC user-plane and control-plane functions with full mobility support features (e.g., access management, session management, mobility management, and charging), and IP multimedia subsystem and TCP optimization VNFs running in a core cloud.

[9] MVO covers the use of technologies and solutions that enable mobile network operators (MNOs) to optimize the delivery of video content through a host of video optimization techniques, prioritization policies, user-based customization, and intelligent management of overall traffic on their data networks. MVO aims at maximizing the quality of video services and enhancing end user experience, while enabling MNOs to monetize the mobile video services, both operators' own and third-party over-the-top services.

- *Massive MTC slice:* Light-duty 5GC without mobility management functions in the core cloud.
- *Mission-critical IoT slice:* 5GC user-plane and the associated servers (e.g., V2X server) are all located in the edge cloud to minimize transport delay.

Dedicated slices are created for services with different requirements and virtualized network functions are assigned to different locations in each slice (i.e., edge cloud or core cloud) depending on the services. Some network functions, such as charging, policy control, might be essential in one slice but unnecessary in other slices. Operators can customize network slices in the most cost-effective manner.

In the RAN side the slicing can be realized based on grouping of physical radio resources or logical radio resources that are abstracted from physical radio resources. RAN slicing can be implemented by mapping a slice identifier to a set of configuration rules applied to the RAN control-plane and user-plane functions. Some network functions, such as mobility management, can be common to several slices. There will also be common control functions that coordinate RAN resource usage among the slices. Radio slices in the RAN may share radio resources (time, frequency, code, power, and space) and the corresponding communication hardware such as digital baseband processing components or analog radio components. The sharing may be done in a dynamic or static manner depending on the configuration rules of the network slice. In the case of dynamic sharing, each slice obtains resources based on its demand and priority using either scheduling (that is, the slice requests resources from a centralized scheduler, which allocates resources based on the overall traffic load or the priority of the slice) or contention. In the case of static sharing, a slice is pre-configured to operate with a dedicated resource throughout its operation time. Static sharing enables guaranteed resource allocation to the slice, whereas dynamic resource sharing allows overall resource usage optimization. Slice-specific configuration rules for the control-plane functions adapt the RAN control-plane functions for each slice. This is because slices may not require all control-plane functions. There is also a need for network slice-specific admission control, so that the system can meet the initial access requirements of various network slices.

3 Key Performance Indicators, Architectural, System, and Service Requirements

Unlike previous generations of 3GPP systems that attempted to provide a general-purpose framework, the 5G systems are expected to provide customized and optimized support for a variety of different services, traffic patterns, and end-user demographics. Several industry-led 5G white papers and in particular the one from Next-generation Mobile Network[10] Alliance [14,25,32,35] described a multi-purpose 5G system capable of simultaneously

[10] Next-Generation Mobile Network Alliance, http://www.ngmn.org/.

supporting various combinations of requirements such as reliability, latency, throughput, positioning, and availability to enable different use cases and deployment scenarios. Such revolutionary systems would be achievable with introduction of new technologies, both in access network and the core network, with configurable, flexible, and scalable assignment of network resources. In addition to increased flexibility and optimization, a 5G system needs to support stringent key performance indicators (KPIs) for latency, reliability, throughput, etc. Enhancements in the radio access network contribute to meeting these KPIs as do core network improvements such as network slicing, in-network caching, and hosting services closer to the end nodes [21]. Flexible network operations are the cornerstone of the 5G systems and the new features that allow this flexibility include NFV, network slicing, SDN, network scalability, and seamless mobility. Other network operational requirements address the indispensable control and data plane resource allocation efficiencies as well as network configurations that optimize service delivery by adopting optimal routing strategies between end users and application servers. Enhanced charging and security mechanisms manage new types of UEs connecting to the network in different ways.

The enhancements in mobile broadband access mechanisms aim to meet a number of new KPIs, which pertain to high data rates, high user density, high user mobility, highly variable data rates, flexible deployment options, and improved coverage. High data rates are driven by the increasing use of data for services such as streaming (e.g., video, music, and user generated content), interactive services (e.g., augmented reality), and some IoT use cases. The availability of these services impose stringent requirements on user experienced data rates and latency in order to meet certain QoS requirements. In addition, increased coverage in densely populated areas such as sport arenas, urban areas, and transportation hubs has become an essential demand for pedestrians and users in vehicles. New KPIs related to traffic and connection density enable transport of high volume of user traffic per area (traffic density) and transport of data for high number of connections (UE density or connection density). Many UEs are expected to support a variety of services which exchange either a very large (e.g., streaming video) or very small (e.g., sensor data burst) amount of data. The 5G systems are required to manage this service variability in a resource efficient manner. All of these cases introduce new deployment requirements for indoor and outdoor, local area connectivity, high user density, wide-area connectivity, and UEs travelling at high-speed scenarios.

Another aspect of 5G KPIs includes requirements for various combinations of latency and reliability, as well as higher accuracy for positioning. These KPIs are driven by support of commercial and public safety services. On the commercial side, industrial control, industrial automation, control of aerial vehicles, and the augmented/virtual reality are examples of such services. Services such as aerial vehicle control will require more precise positioning information that includes altitude, speed, and direction, in addition to planar coordinates. Support for massive connectivity brings many new requirements in addition to those for the enhanced KPIs. The expansion of connected things paradigm introduces need for significant improvements in resource efficiency in all system components (e.g., UEs, IoT devices, access network, and core network). The 5G system also targets enhancement of its capabilities to meet the KPIs that the emerging V2X applications require. For these advanced applications, the requirements, such as data rate, reliability, latency, communication range, and speed, become more stringent. These applications have specific technical requirements that need to be addressed through sophisticated design of the 5G radio interface and the use of appropriate frequency bands in order to ensure consistent satisfaction of their requirements.

3.1 Definition of the Performance Metrics

In order to quantify how certain technical solutions would affect end-user experience or what would be the 5G system performance in a desired use case, specific performance metrics are needed for evaluation. This section provides definitions of 5G main characteristics and KPIs, similar to the ones defined in references [1,14]. The following section provides the definitions of the performance metrics that have been used to measure and benchmark the performance of 5G systems. These metrics are related to the KPIs that describe the main 5G use cases and certain deployment scenarios.

Peak Data Rate: Described in bits/s, this is the maximum (theoretically) achievable data rate under error-free conditions at the physical layer. It denotes the received data bits allocated to a single user terminal, when all assignable radio resources for the corresponding link direction are utilized excluding the overhead resources which include radio resources that are used for physical layer synchronization, reference signals, guard bands, and guard times. The peak data rate is defined for a single mobile station. For a single frequency band, it is related to the peak spectral efficiency in that band. Let W denote the channel bandwidth and SE_p denote the peak spectral efficiency in that band. Then the user peak data rate R_p is given by

$$R_p = SE_p W$$

The peak spectral efficiency and available bandwidth may have different values in different frequency ranges. If the bandwidth is aggregated across multiple bands, the peak data rate will be summed over those bands. Therefore, if bandwidth is aggregated across N bands, then the total peak data rate is defined as follows:

$$R_p = \sum_{n=1}^{N} SE_p^n W_n$$

where W_n and SE_p^n represent the nth component bandwidth and its peak spectral efficiency, respectively.

Average Spectral Efficiency: Let $R_n\ (T)$ denote the number of correctly received bits by the nth user in the downlink or from the nth user in the uplink. It is assumed that there are N users (or devices) and N_{TRP} transmission/reception points (TRPs)[11] in the system within the coverage area. Furthermore, let W denote the channel bandwidth and T the time interval over which the data bits are received. The average spectral efficiency may be estimated by running system-level simulations over N_{drops} (number of drops), where each drop results in certain rate denoted as $R^{(1)}(T), \ldots, R^{(N_{drops})}(T)$, subsequently

$$SE_{avg} = \frac{1}{N_{drops} TWN_{TRP}} \sum_{n=1}^{N_{drops}} R^{(n)}(T) = \frac{1}{N_{drops} TWN_{TRP}} \sum_{n=1}^{N_{drops}} \sum_{t=1}^{N} R_i^{(n)}(T)$$

where SE_{avg} is the estimated average spectral efficiency that approaches the actual average value with increasing number of drops, and $R_i^{(n)}(T)$ is the total number of correctly received bits by the ith user in the nth drop. The average spectral efficiency is evaluated through system-level simulations using the evaluation configuration parameters of Indoor Hotspot-eMBB, Dense Urban-eMBB, and Rural-eMBB test environments defined in references [9,10] while the temporal and spectral resources consumed by layers 1 and 2 overhead are taken into consideration. Examples of layer 1 overhead include the time/frequency resources used by synchronization signals, guard band and DC subcarriers, guard intervals or switching times, reference signals, and cyclic prefixes. Example layer 2 overheads include time/frequency resources utilized to carry common control channels, HARQ ACK/NACK signaling, CSI/CQI feedback, random access channel, payload headers, and CRC. Power allocation and/or boosting should also be accounted for in modeling resource allocation for control channels.

5th Percentile User Spectral Efficiency: This is the 5% point of the cumulative distribution function (CDF)[12] of the normalized user throughput, calculated at all possible user locations. Let's assume that the ith user in the nth drop can correctly decode $R_i^{(n)}(T)$ accumulated bits in the time interval $[1,T]$. For the non-scheduled duration of user i, zero bits are accumulated. During this total time, user i receives accumulated service time of $T_i \leq T$, where the service time is the time duration between the first packet arrival and when the

[11] TRP is defined as physical node, located at a specific geographical location within a cell, with an antenna array that consists of one or more antenna elements that are available for the network operation.

[12] The distribution function $D(x)$, also called the CDF or cumulative distribution function, describes the probability that a random variable X takes on a value less than or equal to a number x. The distribution function is sometimes also denoted as $F_X(x)$. The distribution function is therefore related to a continuous probability density function $p(x)$ by $D(x) = P(X \leq x) = \int_{-\infty}^{x} p(\tau)d\tau$. As a result, $p(x)$ (when it exists) is simply the derivative of the distribution function $p(x) = \partial D(x)/\partial x$. Similarly, the distribution function is related to a discrete probability $p(x)$ by $D(x) = P(X \leq x) = \sum_{X \leq x} p(x)$. There exist distributions that are neither continuous nor discrete.

last packet of the burst is correctly decoded. In the case of full buffer traffic model $T_i = T$. Therefore, the rate normalized by time and bandwidth of user i in drop n is given as

$$r_i^{(n)} = \frac{1}{T_i W} R_i^{(n)}(T)$$

By running the system-level simulations N_{drops} times, $N_{drops}\ N$ values of $r_i^{(n)}$ are obtained of which the lowest 5% point of the CDF is used to estimate the fifth percentile user spectral efficiency. The fifth percentile user spectral efficiency is evaluated by system-level simulations using the evaluation configuration parameters of Indoor Hotspot-eMBB, Dense Urban-eMBB, and Rural-eMBB test environments. It should be noted that the fifth percentile user spectral efficiency is evaluated by system-level simulations only using a single-layer layout configuration even if the test environment comprises multi-layer layout configuration. Furthermore, the fifth percentile user spectral efficiency is evaluated using identical simulation assumptions as the average spectral efficiency for that test environment taking into account the layers 1 and 2 overhead. Note that for time division duplex (TDD) systems, the effective bandwidth is used which is the operating bandwidth normalized by the uplink/downlink ratio.

User Experienced Data Rate: This is the 5% point of the CDF of the user throughput. The user throughput during active state is defined as the number of correctly received bits, that is, the number of bits contained in the service data units (SDUs) delivered to network layer over a certain period of time. In the case of single frequency band and a single TRP layer, the user experienced data rate can be derived from the fifth percentile user spectral efficiency. Let W and SE_{user} denote the system bandwidth and the fifth percentile user spectral efficiency, respectively, then the user experienced data rate denoted by R_{user} is given as $R_{user} = SE_{user} W$. In the case where the system bandwidth is aggregated across multiple bands (one or more TRP layers), the user experienced data rate will be summed over the bands.

The user experienced data rate is evaluated with non-full-buffer traffic assumption. For non-full-buffer traffic, user experienced data rate is fifth percentile of the user throughput. The user throughput during active time is defined as the size of a burst divided by the time between the arrival of the first packet of a burst and the reception of the last packet of the burst. It must be noted that the 5% user spectrum efficiency depends on the number of active users sharing the channel and for a fixed transmit power; it may vary with the system bandwidth. A dense network architecture would increase the 5% user spectrum efficiency and results in fewer users sharing the channel, that is, 5% user spectrum efficiency may vary with the site density. Support of very large system bandwidths in 3GPP 5G systems would further increase the user experienced data rates. The target 5% user spectrum efficiency is three times the IMT-Advanced minimum requirements [8].

Area Traffic Capacity: This metric represents the system throughput over a unit of geographic area and can be evaluated using either a full-buffer traffic model or a non-full-buffer traffic model. The area traffic capacity is described in bits/s/m^2. Both the user experienced data rate and the area traffic capacity must be evaluated at the same time using the same traffic model. The throughput is defined as the number of correctly received bits, that is, the number of bits contained in the SDUs delivered to the network layer over a certain period of time. It can be derived for a particular deployment scenario which utilizes one frequency band and one TRP layer based on the achievable average spectral efficiency, TRP density, and system bandwidth. Let W and $d_{TRP} = N_{TRP}/S$ denote the system bandwidth and the TRP density, respectively, where S is the coverage area. The area traffic capacity C_{area} is proportional to the average spectral efficiency SE_{avg} and can be expressed as follows:

$$C_{area} = SE_{avg} d_{TRP} W$$

In case the system bandwidth is aggregated across multiple frequency bands the area traffic capacity will be accumulated over those bands.

In other words, the area traffic capacity is a measure of traffic a network can deliver over a unit area. It depends on site density, bandwidth, and spectrum efficiency. In the case of full-buffer traffic and a single-layer single-band system, the previous equation may be expressed as

$$\begin{aligned}&\textit{area traffic capacity}\ (\text{bps}/m^2) = \\ &\textit{site density}\ (\text{site}/m^2) \times \textit{bandwidth}\ (\text{Hz}) \times \textit{spectrum efficiency}\ (\text{bps}/\text{Hz}/\text{site})\end{aligned}$$

Several techniques have been utilized in 3GPP NR to achieve high spectrum efficiency and thereby to improve area traffic capacity that will be discussed in the next chapters. To this end, spectrum efficiency improvements in the order of three times of IMT-advanced are targeted [8]. Furthermore, 3GPP has strived to develop standards in order to support large system bandwidths in the excess of 1 GHz (aggregated bandwidth) in mmWave spectrum bands. The available bandwidth and site density, which both have direct impact on the available area capacity, vary in different parts of the world and for different operators.

Connection Density: This metric is defined as the total number of devices served by a network within a geographical area where the service satisfies the requirements of a specific QoS using a limited bandwidth and a limited number of TRPs. The target QoS is set such that the delivery of a payload of a certain size within a certain time and with a certain success probability can be guaranteed. This requirement was defined for the purpose of evaluation in the mMTC use case. More specifically, the connection density represents the total number of terminals that can achieve a target QoS per unit geographical area, where the target QoS is to ensure that the system packet drop rate remains less than 1% at a given packet inter-arrival time and payload size. Connection density is proportional to the connection efficiency, the channel bandwidth and TRP density. Let W denote the channel bandwidth and d_{TRP} the TRP density (i.e., number of TRPs per square kilometer). Then connection density $d_{connection}$ can be expressed as follows:

$$d_{connection} = \alpha d_{TRP} W$$

The connection efficiency α is evaluated with a small packet model. Considering the number of users/devices N in the coverage area with average inter-packet time interval T_{packet}, the average number of packets arrival in 1 second denoted as $N_{packet} = N/T_{packet}$. The system must guarantee that the percentage of packets in outage is less than 1%, where a packet is defined to have experienced outage, if the packet fails to be delivered to the destination within a permissible packet delay bound of 4 milliseconds. The packet delay is measured from the time when a packet arrives at the transmitter buffer to the time when it is successfully decoded at the receiver.

According to the IMT-2020 evaluation methodology [10], in order to compute the connection efficiency, one must run system-level simulations by assuming certain number of devices N, packet arrival rate N_{packet}, number of TRPs N_{TRP} and system bandwidth W. One then generates user traffic packets according to the small packet model specified in reference [10] with packet arrival rate N_{packet}. The system-level simulations are run again to obtain the packet outage rate. The number of devices will be iterated until the number of devices satisfying the packet outage rate of 1% is reached, which is denoted as $N_{capacity}$. The connection efficiency is then calculated as $\alpha = N_{capacity} W/N_{TRP}$.

User-Plane Latency: The one-way user-plane latency is the defined as delay introduced by the radio access network during transmission of an IP packet from the source to the destination. The user-plane latency also known as transport delay is the transmission time between an SDU packet being available at the IP layer in the source UE/BS and the availability of this packet at IP layer in the destination BS/UE depending on the link direction. The user-plane packet delay includes delay introduced by associated protocols and control signaling in either uplink or downlink direction assuming the user terminal is in the connected mode and the network operates in unloaded condition. This requirement is defined for the purpose of evaluation in the eMBB and URLLC use cases. The breakdown of the user-plane latency components is as follows [10]:

1. UE/BS (transmitter) processing delay (typically one subframe)
2. Frame alignment (typically half of subframe duration)
3. Number of transmission time intervals (TTIs)[13] used for data packet transmission, which typically includes UE scheduling request and access grant reception times.

[13] The transmission time interval (TTI) is defined as the duration of the transmission of the physical layer encoded packets over the radio air-interface. In other words, the TTI refers to the length of an independently decodable transmission on the air-link. The data on a transport channel is organized into transport blocks. In each TTI, at most one transport block of variable size is transmitted over the radio air-interface to or received from a terminal in the absence of spatial multiplexing. In the case of spatial multiplexing, up to two transport blocks per TTI are transmitted.

The number of TTIs used for each packet transmission depends on channel quality, allocated frequency resource, and the use of multi-connectivity. In case of user-plane multi-connectivity, this delay component should be derived with respect to different multi-connectivity configurations, that is, whether different data streams are transmitted over different links or multiple links are simply used for redundant data transmission. In 5G, both transmitter and receiver can be user devices considering D2D communication. In D2D communications, the user terminal may need some time for downlink/uplink synchronization.

4. HARQ retransmission time which is typically calculated assuming 10%–30% error rate for transmission over the air prior to HARQ. Both control-plane and user-plane multi-connectivity impact this delay component.
5. BS/UE (receiver) processing delay (typically two to four subframes)

Control-Plane Latency: The control-plane latency is the time for the UE to transition from the IDLE state to the CONNECTED state. In other words, the time that it takes to establish the control-plane and data bearer for data transmission over the access link excludes downlink paging delay and backhaul signaling delay. The control-plane latency includes random access procedure duration, uplink synchronization time, connection establishment and HARQ retransmission interval, and data bearer establishment inclusive of HARQ retransmissions. More specifically, the breakdown of control-plane latency components is as follows [10,39]:

1. UE wakeup time: Wakeup time may significantly depend on the implementation. 3GPP NR has introduced an intermediate state referred to as INACTIVE state in addition to the LTE IDLE and CONNECTED states, for the purpose of control-plane latency reduction and improving device energy consumption. The new NR UE state provides a broadly configurable discontinuous reception (DRX) and thus contributes to different control-plane latencies for different traffic patterns and power saving modes. Since UE can be configured by the network with different DRX cycles in different conditions, this delay component is better characterized through simulation.
2. Downlink scanning and synchronization + broadcast channel acquisition: This step may further require taking into consideration the beam tracking procedures in the terminal side. On the other hand, 5G has introduced different forms of multi-connectivity which would allow skipping this step, for example, the broadcast information for the idle link can be delivered over one of the active links where UE is able to receive. With different configuration of multi-connectivity, broadcast information for the idle link might be delivered in different ways. In the case of control-plane/user-plane decoupling, detection of cells providing the user-plane needs to be taken into account. Note that the periodicity of certain common signals/channels for access may vary over different links.
3. Random access procedure: The delay due to the use of random access (physical channel) preamble for transmission of small payloads needs to be considered. Furthermore,

in case of mMTC traffic, the delay due to possible collision of random access preambles and their retransmissions must be taken into consideration.

4. Uplink synchronization: During the NR feasibility studies, it was observed that some candidate waveforms may relax the requirements for uplink synchronization by allowing asynchronous uplink access. In certain cases, the delay due to capability negotiation procedure and the associated HARQ retransmissions can also avoided, if the UE capability information is already available at the base station. In the case of control-plane/user-plane decoupling in two or more cells, the information about the capabilities of user-plane and control-plane needs to be obtained prior to the uplink access.
5. RRC connection establishment/reconfiguration + HARQ retransmissions: In some new 5G multi-connectivity scenarios, this step is considered as complete when the RRC connection is established with the target radio access network. When data bearers are split/aggregated at the core network, the RRC connection setup is required across all of those radio links.
6. Authorization and authentication/key exchange + HARQ retransmissions: Security information may already be available in the intermediate state introduced by 5G NR. Nevertheless, it is important to consider whether the security context has been discarded during the transition between the UE states.
7. Registration with the anchor gNB + HARQ retransmissions: In the case of user-plane/control-plane split, the UE often registers with the cell that provides the control plane. In that case, when user-plane and control-plane belong to different radio access networks, the UE may also register with both cells. In the case of control-plane establishment in multi-connectivity, the UE may register with multiple cells which provide control-plane functionalities. If the radio access network does not require registration, this component can be omitted, assuming that the context can be retrieved from the previous session.

Mobility: Mobility is the maximum UE speed at which certain QoS can be achieved or sustained. There are four mobility classes defined in IMT-2020 as follows:

- Stationary: 0 km/h
- Pedestrian: 0–10 km/h
- Vehicular: 10–120 km/h
- High-speed vehicular: 120–500 km/h

The upper range of high-speed vehicular mobility class is mainly considered for high-speed trains. A mobility class is supported, if the minimum required data channel link-level spectral efficiency on the uplink can be achieved. This assumes the user is moving at the maximum speed in that mobility class in each of the test environments. The mobility class compliance is evaluated under Indoor Hotspot-eMBB, Dense Urban-eMBB, and Rural-eMBB test environments. In order to verify satisfaction of the mobility requirement, one should run system-level simulations, identical to those for calculating the average spectral efficiency, at the maximum speed of each mobility class, starting with link-level simulations and an appropriate link-to-system mapping (also known as physical layer abstraction), in order to collect overall statistics for uplink SINR values (i.e., SINR distribution or geometry), and to calculate the CDF corresponding to those values for each test environment. The CDF for the test environment(s) is used to find the respective 50th percentile of SINR values. A new link-level simulation in the uplink is run for the selected test environment(s) under either NLoS or LoS propagation conditions with the purpose of obtaining link-level data rate and the residual packet error rate as a function of SINR. The link-level simulation uses air interface configuration(s) supported by the NR while taking into consideration the HARQ retransmissions, channel estimation, and phase noise impairments. The link-level spectral efficiency values (link data rate normalized by channel bandwidth) obtained from the previous step are compared using the set of SINR values obtained earlier for selected test environments. The mobility requirement is satisfied, if the value of the spectral efficiency is larger than or equal to the corresponding threshold value and if the residual decoded packet error rate is less than 1%, for all selected test environments. For the selected test environment, it is sufficient if one of the spectral efficiency values (of either NLoS or LoS propagation model) satisfy the minimum requirement. A similar methodology can be used for the assessment of the mobility requirement in the downlink.

Reliability: Reliability refers to the capability of transmitting certain amount of user traffic within a predetermined time duration with high probability of success. In other words, reliability is the success probability of transmitting a layer 2/3 packet within a permissible time, which is the time it takes to deliver the small payload from the source layer 2 SDU ingress point to the destination layer 2/3 SDU egress point under certain channel conditions. This metric was defined for the purpose of system evaluation in the URLLC usage scenario. The minimum requirement for the reliability is a success probability in the order of $1-10^{-5}$ for transmitting a layer 2 protocol data unit of 32 bytes (i.e., 20 bytes application data + protocol overhead) within 1 ms in channel quality corresponding to the edge of the coverage (cell-edge conditions) for the Urban Macro-URLLC test environment. Note that target coverage and reliability requirement depend on the deployment scenario and type of operation (e.g., V2X with certain average inter-vehicle speed and distance).

In order to verify the compliance, one must run system-level simulations, identical to those for average spectral efficiencies. Using link-level simulations and an appropriate link-to-system mapping for the desired mobility class and the test environment(s), overall statistics for (downlink and/or uplink) SINR values are obtained and the CDF of SINR distribution is calculated for each test environment. Using a new downlink and/or uplink link-level simulation, the residual packet error rate within the maximum delay time as a function of SINR is calculated. The satisfaction of reliability requirement is ensured, if the fraction of messages that are correctly delivered within the permissible delay bound is larger than or equal to the required success probability.

Energy Efficiency: Network energy efficiency is the capability of a network to minimize the energy consumption for radio access (and/or core) network operations without affecting the performance of the network. Device energy efficiency (RAN aspects) is the capability of the network to minimize the power consumed by a user device communication subsystem (modem). Network and the device energy efficiencies are related in the sense that they mutually affect efficient data transmission in a loaded network. The energy consumption of a device can be reduced when there is no data transmission to or from the device. Efficient data transmission in a loaded network is demonstrated by the average spectral efficiency. The device energy consumption can be estimated based on the ratio of sleep and wake cycles of the device. The sleep ratio is the fraction of unoccupied time slots (for the network) or sleeping time (for the device) in a period of time corresponding to the cycle of the control signaling (for the network) or the periods of DRX (for the device) when no user data transfer takes place. Furthermore, the sleep duration, that is, the continuous periods of time with no transmission (for network and device) and reception (for the device), should be sufficiently long.

Network energy efficiency (both quantitative and qualitative aspects) is considered as a basic principle in the NR system design. This is the capability to minimize energy consumption for radio access while providing the highest possible area traffic capacity. The target is to design the system with the ability to efficiently deliver data, and the ability to provide sufficiently granular transmission cycles, so that transmission can be discontinued when there are no data to transmit. It further includes the ability to provide the operator with sufficient flexibility to adapt sleep cycles of base stations depending on the load, services, and coverage area. In order to quantitatively express and compare this KPI, the following network energy efficiency metric in bits per Joule is defined. This definition would allow quantitative comparison of different solutions or energy-saving mechanisms on the basis of efficiency merits, when their impact is not obvious from qualitative analysis. In order to evaluate the overall improvements achieved in 3GPP NR system, the network energy efficiency metric is defined as follows:

$$EE_{global} = \sum\nolimits_{scenario_k} b_k EE_{scenario_k}$$
$$EE_{scenario_l} = \sum\nolimits_{load\ level_l} \frac{a_l V_l}{EC_l}$$

In the above equation b_k refers to the weights of each deployment scenario where the network energy efficiency is evaluated. Furthermore, V_l, EC_l, a_l denote the traffic per second served by a base station (in bits/s); the power consumed by a base station to serve V_l in Watt = Joule/s, and the weight for each traffic load level, respectively. In other words

$$EE_{scenario_k} = \sum_{load\ level_k} a_k \frac{V_k(\text{traffic per second served by a base station})}{EC_k(\text{power consumed by a base station to serve} V_k)} \quad \text{Bits/Joules}$$

The UE battery life can be measured as the total time that a UE can operate with one charge. For mMTC, the UE battery life in extreme coverage is based on the activity of the mobile-originated data transfer consisting of 200 bytes of uplink traffic per day followed by 20 bytes of downlink traffic at maximum coupling loss of 164 dB (a coverage criterion), assuming that the UE battery has a power-rating of 5 Watts-hour. The desired UE battery life for mMTC use cases is between 10 and 15 years. This metric is analytically evaluated. The UE energy efficiency is an indication of how a UE can sustain high mobile broadband data rate while minimizing the UE modem energy consumption.

Coverage: Downlink or uplink coverage in a cellular system is measured by the maximum coupling loss between the transmit antenna of the transmitter and the receive antenna of the receiver. More specifically, the maximum coupling loss in the uplink or downlink between the user device and the base station antenna connectors is defined for a reference data rate of 160 bps, where the data rate is measured at the egress/ingress point of the MAC layers in the uplink and downlink. The target for 3GPP NR coverage is maximum coupling loss of 164 dB. Link budget and/or link-level analysis is used as the evaluation methodology to derive the parameters necessary for calculation of the coverage. In practice, the coupling loss is defined as the total long-term channel loss over the link between the UE antenna ports, and the base station antenna ports and includes antenna gains, path loss, shadowing, and body loss. The maximum coupling loss is the maximum value of the coupling loss at which a minimum SINR can be achieved and a particular service can be delivered. It is defined in the downlink/uplink direction as follows:

$$MaxCL_{UL} = \max(Tx\ \ power_{UL}) - BS\ \ Sensitivity$$
$$MaxCL_{DL} = \max(Tx\ \ power_{DL}) - UE\ \ Sensitivity$$

The parameter *MaxCL* denotes maximum coupling loss and is calculated using link budget analysis and based on the data obtained though link-level simulations. Other parameters including max (*Tx power*) and *BS/UE Sensitivity* depend on implementation and the deployment scenario [14].

For extreme coverage scenario and for a basic mobile broadband service characterized by a downlink data rate of 2 Mbps and an uplink data rate of 60 kbps for stationary users, the target maximum coupling loss is 140 dB. In this case and for most users, a downlink data rate of 384 kbps is acceptable. For a basic eMBB service characterized by a downlink data rate of 1 Mbps and an uplink data rate of 30 kbps for stationary users, the target maximum coupling loss is 143 dB. In this case, the downlink and uplink control channels must satisfy the performance requirements at this coupling loss.

Handover Interruption Time: This metric refers to the shortest time interval during which a user terminal cannot exchange user-plane packets with any base station when traversing boundary of the two or more cells in a cellular network. The handover interruption time includes the time required to execute any radio access network procedure, RRC signaling protocol, or other message exchanges between the mobile station and the radio access network. This benchmark only applies to eMBB and URLLC usage scenarios where effectively a seamless handover interruption time (zero second) is required.

3.2 Test Environments

In the course of development of technical standards, satisfaction of the service and system requirements by candidate technologies are often verified under certain channel propagation conditions, cell layouts, and the technical parameter set pertaining to the deployment scenario under consideration. A test environment is thereby an approximate representation of a practical deployment scenario with predefined channel propagation model, topology, and system configuration. Appropriate channel models are often developed and used in the evaluations of the candidate technologies to allow realistic modeling of the propagation conditions for the radio transmissions in different test environments. The channel models need to cover all required test environments and usage scenarios where the 5G networks are going to be deployed.

Several test environments related to the main use cases of 5G (i.e., eMBB, mMTC, URLLC) have been defined where the candidate 5G radio access technologies will be evaluated, which include the following [32]:

- *Indoor Hotspot-eMBB* test environment represents an isolated indoor environment with very high user density, which is typically found at office buildings and/or in shopping malls populated by stationary and pedestrian users. This scenario is characterized by small coverage areas per site or TRP, high user throughput, high user density, and consistent user experience in an indoor environment. The typical radial coverage in this environment is less than 10 m.
- *Dense Urban-eMBB* test environment exemplifies an urban environment with high user density and large amount of user traffic where pedestrians and users in vehicles are covered in an outdoor and outdoor-to-indoor setting. This interference-limited scenario mainly focuses on overlaying macro-TRPs with or without micro-TRPs, which is typically found in city centers and dense urban areas. The radial coverage in this environment is typically less than 100 m.
- *Rural-eMBB* test environment represents a rural environment with larger and continuous wide-area coverage, supporting pedestrian, vehicular, and high-speed vehicular users. The main feature of this scenario is unspotted wide-area coverage supporting high-speed vehicles. This scenario is noise-limited and/or interference-limited and uses a setup comprising macro-TRPs. The radial coverage in this environment is typically less than 2500 m.

- *Urban Macro-mMTC* test environment characterizes an urban macro environment with blanket coverage over a large cell serving a large number of connected (machine-type) devices. The key characteristics of this interference-limited scenario are continuous and ubiquitous coverage in urban areas comprising macro-TRPs. The typical coverage in this environment is less than 250 m (in sub-GHz frequency bands an inter-site distance of 1732 m is assumed).
- *Urban Macro-URLLC* test environment symbolizes an urban macro environment targeting URLLC services.
- *Extreme Long Distance* deployment scenario is defined to provision services for very large areas with low density of users that could be humans or machines (e.g., distribution of sensors over a large geographical area). The main features of this scenario include macro-cells with very large coverage area supporting basic data rates and voice services, with low-to-moderate user throughput and low user density.

The mapping of the IMT-2020 test environments to the usage scenarios is shown in Table 1. This mapping is based on the similarities between the key features of each test environment and those of the use case. It must be noted that a test environment represents a certain system/geometrical configuration to evaluate the operation of a candidate radio access technology in a particular usage scenario and to verify satisfaction of the requirements for that usage scenario. A test environment typically signifies a practical deployment scenario with limited parameter set that are selected to model the realistic behavior of the radio access technology as accurate as possible. Therefore, the choice of parameters and their values is very important to the accuracy of the modeling of deployment scenarios. In the past two decades, 3GPP and ITU-R technical working groups have made significant efforts to define the test environments and their associated parameter sets to facilitate characterization of the radio access technologies.

The selected parameter sets corresponding to the test environments are shown in Table 2. The complete list of parameters of each test environment is provided in reference [10]. The configurations shown in Table 2 are chosen to ensure fulfillment of the performance requirements in the associated test environment while taking into consideration the services and the practical limitations (e.g., form factor, hardware, and installation) of the gNB/TRP and/or the terminals operating in that environment. Furthermore, DL + UL bandwidth in this table refers to symmetric bandwidth allocations between DL and UL in frequency

Table 1: Mapping of test environments and usage scenarios [10].

Usage Scenarios	eMBB	mMTC	URLLC		
Test environments	Indoor Hotspot-eMBB	Dense Urban-eMBB	Rural-eMBB	Urban macro-mMTC	Urban macro-URLLC

Table 2: Selected parameters of the test environments [10,14].

Deployment Scenario	Frequency Band (GHz)	Maximum System Bandwidth (MHz)	Maximum Number of Tx/Rx Antennas (TRP)	Maximum Number of Tx/Rx Antennas (UE)	Mobility (km/h)
Indoor hotspot	4 30 70	200 (DL + UL) 1000 (DL + UL)	256Tx × 256Rx 256Tx × 256Rx 1024Tx × 1024Rx	8Tx × 8Rx 32Tx × 32R x64Tx × 64Rx	3 (Indoor)
Dense urban	4 30	200 (DL + UL) 1000 (DL + UL)	256Tx × 256Rx 256Tx × 256Rx	8Tx × 8Rx 32Tx × 32Rx	3 (Indoor) 30 (Outdoor)
Rural	0.7 4	20 (DL + UL) 200 (DL + UL)	64Tx × 64Rx 256Tx × 256Rx	4Tx × 4Rx 8Tx × 8Rx	3 (Indoor) 120 (Outdoor)
Urban macro	2 and 4 30	200 (DL + UL) 1000 (DL + UL)	256Tx × 256Rx 256Tx × 256Rx	8Tx × 8Rx 32Tx × 32Rx	3 (Indoor) 30 (Outdoor)
High speed	4 30, 70	200 (DL + UL) 1000 (DL + UL)	256Tx × 256Rx 256Tx × 256Rx	8Tx × 8Rx 32Tx × 32Rx	500 (Outdoor)
Extreme long distance coverage	0.7	40 (DL + UL)	N/A	N/A	160 (Outdoor)
Urban macro-mMTC	3 0.7 (2.1)	10 (50)	64Tx × 4Rx	2Tx × 2Rx	3 (Indoor/Outdoor)

division duplex (FDD) systems and the aggregated system bandwidth used for either DL or UL via switching in time-domain in TDD systems.

3.3 High-Level Architectural Requirements

The desire to support different UE types, services, and technologies in 5G has been driving the development of the NR standards in 3GPP. The key objective of the 5G systems is to support new deployment scenarios and to address the requirements of diverse market segments. As we mentioned earlier, 5G systems have certain key characteristics such as support for myriad of radio access technologies, scalable and customizable network architecture, rigorous KPIs, flexibility and programmability, and resource efficiency (both on the user and control planes) as well as seamless mobility in densely populated heterogeneous environments and support for real-time and non-real-time multimedia services and applications with improved quality of experience (QoE).[14] Thus the architecture of the next-generation

[14] QoE is an important measure of the end-to-end performance at the service level from the user's perspective and an important metric for the design of systems. QoE is related to but differs from QoS, which embodies the notion that hardware and software characteristics can be measured, improved, and guaranteed. In contrast, QoE expresses user satisfaction both objectively and subjectively. It is often used in information technology and consumer electronics. QoE, while not always numerically quantifiable, is the most significant single factor in the real-world evaluation of the user experience. Major factors that affect QoE include cost, reliability, efficiency, privacy, security, interface user-friendliness, and user confidence.

networks is required to address some key design principles in order to meet the above-mentioned requirements. The following is a list of key architectural focus areas and their associated requirements that were identified by system architecture working group in 3GPP in the early stages of 5G standards development [13,21]:

1. Support of the 3GPP NR access, evolution of LTE, legacy 3GPP access, and non-3GPP access such as Wi-Fi networks is a key requirement for the new architecture. As part of non-3GPP access types, trusted and untrusted WLAN access support is necessary for heterogeneous networks. Satellite radio access network needs to be supported in order to enable coverage of rural and remote areas using satellite access nodes. The 5G system supports most of the existing LTE evolved packet system services in addition to the new services that includes seamless handover between the NR and the legacy 3GPP radio access networks. Interworking between the new radio and LTE is required considering the necessity for efficient inter-RAT mobility and aggregation/distribution of data flows via dual connectivity between LTE and the new radio in NSA scenarios. This requirement applies to both collocated and non-collocated site deployments.
2. Unified authentication framework for different access systems is required to simplify access to the next-generation networks via any of their constituent radio access networks. The authentication protocol is responsible for the validation of the user identity that is presented to the network when a UE requests to receive service(s) from the next-generation network.
3. Support of several simultaneous connections of a UE via multiple access technologies served by a single 5GC network is another important requirement. For UEs that can be simultaneously connected to both 3GPP access and non-3GPP access, the next-generation system should be able to exploit the availability of multiple radio links in a way that improves the user experience, optimizes the traffic distribution across various access links, and enables the provision of new high-data-rate services. The new RAN architecture is also required to support operator-controlled side-link (D2D and V2X) operation in both in-coverage and out-of-coverage scenarios. The UE can connect to the network directly (direct network connection), connect via another UE acting as a relay (indirect network connection), or connect using both types of connections. These user terminals can be anything from a wearable, monitoring human biometrics, to non-wearable devices that communicate in a personal area network such as a set of home appliances (e.g., smart thermostat and entry key) or the electronics in an office environment (e.g., smart printers). The relay UE can access the network using 3GPP or non-3GPP access (e.g., WLAN access, fixed broadband access). 3GPP and non-3GPP radio technologies and fixed broadband technologies over licensed bands or unlicensed bands can be supported as connectivity options between the remote UE and the relay UE.
4. Separation of control- and user-plane functions is a key step toward enabling SDN-controlled networks. This feature is also an enabler for the network slicing and the NFV

infrastructure. The next-generation network is built on NFV and software-defined networking to reduce the total cost of ownership and to improve operational efficiency, energy efficiency, and simplicity and flexibility of the network for offering new services.

5. Support of network sharing and network slicing is yet another important requirement. Network slicing allows the operators to provide several customized network services over the same physical network. For example, there can be different requirements on each network slice functionality (e.g., priority, charging, policy control, security, and mobility), differences in performance requirements (e.g., latency, mobility, availability, reliability, and data rates), or each network slice can serve only specific users (e.g., multimedia priority service users, public safety users, enterprise customers, and mobile virtual network operators). A network slice can provide the functionality of a complete network including radio access and core network functions. Each physical network can support one or more network slices [21]. It must be possible to verify if the UE is allowed to access a specific network slice. The management of life cycle of network slice and network function instances is also a key aspect in the overall framework for network slicing in 3GPP.
6. Energy efficiency is a critical issue in 5G systems. The potential to deploy systems in areas without a reliable energy source requires new methods of managing energy consumption not only in the UEs but also through the entire 5G network. The 5G access network is required to support an energy-saving mode which can be activated/deactivated either manually or automatically and the service can be restricted to a group of users (e.g., public safety user, emergency callers). When in energy-saving mode, the inactive UEs transmit power may be reduced or turned off or their latency and jitter constraints may be relaxed with no impact on active users or applications in the network. Small form factor UEs also typically have a small battery, and this not only imposes restrictions on power optimization but also on how the energy is consumed.
7. The unambiguous definition of the functional split between the next-generation core and the access network(s) is an important consideration for the support of heterogeneous access types where new interfaces between the next-generation core and the new 3GPP RAN need to be specified.
8. Mobility management is a key feature of 5G to support UEs with different mobility conditions which include stationary UEs during their entire usable life (e.g., sensors embedded in an infrastructure), stationary UEs during active periods, but nomadic between activations (e.g., fixed access), limited mobility within a constrained area (e.g., robots in a factory), and fully mobile UEs. Some applications require the network to guarantee seamless mobility, transparent to the application layer, in order to avoid service interruption, and to ensure service continuity. The increasing multimedia broadband data traffic necessitates offloading of IP traffic from the 5G network to traditional IP routing networks via an IP anchor node close to the network edge. As the UE

traverses the network, changing the IP anchor node may be needed in order to avoid traffic congestion in highly loaded networks, to reduce end-to-end latency, and to provide better user experience. The flexible nature of the 5G systems support different mobility management options that minimize signaling overhead and optimize network access for different categories of UEs.

3.4 System Performance Requirements

The groundbreaking improvements initiated by 5G are achievable with the introduction of new technologies, both in the access and the core networks, which include flexible and scalable allocation of network resources to various services and applications. In addition to increased flexibility and optimization, a 5G system must support stringent criteria for latency, reliability, throughput, etc. Enhancements in the air-interface contribute to meeting these KPIs as do improvements in the core network, such as network slicing, in-network caching, and hosting services closer to the network edges [21]. 5G systems further support new and emerging business models. Drivers for the 5G KPIs include services such as unmanned aerial vehicle (UAV) remote control, VR/AR,[15] and industrial automation. Increasing network flexibility would allow support of self-contained enterprise networks, which are installed and maintained by network operators while being managed by the enterprise. Enhanced connection modes and improved security enable support of massive IoT use cases, where those are expected to include numerous UEs of different categories sending and receiving data over the 5G network. The essential capabilities for providing this level of flexibility include network slicing, network capability exposure, scalability, and diverse mobility. The requirements for other network operations are meant to address the necessary control and user-plane resource allocation and utilization efficiencies, as well as network configurations that optimize service delivery by minimizing routing between end users and application servers.

Unlike previous generations of wireless systems, the main objective of 5G has been to develop a unified system that can be exclusively configured and optimized for different use cases. Increasing user expectations and diversified and challenging service requirements necessitate a coherent approach to technology development and deployment. In the previous generations of wireless technology, functionality was primarily provided by network equipment, and devices were primarily used to access network services. However, in the case of 5G, it is recognized that the boundary between functionality provided by the network and

[15] VR is the term used to describe a three-dimensional, computer-generated environment which can be explored and interacted with by a person. That person becomes part of this virtual world or is immersed within this environment and whilst there, is able to manipulate objects or perform a series of actions. AR is the integration of digital information with the user's environment in real time. Unlike virtual reality, which creates a totally artificial environment, augmented reality uses the existing environment and overlays new information on top of it.

functionality provided by the device is subtle. Virtualization of network functions, the role of applications, and the ubiquitous availability of increasingly sophisticated smart devices will only accelerate this trend. Nonetheless, there are some cases where the desired functionality is expected to be exclusively provided in the network [45].

The eMBB use cases require a number of new KPIs related to higher data rates, user density, user mobility, and coverage. Higher data rates are driven by the increasing use of data for services such as streaming (e.g., video, music, and user generated content), interactive services, and some IoT applications. These services have stringent requirements for user experienced data rates and end-to-end latency in order to guarantee good user experience. In addition, increased coverage in densely populated areas such as sports arenas, urban areas, and transportation hubs is essential for nomadic and mobile users. The new KPIs on traffic and connection density would enable transport of large amount of user traffic per area (traffic density) and transport of data for a large number of connections (connection density). A large group of UEs are expected to support variety of services which exchange either a very large (e.g., streaming video) or very small (e.g., data burst) amount of data. The 5G systems are designed to manage this variability in an extremely resource and energy efficient manner. All of these scenarios introduce new requirements for indoor and outdoor deployments, local area connectivity, high user density, wide-area connectivity, and UEs travelling at high speeds. Another aspect of 5G KPIs includes requirements for various combinations of latency and reliability, as well as higher position determination accuracy. These KPIs are further driven by support for commercial and public safety services. Support for mMTC presents several new requirements in addition to those for the eMBB use cases. The proliferation of the connected things introduces a need for significant improvements in resource efficiency in all system components (e.g., UEs, IoT devices, radio access network, core network). The 5G systems are intended to extend their capabilities to meet KPIs required by emerging mission-critical applications. For these advanced applications, the requirements, such as data rate, reliability, latency, communication range, and speed, are made more stringent. Fig. 13 illustrates the latency and throughput requirements of the prominent 5G applications (see also Table 3).

3.5 Service Requirements

The 5G systems are required to support the existing LTE network services in addition to the newly introduced services, which means that the existing EPS services can be accessed via the new 5G access technologies. The following is a list of important service requirements for 5G networks:

- *Multimedia Broadcast/Multicast Service:* The proliferation of video streaming services, software delivery over wireless network, group communications, and multicast/broadcast IoT applications have created a demand for flexible and dynamic allocation of

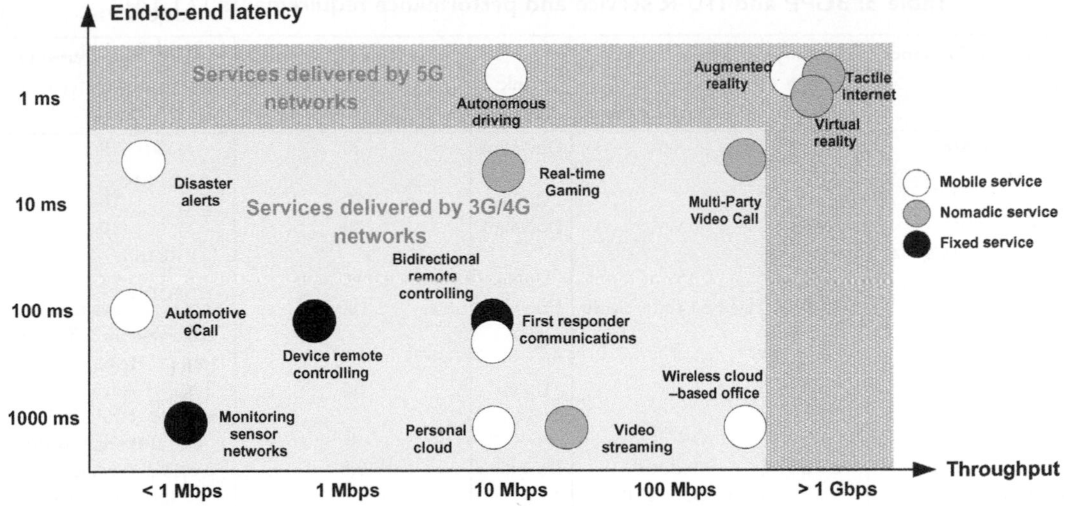

Figure 13

Performance requirements of some 5G applications [44].

radio resources between unicast and multicast services within the network as well as support for a standalone deployment of multicast/broadcast network. Moreover, enabling such a service over a network for a wide range of intersite distances between the radio base stations will enable a more efficient and effective delivery system for real-time and streaming multicast/broadcast content over wide geographical areas as well as in specific geographical areas spanning a limited number of base stations. A flexible multicast/broadcast service will allow the 5G system to efficiently deliver such services.

- *Location/Positioning Service:* The NR systems enable and improve, if necessary, advanced positioning techniques such as RAN-assisted (Cell-ID, E-Cell ID, OTDOA, and UTDOA)[16] and non-RAN-assisted (GNSS, Bluetooth, WLAN, terrestrial beacon systems (TBS),[17] and sensor-based location determination) that have been specified and used in LTE systems. In addition, support of D2D-based positioning techniques has also been considered in NR. The NR positioning schemes exploit high bandwidth, massive antenna systems, new network architecture, and functionalities as well as deployment of massive number of devices to support indoor and outdoor location-based services. Higher precision positioning is characterized by ambitious system requirements for

[16] Cell-ID, E-Cell ID, OTDOA, UTDOA, and GNSS denote cell identifier, enhanced cell identifier, observed time difference of arrival, uplink time difference of arrival, and global navigation satellite system and are commonly used methods for positioning in 3GPP systems.

[17] TBS consist of a network of terrestrial beacons broadcasting signals for positioning purposes. They may use dedicated, unshared spectrum or spectrum shared with other users, including FDD and TDD licensed spectrum.

Table 3: 3GPP and ITU-R service and performance requirements [11,14].

Service/Performance Metric	Category Usage Scenario	Test Environment	Link Direction	ITU-R Requirements	3GPP Requirements
Peak data rate (Gbps)	eMBB	N/A	Downlink	20	20
			Uplink	10	10
Peak spectral efficiency (bits/s/Hz)	eMBB	N/A	Downlink	30	30
			Uplink	15	15
User experienced data rate (Mbps)	eMBB	Dense Urban-eMBB	Downlink	100	3 × IMT-Advanced cell-edge spectral efficiency at 1 GHz Bandwidth
			Uplink	50	3 × IMT-Advanced cell-edge spectral efficiency at 1 GHz bandwidth
5th Percentile user spectral efficiency (bits/s/Hz)	eMBB	Indoor Hotspot-eMBB	Downlink	0.3	3 × IMT-Advanced cell-edge spectral efficiency in InH[a]
			Uplink	0.21	3 × IMT-Advanced Cell-edge Spectral Efficiency in InH
	eMBB	Urban macro	Downlink	N/A	3 × IMT-Advanced cell-edge spectral efficiency in UMa
			Uplink	N/A	3 × IMT-Advanced cell-edge spectral efficiency in UMa
	eMBB	Dense Urban-eMBB	Downlink	0.225	3 × IMT-Advanced cell-edge spectral efficiency in UMi
			Uplink	0.15	3 × IMT-Advanced cell-edge spectral efficiency in UMi
	eMBB	Rural-eMBB	Downlink	0.12	3 × IMT-Advanced cell-edge spectral efficiency in RMa
			Uplink	0.045	3 × IMT-Advanced cell-edge spectral efficiency in RMa
Average spectral efficiency (bits/s/Hz/TRP)	eMBB	Indoor Hotspot-eMBB	Downlink	9	3 × IMT-Advanced cell spectral efficiency in InH
			Uplink	6.75	3 × IMT-Advanced cell spectral efficiency in InH
	eMBB	Dense Urban-eMBB	Downlink	7.8	3 × IMT-Advanced cell spectral efficiency in UMi
			Uplink	5.4	3 × IMT-Advanced cell spectral efficiency in UMi
	eMBB	Rural-eMBB	Downlink	3.3	3 × IMT-Advanced cell spectral efficiency in RMa
			Uplink	1.6	3 × IMT-Advanced cell spectral efficiency in RMa

(*Continued*)

Table 3: (Continued)

Service/Performance Metric	Category Usage Scenario	Test Environment	Link Direction	ITU-R Requirements	3GPP Requirements
Area traffic capacity (Mbit/s/m^2)	eMBB	Indoor Hotspot-eMBB	Downlink	10	No explicit requirement
User-plane latency (ms)	eMBB	N/A	Uplink and downlink	4	4 *The system user plane should support RTT of 600, 180, and 50 ms in the case of GEO, MEO, LEO satellite systems, respectively*
	URLLC	N/A	Uplink and downlink	1	0.5
Control-plane latency (ms)	eMBB	N/A	Uplink and downlink	20	10 *The system control plane should support RTT of 600, 180, and 50 ms in the case of GEO, MEO, and LEO satellite systems, respectively.*
	URLLC	N/A	Uplink and downlink	20	10
Connection density (devices/km^2)	mMTC	Urban Macro-mMTC	Uplink	1000,000	1000,000
Energy efficiency	eMBB	N/A	N/A	Capability to support a high sleep ratio and long sleep duration	No explicit requirement
Reliability	URLLC	Urban Macro-URLLC	Uplink or downlink	$1-10^{-5}$ success probability of transmitting a layer 2 PDU of size 32 B within 1 ms in channel quality of coverage edge	$1-10^{-5}$ for a packet of 32 bytes with a user-plane latency of 1 ms *For eV2X cases (direct communication via side-link or when the packet is relayed via gNB), the requirement is $1-10^{-5}$ for packet size 300 bytes and user-plane latency 3-10 ms.*
Mobility classes	eMBB	Indoor Hotspot-eMBB	Uplink	Stationary, pedestrian	No explicit requirement
	eMBB	Dense Urban-eMBB	Uplink	Stationary, pedestrian, vehicular (up to 30 km/h)	No explicit requirement
	eMBB	Rural-eMBB	Uplink	Pedestrian, vehicular, high-speed vehicular	No explicit requirement
Mobility traffic channel link	eMBB	Indoor Hotspot-eMBB	Uplink	1.5 (10 km/h)	No explicit requirement
Data rates (bits/s/Hz)	eMBB	Dense Urban-eMBB	Uplink	1.12 (30 km/h)	No explicit requirement
	eMBB	Rural-eMBB	Uplink	0.8 (120 km/h) 0.45 (500 km/h)	No explicit requirement No explicit requirement
Handover interruption time (ms)	eMBB and URLLC	N/A	N/A	0	0

(Continued)

Table 3: (Continued)

Service/Performance Metric	Category Usage Scenario	Test Environment	Link Direction	ITU-R Requirements	3GPP Requirements
Bandwidth (MHz/ GHz) and scalability	N/A	N/A	N/A	At least 100 MHz	No explicit requirement
				Up to 1 GHz	No explicit requirement
				Support of multiple different bandwidth values	No explicit requirement

[a]The IMT-Advanced minimum requirements for 5th percentile user spectral efficiencies in InH, UMa, UMi, and RMa test environments can be found in reference [8].

positioning accuracy. One use case where higher precision positioning capability is critically needed is collision avoidance in a busy street where each vehicle must be aware of its own position, the positions of neighboring vehicles, and their expected driving paths and movements, in order to reduce the risk of collisions. In another use case on the factory floor, it is important to locate moving objects such as machinery or parts to be assembled with sufficiently high precision. Depending on the use case, the 5G system is required to support the use of 3GPP and non-3GPP technologies to achieve higher accuracy indoor/outdoor positioning. The corresponding positioning information must be acquired with proper timing, to be reliable, and to be available (e.g., it is possible to unambiguously determine the position). The 5G UEs must be able to share positioning information between each other or with a cloud-based controller if the location information cannot be processed or used locally. Table 4 summarizes the positioning requirements for the latter use case. Another aspect of 5G KPIs includes requirements for various combinations of latency and reliability, as well as higher accuracy for positioning in order to support mission-critical services such as public safety, emergency communications, and public warning/emergency alert systems. These KPIs are driven by support for both commercial and non-commercial public safety services. On the commercial side, industrial control, industrial automation, UAV control, and AR exemplify those services. Services such as UAV control will require more precise positioning information that includes altitude, speed, and direction, in addition to planar coordinates.

- *Service Continuity and Reliability:* Communication service availability is defined as percentage of time where the end-to-end communication service is delivered according to an agreed QoS, divided by the amount of time the system is expected to deliver the an end-to-end service according to the specification in a specific area. Note that the end point in an end-to-end service is assumed to be the communication service interface. The communication service is considered unavailable, if it does not meet the pertinent QoS requirements. The system is considered unavailable in case an expected message is not received within a specified time, which, at minimum, is the sum of end-to-end

Table 4: Performance requirements for higher precision positioning service [21].

Service	Position Acquisition Time (ms)	Survival Time (s)	Availability (%)	Dimension of Service Area (m)	Position Accuracy (m)
Mobile objects on factory floor	500	1	99.99	500 × 500 × 30	0.5

latency, jitter, and the survival time.[18] Reliability expressed in percent is the amount of transmitted network layer packets successfully delivered to a given node within the time constraint required by the targeted service, divided by the total number of sent network layer packets. Communication service availability and reliability are well-defined terms that have used not only within 3GPP but also in vertical industries. Communication service availability addresses the availability of a communication service, in vertical applications in accordance to IEC 61907,[19] whereas reliability is related to the availability of the communication network (see Fig. 14).

As depicted in Fig. 14, reliability covers the communication-related aspects between two nodes which are often the end nodes, while communication service availability addresses the communication-related aspects between two communication service interfaces. In other words, the gap between the two concepts is the communication interface. This might seem to be a small difference, but this difference can lead to conditions where reliability and communication service availability have different values.

- *Priority, QoS, and Policy Control:* The 5G networks are expected to support a wide range of commercial services and regional/national regulatory services with appropriate access prioritization. Some of these services share common QoS characteristics such as latency and packet loss rate but may have different priority requirements. For example, UAV control and air traffic control may have stringent latency and reliability requirements but not necessarily the same priority requirements. In another example voice-based services for multimedia priority service (MPS)[20] and emergency services share

[18] Survival time is the time that an application consuming a communication service may continue without an anticipated message.

[19] IEC 61907 2009 provides guidance on dependability engineering of communication networks. It establishes a generic framework for network dependability performance, provides a process for network dependability implementation, and presents criteria and methodology for network technology designs, performance evaluation, security consideration, and quality of service measurement to achieve network dependability performance objectives (see https://webstore.iec.ch/publication/).

[20] MPS, supported by 3GPP, is a set of services and features which create the ability to deliver calls or complete sessions of a high-priority nature from mobile-to-mobile networks, mobile-to-fixed networks, to fixed-to-mobile networks. MPS provides broadband IP-based multimedia services (IMS-based and non-IMS-based) over wireless networks in support of voice, video, and data services. Network support for MPS will require end-to-end priority treatment in call/session origination/termination including the non-access stratum (NAS) and access stratum (AS) signaling establishment procedures at originating/terminating network side as well as resource allocation in the core and radio networks for bearers. The MPS will also require end-to-end priority treatment in case of roaming if supported by the visiting network and if the roaming user is authorized to receive priority service.

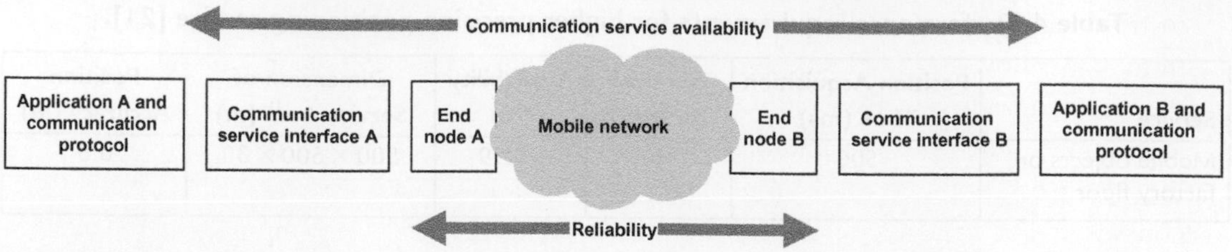

Figure 14
Illustration of the relationship between communication service availability and reliability [21].

common QoS characteristics as applicable for normal public voice communications, yet they may have different priority requirements. The 5G network supports mechanisms that allow decoupling of the priority of a particular communication from the associated QoS characteristics such as latency and reliability to allow flexibility and configurability to support different priority services in operators' networks while adhering to their network policies and the corresponding national/regional regulations. The network needs to support flexible means to make priority decisions based on the state of the network (e.g., during disaster events and network congestion) recognizing that the service priorities may change during a crisis. The priority of any service may be different for each user depending on the operational requirements and regional/national regulations. The 5G systems provide flexible means to prioritize and enforce prioritization among the services (e.g., MPS, emergency, medical, public safety) and among the users of those services. The traffic prioritization may be enforced by adjusting resource utilization or preempting lower priority traffic. The networks are typically capable of providing the required QoS (e.g., reliability, latency, and bandwidth) for a service and are able to prioritize resources when necessary in order to meet the service requirements. The existing QoS and policy frameworks manage latency and improve reliability by traffic engineering. In order to support 5G service requirements, it is necessary for the 5G network to offer QoS and policy control for reliable communication with latency required for a service and enable the resource adaptations as necessary. Also, as 5G network is expected to operate in a heterogeneous environment with multiple access technologies, multiple types of UE, etc., it should support a harmonized QoS and policy framework that can be applied to multiple access networks. Unlike the QoS control in EPS which only covers RAN and core network, the 5G network end-to-end QoS (e.g., RAN, backhaul, core network, network-to-network inter-connect) is required to achieve the 5G user experience (e.g., ultra-low latency, ultra-high bandwidth).

- *V2X Services:* Platooning enables the vehicles to dynamically form a group travelling together. All the vehicles in the platoon receive periodic data from the leading vehicle, in order to carry on platoon operations. This information allows the distance between vehicles to become extremely small, that is, the inter-vehicle distance when translated to time can be very small and in the order of fraction of a second. Platooning applications may allow the participating vehicles to be autonomously driven. Advanced driving enables semi-automated or fully automated driving where longer inter-vehicle distance is assumed. Each vehicle and/or road side unit (RSU)[21] shares data obtained from its local sensors with vehicles in proximity, thus allowing vehicles to coordinate their courses and/or maneuvers. In addition, each vehicle shares its driving plan with vehicles in proximity. The advantage of this feature is safer driving, collision avoidance, and improved traffic efficiency. Extended sensors enable the exchange of raw or processed data gathered through local sensors or live video data among vehicles, road site units, UEs of pedestrians, and V2X application servers. The vehicles can enhance the perception of their environment beyond what their own sensors can detect and have a more holistic view of the local conditions. Remote driving enables a remote driver or a V2X application to operate a remote vehicle for those passengers who cannot drive themselves or a remote vehicle located in dangerous environments. For a case where variation is limited and routes are predictable, such as public transportation, driving based on cloud computing can be used.

Intelligent transport systems (ITS) embrace a wide range of applications that are intended to increase travel safety, minimize environmental impact, improve traffic management, and maximize the benefits of transportation to both commercial users and the general public. Over the years, the emphasis on intelligent vehicle research has turned into cooperative systems in which the traffic participants (vehicles, bicycles, pedestrians, etc.) communicate with each other and/or with the infrastructure. In this context, a cooperative ITS scheme is a technology that allows vehicles to be wirelessly connected to each other, to the infrastructure and other parts of the transport network. In addition to what drivers can immediately see around them, and what vehicle sensors can detect, all parts of the transport system will increasingly be able to share information to improve decision-making in the system. Thus this technology can improve road safety through avoiding collisions and further can assist the transportation network by reducing congestion and improving traffic flows. Once the basic technology is in place as a platform, subsequent applications can be developed. Cooperative ITS can greatly increase the quality and reliability of information available about vehicles, their location and the driving environment. In the future, cars will know the location of road works and the switching phases of traffic lights ahead, and they will be able to react accordingly. This will make travel safer and more convenient. On-board driver assistance, coupled with two-way communication between vehicles and between cars and road infrastructure, can help drivers to better control their vehicle and hence have positive effects in terms of safety and traffic efficiency. RSUs play an important role in this

[21] RSU is a communication/computing device located on the roadside that provides connectivity support to passing vehicles.

technology. Vehicles can also function as sensors reporting weather and road conditions including incidents. In this case, cars can be used as information sources for high-quality information services. RSUs are connected to the traffic control center (TCC) for management and control purposes. The traffic light information and the traffic information obtained from the TCC can be broadcast via the RSUs to the vehicles and the RSUs can further collect vehicle probe data for the TCC. For reliable distribution of data, low-latency and high-capacity connections between RSUs and the TCC are required. This type of application is made possible by setting stringent end-to-end latency requirements for the communication service between RSU and TCC since relayed data needs to be processed in the TCC and the results need to be forwarded to the neighboring RSUs. It must be noted that the availability of the communication service has to be very high in order to compete with existing wired technology and in order to justify the costly deployment and maintenance of RSUs.

- *Security:* IoT introduces new UEs with different characteristics, including IoT devices with no user interface (e.g., embedded sensors), long life spans during which an IoT device may change ownership several times (e.g., consumer goods), and non-provisioned IoT devices (e.g., consumer goods). These applications necessitate secure mechanisms to dynamically establish or refresh credentials and subscriptions. New access technologies, including licensed and unlicensed, 3GPP and non-3GPP, are driving new efforts for creating access independent security mechanisms that are seamlessly available while the IoT device is active. A high level of 5G security is essential for mission-critical communications, for example, in industrial automation, V2X services, industrial IoT, and the smart grid. Expansion into enterprise, vehicular, and public safety markets will drive the efforts toward increased end-user privacy protection. 5G security is attempting to address all of these new requirements while continuing to provide connection/session security consistent with 3GPP legacy systems.

4 ITU-R IMT-2020 Standardization Activities

ITU has a long and reputable history in developing radio interface standards for mobile communications in the form of recommendations and reports. The development of standards for IMT systems, which started in late 1990's, encompassing IMT-2000 and IMT-Advanced [3], helped globalized the 3G and 4G networks and this trend will continue to evolve in to 5G with publication of IMT-2020 recommendations (see Fig. 15). In early 2012, ITU-R[22] initiated a path finding program to develop vision for IMT systems in 2020 and beyond, setting the stage for widespread 5G research activities that subsequently ensued around the world.

[22] ITU toward IMT for 2020 and beyond at http://www.itu.int/.

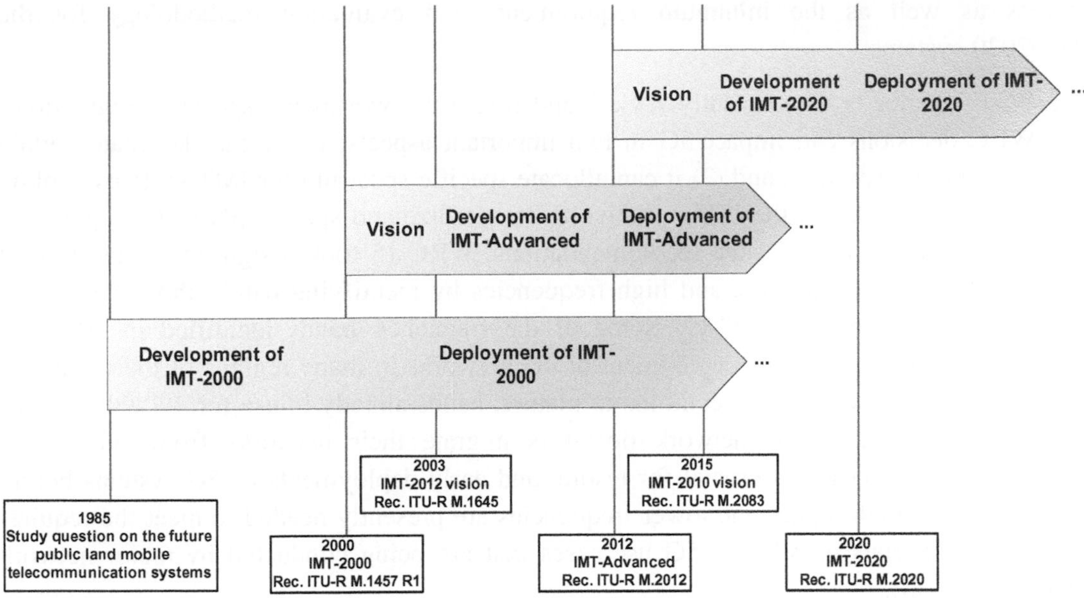

Figure 15
Overview of timelines for IMT standards development and systems deployment [1].

The role of the ITU-R[23] is to ensure efficient and economical use of the worldwide spectrum by all wireless services, including maritime and satellite communications. ITU-R further allocates and regulates the radio spectrum for global deployments of wireless and radio telecommunication technologies. The vision documents published by ITU-R (e.g., recommendations ITU-R M.1645 [2] and ITU-R M.2083 [1]) have been used by standards development organizations as guideline to develop technologies that materialize those visions.

Within ITU-R, there are several study groups that develop the technical bases for decisions taken at world radio-communication conferences and develop global standards in the form of recommendations, reports, and handbooks concerning radio telecommunication topics. The ITU-R members are typically from administrations, the telecommunications industry as a whole, and academic/research organizations throughout the world, who participate in the work of the study groups on topics such as efficient management and use of the spectrum/orbit resources, radio systems characteristics and performance, spectrum monitoring, and emergency radio communications for public safety. Working Party 5D (WP 5D)[24] is part of ITU-R Study Group 5 which is responsible for the overall radio system aspects of IMT systems. WP 5D has already finalized its views on the IMT-2020 timeline and submission

23 ITU-R, https://www.itu.int/en/ITU-R.

24 WP 5D—IMT Systems, https://www.itu.int/en/ITU-R/study-groups/rsg5/rwp5d.

process as well as the minimum requirements and evaluation methodology for the IMT-2020 systems.

The WRC charter is to constantly review and revise the worldwide spectrum regulations. The WRC decisions can impact 5G in two important aspects: (1) it can designate certain bands for mobile services, and (2) it can allocate specific spectrum for IMT systems deployments. The IMT-2020 is the ITU initiative to standardize and specify 5G technologies that are included in the IMT related recommendations. WRC-15 took a significant step toward 5G spectrum at low, medium, and high frequencies by identifying bands above 6 GHz for sharing studies prior to WRC-19. Some of the frequency bands identified in 2015 are already being utilized for the deployment of 4G networks in many regions of the world. As 5G standards and technologies continue to mature, bands already in use for 4G will be reallocated to 5G systems as network operators migrate their networks from 4G to 5G. Availability of spectrum is crucial for testing and early deployments of 5G systems before 2020; therefore both higher and lower frequencies are presently needed to meet the requirements of field trials of relevant 5G use cases that are being conducted by major network operators [29,30].

According to ITU-R vision, IMT-2020 systems are mobile systems which include the new capabilities of IMT systems that go beyond those of IMT-Advanced (see Fig. 16). The IMT-2020 systems support low to high-mobility applications and a wide range of data rates in accordance with user and service requirements in various deployment scenarios. IMT-2020 systems also have capabilities for high-quality multimedia applications within a wide range of services and platforms, providing a significant improvement in performance and quality of service. A broad range of capabilities, tightly coupled with intended use cases and applications for IMT-2020 have been laid out in ITU-R vision in order to support emerging new usage scenarios and applications for 2020+ [1].

While the minimum technical requirements and corresponding evaluation criteria defined by ITU-R based on the envisioned capabilities for IMT-2020 systems could have been met by certain enhancements to the existing IMT systems, true fulfillment of the IMT-2020 system and service requirements necessitated development of groundbreaking technology components and functionalities, and development of a set of new radio access technologies. Nevertheless, IMT-2020 systems continue to interwork with and complement the existing IMT systems and their enhancements [1–12].

IMT-2020 can be considered from multiple aspects including the users, manufacturers, application developers, network operators, and service and content providers as the usage scenarios of IMT-2020 will continue to expand. Therefore, it is expected that the technologies for IMT-2020 can be applied in a broader range of usage scenarios and can support a wider range of environments, diverse service capabilities, and technology options compared to IMT-Advanced and IMT-2000 systems. The task of ITU-R WP 5D is to determine a set

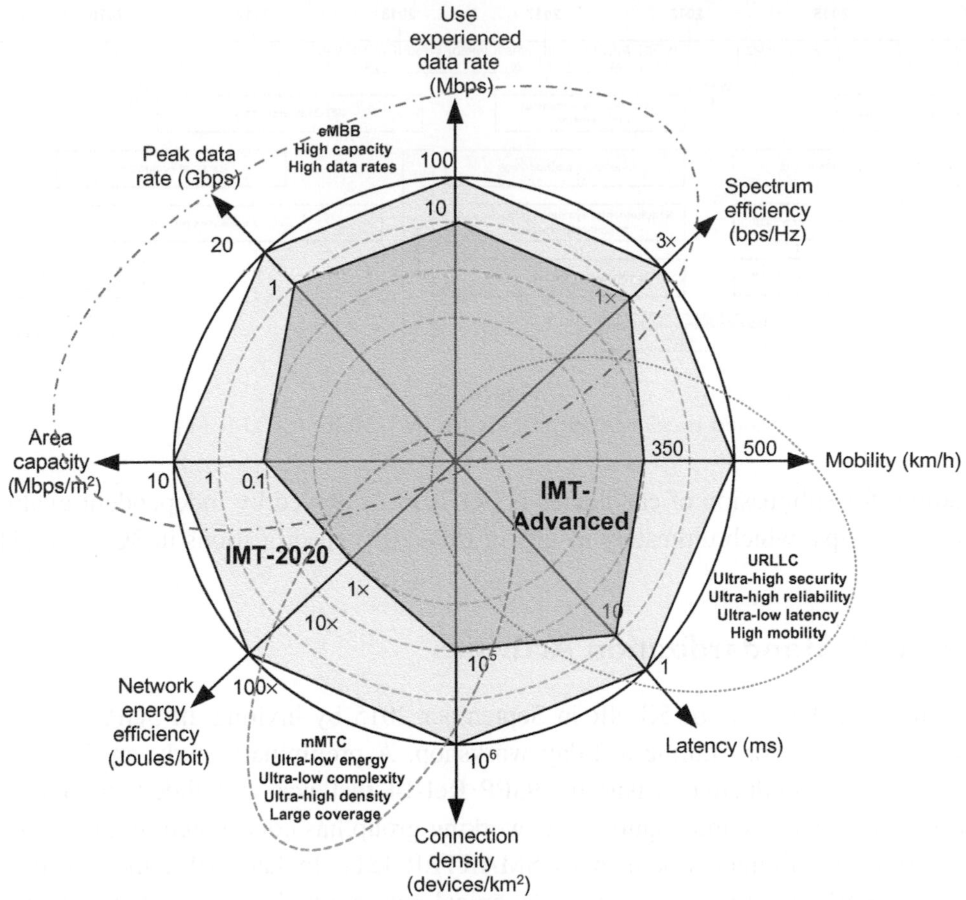

Figure 16

Comparison of IMT-advanced and IMT-2020 features and requirements [1].

of representations of IMT-2020 systems that can satisfy the requirements related to technical performance for IMT-2020 radio interfaces in the aforementioned test environments. Similar to IMT-Advanced process, the proponents could propose a single radio interface technology or a set of radio interface technologies to fulfill the requirements.

The technical characteristics of IMT-2020 systems chosen for evaluation have been explained in detail in Report ITU-R M.[IMT-2020.SUBMISSION], including service and spectrum aspects as well as the requirements related to technical performance, which are based on Report ITU-R M.[IMT-2020.TECH PERF REQ]. These requirements are summarized in Table 3, together with the high-level assessment methods which include system-level and link-level simulations, analytical, and inspection by reviewing the functionalities and supported configurations of the candidates. The process defined by ITU-R for standardization of IMT-2020 systems is illustrated in Fig. 17. According to the IMT-2020 process,

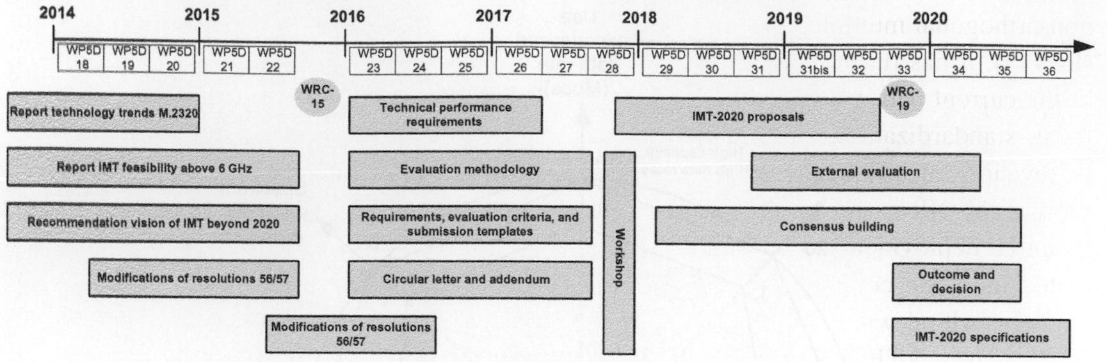

Figure 17
Detailed timeline and process for IMT-2020 in ITU-R [1].

the deadline for submission of candidates is 3Q 2019, followed by independent evaluations by external groups, which ultimately results in IMT-2020 specifications in 3Q 2020 [1].

5 3GPP 5G Standardization Activities

3GPP[25] initiated the study of 5G NR in September 2015 by inviting all 3GPP membership to share their 5G visions during a 2-day workshop. A preliminary study on 5G candidate technologies was conducted as part of 3GPP Rel-14 that was concluded in March 2017 [15]. The 3GPP services and requirements working group has contributed to the study phase for 5G service requirements, known as SMARTER [21]. In late 2015, the 3GPP system architecture working group approved a technical report on the study of architecture for next-generation systems [13] which laid out 3GPP vision for the design of 5G architecture, support of new RAT(s), the evolved LTE, as well as interworking with non-3GPP access networks. The latter study investigated high-level architectural requirements and defined a common terminology for the next-generation networks. At the same time, based on the outcome of the RAN feasibility studies, 3GPP began to develop 5G specifications in two phases. 3GPP Rel-15 is associated with the first phase of 5G specifications that define basic features of 5G systems. 3GPP Rel-16 will specify the second phase of 5G specifications defining additional features of 5G that combined with those of the first phase will satisfy the IMT-2020 requirements. The two phases of 5G specifications development in 3GPP are scheduled to end in September 2018 (already completed in March 2019) and March 2020, respectively [34].

While the prospect of the next-generation networks has inspired extensive and non-backward compatible changes in the air interface design (e.g., channel coding schemes, use of

[25] 3GPP, http://www.3gpp.org/.

non-orthogonal multiple access) and the transport layer [e.g., information centric networking (ICN)[26]], it has further stimulated the development of numerous incremental improvements in the current networks. Many of these technologies are already in the pipeline for development, standardization, and deployment. In fact, what 3GPP calls LTE-Advanced Pro, that is, revisions of LTE technology that build upon 3GPP Rel-8 through 14, along with the first drop of the NR air interface features with Rel-15 in late 2018, will form the basis of 5G in its initial deployments around 2020. Full satisfaction of IMT-2020 requirements is not anticipated until the completion of 3GPP Rel-16 in late 2019. The momentum is clearly and strongly behind what is being provided with many enhancements to LTE in a sequence of new releases. LTE has already been significantly improved since its inception with 3GPP Rel-8 and with further improvements in LTE-Advanced, as initially standardized in 3GPP Rel-10. In the LTE track of 5G in 3GPP, enhancements will continue to enable the standard to support as many 5G requirements and use cases as possible.

In order to increase the performance of the existing systems by an order of magnitude, it would require a combination of new spectrum, improved spectrum efficiency and increased network density. In particular, carrier aggregation which was introduced in 3GPP Rel-10 has been continuously enhanced in conjunction with availability of new spectrum. Among various additional capabilities in 3GPP Rel-11 and Rel-12, the performance of mobile broadband data has been increased to peak data rates of 4 Gbps in the downlink and 1.5 Gbps in the uplink through aggregation of multiple bands. Enhanced MIMO schemes and higher order modulations (i.e., 256QAM and 1024QAM) have further increased the cell spectral efficiency in the latest releases of LTE. For the time being, the evolution of mobile broadband will remain the most viable path because the demand for such services is proven to be strong and profitable for the operators and device/equipment vendors. Commercial implementation of other capabilities introduced by 3GPP in the past releases of LTE has not been as strong as anticipated. Deployment of HetNets with various features coordinating radio resources among large and small cells has been slow. Challenges include the significant difficulties and cost of upgrades and deployment of large number of small cells in urban areas and providing backhaul connection via optical fiber for user data and signaling.

The process of making LTE as of part of 5G standards package requires various enhancements and new features in LTE Rel-14 and Rel-15. The most significant ones are

[26] ICN is an approach to evolve the Internet infrastructure away from a host-centric paradigm based on perpetual connectivity and the end-to-end principle, to a network architecture in which the focus is on content or data [38]. In other words, ICN is an approach to evolve the Internet infrastructure to directly support information distribution by introducing uniquely named data as a core Internet principle. Data becomes independent from location, application, storage, and means of transportation, enabling or enhancing a number of desirable features, such as security, user mobility, multicast, and in- network caching. Mechanisms for realizing these benefits are the subject of ongoing research in Internet Engineering Task Force (IETF) and elsewhere. Current research challenges in ICN include naming, security, routing, system scalability, mobility management, wireless networking, transport services, in-network caching, and network management.

enhancements to user data rates and system capacity with full-dimension MIMO (FD-MIMO), improved support for unlicensed operations [47], and latency reduction in both control and user planes. The enhancements in LTE Rel-14 and Rel-15 are also intended to provide better support for use cases such as massive MTC, mission-critical communications, and intelligent transportation systems. The MIMO enhancement in 3GPP makes it possible to dynamically adapt the transmission both vertically and horizontally by steering a two-dimensional antenna array. The concept of FD-MIMO in the current LTE releases builds on the channel state information feedback mechanisms introduced in LTE Rel-13, in which precoding matrix codebooks support two-dimensional port layouts with up to 16 antenna ports. To enhance both non-precoded and beamformed CSI-RS operation, LTE Rel-14 introduced several new features, including hybrid non-precoded/beamformed CSI mode with optimized feedback; aperiodic triggering of CSI-RS measurements and support for up to 32 antenna ports. Other 3GPP work items include uplink enhancements such as higher order modulation and coding (e.g., 256QAM in the uplink) in baseline LTE systems. While it is theoretically possible to increase uplink peak data rates with improvements such as this in small-cell deployments, commercially available mobile devices are almost invariably use modulation orders limited to 16QAM in the uplink due to practical constraints. The ability to combine unlicensed spectrum with licensed spectrum is a highly attractive opportunity for network operators. This can be achieved by LTE−WLAN link aggregation and/or by carrier aggregation of LTE in both licensed and unlicensed bands through licensed-assisted access. These features are included in 3GPP Rel-13 onward.

There has been significant interest expressed by some mobile operators and broadcasters for enhanced multimedia broadcast multicast service (eMBMS) in the past few years; nevertheless, the use of eMBMS in general has been only nascent and the standardization work is ongoing with further improvements in 3GPP Rel-14.

Reduction in end-to-end latency is required for various applications, including those envisaged for 5G. However, somewhat lower latency than current levels is desirable for more conventional mobile broadband use cases. LTE control-plane latency figures vary quite widely among different networks with figures up to 75 ms quite common. This relatively high latency can result from the routing of data packets through many network elements before they reach a network gateway to the Internet. From that point onward, there will be additional packet-routing delays which are beyond operator's control. 3GPP has been studying the possibility of reducing the LTE TTI of 1 ms and to further identify additional latency reductions at layers 2 and 3 of the LTE protocol stack.

While LTE evolution is mainly focused on eMBB aspects, 3GPP is also trying to fulfill some of the 5G use cases with the specific performance requirements in the LTE path. For example, a downscaled UE Category 0 was introduced in LTE Rel-12 for IoT applications. This and the subsequent LTE-based NB-IoT standardization could satisfy some of the 5G requirements for mMTC use cases.

The scope of NR phase 1 standardized in 3GPP Rel-15 included both NSA and SA operations. In NSA mode, 5G NR uses LTE as an anchor in the control plane. The SA mode allows 5G NR to work independently with full control plane capability. While focusing on eMBB use case, it also provides support for some URLLC services. The LTE and NR share much more than the common 3GPP release schedule shown in Fig. 18. It is inevitable that much of what is being developed for LTE will also significantly contribute to 5G, and there will be significant commonalities and interdependencies, if not compatibilities between the two. For example, 5G might be dependent on low-band LTE for coverage and control channel signaling where such spectrum is not directly available for 5G. However, the transition to 5G also has opened up the possibility of introducing an entirely new air interface and creating network architecture and designs that are not necessarily backward compatible with LTE. There are significant tradeoffs in NR implementation and deployments. For example, dual connectivity of the UE can improve performance, but this requires compromise on

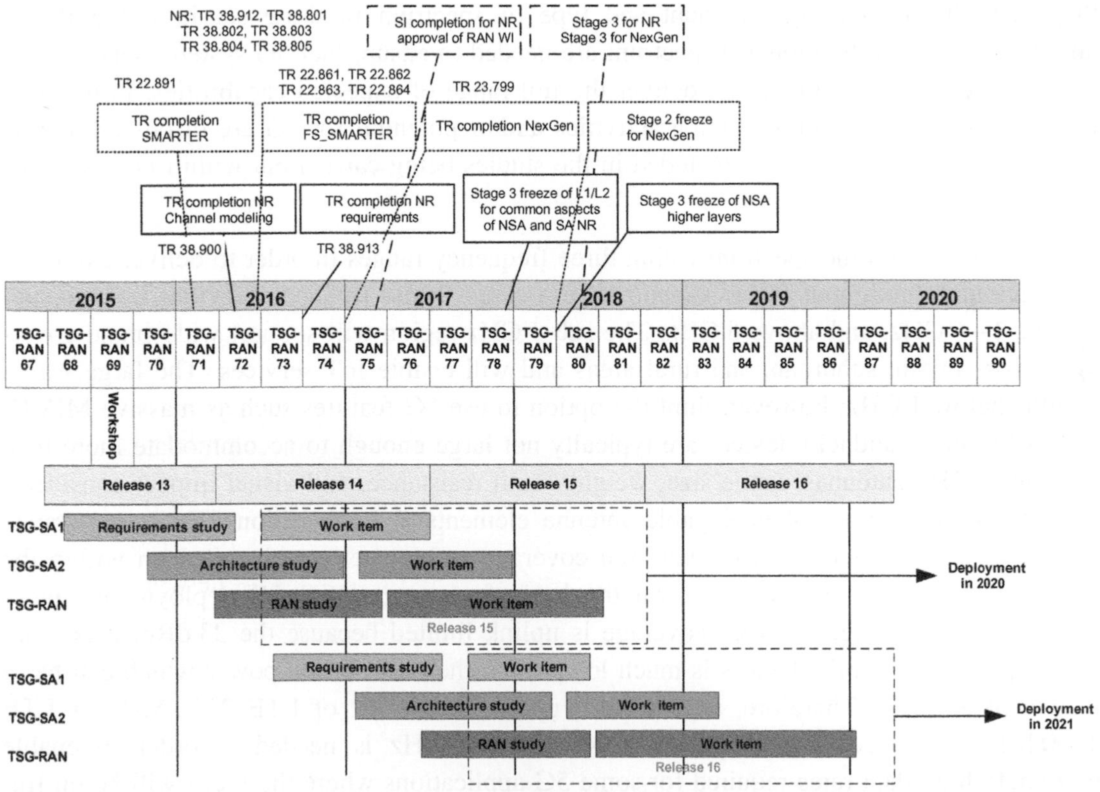

Figure 18

Detailed timeline and process of 5G development in 3GPP (3GPP Work Plan, http://www.3gpp.org/specifications/work-plan).

other aspects. The robustness of LTE control mechanisms (such as mobility, resources management, and scheduling) at lower frequencies makes it very attractive as an anchor technology for the different access techniques used in 5G.

6 Spectrum and Regulations

Spectrum is the key asset for a communications service provider. The available spectrum has a major impact on how a network's maximum capacity and coverage are defined. From VR/AR and autonomous cars to the industrial automation and smart cities, 5G is expected to enable a wide range of new use cases and applications and ultimately define a new framework for future telecommunication systems. Although the mobile industry, academia, and international standards developing organizations have been actively involved in developing the technologies that are central to 5G systems, the success of the 5G services will essentially be heavily reliant on the work of regional regulatory bodies and how quickly they can allocate the right amount and type of spectrum to those services. Significant amount of new and harmonized spectrum are needed to ensure that 5G systems/services can meet the future expectations and deliver the full range of expected capabilities. In this section, frequency bands with ongoing activities and/or potential for 5G are discussed, noting that not all of these bands are included in the studies being carried out within ITU-R in the lead up to WRC-19.

5G systems rely on the spectrum within three frequency ranges in order to deliver extensive coverage and throughput and to support all use cases. Those frequency regions include sub-1 GHz, 1–6 GHz, and above 6 GHz. The sub-1 GHz region will support widespread coverage across urban, suburban, and rural areas and will enable IoT services. The large wavelengths below 1 GHz; however, limit the option to use 5G features such as massive MIMO. That's because handheld devices are typically not large enough to accommodate more than two sub-1 GHz antennas, while size, weight, wind resistance, and visual impact considerations limit the number of deployable antenna elements at base stations. Frequency range 1–6 GHz offers a good tradeoff between coverage and capacity with spectrum within the 3.3–3.8 GHz range expected to form the basis for many initial 5G deployments under 6 GHz [29]. Macrocell network coverage is uplink limited because the 23 dBm maximum output power of mobile devices is much lower than the base station power which can be in excess of 46 dBm. Therefore, a 3500 MHz uplink falls short of LTE 2100 MHz or LTE 1800 MHz coverage. The frequency range above 6 GHz is needed in order to enable extremely high data rates required for some 5G applications where the focus will be on frequency bands above 24 GHz which include growing interest in the 24 GHz and/or 28 GHz bands. There is also some interest in exploring bands in the 6–24 GHz frequency region; however, this range has not been included in WRC-19 sharing studies [30,37].

One of the main objectives of the telecommunication industry is to connect everything and everywhere by more comprehensively and intelligently integrating LTE (in licensed and unlicensed bands), Wi-Fi, and cellular IoT technologies, together with 5G radio interface. This will allow mobile networks to dynamically allocate resources to support the varying needs of an extremely diverse set of network connections ranging from factory automation, sensor networks to self-driving and connected vehicles as well as smartphones and traditional voice services. The increase in the capacity of the 5G radio network needs to be supported by faster and wider bandwidth backhaul that incorporates fast Ethernet, optical fiber, or microwave/mmWave wired/wireless point-to-point links. The use of satellite networks may also provide backhaul for 5G systems despite their limited ability to satisfy 5G's stringent latency and bandwidth requirements.

A central component in the evolution of all mobile technology generations in the past two decades has been the use of continuously increasing bandwidths to support higher peak data rates and larger amounts of user traffic. Ultra-fast 5G services require large amounts of spectrum, compelling the governments and regulators to actively look at significantly higher frequencies that have been traditionally used in mobile services for spectrum allocation. This includes spectrum above 24 GHz where wider bandwidths are more readily available. Without making these higher frequency bands accessible to 5G, it may not be possible to materialize significantly higher data rates and to support rapidly growing mobile data traffic particularly in crowded urban areas.

5G services will struggle to reach beyond urban centers and deep inside the buildings without the sub-1 GHz spectrum. There is existing mobile spectrum in this range which should be refarmed and used in the future. For example, the European Commission has already expressed a desire for the use of 700 MHz band for 5G services in Europe. The United States Federal Communications Commission (FCC)[27] has indicated that the 600 MHz band could be used to drive 5G services in the United States. ITU-R is also considering identifying additional spectrum for mobile broadband from 470 to 694/8 MHz in 2023 which would be well-timed for the large-scale rollout of 5G services, especially if countries prepare to use it quickly after international agreement is reached [30]. The 3.4−3.6 GHz range is practically globally harmonized, which can drive the economies of scale needed for low-cost devices. A number of countries are exploring whether a portion of other bands could be used such as 3.8−4.2 GHz and spectrum in the 4−5 GHz range, in particular 4.8−4.99 GHz region. There are also numerous other mobile bands in the 1−6 GHz range that are currently used for 3G and 4G services which should be gradually repurposed for 5G use. Due to its favorable properties such as radio wave propagation and available bandwidth, the bands in the range of 3300−4200 and 4400−4990 MHz will be the primary spectrum under 6 GHz for the introduction of 5G. Parts of the band between 3300−4200 MHz and 4400−4990 MHz are being considered for the first field trials and

[27] United States FCC, https://www.fcc.gov/.

introduction of 5G services in a number of countries and regions in the world, including Europe (3400–3800 MHz), China (3300–3600, 4400–4500, 4800–4990 MHz), Japan (3600–4200, 4400–4900 MHz), Korea (3400–3700 MHz), and the United States (3100–3550, 3700–4200 MHz).

The above 6 GHz spectrum comprises a combination of licensed and unlicensed mobile bands. There are several mmWave bands including 24.25–27.5, 31.8–33.4, 37–43.5, 45.5–50.2, 50.4–52.6, 66–76, and 81–86 GHz which have been identified for sharing studies by WRC-15 that must be ratified by WRC-19 for worldwide allocation. However, some countries are also investigating other mobile bands above 6 GHz for 5G services, which are not being considered at WRC-19. The 28 GHz band is of particular interest as it has been permitted for 5G use in the United States and is being closely examined by Japan and Korea. This would complement the 24 GHz band, which is being studied at WRC-19 and is supported in the European Union, because the same equipment could easily support both bands thus helping to lower device costs. Fig. 19 illustrates the frequency ranges that are being studied for specification at WRC-19 in different regions of the world [40].

The smaller coverage areas of higher frequencies, when employed terrestrially, provide more frequency reuse opportunities with limited inter-cell interference issues. As such, 5G services in urban areas may be able to occupy the same bands as other wireless services (e.g., satellite and fixed links) which operate in different geographical areas (e.g., rural)

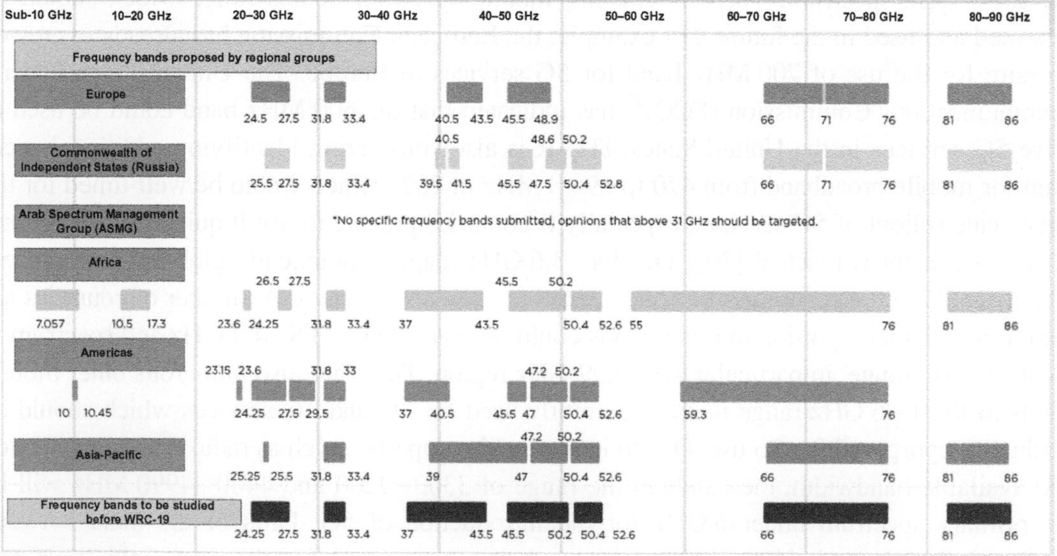

Figure 19
Frequency bands under study for ratification in WRC-19 [40].

when suitable interference mitigation techniques such as beamforming are in place. The potential for spectrum sharing will be explored through sharing studies being conducted by ITU-R, in preparation for WRC-19. Adopting viable sharing methods is particularly important as many of the bands above 24 GHz that are being considered for 5G access are or will be used for 4G/5G mobile backhaul especially in rural areas where fiber links might not be available or practical [30].

As we mentioned earlier, the wide bandwidth requirements for some 5G applications have necessitated comprehensive study of frequency bands above 6 GHz for utilization in various 5G deployment scenarios and use cases. There are ongoing global activities to identify and specify suitable spectrum including frequency bands that can be used in as many countries as possible to enable global roaming and economies of scale. While not very successful for 4G and previous generations, various efforts around the world are underway to achieve spectrum harmonization in 5G networks. In addition to setting different requirements on the network features, new applications will drive a wide variety of deployment scenarios. The different physical characteristics of various spectrum bands (e.g., range, path loss, penetration into structures, propagation) make some applications more suitable in certain frequency bands [46]. While sub-1 GHz spectrum has very good propagation properties that make it the most suitable option for large area coverage, it has limited capacity due to lack of available spectrum and component design considerations. The sub-6 GHz spectrum provides a type of coverage that is more suitable for urban deployment, with increased capacity. The above 6 GHz spectrum is more limited in coverage but can provide very high capacity due to potentially very large bandwidth available at these frequencies.

In late 2015, WRC-15 agreed that ITU-R should conduct sharing and compatibility studies for a number of frequency bands between 24.25 and 86 GHz prior to WRC-19 in late 2019. Some of these frequency bands enable wide contiguous bandwidths, which would allow satisfaction of the stringent requirements of bandwidth-demanding applications. Although the band 27.5–29.5 GHz was not selected for ITU-R studies, the United States and Korea have continued to make progress with trials in this band for 5G which could, when combined with the band 24.25–27.5 GHz (which was selected at WRC-15 for ITU-R studies), provide a good solution for a global implementation of 5G systems. The following frequency bands in sub-6 GHz range have been identified for IMT in ITU-R radio regulations since 1992 [6]:

- 450–470 MHz
- 470–698 MHz
- 694/698–960 MHz
- 1427–1518 MHz
- 1710–2025 MHz
- 2110–2200 MHz

- 2300–2400 MHz
- 2500–2690 MHz
- 3300–3400 MHz
- 3400–3600 MHz
- 3600–3700 MHz
- 4800–4990 MHz

As shown in Table 5, there are two frequency regions which are playing an important role in 5G NR deployments. In the 3–6 GHz band, there is generally globally available spectrum in the range of 3.3–3.8, 3.8–4.2, and 4.4–4.9 GHz. These bands, which are all based on TDD or unpaired spectrum, generally have wider bandwidth than their 4G predecessors. They will be particularly important in user equipment implementations that will use LTE as the anchor (or primary) and the NR as a secondary access scheme (i.e., NSA deployments). 5G systems can also use sub-1 GHz FDD bands to provide wide-area coverage, including deep indoor penetration. The low-band spectrum can take advantage of the new 700 MHz allocation in Europe or 600 MHz allocation in the United States, as well as the refarming of 850/900 MHz released by minimizing the use of legacy 2G/3G spectrum. The aggregation of the different spectrum bands from sub-1 GHz to millimeter waves provides the best combination of coverage, capacity, and user data rates. In addition to the licensed bands in the 3–6 GHz range, there is the possibility of using supplemental unlicensed bands to create even wider operating bandwidth. The 3GPP NR will utilize mmWave spectrum to achieve wider bandwidths and higher data rates. Industry consensus is building around the use of mmWave spectrum for fixed wireless applications due to challenges of mobility management in very high frequencies. Applying mmWave technology to mobile devices represents a significant technological challenge for the near future. Small cells are going to be a critical part of the 5G network deployments. We referred to this trend earlier as network densification. The use of mmWave does leverage large antenna arrays and directional beams in the downlink over a short range, enabling design of high-capacity small cells.

Spectrum harmonization remains an important issue for the development of 5G and even more important for higher frequencies in order to support the development of a new ecosystem as well as the deployment of massive MIMO systems. The United States has adopted new rules to enable rapid development and deployment of 5G technologies and services in licensed spectrum in 28 and 39 GHz. China is also planning to deploy commercial 5G networks to meet the demands for the extremely high peak data rates in the frequency ranges of 26 and 42 GHz. In Europe, the spectrum around 26 GHz has been identified as 5G pioneer band and work is well underway in order to harmonize the frequency bands in Europe for 5G before WRC-19 through adoption of a harmonized band and to promote this band for worldwide use. The use of higher frequency bands for early and commercial deployments of 5G can be summarized as follows [30,40,46]:

Table 5: Candidate spectrum for 5G new radio [48].

Region	Sub-6 GHz Spectrum							mmWave Spectrum			
	New Band		Existing Band		3GPP Band	Bandwidth (MHz)	Total Bandwidth (MHz)	F_{Low} (GHz)	F_{High} (GHz)	Bandwidth (GHz)	Total Bandwidth (GHz)
	F_{Low} (MHz)	F_{High} (MHz)	F_{Low} (MHz)	F_{High} (MHz)							
Korea	3400	3700				300	300	26.50	29.50	3.00	3.00
European	3400	3800	2570	2620	38	50	450	24.25	27.35	3.10	7.70
Union			3400	3800	42 + 43	400		31.80	33.40	1.60	
								40.50	43.50	3.00	
Japan	3600	4200	2496	2690	41	194	1494	27.50	29.50	2.00	2.00
	4400	4900	3400	3600	42	800					
						500					
United	–	–	2496	2690	41	194	344	27.50	28.35	0.85	10.85
States			3550	3700	48	150		37.00	38.60	1.60	
								38.60	40.00	1.40	
								64.00	71.00	7.00	
China	3300	3600	2300	2400	40	100	790	TBD			
	4400	4500	2555	2655	41B	100					
	4800	4990	3400	3600	42	300					
						100					
						190					

- United States: 27.5−28.35 and 37−40 GHz pre-commercial deployments in 2018
- Korea: 26.5−29.5 GHz trials in 2018 and commercial deployments in 2019
- Japan: 27.5−28.28 GHz trials planned from 2017 and potentially commercial deployments in 2020
- China: 24.25−27.5 and 37−43.5 GHz for deployments in 2020 and beyond
- Sweden: 26.5−27.5 GHz awarding trial licenses for use in 2018 and onward
- European Union: 24.25−27.5 GHz for commercial deployments from 2020

Note that the range 24.25−27.5 GHz (26 GHz) is overlapping with the band 26.5−29.5 GHz (28 GHz), which suggests that countries supporting 26 GHz may also benefit from early ecosystem development for the 28 GHz band in other regions. As shown in Table 5, in addition to the bands above 6 GHz, the spectrum in 600, 700, 800, 900, 1500, 2100, 2300, and 2600 MHz may be of particular interest for both legacy and new applications and are key to deliver necessary 5G broadband coverage for IoT, industry automation, and mission-critical usage scenarios.

Extensive research is currently being conducted on the use of much higher frequency bands for 5G systems than are currently used for existing mobile services. The frequencies as high as 300 GHz, according to some reports, have the potential of supporting extremely wideband 5G applications. However, due to their radio propagation characteristics, these bands are likely to be utilized primarily in small cells and may have only limited practical usage outside urban areas. To ensure 5G services provide good coverage that extends beyond small urban hotspots, it will be important to ensure that the suitable spectrum is available in sub-1 GHz region. Significant cooperation among mobile industry leaders, standards developing organizations, and national/regional regulatory bodies is required to pave the way for large-scale commercial 5G deployments in 2020 and beyond. Progressive refarming of existing mobile bands should be possible and permitted to accommodate future 5G services and to ensure spectrum is used as efficiently as possible. The success of refarming is best illustrated by the 1800 MHz band which is now the most common LTE band worldwide and was previously used for 2G services.

A secondary effect of densification of the network with high-data-rate links is the need to improve backhaul transport. Today, optical fiber is the preferred backhaul option. However, as networks grow to become denser with the deployment of more small cells, licensed fixed or point-to-multipoint (PMP) mmWave wireless backhaul may emerge as the most flexible and viable solution. For example, an operator (leveraging guaranteed QoS with licensed PMP) with a 10 Gbps + hub could aggregate backhaul traffic from multiple base stations. The economies of this scenario improve as more base stations are added progressively through network densification. Millimeter wave communications technologies in the 60 GHz and 70−80 GHz frequency range for high capacity in the last mile and pre-aggregation backhaul have been explored in the literature [27,48]. Table 6 provides the existing

Table 6: Mapping of RF components and technologies to spectrum bands [48].

TDD bands	700 MHz–3 GHz	3–6 GHz B42/B43/B46	24.25–29.50 GHz	37–71 GHz
FDD bands	700 MHz–3 GHz	3–6 GHz	-	-
Front-end technology	*Sub-6 GHz frequencies*		*mmWave frequencies*	
Power amplifier	FEMiD[a]/PAMiD[b]/DRx[c] III-V[d]/SiGe[e]/Bulk CMOS[f]	FEMiD/PAMiD/DRx III-V/SiGe/Bulk CMOS	8Tx × 8Rx antenna complete front-end InP/SiGe BiCMOS/Advanced SOI	8Tx × 8Rx antenna complete front-end InP/GaN/SiGe BiCMOS/Advanced SOI
Low-noise amplifier	III-V/SiGe/SOI[g] CMOS	III-V/SiGe/SOI CMOS	Advanced SOI/GaN	SiGe BiCMOS/Advanced SOI
RF switching	SOI CMOS	SOI CMOS	Advanced SOI	Advanced SOI
Filtering	Acoustic/IPD/Ceramic	Acoustic/IPD/Ceramic	IPD/Ceramic	IPD
Antenna integration	N/A	N/A	Yes	Yes
Signal generation method	N/A	N/A	Advanced SOI/SiGe BiCMOS	Advanced SOI/SiGe BiCMOS

DRx, Diversity receive module; *FEMiD*, front-end module with integrated duplexers; *IPD*, integrated passive device; *PAMiD*, power amplifier module integrated duplexer; *SOI*; silicon on insulator.

[a]FEMiD.

[b]PAMiD is a multimode multiband RF front-end module that integrates multiplexers, filters, antenna switches, and low-noise amplifiers allowing highly integrated transmit and receive chains.

[c]DRx combines switches, filters, and low-noise amplifiers into a module, providing a highly integrated solution for implementing high-order diversity receive paths for advanced 4G/5G architectures.

[d]A substance that can act as an electrical conductor or insulator depending on chemical alterations or external conditions. Examples include silicon, germanium, and gallium arsenide. It is called III-V materials since semiconductor elements are in groups III and V of the periodic table of chemical elements.

[e]Silicon Germanium process

[f]Bulk CMOS refers to a chip manufactured on a standard silicon wafer.

[g]SOI technology refers to the use of a layered silicon–insulator–silicon substrate in place of conventional silicon substrates in semiconductor manufacturing, especially microelectronics, to reduce parasitic device capacitance, thereby improving performance.

technology and spectrum, as well as the planned 5G NR spectrum. The table further shows the mapping between those applications and the technologies needed to implement power amplifiers, low-noise amplifiers, RF switches, RF filters, and integrated antennas to enable 5G transceivers to be implemented. One important conclusion is that the entire new spectrum for 5G NR, whether sub-6 GHz or greater than 6 GHz, is mainly TDD bands; thus frequency duplexers are not required in order to implement a front-end solution. Filtering, as needed, is accomplished using band-pass filters. In the 3–6 GHz region, filtering can be accomplished using acoustic, integrated passive devices (IPDs)[28] or ceramic technologies.

[28] IPDs or integrated passive components are attracting an increasing interest due to constant needs of handheld wireless devices to further decrease in size and cost and increase in functionality. Many functional blocks such as impedance matching circuits, harmonic filters, couplers and RF baluns, and power combiner/divider can be realized by IPDs technology. IPDs are generally fabricated using standard wafer fabrication technologies such as thin film and photolithography processing. IPDs can be designed as flip chip mountable or wire bondable components and the substrates for IPDs usually are thin film substrates like silicon, alumina or glass.

For the 5G mmWave fixed wireless applications, requirements for massive MIMO and beamforming mean that transmit and receive functions will most likely be in distributed array formats. As a result, there will be multiple transmit/receive chains to accomplish transceiver functionality for fixed/mobile wireless devices. Filter technology for mmWave 5G is likely to be based on transmission line and waveguide cavity technology. micro-electromechanical systems cavity resonators are also an attractive choice to avoid manual-tuning filters and leveraging silicon wafer-based manufacturing approaches. Multi-pole filters suitable for operation in frequencies of 20–100 GHz have been demonstrated in the literature [27]. User equipment PA technology for 5G sub-6 GHz will leverage traditional GaAs heterojunction bipolar transistor (HBT) technology,[29] but it will require some improvement or innovation in dealing with wide bandwidth signals, especially in biasing. The current skills in IEEE 802.11ac and IEEE 802.11ax implementations (up to 160 MHz bandwidth with extremely challenging error vector magnitude requirements) can be leveraged for development of 5G wideband transceivers. Above 6 GHz, the experience in system-in-package technology will be leveraged to create antenna front-ends on organic laminates.

7 Future Outlook

Users' growing reliance on mobile connectivity continues to present technical and economic challenges for the industry and in particular network operators for provisioning adequate coverage and capacity. Increasing utilization of VR/AR services are examples of changes in user behavior that can generate significantly greater performance demands on the wireless networks. In this chapter we have shown various scenarios for how the 5G technologies can enhance coverage and capacity beyond what can be achieved by legacy networks. Some key aspects of this trend include intrinsic capabilities of 5G systems to provide service in various indoor/outdoor environments particularly in dense urban areas, flexible use, and aggregation of available spectrum and access technologies, including Wi-Fi, LTE-U, and other technologies in the unlicensed bands, use of spectrum above 6 GHz to provide high-capacity short-range connectivity in small cells, and flexible network architectures and models of network operation and coverage.

The industry recognizes that sustaining excellent coverage and capacity is not a transitory event but a continuous process. It must be possible to efficiently reconfigure 5G systems to

[29] HBT is a type of bipolar junction transistor (BJT) which uses different semiconductor materials for the emitter and base regions, creating a heterojunction. The effect is to limit the injection of holes from the base into the emitter region, since the potential barrier in the valence band is higher than in the conduction band. Unlike BJT technology, this allows a high doping density to be used in the base, reducing the base resistance while maintaining gain. The HBT improves on the BJT in that it can process signals of very high frequencies, up to several hundred GHz. It is commonly used in modern ultra-fast electronic circuits, RF systems, and in applications requiring high-power efficiency, such as RF power amplifiers in cellular phones.

account for special events, including planned events such as sports games and unplanned events such as natural disasters. For smart devices, coverage requirements include global roaming and support for services on moving platforms such as cars, planes, high-speed trains, and mass transit. 5G will continue to improve the user and operational experience in these areas.

Leveraging recent advances in storage/memory technologies, communication/connectivity, computing, big-data analytics, artificial intelligence (AI), machine learning (ML), machine vision, and other related areas will enable the fruition of immersive technologies such as augmented and virtual reality. The use of virtual reality is expected to go beyond early adopters such as interactive gaming to enhancing cyber-physical and social experiences such as conversing with family, acquaintances, business meeting, or the disabled individuals. The AR/VR technologies would hypothetically allow walking around a street where everyone is talking in a different language, and interacting with those people in the same language in a fully immersive experience. The growing number of drones, robots, and other self-driving vehicles taking cameras to places that humans could never imagine reaching would allow us to collect new content from attractive points of view around the globe. Ultimately, AR/VR will provide the most personal experience with the closest screen, providing the most connected, most immersive experience witnessed thus far.

The concept of content caching has been recently investigated in great detail, where the idea is to cache essential content at the edge of the network (e.g., a base station), devices, or other intermediate locations. One distinguishes between reactive and proactive caching. While the former serves end users when they request contents, the latter is proactive and anticipates users' requests. Proactive caching depends on the availability of fine-grained spatial–temporal traffic predictions. Miscellaneous information such as user's location, mobility patterns, and social behavior and connections can be further exploited to optimize the services provided to the end user. Storage will play a crucial role in AR/VR where for instance upon the arrival of a task query, the network/server needs to promptly decide whether to store the object if the same request will come in the near future or instead recompute the query from scratch if the arrival rate of the queries will be sparse in the future. Content/media location and delivery will also be important in terms of storing different qualities of the same content at various network locations.

Migrating computationally intensive tasks/processes from AR/VR devices to more resourceful cloud/fog servers is necessary in order to relax some additional computational requirements for the low-cost devices and to extend the battery life of such devices. For this purpose, MEC is used to enable the client on-demand access to the cloud/fog resources (e.g., infrastructures, platforms, and software). While the current solutions allocate cloud-based radio and computing resources in a centralized manner, for optimal AR/VR performance, both radio access and computational resources must be brought closer to the users by binding the availability of small-cell base stations with proximity access to computing/storage/memory resources. Furthermore, the network infrastructure must enable a fully distributed cloud-based immersive experience where intensive computations occur on very powerful servers that are in the cloud/network edge while sharing the sensor data that are being delivered by end-user devices at the client side.

Leveraging short-range communication such as D2D and edge proximity services among co-located AR/VR users can help alleviate network congestion. The idea is to extract, appropriately combine and share relevant contextual information among AR/VR users in terms of views and camera feeds. In the context of self- driving vehicles equipped with ultra-high-definition (UHD) cameras capturing their surrounding area, the task for the vehicle is not only to recognize objects/faces in real time, but also to decide which of the objects should be included in the map and to share the information with neighboring vehicles for richer and more context-aware maps.

The introduction of UHD cameras (e.g., 8K cameras with 360-degree panoramic video recorders) has enriched new video and media experiences. User interaction with media content currently falls at two extreme ends of the spectrum. On the one hand, there are lean-back experiences such as movies and television where consumers are passive and are led through a story by content authors/producers. On the other hand, there are lean-forward experiences in the form of games in which the user is highly engaged and drives the action through an environment created by content authors/producers. The next generation of interactive media where the narrative can be driven by authors/producers will be personalized dynamically to the situation and preferences of end users.

The use of context-aware communication has already been promoted as a means of optimizing complex networks. In order to improve user's connected and immersive experience, the users' personal habits, emotions, and other behavioral and psychological aspects must be factored in their service delivery and optimization. This involves predicting users' disengagement and preventing it by dynamically shifting the content to better match individual's preferences, emotional state, and situation. Since a large amount of users' data in the network can be considered for the big-data processing, ML tools can be exploited to analyze the contextual user information and to react accordingly. Of particular importance is the fact that deep learning models have been recently on the rise in ML applications, due to their human-like behavior in training and good performance in feature extraction.

Previous generations of mobile networks enabled voice, data, video, and other impactful services. In comparison, 5G is going to change our society by opening up the telecommunication ecosystem to vertical industries. 5G will help vertical industries to achieve the IoT vision of ubiquitously connected, seamlessly mobile, highly reliable, ultra-low-latency services for massive number of devices. Service-guaranteed network slicing is one of the essential features for 5G to achieve this vision. Prominent operators, vendors, and vertical industries have to come together in order to establish a common understanding of service-guaranteed network slicing in terms of the vision, requirements, end-to-end solution, key enabling technologies, and the impacts on vertical industries.

The 3GPP standard releases are driven by users and operator demands, with smaller releases launched every few years to enhance features and performance of the status quo. Therefore, one might reasonably expect a beyond 5G release to arrive at some point in the future. There is some debate about what systems beyond 5G would entail and whether indeed it is relevant to consider the term, as industry/user requirements will change greatly in the next 10–20 years. A high-level answer is that systems beyond 5G will explore and include relevant technologies that will be left out from 5G, due to being late, experimental in status or simply outside the defined scope for 5G. Future applications and technologies will be integrated/incorporated when they achieve maturity. For systems beyond 5G, one proposal is to integrate terrestrial wireless with satellite systems, for ubiquitous always-on broadband global network coverage. Many of the cellular devices currently connected are machines rather than people, and with the rise of smart homes, smart buildings, and smart cities, 5G and systems beyond 5G will encounter increased demands for machine-to-machine communications, including robotic interactions and autonomous drone delivery and transport systems. Other trends predicted for systems beyond 5G include ultra-dense cellular networks, reconfigurable and fully programmable hardware, and extended use of mmWave spectrum for user access, enhanced optical-wireless interface and photonics, networked visible light communication (VLC),[30] intelligent networking, and technologies to enable full immersive experience for users [28].

The user demand for greater global coverage, higher network capacities, seamless and always-on connectivity, new Internet services and applications continue to grow, and systems beyond 5G will be expected to deliver even more features and services. One of the potential drivers for systems beyond 5G is the growing prospects of software-defined radio and software-defined networking, which means that the underlying technologies for the systems beyond 5G will be easier to upgrade. This reduces the expensive and disruptive upgrades of previous mobile standards which generally involve replacement of physical infrastructure. Inter-vendor interoperability is also an increasing trend, with disaggregation fueled by open-source/white-box development of systems/technologies.

[30] Networked VLC for ultra-dense wireless networking is based on VLC, which promises quantum step improvements in area spectral efficiency while exploiting existing infrastructures by piggy-backing high speed data communication on existing lighting infrastructures. Networked VLC is also referred to as Li-Fi. The visible light spectrum is unlicensed and 10000 times wider than the range of radio frequencies between 0 and 30 GHz. The use of the visible light spectrum for data communication is enabled by inexpensive and off the-shelf available light emitting diodes which also form the basis for next generation energy efficient lighting. Individual LEDs can be modulated at very high speeds of 3.5 Gbps at approximately 2 m distance have been demonstrated using micro-LEDs with a total optical output power of 5 mW [36].

In the early stage of network slicing deployment there will be only a few network slice instances (NSIs)[31] and the deployment may occur in a semi-automatic mode. As the number of NSIs increases and scenarios, such as, dynamic instantiation of NSIs or run-time adaptation of the deployed NSI emerge, more advanced technologies will be needed to support network slicing and its further evolution [31]. Management functions will become real-time, implying that the difference between management and control will gradually disappear. Some management functions will be tightly integrated with the NSIs as well as the network infrastructure. In current networks, technical domains are normally coordinated via centralized network management system. In 5G, performing real-time cross-domain coordination through distributed lower layer such as control plane would be possible, with potentially unified control logic of different domains. Advanced automation and AI algorithms can be applied in a unified, holistic network manner, which would be scalable and flexible, which might then result in run-time deployment and adaptation of NSIs. The 5G networks are not only envisioned as a support for IoT, but also as means to give rise to an unprecedented scale of emerging industries. IoT requires support for a diverse range of service types, such as e-health, Internet of Vehicles, smart households, industrial control, and environment monitoring. These services will drive the rapid growth of IoT and facilitate numerous devices to connect to the network, leading to the notion of the Internet of Everything especially for vertical industries.

From ITU-R perspective, the IMT systems serve as a communication tool for people and a facilitator which enable the development of other industry sectors, such as medical science, transportation, and education. IMT systems will continue to contribute to the following aspects [1,4,5]:

- Wireless Infrastructure: Broadband wireless connectivity will attain the same level of importance as access to electricity. IMT will continue to play an important role in this context as it will act as one of the key pillars to enable mobile service delivery and information exchange. In the future, private and professional users will be provided with a wide variety of applications and services, ranging from infotainment services to new industrial and professional applications.

[31] NSI is defined as a set of network functions and resources to run network functions, forming a complete instantiated logical network to meet certain network characteristics required by the service instance(s). A network slice instance may be fully or partly, logically and/or physically, isolated from another network slice instance. The resources consist of physical and logical resources. A network slice instance may be composed of subnetwork instances, which as a special case may be shared by multiple network slice instances. The network slice instance is defined by a network slice blueprint. Instance-specific policies and configurations are required when creating a network slice instance. Network characteristics examples are ultra-low-latency, ultra-high reliability, etc. [25,26].

- Integrated Information and Communication Technologies (ICT): The development of future IMT systems is expected to promote the emergence of an integrated ICT industry which constitute a driver for economies around the globe. Some possible areas include the accumulation, aggregation and analysis of big data as well as delivering customized network services for enterprise and social network groups on wireless networks.
- Service Availability and Affordability: IMT systems will continue to help closing the gaps caused by an increasing digital divide. Affordable, sustainable, and easy-to-deploy mobile and wireless communication systems can support this objective while effectively saving energy and maximizing efficiency. IMT systems will enable sharing of any type of content anytime, anywhere through any device. Users will generate more content and share the content without being limited by time and location.
- Energy Efficiency: IMT enables energy efficiency across a range of industries by supporting machine-to-machine communication and solutions such as smart grid, teleconferencing and telepresence, smart logistics, and transportation.
- Education and Culture: IMT systems can change the method of education by providing easy access to digital textbooks or cloud-based storage of knowledge on the Internet, advancing applications such as e-learning, e-health, and e-commerce. IMT systems will support people to create collaborative art works or remotely/virtually participate in group performances or activities.

It is an inevitable fact that future spectrum is going to be highly fragmented, ranging from frequencies below 6 GHz to the mmWave band and THz frequency region. New hardware design and manufacturing paradigms are key to the exploitation of such a fragmented spectrum and in particular for the mmWave part of the spectrum, essential for achieving the data rates required for systems beyond 5G. Electronically steered high-gain 3D antennas and high levels of RF/DSP integration are going to be essential. With unprecedented aggregate data rates in the backbone network, optical fiber connections and wireless backhaul at E and W-bands and beyond will be required. It is an immense challenge to realize hardware (i.e., transceivers, filters, power amplifiers, low-noise amplifiers, mixers, and antennas) at frequencies from 28 to 300 GHz and beyond, with the low manufacturing costs demanded by network operators. Silicon RFIC technology is the key enabler, potentially even up to 1 THz, but considerable advances in RF design techniques are required in order to realize complete beyond 5G subsystems. Furthermore, while mmWave spectrum is the key to providing the expected 10 Gbps+ data rates, it should not be forgotten that the massive IoT presents another set of RF design challenges, such as ultra-low-power operation, energy harvesting, and sensing in ultra-low-cost technologies such as printed and wearable electronics [48].

References

ITU-R Specifications[32]

[1] Recommendation ITU-R M.2083-0, IMT Vision—Framework and Overall Objectives of the Future Development of IMT for 2020 and Beyond, September 2015.
[2] Recommendation ITU-R M.1645—Framework and Overall Objectives of the Future Development of IMT 2000 and Systems Beyond IMT 2000, June 2003.
[3] Recommendation ITU-R M.2012—Detailed Specifications of the Terrestrial Radio Interfaces of International Mobile Telecommunications Advanced (IMT-Advanced), January 2012.
[4] Report ITU-R M.2320—Future Technology Trends of Terrestrial IMT Systems, November 2014.
[5] Report ITU-R M.2370—IMT Traffic Estimates for the Years 2020 to 2030, July 2015.
[6] Report ITU-R M.2376—Technical Feasibility of IMT in Bands Above 6 GHz, July 2015.
[7] Report ITU-R M.2133, Requirements, Evaluation Criteria and Submission Templates for the Development of IMT-Advanced, December 2008.
[8] Report ITU-R M.2134—Requirements Related to Technical Performance for IMT Advanced Radio Interface(s), December 2008.
[9] Report ITU-R M.2135-1, Guidelines for Evaluation of Radio Interface Technologies for IMT-Advanced, December 2009.
[10] Report ITU-R M. [IMT-2020.EVAL], Guidelines for Evaluation of Radio Interface Technologies for IMT-2020, June 2017.
[11] Report ITU-R M. [IMT-2020.TECH PERF REQ], Minimum Requirements Related to Technical Performance for IMT-2020 Radio Interface(s), March 2017.
[12] FG IMT-2020: Report on Standards Gap Analysis, Focus Group on IMT-2020, IMT-O-016, October 2015.

3GPP Specifications[33]

[13] 3GPP TR 23.799: Study on Architecture for Next Generation System (Release 14), December 2016.
[14] 3GPP TR 38.913: Study on Scenarios and Requirements for Next Generation Access Technologies (Release 14), March 2017.
[15] 3GPP TR 38.912: Study on New Radio (NR) Access Technology (Release 14), March 2017.
[16] 3GPP TR 38.801: Study on New Radio Access Technology: Radio Access Architecture and Interfaces (Release 14), March 2017.
[17] 3GPP TS 38.300: NR; NR and NG-RAN Overall Description, Stage 2 (Release 15), April 2019.
[18] 3GPP TR 38.802: Study on New Radio Access Technology, Physical Layer Aspects (Release 14), March 2017.
[19] 3GPP TR 38.803: Study on New Radio Access Technology, RF and Co-existence Aspects (Release 14), March 2017.
[20] 3GPP TR 38.804: Study on New Radio Access Technology, Radio Interface Protocol Aspects (Release 14), March 2017.
[21] 3GPP TS 22.261: Service Requirements for the 5G system, Stage 1 (Release 16), June 2017.
[22] 3GPP TR 38.900: Study on Channel Model for Frequency Spectrum Above 6 GHz (Release 14), December 2016.

Articles, Books, White Papers, and Application Notes

[23] China Mobile Research Institute, C-RAN: The Road Towards Green RAN (Version 3.0), June 2014.
[24] China Mobile Research Institute, NGFI: Next Generation Fronthaul Interface (Version 1.0), June 2015.

[32] ITU-R specifications can be accessed at the following URL: http://www.itu.int/en/ITU-R/study-groups/rsg5/rwp5d/imt-2020/.

[33] 3GPP specifications can be accessed at the following URL: http://www.3gpp.org/ftp/Specs/archive/.

[25] NGMN Alliance, NGMN 5G White Paper v1.0, February 2015.

[26] NGMN, Further Study on Critical C-RAN Technologies, April 2015.

[27] D. Bojic et al., Advanced wireless and optical technologies for small cell mobile backhaul and dynamic software-defined management, IEEE Commun. Mag., 51 (9) (2013).

[28] M. Dohler, et al., Beyond 5G Challenges: White Paper, The University of Sheffield, March 2015.

[29] Global mobile Suppliers Association (GSA), The Future of IMT in the 3300-4200 MHz Frequency Range, June 2017.

[30] GSM Association, 5G Spectrum, Public Policy Position, November 2016.

[31] 5G Service-Guaranteed Network Slicing White Paper, China Mobile Communications Corporation, Huawei Technologies Co., Ltd., Deutsche Telekom AG, Volkswagen, February 2017.

[32] Mobile and Wireless Communications Enablers for the Twenty-Twenty Information Society-II (METIS II): Deliverable D1.1, Refined Scenarios and Requirements, Consolidated Use Cases, and Qualitative Techno-Economic Feasibility Assessment, January 2016.

[33] Mobile and Wireless Communications Enablers for the Twenty-Twenty Information Society (METIS), Deliverable D6.1, Simulation Guidelines, October 2013.

[34] K. Mallinson, The path to 5G: as much evolution as revolution, 3GPP. <http://www.3gpp.org/news-events/3gpp-news/1774-5g_wiseharbour>, May 2016.

[35] 5G Forum, Republic of Korea, 5G vision, requirements, and enabling technologies (V.1.0). <https://www.5gforum.org>, March 2015.

[36] IEEE Std 802.15.7-2011 – IEEE Standard for Local and Metropolitan Area Networks—Part 15.7: Short-Range Wireless Optical Communication Using Visible Light, September 2011.

[37] Evolving LTE to fit the 5G future, Ericsson Technology Review, January 2017.

[38] Wikipedia, Information-centric networking. <https://en.wikipedia.org/wiki/Information-centric_networking>.

[39] 5G-PPP Use Cases and Performance Evaluation Models. <http://www.5g-ppp.eu/>, April 2016.

[40] NTT DoCoMo Technical Journal, ITU Radio-Communication Assembly 2015 (RA-15) Report – Future Mobile Phone Technologies Standardization – 2015 ITU World Radio-communication Conference (WRC-15) Report – Standardization of Mobile Phone Spectrum Vol. 18 No. 1, July 2016.

[41] NMC Consulting Group (NETMANIAS), Timeline of 5G Standardization in ITU-R and 3GPP, January 2017.

[42] NMC Consulting Group (NETMANIAS), Network Architecture Evolution from 4G to 5G, December 2015.

[43] NMC Consulting Group (NETMANIAS), E2E Network Slicing – Key 5G Technology: What Is It? Why Do We Need It? How Do We Implement It?, November 2015.

[44] Ixia, Test Considerations for 5G, White Paper, July 2017.

[45] 5G Reimagined: A North American Perspective, Alliance for Telecommunications Industry Solutions (ATIS), February 2017.

[46] 5G Americas, White Paper on 5G Spectrum Recommendations, April 2017.

[47] Qualcomm Technologies, Inc. LTE-U/LAA, MuLTEfire™ and Wi-Fi; Making Best Use of Unlicensed Spectrum, September 2015.

[48] Skyworks Solutions, Inc., White Paper, 5G in Perspective: a Pragmatic Guide to What's Next, March 2017.

CHAPTER 1

5G Network Architecture

5G network architecture has been designed to support fast and reliable connectivity as well as diverse applications and services, enabling flexible deployments using new concepts, such as network function virtualization (NFV), software-defined networking (SDN), and network slicing. The 5G system supports a service-oriented architecture with modularized network services. The service-oriented 5G core (5GC) network is built on the principle that 5G systems must support wide range of services with different characteristics and performance requirements. The service-oriented architecture and interfaces in the 5G systems make the future networks flexible, customizable, and scalable. Network service providers can leverage service-oriented architecture design in 5G to manage and adapt the network capabilities, for example, by dynamically discovering, adding, and updating network services while preserving the performance and backward compatibility.

The main difference in the service-based architecture is in the control plane where, instead of predefined interfaces between entities, a service model is used in which components request a new network entity to discover and communicate with other entities over application programming interfaces (APIs). This notion is closer to the cloud networking concept and more attractive to the operators that demand flexibility and adaptability in their networks. The challenge is that it is harder to implement using today's cloud platforms and it is likely to be part of future deployments. It is also important for the 5G network to enable each network function (NF) to directly interact with other network functions. The architecture design does not preclude the use of an intermediate function to help route control-plane messages. It is also desirable to minimize dependencies between the access network (AN) and the core network.

There are some new concepts that have fundamentally changed the framework of 5G networks and made them differentiated from the previous generations. 5G network architecture leverages structural separation of hardware and software, as well as the programmability offered by software-defined network (SDN) and network function virtualization (NFV). As such, 5G network architecture is a native SDN/NFV architecture covering mobile/fixed terminals, infrastructure, NFs, enabling new capabilities and management and orchestration (MANO) functions. One of the most innovative concepts that has been incorporated into the design of the next-generation networks is the separation of user-plane and control-plane

5G NR. DOI: https://doi.org/10.1016/B978-0-08-102267-2.00001-4

functions, which allows individually scalable and flexible centralized or distributed network deployments. This concept forms the basis of the SDN. Other schemes include modularized functional design, which enables flexible and efficient network slicing. Having such requirements in mind, third-generation partnership project (3GPP) has developed a flat architecture where the control-plane functions are separated from the user plane in order to allow them to scale independently. Another central idea in the design of 5G networks has been to minimize dependencies between the AN and the core network with a unified access-agnostic core network and a common AN/core network interface which integrates different 3GPP and non-3GPP access types.

In order to further support multi-radio access, the network architecture was required to provision a unified authentication framework. The support of stateless NFs, where the compute resource elements are decoupled from the storage resource elements, is intended to create a disaggregated architecture. To support low-latency services and access to local data networks, the user-plane functions (UPFs) can be deployed close to the edge of the AN which further requires support of capability exposure and concurrent access to local and centralized services.

The combination of SDN and functional virtualization enables dynamic, flexible deployment, and on-demand scaling of NFs, which are necessary for the development of 5G mobile packet core networks. 5G network design requires a common core network associated with one or more ANs to be part of a network slice (e.g., fixed and mobile access within the same network slice). A network slice may include control-plane functions associated with one or more UPFs, and/or service or service-category-specific control-plane and user-plane functional pairs (e.g., user-specific multimedia application session). A device may connect to more than one slice. When a device accesses multiple network slices simultaneously, a control-plane function or a set of control-plane functions should be in common and shared among multiple network slices, and their associated resources. In order to enable different data services and requirements, the elements of the 5GC, also called NFs, have been further simplified with most of them being software-based so that they can be adapted and scaled on a need basis.

Today's static measurements of network and application performance are neither extensible to the dynamic nature of SDN/NFV-based 5G networks and functionalities, nor capable of creating any form of automation to create self-adapting behavior. The pace at which these environments change requires sophisticated analysis of real-time measurements, telemetry data, flow-based information, etc., in combination with user profile and behavior statistics. Creating a dynamic model to analyze the resulting big data requires artificial intelligence/machine learning techniques that will pave the way for migration from today's process-based analytics toward predictive, descriptive, and ultimately cognitive analytics required for self-organizing, self-optimizing, and self-healing networks.

3GPP has been working on the standardization of 5G access and core networks since 2015 with a goal of large-scale commercialization in 2020 + . 3GPP system architecture group finalized the first study items in December 2016 and published the 3GPP TR 23.799 specification as an outcome of the study. The normative specifications of the 5G network architecture and services have been published in numerous 3GPP standards documents [6].

In this chapter we discuss 5G network architecture design principles from 3GPP perspective and the innovative solutions that have formed the foundation of the 5G networks. We will further study the access/core network entities, interfaces, and protocols as well as the quality of service (QoS), security, mobility, and power management in 5G networks.

1.1 Design Principles and Prominent Network Topologies

The 5G system supports a service-based architecture and interfaces with modularized network services, enabling flexible, customizable, and independently deployable networks. Network service providers can leverage 5G service-oriented architecture to manage and customize the network capabilities by dynamically discovering, adding, and updating network services while preserving performance and compatibility with the existing deployments. 5GC and ANs were required to be functionally decoupled to create a radio technology agnostic architecture in order to realize the 5G performance targets for different usage scenarios. As an example, reduction in network transport latency requires placement of computing and storage resources at the edge of the network to enhance service quality and user experience. The tactile Internet is a forward-looking usage scenario under the category of ultra-reliable low-latency communication (URLLC) services. A notable requirement for enabling the tactile Internet is to place the content and context-bearing virtualized infrastructure at the edge of the AN [mobile edge computing (MEC)]. This relocation provides a path for new business opportunities and collaborative models across various service platforms. Improved access to the content, context, and mobility are vital elements to address the demands for reliability, availability, and low latency [62].

While much has been written and speculated about the next-generation radio standards that are going to form the basis of the forthcoming 5G systems, the core network is also an essential piece in achieving the goals set forth for these systems and in helping to ensure the competitiveness and relevance of network operations in the future. With the advent of heterogeneous ANs, that is, deployment of different radio access technologies with different coverage footprints, seamless connectivity, and service continuity can be provided in various mobility classes. The availability of different types of footprints for a given type of wireless access technology characterizes heterogeneous access, for example, a radio network access node such as a base station with large to small coverage area is referred to as macro-, pico-, and femtocell, respectively. A combination of these types of base stations offers the potential to optimize both coverage and capacity by appropriately distributing smaller-size base stations within a larger macro-base station coverage area. Since the radio access technology is common across these different types of base stations, common methods for configuration and operation can be utilized, thereby enhancing integration and operational efficiencies. The diversity of coverage, harnessing of spectrum (e.g., licensed and unlicensed spectrum), and different transmission power levels based on coverage area provide strategies for optimizing the allocation and efficient utilization of radio resources.

The expanding diversity of deployment options while considering the ultra-low latency, high reliability, availability, and mobility requirements demands a significant reduction in the overhead and complexity associated with the frequent setup and teardown of the access/core network bearers and tunneling protocols. However, the changes in the geographical location of point of attachment of a device to the AN, as a result of mobility, would inevitably add more overhead with tunneling, in a functionally virtualized network, which could adversely affect the delay-sensitive service experience. The 5G networks support multiple radio access mechanisms including fixed and mobile access, making fixed/mobile convergence an important consideration. 5G systems further support the use of non-3GPP access for off-loading and maintaining service continuity.

The logical/physical decomposition of radio NFs is required to meet the diverse information transport demands and to align them with the requirements of next-generation services in various use cases. The decomposition of the radio network protocols/functions across layer-1, layer-2, and layer-3 would depend on the degree of centralization or distribution required. It includes placing more functions corresponding to the upper layers of the radio network protocol stack in the distributed entities when high-performance transport (e.g., high bandwidth, high capacity, low latency, low jitter, etc.) is available. Optimized scheduling at a centralized entity is critical for high-performance transport across multiple distributed entities (e.g., base stations, remote radio heads (RRHs),[1] etc.). For relatively low-performance transport options, more functions corresponding to the upper layer of the radio network protocol stack is moved to the central entity to optimize the cost/performance metrics, associated with the distributed entities. The choice of functional split will determine the fronthaul/backhaul capacity requirements and associated latency specifications and

[1] The terms "remote radio head" abbreviated as RRH and "remote radio unit" abbreviated as RRU have the same meaning and will be used interchangeably in this book.

performance. This will impact the network architecture planning, since it determines the placement of nodes and permissible distance between them.

The core network in the 5G systems allows a user to access network services, independent of the type of access technology. The network service provider utilizes a common framework for authentication and billing via a unified customer database to authorize the access to a service independent of the type of access. The 5G system provides termination points or points of attachment to the core network for control-plane and user-plane entities. These points are selected based on location, mobility, and service requirements. They may dynamically change during the lifetime of a service flow, based on the aforementioned requirements. To achieve a unified core network, common mechanisms of attachment are supported for both 3GPP and non-3GPP ANs. The 5G system will allow simultaneous multiple points of attachment to be selected per device on a per-service flow basis. Control-plane functions and UPFs are clearly separated with appropriate open interfaces.

Device types are characterized by a variety of attributes including three broad categories of interfaces, namely human–human (H–H), human–machine (H–M), and machine–machine (M–M). Examples of devices that belong to these categories include smartphones (H–H), robots (H–M) or (M–M), drones (H–M) or (M–M), wearable devices (H–M), smart objects and sensors (M–M), etc. The attributes and capabilities associated with these devices are varying such as high power/low power, energy constraint/non-energy constraint, high cost/low cost, high performance/low performance, delay sensitive/delay tolerant, high reliability, and precision sensitive. These devices are distinguished in terms of diverse media types, such as audio, visual, haptic, vestibular, etc. The devices may be connected to a network either via a wired connection (e.g., Ethernet or optical transport) or a wireless connection (e.g., cellular, Wi-Fi, or Bluetooth). The cloud radio access model includes both composite and heterogeneous types of access, where moving the computational complexity and storage from the device to the edge of the network would enable diverse services using a variety of device types (e.g., H–H, H–M, and M–M) and would enable energy conservation in the devices with limited computing/storage resources.

Flexibility applies not only to network hardware and software but also to network management. An example would be the automation of network instance setup in the context of network slicing that relies on optimization of different NFs to deliver a specific service satisfying certain service requirements. Flexible management will enable future networks to support new types of service offerings that previously would have made no technical or economic sense. Many aspects of the 5G network architecture need to be flexible to allow services to scale. It is likely that networks will need to be deployed using different hardware technologies with different feature sets implemented at different physical locations in the network depending on the use case. In some use cases, the majority of user-plane traffic may require only very simple processing, which can be run on low-cost hardware, whereas other traffic may require more advanced/complex processing. Cost-efficient scaling of the user plane to handle the increasing individual and aggregated bandwidths is a key component of a 5GC network.

As we mentioned earlier, supporting the separation of the control-plane and user-plane functions is one of the most significant principles of the 5GC network architecture. The separation allows control- and user-plane resources to scale independently and supports migration to cloud-based deployments. By separating user- and control-plane resources, the user-plane/control-plane entities may also be implemented/instantiated in different logical/physical locations. For example, the control plane can be implemented in a central site, which makes management and operation less complex and the user plane can be distributed over a number of local sites, moving it closer to the user. This is beneficial, as it shortens the round-trip time between the user and the network service, and reduces the amount of bandwidth required between sites. Content caching is an example of how locating functions on a local site reduces the required bandwidth between sites. Separation of the control and user planes is a fundamental concept of SDN, as the flexibility of 5GC networks will improve significantly by adopting SDN principles. User-plane protocols, which can be seen as a chain of functions, can be deployed to suit a specific use case. Given that the connectivity needs of each use case varies, the most cost-efficient deployment can be uniquely created for each scenario. For example, the connectivity needs for a massive machine type communication (mMTC) service characterized by small payload and low mobility are quite different from the needs of an enhanced mobile broadband (eMBB) service with large payload and high mobility characteristics. An eMBB service can be broken down into several subservices, such as video streaming and web browsing, which can in turn be implemented by separate functional chains within the network slice. Such additional decomposition within the user-plane domain further increases the flexibility of the core network. The separation of the control and user planes enables the use of different processing platforms for each one. Similarly, different user planes can be deployed with different run-time platforms within a user plane, all depending on the cost efficiency of the solution. In the eMBB use case, one functional chain of services may run on general-purpose processors, whereas the service that requires simple user-data processing can be processed on low-cost hardware platforms. It is obvious that enabling future expansions requires greater flexibility in the way that the networks are built. While network slicing is a key enabler to achieving greater flexibility, increasing flexibility may lead to greater complexity at all levels of the system, which in turn tends to increase the cost of operation and delay the deployments.

Traditional radio access network (RAN) architectures consist of several stand-alone base stations, each covering relatively a small area. Each base station processes and transmits/receives its own signal to/from the mobile terminals in its coverage area and forwards the user data payloads from the mobile terminal to the core network via a dedicated backhaul link. Owing to the limited spectral resources, network operators reuse the frequency among different base stations, which can cause interference among neighboring cells.

There are several limitations in the traditional cellular architecture, including the cost and the operation and maintenance of the individual base stations; increased inter-cell interference level due to proximity of the other base stations used to increase network capacity; and variation of the amount of loading and user traffic across different base stations. As a result, the average utilization rate of individual base stations is very low since the radio resources cannot be shared among different base stations. Therefore, all base stations are designed to handle the maximum traffic and not the average traffic, resulting in overprovisioning of radio resources and increasing power consumption at idle times.

In earlier generations of cellular networks, the macro-base stations used an all-in-one architecture, that is, analog circuitry and digital processing hardware were physically co-located. The radio frequency (RF) signal generated by the base station transported over the RF cables up to the antennas on top of a tower or other mounting points. In more recent generations, a distributed base station architecture was introduced where the radio unit, also known as the RRH was separated from the digital unit, or baseband unit (BBU) through a fronthaul transport mechanism such as optical fiber. Complex-valued I/Q samples were carried over fiber using Common Public Radio Interface (CPRI)[2] between the RRH and the BBU. The RRH was installed on top of a tower close to the antenna, reducing the cable loss compared to the traditional base stations where the RF signal has to travel through a long cable from the base station cabinet to the antenna. The fiber link between RRH and BBU also allows more flexibility in network planning and deployment as they can be placed a few hundred meters or a few kilometers away. Most modern base stations now use this decoupled architecture [47].

The cloud-RAN (C-RAN) may be viewed as an architectural evolution of the distributed base station system (Fig. 1.1). It takes advantage of many technological advances in wireless and optical communication systems. For example, it uses the latest CPRI specifications,

[2] Common Public Radio Interface, http://www.cpri.info/.

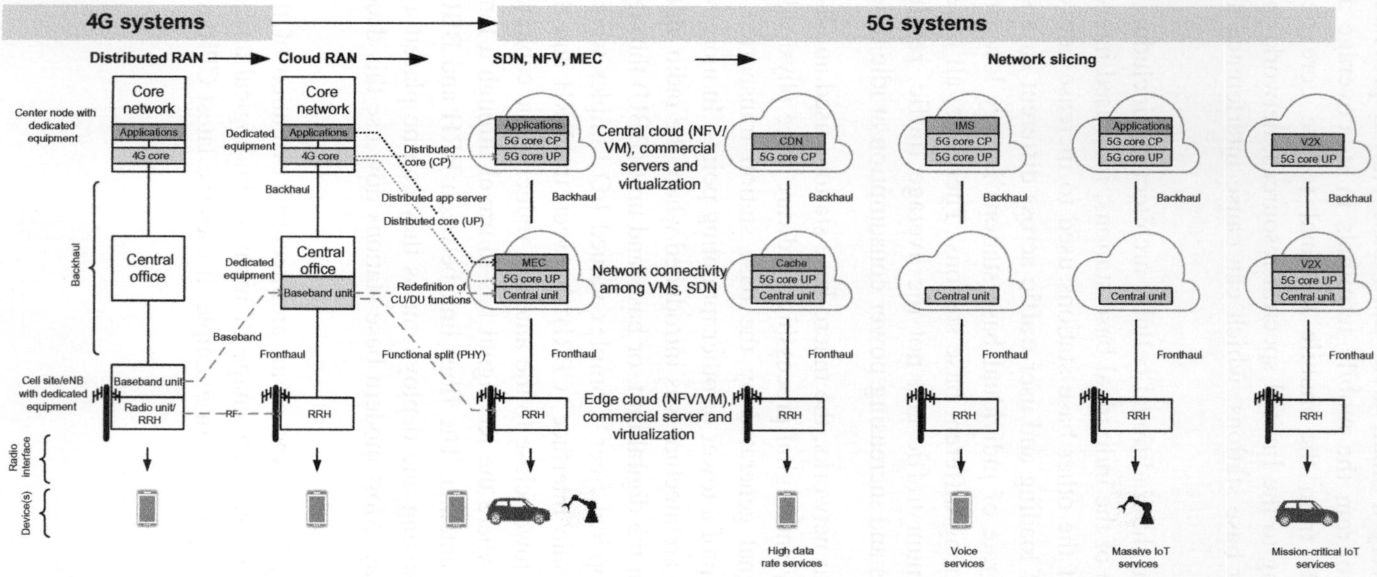

Figure 1.1
Network architecture evolution from 4G to 5G [60].

low-cost coarse/dense wavelength division multiplexing (CWDM/DWDM)[3] technology, and mmWave to allow transmission of baseband signals over long distance, thus achieving large-scale centralized base station deployment. It applies recent data center network technology to allow a low cost, high reliability, low latency, and high bandwidth interconnect network in the BBU pool. In the run up to 5G networks, the C-RAN utilizes open platforms and real-time virtualization technology rooted in cloud computing to achieve dynamic shared resource allocation and support of multi-vendor, multi-technology environments. Fig. 1.1 illustrates the evolution stages of 4G to 5G networks, where the distributed architectures evolved to centralized architectures, NFs have been virtualized, and later the

[3] Wavelength division multiplexing allows different data streams to be sent simultaneously over a single optical fiber network. There are two main types of wavelength division multiplexing technologies in use: Coarse wavelength division multiplexing and dense wavelength division multiplexing. Coarse wavelength division multiplexing allows up to 18 channels to be transported over a single dark fiber, while dense wavelength division multiplexing supports up to 88 channels. Both technologies are independent of transport protocol. The main difference between coarse wavelength division multiplexing and dense wavelength division multiplexing technologies lies in how the transmission channels are spaced along the electromagnetic spectrum. Wavelength division multiplexing technology uses infrared light, which lies beyond the spectrum of visible light. It can use wavelengths between 1260 and 1670 nm. Most fibers are optimized for the two regions 1310 and 1550 nm, which allow effective channels for optical networking. Coarse wavelength division multiplexing is a convenient and low-cost solution for distances up to 70 km. But between 40 and its maximum distance of 70 km, coarse wavelength division multiplexing tends to be limited to eight channels due to a phenomenon called the water peak of the fiber. Coarse wavelength division multiplexing signals cannot be amplified, making the 70 km estimate an absolute maximum. Dense wavelength division multiplexing works on the same principle as coarse wavelength division multiplexing, but in addition to the increased channel capacity, it can also be amplified to support much longer distances. The following figure shows how the dense wavelength division multiplexing channels fit into the wavelength spectrum compared to coarse wavelength division multiplexing channels. Each coarse wavelength division multiplexing channel is spaced 20 nm apart from the adjacent channel (www.Smartoptics.com).

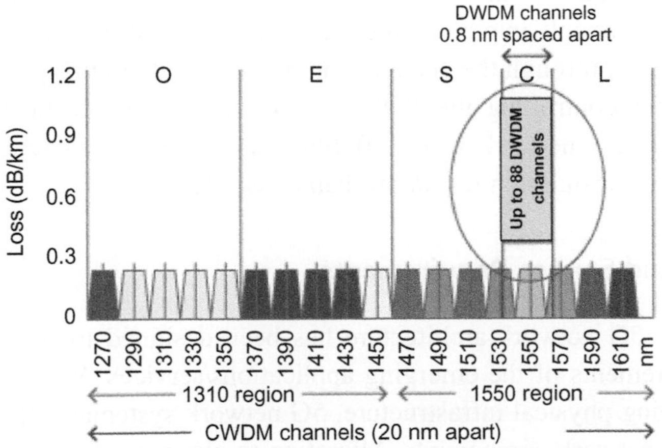

control-plane and user-plane functions were separated, and ultimately network slicing and edge computing have been introduced to further advance the network architectures toward flexibly supporting various 5G use cases and applications.

Having the above principles and requirements in mind, 3GPP 5G system (5GS) architecture has been designed to support data connectivity and new services by enabling deployments to use SDN/NFV methods. The 5GS architecture leverages service-based interactions between control-plane NFs and supports separation of user-plane functions from control-plane functions. The modularized functional design would allow flexible and efficient network slicing. It defines procedures, that is, set of interactions between NFs, as services so that their reuse is possible. It enables each NF to directly interact with other NFs. The 5GS design minimizes dependencies between the access and the core networks. It further supports a unified authentication framework, stateless NFs, where compute resources are decoupled from storage resources, and capability exposure, as well as concurrent access to local and centralized services. The 5GS supports roaming with both home-routed traffic as well as local breakout traffic in the visited network. In 5GS, the interactions between NFs are represented either through a service-based representation, where NFs within the control plane enable other authorized NFs to access their services which may include point-to-point reference points; or a reference-point representation, where the interactions between the NFs are described by point-to-point reference points between any two NFs. The NFs within the 5GC network control plane use service-based interfaces for their interactions [3].

The general principles that guided the definition of 3GPP 5G radio access network (NG-RAN) and 3GPP 5GC network architecture and network interfaces are based on logical separation of signaling and data transport networks, as well as separation of NG-RAN and 5GC functions from the transport functions. As a result, the addressing schemes used in NG-RAN and 5GC are decoupled from the addressing schemes of the transport functions. The protocols over the air interface and the NG interfaces are divided into user-plane protocols, which are the protocols implementing the actual protocol data unit (PDU) session service, carrying user data through the access stratum (AS); and control-plane protocols, which are the protocols for controlling the PDU sessions and the connection between the user equipment (UE) and the network from different aspects, including requesting the service, controlling different transmission resources, handover, etc. [15].

1.1.1 Network and Service Requirements

The service-centric 5G network architecture has been designed to flexibly and efficiently meet diverse requirements of the emerging applications/services. With SDN and NFV supporting the underlying physical infrastructure, 5G network systematically centralizes access, transport, and core network components. Migration to cloud-based architectures is meant to support wide-ranging 5G services and enables key technologies, such as network slicing, on-demand deployment of service anchors, and component-based NFs.

The design principles of 3GPP next-generation system architecture have deviated from those of the long-term evolution (LTE) evolved packet core (EPC) network in order to address the challenging requirements of 5G applications/services. While the design of 5G network started from a clean slate, the requirements for support of the new radio access technologies (RATs) as well as the evolved LTE, legacy systems, and non-3GPP radio access have caused the new design to borrow a large amount of concepts from the predecessor networks. Some of the key tenets of 5G network design include the requirement for logically independent network slicing based on a single network (physical) infrastructure to meet the 5G service requirements; and to provide dual-connectivity-based cloud architecture to support various deployment scenarios. The network design further relies on C-RAN architecture to deploy different radio access technologies in order to provide multi-standard connectivity and to implement on-demand deployment of RAN functions required by 5G services. It simplifies core network architecture to implement on-demand configuration of NFs through control and user-plane separation, component-based functions, and unified database management. It further implements automatic network slicing service generation, maintenance, and termination for various services to reduce operating expenses through agile network operation and management.

3GPP TS 22.261 specification, service requirements for the 5G system, contains performance targets and basic capabilities prescribed for the 5G networks. Among those requirements, one can distinguish support for fixed, mobile, wireless, and satellite access technologies; scalable and customizable network that can be tailored to serve multiple services and vertical markets (e.g., network slicing, NFV); resource efficiency for services ranging from low-rate Internet of things (IoT) services to high-bandwidth multimedia services; energy efficiency and network power optimization; network capability exposure to allow third party Internet service providers and Internet content providers to manage network slices, and deploy applications in the operator's service hosting environment; indirect connectivity from a remote UE via a relay UE to the network; and service continuity between indirect connections and direct connections. 3GPP TS22.261 defines performance targets for different scenarios (e.g., urban macro, rural macro, and indoor hotspot) and applications (e.g., remote control, monitoring, intelligent transport systems, and tactile communications).

The general requirements that led the design of NG-RAN architecture and interfaces included logical separation of signaling and data transport networks and separation of access and core NFs from transport functions, regardless of their possible physical co-location. Other considerations included independence of addressing scheme used in NG-RAN and 5GC from those of transport functions and control of mobility for radio resource control (RRC) connection via NG-RAN. The functional division across the NG-RAN interfaces has limited options and the interfaces are based on a logical model of the entity controlled through the corresponding interface. As was the case with LTE, one physical network element can implement multiple logical nodes.

1.1.2 Virtualization of Network Functions

NFV is an alternative approach to design, deploy, and manage networking services as well as a complement to SDN for network management. While they both manage networks, they rely on different methods. SDN separates the control and forwarding planes to offer a centralized view of the network, whereas NFV primarily focuses on optimizing the network services themselves.

NFV transforms the way that network operators architect networks by evolving standard server-based virtualization technology to consolidate various network equipment types into industrial-grade high-volume servers, switches,[4] and storage, which could be located in data centers, network nodes, and in the end-user premises. The NFV involves implementation of NFs in software that can run on a range of network server hardware that can be moved to or instantiated in various locations in the network as required, without the need for installation of new equipment. The NFV is complementary to SDN, but not dependent on it or vice versa. It can be implemented without an SDN being required, although the two concepts/solutions can be combined to gain potentially greater value. NFV goals can be achieved using non-SDN mechanisms, relying on the techniques currently in use in many data centers. But approaches relying on the separation of the control and data forwarding planes as proposed by SDN can enhance performance, simplify compatibility with the existing deployments, and facilitate operation and maintenance procedures. The NFV is able to support SDN by providing the infrastructure upon which the SDN software can be run. Furthermore, NFV aligns closely with the SDN objectives to use commodity servers and switches. The latter is applicable to any user-plane or control-plane functional processing in mobile and fixed networks. Some example application areas include switching elements, mobile core network nodes, functions contained in home routers and set top boxes, tunneling gateway elements, traffic analysis, test and diagnostics, Internet protocol (IP) multimedia subsystem, authentication, authorization, and accounting (AAA) servers,[5] policy control and charging platforms, and security functions [48].

---

[4] Switch is a device that typically transports traffic between segments of a single local area network. Internal firmware instructs the switch where to forward each packet it receives. Typically, a switch uses the same path for every packet. In a software-defined networking environment, the switches' firmware that dictates the path of packets would be removed from the device and moved to the controller, which would orchestrate the path based on a macro-view of real-time traffic patterns and requirements.

[5] Authentication, authorization, and accounting is a framework for intelligently controlling access to computer resources, enforcing policies, auditing usage, and providing the information necessary to bill for services. These combined processes are considered important for effective network management and security. An authentication, authorization, and accounting server is a server program that handles user requests for access to computer resources and, for an enterprise, provides authentication, authorization, and accounting services. The authentication, authorization, and accounting server typically interacts with network access and gateway servers and with databases and directories containing user information.

NFV leverages modern technologies such as those developed for cloud computing. At the core of these cloud technologies are virtualization mechanisms. Hardware virtualization is realized by means of hypervisors[6] as well as the usage of virtual Ethernet switches (e.g., vSwitch[7]) for connecting traffic between virtual machines (VMs) and physical interfaces. For communication-oriented functions, high-performance packet processing is made possible through high-speed multi-core CPUs with high I/O bandwidth, smart network interface cards for load sharing and transmission control protocol (TCP) offloading, routing packets directly to VM memory, and poll-mode Ethernet drivers (rather than interrupt driven; e.g., Data Plane Development Kit[8]). Cloud infrastructures provide methods to enhance resource availability and usage by means of orchestration and management mechanisms, which is applicable to the automatic instantiation of virtual appliances in the network, management of resources by properly assigning virtual appliances to the CPU cores, memory and interfaces, reinitialization of failed VMs,[9] snapshot of VM states, and the migration of VMs. As shown in Fig. 1.2, containers and VMs are two ways to deploy multiple, isolated services on a single platform [42–44].

The decision whether to use containers or VMs depends on the objective. Virtualization enables workloads to run in environments that are separated from their underlying hardware by a layer of abstraction. This abstraction allows servers to be divided into virtualized machines that can run different operating systems. Container technology offers an alternative method for virtualization, in which a single operating system on a host can run many

[6] A hypervisor is a function which abstracts or isolates the operating systems and applications from the underlying computer hardware. This abstraction allows the underlying host machine hardware to independently operate one or more virtual machines as guests, allowing multiple guest virtual machines to effectively share the system's physical compute resources, such as processor cycles, memory space, and network bandwidth.

[7] A virtual switch is a software program that allows one virtual machine to communicate with another. Similar to its counterpart, the physical Ethernet switch, a virtual switch does more than just forwarding data packets. It can intelligently direct communication on the network by inspecting packets before forwarding them. Some vendors embed virtual switches into their virtualization software, but a virtual switch can also be included in a server's hardware as part of its firmware. Open vSwitch is an open-source implementation of a distributed virtual multilayer switch. The main purpose of Open vSwitch is to provide a switching stack for hardware virtualization environments, while supporting multiple protocols and standards used in computer networks.

[8] Data plane development kit is a set of data-plane libraries and network interface controller drivers for fast packet processing from Intel Corporation. The data plane development kit provides a programming framework for x86-based servers and enables faster development of high-speed data packet networking applications. The data plane development kit framework creates a set of libraries for specific hardware/software environments through the creation of an environment abstraction layer. The environment abstraction layer conceals the environmental-specific parameters and provides a standard programming interface to libraries, available hardware accelerators, and other hardware and operating system (Linux, FreeBSD) elements.

[9] A virtual machine is an operating system or application environment that is installed on software, which imitates dedicated hardware. The end user has the same experience on a virtual machine as they would have on dedicated hardware.

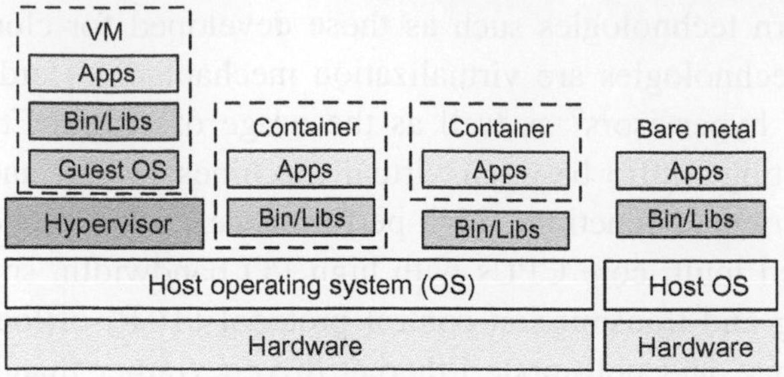

Figure 1.2
NFV software/hardware architecture models.

different applications from the cloud. One way to think of containers versus VMs is that VMs run several different operating systems on one compute node, whereas container technology offers the opportunity to virtualize the operating system itself. A VM is a software-based environment geared to simulate a hardware-based environment, for the benefit of the applications it will host. Conventional applications are designed to be managed by an operating system and executed by a set of processor cores. Such applications can run within a VM without any rearchitecture. On the contrary, container technology has been around for more than a decade and is an approach to software development in which pieces of code are packaged in a standardized way so that they can quickly be plugged in and run on the Linux operating system. This enables portability of code and allows the operating system to be virtualized and share an instance of an operating system in a same way that a VM would divide a server. Therefore instead of virtualizing the hardware like a VM, a container virtualizes at the operating system level. Containers run at a layer on top of the host operating system and they share the kernel. Containers have much lower overhead relative to the VMs and much smaller footprint.

NFV decouples software implementations of NFs from the compute, storage, and networking resources they use. It thereby expands options for both enterprises and service providers, enabling both to create new capabilities and new services for their customers. With new opportunities come new challenges. By tradition, NF implementations are packaged with the infrastructure they utilize; however, this may not be the case anymore. As the physical network is decoupled from the infrastructure and network services, it is necessary to create both new management tools and orchestration solutions for providers to realize the benefits of NFV-based solutions. There are a number of challenges to implement NFV, which need to be addressed by the industry. Some of the main challenges include the following [43]:

- *Portability/interoperability:* This is the ability to load and execute virtual appliances in different but standardized data center environments, which can be provided by different vendors for different operators. The challenge is to define a unified interface which clearly decouples the software instances from the underlying hardware, as represented by VMs and their hypervisors. Portability and interoperability are very important as they create different ecosystems for virtual appliance vendors and data center vendors, while both ecosystems are clearly coupled and depend on each other. Portability also provides the operator with the freedom to optimize the location and required resources of the virtual appliances without constraints.

- *Performance trade-off:* Since the NFV approach is based on conventional hardware as opposed to customized hardware, there could be a possible decrease in performance. The challenge is how to limit the performance degradation by using appropriate hypervisors, hardware accelerators, and advanced software technologies such that the effects on latency, throughput, and processing overhead are minimized.
- *Migration, coexistence, and compatibility with the existing platforms:* Implementations of NFV must coexist with network operators' legacy network equipment, and further it must be compatible with their existing element management systems (EMSs),[10] network management systems (NMSs), operations support system (OSS),[11] and business support system (BSS),[12] and potentially existing IT orchestration systems, if NFV orchestration and IT orchestration need to converge. The NFV architecture must support a migration path from today's proprietary physical network appliance-based solutions to more open standards-based virtual network appliance solutions. In other words, NFV must work in a hybrid network composed of classical physical network appliances and virtual network appliances. Virtual appliances must therefore use existing north-bound interfaces (for management and control) and interwork with physical appliances implementing the same functions.
- *Management and orchestration:* NFV presents an opportunity through the flexibility afforded by software network appliances operating in an open and standardized infrastructure to rapidly align MANO north-bound interfaces to well-defined standards and abstract specifications. Therefore, a consistent MANO architecture is required. This will greatly reduce the cost and time to integrate new virtual appliances into a network

[10] Element management system consists of systems and applications for managing network elements on the network element management layer. An element management system manages a specific type of telecommunications network element. The element management system typically manages the functions and capabilities within each network element but does not manage the traffic between different network elements in the network. To support management of the traffic between network elements, the element management system communicates upward to higher-level network management systems, as described in the telecommunications management network layered model.

[11] Operations support system is a platform used by service providers and network operators to support their network systems. The operations support system can help the operators to maintain network inventory, provision services, configure components, and resolve network issues. It is typically linked with the business support system to improve the overall customer experience.

[12] Business support systems (BSS) are platforms used by service providers, network operator delivery product management, customer management, revenue management (billing) and order management applications that help them run their business operations. Business support system platforms are often linked to operations support system platforms to support the overall delivery of services to customers.

operator's operating environment. The SDN further extends this concept to streamlining the integration of packet and optical switches into the system, for example, a virtual appliance or NFV orchestration system may control the forwarding behavior of physical switches using SDN. Note that NFV will only scale, if all of the functions can be automated.

- *Security and resilience:* Network operators need to be assured that the security, resilience, and availability of their networks are not compromised when VNFs are introduced. The NFV improves network resilience and availability by allowing NFs to be recreated on demand after a failure. A virtual appliance should be as secure as a physical appliance if the infrastructure, particularly the hypervisor and its configuration, is secure. Network operators will be seeking tools to control and verify hypervisor configurations. They will also require security-certified hypervisors and virtual appliances.
- *Network stability:* It is important to ensure that the stability of the network is not impacted when managing and orchestrating a large number of virtual appliances created by different hardware and hypervisor vendors. This is particularly important when virtual functions are relocated or during reconfiguration events (e.g., due to hardware and software failures) or due to a cyber attack. This challenge is not unique to NFV systems and such unsteadiness might also occur in current networks. It should be noted that the occurrence of network instability may have adverse effects on performance parameters or optimized use of resources.
- *Complexity:* It must be ensured that virtualized network platforms will be simpler to operate than those that exist today. A significant focus area for network operators is simplification of the plethora of complex network platforms and support systems which have evolved over decades of network technology evolution, while maintaining continuity to support important revenue-generating services.
- *Integration:* Seamless integration of multiple virtual appliances into existing industrial-grade servers and hypervisors is a major challenge for NFV schemes. Network operators need to be able to combine servers, hypervisors, and virtual appliances from different vendors without incurring significant integration costs. The ecosystem offers integration services and maintenance and third-party application support. It must be possible to resolve integration issues between several suppliers. The ecosystem will require mechanisms to validate new NFV products. Tools must be identified and/or created to address these issues.

1.1.2.1 Architectural Aspects

The NFV initiative began when network operators attempted to accelerate deployment of new network services in order to advance their revenue and growth plans. They found that customized hardware-based equipment limited their ability to achieve these goals. They studied standard IT virtualization technologies and found NFV helped accelerate service innovation and provisioning in that space [48].

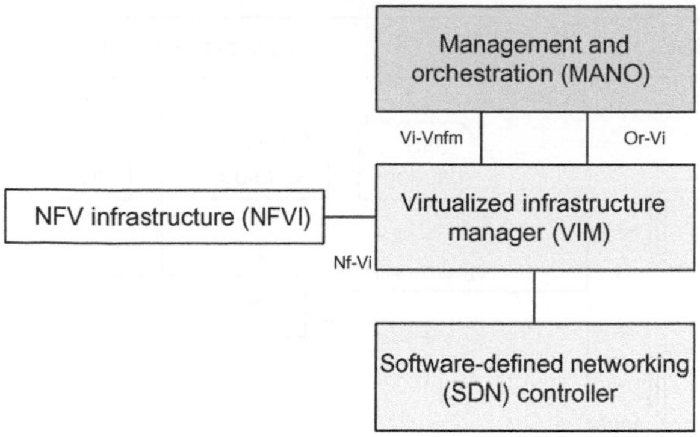

Figure 1.3
High-level concept of a virtualized network.

Fig. 1.3 illustrates the conceptual structure of a virtualized network and its main components. Following conversion of physical NFs to software, that is, virtual network functions (VNFs), the software applications need a platform to run. The NFV infrastructure (NFVI) consists of physical and virtual compute, storage, and networking resources that VNFs need to run. The NFVI layer primarily interacts with two other NFV framework components: VNFs and the virtual infrastructure manager (VIM). As we mentioned earlier, the VNF software runs on NFVI. The VIM, on the other hand, is responsible for provisioning and managing the virtual infrastructure. As shown in Fig. 1.4, the VNF to NFVI interface (Vn–Nf) constitutes a data path through which network traffic traverses, while the NFV to VIM interface (Nf–Vi) creates a control path that is used solely for management but not for any network traffic. The NFVI consists of three distinct layers: physical infrastructure, virtualization layer, and the virtual infrastructure. The VIM manages the NFVI and acts as a conduit for control path interaction between VNFs and NFVI. In general, the VIM provisions, de-provisions, and manages virtual compute, storage, and networking while communicating with the underlying physical resources. The VIM is responsible for operational aspects such as logs, metrics, alerts, analytics, policy enforcement, and service assurance. It is also responsible for interacting with the orchestration layer and SDN controller. Unlike the NFVI which consists of several technologies that can be assembled independently, the VIM comes in the form of complete software stacks. OpenStack[13] is the main VIM software stack which is very common in NFV realization.

[13] OpenStack software controls large pools of compute, storage, and networking resources throughout a data center, managed through a dashboard or via the OpenStack application programming interface. OpenStack works with popular enterprise and open source technologies, making it ideal for heterogeneous infrastructure (https://www.openstack.org/).

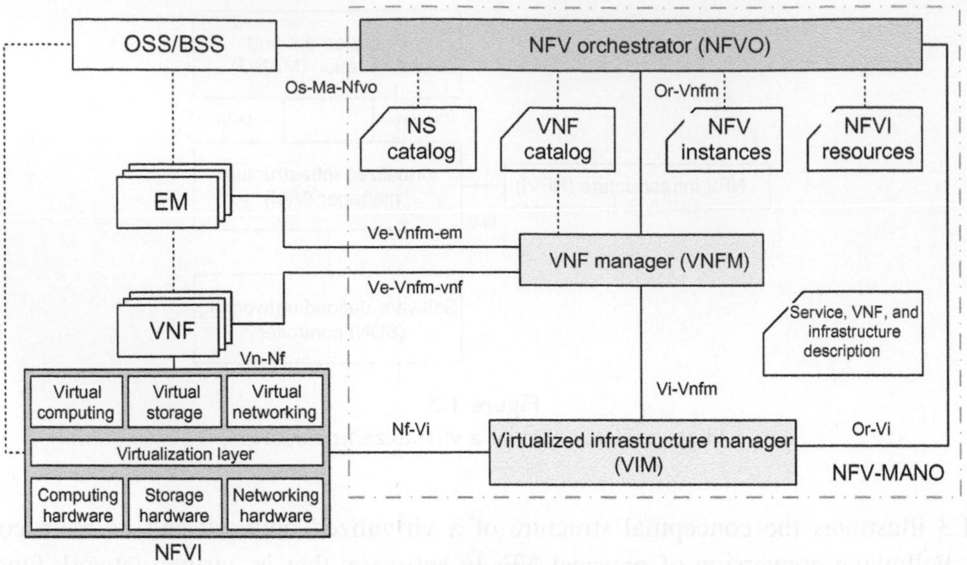

Figure 1.4
NFV architecture [37,40].

As we mentioned earlier, NFV defines standards for compute, storage, and networking resources that can be used to build VNFs. The NFVI is a key component of the NFV architecture that describes the hardware and software components on which virtual networks are built. The NFV leverages the economies of scale of the IT industry. The NVFI is based on widely available and low-cost, standardized computing components. The NFVI works with different types of servers, for example, virtual or bare metal, software, hypervisors, VMs, and VIMs in order to create a platform for VNFs to run. The NFVI standards help increase the interoperability of the components of the VNFs and enable multivendor environments [42–44].

The NFV architecture comprises major components including VNFs, NFV-MANO, and NFVI that work with traditional network components like OSS/BSS. The NFVI further consists of NFVI points-of-presence (NFVI-PoPs[14]), which are the sites at which the VNFs are deployed by the network operator, including resources for computation, storage, and networking. NFVI networks interconnect the computing and storage resources contained in an NFVI-PoP. This may include specific switching and routing devices to allow external connectivity. The NFVI works directly with VNFs and VIMs and in concert with the NFV orchestrator (NFVO). NFV services are instantiated and instructed by the NFVO, which utilizes VIMs that manage the resources from the underlying infrastructure. The NFVI is

[14] Network function virtualization infrastructure points-of-presence is a single geographic location where a number of network function virtualization infrastructure nodes are situated.

critical to realizing the business benefits outlined by the NFV architecture. It delivers the actual physical resources and corresponding software on which VNFs can be deployed. NFVI creates a virtualization layer on top of the hardware and abstracts the hardware resources, so they can be logically partitioned and allocated to the VNF in order to perform their functions. NFVI is also critical to building more complex networks without geographical limitations of traditional network architectures.

A network service can be viewed architecturally as a forwarding graph of NFs interconnected by the supporting network infrastructure. These NFs can be implemented in a single operator network or inter-work between different operator networks. The underlying NF behavior contributes to the behavior of the higher level service. Therefore, the network service behavior is a combination of the behavior of its constituent functional blocks, which can include individual NFs, NF sets, NF forwarding graphs, and/or the infrastructure network. The end points and the NFs of the network service are represented as nodes and correspond to devices, applications, and/or physical server applications. An NF forwarding graph can have NF nodes connected by logical links that can be unidirectional, bidirectional, multicast, and/or broadcast. Fig. 1.4 shows the NFV architectural framework depicting the functional blocks and reference points in the NFV framework. The main reference points and execution reference points are shown by solid lines and are in the scope of European Telecommunications Standards Institute (ETSI) NFV specification [37–41]. The dotted reference points are available in present deployments but may need extensions for handling NFV. However, the dotted reference points are not the main focus of the NFV at present. The illustrated architectural framework focuses on the functionalities necessary for the virtualization and the resulting operation of the network. It does not specify which NFs should be virtualized, as that is solely a decision of the network operator.

1.1.2.2 Functional Aspects

The NFV architectural framework, shown in Fig. 1.4, identifies functional blocks and the main reference points between the blocks. Some of these blocks are already present in current deployments, whereas others might be necessary additions in order to support the virtualization process and the subsequent operation. The functional blocks are as follows [37]:

- VNF is a virtualization of a network function in a legacy non-virtualized network.
- Element management (EM) performs the typical management functionality for one or several VNFs. NFV elements are the discrete hardware and software requirements that are managed in an NFV installation to provide new communication services and application services on commodity-based hardware. NFV services are deployed on commercial off-the-shelf hardware platform, typically run on x86-based or ARM-based computing platform and standard switching hardware. The early model of NFV, ETSI MANO, is a common reference architecture. The NFV architecture developed by ETSI

MANO includes EMSs, which describe how individual VNFs are managed on a commodity hardware platform.

- NFVI represents the entire hardware and software components which create the environment in which VNFs are deployed, managed, and executed. The NFVI can span across several locations, that is, places where NFVI-PoPs are operated. The network providing connectivity between these locations is regarded as part of the NFVI.
- Virtualization layer abstracts the hardware resources and decouples the VNF software from the underlying hardware, thus ensuring a hardware independent life cycle for the VNFs. The virtualization layer is responsible for abstracting and logically partitioning physical resources; enabling the software that implements the VNF to use the underlying virtualized infrastructure; and providing virtualized resources to the VNF. The virtualization layer ensures VNFs are decoupled from hardware resources, thus the software can be deployed on different physical hardware resources. Typically, this type of functionality is provided for computing and storage resources in the form of hypervisors and VMs. A VNF can be deployed in one or several VMs.
- VIM(s) comprises the functionalities that are used to control and manage the interaction of a VNF with computing, storage, and network resources under its authority as well as their virtualization.
- NFVO is in charge of the orchestration and management of NFVI and software resources and realizing network services on NFVI.
- VNF manager(s) is responsible for VNF life cycle management (e.g., instantiation, update, query, scaling, and termination). Multiple VNF managers may be deployed where a VNF manager may be deployed for each VNF or multiple VNFs.
- Service, VNF and infrastructure description is a data set which provides information regarding the VNF deployment template, VNF forwarding graph, service-related information, and NFVI information models. These templates/descriptors are used internally within NFV-MANO. The NFV-MANO functional blocks handle information contained in the templates/descriptors and may expose (subsets of) such information to applicable functional blocks.
- Operations and Business Support Systems (OSS/BSS)

The management and organization working group of the ETSI[15] has defined the NFV-MANO architecture. According to ETSI specification, NFV-MANO comprises three major functional blocks: VIM, VNF manager, and NFVO [37,40]. The VIM is a key component of the NFV-MANO architectural framework. It is responsible for controlling and managing the NFVI compute, storage, and network resources, usually within one operator's infrastructure domain (see Fig. 1.4). These functional blocks help standardize the functions of virtual

[15] European Telecommunications Standards Institute, http://www.etsi.org/.ETSI, network function virtualization specifications are listed and can be found at http://www.etsi.org/technologies-clusters/technologies/nfv.

networking to increase interoperability of SDN elements. The VIMs can also handle hardware in a multi-domain environment or may be optimized for a specific NFVI environment. The VIM is responsible for managing the virtualized infrastructure of an NFV-based solution. The VIM operations include the following:

- It maintains an inventory of the allocation of virtual resources to physical resources. This allows the VIM to orchestrate the allocations, upgrade, release, and retrieval of NFVI resources and optimize their use.
- It supports the management of VNF forwarding graphs by organizing virtual links, networks, subnets, and ports.
- The VIM also manages security group policies to ensure access control.
- It manages a repository of NFVI hardware resources (compute, storage, and networking) and software resources (hypervisors), along with the discovery of the capabilities and features to optimize the use of such resources.
- The VIM performs other functions as well, such as collecting performance and failure information via notifications; managing software images (add, delete, update, query, or copy) as requested by other NFV-MANO functional blocks; and managing catalogs of virtualized resources that can be used by NFVI.

In summary, the VIM is a management layer between the hardware and the software in an NFV domain. VIMs are critical to realizing the business benefits that can be provided by the NFV architecture. They coordinate the physical resources that are necessary to deliver network services. This is particularly noticeable by infrastructure-as-a-service providers, where their servers, networking, and storage resources must work smoothly with the software components running on top of them. They must ensure that resources can be appropriately allocated to fulfill the dynamic service requirements.

The NFVI consists of three distinct layers: physical infrastructure, virtualization layer, and the virtual infrastructure, as shown in Fig. 1.4. The NFVI hardware consists of computing, storage, and networking components. OpenStack is often used in conjunction with NFV technology in data centers to deploy cloud services, especially communication services offered by large service providers and cloud providers. OpenStack is an open source virtualization platform. It enables the service providers to deploy VNFs using commercial off-the-shelf hardware. These applications are hosted in a data center so that they could be accessed via the cloud, which is the underlying model to use NFV. The VIM manages the NFVI and serves as a conduit for control path interaction between VNFs and NFVI. The VIM assigns, provisions, de-provisions, and manages virtual computing, storage, and networking resources while communicating with the underlying physical resources. The VIM is responsible for operational aspects, such as logs, metrics, alerts, root cause analysis, policy enforcement, and service assurance. It is also responsible for interacting with the orchestration layer (MANO) and SDN controller.

NFV is managed by NFV-MANO, which is an ETSI-defined framework for the management and orchestration of all resources in the cloud data center. This includes computing, networking, storage, and VM resources. The main focus of NFV-MANO is to allow flexible on-boarding and to avoid the possible disorder that can arise during transition states of network components. As we mentioned earlier, the NFV-MANO consists of three main components:

1. NFVO is responsible for on-boarding of new network services and VNF packages; network service life cycle management; global resource management; and validation and authorization of NFVI resource requests. The NFVO is a key component of the NFV-MANO architectural framework, which helps standardize the functions of virtual networking to increase interoperability of SDN-controlled elements. Resource management is important to ensure there are adequate compute, storage, and networking resources available to provide network services. To meet that objective, the NFVO can work either with the VIM or directly with NFVI resources, depending on the requirements. It has the ability to coordinate, authorize, release, and engage NFVI resources independent of any specific VIM. It can also control VNF instances sharing resources of the NFVI.
2. VNF manager oversees life cycle management of VNF instances; coordination and adaptation role for configuration; and event reporting between NFVI and EMs.
3. VIM controls and manages the NFVI compute, storage, and networking resources.

1.1.2.3 Operational Aspects

A VNF may be composed of one or multiple VNF components (VNFC). A VNFC may be a software entity deployed in the form of a virtualization container. A VNF realized by a set of one or more VNFCs appears to the outside as a single, integrated system; however, the same VNF may be realized differently by each VNF provider. For example, one VNF developer may implement a VNF as a monolithic, vertically integrated VNFC, and another VNF developer may implement the same VNF using separate VNFCs, for example, one for the control plane, one for the user plane, and one for the EM. VNFCs of a VNF are connected in a graph. For a VNF with only a single VNFC, the internal connectivity graph is a null graph [37].

A VNF can assume a number of internal states to represent the status of the VNF. Transitions between these states provide architectural patterns for some expected VNF functionality. Before a VNF can start its life cycle, it is a prerequisite that the VNF was on-boarded (process of registering the VNF with the NFVO and uploading the VNF descriptor). Fig. 1.5 provides a graphical overview of the VNF states and state transitions. Each VNFC of a VNF is either parallelizable or nonparallelizable. If it is parallelizable, it may be instantiated multiple times per VNF instance, but there may be a constraint on the minimum and maximum number of parallel instances. If it is nonparallelizable, it is instantiated once per VNF instance. Each VNFC of a VNF may need to handle the state information, where it can be either stateful or stateless. A VNFC that does not have to handle state information is

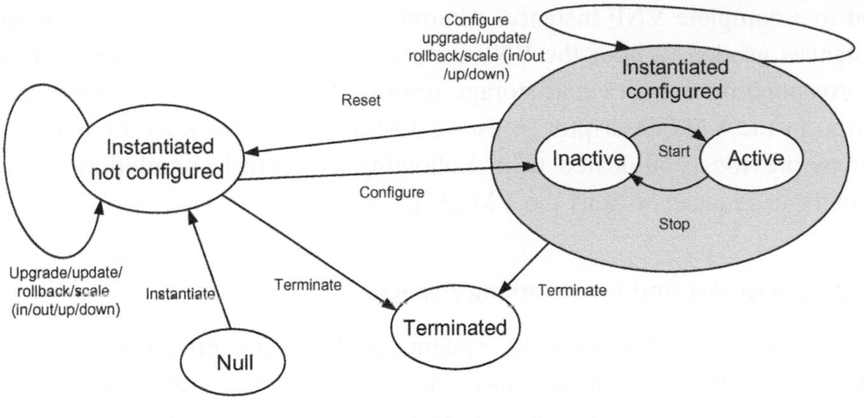

Figure 1.5
VNF instance and state transitions [37].

a stateless VNFC. A VNFC that needs to handle state information may be implemented either as a stateful VNFC or as a stateless VNFC with an external state where the state information is held in a data repository external to the VNFC.

Depending on the type of VNF, the instantiation can be more or less complex. For a simple VNF consisting of a single VNFC, the instantiation based on VNF descriptor is straightforward, while for a complex VNF consisting of several VNFCs connected via virtual networks, the VNF manager may require support from an internal function that is implemented by VNF to facilitate the process of instantiation. As an example, the VNF manager could boot a predefined number of VNFCs, leaving booting of the remaining VNFCs and the configuration of the VNF to an internal VNF provider-specific process that may also involve an EM.

The VNF descriptor is a specification template provided by the VNF developer for describing virtual resource requirements of a VNF. It is used by the NFV-MANO functions to determine how to execute VNF life cycle operations such as instantiation. The NFV-MANO functions consider all VNF descriptor attributes to check the feasibility of instantiating a given VNF. There are several options for how the instances of individual VNFCs can be created, which can be fully or partially loaded virtualization containers; or empty virtualization containers prepared for booting and loading. It is then the responsibility of the VNF MANO functions to instruct the VIM to create an empty virtualization container with an associated interface that is ready for use.

To instantiate a VNF, the VNF manager creates the VNF's set of VNFC instances as defined in the VNF descriptor by executing one or more VNFC instantiation procedures. The VNF descriptor defines which VNFC instances may be created in parallel or sequentially as well as the order of instantiation. The set of created VNFC instances may already correspond to a complete VNF instance. Alternatively, it may contain only a minimal set of VNFC instances needed to boot the VNF instance. The VNF manager requests a new VM and the corresponding network and storage resources for the VNFC instance according to the definition in the VNF descriptor or uses a VM and the corresponding network and storage resources previously allocated to it. Following successful completion of this process, the VNF manager requests to start the VM [37].

1.1.2.4 Legacy Support and Interworking Aspects

In general, the behavior of a complete system can be characterized when the constituent functional blocks and their interconnections are specified. An inherent property of a functional block (in traditional sense) is that its operation is autonomous. The behavior of a functional block is characterized by the static transfer function of the functional block, the dynamic state of the functional block, and the inputs/outputs received/generated at the corresponding reference points. If a functional block is disconnected from an immediately preceding functional block, it will continue to function and generate outputs; however, it will process a null or invalid input. As we mentioned earlier, the objective of NFV is to separate software that defines the NF (the VNF) from the hardware and the generic software that creates the hosting NFVI on which the VNF runs. Therefore, it is a requirement that the VNFs and the NFVI be separately specified. However, this is a requirement that is not immediately satisfied by traditional method of functional blocks and associated interfaces. Fig. 1.6 shows an example where a traditional network comprising three functional blocks is evolved into a hypothetical case where two of the three functional blocks have been virtualized. In each case, the functional block is implemented as a VNF that runs on a host function in the NFVI. However, in this process, there are two important differences with the standard functional block representation that must be noted. The VNF is not a functional block independent of its host function, because the VNF cannot exist autonomously in the way that a functional block can exist. The VNF depends on the host function for its

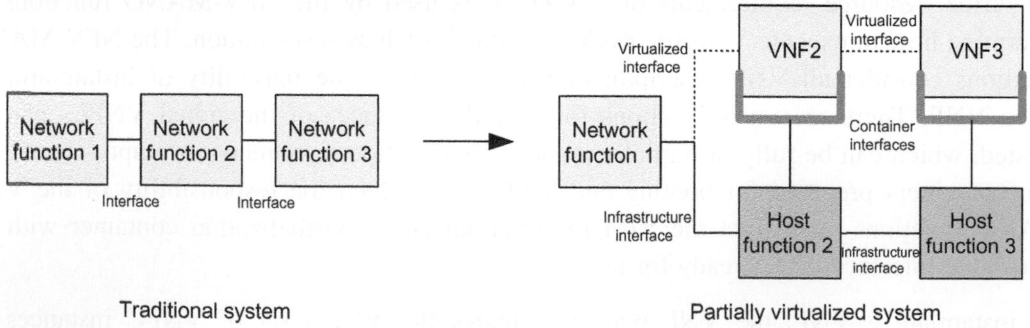

Figure 1.6
Traditional and virtualized network functions [40].

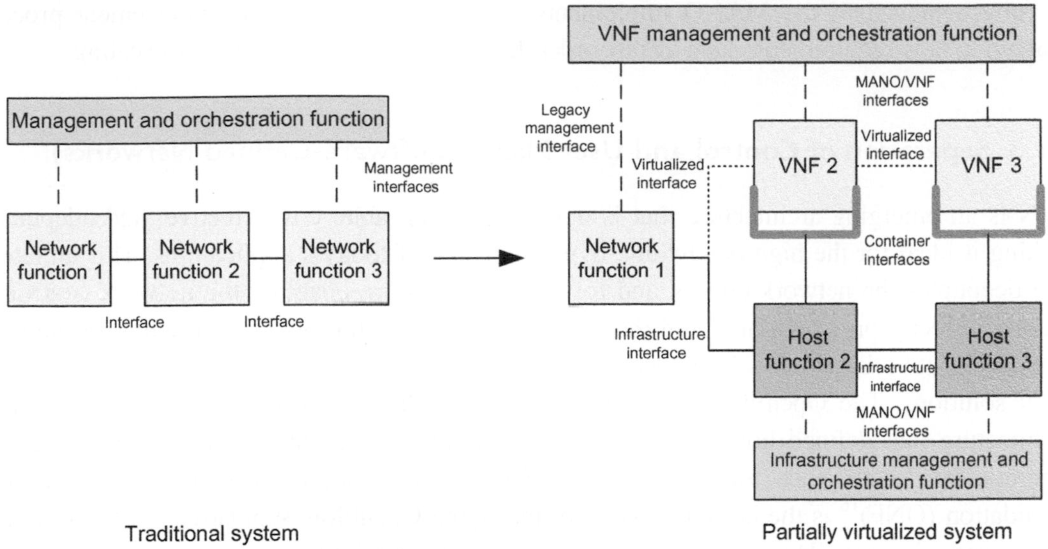

Figure 1.7
MANO of traditional and virtualized network functions [40].

existence and if the host function is interrupted, or disappears, then the VNF will be interrupted or disappear. Similarly, the container interface reflects this existence dependency between a VNF and its host function. The relationship between the VNF and its host function can be described as follows: the VNF is a configuration of the host function and the VNF is an abstract view of the host function when the host function is configured by the VNF. Therefore a host function, when configured with a VNF, has the external appearance as a functional block in traditional sense, implementing the VNF specification. It is the host function that is the functional block, but it externally appears to be the VNF. Equivalently, the VNF is an abstract view of the host function.

In an operator's network, NFs can be remotely configured and managed. For this purpose, the NFs have an interface, often referred to as a north-bound interface, to the MANO function. The MANO function is often complex and includes a large number of distributed components. However, it can be characterized using the same system model comprising functional blocks and their interfaces (see Fig. 1.7). The objective of NFV is to separate the VNFs from the infrastructure including their management. As shown in Fig. 1.7, the MANO functions are divided between the MANO of the NFVI and the MANO of the VNFs.

The MANO of the NFVI is an integral part of the NFV framework. One possible scenario is the management of the existing NFs that are partially virtualized by an NFV deployment. Managing the VNFs using the existing systems can be used for the deployment of NFV in the transition period as illustrated in Fig. 1.7. The removal of the hardware from the VNFs eliminates the requirement of managing hardware aspects. The flexibility provided by NFV can only be fully achieved, if the MANO implements efficient VNF life cycle management process adapted to new requirements such as fast order delivery, fast recovery, and auto-scaling.

1.1.3 Separation of Control and User Planes (Software-Defined Networks)

SDN is an emerging architecture that is dynamic, manageable, cost-effective, and adaptable, making it ideal for the high bandwidth, dynamic nature of today's applications. This architecture decouples the network control and forwarding functions, enabling the network control to become directly programmable and the underlying infrastructure to be abstracted for applications and network services. The OpenFlow protocol is a fundamental element for building SDN solutions. The OpenFlow standard, created in 2008, was recognized as the first SDN architecture that defined how the control- and data-plane elements would be separated and communicate with each other using the OpenFlow protocol. The Open Networking Foundation (ONF)[16] is the body in charge of managing OpenFlow standards, which are open-source specifications. However, there are other standards and open-source organizations with SDN resources, thus OpenFlow is not the only protocol that makes up SDN framework. SDN is a complementary approach to NFV for network management. While they both manage networks, both rely on different methods. SDN offers a centralized view of the network, giving an SDN controller the ability to act as the intelligence of the network. As shown in Fig. 1.8, the SDN controller communicates with switches and routers via south-bound APIs and to the applications with north-bound APIs. In the SDN architecture, the splitting of the control and data forwarding functions is referred to as disaggregation because these components can be sourced separately, rather than deployed as a single integrated system. This architecture provides the applications with more information about the state of the entire network from the controller's perspective compared to the traditional networks where the network is application aware. The SDN architectures generally consist of three functional groups, as follows:

- *SDN applications:* The application plane consists of applications such as routing and load balancing, which communicates with the SDN controller in the control plane through north-bound interfaces (e.g., REST[17] and JSON[18]). SDN applications are programs that communicate behaviors and needed resources with the SDN controller via APIs. In addition, the applications can build an abstracted view of the network by collecting information from the controller for decision-making purposes. These

[16] Open Networking Foundation (https://www.opennetworking.org).

[17] A REST application programming interface, also referred to as a RESTful web service, is based on Representational State Transfer (REST) scheme that is an architectural style and approach to communications often used in web services development. REST-compliant web services allow requesting systems to access and manipulate textual representations of web resources using a uniform and predefined set of stateless operations. A RESTful API is an application program interface that uses HTTP requests to GET, PUT, POST, and DELETE data.

[18] JSON or JavaScript Object Notation, is a minimal, readable format for structuring data. It is used primarily to transmit data between a server and web application, as an alternative to XML.

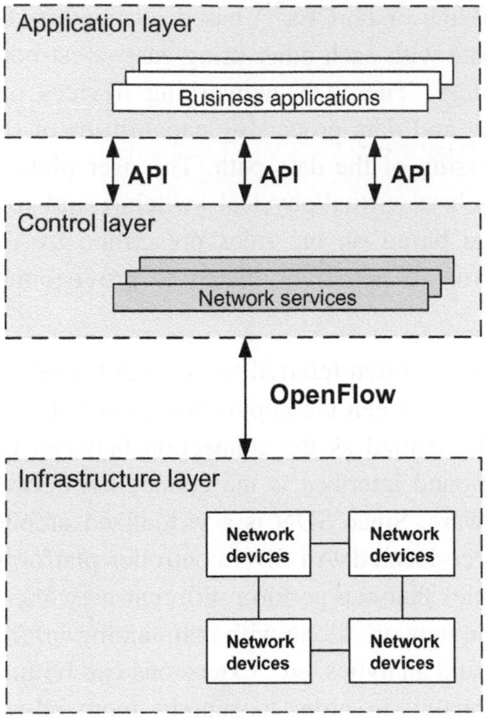

Figure 1.8
Illustration of the SDN concept [49].

applications include network management, analytics, or business applications that are used to run large data centers. For example, an analytics application might be built to recognize suspicious network activity for security purposes.

- *SDN controller:* The SDN controller is a logical entity that receives instructions or requirements from the SDN application layer and relays them to the networking components. The controller also extracts information about the network from the hardware devices and communicates back to the SDN applications with an abstract view of the network, including statistics and events. The control plane consists of one or a set of SDN controllers (e.g., Open Network Operating System[19] or OpenDayLight[20]), which logically maintain a global and dynamic network view, and provide control tasks to manage the network devices in the user plane via south-bound

[19] Open Network Operating System is an open-source software-defined network operating system (https://wiki.-onosproject.org/).

[20] OpenDaylight is an open-source software-defined networking project hosted by Linux Foundation, which was created in order to advance software-defined networking adoption and to create a strong basis for network function virtualization. It was created as a community-led and industry-supported open-source framework. The goal of the OpenDaylight project is to offer a functional software-defined networking platform that can provide the users with directly deployable software-defined networking platform without the need for other components. In addition to this, contributors and vendors can deliver add-ons and other pieces that will offer more value to OpenDaylight (https://www.opendaylight.org/).

interfaces (e.g., OpenFlow or ForCES[21]) based on requests from the applications. The controllers communicate with each other using east–west-bound interfaces.

- *SDN networking devices:* The SDN networking devices on the infrastructure layer control the forwarding and data processing capabilities of the network. This includes forwarding and processing of the data path. The user plane is composed of data forwarding elements, such as virtual/physical switches and routers, which forward and route the data packets based on the rules prescribed by the SDN controllers. This plane is responsible for all activities related to provisioning and monitoring of the networks.

The SDN architecture APIs are often referred to as north-bound and south-bound interfaces, defining the communication between the applications, controllers, and networking systems. A north-bound interface is defined as the connection between the controller and applications, whereas the south-bound interface is the connection between the controller and the physical networking hardware. Since SDN is a virtualized architecture, these elements do not have to be physically co-located. An SDN controller platform typically contains a collection of pluggable modules that can perform different network functions such as tracking device inventory within the network along with maintaining information about device capabilities, network statistics and analytics, etc. Extensions can be inserted in the SDN controller to enhance its functionality in order to support more advanced capabilities such as running algorithms to perform analytics and orchestrating new rules throughout the network.

The centralized, programmable SDN environments can easily adjust to the rapidly evolving needs of enterprise networks. The SDN can lower the cost and limit the uneconomical provisioning and can further provide flexibility in the networks. As already mentioned above, SDN and NFV are set to play key roles for operators as they prepare to migrate from 4G to 5G and to gradually scale their networks. NFVO and NFVI, along with SDN, are critical to the success of 5G rollouts, enabling agile infrastructure that can adapt to network slicing, low latency, and high-capacity requirements of major 5G use cases.

One of the key architectural enhancements in 3GPP's 5G network is control- and user-plane separation (CUPS) of EPC nodes, which enables flexible network deployment and operation

[21] Forwarding and control element separation defines an architectural framework and associated protocols to standardize information exchange between the control plane and the user/forwarding plane in a forwarding and control element separation network element. IETF RFC 3654 and RFC 3746 have defined the forwarding and control element separation requirements and framework, respectively (see https://tools.ietf.org/html/rfc5810).

by distributed or centralized deployment and the independent scaling between control-plane and user-plane functions. In other words, the network equipment is now changing from closed and vendor-specific to open and generic with SDN architectural model, which enables the separation of control and data planes, and allows networks to be programmed through open interfaces. With NFV, network functions that were previously realized in costly customized-hardware platforms are now implemented as software appliances running on low-cost commodity hardware or running in cloud computing environments. By splitting the network entities in this manner (i.e., from serving gateway (SGW) to SGW-C and SGW-U and from packet gateway (PGW) to PGW-C and PGW-U), it is possible to scale these components independently and to enable a range of deployment options. The protocol used between the control and user planes can be either an extension of the existing OpenFlow protocol or new interfaces which have been specified as part of 3GPP CUPS work item [1].

1.1.3.1 Architectural Aspects

Next-generation networks are experiencing an increase in use of very dense deployments where user terminals will be able to simultaneously connect to multiple transmission points. It is a significant advantage for the next-generation access networks to design the architecture on the premises of separation of the control-plane and user-plane functions. This separation would imply allocation of specific control-plane and user-plane functions between different nodes. As we stated earlier, the goal of SDN is to enable cloud and network engineers and administrators to respond quickly to changing business requirements via a centralized control platform. SDN encompasses multiple types of network technologies designed to make the networks more flexible and agile to support the virtualized server and storage infrastructure of the modern data centers. The SDN was originally defined as an approach to designing, building, and managing networks that separates the network's control and forwarding planes, enabling the network control to become directly programmable and the underlying infrastructure to be abstracted for applications and network services. All SDN models have some version of an SDN controller, as well as south-bound APIs and north-bound APIs. As shown in Fig. 1.9, an SDN controller is interfaced with the application layer and the infrastructure layer via north-bound and south-bound APIs, respectively. As the intelligence of the network, SDN controllers provide a centralized view of the overall network and enable network administrators to instruct the underlying systems (e.g., switches and routers) on how the forwarding plane should route/handle network traffic. SDN uses south-bound APIs to relay information to the switches and routers. OpenFlow, considered the first standard in SDN, was the original south-bound API and remains as one of the most commonly used protocols. Despite some belief considering OpenFlow and SDN to be the same, OpenFlow is merely one piece of the larger SDN framework. SDN uses north-bound APIs to communicate with the

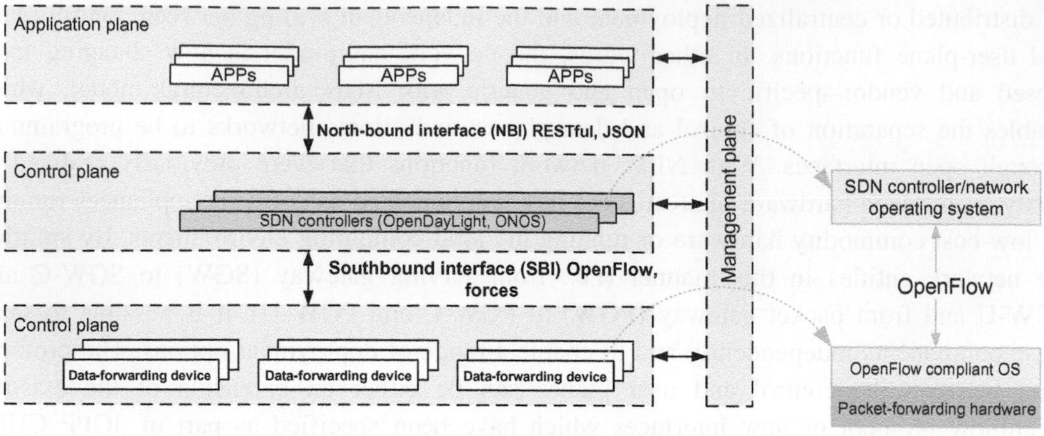

Figure 1.9
SDN framework and its main components [71].

applications in the application layer. These help network administrators to programmatically shape the traffic and deploy services.

In the SDN architecture for 5G networks, shown in Fig. 1.9, there are notably three layers: an infrastructure layer (user plane), a control layer (control plane), and an application layer (application plane). The infrastructure layer mainly consists of forwarding elements (e.g., physical and virtual switches, routers, wireless access points) that comprise the data plane. These devices are mainly responsible for collecting network status, storing them temporarily in local network devices, and sending the stored data to the network controllers and for managing packets based on the rules set by the network controllers or administrators. They allow the SDN architecture to perform packet switching and forwarding via an open interface. The control layer/control plane maintains the link between the application layer and the infrastructure layer through open interfaces. Three communication interfaces allow the controller to interact with other layers: the south-bound interface for interacting with the infrastructure layer, the north-bound interface for interacting with the application layer, and east–west-bound interfaces for communicating with groups of controllers. Their functions may include reporting network status and importing packet-forwarding rules and providing various service access points (SAPs) in various forms. The application layer is designed mainly to fulfill user requirements. It consists of the end-user business applications that utilize network services and resources. The SDN applications are able to control and access switching devices at the data layer through the control-plane interfaces. The SDN applications include network visualization, dynamic access control, security, mobility, cloud computing, and load balancing.

The functionalities of an SDN controller can be classified into four categories: (1) a high-level language for SDN applications to define their network operation policies; (2) a rule update process to install rules generated from those policies; (3) a network status collection process to gather network infrastructure information; and (4) a network status synchronization process to build a global network view using the network status collected by each individual controller. One of the basic functions of the SDN controller is to translate application specifications into packet-forwarding rules. This function uses a protocol to address communication between its application layer and control layer. Therefore, it is imperative to utilize some high-level languages (e.g., C++, Java, and Python) for the development of applications between the interface and the controllers. An SDN controller is accountable for generating packet-forwarding rules as well as clearly describing the policies and installing the rules in relevant devices. The forwarding rules can be updated with policy changes. Furthermore, the controller should maintain consistency for packet forwarding by using either the original rule set/updated rule set or by using the updated rules after the update process is completed. The SDN controllers collect network status to provide a global view of the entire network to the application layer. The network status includes time duration, packet number, data size, and flow bandwidth. Unauthorized control of the centralized controller can degrade controller performance. Generally, this can be overcome by maintaining a consistent global view of all controllers. Moreover, SDN applications play a significant role in ensuring application simplicity and guaranteeing network consistency.

In most SDNs, OpenFlow is used as the south-bound interface. OpenFlow is a flow-oriented protocol and includes switches and port abstraction for flow control. The OpenFlow protocol is currently maintained by ONF and serves as a fundamental element for developing SDN solutions. The OpenFlow, the first standard interface linking the forwarding and controls layers of the SDN architecture, allows management and control of the forwarding plane of network devices (e.g., switches and routers) both physically and virtually. The OpenFlow helps SDN architecture to adapt to the high bandwidth and dynamic nature of user applications, adjust the network to different business needs, and reduce management and maintenance complexity. It must be noted that OpenFlow is not the only protocol available or in development for SDN. To work in an OpenFlow environment, any device that wants to communicate to an SDN controller must support the OpenFlow protocol. The SDN controller sends changes to the switch/router flow table through south-bound interface, allowing network administrators to partition traffic, control flows for optimal performance, and start testing new configurations and applications (see Fig. 1.10).

The OpenFlow features support a number of commonly used data-plane protocols, ranging from layer-2 to layer-4, with packet classification being performed using stateless match tables, and packet processing operations, known as actions or instructions, ranging from header modification, metering, QoS, packet replication (e.g., to implement multicast or link aggregation), and packet encapsulation/de-encapsulation. Various statistics are defined per port, per table, and per table entry. Information can be retrieved on demand or via notifications. OpenFlow is, however, not merely an interface. It also defines the expected behavior of the switch and how the behavior can be customized using the interface. An OpenFlow controller is an SDN controller that uses the OpenFlow protocol to connect and

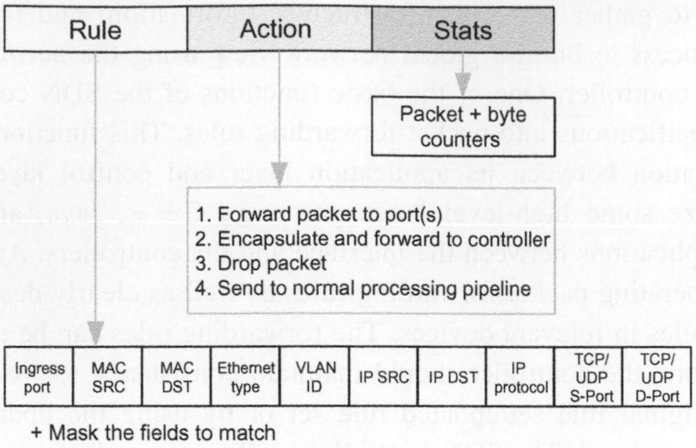

Figure 1.10
Example of OpenFlow flow table entries [48,49].

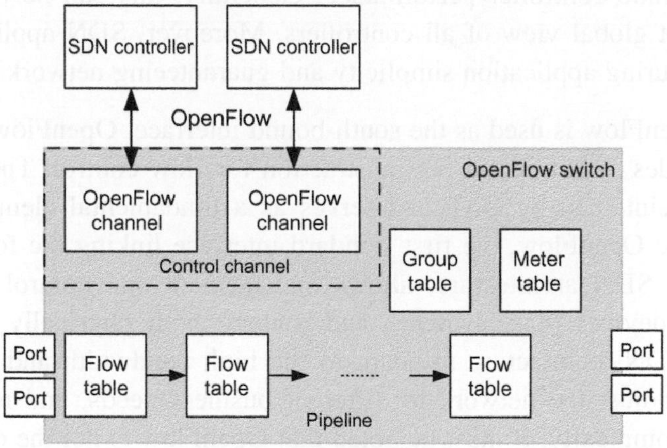

Figure 1.11
Main components of an OpenFlow switch [49].

configure the network devices in order to find the best path for application traffic. OpenFlow controllers create a central control point to manage OpenFlow-enabled network components.

An OpenFlow logical switch (see Fig. 1.11) consists of one or more flow tables and a group table, which perform packet lookups and forwarding, and one or more OpenFlow channels to an external controller. The switch communicates with the controller and the controller manages the switch via the OpenFlow switch protocol. Using the OpenFlow switch protocol, the controller can add, update, and delete flow entries in flow tables, both reactively (in response to packets) and proactively. As shown in Fig. 1.10, each flow table in the switch contains a set of flow entries; each flow entry consists of match fields, counters, and a set of instructions to apply to matching packets. Matching starts at the first flow table and may continue to additional flow tables of the pipeline. Flow entries match packets in priority order, with the first matching entry in each table being used. If a matching entry is found, the instructions associated with the specific flow entry are executed. If no match is found in a flow table, the outcome depends on configuration of the table-miss flow entry, for example, the packet may be forwarded to the controllers over the OpenFlow channel, dropped, or may continue to the next flow table. Instructions associated with each flow entry either contain actions or modify pipeline processing. Actions included in instructions describe packet forwarding, packet modification, and group table processing. Pipeline processing instructions allow packets to be sent to subsequent tables for further processing and allow information, in the form of meta-data, to be communicated between tables. Table pipeline processing stops when the instruction set associated with a matching flow entry does not specify a next table; at this point the packet is usually modified and forwarded [49].

We now change our focus to 3GPP CUPS and discuss the efforts within 3GPP to enable SDN control of networks. 3GPP completed Rel-14 specification of control- and user-plane separation work item in June 2014, which is a key core network feature for many operators. Control- and user-plane separation of EPC nodes provides architecture enhancements for the separation of functionalities in the EPC's SGW, PGW, and traffic detection function (TDF).[22] This enables flexible network deployment and operation, distributed or centralized architecture, as well as independent scaling between control-plane and user-plane functions without affecting the functionality of the existing nodes as a result of the split. The user data traffic in operators' networks have been doubling on an annual basis in recent years due to increasing use of smart devices, proliferation of video streaming, and other broadband applications. At the same time, there is a strong consumer demand for improved user experience, higher throughput, and lower latency. The CUPS scheme allows for reducing the latency of application/service by selecting user-plane nodes which are closer to the RAN or more appropriate for the intended usage type without increasing the number of control-plane nodes. It further supports increase of data traffic by enabling addition of user-plane nodes without changing the number of control-plane nodes (SGW-C, PGW-C, and TDF-C) in the network. The CUPS scheme further allows locating and scaling control-plane and user-plane resources of the EPC nodes independently as well as enabling independent evolution of the control-plane and user-plane functions. The CUPS paradigm is a precursor

[22] Traffic detection function has become an important element in the mobile networks due to the increasing complexities in managing data services, demand for personalization, and service differentiation. Traffic detection function provides communication service providers the opportunity to capitalize on analytics for traffic optimization, charging and content manipulation, working in conjunction with policy management and charging system. Traffic detection function enforces traffic policies based on predetermined rules or dynamically determined rules by the policy and charging rules function on data flows in real time. Traffic detection function was introduced together with Sd reference point as a means for traffic management in the 3GPP Rel-11 specifications using layer-7 traffic identification.

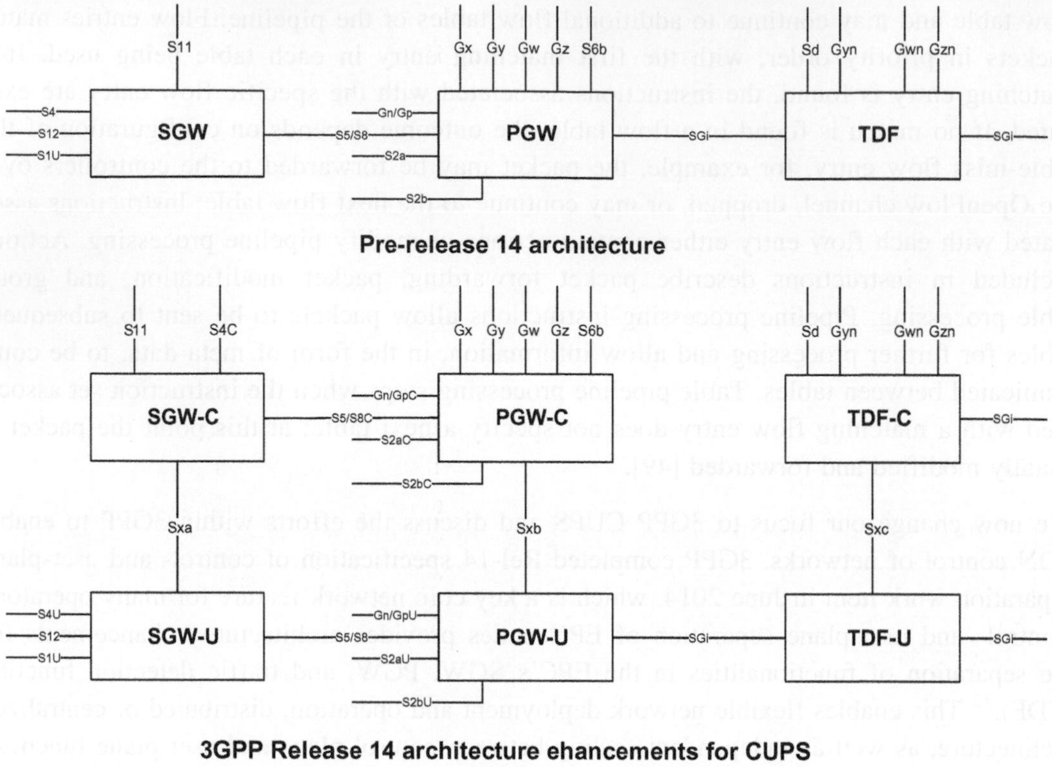

Figure 1.12
Separation of control plane and user plane in EPC [1].

to the use of SDN concept in 3GPP networks. The following high-level principles were incorporated in the CUPS framework [8,36]:

- As shown in Fig. 1.12, the control-plane functions terminate control-plane protocols such as GTP-C, Diameter[23] (Gx, Gy, Gz) and a control-plane function can interface multiple user-plane functions, as well as a user-plane function can be shared by multiple control-plane functions.
- A UE is served by a single SGW-C but multiple SGW-U can be selected for different packet data network (PDN) connections. A user-plane data packet may traverse multiple user-plane functions.

[23] Diameter is an application-layer protocol for authentication authorization and accounting. It is a message-based protocol, where authentication authorization and accounting nodes receive positive or negative acknowledgment for each message exchanged. For message exchange, Diameter uses the transmission control protocol and stream control transmission protocol, which makes it more reliable. Diameter base protocol is specified in IETF RFC 6733 (https://tools.ietf.org/html/rfc6733).

- The control-plane functions control the processing of the packets in the user-plane by provisioning a set of rules in Sx sessions, that is, packet detection, forwarding, QoS enforcement, and usage reporting rules.
- While all 3GPP features impacting the user-plane functions (e.g., policy and charging control, lawful interception, etc.) are supported, the user-plane functions are designed to be 3GPP agnostic as much as possible.
- A legacy SGW, PGW, and TDF can be replaced by a split node without effecting connected legacy nodes.

As shown in Fig. 1.12, CUPS introduces three new interfaces, namely Sxa, Sxb, and Sxc between the control-plane and user-plane functions of the SGW, PGW, and TDF, respectively.

3GPP evaluated candidate protocols such as OpenFlow, ForCES, and Diameter. The criteria identified for the selection process included ease of implementation on simple forwarding devices, no transport blocking, low latency, and capabilities to support the existing 3GPP features, ease of extension and maintenance of the protocols to support 3GPP features, and backward compatibility across releases. Based on these criteria, it was decided to define a new 3GPP native protocol with type-length value-encoded messages over user datagram protocol (UDP)/IP, called packet-forwarding control protocol (PFCP), for Sxa, Sxb, and Sxc interfaces [3gpp]. The protocol stack for the control-plane/user-plane over Sxa, Sxb, Sxc, and combined Sxa/Sxb reference points are depicted in Fig. 1.13.

The PFCP is a new protocol layer, which has the following properties:

- One Sx association is established between a control-plane function and a user-plane function before being able to establish Sx sessions on the user-plane function. The Sx association may be established by the control-plane function or by the user-plane function.
- An Sx session is established in the user-plane function to provision rules instructing the user-plane function on how to process certain traffic. An Sx session may correspond to

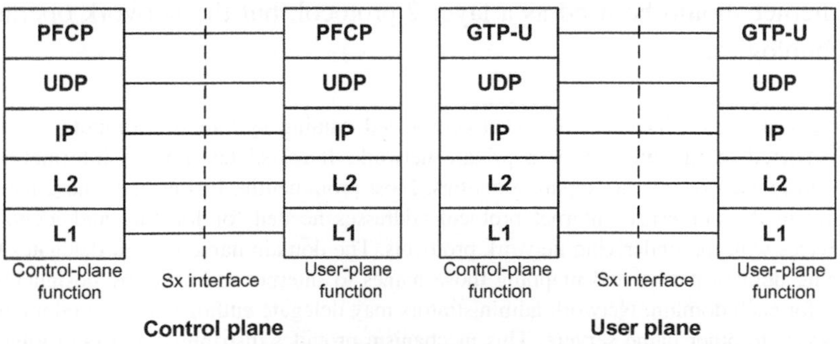

Figure 1.13
Control-plane/user-plane protocol stack over Sxa, Sxb, Sxc, and combined Sxa/Sxb [8].

an individual PDN connection, TDF session, or this can be a standalone session and not tied to any PDN connection/TDF session, for example, for forwarding DHCP/RADIUS/Diameter signaling between the PGW-C and PDN over SGi.

- Sx node–related procedures include Sx association setup/update/release procedures, monitoring peer PFCP, load and overload control procedures to balance loading across user-plane functions, and to reduce signaling toward user-plane function under overload conditions, Sx packet flow description (PFD) management procedure to provision PFDs for one or more application identifiers in the user-plane function.
- Sx session–related procedures include Sx session establishment/modification/deletion procedures; Sx session report procedure to report traffic usage or specific events.

Data forwarding between the control-plane function and user-plane function is supported by GTP-U encapsulation on the user plane and PFCP on the control plane where the latter protocol supports reliable delivery of messages. A set of new domain name system (DNS)[24] procedures are defined for user-plane function selection. The control-plane function selects a user-plane function based on DNS or local configuration, the capabilities of the user-plane function and the overload control information provided by the user-plane function. Figs 1.14 and 1.15 show two [example] deployment scenarios based on higher layer functional split between the central unit and the distributed unit (DU) of the base station reusing Rel-12 dual-connectivity concepts. The control-plane/user-plane separation permits flexibility for different operational scenarios such as moving the PDCP to a central unit while retaining the radio resource management (RRM) in a master cell; moving the RRM to a more central location where it has oversight over multiple cells, allowing independent scalability of user plane and control plane; and centralizing RRM with local breakout of data connections of some UEs closer to the base station site. During the establishment of an Sx session, the control plane function and the user-plane function select and communicate to each other the IP destination address at which they expect to receive subsequent request messages related to that Sx session. The control-plane function and the user-plane function may change this IP address subsequently during an Sx session modification procedure. Typically, Ethernet should be used as a layer-2 protocol, but the network operators may use any other technologies.

---

[24] Domain name system is a hierarchical and decentralized naming system for computers, services, or other resources connected to the Internet or a private network. It associates various information with domain names assigned to each of the participating entities. Most prominently, it translates more readily memorized domain names to the numerical Internet protocol addresses needed for locating and identifying computer hosts and devices with the underlying network protocols. The domain name system delegates the responsibility of assigning domain names and mapping those names to Internet resources by designating authoritative name servers for each domain. Network administrators may delegate authority over subdomains of their allocated name space to other name servers. This mechanism provides distributed and fault-tolerant service and was designed to avoid a single large central database. The domain name system also specifies the technical functionality of the database service that is at its core. It defines the domain name system protocol, a detailed specification of the data structures and data communication exchanges used in the domain name system, as part of the Internet protocol suite.

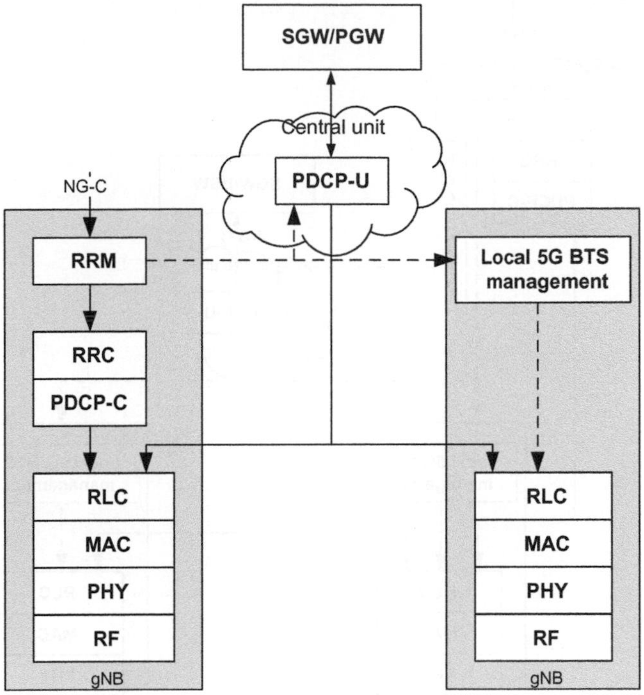

Figure 1.14
Centralized PDCP-U with local RRM [34].

1.1.4 Network Slicing

The combination of SDN and NFV enables dynamic, flexible deployment and on-demand scaling of NFs, which are necessary for the development of the 5G packet core network. Such characteristics have also encouraged the development of network slicing and service function chaining. From a UE perspective, slicing a network is to group devices with similar performance requirements (transmission rate, delay, throughput, etc.) into a slice. From network perspective, slicing a network is to divide an underlying physical network infrastructure into a set of logically isolated virtual networks. This concept is considered as an important feature of a 5G network, which is standardized by 3GPP. Service function chaining (SFC)[25] or network service chaining allows traffic flows to be routed through an ordered

[25] Network service chaining or service function chaining is a capability that uses software-defined networking capabilities to create a service chain of connected network services and connect them in a virtual chain. This capability can be used by network operators to set up suites or catalogs of connected services that enable the use of a single network connection for many services, with different characteristics. The primary advantage of network service chaining is the way virtual network connections can be set up to handle traffic flows for connected services. For example, a software-defined networking controller may use a chain of services and apply them to different traffic flows depending on the source, destination, or type of traffic.

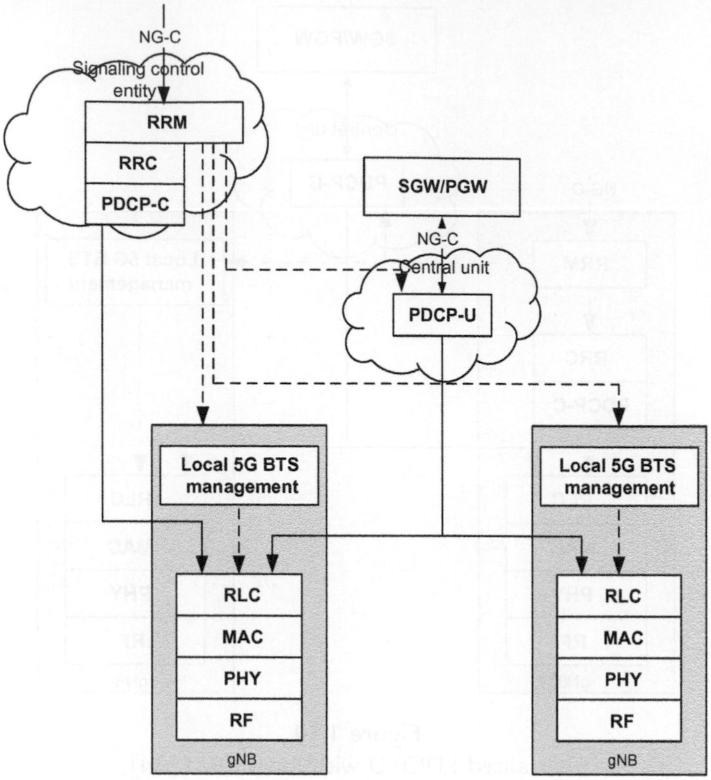

Figure 1.15
Centralized PDCP with centralized RRM in separate platforms [34].

list of NFs (e.g., firewall, load balancers, etc.). The best practical use case of SFC is to chain NFs (i.e., middle boxes in this case) placed in the interface between PGW and the external networks.

As depicted in Fig. 1.16, the network slicing architecture comprises three layers: (1) service instance layer, (2) network slice instance (NSI) layer, and (3) resource layer. The service instance layer represents the services (end-user or business services) which are supported by the network where each service is represented by a service instance. The services are typically provided by network operator or a third party. A service instance can either represent an operator service or a third-party service. A network operator uses a network slice blueprint[26] to create an NSI. An NSI provides the network characteristics which are required by

[26] Network slice blueprint is a complete description of the structure, configuration, and the plans/workflows for how to instantiate and control a network slice instance during its lifecycle. A network slice blueprint enables the instantiation of a network slice, which provides certain network characteristics (e.g., ultralow latency, ultrareliability, and value-added services for enterprises). A network slice blueprint refers to required physical and logical resources and/or to sub-network blueprint(s).

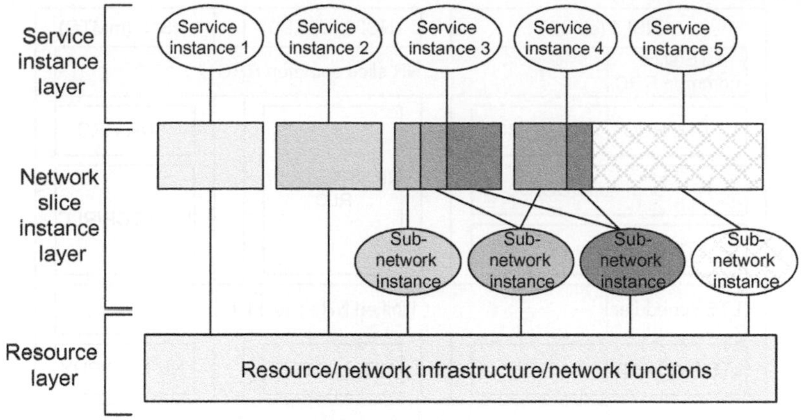

Figure 1.16
Network slicing conceptual architecture [6].

a service instance. An NSI may also be shared across multiple service instances provided by the network operator. The NSI may be composed of zero or more sub-network instances, which may be shared by another NSI. Similarly, the sub-network blueprint[27] is used to create a sub-network instance to form a set of NFs, which run on the physical/logical resources. The sub-network instance is a set of NFs, which run on the physical or logical resources. The network slice is a complete logical network providing telecommunications services and network capabilities. Network slices vary depending on the features of the service they need to support.

Fig. 1.17 shows an example of logical architecture of network slicing in a radio access network. In this example, different NSIs[28] can either share the same functions or have dedicated functions. The new RAT features flexible air interface design and a unified medium access control (MAC) scheduling to support different network slice types. Such a combination allows time-domain and frequency-domain resource isolation without compromising resource efficiency. The protocol stack can be tailored to meet the diverse service requirements from different NSIs. For instance, layer-3 RRC functions can be customized in network slice design phase. Layer-2 can have various configurations for different NSIs to meet specific requirements for radio bearers. In addition, layer-1 uses flexible numerology to support different

[27] Sub-network blueprint: a description of the structure (and contained components) and configuration of the sub-network instances and the plans/workflows for how to instantiate it. A sub-network blueprint refers to physical and logical resources and may refer to other sub-network blueprints.

[28] Network slice instance is the realization of network slicing concept. It is an end-to-end logical network, which comprises a group of network functions, resources, and connections. A network slice instance typically covers multiple technical domains, which includes terminal, access network, transport network, and core network, as well as dual-connectivity domain that hosts third-party applications from vertical industries. Different network slice instances may have different network functions and resources. They may also share some of the network functions and resources.

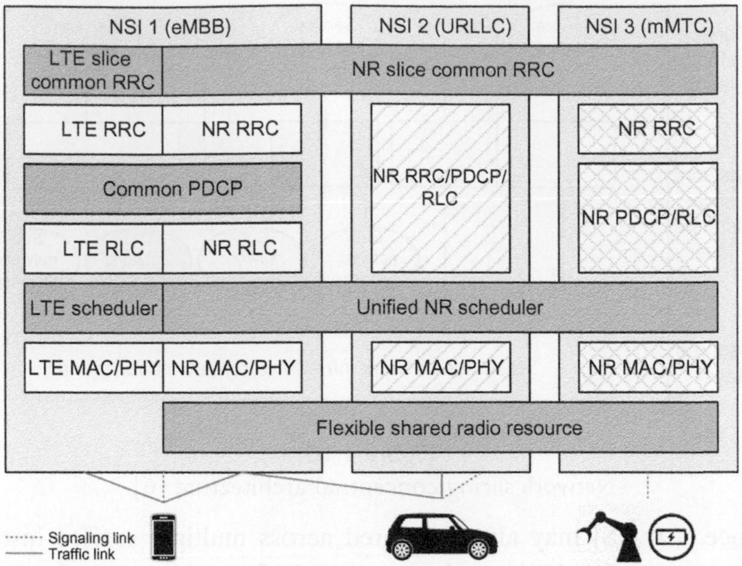

Figure 1.17
Example of RAN architecture with network slicing support [74].

network slice types. An NSI may contain different types of access nodes, such as 3GPP new RAT and a non-3GPP RAT. Consolidating fixed and wireless access in 5G is a desirable approach, which also requires further enhancements in the architecture design [74].

According to 3GPP specifications [3,16], a network slice always consists of an access and a core network part. The support of network slicing relies on the principle that traffic for different slices is handled by different PDU sessions. Network can create different network slices by scheduling and also by providing different L1/L2 configurations. The UE should be able to provide assistance information for network slice selection in an RRC message. While the network can potentially support a large number of slices, the UE does not need to support more than eight slices in parallel. As we mentioned earlier, network slicing is a concept that allows differentiated network services depending on each customer requirements. The mobile network operators can classify customers into different tenant types each having different service requirements that control in terms of what slice types each tenant is authorized to use based on service-level agreement and subscriptions.

Network slices may differ depending on the supported features and optimization of the network functions. An operator may opt to deploy multiple NSIs delivering exactly the same features but for different groups of UEs. A single UE can simultaneously be served by one or more NSIs via NG-RAN. A single UE may be served by a maximum of eight network slices at any time. The access and mobility management function (AMF) instance serving the UE logically belongs to each of the NSIs serving the UE, that is, this AMF instance is common

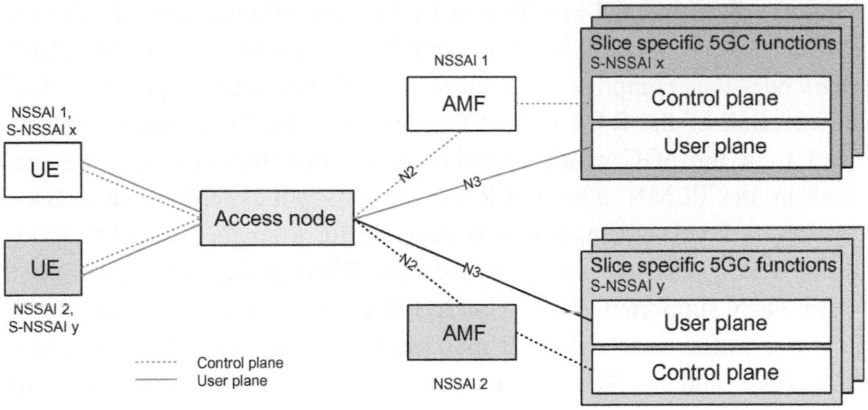

Figure 1.18
Slice selection and identifiers [55].

to the NSIs serving a UE. The selection of the set of NSIs for a UE is triggered by the first associated AMF during the registration procedure typically by interacting with the network slice selection function (NSSF), which may lead to change of AMF.

A network slice is identified by an identifier known as single-network slice selection assistance information (S-NSSAI). The S-NSSAI identity consists of a slice/service type (SST), which refers to the expected network slice behavior in terms of features and services and a slice differentiator (SD), which is an optional information element that complements the SST(s) to differentiate among multiple network slices of the same SST. The support of all standardized SST values by a public land mobile network (PLMN)[29] is not required. The S-NSSAI can have standard values or PLMN-specific values. The S-NSSAI identifiers with PLMN-specific values are associated with the PLMN ID of the PLMN that assigns them. An S-NSSAI cannot be used by the UE in AS procedures in any PLMN other than the one to which the S-NSSAI is associated. Standardized SST values provide a way for establishing global interoperability for slicing so that PLMNs can support the roaming use case more efficiently for the most commonly used SSTs. Currently, the SST values of 1, 2, and 3 are associated with eMBB, URLLC, and mMTC slice types [16]. The NSSAI is a collection of S-NSSAI. There can be at most eight S-NSSAIs in the NSSAI sent in signaling messages between the UE and the network. Each S-NSSAI assists the network in selecting a particular NSI. The same NSI may be selected via different S-NSSAIs. NSSAI includes one or more S-NSSAIs. Each network slice is uniquely identified by an S-NSSAI [3] (see Fig. 1.18).

[29] Public land mobile network is a mobile wireless network that is centrally operated and administrated by an organization and uses land-based RF transceivers or base stations as network hubs. This term is generally used to refer to an operator network.

NG-RAN supports differentiated handling of traffic for different network slices which have been preconfigured. The support of slice capabilities in terms of the NG-RAN functions (i.e., the set of NFs that comprise each slice) is implementation dependent. The NG-RAN supports the selection of the RAN part of the network slice by assistance information provided by the UE or the 5GC which unambiguously identifies one or more preconfigured network slices in the PLMN. The NG-RAN supports policy enforcement between slices according to service-level agreements. It is possible for a single NG-RAN node to support multiple slices. The NG-RAN can apply the best RRM policy depending on the service agreements for each supported slice. The NG-RAN further supports QoS differentiation within a slice. For initial attach, the UE may provide assistance information to support the selection of an AMF and the NG-RAN uses this information for routing the initial NAS to an AMF. If the NG-RAN is unable to select an AMF using this information or the UE does not provide such information, the NG-RAN sends the NAS signaling to a default AMF. For subsequent accesses, the UE provides a Temp ID, which is assigned to the UE by the 5GC, to enable the NG-RAN to route the NAS message to the appropriate AMF as long as the Temp ID is valid. Note that the NG-RAN is aware of and can reach the AMF which is associated with the Temp ID.

The NG-RAN supports resource isolation between slices via RRM policies and protection mechanisms that would avoid conditions such as shortage of shared resources in one slice resulting in under-service issues in another slice. It is possible to fully dedicate the NG-RAN resources to a certain slice. Some slices may be available only in certain parts of the network. Awareness in the NG-RAN of the slices supported in the cells of its neighbors may be beneficial for inter-frequency mobility in the connected mode. It is assumed that the slice configuration does not change within the UE's registration area. The NG-RAN and the 5GC can manage service requests for a slice that may or may not be available in a given area. Admission or rejection of access to a slice may depend on certain factors such as support for the slice, availability of resources, and support of the requested service by other slices. In the case where a UE is simultaneously associated with multiple slices, only one signaling connection is maintained. For intra-frequency cell reselection, the UE always tries to camp on the best cell, whereas for inter-frequency cell reselection, dedicated priorities can be used to control the frequency on which the UE camps. Slice awareness in NG-RAN is introduced at PDU session level by indicating the S-NSSAI corresponding to the PDU session in all signaling containing PDU session resource information. 5GC validates whether the UE is authorized to access a certain network slice. The NG-RAN is informed about all network slices for which resources are being requested during the initial context setup [3,15].

Resource isolation enables specialized customization of network slices and prevents adverse effects of one slice on other slices. Hardware/software resource isolation is up to implementation; nevertheless, RRM procedures and service agreements determine whether each slice may be assigned to shared or dedicated radio resources. To enable differentiated handling of traffic for network slices with different service agreements, NG-RAN is configured with a set of different configurations for different network slices and receives relevant information, indicating which of the configurations applies to each specific network slice. The NG-RAN selects the AMF based on a Temp ID or assistance information provided by the UE. In the event that a Temp ID is not available, the NG-RAN uses the assistance information provided by the UE at RRC connection establishment to select the appropriate AMF instance (i.e., the information is provided after random access procedure). If such information is not available, the NG-RAN routes the UE to a default AMF instance [3,15].

Enabling network slicing in 5G requires native support from the overall system architecture. As shown in Fig. 1.16, the overall architecture consists of three fundamental layers: the infrastructure layer, network slice layer, and network management layer. The infrastructure layer provides the physical and virtualized resources, for instance, computing resource, storage resource, and connectivity. The network slice layer is located above the infrastructure layer and provides necessary NFs, tools and mechanisms to form end-to-end logical networks via NSIs. The network management layer contains the generic BSS/OSS and network slice management (NSM) system, which manages network slicing and ensures satisfaction of the SLA requirements. The overall architecture has the following key features:

- *Common infrastructure:* Network slicing is different than a dedicated network solution that uses physically isolated and static network resources to support tenants. Network slicing promotes the use of a common infrastructure among tenants operated by the same operator. It helps to achieve higher resource utilization efficiency and to reduce the service time to market. Moreover, such design is beneficial for long-term technology evolution as well as for maintaining a dynamic ecosystem.
- *On-demand customization:* Each technical domain in an NSI has different customization capabilities, which are coordinated through the NSM system during the process of network slice template (NST) design, NSI deployment, and operation and management. Each technical domain can perform an independent customization process in terms of design schemes to achieve an effective balance between the simplicity needed by commercial practice and architectural complexity.
- *Isolation:* The overall architecture supports the isolation of NSIs, including resource isolation, operation and management isolation, and security isolation. The NSIs can be either physically or logically isolated at different levels.
- *Guaranteed performance:* Network slicing seamlessly integrates different domains to satisfy industry-defined 5G performance specifications and to accommodate vertical industry requirements.
- *Scalability:* Owing to virtualization, which is one of the key enabling technologies for network slicing, resources occupied by an NSI can dynamically change.

- *Operation and management capability exposure:* Tenants may use dedicated, shared, or partially shared NSIs. Furthermore, different tenants may have independent operation and management demands. The NSM system provides access to a number of operation and management functions of NSIs for the tenants, which for instance allows them to configure NSI-related parameters such as policy.
- *Support for multi-vendor and multi-operator scenarios:* Network slicing allows a single operator to manage multiple technical domains, which may be composed of network elements supplied by different vendors. In addition, the architecture needs to support a scenario, where the services from the tenants may cover different administrative domains owned by different operators.

As we mentioned earlier, an NSI is a managed entity in the operator's network with a life cycle independent of the life cycle of the service instance(s). In particular, service instances are not necessarily active through the entire duration of the run-time phase of the supporting NSI. The NSI life cycle typically includes an instantiation, configuration and activation phase, a run-time phase, and a decommissioning phase. During the NSI life cycle the operator manages the NSI. As shown in Fig. 1.19, the network slice life cycle is described by a number of phases as follows [7]:

- *Preparation:* In this phase, the NSI does not exist. The preparation phase includes the creation and verification of NST(s), on-boarding, preparing the necessary network environment which is used to support the life cycle of NSIs, and any other preparations that are needed in the network.
- *Instantiation, configuration, and activation:* During instantiation/configuration, all shared/dedicated resources associated with the NSI have been created and configured and the NSI is ready for operation. The activation step includes any actions that make the NSI active such as routing traffic to it, provisioning databases (if dedicated to the network slice, otherwise this takes place in the preparation phase), and instantiation, configuration, and activation of other shared and/or non-shared NF(s).
- *Run-time:* In this phase, the NSI is capable of traffic handling and supports certain types of communication services. The run-time phase includes supervision/reporting, as well as activities related to modification. Modification of the workflows related to runtime tasks may include upgrade, reconfiguration, NSI scaling, changes of NSI capacity, changes of NSI topology, and association and disassociation of NFs with NSI.
- *Decommissioning:* This step includes deactivation by taking the NSI out of active state as well as the retrieval of dedicated resources (e.g., termination or reuse of NFs) and configuration of shared/dependent resources. Following this phase, the NSI does not exist anymore.

An NSI is complete in the sense that it includes all functionalities and resources necessary to support certain set of communication services. The NSI contains NFs belonging to the access and the core networks.

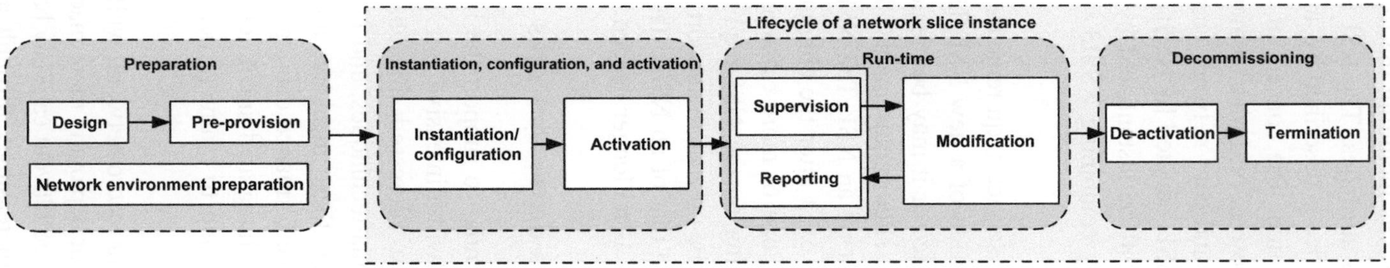

Figure 1.19
Life cycle phases of an NSI [7].

If the NFs are interconnected, the 3GPP management system contains the information relevant to connections between these NFs including the topology of connections, and individual link requirements such as QoS attributes. The NSI is realized via the required physical and logical resources. A network slice is described by an NST. The NSI is created using the NST and instance-specific information. The concept of network slice subnet instance management is introduced for the purpose of NSI management. For example, for instantiation of an NSI that contains radio access and core network components, these components can be defined and instantiated as two NSSIs denoted by NSSI1 (RAT1) in RAN and NSSI3 in the core network. The targeted NSI will be instantiated by combining the NSSI1 and NSSI3. Another NSI can be instantiated by combining NSSI3 with another RAN NSSI denoted by NSSI2 (RAT2).

Depending on the communication service requirements, a communication service can use an existing NSI or trigger the creation of a new NSI. The new NSI may be created exclusively for this communication service or it may be created to support multiple communication services with similar network slice requirements. The life cycle of a communication service is related but not dependent on an NSI. The NSI may exist before the communication service uses the NSI and may exist after the communication service stopped using the NSI. An NSI can be created using one or more existing NSSI(s) or initiate the creation of one or more new NSSI(s) depending on the NSI requirements. The new NSSI(s) may be created just for this NSI or it may be created to support multiple NSIs. The life cycle of an NSI is related but not dependent on that of an NSSI. The NSSI may exist before the NSI is created and may exist after the NSI is no longer needed.

1.1.5 Heterogeneous and Ultra-dense Networks

Effective network planning is essential to support the increasing number of mobile broadband data subscribers and bandwidth-intensive services competing for limited radio resources. Network operators have addressed this challenge by increasing the capacity through new spectrum, multi-antenna techniques, and implementing more efficient modulation and coding schemes. However, these measures are not adequate in highly populated areas and at the cell edges where performance can significantly degrade. In addition to the above remedies, the operators have integrated small cells into their macro-networks to efficiently distribute network loading, and to maintain performance and service quality while reusing spectrum more efficiently.

One solution to expanding an existing macro-network, while maintaining it as homogeneous, is to add more sectors per base station or deploying more macro-base stations. However, reducing the site-to-site distance in the macro-cell layout can only be pursued to a certain extent because finding new macro-sites becomes increasingly difficult and can be expensive, especially in dense urban areas. An alternative is to introduce small cells through addition of low-power access nodes or RRH to the existing [overlaid] macro-cells due to their more economical site acquisition and equipment installation. Small cells are primarily added to increase capacity in hotspots with high user demand and to fill coverage holes in the macro-network in both outdoor and indoor environments. They also improve network performance and service quality by off-loading the overlaid macro-cells. The result is a heterogeneous network topology with large macro-cells in conjunction with small cells providing increased capacity per unit area. Heterogeneous network planning dates back to GSM era where cells were separated through the use of frequency reuse. While this approach could still be taken in LTE, LTE networks primarily use a frequency reuse of one to maximize utilization of the licensed bandwidth. In heterogeneous networks, the cells of different sizes are referred to as macro-cells, micro-cells, pico-cells, and femto-cells in the order of decreasing transmit power. The actual cell size depends not only on the access node power but also on physical antenna positions, as well as the topology of the cells and propagation conditions.

In general, the small cells in an ultra-dense network (UDN) are classified into full-functional base stations (pico-cells and femto-cells) and macro-extension access points (relays and RRHs). A full-functional base station is capable of performing all functions of a macro-cell with a lower power in a smaller coverage area and encompasses the full RAN protocol stack. On the other hand, a macro-extension access node is an extension of a macro-cell to effectively increase the signal coverage, and it performs all or some of the lower protocol layer functions. Moreover, the small cells feature different capabilities, transmission powers, coverage, and deployment scenarios. The UDN deployment scenarios introduce a different coverage environment where any given user would be in close proximity to many cells.

Small-cell architectures using low-power nodes were considered promising to mitigate the substantial increase in network traffic, especially for hotspot deployments in indoor and outdoor scenarios. A low-power node generally means a node whose transmit power is lower than the corresponding macro-node and base station classes, for example, pico-cell and femto-cell access nodes. Small-cell enhancements for LTE focused on additional functionalities for improved performance in hotspot areas for indoor and outdoor using low-power nodes. Network architectures comprising small cells (of various types including non-3GPP access nodes) and macro-cells can be considered as practical realization of the heterogeneous networks. Increasing the density of the 3GPP native or non-indigenous small cells overlaid by macro-cells constitutes what is considered as UDNs.

Small cells can be deployed sparsely or densely with or without overlaid macro-cell coverage as well as in outdoor or indoor environments using ideal or non-ideal backhaul. Small cells deployment scenarios include small-cell access nodes overlaid with one or more macro-cell layer(s)[30] in order to increase the capacity of an already deployed cellular

[30] Note: 3GPP uses the term layer in this context to refer to different radio frequencies or different component carriers.

Table 1.1: Backhaul options for the small cells [12].

Backhaul Type	Backhaul Technology	One-Way Latency (ms)	Throughput (Mbps)
Nonideal	Fiber access 1	10–30	10–10,000
	Fiber access 2	5–10	100–1000
	Fiber access 3	2–5	50–10,000
	DSL access	15–60	10–100
	Cable	25–35	10–100
	Wireless backhaul	5–35	10–100+
Ideal	Fiber access 4	<2.5 μs	Up to 10,000

network. In this context, two scenarios can be considered: (1) UE is simultaneously in coverage of both the macro-cell and the small cell(s); and (2) UE is not simultaneously covered by the macro-cell and small cell(s). The small-cell nodes may be deployed indoors or outdoors, and in either case could provide service to indoor or outdoor UEs, respectively. For an indoor UE, only low UE speeds of 0–3 km/h are considered. For outdoor UEs, in addition to low UE speeds, medium UE speeds up to 30 km/h were targeted. Throughput and mobility (seamless connectivity) were used as performance metrics for low- and medium-speed mobility scenarios. Cell-edge performance (e.g., fifth percentile of user throughput CDF) and network/UE power efficiency were also used as evaluation metrics.

Backhaul is a critical component of heterogeneous networks, especially when the number of small nodes increases. Ideal backhaul, characterized by very high throughput and very low latency link such as dedicated point-to-point connection using optical fiber; and non-ideal backhaul (e.g., xDSL,[31] microwave backhaul, and relaying) for small cells were studied by 3GPP, and performance and cost trade-offs were made. A categorization of ideal and non-ideal backhaul based on operators' data is listed in Table 1.1.

3GPP conducted extensive studies to investigate the interfaces between macro and small cell, as well as between small cells considering the amount and type of information/signaling needed to be exchanged between the nodes in order to achieve the desired performance improvements; and whether a direct interface should be used between macro and small cells or between small cells. In those studies, LTE X2 interface (i.e., inter-eNB interface) was used as a starting point. In small cell enhancements, both sparse and dense small cell deployments were considered. In some scenarios (e.g., indoor/outdoor hotspots), a single or multiple small-cell node(s) are sparsely deployed to cover the hotspot(s), whereas in other

[31] Digital subscriber line is a family of technologies that are used to transmit digital data over telephone lines. The asymmetric digital subscriber line is the most commonly used type of digital subscriber line technology for Internet access. The xDSL service can be delivered simultaneously with wired telephone service on the same telephone line since digital subscriber line uses high-frequency bands for data transmission. On the customer premises, a digital subscriber line filter on each non-digital subscriber line outlet blocks any high-frequency interference to enable simultaneous use of the voice and digital subscriber line services.

scenarios (e.g., dense urban, large shopping mall, etc.), a large number of small-cell nodes are densely deployed to support high-volume traffic over a relatively wide area covered by the small-cell nodes. The coverage of the small-cell layer is generally discontinuous between different hotspot areas. Each hotspot area is typically covered by a group of small cells or a small cell cluster. Furthermore, future extension or scalability of these architectures is an important consideration. For mobility or connectivity performance, both sparse and dense deployments were studied with equal priority.

Network synchronization is an important consideration where both synchronized and unsynchronized scenarios should be considered between small cells as well as between small cells and macro-cell(s). For specific operational features such as interference coordination, carrier aggregation, and coordinated multipoint (CoMP) transmission/reception, small-cell enhancements can benefit from synchronized deployments with respect to small-cell search/measurements and interference/resource management procedures. Small-cell enhancements tried to address deployment scenarios in which different frequency bands are separately assigned to macro-layer and small-cell layer. Small-cell enhancements are applicable to the existing cellular bands and those that might be allocated in the future with special focus on higher frequency bands, for example, the 3.5 GHz band, to take advantage of more available spectrum and wider bandwidths. Small-cell enhancements have considered the possibility of frequency bands that, at least locally, are only used for small cell deployments. The studies further considered cochannel deployment scenarios between macro-layer and small-cell layer. Some example spectrum configurations may include the use of carrier aggregation in the macro-layer with bands X and Y and the use of only band X in the small-cell layer. Other example scenarios may include small cells supporting carrier aggregation bands that are cochannel with the macro-layer; or small cells supporting carrier aggregation bands that are not cochannel with the macro-layer. One potential cochannel scenario may include deployment of dense outdoor cochannel small cells, including low-mobility UEs and non-ideal backhaul. All small cells operate under a macro-cell coverage irrespective of duplex schemes [frequency division duplex/time division duplex (FDD/TDD)] that are used for macro-layer and small-cell layer. Air interface and solutions for small cell enhancement are band-independent [12]. In a small cell deployment, it is likely that the traffic volume and the user distribution are dynamically varying between the small-cell nodes. It is also possible that the traffic is highly asymmetrical and it is either downlink or uplink centric. Both uniform and nonuniform traffic and load distribution in time-domain and spatial-domain are possible. During performance modeling, both non-full-buffer and full-buffer traffic were considered, where non-full buffer traffic was prioritized as it was deemed to be more practical representation of user activity.

Small-cell enhancements target high network energy efficiency and a reasonable system complexity. The small cells can save energy by switching to a dormant mode due to increased likelihood of periods of low or no user activity during operation. The trade-off between user throughput/capacity per unit area and network energy efficiency is an important consideration for small cell deployments. The small cells can further achieve UE energy efficiency considering the small cell's short-range transmission path, resulting in reduced energy/bit for the uplink transmission, mobility measurements, cell identification, and small cell discovery.

Given that some of the heterogeneous network deployments include user-installed access nodes in indoor environments, a self-organizing mechanism for deployment and operation of the small cell without direct operator intervention is required. 3GPP self-organizing networks (SON) solutions aim to configure and optimize the network automatically, so that the intervention of human can be reduced and the capacity of the network can be increased. These solutions can be divided into three categories [85]:

- *Self-configuration:* This is the dynamic plug-and-play configuration of newly deployed access node. As shown in Fig. 1.20, the access node configures its physical cell identity (PCI), transmission frequency, and power, leading to faster cell planning and rollout.

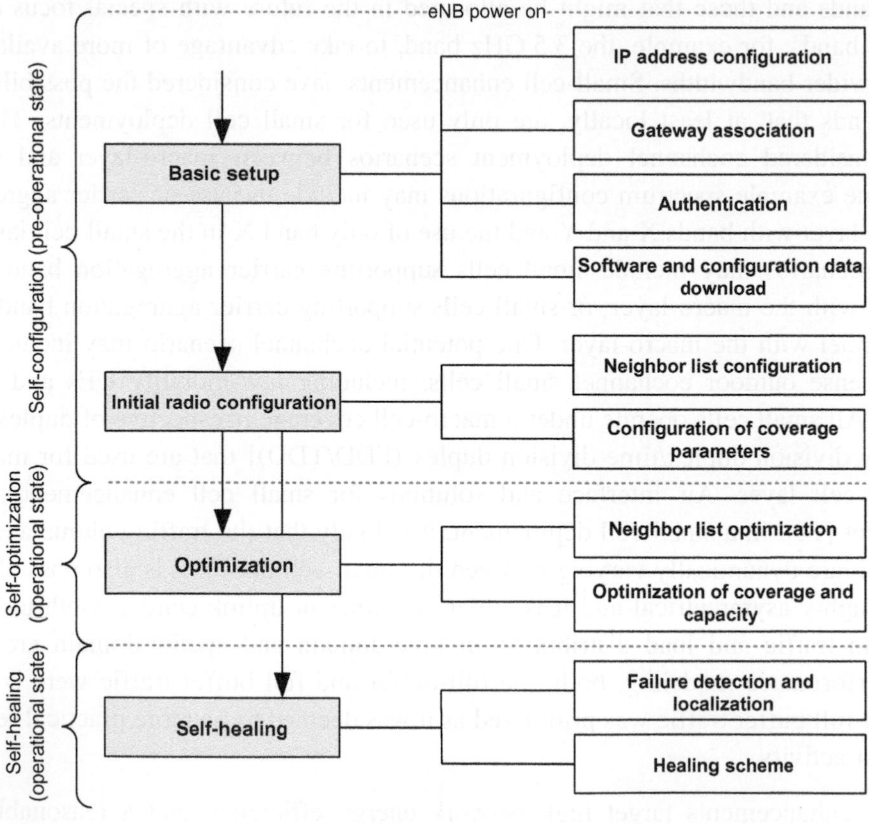

Figure 1.20
SON framework [85].

The network interfaces (e.g., S1 and X2 in the case of LTE) are dynamically configured, and the IP address as well as connection to IP backhaul is established. To reduce manual operation, the automatic neighbor relation (ANR) scheme is used. The ANR configures the neighbor list in newly deployed access nodes and optimizes the list over the course of operation. Dynamic configuration includes the configuration of the physical layer identifier, PCI, and cell global ID (CGID). The PCI mapping attempts to avoid assignment of duplicate identifiers to the access nodes in order to prevent collision. The PCI can be assigned in a centralized or distributed manner. When centralized assignment is used, the operation and management system will have a complete knowledge and control of the PCIs. When the distributed solution is used, the operation and management system assigns a list of possible PCIs to the newly deployed access nodes, but the adoption of the PCI is in control of the eNB. The newly deployed eNB will request a report, sent either by UEs over the air interface or by other eNBs over the X2 interface, including already in-use PCIs. The eNB will randomly select its PCI from the remaining values. The ANR is used to minimize the work required for configuration in newly deployed eNBs as well as to optimize configuration during operation. Correct and up-to-date neighbor lists will increase the number of successful handovers and minimize the number of dropped calls. Before a handover can be executed, the source eNB requires the neighbor information such as PCI and CGID of the target eNB. The PCI is included in normal measurement reports. The mapping between the PCI and CGID parameters can be done by using information from the operation and management or that reported by UEs decoding the target cell CGID on the broadcast channel in the target cell. The capability of decoding CGID is an optional UE feature. A network operator can put a cell on an ANR black list, to block certain handover candidates, for example, from indoor to outdoor cells. 3GPP has also specified LTE inter-frequency and inter-RAT ANR.

- *Self-optimization:* The self-optimization functions were mainly specified in 3GPP Rel-9, which included optimization of coverage, capacity, handover, and interference. Mobility load balancing is a feature where cells experiencing congestion can transfer excess load to other cells which have available resources. Mobility load balancing further allows the eNBs to exchange information about the load level and the available capacity. The report can contain computational load, S1 transport network load, and radio resource availability status. There are separate radio resource status reports for the uplink and downlink, which may include the total allocated resources, guaranteed bit rate (GBR) and non-guaranteed bit rate traffic statistics, the percentage of allocated physical resources relative to total resources, and the percentage of resources available for load balancing. Mobility load balancing can also be used across different radio access technologies. In case of inter-RAT, the load reporting RAN information management protocol will be used to transfer the information via the core network between the base stations of different radio technologies. A cell capacity class value, set by the operation and management system, is used to relatively compare the capacities of different radio access technologies. A handover due to load balancing is performed as a regular handover; however, it may be necessary to set the parameters such that the UE cannot return to the [congested] source cell. Mobility robustness optimization is a solution for automatic detection and correction of errors in the mobility configuration which may cause radio link failure as a result of unsuccessful handover.

- *Self-healing:* Features for automatic detection and removal of failures and automatic adjustment of parameters were mainly specified in 3GPP Rel-10. Coverage and capacity optimization enables automatic correction of capacity problems due to variations of the environment. The minimization of drive tests was a feature that enables normal UEs to provide the same type of information as those collected during the drive tests with the advantage that UEs can further retrieve and report parameters from indoor environments.

The home eNB (HeNB) concept was introduced in LTE Rel-9, which defines a low-power node primarily used to enhance indoor coverage. Home eNBs and particularly femtocells are privately owned and deployed without coordination with the macro-network; as such, if their operating frequency is the same as the frequency used in the macro-cells and access to them is limited, then there is a risk of interference between the femtocell and the surrounding network. The use of different cell sizes with overlapping coverage and creation of a heterogeneous network adds to the complexity of network planning. In a network with a frequency reuse of one, the UE normally camps on the cell with the strongest received signal power, hence the cell edge is located at a point where the received signal strengths are the same in both cells. In homogeneous network deployments, this also typically coincides with the point of equal path loss of the uplink in both cells, whereas in a heterogeneous network, with high-power nodes in the large cells and low-power nodes in the small cells, the point of equal received signal strengths is not necessarily the same as that of equal uplink path loss. Therefore, a challenge in heterogeneous network planning is to ensure that the small cells actually serve certain number of users. This can be done by increasing the area served by the small cell through the use of a positive cell selection offset which is referred to as cell range extension (Fig. 1.21). A drawback of this scheme is the increased interference in the downlink experienced by the UE located in the extended cell region and served by the base station in the small cell. This effect may impact the quality of reception of the downlink control channels. It is important to highlight that indoor small cells (femtocells) operate in three different access modes: open, closed, and hybrid. In open access mode, all subscribers of a given operator can access the node, while in closed access mode, the access is restricted to a closed subscriber group. In hybrid mode, all subscribers can connect to the femtocell with the priority always given to the designated subscribers. A network comprising small cells and

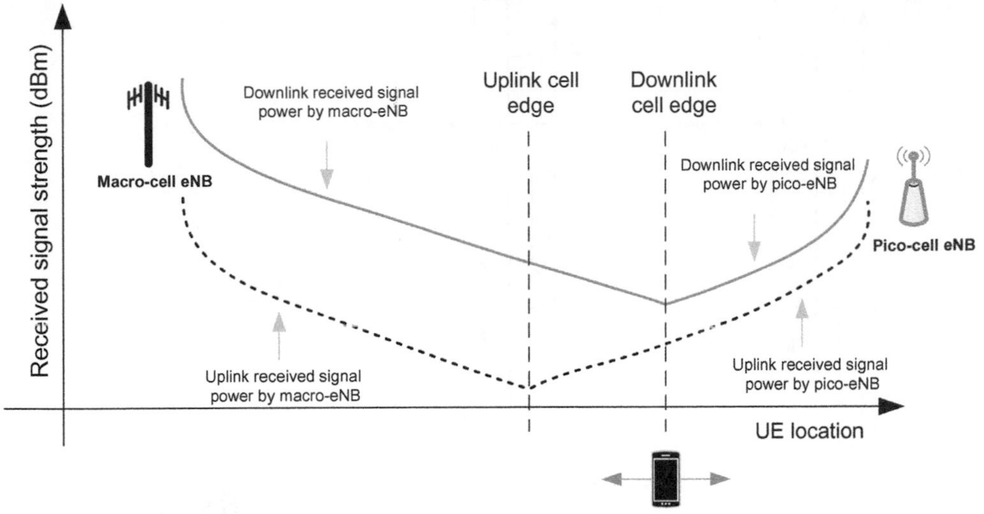

Figure 1.21
Uplink/downlink imbalance issue in HetNet deployments [12].

macro-cells is referred to as HetNet in the literature. HetNets, in general, are considered as a paradigm shift from the classic homogeneous networks.

A number of features were added to the later releases of LTE that can be used to mitigate the inter-cell interference issue in the heterogeneous networks. Inter-cell interference cancelation (ICIC) was introduced in LTE Rel-8, in which the eNBs can coordinate over the X2 interface in order to mitigate inter-cell interference for UEs at the cell edge. Frequency-domain ICIC scheme evolved to enhanced ICIC (eICIC) in LTE Rel-10 where the time-domain ICIC was added through the use of almost blank subframes. Those subframes included only LTE control channels and cell-specific reference signals and no user data, transmitted with reduced power. In that case, the macro-eNB would transmit the almost blank subframes according to a semistatic pattern and the UEs in the extended range of the small cells could better receive downlink control and data channels from the small cell. Further enhancement of ICIC focused on interference handling by the UE through ICIC for control signals, enabling even further cell range extension.

Carrier aggregation (CA) was introduced in LTE Rel-10 to increase the total system bandwidth and the maximum user throughputs. In this scheme, the component carriers (CCs) are aggregated and any CA-capable UE can be allocated resources on all or some component carrier combinations. Cross-carrier scheduling is an important feature in heterogeneous networks supporting CA where the downlink control channels are mapped to different component carriers in the large and small cells (as shown in Fig. 1.22). As an example, when LTE downlink control channel which carries downlink control information along with scheduling

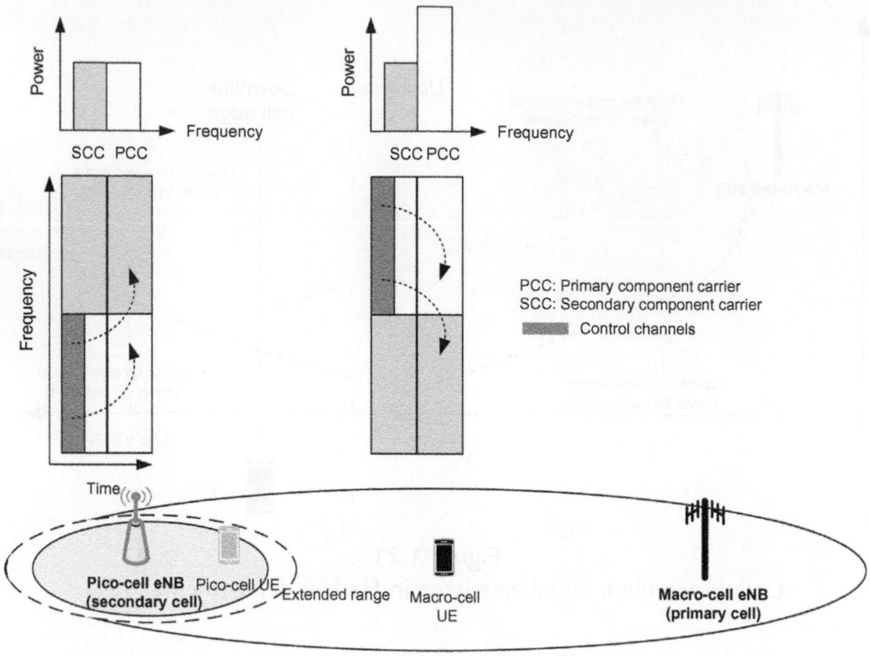

Figure 1.22
Illustration of cross-carrier scheduling and multi-cell operation in LTE [69].

information must be received by the UEs at the cell edge, it may be transmitted with a higher power than the traffic channels. Therefore, using different carriers for the downlink control channels in the large and small cells reduces the risk of inter-cell interference. The latest releases of LTE allow multi-carrier operation with different timing advances, enabling combination of component carriers from macro-eNBs with those of small cells.

The CoMP transmission and reception feature, introduced in LTE Rel-11, is an inter-cell and inter-user interference mitigation scheme that allows coordinated eNBs, relay nodes, or RRHs to simultaneously transmit to or receive from a UE. When CoMP is used in a heterogeneous network, a number of macro-cells and small cells can participate in data transmission to and from a UE. However, this would require that the macro-eNBs and the small cells to be synchronized and coordinated via backhaul links and the content be available almost at the same time at multiple transmission points, which is practically challenging. In the next section, we will see how cloud-RANs with distributed RRHs have made multi-point transmission/reception and data processing practically feasible. In CoMP scheme, a dynamic muting mechanism can be exploited to mute some (radio) resources at certain small cells for the benefit of other small cells. In this manner, the inter-cell coordination function which generates the muting patterns considers an appropriate metric for individual users and a proportional fairness scheduling among all users. The muting technique not only takes into account the first dominant interferer but also the second dominant interference source for optimal interference mitigation.

Owing to the network traffic load fluctuation, switching off the base stations/access nodes in the cells with low or no traffic load is an essential method for UDNs to improve energy efficiency and to reduce inter-cell interference. In practice, the network load fluctuates over different times and locations due to diversity of user behavior and mobility, which is especially true for UDNs that warrants switching off under-utilized base stations. In a sparse network using frequency reuse of one with idle base stations where access node density is less than user density, the average spectral efficiency can still increase linearly with access node density as in the network without idle mode base stations. In a frequency reuse of one UDN with idle mode base stations where access node density is larger than user density, the spectral efficiency only increases logarithmically with access node density [68].

Optimal utilization of large amount of radio resources in a UDN can become increasingly complex. Improper allocation of abounded radio resources in a UDN can lead to higher inter-cell interference, unbalanced load distributions, and higher power consumption. Furthermore, due to inter-cell interference, local radio resource allocation strategies may have a global impact on a UDN operation. In other words, a localized allocation strategy may not work in a UDN environment, which necessitates the use of a centralized RRM that has a holistic view of the UDN, allowing tight interworking across the network. Providing sufficient bandwidth over direct-wired backhaul to each access node in a UDN may not be practical. As a result, in the last decade several schemes have been devised and studied in the literature such as wireless self-backhauling, which consumes valuable radio resources and may cause additional interference and latency. Recall that a UDN is a densified HetNet. The user association in HetNets follows a load-based association rule, where the users are biased to connect to the nearest small cell to offload their traffic. The small cells are usually lightly loaded due to the limited coverage area; hence, the association of a given user to the nearest small cell gives the user a higher data rate privilege. The biasing of users to small cells is performed via virtual extension of their coverage area. Interference management is a challenging task in densified networks. Various types of small cells are deployed with large densities to provide the users with very high throughput connections. The use of inter-cell coordination to mitigate the interference requires increasing signaling overhead due to the large number of deployed small cells. Thus, distributed control is preferred to mitigate the interference in a UDN.

A UDN can be defined as a network where there are more cells than active users. In mathematical terms, $\rho_{BS} \gg \rho_{user}$, where ρ_{BS} denotes the area density of access nodes and ρ_{user} denotes the area density of users. Another definition of UDN can be solely given in terms of the access node density irrespective of the user density. The access nodes in UDN environments are typically low-power small cells with a small footprint, resulting in a small coverage area. Accordingly, the inter-site distance would be in the range of meters or tens of meters. Strong interference between neighboring cells is a limiting factor in UDN. The proximity of the small cells to each other in a UDN environment causes strong interference, thus the use of effective interference management schemes is inevitable to mitigate the interference of neighboring cells. Densification of wireless networks can be realized either by deploying an increasing number of access nodes or by increasing the number of links per unit area. In the first approach, the densification of access nodes can be realized in a distributed manner through deployment of small cells (e.g., pico-cells or femto-cells) or via a centralized scheme using distributed antenna system (DAS)[32] in the form of C-RAN architecture. In small-cell networks, femtocells are typically installed by the subscribers to improve the coverage and capacity in residential areas, and the pico-cells are installed by the operators in hotspots. Thus, in small-cell networks the coordination mechanism is often distributed. Compared to relays, DAS transmits the user signals to the base station via fiber links, while the relays use the wireless spectrum either in the form of in-band or out-of-band.

1.1.6 Cloud-RAN and Virtual-RAN

Operators in quest of more efficient ways to accommodate the increasing use of smartphones and other heavy data-consuming wireless devices in their networks face a dilemma when it comes to expanding network capacity and coverage. Optical fiber is typically the first option which is considered when addressing the problem of exponential traffic growth in the network. However, optical fiber is expensive; it takes a long time to install; and in some locations, it cannot be installed. To improve the network capacity and coverage, operators have several options among them small cells, carrier Wi-Fi,[33] and DASs. These and a host of other solutions are being used by network operators as methods of expanding their network to accommodate the exponential user traffic and new applications.

---

[32] A distributed antenna system is a network of spatially separated antennas which are connected to a common source via a transport mechanism that provides wireless service within a geographic area or structure. Distributed antenna system improves mobile broadband coverage and reliability in areas with heavy traffic and enhances network capacity, alleviating pressure on wireless networks when a large group of people in close proximity are actively using their terminals. Distributed antenna system is an approach to extending outdoor base station signals in indoor environments. It is a network of geographically separated antennas which receive input from a common base station source. Distributed antenna system uses multiple smaller antennas to cover the same area (that otherwise the macro base station would cover) and provides deeper penetration and coverage inside buildings. The RF input to the antennas can be conveyed either by lossy coaxial cables or more expensive optical fiber links. Some in-building distributed antenna systems can support multiple operators and standards at various levels, but advanced equipment is needed to meet a wider range of frequency bands and power outputs. Unwanted signal by-products and interference are serious issues in a shared distributed antenna system environment.

[33] Carrier Wi-Fi provides improved, scalable, robust unlicensed spectrum coverage and is often deployed as a stand-alone solution. It is an easy data offload from the cellular networks with access and policy control capable of supporting large numbers of users. Wi-Fi with new standards, such as Hotspot 2.0, can provide high data rates for users who are continuously streaming content on mobile devices.

In conjunction with the question of network expansion, there are other business imperatives. Mobile data transport architectures must be evaluated based on characteristics such as fastness, time to market, cost-effectiveness, operational and architectural simplicity, expandability, and flexibility. Energy consumption and physical size are also key factors in the deployment of new network architectures considering power and space are expensive and scarce resources at base station sites and central offices (COs).

A centralized-RAN, or C-RAN architecture addresses capacity and coverage issues, while supporting mobile fronthaul and/or backhaul solutions as well as network self-organization, self-optimization, configuration, and adaptation with software control and management through SDN and NFV. Cloud-RAN also provides advantages in controlling ongoing operational costs, improving network security, network controllability, network agility, and flexibility. The application of the C-RAN concept to small-cell architectures provides capacity benefits beyond those achieved through cell virtualization. In a traditional small-cell architecture, each access point provides a fixed amount of capacity within its coverage area. This might work well only if the user traffic is evenly distributed across the coverage area, a condition that is rarely happens in real life. The result is that some access points will be overloaded and others are relatively idle across time. Unlike stand-alone small cells where the addition of the new cells further aggravates the inter-cell interference, C-RAN architectures can be expanded and scaled.

1.1.6.1 Architectural Aspects

Cloud-based processing techniques can be implemented to centralize the baseband processing of multiple small cells and to improve inter-cell mobility and interference management. Small cells can support a variety of applications and services including voice-over-IP and videoconferencing, which can greatly benefit from a centralized architecture. C-RAN architecture comprises distributed RRHs commonly connected to centralized BBUs using optical transport or Ethernet links. The RRHs typically include the radio, the associated RF amplifiers/filters/mixers, and the antenna. The centralized BBU is implemented separately and performs the signal processing functionalities of the RAN protocols. The centralized BBU model enables faster service delivery, cost savings, and improved coordination of radio capabilities across a set of RRHs. Fig. 1.23 shows the migration from distributed RAN architecture to the centralized model. In a V-RAN architecture, the BBU functionalities and services are virtualized in the form of VMs running on general-purpose processor platforms that are located in a centralized BBU pool in the CO that can effectively manage on-demand resource allocation, mobility, and interference control for a large number of interfaces [toward remote radio units (RRUs)] using programmable software layers. The V-RAN architecture benefits from software-defined capacity and scaling limits. It enables selective content caching, which helps to further reduce network deployment and maintenance costs as well as to improve user experience based on its cloud infrastructure.

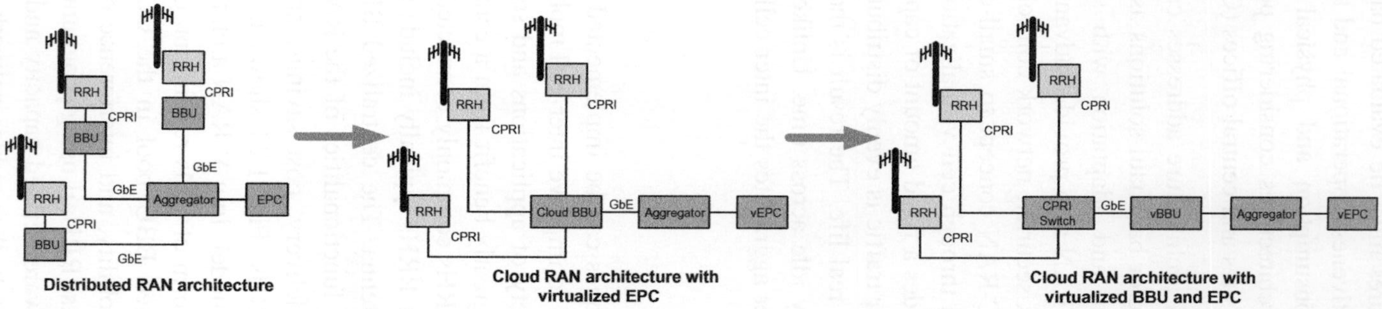

Figure 1.23
Base station architecture evolution and high-level C-RAN architecture [61].

In a traditional cellular network architecture, each physical base station unit encompasses both baseband and radio processing functions. In C-RAN, the baseband processing for a large number of cells is centralized, resulting in improved performance due to the ability to coordinate among multiple cells, and cost reduction as a result of pooling the shared resources. Small cells, when densely deployed across a large indoor environment, create large areas of overlap between neighboring cells. Inter-cell interference occurs at the cell boundaries. Some enterprise small cells use a central service controller to assist in handovers and backhaul aggregation, but it cannot overcome the fact that each cell interferes with its neighbors since some level of interference is inevitable. Creating multiple independent cells further necessitates frequent handovers for mobile terminals, degrading the user experience and creating the potential for handover failures or constant back-and-forth handovers between adjacent cells, a phenomenon known as ping-pong effect. Cell virtualization enables allocation of users' data over the same radio resources but sent to different access nodes for transmission to different users.

In general, a C-RAN architecture consists of three main entities: BBU(s), RRH/units, and the transport network or what is called the fronthaul, as shown in Fig. 1.23. In order to reduce the power consumption across network and to reduce inter-cell interference, the C-RAN architecture allows shutting off idle RRHs. This flexibility provides network adaptation based on traffic profiles that vary temporally and/or geographically that further reduces the interference to neighboring cells in order to optimize overall system performance. From a hardware usage perspective, scheduling baseband resources on-demand to perform communication process for multiple RATs while taking advantage of network virtualization techniques can improve operational efficiency and reduce energy consumption. CoMP, dual connectivity, virtual multiple input and multiple output (MIMO), and coordinated beamforming are effective standard approaches to improve network capacity specified by 3GPP [47]. To support CoMP schemes with joint transmission/processing, the network configures the UEs to measure channel-state information and to periodically report the measurements to a set of collaborating nodes, which results in a large amount of signaling overhead. The logical interface between various RATs in a centralized BBU model would enable signaling exchange among participating nodes and node selection for collaborative transmission and reception.

Since most of the current and the future deployment scenarios are in the form of ultradense heterogeneous networks, multi-RAT coexistence will be a key issue. 3GPP LTE was designed to support inter-RAT handover for GSM, UMTS, and interworking with wireless local area network. As shown in Fig. 1.24, the architecture for interworking between LTE and different RATs requires connection of their core networks for higher layer signaling. This multi-RAT architecture might be feasible but the latency of an inter-RAT handover may be prohibitive for many applications and services. In order to improve the user experience, simplified network architectures have been studied for lower latency based on C-RAN concept. If the BBU

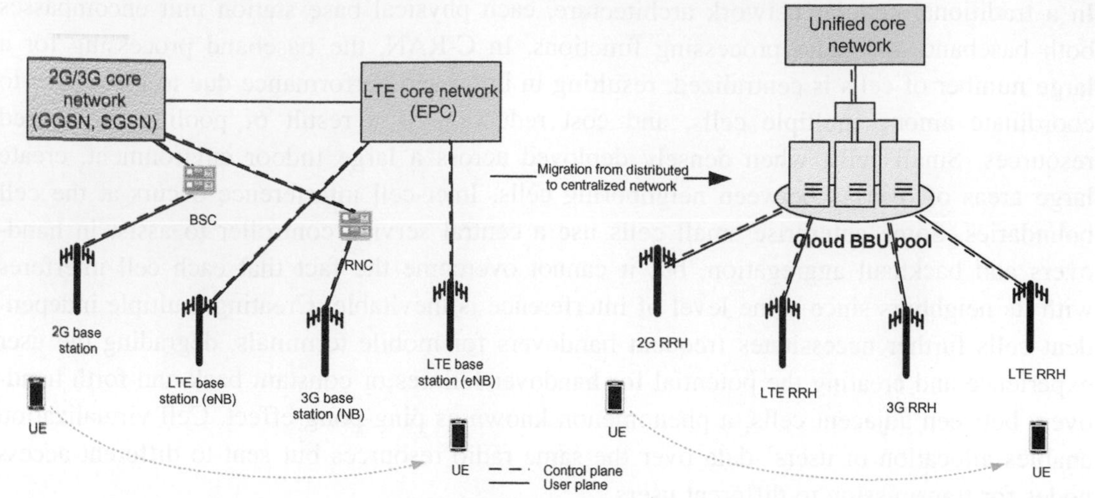

Figure 1.24
Mobility management in multi-RAT scenarios in distributed and centralized RAN architectures [61].

pool, shown in Fig. 1.24, integrates the essential 2G, 3G, and LTE [and later 5G NR (new radio)] network protocols, the inter-RAT handover process can be simplified and the distance between traffic anchor and interface can be reduced, which ultimately reduces the handover interruption time for network services. The inter-RAT handover differs from intra-RAT handover, which is a procedure to select a cell with the strongest received signal. The objective of inter-RAT handover is to select a suitable cell, considering user needs (especially mobility prediction), traffic type, and network property and state, by switching the logical interface between different RATs, which is more manageable in a C-RAN architecture.

Based on the distributed base station architectural model for the C-RAN, all or some of the baseband functions are performed in the centralized unit (CU). The processing resources on the CU can be dynamically managed and allocated. The C-RAN architecture allows improvement of resource utilization rate and energy efficiency as well as support of collaborative techniques. The concept of C-RAN has been evolving in the past decade and numerous architectural and deployment schemes for improving the spectral efficiency, latency, and support of advanced interference mitigation techniques have been studied and trialed by the leading operators. As an example, demarcation of the BBU into a CU and DU(s), functional split, and the next-generation fronthaul interface (NGFI) were introduced under the C-RAN context to meet the 5G requirements. The principle of CU/DU functional split originated from the real-time processing requirements of different applications. As shown in Fig. 1.25, a CU typically hosts non-real-time RAN protocols and functions offloaded from the core network as well as MEC services. Accordingly, a DU is primarily responsible for physical layer processing and real-time processing of layer-2 functions. In order to relax the transport requirements on the fronthaul link between the DU and the RRUs, some physical layer functions can be relocated from the DU to the RRU(s). From the equipment point of view, the CU equipment can be developed based on a general-purpose platform, which supports RAN functions, functions offloaded from core network, and the MEC services, whereas the DU equipment must be (typically) developed using customized platforms, which can support intensive real-time computations. Network function virtualization infrastructure allows system resources, including those in the CU and the DU, to be flexibly orchestrated via MANO, SDN controller, and traditional network operation and maintenance center, supporting fast service rollout. In order to address the transport challenges between CU, DU, and RRU(s), the NGFI standard has been developed where an NGFI switch network is used to connect the C-RAN entities. Using the NGFI standard, the C-RAN entities can be flexibly configured and deployed in various scenarios. In case of ideal fronthaul, the deployment of DU can also be centralized, which could subsequently support cooperative physical layers. In case of non-ideal fronthaul, the DU can be deployed in a fully or partially distributed manner. Therefore C-RAN architecture in conjunction with the NGFI

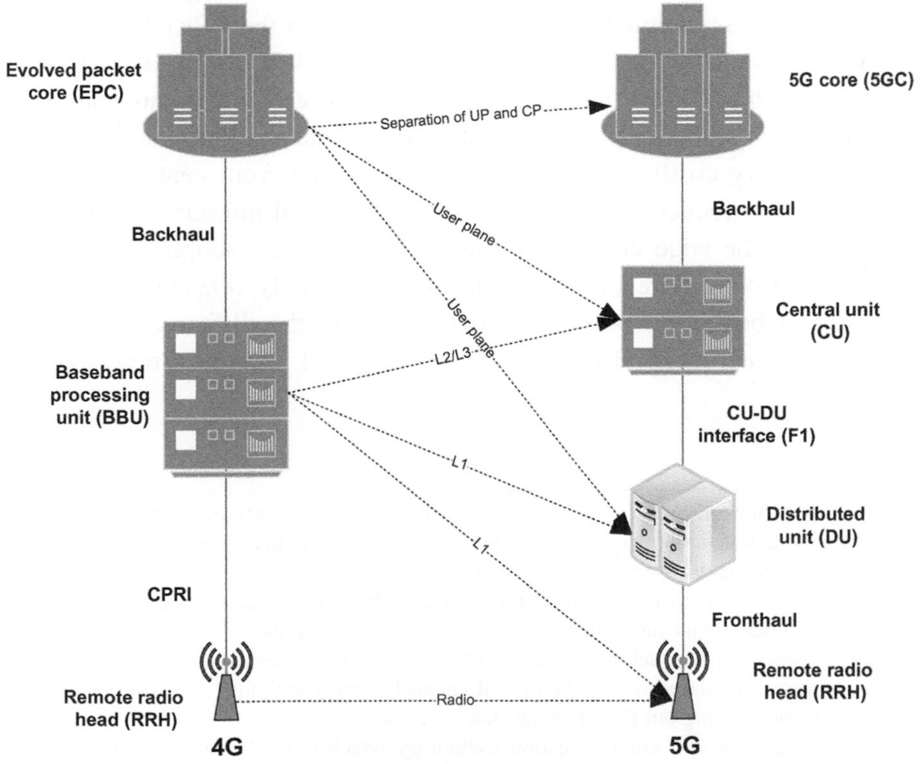

Figure 1.25
BBU architecture and evolution to 5G [47].

standard can support CU and DU deployments [47]. Latency, cost, and distance should be carefully evaluated in determining the proper mode of transport. Some of the available options include dedicated fiber, optical transport network (OTN),[34] passive optical network (PON),[35] microwave, and wavelength-division multiplexing (WDM) schemes. Mobile operators can further leverage low-cost, high-capacity fronthaul solutions using microwave E-band transport as an advanced application of C-RAN architecture. E-band radios are point-to-point, line-of-sight (LoS) microwave radios operating at 71−86 GHz.

1.1.6.2 Fronthaul Transport and Functional Split Options

In a C-RAN architecture, the baseband signal processing is centralized and often moved from individual RRUs to the edge cloud, resulting in a simplified network architecture, smaller form-factor radio units, efficient sharing and use of network resources, reduced costs of equipment installation and site maintenance, and higher spectral efficiency gains from joint processing schemes such as CoMP transmission/reception. However, it is necessary to overcome the constraints of fronthaul network transporting the raw in-phase and quadrature (I/Q) digital radio samples from the radio units to the edge cloud for processing. The fronthaul network is traditionally implemented based on the CPRI standard, which currently supports data rates of up to 24 Gbps per cell, a total fronthaul latency up to 200 μs, low jitter, tight synchronization, and high reliability. These requirements can only be realized with high-capacity fiber or point-to-point wireless links, making the deployment of the fronthaul network very costly, reducing the gains expected from centralization. The most promising approach to reduce the traffic load of the fronthaul interface is through the functional split between the edge cloud BBU and the RRU(s). By adopting only partial radio protocol split, the fronthaul requirements can be significantly relaxed while retaining the main centralization benefits. These functional splits blur the difference between classical fronthaul and backhaul networks, calling for converged transport networks that unify

[34] An optical transport network consists of a set of optical network elements connected by optical fiber links and is able to provide functionality of transport, multiplexing, routing, management, supervision, and survivability of optical channels carrying user signals, according to the requirements given in ITU-T Recommendation G.872. A distinguishing characteristic of the optical transport network is its provision of transport for any digital signal independent of client-specific aspects, that is, client independence. As such, according to the general functional modeling described in ITU-T Recommendation G.805, the optical transport network boundary is placed across the optical channel/client adaptation, in a way to include the server specific processes and leaving out the client-specific processes.

[35] A passive optical network is a communication technology which is used to provide fiber links to the end users. One of the passive optical network's distinguishing features is that it implements a point-to-multipoint architecture where passive fiber-optic splitters are used to enable a single optical fiber to serve multiple endpoints. A passive optical network does not have to provision individual fibers between the hub and customer.

backhaul and fronthaul equipment (i.e., integrated access and backhaul[36]), hence reducing deployment and operational costs. The deployment of these networks can be facilitated by the introduction of NGFI.

Centralizing baseband processing simplifies network management and enables resource pooling and coordination of radio resources. The fronthaul represents the transport network connecting the central site to cell sites when some or the entire baseband functions are hosted in a central site. Point-to-point dark fiber would be the ideal transport medium for fronthaul because of its high bandwidth, low jitter, and low latency. However, dark fiber is not widely available, thus there has been a need to relax the fronthaul requirements in order to enable the use of widely available transport networks such as packet-based Ethernet or E-band microwave. With such transport networks, bandwidth may be limited, jitter may be higher, and latencies may be on the order of several milliseconds. Cloud-RAN architecture is able to support different functional splits. Centralizing only a portion of the baseband functions and leaving the remaining functions at the remote sites would be a way to relax the fronthaul requirements. Depending on which functions are centralized, we will have different bandwidth, latency, and jitter requirements.

In a C-RAN architecture, the RRHs are connected to the BBU pool through high-bandwidth transport links known as fronthaul. There are a few standard interface options between the RRH and BBU including CPRI, radio-over-Ethernet (RoE), and Ethernet. However, CPRI and its most recent version, eCPRI, are currently the most common technologies used by C-RAN equipment vendors. The fronthaul link is responsible for carrying the radio signals, typically over an OTN, using either digitized form based on protocols such as the CPRI, or in analog form through radio-over-fiber technology. The main advantage of digitized transmission is the reduced signal degradation, allowing data transmission over longer distances, offering higher degree of BBU centralization. The common fronthaul solution in C-RAN is to use dedicated fiber. However, centralization requires use of a large number of fiber links which are limited and expensive to deploy. Alternative solutions include the use of other transport technologies such as WDM and OTN, or even the transmission of fronthaul data wirelessly using microwave or mmWave frequency bands. CPRI imposes very strict requirements on the fronthaul network. These requirements make the fronthaul network very expensive to deploy, thereby offsetting the cost saving expected from C-RAN. It can therefore be argued that the fronthaul network could become the bottleneck of 5G mobile networks.

[36] One of the potential technologies targeted to enable future cellular network deployment scenarios and applications is the support for wireless backhaul and relay links enabling flexible and very dense deployment of new radio cells without the need for densifying the transport network proportionately.

The data rates on the fronthaul links are substantially higher compared to the data rates on the radio interface due to CPRI I/Q sampling and additional control information. The centralized BBU and the distributed RRHs exchange uncompressed I/Q samples; therefore, efficient compression schemes are needed to optimize such wideband transmission over capacity-constrained fronthaul links. Possible solutions include digital RF signal sampling rate reduction, use of nonlinear quantization, frequency-domain subcarrier compression, or I/Q data compression. The choice of the most suitable compression scheme is a trade-off between achievable compression ratio, algorithm and design complexity, computational delay, and the signal distortion it introduces, as well as power consumption. Reducing signal sampling rate is a low-complexity scheme with minimal impact on the protocols. Nonlinear quantization improves the signal-to-noise ratio (SNR). Logarithmic encoding algorithms such as μ-law or A-law[37] can also be used to achieve higher transport efficiency on the fronthaul links. Implementation of the orthogonal frequency division multiplexing (OFDM) processing blocks at the RRH allows further reduction in the required fronthaul capacity.

CPRI requires a round-trip latency of 5 μs, excluding propagation delay [75]. More importantly, the total delay including propagation delay is limited by the air-interface hybrid automatic repeat request (HARQ) timing, since HARQ acknowledgments have to be received at the DL/UL transceiver within certain duration and baseband processing would take certain time depending on the air-interface technology. As a result, typically around 200 μs is available for total fronthaul latency. Assuming the speed of light of 200,000 km/s in fiber, CPRI maximum transmission distance is limited to approximately 20 km.

In a fronthaul network which utilizes dark fiber, jitter rarely occurs between BBU and RRH, because this type of optical fiber rarely causes any jitter. On the contrary, in a fronthaul network containing active equipment like WDM or PON, jitter can be introduced to the fronthaul network during signal processing (e.g., mapping/multiplexing in OTN). CPRI I/Q bit streams with such jitter can cause errors in the clock and data recovery process at RRH, subsequently leading to degraded system performance of RRH. Degraded frequency accuracy of the reference clock recovered in RRH can affect the performance of all relevant components that use the reference clock. For example, an inaccurate reference clock may cause errors when converting LTE/NR I/Q samples into analog signals during the digital-to-analog conversion. It can further lead to inaccurate frequency of carrier signals used for radio transmission of the analog signals. Therefore jitter in the fronthaul network can cause significant impacts on the quality of

[37] A-law is a companding algorithm, which is used in European 8-bit PCM digital communications systems to modify the dynamic range of an analog signal for digitization. It is one of the versions of the ITU-T G.711 standard. The μ-law algorithm is another companding algorithm, which is primarily used in 8-bit PCM digital telecommunication systems in North America and Japan. Companding algorithms reduce the dynamic range of an audio signal, resulting in an increase in the signal-to-noise ratio achieved during transmission. In the digital domain, it can reduce the quantization error.

LTE/NR signals transmitted through RRH antennas. Therefore when implementing a fronthaul network for C-RAN, extensive verification is required to ensure jitter introduced by active equipment is maintained within the tolerable range [60].

The stringent latency requirements have kept the fronthaul interface away from packet-switched schemes such as Ethernet. However, with the exponential increase in bandwidth requirements for 5G networks, packet-based transport schemes cannot be disregarded. The economy of scale and the statistical multiplexing gain of Ethernet are essential for the new fronthaul transport. In cooperation with CPRI Forum, IEEE 802.1 took the task of defining a new fronthaul transport standard under the IEEE 802.1cm[38] project. The project was entitled, Time-Sensitive Networking (TSN)[39] for fronthaul, which defines profiles for bridged Ethernet networks that will carry fronthaul payloads in response to requirements contributed by the CPRI forum. The requirements can be divided into three categories:

1. Class 1: I/Q and Control and Management (C&M) data
2. Synchronization
3. Class 2: eCPRI which has been recently added

38 IEEE 802.1CM—time-sensitive networking for fronthaul (http://www.ieee802.org/1/pages/802.1cm.html).

39 Time-sensitive networking is a standard developed by IEEE 802.1Q to provide deterministic messaging on standard Ethernet. Time-sensitive networking scheme is centrally managed with guaranteed delivery and minimized jitter using scheduling for those real-time applications that require deterministic behavior. Time-sensitive networking is a data link layer protocol and as such is part of the Ethernet standard. The forwarding decisions made by the time-sensitive networking bridges use the Ethernet header contents and not the Internet protocol address. The payloads of the Ethernet frames can be anything and are not limited to Internet protocol packets. This means that time-sensitive networking can be used in any environment and can carry the payload of any application. There are five main components in the time-sensitive networking solution as follows [78]:

- Time-sensitive networking flow: The time-critical communication between end devices where each flow has strict timing requirements and each time-sensitive networking flow is uniquely identified by the network devices.
- End devices: These are the source and destination of the time-sensitive networking flows. The end devices are running an application that requires deterministic communication.
- Bridges: Also referred as Ethernet switches, these are special bridges capable of transmitting the Ethernet frames of a time-sensitive networking flow on schedule and receiving Ethernet frames of a time-sensitive networking flow according to a schedule.
- Central network controller: A proxy for the network comprising the time-sensitive networking bridges and their interconnections, and the control applications that require deterministic communication. The central network controller defines the schedule based on which all time-sensitive networking frames are transmitted. Centralized user configuration: An application that communicates with the central network controller and the end devices and represents the control applications and the end devices. The centralized user configuration makes requests to the central network controller for deterministic communication with specific requirements for the flows.

I/Q and C&M data can be transported independently. The round-trip delay for I/Q is limited to 200 μs and maximum frame error rate is 10^{-7}. The C&M data has more relaxed time budgets. Synchronization signals represent an interesting aspect, with a wide range of requirements driven by wireless standards such as 3GPP LTE/NR. Four classes have been defined [75]:

1. Class A+: Strictest class with time error budget of 12.5 ns (one way) for applications such as MIMO and transmit diversity.
2. Class A: Time error budget up to 45 ns for applications including contiguous intra-cell CA.
3. Class B: Budgets up to 110 ns for non-contiguous intra-cell CA.
4. Class C: The least strict class delivers a budget up to 1.5 μs from the primary reference time clock[40] to the end application clock recovery output.

The above synchronization requirements, which continue to become more stringent with the new releases of the standard, pose new challenges for network designers. Traditional backhaul networks mostly rely on GPS receivers at cell sites. It is the simplest solution from the perspective of backhaul network design, but GPS systems have their own vulnerabilities and are not available in certain locations (e.g., deep indoor environments). Therefore, operators around the world have increasingly begun to deploy precision time protocol (PTP)/IEEE 1588v2[41] scheme as a backup mechanism and in some cases as the primary synchronization source in the absence of a viable GPS-based solution.

Standard bodies such as ITU-T have continued to refine and enhance the architectures and metrics for packet-based synchronization networks in parallel with the development of a new fronthaul. The ITU-T G.826x and G.827x[42] series provide a rich set of documents that define the architectures, profiles, and network limits for frequency and time/phase synchronization services. Phase synchronization exhibits an especially interesting challenge for synchronization experts as pointed out by the above fronthaul synchronization requirements. PTP has been defined to synchronize the time and phase of end applications to a primary reference. The PTP protocol continuously measures and attempts to eliminate any offset between the phase of the end application and the primary reference. However, in conventional Ethernet networks, packet delay variation has posed a major challenge to transferring acceptable clock qualities in wireless applications. Ethernet switch manufacturers responded

[40] The primary reference time clock provides a reference time signal traceable to a recognized time standard UTC.

[41] IEEE Std 1588-2008: IEEE Standard for a Precision Clock Synchronization Protocol for Networked Measurement and Control Systems (https://standards.ieee.org/findstds/standard/1588-2008.html).

[42] G.826: End-to-end error performance parameters and objectives for international, constant bit-rate digital paths and connections (https://www.itu.int/rec/T-REC-G.826/en) and G.827: Availability performance parameters and objectives for end-to-end international constant bit-rate digital paths (https://www.itu.int/rec/T-REC-G.827/en).

to this challenge by delivering new classes of PTP-aware nodes such as boundary clocks and transparent clocks. PTP-aware nodes are being increasingly deployed in wireless access networks around the world. While the packet delay variation is not a major concern for these deployments, timing error analysis remains a major point of focus. Timing error defines the difference between the time of a clock at any relevant part of the network and the time of a reference clock such as one delivered by a GPS source at another part of the network. It can result from network asymmetries and node configuration/performance issues [84].

In order to relax the excessive latency and capacity constraints on the fronthaul, the operators and vendors have revisited the concept of C-RAN and considered more flexible distribution of baseband functionality between the RRH and the BBU pool. Instead of centralizing the entire BBU processing on the cloud, by dividing the physical receive and transmit chain in different blocks, it is possible to keep a subset of these blocks in the RRH. By gradually placing more and more BBU processing at the edge of the network, the fronthaul capacity requirement becomes less stringent. Nevertheless, partial centralization has two main drawbacks, both relating to the initially envisioned benefits of C-RAN: (1) RRHs become more complex, and thus more expensive; and (2) de-centralizing the BBU processing reduces the opportunities for multiplexing gains, coordinated signal processing, and advanced interference avoidance schemes. Consequently, flexible or partial centralization is a trade-off between what is gained in terms fronthaul requirements and what is lost in terms of C-RAN features.

Another key question is how the information between the RRH and the BBU is transported over the fronthaul link. A number of fronthaul transmission protocols have been studied since the inception of the C-RAN architecture. However, transport schemes such as CPRI have been predominantly considered for carrying raw I/Q samples in a traditional C-RAN architecture. Considering the potential for various functional splits between the BBU and the RRH, different types of information might need to be transported over the fronthaul link. Given the extensive adoption of Ethernet in the data centers and the core network, RoE could be a generic, cost-effective, off-the-shelf alternative for fronthaul transport. Furthermore, while a single fronthaul link per RRH to the BBU pool has usually been assumed, it is expected that the fronthaul network will evolve to more complex multi-hop topologies, requiring switching and aggregation. This is further facilitated by a standard Ethernet approach. Nevertheless, packetization over the fronthaul introduces some additional concerns related to latency and overhead. As information arriving at the RRH and/or BBU needs to be encapsulated in an Ethernet frame, header-related overhead is introduced per frame. To ensure that this overhead is small, and does not waste the potential bandwidth gains from baseband functional splitting, it would be desirable to fill an Ethernet payload before sending a frame. However, waiting to fill a payload introduces additional latency. Hence, it is important to consider the impact of packetization on the fronthaul bandwidth and latency, in conjunction with possible functional splits between RRH and BBUs, in order to understand the feasibility and potential gains of different approaches.

In the study item for the new radio access technology, 3GPP studied different functional splits between the CU and the DU [34]. Fig. 1.26 shows the possible functional splits between the CU and the DU. After months of discussions between the opponents and proponents of open interfaces, 3GPP initially decided to specify two out of eight possible functional splits, that is, options 2 and 7, but no agreement on option 7 could be reached. Note that option 8 has already been used in LTE and previous generations, where CPRI was the main fronthaul interface transport scheme. As we move from left to right in Fig. 1.26, the split point moves from layer-3 protocols and functions to the lower layer protocols and functions, that is, layers 1 and 2, and ultimately I/Q samples transmission over the fronthaul.

Split option 7 has three variants, as shown in Fig. 1.27, depending on what aspects of physical layer processing are performed in the DU/RRUs. In the following, a detailed description of each functional split and its corresponding advantages and disadvantages are provided [34]:

- *Option 1:* This functional split option is similar to the reference architecture for dual connectivity. In this option, the RRC sublayer is located in the CU. The packet data convergence protocol (PDCP), radio link control (RLC), MAC, physical layer, and RF functional processing are located in the DU(s). This option allows separate user-plane connections (split bearers) while providing centralized RRC and management. It may, in some circumstances, provide benefits in handling edge computing or low-latency use cases where the user data must be stored/processed in the proximity of the transmission point. However, due to the separation of RRC and PDCP, securing the interface in practical deployments may affect performance of this option. Furthermore, the RRUs will be more complex and expensive due to the additional hardware for local processing of layer-1 and layer-2 functions.
- *Option 2:* This functional split is similar to option 1 except PDCP functions are also co-located with the RRC functions in the CU. This option would allow traffic aggregation from NR and LTE transmission points to be centralized. It can further facilitate management of traffic load between NR and LTE links. Note that the PDCP-RLC split was already standardized in LTE under dual-connectivity work item. In addition, this option can be implemented by separating the RRC and PDCP for the control-plane stack and the PDCP for the user-plane stack into different central entities. This option enables centralization of the PDCP sublayer, which may be predominantly affected by user-plane process and may scale with user-plane traffic load.
- *Option 3:* In this option, the lower RLC functions, MAC sublayer, physical layer, and RF functions are located in the DU, whereas PDCP and higher RLC functions are

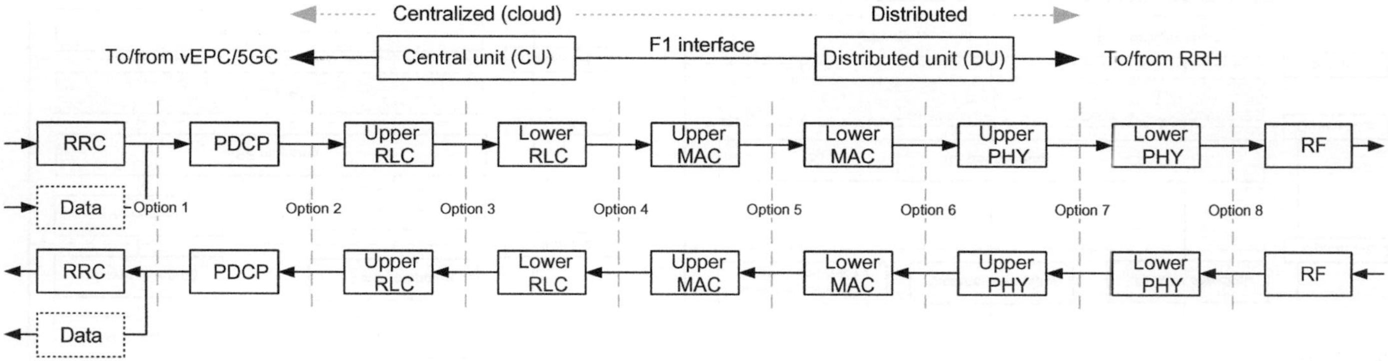

Figure 1.26
Functional split between the CU and the DU [34].

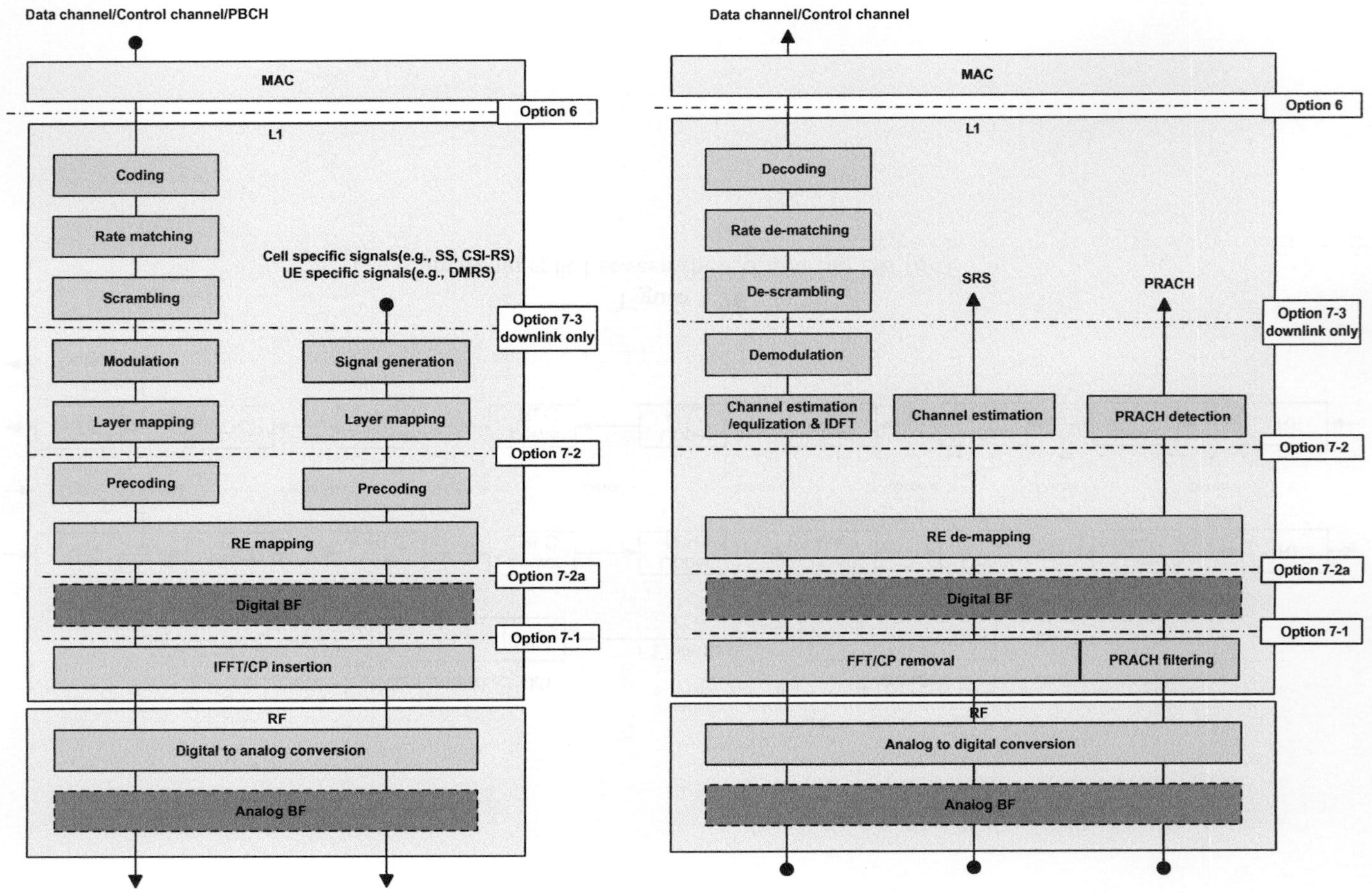

Figure 1.27
Variants of option 7 functional split [34].

implemented in the CU. Depending on real-time/non-real-time RLC processing requirements, the lower RLC may include segmentation functions and the higher RLC may include ARQ and other RLC functions. In this case, the centralized RLC functions will segment the RLC PDUs based on the status reports, while the distributed RLC functions will segment the RLC PDUs into the available MAC PDU resources. This option will allow traffic aggregation from NR and LTE transmission points to be centralized. It can further facilitate the management of traffic load between NR and LTE transmission points and it may have a better flow control across the split. The ARQ functions located in the CU may provide centralization gains. The failure over transport network may also be recovered using the end-to-end ARQ mechanism at the CU. This may provide more protection for critical data and control-plane signaling. The DUs without RLC functions may handle more connected mode UEs as there is no RLC state information stored and hence no need for UE context. Furthermore, this option may facilitate implementation of the integrated access and backhaul to support self-backhauled NR transmission points. It was argued that this option may be more robust under non-ideal transport conditions since the ARQ and packet ordering are performed at the CU. This option may reduce processing and buffering requirements in the DUs due to absence of ARQ protocol. This option may provide an efficient way for implementing intra-gNB mobility. Nonetheless, this option is more prone to latency compared to the option with ARQ in the DUs, since retransmissions are susceptible to transport network latency.

In an alternative implementation, the lower RLC functional group may consist of transmitting side of RLC protocol associated with downlink transmission and the higher RLC functional group may comprise the receiving side of RLC protocol, which are related to uplink transmission. This functional regrouping is not sensitive to the transmission network latency between the CU and the DU and uses interface format inherited from the legacy interfaces of PDCP-RLC and MAC-RLC. Since the receiving side of RLC protocol is located in the CU, there would be no additional transmission delay of PDCP/RLC reestablishment procedure when submitting the RLC SDUs[43] to PDCP. Furthermore, this alternative does not impose any transport constraint, for example,

[43] A service data unit is a specific unit of data that has been passed down from an open-system interconnection layer to a lower layer, which the lower layer has not yet encapsulated into a protocol data unit. A service data unit is a set of data that is sent by a user of the services of a given layer, and is transmitted semantically unchanged to a peer service user. The protocol data unit at a layer N is the service data unit of layer $N-1$. In fact, the service data unit is the payload of a given protocol data unit. That is, the process of changing a service data unit to a protocol data unit consists of an encapsulation process, performed by the lower layer. All data contained in the service data unit becomes encapsulated within the protocol data unit. The layer $N-1$ adds headers/subheaders and padding bits (if necessary to adjust the size) to the service data unit, transforming it into the protocol data unit of layer N. The added headers/subheaders and padding bits are part of the process used to make it possible to send data from a source node to a destination node.

transport network congestion. Nevertheless, due to performing flow control in the CU and the RLC transmit side in the DU, double buffering is needed for transmission.

- *Option 4:* In this case, MAC sublayer, physical layer, and RF functions are processed in the DUs, whereas PDCP and RLC protocols are processed in the CU. No particular advantage was shown for this option.
- *Option 5:* In this option, the RF, physical layer, and some part the MAC sublayer functions (e.g., HARQ protocol) are implemented in the DUs. The upper protocol stack is implemented in the CU. Therefore by splitting the MAC sublayer into two entities (e.g., upper and lower MAC), the services and functions provided by the MAC sublayer will be implemented in the CU and/or the DU. As an example, the centralized scheduling function located in the upper MAC will be in charge of the control of multiple lower MAC sublayers. The inter-cell interference coordination located in the upper MAC will be responsible for interference coordination. Time-critical functions in the lower MAC may include functions with stringent delay requirements (e.g., HARQ protocol) or the functions where performance is proportional to latency (e.g., radio channel and signal measurements at physical layer or random access control). Radio-specific functions in the lower MAC can perform scheduling-related processing and reporting. They can also control activities of the configured UEs and report statistics periodically or on-demand to the upper MAC. This option will allow traffic aggregation/distribution from/to NR and LTE transmission points. Moreover, it can facilitate the management of traffic load between NR and LTE transmission points. In this option, the requirement for fronthaul bandwidth and latency can be relaxed depending on the load across access and core network interface. It allows efficient interference management across multiple cells and enhanced coordinated scheduling schemes such as multipoint transmission/reception.
- *Option 6:* In this option, the physical layer and RF functions are implemented in the DU, whereas the upper protocol layers are located in the CU. The interface between the CU and the DUs carries data, configuration, and scheduling-related information, as well as measurement reports. This option will allow centralized traffic aggregation from NR and LTE transmission points, which can facilitate management of traffic load between NR and LTE access nodes. The fronthaul requirements in terms of throughput are reduced as the payload for this option are transport block bits. Joint transmission and scheduling is also possible in this case since MAC sublayer is centralized. This option may require subframe-level timing synchronization between MAC sublayer in the CU and physical layer in the DUs. Note that round-trip fronthaul delay may affect HARQ timing and scheduling.
- *Option 7:* In this option as shown in Fig. 1.27, the lower physical layer functions and RF circuits are located in the DU(s). The upper protocol layers including the upper physical layer functions reside in the CU. There are multiple realizations of this option including asymmetrical implementation of the option in the downlink and uplink (e.g., option 7-1 in the uplink and option 7-2 in the downlink). A compression technique may

be applied to reduce the required transport bandwidth between the CU and the DU. This option will allow traffic aggregation from NR and LTE transmission points to be centralized and can facilitate the management of traffic load between NR and LTE transmission points. This option can to some extent relax the fronthaul throughput requirements and allows centralized scheduling and joint processing in both transmit and receive sides. However, it may require subframe-level timing synchronization between the fragmented parts of physical layer in the CU and the DUs. The following represent different forms where this option can be implemented:

- *Option 7-1:* In this variant, the fast Fourier transform (FFT), CP removal (OFDM processing), and possibly PRACH processing is implemented in the uplink and in the DUs, and the remaining physical layer functions reside in the CU. In the downlink, inverse FFT (IFFT) and CP insertion blocks (OFDM processing) reside in the DUs and the rest of physical layer functions will be performed in the CU. This variant would allow implementation of advanced receivers.
- *Option 7-2:* In this variant, FFT, CP removal, resource de-mapping, and possibly MIMO decoding functions are implemented in the DU in the uplink and the remaining physical layer functional processing are performed in the CU. In the downlink, IFFT, CP addition, resource mapping, and MIMO precoding functions are performed in the DU, and the rest of physical layer processing is performed in the CU. This variant also allows the use of advanced receivers for enhanced performance.
- *Option 7-3:* This downlink only option implements the channel encoder in the CU, and the rest of physical layer functions are performed in the DU(s). This option can reduce the fronthaul throughput requirements as the payloads consist of the encoded bits.

- *Option 8:* In this option, RF functionality is in the DU and the entire upper layer functions are located in the CU. Option 8 allows for separation of RF and the physical layer and further facilitates centralization of processes at all protocol layer levels, resulting in very tight coordination of the RAN and efficient support of features such as CoMP, MIMO, load balancing, and mobility. This option will allow traffic aggregation from NR and LTE transmission points to be centralized. Moreover, it can facilitate the management of traffic load between NR and LTE transmission points, yielding high degree of centralization and coordination across the entire protocol stack, which enables more efficient resource management and radio performance. Separation between RF and physical layer decouples RF components from physical layer updates, which may improve their respective scalability. Separation of RF and physical layer allows reuse of the RF components to serve physical layers of different radio access technologies and allows pooling of physical layer resources, which may enable cost-efficient dimensioning of the physical layer. It further allows operators to share RF components, which may reduce system and site development/maintenance costs. However, it results in more stringent requirements on fronthaul latency, which may cause constraints on network deployments with respect to network topology and available transport options, as well as rigorous requirements on fronthaul bandwidth, which may imply higher resource consumption and costs in transport mechanisms.

There are strict timing, frequency, and synchronization requirements for the fronthaul links. In particular, 3GPP has imposed stringent latency requirements on transporting I/Q signals over the fronthaul, which pose certain challenges for system designers [11,14]. CPRI transport for fronthaul requires a low latency link and 200 μs is a generally accepted value for the round-trip latency, which limits the length of the fronthaul links to about 20 km. CPRI implementations require tight frequency and timing synchronization and accurate time of day (ToD) clock synchronization. A frequency precision of 16 ppb and ToD accuracy within 1.5 μs are required for CPRI transport.

The advantage of any functional split mainly depends on the availability of an ideal or non-ideal transport network. In a non-ideal fronthaul transport case, the functional split needs to occur at a higher level in the protocol stack, which reduces the level of centralization that can be achieved through C-RAN. In this case, the synchronization and bandwidth requirements can be relaxed at the expense of some of these 4G/5G RAN features such as massive MIMO, CA, and multipoint joint processing. The following solutions can be used to help overcome these obstacles. White Rabbit technology[44] is a combination of physical layer and PTP timing. White Rabbit introduces the technique of measuring and compensation for asymmetry to mitigate time and phase transfer errors. White Rabbit provides sub-nanosecond timestamp accuracy and pico-second precision of synchronization for large distributed systems and allows for deterministic and reliable data delivery. To achieve sub-nanosecond synchronization, White Rabbit utilizes synchronous Ethernet (SyncE)[45] to achieve synchronization and IEEE 1588v2. White Rabbit uses the PTP to achieve sub-nanosecond accuracy. A two-way exchange of the PTP synchronization messages allows precise adjustment of clock phase and offset. The link delay is known precisely via accurate hardware timestamps and the calculation of delay asymmetry. Alternatively, partial timing support compatible with ITU-T G.8275.2 standard can be used where the position of a grandmaster clock is moved closer to the PTP slaves in the RRHs. This is an excellent alternative to full on-path support White Rabbit for those operators who are not willing or cannot upgrade their networks for White Rabbit physical layer support.

---

44 White Rabbit is a multidisciplinary project for development of a new Ethernet-based technology which ensures sub-nanosecond synchronization and deterministic data transfer. The project uses an open-source paradigm for the development of its hardware and software components (https://www.ohwr.org/projects/white-rabbit/wiki/).

45 Synchronous Ethernet is an ITU-T standard for computer networking that facilitates the transfer of clock signals over the Ethernet physical layer. This signal can then be made traceable to an external clock.

CPRI Transport The CPRI is an industry forum defining a publicly available specification for the interface between a radio equipment control (REC) and a radio equipment (RE) in wireless networks. CPRI specifies a digitized serial interface between a base station referred to as REC in CPRI terminology and an RRH or RE. The specifications cover the user-plane, the control-plane transport mechanisms, as well as the synchronization schemes. The specification supports both electrical and optical interfaces as well as point-to-point, star, ring, daisy-chain topologies. The CPRI interface provides a physical connection for I/Q samples transport as well as radio unit management, control signaling, and synchronization such as clock frequency and timing synchronization [75].

CPRI transports I/Q samples to/from a particular antenna port and RF carrier. This is called an antenna-carrier (AxC) and is the amount of digital baseband (I/Q) user-plane data necessary for either reception or transmission of only one carrier at one independent antenna element. An AxC group is an aggregation of multiple AxC streams with the same sample rate, the same sample width, and the same destination. An AxC container consists of a number of AxCs and is a part of a basic CPRI frame (see Fig. 1.28). Data is organized into basic frames of 16 words. The first word of each basic frame is the control word. Each word can be 8, 16, or 32 bits, depending on the width of the I/Q samples. The width of the word depends on the CPRI line rate. For example, in an LTE system, if I = 16 bits and Q = 16 bits, then one AxC is 32 bits. Each 256 basic frames make up a hyperframe and 150 hyperframes are needed to transport an LTE 10 ms frame. Data in a basic frame is encoded with 8B/10B encoding, that is, 8 bits of data are encoded in 10 bits. The extra bits are used to detect link failures. Some of the CPRI rates support 64B/66B encoding scheme and this

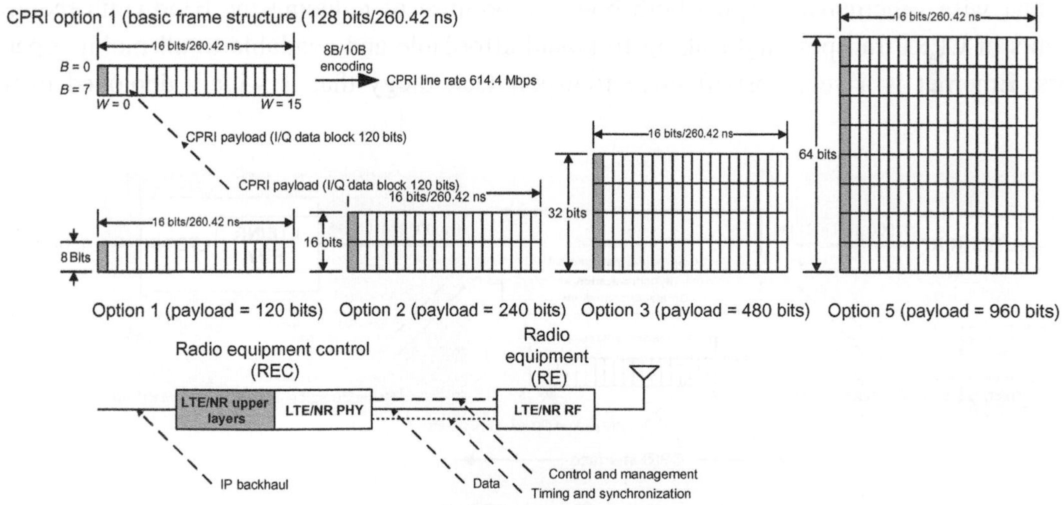

Figure 1.28
CPRI frame structure [60].

extension is used to detect sync header impairments and link failures. Note that 8B/10B and 64B/66B encodings incur 20% and 3% overhead, thus the latter would be a significant improvement in overhead reduction.

The CPRI specification specifies the maximum allowed effect of the fronthaul jitter on the frequency accuracy of the clock recovered at the RRH relative to a master reference clock at the BBU. One of the CPRI technical requirements defines the clock frequency accuracy of RRH as ± 0.002 ppm. This requirement states that the maximum impact of jitter from the CPRI fronthaul on the frequency accuracy of RRH should be less than ± 0.002 ppm. In addition to jitter, there are other factors that may affect the accuracy of clock frequency. CPRI requires very high reliability with bit error rates (BERs) of $\leq 10^{-12}$.

The CPRI framing process is illustrated in Fig. 1.29. User data is transported as baseband digital I/Q stream in a data block of a CPRI basic frame. The RRH, upon receiving the data, converts it into an analog signal, amplifies it, and then radiates the signal over the air. Control and management data and synchronization information are delivered through CPRI subchannels, more specifically through control words in the CPRI basic frames. This information is only used by the REC (on the BBU side) and the RE (on the RRH side). CPRI subchannels are created per CPRI hyper-frame, which is 66.67 μs and a hyperframe consists of 256 basic frames (260.42 ns). Each basic frame has one byte of a control word and 15 bytes of payload. A group of 256 control words in one hyperframe collectively constitute 64 subchannels. Fig. 1.29 shows a CPRI subframe and how the control and management data and synchronization information are mapped and transported [75].

eCPRI Transport The concepts of CPRI-over-Ethernet and replacing the TDM-like CPRI format with Ethernet messaging both hold the promise of reducing the bandwidth requirements of CPRI transport and making fronthaul affordable and available to all mobile operators. Ethernet is a very cost-effective transport technology that is widely deployed in the

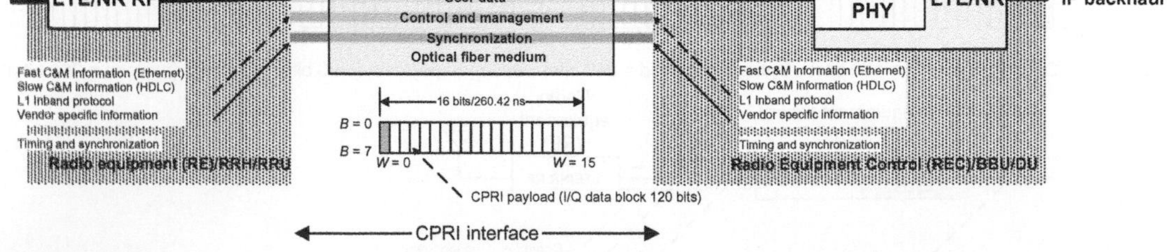

Figure 1.29
Illustration of CPRI framing [60].

backhaul transport network. However, it is also an asynchronous best effort technology that has not been originally designed to meet the low latency, low jitter, and tight synchronization requirements of baseband signal transmission. The new specification, known as eCPRI, introduces improved transport efficiency to match the speed and bandwidth requirements of 5G fronthaul networks. The eCPRI specification was released in August 2017 that supports partitioning of base station functions. The main advantages of the eCPRI protocol include support of functional split option 7, flexible bandwidth scaling according to user-plane traffic, and the use of mainstream transport technologies, which makes it possible carrying eCPRI and other traffic simultaneously in the same switched network.

The main difference between eCPRI and CPRI v7.0 can be summarized by looking at their respective characteristics [76].

- CPRI characteristics
  - It is intrinsically a point-to-point interface.
  - There is a master port and a slave port connected directly by optical/electrical cable (s) as a hop.
  - Networking functions are application layer functions and not supported by the CPRI interface itself.
  - Supported topologies depend on REC/RE functions.
  - Supported logical connections include point-to-point (one REC ↔ one RE) and point-to-multipoint (one REC ↔ several REs).
  - Redundancy, QoS, security, etc. are REC/RE functions.
- eCPRI characteristics
  - An eCPRI network consists of eCPRI nodes (eRECs and eREs), transport network, as well as other network elements including grand master for timing and EMS/NMS for management.
  - There is no longer a master port/slave port classification at physical level. SAP_S: master of PTP and synchronous Ethernet is not an eREC entity in general. SAP_{CM}: some of management-plane entities may be managed by EMS/NMS.
  - The eCPRI layer is above the transport networking layer.
  - The eCPRI layer does not depend on a specific transport network layer (TNL) topology.
  - The transport network may include local network and local switches provided by the eREC/eRE vendors.
  - Supported logical connections include point-to-point (one eREC ↔ one eRE), point-to-multipoint (one eREC ↔ several eREs), multipoint-to-multipoint (eRECs ↔ eREs, eRECs ↔ eRECs, eREs ↔ eREs).
  - Redundancy, QoS, security, etc. are mainly transport network functions; eCPRI nodes need to implement proper TNL protocols to support these capabilities.

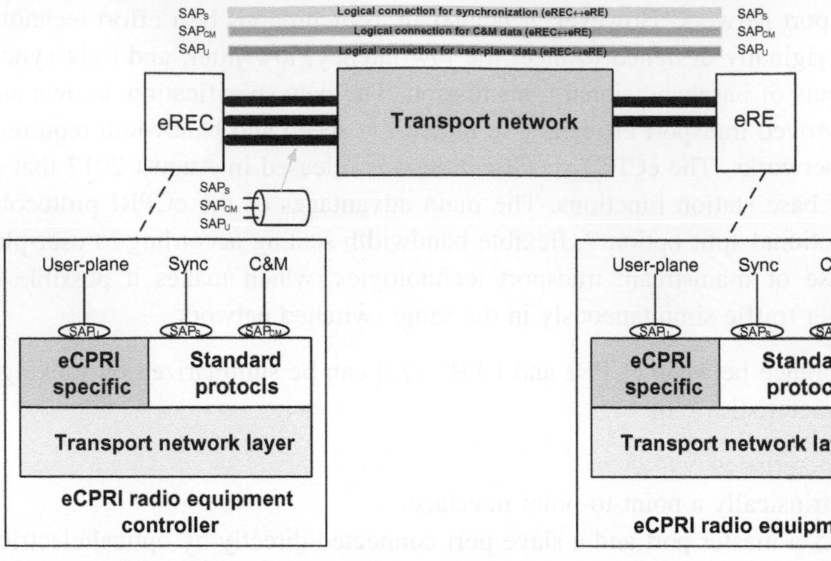

Figure 1.30
eCPRI system and interface definition [76].

As shown in Fig. 1.30, in eCPRI, the radio base station is divided into two building blocks: eCPRI radio equipment control (eREC) and eCPRI radio equipment (eRE), which are physically separated and are connected via a transport network. The eREC implements part of the physical layer functions and higher layer functions of the air interface, whereas the eRE contains the remaining part of the physical layer functions and the analog RF functions. User-plane data, control and management, and synchronization signals (i.e., synchronization data used for frame and timing alignment) are packetized, multiplexed, and transferred over the transport network which connects eREC(s) and eRE(s). The eCPRI does not rely on specific transport network and data-link-layer protocols, thus any type of network can be used for eCPRI provided that eCPRI requirements are fulfilled (see Fig. 1.31).

Fig. 1.32 shows high-level protocol stack and physical layer processing of an LTE eNB or NR gNB. The eCPRI specification defines five functional splits identified as A to E splits. An additional set of intra-PHY functional divisions identified as I_D, II_D, and I_U are also defined. It is understood that the CPRI specification supports only functional split E. The physical layer processing stages shown in Fig. 1.32 are consistent with those of the NR. The eCPRI specification focuses on three different reference splits, two splits in the downlink and one split in the uplink. Any combination of the different DL/UL splits is also possible. Other functional splits within the physical layer and/or upper layers are not precluded by eCPRI specification. The information flows for the eCPRI interface are defined as user plane including user data, real-time control data, and other eCPRI services; control and

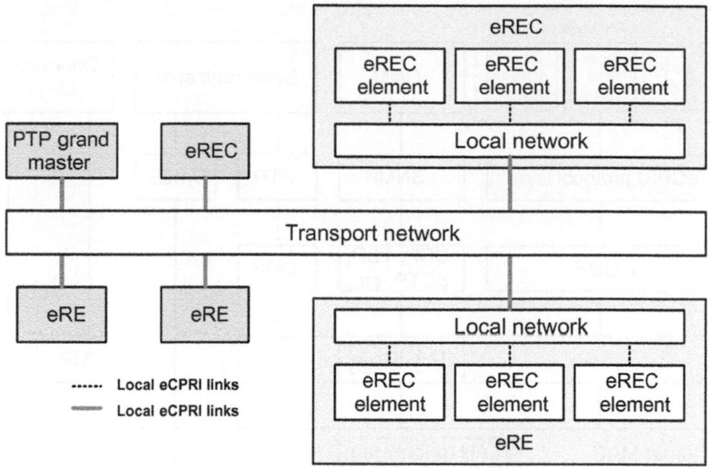

Figure 1.31
eCPRI example system architecture [76].

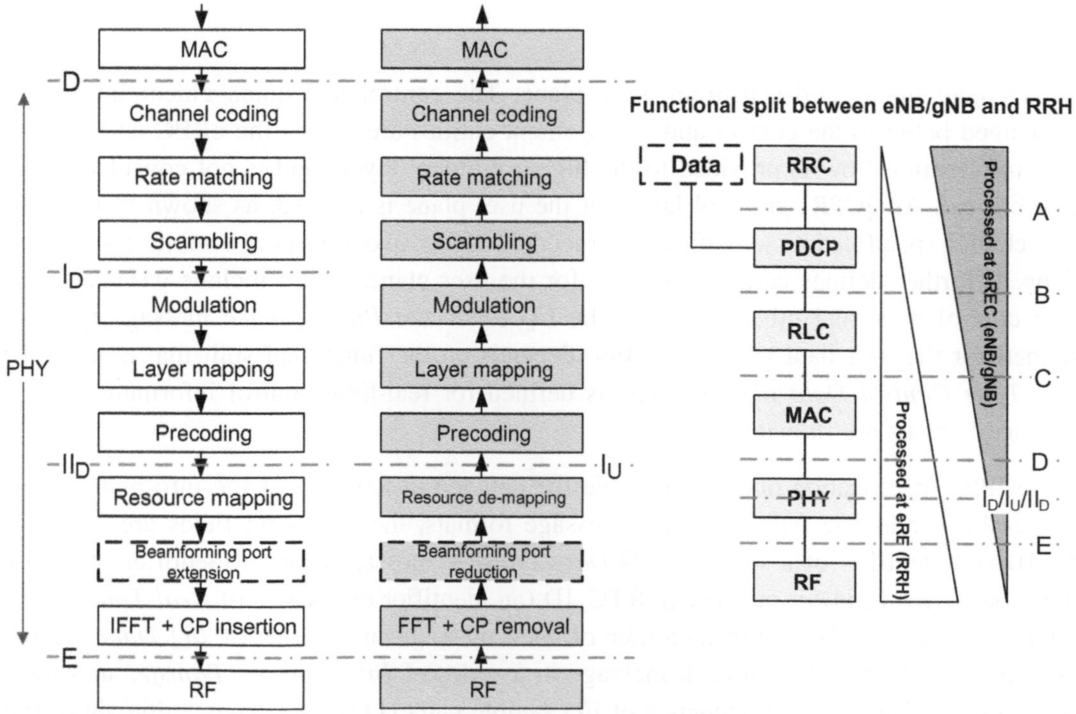

Figure 1.32
Processing stages of the physical layer shown in eCPRI specification [76].

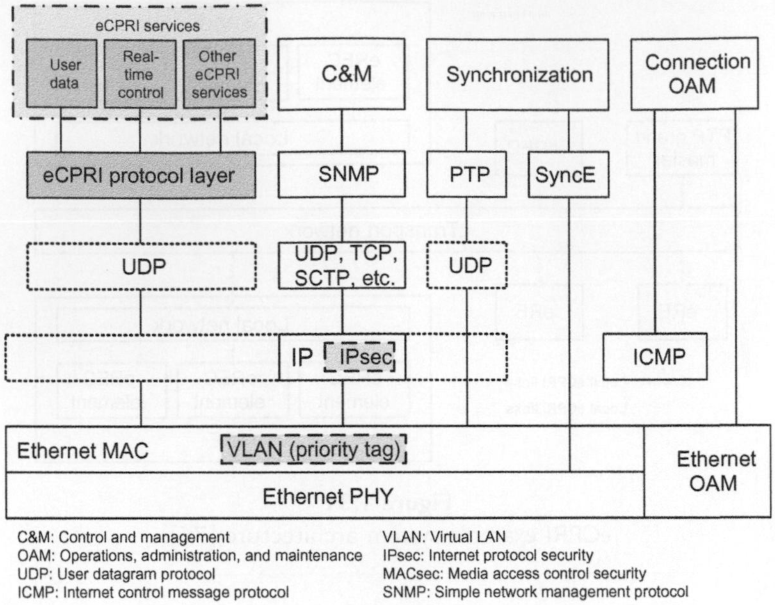

Figure 1.33
eCPRI protocol stack over IP or Ethernet [76].

management plane; and synchronization plane. The control and management information exchanged between the control and management entities are within the eREC and the eRE. This information flow is provided to the higher protocol layers and is not considered to be time critical. An eCPRI protocol layer for the user plane is defined, as shown in Fig. 1.33. The eCPRI specification identifies Ethernet and IP as two transport options for the user plane. It further defines certain messages for the user plane, which include user data, real-time control data, and other services. The *I/Q Data* and *Bit Sequence* message types are defined for the user data whose selection depends on the functional split that is used. The *Real-Time Control Data* message type is defined for real-time control information. These message formats are shown in Fig. 1.34.

The eCPRI specification does not provide the detailed description of the information fields of the above message types. In these message formats, the following fields are identified: PC_ID (an identifier of a series of *I/Q Data Transfer* messages; or an identifier of a series of *Bit Sequence Transfer* messages), RTC_ID (an identifier of a series of *Real-Time Control Data* messages), SEQ_ID (an identifier of each message in a series of *I/Q Data Transfer* messages; or an identifier of each message in a series of *Bit Sequence Transfer* messages), I/Q samples of user data (a sequence of I/Q sample pairs (I,Q) in frequency domain or time domain and associated control information, if necessary), bit sequence of user data, real-time control data, whose interpretation is left to implementation. Therefore, the details of

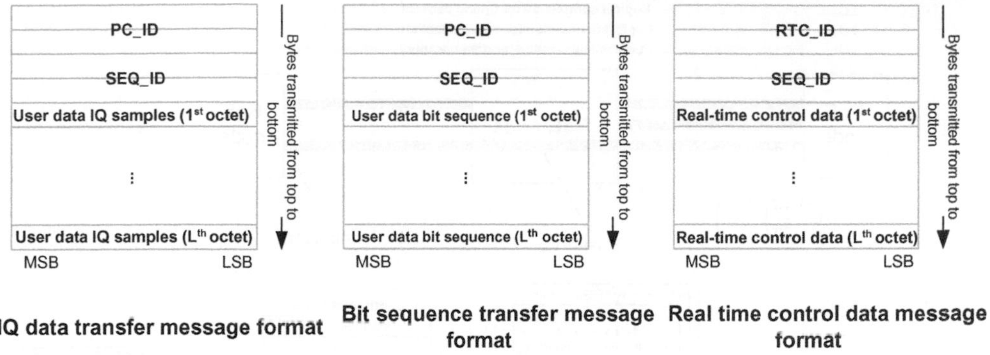

Figure 1.34
eCPRI user-plane message formats [76].

information flow are out of the scope of the eCPRI specification. This flexibility implies that there is additional work required to realize multi-vendor interoperable solutions.

IEEE 1914.3 Radio-Over-Ethernet Transport As we mentioned earlier, the TSN has been designed to ensure timely transport of delay-sensitive packetized streams such as CPRI-over-Ethernet; however, it does not deal with the encapsulation of various fronthaul transport payloads. RoE is a standard for radio transport over Ethernet including specification of encapsulations and mappings developed by IEEE 1914.3 working group.[46] It enables transport of I/Q data over Ethernet (i.e., native RoE packet mapper) as well as support of structure-aware and structure-agnostic mappers for CPRI and other data formats. The work targeted among others definition of a native RoE encapsulation and mapper transport format for both digitized radio payload (I/Q data) and management and control data. The Ethernet packet format itself is not changed and neither is the MAC protocol, as shown in Fig. 1.35.

The RoE defines a native encapsulation header format for transporting time-sensitive radio data and control information. The definition of protocol primitives allows multiplexing of independent streams, for example, antenna and carriers; time stamping or sequence numbering to enable time synchronization of RoE packets and timing alignment of the streams; control protocol for auxiliary non-data streams and for link and RoE endpoint management; and mapper(s) for existing CPRI framing standards to a native RoE encapsulation and transport. There is also focus on enabling support for non-native radio data where the transport structure is simply a container for the data (see Fig. 1.35). IEEE 1914.3 has further

[46] IEEE 1914.3: Standard for radio-over-Ethernet encapsulations and mappings (http://sites.ieee.org/sagroups-1914/p1914-3/).

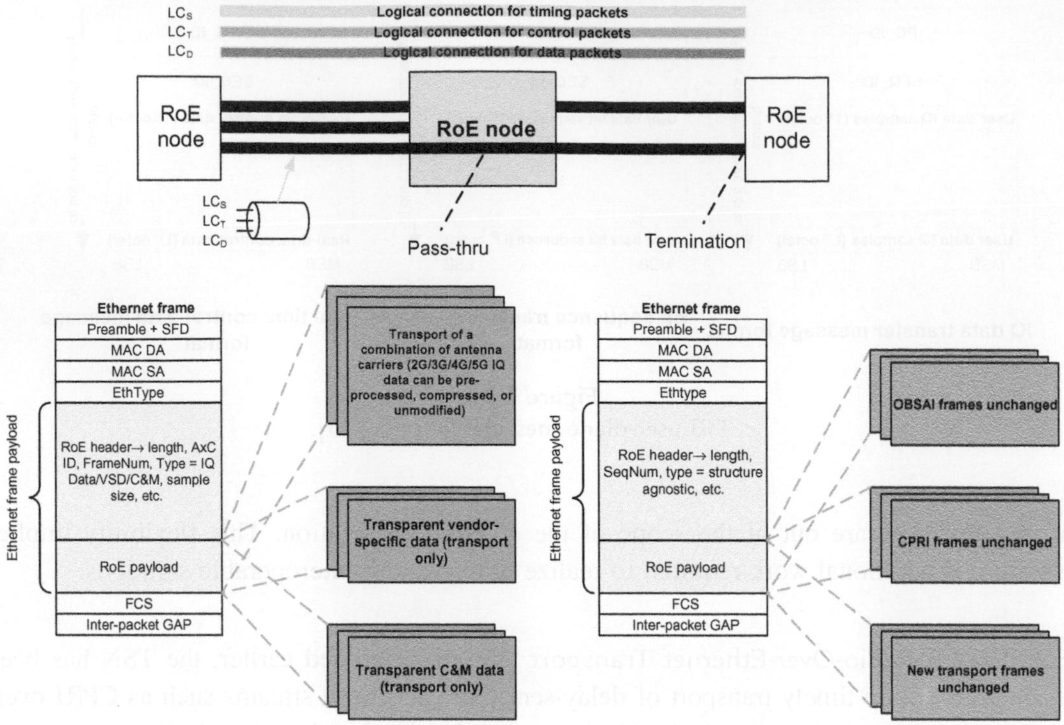

Figure 1.35
RoE encapsulation: RoE structure aware and agnostic mappers [77].

considered implementing a structure-aware mapper for CPRI with well-defined encapsulation procedure. One can therefore distinguish between a simple tunneling mapper for the former use case and the structure-aware mapper for CPRI. There is also a structure-agnostic mapper in the specs which offers a middle ground for efficiency and complexity relative to the other mappers. Several use cases of RoE can be considered including aggregation of multiple CPRI streams from a number of RRHs to a single RoE link to the BBU pool, or a native edge-to-edge RoE connection from the RRH directly to the BBU pool. The RoE will logically add a new switching/aggregation node between the baseband pool and the radio resources.

IEEE 1914.1 Next-Generation Fronthaul Interface The fronthaul packet transport enables implementation of critical 4G/5G technologies such as massive MIMO, CoMP transmission and reception, and scalable centralized/virtual RAN functions. Current network deployment models and practices based on traditional backhaul or fronthaul requirements are likely to be unsustainable and expensive for 5G deployments as the need for integrated access and backhaul in heterogeneous networks becomes more compelling. The NGFI is the new

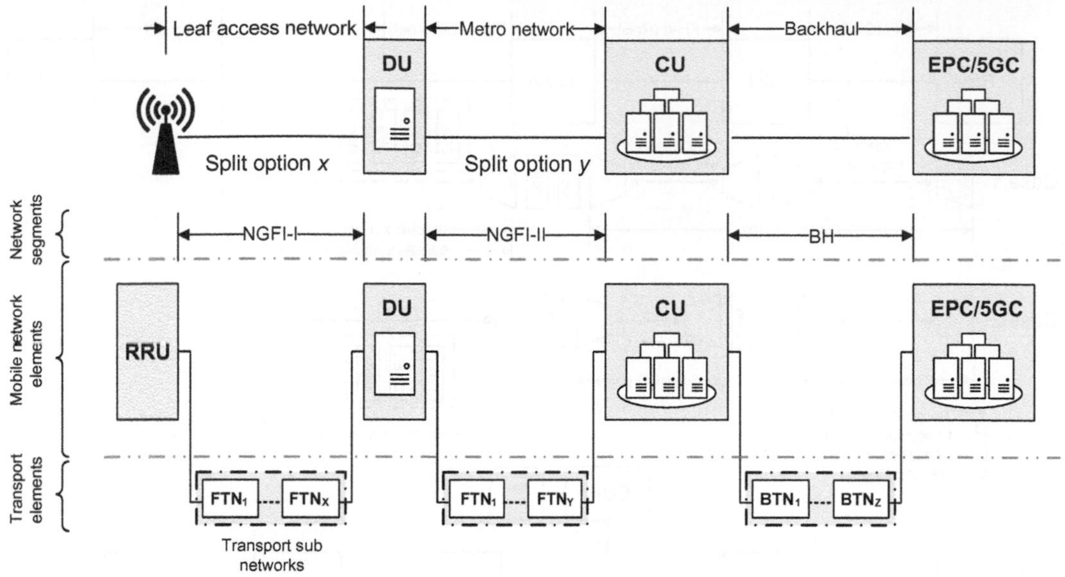

Figure 1.36
Example of RAN aggregation node with CU and DUs [78].

packet-based fronthaul interface specified by IEEE 1914.1[47] in order to provide bandwidth efficiency and to achieve scalability in transport networks as the CPRI standard cannot scale with the bandwidth requirements of 5G. IEEE 1914.1 standard simplifies network design and operation, increases network flexibility and resource utilization, and lowers cost by leveraging the existing and mature Ethernet-based solutions for critical functions such as QoS, synchronization, and data security. The fronthaul architecture provides unified management and control mechanisms, common networking protocols, and universal network elements, thus facilitating migration to cloud-based mobile networks.

As shown in Fig. 1.36, the high-level architecture for transport network for the next-generation mobile systems is typically hierarchical and rigorously follows the physical OTN topology. In the reference architecture, the core network is where the packet core gateways are located. The metro aggregation network aggregates one or more metro ANs, which again aggregates one or more cell site or leaf transport networks. Ring network topologies are common due to their resilience properties; however, other topologies are also conceivable. The leaf networks are often point-to-point topologies. There is no single location for central units and BBU pools. Each of the larger transport network domains may have their own CU/BBU pool sites. The same also applies to DUs. For instance, a DU may be located in an evolved RRH, in an aggregation node connecting to multiple RRHs, or in a central

[47] IEEE 1914.1: Standard for packet-based fronthaul transport networks (http://sites.ieee.org/sagroups-1914/p1914-1/).

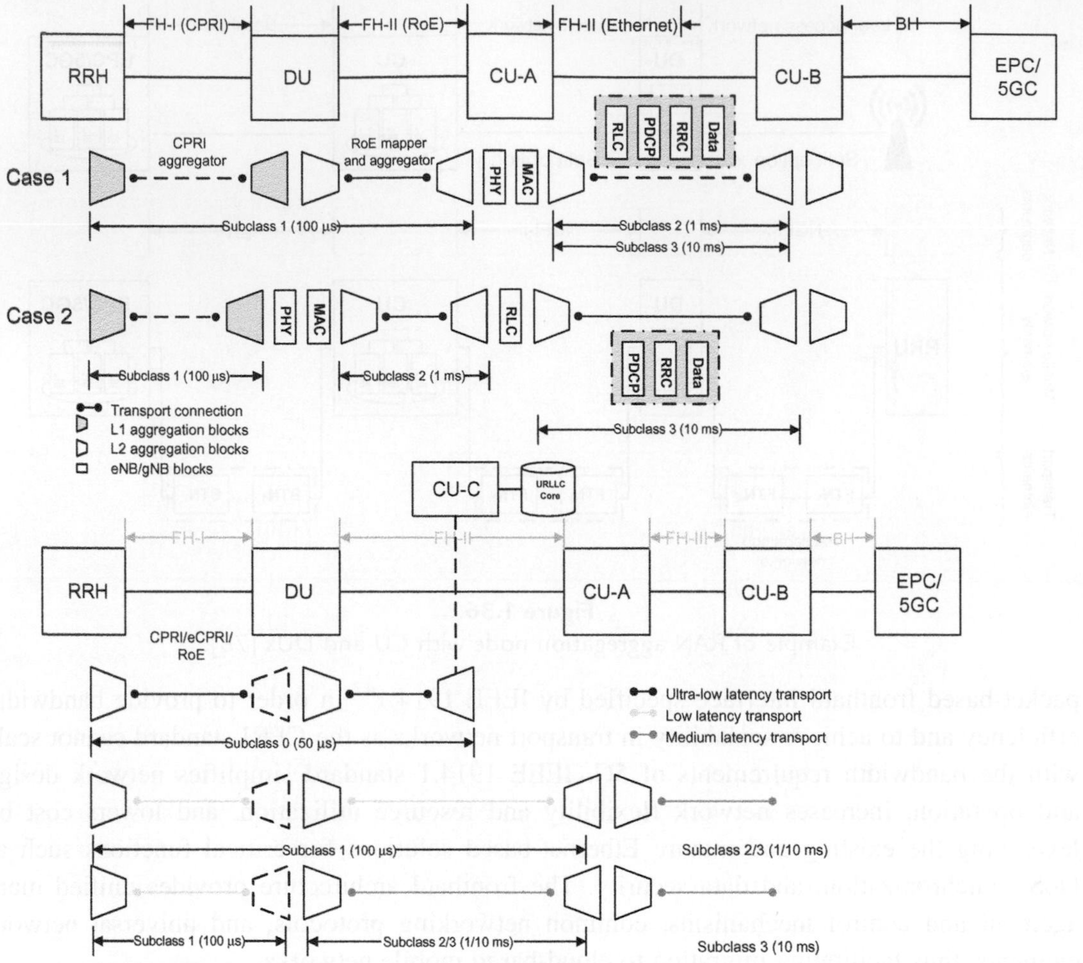

Figure 1.37
Next-generation fronthaul: CPRI-over-Ethernet for LTE and NR [78].

office. These networks must be able to transport 5G flows with heterogeneous traffic profiles. The traffic profiles may consist of traditional backhaul IP traffic, several different 3GPP functional split options traffic profiles, application traffic with varying latency needs, or non-IP CPRI TDM traffic. The transport network has to be able to serve all traffic profiles with vastly different service-level requirements in the same transport network infrastructure [78].

The transport network dedicated to 5G services is hierarchical, which means it comprises different domains and progressively aggregates signals from RRHs, at one end and up to the packet core at the other end. The support of multi-level functional split, results in a logical partitioning where more fronthaul segments and a backhaul segments may be identified. Fig. 1.37 depicts a generic model for converged fronthaul/backhaul network, where the following segments are identified: Fronthaul-I (NGFI-I), connecting the RU to a DU; Fronthaul-II (NGFI-II), connecting the DU to a CU; and Backhaul (BH), connecting the CU to the packet core elements. The NGFI reference architecture assumes all network deployment scenarios can leverage the same transport network infrastructure. Fig. 1.37 illustrates a deployment scenario where both new and legacy radio technologies coexist in the same cell site. In essence, the same transport infrastructure has to serve legacy backhaul, legacy non-packet CPRI fronthaul, and the highly versatile 5G fronthaul incorporating multiple functional split options with varying traffic profiles. It is also possible to deploy multi-level functional splits between an RRH and a BBU pool, resulting in multiple fronthaul transport domains in the network. However, practical limitations limit the number of splits that can be supported. For instance, an RRH to the aggregating DU node connection may implement 3GPP functional split option 7, and the DU to the CU connection may subsequently implement 3GPP functional split option 2. Fig. 1.37 illustrates an example deployment with two functional splits between the RRH and the CU. Typically, the DU would be an aggregation node close to the radio access network edge, and possibly in charge of time- and latency-sensitive coordinated scheduling and MAC-level retransmission functions.

IEEE 1914.1 standard specifies a number of service classes corresponding to the characteristics of traffic being transported over the backhaul/fronthaul links in a C-RAN, which includes control information or user traffic. In particular, it defines dedicated subclasses providing requirements for maximum tolerable latency for mobile control, transport control, and synchronization signals. The data-plane (or user traffic) class is further divided into five different subclasses, each addressing a specific network application (e.g., 5G service transport or functional split options implemented between RRH, DU, and CU). More specifically, the very low-latency subclass addresses network segment supporting URLLC service; the low-latency subclass addresses network segment where 3GPP split options 6, 7, 8 are implemented, the medium-latency subclass addresses network segment where 3GPP split options 2, 3, 4, 5 are implemented, the high-latency subclass applies to functional split options 2 and 3 with longer transport distances, and the very high-latency subclass is applicable to functional split option 1 or legacy transport that connects the access and core networks with much longer transport distance.

Network slicing enables next-generation network to provide multiple type of services and applications with different service requirements under a common and shared physical network infrastructure. An NGFI-compliant transport network supports transport of the network slicing traffic and serves as a sub-network instance and resource manageable by the network slicing orchestrator. From architectural perspective, two operation modes may be defined depending on the level of transparency to the network slicing.

In slicing-agnostic transport mode, the NGFI deployment scenarios may coexist across multiple instances in the same transport network segment, potentially owing to variations in RAN node (RRU, DU, CU) locations that resulted from network slicing operation.

Therefore, it is possible that multiple classes of service be simultaneously enacted in a transport connection between two network nodes. The combined classes of service are logically presented in Fig. 1.38, which are supported by a slicing-agnostic transport network. The slicing-agnostic transport is able to provide multiple levels of transport service and to maintain different transport QoS based on the class of service requirement. It also performs transport key performance indicator (KPI) monitoring and reports the results to NGFI operation and maintenance (OAM), on the basis of classes of service. The transport network in this mode is not directly aware of the network slicing operation. Instead, it simply makes the classes of service available via its interface to the other network entities, where a slicing-to-classes of service mapping needs to be performed by aggregating the slicing traffic with similar KPI requirements. If RAN nodes that support the sliced services are not geographically co-located, this class of service may be routable to multiple destinations. Upon addition, deletion, or modification of the slicing operations, the slicing-agnostic transport is not expected to adapt to the change and to perform any reconfiguration process for optimization of the transport network.

In slicing-aware transport mode, while the user plane remains class of service based and the transport QoS is maintained at class of service level, a slicing-aware transport network interfaces with the network slicing orchestrator which is aware of the network slicing operation. Thus, the transport operation can be controlled and managed via the NGFI OAM configuration and provisioning functions for the purpose of network slicing optimization. As depicted in Fig. 1.38, the slice-to-class of service mapping is required to support class of service-based transport operation and is realized within the transport network and controlled by OAM configuration function that communicates with the network slicing orchestrator. This slice-to-class of service mapping correspondence is flexible by nature and should be dynamically or semi-dynamically reconfigured without service interruption. Furthermore, the transport OAM monitoring function reports the transport KPIs to the network slicing orchestrator, providing a means of feedback for integrated network optimization. Upon deletion, addition, or any modification of the network slicing services, the slice-aware transport adapts to the changes and performs seamless reconfiguration over the transport network for the purpose of overall network optimization. A slice-based transport that is fully optimized for network slicing should not be class of service-based where each slice traffic should be individually identified, labeled, and transported according to its own QoS requirements [78].

1.1.6.3 Backhaul Transport Options

The increasing demand for broadband wireless applications has a significant effect on the entire mobile infrastructure including the mobile backhaul network. In an operator's

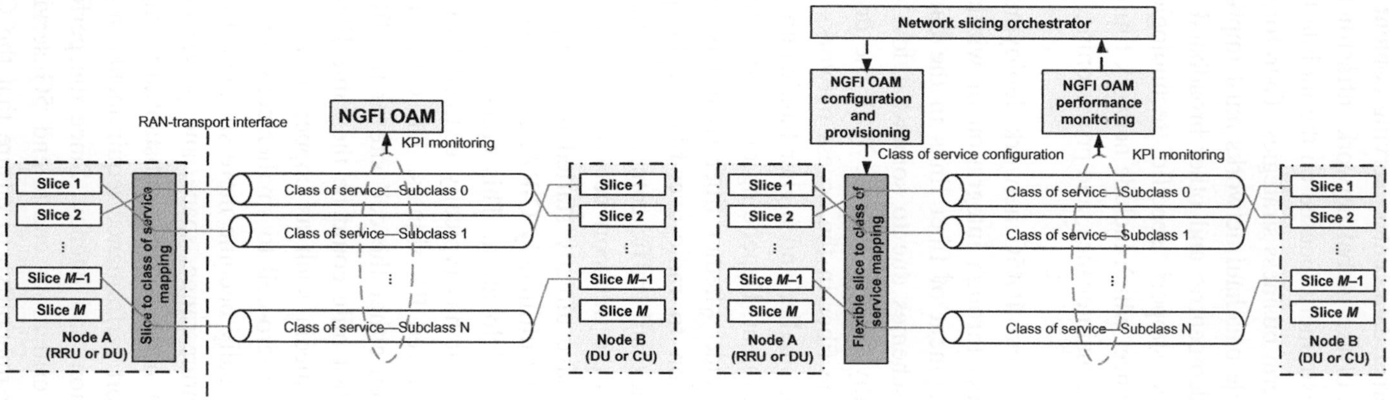

Figure 1.38
Comparison of slice-agnostic and slice-aware transports [78].

network, the mobile backhaul connects small and macro-cell sites to the core network that is further connected to external data centers serving content and applications. The RAN is an increasingly critical part of the global network infrastructure and is the primary reason that network operators are extremely focused on the mobile backhaul network as a key element of their short to long-term business strategies. Therefore the capacity, latency, reliability, and availability of mobile backhaul networks must improve as the wireless access data rates increase to enable video-centric and other broadband user applications. In addition, mobile backhaul networks support specific technologies that together ensure an acceptable quality of experience, which include network timing and synchronization as well as operations, administration, maintenance, and provisioning.

Recent technology advances and 4G/5G network deployments have created a new landscape in the access networks through integration of wireless and fiber technologies. In the past, the use and deployment of fiber links in the backhaul was slow compared to that of wireless backhaul schemes due to somewhat low data rate requirements of the backhaul supporting typically large cell sites. With the decreasing cost of fiber deployments and penetration of the fiber in the access networks as well as the demand of the latest wireless standards for smaller and higher bandwidth cells, the use of fiber connectivity has become more prevalent. Depending on the demarcation point between key network elements, one can decide whether fiber should be used only as a high data rate backhaul path or a transition to radio-over-fiber techniques can be afforded for the fronthaul links, as well. Backhaul traffic comprises a number of components in addition to the user-plane traffic. As the transport networks evolve, network operators are evaluating and, in some cases, have already started to deploy mobile fronthaul networks to support centralized RANs and ultimately C-RANs. In these networks, the BBU is moved from the cell site to a central location. This creates a new fronthaul network between the BBU and the cell site that has typically utilized CPRI protocol until just recently, which effectively carry digitized RF signals. These networks require fiber-based backhaul of Ethernet traffic from either the cell site or the BBU location, regardless of the last mile technology. The last mile could be the same Ethernet-over-fiber connection or Ethernet over some other media such as copper or microwave. As the cell topology changes from a traditional macro-cell to smaller cells, the requirements for the endpoint of the backhaul service typically become more stringent to meet certain deployment specifications such as temperature range, space, and power consumption. The performance of the underlying transport network must substantially improve in order to support the considerably tighter transport performance requirements in terms of frequency synchronization, phase synchronization, and latency. Since the performance of the backhaul networks is becoming more critical to end-to-end 5G services/applications, performance monitoring capabilities are imperative to ensure that the QoS requirements and service-level agreements are met (see Fig. 1.39).

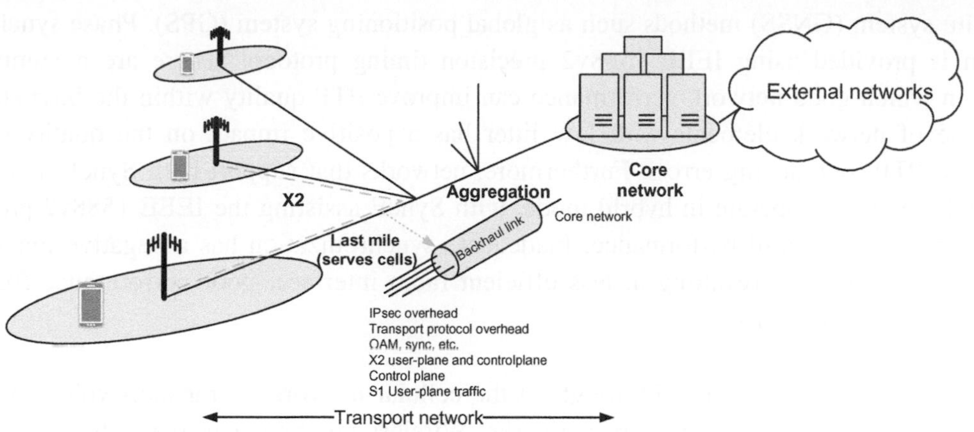

Figure 1.39
Illustration of backhaul components in a typical cellular network [66].

Mobile backhaul is an example application area where packet optical technology[48] enables the paradigm shift from traditional TDM-based backhaul to high performance, low cost, and scalable solutions that are currently required. Modern packet optical solutions enable the optimization of IP traffic between cell site gateways and core routers, avoiding unnecessary router hops. New generation of mobile backhaul supports Ethernet transport for all cell types and all locations regardless of last mile technology, whether CPRI-based fronthaul, DASs, fiber-connected small-cells or macro-cells, or fiber aggregation points supporting microwave backhaul in non-fiber environments are utilized. The use of frequency-division duplex scheme over the air-interface requires only simple frequency synchronization using SyncE or IEEE 1588v2 packet-based synchronization schemes. However, time-division duplex scheme requires phase synchronization, in which the network needs to track the phase of the synchronization signal and to receive accurate time-of-day timestamps. The more complex radio-access network features such as CoMP transmission and reception and eICIC further require phase synchronization.

Frequency synchronization is provided through a number of methods, where the most common solution is SyncE. It can also be provided in some regions using the global navigation

[48] Packet optical transport covers technologies and architectures that enable the transport of Internet protocol packets on both fixed and mobile optical networks. Converged products include the functional switching capability of wavelength division multiplexing schemes, Ethernet switching via various protocols, as well as time division multiplexing and optical transport network switching. The technologies include a combination of optical networking products that operate separately or within a single converged platform called packet optical transport system such as reconfigurable optical add-drop multiplexers, time division multiplexing, and carrier Ethernet switching products. These platforms reside in the metro edge, metro core, and the long-haul networks of major service providers.

satellite system (GNSS) methods such as global positioning system (GPS). Phase synchronization is provided using IEEE 1588v2 precision timing protocol. There are a number of ways in which good network performance can improve PTP quality within the base station. The use of network elements with low jitter has a positive impact on the quality of the received PTP by reducing errors. Furthermore, networks that support both SyncE and IEEE 1588v2 are able to operate in hybrid mode, with SyncE assisting the IEEE 1588v2 protocol for an improved overall performance. Inadequate synchronization has a negative impact on network performance, resulting in less efficient radio interface, poor performance for data traffic, and dropped calls.

Mobile backhaul is provisioned throughout the cellular network to transport voice and data traffic between the access and core networks. Wireless equipment in the radio access network includes macro-cell base stations, small-cell access points, and DASs. Wired and wireless transport mechanisms are the two types of mobile backhaul deployed across the RANs. With the emergence of heterogeneous networks, mobile backhaul has become a critical component in the 4G and 5G networks.

The cellular networks are evolving toward a heterogeneous architecture where different classes of small-cell base stations and DAS installations are coordinated, cooperating with macro-site base stations. A HetNet topology improves cellular network capacity and coverage to support the exponential growth in mobile traffic. Subsequently, HetNets can deliver ubiquitous connectivity to the mobile users with exceptional quality of experience. Backhaul is the confluence of mobile broadband users, small cells, DAS, macro-cells, and the core network. The emerging HetNet topologies have created a need for diverse wired and wireless mobile backhaul solutions. Small cells are being deployed in indoor and outdoor environments, on utility poles, and other urban structures. The sites can be located in private or public locations. Depending on the use case, each small-cell site has specific requirements for power sourcing, power budget, and backhaul transport. Meanwhile, the conventional macro-cell base stations will continue to increase network capacity, further driving demand for high-throughput backhaul.

Wireless backhaul, emerged as a cost-effective connectivity option, has many advantages relative to wired technologies, but wireless solutions also present unique design challenges among which are the need for spectrally efficient radio links, low operating power, small form factor, and environmentally resilient high-reliability equipment. RF analog integration also plays an important role. As an example, a typical microwave radio transmitter relies on RF analog integration and RF building blocks to reduce the size, lower the power, and improve the dynamic performance. A wireless backhaul equipment requires four key components: antennas, radio transceiver, modem, and interface. The antenna transmits and receives electromagnetic waves. The radio transceiver handles RF carrier frequency transformation to and from the baseband.

The modem performs channel coding/decoding and modulation/demodulation of the baseband signals, and the interface transports information between the radio and TDM/IP transport.

As shown in Fig. 1.40, wired and wireless backhaul solutions employ broadband technologies that vary in terms of physical media and access method. Wired backhaul physical media include copper wire, hybrid fiber-coaxial (HFC)[49] cable, and single-mode and multimode optical fiber. Transport access technologies are fractional-T/E carrier (T1/E1), digital subscriber line (DSL), pseudo-wire, Ethernet, WDM, and gigabit passive optical network.[50] The choice of wired backhaul is driven by the availability of physical media, cost, and capacity requirements. Wireless backhaul is needed when base stations do not have access to copper, HFC, or fiber transport. Wireless backhaul is also attractive when time to deployment is critical or when leasing costs are prohibitive. It is estimated that nearly 70% of worldwide LTE base station installations use wireless backhaul [80]. The methods of delivering wireless backhaul transport are LoS microwave, LoS millimeter wave, non-LoS (NLoS) sub-6 GHz microwave, broadband satellite links, and in-band or out-of-band relay nodes.

The HetNet architecture comprises four general classes of base station: (1) macro-cell, (2) metro-cell, (3) pico-cell, and (4) femto-cell. Table 1.2 compares the types of base station, deployment scenarios, and the set of possible wireless backhaul solutions. Wireless backhaul radios operate over a wide spectrum of licensed and unlicensed RF bands extending to 80 GHz (see Fig. 1.41). The RF spectrum for wireless backhaul ranges from sub-6 GHz NLoS to C/Ka/Ku-band microwave LoS, and Q/V/E-band mmWave LoS. Each RF band has spectrum restrictions, channel bandwidth limitations, and specific propagation characteristics. Channel bandwidth can vary from 5 to 160 MHz in NLoS systems; from 3.5 to 56 MHz in microwave LoS systems; or from 28 to 112 MHz and 250 MHz to 5 GHz in mmWave systems. All these specifications impact the type of modulation and carrier-to-noise ratio, and thereby mandating certain capacity trade-offs and maximum link distance.

Table 1.2 shows that backhaul throughput for each base station class must match the respective cell-site capacity. An optimal wireless backhaul solution further adapts the performance requirement with a particular deployment scenario. The method of wireless backhaul determines frequency band operation, radio design specifications, and the radio architecture.

[49] A hybrid fiber coaxial network is a telecommunication technology in which optical fiber cable and coaxial cable are used in different portions of a network to carry broadband content. An advantage of hybrid fiber coaxial is that some of the characteristics of fiber-optic cable (high bandwidth and low noise and interference susceptibility) can be brought close to the user without having to replace the existing coaxial cable that is installed at home or business.

[50] Gigabit passive optical network is a point-to-multipoint access network. Its main characteristic is the use of passive splitters in the fiber distribution network, enabling one single feeding fiber from the provider to serve multiple homes and small businesses. Gigabit passive optical network has a downstream capacity of 2.488 Gbps and an upstream capacity of 1.244 Gbps that is shared among users. Encryption is used to keep each user's data secured and private from other users [ITU-T G.984].

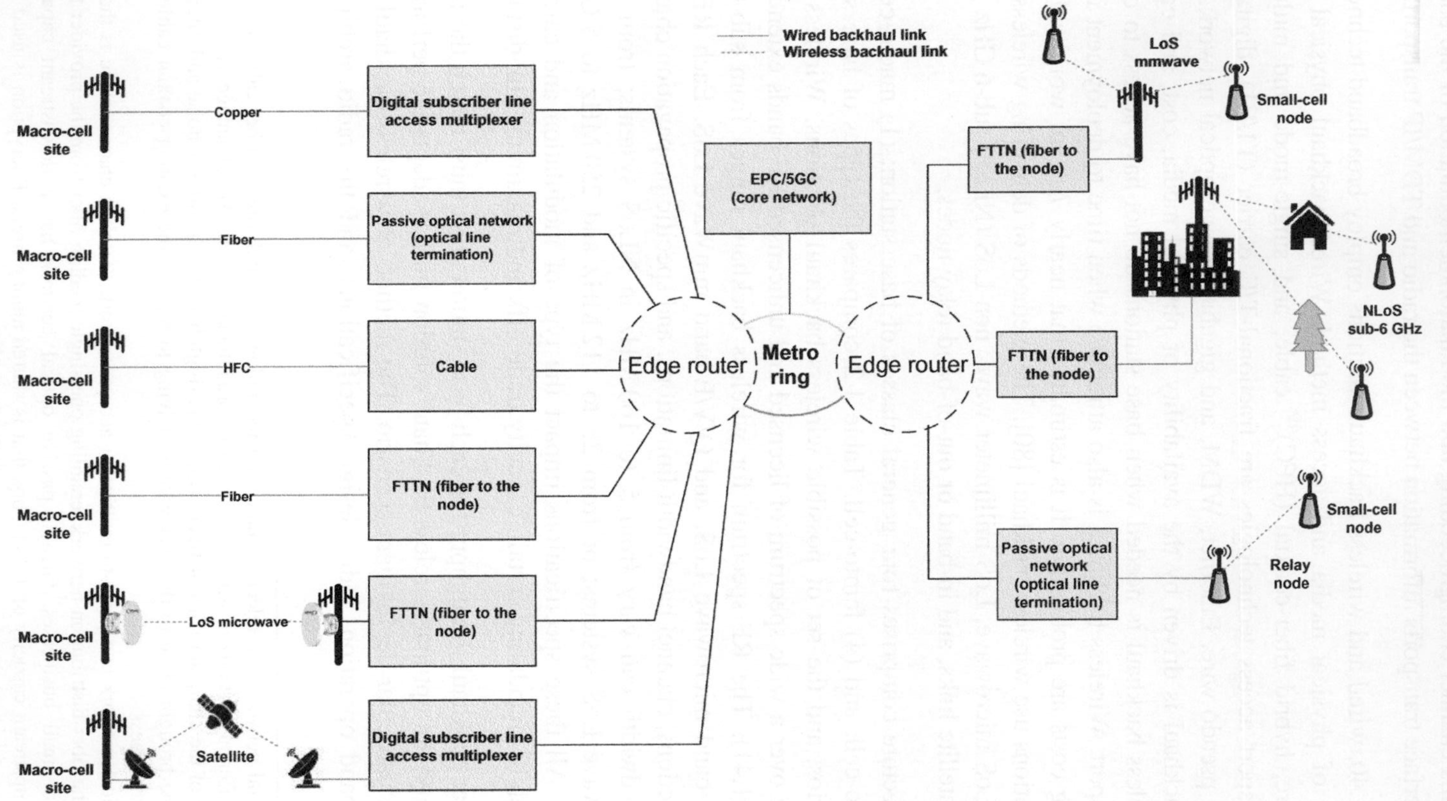

Figure 1.40

Illustration of mobile backhaul network physical media and access layer with macro-cell and small-cell base station [79].

Table 1.2: Example of LTE-advanced base stations and wireless backhaul solutions [79].

Base Station Type	Deployment Scenario	Number of Active Users	Cell Radius (km)	Power Amplifier Output Power (dBm)	Signal Bandwidth (MHz)	Number of Sectors/ Number of Antennas	Total Base Station Theoretical Capacity (Gbps)	Wireless Backhaul Scheme
Macro-cell	Outdoor	200–1000	> 1	50	100	3 4 × 4	9	Microware, mmWave, VSAT
Metro-cell	Outdoor	200	< 1	38	40	2 2 × 2	1.6	Microware, mmWave, VSAT
Pico-cell	Indoor/ outdoor	32–100	< 0.3	24	20	1 2 × 2	0.8	mmWave, sub-6 GHz, relay
Femto-cell	Indoor	4–16	< 0.1	20	10	1 2 × 2	0.4	Sub-6 GHz, relay

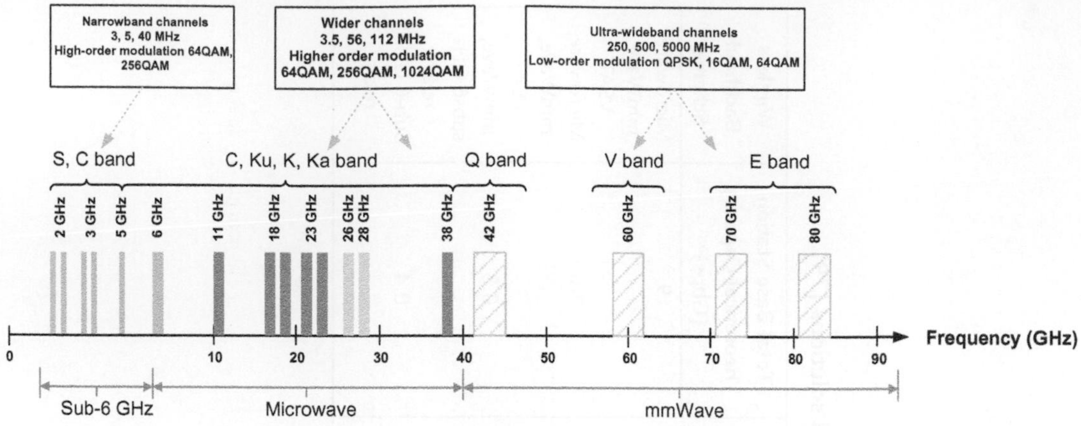

Figure 1.41
Frequency spectrum, channel bandwidth, and modulation considerations for RF bands [79].

Table 1.3 summarizes different characteristics for each wireless backhaul method and further outlines the key factors to be considered in backhaul selection.

Low-order modulation schemes such as QPSK can be used with wideband channels or for operation in poor atmospheric conditions with low SNR channel conditions. On the other hand, high-order modulation up to 2048 QAM can be used with narrowband channels or for operation in good channel quality and in clear atmospheric conditions. Depending on the link capacity requirements, deployment scenario, SNR, and atmospheric conditions, the data throughput can range from 100 Mbps to 10 Gbps. For macro-cell base station applications, the migration from indoor units to full outdoor units lowers the power, improves signal quality, and lowers OPEX. Macro-cells take advantage of full outdoor unit partition because RF loss in the waveguide or coaxial cable is minimized, or eliminated, which lowers RF power output and improves receiver input sensitivity. In small-cell base stations, the adoption of an integrated configuration means that systems can achieve a small footprint with lower equipment cost. Small cells benefit from an embedded partition because a single unit houses the wide and local area network connectivity, and wireless backhaul radio functions, resulting in reduced system size and simplified installation. Furthermore, since many small-cell deployments rely on E-band/V-band backhaul operating at 60–80 GHz, the RF losses associated with radio and antenna connections are significantly reduced.

Conventional point-to-point, LoS microwave systems operate in the licensed spectrum from C-band up to Ka-band. Common operating band frequencies are 6, 11, 18, 23, 26, and 38 GHz. These systems require unobstructed propagation. Fig. 1.42 illustrates a point-to-point microwave backhaul link connecting a macro-cell base station node to an aggregation node. For small-cell base station applications some microwave equipment vendors have demonstrated NLoS operation using conventional LoS microwave bands, leveraging high antenna gains with

Table 1.3: Wireless backhaul options and associated parameters [79].

Backhaul Type	NLoS/ LoS	PTP/ PMP	Frequency Band	Licensed/ Unlicensed	Channel Bandwidth (MHz)	Modulation	Range (km)	Latency (ms)	Single-Channel Capacity (Gbps)	Application	Notes
Microwave	LoS	PTP/ PMP	C, Ku, Ka	Licensed	3.5/7/14/ 28/56 10/20/30/ 40/50	QPSK 16−2048QAM	<10	<0.2	0.5	Macrocell small-cell aggregation	High reliability and high capacity links
mmWave	LoS	PTP/ PMP PTP PTP	Q E V	Licensed Unlicensed Lightly licensed	28/56/112 >250 250−5000	QPSK 16−2048QAM QPSK 16QAM, 64QAM	– <4 ≪1	<0.05	0.3−10	Macrocell small-cell aggregation	Narrow beams and oxygen absorption to improve frequency reuse in V band
VSAT	LoS	PTP	C, Ku, Ka	Licensed	26/33/50/ 72/500	QPSK, 8PSK, 32ASK	>10	<120 (medium earth orbit) <330 (GEO)	<1	Small cell	Remote and rural areas with no infrastructure
Sub-6 GHz	NLoS/ LoS	PTP/ PMP	S, C	Licensed/ unlicensed	5/10/20/ 40/80/160	64QAM,256QAM	1 (NLoS) 10 (LoS)	<12	0.5	Metro-cell, picocell, femtocell	Fast deployment and unpredictable capacity
In-band relay	NLoS	PTP	LTE bands 1−43	Licensed	5/10/15/ 20/40	64QAM	<10	>10	–	Picocell, femtocell	Occupies cellular spectrum and adds latency

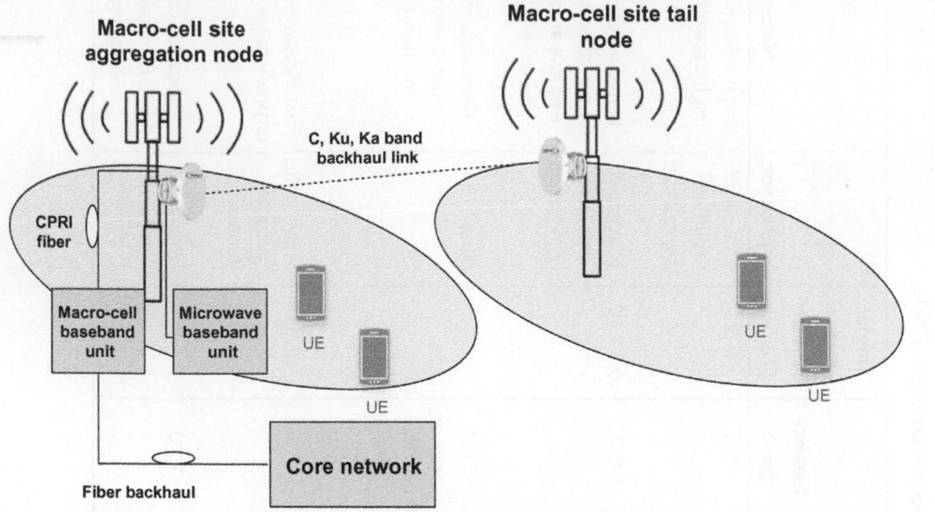

Figure 1.42
A point-to-point LoS microwave link with a relay node and aggregation node [79].

operating guidelines for the electromagnetic wave propagation effects such as diffraction, reflection, and penetration to overcome the additional path loss as a result of NLoS operation.

In clear weather conditions when radio-link SNR is high, the spectral efficiency and throughput are increased by employing higher order QAM constellations such as 256QAM, 1024QAM, or even 4096QAM. In poor weather conditions as SNR degrades, the modulation order can be lowered to 16QAM or QPSK to ensure operational link for high-priority data but at reduced throughput. However, shifting to higher order QAM constellations yields a point of diminishing returns in terms of throughput gained versus added cost, RF transmitter power utilized, required RF signal-chain linearity, and higher dynamic range. With each increase in modulation order, 3–4 dB increase in SNR or transmitter power is needed; however, with each increase in modulation order the throughput only improves by about 10%.

Co-channel dual polarization (CCDP) utilizes cross-polarization interference cancelation (XPIC)[51] to double the link capacity over the same channel. CCDP-XPIC allows simultaneous transmission of two separate data streams on the same frequency. Data is transmitted

[51] Cross-polarization interference cancelation is an algorithm to suppress mutual interference between two received streams in a polarization-division multiplexing communication system. The cross-polarization interference canceler is a signal processing technique implemented on the demodulated received signals at the baseband level. This technique is typically necessary in polarization-division multiplexing systems where the information to be transmitted is encoded and modulated at the system's symbol rate and upconverted to an RF carrier frequency, generating two (orthogonally polarized) radio streams radiated by a single dual-polarized antenna. A corresponding dual-polarized antenna is located at the remote receiver site and connected to two RF receivers, which down-convert and later combine the radio streams into the baseband signal.

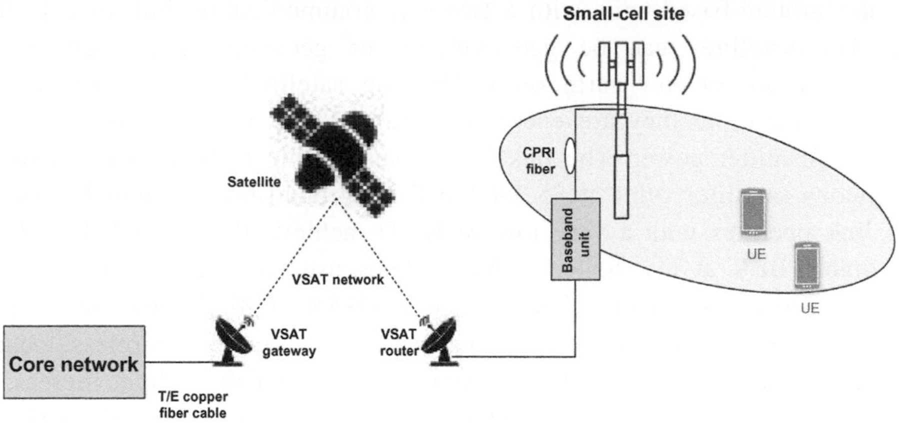

Figure 1.43
Illustration of an example satellite backhaul [79].

on orthogonal antenna polarizations (vertical and horizontal) and cross-polarization interference is canceled using digital signal processing. Spatial multiplexing significantly improves spectral efficiency and uses multi-antenna techniques to send multiple data streams over the same RF channel. A 2×2 MIMO link can ideally double the capacity. Spatial multiplexing has been used in many wireless access technologies including LTE/NR and IEEE 802.11n/ac/ax, which relies on multipath interference and exploits spatial propagation paths caused by reflections. However, an LoS microwave link does not exhibit multipath phenomenon, thus a multipath condition is simulated by deliberate separation of the antennas, thereby creating a pseudo multipath condition. While it is evident that high-density RF analog integration is important to reduce the size and lower the component count, there are still many radio functions that rely on [discrete] RF building blocks. The IF circuitry and frequency up/down conversion require several key analog functions including phase-locked loop (PLL)[52] frequency synthesizer, and the variable gain amplifier (VGA).

Commercial satellite systems use very small aperture terminals (VSAT) for cost-effective delivery of telephony, broadband access, and video content. VSAT systems are deployed in enterprise-grade private networks, consumer broadband services, and cellular base station backhaul. In cellular base station backhaul applications, VSAT systems are ideal for remotely located small-cell sites. Fig. 1.43 shows a typical VSAT system used in a base station backhaul application. Router and gateway VSAT

[52] A phase-locked loop is a control system that generates an output signal whose phase is related to the phase of an input reference signal. There are several different phase-locked loop types, but the simplest form is an electronic circuit consisting of a variable frequency oscillator and a phase detector in a feedback loop. The oscillator generates a periodic signal, and the phase detector compares the phase of that signal with the phase of the input reference signal, adjusting the oscillator frequency to maintain phase matching.

terminals are ground-based units with a two-way communication link to C/Ka/Ku-band satellites. The satellites can be geostationary or geosynchronous equatorial orbit, medium earth orbit, or low earth orbit. Orbiting satellites are powered by a finite energy source; therefore they are energy-constrained and a satellite downlink channel has limited transmitter power. The link is also susceptible to atmospheric loss because geosynchronous satellites orbit at 35,786 km from ground-level terminals. As a result, the radio link operates with a very low SNR. To achieve the desired data throughput with acceptable BER at low SNR, VSAT systems use wide-channel bandwidths with relatively low-order modulation schemes such as QPSK or 8PSK and low coding rates. Microwave and broadband satellite backhaul are two common wireless technologies deployed across the RAN. As the base station capacity and throughput increase to support growing mobile data demand, backhaul capacity must increase. Similarly, as base station size and power decrease, the backhaul solutions must become smaller and more efficient. As such, wireless backhaul systems will continue relying on RF analog integration and RF building block solutions to achieve high spectral efficiency, smaller form factor, and lower operating power.

1.1.7 Mobile Edge Computing

MEC or multi-access edge computing is an emerging 5G technology which enables provisioning of cloud-based network resources and services (e.g., processing, storage, and networking) at the edge of the network and in the proximity of the users. The edge may refer to the base stations themselves and/or data centers close to the radio network possibly located at the aggregation points. The end-to-end latency perceived by the mobile user can be significantly reduced using the MEC platform, which is a key enabling factor for many 5G services such as the tactile Internet. MEC supports different deployment scenarios, and the MEC servers can be located at different locations within the radio access network depending on technical and business requirements. Applications and analytics at the MEC server will not be impacted by congestion in other parts of the network. By performing analytics or caching content at the MEC server, the volume of data transmitted to the core network for processing is reduced, and the impact on the data traffic through the network is minimized, resulting in more efficient use of existing network bandwidth. This establishes an ultra-low-latency environment capable of providing mission-critical and real-time services. The user location information can be used by applications and services hosted by the MEC server to offer location/context-related services to the subscribers. Since these applications and services are found at the edge of the network instead of within a centralized cloud, responsiveness can be improved, resulting in an enriched quality of experience for the user.

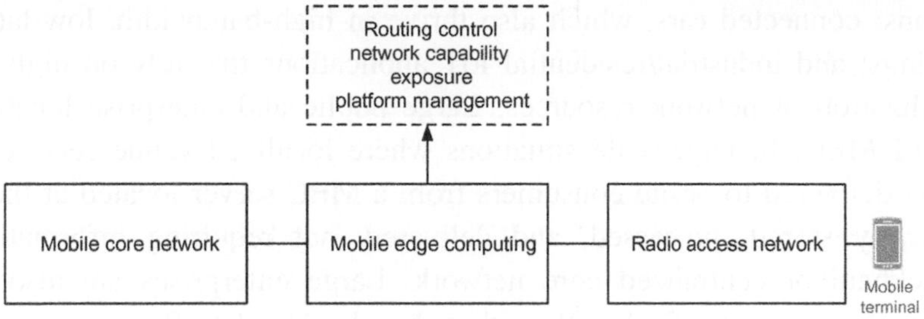

Figure 1.44
Conceptual MEC architecture [48].

Established in December 2014, the ETSI Industry Specification Group (ISG)[53] on MEC has produced normative specifications that enable the hosting of third-party applications in a multi-vendor environment. The initial scope of the ISG MEC was to focus on use cases, and to specify the requirements and the reference architecture, including the components and functional elements that were the key enablers for MEC solutions. The work has continued on platform services, APIs, and interfaces. The MEC platform API is application-agnostic and allows smooth porting of value-creating applications on every mobile-edge server with guaranteed service-level agreement.

The main functions in the MEC platform include routing, network capability exposure, and management (see Fig. 1.44). The routing entity is responsible for packet forwarding among the MEC platform, radio access network, and the mobile core network, as well as within the MEC platform. The network capability exposure entity enables the authorized exposure of the radio network information service and the RRM. The management entity supports the authentication, authorization, and accounting and management of the third-party applications in the MEC platform. This section presents the architectural description of the MEC platform as well as main applications and key functionalities enabling this technology.

1.1.7.1 Service and Deployment Scenarios

The primary objective of multi-access edge computing is to reduce network congestion and to improve application performance by processing the corresponding tasks in the proximity of the users. Furthermore, it intends to improve the type and delivery of the content and applications to those users. The use cases already being realized include augmented reality and virtual reality, which both benefit from fast response times and low-latency

[53] ETSI multi-access edge computing (http://www.etsi.org/index.php/technologies-clusters/technologies/multi-access-edge-computing).

communications; connected cars, which also thrive in high-bandwidth, low-latency, highly available settings; and industrial/residential IoT applications that rely on high performance and smart utilization of network resources. Large public and enterprise locations are also beneficiaries of MEC. In large-scale situations where localized venue services are important, content is delivered to onsite consumers from a MEC server located at the venue. The content is locally stored, processed, and delivered, not requiring information transport through a backhaul or centralized core network. Large enterprises are also increasingly motivated to process user traffic locally rather than backhaul traffic to a central network, using small-cell networks instead [48]. As shown in Fig. 1.45, edge computing use cases may be classified into two major categories, that is, third-party applications and operator applications.

The MEC ecosystem is likely to bring significant benefit to the mobile operators and other industries as well as to the consumers that will be able to experience services which need to rely on very accurate localization or high performance in terms of latency and throughput. In other words, MEC will be able to support new IoT services that would not be technically or economically feasible without 5G networks. Several use cases are addressed by MEC and also currently considered as part of next-generation mobile networks. Security and safety have been among the most important verticals for IoT. The advances in technology with an ever-increasing amount of information collected from sensors and high-resolution video cameras create the need for a scalable, flexible, and cost-effective solution to analyze the content in real time. MEC can host the analytics applications close to the source and enable increased reliability, better performance, and significant savings by locally processing large volume of data.

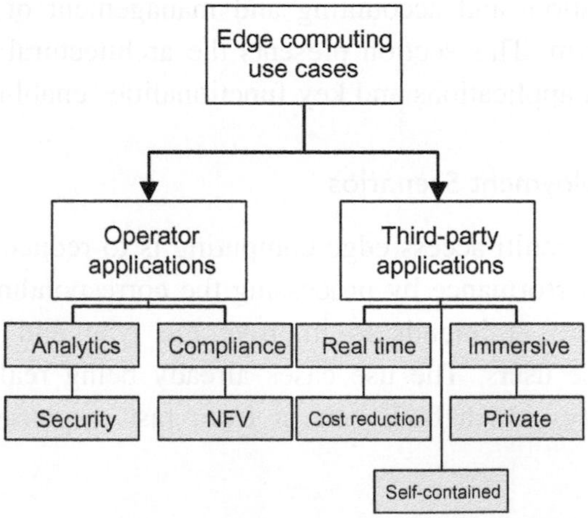

Figure 1.45
MEC use cases [45,48].

The automotive sector is another area where the new technologies are revolutionizing the industry. Self-driving cars have been already demonstrated by both traditional automotive and new Internet players, and it is anticipated to make the first autonomous cars commercially available in 2020+. While the work on future 5G systems is currently being conducted by various organizations around the globe, the digitalization in the automotive industry is clearly reflected in the use cases and the requirements coming from this sector. The IoT is a key driver for the next-generation technology and most of the use cases appear to focus on the connected cars. With the next-generation system, the latency requirements are set to less than 1 ms to empower a wide range of use cases. MEC is the ideal solution and has been identified as a key component to support these ultra-low latency scenarios as it enables hosting applications close to the users at the edge cloud and therefore provides the shortest path between the applications and the servers.

Computationally intensive applications running on mobile terminals may be offloaded to the cloud for various reasons, such as availability of more computing power or of specific hardware capabilities, reliability, joint use of the resources in collaborative applications, or saving network capacity. The computation offload is particularly suitable for IoT applications and scenarios where terminals have limited computing capabilities, to guarantee longer battery life. Such offload may happen statically (server components are deployed by the service provider proactively in advance) or dynamically (server components are deployed on demand by request from the UE). In this case, applications benefit from the low delay provided by the MEC.

1.1.7.2 Architectural Aspects

MEC provides a highly distributed computing environment that can be used to deploy applications and services as well as to store and process content in close proximity of the mobile users. Applications can also be exposed to real-time radio and network information and can offer a personalized and contextualized experience to the mobile subscriber. This translates into a mobile broadband experience that is not only more responsive but also opens up new business opportunities and creates an ecosystem where new services can be developed in and around the base station.

Fig. 1.46 shows the high-level functional entities in the MEC framework, which are further grouped into the system-level, host-level, and network-level entities. The host-level group consists of the MEC host and the corresponding MEC host-level management entity. The MEC host is further split to include the MEC platform, applications, and the virtualization infrastructure. The network-level group consists of the corresponding external entities that are 3GPP radio access networks, the local networks, and the external networks. This layer represents the connectivity to local area networks, cellular networks, and external networks such as the Internet. Above everything is the MEC system-level management that by

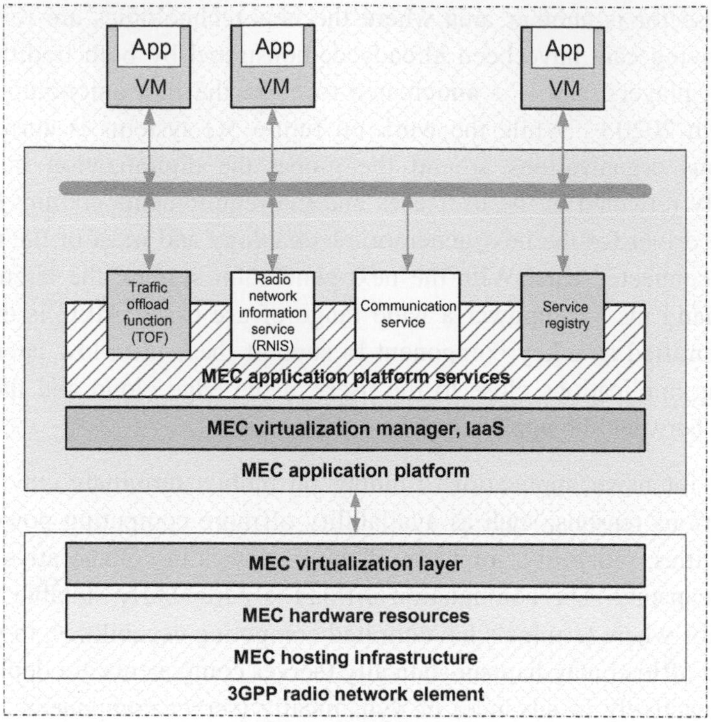

Figure 1.46
Overview of ETSI MEC [46].

definition has the overall visibility to the entire MEC system. The MEC system consists of the MEC hosts and the MEC management necessary to run MEC applications within an operator network or a subset of an operator network [53].

An in-depth understanding of the MEC systems can be attained from the reference architecture depicted in Fig. 1.47, which defines the functional entities and their relationships. The reference architecture follows the earlier described functional grouping of the general framework and includes system-level and host-level functions; however, the network-level functional group is not visible because there are no MEC-specific reference points needed to access those entities. The MEC host is an entity consisting of the MEC platform and the virtualization infrastructure that provides computing, storage, and network resources for the MEC applications. In addition, the MEC host can provide storage and ToD information for the applications. The virtualization infrastructure includes a data plane that executes the forwarding rules received by the MEC platform and routes the traffic between the applications, services, and networks. The MEC server provides computing resources, storage capacity, connectivity, and access to user traffic and radio and network information.

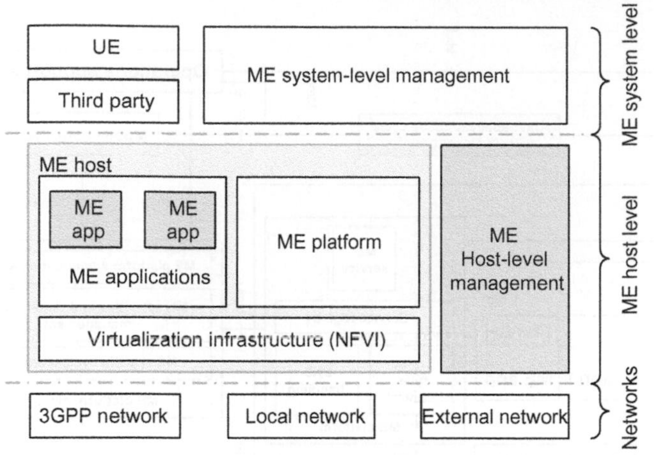

Figure 1.47
MEC general framework [46].

The MEC platform represents a collection of baseline functionalities that are required to run applications on a particular MEC host and to enable MEC applications to discover, advertise, offer, and exploit the MEC services. MEC services can be provided by the platform and by the applications, where both the platform and applications may utilize MEC services. The baseline functionalities of the MEC platform are needed to navigate the traffic between the applications, services, and networks. The MEC platform receives the traffic forwarding rules from the MEC platform manager, MEC applications, and MEC services, and based on those criteria, as well as policies, it provides the instructions to the forwarding plane.

The reference architecture shows the functional elements that comprise the mobile edge (ME) system and the reference points between them. Fig. 1.48 depicts the ME system reference architecture. There are three groups of reference points defined between the system entities as follows [46]:

1. Reference points related to ME platform functionality (Mp)
2. Management reference points (Mm)
3. Reference points interfacing with external entities (Mx)

The MEC applications run as VMs on top of virtualization infrastructure provided by the MEC host. The applications interact with the MEC platform over an Mp1 reference point to utilize the services offered by the platform. The MEC platform manager is a host-level entity that is further split into MEC platform element management, MEC application life cycle management, and MEC application rules and requirements management functions. The application life cycle management consists of application instantiation and termination procedures as well as providing indication to the MEC orchestrator on application-related

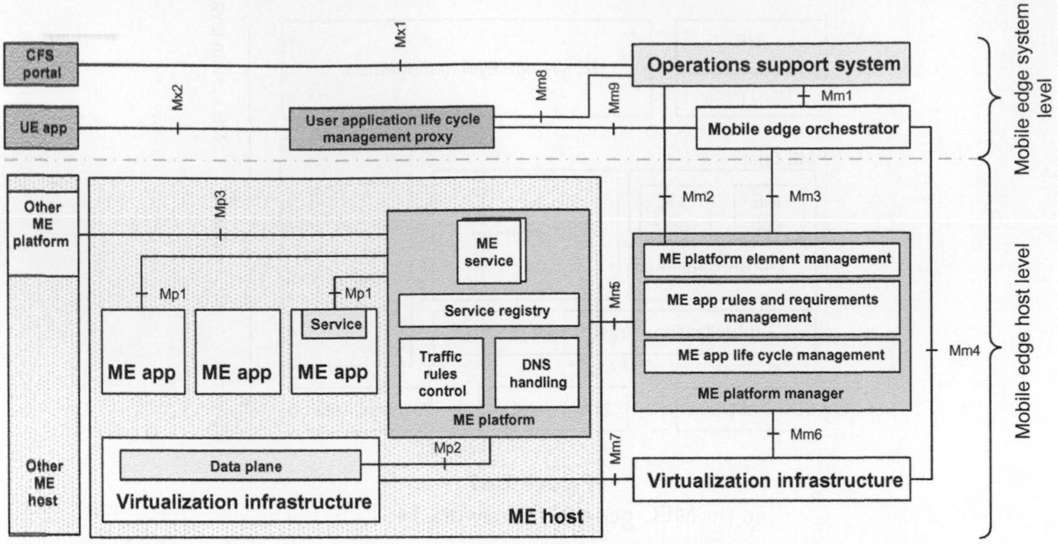

Figure 1.48
MEC reference architecture [46].

events. The MEC orchestrator is the central function in the MEC system that has visibility over the resources and capabilities of the entire MEC network. The MEC orchestrator maintains information on the entire MEC system, the services and resources available in each host, the applications that are instantiated, and the topology of the network. The orchestrator is also responsible for managing the MEC applications and the related procedures by integrating the applications, checking the integrity and authenticity of the application, validating the policies for the applications, and maintaining a catalog of the applications that are available.

The MEC applications may indicate their requirements for the resources, services, location, and performance, such as maximum allowed latency, and it is the MEC orchestrator's responsibility to ensure that their requirements are satisfied. The orchestrator uses the requirements received from the applications in the selection process for the target MEC host. The reference point toward the VIM is used to manage the virtualized resources of the MEC host and to manage the application images that are provided for instantiation. It is further used for maintaining status information on available resources. The operations support system of an operator is a function that is widely used to manage various services and subsystems in the operators' network. The reference point between the MEC orchestrator and VIM is used for management of the application images and the virtualized resources as well as for monitoring the availability of the resources.

The customer-facing service (CFS) acts as an entry point for a third-party application. This portal can be used for operations to manage the provisioning, selection, and ordering of the

MEC applications. The user application life cycle management proxy is a function that the MEC-related clients and applications use to request services related to onboarding, instantiation, and termination of the applications. This proxy can be used to request transfer of the application from the MEC system to the external cloud or to the MEC system from the external cloud. More specifically, the ME system-level management includes the ME orchestrator as its core component, which has an overall view of the complete ME system. The ME host-level management comprises the ME platform manager and the VIM, which handles the management of the ME-specific functionality of a particular ME host and the applications running on it. The ME host is an entity that contains the ME platform and a virtualization infrastructure which provides compute, storage, and network resources for the ME applications. The virtualization infrastructure includes a data plane that executes the traffic rules received by the ME platform, and routes the traffic among applications, services, DNS server/proxy,[54] 3GPP network, local networks, and external networks.

The ME platform is responsible for offering an environment where the ME applications can discover, advertise, consume, and offer ME services, including ME services available via other platforms; receiving traffic rules from the ME platform manager, applications, or services; and instructing the data plane. This includes the translation of tokens representing UEs in the traffic rules into specific IP addresses, receiving DNS records from the ME platform manager, configuring a DNS proxy/server, hosting ME services, and providing access to persistent storage and ToD information. The ME applications run as VMs on top of the virtualization infrastructure provided by the ME host, and can interact with the ME platform to utilize and provide ME services. Under certain conditions, the ME applications can also interact with the ME platform to perform certain support procedures related to the life cycle of the application, such as indicating availability, preparing relocation of user state, etc. The ME applications can have a certain number of rules and requirements associated with them such as required resources, maximum latency, required or useful services, etc. These requirements are validated by the ME system-level management and can be assigned to default values if not provided.

The ME orchestrator is the core functionality in ME system-level management. The ME orchestrator is responsible for maintaining an overall view of the ME system based on deployed ME hosts, available resources, available ME services, and topology; on-boarding

[54] Domain name system is a hierarchical decentralized naming system for computers, services, or other resources connected to the Internet or a private network. It associates various information with domain names assigned to each of the participating entities. It translates readily memorized domain names to the numerical IP addresses needed for locating and identifying computer services and devices with the underlying network protocols. A domain name system proxy improves domain lookup performance by caching previous lookups. A typical domain name system proxy processes domain name system queries by issuing a new domain name system resolution query to each name server that it has detected until the hostname is resolved.

of application packages, including checking the integrity and authenticity of the packages, validating application rules and requirements, and if necessary adjusting them to comply with operator policies, keeping a record of on-boarded packages, and preparing the VIM(s) to handle the applications; selecting appropriate ME host(s) for application instantiation based on constraints such as latency, available resources, and available services; triggering application instantiation and termination; and triggering application relocation as needed when supported.

The operations support system shown in Fig. 1.48 refers to the OSS of an operator. It receives requests via the CFS portal and from UE applications for instantiation or termination of applications and decides whether to grant these requests. Granted requests are forwarded to the ME orchestrator for further processing. The OSS also receives requests from UE applications for relocating applications between external clouds and the ME system. A user application is an ME application that is instantiated in the ME system in response to a request of a user via an application running in the UE. The user application life cycle management proxy allows UE applications to request on-boarding, instantiation, termination of user applications, and relocation of user applications in and out of the ME system. It also allows informing the UE applications about the state of the user applications. The user application life cycle management proxy authorizes requests from UE applications in the UE and interacts with the OSS and the ME orchestrator for further processing of these requests. The user application life cycle management proxy is only accessible from within the mobile network. It is only available when supported by the ME system.

The VIM is responsible for managing the virtualized resources for the ME applications. The management tasks consist of allocating and releasing virtualized computing, storage, and network resources provided by the virtualization infrastructure. The VIM also prepares the virtualization infrastructure to run software images, which can also be stored by the VIM for a faster application instantiation. Since it is possible for virtualized resources to run out of capacity or to fail in operation, it is important to closely monitor them. The VIM provides support for fault and performance monitoring by collecting and reporting information on virtualized resources and providing the information to server and system-level management entities. The VIM has a reference point toward the virtualization infrastructure to manage the virtualized resources.

The ME reference architecture shown in Fig. 1.48 incorporates the following reference points [46]:

- Mp1 is a reference point between the ME platform and the ME applications that provides service registration, service discovery, and communication support for services. It also enables other functionalities such as application availability, session state relocation support procedures, traffic rules and DNS rules activation, access to persistent storage and ToD information, etc.

- Mp2 is a reference point between the ME platform and the data plane of the virtualization infrastructure and is used to instruct the data plane on how to route traffic among applications, networks, services, etc.
- Mp3 is a reference point between the ME platforms and is used for controlling communication between ME platforms.

Reference points related to the ME management

- Mm1 is a reference point between the ME orchestrator and the OSS that is used for triggering the instantiation and the termination of ME applications in the ME system.
- Mm2 is a reference point between the OSS and the ME platform manager that is used for the ME platform configuration, fault detection, and performance management.
- Mm3 is a reference point between the ME orchestrator and the ME platform manager and is used for the management of the application life cycle, application rules and requirements, and keeping track of available ME services.
- Mm4 is a reference point between the ME orchestrator and the VIM which is used to manage virtualized resources of the ME host including maintaining track of available resource capacity and managing application images.
- Mm5 is a reference point between the ME platform manager and the ME platform and is used to perform platform configuration, configuration of the application rules and requirements, application life cycle support procedures, management of application relocation, etc.
- Mm6 is a reference point between the ME platform manager and the VIM which is used to manage virtualized resources and to realize the application life cycle management.
- Mm7 is a reference point between the VIM and the virtualization infrastructure that is used to manage the virtualization infrastructure.
- Mm8 is a reference point between the user application life cycle management proxy and the OSS and is used to handle UE requests for running applications in the ME system.
- Mm9 is a reference point between the user application life cycle management proxy and the ME orchestrator of the ME system and is used to manage ME applications requested by UE application.

Reference points related to external entities

- Mx1 is a reference point between the OSS and the CFS portal and is used by a third party to request the ME system to run applications in the ME system.
- Mx2 is a reference point between the user application life cycle management proxy and the UE application, which is used by a UE application to request the ME system to run an application in the ME system, or to move an application in or out of the ME system. This reference point is only accessible within the mobile network. It is only available when supported by the ME system.

The ME computing and NFV are complementary concepts that can exist independently. The ME architecture has been designed in such a way that a number of different deployment options of ME systems are possible. An ME system can be realized independent of an NFV environment in the same network, or can coexist with that environment. Since both MEC and NFV are based on the use of virtualization concept, the MEC applications and VNFs can be fully or partially instantiated over the same virtualization infrastructure. The MEC reference architecture reuses the concept of a VIM similar to that of the VIM of NFV framework. Multiple scenarios for deployments are possible, depending on operators' preferences for their networks and their network migration plans. The relationship between MEC and NFV-MANO components is an important aspect of integrated MEC/NFV deployments.

In 5G networks, there are three types of session and service continuity (SSC) modes. Different SSC modes can guarantee different levels of service continuity. As shown in Fig. 1.49, SSC mode 1 maintains the same UPF. In SSC mode 2, the network may trigger the release of the PDU session and instruct the UE to establish a new PDU session to the same data network immediately. Upon establishment of the new PDU session, a new UPF acting as PDU session anchor can be selected. In SSC mode 3, the network allows the establishment of UE connectivity via a new UPF to the same application server before connectivity between the UE and the previous UPF is terminated. Different applications have different service and session continuity requirements. Therefore, in order to achieve efficient control of MEC APPs with different SSC mode requirements in 5G network, the coordination between ME APPs on the ME host and the 5G network, for example, how to indicate the SSC mode requirement of an ME APP to the 5G network, must be carefully considered.

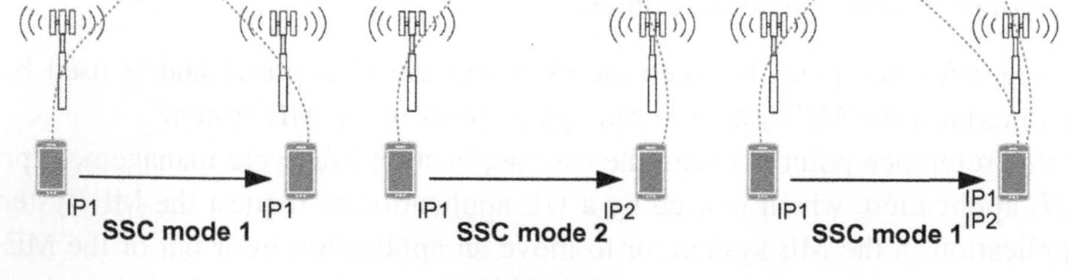

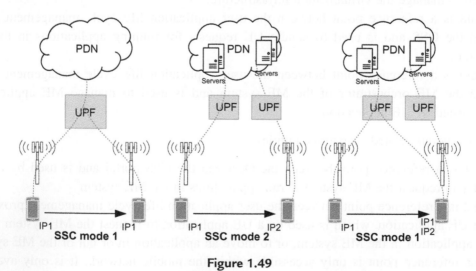

Figure 1.49
SSC modes [3,4].

A large part of the functionality providing data connectivity is for supporting flexible deployment of application functions (AFs) in the network topology as needed for edge computing, which is supported via the three different SSC modes or via the functionality of uplink classifiers and branching points. The SSC modes include the traditional SSC 1 mode, where the IP anchor remains stable to provide continuous support of applications and to maintain the path toward the UE as its location is updated. The new modes allow relocating the IP anchor. There are two options, make-before-break (SSC mode 3) and break-before-make (SSC mode 2). The architecture enables applications to influence selection of suitable data service characteristics and SSC modes. Given that 5G network deployments are expected to serve extremely large amount of mobile data traffic, an efficient user-plane path management is critical. The system architecture defines in addition to the SSC modes the functionality of uplink classifiers and branching points to allow breaking out and injecting traffic selectively to and from AFs on the user plane path before the IP anchor. Also, as permitted by policies, AFs may coordinate with the network by providing information relevant for optimizing the traffic route or may subscribe to 5G system events that may be relevant for applications [3,4].

1.1.8 Network Sharing

A network sharing architecture allows multiple participating operators to share resources of a single shared network according to agreed allocation terms. The shared network includes a RAN. The shared resources include radio resources of that network. The shared network operator allocates shared resources to the participating operators based on their plans and current needs and according to service-level agreements. A UE that has a subscription to a participating network operator can select the participating network operator while within the coverage area of the shared network and to receive subscribed services from the participating network operator. 3GPP laid out two approaches to sharing a RAN, which are illustrated in Fig. 1.50, where they primarily differ in the core network aspects. In multi-operator core network (MOCN) approach, each network operator has its own core network. Maintaining a strict separation between the core network and the radio network has a number of benefits related to service differentiation, interworking with legacy networks, fall back to circuit-switched voice services, and the support of roaming. Alternatively, in the gateway core network (GWCN) approach, the network operators also share the mobility management entity of the core network, which is responsible for bearer management and connection management between the mobile terminal and the network. The GWCN approach enables additional cost savings compared to the MOCN approach, but it is relatively less flexible, potentially reducing the level of differentiation among the participating operators. 3GPP Rel-15 supports MOCN network sharing architecture, in which only the RAN is shared in the 5G system. However, RAN and AMF support for operators that use more than one PLMN ID is required [3].

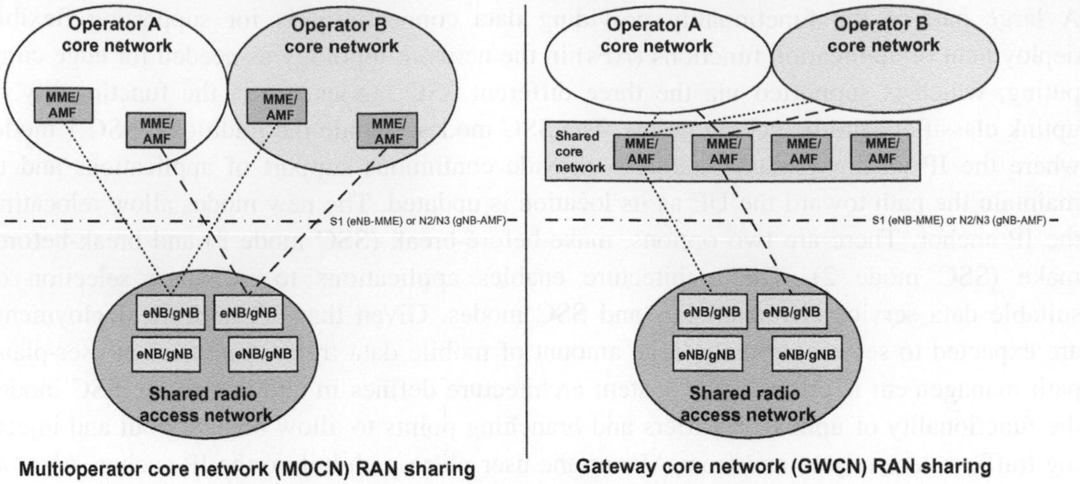

Figure 1.50
3GPP-defined network sharing architecture options [34].

In each case, the network broadcasts system information and supports signaling exchanges that allow the UE to distinguish between up to six different sharing network operators, to obtain service or to perform handover, with no consideration of the underlying network sharing arrangement. As network sharing becomes a central feature of mobile network operation, there is a need to address a wide variety of technical, commercial, and regulatory requirements. Among other things, there is an interest in pooling spectrum, sharing resources asymmetrically and dynamically based on financial considerations and load, and the ability for participating operators to manage and control the use of resources independently. If a shared NG-RAN is configured to indicate the available core network operators to the UEs for selection, each cell in the shared RAN would then include the PLMNs related to the available core network operators in the shared network in the broadcast system information. The broadcast system information provides a set of PLMN IDs and optionally one or more additional set of parameters per PLMN such as cell-ID, tracking areas (TAs), etc. All UEs compliant with the 5G system attempting to connect to NG-RAN must support reception of the basic and additional set of PLMN IDs. The available core network operators must be the same for all cells of a TA in a shared NG-RAN network. The UE decodes the broadcast system information and takes the information concerning the available PLMN IDs into account in the network and cell (re-)selection procedures.

A UE that has a subscription to a participating operator in a network sharing scenario must be able to select this participating network operator while present within the coverage area of the shared network and to receive subscribed services from the participating network operator. Each cell in the shared NG-RAN must include the PLMN-IDs corresponding to the available core network operators in the shared network in the broadcast system information. When a UE performs an initial access to a shared network, one of the available PLMNs is selected to serve the UE. The UE uses all received broadcast PLMN-IDs in its PLMN (re)selection processes and informs the NG-RAN of the selected PLMN so that the NG-RAN can properly route its traffic. After initial access to the shared network, the UE does not switch to another available PLMN as long as the selected PLMN is available to serve the UE at its present location. The network does not move the UE to another available PLMN by handover as long as the selected PLMN is available to serve the UE.

The NG-RAN uses the selected PLMN information, which is provided by the UE at RRC establishment or provided by the AMF/source NG-RAN at N2/Xn handover, to select target cells for future handovers and allocation of radio resources. In case of handover to a shared network, the NG-RAN selects a target PLMN based on either PLMN in use, preset configuration, or the equivalent PLMN list in the handover restriction list provided by the AMF. For Xn-based handover procedure, the source NG-RAN indicates the selected PLMN ID to the target NG-RAN by using target cell ID. For N2-based handover procedure, the NG-RAN indicates the selected PLMN ID to the AMF as part of the tracking area identity (TAI) sent in the handover required message. The source AMF uses the TAI information supplied by the source NG-RAN to select the target AMF/MME and to forward the selected PLMN ID to the target AMF/MME. The selected PLMN ID is signaled to the target NG-RAN/eNB so that it can select target cells for future handovers. In a network slicing scenario, a network slice is defined within a PLMN. Network sharing is performed among different PLMNs and each PLMN sharing the NG-RAN defines and supports its PLMN-specific set of slices that are supported by the common NG-RAN [3].

1.2 Reference Architectures

5G systems have been designed to support seamless user connectivity and to render new services/applications which would require deployment of networks that exploit innovative techniques such as NFV and SDN. A distinct feature of 5G system architecture is network slicing. The previous generation supported certain aspects of this with the functionality for dedicated core networks. In the context of 3GPP 5G system architecture, a network slice refers to the set of 3GPP defined features and functionalities that together form a complete PLMN for providing services to UEs. Network slicing allows controlled composition of a PLMN from the specified NFs with their specifics and provides services that are required for a specific usage scenario. The need for these new techniques is increasing due to the versatility of data services that are supported by 5G networks. Mobile networks were traditionally designed as voice-centric and later data-centric systems; however, with 5G this design philosophy has changed as a result of proliferation of new use cases and applications.

Having such requirements in mind, 3GPP attempted to maintain the premises of flat architecture where the control-plane functions are separated from the user plane in order to make them scale independently, allowing the operators to exploit logical functional split for dimensioning in deploying and adapting their networks. Another central idea in the design of 5G network was to minimize dependencies between the access network and the core network with a converged access-agnostic core network with a common interface which integrates different 3GPP and non-3GPP access types.

5G is a service-centric architecture which strives to deliver the entire network as a service. 3GPP system architecture group took the approach to rearchitect the LTE core network based on a service-oriented framework. This involves breaking everything down to more functional granularity. The MME no longer exists and its functionality has been redistributed between mobility management and session management NFs. As such, registration, reachability, mobility management, and connection management are now considered new services offered by a new general NF labeled as AMF. Session establishment and session management, also formerly part of the MME, are now new services offered by a new NF called session management function (SMF). Furthermore, packet routing and forwarding functions, currently performed by the SGW and PGW in 4G, will now be realized as services rendered through a new NF called UPF. The main reason for this new architectural approach is to enable a flexible network as a service solution. By standardizing a modularized set of services, various deployment options such as centralized, distributed, or mixed configurations will be enabled for different users or applications. The dynamic service chaining lay the groundwork for network slicing which is an important concept in 5G to satisfy the diverse user and application demands, shifting the design emphasis on software rather than hardware. The physical boxes where these software services are instantiated could be in the cloud or on any targeted general-purpose hardware in the system.

In the following sections, we will discuss the reference architecture, network entities, and interfaces of the 3GPP 5G access and the core networks. The user-plane and control-plane protocols will be further discussed.

1.2.1 Access Network

1.2.1.1 Reference Architecture: Network Entities and Interfaces

The overall reference architecture of next-generation network comprises the entities associated with the radio access network (NG-RAN) and the core network (5GC) and their corresponding interfaces that terminate the protocols. In this section, we will describe the access network reference architecture. The NG-RAN architecture consists of a set of gNBs connected to the 5GC through the NG interface (see Fig. 1.51). The interface between the NG-RAN and 5GC is referred to as N2 and N3 depending on the termination point in 3GPP

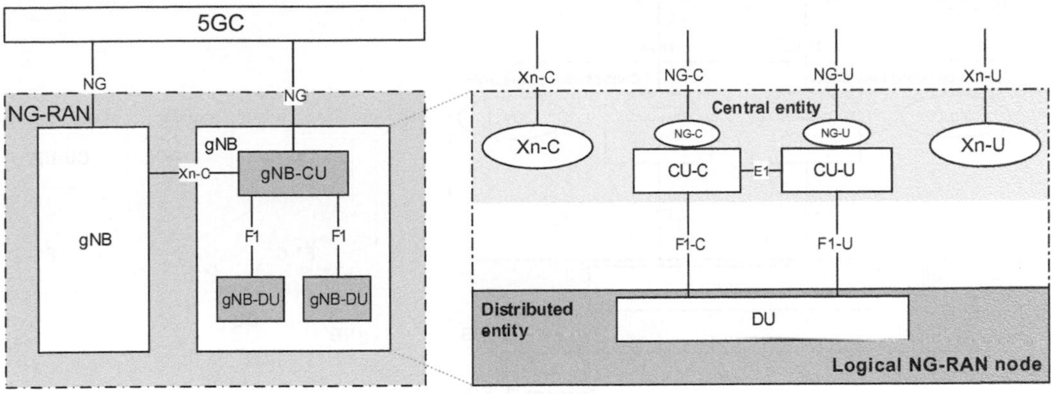

Figure 1.51
Overall NG-RAN architecture [15].

system architecture specifications. Furthermore, gNBs are interconnected through the Xn interface. In a C-RAN architecture, a gNB may be further disaggregated such that some lower layer protocol functions are implemented in the DUs and the remaining upper layer protocol functions are implemented in the edge cloud and as part of the CU(s). In that case, the gNB would consist of a gNB-CU and gNB-DUs as shown in Fig. 1.51. The gNB-CU and gNB-DU entities are connected via F1 logical interface. Note that one gNB-DU is connected to only one gNB-CU. In some deployment scenarios, one gNB-DU may be connected to multiple gNB-CUs. In NG-RAN reference architecture, NG, Xn and F1 are logical interfaces [16].

The traditional architecture of a base station where all protocol layers and functionalities were concentrated in a single logical RAN entity has evolved into a disaggregated model in 5G network architecture. In a disaggregated gNB architecture, the NG and Xn-C interfaces are terminated at the gNB-CU. In an LTE–NR dual connectivity (EN-DC) scenario, the S1-U and X2-C interfaces for a disaggregated gNB terminate at the gNB-CU. It must be noted that the gNB-CU and its associated gNB-DUs are seen as a gNB to other gNBs and the 5GC. The NG-RAN comprises a radio network layer (RNL) and a TNL. The NG-RAN architecture, that is, the NG-RAN logical nodes and the corresponding interfaces, is defined as part of the RNL, whereas the NG-RAN interfaces (NG, Xn, F1) are specified as part of TNL protocols and functionalities. The TNL provides services for user-plane and signaling transport. The protocols over Uu (i.e., the radio air-interface) and NG interfaces are divided into two classes: user-plane protocols, which are the protocols implementing the actual PDU carrying user data through the AS; and control-plane protocols, which are the protocols for controlling the PDU sessions and the connection between the UE and the network from different aspects including requesting the service, controlling different transmission resources, handover, etc. as well as a mechanism for transparent transfer of NAS messages

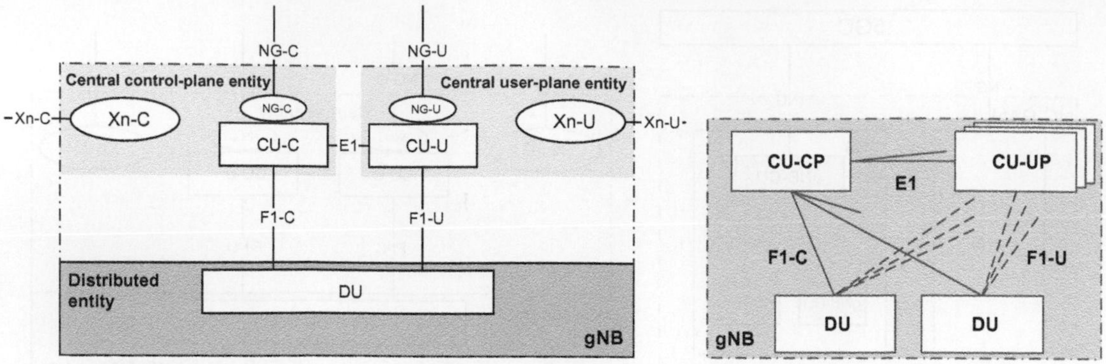

Figure 1.52
Disaggregated gNB model [15].

via encapsulation in RRC messages. The NG interface, comprising a user-plane interface (NG-U) and a control-plane interface (NG-C), connects the 5GC and the NG-RAN. In this architecture, the NG-RAN termination is an NG-RAN node which can be either an ng-eNB or a gNB and the 5GC termination is either the control-plane AMF logical node or the user-plane UPF logical node. There may be multiple NG-C logical interfaces toward 5GC from any NG-RAN nodes which are selected by the NAS node selection function. Likewise, there may be multiple NG-U logical interfaces toward 5GC from any NG-RAN node which can be selected within the 5GC and signaled to the NG-RAN node by the AMF [15,17] (see Fig. 1.52).

The NG interface supports procedures to establish, maintain, and release NG-RAN part of PDU sessions; to perform intra-RAT handover and inter-RAT handover; the separation of each UE on the protocol level for user-specific signaling management; the transfer of NAS signaling messages between UE and AMF; and mechanisms for resource reservation for packet data streams. The functions supported over the NG interface include the following [17]:

- Paging function which supports transmission of paging requests to the NG-RAN nodes that are part of the paging area to which the UE is registered.
- UE context management function which allows the AMF to establish, modify, or release the UE context in the AMF and the NG-RAN node in order to support user-specific signaling on NG interface.
- Mobility function for UEs in ECM-CONNECTED, which includes the intra-system handover function to support mobility within NG-RAN and inter-system handover function to support mobility from/to EPS system including the preparation, execution, and completion of handover via the NG interface.

- PDU session function that is responsible for establishing, modifying, and releasing the involved PDU sessions NG-RAN resources for user data transport once the UE context is available in the NG-RAN node.
- NAS signaling transport function provides means to transport or re-route a NAS message for a specific UE over the NG interface.
- NG-interface management functions which provide mechanisms to ensure a default start of NG-interface operation and handling different versions of application part implementations and protocol errors.

The interconnection of NG-RAN nodes to multiple AMFs is supported in the 5G system architecture. Therefore, a NAS node selection function is located in the NG-RAN node to determine the AMF association of the UE, based on the UE's temporary identifier, which is assigned to the UE by the AMF. If the UE's temporary identifier has not been assigned or is no longer valid, the NG-RAN node may consider the slicing information to determine the AMF. This functionality is located in the NG-RAN node and enables proper routing via the NG interface.

As shown in Fig. 1.53, each NG-RAN node is either a gNB, that is, an NR base station, providing NR user-plane and control-plane protocol terminations toward the UE or an ng-eNB, that is, a Rel-15/16 LTE base station, terminating LTE user-plane and control-plane protocols to/from the UE. The gNBs and ng-eNBs are interconnected through Xn interface. The gNBs and ng-eNBs are also connected via NG interfaces to the 5GC. More specifically, NG-C interfaces NG-RAN nodes to the AMF and NG-U interfaces the NG-RAN nodes to the UPF [16].

In 5G network architecture, the gNB entity is responsible for performing functions corresponding to RRM including radio bearer control, radio admission control, connection mobility control, dynamic allocation of resources to UEs in both uplink and downlink (scheduling). The gNB further performs IP header compression and encryption and integrity protection of data as well as selection of an AMF upon UE attachment when no routing to an AMF can be determined from the information provided by the UE. It also provides routing of user-plane data toward UPF(s) along with routing of control-plane information toward AMF; the connection setup and release; scheduling and transmission of paging messages originated from the AMF; scheduling and transmission of system information originated from the AMF or network operation and management; measurement and reporting configuration for mobility and scheduling; session management; support of network slicing; QoS flow management and mapping to data radio bearers (DRBs); support of UEs in RRC_INACTIVE state; as well as RAN sharing and DC [16].

In NG-RAN architecture, the role of AMF is similar to that of MME in the EPC which hosts core network control-plane functions such as terminating NAS signaling and security; AS security control; internetwork signaling for mobility between 3GPP access networks;

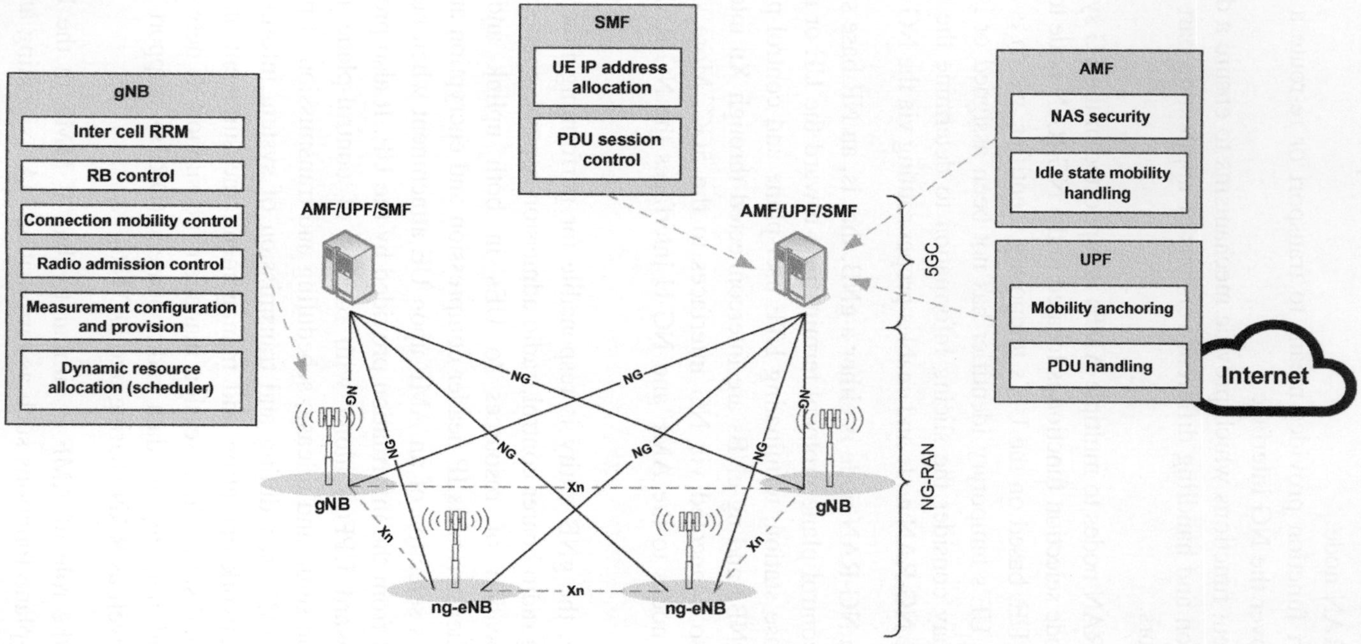

Figure 1.53
Overall access network architecture [16].

idle-mode UE reachability which includes paging control, registration area management; support of intra-system and inter-system mobility; access authentication and authorization; mobility management control; support of network slicing; and SMF selection.

The role of SMF in the new architecture is similar to that of EPC PGW entity hosting user-plane functions such as session management; allocating and managing UE IP address; selecting and controlling of UPFs; configuring traffic steering at UPF in order to route traffic toward proper destination; controlling part of policy enforcement; and the QoS.

The new architecture further features a UPF entity which is similar to the EPC SGW user-plane entity. This entity hosts the UPFs such as anchor point for intra- or inter-RAT mobility, external PDU session point of interconnect to data network, packet routing and forwarding, packet inspection and user plane part of policy rule enforcement, traffic usage reporting, uplink classifier to support routing traffic flows to a data network, QoS handling for user plane which includes packet filtering, gating, UL/DL rate enforcement, uplink traffic verification, that is, service data flow (SDF)[55] to QoS flow mapping, and downlink packet buffering and downlink data notification triggering [16].

1.2.1.1.1 Xn Control-Plane/User-Plane Functions and Procedures

Xn is a reference point connecting the NG-RAN access nodes, supporting the transfer of signaling information and forwarding of user traffic between those nodes. Xn is a logical point-to-point interface between two NG-RAN access nodes which establishes a logical connection between the two nodes even when a direct physical connection does not exist between the two nodes (which is often the case). It comprises both user-plane and control--plane protocols.

The user-plane protocol stack of Xn interface is shown in Fig. 1.54. The TNL uses GTP-U[56] over UDP/IP to transfer the user-plane PDUs. As a result, Xn-U interface can

[55] User traffic using different services or applications has different quality of service classes. A service data flow is an IP flow or an aggregation of IP flows of user traffic classified by the type of the service in use. Different service data flows have different quality of service class attributes, thereby a service data flow serves as a unit by which quality of service rules are applied in accordance with network policy and charging rules.

[56] GPRS tunneling protocol is a protocol that is used over various interfaces within the packet core in 3GPP networks to allow the user terminals to maintain a connection to a packet data network while on the move. The protocol uses tunnels to allow two GPRS support nodes to communicate over a GPRS tunneling protocol-based interface and separates traffic into different communication flows. GPRS tunneling protocol creates, modifies, and deletes tunnels for transporting IP payloads between the user equipment, the GPRS support node in the core network and the Internet. GPRS tunneling protocol comprises three types of traffic, namely control-plane (GTP-C), user-plane (GTP-U), and charging. GTP-C protocol supports exchange of control information for creation, modification, and termination of GPRS tunneling protocol tunnels. It creates data forwarding tunnels in case of handover. GTP-U protocol is used to forward user IP packets over certain network interfaces. When a GTP tunnel is established for data forwarding during LTE handover, an End Marker packet is transferred as the last packet over the GPRS tunneling protocol tunnel.

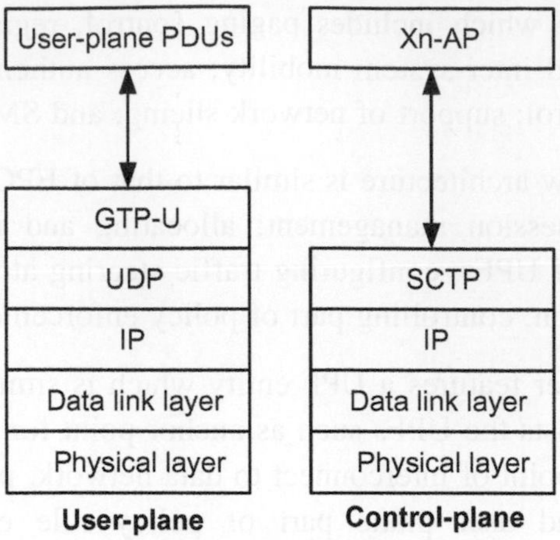

Figure 1.54
Xn-U and Xn-C protocol stacks [16].

only provide non-guaranteed delivery of user-plane PDUs, but it supports data forwarding and flow control. The Xn interface specifications facilitate interconnection of NG-RAN nodes manufactured by different vendors as well as separation of Xn interface radio network and transport network functionalities in order to allow future extensions. The Xn interface supports intra-NG-RAN mobility and DC procedures.

The control-plane protocol stack of Xn interface is also illustrated in Fig. 1.54. The TNL uses stream control transmission protocol (SCTP)[57] over IP. The application layer signaling protocol is referred to as Xn application protocol (Xn-AP). The SCTP layer provides guaranteed delivery of application layer messages. In the transport IP layer, point-to-point transmission is used to deliver the signaling PDUs. The Xn-C interface supports Xn interface management and error handling (the functionality to manage the Xn-C interface), mobility support for UE in CM-CONNECTED (the functionality to manage the UE mobility for connected mode between nodes in the NG-RAN), context transfer from the current serving NG-RAN node to the new serving NG-RAN node, and control of user-plane tunnels between the existing serving NG-RAN node and the new serving NG-RAN node. It further supports DC (the functionality to enable usage of additional resources in a secondary node in the NG-RAN) as well as support of paging (sending of paging messages via the last serving NG-RAN node toward other nodes in the RAN-based notification area to a UE in RRC_INACTIVE state).

57 Stream control transmission protocol is a transport layer protocol for transmitting multiple streams of data at the same time between two end points that have established a connection in a network. It is message-oriented protocol like user data protocol and ensures reliable, in-sequence transport of messages with congestion control like transmission control protocol. Stream control transmission protocol supports multihoming and redundant paths to increase resilience and reliability.

1.2.1.1.2 F1 Control-Plane/User-Plane Functions and Procedures

As we described earlier, the NG-RAN gNB functions may be implemented in one gNB-CU and one or more gNB-DUs. The gNB-CU and gNB-DU are connected via an F1 logical interface, which is a standardized interface that supports interchange of signaling information and data transmission between the termination points. From a logical point of view, F1 is a point-to-point interface between the termination points, which may be established irrespective of the existence of a physical connectivity between the two points. F1 interface supports control-plane and user-plane separation and separates radio network and TNLs. This interface enables exchange of UE and non-UE associated information. Other nodes in the network view the gNB-CU and a set of gNB-DUs as a gNB. The gNB terminates X2, Xn, NG, and S1-U interfaces [28].

The RRM functions ensure the efficient use of the available network resources. In a gNB-CU/gNB-DU disaggregated model, different RRM functions may be located at different locations. As an example, radio bearer control function (for establishment, maintenance, and release of radio bearers) can be located either at gNB-CU or gNB-DUs, whereas inter-RAT RRM function (for management of radio resources in conjunction with inter-RAT mobility) and dynamic resource allocation and packet scheduling functions (for allocation and de-allocation of resources to user and control-plane packets) are exclusively located at gNB-CU and gNB-DU(s), respectively. Fig. 1.55 shows F1 control-plane and user-plane protocol structures. In the control plane, the TNL is based on IP transport using SCTP over IP for transfer of control messages. The application layer signaling protocol is referred to as F1 application protocol (F1AP). In the user plane, the IP-based TNL uses GTP-U over UDP/IP for transfer of data packets.

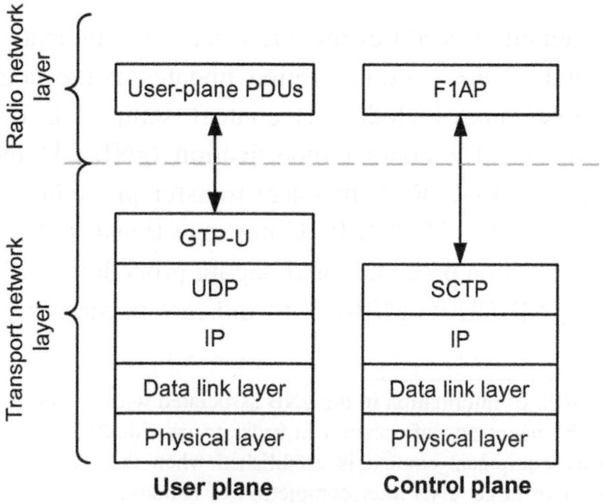

Figure 1.55
User-plane and control-plane protocol structure of F1 interface [28].

F1AP provides a signaling conduit between gNB-DU and the gNB-CU whose services are divided into non-UE-associated (related to the entire F1 interface instance between the gNB-DU and gNB-CU utilizing a non-UE-associated signaling connection); and UE-associated (F1AP functions that provide these services are associated with a UE-associated signaling connection that is maintained for a specific UE) services [31]. F1AP consists of elementary procedures (EPs), where an EP is defined as the unit of interaction between gNB-CU and gNB-DU over F1. These EPs are defined separately and are used to create complete sequences in a flexible manner. Unless otherwise stated by the restrictions, these EPs may be invoked independent of each other as stand-alone procedures, which can be active in parallel. An EP consists of an initiating message and possibly its response message. Two types of EPs, referred to as class 1 and class 2, are used, where the former consists of elementary procedures with response (success and/or failure) and the latter comprises EPs without response.

A gNB-CU UE F1AP ID is allocated to uniquely identify the UE over the F1 interface within a gNB-CU. The gNB-DU stores the received gNB-CU UE F1AP ID for the duration of time that the UE-associated logical F1-connection is valid. A gNB-DU UE F1AP ID is assigned to uniquely identify the UE over the F1 interface within a gNB-DU. When a gNB-CU receives a gNB-DU UE F1AP ID, it stores it for the period of time that the UE-associated logical F1-connection for the UE remains valid. The UE-associated signaling is used when F1AP messages corresponding to a UE utilize the UE-associated logical F1-connection for association of the message to the UE in gNB-DU and gNB-CU. A UE-associated logical F1-connection uses unique identities that are used to identify UE F1AP messages by the gNB-CU and gNB-DU. A UE-associated logical F1-connection may exist before the F1 UE context[58] is setup in the gNB-DU.

The F1 interface management procedures include reset, error indication, F1 setup, gNB-DU configuration update, and gNB-CU configuration update on the control plane. The F1-C context management procedures include UE context setup, UE context release request (gNB-DU/gNB-CU initiated), UE context modification (gNB-CU/gNB-DU initiated), and UE mobility command. The F1-C RRC message transfer procedures include initial uplink RRC message transfer as well as UL/DL RRC message transfer. The F1 control plane further includes system information procedure and paging procedures [28]. The error indication function is used by the gNB-DU or gNB-CU to indicate occurrence of an error. The reset

[58] A gNB UE context is a block of information in the gNB associated with an active user equipment. The block of information contains the necessary information in order to provide NG-RAN services to the active user equipment. The gNB user equipment context is established when the transition to active state for a user equipment is completed or in target gNB after completion of handover resource allocation during handover preparation, where in each case the user equipment state information, security information, user equipment capability information, and the identities of the user equipment-associated logical NG connection are stored in the gNB user equipment context.

function is used to initialize the peer entity after node setup and after a failure event. This procedure can be used by both the gNB-DU and gNB-CU. The F1 setup function, initiated by gNB-DU, allows exchange of application-level data between gNB-DU and gNB-CU while ensuring proper interoperability over the F1 interface. The gNB-CU and gNB-DU configuration update functions facilitate updating application-level configuration data between gNB-CU and gNB-DU to properly interoperate over the F1 interface, and activate or deactivate the cells. Scheduling of broadcast system information is performed in the gNB-DU. The gNB-DU is responsible for encoding of NR-MIB (master information block portion of NR system information). In case broadcast of RMSI (remaining system information) or other SI messages is needed, the gNB-DU will be responsible for encoding of RMSI and the gNB-CU is responsible for encoding of other SI messages. The gNB-DU and gNB-CU measurement reporting functions are used to report the measurements of gNB-DU and gNB-CU, respectively.

The gNB-DU is further responsible for transmitting paging information according to the scheduling parameters. The gNB-CU provides paging information to enable gNB-DU to calculate the exact paging occasion and paging frame. The gNB-CU is responsible for calculating paging area. The gNB-DU combines the paging records for a particular paging occasion, frame, and area and further encodes the RRC message and broadcasts the paging message. The F1 UE context management function supports the establishment and modification of the necessary overall initial UE context. The mapping between QoS flows and radio bearers is performed by gNB-CU where the granularity of bearer related management over F1 is at radio-bearer level. To support PDCP duplication for intra-DU carrier aggregation, one DRB should be configured with two GTP-U tunnels between gNB-CU and gNB-DU.

1.2.1.1.3 E1 Control-Plane Functions and Procedures

As we discussed earlier, the disaggregated gNB model was introduced to enable separation of control-plane and user-plane functions in addition to the functional split of the protocol stack in gNB which facilitates design and development of new-generation gNBs based on the SDN/NFV concepts. As such, a new interface between the gNB-CU-CP and gNB-CU-UP components has been defined by 3GPP to support exchange of signaling information between these entities (see Fig. 1.52). E1 is an open interface which would allow multi-vendor implementation of the control-plane and data-plane components of the CU. E1 establishes a logical point-to-point interface between a gNB-CU-CP and a gNB-CU-UP even in the absence of a direct physical connection between the endpoints. The E1 interface separates radio network and TNLs and enables exchange of UE associated/non-UE associated information. The E1 interface is a control interface and is not used for user data forwarding. As shown in Fig. 1.52, a gNB may consist of one gNB-CU-CP, multiple gNB-CU-UPs, and multiple gNB-DUs. One gNB-DU can be connected to multiple gNB-CU-UPs under the

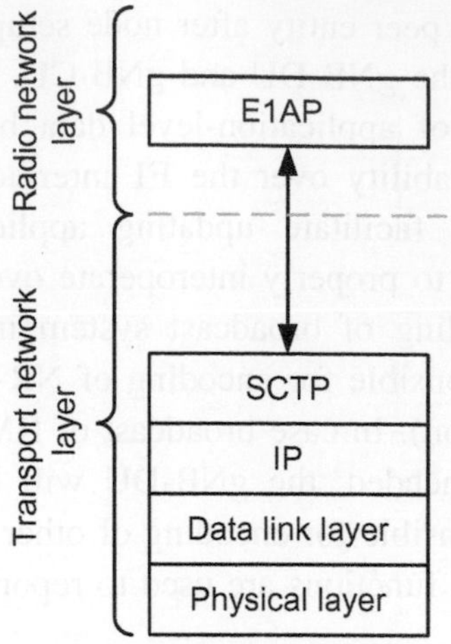

Figure 1.56
Protocol stack for E1 interface [36].

control of the same gNB-CU-CP and one gNB-CU-UP can be connected to multiple DUs under the control of the same gNB-CU-CP. The connectivity between a gNB-CU-UP and a gNB-DU is established by the gNB-CU-CP using bearer or UE context management functions. The gNB-CU-CP further selects the appropriate gNB-CU-UP(s) for the requested UE services. The E1 interface would support independent virtualization of the control and UPFs. It would also enable more flexible allocation of the functions of the central unit. It allows for energy and cost-efficient central processing and resource pooling for the user plane. Several functions such as security, packet inspection, header compression, and data mining could benefit from centralization. Furthermore, it would provide optimum routing of packets in case of multi-connectivity and interworking with other systems. The protocol structure for E1 is shown in Fig. 1.56. The TNL uses IP transport with SCTP over IP. The application layer signaling protocol is referred to as E1 application protocol (E1AP).

1.2.1.1.4 NG Control-Plane/User-Plane Functions and Procedures

The NG control-plane interface (NG-C) is defined between the gNB/ng-eNB and the AMF. The control-plane protocol stack of NG interface is shown in Fig. 1.57. The TNL relies on IP transport; however, for more reliable transmission of signaling messages, the SCTP protocol is used over IP. The application layer signaling protocol is referred to as NG application protocol (NGAP). The SCTP layer provides guaranteed delivery of application layer messages. IP layer point-to-point transmission is used to deliver the signaling PDUs. The NG-C interface further enables interface management (the functionality to manage the NG-C interface), UE context management (the functionality to manage the UE context between NG-RAN and 5GC), UE mobility

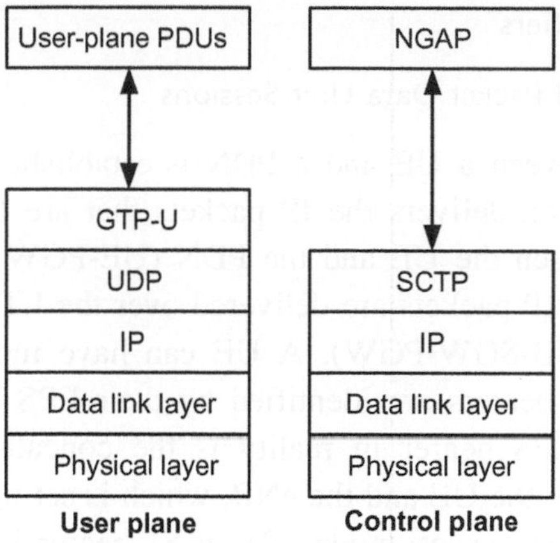

Figure 1.57
NG-U and NG-C protocol stack [16].

management (the functionality to manage the UE mobility in connected mode between NG-RAN and 5GC), transport of NAS messages (procedures to transfer NAS messages between 5GC and UE), and paging (the functionality to enable 5GC to generate paging messages sent to NG-RAN to allow NG-RAN to page the UE in RRC_IDLE state). It further enables PDU session management (the functionality to establish, manage, and remove PDU sessions and respective NG-RAN resources that are made of data flows carrying user-plane traffic) as well as configuration transfer (the functionality to transfer the NG-RAN configuration information, e.g., transport layer addresses for establishment of Xn interface, between two NG-RAN nodes via 5GC). The NGAP protocol consists of elementary procedures where an elementary procedure is a unit of interaction between the NG-RAN node and the AMF. These elementary procedures are defined separately and are used to create complete sequences in a flexible manner. Unless otherwise stated by the restrictions, the EPs may be invoked independent of each other as stand-alone procedures, which can be active in parallel. An EP consists of an initiating message and possibly the corresponding response message. Two types of EPs referred to as class 1 and class 2 are defined, where class 1 EP consists of elementary procedures with response (success and/or failure) and class 2 EP comprises elementary procedures without response.

The NG user-plane (NG-U) interface is defined between a NG-RAN node and a UPF. The NG-U interface provides non-guaranteed delivery of user-plane PDUs between the two nodes. The user-plane protocol stack of the NG interface is also shown in Fig. 1.57. The TNL relies on IP transport and use of GTP-U over UDP/IP to carry the user-plane PDUs between the NG-RAN node and the UPF.

1.2.1.2 Bearers and Identifiers

1.2.1.2.1 Radio Bearers and Packet Data Unit Sessions

In LTE, IP connection between a UE and a PDN is established via a PDN connection or EPS session. An EPS session delivers the IP packets that are labeled with UE IP address through logical paths between the UE and the PDN (UE-PGW-PDN). An EPS bearer is a logical pipe through which IP packets are delivered over the LTE network, that is, between a UE and a PGW (UE-eNB-SGW-PGW). A UE can have multiple EPS bearers concurrently. Thus different EPS bearers are identified by their EPS bearer IDs, which are allocated by the MME. An EPS bearer in reality is the concatenation of three underlying bearers: (1) A DRB between the UE and the eNB, which is set up upon user-plane establishment between the UE and the access node; (2) an S1 bearer between the eNB and SGW, which is set up upon establishment of EPS connection between the UE and the EPC; and (3) an S5/S8 bearer between the SGW and PGW, which is set up upon establishment of PDN connection between the UE and the PGW. An E-URAN random access bearer (E-RAB) is a bearer with two endpoints at the UE and at the SGW, which consists of a DRB and an S1 bearer. As such, E-RAB is a concatenation of a DRB and an S1 bearer, which logically connects the UE to the SGW. LTE QoS architecture describes how a network operator could create and configure different bearer types in order to map various IP traffic categories (user services) to the appropriate bearers according to the service QoS requirements. In that context, a bearer is an encapsulation mechanism or a tunnel that is created per user for transporting various traffic flows with specific QoS requirements within the LTE access and core networks.

In LTE and NR, there are two types of radio bearers, namely signaling radio bearers (SRB) and DRB. The SRBs are radio bearers that are only used for the transmission of RRC and NAS messages, whereas the DRBs are radio bearers that are used to transport user-plane traffic. The RRC messages are used as signaling between UE and the access node (i.e., eNB or gNB). The NAS messages are used for signaling between the UE and the core network. The RRC messages are used to encapsulate the NAS messages for transfer between the UE and the core network through the access node (transparent to the access node). The SRBs are further classified into three types: SRB0, SRB1, and SRB2, where SRB0 is used to transfer RRC messages which use common control channel, SRB1 is used to transfer RRC messages which use dedicated control channel, and SRB2 is used to transfer RRC messages which use dedicated control channel and encapsulate NAS messages. The SRB1 can be used to encapsulate NAS message, if SRB2 has not been configured. The SRB2 has lower priority than SRB1 and it is always configured after security activation. The SRB0 uses RLC transparent mode, while SRB1 and SRB2 use RLC acknowledged mode.

In LTE, upon connection establishment and at the beginning of an EPS session, a default EPS bearer, with no guaranteed bit rate and best effort QoS characteristics, is established.

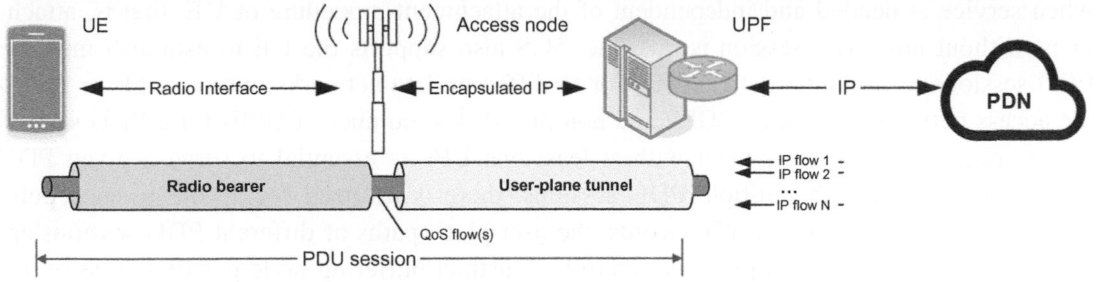

Figure 1.58
High-level illustration of 3GPP bearer architecture [16].

Dedicated EPS bearers with guaranteed bit rate and specific QoS attributes for various services can be established through negotiation and connection reconfiguration for active users. In LTE, the QoS control is performed per EPS bearer such that EPS bearers and radio bearers have a one-to-one relationship. There are two types of end-to-end bearers. The default bearer is established during the attach procedure and after an IP address is allocated to the UE and has best effort QoS characteristics. The dedicated bearer is typically established during the call setup after transition from the idle mode but can also be established during the attach procedure and supports various QoS characteristics with guaranteed bit rate. Bearer establishment negotiations are performed between the UE and the access point name (APN)[59] in the core network which maps the bearers to an external network such as the Internet or an IP multimedia subsystem (IMS).[60] However, it is important to note that the availability and provisioning of bearers is strictly controlled by operator configuration, as well as the association between the UE and the PDN that provides PDU connectivity service. The PDU sessions can be based on IPv4, IPv6, Ethernet or unstructured.

In NR and as shown in Fig. 1.58, the UE receives services through a PDU session, which is a logical connection between the UE and the data network. Various PDU session types are supported, for example, IPv4, IPv6, and Ethernet. Unlike the EPS, where at least one default bearer is always created when the UE attaches to the network, 5GS can establish a session

[59] Access point name is a gateway or anchor node between a mobile network and another IP network such as the Internet. A mobile device attempting a data connection must be configured with an access point name to present to the carrier. The carrier will then examine this identifier in order to determine what type of network connection should be created, which IP addresses should be assigned to the wireless device, and which security methods should be used. Therefore, the access point name identifies the packet data network with which a mobile data user communicates.

[60] The IP multimedia subsystem is an architectural framework developed by 3GPP for delivering IP multimedia services. Mobile networks originally provided voice services through circuit-switched networks and later migrated to all-IP network architectures. IP multimedia subsystem provides real-time multimedia sessions (voice over IP, video, teleconferencing, etc.) and nonreal-time multimedia sessions (push to talk, presence, and instant messaging) over an all-IP network. IP multimedia subsystem enables convergence of services provided by different types of networks which include fixed, mobile, and Internet.

when service is needed and independent of the attachment procedure of UE, that is, attachment without any PDU session is possible. 5GS also supports the UE to establish multiple PDU sessions to the same data network or to different data networks over a single or multiple access networks including 3GPP and non-3GPP. The number of UPFs for a PDU session is not specified. The deployment with at least one UPF is essential to serve a given PDU session. For a UE with multiple PDU sessions, there is no need for a convergence point such as SGW in the EPC. In other words, the user-plane paths of different PDU sessions are completely disjoint. This implies that there is a distinct buffering node per PDU session for the UE in the RRC_IDLE state.

5G QoS framework has been designed to allow detection and differentiation of sub-service flows in order to provide improved quality of experience relative to the previous generations of 3GPP radio interface standards. Since LTE bearer framework was insufficient to address certain 5G service requirements, a refined QoS model based on the concept of QoS flow was introduced in 5G. The QoS flow is the finest granularity for QoS enforcement in 3GPP 5G systems. All traffic mapped to the same 5G QoS flow receives the same forwarding treatment. Providing different QoS forwarding treatment requires the use of different 5G QoS flows. Fig. 1.59 illustrates the comparison between 4G and 5G QoS models. It is shown that the 5G concept allows flexible mapping of the 5G QoS flows to radio bearers, for example, the first 5G QoS flow is transported over the first 5G radio bearer while the second and third 5G QoS flows are transported together in the second 5G radio bearer. In order to support 5G QoS flows, either the existing protocols (e.g., PDCP) had to be enhanced or a new (layer-2) sublayer known as service data adaptation protocol (SDAP) had to be introduced. The main services and functions of SDAP include mapping between a

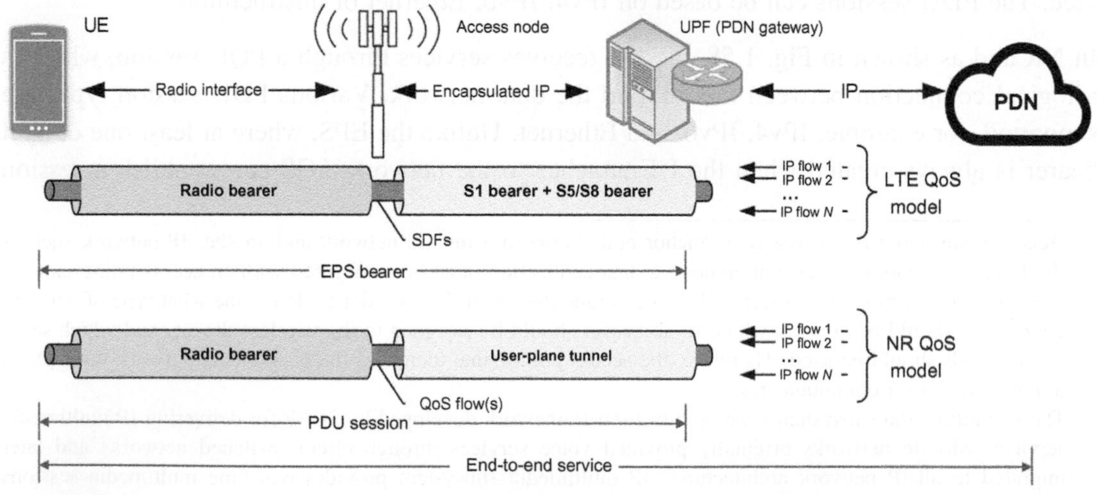

Figure 1.59
Comparison of QoS models in 4G LTE and 5G NR.

QoS flow and a data radio bearer, marking QoS flow identifier (QFI) in the downlink and uplink packets. For each DRB, the UE monitors the QoS flow ID(s) of the downlink packets and applies the same mapping in the uplink, that is, for a DRB, the UE maps the uplink packets belonging to the QoS flows(s) corresponding to the QoS flow ID(s) and PDU session in the downlink packets for that DRB.

In order to establish a PDU session, a PDU session establishment message is sent by 5GC to the gNB serving the UE, which includes the NAS message to be transferred to the UE containing the QoS-related information. The gNB sends a DRB setup request message to the UE and includes the DRB parameters and the NAS message that it received earlier. The UE establishes at least a default DRB associated with the new PDU session. It further creates the QFI to DRB mapping and sends an RRC DRB setup complete message to the gNB. The gNB sends PDU session establishment acknowledgment message to 5GC, indicating successful establishment of the PDU session. Data is sent over the N3 tunnel to the gNB and then over the DRB to the UE. The data packets may optionally include a QoS marking (same as or corresponding to QFI) in their SDAP header. The UE sends uplink packets over the DRB to the gNB. The uplink data packets include a QoS marking (same as or corresponding to QFI) in the SDAP header [16].

Dual-connectivity was introduced as part of LTE Rel-12. In LTE DC, the radio protocol architecture that a particular bearer uses depends on how the bearer is setup. Three bearer types have been defined: master cell group (MCG), secondary cell group (SCG), and split bearer. The three bearer types are illustrated in Figs. 1.60–1.62. The RRC layer is always located in the master node and signaling bearers are mapped to MCG bearer; therefore, they only use the radio resources provided by the master node. The MCG bearer can be seen as the legacy bearer that transports both data and signaling. Split bearer and SCG bearer are data only bearers. The main difference between the two is that for split bearer,

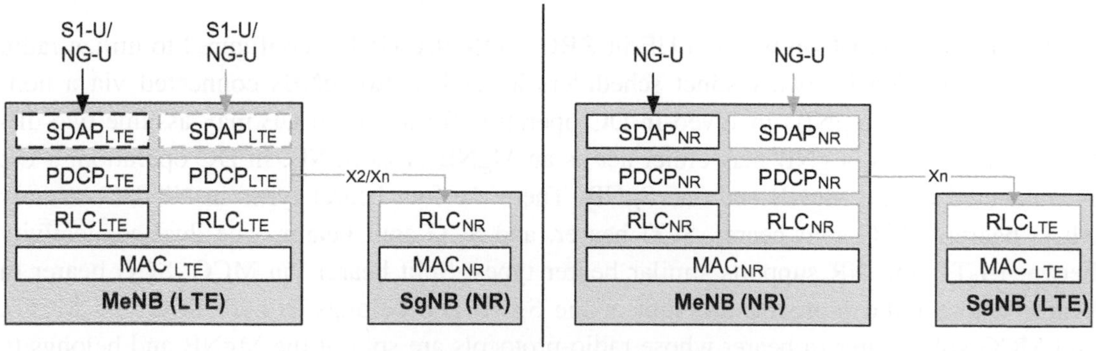

Figure 1.60
Illustration of split bearer via MCG [13,16].

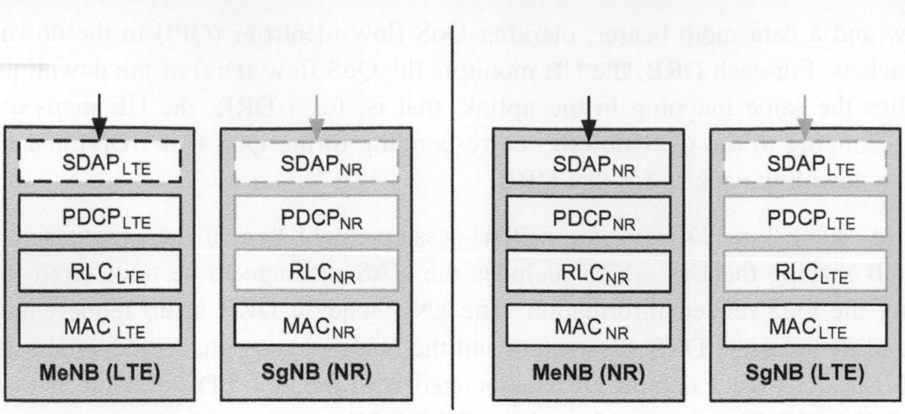

Figure 1.61
Illustration of SCG bearer [13].

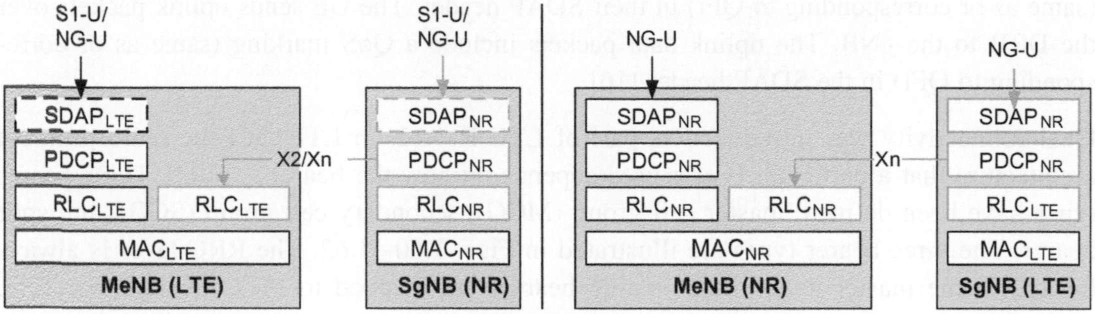

Figure 1.62
Illustration of split bearer via SCG [13].

the S1-U/NG-U interface terminates in the master node, whereas for SCG bearer, the S1-U/NG-U interface terminates at the secondary node [13].

3GPP NR supports DC in which a UE in RRC_CONNECTED is configured to utilize radio resources provided by two distinct schedulers located in two gNBs connected via a non-ideal backhaul. The gNBs involved in DC operation for a certain UE may assume two different roles, that is, a gNB may either act as an MgNB or as SgNB. In DC operation, a UE is connected to one MgNB and one SgNB. There are four bearer types in NR DC, namely MCG bearer, MCG split bearer, SCG bearer, and SCG split bearer. The dual-connectivity between LTE and NR supports similar bearer types. Split bearer via MCG, SCG bearer (a bearer whose radio protocols are split at the SgNB and belongs to both SCG and MCG), and MCG split bearer (a bearer whose radio protocols are split at the MgNB and belongs to both MCG and SCG). The MCG bearer and one SCG bearer are used for two different QoS flows [16].

In the downlink, the incoming data packets are classified by the UPF based on SDF[61] templates according to their precedence (without initiating additional N4 signaling). The UPF conveys the classification of user-plane traffic associated with a QoS flow through an N3 (and N9) user-plane marking using a QFI. The AN binds QoS flows to AN resources (i.e., data radio bearers). There is no one-to-one relationship between QoS flows and AN resources and it is the responsibility of the AN to establish the necessary resources for mapping to the QoS flows [13].

In 3GPP NR, the DRB defines the packet treatment on the radio interface. A DRB serves packets with the same packet forwarding treatment. Separate DRBs may be established for QoS flows requiring different packet handling. In the downlink, the RAN maps QoS flows to DRBs based on QoS flow ID and the associated QoS profiles. In the uplink, the UE marks uplink packets over the radio air-interface with the QoS flow ID for the purpose of marking forwarded packets to the core network. Downlink traffic is marked to enable prioritization in the IP network and in the access node. Similar traffic marking may be used for uplink traffic, according to the operator's configuration. Standardized packet marking informs the QoS enforcement functions of what QoS to provide without any QoS signaling, although the option with QoS signaling offers more flexibility and QoS granularity.

1.2.1.2.2 Radio Network Identifiers

Each entity and bearer in NG-RAN and 5GC are identified with a unique identifier. The identifiers are either permanently provisioned, such as the International Mobile Subscriber Identity (IMSI) and International Mobile Equipment Identity (IMEI) that are assigned by an operator to the UE, or they are assigned during the lifetime of UE operation in an operator's network. Fig. 1.56 illustrates the identities that are either provisioned or dynamically/semi-statically assigned to the bearers and the entities in 5GS. This section describes the bearers and identifiers that have been introduced in 3GPP 5G specifications. The direction of the arrows in the figure indicates the entity which assigns the identifier and the entity to which the identifier is assigned. The bearers and various bearer identifiers depending on the network interface and the flow direction are also shown in the figure. Note that the protocols over NR-Uu and NG interfaces are divided into two categories: user-plane protocols, which

[61] Service data flow is a fundamental concept in the 3GPP definition of QoS and policy management. Service data flows represent the IP packets related to a user service (web browsing, e-mail, etc.). Service data flows are bound to specific bearers based on policies defined by the network operator. The traffic detection filters, for example, IP packet filter, required in the user-plane function can be configured either in the SMF and provided to the UPF, as service data flow filter(s), or be configured in the UPF, as the application detection filter identified by an application identifier. In the latter case, the application identifier has to be configured in the SMF and the UPF. In this context, service data flow filter is a set of packet flow header parameter values/ranges used to identify one or more of the packet (IP or Ethernet) flows constituting a service data flow. service data flow template is the set of service data flow filters in a policy rule or an application identifier in a policy rule referring to an application detection filter, required for defining a service data flow.

are the protocols implementing the actual PDU session carrying user data through the AS; and the control-plane protocols that are the protocols for controlling the PDU session and the connection between the UE and the network from different aspects including requesting the service, controlling different transmission resources, handover, etc.

In LTE, several radio network temporary identifiers (RNTIs) were used to identify a connected-mode UE within a cell, a specific physical channel, a group of UEs in case of paging, a group of UEs for which power control command is issued by the eNB, system information transmitted for all UEs by the eNB, etc. In general, the RNTIs are used to scramble the CRC part of the radio channel messages. This implies that if the UE does not know the exact RNTI values for each of the cases, it cannot decode the radio channel messages even though the message reaches to the UE. The radio network and UE identifiers in NR, while similar to those of LTE, have been adapted to support new features and functionalities of NR and the 5GC such as multi-connectivity and network slicing. When an NR-compliant UE is connected to 5GC, the following identities are used at cell level to uniquely identify the UE (Fig. 1.63):

- Cell Radio Network Temporary Identifier (C-RNTI) is a unique identification, which is used as an identifier of the RRC connection and for scheduling purposes. In DC scenarios, two C-RNTIs are independently allocated to the UE, one for MCG and one for SCG.
- Temporary C-RNTI, which is used during the random-access procedure, is a random value for contention resolution which during some transient states, the UE is temporarily identified with a random value used for contention resolution purposes.
- Inactive RNTI (I-RNTI) is used to identify the UE context for RRC_INACTIVE.

The following identities are used in NG-RAN for identifying a specific network entity:

- AMF Name is used to identify an AMF. The AMF Name fully-qualified domain name (FQDN) uniquely identifies an AMF, where FQDN consists of one or more labels. Each label is coded as a one octet-length field followed by that number of octets coded as 8 bit ASCII characters. An AMF Set within an operator's network is identified by its AMF Set ID, AMF Region ID, mobile country code (MCC), and mobile network code (MNC).
- NR cell global identifier (NCGI) is used to globally identify the NR cells. The NCGI is constructed from the PLMN identity to which the cell belongs and the NR cell identity (NCI) of the cell. It can be assumed that it is equivalent to CGI in LTE system.
- gNB Identifier (gNB ID) is used to identify gNBs within a PLMN. The gNB ID is contained within the NCI of its cells.
- Global gNB ID is used to globally identify the gNBs, which is constructed from the PLMN identity to which the gNB belongs and the gNB ID. The MCC and MNC are the same as included in the NCGI.
- TAI is used to identify TAs. The TAI is constructed from the PLMN identity to which the TA belongs and the tracking area code (TAC) of the TA.
- S-NSSAI is used to identify a network slice.

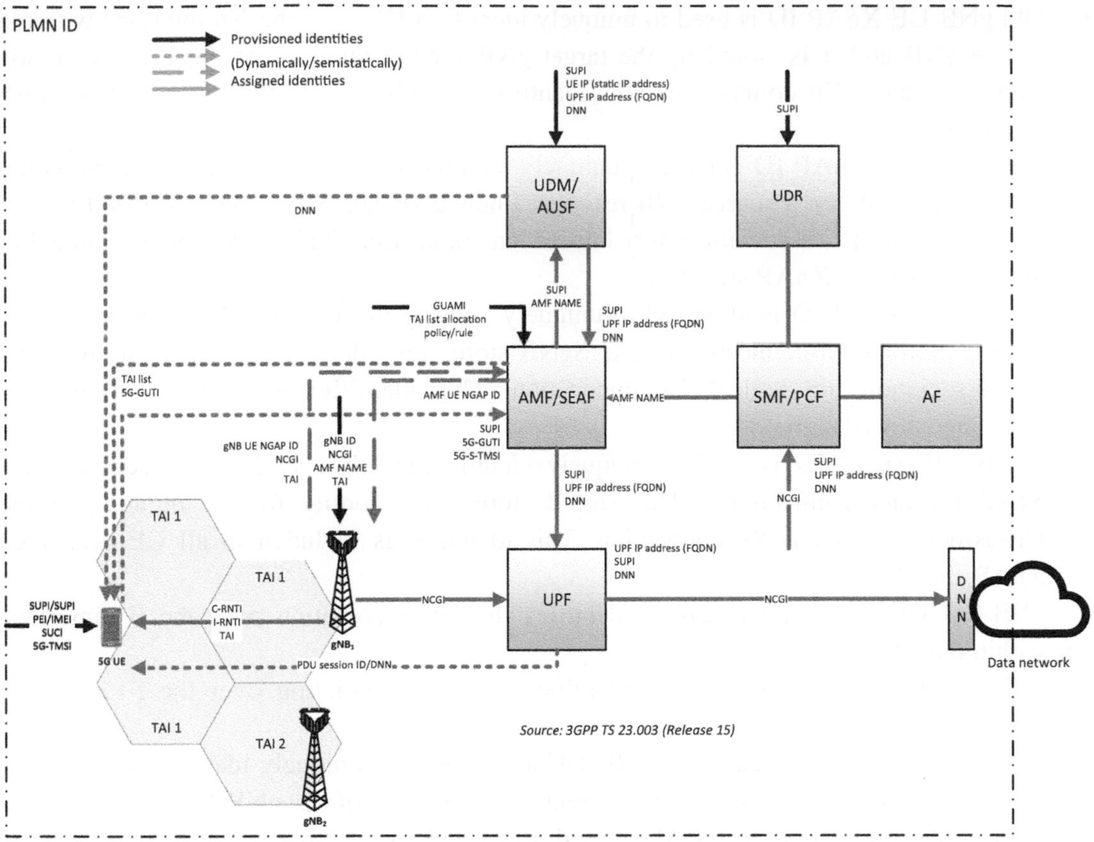

Figure 1.63
NG-RAN/5GC entities and bearers and identifiers.

An application protocol identity (AP ID) is assigned when a new UE-associated logical connection is created in either a gNB or an AMF. An AP ID uniquely identifies a logical connection with a UE over the NG interface or Xn interface within a node (gNB or AMF). Upon receipt of a message that has a new AP ID from the originating node, the receiving node stores the corresponding AP ID for the duration of the logical connection. The definition of AP IDs used over the NG, Xn, or F1 interface are as follows [15]:

- gNB UE NG application protocol (NGAP) ID is used to uniquely identify a UE over the NG interface within a gNB. This identifier is stored by the AMF for the duration of the UE association through the logical NG connection. This identifier is included in all UE associated NGAP signaling.
- AMF UE NGAP ID is allocated to uniquely identify the UE over the NG interface within an AMF. This identifier is stored for the duration of the UE-associated logical NG connection by the gNB. This identifier is included in all UE-associated NGAP signaling once known to the gNB.

- Old gNB UE XnAP ID is used to uniquely identify a UE over the Xn interface within a source gNB and it is stored by the target gNB for the duration of the UE association over the logical Xn connection. This identifier is included in all UE-associated XnAP signaling.
- New gNB UE XnAP ID is used to uniquely identify a UE over the Xn interface within a target gNB. When a source gNB receives a new gNB UE XnAP ID, it stores it for the duration of the UE association over logical Xn connection. This identifier is included in all UE-associated XnAP signaling.
- MgNB UE XnAP ID is allocated to uniquely identify the UE over Xn interface within an MgNB for dual-connectivity. The SgNB stores this identity for the duration of the UE association via logical Xn connection. This identifier is included in all UE-associated XnAP signaling.
- SgNB UE XnAP ID is used to uniquely identify the UE over Xn interface within a SgNB for dual-connectivity. The MgNB stores this identity for the duration of the UE-associated logical Xn connection. This identifier is included in all UE-associated XnAP signaling.
- gNB-CU UE F1AP ID uniquely identifies the UE association over the F1 interface within the gNB-CU.
- gNB-DU UE F1AP ID uniquely identifies the UE association over the F1 interface within the gNB-DU.
- gNB-DU ID is configured at the gNB-DU and is used to uniquely identify the gNB-DU within a gNB-CU. The gNB-DU informs the gNB-CU of its gNB-DU ID during F1 setup procedure. The gNB-DU ID is used over F1AP procedures.

1.2.1.3 User-Plane and Control-Plane Protocol Stacks

This section describes an overview of the NG-RAN protocol structure, protocol layer terminations at various access and core network nodes, as well as the functional split between the NG-RAN and the 5GC. The NG-RAN radio protocols can be divided into control-plane and user-plane categories, where the user-plane protocols are typically responsible for carrying user data and the control-plane protocols are used to transfer signaling and control information.

In general, a communication protocol is a set of rules for message exchange and/or sending blocks of data known as PDUs between network nodes. A protocol may define the packet structure of the data transmitted and/or the control commands that manage the session. A protocol suite consists of several levels of functionality. This modularity facilitates the design and evaluation of protocols. Since each protocol layer usually (logically or physically) communicates with its peer entity across a communication link, they are commonly seen as layers in a stack of protocols, where the lowest protocol layer always deals with physical interaction of the hardware across the communication link. Each higher layer protocol adds more features or functionalities. User applications usually deal with the top-most layers.

In the context of protocol structure, we will frequently use the terms service and protocol. It must be noted that services and protocols are distinct concepts. A service is a set of primitives or operations that a layer provides to the layer(s) to which it is logically/physically connected. The service defines what operations a layer performs without specifying how the operations are implemented. It is further related to the interface between two adjacent layers. A protocol, in contrast, is a set of rules presiding over the format and interpretation of the information/messages that are exchanged by peer entities within a layer. The entities use protocols to implement their service definitions. Thus a protocol is related to the implementation of a service. The protocols and functional elements defined by 3GPP standards correspond to all layers of the open system interconnection (OSI), that is, the seven-layer network reference model.[62] As shown in Fig. 1.64, what 3GPP considers as layer-2 and layer-3 protocols is mapped to the OSI data link layer. The higher layer protocols in the 3GPP stack are the application and transport layers. The presentation and session layers are often abstracted in practice.

The NG-RAN protocol structure is depicted in Fig. 1.65 for UEs and gNBs in the user plane and control plane. In the control plane, the NAS functional block is used for network attachment, authentication, setting up bearers, and mobility management. All NAS messages are ciphered and integrity protected by the AMF and the UE. There is also a mechanism for transparent transfer of NAS messages. As shown in Fig. 1.66, the layer-2 of NR is divided into MAC, RLC, PDCP, and SDAP sublayers. The SAP or the interface between two adjacent protocol layers is marked with a circle at the interface between the sublayers in the figure. The SAP between the physical layer and the MAC sublayer provides the transport channels. The SAP between the MAC sublayer and the RLC sublayer provides the logical channels. The physical layer provides transport channels to the MAC sublayer. From the physical layer perspective, the MAC sublayer provides and receives services in the form of transport channels. The data in a transport channel is organized into transport blocks. By

[62] Open systems interconnection is a standard description or a reference model for the computer networks which describes how messages should be transmitted between any two nodes in the network. Its original purpose was to guide product implementations to ensure consistency and interoperability between products from different vendors. This reference model defines seven layers of functions that take place at each end of a communication link. Although open systems interconnection is not always strictly adhered to in terms of grouping, the related functions together in a well-defined layer, many if not most products involved in telecommunication make an attempt to describe themselves in relation to the open systems interconnection model. Open systems interconnection was officially adopted as an international standard by the International Organization of Standards and it is presently known as Recommendation X.200 from ITU-T. The layers of open systems interconnection model are classified into two groups. The upper four layers are used whenever a message passes from or to a user. The lower three layers (up to the network layer) are used when any message passes through the host computer.

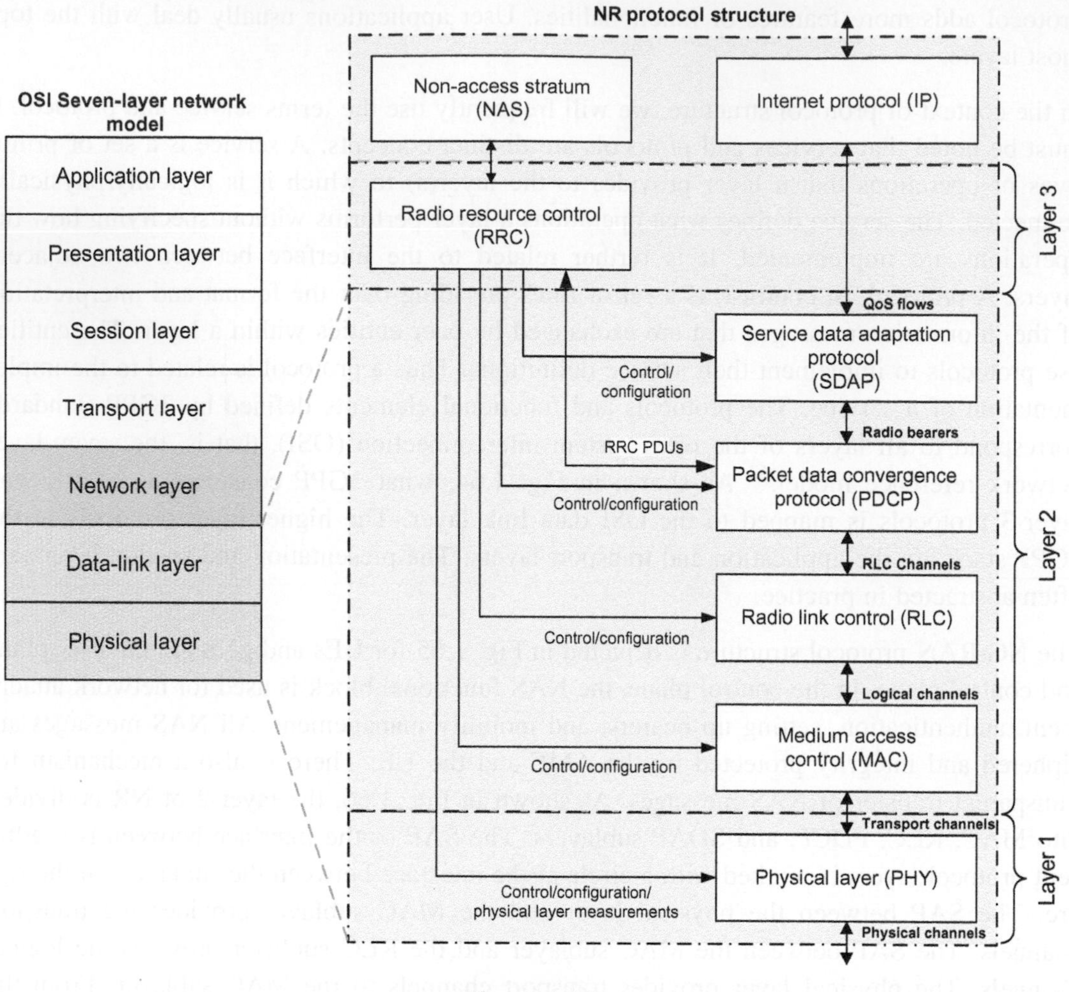

Figure 1.64
Mapping of lower NR protocol layers to OSI network reference model.

varying the transmission format of the transport blocks, the MAC sublayer can realize different data rates and reliability levels. The MAC sublayer receives the RLC SDUs mapped to various logical channels in the downlink, and generates the MAC PDUs that further become the transport blocks in the physical layer. The RLC sublayer provides RLC channels to the PDCP sublayer and the latter provides radio bearers to the SDAP sublayer. As we mentioned earlier, radio bearers are classified into data radio bearers for user-plane data and SRBs for control-plane information. The SDAP sublayer is configured by RRC and maps QoS flows to DRBs where one or more QoS flows may be mapped into one DRB in the downlink; however, one QoS flow is mapped into only one DRB at a time in the uplink

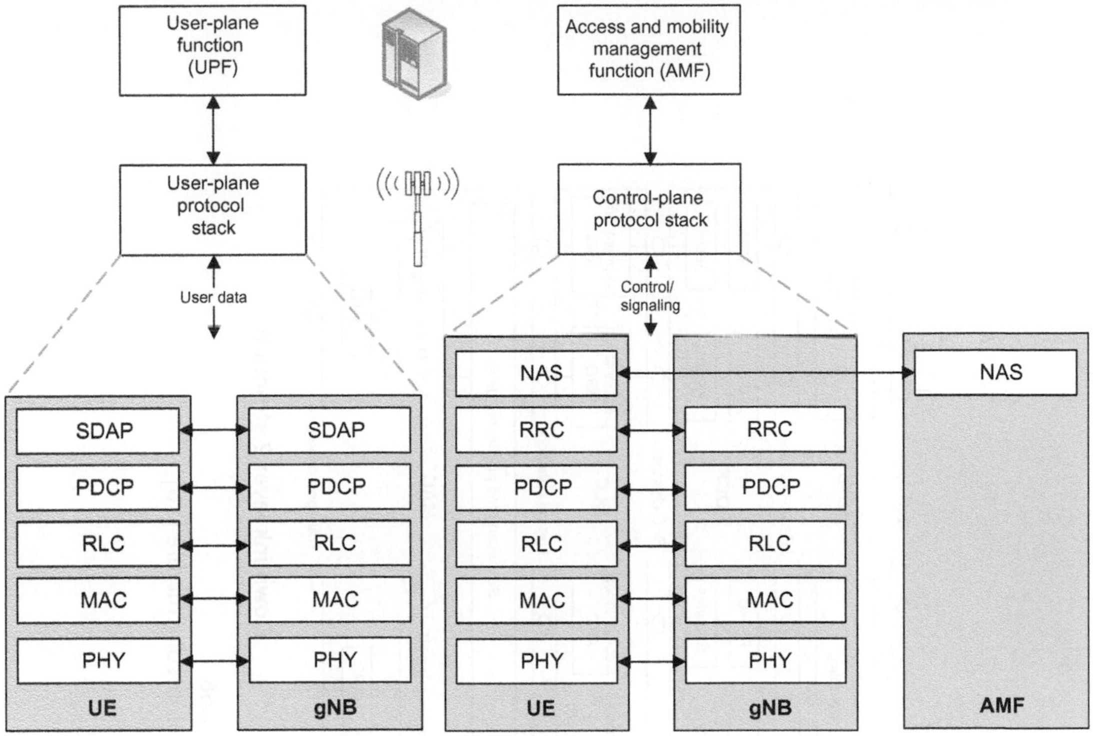

Figure 1.65
NG-RAN protocol stack [16].

[16]. The introduction of a new sublayer in NR layer-2 was meant to support the improved flow-based QoS model in NR as opposed to bearer-based QoS model in LTE.

MAC sublayer is responsible for mapping between logical channels and transport channels and multiplexing/de-multiplexing of MAC SDUs belonging to one or different logical channels to transport blocks which are delivered to or received from the physical layer through the transport channels. The MAC sublayer further handles scheduling and UE measurements/reporting procedures as well as error correction through HARQ (one HARQ entity per carrier in case of CA). It further manages user prioritization via dynamic scheduling, priority handling among logical channels of one UE through logical channel prioritization. A single MAC instantiation can support one or more OFDM numerologies, transmission timings as well as mapping restrictions on logical channels. In case of CA, the multi-carrier properties of the physical layer are only exposed to the MAC sublayer. In that case, one HARQ entity is required per serving cell. In both uplink and downlink, there is one independent HARQ entity per serving cell and one transport block is generated per transmission time interval per serving cell in the absence of spatial multiplexing. Each transport block and the associated HARQ retransmissions are mapped to a single serving cell [16].

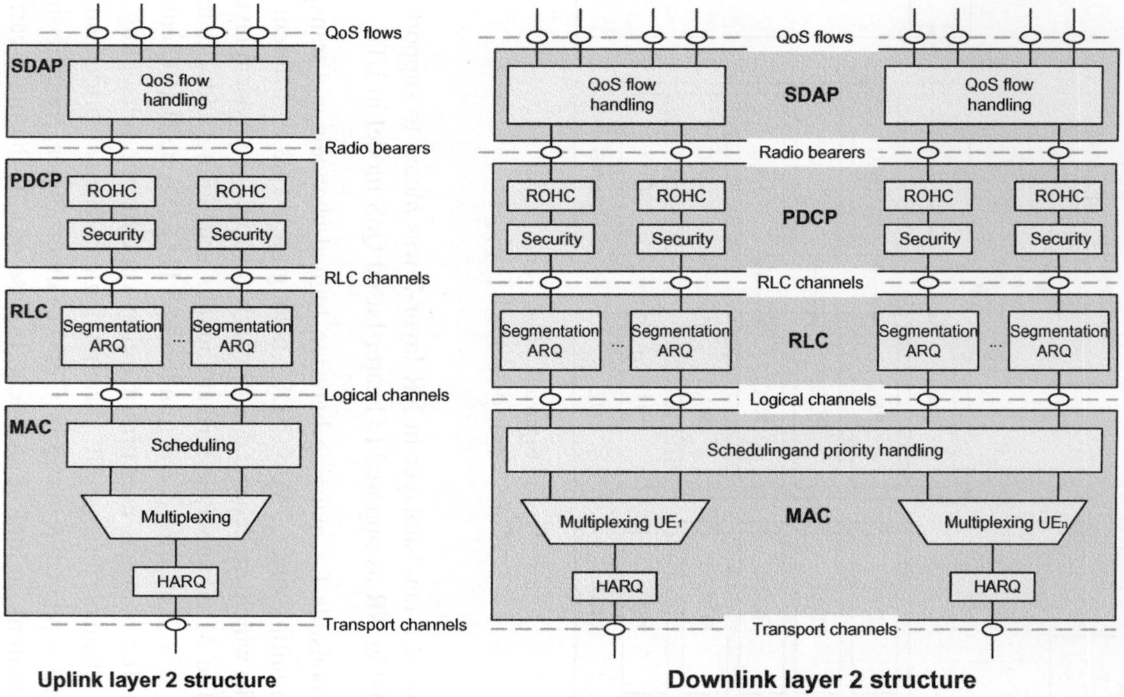

Figure 1.66
3GPP NR DL/UL layer-2 protocol structure [16].

MAC sublayer provides different type of data transfer services through mapping of logical channels to transport channels. Each logical channel type is defined by the type of information being transferred. Logical channels are classified into two groups of control and traffic channels. In NR, the control channels are used for transport of control-plane information and are of the following types (see Fig. 1.67):

- Broadcast control channel (BCCH) is a downlink channel for broadcasting system control information.
- Paging control channel (PCCH) is a downlink channel that transports paging and system information change notifications.
- Common control channel (CCCH) is a logical channel for transmitting control information between UEs and the network when the UEs have no RRC connection with the network.
- Dedicated control channel (DCCH) is a point-to-point bidirectional channel that transmits UE-specific control information between a UE and the network and is used by the UEs that have established RRC connection with the network.

Traffic channels are used for the transfer of user-plane information and are of the following type:

- Dedicated traffic channel (DTCH) is a UE-specific point-to-point channel for transport of user information which can exist in both uplink and downlink.

The physical layer provides information transfer services to the MAC and higher layers. The physical layer transport services are described by how and with what characteristics data is transferred over the radio interface. This should be clearly distinguished from the classification of what is transported which relates to the concept of logical channels in the

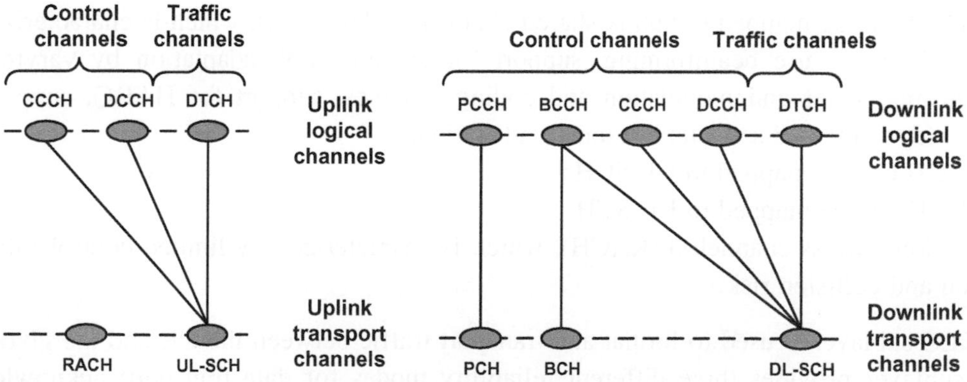

Figure 1.67
Mapping of logical and transport channels [16].

MAC sublayer. In the downlink, the logical channels and transport channels are mapped as follows (see Fig. 1.67):

- BCCH can be mapped to broadcast channel (BCH), which is characterized by fixed, predefined transport format, and is required to be broadcast in the entire coverage area of the cell. A DL-SCH may support receptions using different numerologies and/or TTI duration within the MAC entity. An UL-SCH may also support transmissions using different numerologies and/or TTI duration within the MAC entity.
- BCCH can be mapped to downlink shared channel(s) (DL-SCH), which is characterized by support for HARQ protocol, dynamic link adaptation by varying modulation, coding, and transmit power, possibility for broadcast in the entire cell, possibility to use beamforming, dynamic and semistatic resource allocation, and UE discontinuous reception (DRX) to enable power saving.
- PCCH can be mapped to paging channel (PCH), which is characterized by support for UE DRX in order to enable power saving, requirement for broadcast in the entire coverage area of the cell, and is mapped to physical resources which can also be used dynamically for traffic or other control channels. This channel is used for paging when the network does not know the location of the UE.
- Common control channel (CCCH) can be mapped to DL-SCH and represents a logical channel for transmitting control information between UEs and gNBs. This channel is used for UEs that have no RRC connection with the network.
- Dedicated control channel (DCCH) can be mapped to DL-SCH and is a point-to-point bidirectional channel that transmits dedicated control information between a UE and the network. It is used by UEs that have already established RRC connection.
- Dedicated traffic channel (DTCH) can be mapped to DL-SCH and represents a point-to-point bidirectional channel dedicated to a single UE for the transfer of user information.

In the uplink, the logical channels and the transport channels are mapped as follows:

- CCCH can be mapped to uplink shared channel(s) (UL-SCH), which is characterized by possibility to use beamforming, support for dynamic link adaptation by varying the transmit power and modulation and coding schemes, support for HARQ, support for both dynamic and semistatic resource allocation.
- DCCH can be mapped to UL-SCH.
- DTCH can be mapped to UL-SCH.
- Random access channel(s) (RACH), which is characterized by limited control information and collision risk.

The RLC sublayer is used to format and transport traffic between the UE and the gNB. The RLC sublayer provides three different reliability modes for data transport: acknowledged mode (AM), unacknowledged mode (UM), and transparent mode (TM). The UM is suitable for transport of real-time services since such services are delay-sensitive and cannot tolerate delay due to ARQ retransmissions. The acknowledged mode is appropriate for non-real-time services such as file transfers. The transparent mode is used when the size of SDUs are known in advance such as for broadcasting system information. The RLC sublayer also provides sequential delivery of SDUs to the upper layers and eliminates duplicate packets from being delivered to the upper layers. It may also segment the SDUs. The RLC configuration is defined per logical channel with no dependency on numerologies and/or TTI durations, and ARQ can operate with any of the numerologies and/or TTI durations for which the logical channel is configured. For SRB0, paging and broadcast system information, RLC TM mode is used. For other SRBs, RLC AM mode is used. For DRBs, either RLC UM or AM mode is used.

The services and functions provided by the PDCP sublayer in the user plane include header compression/decompression of IP packets; transfer of user data between NAS and RLC sublayer; sequential delivery of upper layer PDUs and duplicate detection of lower layer SDUs following a handover in RLC acknowledged mode; retransmission of PDCP SDUs following a handover in RLC acknowledged mode; and ciphering/deciphering and integrity protection. The services and functions provided by the PDCP for the control plane include ciphering and integrity protection and transfer of control-plane data where PDCP receives PDCP SDUs from RRC and forwards them to the RLC sublayer and vice versa.

The main services and functions provided by SDAP sublayer include mapping between a QoS flow and a DRB and marking QoS flow IDs in downlink and uplink packets. A single instantiation of SDAP protocol is configured for each individual PDU session, with the exception of dual-connectivity mode, where two entities can be configured.

The RRC sublayer in the gNB makes handover decisions based on neighbor cell measurements reported by the UE; performs paging of the users over the air-interface; broadcasts system information; controls UE measurement and reporting functions such as the periodicity of channel quality indicator reports; and further allocates cell-level temporary identifiers to the active users. It also executes transfer of UE context from the serving gNB to the target-gNB during handover and performs integrity protection of RRC messages. The RRC sublayer is responsible for setting up and maintenance of radio bearers. Note that the RRC sublayer in 3GPP protocol hierarchy is considered as layer-3 protocol [16].

Fig. 1.68 shows an example of layer-2 data flow and packet processing, where a transport block is generated by MAC sublayer by concatenating two RLC PDUs from the radio bearer RB_x and one RLC PDU from the radio bearer RB_y. In this figure, H denotes the layer-specific headers or subheaders of each sublayer. The two RLC PDUs from RB_x each corresponds to one of the IP packets n and $n + 1$, while the RLC PDU from RB_y is a segment of the IP packet m.

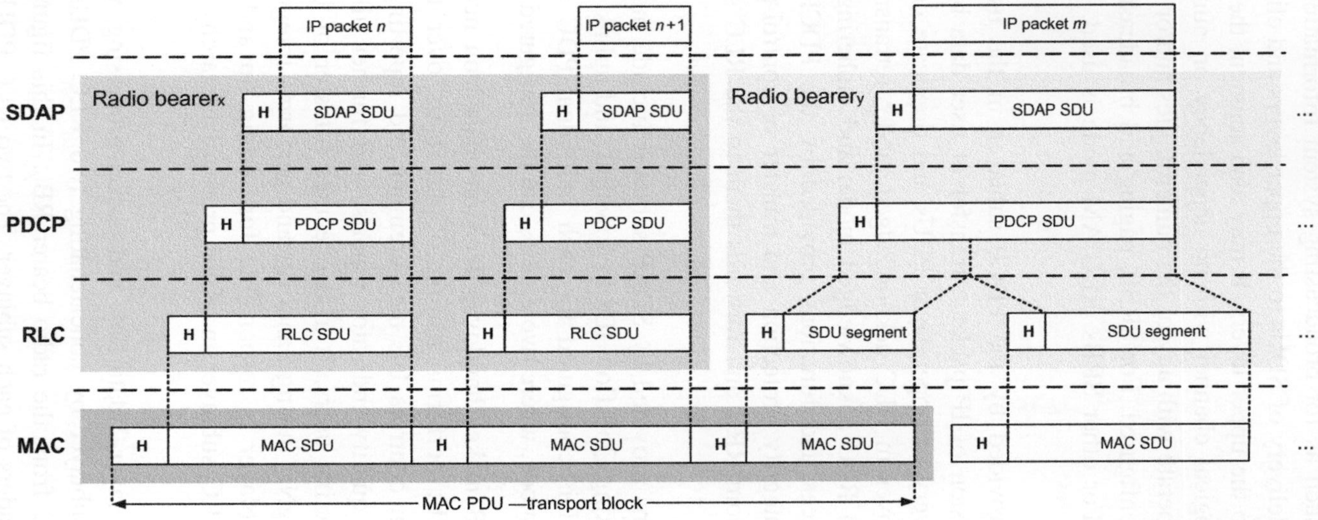

Figure 1.68
Example of layer-2 packet processing in NR [16].

1.2.2 Core Network

The design of 5G core network architecture in 3GPP was based on the following design principles in order to allow efficient support of new service networks such as information-centric networking (ICN)[63] [81]:

- Control- and user-plane separation (CUPS): This was a departure from LTE's vertically integrated control/user-plane network design to the one adopting NFV framework with modular NFs decoupled from the hardware for service centricity, flexibility, and programmability. In doing so, NFs are going to be implemented both physically and virtually, while allowing each to be customized and scaled based on their individual requirements, also allowing the realization of multi-slice coexistence. This further allows the introduction of UPF with new control functions, or reusing/extending the existing ones, to manage the new user-plane realizations.
- Decoupling of RAT and core network: Unlike LTE's unified control plane for access and the core networks, 5GC offers control-plane separation of RAN from the core network. This allows introduction of new radio access technologies and mapping of multiple heterogeneous RAN sessions to arbitrary core network slices based on service requirements.
- Non-IP PDU session support: A PDU session is defined as the logical connection between the UE and the data network. The PDU session establishment in 5GC supports both IP and non-IP PDUs (known as unstructured payloads), and this feature can potentially allow the support for ICN PDUs by extending or reusing the existing control functions.
- Service-centric design: 5GC service orchestration and control functions, such as naming, addressing, registration/authentication, and mobility, will utilize cloud-based service APIs. This enables open interfaces for authorized service function interaction and creating service-level extensions to support new network architectures. These APIs include widely used approaches, while not precluding the use of procedural approach between functional units.

Compared to LTE core network, where PDU session states in RAN and core were synchronized from session management perspective, 5GC decouples those states by allowing PDU

[63] Information-centric networking is an approach to evolve the Internet infrastructure away from a host-centric paradigm based on perpetual connectivity and the end-to-end principle, to a network architecture in which the focus is on content or data. In other words, information-centric networking is an approach to evolve the Internet infrastructure to directly support information distribution by introducing uniquely named data as a core Internet principle. Data becomes independent from location, application, storage, and means of transportation, enabling or enhancing a number of desirable features, such as security, user mobility, multicast, and in-network caching. Mechanisms for realizing these benefits is the subject of ongoing research in Internet Engineering Task Force and elsewhere. Current research challenges in information-centric networking includes naming, security, routing, system scalability, mobility management, wireless networking, transport services, in-network caching, and network management.

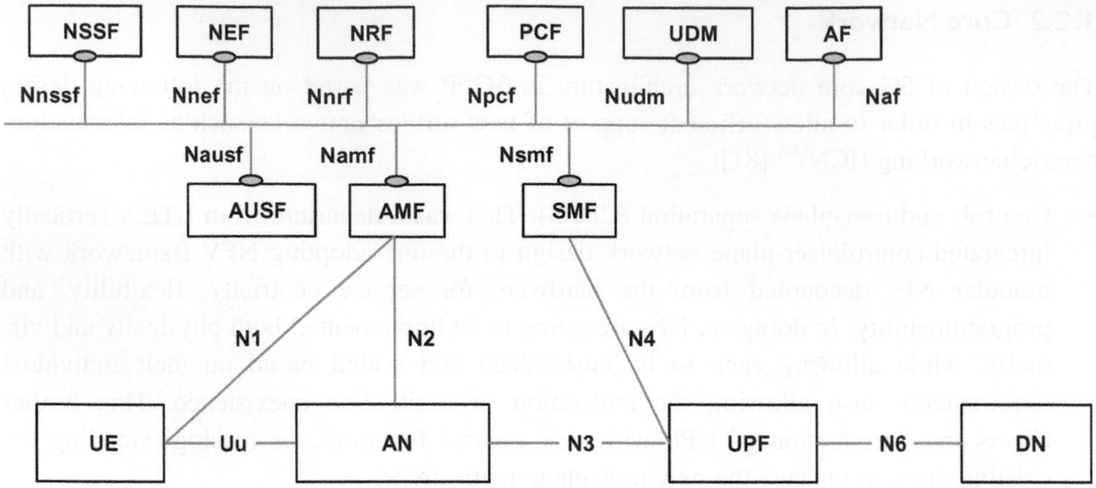

Figure 1.69
5G system service-based reference architecture [3].

sessions to be defined prior to a PDU session request by a UE. This de-coupling allows dynamic and policy-based interconnection of UE flows with slices provisioned in the core network. The SMF is used to handle IP anchor point selection and addressing functionality, management of the user-plane state in the UPFs such as in uplink classifier and branching point functions during PDU session establishment, modification and termination, and interaction with RAN to allow PDU session forwarding in UL/DL to the respective data networks. In the user plane, UE's PDUs are tunneled to the RAN using the 5G RAN protocols. From the RAN perspective, the PDU's five-tuple header information (IP source/destination, port, protocol, etc.) is used to map the flow to an appropriate tunnel from RAN to UPF.

1.2.2.1 Reference Architecture: Network Entities and Interfaces

The 5G core network architecture comprises a number of NFs,[64] some of which have newly been introduced. In this section, we provide a brief functional description of these NFs. Fig. 1.69 illustrates the 5GC service-based reference architecture. Service-based interfaces are used within the control plane. The main 5GC network entities include the following [3]:

Fig. 1.70 depicts the 5G system architecture in the non-roaming case, using the reference-point representation showing how various NFs interact with each other. Note that the

[64] NF is a 3GPP-adopted processing function in next generation network that has both functional behavior and interface. An NF can be implemented either as a network element on a dedicated hardware, as a software instance running on a dedicated hardware, or as a virtualized function instantiated on an appropriate platform such as cloud infrastructure.

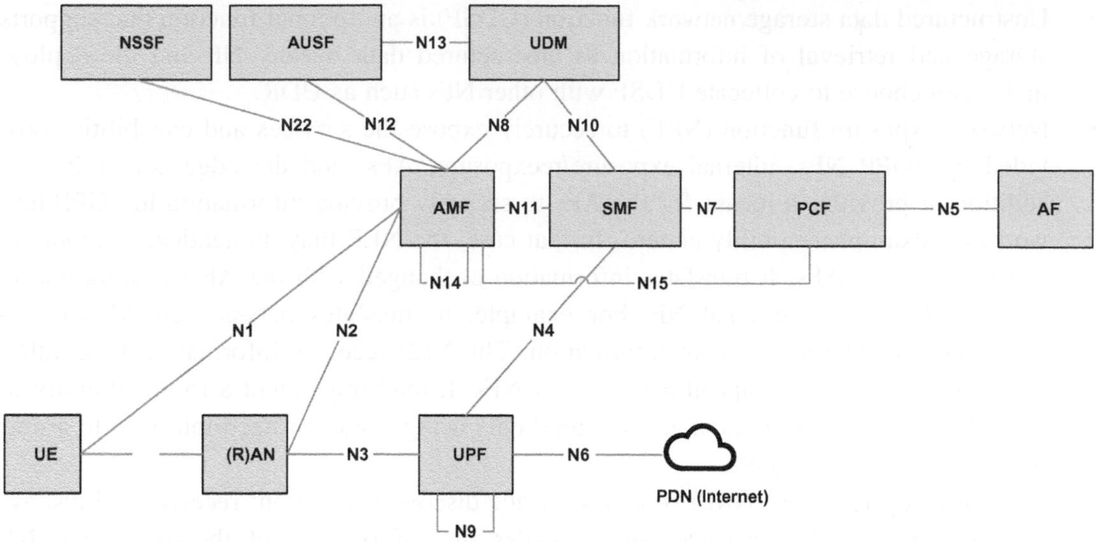

Figure 1.70
Non-roaming reference-point representation of 5G system architectures [3].

service-based and reference-point representations are two different representations of the 5GC which make it distinguished from LTE EPC. Service-based interfaces and reference points are two different ways to model interactions between architectural entities. A reference point is a conceptual point at the conjunction of two non-overlapping functional groups. A reference point can be replaced by one or more service-based interfaces that provide equivalent functionality. A unique reference point exists between two NFs, which means even if the functionality of two reference points is the same with different NFs, different reference point names must be assigned. However, when service-based interface representation is used, the same service-based interface is assigned if the functionality is equal on each interface.The functional description of the network functions is as follows [3]:

- Authentication server function (AUSF) is responsible for performing authentication process with the user terminals.
- AMF is responsible for termination of RAN control-plane interface (N2) and NAS (N1); ciphering and integrity protection of NAS messages; registration management; connection management; mobility management; lawful interception; transport of session management messages between UE and SMF; transparent proxy for routing session management messages; access authentication and authorization; and security anchor function (SEAF) and security context management (SCM). In addition to the functionalities described earlier, the AMF may support functions associated with non-3GPP ANs.
- Data network (DN) comprises operator's services, Internet access, or other services.

- Unstructured data storage network function (UDSF) is an optional function that supports storage and retrieval of information as unstructured data by any NF and the deployments can choose to collocate UDSF with other NFs such as UDR.
- Network exposure function (NEF) to securely expose the services and capabilities provided by 3GPP NFs, internal exposure/reexposure, AFs, and the edge computing. In addition, it provides a means for the AFs to securely provide information to 3GPP network, for example, mobility pattern. In that case, the NEF may authenticate, authorize, and regulate the AFs. It translates information exchanged with the AF and information exchanged with the internal NF. For example, it translates between an AF-service-identifier and internal 5G core information. The NEF receives information from other NFs based on exposed capabilities of other NFs. It may implement a frontend entity to store the received information as structured data using a standardized interface to a unified data repository (UDR).
- NF repository function (NRF) supports service discovery function; receives NF discovery request from NF instance; and provides the information of the discovered NF instances to the NF instance. It further maintains the NF profile of available NF instances and their supported services.
- NSSF selects the set of NSIs to serve a UE and to determine the allowed NSSAI and to determine the AMF set to serve the UE or depending on the configuration, a list of candidate AMF(s).
- Policy control function (PCF) supports interactions with the access and mobility policy enforcement in the AMF through service-based interfaces and further provides access and mobility management-related policies to the AMF.
- SMF handles session management (session establishment, modification, and release); UE IP address allocation and management; selection and control of UPF; traffic steering configuration at UPF to route traffic to the proper destination; termination of interfaces toward PCFs; control part of policy enforcement and QoS; and lawful interception among other functions.
- Unified data management (UDM) supports generation of 3GPP authentication and key agreement (AKA)[65] authentication credentials; user identification handling; access authorization based on subscription data; UE's serving NF registration management; service/session continuity; lawful interception functionality; subscription management; and SMS management. To provide these functions, the UDM uses subscription data (including authentication data) that may be stored in the UDR, in that case the UDM implements the application logic and does not require an internal user data storage, thus

[65] Authentication and key agreement is a mechanism which performs authentication and session key distribution in UMTS networks. Authentication and key agreement is a challenge-response-based mechanism that uses symmetric cryptography. Authentication and key agreement is typically run on a UMTS IP multimedia services identity module, which resides on a smart card device that also provides tamper-resistant storage of shared secrets. Authentication and key agreement is defined in IETF RFC 3310.

different UDMs may serve the same user in different transactions. The UDM is located in the home PLMN of the subscribers which it serves and accesses the information of the UDR located in the same PLMN.

- UDR supports storage and retrieval of subscription data by the UDM; storage and retrieval of policy data by the PCF; storage and retrieval of structured data for exposure; and application data by the NEF, including packet flow descriptions for application detection, and application request information for multiple UEs. During the deployments, the operators can opt to co-locate UDR with UDSF.
- Non-3GPP interworking function (N3IWF) supports untrusted non-3GPP access to 5GC. It further supports IPsec tunnel establishment with the UE; terminates the IKEv2[66] or IPsec[67] protocols with the UE over NWu and relays over N2 the information needed to authenticate the UE and authorize its access to the 5G core network as well as termination of N2 and N3 interfaces to 5G core network for control plane and user plane, respectively; relaying uplink and downlink control-plane NAS (N1) signaling between the UE and AMF; handling of N2 signaling from SMF related to PDU sessions and QoS; and establishment of IPsec security association (IPsec SA) to support PDU session traffic.
- UPF acts as the anchor point for intra-RAT or inter-RAT mobility; external PDU session point of interconnect to the data network; packet routing and forwarding; packet inspection and user-plane part of policy rule enforcement; lawful interception; traffic usage reporting; uplink classifier to support routing traffic flows to a data network; branching point to support multi-homed PDU session; QoS handling for user plane (packet filtering, gating, and UL/DL rate enforcement); uplink traffic verification (SDF to QoS flow mapping); transport-level packet marking in the uplink and downlink; and downlink packet buffering and downlink data notification triggering.
- AF is responsible for interacting with the 3GPP core network in order to support application influence on traffic routing; accessing network exposure function; and interacting with the policy framework for policy control. Based on operator deployment, the AF is considered to be trusted by the operator and can be allowed to interact directly with relevant NFs.

---

[66] IKE or IKEv2 is the protocol used to set up a security association in the IPsec protocol suite. The IKE protocol is based on a key-agreement protocol and the Internet security association and key management protocol which is a protocol defined by IETF RFC 2408 for establishing security associations and cryptographic keys in an Internet environment. The IKE uses X.509 certificates for authentication which are either pre-shared or distributed to set up a shared session secret from which cryptographic keys are derived. In addition, a security policy for every peer which will connect must be manually maintained.

[67] IPsec is a set of protocols for securing IP-based communications by authenticating and encrypting each IP packet of a data stream. IPsec also includes protocols for establishing mutual authentication between agents at the beginning of the session and negotiation of cryptographic keys to be used during the session. IPsec can be used to protect data flows between a pair of hosts, between a pair of security gateways, or between a security gateway and a host.

- Security edge protection proxy is a non-transparent proxy which supports message filtering and policing inter-PLMN control-plane interfaces and topology hiding.
- User equipment
- Access network

Reference-point representation of the architecture can be used to develop detailed call flows in the normative standardization. N1 is defined to carry signaling between UE and AMF. The reference points for connecting AN and AMF and AN and UPF are defined as N2 and N3, respectively. There is no reference point between AN and SMF, but there is a reference point, N11, between AMF and SMF. Therefore, the SMF is controlled by AMF. N4 is used by SMF and UPF so that the UPF can be configured using the control information generated by the SMF, and the UPF can report its state to the SMF. N9 is the reference point for the connection between different UPFs, and N14 is the reference point connecting different AMFs. N15 and N7 are defined for the PCF to apply policies to AMF and SMF, respectively. N12 is required for the AMF to perform authentication of the UE. N8 and N10 are defined to provide the UE subscription data to AMF and SMF.

The 5G core network supports UE connectivity via untrusted non-3GPP access networks such as Wi-Fi. Non-3GPP access networks can connect to the 5G core network via a non-3GPP interworking function (N3IWF). The N3IWF interfaces the 5G core network control plane and UPFs via N2 and N3 interfaces, respectively, with the external network. The N2 and N3 reference points are used to connect stand-alone non-3GPP access networks to 5G core network control plane function and UPF, correspondingly. A UE that attempts to access the 5G core network over a stand-alone non-3GPP access, after UE attachment, must support NAS signaling with 5G core network control-plane functions using N1 reference point. When the UE is connected via a NG-RAN and via a stand-alone non-3GPP access, multiple N1 instances could exist for the UE, that is, there is one N1 instance over NG-RAN and one N1 instance over non-3GPP access [3].

1.2.2.2 PDN Sessions and 5GC Identifiers

In an LTE network, once a UE connects to a PDN using the IP address assigned to it upon successful initial attach to the network, the IP connection remains in place after a default EPS bearer is established over the LTE network and until the UE detaches from the LTE network (i.e., the PDN connection is terminated). Even when there is no user traffic to send, the default EPS bearer always stays activated and ready for possible incoming user traffic. Additional EPS bearers can be established, if the best effort QoS attributes of the default EPS bearer do not satisfy the service requirements. The additional EPS bearer is called a dedicated EPS bearer, where multiple dedicated bearers can be created, if required by the user or the network. When there is no user traffic, the dedicated EPS bearers can be removed. Dedicated EPS bearers are linked to a default EPS bearer. Therefore, IP traffic from or to a UE is delivered through an EPS bearer depending on the required QoS class over the LTE network. Uplink IP traffic is mapped from a UE to the EPS bearer while downlink IP traffic is mapped from a PGW to the EPS bearer. Each E-RAB is associated with a QCI and an allocation and retention priority (ARP), where each QCI is characterized by priority, packet delay budget (PDB), and acceptable packet loss rate.

The 5G core network supports PDU connectivity service, that is, a service that provides exchange of PDUs between a UE and a data network. The PDU connectivity service is supported via PDU sessions that are established upon request from the UE. The PDU sessions are established (upon UE request), modified (upon UE and 5GC request), and released (upon UE and 5GC request) using NAS session management signaling over N1 between the UE and the SMF. Upon request from an application server, 5GC is able to trigger a specific application in the UE. The UE conveys the message to the application upon receiving the trigger message. Note that unlike LTE, 3GPP NR Rel-15 does not support dual-stack PDU session. The 5GC supports dual-stack UEs using separate PDU sessions for IPv4 and IPv6. In 3GPP NR, the QoS granularity is refined further to the flow level. In a typical case, multiple applications will be running on a UE; however, in LTE eNB, each E-RAB does not have an associated QCI or an ARP, whereas in NR, the SDAP sublayer can be configured by RRC sublayer to map QoS flows to DRBs. One or more QoS flows may be mapped to one DRB. Thus, QFI is used to identify a QoS flow within the 5G system. User-plane traffic with the same QFI within a PDU session receives the same traffic forwarding treatment (e.g., scheduling, admission threshold, etc.). The QFI is carried in an encapsulation header on N3 and is unique within a PDU session.

The 5GC is access-agnostic and allows running the N1 reference point on non-3GPP radio access schemes such as Wi-Fi. The UE can also send NAS messages for session and mobility management to the 5GC via a non-3GPP access, which was not possible in the previous 3GPP radio access standards. Non-access stratum is the signaling protocol of the UE for mobility and session-related control messages, which requires a new security procedure in order to authenticate the UE over the non-3GPP access with the AMF in the 3GPP network. Non-3GPP access networks must be connected to the 5G core network via N3IWF entity. The N3IWF interfaces the 5G core network control-plane function and UPF via N2 and N3 interfaces, respectively. When a UE is connected via an NG-RAN and via a stand-alone non-3GPP access, multiple N1 instances, that is, one N1 instance over NG-RAN and one N1 instance over non-3GPP access, will be created. The UE is simultaneously connected to the same 5G core network of a PLMN over a 3GPP access and a non-3GPP access and is served by a single AMF provided that the selected N3IWF is located in the same PLMN as the 3GPP access. However, if the UE is connected to the 3GPP access network of a PLMN and if it selects an N3IWF which is located in a different PLMN, then the UE will be served separately by two PLMNs. The UE is registered with two separate AMFs. The PDU sessions over 3GPP access are served by the visiting SMFs which are different from the ones serving the PDU sessions over the non-3GPP access. The UE establishes an IPsec tunnel with N3IWF in order to attach to the 5G core network over the untrusted non-3GPP access and is authenticated by and attached to the 5G core network during the IPsec tunnel establishment procedure [3].

The network identifiers in 5G system are divided into subscriber identifiers and UE identifiers. Each subscriber in the 5G system is assigned a 5G subscription permanent identifier (SUPI) to use within the 3GPP system. The 5G system treats subscription identification independent of the UE identification. In that sense, each UE accessing the 5G system is assigned a permanent equipment identifier (PEI). The 5G system assigns a temporary identifier (5G-GUTI) to the UE in order to protect user confidentiality. The 5G network identifiers can be summarized as follows [3]:

- 5G Subscription Permanent Identifier is a global unique identifier that is assigned to each subscriber in the 5G system, which is provisioned in the UDM/UDR. The SUPI is used only within 3GPP system. The previous generations' IMSI[68] and network access identifier (NAI)[69] can still be used in 3GPP Rel-15 as SUPI. The use of generic NAI makes the use of non-IMSI-based SUPIs possible. The SUPI must contain the address of the home network in order to enable roaming scenarios. For interworking with the EPC, the SUPI allocated to the 3GPP UE is based on the IMSI. Furthermore, 5GS defines a subscription concealed identifier (SUCI) which is a privacy preserving identifier containing the concealed SUPI.
- Permanent Equipment Identifier is defined for a 3GPP UE accessing the 5G system. The PEI can assume different formats for different UE types and use cases. The UE presents the PEI to the network along with an indication of the PEI format being used. If the UE supports at least one 3GPP access technology, the UE must be allocated a PEI in the IMEI format. In 3GPP Rel-15, the only format supported for the PEI parameter is an IMEI.
- 5G Globally Unique Temporary Identifier is allocated by an AMF to the UE that is common in both 3GPP and non-3GPP access. A UE can use the same 5G-GUTI for

68 International Mobile Subscriber Identity (IMSI) is used as a unique identification of mobile subscriber in the 3GPP networks. The IMSI consists of three parts: (1) mobile country code consisting of three digits. The mobile country code uniquely identifies the country of residence of the mobile subscriber; (2) mobile network code consisting of two or three digits for 3GPP applications. The mobile network code identifies the home public land mobile network of the mobile subscriber. The length of the mobile network code (two or three digits) depends on the value of the mobile country code; and (3) mobile subscriber identification number identifying the mobile subscriber within a public land mobile network.

69 Network access identifier defined by IETF RFC 7542 is a common format for user identifiers submitted by a client during authentication. The purpose of the network access identifier is to allow a user to be associated with an account name, as well as to assist in the routing of the authentication request across multiple domains. Note that the network access identifier may not necessarily be the same as the user's email address or the user identifier submitted in an application-layer authentication.

accessing 3GPP access and non-3GPP access security context within the AMF. The AMF may assign a new 5G-GUTI to the UE at any time. The AMF may delay updating the UE with its new 5G-GUTI until the next NAS signaling exchange. The 5G-GUTI comprises a GUAMI and a 5G-TMSI, where GUAMI identifies the assigned AMF and 5G-TMSI identifies the UE uniquely within the AMF. The 5G-S-TMSI is the shortened form of the GUTI to enable more efficient radio signaling procedures.

- AMF Name is used to identify an AMF. It can be configured with one or more GUAMIs. At any given time, the GUAMI value is exclusively associated to one AMF name.
- Data Network Name (DNN) is equivalent to an APN which may be used to select an SMF and UPF(s) for a PDU session, to select N6 interface(s) for a PDU session, or to determine policies that are applied to a PDU session.
- Internal-Group Identifier is used to identify a group as the subscription data for a UE in UDM may associate the subscriber with different groups. A UE can belong to a limited number of groups. The group identifiers corresponding to a UE are provided by the UDM to the SMF and when PCC applies to a PDU session by the SMF to the PCF. The SMF may use this information to apply local policies and to store this information in charging data record.
- Generic Public Subscription Identifier (GPSI) is used for addressing a 3GPP subscription in different data networks outside of the 3GPP system. The 3GPP system stores within the subscription data the association between the GPSI and the corresponding SUPI. GPSIs are public identifiers used both inside and outside of the 3GPP system. The GPSI is either a mobile subscriber ISDN number (MSISDN) or an external identifier. If MSISDN is included in the subscription data, it will be possible that the same MSISDN value is supported in both 5GS and EPS. There is no one-to-one relationship between GPSI and SUPI.

1.2.2.3 User-Plane and Control-Plane Protocol Stacks

1.2.2.3.1 Control-Plane Protocol Stacks

The 5GC supports PDU connectivity service which provides exchange of PDUs between a UE and a data network. The PDU connectivity service is supported through PDU sessions that are established upon request from the UE. In order to establish a PDU session and access to the PDN, the UE must establish user plane and control plane over the NG-RAN and the 5GC network interfaces to the PDN. Connection management comprises establishing and releasing a signaling connection between a UE and the AMF over N1. This signaling connection is used to enable NAS signaling exchange between the UE and the core network, which includes both access network signaling connection between the UE and the access node (RRC connection over 3GPP access or UE-N3IWF connection over non-3GPP access) and the N2 connection for this UE between the access node and the AMF. A NAS connection over N1 is used to connect a UE to the AMF. This NAS connection is used for registration management and connection management functions as well as for transport of session management messages and procedures for the UE. The NAS protocol over N1 comprises NAS mobility and session management (NAS-MM and NAS-SM) components. There are several protocol information that need to be transported over N1 using NAS-MM protocol between a UE and a core NF besides the AMF (e.g., session management signaling). Note that in 5G systems, registration/connection management NAS messages and other types of NAS messages as well as the corresponding procedures are decoupled. The NAS-MM supports NAS procedures that terminate at the AMF such as handling registration and connection management state machines and procedures of the UE, including NAS transport. There is a single NAS protocol that applies to both 3GPP and non-3GPP access. When a UE is served by a single AMF while it is connected through multiple (3GPP and/or non-3GPP) access schemes, there would be one N1 NAS connection per access link. The security for the NAS messages is provided based on the security context established between the UE and the AMF. It is possible to transmit the other types of NAS messages (e.g., NAS SM) along with RM/CM NAS messages by supporting NAS transport of different types of payload or messages that do not terminate at the AMF. This includes information about the payload type, information for forwarding purposes, and the SM message in case of SM signaling.

The NAS-SM messages control the session management functions between the UE and the SMF. The session management message is created and processed in the NAS-SM layer of UE and the SMF (see Fig. 1.71). The content of the NAS-SM message is transparent to the AMF. The NAS-MM layer creates a NAS-MM message, including security header, indicating NAS transport of SM signaling, as well as additional information for the receiving NAS mobility management (NAS-MM) entity to determine how and where to forward the SM signaling message. The receiving NAS-MM layer performs integrity check and interpretation of NAS message content. Fig. 1.71 further depicts the NAS-MM layer, which is a NAS protocol for mobility management; support of registration management; connection

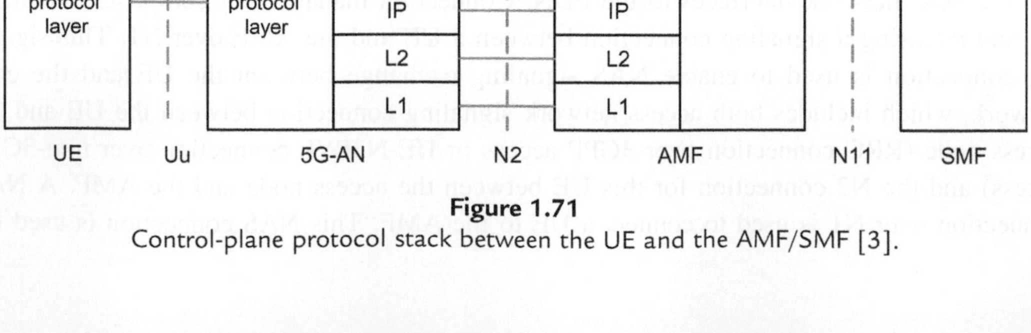

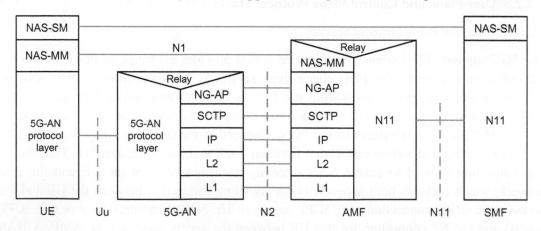

Figure 1.71
Control-plane protocol stack between the UE and the AMF/SMF [3].

management and user-plane connection activation and deactivation functions. It is also responsible for ciphering and integrity protection of NAS signaling.

The UE and AMF support NAS signaling connection setup function, which is used to establish a NAS signaling connection for a UE in CM-IDLE state. It will be explained in the next chapter that two connection management states are used to reflect the NAS signaling connectivity between the UE and the AMF, that is, CM-IDLE and CM-CONNECTED. The CM state for 3GPP access and non-3GPP access are independent of each other, that is, one can be in CM-IDLE state while the other is in CM-CONNECTED state [3].

A UE must register with the network to be authorized to use network services, to assist mobility tracking, and to be reachable. The registration procedure is used when the UE needs to perform initial registration with the 5GS; location update upon entering a new tracking area outside of the UE's registration area in CM-CONNECTED and CM-IDLE modes; when the UE performs a periodic registration update (due to a predefined inactivity time interval); and additionally when the UE needs to update its capabilities or protocol parameters that were negotiated during registration procedure. The AMF provides a list of recommended cells/TAs/NG-RAN node identifiers for paging.

The N2 interface supports management procedures which are not UE-specific, rather for configuration or reset of the N2 interface. These procedures are applicable to any access scheme and access-specific messages that carry some information for a particular access scheme such as information on the default paging DRX cycle that is used only for 3GPP access. The N2 interface further supports UE-specific and NAS-transport procedures. These procedures are in general access-agnostic, but they may also correspond to uplink NAS transport messages that carry some access-dependent information such as user location information. The N2 interface also supports procedures related to UE context management and resources for PDU sessions. These messages carry information on N3 addressing and QoS requirements that should be transparently forwarded by the AMF between the 5G access node and the SMF. The N2 interface further enables procedures related to handover management for 3GPP access.

The control-plane interface between a 5G access node and the 5G core supports connection of different types of 5G access nodes to the 5GC via a unique control-plane protocol. A single NGAP protocol is used for 3GPP and non-3GPP access schemes. There is a unique N2 termination point at the AMF for a given UE (for each access node used by the UE) regardless of the number of PDU sessions of the UE. The N2 control plane supports separation of AMF and other functions such as SMF that may need to control the services supported by 5G access nodes, where in this case, AMF transparently forwards NGAP messages between the 5G access node and the SMF. The N2 session management information (i.e., a subset of NGAP information that AMF transparently relays between an access node and SMF) is exchanged between the SMF and the 5G access node which is transparent to the AMF. The NG application protocol enables message exchange between a 5G access node and the

AMF over N2 interface using SCTP protocol, which guarantees delivery of signaling messages between AMF and 5G access node.

1.2.2.3.2 User-Plane Protocol Stacks

The protocol stack for the user-plane transport related to a PDU session is illustrated in Fig. 1.72. The PDU layer corresponds to the PDU that is transported between the UE and the PDN during a PDU session. The PDU session type can be IPv6 or Ethernet for transporting IP packets or Ethernet frames. The GPRS tunneling protocol for the user plane (GTP-U) supports multiplexing of the traffic from different PDU sessions by tunneling user data over N3 interface (i.e., between the 5G access node and the UPF) in the core network. GTP encapsulates all end-user PDUs and provides encapsulation per-PDU-session. This layer also transports the marking associated with a QoS flow [3].

The 5G encapsulation layer supports multiplexing the traffic from different PDU sessions over N9 interface (i.e., an interface between different UPFs). It provides encapsulation per PDU session and carries the marking associated with the QoS flows. The 5G access node protocol stack is a set of protocols/layers which are related to the access network as described in the previous sections. The number of UPF entities in the data path is not constrained by the 3GPP specifications; therefore, there could be none or more than one UPF entities in the data path of a PDU session that may not support PDU session anchor functionality for that PDU session. In certain cases, there is an uplink classifier or a branching point in the data path of a PDU session, which does not act as the non-PDU session anchor UPF. In that case, there are multiple N9 interfaces branching out of the uplink classifier/branching point, each leading to different PDU session anchors.

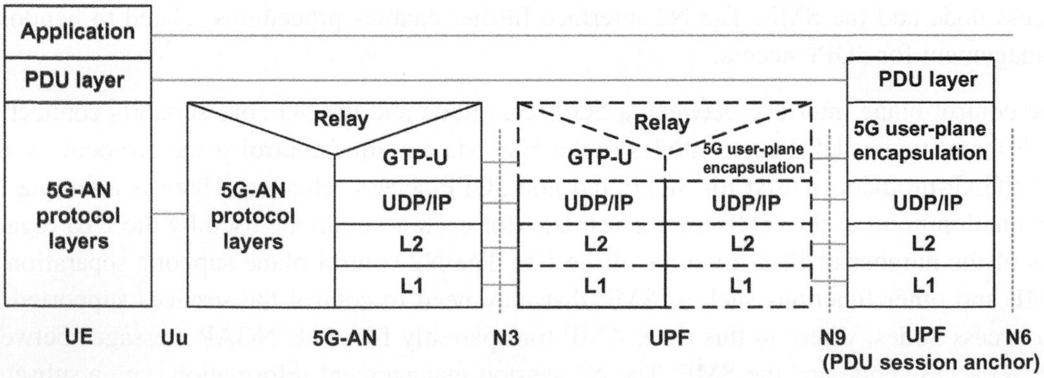

Figure 1.72
User-plane protocol stack between UE and UPF [3].

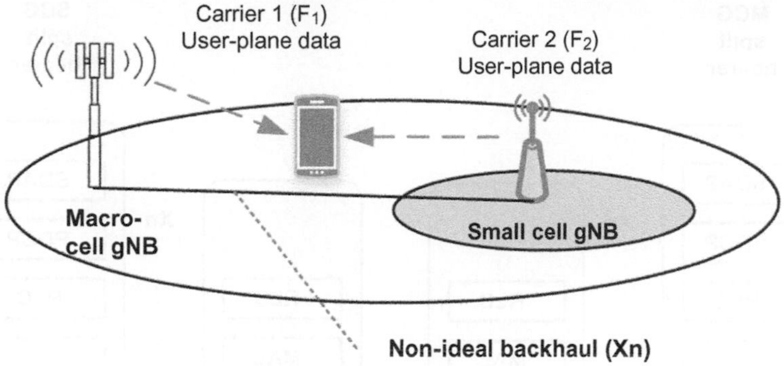

Figure 1.73
Illustration of inter-node radio resource aggregation (dual connectivity concept).

1.3 Dual Connectivity and Multi-connectivity Schemes

Dual connectivity (or multi-connectivity) is a term that is used to refer to an operation where a given UE is allocated radio resources provided by at least two different network nodes connected with non-ideal backhaul (see Fig. 1.73). Each access node involved in dual connectivity for a UE may assume different roles. Those roles do not necessarily depend on the access node's power class and can vary for different UEs. To support tight interworking between LTE and NR, where both LTE eNB and NR gNB can act as a master node. It is assumed that the dual connectivity between LTE and NR supports the deployment scenario where LTE eNB is not required to be synchronized with NR gNB.

3GPP NR supports dual connectivity operation in which a UE in the connected mode is configured to utilize radio resources provided by two distinct schedulers, located in two gNBs connected via a non-ideal backhaul as shown in Fig. 1.73. The gNBs involved in the dual connectivity for a certain UE may assume two different roles, that is, a gNB may either act as a master gNB (MgNB) or as a secondary gNB (SgNB). Under this condition, a UE is connected to one MgNB and one SgNB. Under dual connectivity framework, the radio protocol stack that a radio bearer uses depends on how the radio bearer is setup. There are four bearer types in dual connectivity framework, namely MCG[70] bearer, MCG split bearer, SCG[71] bearer, and SCG split bearer as depicted in Fig. 1.74.

Under dual connectivity framework, the UE is configured with two MAC entities, one for the MCG and one for the SCG. For a split bearer,[72] the UE is configured over one of the

[70] Master cell group in a multi-RAT dual connectivity scheme refers to a group of serving cells associated with the master node, comprising the primary cell and optionally one or more secondary cells.

[71] Secondary cell group in a multi-RAT dual connectivity scheme refers to a group of serving cells associated with the secondary node, comprising primary cell and optionally one or more secondary cells.

[72] Split bearer in multi-RAT dual connectivity is a bearer whose radio protocols are split either at the master node or at the secondary node and belongs to both secondary cell group and master cell group.

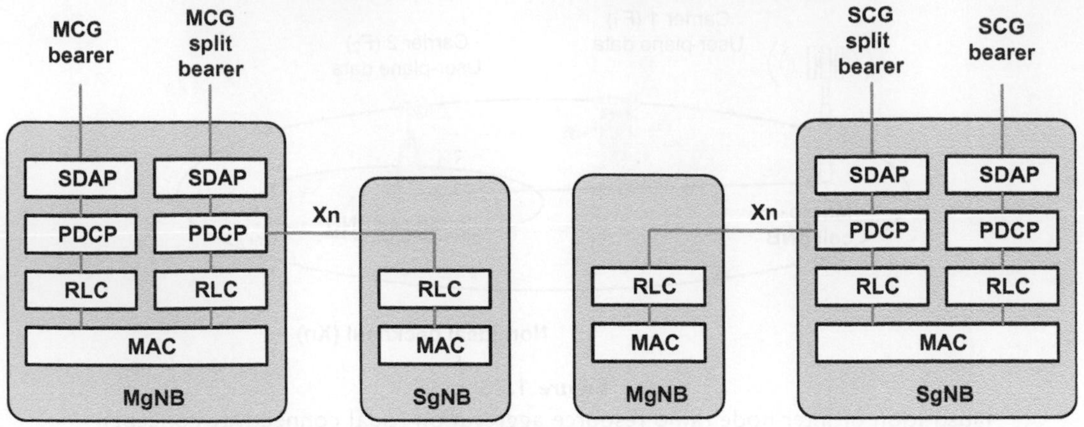

Figure 1.74
MgNB and SgNB bearers for dual connectivity [16].

links (or both) that it transmits uplink PDCP PDUs. The RLC sublayer only transmits corresponding feedback for the downlink data over the link which is not used for transmitting PDCP PDUs.

Multi-RAT DC (MR-DC) is a generalization of the intra-LTE dual connectivity, where a UE with multiple RF transceivers may be configured to utilize radio resources provided by two distinct schedulers located in two different nodes and connected via non-ideal backhaul, one providing LTE access and the other one providing NR connectivity. One scheduler is located in the master node (MN)[73] and the other in the secondary node (SN).[74] The MN and SN entities are connected via a network interface and the MN is typically connected to the core network. LTE supports MR-DC via E-UTRA-NR dual connectivity (EN-DC), in which a UE is connected to one eNB that acts as the MN and one gNB that acts as the SN. The eNB is connected to the EPC and the gNB is connected to the eNB via the X2 interface (i.e., the logical interface between LTE eNBs).

In MR-DC scenarios, the UE has a single RRC state, based on the MN RRC state and a single control-plane connection toward the core network. Fig. 1.75 shows the control-plane architecture for MR-DC. Each radio node has its own RRC entity, which can generate RRC PDUs to be sent to the UE. The RRC PDUs generated by the SN can be transported via the MN to the UE. The MN always sends the initial SN RRC configuration via MCG SRB, for example, SRB1, but subsequent reconfigurations may be sent via the MN or the SN entities. When transporting RRC PDU from the SN, the MN does not modify the UE configuration

[73] Master node in a multi-RAT dual connectivity architecture is either a master eNB or a master gNB.
[74] Secondary node in multi-RAT dual connectivity architecture is either a secondary eNB or a secondary gNB.

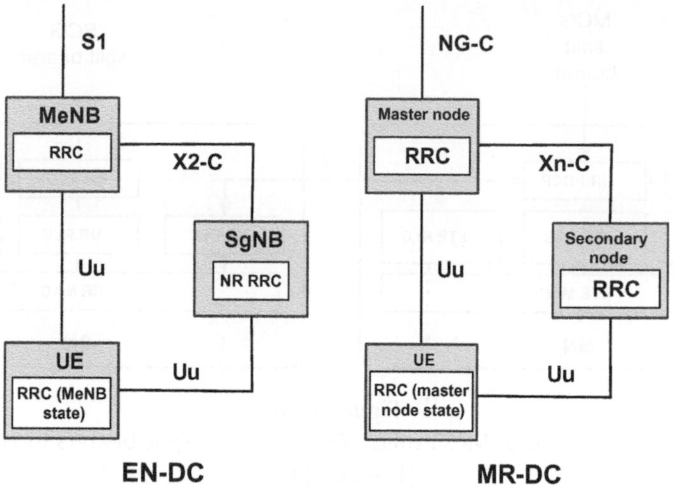

Figure 1.75
Control-plane architecture for EN-DC and MR-DC with 5GC [13].

provided by the SN. In EN-DC and NG-RAN E-UTRA-NR dual connectivity (NGEN-DC[75]) scenarios, during initial connection establishment SRB1 uses LTE PDCP; however, after initial connection establishment MCG SRB (SRB1 and SRB2) can be configured by the network to use either LTE PDCP or NR PDCP. The PDCP version change (release of old PDCP and establishment of new PDCP) of SRBs can be supported via a handover procedure (reconfiguration with mobility) or through a reconfiguration without mobility, when the network is aware that there is no uplink data in the UE buffer. For EN-DC capable UEs, NR PDCP can be configured for DRBs and SRBs before EN-DC is configured. If the SN is a gNB (i.e., for EN-DC and NGEN-DC), the UE can be configured to establish an SRB with the SN (e.g., SRB3[76]) to enable RRC PDUs for the SN to be sent directly between the UE and the SN. The RRC PDUs for the SN can only be sent directly to the UE for SN RRC reconfiguration without any coordination with the MN. Measurement reporting for mobility within the SN can be conducted directly from the UE to the SN, if SRB3 is configured. The MCG split SRB is supported for all MR-DC cases, allowing duplication of RRC PDUs generated by the MN, via the direct path and through the SN. The MCG split SRB uses NR PDCP. The SCG split SRB is not currently supported in 3GPP specifications [13].

---

[75] NG-RAN supports NGEN-DC, in which a UE is connected to one ng-eNB that acts as a master node and one gNB that acts as a secondary node. The ng-eNB is connected to the 5GC and the gNB is connected to the ng-eNB via the Xn interface.

[76] SRB3 in EN-DC and NGEN-DC represents a direct signaling radio bearer between the secondary node and the user equipment.

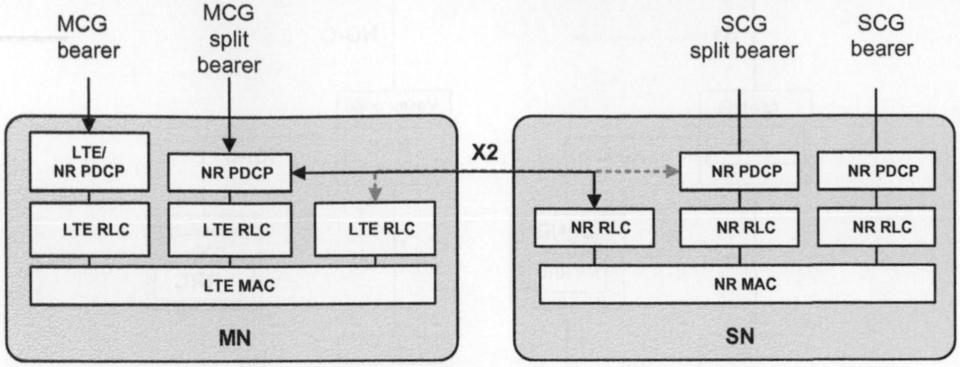

Figure 1.76
Radio protocol stack for MCG, MCG split, SCG, and SCG split bearers in MR-DC with EPC (EN-DC) [13].

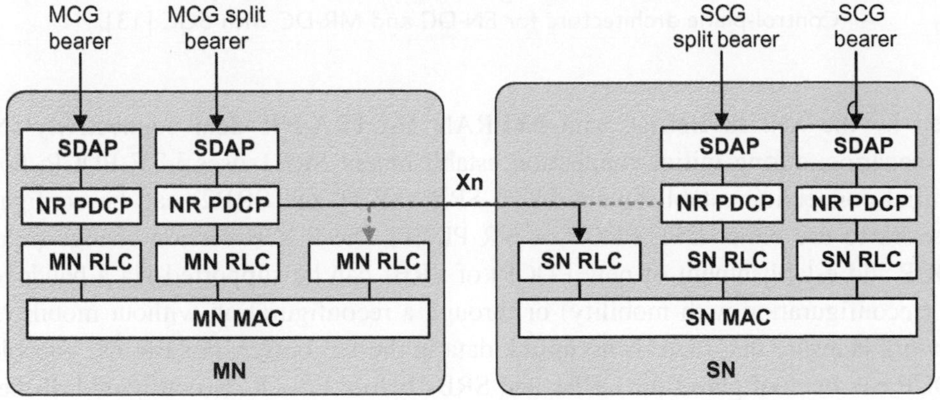

Figure 1.77
Radio protocol stack for MGC, MCG Split, SCG, and SCG split bearers in MR-DC with 5GC (NGEN-DC, NE-DC) [13].

As we mentioned earlier, there are four bearer types identified as MCG bearer, MCG split bearer, SCG bearer, and SCG split bearer in MR-DC scenarios. These four bearer types are depicted in Fig. 1.76 for MR-DC with EPC (EN-DC) and in Fig. 1.77 for MR-DC with 5GC (NGEN-DC, NE-DC). For EN-DC, the network can configure either LTE PDCP or NR PDCP for MCG bearers while NR PDCP is always used for SCG bearers. For split bearers, NR PDCP is always used and from the UE perspective there is no difference between MCG and SCG split bearers. In MR-DC with 5GC, NR PDCP is always used for all bearer types.

From system architecture point of view, in MR-DC, there is an interface between the MN and the SN entities to facilitate control-plane signaling and coordination. For each MR-DC

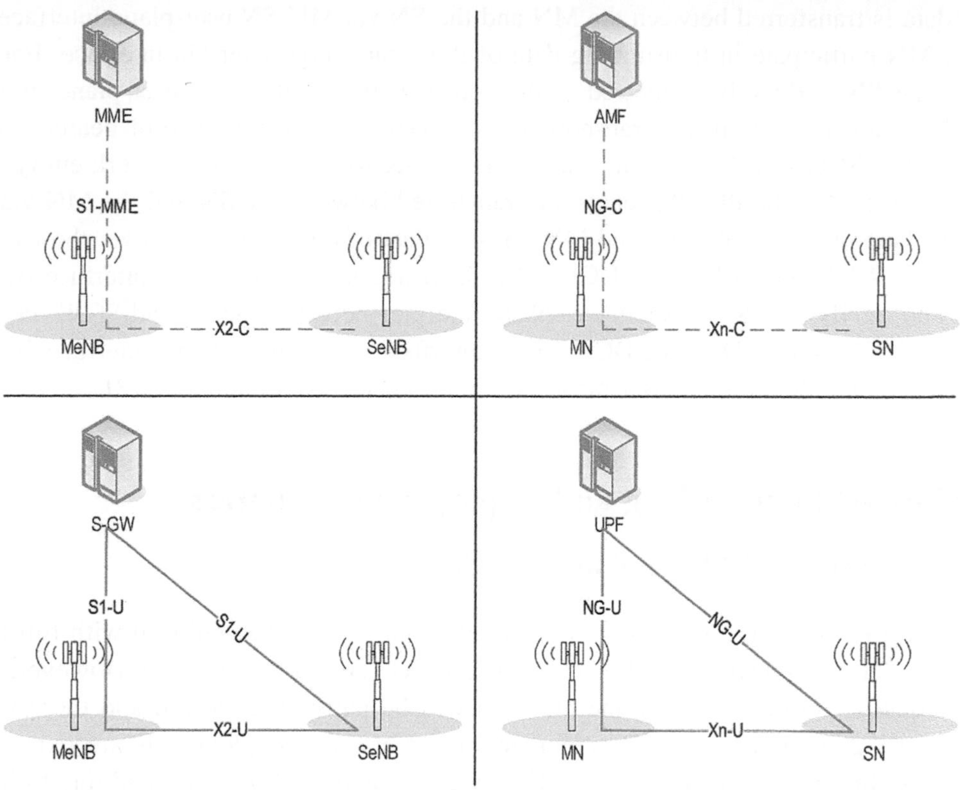

Figure 1.78
Control-plane and user-plane architecture for EN-DC and MR-DC with 5GC [13].

capable UE, there is also one control-plane connection between the MN and a corresponding core network entity. The MN and the SN entities involved in MR-DC operation for a UE control their own radio resources and are primarily responsible for allocating radio resources of their cells. Fig. 1.78 shows control-plane connectivity of an MN and SN involved in MR-DC with a UE. In MR-DC with EPC (EN-DC) scenario, the involved core network entity is the MME. S1-MME is terminated in MeNB and the MeNB and the SgNB are interconnected via X2-C. In MR-DC with 5GC (NGEN-DC, NE-DC) scenario, the terminating core network entity is the AMF. The NG-C interface is terminated at the MN and the MN and the SN are interconnected via Xn-C interface [13].

There are different user-plane connectivity options for the MN and SN involved in MR-DC operation with a certain UE, as shown in Fig. 1.78. The user-plane connectivity depends on the configured bearer type. For MCG bearers, the user-plane connection to the core network entity is terminated at the MN. The SN is not involved in the transport of user-plane data for this type of bearer over the Uu interface (i.e., the radio interface). For MCG split bearers, the user-plane connection to the core network entity is terminated in the MN.

PDCP data is transferred between the MN and the SN via MN-SN user-plane interface. The SN and MN participate in transmitting data of this bearer type over Uu interface. For SCG bearers, the SN is directly connected to the core network entity via a user-plane interface. The MN is not involved in the transport of user-plane data for this type of bearer over Uu interface. For SCG split bearers, the user-plane connection to the core network entity is terminated at the SN. The PDCP packets are transferred between the SN and the MN via MN/SN user-plane interface. The SN and MN transmit data of this bearer type over the Uu interface. For MR-DC with EPC (EN-DC), X2-U interface is the user-plane interface between MeNB and SgNB, and S1-U is the user-plane interface between MeNB and SGW. For MR-DC with 5GC (NGEN-DC, NE-DC), Xn-U interface is the user-plane interface between MN and SN, and NG-U is the user-plane interface between MN and UPF [13].

1.4 LTE-NR Interworking and Deployment Scenarios

1.4.1 RAN-Level and CN-Level Interworking

To provide full 5G services to the users, 5G cells will have to be deployed with full coverage and all UEs will have to be able to connect to 5G network everywhere. However, in the early phases of 5G deployments, the 5G cells will be partially deployed and there will be 5G coverage gaps. Large-scale 5G commercial services and deployments are expected in 2020 + and the initial investments for 5G service are expected to be limited due to lack of 5G user equipment. As a result, the 5G networks need to be able to interwork with the existing LTE networks. The interworking solution can provide seamless service to the users. In this section, we discuss the solutions for LTE–NR interworking and we will compare the solutions in terms of performance, features and ease of migration to a full 5G network. If 5G cells are not deployed with full coverage, a seamless service can be provided to the users by interworking with the existing LTE networks, which are already deployed with full coverage. For LTE–NR interworking, two types of solutions namely access-network-level interworking and core-network-level interworking, have been studied in 3GPP.

In RAN-level interworking solutions, the interworking service between LTE and 5G is made possible by using a direct interface between LTE eNB and NR gNB. The RRC messages are transmitted over the LTE radio interface, thus the connection and the mobility of UE are controlled by the LTE eNB. The user traffic is simultaneously transmitted through LTE eNB and NR gNB either by PDCP aggregation or using NR gNB split bearer. Although the RRC messages can be processed by the LTE eNBs that provide larger coverage than NR gNBs, LTE radio interface always remains connected, even though user traffic is transmitted over the NR. RAN-level interworking is necessary in non-stand-alone architecture, where the NR cannot be used without overlaid LTE network. Note that when the LTE EPC is used, only EPC-based services can be provided, even though 5G radio technology is used. Two different core networks can be used for RAN-level interworking, as shown in Fig. 1.79. LTE and NR interworking can be achieved by upgrading some LTE eNBs connected to NR gNBs and by increasing the gateway capacity in EPC. The new 5G core network, 5GC, has been designed to support RAN-level interworking. In this solution, the new 5GC network slicing feature can be used to separate 5G services from the LTE services. In this case, all LTE eNBs will have to be upgraded to ng-eNBs so that they can be connected to 5GC [16].

In core-network-level interworking, a direct interface between the LTE eNB and the NR gNB is not required and the EPC SGW is connected to the 5GC UPF. The UE manages LTE and NR interface connection independently, and can be connected to a single network, either LTE or 5G. When the UE is located in 5G coverage, it can only connect to the 5G network and receive 5G service. When the UE moves out of 5G coverage, it releases NR interface connection and establishes LTE radio interface connection. Although the network to which the UE connects changes, the IP address assigned to the UE stays the same and seamless service can be provided to the user. The core-network-level interworking is necessary in stand-alone architecture models, where NR can be used without relying on LTE network. In this case, single registration or dual registration is possible, as shown in Fig. 1.79. With the single registration, the UE registers with either LTE EPC or 5G networks at any time, and the UE context can be transferred via the control interface between the EPC MME and 5GC AMF when the UE's network association changes. In order to support the single registration, the MME will have to be upgraded in order to support the MME-AMF interface and the SGW needs to be connected to UPF in 5GC. LTE eNB must also be upgraded to support the mobility between the LTE and the 5G networks. N26 interface is an inter-CN interface between the MME and AMF in order to enable interworking between EPC and the 5G core. Support of N26 interface in the network is optional for interworking. N26 supports a subset of the functionalities that are essential for interworking. Networks that support interworking with EPC may support interworking procedures that use the N26 interface or interworking procedures that do not use the N26 interface. Interworking procedures with N26 support provide IP address continuity during inter-system mobility to UEs that support 5GC NAS and EPC NAS. Networks that support interworking procedures without N26 must support procedures to provide IP address continuity during inter-system mobility to the UEs that operate in both single-registration and dual-registration modes [3]. Interworking procedures using N26 interface enable the exchange of MM and SM states between the source and target networks. Handover procedures are supported through N26 interface. When interworking procedures with N26 is used, the UE operates in single-registration mode. The network retains only one valid MM state for the UE, either in the AMF or MME. Either the AMF or the MME is registered in the HSS + UDM. The support for N26 interface between AMF in 5GC and MME in EPC is required to enable seamless session continuity (e.g., for voice services) for inter-system handover. When the UE moves

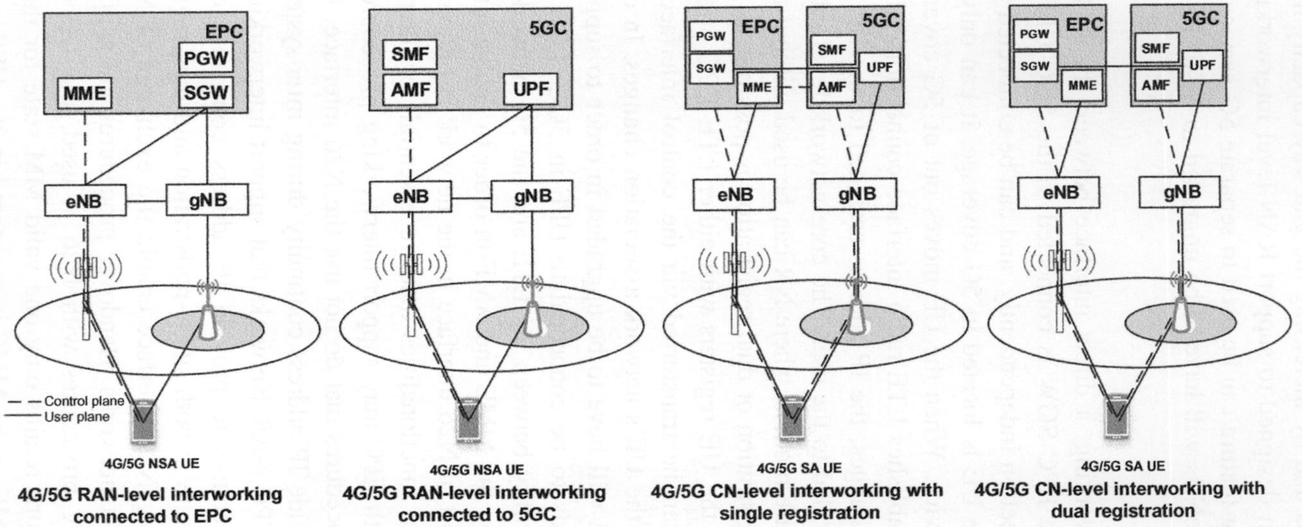

Figure 1.79
RAN-level and core-network-level interworking models [82].

from 5GC to EPC, the SMF determines which PDU sessions can be relocated to the target EPS. The SMF can release the PDU sessions that cannot be transferred as part of the handover. However, the target EPS determines if the PDU session can be successfully moved to the target network.

The dual registration approach requires the UE to register separately with the EPC and the 5GC. Thus it does not need to forward the UE context between MME and AMF, and the interface between MME and AMF is not required. The handover between LTE and 5G systems is decided by the UE. The UE performs normal access procedures after moving to the other network. Therefore, the solution can be supported by LTE eNBs with minimal changes. Furthermore, the impact on EPC to support the dual registration is minimized. However, in order to improve the mobility performance between LTE and 5G, it is necessary to temporarily support dual radio transmission when moving to the other network, although the UE can support dual registration solution even with single radio transmission capability. Deployments based on different 3GPP architecture options (i.e., EPC based or 5GC based) and UEs with different capabilities (EPC NAS and 5GC NAS) may coexist at the same time within one PLMN.

In order to interwork with EPC, the UE that supports both 5GC and EPC NAS can operate in single- or dual-registration mode. In single-registration mode, the UE has only one active MM state (either RM state in 5GC or EMM state in EPC) and it is either in 5GC NAS mode or in EPC NAS mode (when connected to 5GC or EPC, respectively). The UE maintains a single coordinated registration for 5GC and EPC. Accordingly, the UE maps the EPS-GUTI to 5G-GUTI during mobility between EPC and 5GC and vice versa. In dual-registration mode, the UE can handle independent registrations for 5GC and EPC. In this mode, UE maintains 5G-GUTI and EPS-GUTI independently. In this mode, the UE provides native 5G-GUTI, if previously allocated by 5GC, for registration with 5GC and it provides native EPS-GUTI, if previously allocated by EPC, for Attach/TAU with EPC. In this mode, the UE may be registered to 5GC only, EPC only, or to both 5GC and EPC. The support of single registration mode is mandatory for UEs that support both 5GC and EPC NAS. During LTE initial attach procedure, the UE supporting both 5GC and EPC NAS indicates its support of 5G NAS in UE network capability [3].

The Ethernet and unstructured PDU session types are transferred to EPC as non-IP PDN type (when supported by UE and network). The UE sets the PDN type to non-IP when it moves from 5GS to EPS. After the transfer to EPS, the UE and the SMF maintain information about the PDU session type used in 5GS, that is, the information indicating that the PDN connection with non-IP PDN type corresponds to PDU session type of Ethernet or unstructured, respectively. This is to ensure that the appropriate PDU session type will be used if the UE transfers back to the 5GS.

1.4.2 5G Deployments Scenarios and Architecture Options

This section describes 5G network architecture deployment options supporting stand-alone and non-stand-alone mode of operation. These deployment options were extensively discussed in 3GPP and have been prioritized based on their viability and practicality. Each option provides NR access to sufficiently capable UEs either through direct access to gNB/5GC or indirectly via LTE interworking. 3GPP defined both stand-alone and non-stand-alone deployment configurations for NR in 3GPP Rel-15. A stand-alone NR deployment would not require an associated LTE network. The NR-capable UE could use random access to directly establish a radio link with a gNB and attach to the 5GC in order to use network services. The stand-alone NR deployment required a complete set of specifications from 3GPP for all entities and interfaces in the network, which was subsequently defined in 3GPP Rel-15 specifications.

In stand-alone operation (shown as option 2 in Fig. 1.80), the network access procedures closely follow the LTE counterparts. The additional requirements mainly include broadcast of NR system information, which includes a minimum set of parameters and extended set of parameters where the former is periodically broadcast and comprises basic information required for initial access and the scheduling information for other system information; and the latter encompasses other system information that are not transmitted via the broadcast channel, which may either be broadcast or provisioned in a dedicated manner which can be triggered by the network configuration or upon request from the UE. In comparison to LTE system information broadcasting scheme, on-demand broadcasting is a new mechanism introduced in NR to deliver other system information by UE request. For UEs in RRC_CONNECTED state, dedicated RRC signaling is used for the request and delivery of the other system information. For UEs in RRC_IDLE and RRC_INACTIVE states, making the request will trigger a random access procedure [16].

As an interim step for NR deployments, 3GPP has defined a set of non-stand-alone deployment configurations using dual connectivity between NR gNB and LTE eNB (or ng-eNBs). Since initial NR networks will not have full coverage, dual connectivity can be used to combine the coverage advantage of the existing LTE networks with the throughput and latency advantages of the NR. However, it requires more complex UE implementations to allow simultaneous connections with both LTE and NR networks, potentially increasing the cost of the UEs. This will require more complex UE radio capabilities including the ability to simultaneously receive downlink transmissions from NR and LTE base stations on separate frequency bands. The non-stand-alone NR deployments use architectures where NR gNBs are associated with LTE eNBs and do not require separate signaling connections to the 5GC. These architectures are classified based on the control-plane and user-plane connections used between eNB, gNB, EPC, and 5GC.

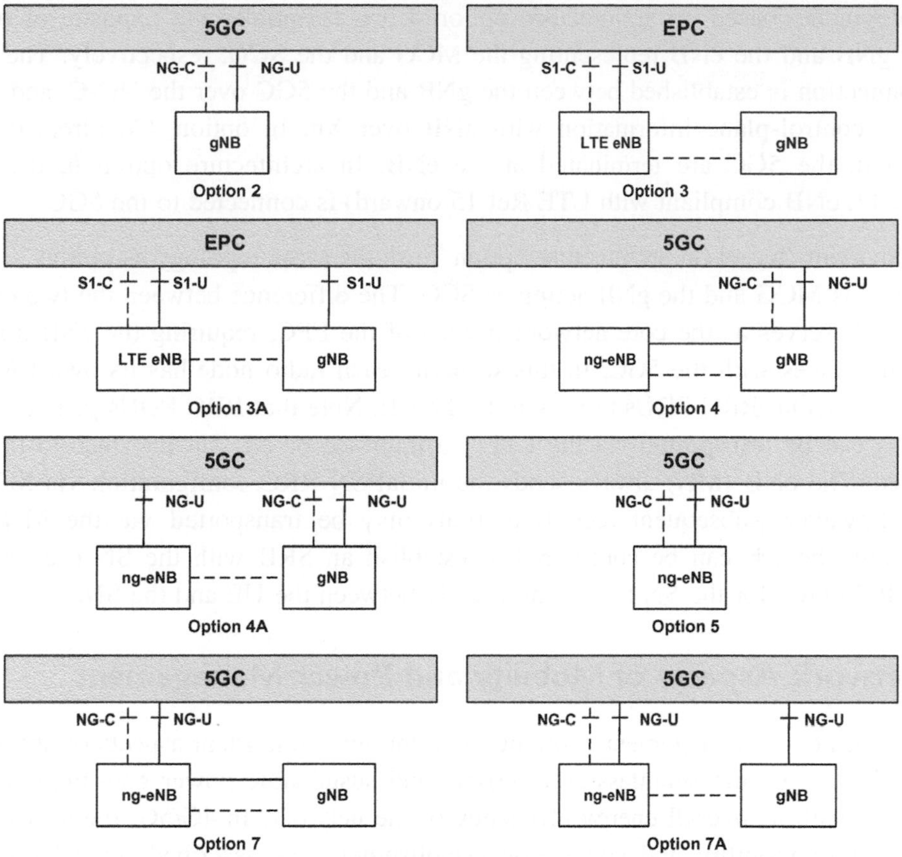

Figure 1.80
NR deployment options [34].

The deployments based on architecture option 3 use EPC as the core network as shown in Fig. 1.80. In this case, S1-C control-plane interface for the UE is established between the LTE eNB and the EPC. The gNB acts as a secondary node connected to the master node represented by the eNB. Control-plane information is exchanged between the LTE eNB and the NR gNB, and no direct control-plane interface exists between the gNB and the EPC. User-plane bearers are supported between eNB and EPC over S1-U. In option 3A, the gNB also terminates user-plane bearers with the EPC. In option 3A, those gNB terminated S1-U bearers may further split and carried over the X2/Xn interface to the eNB and over the LTE air interface. The deployments based on architecture option 3 do not require interface with the 5GC, and allow service over the NR air interface with Uu (i.e., the interface between the UE and the serving gNB) and X2/Xn interfaces defined. As such, this is seen as the most likely architectural scenario for early NR deployments. From control-plane perspective, there is only one RRC state in the UE, which is based on the LTE RRC protocols and there is only one control-plane connection toward the core network.

The deployments based on architecture option 4 are essentially the opposite of option 3, with the gNB and the eNB representing the MCG and the SCG, respectively. The control-plane connection is established between the gNB and the 5GC over the NG-C, and the eNB exchanges control-plane information with gNB over Xn. In option 4A, direct user-plane bearers with the 5GC are terminated at the eNB. In architecture option 5, the ng-eNB (i.e., an LTE eNB compliant with LTE Rel-15 onward) is connected to the 5GC.

The deployments based on architecture option 7 use the same topology as option 3, with the eNB acting as MCG and the gNB acting as SCG. The difference between the two options is that the 5GC serves as the core network instead of the EPC, requiring the eNB upgrade to ng-eNB interfaces with the 5GC. In this scenario, each radio node has its own RRC entity which can generate RRC PDUs to be sent to the UE. Note that RRC PDUs generated by the gNB (SN) can be transported via the LTE Uu interface or NR Uu interface to the UE, if configured. The eNB (MN) always sends the initial SN RRC configuration via MCG SRB (SRB1); however, subsequent reconfigurations may be transported via the MN or SN. Furthermore, the UE can be configured to establish an SRB with the SN (i.e., SRB3) to enable RRC PDUs for the SN to be sent directly between the UE and the SN.

1.5 Network Aspects of Mobility and Power Management

Mobility and power management continue to be the most important aspects of any new cellular standard to ensure seamless connectivity and sustainable power consumption of user terminals as well as overall energy efficiency of the network. In 4G/5G, the states of a UE with respect to mobility and connection establishment are described by NAS and RRC states. Fig. 1.81 shows and compares the EPS and 5GS/NG-RAN NAS and RRC states. There are three states shown in the EPS model, that is, EPS mobility management (EMM),

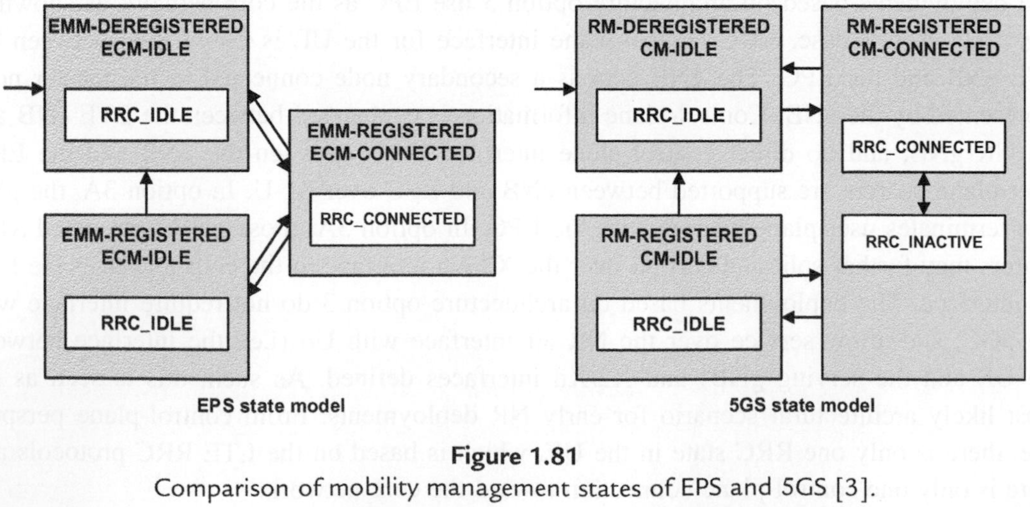

Figure 1.81
Comparison of mobility management states of EPS and 5GS [3].

EPS connection management (ECM), and RRC. The EMM and ECM states are managed by the core network, where the EMM state represents whether a UE is registered in the EPC and the ECM state indicates if NAS signaling connection between the UE and the MME is established. On the other hand, the RRC state is managed by E-UTRAN, and it represents whether there is a connection between the UE and the serving eNB. A UE in the ECM-CONNECTED state needs to be in the RRC_CONNECTED state, because a radio link connection is required in order to establish NAS signaling [3,16].

1.5.1 Mobility Management

In 5GS, the mobility management state of a UE can be either RM-REGISTERED or RM-DEREGISTERED depending on whether the UE is registered in 5GC, which is very similar to EMM-REGISTERED and EMM-DEREGISTERED states in EPC. Two connection management states are used to reflect the NAS signaling connectivity of the UE with the AMF namely CM-IDLE and CM-CONNECTED. A UE in CM-IDLE state has no NAS signaling connection established with the AMF over N1. A UE in CM-CONNECTED state has a NAS signaling connection with the AMF over N1. A NAS signaling connection uses an RRC connection between the UE and the NG-RAN and an NGAP UE association between the access network and the AMF for 3GPP access. NR RRC protocol states consist of three states, where in addition to RRC_IDLE and RRC_CONNECTED states, a third state has been introduced, RRC_INACTIVE, as an intermediate state prior to transition to RRC_IDLE state in order to save UE power and to allow fast connection setup [3,16].

As shown in Fig. 1.81, in EPS, when a UE is in the RRC_CONNECTED state, the serving eNB evaluates the received signal strength measurements from the UE and initiates a handover procedure when the UE's received signal strength goes below a threshold. However, in the RRC_IDLE state, where the eNB is not aware of the UE's location, the UE decides whether to camp on the current cell or to reselect a neighboring target cell based on received signal strength measurements. This procedure is referred to as cell reselection. In EPS, the mobility procedures, that is, as handover in the RRC_CONNECTED state and cell reselection in the RRC_IDLE state, are not flexible, whereas in 5GS, the core network is able to flexibly control whether to perform a handover or cell reselection for a UE in CM-CONNECTED state [3,16].

In EPC, the location of a UE is tracked by the MME. The granularity of location tracking is different depending on the RRC state of the UE. In the RRC_CONNECTED state, the UE's location is tracked at the cell level, whereas in the RRC_IDLE state, its location is tracked at the tracking area level, which is a set of cells belonging to a paging group that simultaneously transmit paging messages. Similarly, 5GC can track the location of a UE at the tracking area level in the CM-IDLE state, whereas the UE's location is known at the level of the serving cell to the core network in the CM-CONNECTED state. In 5GC, when a UE registers with the network over the 3GPP access, the AMF allocates a set of tracking areas in TAI list to the UE. The AMF takes into account various information (e.g., mobility pattern and allowed/non-allowed area, etc.), when it allocates registration area to the UE, that is, the set of tracking areas in TAI list. In 5GS; however, NG-RAN also supports the location tracking for the UEs in RRC_INACTIVE state. In that state, the core network knows that the UE is somewhere within the NG-RAN, but the NG-RAN node needs a new location tracking functionality to determine the exact location of the UE because the connection between the UE and the NG-RAN node is not active.

In EPS, when downlink traffic for a UE in the RRC_IDLE state arrives at the SGW, the MME performs a paging procedure based on the detected location of the UE. However, 5GS supports two types of paging, namely core-network-initiated paging and access-network-initiated paging. The UE in RRC_IDLE and RRC_INACTIVE states may use DRX in order to reduce power consumption. While in RRC_IDLE, the UE monitors 5GC-initiated paging, in RRC_INACTIVE, the UE is reachable via RAN-initiated paging and 5GC-initiated paging. The RAN and 5GC paging occasions overlap and the same paging mechanism is used. In core-network-initiated paging, the default paging procedure is requested by the core network when the UE is in the CM-IDLE state. The newly introduced RAN-initiated paging mode is used for the UEs in the RRC_INACTIVE state. Since a UE in the RRC_INACTIVE state is in the CM-CONNECTED state (see Fig. 1.81), the core network simply forwards the data or the signaling messages to the corresponding RAN when data or signaling messages arrive for the UE. Therefore, RAN itself generates the paging message and performs paging to find the updated location of the UE, and then sends the data or signaling messages to the UE. The 5GC can transmit additional assistance information for RAN paging [16].

Service area restriction also known as mobility-on-demand, which defines areas where the UE may or may not initiate communication with the network, is a concept to selectively support mobility of devices on a need basis. It includes supporting UE's mobility at a certain level classified as mobility restriction and mobility pattern (or mobility level). The former addresses mobility restriction in terms of allowed, non-allowed, and forbidden areas. The minimum granularity of the area is at tracking area level. In the allowed area, UE can communicate through the control or user planes. The UE cannot send service request and session management signaling in the non-allowed area. However, periodic registration update is possible. It can also respond to the paging messages from the core network. Furthermore, emergency calls or multimedia priority services are allowed. In the forbidden area, UE is not allowed to have any communication with the network except for the emergency services. The mobility pattern is used as a concept to describe the expected mobility of UE in 5GC. Mobility pattern may be used by the AMF to characterize and optimize the UE mobility. The AMF determines and updates mobility pattern of the UE based on subscription of the UE, statistics of the UE mobility, network local policy, and the UE-assisted information, or any combination of these parameters. The statistics of the UE mobility can be history-based or expected UE moving trajectory. The UE mobility pattern can be used by the AMF to optimize mobility support provided to the UE [16]. This procedure is used for the case where the UE moves from one gNB-DU to another gNB-DU within the same gNB-CU during NR operation.

The Internet of things is an important 5G service category. IoT devices mostly send mobile-originated data. For this type of devices, the mobile-originated only mode is defined in 5GS where the core network determines whether to apply the mobile-originated only mode to a UE during the registration procedure based on the UE subscription data and the network policy. The mobile-originated only mode is allocated to a UE, which does not require mobile-terminated traffic. Therefore, the UE in mobile-originated-only mode does not listen to the paging messages. The core network does not need to manage the UE's location while it is registered in the 5GC. For optimization, the core network may decide to deregister the UE after the mobile-originated data communication is finished, without transferring the UE's state into the CM-IDLE state in the RM-REGISTERED state, because most functions supported in the CM-IDLE state are not relevant to the UE in mobile-originated only mode, for example, UE location tracking and reachability management. In such cases, the UE needs to perform attach procedure whenever the mobile-originated data transmission is necessary to communicate with the core network.

A UE receives services through a PDU session, which is a logical connection between the UE and the data network. In 5GS, various PDU session types are supported, for example, IPv4, IPv6, Ethernet, etc. Unlike EPS, where at least one default session (i.e., default EPS bearer) is always created while the UE attaches to the network, 5GS can establish a session when service is needed irrespective of the attachment procedure of UE, that is, attachment without any PDU session is possible. 5GS also supports UE establishing multiple PDU sessions to the same data network or to different data networks over a single or multiple access networks including 3GPP and non-3GPP access. The number of UPFs for a PDU session is not specified. The deployment with at least one UPF is essential to serve a given PDU session. For a UE with multiple PDU sessions, there is no need for a convergence point like SGW in the EPC. In other words, the user-plane paths of different PDU sessions are completely disjoint. This implies that there is a distinct buffering node per PDU session for the UE in the RRC_IDLE state. In order to ensure slice-aware mobility management when network slicing is supported, a slice ID is introduced as part of the PDU session information that is transferred during mobility signaling. This enables slice-aware admission and congestion control [34].

LTE–NR interworking is important for stand-alone mode of operation between LTE and NR unlike for dual connectivity where there is simultaneous transmission across both RATs most of the time. Interworking between LTE and NR is not expected to be significantly

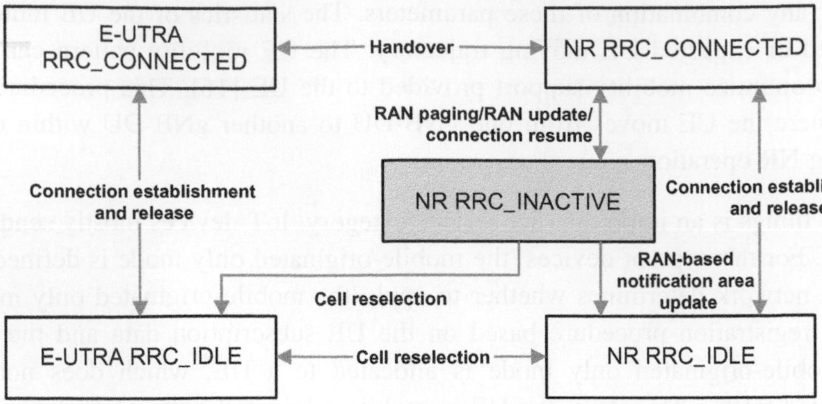

Figure 1.82
LTE–NR mobility state diagram [16,35].

different from what is defined in LTE specifications for interworking with other 3G networks. The inter-RAT mobility is expected to be supported both in idle mode as well as in the connected mode. Fig. 1.82 illustrates possible mobility scenarios across LTE and NR.

RRC_INACTIVE is a new RRC state in NR, in addition to RRC_IDLE and RRC_CONNECTED. It is a state where a UE remains in CM-CONNECTED and is able to move within an area configured by NG-RAN (i.e., RAN-based notification area or RNA) without notifying NG-RAN. The RNA can cover a single cell or multiple cells. In RRC_INACTIVE, the last serving NG-RAN node retains the UE context and the UE-associated NG connection with the serving AMF and UPF. The UE notifies the network via RAN-based notification area update procedure, if it has moved out of the configured RNA. If the last serving gNB receives downlink data from the UPF or downlink signaling from the AMF while the UE is in RRC_INACTIVE state, it pages the UE in the cells corresponding to the RNA and may forward the paging message to the neighbor gNB(s), if the RNA includes cells of neighbor gNB(s) [16]. Connected mode mobility is enabled using Xn interface between LTE eNB and NR gNB where both eNB and gNB are connected to 5G core network. S1/N2-based handover is supported when LTE eNB is connected to EPC and NR gNB is connected to 5GC. Xn and core network handover in 5GC, when both eNB and gNB are connected to 5GC, is transparent to the UE. Seamless handover is possible based on tight interworking between radio access technologies when anchored at 5GC.

1.5.2 Network-Controlled Power Management

Power management schemes are important to sustaining UE services and prolonged UE battery life. In this section, we discuss the power management schemes and strategies used by 5G networks to optimize UE power consumption while satisfying user expectations and

QoS requirements of the applications. During the normal operation, a 5G UE can be in one of the three states of connected, inactive, or idle as explained in the previous section. If there is no data to be transmitted/received, the UE stays in the energy-efficient idle state. In contrast, the connected state is the energy-consuming state, as the UE needs to continuously monitor the link quality of the serving and neighboring cells and to provide periodic reports on the quality and status of the radio link.

Mobile devices in connected mode may be configured with discontinuous reception (DRX) for power saving purposes. The parameters of DRX configuration can be optimized to either maximize power saving or minimize latency performance based on the UE's active applications/services. The DRX cycles are broadly configurable to support a wide range of services with different requirements in terms of power consumption and accessibility delays. The implementation of DRX allows the UE to avoid frequently monitoring the physical downlink control channel, except during specific time intervals configured by higher layers. A typical DRX cycle can be divided into an active time interval and an inactive period. During active time interval, the UE is awake and performs continuous reception, while the inactivity timer has not expired. During this time, the UE is performing continuous reception while waiting for a downlink transmission. The UE then enters the inactive period, if there is no traffic activity longer than the inactivity timer duration, after expiration of on-duration, or if the UE receives a MAC control message and is instructed to enter the DRX mode. When the DRX is configured and the UE is in an active state while in the on-duration interval (see Fig. 1.83), the UE monitors the relevant control channels to detect any downlink allocations and to receive pending transmissions from the serving base station. If no allocations including paging messages are detected within the on-duration, the UE will enter the inactive interval and will follow the DRX configuration to wake up in the next DRX cycle. During inactivity period, the UE first follows the short DRX cycle, if configured, and starts the DRX short cycle timer. After the short cycle timer expires, the UE follows the long DRX cycle. If no short DRX cycle is configured, the UE directly enters the long DRX cycle.

It is necessary to optimize the DRX parameters according to the QoS requirements of various services such as VoIP, web browsing and video streaming, where each has a different traffic model and specific QoS requirements, which may significantly impact the configuration of DRX parameters. The services with low delay requirement can be dealt with by activating the UE more frequently to monitor downlink allocations with short DRX cycle setting. While delay-tolerant services can achieve high energy-saving gains by waking up the UE to monitor downlink allocations with long DRX cycle configuration. Therefore, the best trade-off between power saving and responsiveness of the UEs should be made to optimally configure DRX parameters. 3GPP releases define the DRX configuration per UE. When multiple data bearers are established, DRX is enabled only when all the data bearers met their corresponding DRX inactivity timer condition, and the shortest DRX cycles

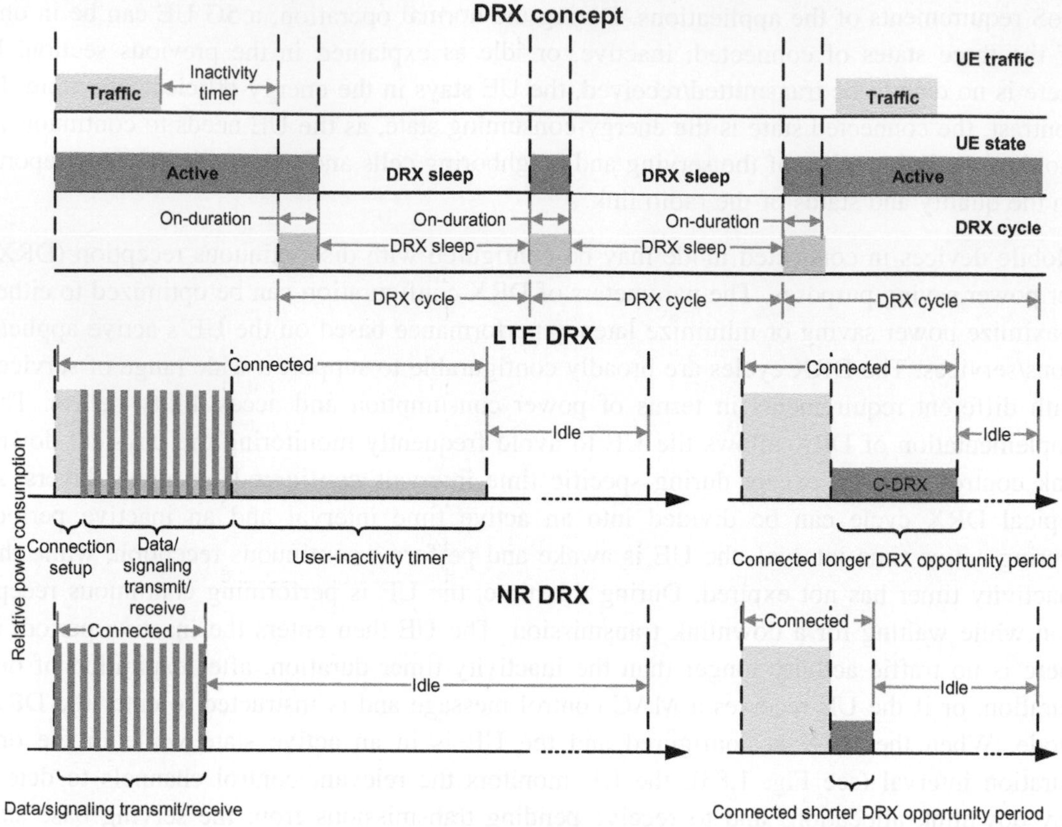

Figure 1.83
DRX concept and comparison of LTE and NR DRX schemes (example) [16,82].

among all the data bearers are followed. This solution is simple and effective for non-CA scenarios. In CA scenarios, the UE may operate over several component carriers and supports separate RF transceivers for each RF carrier. The baseline UE-specific DRX mechanism is no longer suitable to achieve higher energy-saving gains since the same DRX setting has to be configured for all component carriers according to the delay requirement of all applications running on the UE. The UE has to wake up the RF circuitries for all RF carriers at the same time to monitor possible downlink allocations. When bandwidth adaptation is configured in NR, the UE only has to monitor the downlink control channel on the [single] active bandwidth part, thus it does not have to monitor downlink control channels on the entire downlink frequency bands used in the cell. A bandwidth part inactivity timer (independent of DRX inactivity-timer described earlier) is used to switch the active bandwidth part to the default one, that is, the timer is restarted upon successful downlink control channel decoding and switching to the default bandwidth part occurs upon its expiration (see Fig. 1.83).

The percentage of time spent in the connected and idle states depends on a number of parameters controlled by the network including paging occasions, DRX cycles, user-inactivity timer, etc. The user-inactivity timer determines how long the UE stays in the connected state after it receives or transmits the last data packet. When the timer expires, the gNB releases the RRC connection and the UE immediately transitions to the idle state. The shorter the user-inactivity timer, the longer the UE battery life. However, if a new packet arrives the gNB queue shortly after the UE transitions to the idle state, the core network has to page the UE with network and radio signaling, causing extra service latency to transition to the connected state. In other words, the length of the user-inactivity timer determines a trade-off between UE energy consumption and connection latency as well as network control signaling overhead. Whenever the latency requirement can be relaxed, the DRX can provide further power savings. This will reduce the active duty cycle for downlink control channel monitoring, and if cross-slot scheduling is configured, then data channel reception is only needed when UE data is present. With this configuration the UE modem/transceivers may spend approximately 90% of time in a low-power sleep mode, but the penalty would be the substantially increased latency. The UE in RRC_IDLE and RRC_INACTIVE states may use DRX in order to reduce power consumption. While in RRC_IDLE, the UE monitors 5GC-initiated paging, in RRC_INACTIVE the UE is reachable via RAN-initiated paging and 5GC-initiated paging. The RAN and 5GC paging occasions overlap and same paging mechanism is used. The UE monitors one paging occasion per DRX cycle to receive the paging message. Paging DRX cycle length is configurable and a default DRX cycle for core-network-initiated paging is sent via system information. A UE-specific DRX cycle for core-network-initiated paging can be configured via UE dedicated signaling. The NG-RAN can configure a UE with a DRX cycle for RAN-initiated paging, which can be UE specific. The number of paging occasions in a DRX cycle is configured and signaled via system information. A network may assign UEs to the paging occasions based on UE identities when multiple paging occasions are configured in the DRX cycle. When DRX is configured, the UE does not have to continuously monitor downlink control channels. If the UE detects a relevant downlink control channel, it stays awake and starts the inactivity timer.

The DRX mechanism is characterized by several parameters such as *on-duration*, which is the time interval that the UE waits for, after waking up, to receive possible downlink control channels; *inactivity-timer*, which measures the duration of time from the last successful detection that the UE waits to successfully decode a downlink control channel. If the UE fails to detect a relevant downlink allocation, it goes back to sleep mode; *retransmission-timer* measures the duration of time when a HARQ retransmission is expected; and *cycle*, which specifies the periodic repetition of the on-duration followed by a possible period of inactivity [16].

1.6 Quality-of-Service Framework

We begin this section with a review of QoS framework in LTE to set the stage for the 5G QoS framework. As shown in Fig. 1.84, in EPC, the user traffic is classified into different SDFs each associated with different QoS classes based on the type of the service that is being provided through the SDFs. Different QoS rules are then applied to each SDF. Since SDFs are delivered through EPS bearers in an LTE network, the EPS bearer QoS has to be controlled in a way that SDF QoS is maintained. In an LTE network, the user traffic (IP flows or IP packets) is classified into SDF traffic and EPS bearer traffic. An SDF refers to a group of IP flows associated with a service that a user is utilizing, whereas an EPS bearer refers to IP flows of aggregated SDFs that have the same QoS class. The SDF and EPS bearers are detected by matching the IP flows against the packet filters, that is, SDF templates for SDFs or traffic flow templates (TFTs) for EPS bearers. These packet filters are preconfigured by network operators in accordance with their policy and each of them typically consists of 5-tuple (source IP address, destination IP address, source port number, destination port number, and protocol ID). In other words, in LTE network, IP flows with the same service characteristics that match the packet filters of an SDF template are designated to an SDF. SDFs that match the packet filters of a TFT are mapped to an EPS bearer, in order to be delivered to the UE. SDFs with the same QoS class are delivered, as aggregated, through an EPS bearer, whereas the ones with different QoS class are delivered through different EPS bearers. In LTE, there are two types of EPS bearers, default and dedicated. When a UE attaches to the LTE network, an IP address is assigned for PDN connection and a default EPS bearer is established at the same time [58,59].

In an LTE network, the QoS parameters are defined at service and bearer levels. SDF QoS parameters are service-level QoS parameters, whereas EPS bearer QoS parameters are bearer-level QoS parameters. Service level and bearer level are also called as SDF level and SDF aggregate level, respectively. An SDF aggregate refers to a group of SDFs which have

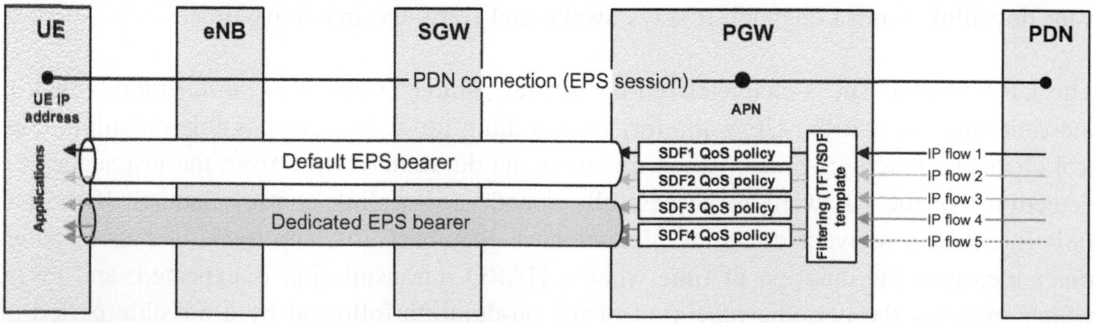

Figure 1.84
QoS architecture and process in LTE [59].

the same QCI and ARP values and belong to one EPS session. Both QCI and ARP are the basic QoS parameters applied to all SDFs and EPS bearers. The QoS class identifier (QCI) is particularly important because it serves as a reference that indicates the performance characteristics of SDFs and EPS bearers. In addition to these two basic parameters, there are other QoS parameters, such as GBR, maximum bit rate (MBR), and aggregated maximum bit rate (AMBR) that specify the bandwidth (or bit rate) characteristics of SDFs and EPS bearers. The SDF and EPS bearer QoS parameters are as follows: SDF QoS parameters (QCI, ARP, GBR, and MBR) and EPS bearer QoS parameters (QCI, ARP, GBR, MBR, APN-AMBR, and UE-AMBR). The QCI and ARP are applied to all EPS bearers. An EPS bearer is classified as a GBR bearer or a non-GBR bearer depending on the resource type specified by its QCI. A default bearer must be non-GBR, while a dedicated bearer can be either GBR or non-GBR. Other than QCI and ARP, there are other QoS parameters for EPS bearers including MBR and GBR indicating the bandwidth (or bit rate) of an EPS bearer, and AMBR indicating the total bandwidth of multiple EPS bearers. The MBR and GBR are the maximum and the guaranteed bandwidths of an EPS bearer, respectively, and AMBR is the maximum total bandwidth of multiple EPS bearers [16,59].

Fig. 1.85 illustrates the QoS parameters applied to SDFs and EPS bearers. In this figure, the UE is connected to two PDNs. The UE has two IP addresses: IP address 1 assigned by PGW-1 for use in PDN-1, and IP address 2 assigned by PGW-2 for use in PDN-2. The UE has one default bearer and two dedicated bearers established for each PDN. The IP flows (user traffic) are filtered into SDFs in the PGW by using SDF templates. There are two groups of SDFs, each received from PDN-1 and PDN-2. For these SDFs, network resources are allocated and packet forwarding is treated according to the QoS rules set in the PGW. The SDFs are then mapped to EPS bearers based on their specified QCI and ARP. In case of PDN-1, as shown in the figure, the SDFs 1 and 2 are mapped to the default bearer, SDFs 3 and 4 are mapped to the non-GBR dedicated bearer, and SDF 5 is mapped to the GBR dedicated bearer, all forwarded to the UE. Such traffic mapped from SDF to EPS bearer is defined by using traffic filter template. All user traffic is subject to the EPS bearer QoS while being delivered through the EPS bearers. All non-GBR bearers associated with a PDN are controlled by the maximum APN-AMBR that they share while the ones associated with a UE are controlled by the maximum UE-AMBR that they share. In LTE, all QoS parameters for SDFs are provisioned by policy and charging rules function of the EPC [59].

The NG-RAN general QoS framework, both for NR connected to 5GC and for LTE connected to 5GC scenarios, is depicted in Fig. 1.86. For each UE, 5GC establishes one or more PDU sessions and the NG-RAN establishes one or more data radio bearers per PDU session. The NG-RAN maps packets belonging to different PDU sessions to different DRBs and establishes at least one default DRB for each PDU session. The NAS-level packet filters in the UE and in the 5GC associate uplink and downlink packets with QoS flows.

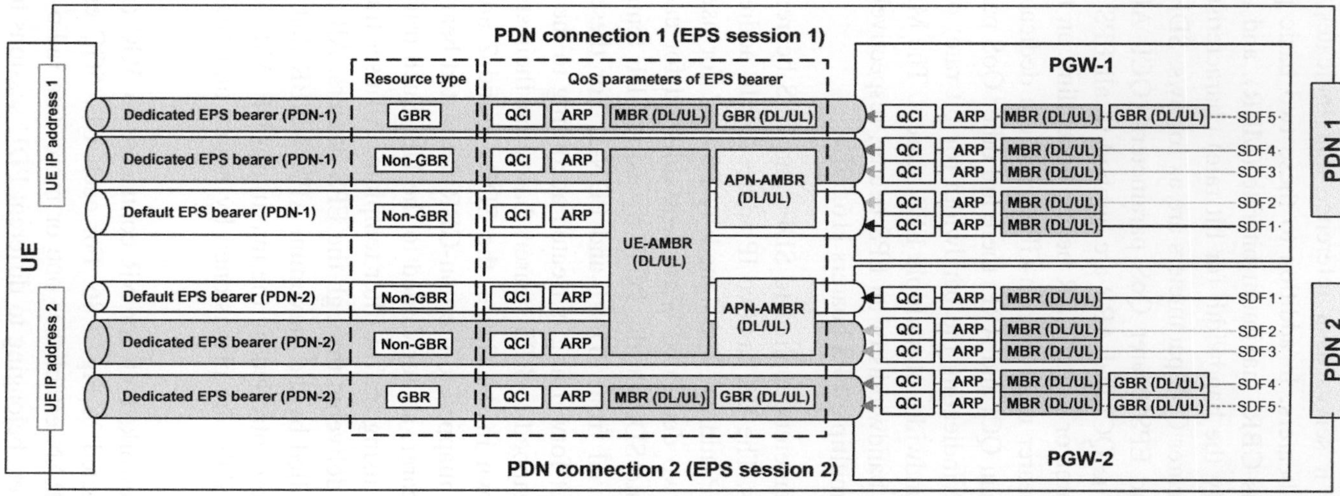

Figure 1.85
QoS parameters in LTE [59].

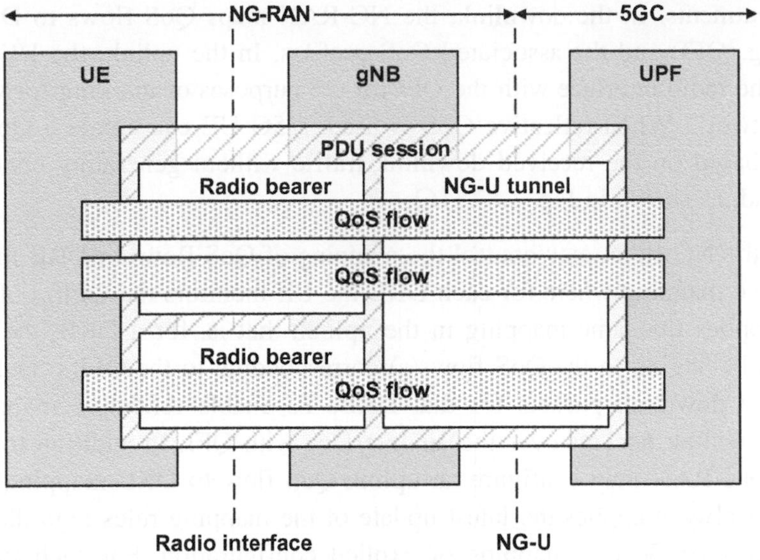

Figure 1.86
QoS framework in NR [16].

The AS-level mapping rules in the UE and in the NG-RAN associate uplink and downlink QoS flows with DRBs.

At NAS level, the QoS flow is the finest granularity for QoS differentiation in a PDU session. A QoS flow is identified within a PDU session by a QFI transferred in encapsulated format over NG-U. NG-RAN and 5GC ensure quality of service (e.g., reliability and maximum tolerable delay) by mapping packets to appropriate QoS flows and DRBs. There is a two-step mapping of IP-flows to QoS flows (NAS level) and from QoS flows to DRBs (AS level). At NAS level, a QoS flow is characterized by a QoS profile which is provided by 5GC to NG-RAN as well as a set of QoS rule(s) which are provided by 5GC to the UE. The QoS profile is used by NG-RAN to determine the treatment on the radio interface while the QoS rules define the mapping between uplink user-plane traffic and QoS flows in the UE. A QoS flow may either be GBR or non-GBR depending on its profile.

The QoS profile of a QoS flow contains QoS parameters, that is, a 5G QoS identifier (5QI) and an ARP. In case of a GBR QoS flow, the QoS parameters are guaranteed flow bit rate (GFBR) and maximum flow bit rate (MFBR) for uplink and downlink. In case of non-GBR, the QoS parameters include the newly defined reflective QoS attribute (RQA). The RQA, when included, indicates that some and not necessarily all traffic carried on this QoS flow is subject to reflective QoS at NAS level. At AS level, the DRB defines the packet treatment on the radio interface. A DRB serves packets with the same packet forwarding treatment. Separate DRBs may be established for QoS flows requiring different packet forwarding treatments. In the downlink, the NG-RAN maps QoS flows to DRBs based on NG-U marking (QFI) and the associated QoS profiles. In the uplink, the UE marks uplink packets over the radio interface with the QFI for the purposes of marking forwarded packets to the core network. When reflective QoS is used, a 5G UE can create a QoS rule for the uplink traffic based on the received downlink traffic without generating control-plane signaling overhead, as shown in Fig. 1.87 [16].

In the uplink, the NG-RAN may control the mapping of QoS flows to DRB in two different ways. Reflective mapping where for each DRB the UE monitors the QFI(s) of the downlink packets and applies the same mapping in the uplink, that is, for a DRB, the UE maps the uplink packets belonging to the QoS flows(s) corresponding to the QFI(s) and PDU session observed in the downlink packets for that DRB. To enable reflective mapping, the NG-RAN marks downlink packets over the air interface with QFI. In addition to the reflective mapping, the NG-RAN may configure an uplink QoS flow to DRB mapping via RRC signaling. The UE always applies the latest update of the mapping rules regardless of whether it is performed via reflective mapping or explicit configuration. For each PDU session, a default DRB is configured. If an incoming uplink packet matches neither an RRC configured nor a reflective QoS flow ID to DRB mapping, the UE maps that packet to the default DRB of the PDU session. Within each PDU session, it is up to NG-RAN to decide how to map multiple QoS flows to a DRB. The NG-RAN may map a GBR flow and a non-GBR flow, or more than one GBR flow to the same DRB. The time when a non-default DRB between NG-RAN and UE is established for QoS flow can be different from the time when the PDU session is established. It is up to NG-RAN to decide when non-default DRBs are established. In dual connectivity scenarios, the QoS flows belonging to the same PDU session can be mapped to different bearer types and, consequently, there can be two different SDAP entities configured for the same PDU session, that is, one for MCG and another one for SCG [16].

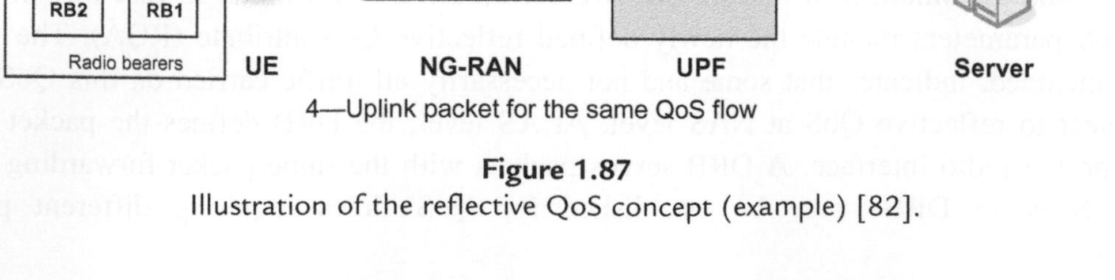

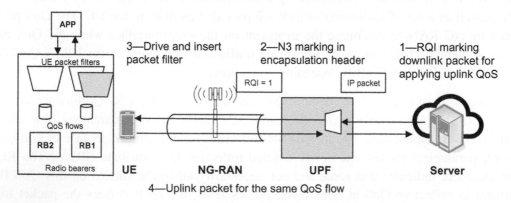

Figure 1.87
Illustration of the reflective QoS concept (example) [82].

As we mentioned earlier, the 5G QoS model is based on QoS flows. The 5G QoS model supports both QoS flows that require GBFR QoS flows and QoS flows that do not require guaranteed flow bit rate (non-GBFR QoS flows). The 5G QoS model also supports reflective QoS. The QoS flow is the finest granularity to differentiate QoS classes in the PDU session. A QFI is used to identify a QoS flow in the 5G system. User-plane traffic with the same QFI within a PDU session receives the same traffic forwarding treatment (e.g., scheduling and admission control). The QFI is carried in an encapsulated header format on N3 (and N9), that is, without any changes to the end-to-end packet header. The QFI is used for all PDU session types and is unique within a PDU session. The QFI may be dynamically assigned or may be equal to the 5QI. Monitoring of user-plane traffic (e.g., MFBR enforcement) is not considered as QoS differentiation and is done by UPFs on an SDF-level basis. Within the 5GS, a QoS flow is controlled by the SMF and may be preconfigured, established via the PDU session establishment procedure, or the PDU session modification procedures. Any QoS flow is characterized by a QoS profile provided by the SMF to the AN via the AMF over N2 reference point or is preconfigured in the AN; one or more QoS rule(s) which can be provided by the SMF to the UE via the AMF over N1 reference point and/or derived by the UE by applying reflective QoS control; and one or more SDF templates provided by the SMF to the UPF. In 5GS, the QoS flow of the default QoS rule is required to be established for a PDU session and to remain active throughout the lifetime of the PDU session. The QoS flow of the default QoS rule is a non-GBR QoS flow [3].

A QoS flow may be either GBR or non-GBR depending on its QoS profile, which contains the corresponding QoS parameters. For each QoS flow, the QoS profile includes the following parameters [3]:

- 5G QoS identifier (5QI)
- ARP
- For each non-GBR QoS flow, the QoS profile may include RQA
- For each GBR QoS flow, the QoS profile includes GFBR and MFBR for uplink and downlink
- In case of a GBR QoS flow only, the QoS parameters may include notification control and maximum packet loss rate for uplink and downlink

Each QoS profile has one corresponding QFI, which is not included in the QoS profile itself. The 5QI value may indicate that a QoS flow has signaled QoS characteristics, and if so, the QoS characteristics are included in the QoS profile.

The UE performs classification and marking of uplink user-plane traffic, that is, the association of uplink traffic to QoS flows based on the QoS rules. These QoS rules may be explicitly provided to the UE via the PDU session establishment/modification procedure, preconfigured in the UE or implicitly derived by the UE by applying reflective QoS. A QoS rule contains a QoS rule identifier which is unique within the PDU session, the QFI of the associated QoS flow and except for the default QoS rule a packet filter set[77] for uplink and optionally for downlink and a precedence value.[78] Furthermore, for a dynamically assigned QFI, the QoS rule contains the QoS parameters relevant to the UE (e.g., 5QI, GBR and MBR, and the averaging window[79]). There are more than one QoS rule associated with the same QoS flow. A default QoS rule is required for each PDU session and associated with the QoS flow of the default QoS rule. The default QoS rule is the only QoS rule of a PDU session that may contain no packet filter set in which case, the highest precedence value is used. If the default QoS rule does not contain a packet filter set, the default QoS rule defines the treatment of packets that do not match any other QoS rules in a PDU session. If the default QoS rule does not contain a packet filter, the reflective QoS is not applied to the QoS flow of the default QoS rule.

The SMF performs binding of SDFs to QoS flows based on the QoS and service requirements. The SMF assigns the QFI for a new QoS flow and derives its QoS profile from the information provided by the PCF.[80] The SMF provides the QFI along with the QoS profile and a transport-level packet marking value for uplink traffic to the AN. The SMF further provides the SDF template, that is, the packet filter set associated with the SDF received from the PCF together with the SDF precedence value, the QoS-related information, and the corresponding packet marking information, that is, the QFI, the transport level packet marking value for downlink traffic and optionally the reflective QoS indication to the UPF enabling classification, bandwidth enforcement and marking of user-plane traffic. For each SDF, when applicable, the SMF generates a QoS rule. Each of these QoS rules contain the QoS rule identifier, the QFI of the QoS flow, the packet filter set of the uplink part of the SDF template, and optionally the packet filter set for the downlink part of the SDF template, as well as the QoS rule priority value set to the SDF precedence value. The QoS rules are then provided to the UE. The principle of classification and marking of user-plane traffic and mapping of QoS flows to AN resources is illustrated in Fig. 1.88 [3].

---

[77] A packet filter set is used in the quality of service rules or service data flow template to identify a quality of service flow. The packet filter set may contain packet filters for the downlink direction, the uplink direction, or packet filters that are applicable to both directions. There are two types of packet filter set, that is, IP packet filter set and Ethernet packet filter set, corresponding to those protocol data unit session types.

[78] The quality of service rule precedence value and the service data flow template precedence value determine the order in which a quality of service rule or a service data flow template, respectively, is evaluated. The evaluation of the quality of service rules or service data flow templates is performed in the increasing order of their precedence value.

[79] The averaging window is defined only for guaranteed bit rate quality of service flows and represents the duration over which the guaranteed flow bit rate and maximum flow bit rate are calculated. The averaging window may be signaled along with 5 QoS identifiers to the access network and user-plane function, and if it is not received, a predefined value will be applied.

[80] Policy control function provides policy framework incorporating network slicing, roaming and mobility management and is equivalent to policy and charging rules function in evolved packet core.

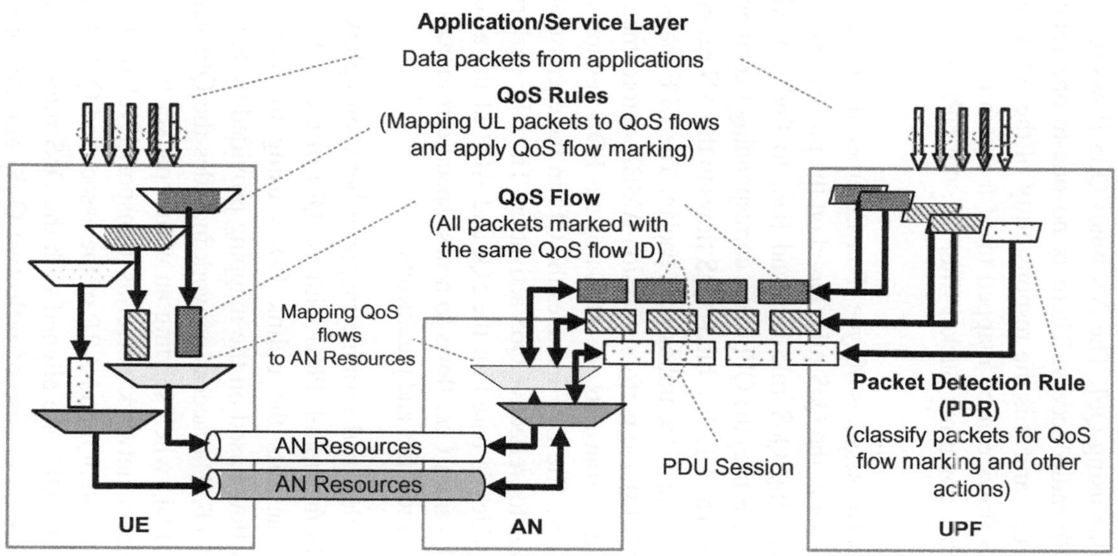

Figure 1.88
Classification and user-plane marking of QoS flows and mapping to access network resources [3].

In the downlink direction, the incoming data packets are classified by the UPF based on the Packet Filter Sets of the packet detection rules (PDRs) in the order of their precedence (without initiating additional N4 signaling). Note that packet detection rules contain information that is necessary to classify the PDU(s) arriving at the UPF. The UPF conveys the classification information of the user-plane traffic, corresponding to a QoS flow, through N3 (and N9) user-plane marking using QFI. The AN binds QoS flows to AN resources, that is, data radio bearers in 3GPP radio access. There is no one-to-one relationship between QoS flows and the AN resources, and it is the responsibility of the AN to establish the necessary AN resources that QoS flows can be mapped to. If a match cannot be found and all QoS flows are associated with a downlink packet filter set, the UPF will discard the downlink data packet.

In the uplink direction and for PDU sessions of type IP or Ethernet, the UE evaluates the uplink packets against the packet filter set in the QoS rules based on the precedence value of QoS rules in increasing order until a matching QoS rule is found (i.e., to find out which packet filter matches the uplink packet). The UE uses the QFI in the corresponding matching QoS rule to bind the uplink packet to a QoS flow. The UE then binds QoS flows to the AN resources. If no matching QoS rule is found, the UE discards the uplink data packet. The UPF maps user-plane traffic to QoS flows based on the PDRs [3] and performs session-AMBR enforcement and PDU counting for charging. The UPF further transmits the PDUs of the PDU session in a single tunnel between 5GC and access network. The UPF includes the QFI in the encapsulation header and it may further include an indication for reflective QoS activation. The UPF performs transport-level packet marking in downlink, which is based on the 5QI and ARP of the associated QoS flow. The access network maps PDUs from QoS flows to access-specific resources based on the QFI and the associated 5G QoS characteristics and parameters.

The UE uses the stored QoS rules to determine mapping between uplink user-plane traffic and QoS flows. The UE marks the uplink PDU with the QFI of the QoS rule containing the matching packet filter and transmits the uplink PDUs using the corresponding access-specific resource for the QoS flow based on the mapping provided by RAN. The access network transmits the PDUs to UPF. The access network includes the QFI value, in the encapsulation header of the uplink PDU when sending an uplink packet from the access network to the core network. The access network performs transport-level packet marking in the uplink, which may be based on the 5QI and ARP of the associated QoS flow. The UPF verifies whether QFIs in the uplink PDUs are aligned with the QoS rules provided to the UE or implicitly derived by the UE (e.g., in case of reflective QoS) and performs session-AMBR enforcement and counting of packets for charging.

The 5G QoS parameters can be further described in detail as follows [3]:

- 5QI is a scalar that is used as a reference to 5G QoS characteristics, that is, access node-specific parameters that control QoS forwarding treatment for the QoS flow (e.g., scheduling weights, admission thresholds, queue management thresholds, link layer

Table 1.4: 5QI to QoS characteristics mapping [3].

5QI Value	Resource Type	Default Priority Level	Packet Delay Budget (ms)	Packet Error Rate	Default Maximum Data Burst Volume (Bytes)	Default Averaging Window (ms)	Example Services
1	GBR	20	100	10^{-2}	N/A	2000	Conversational Voice
2		40	150	10^{-3}	N/A	2000	Conversational Video (Live Streaming)
3		30	50	10^{-3}	N/A	2000	Real-time Gaming, V2X Messages, Electricity Distribution, Process Automation
4		50	300	10^{-6}	N/A	2000	Non-Conversational Video (Buffered Streaming)
65		7	75	10^{-2}	N/A	2000	Mission-critical User-plane Push-to-Talk Voice
66		20	100	10^{-2}	N/A	2000	Non-mission-critical User-plane Push-to-Talk Voice
67		15	100	10^{-3}	N/A	2000	Mission-critical Video
75		—	—	—	—	—	—
5	Non-GBR	10	100	10^{-6}	N/A	N/A	IMS Signaling
6		60	300	10^{-6}	N/A	N/A	Video (Buffered Streaming) TCP-based (e.g., www, e-mail, chat, ftp, p2p file sharing, progressive video, etc.)
7		70	100	10^{-3}	N/A	N/A	Voice, Video (Live Streaming) Interactive Gaming
8		80	300	10^{-6}	N/A	N/A	Video (Buffered Streaming) TCP-based (e.g., www, e-mail, chat, ftp, p2p file sharing, progressive video, etc.)
9		90					
69		5	60	10^{-6}	N/A	N/A	Mission-critical Delay Sensitive Signaling
70		55	200	10^{-6}	N/A	N/A	Mission-critical Data
79		65	50	10^{-2}	N/A	N/A	V2X Messages
80		68	10	10^{-6}	N/A	N/A	Low-latency eMBB Applications, Augmented Reality
82	Delay Critical GBR	19	10	10^{-4}	255	2000	Discrete Automation
83		22	10	10^{-4}	1354	2000	Discrete Automation
84		24	30	10^{-5}	1354	2000	Intelligent Transport Systems
85		21	5	10^{-5}	255	2000	Electricity Distribution

protocol configuration, etc.). The 5QI values have one-to-one mapping to standardized combination of 5G QoS attributes, as shown in Table 1.4. The 5G QoS characteristics for preconfigured 5QI values are preset in the AN, whereas the dynamically assigned 5QI values are signaled as part of the QoS profile.

- The ARP contains information about the priority level, the preemption capability and the preemption susceptibility. The priority level defines the relative importance of a resource request. This allows deciding whether a new QoS flow may be accepted or should be rejected in case of resource limitation, which is typically used for admission control of GBR traffic. It may also be used to decide which of the existing QoS flows to preempt in limited resource scenarios. The range of the ARP priority level is from 1 to 15, with 1 as the highest priority level. The preemption capability information defines if an SDF may use the resources that were already assigned to another SDF with a lower priority level. The preemption susceptibility information defines whether an SDF may lose the resources assigned to it in order to admit an SDF with higher priority level.
- The RQA is an optional parameter, which indicates that certain traffic carried in this QoS flow is subject to reflective QoS. The access network enables the transfer of the RQI for AN resource corresponding to this QoS flow when RQA is signaled. The RQA may be signaled to the NG-RAN via N2 reference point upon UE context establishment in NG-RAN and upon QoS flow establishment or modification.
- The notification control indicates whether notifications are requested from the RAN when the GFBR requirement can no longer be satisfied for a QoS flow during its lifetime. If the notification control is enabled for a given GBR QoS flow and the NG-RAN determines that the GFBR cannot be satisfied, the AN sends a notification to the SMF, where 5GC upon receiving the notification may initiate an N2 signaling to modify or remove the QoS flow.
- Each PDU session of a UE is associated with a session aggregate maximum bit rate (session-AMBR). The subscribed session-AMBR is a subscription parameter which is retrieved by the SMF from UDM. The SMF may use the subscribed session-AMBR or modify it based on local policy or use the authorized session-AMBR received from PCF. The session-AMBR limits the aggregate bit rate that can be expected across all non-GBR QoS flows for a specific PDU session.
- For GBR QoS flows, the 5G QoS profile includes DL/UL GFBR and DL/UL MFBR. The GFBR denotes the bit rate that may be expected to be provided by a GBR QoS flow. The MFBR limits the bit rate that may be expected to be provided by a GBR QoS flow, which means that the excess traffic may be discarded by a rate shaping function. The GFBR and MFBR parameters are signaled to the AN in the QoS profile and signaled to the UE as QoS flow level for each individual QoS flow.
- For each PDU session, the SMF retrieves the default 5QI and ARP from UDM. The SMF may change the default 5QI/ARP based on local configuration or interaction with the PCF. The default 5QI is derived from the standardized range of values for non-GBR 5QIs.

- The DL/UL maximum packet loss rate indicates the maximum rate for lost packets of the QoS flow that can be tolerated in the downlink or uplink, which is provided to the QoS flow, if it is compliant to GFBR.

The 5G QoS characteristics describe the packet forwarding treatment that a QoS flow receives between the UE and the UPF, which are described in terms of the following performance metrics [3]:

- Resource type (GBR, delay critical GBR, or non-GBR) determines whether dedicated network resources related to QoS flow-level GFBR value are permanently allocated, for example, by an admission control function in a base station. The GBR QoS flow is often dynamically authorized, which requires dynamic policy and charging control. A non-GBR QoS flow may be pre-authorized through static policy and charging control. There are two types of GBR resource types, GBR and delay critical GBR, where both resource types are treated in the same manner, except that the definition of PDB and packet error rate (PER) are different.
- Priority level indicates the resource scheduling priority among QoS flows. The priority levels are used to differentiate between QoS flows of the same UE and they are also used to differentiate between QoS flows from different UEs. Once all QoS requirements are satisfied for the GBR QoS flows, additional resources can be used for any remaining traffic in an implementation-specific manner. In addition, the scheduler may prioritize QoS flows based on other parameters such as resource type, radio condition, etc. in order to optimize application performance and network capacity.
- The packet delay budget defines an upper bound for the time that a packet may be delayed between the UE and the UPF that terminates N6 interface. The value of the PDB is the same in downlink and uplink for a certain 5QI. The PDB is used to support the configuration of scheduling and link layer functions (e.g., configuration of scheduling priority weights and HARQ target operating points). For a delay-sensitive GBR flow, a packet delayed more than PDB is considered as a lost packet, if the data burst is not exceeding the MDBV within the period of PDB and the QoS flow is not exceeding the GFBR. For GBR QoS flows with GBR resource type, the PDB is interpreted as a maximum delay with a confidence interval of 98%.
- The packet error rate defines an upper bound for the rate of PDUs or IP packets that have been processed by the sender of a link layer protocol, but are not successfully delivered by the corresponding receiver to the upper layers. Therefore, the PER defines an upper bound for non-congestion related packet losses. The purpose of the PER is to find appropriate link layer protocol configurations. For some 5QI values, the target PER is the same in the downlink and uplink. For QoS flows with delay-sensitive GBR resource type, a packet which is delayed more than PDB is dropped and included in the PER calculation, unless the data burst is exceeding the MDBV within the period of PDB or the QoS flow is exceeding the GFBR.

- Averaging window is defined only for GBR QoS flows and denotes the time interval over which the GFBR and MFBR are calculated. The averaging window may be signaled with 5QIs to the AN and the UPF and if it is not received a predefined value is used.
- The maximum data burst volume (MDBV) is associated with each GBR QoS flow with delay-sensitive resource type. The MDBV denotes the largest amount of data that a 5G-AN is required to serve within a period of 5G-AN PDB (i.e., 5G-AN part of the PDB). Each standardized 5QI of delay-sensitive GBR resource type is associated with a default value for the MDBV. The MDBV may also be signaled together with a standardized or pre-configured 5QI to the AN.

Reflective QoS enables a UE to map uplink user-plane traffic to QoS flows without SMF provided QoS rules, which is applied to IP and Ethernet PDU sessions. The support of reflective QoS over the access network is controlled by 5GC. The UE derives the reflective QoS rule from the received downlink traffic. It must be noted that it is possible to apply reflective QoS and non-reflective QoS concurrently within the same PDU session. For user traffic that is subject to reflective QoS, the uplink packets are assigned the same QoS marking as the reflected downlink packets. Reflective QoS is controlled on per-packet basis using the RQI in the encapsulation header on N3 reference point together with the QFI and a reflective QoS timer (RQ timer) value that is either signaled to the UE upon PDU session establishment or set to a default value.

To summarize this section, as we discussed 5G session management supports a PDU connectivity service that provides PDU exchange between a UE and a data network. In 5GC, the SMF is responsible for handling session management procedures. There is a notable difference in session management between EPC and 5GC. In EPC, the entire session is maintained by a single MME in a centralized manner, so that the user-plane path is established via a centralized PGW. This potentially results in congestion of backhaul traffic at the PGW, whereas in 5GC, different PDU sessions can be maintained by conceivably different SMFs, and their user-plane paths are established via multiple UPFs. This can distribute the cellular operator's backhaul traffic within the 5GC and reduce the perceived latency by the user. Compared to LTE's QoS framework, which is bearer-based and uses control-plane signaling, the 5G system adopts the QoS flow-based framework, and uses both control-plane and user-plane (i.e., reflective QoS) signaling in order to satisfy various application/service QoS requirements. The QoS flow-based framework enables flexible mapping of QoS flows to DRB(s) by decoupling the QoS flow and the radio bearer, allowing more flexible QoS characteristics. When reflective QoS is used, the 5G UE can create a QoS rule for the uplink traffic based on the received downlink traffic without generating control-plane signaling overhead. Table 1.5 summarizes and compares the EPS and 5GS QoS and session management features.

Table 1.5: Comparison of LTE and NR QoS and session management characteristics [71].

	RAN-Level Interworking		5G SA With Core-Network-Level Interworking
	With EPC	With 5GC	
Network slicing	Per device (dedicated core)	Per service (enabling third-party service)	Per service (enabling third-party service)
Session management	Limited and centralized	Flexible and distributed (lower cost, lower latency)	Flexible and distributed (lower cost, lower latency)
QoS	Per-bearer Network-initiated	Per-flow UE/network-initiated (dynamic QoS)	Per-flow UE/network-initiated (dynamic QoS)

1.7 Security Framework

We begin this section with a review of security framework in LTE to set the stage for the 5G security framework discussion. Fig. 1.89 shows the scope and overall concept of the LTE security architecture. In LTE, the authentication function performs mutual authentication between the UE and the network. The NAS security performs integrity protection/verification and ciphering (encryption/decryption) of NAS signaling between the UE and the MME. The AS security is responsible for integrity protection/verification and ciphering of RRC signaling between the UE and the eNB and further performs ciphering of user traffic between the UE and the eNB.

In 3GPP networks, authentication refers to the process of determining whether a user is an authorized subscriber to the network that it is trying to access. Among various authentication procedures, EPS AKA procedure is used in LTE networks for mutual authentication between users and networks. The EPS AKA procedure consists of two steps: (1) the home subscriber server (HSS) generates EPS authentication vector(s) (RAND, AUTN, XRES, K_{ASME}) and delivers them to the MME and (2) the MME selects one of the authentication vectors and uses it for mutual authentication with the UE and shares the same authentication key (K_{ASME}). Mutual authentication is the process in which a network and a user authenticate each other. In LTE networks, since the identification of the user's serving network is required when generating authentication vectors, authentication of the network by the user is performed in addition to authentication of the user by the network. Access security management entity (ASME) is an entity that receives top-level key(s) from the HSS to be used in an AN. In EPS, the MME serves as the ASME and K_{ASME} is used as the top-level key in the AN. The MME conducts mutual authentication with the UE on behalf of the HSS using K_{ASME}. Once mutually authenticated, the UE and MME share the same K_{ASME} as an authentication key.

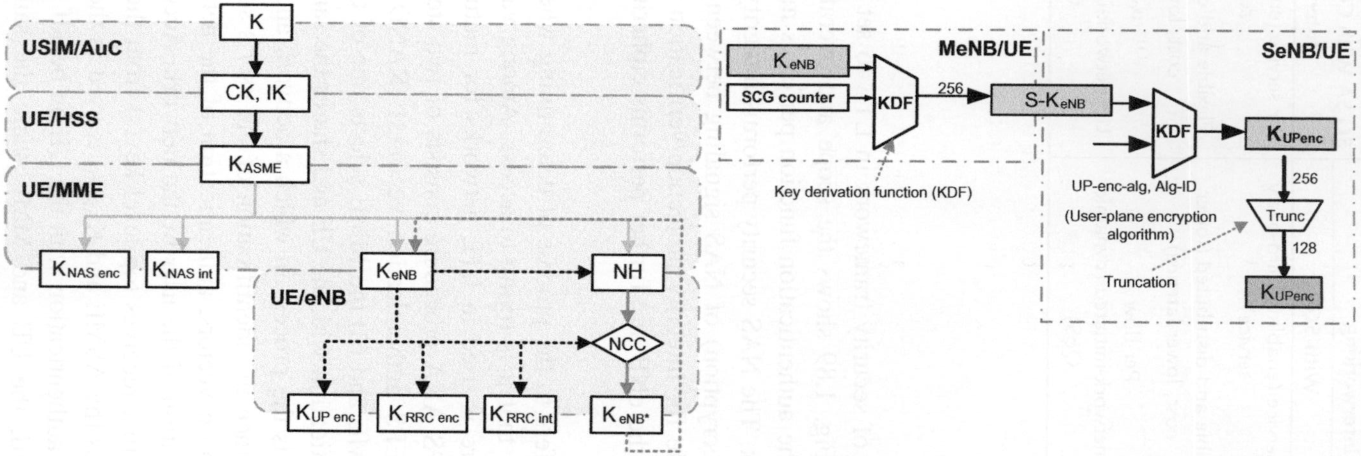

Figure 1.89
Key derivation in LTE and dual connectivity [56,57].

NAS security is designed to securely deliver signaling messages between the UE and the MME over the radio link and to perform integrity protection/verification as well as ciphering of NAS signaling messages. Different keys are used for integrity verification and ciphering. While integrity verification is a mandatory function, ciphering is an optional function. The NAS security keys, such as integrity key (K_{NASint}) and ciphering key (K_{NASenc}), are derived by the UEs and the MMEs from K_{ASME} (see Fig. 1.89). In Fig. 1.89, next hop (NH) key is used by the UE and eNB in the derivation of K_{eNB^*} for provisioning forward security. The NH is derived by UE and MME from K_{ASME} and K_{eNB} when the security context is established, or from K_{ASME} and the previous NH. The NH chaining count (NCC) is a counter related to NH, that is, the number of key chaining that has been performed, which allows the UE to be synchronized with the eNB and to determine whether the next K_{eNB^*} needs to be based on the current K_{eNB} or a fresh NH value [56,57].

AS security is used to ensure secure delivery of data between a UE and an eNB over the radio interface. It includes both integrity check and ciphering of RRC signaling messages over the control plane, and only ciphering of IP packets over the user plane. Different keys are used for integrity check/ciphering of RRC signaling messages and ciphering of IP packets. Integrity verification is mandatory, but ciphering is optional. AS security keys, such as K_{RRCint}, K_{RRCenc}, and K_{UPenc}, are derived from K_{eNB} by a UE and an eNB. K_{RRCint} and K_{RRCenc} are used for integrity check and ciphering of control-plane information (i.e., RRC signaling messages), and K_{UPenc} is used for ciphering of user-plane data (i.e., IP packets). Integrity verification and ciphering are performed at the PDCP sublayer.

Key derivation for dual connectivity SCG bearers is depicted in Fig. 1.89, where SCG counter is a counter used as freshness input into S-K_{eNB} derivations. For SCG bearers in dual connectivity, the user-plane keys are updated upon SCG change by conveying the value of the SCG counter to be used in key derivation to the UE via RRC signaling. When K_{eNB} is refreshed, SCG counter is reset and S-K_{eNB} is derived from the K_{eNB}. The SCG bearers in dual connectivity scenarios share a common pool of radio bearer identities (DRB IDs) with the MCG bearers. When no new DRB ID can be allocated for an SCG bearer without guaranteeing COUNT reuse avoidance, the MeNB derives a new S-K_{eNB}. The SeNB informs MeNB when the uplink or downlink PDCP COUNTs are about to wrap around. In that case, the MeNB updates the S-K_{eNB}. To update the S-K_{eNB}, the MeNB increases the SCG counter and uses it to derive a new S-K_{eNB} from the currently active K_{eNB} in the MeNB. The MeNB sends the freshly derived S-K_{eNB} to the SeNB. The newly derived S-K_{eNB} is then used by the SeNB in computing a new encryption key K_{UPenc} which is used for all DRBs in the SeNB for the target UE. Furthermore, when the SCG counter approaches its maximum value, the MeNB refreshes the currently active K_{eNB}, before any further S-K_{eNB} is derived [56,57].

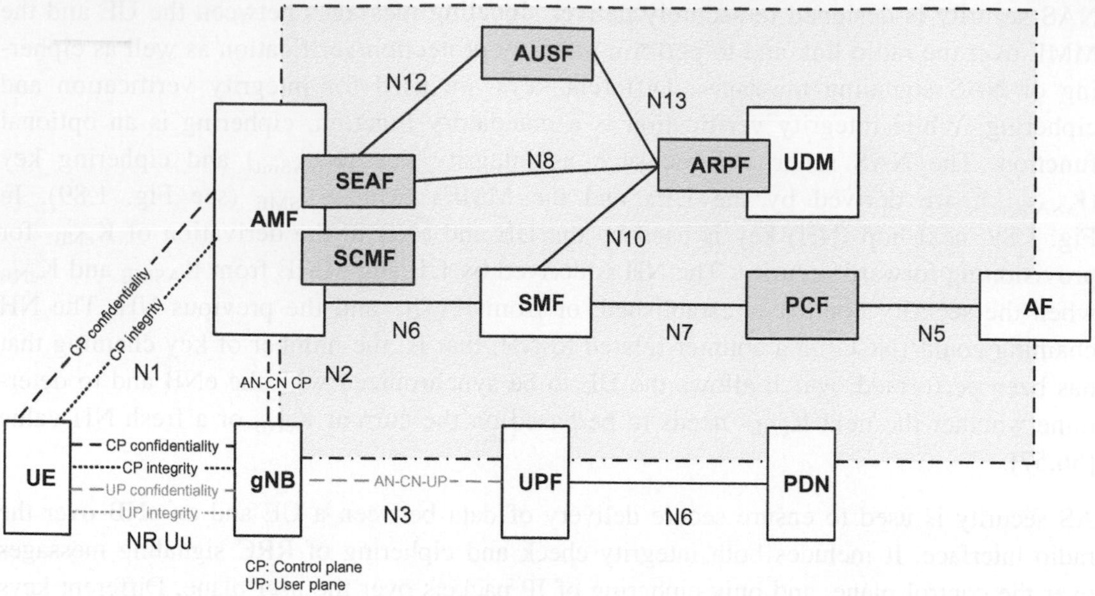

Figure 1.90
5GS security architecture and termination points [54].

Fig. 1.90 shows 5G security architecture including the new security entities: SEAF, AUSF, authentication credential repository and processing function (ARPF), security context management function (SCMF), and security policy control function (SPCF).

5GC has introduced a new security anchor called SEAF, which may be co-located with the AMF. The SEAF will create, for the primary authentication, a unified anchor key K_{SEAF} (common for all access links) that can be used by the UE and the serving network to protect the subsequent communications. It is possible to generate two anchor keys for certain scenarios where a UE is connected to 3GPP access (visited network) and to a non-3GPP access (home network). For normal roaming scenarios, the SEAF is located in the visited network. The AUSF terminates requests from the SEAF and further interacts with the ARPF. Depending on how the authentication functionality is split, the AUSF and the ARPF may be co-located, but an interface similar to SWx interface is defined for EAP-AKA and EAP-AKA′. The ARPF is co-located with the UDM and stores long-term security credentials such as the key K in EPS AKA or EAP-AKA for authentication. It can run cryptographic algorithms using long-term security credentials as input and can create the authentication vectors. Another new functional entity is the SCMF, which may be co-located with the SEAF in the AMF and retrieve a key from the SEAF, which is used to derive further access network specific keys. The SPCF provides the security policy to the network entities (e.g., SMF, AMF) and/or to the UE depending on the application-level input from the AF and may be stand-alone or co-located with the PCF. The security policy may include

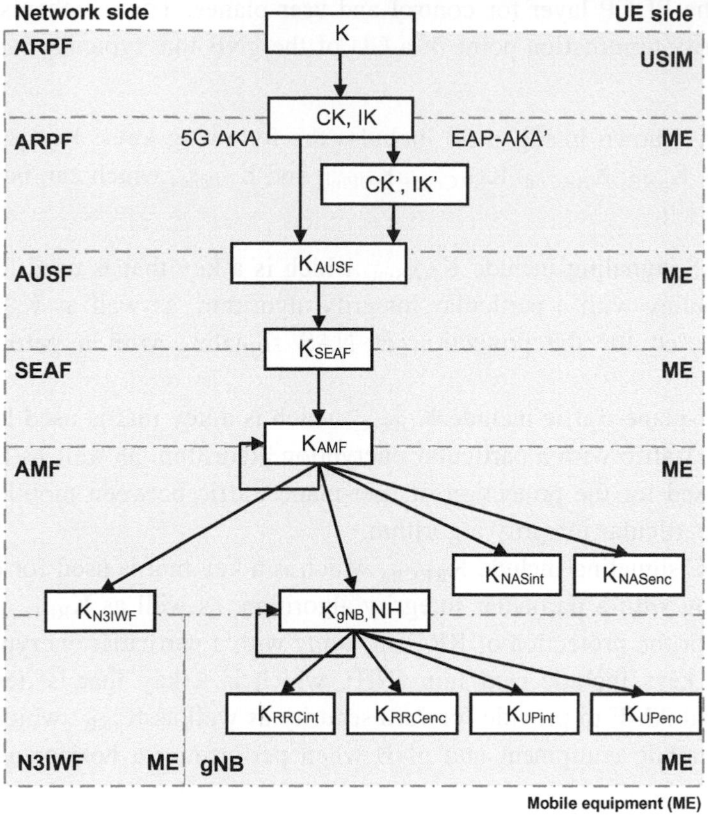

Figure 1.91
Key hierarchy in 5GS [9].

information about AUSF selection, confidentiality protection algorithm, integrity protection algorithm, key length and key life cycle. Fig. 1.91 shows the key hierarchy in 5GS [54].

The new security anchor key K_{SEAF} is used to further derive the access network key K_{AN} and the NAS keys K_{NAS}. There is only one NAS security termination entity, which is the AMF. The user-plane data on the radio bearer can be secured on a per session basis with the key K_{UP}. A session can belong to the same or different network slices. All key sets for NAS, RRC, and user-plane consist of an integrity key and a confidentiality key for encryption. Recall that the termination point of the user-plane security is at eNB in LTE networks. However, the gateway location may vary in order to provide different type of services, and the gNB may be located at the edge, that is, an exposed environment. Thus, the termination point of user-plane security should be reconsidered with the principle that security termination is in the entity where the traffic terminates. The user-plane security terminates at PDCP sublayer of gNB. This is aligned with LTE security framework that radio interface security is provided by the PDCP layer for control and user planes. This mechanism allows provisioning the security termination point in a CU of the gNB that typically resides in a secure location.

The key hierarchy shown in Fig. 1.90 includes the following keys: K_{SEAF}, K_{AMF}, K_{NASint}, K_{NASenc}, K_{N3IWF}, K_{gNB}, K_{RRCint}, K_{RRCenc}, K_{UPint}, and K_{UPenc}, which can be divided into the following groups [9]:

- Keys for NAS signaling include K_{NASint}, which is a key that is used for the protection of NAS signaling with a particular integrity algorithm, as well as K_{NASenc} which is a key that is used for the protection of NAS signaling with a particular encryption algorithm.
- Keys for user-plane traffic include K_{UPenc}, which is a key that is used for the protection of user-plane traffic with a particular encryption algorithm, as well as K_{UPint} which is a key that is used for the protection of user-plane traffic between mobile equipment and gNB with a particular integrity algorithm.
- Keys for RRC signaling include K_{RRCint} which is a key that is used for the protection of RRC signaling with a particular integrity algorithm, as well as K_{RRCenc} which is a key that is used for the protection of RRC signaling with a particular encryption algorithm.
- Intermediate keys include next hop (NH), which is a key that is derived by mobile equipment and AMF to provide forward security as well as $K_{gNB}{}^*$ which is a key that is derived by mobile equipment and gNB when performing a horizontal or vertical key derivation.

When a UE obtains services in RRC idle mode, it does not validate the eNB, which may result in camping on a wrong base station, ultimately leading to denial of service attack. In current LTE systems, the RAN security has been focused on RRC_CONNECTED state, which has been improved in 5G security framework.

The purpose of the primary authentication and key agreement procedures is to enable mutual authentication between the UE and the network and to provide key derivation material that can be used between the UE and network in subsequent security procedures. The keying material generated by the primary authentication and key agreement procedure results in an anchor key called the K_{SEAF} provided by the AUSF to SEAF. Keys for more than one security context can be derived from the K_{SEAF} with no need for a new authentication.

In 5G systems, the storage of credentials and identities for both human and machine type devices is required in the UE. The credentials and identities can be stolen through hardware/software attacks. Such security threats can impact the subscriber and/or the network. 3GPP UE security framework provides the following features for storage of UE credentials: integrity protection of the subscription credential(s); confidentiality protection of the long-term key(s) of the subscription credential(s); and execution of the authentication algorithm(s) that make use of the subscription credentials. These features must be implemented in the UE, with using a tamper resistant secure hardware component. The subscriber identity module (SIM) functions for 5G are inherited from previous standards. Similar to LTE systems, the 5G USIM will be able to generate symmetric keys. It may also be able to generate new asymmetric key pairs and even new trusted public keys.

Network slicing requires basic security from the UE side when accessing a slice; however, this is not trivial and there are new security challenges. The slice isolation must be ensured for network slices, without which attackers who access to one slice may attempt an attack to other slices. Proper isolation will enable integrity and confidentiality protection. Moreover, it should be ensured that resources of the network infrastructure or an NSI are not impacted from another slice instance, to minimize attacks and to provide availability. 5G UE can simultaneously access to different network slices for multiple services. Such access can be via various type of RANs including both 3GPP and non-3GPP access. When the network slice selection data is tampered, unauthorized UEs may use such information to establish connection with the network slice and consume network resources. On the other hand, the advantage of network slicing is that operators are able to provide customized security for each slice. Different access authentication and authorization can be provided within different network slice tenants that can be extended to host applications. In order to support network-controlled privacy of slice information for the slices that the UE can access, the UE must be made aware or configured with privacy considerations that apply to NSSAI. The UE must not include NSSAI in NAS signaling or unprotected RRC signaling unless it has a NAS security context setup [3].

References

3GPP Specifications[81]

[1] 3GPP TS 23.214, Architecture enhancements for control and user plane separation of EPC nodes, Stage 2 (Release 15), September 2017.
[2] 3GPP TS 23.402, Architecture enhancements for non-3GPP accesses (Release 15), March 2018.
[3] 3GPP TS 23.501, System architecture for the 5G system (Release 15), December 2018.
[4] 3GPP TS 23.502, Procedures for the 5G system (Release 15), April 2019.
[5] 3GPP TR 23.714, Study on control and user plane separation of EPC nodes (Release 14), June 2016.
[6] 3GPP TR 23.799, Study on architecture for next generation system (Release 14), December 2016.
[7] 3GPP TR 28.801, Study on management and orchestration of network slicing for next generation network (Release 15), January 2018.
[8] 3GPP TS 29.244, Interface between the control plane and the user plane of EPC nodes, Stage 3 (Release 15), March 2019.
[9] 3GPP TS 33.501, Security architecture and procedures for 5G system (Release 15), December 2018.
[10] 3GPP TR 33.899, Study on the security aspects of the next generation system (Release 14), August 2017.

[81] 3GPP specifications can be accessed at the following URL: http://www.3gpp.org/ftp/Specs/archive/.

[11] 3GPP TS 36.104, Evolved universal terrestrial radio access (E-UTRA), Base Station (BS) Radio Transmission and Reception (Release 15), June 2018.
[12] 3GPP TR 36.932, Scenarios and requirements for small cell enhancements for E-UTRA and E-UTRAN (Release 14), March 2017.
[13] 3GPP TS 37.340, Multi-connectivity, Stage 2 (Release 15), January 2018.
[14] 3GPP TS 38.104, NR; base station (BS) radio transmission and reception (Release 15), January 2018.
[15] 3GPP TS 38.401, NG-RAN, architecture description (Release 15), December 2018.
[16] 3GPP TS 38.300 NR, Overall description, Stage-2 (Release 15), December 2018.
[17] 3GPP TS 38.410 NG-RAN, NG general aspects and principles (Release 15), December 2018.
[18] 3GPP TS 38.411 NG-RAN, NG layer 1 (Release 15), December 2018.
[19] 3GPP TS 38.412 NG-RAN, NG signaling transport (Release 15), December 2018.
[20] 3GPP TS 38.413 NG-RAN, NG application protocol (NGAP) (Release 15), December 2018.
[21] 3GPP TS 38.414 NG-RAN, NG data transport (Release 15), December 2018.
[22] 3GPP TS 38.420 NG-RAN, Xn general aspects and principles (Release 15), December 2018.
[23] 3GPP TS 38.421 NG-RAN, Xn layer 1 (Release 15), December 2018.
[24] 3GPP TS 38.422 NG-RAN, Xn signaling transport (Release 15), December 2018.
[25] 3GPP TS 38.423 NG-RAN, Xn application protocol (XnAP) (Release 15), December 2018.
[26] 3GPP TS 38.424 NG-RAN, Xn data transport (Release 15), December 2018.
[27] 3GPP TS 38.425 NG-RAN, Xn interface user plane protocol (Release 15), December 2018.
[28] 3GPP TS 38.470 NG-RAN, F1 general aspects and principles (Release 15), December 2018.
[29] 3GPP TS 38.471 NG-RAN, F1 layer 1 (Release 15), December 2018.
[30] 3GPP TS 38.472 NG-RAN, F1 signaling transport (Release 15), December 2018.
[31] 3GPP TS 38.473 NG-RAN, F1 application protocol (XnAP) (Release 15), December 2018.
[32] 3GPP TS 38.474 NG-RAN, F1 data transport (Release 15), December 2018.
[33] 3GPP TS 38.475 NG-RAN, F1 interface user plane protocol (Release 15), December 2018.
[34] 3GPP TR 38.801, Study on new radio access technology: radio access architecture and interfaces (Release 14), March 2017.
[35] 3GPP TR 38.804, Study on new radio access technology radio interface protocol aspects (Release 14), March 2017.
[36] 3GPP TR 38.806, Study of separation of NR control plane (CP) and user plane (UP) for split option 2 (Release 15), December 2017.

ETSI Specifications[82]

[37] ETSI GS NFV-SWA 001, Network functions virtualization (NFV), virtual network functions architecture, December 2014.
[38] ETSI GS NFV-IFA 001, Network functions virtualization (NFV), acceleration technologies, report on acceleration technologies & use cases, December 2015.
[39] ETSI GS NFV-IFA 002, Network functions virtualization (NFV) Release 2, Acceleration Technologies, VNF Interfaces Specification, August 2017.
[40] ETSI GS NFV-INF 001, Network functions virtualization (NFV), Infrastructure Overview, January 2015.
[41] ETSI GS NFV 002, Network functions virtualization (NFV), Architectural Framework, December 2014.
[42] ETSI, Network functions virtualization, White Paper on NFV Priorities for 5G, February 2017.
[43] ETSI, Network functions virtualization, Introductory White Paper, October 2012.
[44] ETSI, Network functions virtualization, White Paper, October 2014.
[45] ETSI GS MEC-IEG 004, Mobile-edge computing (MEC), Service Scenarios, November 2015.
[46] ETSI GS MEC 003, Mobile edge computing (MEC), Framework and Reference Architecture, March 2016.

[82] ETSI specifications can be accessed at the following URL: http://www.etsi.org/deliver/.

Articles, Books, White Papers, and Application Notes

[47] China Mobile Research Institute, Toward 5G C-RAN: Requirements, Architecture and Challenges, November 2016.
[48] SDN, NFV, and MEC on SDxCentral. available at: <https://www.sdxcentral.com/>.
[49] Open Networking Foundation, OpenFlow Switch Specification, version 1.5.1, March 2015.
[50] Y. Chao Hu, et al., Mobile edge computing: a key technology towards 5G, ETSI White Paper No. 11, September 2015.
[51] 5G network architecture, a high-level perspective, Huawei Technologies Co., Ltd., 2016.
[52] K. Miyamoto, et al., Analysis of mobile fronthaul bandwidth and wireless transmission performance in split-PHY processing architecture, Optics Express 24 (2) (2016) 1261–1268.
[53] D. Sabella, et al., Mobile-edge computing architecture, the role of MEC in the Internet of Things, IEEE Consumer Electron Mag. 5 (4) (2016).
[54] X. Zhang, et al., Overview of 5G security in 3GPP, in: IEEE Conference on Standards for Communications and Networking (CSCN), September 2017.
[55] J. Kim, et al., 3GPP SA2 architecture and functions for 5G mobile communication system, The Korean Institute of Communications Information Sciences, 2017.
[56] LTE security I: LTE security concept and LTE authentication, NMC Consulting Group, July 2013.
[57] LTE security II: NAS and AS security, NMC Consulting Group, July 2013.
[58] LTE network architecture, NMC Consulting Group, July 2013.
[59] LTE QoS-SDF and EPS bearer QoS, NMC Consulting Group, September 2013.
[60] Emergence of C-RAN: separation of baseband and radio, and baseband centralization, NMC Consulting Group, March 2014.
[61] The benefits of cloud-RAN architecture in mobile network expansion, Fujitsu Network Communications Inc., 2014.
[62] NGMN Alliance, 5G end-to-end architecture framework, October 2017.
[63] NGMN Alliance, Service-based architecture in 5G, January 2018.
[64] NGMN Alliance, Update to NGMN description of network slicing concept, October 2016.
[65] NGMN Alliance, NGMN paper on edge computing, October 2016.
[66] NGMN Alliance, Backhaul provisioning for LTE-advanced & small cells, October 2015.
[67] NGMN Alliance, Project RAN evolution: further study on critical C-RAN technologies, March 2015.
[68] J. Liu, et al., Ultra-dense networks (UDNs) for 5G, IEEE 5G Tech Focus 1 (1) (2017) 6.
[69] J. Wannstrom, et al., HetNet/small cells. Available from: <http://www.3gpp.org/hetnet>.
[70] N.T. Le, et al., Survey of promising technologies for 5G networks, Mobile Information Systems 2016 (2016) 6 pp.
[71] V.G. Nguyen, K.J. Grinnemo, SDN/NFV-based mobile packet core network architectures: a survey, IEEE Commun. Surv. Tutor 19 (3) (2017) 1567–1602.
[72] J.E. Mitchell, Integrated wireless backhaul over optical access networks, J Lightw. Technol. 32 (20) (2014) 3373–3382.
[73] R. Trivisonno, et al., Network slicing for 5G systems, in: IEEE Conference on Standards for Communications and Networking (CSCN), 2017.
[74] 5G Service-Guaranteed Network Slicing, White Paper, Huawei Technologies Co., Ltd., February 2017.
[75] CPRI Specification V7.0, Common Public Radio Interface (CPRI): Interface Specification, October 2015.
[76] eCPRI Specification V1.2, Common Public Radio Interface: eCPRI Interface Specification, June 2018.
[77] IEEE Std 1914.3-2018, Standard for radio over Ethernet encapsulations and mappings, September 2018.
[78] IEEE P1914.1/D4.1, Draft standard for packet-based fronthaul transport networks, April 2019.
[79] D. Anzaldo, Backhaul alternatives for HetNet small cells, Part 1 and 2, Microwaves & RF, September 2015.
[80] Wireless Backhaul Spectrum Policy Recommendations and Analysis Report, GSMA, November 2014.
[81] R. Ravindran, et al., Realizing ICN in 3GPP's 5G NextGen core architecture, Cornell University Library, November 2017.

[82] 4G-5G Interworking, RAN-level and CN-level Interworking, Samsung, June 2017.
[83] Nomor Research, 5G RAN Architecture Interfaces and eCPRI, September 2017.
[84] R. Vaez-Ghaemi, The evolution of fronthaul networks, Viavi Solutions, June 2017.
[85] Sujuan Feng and Eiko Seidel, Self-Organizing Networks (SON) in 3GPP Long Term Evolution, Nomor Research GmbH, May 2008.

CHAPTER 2

New Radio Access Layer 2/3 Aspects and System Operation

2.1 Overview of Layer 2 and Layer 3 Functions

The NR radio interface protocols (alternatively referred to as layer-1, layer-2, and layer-3 protocols) operate between the NG-RAN and the UE and consist of user-plane protocols, for transfer of user data (IP packets) between the network and the UE, and control-plane protocols, for transporting control signaling information between the NG-RAN and the UE. The non-access stratum (NAS) protocols terminate in the UE and the AMF entity of the 5G core network and are used for core network related functions and signaling including registration, authentication, location update and session management. In other words, the protocols over Uu and NG interfaces are categorized into user plane and control plane protocols. User plane protocols implement the actual PDU Session service which carries user data through the access stratum. Control plane protocols control PDU Sessions and the connection between UE and the network from various aspects which include requesting the service, controlling different transmission resources, handover etc. The mechanism for transparent transfer of NAS messages is also included. The layer 2 of the new radio protocol stack is split into four sublayers: medium access control (MAC), radio link control (RLC), packet data convergence protocol (PDCP), and service data adaptation protocol (SDAP), where each sublayer hosts a number of functions and performs certain functionalities that are configurable via radio resource control (RRC) at layer 3. Fig. 2.1 shows the OSI protocol layers and how they map to 3GPP radio protocol architecture. As shown in the figure (*dark-shaded boxes*), the data link layer of OSI network model maps to these four sublayers that constitute layer 2 of the 3GPP new radio protocols. The layer 2 protocols of NR have some similarities with the corresponding LTE protocols; however, the NR has added more configuration flexibility and more functionalities to support the new features such as beam management that have no counterpart in LTE. A significant difference in NR RRC compared to LTE RRC is the introduction of a 3-state UE behavior with the addition of the RRC_INACTIVE state. The RRC_INACTIVE provides a state with battery efficiency similar to RRC_IDLE while storing the UE context within the NG-RAN so that the transitions to/from RRC_CONNECTED are faster and incur less signaling overhead. The other significant improvements relative to LTE RRC are the support of on-demand system information

5G NR. DOI: https://doi.org/10.1016/B978-0-08-102267-2.00002-6

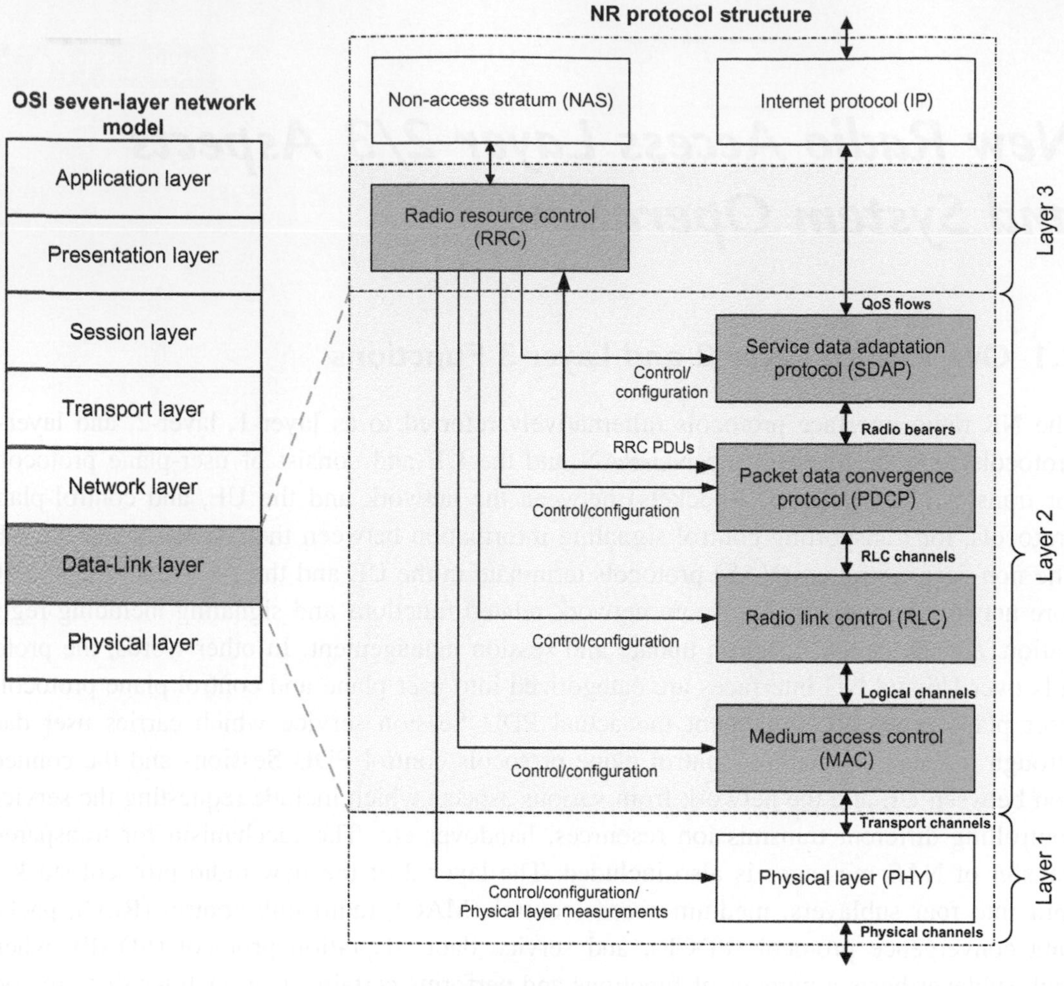

Figure 2.1
The layer 2/3 protocols in NR protocol stack [8].

transmission that enables the UE to request when specific system information is required instead of NG-RAN consuming radio resources to periodically broadcast system information, and the extension of the measurement reporting framework to support beam measurements for handover in a beamformed operation [8]. In NR layer 2 protocol structure, each sublayer provides certain services to the immediately adjacent layers by processing the incoming service data units (SDUs) and generating the proper protocol data units (PDUs). The functional processing of the protocol layers is further classified into user-plane and control-plane protocols (Fig. 2.2), where the reliability requirement of the control-plane information is much higher than that of the user-plane data.

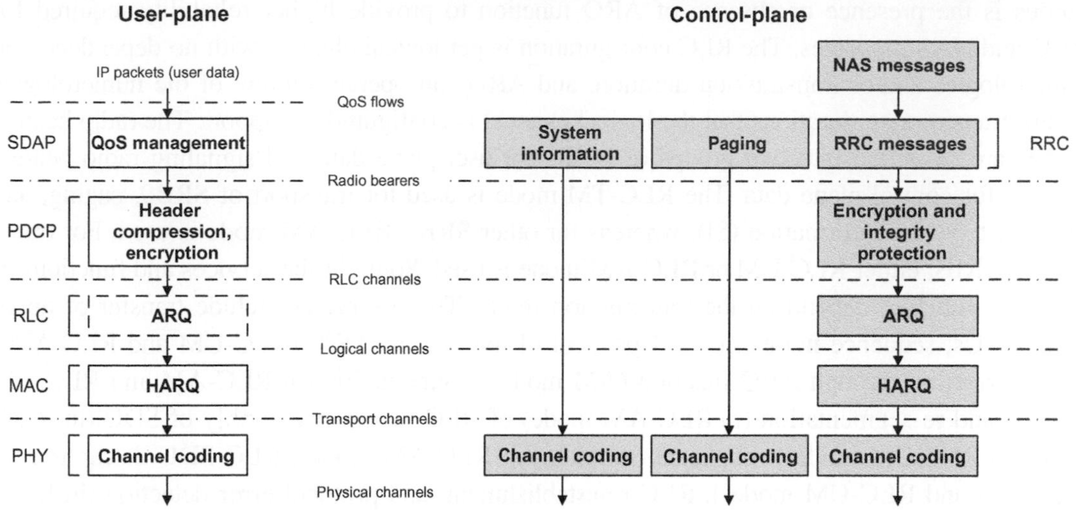

Figure 2.2
NR layer 2 user-plane and control-plane functional mapping and processing.

The service data adaptation protocol is a new sublayer in layer 2 which immediately interfaces with the network layer and provides a mapping between the QoS flows and data radio bearers (DRBs). It also marks the QoS flow identifiers (QFIs) in the downlink and uplink packets. A single-protocol entity of SDAP is configured for each individual PDU Session.

The services and functions of the PDCP sublayer on the user-plane include sequence numbering; header compression and decompression [only supports Robust Header Compression (ROHC) protocol]; transfer of user data; reordering and duplicate detection; in-sequence delivery; PDCP PDU routing (in case of split bearers); retransmission of PDCP SDUs; ciphering, deciphering, and integrity protection; PDCP SDU discard; PDCP re-establishment and data recovery for RLC acknowledged mode (AM); PDCP status reporting for RLC-AM; duplication of PDCP PDUs and duplicate discard indication to lower layers.

The main services and functions of the PDCP sublayer on the control-plane consist of sequence numbering; ciphering, deciphering, and integrity protection; transfer of control-plane data; reordering and duplicate detection; in-sequence delivery; duplication of PDCP PDUs and duplicate discard indication to lower layers [8]. The PDCP protocol performs (optional) IP-header compression, followed by ciphering, for each radio bearer. A PDCP header is added, carrying information required for deciphering in the other end, as well as a sequence number used for retransmission and in-sequence delivery.

The RLC sublayer supports three transmission modes, which include transparent mode (TM), unacknowledged mode (UM), and the acknowledged mode. The main difference between these modes is the presence or absence of ARQ function to provide higher reliability required for RRC and NAS messages. The RLC configuration is per logical channel with no dependency on numerologies and/or transmission duration, and ARQ can operate on any of the numerologies and/or transmission durations that the logical channel is configured to support. The radio bearers in NR are classified into two groups of DRBs for user-plane data and signaling radio bearers (SRBs) for control-plane data. The RLC-TM mode is used for transport of SRB0, paging, and broadcast system information (SI), whereas for other SRBs, RLC-AM mode is used. For transport of DRBs, either RLC-UM or RLC-AM mode is used. Some of the services and functions of the RLC sublayer depend on the transmission mode. Those services include transfer of upper layer PDUs; sequence numbering independent of the one in PDCP (RLC-UM and RLC-AM); error correction through ARQ function (AM mode); segmentation (in RLC-AM and RLC-UM modes) and re-segmentation (in RLC-AM mode) of RLC SDUs; reassembly of SDU (in RLC-AM and RLC-UM modes); duplicate detection (in RLC-AM mode); RLC SDU discarding (in RLC-AM and RLC-UM modes); RLC re-establishment; and protocol error detection (in RLC-AM mode). The ARQ function within the RLC sublayer performs ARQ retransmission of RLC SDUs or SDU segments based on the RLC status reports. It further sends requests for RLC status reports when needed, and triggers an RLC status report after detecting a missing RLC SDU or SDU segment [8]. The RLC protocol performs segmentation of the PDCP PDUs, if necessary, and adds an RLC header containing a sequence number used for handling retransmissions. Unlike LTE, the NR RLC is not providing in-sequence delivery of data to higher layers due to the additional delay incurred by the reordering mechanism. In-sequence delivery can be provided by the PDCP layer, if necessary.

The services and functions of the MAC sublayer include mapping between logical and transport channels, as well as multiplexing/demultiplexing of MAC SDUs belonging to one or different logical channels on the transport blocks, which are mapped to the transport channels depending on the required physical layer processing. The MAC sublayer also performs scheduling of measurement reporting; error correction through HARQ [one HARQ entity per cell when carrier aggregation is utilized]; priority handling between user equipments through dynamic scheduling; and priority handling between logical channels of one UE through logical channel prioritization. A single MAC instantiation can support multiple numerologies, transmission timings, and cells. The mapping restrictions in logical channel prioritization can control which numerology(ies), cell(s), and transmission timing(s) a logical channel can use. The main difference between MAC and RRC control lies in the signaling reliability. The signaling corresponding to state transitions and radio bearer configurations should be performed by the RRC sublayer due to higher signaling reliability requirement. The RLC PDUs are delivered to the MAC sublayer, which multiplexes a number of RLC PDUs and adds a MAC header to form a transport block. It must be noted that in NR, the MAC headers are distributed across the MAC PDU, such that the MAC header corresponding to a particular RLC PDU is located immediately prior to it (see Fig. 2.3).

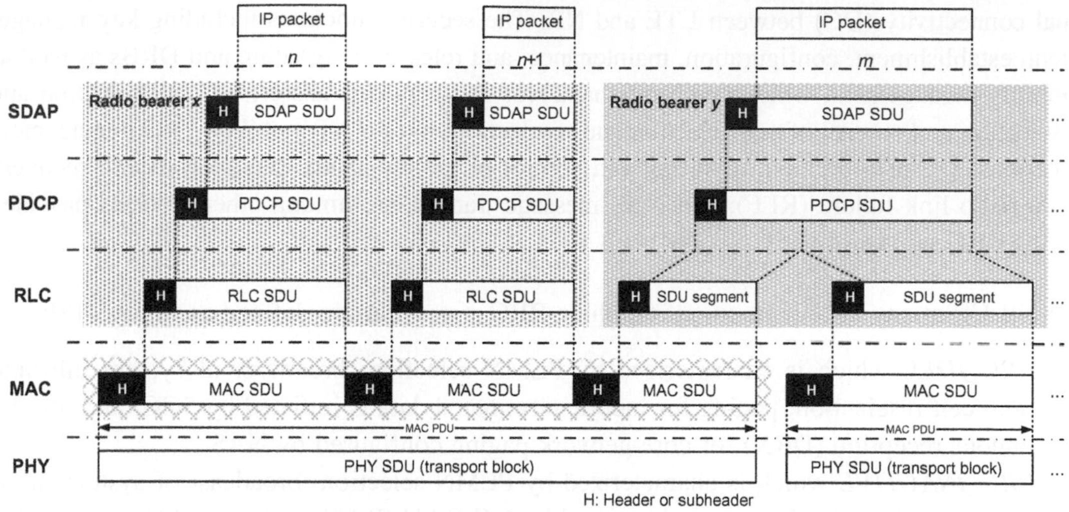

Figure 2.3
Layer 2 packet processing [8].

This is different compared to LTE, in which all header information are located at the beginning of the MAC PDU. The NR MAC PDUs can be assembled as soon as the RLC PDUs become available; thus there is no need to assemble the full MAC PDU before the header fields can be computed. This reduces the processing time and the overall latency [8,18].

The physical layer provides information transfer services to the MAC and higher layers. The physical layer transport services are described by how and with what characteristics data is transferred over the radio interface. This should be clearly distinguished from the classification of what is transported which relates to the concept of logical channels at the MAC sublayer.

It was mentioned earlier that each sublayer in layer 2 radio protocols receives SDUs from the previous layer, processes the information according to the configured functions and parameters of the sublayer, and generates PDUs that are delivered to the next layer. In this process, a unique header or subheader is attached to the SDU by each sublayer. An example is shown in Fig. 2.3, where a transport block is generated by MAC sublayer by means of concatenating two RLC PDUs from radio bearer *x* and one RLC PDU from radio bearer *y*. The RLC PDUs from radio bearer *x* each corresponds to one IP packet (n and $n + 1$) while the RLC PDU from radio bearer *y* is a segment of an IP packet (m) [8].

The services and functions of the RRC sublayer include broadcast of SI related to access stratum (AS) and NAS, as well as paging initiated by 5GC or NG-RAN (paging initiated by NG-RAN is a new NR feature). The RRC sublayer services further includes establishment, maintenance, and release of an RRC connection between the UE and NG-RAN that consist of addition, modification, and release of carrier aggregation; addition, modification, and release of dual connectivity (DC) between LTE and NR. The security functions including key management, establishment, configuration, maintenance, and release of signaling and DRBs as well as mobility management, which comprises handover and context transfer; UE cell selection and reselection and control of cell selection and reselection; inter-RAT mobility; QoS management functions; UE measurement reporting and control of the reporting; detection of and recovery from radio link failure (RLF); and NAS message transfer are among other services provided by the RRC sublayer [8,15].

An NR UE at any time is in one of the three RRC states that are defined as follows [8]:

- *RRC_IDLE* which is characterized by PLMN selection; broadcast of system information; cell reselection; paging for mobile terminated data is initiated by 5GC; and discontinuous reception (DRX) for core-network paging configured by NAS.
- *RRC_INACTIVE* which is characterized by PLMN selection; broadcast of system information; cell reselection; paging initiated by NG-RAN (RAN paging); RAN-based notification area (RNA) is managed by NG-RAN; DRX for RAN paging configured by NG-RAN; 5GC and NG-RAN connection (both control and user-planes) establishment for the UEs; storage of UE AS context in NG-RAN and the UE; and NG-RAN knowledge of UE location at RNA-level.
- *RRC_CONNECTED* which is characterized by 5GC and NG-RAN connection (both control and user-planes) establishment for the UEs; storage of the UE AS context in NG-RAN and the UE; NG-RAN knowledge of UE location at cell level; transfer of unicast data between the UE and gNB; and network-controlled mobility including measurements.

The SI consists of a master information block (MIB) and a number of system information blocks (SIBs), which are divided into minimum SI and other SI. The minimum SI comprises basic information required for initial access and information for acquiring any other SI. The other SI encompasses all SIBs that are not broadcast in the minimum SI. Those SIBs can either be periodically broadcast on downlink shared channel (DL-SCH), broadcast on-demand on DL-SCH upon request from UEs in RRC_IDLE or RRC_INACTIVE states or sent in a dedicated manner on DL-SCH to UEs in RRC_CONNECTED state.

A UE is not required to acquire the content of the minimum SI of a cell/frequency that is considered for camping from another cell/frequency. This does not preclude the case that the UE applies stored SI from previously visited cell(s). A cell is barred, if a UE cannot determine/receive the full content of the minimum SI of that cell. In case of bandwidth adaptation the UE only acquires SI on the active BWP.

In the next sections, we will discuss layers 2 and 3 functions and procedures in more detail. We will further discuss UE states, state transitions, and important procedures such as idle, inactive, and connected mode procedures, random-access procedure, mobility and power management, UE capability, and carrier aggregation.

2.2 Layer 2 Functions and Services

2.2.1 Medium Access Control Sublayer

The MAC sublayer performs logical channel multiplexing and controls HARQ retransmissions. It also handles scheduling functions and is responsible for multiplexing/demultiplexing data packets across multiple component carriers when carrier aggregation is configured. The services of MAC sublayer to the RLC sublayer are in the form of logical channels. A logical channel is defined by the type of information it carries and it is generally classified either as a control channel, for transmission of control and configuration information, or a traffic channel, for transmission of user data. The MAC sublayer provides services to the physical layer in the form of transport channels (see Fig. 2.4). A transport channel is defined by how and with what characteristics the information is transmitted over the radio interface. The information traversing a transport channel is organized in the form of transport blocks. In each physical layer transmission time interval (TTI), one transport block with dynamic size is transmitted over the radio interface to a device. In the case of spatial

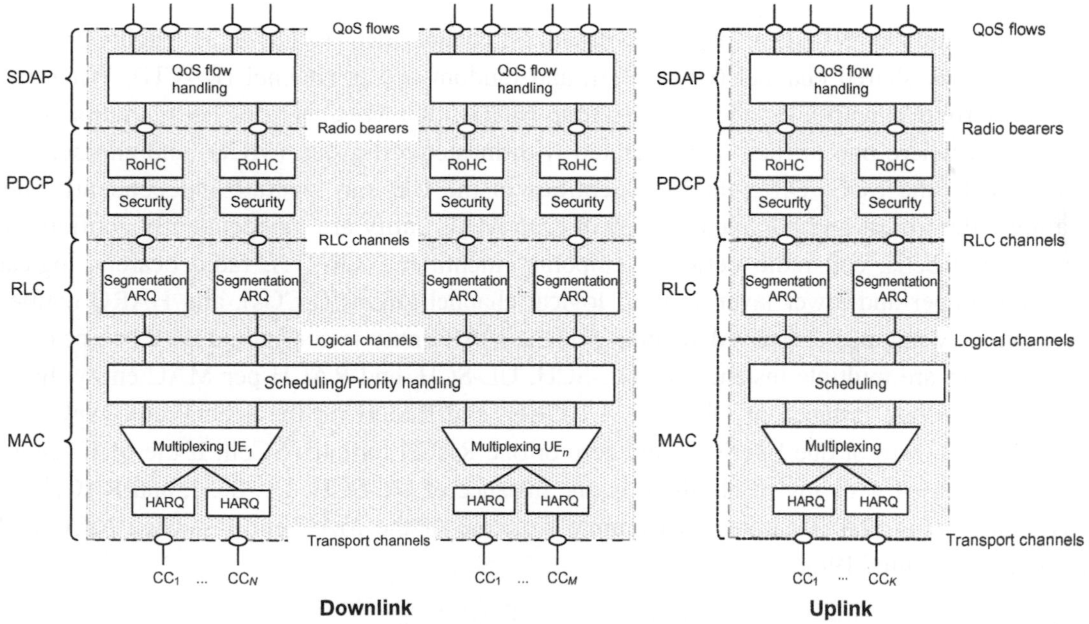

Figure 2.4
NR layer 2 functions in the downlink and uplink [8].

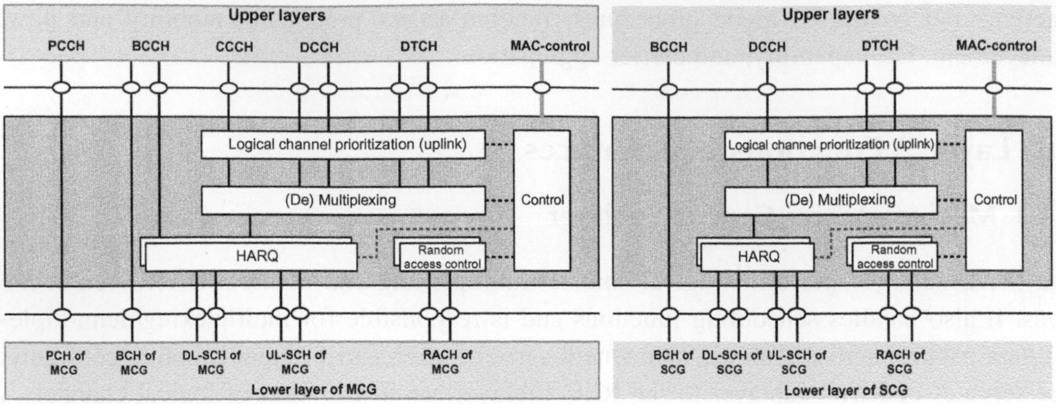

Figure 2.5
Example MAC structure with two MAC entities [9].

multiplexing when more than four layers are configured, there are two transport blocks per TTI. There is a transport format associated with each transport block, specifying how the transport block is transmitted over the radio interface. The transport format includes information about the transport block size, the modulation and coding scheme, and the antenna mapping. By varying the transport format, the MAC sublayer can realize different data rates, which is known as transport format selection.

The MAC entity in the UE manages the broadcast channel (BCH), DL-SCH, paging channel (PCH), uplink shared channel (UL-SCH)], and random-access channel (RACH). When the UE is configured with a secondary cell group (SCG), two MAC entities (instantiations) are configured where one of them is associated with master cell group (MCG) and another one is related to the SCG. Different MAC entities in the UE may operate independently, for example, the timers and parameters used in each MAC entity are independently configured. The serving cells, cell radio network temporary identifier (C-RNTI), radio bearers, logical channels, upper and lower layer entities, logical channel groups (LCGs), and HARQ entities are separately mapped to each MAC entity. If the MAC entity is configured with one or more SCells, there are multiple instances of DL-SCH, UL-SCH, and RACH per MAC entity; however, there is one instance of DL-SCH, UL-SCH, and RACH on the special cell (SpCell),[1] one DL-SCH, zero or one UL-SCH, and zero or one RACH for each SCell. If the MAC entity is not configured with any SCell, there is one instance of DL-SCH, UL-SCH, and RACH per MAC entity. Fig. 2.5 illustrates an example structure of the MAC entities when MCG and SCG are configured [9].

[1] In the context of dual connectivity, the SpCell refers to the primary cell of the MCG or the primary SCell of the SCG depending on whether the MAC entity is associated with the MCG or the SCG. Otherwise, the SpCell refers to the PCell. A SpCell supports PUCCH transmission and contention-based random-access procedure and it is always activated [9].

The MAC sublayer provides data transfer and radio resource allocation services to the upper layers and at the same time receives certain services from the physical layer which include data transfer, signaling of HARQ feedback, and scheduling request (SR), as well as conducting certain link-level measurements. The MAC sublayer provides a mapping between logical channels and transport channels; multiplexing of MAC SDUs from one or different logical channels to transport blocks to be delivered to the physical layer on transport channels; demultiplexing of MAC SDUs to one or different logical channels from transport blocks delivered from the physical layer on transport channels; scheduling measurement reporting; forward error correction through HARQ; and logical channel prioritization. The MAC sublayer further provides data transfer services via logical channels. To accommodate different types of data transfer services, various logical channels are defined, each supporting transfer of a particular type of information. Each logical channel is defined by the type of information which is transferred [9]. The MAC entity maps logical channels to transport channels in the uplink and downlink. This mapping depends on the multiplexing that is configured by RRC sublayer.

The priority handling among multiple logical channels, where each logical channel has its own RLC entity, is supported by multiplexing the logical channels to one transport channel. The MAC entity at the receiving side handles the corresponding demultiplexing and forwards the RLC PDUs to their respective RLC entity. To enable the demultiplexing function at the receiver, a MAC header is used. In NR, instead of putting the entire MAC header information at the beginning of a MAC PDU as LTE does, which implies that the assembly of a MAC PDU cannot start until the scheduling decision is available, the subheader corresponding to a certain MAC SDU is placed immediately in front of the SDU, as shown in Fig. 2.6. This allows the PDUs to be processed before a scheduling decision is received. If necessary, padding can be used to align the transport block size with those supported in NR.

A MAC subheader contains the identity of the logical channel (LCID) from which the RLC PDU originated, and the length of the PDU in bytes. There is also a flag indicating the size of the length indicator, as well as a reserved bit for future extension. In addition to multiplexing of different logical channels, the MAC sublayer can insert MAC control elements (CEs) in the transport blocks to be transmitted over the transport channels. A MAC CE is a form of in-band control signaling and is identified with reserved values in the LCID field, where the LCID value indicates the type of control information. Both fixed-length and variable-length MAC CEs are supported, depending on the use case. For downlink transmissions, MAC CEs are located at the beginning of the MAC PDU, whereas for uplink transmissions, the MAC CEs are located at the end, immediately before the padding (see Fig. 2.6). In some cases the size of padding can be zero. A MAC CE provides a faster way to send control signaling than RLC, without having to resort to the restrictions in terms of payload sizes and reliability offered by L1/L2 control signaling.

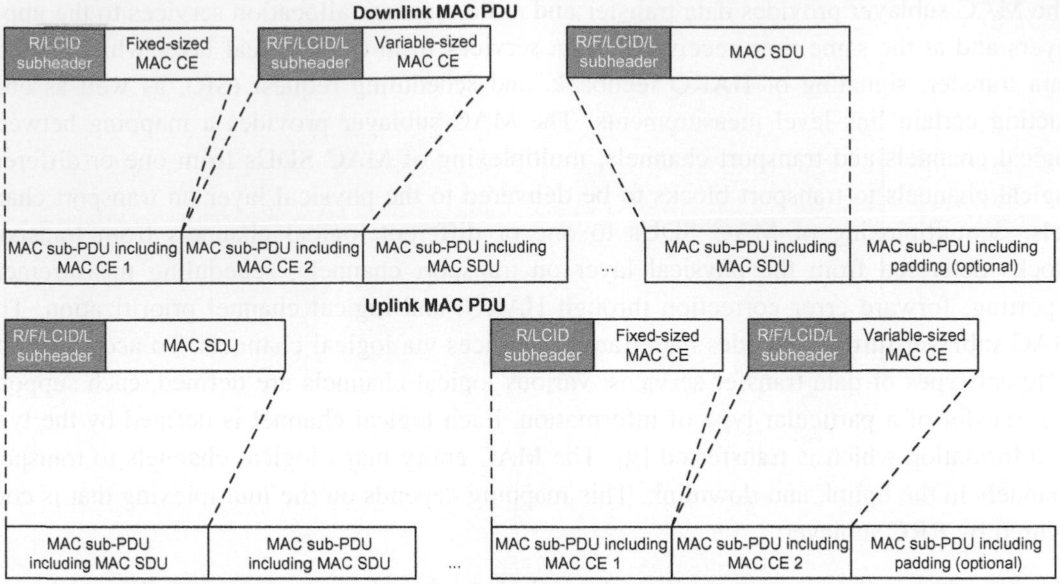

Figure 2.6
Example structures of downlink/uplink MAC PDUs [9].

A MAC PDU is a bit string that is octet-aligned and consists of one or more MAC sub-PDUs. Each MAC sub-PDU may consist of a subheader only (including padding); a subheader and a MAC SDU; a subheader and a MAC CE; or a subheader and padding. The MAC SDUs have variable sizes, where each MAC subheader corresponds to either a MAC SDU, a MAC CE, or padding. A MAC subheader typically consists of four header fields R/F/LCID/L (see Fig. 2.7). However, a MAC subheader for fixed-sized MAC CE, padding, and a MAC SDU containing uplink common control channel (CCCH), consists of two header fields R/LCID. MAC control elements are placed together. The downlink MAC subPDU(s) with MAC CE(s) is placed before any MAC subPDU with MAC SDU and MAC subPDU with padding. The uplink MAC subPDU(s) with MAC CE(s) is placed after all MAC subPDU(s) with MAC SDU and before the MAC subPDU with padding in the MAC PDU. Note that the size of padding can be zero. At most one MAC PDU can be transmitted per transport block per MAC entity. The aforementioned subheader fields are defined as follows [9]:

- *LCID*: The LCID field identifies the logical channel instance of the corresponding MAC SDU or the type of the corresponding MAC CE or padding. There is one LCID field per MAC subheader whose size is 6 bits.
- *L*: The length field indicates the length of the corresponding MAC SDU or variable-sized MAC CE in bytes. There is one L field per MAC subheader except for subheaders corresponding to fixed-sized MAC CEs, padding, and MAC SDUs containing uplink CCCH. The size of the L field is indicated by the F field.

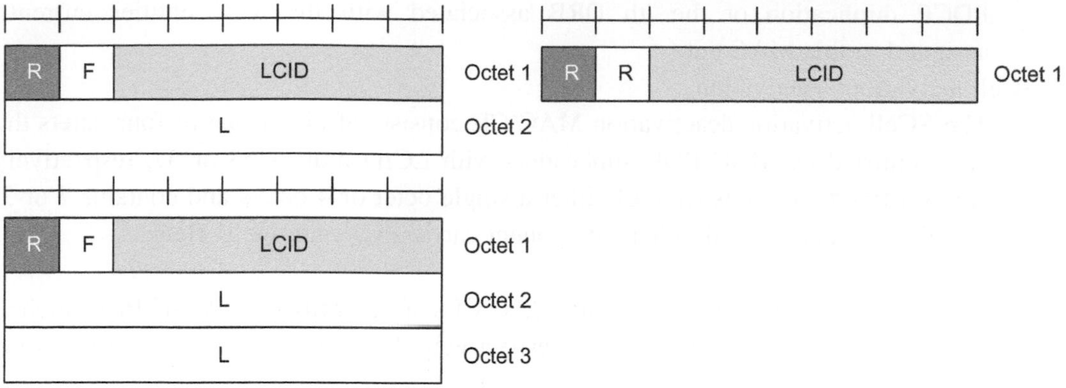

Figure 2.7
Structure of various MAC subheaders [9].

- *F*: The format field indicates the size of the length field. There is one F field per MAC subheader except for subheaders corresponding to fixed-sized MAC CEs, padding, and MAC SDUs containing uplink CCCH. The size of the F field is 1 bit, where the 8-bit and 16-bit Length fields are indicated by $F=0$ and $F=1$, respectively.
- *R*: Reserved bit, which is set to zero.

The MAC subheaders consist of the following fields:

- *E*: The extension field is a flag indicating whether the MAC sub-PDU, including the MAC subheader is the last MAC sub-PDU in the MAC PDU. It is set to one to indicate at least another MAC sub-PDU follows; otherwise, it is set to zero.
- *T*: The type field is a flag indicating whether the MAC subheader contains a random-access preamble identifier (RAPID) or a backoff indicator (BI). It is set to zero to indicate the presence of a BI field in the subheader; otherwise it is set to one to indicate the presence of a RAPID field in the subheader.
- *R*: Reserved bit, which is set to zero.
- *BI*: The BI field identifies the overload condition in the cell. The size of the BI field is 4 bit.
- *RAPID*: The RAPID field identifies the transmitted random-access preamble. The size of the RAPID field is 6 bits. If the RAPID in the MAC subheader of a MAC sub-PDU corresponds to one of the random-access preambles configured for SI request, MAC random-access response (RAR) is not included in the MAC sub-PDU. The MAC subheader is octet aligned.

There are a number of MAC CEs that are specified for the following purposes [9]:

- Duplication activation/deactivation
  - The duplication activation/deactivation MAC CE consists of one octet that is identified by a MAC PDU subheader with LCID index 56. It has a fixed size and contains eight D-fields, where the D_i field indicates the activation/deactivation status of the

PDCP duplication of the *i*th DRB associated with the RLC entities currently assigned to this MAC entity.

- SCell activation/deactivation
  - The SCell activation/deactivation MAC CE consists of either one or four octets that are identified by MAC PDU subheaders with LCID indices 58 or 57, respectively. It has a fixed size consisting of either a single octet or 4 octets and contains 7 or 31 C-fields (mapped to individual component carriers C_i) and one R-field.
- DRX
  - The DRX and long DRX command MAC CEs are identified by MAC PDU subheaders with LCID indices 60 and 59, respectively, both of which have a fixed size of zero bits.
- Timing advance command
  - The timing advance command MAC CE is identified by a MAC PDU subheader with LCID index 61. It has a fixed size of one octet and includes timing advance group (TAG) identity (TAG ID) which indicates the TAG ID of the addressed TAG (2 bit); and timing advance command which indicates the index value $TA = (0, 1, 2, \ldots, 63)$ that is used to control the amount of timing adjustment that MAC entity has to apply (6 bits).
- UE contention resolution identity
  - The UE contention resolution identity MAC CE is identified by a MAC PDU subheader with LCID index 62. It has a fixed 48 bit size and consists of a single field containing UE contention resolution identities.
- Semi-persistent (SP) CSI-RS/CSI-IM resource set activation/deactivation
  - The network may activate and deactivate the configured SP CSI-RS/CSI-IM resource sets of a serving cell by sending the SP CSI-RS/CSI-IM resource set activation/deactivation MAC CE. The configured SP CSI-RS/CSI-IM resource sets are initially deactivated upon configuration and after a handover.
- Aperiodic CSI trigger state sub-selection
  - The network may select among the configured aperiodic CSI trigger states of a serving cell by sending the aperiodic CSI trigger state sub-selection MAC CE.
- Transmission configuration indicator (TCI) states activation/deactivation for UE-specific PDSCH
  - The network may activate and deactivate the configured TCI states for physical downlink shared channel (PDSCH) of a serving cell by sending the TCI states activation/deactivation for UE-specific PDSCH MAC CE. The configured TCI states for PDSCH are initially deactivated upon configuration and after a handover.
- TCI state indication for UE-specific PDCCH
  - The network may indicate a TCI state for physical downlink control (PDCCH) reception for a CORESET of a serving cell by sending the TCI state indication for UE-specific PDCCH MAC CE.

- SP CSI reporting on physical uplink control channel (PUCCH) activation/deactivation
  - The network may activate and deactivate the configured SP CSI reporting on PUCCH of a serving cell by sending the SP CSI reporting on PUCCH activation/deactivation MAC CE. The configured SP CSI reporting on PUCCH is initially deactivated upon configuration and after a handover.
- SP SRS activation/deactivation
  - The network may activate and deactivate the configured SP SRS resource sets of a serving cell by sending the SP SRS activation/deactivation MAC CE. The configured SP SRS resource sets are initially deactivated upon configuration and after a handover.
- PUCCH spatial relation activation/deactivation
  - The network may activate and deactivate a spatial relation for a PUCCH resource of a serving cell by sending the PUCCH spatial relation activation/deactivation MAC CE.
- SP zero power (ZP) CSI-RS resource set activation/deactivation
  - The network may activate and deactivate the configured SP ZP CSI-RS resource set of a serving cell by sending the SP ZP CSI-RS resource set activation/deactivation MAC CE. The configured SP ZP CSI-RS resource sets are initially deactivated upon configuration and after a handover.
- Recommended bit rate
  - The recommended bit rate procedure is used to provide the MAC entity with information about the physical layer bit rate which the gNB recommends. An averaging window with default size of 2 seconds is applied. The gNB may transmit the recommended bit rate MAC CE to the receiver-side MAC entity to indicate the recommended bit rate for the UE for a specific logical channel and downlink/uplink direction.
- Single and multiple-entry power headroom report (PHR)
  - The single-entry PHR MAC CE is identified by a MAC PDU subheader with LCID index 57. It has a fixed size of two octets, including power headroom which indicates the PH level (6 bits) and $P_{CMAX_{lm}}$ denoting the maximum permissible UE transmit power for carrier l and serving cell m.
  - The multiple-entry PHR MAC CE is identified by a MAC PDU subheader with LCID index 56. It has a variable size and may include a bitmap, a Type 2 PH field and an octet containing the associated $P_{CMAX_{lm}}$ field for the SpCell of the other MAC entity, as well as a Type 1 PH field and an octet containing the associated $P_{CMAX_{lm}}$ for the primary cell (PCell). It may further include one or more Type x PH fields (x is either 1 or 3) and octets containing the associated $P_{CMAX_{lm}}$ fields for serving cells other than PCell indicated in the bitmap. The MAC entity determines whether power headroom value for an activated serving cell is based on a real transmission or a reference format by considering the configured grant(s) and the downlink control information (DCI) that has been received prior to the PDCCH occasion

in which the first uplink grant for a new transmission is received after a PHR has been triggered.

- Configured grant confirmation
  - The configured grant confirmation MAC CE is identified by a MAC PDU subheader with LCID index 55, which has a fixed size of zero bits.
- C-RNTI
  - The C-RNTI MAC CE is identified by a MAC PDU subheader with LCID index 58. It has a fixed size and consists of a single field containing the C-RNTI. The length of the field is 16 bits.
- Buffer status report (BSR)
  - The BSR MAC CEs consist of short BSR format (fixed size), long BSR format (variable size), short truncated BSR format (fixed size), and long truncated BSR format (variable size). The BSR formats are identified by MAC PDU subheaders with LCID indices 59, 60, 61, and 62 corresponding to short truncated BSR, long truncated BSR, short BSR, and long BSR MAC CEs, respectively. The fields in the BSR MAC CE contain an LCG ID, denoting the LCG ID which identifies the group of logical channels whose buffer status are being reported, and one or several LCG_i for the long BSR format, which indicate the presence of the buffer size field for the *i*th LCG. For the long truncated BSR format, this field indicates whether the *i*th LCG has data available. The buffer size field identifies the total amount of data available according to the data volume calculation procedure in references [10,11] across all logical channels of an LCG after the MAC PDU has been created, that is, after the logical channel prioritization procedure, which may result in setting the value of the buffer size field to zero. The amount of data is indicated in number of bytes. The size of the RLC and MAC headers are not considered in the buffer size calculation. The length of this field for the short BSR format and the short truncated BSR format is 5 bits and for the long BSR format and the long truncated BSR format is 8 bits. For the long BSR format and the long truncated BSR format, the buffer size fields are included in ascending order based on the LCG_i. For the long truncated BSR format the number of buffer size fields included is maximized, as long as it does not exceed the number of padding bits.

The structure and content of some MAC CEs are shown in Fig. 2.8.

As we mentioned earlier, a [transparent] MAC PDU may only consist of a MAC SDU whose size is aligned to a permissible transport block. This MAC PDU is used for transmissions on PCH, BCH, and DL-SCH including BCCH. Furthermore, a random-access response MAC PDU format is defined which consists of one or more MAC sub-PDUs and may contain some padding bits. Each MAC sub-PDU consists of a MAC sub-header with backoff indicator; a MAC sub-header with random access preamble identifier (RAPID), which is an acknowledgment for SI request; or a MAC sub-header with RAPID and MAC

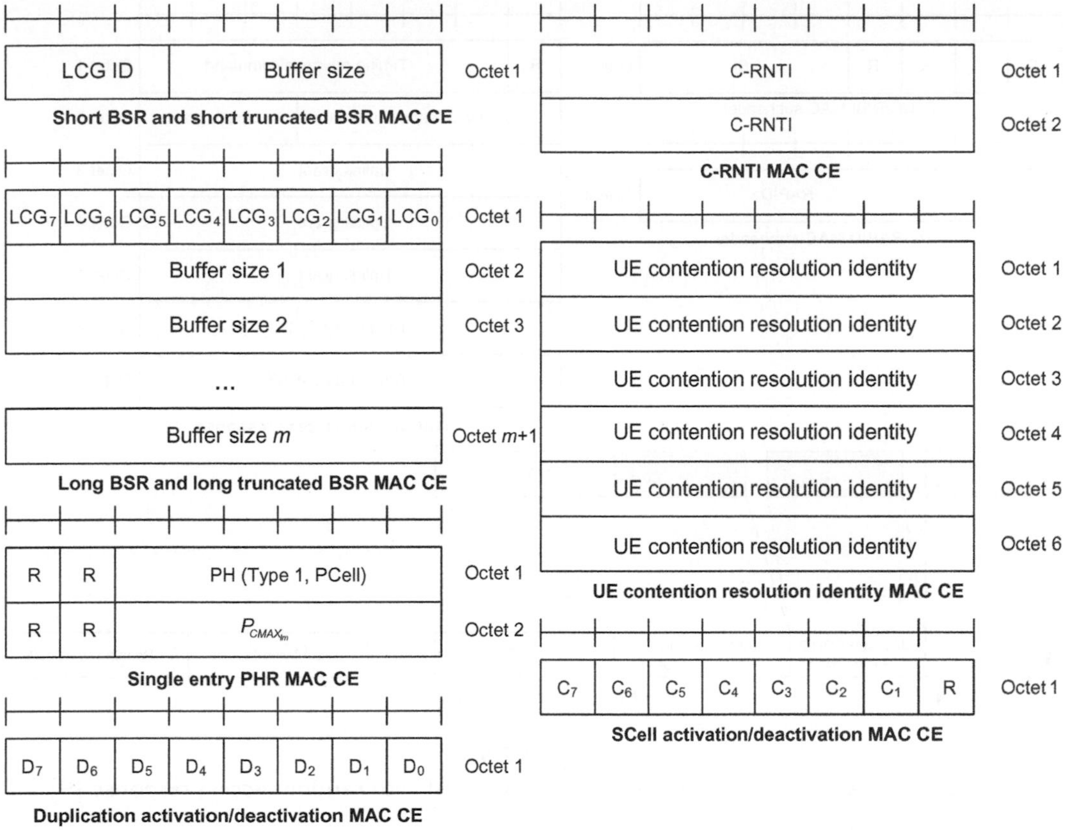

Figure 2.8
Some MAC control element formats [9].

RAR. A MAC sub-header with backoff indicator consists of five header fields E/T/R/R/BI (see Fig. 2.9). A MAC sub-PDU with backoff indicator is placed at the beginning of the MAC PDU, if included. The 'MAC sub-PDU(s) with RAPID' and 'MAC sub-PDU(s) with RAPID and MAC RAR' can be placed anywhere between MAC sub-PDU with backoff indicator and the padding bits. A MAC sub-header with RAPID consists of three header fields E/T/RAPID. Padding bits are placed at the end of the MAC PDU, if necessary. The presence and the length of the padding bits are based on the transport block size and the size of the MAC sub-PDU(s).

The MAC entity is further responsible for distributing IP packets corresponding to each flow across different component carriers or cells, when carrier aggregation is utilized. The carrier aggregation relies on independent processing of the component carriers in the physical layer, including control signaling, scheduling, and HARQ retransmissions. The MAC sublayer makes multi-carrier operation transparent to the upper sublayers/layers. The logical

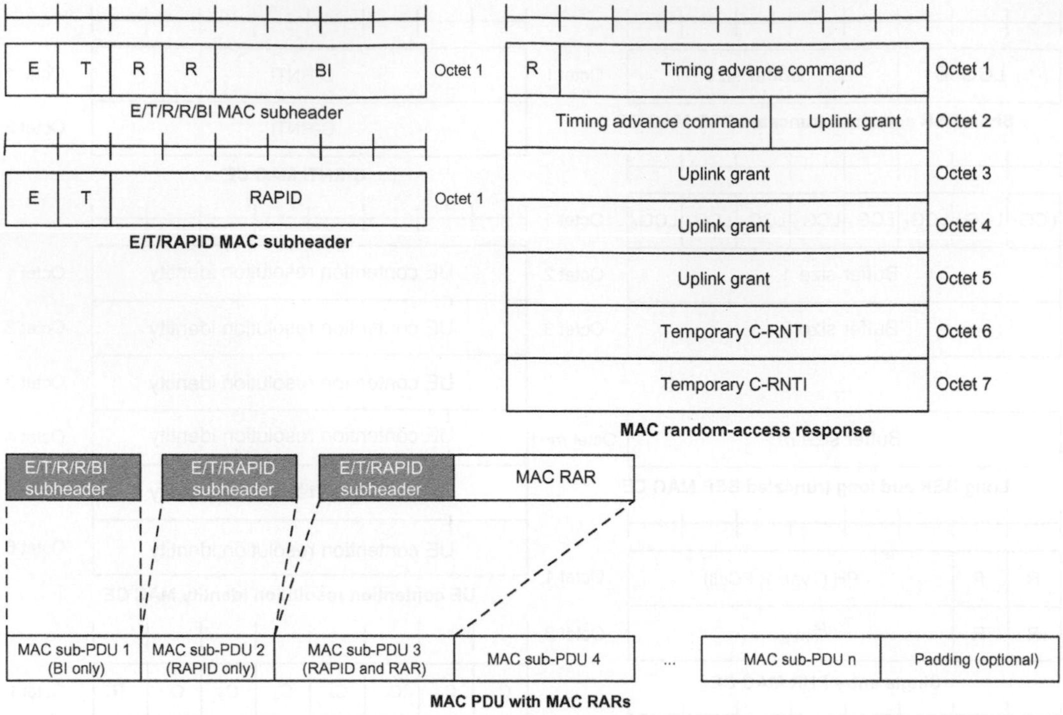

Figure 2.9
Structure of MAC subheaders and example MAC PDU with random-access response [9].

channels, including any MAC CEs, are multiplexed to form transport blocks per component carrier, where each component carrier has its own HARQ entity.

In order to efficiently utilize the radio resources, the MAC sublayer in gNB includes schedulers that allocate radio resources in the downlink/uplink to the active users in the cell. The default mode of operation of a gNB scheduler is dynamic scheduling, in which the gNB makes scheduling decisions, typically once per slot, and sends the scheduling information to a group of devices. While slot-based scheduling is a common practice, neither the scheduling decisions, nor the actual data transmission are required to start or end at the slot boundaries, which is important for low-latency applications and unlicensed spectrum operation. The downlink/uplink scheduling are independent and scheduling decisions can be independently made.

The downlink scheduler dynamically controls the radio resources allocated to active devices in order to efficiently share the DL-SCH. The selection of transport format which includes transport block size, modulation and coding scheme, and antenna mapping, as well as logical channel multiplexing for downlink transmissions are all managed by the gNB scheduler.

The uplink scheduler serves a similar purpose, that is, dynamically controlling the transmission opportunities of the active UEs and efficiently sharing of UL-SCH. The scheduling strategy of a gNB is vendor-dependent and implementation specific and is not specified by 3GPP. In general, the ultimate goal of all schedulers is to take advantage of the channel variations experienced by various devices and to schedule the transmissions/receptions to/from each device on the radio resources with advantageous channel conditions in time and frequency domains, that is, channel-dependent scheduling [8].

The downlink channel-dependent scheduling is supported through periodic/aperiodic channel state information reports sent by the devices to the gNB, which provide information on the instantaneous downlink channel quality in time and frequency domains, as well as information to determine an appropriate antenna/beam configuration. In the uplink, the channel state information required for channel-dependent scheduling can be obtained from the sounding reference signals transmitted by each device to the gNB. In order to assist the uplink scheduler, the device can further transmit BSR (measuring the data that is buffered in the logical channel queues in the UE) and PHR (measuring the difference between the nominal UE maximum transmit power and the estimated power for uplink transmission) to the gNB using MAC CEs. This information can only be transmitted, if the device has been given a valid scheduling grant. While dynamic scheduling is the default operation mode of many base station schedulers, there are cases where semi-persistent scheduling (SPS) is preferred due to reduced signaling overhead [17].

In the downlink, the gNB can dynamically allocate resources to the active UEs and identify them by their C-RNTIs on PDCCH(s). A UE always monitors the PDCCH(s) in order to find possible assignments according to its configured DRX cycles. When carrier aggregation is configured, the same C-RNTI applies to all serving cells. The gNB may preempt an ongoing PDSCH transmission to one UE with a delay-sensitive transmission to another UE. The gNB can configure the UEs to monitor interrupted transmission indications using INT-RNTI on a PDCCH. If a UE receives the interrupted transmission indication, it may assume that no useful information was intended for it by the resource elements included in the indication, even if some of those resource elements were already scheduled for that UE. Furthermore, the gNB can allocate downlink resources for the initial HARQ transmissions to the UEs configured with SPS. In that case, the RRC signaling defines the periodicity of the configured downlink assignments while PDCCH addressed to CS-RNTI can either signal and activate the configured downlink assignment or deactivate it, that is, a PDCCH addressed to CS-RNTI indicates that the downlink assignment can be implicitly reused according to the periodicity defined by the RRC signaling until it is deactivated [8].

The dynamically allocated downlink reception overrides the configured downlink assignment in the same serving cell, if they overlap in time. When carrier aggregation or bandwidth adaptation is configured, one configured downlink assignment can be signaled per serving cell or per BWP, respectively. In each serving cell, the gNB can only configure one active downlink assignment for a UE at a time; however, it can simultaneously configure multiple active downlink assignments on different serving cells. Activation and deactivation of configured downlink assignments are independent among the serving cells [8].

In the uplink, the gNB can dynamically allocate resources to the active UEs and identify them by their C-RNTI on PDCCH(s). A UE always monitors the PDCCH(s) in order to find possible grants for uplink transmission according to its configured DRX cycles. When carrier aggregation is configured, the same C-RNTI applies to all serving cells. In addition, the gNB can allocate uplink resources for the initial HARQ transmissions to the UEs with the configured grants (alternatively known as semi-persistent scheduling). Two types of configured uplink grants are specified in NR. In Type 1 configured grant the RRC signaling directly provides the configured uplink grant (including the periodicity), whereas in Type 2 configured grant, the RRC signaling defines the periodicity of the configured uplink grant, while PDCCH addressed to CS-RNTI can either signal and activate the configured uplink grant or deactivate it. Therefore, a PDCCH addressed to CS-RNTI indicates that the uplink grant can be implicitly reused according to the periodicity defined through RRC signaling until it is deactivated. The dynamically allocated uplink transmission overrides the configured uplink grant in the same serving cell, if they overlap in time. The retransmissions other than repetitions are required to be explicitly allocated via PDCCH(s) [8]. If carrier aggregation or bandwidth adaptation is configured, one configured uplink grant can be signaled per serving cell or per BWP, respectively. In each serving cell, there is only one active configured uplink grant at a time. A configured grant for a serving cell in the uplink can either be Type 1 or Type 2. The activation and deactivation of configured grants for the uplink are independent among the serving cells for Type 2. In the case of supplemental uplink (SUL), a configured grant can only be signaled for one of the two uplink carriers of the cell [8].

When a downlink assignment is configured for SPS, the UE MAC entity will assume that the Nth downlink assignment occurs in the slot number that satisfies the following criterion [9]:

$$\left(N_{frame}^{slot} SFN + n_{slot}\right) = \left[\left(N_{frame}^{slot} SFN_{start-time} + slot_{start-time}\right) + NT_{SPS} N_{frame}^{slot}/10\right] \bmod \left(1024 N_{frame}^{slot}\right)$$

where $SFN_{start-time}$ and $slot_{start-time}$ are the system frame number (SFN) and slot of the first transmission of PDSCH where the configured downlink assignment was (re)initialized.

When a uplink grant is configured for a configured grant Type 1, the UE MAC entity will assume that the uplink grant repeats at symbols for which the following criterion is met [9]:

$$\left[\left(N_{frame}^{slot}N_{slot}^{symbol}SFN\right)+\left(n_{slot}N_{slot}^{symbol}\right)+n_{symbol}\right]=\left(N_{slot}^{symbol}T_{offset}+S+NT_{SPS}\right)\times \bmod\left(1024N_{frame}^{slot}N_{slot}^{symbol}\right)\forall N\geq 0$$

When a uplink grant is configured for a configured grant Type 2, the UE MAC will assume that the uplink grant repeats with each symbol for which the following criterion is satisfied [9]:

$$\left[\left(N_{frame}^{slot}N_{slot}^{symbol}SFN\right)+\left(n_{slot}N_{slot}^{symbol}\right)+n_{symbol}\right]=\left[\left(N_{frame}^{slot}N_{slot}^{symbol}SFN_{start-time}+slot_{start-time}N_{slot}^{symbol}+symbol_{start-time}\right)+NT_{SPS}\right]\times \bmod\left(1024N_{frame}^{slot}N_{slot}^{symbol}\right)\forall N\geq 0$$

where $SFN_{start-time}$, $slot_{start-time}$, and $symbol_{start-time}$ are the SFN, slot, and symbol corresponding to the first transmission opportunity of physical uplink shared channel (PUSCH), where the configured uplink grant was (re)initialized.

Measurement reports are required to enable the MAC scheduler to make scheduling decisions in the downlink and uplink. These reports include transport capabilities and measurements of UEs instantaneous radio conditions. The uplink BSRs are needed to support QoS-aware packet scheduling. In NR, uplink BSRs refer to the data that is buffered for an LCG in the UE. There are eight LCGs and two reporting formats in the uplink: a short format to report only one BSR (of one LCG) and a flexible long format to report several BSRs (up to eight LCGs). The uplink BSRs are transmitted using MAC CEs. When a BSR is triggered upon arrival of data in the transmission buffers of the UE, an SR is transmitted by the UE. The PHRs are needed to support power-aware packet scheduling. In NR, there are three types of reporting, that is, for PUSCH transmission, PUSCH and PUCCH transmission, and SRS transmission. In case of carrier aggregation, when no transmission takes place on an activated SCell, a reference power is used to provide a virtual report. The PHRs are transmitted using MAC CEs [8].

The HARQ functionality in the MAC sublayer ensures reliable transport of MAC PDUs between peer entities over the physical layer. A HARQ mechanism along with soft combining can provide robustness against transmission errors. A single HARQ process supports one or multiple transport blocks depending on whether the physical layer is configured for downlink/uplink spatial multiplexing. An asynchronous incremental redundancy HARQ protocol is supported in the downlink and uplink. The gNB provides the UE with the HARQ-ACK feedback timing either dynamically via DCI or semi-statically through an RRC configuration message. The UE may be configured to receive code-block-group–based transmissions where retransmissions may be scheduled to carry a subset of the code blocks included in a transport block. The gNB schedules each uplink transmission and retransmission using the uplink grant on DCI [8].

The HARQ protocol in NR uses multiple parallel stop-and-wait processes. When a transport block is received at the receiver, it attempts to decode the packet and to inform the transmitter about the outcome of the decoding process through an ACK bit indicating whether the decoding process was successful or requesting the retransmission of the transport block. The HARQ-ACKs are sent by the receiver based on a specific timing relationship between UL HARQ-ACKs and DL HARQ processes or based on the position of the ACK bit in the HARQ-ACK codebook when multiple ACKs are transmitted simultaneously. An asynchronous HARQ protocol is used for both downlink and uplink, that is, an explicit HARQ process number is used to identify a particular process, since the retransmissions are scheduled in the same way as the initial transmission (explicit scheduling). The use of an asynchronous UL HARQ protocol, instead of a synchronous one that was used in LTE, was deemed to be necessary to support dynamic TDD where there is no fixed UL/DL allocation. It also provides more flexibility in terms of prioritization of data flows and devices and is further useful for operation in unlicensed spectrum. The NR supports up to 16 HARQ processes. The larger number of HARQ processes (compared to LTE) was motivated by the disaggregated RAN architectures and consideration for remote radio heads and fronthaul transport delay, as well as the use of shorter slot durations at high-frequency bands. It must be noted that the larger number of maximum HARQ processes does not imply a longer roundtrip delay, since the decoding will typically succeed after a few retransmissions depending on the channel conditions. Note that the PDCP sublayer can provide in-sequence delivery; thus this function is not provided by the RLC sublayer in order to reduce the latency.

A new feature of the NR HARQ protocol (compared to LTE) is the possibility for retransmission of code block groups, which is useful for very large transport blocks or when a transport block is partially preempted by another transmission. As part of the channel coding operation in the physical layer, a transport block is split into one or more code blocks with channel coding applied to each of the code blocks of up to 8448 bits in order to maintain a reasonable complexity. In practice and in the presence of burst errors, only a few code blocks in the transport block may be corrupted and the majority of code blocks are correctly received. In order to correctly deliver the transport blocks to the destination MAC sublayer, it is sufficient to only retransmit the erroneous code blocks. Furthermore, to avoid the excessive control signaling overhead due to individual code block addressing by HARQ mechanism, code block groups have been defined. If per-CBG retransmission is configured, feedback is provided per-CBG and only the erroneously received code block groups are retransmitted. The CBG-based retransmissions are transparent to the MAC sublayer and are handled in the physical layer. From MAC sublayer perspective, the transport block is not correctly received until all CBGs are correctly received. It is not possible to mix the CBGs belonging to another transport block with retransmissions of CBGs belonging to the incorrectly received transport block in the same HARQ process [17].

2.2.2 Radio Link Control Sublayer

The RLC sublayer is located between PDCP and MAC sublayers as shown in Fig. 2.1. The RRC sublayer controls and configures the RLC functions. The RLC functions are performed by the RLC entities. For an RLC entity configured at the gNB, there is a peer RLC entity configured at the UE and vice versa. An RLC entity receives/delivers RLC SDUs from/to the upper layer and sends/receives RLC PDUs to/from its peer RLC entity via lower layers. The RLC sublayer can operate in one of the three modes of operation defined as transparent mode, unacknowledged mode, and acknowledged mode. Depending on the mode of operation, the RLC entity controls the usage of error correction, segmentation, resegmentation, reassembly, and duplicate detection of SDUs. An RLC entity in any mode can be configured either as a transmitting or a receiving entity. The transmitting RLC entity receives RLC SDUs from the upper layer and sends RLC PDUs to its peer receiving RLC entity via lower layers. The receiving RLC entity delivers RLC SDUs to the upper layer and receives RLC PDUs from its peer transmitting RLC entity via lower layers. Fig. 2.10 illustrates the

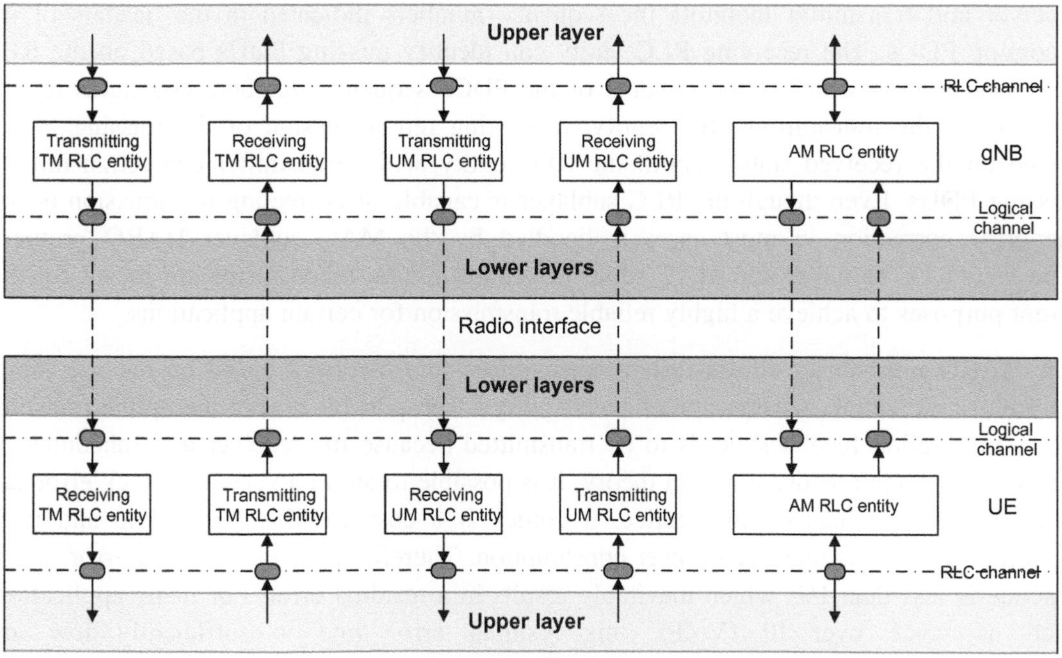

Figure 2.10
High-level architecture of RLC sublayer [10].

high-level architecture of the RLC sublayer. The [octet-aligned] RLC SDUs of variable sizes are supported for RLC entities. Each RLC SDU is used to construct an RLC PDU without waiting for notification from the MAC sublayer for a transmission opportunity. In the case of RLC-UM and RLC-AM entities, an RLC SDU may be segmented and transported using two or more RLC PDUs based on the notification(s) from the lower layer [10].

In RLC-TM, no RLC headers are added. The RLC-UM supports segmentation and duplicate detection, while the RLC-AM supports retransmission of erroneous packets. A key difference with LTE is that the NR RLC sublayer does not handle in-sequence delivery of SDUs to the upper layers. Eliminating the in-sequence delivery function from the RLC sublayer reduces the overall latency since the correctly received packets do not have to wait for retransmission of an earlier (missing) packet before being delivered to the higher layers. Another difference is the removal of concatenation function from the RLC protocol to allow RLC PDUs to be assembled prior to receiving the uplink scheduling grant. By eliminating the concatenation from RLC functions the RLC PDUs can be assembled in advance and upon receipt of the scheduling decision the device can forward a number of RLC PDUs to the MAC sublayer depending on the scheduled transport block size.

The RLC retransmission mechanism is used to provide error correction of data delivered to higher layers. The retransmission protocol operating between the RLC entities in the receiver and transmitter monitors the sequence numbers indicated in the headers of the incoming PDUs. The receiving RLC entity can identify missing PDUs based on the RLC sequence number which is independent of the PDCP sequence number. The status reports are sent to the transmitting RLC entity, requesting retransmission of the missing PDUs. Based on the received status report, the RLC entity at the transmitter can retransmit the missing PDUs. Even though the RLC sublayer is capable of correcting transmission errors, the error correction in most cases is handled by the MAC sublayer HARQ protocol. However, RLC sublayer and MAC sublayer retransmission mechanisms are meant for different purposes to achieve a highly reliable transmission for certain applications.

The HARQ mechanism allows fast retransmissions and feedback based on the success or failure of a downlink transmission after receiving a transport block. For the uplink transmissions, no explicit feedback needs to be transmitted because the receiver and scheduler are located in the same node. While in theory, it is possible to attain a very low block error rate using HARQ mechanism, in practice, it comes at a cost additional signaling and radio resources as well as increased power consumption. Therefore, the target block error rate in practice is less than 1%, which inevitably results in a residual error. For many applications such as voice over IP (VoIP), this residual error rate is sufficiently low and acceptable (because voice decoders can tolerate some level of frame erasure); however, there are use cases where lower block error rates are required, for example, transmission of RRC and NAS messages.

A sufficiently low block error rate is not only required for URLLC services, but it also is important from a system-level perspective in terms of sustaining data rate performance. The TCP protocol (at transport layer) requires virtually error-free delivery of packets to the peer TCP layer. As an example, to obtain sustained data rates in the excess of 100 Mbps in TCP/IP applications, a packet error rate of less than 10^{-5} is required, because the TCP protocol would consider packet errors due to network congestion and would trigger the congestion-avoidance mechanism to decrease the data rate, causing reduction of the overall data rate performance of the system. It must be noted that the infrequent transmission of RLC status reports compared to more frequent HARQ retransmission makes obtaining a reliability level of 10^{-5} using RLC ARQ retransmissions more practical [17]. Therefore, complementing MAC sublayer HARQ protocol with RLC sublayer ARQ mechanism would achieve relatively lower latency and reasonable feedback overhead to satisfy the stringent requirements of URLLC applications. The RLC sublayer ARQ mechanism retransmits RLC SDUs or RLC SDU segments based on RLC status reports, where polling may be used to request RLC status reports. An RLC receiver can also trigger RLC status report after detecting a missing RLC SDU or RLC SDU segment [8]. The RLC-TM entity is configured to transmit/receive RLC PDUs through BCCH, DL/UL CCCH, and paging control channel logical channels. The RLC-UM entity is configured to transmit/receive RLC PDUs via downlink or uplink dedicated traffic channel (DTCH). The RLC-AM entity can be configured to transmit/receive RLC PDUs through downlink or uplink dedicated control channel (DCCH) or DL/UL DTCH logical channels. Functional models of RLC-AM and RLC-UM entities are illustrated in Fig. 2.11, where the functions that are only related to RLC-AM mode (ARQ functions) are marked with a dark color.

The RLC PDUs and SDUs are bit strings that are octet-aligned. The TM data (TMD PDU) consists only of a data field and does not include any RLC headers. In RLC-AM and RLC-UM modes, a sequence number is generated and attached to the incoming SDUs using 6 or 12 bits for the RLC-UM and 12 or 18 bits for the RLC-AM. The sequence number is included in the RLC PDU header as shown in Fig. 2.12. If the SDU is not segmented, the RLC PDU consists of the RLC SDU and a header, which allows the RLC PDUs to be generated in advance as the header does not depend on the transport block size. However, depending on the transport block size after multiplexing at MAC sublayer, the size of the last RLC PDU in a transport block may not match the RLC SDU size, thus requiring dividing the SDU into multiple segments. If no segmentation is done, padding need to be used which would adversely impact the spectral efficiency. As a result, dynamically varying the number of RLC PDUs in a transport block along with segmentation to adjust the size of the last RLC PDU, ensures that the transport block is efficiently utilized. Segmentation is done by dividing the last preprocessed RLC SDU into two segments, the header of the first segment is updated, and a new header is added to the second segment as shown in Fig. 2.3. Each RLC SDU segment carries the same sequence number as the original SDU and the

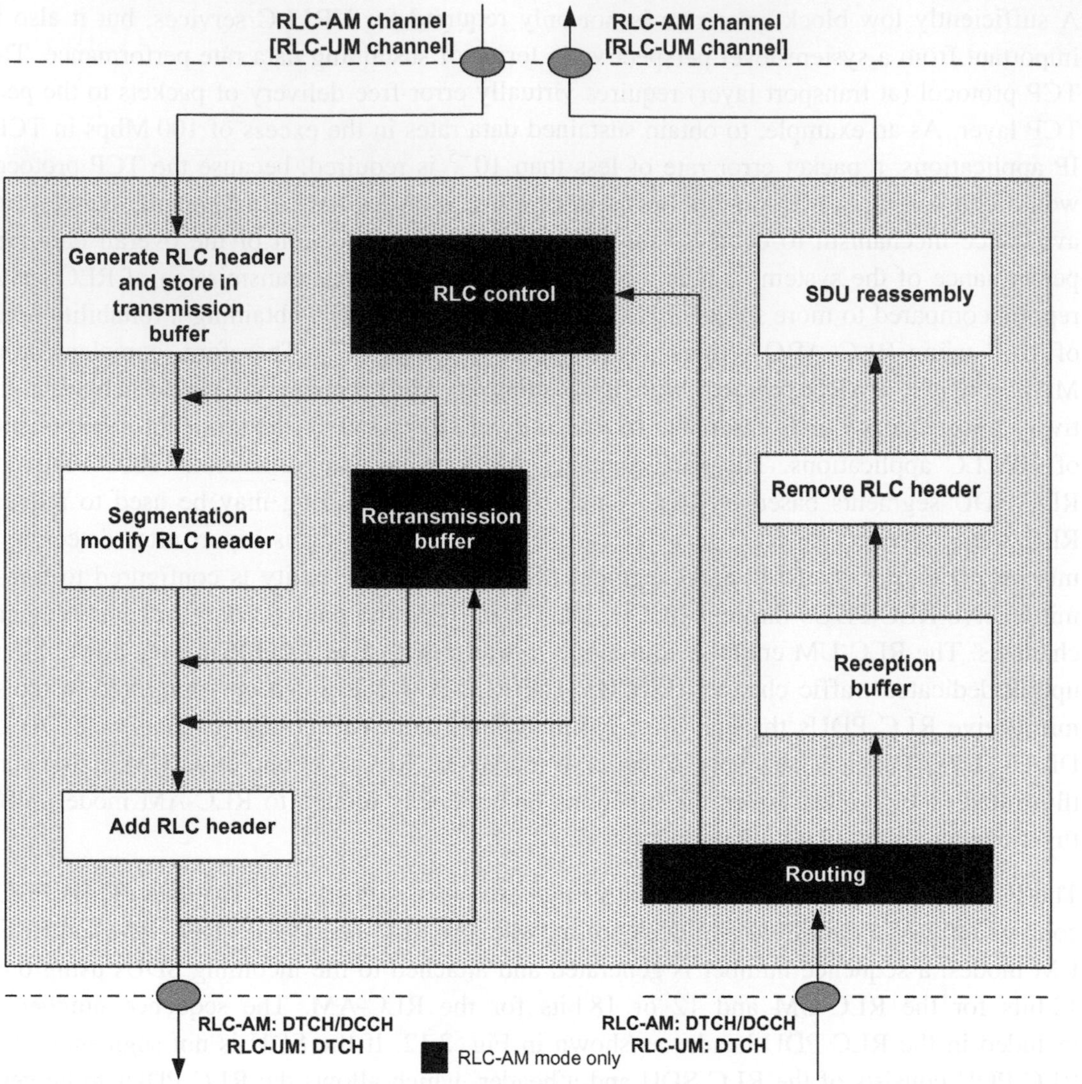

Figure 2.11
RLC-AM and RLC-UM entity models [10].

sequence number is part of the RLC header. To distinguish an unsegmented RLC PDU from a segmented one, a segmentation information field is included in the RLC header, indicating whether the PDU is a complete SDU, the first segment of the SDU, the last segment of the SDU, or a segment between the first and last segments of the SDU. Furthermore, in the case of a segmented SDU, a 16 bit segmentation offset (SO) is included in all segments except the first one to indicate which byte of the SDU is represented by the segment. The RLC header may further include a poll (P) bit, which is used to request status

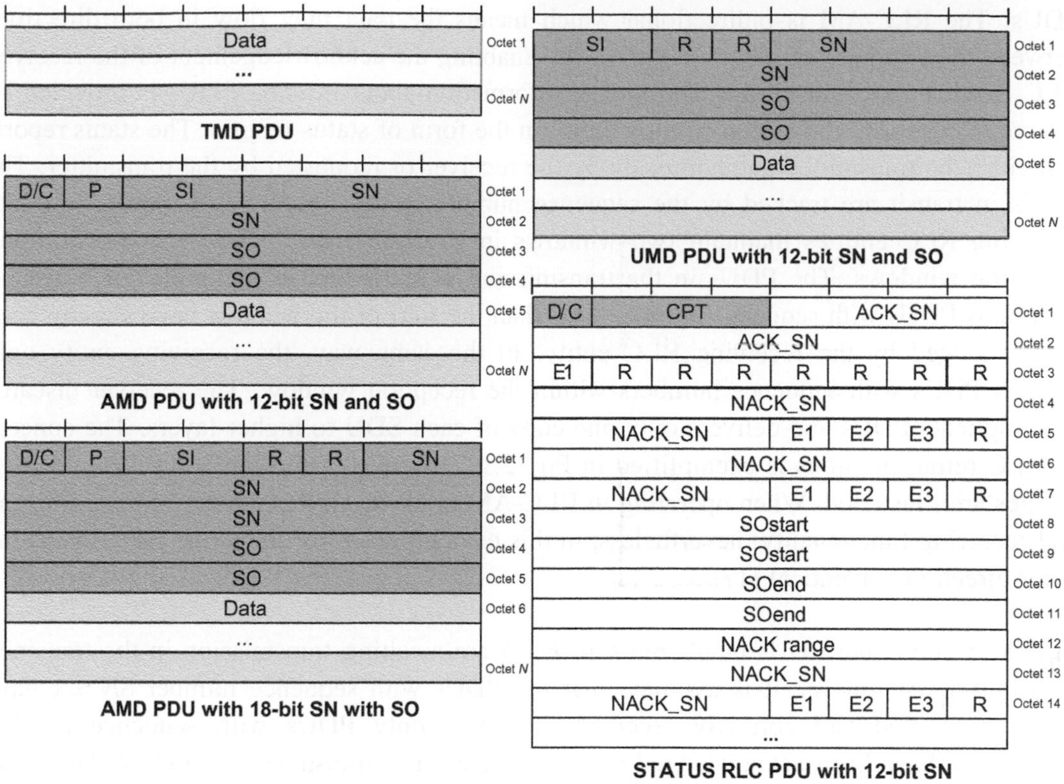

Figure 2.12
Example RLC PDU formats[10].

report for RLC-AM, and a data/control (D/C) indicator, indicating whether the RLC PDU contains data to/from a logical channel or control information required for RLC operation [10,17].

The retransmission of missing RLC PDUs is one of the main services of the RLC-AM entity. While MAC sublayer HARQ protocol provides some level of error correction, when combined with the ARQ retransmission mechanism, they can provide a higher level of reliability. The missing RLC PDUs can be detected by inspecting the sequence numbers of the received PDUs, and a retransmission requested from the transmitting side. The RLC-AM mode in NR is similar to LTE except the in-sequence delivery service is not supported in NR. As we mentioned earlier, eliminating the in-sequence delivery from the RLC helped reduce the overall latency. It further reduces the buffering requirements at the RLC sublayer. In LTE, which supports the in-sequence delivery service at RLC sublayer, an RLC SDU cannot be forwarded to higher layers unless all previous SDUs have been correctly received. A single missing SDU can significantly delay the delivery of the subsequent

SDUs. The RLC-AM is bidirectional which means the data may flow in both directions between the two peer entities (see Fig. 2.11), enabling the acknowledgement of the received RLC PDUs to the transmitting entity. The information about missing PDUs is provided by the receiving entity to the transmitting entity in the form of status reports. The status reports can either be transmitted autonomously by the receiver or requested by the transmitter. The PDUs in transit are tracked by the sequence number in the header. The transmitting and receiving RLC entities maintain two windows in RLC-AM, that is, the transmission and reception windows. The PDUs in the transmission window are only eligible for transmission, thus PDUs with sequence numbers less than the start of the window have already been acknowledged by the receiving RLC entity. In the same way, the receiving entity only accepts PDUs with sequence numbers within the reception window. The receiver discards any duplicate PDUs and delivers only one copy of each SDU to higher layers. The concept of RLC retransmissions is exemplified in Fig. 2.13, where transmitting and receiving RLC entities are illustrated. When operating in RLC-AM mode, each RLC entity has transmitting and receiving functionality; nevertheless, in this example we only show one direction as the other direction is identical [17].

In this example, the PDUs numbered k to $k+4$ are awaiting transmission in the transmission buffer. At time t_0, it is assumed that the PDUs with sequence number $SN \leq k$ have been transmitted and correctly received; however, only PDUs with sequence number $SN \leq k-1$ have been acknowledged by the receiver. The transmission window starts at k, that is, the first unacknowledged PDU, while the reception window starts at $k+1$, that is, the next PDU expected to be received. Upon reception of kth PDU, the SDU is reassembled and delivered to the higher layers. For a PDU containing an unsegmented SDU the reassembly function only involves header removal, but in the case of a segmented SDU, the SDU cannot be delivered to upper layers until the PDUs carrying all segments of the SDU arrive at the receiver. The transmission of PDUs continues such that at time t_1, PDUs $k+1$ and $k+2$ are transmitted but, at the receiving end, only PDU $k+2$ has arrived. As soon as a complete SDU is received, it is delivered to the higher layers; thus PDU $k+2$ is forwarded to the higher sublayer without waiting for the missing PDU $k+1$, which could be undergoing retransmission by the HARQ protocol. Therefore, the transmission window remains unchanged, since none of the PDUs with $SN \geq k$ have been acknowledged by the receiver. This could result in retransmission of these PDUs given that the transmitter is not aware of whether they have been correctly received. The reception window is not updated when PDU $k+2$ arrives because of the missing PDU $k+1$. At this point, the receiver starts the *t-Reassembly* timer. If the missing PDU $k+1$ is not received before the timer expires, a retransmission is requested. If the missing PDU arrives at time t_2 before the timer expires, the reception window is advanced, the reassembly timer is stopped, and PDU $k+1$ is delivered for reassembly with SDU $k+1$. The RLC sublayer is also responsible for duplicate detection using the same sequence number that is used for retransmission management.

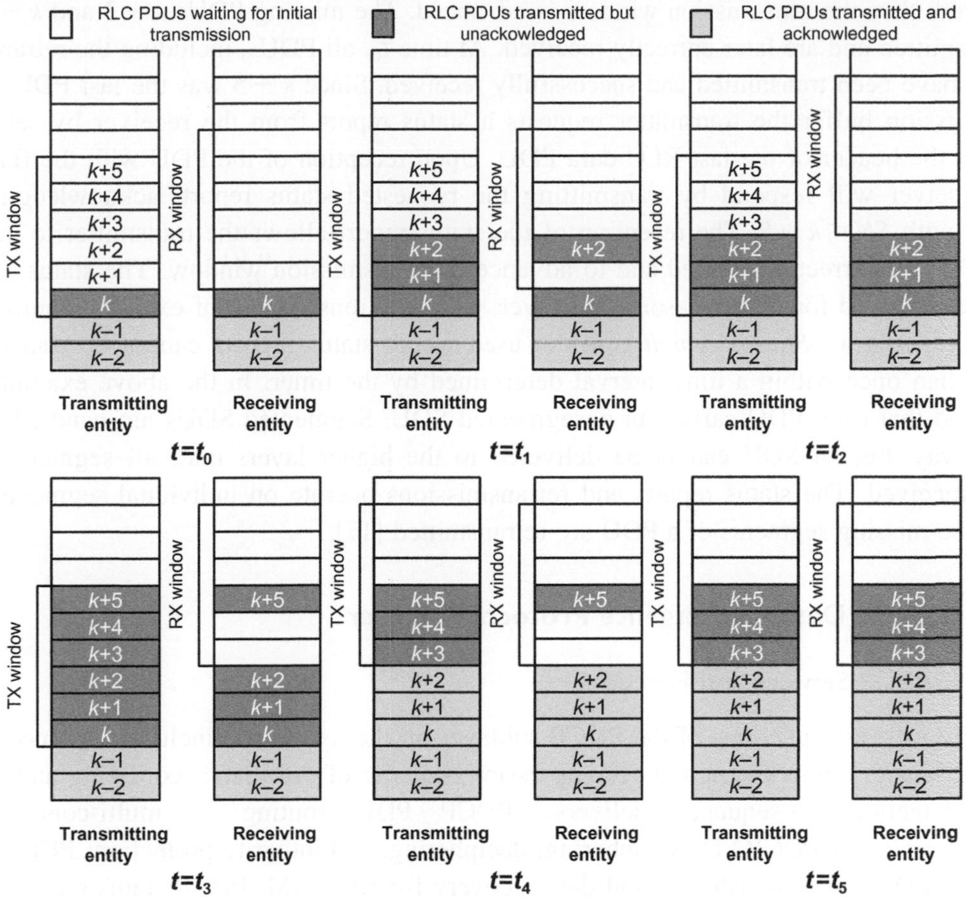

Figure 2.13
Example RLC-AM operation [17].

If PDU $k+2$ arrives again within the reception window, it will be discarded due to existence of another copy.

The transmission continues with PDUs $k+3$, $k+4$, and $k+5$, as shown in Fig. 2.13. At time t_3, PDUs with $SN \leq k+5$ have been transmitted, but only PDU $k+5$ has arrived and PDUs $k+3$ and $k+4$ are missing. Similar to the previous case, this causes the *t-Reassembly* timer to start. However, in this example no PDUs arrive prior to the expiration of the timer. The expiration of the timer at time t_4 triggers the receiver to send a control PDU containing a status report indicating the missing PDUs to its peer entity. The control PDUs have higher priority than data PDUs to avoid delayed transmission of the status reports, which would adversely impact the retransmission delay. Upon receipt of the status report at time t_5, the transmitter is informed that PDUs up to $k+2$ have been correctly received; thus the transmission window is advanced. The missing PDUs $k+3$ and $k+4$ are retransmitted and are later correctly received. At time t_6, all PDUs, including the retransmissions, have been transmitted and successfully received. Since $k+5$ was the last PDU in the transmission buffer the transmitter requests a status report from the receiver by setting a flag in the header of the last RLC data PDU. Upon reception of the PDU with the flag set, the receiver will respond by transmitting the requested status report, acknowledging all PDUs with $SN \leq k+5$. The reception of the status report allows the transmitter to declare all PDUs as correctly received and to advance the transmission window. The status reports can be triggered for many reasons. However, to avoid transmission of excessive number of status reports, a *t-StatusProhibit* timer is used where status reports cannot be transmitted more than once within a time interval determined by the timer. In the above example, we assumed that each PDU carries an unsegmented SDU. Segmented SDUs are handled in the same way, but an SDU cannot be delivered to the higher layers until all segments have been received. The status reports and retransmissions operate on individual segments, and only the missing segments of a PDU are retransmitted [17].

2.2.3 Packet Data Convergence Protocol Sublayer

2.2.3.1 PDCP Services and Functions

The services and functions of the PDCP sublayer on the user-plane include sequence numbering; header compression and decompression; transfer of user data; reordering and duplicate detection; in-sequence delivery; PDCP PDU routing in multi-connectivity; retransmission of PDCP SDUs; ciphering, deciphering, and integrity protection; PDCP SDU discard; PDCP re-establishment and data recovery for RLC-AM; PDCP status reporting for RLC-AM; duplication of PDCP PDUs and duplicate discard indication to lower layers. The main services and functions of the PDCP sublayer on the control-plane consist of sequence numbering; ciphering, deciphering, and integrity protection; transfer of control-plane data; reordering and duplicate detection; in-sequence delivery; duplication of PDCP PDUs and duplicate discard indication to lower layers. The PDCP protocol performs (optional) IP-header compression, followed by ciphering, for each radio bearer. A PDCP header is added, carrying information required for deciphering in the other end, as well as a sequence number used for retransmission and in-sequence delivery [8].

In some services and applications such as VoIP, interactive gaming, and multimedia messaging, the data payload of the IP packet is almost the same size or even smaller than the header itself. Over the end-to-end connection comprising multiple hops, these protocol headers are extremely important, but over a single point-to-point link, these headers serve no useful purpose. It is possible to compress these headers, and thus save the bandwidth and use the expensive radio resources more efficiently. The header compression also provides other important benefits, such as reduction in packet loss and improved interactive response time. The payload header compression is the process of suppressing the repetitive portion of payload headers at the sender side and restoring them at the receiver side of a low-bandwidth/capacity-limited link. The use of header compression has a well-established history in transport of IP-based payloads over capacity-limited wireless links where more bandwidth efficient transport methods are required. The Internet Engineering Task Force (IETF)[2] has developed several header compression protocols that are widely used in telecommunication systems. The header compression mechanism used in 3GPP LTE and NR standards is based on IETF RFC 5795, ROHC framework. The IP together with transport protocols such as TCP or UDP and application layer protocols (e.g., RTP) are described in the form of payload headers. The information carried in the header helps the applications to communicate over large distances connected by multiple links or hops in the network. This information consists of source and destination addresses, ports, protocol identifiers, sequence numbers, error checksums, etc. Under nominal conditions, most of the information carried in packet-headers remains the same or changes in specific patterns. By observing the fields that remain constant or change in specific patterns, it is possible either not to send them in each packet, or to represent them in a smaller number of bits than would have been originally required. This process is referred to as header compression.

The PDCP sublayer, at the transmit side, performs encryption of IP packets to protect user privacy, and additionally for the control-plane messages, performs integrity protection to ensure that control messages originate from the correct source and can be authenticated. At the receiver side the PDCP performs the corresponding decryption and decompression operation. The PDCP further discards duplicate PDUs and performs in-sequence delivery of the packets. Upon handover, the undelivered downlink PDUs will be forwarded by the PDCP sublayer of the source gNB to peer entity of the target gNB for delivery to the UE. The PDCP entity in the device will also handle retransmission of all uplink packets that were not delivered to the gNB given that the HARQ buffers are flushed upon handover. In this case, some PDUs may be received in duplicate form, possibly from both the source and the target gNBs. In this case, the PDCP will remove any duplicates. The PDCP entity can also be configured to perform reordering to ensure in-sequence delivery of SDUs to higher layer protocols. In some cases, the PDUs can be duplicated and transmitted on multiple cells, increasing the likelihood of their correct reception at the receiving end, which can be useful for services requiring very high reliability. At the receiving end, the PDCP duplicate removal functionality removes any duplicates. The PDCP plays an important role in dual connectivity, where a device is connected to two cells, that is, an MCG and an SCG. The two cell groups can be handled by different gNBs. A radio bearer is typically handled by one of the cell groups, but there is also the possibility for split bearers, in which case one

[2] IETF: https://www.ietf.org/.

radio bearer is handled by both cell groups. In this case the PDCP entity is responsible for distributing the data between the MCG and the SCG [17].

The PDCP entities are located in the PDCP sublayer. Several PDCP entities may be defined for a UE. Each PDCP entity carries the data of one radio bearer, which is associated with either the control-plane or the user-plane. Fig. 2.14 illustrates the functional block diagram of a PDCP entity. For split bearers, the routing function is performed in the transmitting PDCP entity. The PDCP sublayer provides services to RRC and SDAP sublayers. The PDCP provides transfer of user-plane and control-plane data; header compression; ciphering; integrity protection services to the higher layer protocols. The maximum size of a data or control PDCP SDU supported in NR is 9000 bytes [8]. It must be noted that an NR system provides protection against eavesdropping and modification attacks. Signaling traffic (RRC messages) is encrypted and integrity protected. User-plane traffic (IP packets) is encrypted and can be integrity protected. User-plane integrity protection is a new feature (relative to LTE) that is useful for small-data transmissions, and particularly

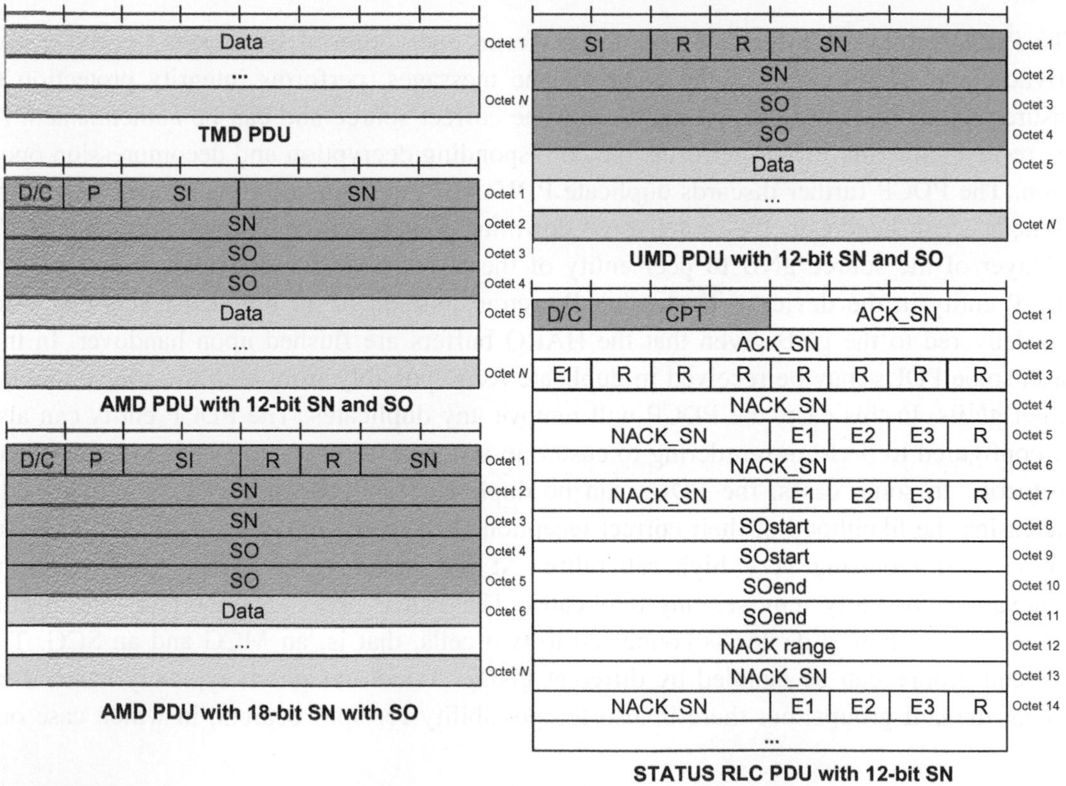

Figure 2.14
Functional block diagram of the PDCP entities [11].

for constrained IoT devices. Data traffic including voice calls, Internet traffic, and text messages are protected using encryption. The device and the network mutually authenticate each other and use integrity-protected signaling. This setup makes nearly impossible for an unauthorized party to decrypt and read the information that is communicated over the air. Although integrity protection of user-plane data is supported in 5G networks, this feature is not used in E-UTRA-NR-DC (EN-DC) scenarios since LTE systems only provide integrity protection of control-plane messages.

When duplication is configured for a radio bearer by RRC, a secondary RLC entity is added to the radio bearer to handle the duplicated PDCP PDUs, where the logical channels corresponding to the primary and the secondary RLC entities are referred to as the *primary logical channel* and the *secondary logical channel*, respectively. Therefore, the duplication function at PDCP sublayer consists of submitting the same PDCP PDUs twice, that is, to the primary RLC entity and to the secondary RLC entity. With two independent transmission paths, packet duplication would improve the transmission reliability and reduce the latency, which is especially advantageous to URLLC services [8]. The PDCP control PDUs are not duplicated and are always submitted to the primary RLC entity. When configuring duplication for a DRB, the RRC sublayer sets the initial state to be either activated or deactivated. After the configuration, the state can be dynamically controlled by means of a MAC CE. In dual connectivity, the UE applies the MAC CE commands regardless of their origin (MCG or SCG). When duplication is configured for an SRB, the state is always active and cannot be dynamically controlled. When activating duplication for a DRB, NG-RAN ensures that at least one serving cell is activated for each logical channel of the DRB. When the deactivation of SCells leaves no activated serving cells for the logical channels of the DRB, NG-RAN ensures that duplication is also deactivated [8].

When duplication is activated, the original PDCP PDU and the corresponding duplicate are not transmitted on the same carrier. The primary and secondary logical channels can either belong to the same MAC entity (referred to as CA duplication) or to different ones (referred to as DC duplication). In CA duplication, logical channel mapping restrictions are applied to the MAC sublayer to ensure that the primary and secondary logical channels are not sent on the same carrier. When duplication is deactivated for a DRB, the secondary RLC entity is not reestablished, the HARQ buffers are not flushed, and the transmitting PDCP entity should indicate to the secondary RLC entity to discard all duplicated PDCP PDUs. In addition, in case of CA duplication, the logical channel mapping restrictions of the primary and secondary logical channels are relaxed as long as duplication remains deactivated. When an RLC entity acknowledges the transmission of a PDCP PDU, the PDCP entity notifies the other RLC entity to discard that PDU. When the secondary RLC entity

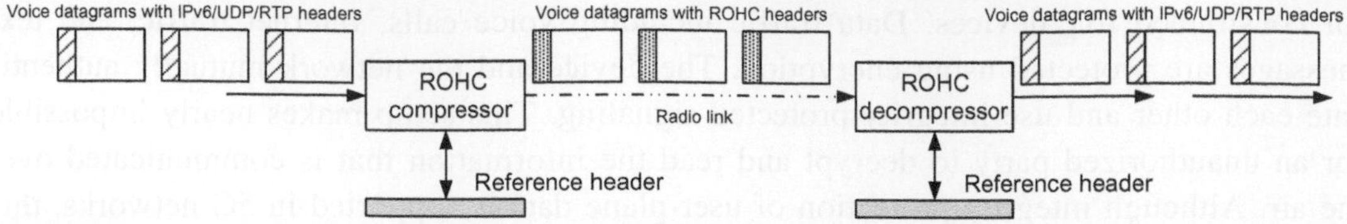

Figure 2.15
Example ROHC compression/decompression of RTP/UDP/IP headers for communication over a radio link [16].

reaches the maximum number of retransmissions for a PDCP PDU, the UE informs the gNB, but does not trigger RLF[3] [8].

2.2.3.2 Header Compression Function

There are multiple header compression algorithms referred to as profiles in 3GPP specifications [11]. Each profile is specific to a particular network layer, transport layer, or upper layer protocol combination, for example, TCP/IP or RTP/UDP/IP. The PDCP entities associated with DRBs carrying user-plane data can be configured by upper layers to use header compression. Each PDCP entity uses at most one ROHC compressor instance and at most one ROHC decompressor instance.

As we mentioned earlier, the ROHC algorithm reduces the size of transmitted RTP/UDP/IP header by removing the redundancies. This mechanism starts by classifying header fields into different classes according to their variation pattern. The fields that are classified as *inferred* are not sent. The *static* fields are sent initially and then are not sent anymore and the fields with varying information are always sent. The ROHC mechanism is based on a context,[4] which is maintained, by both ends, that is, the compressor and the decompressor (see Fig. 2.15). The context encompasses the entire header and ROHC information. Each context has a context ID, which identifies the flows. The ROHC scheme operates in one of the following three operation modes [16]:

---

3 The UE declares a RLF when one of the following criteria is met: (1) Expiry of a timer started after indication of radio problems from the physical layer (if radio problems are recovered before the timer expires, the UE stops the timer), random-access procedure failure, or RLC failure. After an RLF is declared, the UE stays in RRC_CONNECTED and selects a suitable cell and then initiates RRC connection reestablishment. The UE further enters RRC_IDLE, if a suitable cell cannot be found within a certain time after RLF was declared [8].

4 The context of the compressor is the state it uses to compress a header. The context of the decompressor is the state it uses to decompress a header. Either of these or the combinations of these two is usually referred to as "context." The context contains relevant information from previous headers in the packet stream, such as static fields and possible reference values for compression and decompression. Moreover, additional information describing the packet stream is also part of the context, for example, information about how the IP identifier field changes and the typical interpacket increase in sequence numbers or timestamps [16].

1. *Unidirectional mode (U)*, where the packets are only sent in one direction from compressor to decompressor. This mode makes ROHC usable over links where a return path from decompressor to compressor is unavailable or undesirable.
2. *Optimistic mode (O)* is a bidirectional mode similar to the unidirectional mode, except that a feedback channel is used to send error recovery requests and (optionally) acknowledgements of significant context updates from the decompressor to compressor. The O-mode aims to maximize compression efficiency and sparse usage of the feedback channel.
3. *Reliable mode (R)* is a bidirectional mode which differs in many ways from the previous two modes. The most important differences include intensive use of feedback channel and a strict logic at both compressor and decompressor that prevents loss of context synchronization between compressor and decompressor except for very high residual bit error rates.

The U-mode is used when the link is unidirectional or when feedback is not possible. For bidirectional links, O-mode uses positive feedback packets (ACK) and R-mode use positive and negative feedback packets [ACK and NACK]. The ROHC mechanism always starts header compression using U-mode even if it is used over a bidirectional link and it does not send retransmissions when an error occurs; thus the erroneous packet is dropped. The ROHC feedback is used only to indicate to the compressor side that there was an error and probably the context is damaged. After receiving a negative feedback the compressor always reduces its compression level.

The ROHC compressor has three compression states defined as follows [16]:

1. *Initialization and refresh (IR)*, where the compressor has been just created or reset and full packet-headers are sent.
2. *First order (FO)*, where the compressor has detected and stored the static fields such as IP addresses and port numbers on both sides of the connection.
3. *Second order (SO)*, where the compressor is suppressing all dynamic fields such as RTP sequence numbers and sending only a logical sequence number and partial checksum to make the other side generate the headers based on prediction and verify the headers of the next expected packet.

Each compression state uses a different header format in order to send the header information. The IR compression state establishes the context, which contains static and dynamic header information. The FO compression state provides the change pattern of dynamic fields. The SO compression state sends encoded values of *sequence number* (SN) and *timestamp* (TS), forming the minimal size packets (Figs. 2.6–2.15). Using this header format, all header fields can be generated at the other end of the radio link using the previously established change pattern. When some updates or errors occur, the compressor returns to

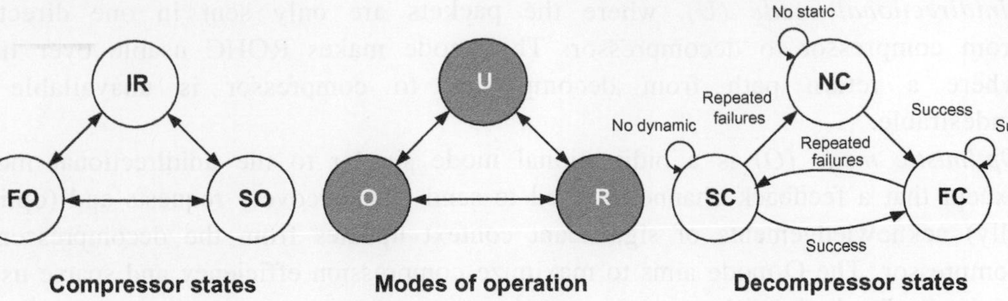

Figure 2.16
ROHC state machines [16].

upper compression states. It only transitions to the SO compression state after retransmitting the updated information and reestablishing the change pattern in the decompressor.

In the U-mode the feedback channel is not used. To increase the compression level an optimistic approach is used for compressor to ensure that the context has been correctly established at decompressor side. This means that compressor uses the same header format for a number of packets. Since the compressor does not know whether the context is lost, it also uses two timers, to be able to return to the FO and IR compression states. The decompressor works at the receiving end of the link and decompresses the headers based on the header fields' information of the context. Both the compressor and the decompressor use a context to store all the information about the header fields. To ensure correct decompression the context should be always synchronized. The decompressor has three states as follows: (1) no context (NC) where there is NC synchronization, (2) static context (SC) where the dynamic information of the context has been lost, and (3) full context (FC) when the decompressor has all the information about header fields. In FC state the decompressor transitions to the initial states as soon as it detects corruption of the context. The decompressor uses the "*k* out of *n*" rule by looking at the last *n* packets with CRC failures. If *k* CRC failures have occurred, it assumes the context has been corrupted and transitions to an initial state (SC or NC). The decompressor also sends feedback according to the operation mode (Fig. 2.16).

The values of the ROHC compression parameters that determine the efficiency and robustness are not defined in ROHC specification and are not negotiated initially but are stated as implementation dependent. The values of these parameters remain fixed during the compression process. The compression parameters are defined as follows [16]:

- *L*: In U-mode and O-mode, the ROHC compressor uses a confidence variable *L* in order to ensure the correct transmission of header information.
- *Timer_1* (*IR_TIMEOUT*): In U-mode, the compressor uses this timer to return to the IR compression level and periodically resends static information.

- *Timer_2* (*FO_TIMEOUT*): The compressor also uses another timer in U-mode, and this timer is used to go downward to FO compression level, if the compressor is working in SO compression level.
- *Sliding window width (SWW)*: The compressor, while compressing header fields such as *sequence number* (SN) and *TS*, utilizes window-based least significant bit (*W_LSB)* encoding that uses a sliding window of width equal to *SWW*.
- *W_LSB* encoding is used to compress those header fields whose change pattern is known. When using this encoding, the compressor sends only the least significant bits. The dccompressor uses these bits to construct the original value of the encoding fields.
- *k* and *n*: The ROHC decompressor uses a "*k* out of *n*" failure rule, where *k* is the number of packets received with an error in the last *n* transmitted packets. This rule is used in the state machine of the decompressor to assume the damage of context and move downwards to a state after sending a NACK to the compressor, if bidirectional link is used. The decompressor does not assume context corruption and remains in the current state until *k* packets arrive with error in the last *n* packets.

2.2.3.3 Ciphering and Integrity Protection Functions

The RRC confidentiality protection is provided by the PDCP sublayer between a UE and the serving gNB. The user-plane security policy indicates whether the user-plane confidentiality and/or user-plane integrity protection is activated for all DRBs belonging to the PDU session. The input parameters to the 128-bit NR encryption algorithm (NEA) (or alternatively encryption algorithm for 5G), which is used for ciphering, are 128-bit cipher key referred to as KEY (K_{RRCenc}), 32-bit COUNT (PDCP COUNT), 5-bit radio bearer identity BEARER, 1-bit direction of the transmission, that is, DIRECTION, and the length of the keystream required identified as LENGTH. The DIRECTION bit is set to zero for uplink and one for downlink. Fig. 2.17 illustrates the use of the ciphering algorithm NEA to encrypt plain text by applying a keystream using a bit-wise binary addition of the *plaintext* block and the *keystream* block. The *plaintext* block may be recovered by generating the same *keystream* block using the same input parameters and applying a bit-wise binary addition with the *ciphertext* block. Based on the input parameters, the algorithm generates the output keystream block *keystream* which is used to encrypt the input *plaintext* block to produce the output *ciphertext* block. The input parameter LENGTH only denotes the length of the *keystream* block and not its content [2,11]. The ciphering algorithm and key to be used by the PDCP entity are configured by upper layers and the ciphering method is applied according to the security architecture of 3GPP system architecture evolution (SAE, which is the LTE system architecture). The ciphering function is activated by upper layers. After security activation, the ciphering function is applied to all PDCP PDUs indicated by upper layers for downlink and uplink transmissions. The COUNT value is composed of a hyper-frame number (HFN) and the PDCP SN. The size of the HFN part in bits is equal to 32

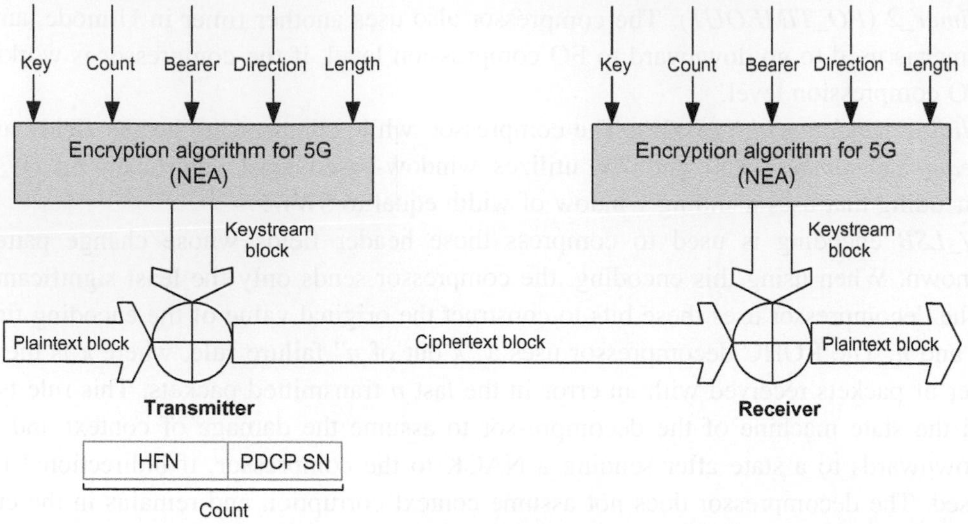

Figure 2.17
Illustration of the ciphering and deciphering procedures [2,11].

minus the length of the PDCP SN. The PDCP does not allow COUNT to wrap around in the downlink and uplink; thus it is up to the network to prevent it from happening [2,8,11].

As shown in Fig. 2.18, the input parameters to the 128-bit NR integrity protection algorithm (NIA) (or alternatively integrity protection algorithm for 5G) are the RRC messages denoted as MESSAGE, 128-bit integrity key K_{RRCint} referred to as KEY, 5-bit bearer identity BEARER, 1-bit direction of transmission denoted as DIRECTION, and a bearer specific direction-dependent 32-bit input COUNT which corresponds to the 32-bit PDCP COUNT. The RRC integrity checks are performed both in the UE and the gNB. If the gNB or the UE receives a PDCP PDU which fails the integrity check with faulty or missing message authentication code (MAC-I) after the start of integrity protection, the PDU will be discarded. The DIRECTION bit set to zero for uplink and one for downlink. The bit length of the MESSAGE is LENGTH. Based on these input parameters, the sender computes a 32-bit message authentication code (MAC-I/NAS-MAC)[5] using the integrity protection algorithm

[5] In cryptography a MAC-I is a cryptographic checksum on data that uses a session key to detect both accidental and intentional modifications of the data. A MAC (not to be confused with medium access control MAC) requires two inputs: a message and a secret key known only to the originator of the message and its intended recipient(s). This allows the recipient of the message to verify the integrity of the message and authenticate that the message sender has the shared secret key. Any mismatch between the sender's and receiver's calculated MAC-I values would invalidate the message. There are four types of message authentication codes: unconditionally secure, hash function-based, stream cipher-based, and block cipher-based. In the past, the most common approach to creating a message authentication code was to use block ciphers; however, hash-based MACs which use a secret key in conjunction with a cryptographic hash function to produce a hash, have become more widely used.

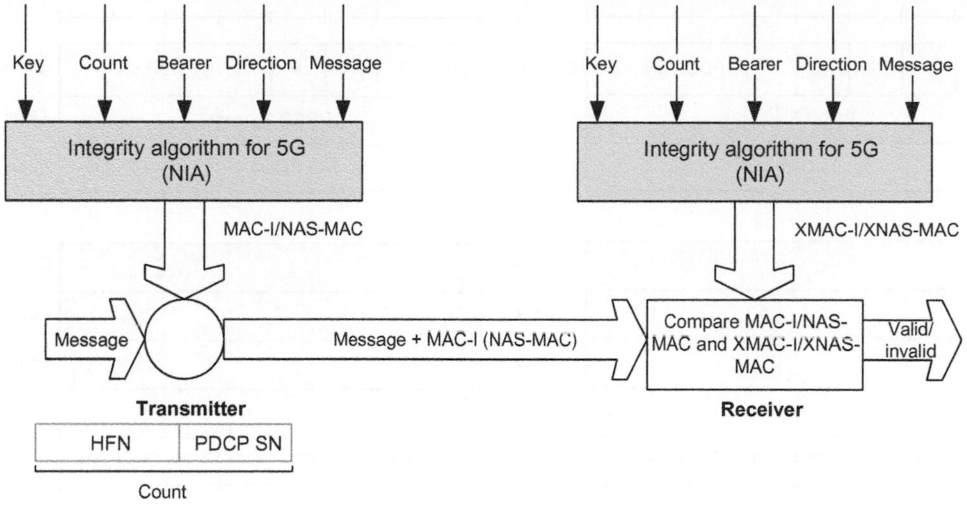

Figure 2.18
Integrity protection and verification procedures [2,11].

NIA. The message authentication code is then appended to the message when sent. For integrity protection algorithms, the receiver computes the expected message authentication code (XMAC-I/XNAS-MAC) on the message received in the same way that the sender computed its message authentication code on the message sent and verifies the integrity of the message by comparing it to the received message authentication code, that is, MAC-I/NAS-MAC. The integrity protection algorithm and key to be used by the PDCP entity are configured by upper layers and the integrity protection method is applied according to security architecture of 3GPP SAE [2]. The integrity protection function is activated by upper layers. Following the security activation, the integrity protection function is applied to all PDUs including and subsequent to the PDU indicated by upper layers for downlink and uplink transmissions. As the RRC message which activates the integrity protection function is itself integrity protected with the configuration included in that RRC message, the message must be decoded by RRC before the integrity protection verification can be performed for the PDU in which the message was received. The parameters that are required by PDCP for integrity protection are defined in reference [11] and are input to the integrity protection algorithm.

The PDCP data PDU is used to convey user-plane and control-plane data, as well as MAC-I in addition to the PDU header. The PDCP control PDU is used to transport PDCP status report and/or interspersed ROHC feedback in addition to the PDU header. A PDCP SDUs and PDUs are octet-aligned bit strings. A compressed or uncompressed SDU is included in a PDCP data PDU. Fig. 2.19 shows the format of the PDCP data PDU with 12-bit SN, which is applicable to SRBs. The figure further shows the format of the PDCP data PDU

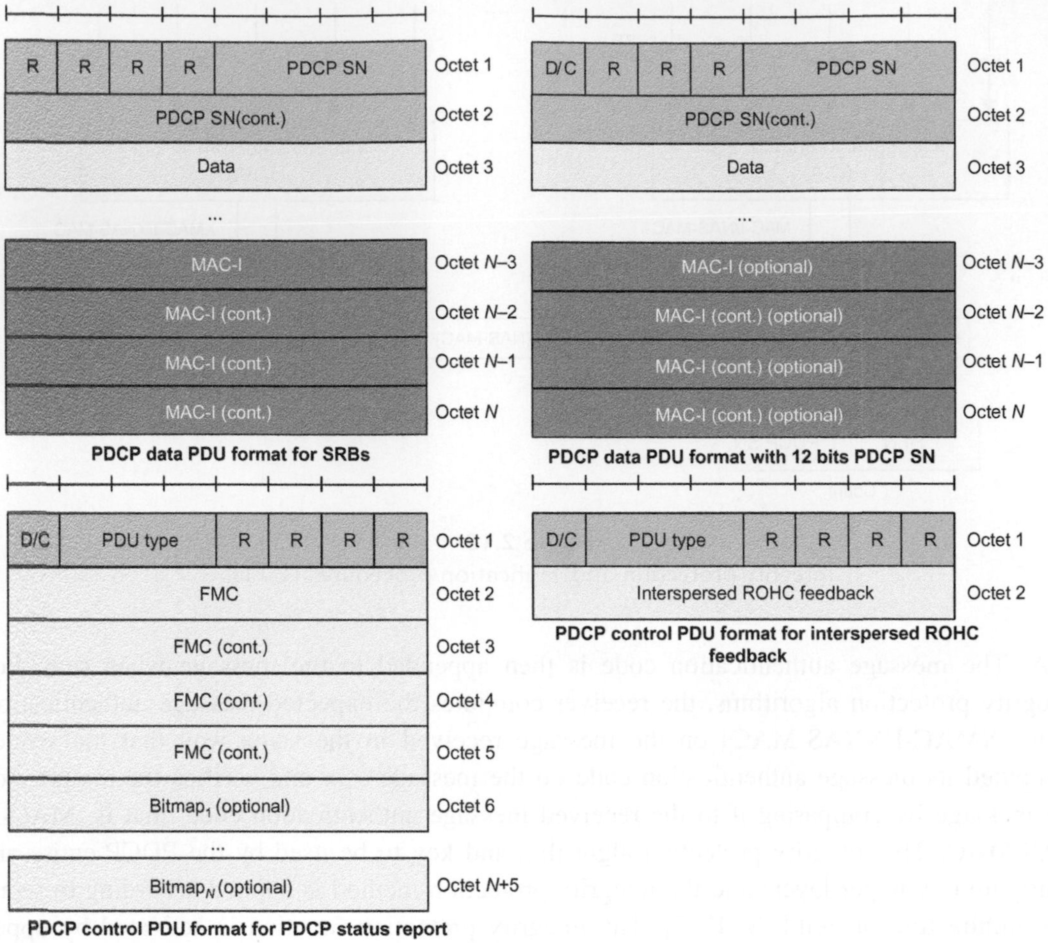

Figure 2.19
PDCP PDU formats [11].

with 12-bit SN for RLC-UM and RLC-AM DRBs. The structure of PDCP control PDU carrying one PDCP status report, which is applicable to RLC-AM DRBs, as well as the structure of PDCP control PDU transporting one interspersed ROHC feedback, which is applicable to RLC-UM and AM DRBs are shown in Fig. 2.19.

In PDCP PDU formats, the sequence number (SN) is a 12- or 18-bit number which is configured by RRC. The data field is a variable-size field which includes uncompressed user-plane/control-plane data or compressed user-plane data. As we stated earlier, the header compression only applies to user-plane data. The MAC-I field carries a message authentication code. For SRBs the MAC-I field is always present; however, if integrity protection is not configured, the MAC-I field is still present in PDCP PDU but is padded with zeros.

For DRBs, the MAC-I field is present only when the DRB is configured with integrity protection, which is unique to NR. The D/C field indicates whether the corresponding PDCP PDU is a PDCP data PDU or a PDCP control PDU. The PDU Type identifies the type of control information included in the corresponding PDCP control PDU, which can be a status report, interspersed ROHC feedback, or reserved. The first missing COUNT indicates the COUNT value of the first missing PDCP SDU within the reordering window. The Bitmap field indicates which SDUs are missing and which SDUs have been correctly received in the receiving entity. The interspersed ROHC feedback has a variable length and contains one ROHC packet with only feedback, that is, a ROHC packet which is not associated with a PDCP SDU. When an interspersed ROHC feedback is generated by the header compression protocol, the transmitting PDCP sends the corresponding PDCP control PDU to the lower layers without associating a PDCP SN or performing ciphering. The receiving PDCP entity delivers the corresponding interspersed ROHC feedback to the header compression protocol without performing deciphering [11].

2.2.4 Service Data Adaptation Protocol Sublayer

The main services and functions of SDAP sublayer include mapping between a QoS flow and a DRB and marking QFI in downlink and uplink IP packets. A single-protocol entity of SDAP is configured for each individual PDU session. The SDAP sublayer was introduced in NR because of the new QoS framework compared to LTE QoS management when connected to the 5G core. However, if the gNB is connected to the EPC, which is the case for non-standalone deployments, the SDAP service/functionality is not used. As we mentioned earlier, the NG-RAN architecture supports disaggregated gNB where gNB functions are split into a central unit (gNB-CU) and one or more distributed units (gNB-DU) connected via F1 interface. In the case of a split gNB, the RRC, PDCP, and SDAP protocols, described in more detail below, reside in the gNB-CU and the remaining protocol entities (RLC, MAC, and PHY) will be located in the gNB-DU. The interface between the gNB (or the gNB-DU) and the device is denoted as the Uu interface. In the example shown in Fig. 2.4, the SDAP protocol maps the IP packets to different radio bearers, that is, IP packets n and $n+1$ are mapped to radio bearer x and IP packet m is mapped to radio bearer y. The SDAP mapping function between a QoS flow and a DRB is due to the new QoS framework which is used in the new radio. The SDAP further marks the QFIs in the downlink due to the use of reflective QoS[6] and in the uplink due to the use of new QoS framework. A single SDAP entity (as shown in Fig. 2.20) is configured for each individual PDU session, except for the dual connectivity scenario where two entities can be configured.

[6] Reflective QoS flow to DRB mapping is a QoS flow to DRB mapping scheme where a UE monitors the QoS flow to DRB mapping rule in the downlink and applies it to in the uplink [3].

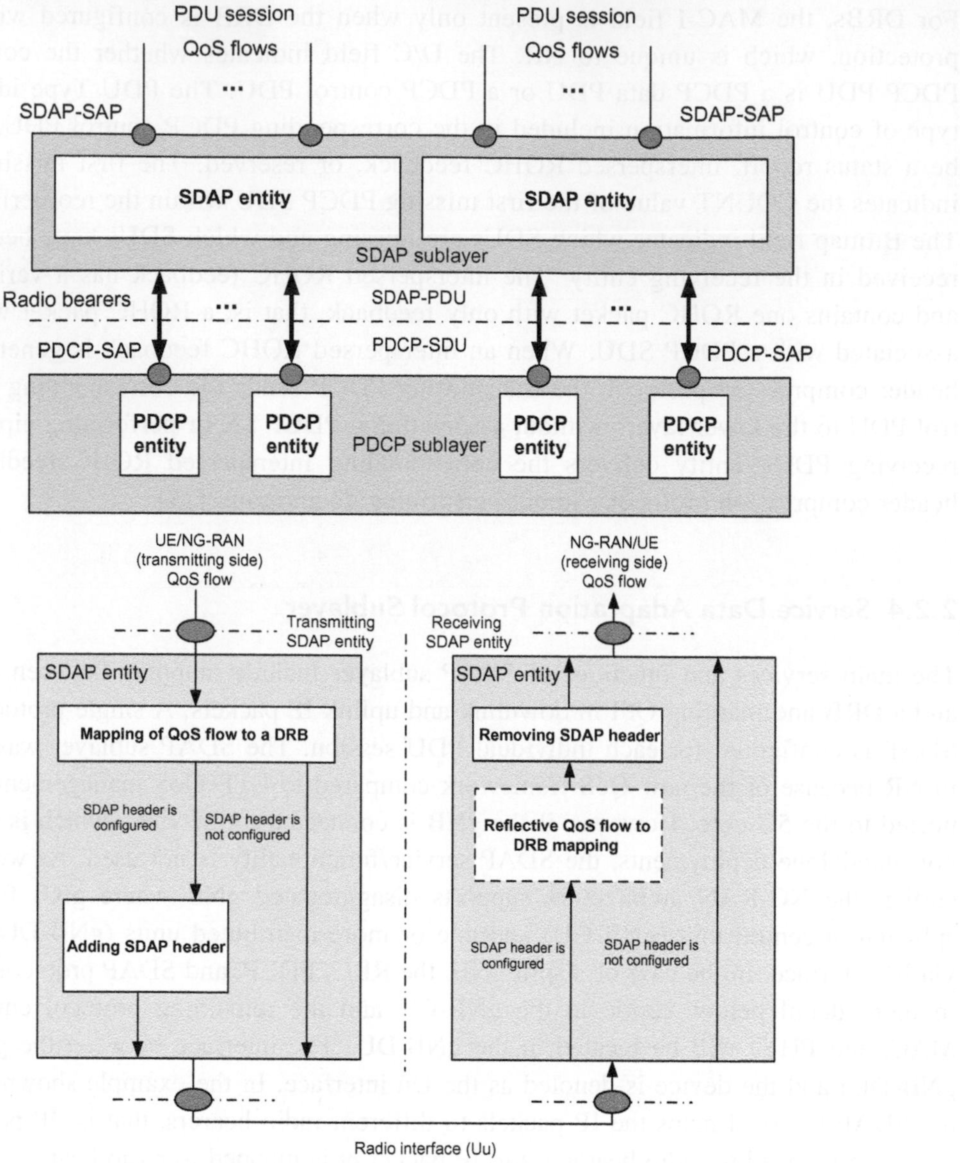

Figure 2.20
High-level SDAP sublayer functional architecture [3].

Fig. 2.20 illustrates one possible structure for the SDAP sublayer; however, the actual implementations may vary. The SDAP sublayer is configured by RRC. It maps the QoS flows to DRBs. One or more QoS flows may be mapped onto one DRB. However, in the uplink, one QoS flow is mapped to only one DRB at a time. The SDAP entities are located in the SDAP sublayer. Several SDAP entities may be defined for a UE. One SDAP entity is

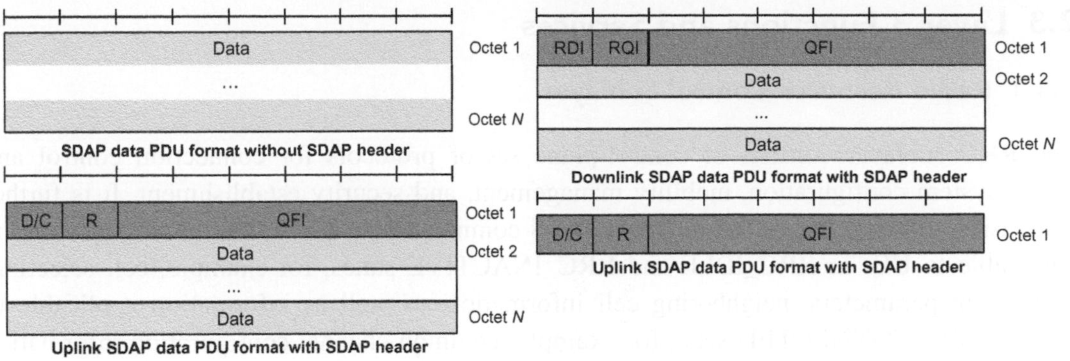

Figure 2.21
SDAP PDU formats [3].

configured for each individual PDU session. An SDAP entity receives/delivers SDAP SDUs from/to upper layers and transmits/receives SDAP data PDUs to/from its peer SDAP entity via lower layers. In the transmitting side, when an SDAP entity receives an SDAP SDU from upper layers, it constructs the corresponding SDAP data PDU and submits it to lower layers. In the receiving side, when an SDAP entity receives an SDAP data PDU from lower layers, it retrieves the corresponding SDAP SDU and delivers it to upper layers. Fig. 2.20 illustrates the functional block diagram of the SDAP entity for the SDAP sublayer [3]. Note that the reflective QoS flow to DRB mapping is performed at UE, if downlink SDAP header is configured. The SDAP sublayer transfers user-plane data and exclusively provides services to the user-plane upper layers. As shown in Fig. 2.20, the SDAP sublayer supports transfer of user-plane data, mapping between a QoS flow and a DRB for both downlink and uplink; marking QFI in both downlink and uplink packets; and reflective QoS flow to DRB mapping for the uplink SDAP data PDUs.

The SDAP data PDU is used to convey SDAP header and user-plane data. An SDAP PDU or SDU is a bit string that is octet-aligned (Fig. 2.21). An SDAP SDU is included in an SDAP PDU. An SDAP data PDU may only consist of a data field with no SDAP header. As shown in Fig. 2.21, the SDAP data PDU in the downlink or uplink consists of a header and data fields, where the headers for the downlink and uplink are different. For each downlink SDAP data PDU, in which reflective QoS indication (RQI) is set to one, the SDAP entity would inform the NAS layer of the RQI and QFI. The end-marker control PDU is used by the SDAP entity at UE to inform the gNB that the SDAP SDU QoS flow mapping to the DRB on which the end-marker PDU is transmitted, has stopped. The D/C bit specifies whether the SDAP PDU is an SDAP data PDU or an SDAP control PDU. The QFI field identifies the QFI to which the SDAP PDU belongs. The RQI bit indicates whether NAS should be informed of the updated SDF to QoS flow mapping rules and the reflective QoS flow to DRB mapping indication bit implies whether QoS flow to DRB mapping rule should be updated [3].

2.3 Layer 3 Functions and Services

2.3.1 Radio Resource Control Sublayer

The RRC sublayer consists of control-plane set of protocols for connection control and setup, system configuration, mobility management, and security establishment. It is further responsible for broadcast of SI including NAS common control information and information applicable to UEs in RRC_IDLE and RRC_INACTIVE states, for example, cell selection/reselection parameters, neighboring cell information, as well as information applicable to UEs in RRC_CONNECTED state, for example, common channel configuration information [14]. The RRC connection control functions include paging; establishment/modification/suspension/resumption/release of RRC connections including assignment/modification of UE identity; establishment/modification/suspension/resumption/release of SRBs except SRB0; access barring; initial security activation including initial configuration of AS integrity protection (SRBs, DRBs), and AS ciphering (SRBs, DRBs); mobility management including intra-frequency and inter-frequency handover, associated security handling such as key or algorithm change, specification of RRC context information transferred between network nodes; establishment/modification/suspension/resumption/release of radio bearers carrying user data (DRBs); radio configuration control including assignment/modification of ARQ configuration, HARQ configuration, and DRX configuration.

In case of dual connectivity, the RRC sublayer provides cell management functions including change of primary second cell (PSCell), addition/modification/release of SCG cell(s). In case of carrier aggregation, RRC sublayer provides cell management functions including addition/modification/release of SCell(s). The RRC sublayer further provides QoS control including assignment/modification of SPS configuration and DL/UL configured grant configuration, assignment/modification of parameters for uplink rate-control in the UE, that is, allocation of a priority and a prioritized bit rate for each resource block. The RRC sublayer also handles recovery from RLF condition; inter-RAT mobility including security activation, transfer of RRC context information; measurement configuration and reporting which includes establishment/modification/release of measurement configuration (e.g., intra-frequency, inter-frequency, and inter-RAT measurements); setup and release of measurement gaps; and measurement reporting [14].

The RRC messages are sent to the UEs using SRBs, based on the same set of protocol layers that are used for user-plane packet processing except the SDAP sublayer. The SRBs are mapped to the CCCH during connection setup and to the DCCH once the connection is established. The control-plane and user-plane data can be multiplexed in the MAC sublayer and transmitted to the device within the same TTI. The MAC CEs can be used for control of radio resources in some specific cases where low latency is more important than ciphering, integrity protection, and reliable transport of data.

Table 2.1: Radio resource control (RRC) functions in standalone and non-standalone NR operation.

Services	Functions	Non-standalone Architecture	Standalone Architecture	Differences With LTE RRC
System information	Broadcast of minimum system information	√	√	–
	Broadcast of other system information		√	Introduction of on-demand and area provision
Connection control	Bearer and cell settings	√	√	Introduction of split SRB[a] and direct SRB[b]
	Connection establishment with the core network		√	Introduction of RRC_INACTIVE state
	Paging		√	Introduction of RAN level paging
	Access control		√	Introduction of unified access control
Mobility	Handover		√	–
	Cell selection/reselection		√	–
Measurement	Downlink quality measurements/reporting	√	√	Introduction of beam measurements
	Cell identifier measurement/reporting	√	√	–

[a]Split SRB is a bearer for duplicating RRC messages generated by the master node for terminals in dual connectivity scenarios and transmitting via the secondary node.
[b]Direct SRB is a bearer whereby the secondary node can send RRC messages directly to the terminals in dual connectivity scenarios.

Table 2.1 shows the functional classification of NR RRC, the relevance of each function to standalone and non-standalone operation, and the similarities and differences with the LTE RRC functions.

SRBs are defined as radio bearers that are used only for the transmission of RRC and NAS messages. More specifically, the new radio has specified four types of SRBs which includes SRB0 for RRC messages using the CCCH logical channel; SRB1 for RRC messages which may include a piggybacked NAS message, as well as for NAS messages prior to the establishment of SRB2 using DCCH logical channel; SRB2 for NAS messages using DCCH logical channel (SRB2 has a lower priority than SRB1 and may be configured by the network after security activation); and SRB3 for specific RRC messages when UE is in EN-DC mode using DCCH logical channel. In the downlink, piggybacking of NAS messages is used only for bearer establishment/modification/release. In the uplink, piggybacking of NAS messages is used only for transferring the initial NAS messages during connection setup and connection resume. The NAS messages transferred via SRB2 are also contained in RRC messages, which do not carry any RRC protocol control information. Once security is activated, all RRC messages on SRB1, SRB2, and SRB3, including those containing NAS messages, are integrity protected and ciphered by PDCP sublayer. The NAS independently applies integrity protection and ciphering to the NAS messages [14].

2.3.2 System Information

The system information consists of an MIB and a number of SIBs, which are divided into minimum SI and other SI. The minimum SI comprises basic information required by the UEs for initial access and acquiring any other SI. The minimum SI itself consists of MIB which contains cell barred status information and essential physical layer information of the cell required for the UEs to receive further SI, for example, CORESET#0 configuration. The MIB is periodically broadcast on BCH. The minimum SI further includes SIB1 which defines the scheduling of other SIBs and contains information required for initial access. The SIB1 is also referred to as remaining minimum system information (RMSI) and is periodically broadcast on DL-SCH or sent in a dedicated manner on DL-SCH to UEs in RRC_CONNECTED state. The other SI encompasses all SIBs that are not broadcast as part of minimum SI. Those SIBs can either be periodically broadcast on DL-SCH, broadcast on-demand on DL-SCH, that is, upon request from the UEs in RRC_IDLE or RRC_INACTIVE state or sent in a dedicated manner on DL-SCH to the UEs in RRC_CONNECTED state. The other SI is divided into the following SIBs [8]:

- SIB2 contains cell reselection information related to the serving cell.
- SIB3 contains information about the serving frequency and intra-frequency neighbor cells relevant for cell reselection, including cell reselection parameters common for a frequency as well as cell-specific reselection parameters.
- SIB4 contains information about other NR frequencies and inter-frequency neighbor cells relevant for cell reselection including cell reselection parameters common for a frequency as well as cell-specific reselection parameters.
- SIB5 contains information about E-UTRA frequencies and E-UTRA neighbor cells relevant for cell reselection, including cell reselection parameters common for a frequency as well as cell-specific reselection parameters.
- SIB6 contains an Earthquake and Tsunami Warning System (ETWS)[7] primary notification.
- SIB7 contains an ETWS secondary notification.

[7] ETWS is a public warning system developed to satisfy the regulatory requirements for warning notifications related to earthquake and/or tsunami events. The ETWS warning notifications can either be a primary notification (short notification) or secondary notification (providing detailed information) [8].

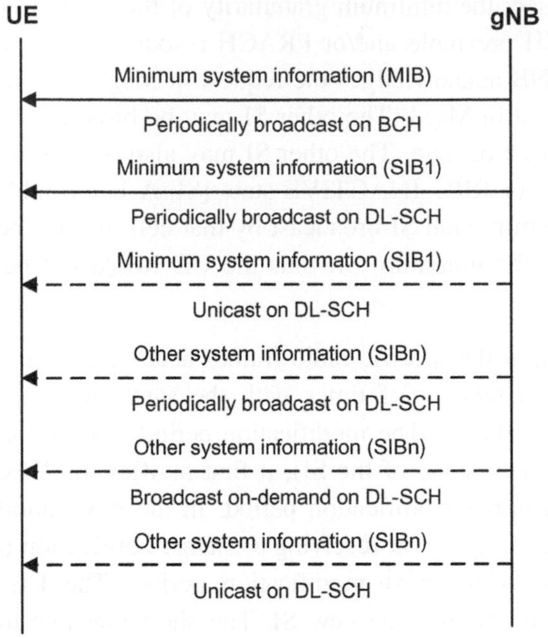

Figure 2.22
System information provisioning [8].

- SIB8 contains a Commercial Mobile Alert System (CMAS)[8] warning notification.
- SIB9 contains information related to Global Positioning System (GPS) time and coordinated universal time.

Fig. 2.22 summarizes SI provisioning. For a cell/frequency that is considered for camping by the UE, the UE is not required to acquire the contents of the minimum SI of that cell/frequency from another cell/frequency layer. This does not preclude the case where the UE applies stored SI from previously visited cell(s). The UE would consider a cell as barred, if it cannot determine the full content of the minimum SI broadcast by that cell. The UE only acquires SI on the active BWP, when using bandwidth adaptation.

The MIB is mapped to BCCH and is exclusively carried on BCH; however, other SI messages are mapped to BCCH and are dynamically carried on DL-SCH. The scheduling of SI messages is part of other SI and is signaled via SIB1. The UEs in RRC_IDLE or RRC_INACTIVE state may request other SI which would trigger a random-access procedure, wherein the corresponding Msg3 includes the SI request message unless the requested SI is associated with a subset of physical RACH (PRACH) resources, in that case Msg1 is

[8] CMAS is a public warning system developed for the delivery of multiple, concurrent warning notifications [8].

used. When Msg1 is used, the minimum granularity of the request is one SI message (i.e., a set of SIBs), one RACH preamble and/or PRACH resource can be used to request multiple SI messages and the gNB acknowledges the request in Msg2. When Msg3 is used, the gNB acknowledges the request in Msg4. The other SI may be broadcast at a configurable periodicity for a certain duration of time. The other SI may also be broadcast when it is requested by a UE in RRC_IDLE or RRC_INACTIVE state [8]. A UE would be allowed to camp on a cell, if it acquires the minimum SI broadcast by that cell. It must be noted that not all cells in a network broadcast the minimum SI; thus the UE would not be able to camp on those cells.

The SI may be changed at the specific radio frames according to a modification period. The SI may be transmitted a number of times with the same content within the modification period defined by its scheduling. The modification period is configured by the SI. When the network parameters change (some of the SI), it first notifies the UEs about this change, that is, this may be done within a modification period. In the next modification period, the network transmits the updated SI. Upon receiving a change notification the UE acquires the new SI from the beginning of the next modification period. The UE applies the previously acquired SI until the UE acquires the new SI. The short message transmitted with P-RNTI via DCI on PDCCH is used to inform UEs in RRC_IDLE, RRC_INACTIVE, or RRC_CONNECTED state about an SI change. If the UE receives a short message with SI change indication, it means that the SI will change at the next modification period boundary [8].

As we mentioned earlier, the SI is divided into the MIB and a number of SIBs. The MIB is always transmitted on the BCH with a periodicity of 80 ms and is repeated within the 80 ms. The MIB includes parameters that are needed to acquire SIB1 from the cell.

The SIB1 is transmitted on the DL-SCH with a periodicity of 160 ms and variable transmission repetition periodicity within 160 ms. The default repetition period of SIB1 is 20 ms; however, the actual repetition periodicity is up to network implementation. For synchronization signal/PBCH block (SSB) and CORESET multiplexing pattern 1, SIB1 repetition transmission period is 20 ms. For SSB and CORESET multiplexing pattern 2/3, SIB1 repetition period is the same as the SSB period. The SIB1 includes information regarding the availability and scheduling (e.g., mapping of SIBs to SI message, periodicity, and SI-window size) of other SIBs with an indication whether the SIBs are only provided on-demand, and in that case the configuration needed by the UE to perform the SI request. The SIB1 is cell-specific.

Other SIBs are carried in SI messages, which are transmitted on the DL-SCH. The SIBs with the same periodicity can only be mapped to the same SI message. Each SI message is transmitted within periodically occurring time domain windows referred to as SI-windows with same length for all SI messages. Each SI message is associated with an SI-window and the SI-windows of different SI messages do not overlap. That is, within one SI-window only the corresponding SI message is transmitted. Any SIB except SIB1 can be configured to be cell-specific or area-specific, using an indication in SIB1. The cell-specific SIB is applicable only within the cell that provides the SIB, while the area-specific SIB is applicable within an area referred to as SI area, which consists of one or several cells and is identified by *systemInformationAreaID*. For a UE in RRC_CONNECTED state, the network can provide SI through dedicated signaling using the *RRCReconfiguration* message, for example, if the UE has an active BWP with no common search space configured to monitor SI or paging. For PSCell and SCells, the network provides the required SI by dedicated signaling, that is, within an *RRCReconfiguration* message. Nevertheless, the UE acquires the MIB of the PSCell in order to obtain the SFN timing of the SCG which may be different from that of MCG. Upon change of the relevant SI for SCell, NG-RAN releases and adds the concerned SCell. The physical layer imposes a limit on the maximum size of a SIB. The maximum SIB1 or *SI message* size is 2976 bits [14].

We will explain in Chapter 3 that the PDCCH (physical downlink control channel) monitoring occasions for SI message are determined based on the search space indicated by *searchSpaceOtherSystemInformation* parameter, if the latter parameter is not set to zero. The PDCCH monitoring occasions for SI message, which are not overlapping with uplink symbols (determined according to *tdd-UL-DL-ConfigurationCommon*) are sequentially numbered from one in the SI-window. The PDCCH monitoring occasion(s) $[xN + K]$ for SI message in SI-window correspond to the Kth transmitted synchronization signal block [see Chapter 3 for description], where $x = 0, 1, \ldots, X - 1$, $K = 1, 2, \ldots, N$, N is the number of actual transmitted synchronization signal blocks determined according to *ssb-PositionsInBurst* in SIB1 and $X = \lfloor Number\ of\ PDCCH\ monitoring\ occasions\ in\ SI-window/N \rfloor$ [14].

2.3.3 User Equipment States and State Transitions

The operation of the RRC sublayer is guided by a state machine which defines the states that a UE may be present in at any time during its operation in the network. Apart from RRC_CONNECTED and RRC_IDLE states, which are similar to those of LTE, the NR has introduced a new RRC state referred to as RRC_INACTIVE state. As shown in Fig. 2.23, when a UE is powered up, it is in disconnected and idle mode. However, the UE can transition to the connected mode with initial access and RRC connection establishment. If there is no UE activity for a short period of time, the UE can suspend its active session and transition to the inactive mode. Nevertheless, it can resume its session by moving to the connected mode. 5G applications and services have different characteristics. To meet the requirements of different services, it was imperative to reduce the control-plane latency by introducing a new RRC state machine and a dormant state. The URLLC services are

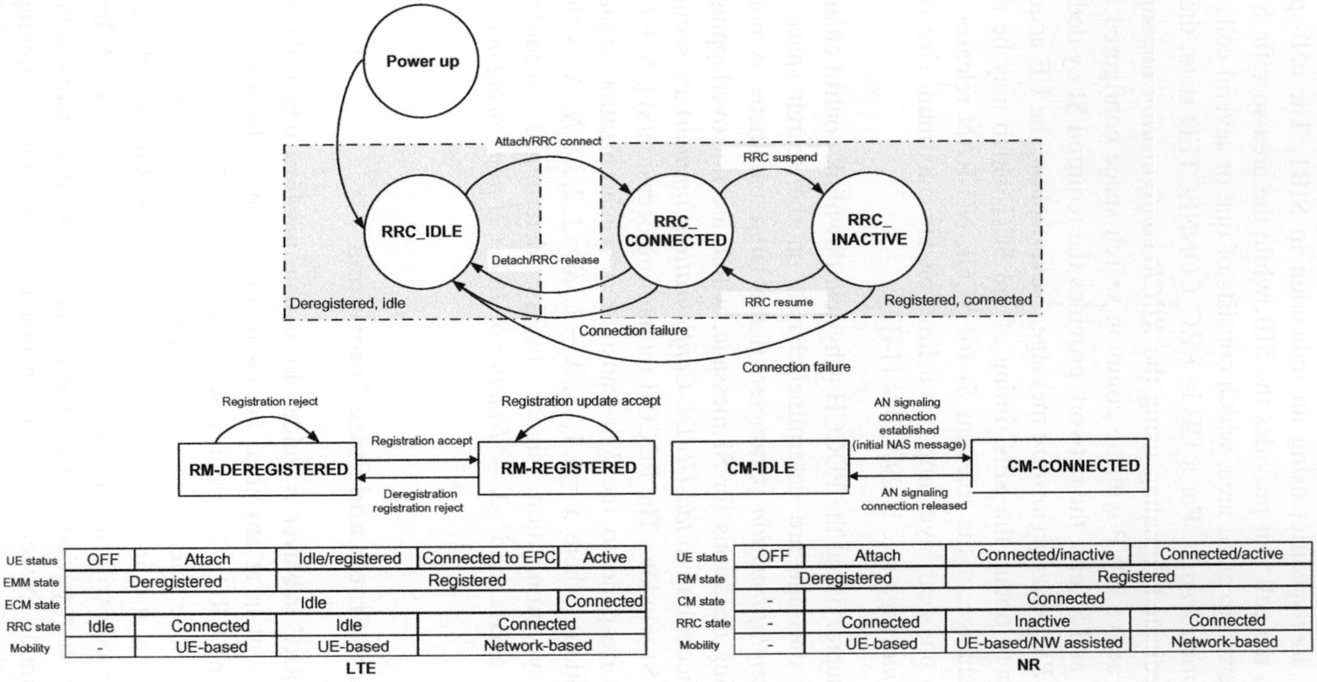

Figure 2.23
NR UE states and comparison to LTE [8,14].

characterized by transmission of frequent/infrequent small packets that require very low latency and high reliability; thus the devices must stay in a low-activity state, and intermittently transmit uplink data and/or status reports with small payloads to the network. There is further a need for periodic/aperiodic downlink small packet transmissions. A UE can move to RRC idle mode from RRC connected or RRC inactive state.

A UE needs to register with the network to receive services that requires registration. Once registered, the UE may need to update its registration with the network either periodically, in order to remain reachable (periodic registration update); or upon mobility (mobility registration update); or to update its capabilities or renegotiate protocol parameters (mobility registration update). As we discussed in Chapter 1, the mobility state of a UE in 5G core network (CN) can be either RM-REGISTERED or RM-DEREGISTERED depending on whether the UE is registered with 5GC. The registration management (RM) states are used in the UE and the Access and Mobility Management Function (AMF) to reflect the registration status of the UE in the selected PLMN. In the RM-DEREGISTERED state, the UE is not registered with the network. The UE context in AMF holds no valid location or routing information for the UE; thus the UE is not reachable by the AMF. However, some parts of UE context may still be stored in the UE and the AMF to avoid performing an authentication procedure in each registration procedure. In the RM-REGISTERED state, the UE is registered with the network and can receive services that require registration with the network [1]. Fig. 2.23 shows the UE RM states and transition between the two states.

Two connection management states are used to reflect the NAS signaling connectivity status of the UE with the CN, namely, CM-IDLE and CM-CONNECTED. A UE in CM-IDLE state has no NAS signaling connection established with the AMF over N1, whereas in CM-CONNECTED state the UE has a NAS signaling connection with the AMF over N1. A NAS signaling connection uses the RRC connection between the UE and the NG-RAN to encapsulate NAS messages exchanged between the UE and the CN in the RRC messages. The NR RRC protocol states consist of three states, where in addition to RRC_IDLE and RRC_CONNECTED states, a third state has been introduced, RRC_INACTIVE, as a primary sleeping state prior to transition to RRC_IDLE state in order to save UE power and to allow fast connection setup [1,8].

In RRC_IDLE state there is no UE context, that is, the parameters necessary for communication between the device and the network, in the radio access network, and the device is not registered to a specific cell. From the CN perspective, the device is in the CM-IDLE state where no data transfer may be performed since the device is in a sleep mode to conserve the battery. The idle mode UEs periodically wake up to receive paging messages, if any, from the network. In this mode, the mobility is managed by the device through cell reselection. The uplink synchronization is not maintained in the idle mode, and thus the UE is required to perform a random-access procedure in order to transition to the connected mode. As part of transitioning to the RRC_CONNECTED state, the UE context is established in the device and the network. From the CN perspective the device is in the CM-CONNECTED state and registered with the network. The cell to which the device belongs is known and a temporary identity for the device, that is, C-RNTI is used to identify the UE in NG-RAN while in the connected mode. The connected mode is intended for data transfer to/from the device; however, a DRX cycle can be configured during inactive times to reduce the UE power consumption. Since there is an already-established UE context in the gNB in the connected mode, transition from DRX mode and starting data transfer is relatively fast, requiring no connection setup. In this mode, the mobility is managed by the network. The device provides neighbor cell measurements to the network, and the network would instruct the device to perform a handover when necessary. The UE may lose uplink synchronization and may need to perform random-access procedure to be synchronized.

The LTE system only supports idle and connected modes. In practice, the idle mode serves as the primary means to reduce the device power consumption during operation in the network. However, intermittent transmission of small packets by some delay-sensitive applications results in frequent state transitions which would cause signaling overhead and additional delays. Therefore to reduce the signaling overhead and the latency, a third state has defined in NR. As shown in Fig. 2.23, in RRC_INACTIVE state, the UE context is maintained in the device and the gNB and the CN connection is preserved (i.e., the device is in CM-CONNECTED state). As a result, the transitions to or from the RRC_CONNECTED state for data transfer becomes more efficient and faster. In this mode, the UE is configured with sleep periods similar to the idle mode, and mobility is controlled through cell reselection without involvement of the network. The characteristics of NR UE states are summarized in Table 2.2. One important difference between the UE states in NR is the way that the mobility is handled. In the idle and inactive states, the mobility is

Table 2.2: Characteristics of RRC states in NR [19].

<table>
<tr><th>RRC_IDLE</th><th>RRC_INACTIVE</th><th>RRC_CONNECTED</th></tr>
<tr><td colspan="2">UE-controlled mobility based on network configuration (cell reselection)</td><td>Network-controlled mobility within NR and to/from LTE</td></tr>
<tr><td>DRX configured by NAS</td><td>DRX configured by NAS or gNB</td><td>DRX configured by gNB</td></tr>
<tr><td colspan="2">Broadcast of system information</td><td>Neighbor cell measurements</td></tr>
<tr><td>Paging (CN-initiated)</td><td>Paging (CN-initiated or NG-RAN-initiated)</td><td>Network can transmit and/or receive data to/from UE</td></tr>
<tr><td>UE has an CN ID that uniquely identifies it w/in a tracking area</td><td>NG-RAN knows the RNA which the UE belongs to</td><td>NG-RAN knows the cell which the UE belongs to</td></tr>
<tr><td>No UE context stored in gNB</td><td colspan="2">UE and NG-RAN have the UE AS context stored, and the 5GC-NG-RAN connection (both control/user-planes) is established for the UE</td></tr>
</table>

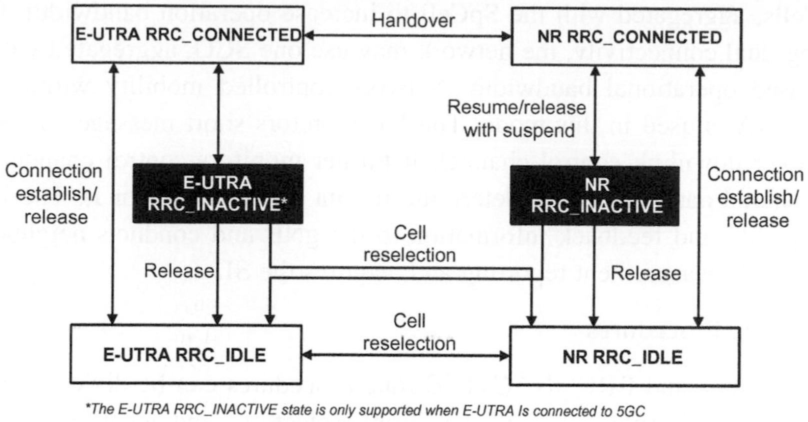

Figure 2.24
UE state machine and state transitions between NR/5GC, LTE/EPC, and LTE/5GC [1].

handled by the device through cell reselection, while in the connected state, the mobility is managed by the network based on measurements.

Fig. 2.24 illustrates an overview of UE RRC state machine and state transitions in NR. A UE has only one RRC state in NR at a given time. The UE is either in RRC_CONNECTED state or in RRC_INACTIVE state when an RRC connection has been established; otherwise, if no RRC connection is established, the UE is in RRC_IDLE state. The RRC states can further be characterized as follows (see the summary in Table 2.2):

- *RRC_IDLE*: In this state, a UE-specific DRX may be configured by upper layers. A UE-controlled mobility based on network configuration will be used. The UE monitors short messages transmitted with P-RNTI over downlink control channel. It also monitors the PCH for CN paging using 5G-S-TMSI and performs neighboring cell measurements and cell (re)selection. It further acquires SI and can send SI request (if configured).
- *RRC_INACTIVE*: In this state, a UE-specific DRX may be configured by upper layers or by RRC layer. A UE-controlled mobility based on network configuration is used. The UE stores the inactive AS context and an RNA is configured by the RRC sublayer. The UE monitors short messages transmitted with P-RNTI over downlink control channel. It monitors the PCH for CN paging using 5G-S-TMSI and RAN paging using full I-RNTI and performs neighboring cell measurements and cell (re)selection. The UE further periodically performs RNA updates (RNAUs) when moving outside the configured RNA. It also acquires SI and can send SI request (if configured).
- *RRC_CONNECTED*: In this state, the UE stores the AS context. The network transfers unicast data to/from the UE. At the lower layers, the UE may be configured with a UE-specific DRX. For carrier-aggregation-capable UEs, the network may use one or

more SCells, aggregated with the SpCell to increase operation bandwidth. For the UEs supporting dual connectivity, the network may use one SCG, aggregated with the MCG, for increased operational bandwidth. Network-controlled mobility within NR and to/from E-UTRA is used in this mode. The UE monitors short messages transmitted with P-RNTI over downlink control channel. It further monitors control channels associated with the shared data channel to determine if data is scheduled for it. The UE provides channel quality and feedback information to the gNB and conducts neighbor cell measurements and measurement reporting and acquires the SI.

2.3.3.1 Idle Mode Procedures

The RRC_IDLE state and RRC_INACTIVE state procedures can be divided into three processes, namely, PLMN selection, cell selection/reselection, and location registration and RNA update. The PLMN selection, cell reselection procedures, and location registration are common for both RRC_IDLE state and RRC_INACTIVE state, whereas RNA update is only applicable to RRC_INACTIVE state. When the UE selects a new PLMN, it transitions from RRC_INACTIVE to RRC_IDLE state. When a UE is powered on, a PLMN is selected by NAS and a number of RATs associated with the selected PLMN are identified for cell selection. The NAS provides a list of equivalent PLMNs that the AS must use for cell selection/reselection.

During cell selection the UE searches for a suitable cell within the selected PLMN. The UE would select a cell, if certain criteria are met. This procedure is known as camping on the cell in 3GPP terminology. The UE then registers with the network by means of NAS registration procedure in the tracking area of the selected cell. A successful location registration would make the selected PLMN as the registered PLMN. If the UE finds a more suitable cell, based on the cell reselection criteria, it reselects that cell and camps on it. If the new cell does not belong to at least one tracking area to which the UE is registered, another location registration is performed. In RRC_INACTIVE state, if the new cell does not belong to the configured RNA, an RNA update procedure is performed. The UE usually searches for higher priority PLMNs at regular time intervals and continues to search for a more suitable cell, if another PLMN has been selected by NAS. If the UE loses the coverage of the registered PLMN, either a new PLMN is automatically selected (automatic mode), or the user is notified of the available PLMNs so that a manual selection can be performed (manual mode).

The purpose of camping on a cell in RRC_IDLE or RRC_INACTIVE state is to enable the UE to receive SI from the PLMN, when registered. If the network needs to send a message or deliver data to the registered UE, it knows the set of tracking areas (in RRC_IDLE state) or RNA (in RRC_INACTIVE state) in which the UE is camped. The network can then send a paging message to the UE on the control channels of all cells in the corresponding set of (tracking) areas. It further enables the UE to receive ETWS and CMAS notifications. During the PLMN selection, the UE scans all RF channels in the NR bands according to its capabilities to find available PLMNs. On each carrier, the UE searches for the strongest cell and acquires the SI in order to find which PLMN(s) the cell belongs to. If the UE can detect one or several PLMN identities in the strongest cell, each detected PLMN is reported to the NAS as a high-quality PLMN provided that the measured reference signal received power value is greater than or equal to −110 dBm [12].

As we mentioned earlier, the PLMN selection in NR is based on the 3GPP PLMN selection, and that is, cell selection is required upon transition from RM-DEREGISTERED to RM-REGISTERED, from CM-IDLE to CM-CONNECTED, and from CM-CONNECTED to CM-IDLE based on the following principles [8]:

- The UE NAS layer identifies a PLMN and equivalent PLMNs.
- Cell selection is always based on cell-defining SSBs (CD-SSBs)[9] located on the synchronization raster.
- The UE searches the designated NR frequency bands and for each carrier frequency identifies the strongest cell consistent with the CD-SSB. It then detects the SI broadcast from that cell to identify its PLMN(s).
- The UE may conduct a full search in initial cell selection or make use of the stored information to shorten the search, that is, stored information cell selection.
- The UE searches for a suitable cell, and if it is not able to identify a suitable cell, it may proceed with an acceptable cell. When a suitable cell or an acceptable cell is found, the UE camps on that cell and starts the cell reselection procedure.
- A suitable cell is one for which the measured cell attributes satisfy the cell selection criteria; the cell PLMN is the selected PLMN, registered, or an equivalent PLMN; the cell is not barred or reserved, and the cell is not part of a tracking area which is in the list of *forbidden tracking areas for roaming*. An acceptable cell is one for which the measured cell attributes satisfy the cell selection criteria and the cell is not barred.
- Upon transition from RRC_CONNECTED or RRC_INACTIVE to RRC_IDLE, the UE should camp on a cell following cell selection depending on the frequency assigned by the RRC sublayer in the state transition message.
- The UE should attempt to find a suitable cell in the manner described for stored information or initial cell selection. If no suitable cell is found on any frequency or RAT, the UE should attempt to find an acceptable cell.
- In multi-beam operation, the cell quality is derived amongst the beams corresponding to the same cell.

[9] Within the frequency span of a carrier, multiple SSBs can be transmitted. The PCIs of SSBs transmitted in different frequency locations do not have to be unique, that is, different SSBs in the frequency domain can have different PCIs. However, when an SSB is associated with an RMSI, the SSB corresponds to an individual cell, which has a unique NR cell global identifier (NCGI). Such an SSB is referred to as CD-SSB. A PCell is always associated with a CD-SSB located on the synchronization raster.

A UE in RRC_IDLE may perform cell reselection according to the following procedure [8]:

- Cell reselection is always based on CD-SSBs located on the synchronization raster.
- The UE measures the attributes of the serving and neighbor cells to facilitate the reselection process.
- The UE would only need information on the carrier frequencies of the inter-frequency neighbor cells, when conducting search and measurement.
- Cell reselection identifies the cell that the UE should camp on. It is based on cell reselection criteria which involves measurements conducted on the serving and the neighbor cells. It must be noted that intra-frequency cell reselection is based on ranking of the cells and inter-frequency cell reselection is based on absolute priorities, where a UE would camp on the highest priority frequency available.
- A neighbor cell list (NCL) can be provided by the serving cell to facilitate cell selection in specific cases for intra- and inter-frequency neighbor cells. The NCL contains cell-specific cell reselection parameters (e.g., cell-specific offset) for specific neighbor cells. Black lists can be provided to prevent the UE from reselecting to specific intra- and inter-frequency neighboring cells.
- In multi-beam operations, the cell quality is derived amongst the beams corresponding to the same cell.

2.3.3.2 Inactive Mode Procedures

RRC_INACTIVE is a state where a UE remains in CM-CONNECTED state while roaming within an area configured by NG-RAN known as RNA without notifying NG-RAN. In RRC_INACTIVE state, the last serving gNB maintains the UE context and the UE-associated NG connection with the serving AMF and user-plane function (UPF). If the last serving gNB receives downlink data from the UPF or downlink UE-associated signaling from the AMF (except the *UE Context Release Command* message) while the UE is in RRC_INACTIVE state, it pages the UE in the cells corresponding to the RNA and may send an XnAP RAN Paging[10] to neighbor gNB(s), if the RNA includes cells of neighboring gNB(s). Upon receiving the *UE Context Release Command* message while the UE is in RRC_INACTIVE state, the last serving gNB may send the paging message in the cells corresponding to the RNA and may send an XnAP RAN Paging to neighbor gNB(s), if the RNA includes the cells of neighbor gNB(s). Upon receiving the *NG RESET* message while the UE is in RRC_INACTIVE state, the last serving gNB may page the involved UE(s) in

[10] The purpose of the RAN Paging procedure is to enable the NG-RAN node$_1$ to request paging of a UE in the NG-RAN node$_2$. The procedure uses non-UE-associated signaling. The RAN paging procedure is triggered by the NG-RAN node$_1$ by sending the RAN paging message to the NG-RAN node$_2$, in which the necessary information such as UE RAN Paging identity is provided.

the cells corresponding to the RNA and may send an XnAP RAN Paging to neighbor gNB (s), if the RNA includes the cells of neighbor gNB(s) [8].

The AMF provides to the NG-RAN node the *CN assistance information* to assist the NG-RAN node's decision on whether the UE can be moved to the RRC_INACTIVE state. The *CN assistance information* includes the registration area configured for the UE, the *periodic registration update* timer, and the *UE identity index* value, and may further include the UE-specific DRX, an indication if the UE is configured with mobile initiated connection only mode by the AMF, and the *expected UE behavior*. The UE registration area is considered by the NG-RAN node, when configuring the RNA. The UE-specific DRX and *UE identity index* value are used by the NG-RAN node for RAN paging. The *periodic registration update* timer is taken into account by the NG-RAN node to configure *periodic RNA update* timer. The NG-RAN node further considers the *expected UE behavior* to assist the UE RRC state transition decision [8].

During transition to RRC_INACTIVE state, the NG-RAN node may configure the UE with a periodic *RNA update* timer value. If the UE attempts to access a gNB other than the last serving gNB, the receiving gNB triggers the *XnAP retrieve UE context* procedure to obtain the UE context from the last serving gNB and may also trigger a data forwarding procedure including tunnel information for recovery of data from the last serving gNB. Upon successful UE context retrieval, the receiving gNB performs the slice-aware admission control in case of receiving slice information and becomes the serving gNB and further triggers the *NGAP path switch request* and applicable RRC procedures. After the path switch procedure, the serving gNB triggers release of the UE context at the last serving gNB by means of the *XnAP UE context release* procedure. If the UE attempts to access a gNB other than the last serving gNB and the receiving gNB does not find a valid UE context, the receiving gNB can establish a new RRC connection instead of resumption of the previous RRC connection.

A UE in the RRC_INACTIVE state is required to initiate RNAU procedure, when it moves out of the configured RNA. When receiving RNAU request from the UE, the receiving gNB triggers the XnAP retrieve UE context procedure to obtain the UE context from the last serving gNB and may decide to move the UE back to RRC_INACTIVE state, RRC_CONNECTED state, or RRC_IDLE state. In case of periodic RNA update, if the last serving gNB decides not to relocate the UE context, it fails the retrieve UE context procedure and directly moves the UE back to RRC_INACTIVE state or to RRC_IDLE state by an encapsulated *RRCRelease* message. Table 2.3 provides the functional split between the UE NAS and AS procedures in RRC_IDLE and RRC_INACTIVE states.

The UE may use DRX in RRC_IDLE and RRC_INACTIVE states in order to reduce power consumption. The UE monitors one paging occasion (PO) per DRX cycle. A PO is a set of

Table 2.3: Functional split between access stratum and non-access stratum in RRC_IDLE and RRC_INACTIVE states [12].

RRC_IDLE and RRC_INACTIVE State Procedure	UE Non-access Stratum	UE Access Stratum
PLMN selection	• Maintain a prioritized list of PLMNs • Select a PLMN using automatic or manual mode • Request AS to select a cell belonging to this PLMN. For each PLMN, associated RAT(s) may be set • Evaluate reports of available PLMNs from AS for PLMN selection • Maintain a list of equivalent PLMN identities	• Search for available PLMNs, if the associated RAT(s) are set for the PLMN, search among those RAT(s) • Perform measurements to support PLMN selection • Synchronize to a broadcast channel to identify PLMNs • Report available PLMNs with the associated RAT(s) to NAS on request from NAS or autonomously
Cell selection	• Control cell selection by indicating RAT(s) associated with the selected PLMN to be used initially in the search of a cell in the cell selection process • Maintain a list of *forbidden tracking areas* and provide the list to AS	• Perform the required measurements to support cell selection • Detect and synchronize to a broadcast channel • Receive and process broadcast information • Forward NAS system information to NAS • Search for a suitable cell. The cells broadcast one or more PLMN identity in the system information. Search among the associated RATs for that PLMN • Respond to NAS whether such cell is found • If a cell is found which satisfies cell selection criteria, camp on that cell
Cell reselection	• Maintain a list of equivalent PLMN identities and provide the list to AS • Maintain a list of *forbidden tracking areas* and provide the list to AS	• Perform the required measurements to support cell reselection • Detect and synchronize to a broadcast channel • Receive and process broadcast information • Forward NAS system information to NAS • Change cell if a more suitable cell is found
Location registration	• Register the UE as active after power up • Register the UE's presence in a registration area regularly or when entering a new tracking area • De-register UE when shutting down • Maintain a list of *forbidden tracking areas*	• Report registration area information to NAS
RAN notification area update	N/A	• Register the UE's presence in a RAN-based notification area and periodically or when entering a new RNA

PDCCH monitoring occasions and can consist of multiple time slots (e.g., subframe or OFDM symbols) where paging DCI can be sent. One paging frame (PF) is one radio frame and may contain one or multiple PO(s) or starting point of a PO. In multi-beam operations, the UE can assume that the same paging message is repeated in all transmitted beams, and thus the selection of the beam(s) for the reception of the paging message is up to UE implementation. The paging message is the same for both RAN-initiated paging and CN-initiated paging. The UE initiates *RRC Connection Resume* procedure upon receiving RAN-initiated paging. If the UE receives a CN-initiated paging in RRC_INACTIVE state, the UE transitions to RRC_IDLE state and informs the NAS. The PF and PO for paging are determined by the following expressions. The *SFN* for the PF is determined by $(SFN + PF\_offset)\text{mod}T = (T/N)(UE\_ID\text{mod}N)$ where index i_s indicating the index of the PO is determined by $i_s = \lfloor UE\_ID/N \rfloor \text{mod}N_s$. The parameters of the latter equations are defined as follows: T denotes the DRX cycle of the UE where T is determined by the shortest of the UE-specific DRX value, if configured by RRC or upper layers and a default DRX value broadcast in SI. If UE-specific DRX is not configured by RRC or by upper layers, the default value is applied; N is the number of total PFs in T; N_s denotes the number of POs for a PF; PF_offset is the offset used for PF determination; and $UE\_ID = 5G{-}S{-}TMSI\text{mod}1024$ [12].

A UE in RRC_INACTIVE state performs cell reselection similar to the procedure earlier defined for the RRC_IDLE state. The UE in the RRC_INACTIVE state can be configured by the last serving NG-RAN node with an RNA, where the RNA can cover a single or multiple cells and is contained within the CN registration area, as well as an RNA update that is periodically sent by the UE and is also sent when the cell reselection procedure of the UE selects a cell that does not belong to the configured RNA [8].

2.3.3.3 Connected Mode Procedures

In the RRC_CONNECTED state, the device has a connection established to the network. The goal of connected-state mobility is to ensure that this connectivity is sustained without interruption or noticeable degradation as the device moves across the network. To satisfy this goal, the device continuously searches for and conducts measurements on new cells both at the current carrier frequency (intra-frequency measurements) and at different carrier frequencies (inter-frequency measurements) that the device has been configured to do. Such measurements are conducted on the SSB in the same way as for initial access and cell selection/reselection in idle and inactive modes. However, the measurements can also be conducted on configured CSI-RS. In the connected mode, the handover is network-controlled, and the UE does not make any decision on handover to a different cell. Based on different triggering conditions such as the relative power of a measured SSB relative to that of the current cell, the device reports the result of the measurements to the network. The network then makes a decision as to whether the device has to be handed-over to a new cell. It should be noted that the reporting is provided through RRC signaling and not layer 1 measurement and reporting framework used for beam management. Apart from some cases in small cell network architectures where the cells are relatively synchronized, the device must perform a new uplink synchronization with respect to the target cell prior to handover. To obtain synchronization to a new cell, the UE has to perform a contention-free random-access procedure using resources specifically assigned to the device with no risk of collision with the goal of attaining synchronization to the target cell. Thus, only first two steps of the random-access procedure are needed which includes the preamble transmission and the corresponding random-access response providing the device with updated transmission timing [17].

In RRC_CONNECTED state, a network-controlled mobility scheme is used which has two variants: cell-level mobility and beam-level mobility.

In *cell-level mobility* explicit RRC signaling is used to trigger a handover. For inter-gNB handover, the signaling procedures consist of four components as follows [8]:

1. The source gNB initiates handover and issues a *Handover Request* message over Xn interface.
2. The target gNB performs admission control and provides the RRC configuration as part of the *Handover ACK* message.
3. The source gNB provides the RRC configuration to the UE in the *Handover Command*. The *Handover Command* message includes at least the cell ID and all information required to access the target cell so that the UE can access the target cell without detecting its SI. In some cases, the information required for contention-based and contention-free random-access procedure can be included in the *Handover Command* message. The access information to the target cell may include beam-specific information.
4. The UE moves the RRC connection to the target gNB and replies with a *Handover Complete* message.

The handover mechanism triggered by RRC signaling requires the UE to at least reset the MAC entity and reestablish RLC entity. The NR supports RRC-managed handovers with and without PDCP entity reestablishment. For DRBs using RLC-AM mode, PDCP can either be reestablished along with security key change or initiate a data recovery procedure without key change. For DRBs using RLC-UM mode and for SRBs the PDCP entity can be reestablished either together with security key change or to remain as it is without key change. Data forwarding, in-sequence delivery, and duplication avoidance during handover can be guaranteed, when the target gNB uses the same DRB configuration as the source gNB. The NR further supports timer-based handover failure procedure where the RRC connection reestablishment procedure is used for recovering from handover failure.

In *beam-level mobility*, explicit RRC signaling is not required to trigger handover. The gNB provides the UE via RRC signaling the measurement configuration containing configurations of SSB/CSI-RS resources and resource sets, reports, and trigger states for triggering channel and interference measurements and reports. The beam-level mobility is performed at lower layers by means of physical layer and MAC layer control signaling, and the RRC is not required to know which beam is being used at any given time. The SSB-based beam-level mobility is based on the SSB associated with the initial DL BWP and can only be configured for the initial DL BWPs and for DL BWPs containing the SSB associated with the initial DL BWP. For other DL BWPs, the beam-level mobility can only be performed based on CSI-RS.

2.3.4 User Equipment Capability

The UE capabilities in NR do not rely on UE categories. Unlike LTE, the NR UE categories are associated with fixed peak data rates and defined for marketing purposes; thus they are not signaled to the network. Instead, the network determines the uplink and downlink data rate supported by a UE from the supported band combinations and from the baseband capabilities such as modulation scheme, and the number of MIMO layers. In order to limit signaling overhead, the gNB can ask the UE to provide NR capabilities for a restricted set of bands. When responding, the UE can skip a subset of the requested band combinations when the corresponding UE capabilities are the same.

The NR defines an approximate (peak) data rate for a given number of aggregated carriers in a band or band combination as follows [13]:

$$D_{NR} = 10^{-6} \sum_{j=1}^{J} \left[v_{layer}^{(j)} Q_m^{(j)} f^{(j)} R_{\max} \frac{12 N_{PRB}^{BW(j)}(\mu)}{T_s(\mu)} \left(1 - \alpha^{(j)}\right) \right] \text{ (Mbps)}$$

where J is the number of aggregated component carriers in a band or band combination; $R_{\max} = 948/1024$; $v_{layer}^{(j)}$ is the maximum number of supported layers given by higher layer parameter *maxNumberMIMO-LayersPDSCH* for the downlink and higher layer parameters *maxNumberMIMO-LayersCB-PUSCH* and *maxNumberMIMO-LayersNonCB-PUSCH* for the uplink; $Q_m^{(j)}$ is the maximum supported modulation order given by higher layer parameter *supportedModulationOrderDL* for the downlink and higher layer parameter *supportedModulationOrderUL* for the uplink; $f^{(j)}$ is the scaling factor given by higher layer parameter *scalingFactor* which can take the values of 1, 0.8, 0.75, and 0.4; μ is the numerology (an OFDM parameter); $T_s(\mu)$ is the average OFDM symbol duration in a subframe for numerology μ which is given as $T_s(\mu) = 0.0142^{\mu}$ for normal cyclic prefix; $N_{PRB}^{BW(j)}(\mu)$ is the maximum resource block allocation in bandwidth $BW^{(j)}$ with numerology μ where $BW^{(j)}$ is the UE supported maximum bandwidth in the given band or band combination; and $\alpha^{(j)}$ is the estimated overhead which takes the values 0.14 [for downlink in frequency range (FR) 1], 0.18 (for downlink in FR2), 0.08 (for uplink in FR1), and 0.10 (for uplink in FR2). Note that only one of the uplink or supplemental uplink carriers with the higher data rate is counted for a cell operating SUL.

The approximate maximum data rate can be computed as the maximum of the approximate data rates computed using the above expression for each of the supported band or band combinations. For LTE in the case of dual connectivity, the approximate data rate for a given number of aggregated carriers in a band or band combination is computed as $D_{MR-DC} = 10^{-3}\sum_{j=1}^{J} TBS_j$ (Mbps), where J is the number of aggregated LTE component carriers in multi-radio dual connectivity (MR-DC) band combination and TBS_j is the total maximum number of DL-SCH transport block bits received within a 1 ms TTI for the jth component carrier based on the UE supported maximum MIMO layers for the jth carrier, and based on the modulation order and the number of physical resource blocks (PRBs) in the bandwidth of the jth carrier. The approximate maximum data rate can be calculated as the maximum of the approximate data rates computed using the latter equation for each of the supported band or band combinations. For MR-DC, the approximate maximum data rate is computed as the sum of the approximate maximum data rates from NR and LTE [13].

The total layer 2 buffer size is another UE capability attribute that is defined as the sum of the number of bytes that the UE is capable of storing in the RLC transmission windows and RLC reception and reordering windows and also in PDCP reordering windows for all radio bearers. The total layer 2 buffer size in MR-DC and NR-DC scenario is the maximum of the calculated values based on the following equations [13]:

$$MaxULDataRate\_MN \times RLCRTT\_MN + MaxULDataRate\_SN \times RLCRTT\_SN + MaxDLDataRate\_SN \times RLCRTT\_SN + MaxDLDataRate\_MN \times \left(RLCRTT\_SN + X2/Xndelay + QueuinginSN\right)$$

$$MaxULDataRate\_MN \times RLCRTT\_MN + MaxULDataRate\_SN \times RLCRTT\_SN + MaxDLDataRate\_MN \times RLCRTT\_MN + MaxDLDataRate\_SN \times \left(RLCRTT\_MN + X2/Xndelay + QueuinginMN\right)$$

In other scenarios, the total layer 2 buffer size is calculated as $MaxDLDataRate \times RLCRTT + MaxULDataRate \times RLCRTT$. It must be noted that the additional layer 2 buffer required for preprocessing of data is not taken into account in above formula. The total layer 2 buffer size is determined as the maximum layer 2 buffer size of all calculated ones for each band combination and the applicable Feature Set combination in the supported MR-DC or NR band combinations. The RLC RTT for NR cell group corresponds to the smallest subcarrier spacing (SCS) numerology supported in the band combination and the applicable Feature Set combination. The NR specifications specify *X2/Xn delay + Queuing in SN =* , if SCG is NR, and 55 ms if SCG is LTE. The NR specifications define *X2/Xn delay + Queuing in MN = 25 ms*, if MCG is NR, and 55 ms if MCG is LTE. They further specify RLC RTT for LTE cell group as 75 ms and the RLC RTT for NR cell group ranging from 20 to 50 ms depending on the subcarrier spacing [13].

The maximum supported data rate for integrity-protected DRBs (see Section 2.2.3.3) is a UE capability indicated at NAS layer, with a minimum value of 64 kbps and a maximum value of the highest data rate supported by the UE. In case of failed integrity check (i.e., due to a faulty or missing MAC-I) the corresponding PDU is discarded by the receiving PDCP entity.

2.4 Discontinuous Reception and Power-Saving Schemes

The UE monitors physical downlink control channel (PDCCH) while in RRC_CONNECTED state. This activity is controlled by the DRX and bandwidth adaptation schemes configured for the UE. When bandwidth adaptation is configured, the UE only has to monitor PDCCH on the active BWP, that is, it does not have to monitor PDCCH on the entire downlink frequency of the cell. A BWP inactivity timer (independent from the DRX inactivity timer) is used to switch the active BWP to the default one. The latter timer is restarted upon successful PDCCH decoding and the switching to the default BWP happens when it expires. When DRX is configured, the UE is not required to continuously monitor the PDCCH. The DRX mechanism is characterized by the following parameters [8]:

- *On-duration*: The time interval during which the UE would expect to receive the PDCCH. If the UE successfully decodes the PDCCH, it stays awake and starts the inactivity timer.
- *Inactivity timer*: The time interval during which the UE waits for successful decoding of the PDCCH, starting from the last successful decoding of a PDCCH. If the decoding fails, the UE can go back to sleep. The UE restarts the inactivity timer following a single successful decoding of a PDCCH for the first transmission only (i.e., not for retransmissions).
- *Retransmission-timer*: The time interval until a retransmission can be expected.
- *Cycle*: It specifies the periodic repetition of the on-duration followed by a possible period of inactivity.
- *Active-time*: The total time duration that the UE monitors PDCCH. This includes the on-duration of the DRX cycle, the time that the UE is performing continuous reception while the inactivity timer is running, and the time when the UE is performing continuous reception while awaiting a retransmission opportunity.

Due to bursty nature of the packet data traffic, which is characterized by intermittent periods of transmission activity followed by longer periods of inactivity, and to reduce the UE power consumption, NR supports a DRX scheme similar to that of LTE. Bandwidth adaptation and dynamic carrier activation/deactivation are two other power-saving mechanisms supported in NR. The underlying mechanism for DRX is a configurable DRX cycle in the device. When a DRX cycle is configured, the device monitors the downlink control channel only during the active-time and sleeps, with its receiver circuitry switched off, during the inactivity time, leading to a significant reduction in UE power consumption. The longer the DRX inactive time, the lower the power consumption. However, this would have certain implications for the scheduler, since the device is only reachable when it is active according to the DRX cycle configured for it. In many cases, if the device has been scheduled and is engaged in receiving or transmitting data, it is likely that it will be scheduled again soon; thus waiting until the next activity period according to the DRX cycle would result in additional delays. Therefore, to reduce the delays, the device remains in the active state for a configurable period of time after being scheduled. This is realized by an inactivity timer started by the UE every time that it is scheduled where the UE remains awake until the time expires (Fig. 2.25). Since NR supports multiple numerologies the time unit of the DRX timers is specified in milliseconds in order to avoid associating the DRX periodicity to a certain numerology (Fig. 2.26).

The NR HARQ retransmissions are asynchronous in both downlink and uplink. If the device has been scheduled a transmission in the downlink that it cannot decode, a typical gNB behavior is to retransmit the data at a later time. In practice, the DRX scheme has a configurable timer which is started after an erroneously received transport block and is used to wake up the UE receiver when it is likely for the gNB to schedule a retransmission. The value of the timer is preferably set to match the (implementation-specific) roundtrip time in the HARQ protocol. The above mechanism is a (long) DRX cycle in conjunction with the device remaining awake for a period of time after being scheduled. However, in some services such voice over IP, which is characterized by periods of regular transmission, followed by periods of no activity, a second (short) DRX cycle can be optionally configured in addition to the long DRX cycle.

The RRC entity controls the DRX operation by configuring the following parameters [14]:

- *drx-onDurationTimer*: The duration at the beginning of a DRX cycle.
- *drx-SlotOffset*: The delay before starting the *drx-onDurationTimer*.
- *drx-InactivityTimer*: The duration after the PDCCH occasion in which a PDCCH indicates a new uplink/downlink transmission for the MAC entity.
- *drx-RetransmissionTimerDL* (per-DL HARQ process except for the broadcast process): The maximum duration until a downlink retransmission is received.

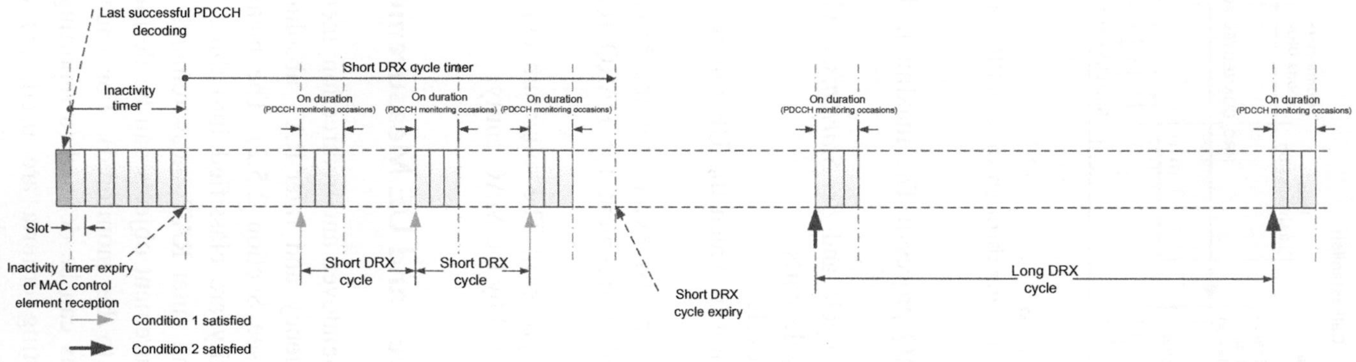

Figure 2.25
Illustration of DRX mechanism [8].

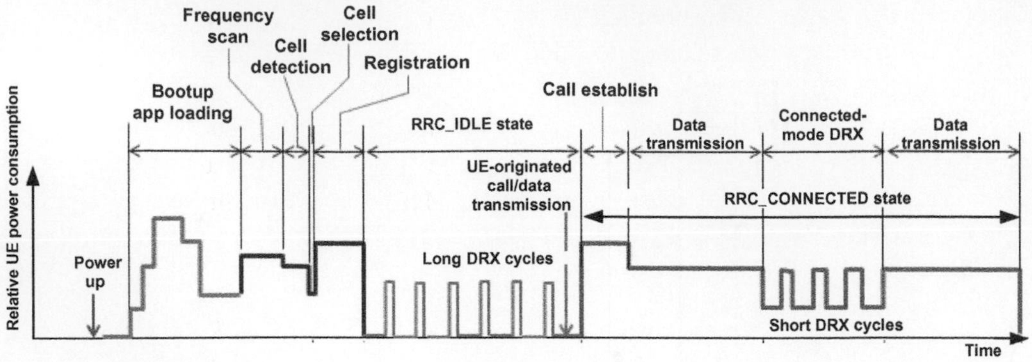

Figure 2.26
Example UE power consumption when transitioning through various RRC states [20].

- *drx-RetransmissionTimerUL* (per-UL HARQ process): The maximum duration until a grant for uplink retransmission is received.
- *drx-LongCycleStartOffset*: The long DRX cycle and *drx-StartOffset* which define the subframe where the long and short DRX cycle starts.
- *drx-ShortCycle* (optional): The short DRX cycle.
- *drx-ShortCycleTimer* (optional): The duration in which the UE follows the short DRX cycle.
- *drx-HARQ-RTT-TimerDL* (per-DL HARQ process except for the broadcast process): The minimum duration before a downlink assignment for HARQ retransmission is expected by the MAC entity.
- *drx-HARQ-RTT-TimerUL* (per-UL HARQ process): The minimum duration before an uplink HARQ retransmission grant is expected by the MAC entity.

2.5 Mobility Management, Handover, and UE Measurements

The NR performs load balancing through handover and redirection mechanisms upon RRC release and through use of inter-frequency and inter-RAT absolute priorities as well as inter-frequency *Qoffset* parameters (see Section 2.5.2). The measurements performed by a UE for connected mode mobility are classified into three types, namely, intra-frequency NR, inter-frequency NR, and inter-RAT measurements for LTE. For each measurement type, one or several measurement objects can be defined (a measurement object defines the carrier frequency to be monitored). For each measurement object, one or several reporting configurations can be defined (a reporting configuration defines the reporting criteria). Three reporting criteria are used: (1) event-triggered reporting, (2) periodic reporting, and (3) event-triggered periodic reporting. The association between a measurement object and a reporting configuration is created by a measurement identity (a measurement identity associates one measurement object and one reporting configuration of the same RAT). By using several measurement identities (one for each measurement object, reporting configuration pair), it is possible to associate several reporting configurations to one measurement object and to associate one reporting configuration to several measurement objects [8]. The measurements identity is used when reporting results of the measurements. Measurement quantities are considered separately for each RAT.

Measurement commands are used by NG-RAN to instruct the UE to start, modify, or stop measurements. Handover can be performed within the same RAT and/or CN, or it can involve a change of the RAT and/or CN. Inter-system fallback toward LTE RAN is performed for load balancing when 5GC does not support emergency services or voice services. Depending on certain criteria such as CN interface availability, network configuration, and radio conditions, the fallback procedure results in either connected-state mobility (handover procedure) or idle state mobility (redirection) [8].

2.5.1 Network-Controlled Mobility

The mobility of the UEs in RRC_CONNECTED state is controlled by the network (network-controlled mobility), which is classified into two types of mobility, namely, cell-level mobility and beam-level mobility. The *cell-level mobility* requires explicit RRC signaling in order to be triggered, which results in handover. The main steps of the inter-gNB handover signaling procedures are illustrated in Fig. 2.27. The inter-gNB handover comprises the following steps [8]:

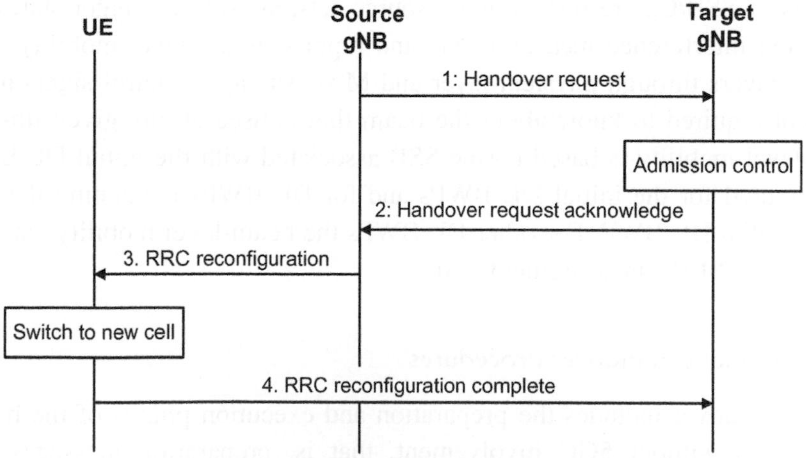

Figure 2.27
Inter-gNB handover procedure [8].

1. The source gNB initiates handover and issues a *Handover Request* over Xn interface.
2. The target gNB performs admission control and provides the RRC configuration as part of the *Handover ACK*.
3. The source gNB provides the RRC configuration to the UE in the *Handover Command* message, which includes the cell ID and all information required to access the target cell, so that the UE can access the target cell without detecting that cell's SI. In some cases, the information required for contention-based and contention-free random-access procedure can be included in the *Handover Command* message. The access information to the target cell may include beam-specific information.
4. The UE moves the RRC connection to the target gNB and replies with the *Handover Complete* message.

The user data can be sent in step 4, if the grant allows. The handover mechanism triggered by RRC signaling requires the UE to reset the MAC entity and reestablish RLC entity. The RRC-triggered handovers with and without PDCP entity reestablishment are both supported in NR. For DRBs using RLC-AM mode, the PDCP entity can either be reestablished along with a security key change or initiate a data recovery procedure without a key change. For DRBs using RLC-UM mode and for SRBs, the PDCP entity can either be reestablished in conjunction with a security key change or to remain as it is without a key change. Data forwarding, in-sequence delivery, and duplication avoidance at handover can be guaranteed when the target gNB uses the same DRB configuration as the source gNB. Timer-based handover failure procedure is supported in NR where an RRC connection reestablishment procedure is used for recovering from handover failure [8].

The *beam-level mobility* does not require explicit RRC signaling in order to be triggered. The gNB provides the UE, via RRC signaling, with measurement configuration containing configurations of SSB/CSI resources and resource sets, as well as trigger states for triggering channel and interference measurements and reports. Beam-level mobility is then managed at lower layers through physical layer and MAC sublayer control signaling. The RRC sublayer is not required to know about the beam that is used at any given time. The SSB-based beam-level mobility is based on the SSB associated with the initial DL BWP and can only be configured for the initial DL BWPs and for DL BWPs containing the SSB associated with the initial DL BWP. For other DL BWPs the beam-level mobility can only be performed based on CSI-RS measurements [8].

2.5.1.1 Control-Plane Handover Procedures

The intra-NR handover includes the preparation and execution phases of the handover procedure performed without 5GC involvement, that is, preparation messages are directly exchanged between the gNBs. The release of the resources at the source gNB during the handover completion phase is triggered by the target gNB. Fig. 2.28 shows the basic

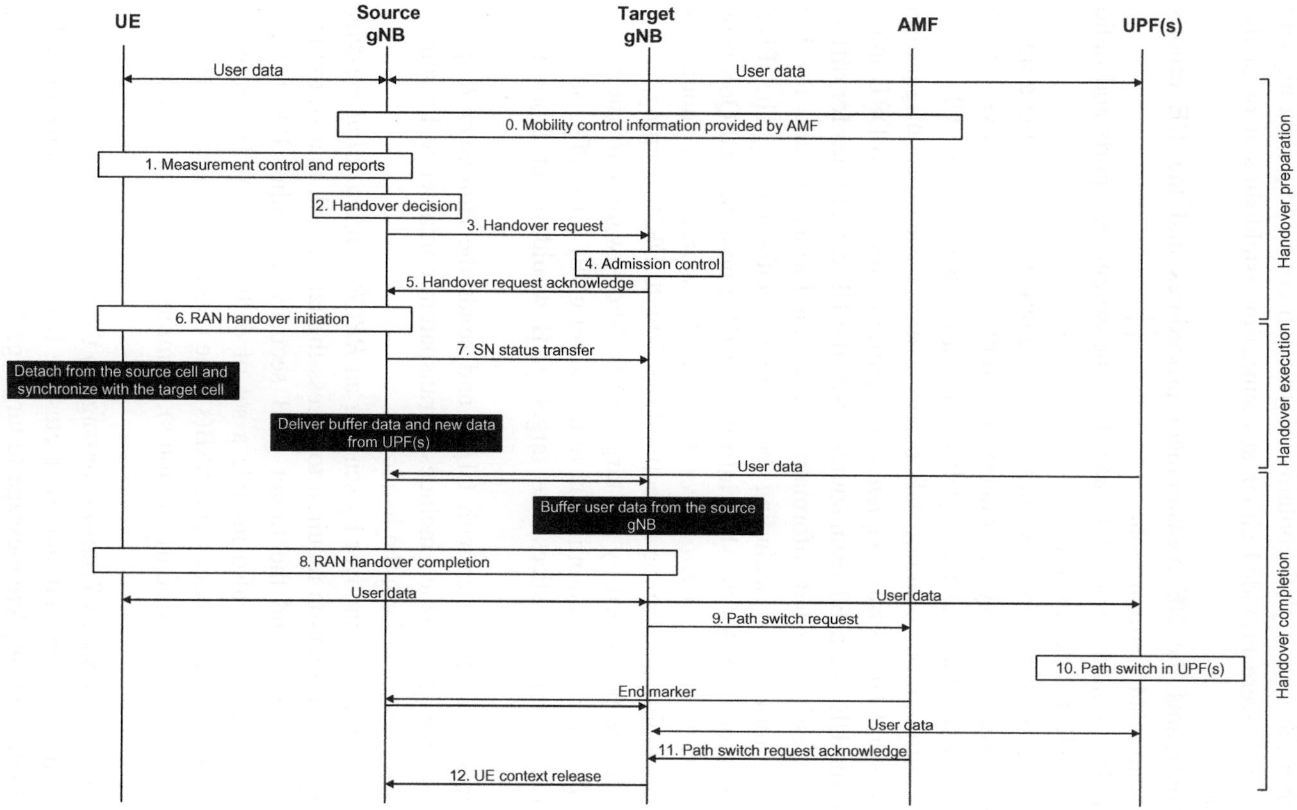

Figure 2.28
Intra-AMF/UPF handover in NR [8].

handover procedure where the AMF and the UPF entities do not change. The processing steps of this handover are as follows [8]:

1. The UE context within the source gNB contains information concerning roaming and access restrictions, which were provided either at connection establishment or at the last tracking area update.
2. The source gNB configures the UE measurement procedures and the UE reports according to the measurement configuration.
3. The source gNB decides to handover the UE, based on measurement reports and radio resource management (RRM) information.
4. The source gNB issues a *Handover Request* message to the target gNB passing a transparent RRC container with necessary information to prepare for the handover at the target gNB. The information includes target cell ID, K_{gNB^*}, C-RNTI of the UE in the source gNB, RRM configuration including UE inactive time, basic AS-configuration including *antenna info and DL carrier frequency*, the current QoS flow to DRB mapping rules applied to the UE, the SIB1 from source gNB, the UE capabilities for different RATs, and PDU Session related information, and can further include the UE reported measurement information including beam-related information. The PDU Session related information includes the slice information (if supported) and QoS flow level QoS profile(s). After issuing a *Handover Request,* the source gNB should not reconfigure the UE, including performing reflective QoS flow to DRB mapping.
5. Admission control may be performed by the target gNB. Slice-aware admission control is performed, if the slice information is sent to the target gNB. If the PDU sessions are associated with non-supported slices, the target gNB would reject those PDU sessions.
6. The target gNB prepares the handover with L1/L2 and sends the *Handover Request Acknowledge* to the source gNB, which includes a transparent container to be sent to the UE as an RRC message to perform the handover.
7. The source gNB triggers the Uu handover by sending an *RRCReconfiguration* message to the UE, containing the information required to access the target cell, that is, the target cell ID, the new C-RNTI, and the target gNB security algorithm IDs for the selected security algorithms. It can also include a set of dedicated RACH resources, the association between RACH resources and SSB(s), the association between RACH resources and UE-specific CSI-RS configuration(s), common RACH resources, and the SI of the target cell.
8. The source gNB sends the *SN Status Transfer* message to the target gNB.
9. The UE synchronizes to the target cell and completes the RRC handover procedure by sending *RRCReconfigurationComplete* message to target gNB.

10. The target gNB sends a *Path Switch Request* message to the AMF to trigger 5GC to switch the downlink data path toward the target gNB and to establish an NG-C interface instance toward the target gNB.
11. 5GC switches the downlink data path toward the target gNB. The UPF sends one or more end-marker packets on the old path to the source gNB per PDU session/tunnel and then can release any user-plane/TNL resources toward the source gNB.
12. The AMF confirms the *Path Switch Request* message with the *Path Switch Request Acknowledge* message.
13. Upon reception of the *Path Switch Request Acknowledge* message from the AMF the target gNB sends the *UE Context Release* message to inform the source gNB about the success of the handover. The source gNB can then release radio and control-plane-related resources associated to the UE context. Any ongoing data forwarding may continue.

The RRM configuration can include both beam measurement information (for layer 3 mobility) associated to SSB(s) and CSI-RS(s) for the reported cell(s), if both types of measurements are available. Also, if carrier aggregation is configured, the RRM configuration can include the list of best cells on each frequency for which measurement information is available. The RRM measurement information can also include the beam measurement for the listed cells that belong to the target gNB.

The common RACH configuration for beams in the target cell is only associated to the SSB(s). The network can have dedicated RACH configurations associated to the SSB(s) and/or have dedicated RACH configurations associated to CSI-RS(s) within a cell. The target gNB can only include one of the following RACH configurations in the handover command to enable the UE to access the target cell:

1. Common RACH configuration
2. Common RACH configuration + dedicated RACH configuration associated with SSB
3. Common RACH configuration + dedicated RACH configuration associated with CSI-RS

The dedicated RACH configuration allocates RACH resource(s) along with a quality threshold to use them. When dedicated RACH resources are provided, they are prioritized by the UE, and the UE does not switch to contention-based RACH resources as long as the quality threshold of those dedicated resources is met. The order to access the dedicated RACH resources is up to UE implementation.

2.5.1.2 User-Plane Handover Procedures

The user-plane aspects of intra-NR handover for the UEs in RRC_CONNECTED state include the following principles to avoid loss of user data during handover [8]. During handover preparation, the user-plane tunnels can be established between the source gNB and the target gNB. During handover execution, the user data can be forwarded from the source gNB to the target gNB. Packet forwarding should be done in order, as long as the packets are received at the source gNB from the UPF or the source gNB buffer has not been emptied. During handover completion, the target gNB sends a path switch request message to the AMF to inform it that the UE has been granted access and the AMF then triggers path switch related 5GC internal signaling and actual path switch of the source gNB to the target gNB in UPF. The source gNB should continue forwarding data, as long as packets are received at the source gNB from the UPF or the source gNB buffer has not been emptied.

For RLC-AM bearers, in-sequence delivery, and duplication avoidance, the PDCP SN is maintained on a per DRB basis and the source gNB informs the target gNB about the next downlink PDCP SN to allocate to a packet which does not have a PDCP sequence number yet, neither from source gNB nor from the UPF. For security synchronization, the HFN is maintained and the source gNB provides the target gNB one reference HFN for the uplink and one for the downlink, that is, HFN and corresponding SN. In both UE and target gNB, a window-based mechanism is used for duplication detection and reordering. The occurrence of duplicates over the air interface in the target gNB is minimized by means of PDCP-SN-based reporting at the target gNB by the UE. In the uplink, the reporting is optionally configured on as per DRB basis by the gNB and the UE initially starts by transmitting those reports when granted resources are in the target gNB. In the downlink, the gNB can decide when and for which bearers a report is sent, and the UE does not wait for the report to resume UL transmission [8].

The target gNB retransmits and prioritizes all downlink data forwarded by the source gNB excluding the PDCP SDUs for which the reception was acknowledged through PDCP-SN-based reporting by the UE, that is, the target gNB should initially send all forwarded PDCP SDUs with PDCP SNs, then all forwarded downlink PDCP SDUs without SNs before sending new data from 5GC. Lossless delivery, when a QoS flow is mapped to a different DRB at handover, requires that the old DRB to be configured in the target cell. For in-order delivery in the downlink, the target gNB should first transmit the forwarded PDCP SDUs on the old DRB before transmitting new data from 5G CN on the new DRB. In the uplink, the target gNB should not deliver data of the QoS flow from the new DRB to 5G CN before receiving the end-marker on the old DRB from the UE [8].

The UE retransmits in the target gNB all uplink PDCP SDUs starting from the oldest PDCP SDU that has not been acknowledged at RLC sublayer in the source, excluding PDCP SDUs for which the reception was acknowledged through PDCP-SN-based reporting by the target. For RLC-UM bearers the PDCP SN and HFN are reset in the target gNB; no PDCP SDUs are retransmitted in the target gNB; and the target gNB prioritizes all downlink SDAP SDUs forwarded by the source gNB over the data from the CN. To minimize the losses when a QoS flow is mapped to a different DRB at handover, the old DRB needs to be configured in the target cell. For in-order delivery in the downlink, the target gNB should first transmit the forwarded PDCP SDUs on the old DRB before transmitting new data from 5G CN on the new DRB. In the uplink, the target gNB should not deliver data of the QoS flow from the new DRB to 5G CN before receiving the end-marker on the old DRB from the UE. The UE does not retransmit any PDCP SDU in the target cell for which transmission had been completed in the source cell.

The source NG-RAN node may request downlink data forwarding per QoS flow to be established for a PDU session and may provide information on how it maps QoS flows to DRBs. The target NG-RAN node decides whether data forwarding per QoS flow should be established for a PDU session. If lossless handover is desired and the QoS flow to DRB mapping, applied at the target NG-RAN node, allows employing data forwarding with the same QoS flow to DRB mapping that was used in the source NG-RAN node for a DRB and if all QoS flows mapped to that DRB are accepted for data forwarding, the target NG-RAN node establishes a downlink forwarding tunnel for that DRB. For a DRB for which SN status preservation is important, the target NG-RAN node may decide to establish an uplink data forwarding tunnel.

The target NG-RAN node may also decide to establish a downlink forwarding tunnel for each PDU session. In this case, the target NG-RAN node provides information related to the QoS flows for which data forwarding has been accepted and the corresponding uplink TNL information for data forwarding tunnels to be established between the source and the target NG-RAN nodes [8].

2.5.2 UE-Based Mobility

The PLMN selection in NR is based on 3GPP PLMN selection rules. Cell selection is required upon transition from RM-DEREGISTERED to RM-REGISTERED, from CM-IDLE to CM-CONNECTED, and from CM-CONNECTED to CM-IDLE. The UE NAS layer identifies a selected PLMN and equivalent PLMNs. Cell selection is always based on CD-SSBs located on the synchronization raster. The UE scans the NR frequency bands and for each carrier frequency identifies the strongest cell and the associated CD-SSB. It then detects broadcast SI of the cell to identify its PLMN(s). The UE may scan each carrier in certain order during initial cell selection or take advantage of stored information to expedite the search during stored information cell selection. The UE then tries to identify a suitable cell. If it is not able to identify a suitable cell, it then tries to identify an acceptable cell. When a suitable cell or an acceptable cell is found, the UE camps on it and begins the cell reselection procedure. A suitable cell is a cell whose measured cell attributes satisfy the cell selection criteria; the cell PLMN is the selected PLMN, registered or an equivalent PLMN; the cell is not barred or reserved, and the cell is not part of a tracking area which is in the list of forbidden tracking areas for roaming. An acceptable cell is a cell whose measured cell attributes satisfy the cell selection criteria, and the cell is not barred.

Upon transition from RRC_CONNECTED or RRC_INACTIVE to RRC_IDLE state, a UE may camp on a cell as a result of cell selection according to the frequency assigned to the UE via RRC signaling in the state transition message. The UE may attempt to find a suitable cell in the above manner described for stored information or initial cell selection. If no suitable cell is found on any frequency or RAT, the UE may attempt to find an acceptable cell.

In multi-beam operations, the cell quality is derived among the beams that are corresponding to the same cell [8].

The cell selection criterion S is considered fulfilled, if the following criteria are satisfied: $Srxlev > 0$ AND $Squal > 0$ where $Srxlev = Q_{rxlevmeas} - (Q_{rxlevmin} + Q_{rxlevminoffset}) - P_{compensation} - Qoffset_{temp}$ and $Squal = Q_{qualmeas} - (Q_{qualmin} + Q_{qualminoffset}) - Qoffset_{temp}$. The signaled values $Q_{rxlevminoffset}$ and $Q_{qualminoffset}$ are only applied when a cell is evaluated for cell selection as a result of a periodic search for a higher priority PLMN, while camped normally in a visited PLMN (VPLMN). During the periodic search for higher priority PLMN, the UE may check the S criterion of a cell using stored parameter values that were obtained from a different cell within the higher priority PLMN [12]. The cell selection parameters are described in Table 2.4.

The following rules are used by the UE to limit the required measurements. If the serving cell fulfills $Srxlev > S_{IntraSearchP}$ AND $Squal > S_{IntraSearchQ}$, the UE may skip intra-frequency measurements; otherwise, the UE must perform intra-frequency measurements. The UE must apply the following rules for NR inter-frequency and inter-RAT frequency measurements, which are identified in the SI and for which the UE has priority. For an NR inter-frequency or inter-RAT frequency with a reselection priority higher than the reselection priority of the current NR frequency, the UE is required to perform measurements on higher priority NR inter-frequency or inter-RAT frequencies [6]. For an NR inter-frequency with an equal or lower reselection priority than the reselection priority of the current NR frequency and for inter-RAT frequency with lower reselection priority than the reselection priority of the current NR frequency, if the serving cell fulfills $Srxlev > S_{nonIntraSearchP}$ AND $Squal > S_{nonIntraSearchQ}$ criterion, the UE may skip measurements of NR inter-frequencies or inter-RAT frequency cells of equal or lower priority; otherwise, the UE is required to perform measurements on NR inter-frequencies or inter-RAT frequency cells of equal or lower priority [6,8].

Table 2.4: Cell selection parameters [7,12].

Parameter	Description
Srxlev	Cell selection receive-level value (dB)
Squal	Cell selection quality value (dB)
$Qoffset_{temp}$	An offset temporarily applied to a cell (dB)
$Q_{rxlevmeas}$	Measured cell receive-level value (RSRP)
$Q_{qualmeas}$	Measured cell quality value (RSRQ)
$Q_{rxlevmin}$	Minimum required receive-level in the cell (dBm). If the UE supports SUL frequency for this cell, *Qrxlevmin* is obtained from *RxLevMinSUL*, if present, in *SIB1*, *SIB2*, and *SIB4*. If $Q_{rxlevminoffsetcellSUL}$ is present in *SIB3* and *SIB4* for the candidate cell, this cell-specific offset is added to the corresponding *Qrxlevmin* to achieve the required minimum receive-level in the candidate cell; otherwise, *Qrxlevmin* is obtained from *q-RxLevMin* in *SIB1*, *SIB2*, and *SIB4*. If $Q_{rxlevminoffsetcell}$ is present in *SIB3* and *SIB4* for the candidate cell, this cell-specific offset is added to the corresponding *Qrxlevmin* to achieve the required minimum receive-level in the candidate cell
$Q_{qualmin}$	Minimum required quality level in the cell (dB). If $Q_{qualminoffsetcell}$ is signaled for the concerned cell, this cell-specific offset is added to achieve the required minimum quality level in the candidate cell
$Q_{rxlevminoffset}$	An offset to the signaled $Q_{rxlevmin}$ taken into account in the *Srxlev* evaluation as a result of periodic search for a higher priority PLMN while camped normally in a VPLMN
$Q_{qualminoffset}$	Offset to the signaled $Q_{qualmin}$ taken into account in the *Squal* evaluation as a result of periodic search for a higher priority PLMN while camped normally in a VPLMN
$P_{compensation}$	If the UE supports the additional P_{max} in the *NR-NS-PmaxList*, if present, in *SIB1*, *SIB2*, and *SIB4*, $\max(P_{EMAX1} - P_{PowerClass}, 0) - [\min(P_{EMAX2}, P_{PowerClass}) - \min(P_{EMAX1}, P_{PowerClass})]$(dB) else $\max(P_{EMAX1} - P_{PowerClass}, 0)$(dB)
P_{EMAX1}, P_{EMAX2}	Maximum transmit-power level that a UE may use when transmitting in the uplink in the cell (dBm) is defined as P_{EMAX}. If the UE supports SUL frequency for this cell, P_{EMAX1} and P_{EMAX2} are obtained from the *p-Max* for SUL in *SIB1* and *NR-NS-PmaxList* for SUL, respectively, in *SIB1*, *SIB2*, and *SIB4*; otherwise, P_{EMAX1} and P_{EMAX2} are obtained from the *p-Max* and *NR-NS-PmaxList*, respectively, in *SIB1*, *SIB2*, and *SIB4* for regular uplink
$P_{PowerClass}$	Maximum RF output power of the UE (dBm) according to the UE power class

When evaluating *Srxlev* and *Squal* of non-serving cells for reselection evaluation purposes, the UE must use the parameters that are provided by the serving cell and for the verification of cell selection criterion, the UE must use the parameters provided by the target cell for cell reselection. The NAS can control the RAT(s) in which the cell selection should be performed, for instance by indicating RAT(s) associated with the selected PLMN, and by maintaining a list of forbidden registration area(s) and a list of equivalent PLMNs. The UE must select a suitable cell based on RRC_IDLE or RRC_INACTIVE state measurements and cell selection criteria. In order to expedite the cell selection process, the previously stored information for other RATs may be used by the UE. After camping on a cell, the UE is required to regularly search for a better cell according to the cell reselection criteria. If a better cell is found, the UE proceed to select that cell. The change of cell may imply a change of RAT.

The UE mobility state is determined, if the parameters ($T_{CRmax}, N_{CR\_H}, N_{CR\_M}$, and $T_{CRmaxHyst}$) are broadcast in SI for the serving cell. The **state detection criteria** are based on the following principles [12]:

- Normal-mobility state criterion: If the number of cell reselections during time period T_{CRmax} is less than N_{CR_M}.
- Medium-mobility state criterion: If the number of cell reselections during time period T_{CRmax} is greater than or equal to N_{CR_M} but less than or equal to N_{CR_M}.
- High-mobility state criterion: If the number of cell reselections during time period T_{CRmax} is greater than N_{CR_M}.

The UE must not attempt to make consecutive reselections, where a cell is reselected again immediately after one reselection for mobility state detection criteria.

The **state transitions** of the UE are determined based on the following criteria [12]:

- If the criteria for high-mobility state is detected, the UE must transition to the high-mobility state.
- If the criteria for medium-mobility state is detected, the UE must transition to medium--mobility state.
- If criteria for either medium- or high-mobility state is not detected during time period $T_{CRmaxHyst}$, the UE must transition to normal-mobility state.

If the UE is in high- or medium-mobility state, it must apply the speed-dependent scaling rules [12].

2.5.3 Paging

Paging allows the network to reach UEs in RRC_IDLE and RRC_INACTIVE states, and to notify the UEs in RRC_IDLE, RRC_INACTIVE, and RRC_CONNECTED states of SI change, as well as to send ETWS/CMAS notifications. While in RRC_IDLE state, the UE monitors the PCHs for CN-initiated paging. The UEs in RRC_INACTIVE state also monitor PCHs for RAN-initiated paging. To ensure limiting the adverse impact of paging procedure on the battery consumption, a UE does not need to continuously monitor the PCHs since NR defines a UE-specific paging DRX. The UE in RRC_IDLE or RRC_INACTIVE state is only required to monitor PCHs during paging occasions in a DRX cycle. The paging DRX cycles are configured by the network as follows [8]. For CN-initiated paging, a default cycle is broadcast in SI and also a UE-specific cycle is configured via NAS signaling. For RAN-initiated paging, a UE-specific cycle is configured via RRC signaling. The UE applies the shortest of the DRX cycles that are configured for it, that is, a UE in RRC_IDLE uses the shortest of the CN-initiated paging cycles, whereas a UE in RRC_INACTIVE applies the shortest of CN-initiated and RAN-initiated paging cycles (see Fig. 2.29).

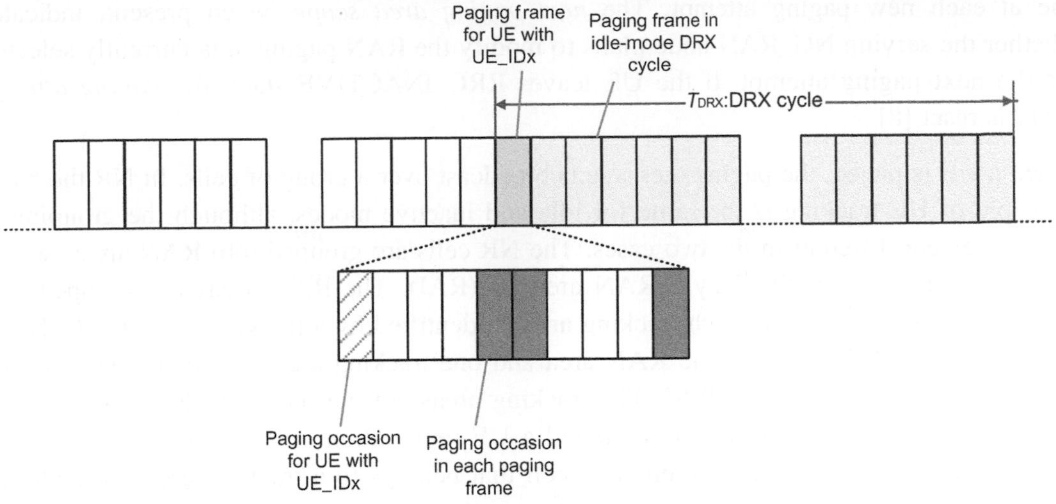

Figure 2.29
Example illustration of NR paging frames and paging occasions [12].

In CN-initiated and RAN-initiated paging, the POs of a UE are based on the same UE_ID, resulting in overlapping POs for both cases. The number of different POs in a DRX cycle is configurable via SI and a network may distribute UEs into those POs based on their UE_IDs. When in RRC_CONNECTED state, the UE monitors the PCHs in any PO signaled in the SI for SI change indication and public warning system notification. In the case of bandwidth adaptation, a UE in RRC_CONNECTED state only monitors PCHs on the active BWP with common search space configured (see Chapter 4).

To optimize the paging procedure for the UEs in CM-IDLE state, upon UE context release, the NG-RAN node may provide the AMF with a list of recommended cells and NG-RAN nodes as assistance information for subsequent paging of the UE. The AMF may further provide *paging attempt information* consisting of a *paging attempt count* and the *intended number of paging attempts* and possibly the *next paging area scope*. If *paging attempt information* is included in the paging message, each paged NG-RAN node receives the same information during a paging attempt. The *paging attempt count* is increased by one at each new paging attempt. The *next paging area scope*, when present, indicates whether the AMF intends to modify the paging area currently selected at the next paging attempt. If the UE has changed its state to CM-CONNECTED, the *paging attempt count* is reset [8].

To optimize the paging procedure for the UEs in RRC_INACTIVE state, upon RAN-initiated paging, the serving NG-RAN node provides the RAN paging area information. The serving NG-RAN node may also provide the RAN paging attempt information. Each paged NG-RAN node receives the same RAN paging attempt information during a paging attempt with the following content: *paging attempt count*, the intended number of paging attempts and the *next paging area scope*. The *paging attempt count* is increased by one at each new paging attempt. The *next paging area scope*, when present, indicates whether the serving NG_RAN node plans to modify the RAN paging area currently selected for the next paging attempt. If the UE leaves RRC_INACTIVE state, the *paging attempt count* is reset [8].

When a UE is paged, the paging message is broadcast over a group of cells. In NR the basic principle of UE tracking is the same for idle and inactive modes, although the grouping is to some extent different in the two cases. The NR cells are grouped into RAN areas, where each RAN area is identified by a RAN area ID (RAI). The RAN areas are grouped into larger tracking areas, where each tracking area is identified by a tracking area ID (TAI). As a result, each cell belongs to one RAN area and one tracking area, the identities of which are provided as part of the cell SI. The tracking areas are the basis for device tracking at core-network level. Each device is assigned a UE registration area by the CN, consisting of a list of TAIs. When a device enters a cell that belongs to a tracking area not included in the assigned UE registration area, it accesses the network, including the CN, and performs a NAS registration update. The CN registers the device location and updates the device registration area and provides the device with a new TAI list that includes the new TAI. The reason that the device is assigned a set of TAIs is to avoid repeated NAS registration updates, every time that the device crosses the border of two neighbor tracking areas. By keeping the old TAI within the updated UE registration area, no new update is needed, if the device moves back to the old TAI. The RAN area is the basis for device tracking on RAN level. The UEs in the inactive mode can be assigned an RNA that consists of either a list of cell identities; a list of RAIs; or a list of TAIs [17].

The procedure for RNA update is similar to the update of the UE registration area. When a device enters a cell that is not included in its RNA, it accesses the network and performs an RNA update. The radio network registers the device location and updates the device RNA. Since the change of tracking area always implies the change of the device RAN area, an RNA update is implicitly performed every time a device performs a UE registration update. In order to track its movement within the network, the device searches for and measures SSBs similar to the initial cell search procedure. Once the device detects an SSB with a received power that exceeds the received power of its current SSB by a certain threshold, it detects the SIB1 of the new cell in order to acquire information about the tracking and RAN areas.

2.5.4 Measurements

The UE in RRC_CONNECTED state conducts measurements on multiple beams (at least one) of a cell and averages the measurement results mainly in the form of power values, in order to derive the cell quality. Therefore, the UE is configured to consider a subset of the detected beams. Filtering is applied at two different levels namely at the physical

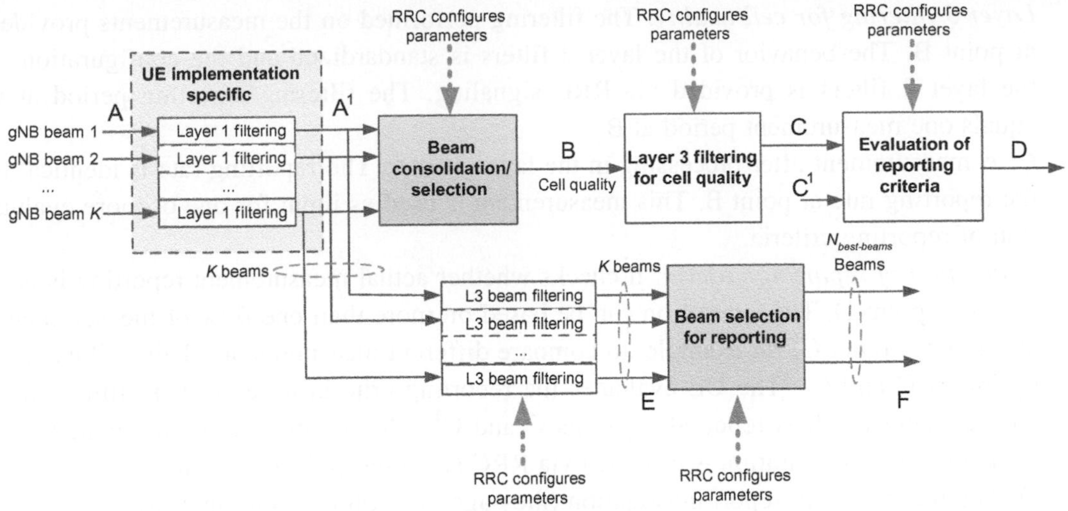

Figure 2.30
NR RRM measurement model [8].

layer to derive beam quality and then at RRC level to derive cell quality from multiple beams. The cell quality from beam measurements is derived in the same way for the serving cell(s) and for the non-serving cell(s). The measurement reports may contain the measurement results of the $N_{best-beams}$ best beams, if the UE is configured by the gNB. The K beams correspond to the measurements on SSB or CSI-RS resources configured for layer 3 mobility by gNB and detected by UE at layer 1.

The corresponding high-level measurement model is illustrated in Fig. 2.30 where the parameters can be described as follows [8]:

- A_i: The measurements (beam-specific samples) internal to the physical layer.
- *Layer 1 filtering*: Internal layer 1 filtering of the inputs measured at point A. The exact filtering function is implementation specific and the way that the measurements are conducted in the physical layer by an implementation (inputs A_i and layer 1 filtering) in not specified by the standard.
- $A_i^{\,1}$: The measurements (i.e., beam-specific measurements) reported by layer 1 to layer 3 after layer 1 filtering.
- *Beam consolidation/selection*: The beam-specific measurements are consolidated to derive cell quality. The behavior of the beam consolidation/selection is standardized, and the configuration of this module is provided via RRC signaling. The reporting period at B equals one measurement period at A^1.
- *B*: A measurement (i.e., cell quality) derived from beam-specific measurements reported to layer 3 after beam consolidation/selection.

- *Layer 3 filtering for cell quality*: The filtering performed on the measurements provided at point B. The behavior of the layer 3 filters is standardized and the configuration of the layer 3 filters is provided via RRC signaling. The filtering reporting period at C equals one measurement period at B.
- *C*: A measurement after processing in the layer 3 filter. The reporting rate is identical to the reporting rate at point B. This measurement is used as input for one or more evaluation of reporting criteria.
- *Evaluation of reporting criteria*: It checks whether actual measurement reporting is necessary at point D. The evaluation can be based on more than one flow of measurements at reference point C, for example, to compare different measurements. This is illustrated by inputs C and C^1. The UE evaluates the reporting criteria at least every time a new measurement result is reported at points C and C^1. The reporting criteria are standardized, and the configuration is provided via RRC signaling (UE measurements).
- *D*: The measurement report information (message) sent on the radio interface.
- *L3 beam filtering*: The filtering performed on the measurements (i.e., beam-specific measurements) provided at point A^1. The behavior of the beam filters is standardized, and the configuration of the beam filters is provided via RRC signaling. The filtering reporting period at E equals one measurement period at A^1.
- *E*: A measurement (i.e., beam-specific measurement) after processing in the beam filter. The reporting rate is identical to the reporting rate at point A^1. This measurement is used as input for selecting the $N_{best-beams}$ measurements to be reported.
- *Beam selection for beam reporting*: It selects the $N_{best-beams}$ measurements from the measurements provided at point E. The behavior of the beam selection is standardized, and the configuration of this module is provided via RRC signaling.
- *F*: The beam measurement information included in measurement report sent over the radio interface.

Layer 1 filtering introduces a certain level of measurement averaging and the manner through which the UE performs the required measurements is implementation specific to the point that the output at B fulfills the performance requirements. The layer 3 filtering function for cell quality and the related parameters do not introduce any delay in the sample availability at points B and C in Fig. 2.30. The measurements at points C and C^1 are the input used in the event evaluation. The L3 beam filtering and the related parameters do not cause any delay in the sample availability at points E and F. The measurement reports are characterized by the following criteria [8]:

- Measurement reports include the measurement identity of the associated measurement configuration that triggered the reporting.
- Cell and beam measurement quantities to be included in measurement reports are configured by the network.

- The number of non-serving cells to be reported can be limited through configuration by the network.
- Cells belonging to a blacklist configured by the network are not used in event evaluation and reporting, and conversely when a whitelist is configured by the network, only the cells belonging to the whitelist are used in event evaluation and reporting.
- Beam measurements to be included in measurement reports are configured by the network (beam ID only, measurement result and beam ID, or no beam reporting).

The intra-frequency neighbor (cell) measurements and inter-frequency neighbor (cell) measurements are defined as follows [8]:

- *SSB-based intra-frequency measurement*: A measurement is defined as an SSB-based intra-frequency measurement provided that the center frequency of the SSB of the serving cell and the center frequency of the SSB of the neighbor cell are the same, and the subcarrier spacing of the two SSBs are also the same.
- *SSB-based inter-frequency measurement*: A measurement is defined as an SSB-based inter-frequency measurement provided that the center frequency of the SSB of the serving cell and the center frequency of the SSB of the neighbor cell are different, or the subcarrier spacing of the two SSBs are different. It must be noted that for SSB-based measurements, one measurement object corresponds to one SSB and the UE considers different SSBs as different cells.
- *CSI-RS-based intra-frequency measurement*: A measurement is defined as a CSI-RS-based intra-frequency measurement, if the bandwidth of the CSI-RS resource on the neighbor cell configured for measurement is within the bandwidth of the CSI-RS resource on the serving cell configured for measurement, and the subcarrier spacing of the two CSI-RS resources is the same.
- *CSI-RS-based inter-frequency measurement*: A measurement is defined as a CSI-RS-based inter-frequency measurement, if the bandwidth of the CSI-RS resource on the neighbor cell configured for measurement is not within the bandwidth of the CSI-RS resource on the serving cell configured for measurement, or the subcarrier spacing of the two CSI-RS resources are different.

A measurement can be non-gap-assisted or gap-assisted depending on the capability of the UE, the active BWP of the UE and the current operating frequency, described as follows [8]:

- For an SSB-based inter-frequency measurement, a measurement gap configuration is always provided if the UE only supports per-UE measurement gaps or if the UE supports per-FR measurement gaps and any of the configured BWP frequencies of any of the serving cells are in the same FR of the measurement object.

- For an SSB-based intra-frequency measurement, a measurement gap configuration is always provided in the case where, other than the initial BWP, if any of the UE configured BWPs do not contain the frequency domain resources of the SSB associated with the initial DL BWP.

In non-gap-assisted scenarios, the UE must be able to conduct measurements without measurement gaps. In gap-assisted scenarios, the UE cannot be assumed to be able to conduct measurements without measurement gaps.

2.6 UE and Network Identifiers

An NR UE in the connected mode uses a number of network-assigned temporary identifiers in order to communicate to gNB and 5GC. Those identifiers, their descriptions, and their usage are summarized as follows [8]:

- C-RNTI: A unique UE identification used as an identifier of the RRC connection and for scheduling purposes.
- CS-RNTI: A unique UE identification used for SPS in the downlink or configured grant in the uplink.
- INT-RNTI: An identification of preemption in the downlink.
- P-RNTI: An identification of paging and SI change notification in the downlink.
- SI-RNTI: An identification of broadcast and SI in the downlink.
- SP-CSI-RNTI: Unique UE identification used for semi-persistent CSI reporting on PUSCH.

The following identifiers are used for power and slot format control [8]:

- SFI-RNTI: An identification of slot format.
- TPC-PUCCH-RNTI: A unique UE identification to control the power of PUCCH.
- TPC-PUSCH-RNTI: A unique UE identification to control the power of PUSCH.
- TPC-SRS-RNTI: A unique UE identification to control the power of SRS.

The following identities are used during random-access procedure [8]:

- RA-RNTI: An identification of the RAR message in the downlink.
- Temporary C-RNTI: A UE identification temporarily used for scheduling during the random-access procedure.
- Random value for contention resolution: A UE identification temporarily used for contention resolution purposes during the random-access procedure.

The following identities are used at NG-RAN level by an NR UE connected to 5GC:

- I-RNTI: An ID used to identify the UE context in RRC_INACTIVE state.

Table 2.5: Radio network temporary identifiers (RNTIs) in NR and their usage [9].

RNTI	Usage	Transport Channel	Logical Channel
P-RNTI	Paging and system information change notification	PCH	PCCH
SI-RNTI	Broadcast of system information	DL-SCH	BCCH
RA-RNTI	Random-access response	DL-SCH	N/A
Temporary C-RNTI	Contention resolution (when no valid C-RNTI is available)	DL-SCH	CCCH
Temporary C-RNTI	Msg3 transmission	UL-SCH	CCCH, DCCH, DTCH
C-RNTI, MCS-C-RNTI	Dynamically scheduled unicast transmission	UL-SCH	DCCH, DTCH
C-RNTI	Dynamically scheduled unicast transmission	DL-SCH	CCCH, DCCH, DTCH
MCS-C-RNTI	Dynamically scheduled unicast transmission	DL-SCH	DCCH, DTCH
C-RNTI	Triggering of PDCCH ordered random access	N/A	N/A
CS-RNTI	Configured scheduled unicast transmission (activation, reactivation, and retransmission)	DL-SCH, UL-SCH	DCCH, DTCH
CS-RNTI	Configured scheduled unicast transmission (deactivation)	N/A	N/A
TPC-PUCCH-RNTI	PUCCH power control	N/A	N/A
TPC-PUSCH-RNTI	PUSCH power control	N/A	N/A
TPC-SRS-RNTI	SRS trigger and power control	N/A	N/A
INT-RNTI	Indication of preemption in the downlink	N/A	N/A
SFI-RNTI	Slot format indication in a given cell	N/A	N/A
SP-CSI-RNTI	Activation of semi-persistent CSI reporting on PUSCH	N/A	N/A

A complete list of UE RNTIs is provided in Table 2.5.

The following identities are used in NG-RAN for identifying a specific network entity [8]:

- AMF name: This is used to identify an AMF.
- NCGI: This is used to globally identify the NR cells. The NCGI is constructed from the PLMN identity to which the cell belongs to and the NR cell identity (NCI) of the cell.
- gNB ID: This is used to identify the gNBs within a PLMN. The gNB ID is contained within the NCI of its cells.
- Global gNB ID: This is used to globally identify the gNBs. The global gNB ID is constructed from the PLMN identity to which the gNB belongs and the gNB ID. The MCC and MNC are the same as included in the NCGI.
- Tracking area identity (TAI): This is used to identify tracking areas. The TAI is constructed from the PLMN identity the tracking area belongs to and the tracking area code of the tracking area.

- Single network slice selection assistance information (S-NSSAI): This is used to identify a network slice.

2.7 Random-Access Procedure (L2/L3 Aspects)

The random-access procedure is triggered by a number of events including initial access from RRC_IDLE state; RRC connection reestablishment procedure, handover, downlink/uplink data arrival when in RRC_CONNECTED state and if uplink synchronization status is *non-synchronized*; uplink data arrival while in RRC_CONNECTED state and when there are no available PUCCH resources for SR; SR failure; request by RRC upon synchronous reconfiguration; transition from RRC_INACTIVE state; establishing timing alignment upon SCell addition; request for other SI; and beam failure recovery. Furthermore, the random-access procedure takes two distinct forms: contention-based random access and contention-free random access as shown in Fig. 2.31. For random access in a cell configured with SUL, the network can explicitly signal which carrier to use (uplink or SUL); otherwise, the UE selects the SUL carrier, if the measured quality of the downlink is lower than a broadcast threshold. Once started, all uplink transmissions of the random-access procedure remain on the selected carrier [8]. The complete description of the random-access procedure including the physical layer aspects can be found in Chapter 3.

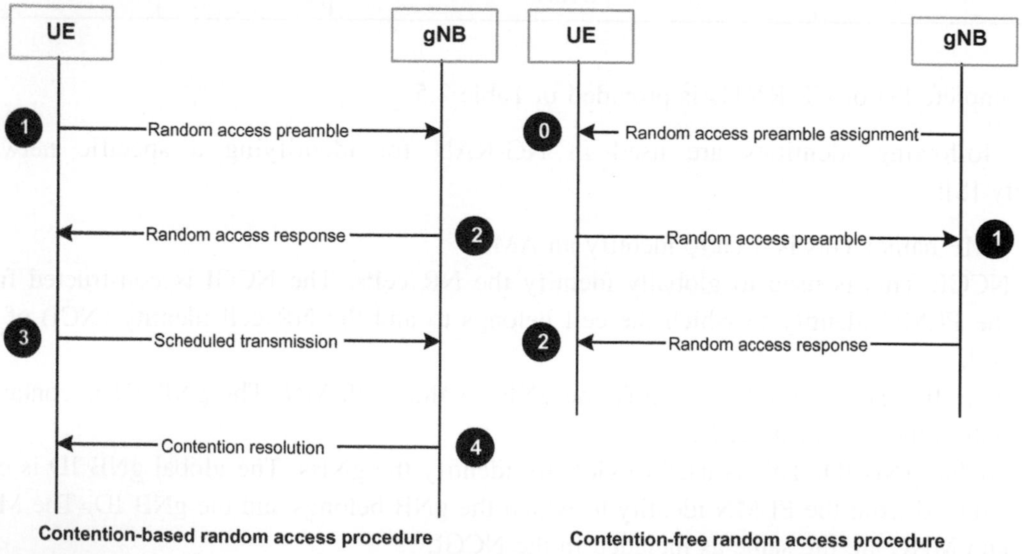

Figure 2.31
Contention-based and contention-free random-access procedures [8].

2.8 Multi-radio Dual Connectivity (L2/L3 Aspects)

MR-DC is a generalization of the intra-E-UTRA dual connectivity. The NG-RAN supports MR-DC operation wherein a UE in RRC_CONNECTED state is configured to utilize radio resources provided by (at least) two distinct schedulers, located in two different NG-RAN nodes connected via a non-ideal backhaul, one providing the NR access and the other providing either E-UTRA or NR access. In MR-DC, one network node acts as the master node (MN) and the other as the secondary node (SN). The MN and SN entities are connected via a network interface and the MN is connected to the CN. The NR MR-DC scheme is designed based on the assumption of non-ideal backhaul between different nodes but can also be used in the case of ideal backhaul. The LTE network supports MR-DC via EN-DC, in which a UE is connected to one eNB that acts as the MN and one en-gNB[11] that acts as a SN. The eNB is connected to the EPC via the S1 interface and to the en-gNB via the X2 interface. The en-gNB may also be connected to the EPC via the S1-U interface and other en-gNBs via the X2-U interface [4].

The NG-RAN supports NG-RAN EN-DC (NGEN-DC), in which a UE is connected to one ng-eNB that acts as the MN and one gNB that acts as an SN. The ng-eNB is connected to the 5GC and the gNB is connected to the ng-eNB via the Xn interface. The NG-RAN further supports NR-E-UTRA DC (NE-DC), in which a UE is connected to one gNB that acts as the MN and one ng-eNB that acts as an SN. The gNB is connected to 5GC and the ng-eNB is connected to the gNB via the Xn interface. Another important scenario is NR–NR dual connectivity (NR-DC), in which a UE is connected to one gNB that acts as the MN and another gNB that acts as the SN. The master gNB is connected to the 5GC via the NG interface and to the secondary gNB via the Xn interface. The secondary gNB may also be connected to the 5GC via the NG-U interface. In addition, NR-DC can also be used when a UE is connected to two gNB-DUs, one serving the MCG and the other serving the SCG, connected to the same gNB-CU, acting both as the MN and the SN. When the UE is configured with SCG, it is configured with two MAC entities: one MAC entity for the MCG and one MAC entity for the SCG [4].

In MR-DC, the UE has a single RRC state, based on the MN RRC and a single control-plane connection toward the CN. Each radio node has its own RRC entity (LTE version, if the node is an eNB or NR version if the node is a gNB) which can generate RRC PDUs to be sent to the UE. The RRC PDUs generated by the SN can be transported via the MN to the UE. The MN always sends the initial SN RRC configuration via MCG SRB (SRB1); however, subsequent reconfigurations may be transported via MN or SN. When

[11] en-gNB is a node providing NR user-plane and control-plane protocol terminations toward the UE and acts as a secondary node in EN-DC. A secondary node in MR-DC is a radio access node, with no control-plane connection to the core network, providing only additional radio resources to the UE. It may be an en-gNB (in EN-DC), a secondary ng-eNB (in NE-DC), or a secondary gNB (in NR-DC and NGEN-DC) [4].

transporting RRC PDU from the SN, the MN does not modify the UE configuration provided by the SN [4].

When an LTE node is connected to the EPC, upon initial connection establishment, SRB1 uses LTE PDCP. If the UE supports EN-DC, regardless of whether EN-DC is configured, after initial connection establishment, the MCG SRBs (SRB1 and SRB2) can be configured by the network to use either LTE PDCP or NR PDCP (either SRB1 and SRB2 are both configured with LTE PDCP, or they are both configured with NR PDCP). A change from LTE PDCP to NR PDCP (or vice versa) is supported via a handover procedure (reconfiguration with mobility) or, for the initial change of SRB1 from LTE PDCP to NR PDCP, with a reconfiguration without mobility before the initial security activation.

If the SN is a gNB (i.e., the case for EN-DC, NGEN-DC, and NR-DC scenario), the UE can be configured to establish an SRB with the SN (SRB3) to enable RRC PDUs for the SN to be directly transferred between the UE and the SN. The RRC PDUs for the SN can only be transported directly to the UE for SN RRC reconfiguration without any coordination with the MN. Measurement reporting for mobility within the SN can be sent directly from the UE to the SN, if SRB3 is configured. The split SRB is supported for all MR-DC options, allowing duplication of RRC PDUs generated by the MN via the direct path and via the SN. The split SRB utilizes the NR PDCP. The NR Rel-15 specifications do not support duplication of RRC PDUs generated by the SN via the MN and SN paths. In EN-DC, the SCG configuration is maintained in the UE during suspension. The UE releases the SCG configuration (but not the radio bearer configuration) during resumption initiation (see Fig. 2.24). In MR-DC with 5GC, the UE stores the PDCP/SDAP configuration when moving to RRC_INACTIVE state, but it releases the SCG configuration [4].

There are three bearer types in MR-DC from a UE perspective: MCG bearer, SCG bearer, and split bearer. For EN-DC, the network can configure either LTE PDCP or NR PDCP for the MN-terminated MCG bearers, while NR PDCP is always used for all other bearers. In MR-DC with 5GC, NR PDCP is always used for all bearer types. In NGEN-DC, LTE RLC/MAC is used in the MN, while NR RLC/MAC is used in the SN. In EN-DC, NR RLC/MAC is used in the MN while LTE RLC/MAC is used in the SN. In NR-DC, NR RLC/MAC is used in both MN and SN. From the network perspective, each bearer (MCG, SCG, and split bearer) can be terminated either in MN or in SN. If only SCG bearers are configured for a UE, for SRB1 and SRB2, the logical channels are always configured at least in the MCG, that is, this is still an MR-DC configuration and a PCell always exists. If only MCG bearers are configured for a UE, that is, there is no SCG, this is still considered an MR-DC configuration, if at least one of the bearers is terminated in the SN [4].

In MR-DC, two or more component carriers may be aggregated over two cell groups. A UE may simultaneously receive or transmit on multiple component carriers depending on its capabilities. The maximum number of configured component carriers for a UE is 32 for downlink and uplink. Depending on UE's capabilities, up to 31 component carriers can be configured for an LTE cell group when the NR cell group is configured. For the NR cell group, the maximum number of configured component carriers for a UE is 16 for downlink and 16 for uplink. A gNB may configure the same physical cell IDs (PCIs) for several NR cells that it serves. To avoid PCI confusion for MR-DC, the NR PCIs may be allocated in a way that an NR cell is uniquely identifiable by a PCell ID. This PCell is in the coverage area of an NR cell included in the MR-DC operation. In addition, the NR PCIs may only be reused in NR cells on the same SSB frequency sufficiently apart from each other. An X2-C/Xn-C signaling can be used to help identify NR PCIs by including the cell global identifier (CGI) of the PCell in the respective X2AP/XnAP messages and by providing neighbor cell relationship via non-UE-associated signaling [4].

In MR-DC, the UE is configured with two MAC entities: one MAC entity for the MCG and one MAC entity for the SCG. In MR-DC, SPS resources can be configured on both PCell and PSCell. In MR-DC, the BSR configuration, triggering, and reporting are independently performed per cell group. For split bearers, the PDCP data is considered in BSR in the cell group(s) configured by RRC signaling. In EN-DC, separate DRX configurations are provided for MCG and SCG. Both RLC-AM and RLC-UM can be configured in MR-DC for all bearer types (i.e., MCG, SCG, and split bearers). In EN-DC, packet duplication can be applied to carrier-aggregation in the MN and in the SN; however, MCG bearer carrier-aggregation packet duplication can be configured only in combination with LTE PDCP; and MCG DRB carrier-aggregation duplication can be configured only if dual-connectivity packet duplication is not configured for any split DRB. In NGEN-DC, carrier-aggregation packet duplication can only be configured for SCG bearer. In NE-DC, carrier-aggregation packet duplication can only be configured for MCG bearer. In NR-DC, carrier-aggregation packet duplication can be configured for both MCG and SCG bearers. In EN-DC, ROHC can be configured for all bearer types. In MR-DC with 5GC, the network may host up to two SDAP protocol entities for each PDU session, one for MN and the other one for SN. The UE is configured with one SDAP protocol entity per PDU session [4].

In MR-DC, the SN is not required to broadcast SI other than for radio frame timing and SFN. The SI for initial configuration is provided to the UE by dedicated RRC signaling via the MN. The UE acquires radio frame timing and SFN of SCG from the LTE primary and secondary synchronization signals and MIB (if the SN is an eNB) and from the NR primary and secondary synchronization signals and MIB (if the SN is a gNB) of the PSCell. Moreover, upon change of the relevant SI of a configured SCell, the network releases and subsequently adds the corresponding SCell (with updated SI), via one or more RRC reconfiguration messages sent on SRB1 or SRB3. If the measurement is configured for the UE in preparation for the secondary node addition procedure, the MN may configure the measurement for the UE. In the case of the intra-secondary node mobility, the SN may configure the measurement for the UE in coordination with the MN [4].

The secondary node change procedure can be triggered by both MN (only for inter-frequency secondary node change) and SN. For secondary node changes triggered by the SN, the RRM measurement configuration is maintained by the SN which also processes the measurement reporting, without providing the measurement results to the MN. Measurements can be configured independently by the MN and by the SN (intra-RAT measurements on serving and non-serving frequencies). The MN indicates the maximum number of frequency layers and measurement identities that can be used in the SN to ensure that UE capabilities are not exceeded. If MN and SN both configure measurements on the same carrier frequency then those configurations must be consistent. Each node (MN or SN) can independently configure a threshold for the SpCell quality. When the PCell quality is above the threshold configured by the MN, the UE is still required to perform inter-RAT measurements configured by the MN on the SN RAT. When SpCell quality is above the threshold configured by the SN, the UE is not required to perform measurements configured by the SN [4].

The measurement reports, configured by the SN, are sent on SRB1 when SRB3 is not configured; otherwise, the measurement reports are sent over SRB3. The measurement results related to the target SN can be provided by MN to target SN at MN-initiated SN change procedure. The measurement results of target SN can be forwarded from the source SN to the target SN via MN at SN-initiated SN change procedure. The measurement results corresponding to the target SN can be provided by the source MN to the target MN at inter-MN handover with/without SN change procedure [4].

Per-UE or per-FR measurement gaps can be configured, depending on UE capability to support independent FR measurement and network preference. Per-UE gap applies to both FR1 (LTE and NR) and FR2 (NR) bands. For per-FR gap, two independent gap patterns (i.e., FR1 gap and FR2 gap) are configured for FR1 and FR2. The UE may also be configured with a per-UE gap sharing configuration (applying to per-UE gap) or with two separate gap sharing configurations (applying to FR1 and FR2 measurement gaps, respectively). A measurement gap configuration is always provided in the following scenarios: for UEs configured with LTE inter-frequency measurements; and for UEs that support either per-UE or per-FR gaps, when the conditions to measure SSB-based inter-frequency measurement or SSB-based intra-frequency measurement are
satisfied [4].

2.9 Carrier Aggregation (L2/L3 Aspects)

Multiple NR component carriers can be aggregated and simultaneously transmitted to a UE in the downlink or from a UE in the uplink, allowing an increased operating bandwidth and correspondingly higher link data rates. The component carriers do not need to be contiguous in the frequency domain and can be in the same frequency band or different frequency bands, resulting in three scenarios: intra-band carrier aggregation with frequency-contiguous component carriers, intra-band carrier aggregation with non-contiguous component carriers, and inter-band carrier aggregation with non-contiguous component carriers. While the system-level operation for the three scenarios is the same, the architecture and complexity of RF transceivers can be very different. The NR supports up to 16 downlink/uplink carriers of different bandwidths and different duplex schemes, with the minimum and maximum contiguous bandwidth of 5 and 400 MHz per component carrier, respectively [5,21].

A UE capable of carrier aggregation may receive or transmit simultaneously on multiple component carriers, while a device not capable of carrier aggregation can access one of the component carriers at any given time. In the case of inter-band carrier aggregation of multiple half-duplex TDD carriers (supplemental uplink or downlink), the transmission direction of different carriers does not necessarily have to be the same, which implies that a carrier-aggregation-capable TDD device may need a front-end duplexer, unlike a typical carrier-aggregation-incapable TDD device that does not include a duplexer. In the LTE and NR specifications, the carrier aggregation is treated as a cell, that is, a carrier-aggregation-capable UE is said to able transmit/receive to/from multiple cells. One of these cells is referred to as the PCell that is the cell which the device initially selects and connects to. Once connected to the gNB, one or more SCells can be configured. The SCells can be activated or deceived to meet various application requirements. Different UEs may have different designated cells as their PCell, meaning that the configuration of the PCell is UE-specific. Furthermore, the number of carriers (or cells) does not have to be the same in uplink and downlink. In a typical scenario, there are more downlink carriers than uplink carriers, since there is often more traffic in the downlink than in the uplink. Furthermore, the RF implementation complexity and cost of operating multiple simultaneously active uplink carriers are often higher than the corresponding complexity/cost of the downlink. The scheduling grants and radio resource assignments can be transmitted on either the same cell as the corresponding data, referred to as self-scheduling, or on a different cell than the corresponding data, referred to as cross-carrier scheduling.

The NR carrier aggregation uses L1/L2 control signaling for scheduling the UE in the downlink, and uplink control signaling to transmit HARQ-ACKs. The uplink feedback is typically transmitted on the PCell to allow asymmetric carrier aggregation. In certain use cases where there are a large number of downlink component carriers and a single uplink component carrier, the uplink carrier would be overloaded with a large number of feedback information. To avoid overloading a single carrier, it is possible to configure two PUCCH groups where the feedback corresponding to the first group is transmitted in the uplink of the PCell and the feedback corresponding to the other group of carriers is transmitted on the PSCell. If carrier aggregation is enabled, the UE may receive and transmit on multiple carriers, but operating multiple carriers is only needed for high data rates, thus is advantageous to deactivate unused carriers. Activation and deactivation of component carriers can be done through MAC CEs (see Section 2.2.1), where a bitmap is used to indicate whether a configured SCell should be activated or deactivated.

As we mentioned earlier, to ensure reasonable UE power consumption when carrier aggregation is configured, an activation/deactivation mechanism of cells is supported. When an SCell is deactivated, the UE no longer needs to receive the corresponding PDCCH or PDSCH, it cannot transmit in the corresponding uplink, it is not required to perform CQI measurements on that cell. On the other hand, when an SCell is activated, the UE receives PDSCH and PDCCH (if the UE is configured to monitor PDCCH on this SCell) and is expected to be able to perform CQI measurements on that cell. The NG-RAN ensures that while PUCCH SCell (i.e., an SCell configured with PUCCH) is deactivated, SCells of the secondary PUCCH group (i.e., a group of SCells whose PUCCH signaling is associated with the PUCCH on the PUCCH SCell) are activated. The NG-RAN further ensures that SCells mapped to PUCCH SCell are deactivated before the PUCCH SCell is changed or removed. When reconfiguring the set of serving cells, SCells added to the set are initially deactivated and SCells which remain in the set (either unchanged or reconfigured) do not change their activation status. During handover, the SCells are deactivated. When bandwidth adaptation is configured, only one UL BWP for each uplink carrier and one DL BWP or only one DL/UL BWP pair can be active at any given time in an active serving cell, all other BWPs that the UE is configured with will be deactivated. The UE does not monitor the PDCCH and does not transmit on PUCCH, PRACH, and UL-SCH of the deactivated BWPs [8].

The SCell activation/deactivation is an efficient mechanism to reduce UE power consumption in addition to DRX. On a deactivated SCell, the UE neither receives downlink signals nor transmits any uplink signal. The UE is also not required to perform measurements on a deactivated SCell. Deactivated SCells can be used as pathloss reference for measurements in uplink power control. It is assumed that these measurements would be less frequent while the SCell is deactivated in order to conserve the UE power. On the other hand, for an activated SCell, the UE performs normal activities for downlink reception and uplink transmission. Activation and deactivation of SCells is controlled by the gNB. As shown in Fig. 2.32, the SCell activation/deactivation is performed when the gNB sends an activation/deactivation command in the form of a MAC CE. A timer may also be used for automatic deactivation, if no data or PDCCH messages are received on a SCell for a certain period of time. This is the only case in which deactivation can be executed autonomously by the UE. Serving cell activation/deactivation is performed independently for each SCell, allowing the UE to be activated only on a particular set of SCells. Activation/deactivation is not applicable to the PCell because it is required to always remain activated when the UE has an RRC connection to the network [8].

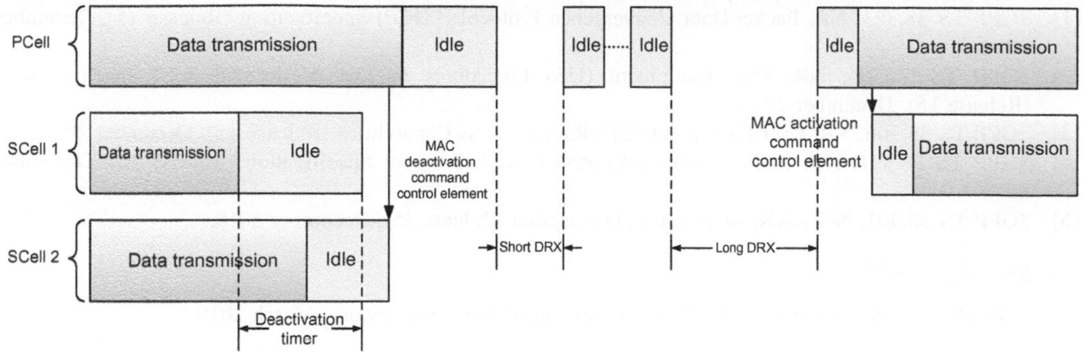

Figure 2.32
Illustration of SCell activation/deactivation procedure [9].

As already mentioned, if the UE is configured with one or more SCells, the network may activate and deactivate the configured SCells. The PCell is always activated. The network activates and deactivates the SCell(s) by sending an activation/deactivation MAC CE described in Section 2.2.1. Furthermore, the UE maintains a *sCellDeactivationTimer* timer per configured SCell (except the SCell configured with PUCCH) and deactivates the associated SCell upon its expiration. The same initial timer value is applied to each instance of the *sCellDeactivationTimer* and it is configured by RRC signaling. The configured SCells are initially deactivated upon addition and after a handover. The HARQ feedback for the MAC PDU containing SCell activation/deactivation MAC CE is not impacted by PCell, PSCell, and PUCCH SCell interruptions due to SCell activation/deactivation [9].

References

3GPP Specifications[12]

[1] 3GPP TS 23.501, System Architecture for the 5G System (Release 15), December 2018.
[2] 3GPP TS 33.501, Security Architecture and Procedures for 5G System (Release 15), December 2018.
[3] 3GPP TS 37.324, NR, Service Data Adaptation Protocol (SDAP) Specification (Release 15), September 2018.
[4] 3GPP TS 37.340, Multi-connectivity, Stage 2 (Release 15), December 2018.
[5] 3GPP TS 38.104, NR, Base Station (BS) Radio Transmission and Reception (Release 15), December 2018.
[6] 3GPP TS 38.133, NR, Requirements for Support of Radio Resource Management (Release 15), December 2018.
[7] 3GPP TS 38.215, NR, Physical Layer Measurements (Release 15), March 2018.
[8] 3GPP TS 38.300, NR, Overall Description, Stage-2 (Release 15), December 2018.
[9] 3GPP TS 38.321, NR, Medium Access Control (MAC) Protocol Specification (Release 15), December 2018.
[10] 3GPP TS 38.322, NR, Radio Link Control (RLC) Protocol Specification (Release 15), December 2018.

[12] 3GPP specifications can be accessed at http://www.3gpp.org/ftp/Specs/archive/.

[11] 3GPP TS 38.323, NR, Packet Data Convergence Protocol (PDCP) Specification (Release 15), December 2018.
[12] 3GPP TS 38.304, NR, User Equipment (UE) Procedures in Idle Mode and RRC Inactive State (Release 15), December 2018.
[13] 3GPP TS 38.306, NR, User Equipment (UE) Radio Access Capabilities (Release 15), December 2018.
[14] 3GPP TS 38.331, NR, Radio Resource Control (RRC); Protocol Specification (Release 15), December 2018.
[15] 3GPP TS 38.401, NG-RAN, Architecture Description (Release 15), December 2018.

IETF Specifications[13]

[16] IETF RFC 5795, The RObust Header Compression (ROHC) Framework, March 2010.

Articles, Books, White Papers, and Application Notes

[17] E. Dahlman, S. Parkvall, 5G NR: The Next Generation Wireless Access Technology, Academic Press, August 2018.
[18] S. Ahmadi, LTE-Advanced: A Practical Systems Approach to Understanding 3GPP LTE Releases 10 and 11 Radio Access Technologies, Academic Press, November 2013.
[19] 3GPP RWS-180010, NR Radio Interface Protocols, Workshop on 3GPP Submission Towards IMT-2020, Brussels, Belgium, October 2018.
[20] 5G New Radio, ShareTechNote. <http://www.sharetechnote.com>.
[21] MediaTek, A New Era for Enhanced Mobile Broadband, White Paper, March 2018.

[13] IETF specifications can be accessed at https://datatracker.ietf.org/.

CHAPTER 3

New Radio Access Physical Layer Aspects (Part 1)

This chapter describes the theoretical and practical aspects of physical layer protocols and functional processing in 3GPP new radio. As shown in Fig. 3.1, the physical layer is the lowest protocol layer in baseband signal processing that interfaces with the digital and the analog radio frontends and the physical media (in this case air interface) through which the signal is transmitted and received. The physical layer further interfaces with the medium access control (MAC) sublayer and receives MAC PDUs and processes the transport blocks through channel coding, rate matching, interleaving/scrambling, baseband modulation, layer mapping for multi-antenna transmission, digital precoding, resource element mapping, orthogonal frequency division multiplexing (OFDM) modulation, and antenna mapping. The choice of appropriate modulation and coding scheme as well as multi-antenna transmission mode is critical to achieve the desired reliability/robustness (coverage) and system/user throughput in mobile communications. Typical mobile radio channels tend to be dispersive and time variant and exhibit severe Doppler effects, multipath delay variation, intra-cell and inter-cell interference, and fading.

A good and robust design of the physical layer ensures that the system can robustly operate and overcome the above deleterious effects and can provide the maximum throughput and lowest latency under various operating conditions. Chapters 3 and 4 on physical layer in this book are dedicated to systematic design of physical layer protocols and functional blocks of 5G systems, the theoretical background on physical layer procedures, and performance evaluation of physical layer components. The theoretical background is provided to make the chapter self-contained and to ensure that the reader understands the underlying theory governing the operation of various functional blocks and procedures. While the focus is mainly on the techniques that were incorporated in the design of 3GPP NR physical layer, the author has attempted to take a more generic and systematic approach to the design of physical layer for the IMT-2020 wireless systems so that the reader can understand and apply the learnings to the design and implementation of any OFDM-based physical layer irrespective of the radio access technology.

5G NR. DOI: https://doi.org/10.1016/B978-0-08-102267-2.00003-8

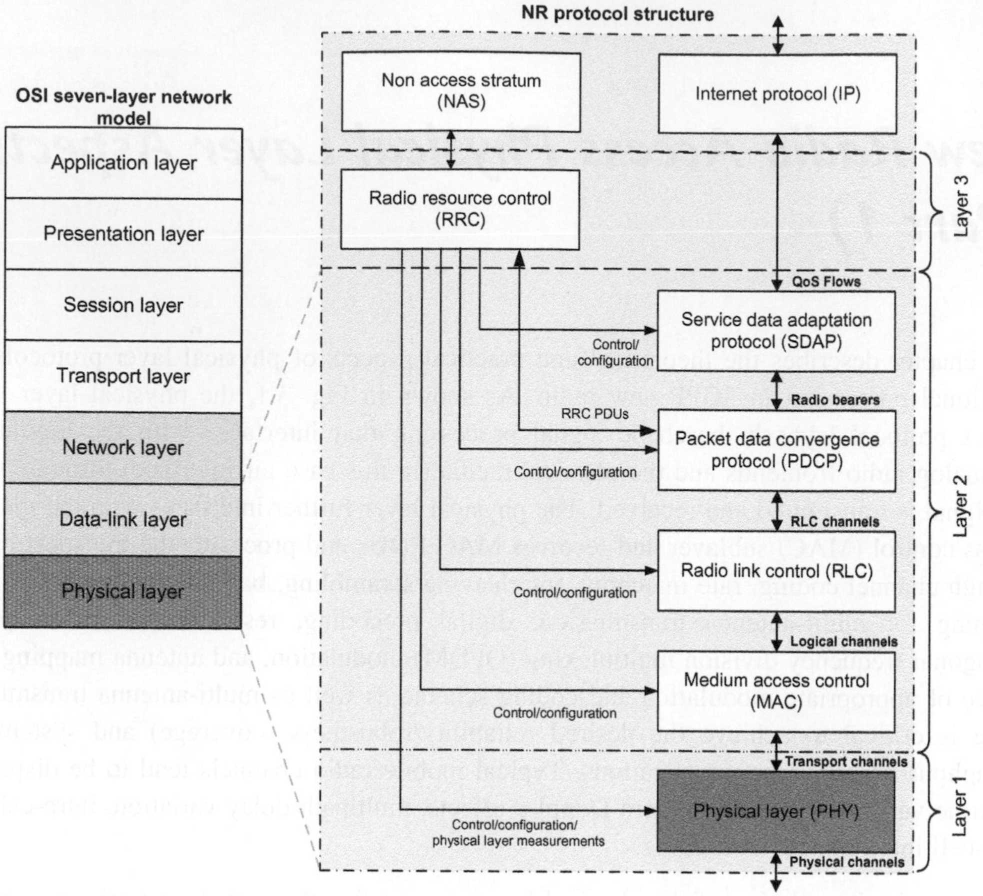

Figure 3.1
The physical layer in NR protocol stack [19].

In this chapter, we start with the study of the fundamental concepts and common features/functions in downlink and uplink of the new radio, which include review of the characteristics of wireless channels in sub-6 GHz and mmWave frequency regions as well as the analysis of two-dimensional (2D) and three-dimensional (3D) channel models and propagation effects. We will then begin our top-down approach to physical layer protocols starting with waveforms, orthogonal and non-orthogonal multiple-access schemes, and duplex schemes as well as the operating frequencies of the new radio. The frame structure, OFDM numerologies, time-frequency resources, and resource allocation techniques will be discussed and analyzed from theoretical and practical point of views.

3.1 Channel Models and Propagation Characteristics

3.1.1 Characteristics of Wireless Channels

In a wireless communication system, a signal can travel from the transmitter to the receiver over multiple paths. This phenomenon is referred to as multipath propagation where signal attenuation varies on different paths. This effect also known as multipath fading can cause stochastic fluctuations in the received signal's magnitude, phase, and angle of arrival (AoA). The propagation over different paths is caused by scattering, reflection, diffraction, and refraction of the radio waves by static and moving objects as well as the transmission medium. It is obvious that different propagation mechanisms result in different channel and path loss models. As a result of wave propagation over multipath fading channels, the radio signal is attenuated due to mean path loss as well as macroscopic and microscopic fading.

A detailed channel model for the frequency range (FR) from 6 to 100 GHz was developed in 3GPP [24]. It is applicable to bandwidth up to 10% of the carrier frequency (with a limit of 2 GHz), which accounts for the mobility of one of the two terminals (i.e., in a typical cellular network, the base station is fixed and the user terminal moves). It further provides several optional features that can be plugged in to the basic model, in order to simulate spatial consistency (i.e., the radio environment conditions of nearby users are correlated), blockage, and oxygen absorption. This model supports different IMT-2020 test environments including urban microcell (UMi), urban macrocell (UMa), rural macrocell (RMa), indoor hotspot (InH), and outdoor to indoor, which must be chosen when setting the simulation parameters [4,5,8,24,25].

3.1.1.1 Path Loss Models

In ideal free-space propagation model, the attenuation of RF signal energy between the transmitter and receiver follows inverse-square law. The received power expressed as a function of the transmitted power is attenuated proportional to the inverse of $L_s(d)$, which is called free-space path loss. When the receiving antenna is isotropic, the received signal power can be expressed as follows [32,33]:

$$P_{RX} = P_{TX}G_{TX}G_{RX}\left(\frac{\lambda}{4\pi d}\right)^2 = \frac{P_{TX}G_{TX}G_{RX}}{L_s(d)}$$

where P_{TX} and P_{RX} denote the transmitted and the received signal power, G_{TX} and G_{RX} denote the transmitting and receiving antenna gains, d is the distance between the transmitter and the receiver, and λ is the wavelength of the RF signal. In mmWave bands, the smaller wavelength translates into smaller captured energy at the receive antenna. For example, in the frequency range 3–30 GHz, an additional 20 dB path loss is added due to effective

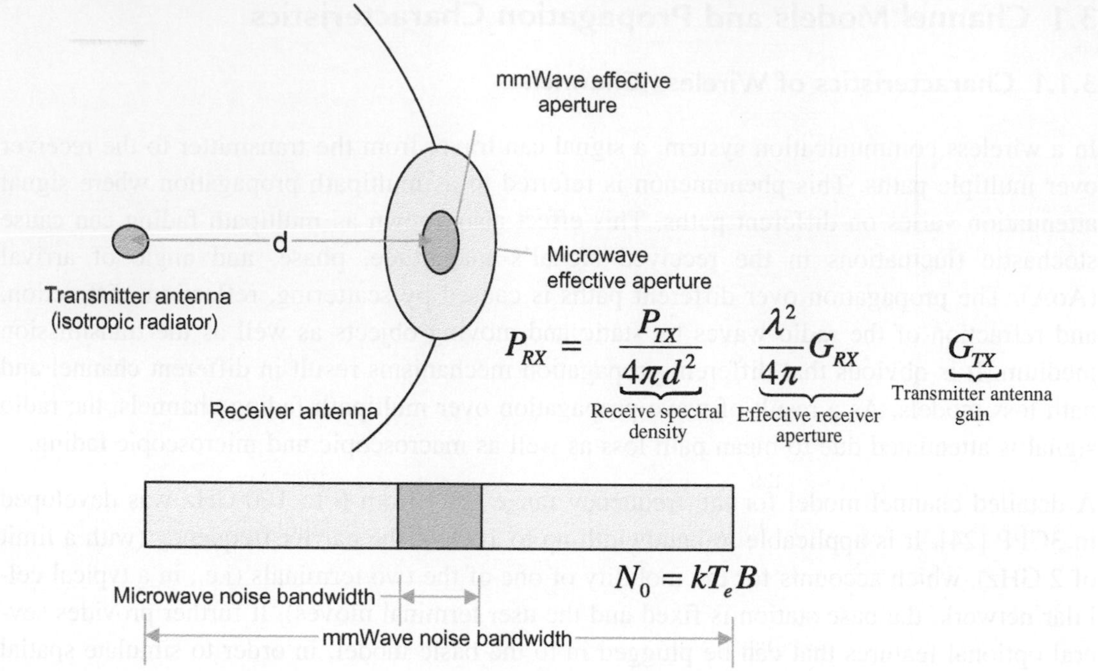

Figure 3.2
Definition of gain and aperture in mmWave [32].

aperture reduction. On the other hand, larger bandwidths in mmWave mean higher noise power and lower signal-to-noise ratios (SNRs), for example, from 50 to 500 MHz bandwidth, the noise power is increased by 10 dB (see Fig. 3.2).

Free-space conditions require a direct line of sight (LoS) between the two antennas involved. Consequently, no obstacles must exist in the path between the antennas at both ends. Furthermore, in order to avoid the majority of effects caused by superposition of direct and reflected signals, it is necessary that the first Fresnel zone[1] is completely free of obstacles. The first Fresnel ellipsoid is defined as a rotational ellipsoid with the two antennas at its focal points. Within this ellipsoid, the phase difference between two potential paths is less than half a wavelength. The radius b at the center of the ellipsoid can be calculated as $b = 8.66\sqrt{d/f}$ where b is the radius in meters, d is the distance between the receiver and the transmitter in kilometers, and f is the frequency in GHz [26,27,32].

[1] A Fresnel zone is one of a series of confocal prolate ellipsoidal regions of space between and around a transmitting antenna and a receiving antenna system. The regions are used to understand and compute the strength of waves (such as sound or radio waves) propagating between a transmitter and a receiver, as well as to predict whether obstructions near the line joining the transmitter and receiver will cause significant interference.

It must be noted that all antenna properties and attributes that we consider in this section assumes far-field patterns. The distance from an antenna, where far-field conditions are met, depends on the dimensions of the antenna relative to the wavelength. For smaller antennas, the wave fronts radiated from the antenna become almost parallel at much closer distance compared to electrically large antennas. A good approximation for small antennas is that far-field conditions are reached at $r \approx 2\lambda$. For larger antennas, that is, reflector antennas or array antennas where the dimensions of the antenna L are significantly larger compared to the wavelength $L \gg \lambda$, the far-field distance is approximated as $r \approx 2L^2/\lambda$ [26,33].

Macroscopic fading is caused by shadowing effects of buildings and natural obstructions and is modeled by the local mean of a fast fading signal. The mean path loss $\overline{L}_p(d)$ as a function of distance d between the transmitter and receiver is proportional to an nth power of d relative to a reference distance d_0. In logarithmic scale, it can be expressed as follows:

$$\overline{L}_p(d) = L_s(d_0) + 10n \log\left(\frac{d}{d_0}\right) \quad \text{(dB)}$$

The reference distance d_0 corresponds to a point located in the far-field of the antenna typically 1 km for large cells, 100 m for microcells, and 1 m for indoor channels. In the above equation, $\overline{L}_p(d)$ is the mean path loss which is typically $10n$ dB per decade attenuation for $d \gg d_0$. The value of n depends on the frequency, antenna heights, and propagation environment that is equal to 2 in free space. The studies show that the path loss $L_p(d)$ is a random variable with log-normal distribution about the mean path loss $\overline{L}_p(d)$. Let $X \sim \mathbf{N}(0, \sigma^2)$ denote a zero-mean Gaussian random variable with standard deviation σ when measured in decibels, then

$$L_p(d) = L_s(d_0) + 10n \log\left(\frac{d}{d_0}\right) + X \quad \text{(dB)}$$

The value of X is often derived empirically based on measurements. A typical value for σ is 8 dB. The parameters that statistically describe path loss due to large-scale fading (macroscopic fading) for an arbitrary location with a specific transmitting-receiving antenna separation include the reference distance d_0, the path loss exponent n, and the standard deviation σ of X [33].

Microscopic fading refers to the rapid fluctuations of the received signal in time and frequency and is caused by scattering objects between the transmitting and receiving antennas. When the received RF signal is a superposition of independent scattered components plus an LoS component, the envelope of the received signal $r(t)$ has a Rician Probability Distribution Function (PDF) that is referred to as Rician fading. As the magnitude of the LoS component approaches zero, the Rician PDF approaches a Rayleigh PDF. Thus

$$f(r) = \frac{r(K+1)}{\sigma^2}\exp\left[-K - \frac{(K+1)r^2}{2\sigma^2}\right] I_0\left(\frac{2r}{\sigma}\sqrt{K\frac{(K+1)}{2}}\right), \quad r \geq 0$$

where K and $I_0(r)$ denote the Rician factor and zero-order modified Bessel function of the first kind.[2] In the absence of LoS path ($K = 0$), the Rician PDF reduces to Rayleigh distribution.

One of the challenges of mobile communications in the higher frequency bands for outdoor access has been to overcome the difficulties in highly varying propagation conditions. Understanding the propagation conditions will be critical to designing an appropriate air interface and determining the type of hardware (particularly the array size) needed for reliable communications. Extensive measurements over a wide range of frequencies were performed by a large number of academic institutions and the industry. Since maintaining link budgets at higher frequencies are challenging, there are few measurements at larger distances for 5G deployment scenarios of interest. Based on the results of the measurements, some important observations were made that helped development of the new channel models. The most notable signal degradation is due to the higher path loss of the bands above 6 GHz relative to sub-6 GHz bands. The additional path loss as result of increasing frequency need to be compensated by some means such as larger antenna array sizes with higher array gains and MIMO schemes to ensure sufficient and robust links. Due to large variation of propagation characteristics in bands above 6 GHz, the propagation characteristics of different frequency bands were independently investigated. The combined effects of all contributors to propagation loss can be expressed via the path loss exponent. In UMi test environment, the LoS path loss in the bands of interest appears to closely follow the free-space path loss model. In lower bands, a higher path loss exponent was observed in NLoS scenarios. The shadow fading in the measurements appears to be similar to lower frequency bands, while ray-tracing results show a much higher shadow fading (>10 dB) than measurements, due to the larger dynamic range allowed in some ray-tracing experiments. In sub-6 GHz NLoS scenarios, the root mean square (RMS) delay spread is typically modeled in the range of 50–500 ns, the RMS azimuth angle spread of departure (from the access point) at around 10–30 degrees, and the RMS azimuth angle spread of arrival (at the UE) at around 50–80 degrees [7,8]. There are measurements of the delay spread above 6 GHz which indicate somewhat smaller ranges as the frequency increases, and some measurements show the millimeter wave omnidirectional channel to be highly directional in nature.

[2] A modified Bessel function of the first kind is a function $I_n(x)$ which is one of the solutions to the modified Bessel differential equation and is closely related to the Bessel function of the first kind $J_n(x)$. The modified Bessel function of the first kind $I_n(z)$ can be defined by the contour integral $I_n(z) = \oint e^{z/2(t+1)/t} t^{-n-1} dt$ where the contour encloses the origin and is traversed in a counterclockwise direction. In terms of $J_n(x)$, $I_n(x) \triangleq j^{-n} J_n(jx) = e^{-jn\pi/2} J_n(xe^{j\pi/2})$.

In UMa test environments, the behavior of LoS path loss is similar to free-space path loss and the NLoS path loss behavior appears not following a certain model over a wide range of frequencies. It was observed that the rate at which the path loss increased with frequency was not linear, as the rate is higher in the lower parts of the spectrum, which can be possibly explained as due to diffraction effect that is frequency-dependent and a more dominating component in lower frequencies. However, in higher frequencies, reflections and scattering are relatively the predominant components. From preliminary ray-tracing studies, the channel spreads in terms of delay and angle appear to be weakly dependent on the frequency and are generally 2−5 times smaller than the values reported in [7]. The cross-polar scattering in the ray-tracing results tends to increase with increasing frequency due to diffuse scattering.[3]

In InH deployment scenarios, under LoS conditions, multiple reflections from walls, floor, and ceiling could give rise to wave-guiding effect. The measurements conducted in office scenarios suggest that path loss exponent, based on a 1 m free-space reference distance is typically below 2, leading to relatively less path loss than predicted by the free-space path loss formula. The strength of the wave-guiding effect is variable and the path loss exponent appears to increase slightly with increasing frequency, possibly due to the relation between the wavelength and surface roughness. Measurements of the small-scale channel properties such as angular spread (AS) and delay spread have shown similarities between channels over a very wide frequency range, where the results suggest that the main multipath components are present at all frequencies with some small variations in magnitudes. Recent studies have shown that polarization discrimination ranges between 15 and 25 dB for indoor mmWave channels with greater polarization discrimination at 73 GHz than at 28 GHz [1−3].

Cross-polarized signal components are generated by reflection and diffraction. It is widely known that the fading correlation characteristic between orthogonally polarized antennas has a very low correlation coefficient. Polarization diversity techniques and MIMO systems with orthogonally polarized antennas are developed that employ this fading characteristic. Employing the polarization diversity technique is one solution to improving the received power, and the effect of the technique is heavily dependent on the cross-polarization discrimination (XPD) ratio characteristic. Moreover, the channel capacity can be improved by appropriately using the cross-polarization components in MIMO systems. Thus the communication quality can be improved by effectively using the information regarding the cross-polarized waves in a wireless system. XPD is an important characteristic, particularly in

[3] Diffuse scattering is a form of scattering that arises from deviation of material structure from that of a perfectly regular lattice, which appears in experimental data as scattering spread over a wide $q-$ range (diffuse). Diffuse scattering is generally difficult to quantify and is closely related to Bragg diffraction, which occurs when scattering amplitudes add constructively. The defect in a crystal lattice results in reduction of the amplitude of the Bragg peak.

dual-polarized systems, where cross-talk between polarizations can prevent the system to achieve its quality objectives. Radios can use cross-polarization interference cancellation (XPIC) to isolate polarizations and compensate for any link or propagation induced coupling. However, good antenna polarization is important to allow the XPICs maximum flexibility to compensate for these dynamic variations.

The following equations provide one of the path loss models for UMa test environment for LoS and NLoS scenarios in the frequency range of 0.5–100 GHz. These models were developed based on measurement results conducted by independent academic and industry organizations and published in the literature [7,24,34–36].

- LoS path loss for $0.5\ \text{GHz} \leq f_c \leq 100\ \text{GHz}$

$$PL_{UMa-LoS} = \begin{cases} PL_1 & 10\ \text{m} \leq d_{2D} \leq d'_{BP} \\ PL_2 & d'_{BP} \leq d_{2D} \leq 5\ \text{km} \end{cases}$$

$$PL_1 = 32.4 + 20\log_{10}(d_{3D}) + 20\log_{10}(f_c)$$

$$PL_2 = 32.4 + 40\log_{10}(d_{3D}) + 20\log_{10}(f_c) - 10\log_{10}\left[(d'_{BP})^2 + (h_{gNB} - h_{UE})^2\right]$$

$$h_{gNB} = 25\ \text{m},\quad 1.5\ \text{m} \leq h_{UE} \leq 22.5\ \text{m},\quad \sigma_{SF} = 4\ \text{dB}$$

- NLoS path loss for $6\ \text{GHz} \leq f_c \leq 100\ \text{GHz}$

$$PL_{UMa-NLoS} = \max\left(PL_{UMa-LoS}, PL'_{UMa-NLoS}\right)$$

$$10\ \text{m} \leq d_{2D} \leq 5\ \text{km}$$

$$PL'_{UMa-NLoS} = 13.54 + 39.08\log_{10}(d_{3D}) + 20\log_{10}(f_c) - 0.6(h_{UE} - 1.5)$$

$$1.5\ \text{m} \leq h_{UE} \leq 22.5\ \text{m}, h_{gNB} = 25\ \text{m}, \sigma_{SF} = 6$$

3.1.1.2 Delay Spread

Time-varying fading due to scattering objects or transmitter/receiver motion results in Doppler spread. The time spreading effect of small-scale or microscopic fading is manifested in the time domain as multipath delay spread and in the frequency domain as channel coherence bandwidth. Similarly, the time variation of the channel is characterized in the time domain as channel coherence time and in the frequency domain as Doppler spread. In a fading channel, the relationship between maximum excess delay time τ_m and symbol time τ_s can be viewed in terms of two different degradation effects, that is, frequency-selective fading and frequency non-selective or flat fading. A channel is said to exhibit frequency-selective fading if $\tau_m > \tau_s$. This condition occurs whenever the received multipath components of a symbol extend beyond the symbol's time duration. Such multipath dispersion of the signal results in inter-symbol interference (ISI) distortion. In the case of frequency-selective fading, mitigating the distortion is possible because many of the multipath components are separable by the receiver. A channel is said to exhibit frequency non-selective or flat fading if $\tau_m < \tau_S$. In this case, all multipath components of a symbol arrive at the receiver within the symbol time duration; therefore, the components are not resolvable. In this case, there is no channel-induced ISI distortion, since the signal time spreading does not result in significant overlap among adjacent received symbols. There is still performance degradation because the irresolvable phasor components can add up destructively to yield a substantial reduction in SNR. Also, signals that are classified as exhibiting flat fading can sometimes experience frequency-selective distortion [33].

Fig. 3.3 illustrates multipath-intensity profile $\Lambda(\tau)$ versus delay τ where the term delay refers to the excess delay. It represents the signal's propagation delay that exceeds the delay of the first signal component arrival at the receiver. For a typical wireless communication

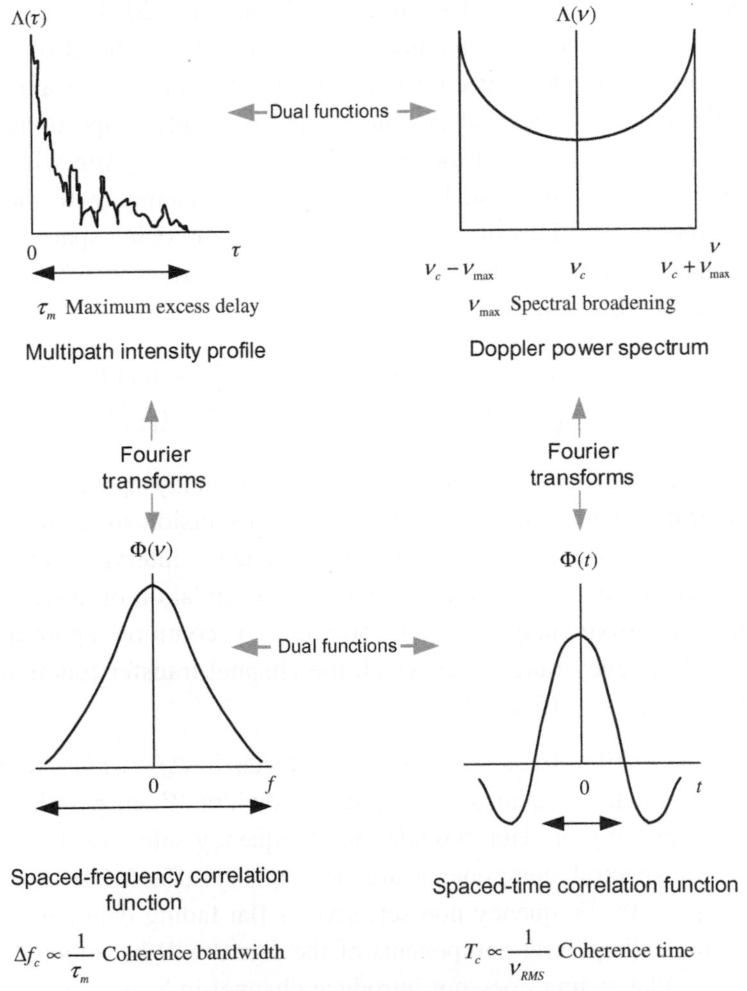

Figure 3.3
Illustration of the duality principle in time and frequency domains [33].

channel, the received signal usually consists of several discrete multipath components. The received signals are composed of a continuum of multipath components in some channels such as the tropospheric channel. In order to perform measurements of the multipath-intensity profile, a wideband signal, that is, a unit impulse or Dirac delta function, is used. For a single transmitted impulse, the time τ_m between the first and last received component is defined as the maximum excess delay during which the multipath signal power typically falls to some level 10–20 dB below that of the strongest component. Note that for an ideal system with zero excess delay, the function $\Lambda(\tau)$ would consist of an ideal impulse with weight equal to the total average received signal power. In the literature, the Fourier transform of $\Lambda(\tau)$ is referred to as spaced-frequency correlation function $\Phi(\upsilon)$. The spaced-frequency correlation function $\Phi(\upsilon)$ represents the channel's response to a pair of sinusoidal signals separated in frequency by υ. The coherence bandwidth Δf_c is a measure of the frequency range over which spectral components have a strong likelihood of amplitude correlation. In other words, a signal's spectral components over this range are affected by the channel in a similar manner. Note that Δf_c and τ_m are inversely proportional $\left(\Delta f_c \propto 1/\tau_m\right)$. The maximum excess delay τ_m is not the best indicator of how a given wireless system will perform over a communication channel because different channels with the same value of τ_m can exhibit different variations of signal intensity over the delay span. The delay spread is often characterized in terms of the RMS delay spread τ_{RMS} in which the average multipath delay $\overline{\tau}$ is calculated as follows:

$$\tau_{RMS} = \sqrt{\frac{\int_0^{\tau_m} (\tau - \overline{\tau})^2 \Lambda(\tau) d\tau}{\int_0^{\tau_m} \Lambda(\tau) d\tau}}, \quad \overline{\tau} = \frac{\int_0^{\tau_m} \tau \Lambda(\tau) d\tau}{\int_0^{\tau_m} \Lambda(\tau) d\tau}$$

An exact relationship between coherence bandwidth and delay spread does not exist and must be derived from signal analysis of actual signal dispersion measurements in specific channels. If coherence bandwidth is defined as the frequency interval over which the channel's complex-valued frequency transfer function has a correlation of at least 0.9, the coherence bandwidth is approximately $\Delta f_c \approx 1/(50\tau_{RMS})$. A common approximation of Δf_c corresponding to a frequency range over which the channel transfer function has a correlation of at least 0.5 is $\Delta f_c \approx 1/(5\tau_{RMS})$.

A channel is said to exhibit frequency-selective effects, if $\Delta f_c < 1/\tau_s$ where the inverse symbol rate is approximately equal to the signal bandwidth W. In practice, W may differ from $1/\tau_s$ due to filtering or data modulation. Frequency-selective fading effects arise whenever a signal's spectral components are not equally affected by the channel. This occurs whenever $\Delta f_c < W$. Frequency non-selective or flat fading degradation occurs whenever $\Delta f_c > W$. Hence, all spectral components of the signal will be affected by the channel in a similar manner. Flat fading does not introduce channel-induced ISI distortion, but performance degradation can still be expected due to loss in SNR, whenever the signal experiences fading. In order to avoid channel-induced ISI distortion, the channel is required to exhibit flat-fading by ensuring that $\Delta f_c > W$. Therefore, the channel coherence bandwidth Δf_c sets an upper limit on the transmission rate that can be used without incorporating an equalizer in the receiver [33].

3.1.1.3 Doppler Spread

Fig. 3.3 shows another function $\Phi(t)$ known as spaced-time correlation function, which is the autocorrelation function of the channel's response to a sinusoid. This function specifies the extent to which there is correlation between the channel's response to a sinusoid sent at time t_1 and the response to a similar sinusoid sent at time t_2, where $\Delta t = t_2 - t_1$. The coherence time is a measure of the expected time duration over which the channel's response is essentially invariant. To estimate $\Phi(t)$, a sinusoidal signal is transmitted through the channel and the autocorrelation function of the channel output is calculated. The function $\Phi(t)$ and the coherence time T_c provide information about the rate of fading channel variation. Note that for an ideal time-invariant channel, the channel's response would be highly correlated for all values of Δt and $\Phi(t)$ would be a constant function. If one ideally assumes uniformly distributed scattering around a mobile station with linearly polarized antennas, the Doppler power spectrum (i.e., the inverse Fourier transform of spaced-time correlation function $\Lambda(\nu)$) has a U-shaped distribution as shown in Fig. 3.3. In a time-varying fading channel, the channel response to a pure sinusoidal tone spreads over a finite frequency range $\nu_c - \nu_{\max} < \nu < \nu_c + \nu_{\max}$, where ν_c and $\nu_{\max}$ denote the frequency of the sinusoidal tone and the maximum Doppler spread, respectively. The RMS bandwidth of $\Lambda(\nu)$ is referred to as Doppler spread and is denoted by ν_{RMS} that can be estimated as follows:

$$\nu_{RMS} = \sqrt{\frac{\int_{\nu_c-\nu_{\max}}^{\nu_c+\nu_{\max}} (\nu-\overline{\nu})^2 \Lambda(\nu) d\nu}{\int_{\nu_c-\nu_{\max}}^{\nu_c+\nu_{\max}} \Lambda(\nu) d\nu}} \qquad \overline{\nu} = \frac{\int_{\nu_c-\nu_{\max}}^{\nu_c+\nu_{\max}} \nu \Lambda(\nu) d\nu}{\int_{\nu_c-\nu_{\max}}^{\nu_c+\nu_{\max}} \Lambda(\nu) d\nu}$$

The coherence time is typically defined as the time lag for which the signal autocorrelation coefficient reduces to 0.7. The coherence time is inversely proportional to Doppler spread $T_c \approx 1/\nu_{RMS}$. A common approximation for the value of coherence time as a function of Doppler spread is $T_c = 0.423/\nu_{RMS}$. It can be observed that the functions on the right side of Fig. 3.3 are dual of the functions on the left side (duality principle).

3.1.1.4 Angular Spread

The angle spread refers to the spread in AoA of the multipath components at the receiver antenna array. At the transmitter, on the other hand, the angle of spread refers to the spread in the angle of departure (AoD) of the multipath components that leave the transmit antennas. If the angle spectrum function $\Theta(\theta)$ denotes the average power as function of AoA, then the RMS angle spread can be estimated as follows:

$$\theta_{RMS} = \sqrt{\frac{\int_{-\pi}^{+\pi} (\theta-\overline{\theta})^2 \Theta(\theta) d\theta}{\int_{-\pi}^{+\pi} \Theta(\theta) d\theta}} \quad \overline{\theta} = \frac{\int_{-\pi}^{+\pi} \Theta(\theta)\theta d\theta}{\int_{-\pi}^{+\pi} \Theta(\theta) d\theta}$$

The angle spread causes space-selective fading, which manifests itself as variation of signal amplitude according to the location of antennas. The space-selective fading is characterized by the coherence distance D_c which is the spatial separation for which the autocorrelation coefficient of the spatial fading reduces to 0.7. The coherence distance is inversely proportional to the angle spread $D_c \propto 1/\theta_{RMS}$. In Fig. 3.3 a duality between multipath-intensity function $\Lambda(\tau)$ and Doppler power spectrum $\Lambda(\nu)$ is shown, which means that the two functions exhibit similar behavior across time domain and frequency domain. As the $\Lambda(\tau)$ function identifies expected power of the received signal as a function of delay, $\Lambda(\nu)$ identifies expected power of the received signal as a function of frequency. Similarly, spaced-frequency correlation function $\Phi(f)$ and spaced-time correlation function $\Phi(t)$ are dual functions. It implies that as $\Phi(f)$ represents channel correlation in frequency, $\Phi(t)$ corresponds to channel correlation function in time in a similar manner [33].

The AS in radians can be expressed based on the circular standard deviation in directional statistics using the following expression [24,25]:

$$\mathrm{AS} = \sqrt{-2\log\left(\left|\frac{\sum_{n=1}^{N}\sum_{m=1}^{M}\left(e^{j\varphi_{nm}}P_{nm}\right)}{\sum_{n=1}^{N}\sum_{m=1}^{M}P_{nm}}\right|\right)}$$

where P_{nm} denotes the power of the mth subpath of the nth path and φ_{nm} is the subpaths angle (either AoA, AoD, elevation angle of arrival [EoA], elevation angle of departure [EoD]). In order to model large signal bandwidths and large antenna arrays, the channel models have been specified with sufficiently high resolution in the delay and angular domains. There are two important aspects related to large antenna arrays. One is the very large size of the antenna array and the other is the large number of antenna elements. These features require high angular resolution in channel modeling, which means more accurate modeling of AoA/AoD, and possibly higher number of multipath components.

3.1.1.5 Blockage

The blockage model describes a phenomenon where the stationary or moving objects standing between the transmitter and receiver dramatically change the channel characteristics and in some cases may block the signal, especially in high-frequency bands, since the signal in mmWave does not effectively penetrate or diffract around human bodies and other objects. Shadowing by these objects is an important factor in the link budget calculations and the time variation of the channel, and such dynamic blocking may be important to capture in evaluations of technologies that include beam-finding and beam-tracking capabilities. The effect of the blockage is considered not only on the total received power, but also on the angle or power of multipath due to different size, location, and direction of the blocker. There are two categories of blockage: (1) dynamic blockage and (2) geometry-induced blockage. Dynamic blockage is caused by the moving objects in the communication environment. The effect is additional transient loss on the paths that intercept the moving objects. Geometry-induced blockage, on the other hand, is a static property of the environment. It is caused by objects in the map environment that block the signal paths. The propagation channels in geometry-induced blockage locations are dominated by diffraction and sometimes by diffuse scattering. The effect is an additional loss beyond the normal path loss and shadow fading. Compared to shadow fading caused by reflections, diffraction-dominated shadow fading may have different statistics (e.g., different mean, variance, and coherence distance) [33]. Radio waves are attenuated by foliage, and this effect increases with frequency. The main propagation phenomena involved are attenuation of the radiation through the foliage, diffraction above/below and sideways around the canopy, and diffuse scattering by the leaves. The vegetation effects are captured implicitly in the path loss equations.

A stochastic method for capturing human and vehicular blocking in mmWave frequency regions can be used (among other methods) to model the blockage effect. In this case, the number of blockers must be first determined. For this purpose, multiple 2D angular blocking regions, in terms of center angle, azimuth, and elevation angular span are generated around the UE. There is one self-blocking region and $K=4$ non-self-blocking regions, where K may be changed for certain scenarios such as higher blocker density. Note that the self-blocking component of the model is important in capturing the effects of human body blocking. In the next step, the size and location of each blocker must be generated. For self-blocking, the blocking region in the UE local coordinate system[4] is defined in terms of elevation and azimuth angles $(\theta'_{sb}, \varphi'_{sb})$ and azimuth and elevation angular span (x_{sb}, y_{sb}) [4,5].

$$\left\{(\theta', \varphi') \middle| \left(\theta'_{sb} - \frac{y_{sb}}{2} \le \theta' \le \theta'_{sb} + \frac{y_{sb}}{2}, \quad \varphi'_{sb} - \frac{x_{sb}}{2} \le \varphi' \le \varphi'_{sb} + \frac{x_{sb}}{2}\right)\right\}$$

where the parameters of the above equation are described in Table 3.1. For non-self-blocking $k=1,2,3,4$, the blocking region in global coordinate system is defined as follows [4,5]:

$$\left\{(\theta, \varphi) \middle| \left(\theta_k - \frac{y_k}{2} \le \theta \le \theta_k + \frac{y_k}{2}, \quad \varphi_k - \frac{x_k}{2} \le \varphi \le \varphi_k + \frac{x_k}{2}\right)\right\}$$

where d is the distance between the UE and the blocker; other parameters are given in Table 3.1.

[4] Global and local coordinate systems are used to locate geometric items in space. By default, a node coordinates are defined in the global Cartesian system.

Table 3.1: Blocking region parameters [4,5].

Self-Blocking Region Parameters					
Mode	φ'_{sb}	x_{sb}	θ'_{sb}	y_{sb}	
Portrait (degree)	260	120	100	80	
Landscape (degree)	40	160	110	75	
Blocking Region Parameters					
Blocker Index $k = 1,2,3,4$	φ_k	x_k	θ_k	y_k	d
InH scenario	Uniform in [0°, 360°]	Uniform in [15°, 45°]	90°	Uniform in [5°,15°]	2 m
UMi_x, Uma_x, RMa_x scenarios	Uniform in [0°, 360°]	Uniform in [5°,15°]	90°	5°	10 m

The attenuation of each cluster due to self-blocking corresponding to the center angle pair $(\theta'_{sb}, \varphi'_{sb})$ $(\theta'_{sb}, \varphi'_{sb})$ is 30 dB provided that $|\varphi'_{AOA} - \varphi'_{sb}| < x_{sb}/2$ and $|\theta'_{ZOA} - \theta'_{sb}| < y_{sb}/2$; otherwise, the attenuation is 0 dB. Note that ZOA denotes the Zenith angle of arrival. The attenuation of each cluster due to the non-self-blocking regions $k = 1,2,3,4$ is given as $L_{dB} = -20\ \log_{10}\left[1 - (F_{A_1} + F_{A_2})(F_{Z_1} + F_{Z_2})\right]$ provided that $|\varphi_{AOA} - \varphi_k| < x_k$ and $|\theta_{ZOA} - \theta_k| < y_k$; otherwise, the attenuation is 0 dB. The terms $F_{A_1|A_2|Z_1|Z_2}$ in the previous equation are defined as follows [4,5]:

$$F_{A_1|A_2|Z_1|Z_2} = \frac{1}{\pi}\tan^{-1}\left[\pm\frac{\pi}{2}\sqrt{\frac{\pi}{\lambda}d\left(\frac{1}{\cos(A_1|A_2|Z_1|Z_2)} - 1\right)}\right]$$

where $A_1 = \varphi_{AOA} - (\varphi_k + x_k/2)$, $A_2 = \varphi_{AOA} - (\varphi_k - x_k/2)$, $Z_1 = \theta_{ZOA} - (\theta_k + y_k/2)$, and $Z_2 = \theta_{ZOA} - (\theta_k - y_k/2)$. The center of the blocker is generated based on a uniformly distributed random variable, which is temporally and spatially consistent. The 2D autocorrelation function $R(\Delta_x, \Delta_t)$ can be described with sufficient accuracy by the exponential function $R(\Delta_x, \Delta_t) = \exp\left[-\left(|\Delta_x|/d_{corr} + |\Delta_t|/t_{corr}\right)\right]$ where d_{corr} denotes the spatial correlation distance or the random variable determining the center of the blocker and t_{corr} is the correlation time given as $t_{corr} = d_{corr}/v$, in which v is the speed of moving blocker [4,5].

3.1.1.6 Oxygen Absorption

The electromagnetic wave may be partially or totally attenuated by an absorbing medium due to atomic and molecular interactions. This gaseous absorption causes additional loss to the radio wave propagation. For frequencies around 60 GHz, additional loss due to oxygen absorption is applied to the cluster responses for different center frequency and bandwidth correspondingly. The additional loss $OL_n(f_c)$ for cluster n at center frequency f_c is given as follows [4]:

$$OL_n(f_c) = \frac{\alpha(f_c)[d_{3D} + c(\tau_n + \tau_\Delta)]}{1000}\ \text{[dB]}$$

Table 3.2: Oxygen attenuation $\alpha(f_c)$ as a function of frequency [4,5].

Frequency (GHz)	0–52	53	54	55	56	57	58	59	60	61	62	63	64	65	66	67	68–100
$\alpha(f_c)$(dB/km)	0	1	2.2	4	6.6	9.7	12.6	14.6	15	14.6	14.3	10.5	6.8	3.9	1.9	1	0

where $\alpha(f_c)$ denotes the frequency-dependent oxygen absorption loss in dB/km whose sample values at some frequencies are shown in Table 3.2, c is the speed of light in m/s, d_{3D} is the 3D distance in meters between the receive and transmit antennas, τ_n is the nth cluster delay in seconds, and $\tau_\Delta = 0$ in the LoS scenarios. For center frequencies not shown in Table 3.2, the frequency-dependent oxygen absorption loss $\alpha(f_c)$ is obtained from a linear interpolation of the values corresponding to the two adjacent frequencies [4,5].

For wideband channels, the time-domain channel response of each cluster (all rays within one cluster share common oxygen absorption loss) are transformed into frequency-domain channel response and the oxygen absorption loss is applied to the cluster's frequency-domain channel response for frequency $f_c + \Delta f$ within the channel bandwidth W. The oxygen loss $OL_n(f_c + \Delta f)$ for cluster n at frequency $f_c + \Delta f$, where $-W/2 \leq \Delta f \leq W/2$ is given as follows:

$$OL_n(f_c + \Delta f) = \frac{\alpha(f_c + \Delta f)[d_{3D} + c(\tau_n + \tau_\Delta)]}{1000} \quad \text{[dB]}$$

where $\alpha(f_c + \Delta f)$ is the oxygen absorption loss in dB/km at frequency $f_c + \Delta f$. The final frequency-domain channel response is obtained by the summation of frequency-domain channel responses of all clusters. Time-domain channel response is obtained by the reverse transform from the obtained frequency-domain channel response [4,5].

Measurements in mmWave frequency bands have shown that ground reflection in mmWave has significant effect which can produce a strong propagation path that superimposes with the direct LoS path and induces severe fading effects. When ground reflection is considered, the randomly generated shadow fading is largely replaced by deterministic fluctuations in terms of distance. As a result, the standard deviation of shadow fading, when ground reflection is considered, is set to 1 dB. The value of 1 dB was obtained via simulations in order to maintain a similar level of random channel fluctuations without ground reflection [4,5].

The mmWave channels are sparse, that is, they have few entries in the delay angle bins, although experimental verification of this may be limited due to the resolution of rotating horn antennas used for such measurements. However, a lower bound on the channel sparsity can still be established from existing measurements, and in many environments, the percentage of delay/angle bins with significant energy is rather low but not necessarily lower than at centimeter-wave frequencies.

Molecular oxygen absorption around 60 GHz is particularly high (~13 dB/km, depending on the altitude) but decreases rapidly away from the oxygen resonance frequency to below 1 dB/km. While these absorption values are considered high for macrocell links, cell densification has already reduced the required link distance to a substantially smaller range in urban areas, and the densification process will continue to reduce cell sizes. For a link distance of approximately 200 m, the path loss of a 60 GHz link in heavy rain condition is less than 3 dB. Fig. 3.4 illustrates the impact of rainfall and oxygen absorption throughout the mmWave transmission. A more serious issue than free-space loss for mmWave signals is their limited penetration through materials and limited diffraction. In the urban environment, coverage for large cells could be particularly challenging; however, cell densification can be used to achieve the ambitious 5G capacity goals.

Atmospheric effects such as oxygen and water vapor absorption as well as fog and precipitation can scale exponentially with the link distance. Limiting to distances below 1 km, attenuation caused by atmospheric gases can be neglected up to 50 GHz, as shown in Fig. 3.4. However, above 50 GHz, it becomes important to consider the oxygen absorption peak of approximately 13 dB/km at 60 GHz and the water vapor resonance peak at 183 GHz of approximately 29 dB/km for relative humidity of 44% under standard

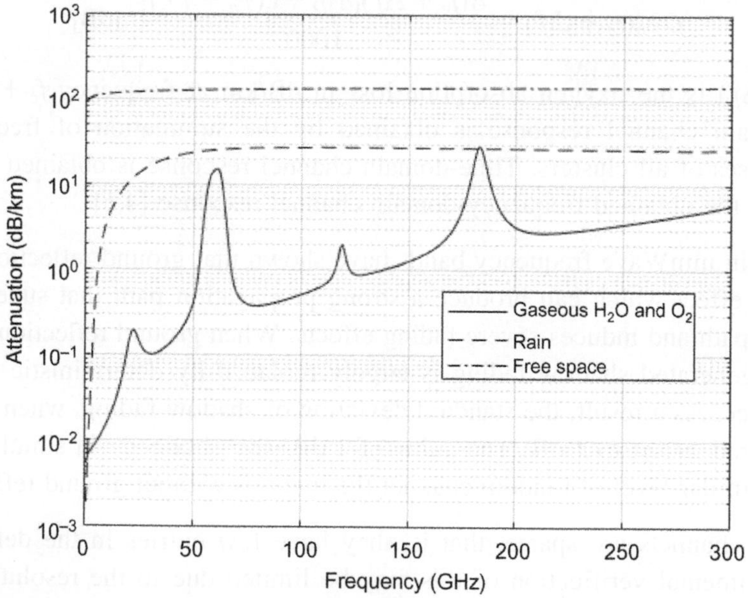

Figure 3.4

Comparison of gaseous $H_2O + O_2$, rain, and free-space attenuation (propagation distance 1 km, rain rate of 95 mm/h, and dry air pressure is 1013 hPa and the water vapor density is 7.5 g/m^3).

conditions. Note that neither fog nor rain is relevant for frequencies below 6 GHz. At frequencies above 80 GHz, dense fog related to a visibility of less than 70 m has a noticeable impact ($>$ 3 dB/km) and becomes severe above 200 GHz ($>$ 10 dB/km). Drizzle and steady rain are not a substantial issue for distances up to 1 km (3.0−4.4 dB/km above 70 GHz). However, as shown in Fig. 3.4, heavy rain attenuation increases up to 10−15 dB/km, and for downpours, up to 40 dB/km can be experienced. In summary, atmospheric effects, especially under bad weather conditions, are relevant for mmWave links over distances greater than 100 m and a crucial issue for longer distances of 1 km and farther [32].

3.1.1.7 LoS Path Loss Probability

It is shown in the literature that a height-dependent path loss for an indoor UE associated with a LoS condition can be modeled considering the dimensions of the building and the location of the UE inside the building. 3GPP has modeled the LoS path loss by using the 3D distance between the gNB and the UE along with the coefficients given by the ITU-R LoS path loss equations for 3D-UMa and 3D-UMi [4,24]. This provides a reasonable approximation to the more accurate models and can be determined without explicitly modeling the building dimensions. The ITU-R LoS path loss model assumes a two-ray model resulting in a path loss equation transitioning from a 22 dB/decade slope to a steeper slope at a break point depending on the environment height, which represents the height of a dominant reflection from the ground (or a moving platform) that can add constructively or destructively to the direct ray received at a UE located at the street level. In the 3D-UMa scenario, it is likely that such a dominant reflection path may come from the street level for indoor UEs associated with a type-1 LoS condition. Therefore, the environmental height is fixed at 1 m for a UE associated with a type-1 LoS condition. In the case of a UE associated with a type-2 LoS condition a dominant path can be likely created by reflection from the rooftop of a neighboring building. Note that a rooftop is at least 12 m in height; in this case the environmental height is randomly determined from a discrete uniform distribution between the UE height in meter and 12 [4,24].

For NLoS path loss modeling, which is the primary radio propagation mechanism in a 3D-UMa scenario, the dominant propagation paths experience multiple diffractions over rooftops followed by diffraction at the edge of the building. The path loss attenuation increases with the diffraction angle as a UE transitions from a high floor to a lower floor. In order to model this phenomenon, a linear height gain term given by $-\alpha(h-1.5)$ is introduced, where α in dB/m is the gain coefficient. A range of values between 0.6 and 1.5 were observed in different studies based on field measurements and ray-tracing simulations and a nominal value of 0.6 dB/m was chosen. In 3D-UMi test environments, the dominant propagation paths travel through and around the buildings. The UE may also receive a small amount of energy from propagation above rooftops. In order to simplify the model, a linear height gain is also applied to the 3D-UMi NLoS path loss with 0.3 dB/m gain coefficient based on the results from multiple studies. In addition, in both 3D-UMa and 3D-UMi scenarios the NLoS path loss is lower bounded by the corresponding LoS path loss because the path loss in a NLoS environment is in principle larger than that in a LoS environments. Measurement results and ray-tracing data have indicated that the marginal distribution of the composite power angular spectrum in zenith statistically has a Laplacian distribution and its conditional distribution given a certain link distance and UE height can also be approximated by Laplacian distribution. To incorporate these observations, zenith angle of departure (ZoD) and zenith angle of arrival (ZoA) are modeled by inverse Laplacian functions. It is also observed that the zenith angle spread of departure (ZSD) decreases significantly as the UE moves further away from the gNB. An intuitive explanation is that the angle subtended by a fixed local ring of scatterers at the UE to the gNB decreases as the UE moves away from the gNB. The ZSD is also observed to slightly change as an indoor UE moves up to higher floors [34–36].

We use the concept of the LoS probability to distinguish between the LoS and NLoS links. A link of length d is LoS with probability $p_{LoS}(d)$. The LoS probability is a non-increasing function of the link length. The LoS probability is obtained based on the certain building models, that is, either we use stochastic models from random shape theory or we use site-specific maps from geographical information system database. In this section, the LoS condition is determined based on a map-based approach, that is, by considering the transmitter and receiver positions and whether any buildings or walls are blocking the direct path between them. The impact of in-between objects not represented in the map is modeled separately through shadowing or blocking path loss components. It is noteworthy that this LoS definition is frequency independent, due to the fact that only buildings and walls are considered in the definition. 3GPP and ITU-R define the UMa LoS probability as follows:

$$p_{LoS}(d) = \min\left(\frac{d_1}{d_{2D}}, 1\right)\left(1 - e^{-d_{2D}/d_2}\right) + e^{-d_{2D}/d_2}$$

where d_{2D} is the 2D distance in meters and d_1 and d_2 can be optimized to fit a set of measurement data in the test environments/scenarios under consideration. For UMi test environments, it was observed that the above LoS probability is sufficient for frequencies above 6 GHz. The fitted d_1/d_2 model provides more consistency with measured data and the error between the measured data and the 3GPP LoS probability model over all distances are small. Note that the 3GPP UMi LoS probability model is not a function of UE height unlike the UMa LoS probability model.

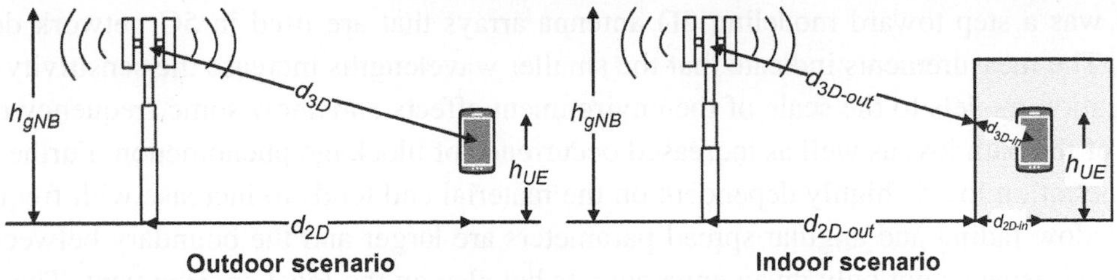

Figure 3.5
Definition of 2D and 3D distances in outdoor/indoor environments [4].

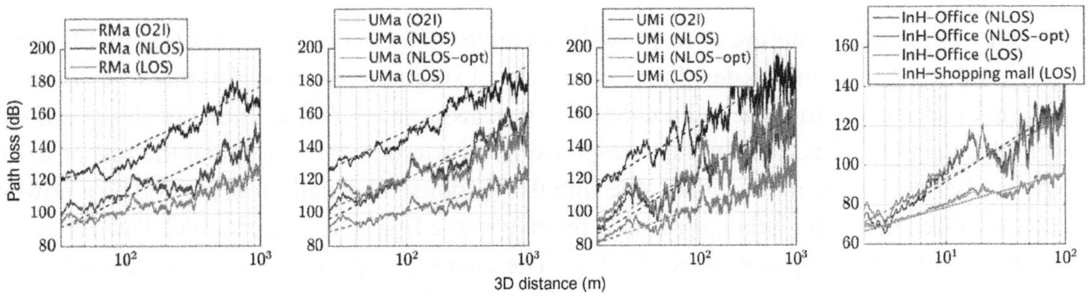

Figure 3.6
Path loss with (*solid line*) and without (*dashed line*) shadowing for various scenarios [37].

In the path loss models we often refer to 2D and 3D distances. Those distances are defined as follows (see Fig. 3.5):

$$d_{3D} = d_{3D-out} + d_{3D-in} = \sqrt{(d_{2D-out} + d_{2D-in})^2 + (h_{gNB} - h_{UE})^2}$$

Fig. 3.6 shows the path loss in dB for the 3D distance from the smallest value supported in each scenario to 10^3 m for the outdoor scenarios and 10^2 m for the indoor cases. In this figure, O2I denotes outdoor-to-indoor loss in various test environments.

3.1.2 Two- and Three-Dimensional Channel Models

The 5G cellular systems are expected to operate over a wide range of frequencies from 450 MHz to 100 GHz. For the development and standardization of the new 5G systems operating in frequency bands above 6 GHz, there was a serious need to accurately model radio signal propagation in these frequency bands, which could not be fully characterized by the existing channel models, because the previous generations of channel models were designed and evaluated for sub-6 GHz frequencies. The development of 3GPP 3D channel model was a step toward modeling 2D antenna arrays that are used in 5G network deployments. The measurements indicate that the smaller wavelengths increase the sensitivity of the propagation models to the scale of the environment effects and show some frequency dependence of the path loss as well as increased occurrence of blockage phenomenon. Furthermore, the penetration loss is highly dependent on the material and tends to increase with frequency. The shadow fading and angular spread parameters are larger and the boundary between LoS and NLoS depends not only on antenna heights but also on the local environment. The small-scale characteristics of the channel such as delay-spread and angular-spread and the multipath richness is somewhat similar over the, which was a good reason for extending the existing 3GPP models to the wider frequency range [24,25].

The goal of channel modeling is to provide accurate mathematical representations of radio propagation to be used in link-level and system-level simulations corresponding to a specific deployment scenario. Since the radio channel can be assumed as linear, it can be described by its impulse response. Once the impulse response is known, one can determine the response of the radio channel to any input signal. The impulse response is usually represented as a power density function of excess delay, measured relative to the first detectable signal. This function is often referred to as a power delay profile. The channel impulse response varies with the position of the receiver and may also vary with time. Therefore, it is usually measured and reported as an average of profiles measured over one wavelength to reduce noise effects, or over several wavelengths to determine a spatial average. It is important to clarify which average is meant, and how the averaging was performed.

The propagation effects of a wireless channel can be modeled with a large-scale propagation model combined with a small-scale fading model, where the former models long-term slow-fading characteristics of the wireless channel, such as path loss and shadowing, while the small-scale fading model provides rapid fluctuation behavior of the wireless channel due to multipath and Doppler spread. For a wireless channel with multiple antennas, static beamforming gain such as the sectorization beam pattern also contributes to the long-term propagation characteristic of the wireless channel and can be modeled as part of the large-scale propagation model. As for the small-scale fading model of a MIMO channel, the correlation of signals between antenna elements also needs to be considered and can be modeled by a spatial channel model that was used to evaluate performance of the previous generations of cellular standards. Although electromagnetic beam patterns generated by base station antenna arrays are 3D in nature, they were usually modeled as linear horizontal arrays, and elevation angles of signal paths have been ignored for simplicity. The 3D spatial channel models non-zero elevation angles associated with signal paths as well as azimuth angles so that the small-scale fading effect on each antenna element of the 2D antenna grid and the correlation between any two pair of antenna elements on the 2D antenna grid can be modeled.

While the increase in average mutual information (capacity) of a MIMO channel with the number of antennas is well understood, it appears that the variance of the mutual information can grow very slowly or even shrink as the number of antennas increases. This phenomenon is referred to as channel hardening in the literature, which has certain implications for control and data transmission [31].

For a three-sector macrocell, the horizontal and vertical antenna patterns are commonly modeled with a 3 dB beam-width of 70 and 10 degrees, respectively, with an antenna gain of 17 dBi. The vertical antenna pattern is also a function of electrical and mechanical antenna downtilt, where the electrical downtilt is a result of vertical analog beamforming, generated by applying common and static phase shifts to each vertical array of 2D antenna elements. For instance, the electrical downtilt is set to 15 and 6 degrees for a macrocell deployment scenario with inter-site distance of 500 m and 1732 m, respectively, in 3GPP evaluation methodology [4].

In order to meet the technical requirements of IMT-2020, new features are captured in 5G channel models such as support of frequencies up to 100 GHz and large bandwidth, 3D modeling, support of large antenna arrays, blockage modeling, and spatial consistency [4,5]. The 3D modeling describes the channel propagation in azimuth and elevation directions between the transmitting and the receiving antennas. It is more complete and accurate relative to the 2D modeling which only considers the propagation characteristics in the azimuth direction. Multi-antenna techniques capable of exploiting the elevation dimension have been developed for LTE since Rel-13 and are considered very important in 5G, which include modeling the elevation angles of departure and arrival, and their correlation with other parameters [7]. For a base station (access node) equipped with columns of active antenna arrays, vertical analog beamforming can be applied prior to the power amplifier (PA); thus the electrical downtilt can be dynamically adjusted over time. This feature allows the base station to dynamically adjust its cell coverage depending on the user distribution in the cell, for example, analog beamforming can be directed toward UEs congregated at the same location. Nonetheless, such beamforming is still fundamentally cell specific (i.e., spatial separation of individual UE is not possible) and operate on a long timescale, although time-varying base station antennas are modeled having one or multiple antenna panels, where an antenna panel has one or multiple antenna elements placed vertically, horizontally or in a 2D array within each panel. As a result, 3D channel modeling is required for performance evaluation of full-dimension MIMO. The full-dimension MIMO involves precoding/beamforming exploiting both horizontal and vertical degrees of freedom on a small timescale in a frequency-selective manner.

We start our study with the 3D antenna models in mmWave bands where implementation of large antenna arrays is feasible in the gNB or the UE. Let's assume that the gNB and/or

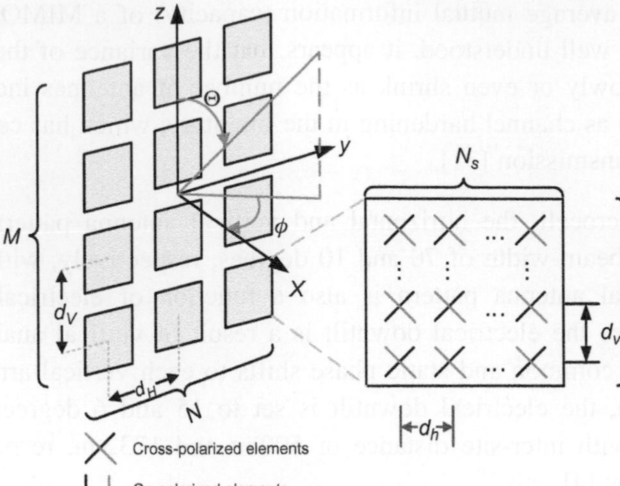

Figure 3.7
3D base station antenna model [4].

the UE each has a 2D planar antenna subarray comprising $M_s \times N_s$ antenna elements, where N_s denotes the number of columns and M_s is the number of antenna elements with the same polarization in each column (see Fig. 3.7). The antenna elements are uniformly spaced with a center-to-center spacing of d_h and d_v in the horizontal and vertical directions, respectively. The $M_s \times N_s$ elements may either be single polarized or dual polarized. A uniform rectangular array is formed comprising $M \times N$ antenna panels where M is the number of panels in a column and N is the number of panels in a row. Antenna panels are uniformly spaced with a center-to-center spacing of d_H and d_V in the horizontal and vertical directions, respectively. The 3GPP 3D channel model allows modeling 2D planar antenna arrays. The antenna elements can either be linearly polarized or cross polarized, as shown in Fig. 3.7. In this regard, the model represents a compromise between practicality and precision as it does not include the mutual coupling effect as well as different propagation effects of horizontally and vertically polarized waves [7].

For each antenna element, the general form of antenna element horizontal radiation pattern can be specified as $A_{E,H}(\phi) = -\min\left(12(\phi/\phi_{3\mathrm{dB}})^2, \alpha_m\right)$ where $-180^\circ \le \phi \le 180^\circ$, $\phi_{3\mathrm{dB}}$ denotes the horizontal 3 dB beam-width, and the parameter α_m is the maximum side-lobe level attenuation.

The general form of antenna element vertical radiation pattern is specified as $A_{E,V}(\theta) = -\min\left(12\left[(\theta - \theta_{tilt})/\theta_{3\mathrm{dB}}\right]^2, \alpha_m\right)$ where $-180^\circ \le \theta \le 180^\circ$, $\theta_{3\mathrm{dB}}$ denotes the vertical 3 dB beam-width, and θ_{tilt} is the tilt angle. It must be noted that $\theta = 0$ points to the zenith and $\theta = 90^\circ$ points to the horizon. The combined vertical and horizontal antenna element pattern is then given as $A(\theta,\phi) = -\min\left(-\lfloor A_{E,V}(\theta) + A_{E,H}(\phi)\rfloor, \alpha_m\right)$ where $A(\theta,\phi)$ is the relative antenna gain (dB) of an antenna element in the direction (θ,ϕ) [4,7,24].

The above concepts can also be applied to the UE antenna arrays. In this case $M \times N$ antenna panels may have different orientations. Let $\left(\Omega_{m_g,n_g}, \Theta_{m_g,n_g}\right)$ denote the orientation angles of the panel $\left(m_g, n_g\right)$ $0 \leq m_g < M_g, 0 \leq n_g < N_g$, where the orientation of the first panel $\left(\Omega_{0,0}, \Theta_{0,0}\right)$ is defined as the UE orientation, Ω_{m_g,n_g} is the array bearing angle, and Θ_{m_g,n_g} is the array downtilt angle. The antenna bearing is defined as the angle between the main antenna lobe center and a line directed toward east. The bearing angle increases in a clockwise direction. The parameters of the base station antenna array pattern for Dense-urban-eMBB (macro-TRP), Rural-eMBB, Urban-macro-mMTC, and Urban-macro-URLLC test environments (i.e., deployment scenarios studied in 3GPP and ITU-R[5]) are defined as $\theta_{3\text{dB}} = \phi_{3\text{dB}} = 65^{\text{o}}$, $\theta_{tilt} = 90^{\text{o}}$, $\alpha_m = 30$ dB. The parameters of the UE antenna pattern for frequencies above 30 GHz are given as $\theta_{3\text{dB}} = \phi_{3\text{dB}} = 90^\circ$, $\theta_{tilt} = 90^\circ$, *and* $\alpha_m = 25$ dB.

The IMT-2020 3D channel model developed in ITU-R is a geometry-based stochastic channel model. It does not explicitly specify the location of the scatterers, rather the directions of the rays. Geometry-based modeling of the radio channels enables separation of propagation parameters and antennas. The channel parameters for individual snapshots are determined stochastically based on statistical distributions obtained from channel measurements. Channel realizations are generated through the application of the geometrical principle by summing contributions of rays with specific small-scale parameters such as delay, power, azimuth angles of arrival and departure, and elevation angles of arrival and departure. The results are further combined with transmit/receive antenna correlations and temporal fading with geometry-dependent Doppler spectrum effects [4]. A single-link channel model is shown in Fig. 3.8 for the downlink direction. Each circle with several dots represents scattering region creating one cluster, each cluster is constituted by M rays, where a total of N clusters are assumed. We assume that there are N_{tx} antennas at the transmitter and N_{rx} antennas at the receiver. The small-scale parameters such as delay $\tau_{n,m}$, azimuth AoA

[5] Test environment reflects geographic environment and usage scenario which is used for the evaluation process; however, it has a direct relevance to the deployment scenario. The test environments defined in 3GPP and ITU-R are as follows:

- Indoor hotspot-eMBB is an indoor isolated environment at offices and/or in shopping malls based on stationary and pedestrian users with very high user density.
- Dense-urban-eMBB is an urban environment with high user density and traffic loads focusing on pedestrian and vehicular users.
- Rural-eMBB is a rural environment with larger and continuous wide area coverage, supporting pedestrian, vehicular, and high speed vehicular users.
- Urban macro-mMTC is an urban macro-environment targeting continuous coverage focusing on a high number of connected machine type devices.
- Urban macro-URLLC is an urban macro-environment targeting ultra-reliable and low-latency communications.

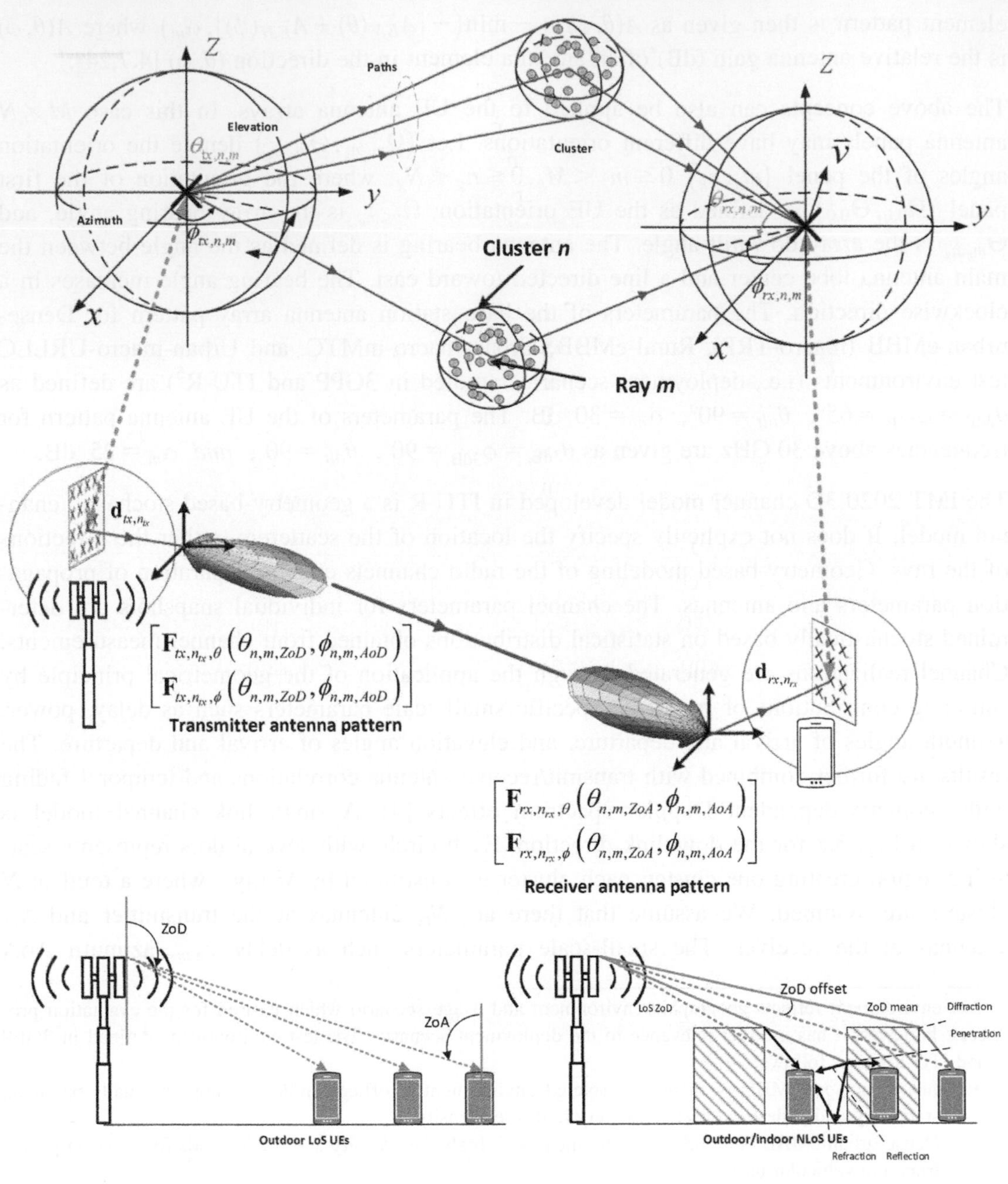

Figure 3.8
Illustration of 3D MIMO channel model and its parameters [4].

$\phi_{rx,n,m}$, elevation AoA $\theta_{rx,n,m}$, azimuth AoD $\phi_{tx,n,m}$ and elevation AoD $\theta_{tx,n,m}$ are assumed to be different for each ray. In the primary module of 3GPP 3D channel model, the number of clusters are fixed and frequency independent. The typical number of clusters reported in the literature is often small, random, and can be modeled as a Poisson distribution. By choosing an appropriate mean value of the Poisson distribution, the events with a larger number of clusters with a low probability may also be produced.

The 3GPP 3D channel model is a geometric stochastic model, describing the scattering environment between base station sector and the UE in both azimuth and elevation dimensions. The scatterers are represented by statistical parameters without having a real physical location. It specifies three propagation conditions, LoS, non-LoS, and outdoor-to-indoor. In each scenario it defines different parameters for mean propagation path loss, macroscopic fading, and microscopic fading. The probability of being in LoS is determined separately for indoor and outdoor UEs depending on the height of the UE as well as the break point distance. The break point distance characterizes the gap between transmitter and receiver at which the Fresnel zone is barely broken for the first time. For an indoor UE, LoS refers to the signal propagation outside of the building in which the UE is located. For each UE location, large-scale parameters are generated according to its geographic position as well as the propagation conditions at this location. The large-scale parameters incorporate shadow fading, the Rician K-factor (only in the LoS case), delay spread, azimuth angle spread of departure and arrival, as well as azimuth angle-spread of departure and arrival. The time-variant impulse response matrix of $N_{rx} \times N_{tx}$ MIMO channel is given by the following expression [4]:

$$\mathbf{H}(t;\tau) = \sum_{n=1}^{N} \mathbf{H}_n(t;\tau)$$

where $t, \tau, N,$ and n denote time, delay, number of clusters, and cluster index, respectively. The above channel impulse response is composed of the antenna array impulse response matrices $\mathbf{F}_{tx}$ and $\mathbf{F}_{rx}$ for the transmitter and the receiver sides, respectively, as well as the dual-polarized propagation channel response matrix. The channel from the transmitter antenna element n_{tx} to the receiver antenna element n_{rx} for the cluster n is given by the following expression [4]:

$$\begin{aligned}
&\mathbf{H}_{n_{rx},n_{tx},n,m}(t;\tau) \\
&= \sqrt{\frac{P_n}{M}} \sum_{m=1}^{M} \begin{bmatrix} \mathbf{F}_{rx,n_{rx},\theta}\left(\theta_{n,m,ZoA}, \phi_{n,m,AoA}\right) \\ \mathbf{F}_{rx,n_{rx},\phi}\left(\theta_{n,m,ZoA}, \phi_{n,m,AoA}\right) \end{bmatrix}^T \begin{bmatrix} \exp\left(j\Phi_{n,m}^{\theta\theta}\right) & \sqrt{\kappa_{n,m}{}^{-1}}\exp\left(j\Phi_{n,m}^{\theta\phi}\right) \\ \sqrt{\kappa_{n,m}{}^{-1}}\exp\left(j\Phi_{n,m}^{\phi\theta}\right) & \exp\left(j\Phi_{n,m}^{\phi\phi}\right) \end{bmatrix} \\
&\times \begin{bmatrix} \mathbf{F}_{tx,n_{tx},\theta}\left(\theta_{n,m,ZoD}, \phi_{n,m,AoD}\right) \\ \mathbf{F}_{tx,n_{tx},\phi}\left(\theta_{n,m,ZoD}, \phi_{n,m,AoD}\right) \end{bmatrix} \exp\left(j2\pi\lambda_0^{-1}\left(\mathbf{r}_{rx,n,m}^T \cdot \mathbf{d}_{rx,n_{rx}}\right)\right) \exp\left(j2\pi\lambda_0^{-1}\left(\mathbf{r}_{tx,n,m}^T \cdot \mathbf{d}_{tx,n_{tx}}\right)\right) \\
&\times \exp\left(j2\pi v_{n,m} t\right) \delta(\tau - \tau_{n,m})
\end{aligned}$$

where P_n is the power of nth path and the other parameters are defined as follows [4]:

- $\Phi^{\theta\theta}_{n,m}$, $\Phi^{\theta\phi}_{n,m}$, $\Phi^{\phi\theta}_{n,m}$, and $\Phi^{\phi\phi}_{n,m}$ denote random initial phase for each ray m of each cluster n and for four different polarization combinations.
- $\mathbf{F}_{rx,n_{rx},\theta}$ and $\mathbf{F}_{rx,n_{rx},\phi}$ represent receive antenna element n_{rx} field patterns in direction of the spherical basis vectors Θ and Φ.
- $\mathbf{F}_{tx,n_{tx},\theta}$ and $\mathbf{F}_{tx,n_{tx},\phi}$ denote transmit antenna element n_{tx} field patterns in direction of the spherical basis vectors Θ and Φ.
- $\mathbf{r}_{rx,n,m}$ and $\mathbf{r}_{tx,n,m}$ denote spherical unit vector with azimuth arrival angle $\phi_{n,m,AoA}$, elevation arrival angle $\theta_{n,m,ZoA}$ and azimuth departure angle $\phi_{n,m,AoD}$, and elevation departure angle $\theta_{n,m,ZoD}$.
- $\mathbf{d}_{rx,n_{rx}}$ and $\mathbf{d}_{tx,n_{tx}}$ represent location vector of receive antenna element n_{rx} and transmit antenna element n_{tx}.
- $\kappa_{n,m}$ is the cross-polarization power ratio in linear scale.
- λ_0 denotes the wavelength of the carrier frequency.
- $v_{n,m}$ represents the Doppler frequency component of ray n, m.

If the radio channel is dynamically modeled, the above small-scale parameters would become time variant. The channel impulse response describes the channel from a transmit antenna element to a receive antenna element. The spherical unit vectors are defined as follows [24]:

$$\hat{\mathbf{r}}_{rx,n,m} = \begin{bmatrix} \sin\theta_{n,m,ZoA} \cos\varphi_{n,m,AoA} \\ \sin\theta_{n,m,ZoA} \sin\varphi_{n,m,AoA} \\ \cos\theta_{n,m,ZoA} \end{bmatrix}, \quad \hat{\mathbf{r}}_{tx,n,m} = \begin{bmatrix} \sin\theta_{n,m,ZoD} \cos\varphi_{n,m,AoD} \\ \sin\theta_{n,m,ZoD} \sin\varphi_{n,m,AoD} \\ \cos\theta_{n,m,ZoD} \end{bmatrix}$$

The Doppler frequency component $v_{n,m}$ depends on the arrival angles (AoA, ZoA), and the UE velocity vector $\bar{\mathbf{v}}$ with speed v, travel azimuth angle ϕ_v, elevation angle θ_v and is given by

$$v_{n,m} = \frac{\hat{\mathbf{r}}^T_{rx,n,m}.\bar{\mathbf{v}}}{\lambda_0}, \quad \bar{\mathbf{v}} = v\left[\sin\theta_v \cos\varphi_v \ \ \sin\theta_v \sin\varphi_v \ \ \cos\theta_v\right]^T$$

The parameters $\mathbf{d}_{rx,n_{rx}}$ and $\mathbf{d}_{tx,n_{tx}}$ are the location vectors of receive and transmit antenna elements, respectively. Considering a base station sector with coordinates (s_x, s_y, s_z) and a planar antenna array, the location vector per antenna element is defined as $\mathbf{d}_{tx,n_{tx}} = \begin{pmatrix} s_x & s_y & s_z \end{pmatrix}^T + \begin{pmatrix} 0 & (k-1)d_H & (l-1)d_V \end{pmatrix}^T$ where $k = 1, 2, \ldots, N$ and $l = 1, 2, \ldots, M$ (see Fig. 3.7). The parameters N and d_H denote the number of antenna elements and the inter-element spacing in the horizontal direction, respectively, while M and d_V represent the number of antenna elements and the inter-element spacing in the vertical direction, respectively. The Doppler frequency component of the UE moving at velocity $\mathbf{v}$ is represented by parameter $v_{n,m}$.

When antenna arrays are deployed at the transmitter and receiver, the impulse response of such arrangement results in a vector channel. An example of this configuration is given below for the case of a 2D antenna array [4]:

$$\mathbf{H}_n^{NLoS}(t) = \sqrt{\frac{P_n}{M}} \sum_{m=1}^{M} \begin{bmatrix} F_{rx,\theta}(\theta_{n,m,ZoA}, \phi_{n,m,AoA}) \\ F_{rx,\phi}(\theta_{n,m,ZoA}, \phi_{n,m,AoA}) \end{bmatrix}^T \begin{bmatrix} \exp\left(j\Phi_{n,m}^{\theta\theta}\right) & \sqrt{\kappa_{n,m}}^{-1}\exp\left(j\Phi_{n,m}^{\theta\phi}\right) \\ \sqrt{\kappa_{n,m}}^{-1}\exp\left(j\Phi_{n,m}^{\phi\theta}\right) & \exp\left(j\Phi_{n,m}^{\phi\phi}\right) \end{bmatrix}$$

$$\times \begin{bmatrix} F_{tx,\theta}(\theta_{n,m,ZoD}, \phi_{n,m,AoD}) \\ F_{tx,\phi}(\theta_{n,m,ZoD}, \phi_{n,m,AoD}) \end{bmatrix} \mathbf{a}_{rx}(\theta_{n,m,ZoA}, \phi_{n,m,AoA}) \mathbf{a}_{tx}^H(\theta_{n,m,ZoD}, \phi_{n,m,AoD}) \exp(j2\pi\nu_{n,m}t)$$

$$\mathbf{H}_n^{LoS}(t) = \begin{bmatrix} F_{rx,\theta}(\theta_{LoS,ZoA}, \phi_{LoS,AoA}) \\ F_{rx,\phi}(\theta_{LoS,ZoA}, \phi_{LoS,AoA}) \end{bmatrix}^T \begin{bmatrix} \exp(j\Phi_{LoS}) & 0 \\ 0 & \exp(j\Phi_{LoS}) \end{bmatrix}$$

$$\times \begin{bmatrix} F_{tx,\theta}(\theta_{LoS,ZoD}, \phi_{LoS,AoD}) \\ F_{tx,\phi}(\theta_{LoS,ZoD}, \phi_{LoS,AoD}) \end{bmatrix} \mathbf{a}_{rx}(\theta_{LoS,ZoA}, \phi_{LoS,AoA}) \mathbf{a}_{tx}^H(\theta_{LoS,ZoD}, \phi_{LoS,AoD}) \exp(j2\pi\nu_{LoS}t)$$

where $\mathbf{a}_{tx}(\theta_{n,m,ZoD}, \phi_{n,m,AoD})$ and $\mathbf{a}_{rx}(\theta_{n,m,ZoA}, \phi_{n,m,AoA})$ are the transmit and receive antenna array impulse response vectors, respectively, corresponding to ray $m \in \{1, \ldots, M\}$ in cluster $n \in \{1, \ldots, N\}$ given by the following expression:

$$\mathbf{a}_{tx}(\theta_{n,m,ZoD}, \phi_{n,m,AoD}) = \exp\left(j\frac{2\pi}{\lambda_0}\left[\mathbf{W}_{tx}\mathbf{r}_{tx}(\theta_{n,m,ZoD}, \phi_{n,m,AoD})\right]\right); \quad \forall n, m$$

$$\mathbf{a}_{rx}(\theta_{n,m,ZoA}, \phi_{n,m,AoA}) = \exp\left(j\frac{2\pi}{\lambda_0}\left[\mathbf{W}_{rx}\mathbf{r}_{rx}(\theta_{n,m,ZoA}, \phi_{n,m,AoA})\theta\right]\right); \quad \forall n, m$$

where λ_0 is the wavelength of carrier frequency f_0; $\mathbf{r}_{tx}(\theta_{n,m,ZoD}, \phi_{n,m,AoD})$ and $\mathbf{r}_{rx}(\theta_{n,m,ZoA}, \phi_{n,m,AoA})$ are the corresponding angular 3D spherical unit vectors of the transit and receive, respectively; $\mathbf{W}_{tx}$ and $\mathbf{W}_{rx}$ denote the location matrices of the transmit and receive antenna elements in 3D Cartesian coordinates. The location matrices in the vectored impulse response are provided for an antenna configuration that is a uniform rectangular array consisting of cross-polarized antenna elements shown in Fig. 3.4.

The Doppler shift generally depends on the time-variance of the channel as it is defined as the derivative of the channel phase over time. It can result from transmitter, receiver, or scatterers movement. The general form of the exponential Doppler component is given as follows [24]:

$$\nu_{n,m} = \exp\left(j2\pi \int_{t_0}^{t} \frac{\hat{\mathbf{r}}_{rx,n,m}^T(\tilde{t}) \cdot \mathbf{v}(\tilde{t})}{\lambda_0} d\tilde{t}\right)$$

where $\hat{\mathbf{r}}_{rx,n,m}(t)$ denotes the normalized vector that points to the direction of the incoming wave as seen from the receiver at time t. The velocity vector of the receiver at time t is denoted by $\mathbf{v}(t)$ while t_0 denotes a reference point in time that defines the initial phase $t_0 = 0$. The above expression is only valid for time-invariant Doppler shift, satisfying $\hat{\mathbf{r}}^T_{rx,n,m}(t) \cdot \mathbf{v}(t) = \hat{\mathbf{r}}^T_{rx,n,m} \cdot \mathbf{v}$.

Spatial correlation is often said to degrade the performance of multi-antenna systems and imposes a limit on the number of antennas that can be effectively fit in a small mobile device. This seems intuitive as the spatial correlation decreases the number of independent channels that can be created by precoding. When modeling spatial correlation, it is useful to employ the Kronecker model, where the correlation between transmit and receive antennas are assumed independent and separable. This model is reasonable when the main scattering appears close to the antenna arrays and has been validated by outdoor and indoor measurements. Assuming Rayleigh fading, the Kronecker model means that the channel matrix can be represented as $\mathbf{H} = \mathbf{R}_{RX}^{1/2}\mathbf{H}_{channel}\left(\mathbf{R}_{TX}^{1/2}\right)^T$ where the elements of $\mathbf{H}_{channel}$ are independent and identically distributed as circular symmetric complex Gaussian random variables with zero-mean and unit variance. The important part of the model is that $\mathbf{H}_{channel}$ is pre-multiplied by the receive-side spatial correlation matrix $\mathbf{R}_{RX}$ and post-multiplied by transmit-side spatial correlation matrix $\mathbf{R}_{TX}$. Equivalently, the channel matrix can be expressed as $\mathbf{H} \sim \mathbb{C}(0, \mathbf{R}_{TX} \otimes \mathbf{R}_{RX})$ where $\otimes$ denotes the Kronecker product [9].

Using the Kronecker model, the spatial correlation depends directly on the eigenvalue distributions of the correlation matrices $\mathbf{R}_{RX}$ and $\mathbf{R}_{TX}$. Each eigenvector represents a spatial direction of the channel and its corresponding eigenvalue describes the average channel/signal gain in that direction. For the transmit-side, the correlation matrix $\mathbf{R}_{TX}$ describes the average gain in a spatial transmit direction, while receive-side correlation matrix $\mathbf{R}_{RX}$ describes a spatial receive direction. High spatial correlation is represented by large eigenvalue spread in $\mathbf{R}_{TX}$ or $\mathbf{R}_{RX}$, implying that some spatial directions are statistically stronger than the others. On the other hand, low spatial correlation is represented by small eigenvalue spread in $\mathbf{R}_{TX}$ or $\mathbf{R}_{RX}$, which implies that almost the same signal gain can be expected from all spatial directions [9].

Let's now focus on a specific case of downlink transmission from a base station to a mobile station. Denoting the nth snapshot of the spatial correlation matrices at the gNB and the UE by $\mathbf{R}_{gNB,n}$ and $\mathbf{R}_{UE,n}$, the per-tap spatial correlation is determined as the Kronecker product[6] of the gNB and UE's antenna correlation matrices as $\mathbf{R}_n = \mathbf{R}_{gNB,n} \otimes \mathbf{R}_{UE,n}$. We denote the number of receive antennas by N_{RX} and the number of transmit antennas by N_{TX}. If

[6] Given a $m \times n$ matrix $\mathbf{A}$ and a $p \times q$ matrix $\mathbf{B}$, the Kronecker product $\mathbf{C} = \mathbf{A} \otimes \mathbf{B}$, also called matrix direct product, is an $mp \times nq$ matrix with elements defined by $c_{\alpha\beta} = a_{ij}b_{kl}$ where $\alpha \equiv p(i-1)+k$ and $\beta \equiv q(j-1)+l$. The matrix direct product provides the matrix of the linear transformation induced by the vector space tensor product of the original vector spaces. More precisely, suppose that operators $S{:}V_1 \rightarrow W_1$ and $T{:}V_2 \rightarrow W_2$ are given by $S(x) = Ax$ and $T(y) = By$, then $S \otimes T{:}V_1 \otimes V_2 \rightarrow W_1 \otimes W_2$ is determined by $S \otimes T(x \otimes y) = (Ax) \otimes (By) = (A \otimes B)(x \otimes y)$.

Table 3.3: gNB/UE correlation matrix [9].

Entity	Number of Antennas: 1	2	4
gNB	$\mathbf{R}_{gNB}=1$	$\mathbf{R}_{gNB}=\begin{pmatrix}1 & \alpha\\ \alpha^* & 1\end{pmatrix}$	$\mathbf{R}_{gNB}=\begin{pmatrix}1 & \alpha^{1/9} & \alpha^{4/9} & \alpha\\ \alpha^{1/9*} & 1 & \alpha^{1/9} & \alpha^{4/9}\\ \alpha^{4/9*} & \alpha^{1/9*} & 1 & \alpha^{1/9}\\ \alpha^* & \alpha^{4/9*} & \alpha^{1/9*} & 1\end{pmatrix}$
UE	$\mathbf{R}_{UE}=1$	$\mathbf{R}_{UE}=\begin{pmatrix}1 & \beta\\ \beta^* & 1\end{pmatrix}$	$\mathbf{R}_{UE}=\begin{pmatrix}1 & \beta^{1/9} & \beta^{4/9} & \beta\\ \beta^{1/9*} & 1 & \beta^{1/9} & \beta^{4/9}\\ \beta^{4/9*} & \beta^{1/9*} & 1 & \beta^{1/9}\\ \beta^* & \beta^{4/9*} & \beta^{1/9*} & 1\end{pmatrix}$

Table 3.4: Values of α and β for different antenna correlations [9].

Low Correlation		Medium Correlation		High Correlation	
α	β	α	β	α	β
0	0	0.9	0.3	0.9	0.9

cross-polarized antennas are present at the receiver, it is assumed that $N_{RX}/2$ receive antennas have the same polarization, while the remaining $N_{RX}/2$ receive antennas have orthogonal polarization. Likewise, if cross-polarized antennas are present at the transmitter, it is assumed that $N_{TX}/2$ transmit antennas have the same polarization, while the remaining $N_{TX}/2$ transmit antennas have orthogonal polarization. It is further assumed that the antenna arrays are composed of pairs of co-located antennas with orthogonal polarization. Under these assumptions, the per-tap channel correlation is determined as $\mathbf{R}_n = \mathbf{R}_{gNB,n} \otimes \boldsymbol{\Gamma} \otimes \mathbf{R}_{UE,n}$ where $\mathbf{R}_{UE,n}$ is an $N_{RX} \times N_{RX}$ matrix, if all receive antennas have the same polarization, or an $N_{RX}/2 \times N_{RX}/2$ matrix, if the receive antennas are cross-polarized. Likewise, $\mathbf{R}_{gNB,n}$ is an $N_{TX} \times N_{TX}$ matrix, if all transmit antennas have the same polarization, or a $N_{TX}/2 \times N_{TX}/2$ matrix, if the transmit antennas are cross-polarized. Matrix $\boldsymbol{\Gamma}$ is a cross-polarization matrix based on the cross-polarization defined in the cluster-delay-line models. Matrix $\boldsymbol{\Gamma}$ is a 2×2 matrix, if cross-polarized antennas are used at the transmitter or at the receiver. It is a 4×4 matrix if cross-polarized antennas are used at both the transmitter and the receiver [9]. Table 3.3 defines the correlation matrices for the gNB and UE in NR with different number of transmit/receive antennas at each entity [9].

For the scenarios with more antennas at either gNB, UE, or both, the channel spatial correlation matrix can be expressed as the Kronecker product of $\mathbf{R}_{gNB}$ and $\mathbf{R}_{UE}$ according to $\mathbf{R}_{spatial} = \mathbf{R}_{UE} \otimes \mathbf{R}_{gNB}$. The parameters α and β for different antenna correlation types are given in Table 3.4. The 3D channel model parameters and model generation procedure are summarized in Fig. 3.9.

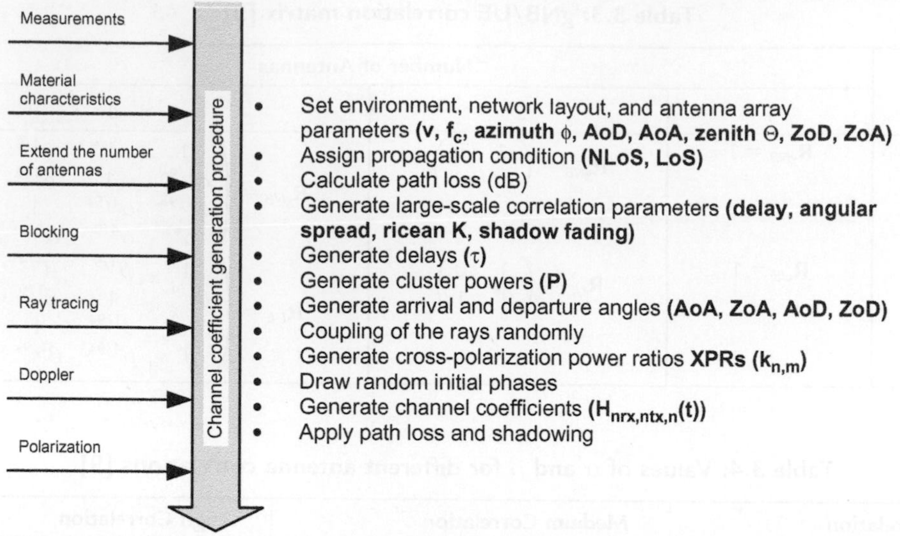

Figure 3.9
Summary of 3D channel model parameters and model generation procedure [7].

3.2 Waveforms

There have been considerable discussions on whether a new type of transmission waveform besides the incumbent cyclic prefix (CP)-OFDM should be adopted for NR [10,11]. Several alternative OFDM-based waveforms, including filter bank multicarrier and generalized frequency division multiplexing, were studied. Many of them claimed advantages in terms of increased bandwidth efficiency, relaxed synchronization requirements, reduced inter-user interference, reduced out-of-band (OOB) emissions, and so on, but at the same time created challenges in terms of increased transceiver complexity, difficulty in MIMO integration, and significant specification impacts. This section describes some of the prominent 5G waveform candidates and their characteristics as well as the reasons that the status-quo OFDM waveform continued to be supported in the new radio. Furthermore, the waveform, numerology, and frame structure should be chosen to enable efficient time/frequency utilization for frequency division duplex (FDD) and time division duplex (TDD) deployments, respectively. Table 3.5 summarizes the design requirements concerning the choice of the waveform, which were used to examine the waveform candidates for the NR.

Gabor's theory of communication suggests that ideally a multicarrier system such as OFDM must satisfy the following requirements [45]:

- The subcarriers are mutually orthogonal in time and frequency to make the receiver as simple as possible and to maintain the inter-carrier interference as low as possible.

Table 3.5: Summary of design targets for the waveform [45,47].

Design Criteria	Remarks
Higher spectral efficiency and scalability	High spectral efficiency for high data rates and efficient use of the available spectrum Ability to efficiently support MIMO and multipath robustness Low latency
Lower in-band and out-of-band emissions	Reduce interference among users within allocated band and reduce interference among neighbor operators
Enables asynchronous multiple access	Support a higher number of small-cell data burst devices with minimal scheduling overhead through asynchronous operations and enables lower power operation
Lower power consumption	Low peak-to-average power ratio allowing efficient power amplifier design
Lower implementation complexity	Reasonable transmitter and receiver complexity and additional complexity must be justified by significant performance improvements
Coexistence with legacy and mobility Support	Simplify LTE coexistence Robust against Doppler shift to allow high mobility

- The transmission waveform is well localized in time and frequency. This provides immunity to ISI from multipath propagation or delay spread and to ICI from Doppler spread. A good time localization is required to enable low latency.
- Maximal spectral efficiency, that is, $\rho = (\delta_T \delta_F)^{-1}$ with ρ denoting the spectral efficiency in data symbols per second per Hertz.

However, it is shown in the literature [38,46], that it is not possible to satisfy these three requirements at the same time and certain tradeoffs are necessary. This conclusion has an impact on the waveform selection in wireless communication systems.

3.2.1 OFDM Basics and Transmission Characteristics

OFDM was selected in LTE/LTE-A due to its efficiency and simplicity using baseband modulation and demodulation stages based on FFT. Mathematically, the nth OFDM symbol can be described as

$$x_n(t) = \sum_{k=0}^{N_{FFT}-1} s_{k,n} \mathrm{rect}(t - nT_u) e^{j(2\pi/T_u)kt}$$

where k denotes the subcarrier index and n denotes symbol index, N_{FFT} is the total number of subcarriers, rect(.) is a rectangular pulse with the symbol period of T_u, and $s_{k,n}$ is the data symbol [e.g., quadrature amplitude modulation (QAM) symbol] of the kth subcarrier at the nth time instant. A CP is appended to the beginning of each symbol. The inserted CP serves as a guard time between symbols which protects against ISI. In addition, it preserves the orthogonality between subcarriers after passing through a channel provided that the CP duration is longer than the channel RMS delay spread. The CP acts as a buffer region where delayed information from the previous symbols can be stored. The receiver must exclude samples from the CP which might be corrupted by the previous symbol when choosing the samples for an OFDM symbol. When demodulating the received symbol, the receiver can choose T_u/T_s samples from a region which is not affected by the previous symbol.

In a conventional serial data transmission system the information bearing symbols are transmitted sequentially, with the frequency spectrum of each symbol occupying the entire available bandwidth. An unfiltered QAM signal spectrum can be described in the form of $\sin(\pi f T_u)/\pi f T_u$ with zero-crossing points at integer multiples of $1/T_u$, where T_u is the QAM symbol period. The concept of OFDM is to transmit the data bits in parallel QAM-modulated subcarriers using frequency division multiplexing. The carrier spacing is carefully selected so that each subcarrier is located on other subcarriers' zero-crossing points in the frequency domain. Although there are spectral overlaps among subcarriers, they do not interfere with each other, if they are sampled at the subcarrier frequencies. In other words, they maintain spectral orthogonality. The OFDM signal in frequency domain is generated through aggregation of N_{FFT} parallel QAM-modulated subcarriers where adjacent subcarriers are separated by subcarrier spacing $1/T_u$. Since an OFDM signal consists of many parallel QAM subcarriers, the mathematical expression of the signal in time domain can be expressed as follows:

$$x(t) = Re\left\{ e^{j\omega_c t} \sum_{k=-(N_{FFT}-1)/2}^{(N_{FFT}-1)/2} s_k e^{j2\pi k(t-t_g)/T_u} \right\}, \quad nT_u \le t \le (n+1)T_u$$

where $x(t)$ denotes the OFDM signal in time domain, s_k is the complex-valued data that is QAM-modulated and transmitted over subcarrier k, N_{FFT} is the number of subcarriers in frequency domain, ω_c is the RF carrier frequency, and T_g is the guard interval or the CP length. For a large number of subcarriers, direct generation and demodulation of the OFDM signal would require arrays of coherent sinusoidal generators which can become excessively complex and expensive. However, one can notice that the OFDM signal is actually the real part of the inverse discrete Fourier transform (IDFT) of the original complex-valued data symbols $\{s_k | k = -(N_{FFT}-1)/2, \ldots, (N_{FFT}-1)/2\}$. It can be observed that there are $N < N_{FFT}$ subcarriers each carrying the corresponding data α_k. The inverse of the subcarrier spacing $\Delta f = 1/T_u$ is defined as the OFDM useful symbol duration T_u, which is N_{FFT} times longer than that of the original input data symbol duration.

3.2.1.1 Cyclic Prefix

The inclusion of CP in OFDM makes it robust to timing synchronization errors. Robustness to synchronization errors is relevant when synchronization is hard to achieve such as over the sidelink. It can also be relevant if asynchronous transmissions are allowed in the uplink. The inclusion of he cyclic prefix adds redundancy to the transmission since the same content is transmitted twice as the CP is a copy of the tail of a symbol placed at its beginning. This overhead can be expressed as a function of symbol duration and duration of the CP as $\mathrm{OH} = T_{CP}/(T_{CP} + T_u)$. OFDM is a flexible waveform that can support diverse services in a wide range of frequencies when properly selecting subcarrier spacing and CP. Further discussion on OFDM numerology design that fulfills a wide range of requirements is given in the next section.

Since IDFT is used in the OFDM modulator, the original data are defined in the frequency domain, while the OFDM signal $s(t)$ is defined in the time domain. The IDFT can be implemented via a computationally efficient FFT algorithm. The orthogonality of subcarriers in OFDM can be maintained and individual subcarriers can be completely separated and demodulated by an FFT at the receiver when there is no ISI introduced by communication channel. In practice, linear distortions such as multipath delay cause ISI between OFDM symbols, resulting in loss of orthogonality and an effect that is similar to cochannel interference. However, when delay spread is small, that is, within a fraction of the OFDM useful symbol length, the impact of ISI is negligible, although it depends on the order of modulation implemented by the subcarriers (see Fig. 3.10). A simple solution to mitigate multipath delay is to increase the OFDM effective symbol duration such that it is much larger than the delay spread; however, when the delay spread is large, it requires a large number of subcarriers and a large FFT size. Meanwhile, the system might become sensitive to Doppler shift and carrier frequency offset. An alternative approach to mitigate multipath distortion is to generate a cyclically extended guard interval, where each OFDM symbol is prefixed with a periodic extension of the signal itself, as shown in Fig. 3.10 where the tail of the symbol is copied to the beginning of the symbol. The OFDM symbol duration then is defined as $T_s = T_u + T_g$, where T_g is the guard interval or CP. When the guard interval is longer than the channel impulse response or the multipath delay, the ISI can be effectively eliminated. The ratio of the guard interval to useful OFDM symbol duration depends on the deployment scenario and the frequency band. Since the insertion of the guard intervals will reduce the system throughput, T_g is usually selected less than $T_u/4$. The CP should absorb most of the signal energy dispersed by the multipath channel. The entire the ISI energy is contained within the CP, if its length is greater than that of the channel RMS delay spread $(T_g > \tau_{RMS})$. In general it is sufficient to have most of the delay spread energy absorbed by the guard interval, considering the inherent robustness of large OFDM symbols to time dispersion. Fig. 3.11 illustrate the OFDM modulation and demodulation process in the transmitter and the receiver, respectively. In practice, a windowing or filtering scheme is utilized in the OFDM transmitter side to reduce the OOB emissions of the OFDM signal (see Fig. 3.10).

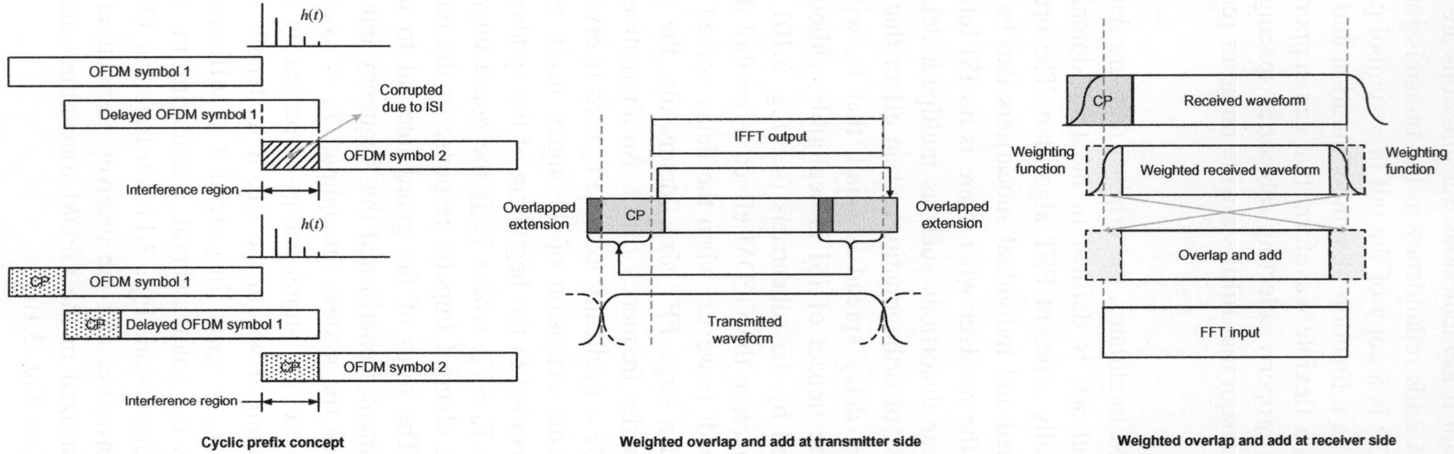

Figure 3.10

Illustration of the effect of cyclic prefix for eliminating ISI ($h(t)$ is the hypothetical channel impulse response) and practical implementation of CP insertion and removal [47].

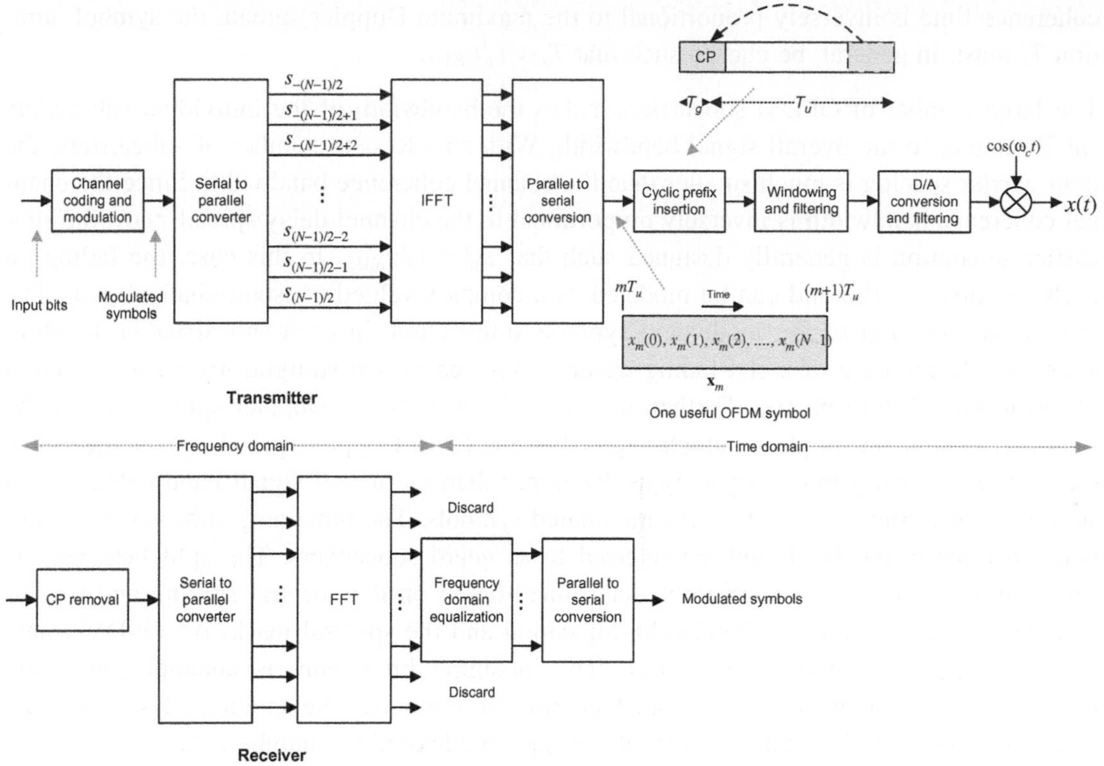

Figure 3.11
OFDM signal generation and reception process.

The mapping of the modulated data symbol into multiple subcarriers also allows an increase in the symbol duration. The symbol duration obtained through an OFDM scheme is much larger than that of a single-carrier modulation technique with a similar transmission bandwidth. In general, when the channel delay spread exceeds the guard time, the energy contained in the ISI will be much smaller with respect to the useful OFDM symbol energy, as long as the symbol duration is much larger than the channel delay spread, that is, $T_S \gg \tau_{RMS}$. Although large OFDM symbol duration is desirable to mitigate the ISI effects caused by time dispersion, large OFDM symbol duration can further reduce the ability to alleviate the effects of fast fading, particularly, if the symbol period is large compared to the channel coherence time, then the channel can no longer be considered as time-invariant over the OFDM symbol duration; therefore, this will introduce the inter-subcarrier orthogonality loss. This can affect the performance in fast fading conditions. Hence, the symbol duration should be kept smaller than the minimum channel coherence time. Since the channel coherence time is inversely proportional to the maximum Doppler spread, the symbol duration T_s must, in general, be chosen such that $T_s \ll 1/\nu_{RMS}$.

The large number of OFDM subcarriers makes the bandwidth of the individual subcarriers small relative to the overall signal bandwidth. With an adequate number of subcarriers, the inter-carrier spacing is much smaller than the channel coherence bandwidth. Since the channel coherence bandwidth is inversely proportional to the channel delay spread τ_{RMS}, the subcarrier separation is generally designed such that $\Delta f \ll 1/\tau_{RMS}$. In this case, the fading on each subcarrier is flat and can be modeled as a complex-valued constant channel gain. The individual reception of the modulated symbols transmitted on each subcarrier is therefore simplified to the case of a flat-fading channel. This enables a straightforward introduction of advanced MIMO schemes. Furthermore, in order to mitigate Doppler spread effects, the inter-carrier spacing should be much larger than the RMS Doppler spread $\Delta f \gg \nu_{RMS}$. Since the OFDM sampling frequency is typically larger than the actual signal bandwidth, only a subset of subcarriers is used to carry modulated symbols. The remaining subcarriers are left inactive prior to the IFFT and are referred to as guard subcarriers. The split between the active and the inactive subcarriers is determined based on the spectral sharing and regulatory constraints, such as the bandwidth allocation and the spectral mask. An OFDM transmitter diagram is shown in Fig. 3.11. The incoming bit stream is channel coded and modulated to form the complex-valued modulated symbols. The modulated symbols are converted from serial to parallel with $N < N_{FFT}$ complex-valued numbers per block, where N_{FFT} is the size of FFT/IFFT operation. Each block is processed by an IFFT and the output of the IFFT forms an OFDM symbol, which is converted back to serial data for transmission. A guard interval or CP is inserted between symbols to eliminate ISI effects caused by multipath distortion. The discrete symbols are windowed/filtered and converted to an analog signal for RF upconversion. The reverse process is performed at the receiver. A one-tap equalizer is usually used for each subcarrier to correct channel distortion. The tap coefficients are calculated based on channel information.

When there is multipath distortion, a conventional single-carrier wideband transmission system suffers from frequency-selective fading. A complex adaptive equalizer must be used to equalize the in-band fading. The number of taps required for the equalizer is proportional to the symbol rate and the multipath delay. For an OFDM system, if the guard interval is larger than the multipath delay, the ISI can be eliminated and orthogonality can be maintained among subcarriers. Since each OFDM subcarrier occupies a very narrow spectrum, in the order of a few kHz, even under severe multipath distortion, they are only subject to flat fading. In other words, the OFDM converts a wideband frequency-selective fading channel to a series of narrowband frequency non-selective fading subchannels by using the parallel multicarrier transmission scheme. Since OFDM data subcarriers are statistically independent and identically distributed, based on the central-limit theorem, when the number of subcarriers N_{FFT} is large, the OFDM signal distribution tends to be Gaussian.

3.2.1.2 Pre- and Post-processing Signal-to-Noise Ratio

In an OFDM system, the SNR is a measure of channel quality and is a key factor of link-level error assessment. There are different methods for calculation of SNR in single-antenna and multi-antenna transmission systems. For single-input/single-output systems, the SNR can be viewed as the received SNR, that is, the received SNR before the detector. The post-processing SNR is often used for MIMO links and represents the SNR after combining in the receiver and measures the likelihood that a coded message is decoded successfully. In link-level simulations, the SNR γ is typically calculated using the following method. Let's vector $\mathbf{x}=(x_1,x_2,\ldots,x_{N_{TX}})^H\in\mathbb{C}^{N_{TX}\times 1}$ denotes the transmit signal where $x_k\in\mathbb{C}\forall k=1,2,\ldots,N_{TX}$ is the complex-valued transmitted symbols from the kth transmit antenna. Note that N_{TX} is the number of transmit antennas. It can be shown that the total transmit signal power can be obtained as $\sigma_{\mathbf{x}}^2=\text{trace}(\mathbf{R}_{\mathbf{x}})$ where $\mathbf{R}_{\mathbf{x}}$ denotes the autocorrelation matrix of the transmitted signal. The transmit power from the kth antenna is given as $\sigma_{x_k}^2=E\{|x_k|^2\}=1/N_{TX}$ (uniformly distributed power). If $\mathbf{H}$ represents the channel matrix with $\|\mathbf{H}\|_F^2=N_{TX}N_{RX}$[7] in which N_{TX} and N_{RX} denote the number of transmit and receive antennas, respectively, the received signal vector $\mathbf{y}$ can be calculated as $\mathbf{y}=\mathbf{Hx}+\mathbf{v}$ where complex-valued Gaussian-distributed noise vector $\mathbf{v}\sim\mathbb{C}(\mathbf{0},\sigma_{\mathbf{v}}^2\mathbf{I})$ denotes the noise vector with respect to the size of the FFT N_{FFT} and the number of used subcarriers before the detector N_{used}. We define the complex-valued Gaussian-distributed noise vector $\mathbf{n}\sim\mathbb{C}(\mathbf{0},\sigma_{\mathbf{n}}^2\mathbf{I})$ to be noise after the FFT operation. The receive SNR before the detector is given as $\gamma_{pre-FFT}=\left(\|\mathbf{Hx}\|_F^2/N_{RX}\sigma_{\mathbf{v}}^2\right)=\sigma_{\mathbf{v}}^{-2}$, whereas the SNR after the FFT operation is given as $\gamma_{post-FFT}=\left(\|\mathbf{Hx}\|_F^2/N_{RX}\sigma_{\mathbf{n}}^2\right)=\sigma_{\mathbf{n}}^{-2}$. It can be observed that the difference between pre-FFT and post-FFT SNRs $\left(\gamma_{pre-FFT}/\gamma_{post-FFT}\right)=(\sigma_{\mathbf{v}}^2/\sigma_{\mathbf{n}}^2)=(N_{FFT}/N_{used})$; $N_{used}\le N_{FFT}$ is always positive, implying that the FFT operation suppresses the noise and enhances the SNR.

3.2.1.3 Peak-to-Average Power Ratio

The peak-to-average power ratio (PAPR) for a single-carrier modulation signal depends on its constellation and the pulse-shaping filter roll-off factor. For a Gaussian distributed

[7] The p-norm of matrix $\mathbf{H}$ is defined as $\|\mathbf{H}\|_p=\left(\sum_{i=0}^{M-1}\sum_{j=0}^{N-1}|h_{ij}|^p\right)^{1/p}$. In the special case when $p=2$ the norm is called the Frobenius norm and for $p=\infty$ is called the maximum norm. The Frobenius norm or Hilbert–Schmidt norm of matrix $\mathbf{H}$ is similar, although the latter term is often reserved for the operators on a Hilbert space. In general, this norm can be defined in the following forms: $\|\mathbf{H}\|_p=\sqrt{\sum_{i=0}^{M-1}\sum_{j=0}^{N-1}|h_{ij}|^2}=\sqrt{\text{trace}(\mathbf{H}^H\mathbf{H})}=\sqrt{\sum_{i=0}^{\min(M,N)-1}\varsigma_i^2}$ where $\mathbf{H}^H$ denotes the conjugate transpose of matrix $\mathbf{H}$ and ς_i are the singular values of matrix $\mathbf{H}$. The Frobenius norm is further similar to the Euclidean norm on $\mathbb{R}^N$ and is obtained from an inner product on the space of all matrices. The Frobenius norm is submultiplicative and is very useful for numerical linear algebra.

OFDM signal, the cumulative distribution function (CDF)[8] of PAPR for 99.0%, 99.9%, and 99.99% are approximately 8.3, 10.3, and 11.8 dB, respectively. Since the OFDM signal has a high PAPR, it could be clipped in the transmitter PA, because of its limited dynamic range or nonlinearity. Higher output back-off is required to prevent performance degradation and intermodulation products spilling into adjacent channels. Therefore, RF PAs should be operated in a very large linear region. Otherwise, the signal peaks leak into nonlinear region of the PA causing signal distortion. This signal distortion introduces intermodulation among the subcarriers and OOB emission. Thus the PA should be operated with large power back-offs. On the other hand, this leads to very inefficient amplification and an expensive transmitter. Thus, it is highly desirable to reduce the PAPR. In addition to inefficient operation of the PA, a high PAPR requires larger dynamic range for the receiver analog to digital converter (ADC). To reduce the PAPR, several techniques have been proposed and used such as clipping,[9] channel coding, temporal windowing, tone reservation,[10] and tone injection. However, most of these methods are unable to achieve simultaneously a large reduction in PAPR with low complexity and without performance degradation. The PAPR ξ of an OFDM signal is defined as follows:

$$\xi = \frac{\max|x(t)|^2}{E\{|x(t)|^2\}}\Bigg|_{nT_u \le t \le (n+1)T_u}$$

In the above equation $E\{.\}$ denotes the expectation operator and n is an integer. From the central-limit theorem, for large values of N_{FFT}, the real and imaginary values of OFDM signal $x(t)$ would have Gaussian distribution. Consequently, the amplitude of the OFDM signal has a Rayleigh distribution with zero mean and a variance of N_{FFT} times the variance of one complex sinusoid. Assuming the samples to be mutually uncorrelated, the CDF for the peak power per OFDM symbol is given by

$$P(\xi > \gamma) = \left[1 - (1 - e^{-\gamma})^{N_{FFT}}\right]$$

From the above equation it can be seen that large PAPR occurs only infrequently due to relatively large values of N_{FFT} used in practice.

---

8 The CDF of the real-valued random variable X is defined as $x \rightarrow F_X(x) = P(X \le x), \forall x \in \mathbb{R}$, where the right-hand side represents the probability that random variable X takes on a real value less than or equal to x. The CDF of X can be defined in terms of the probability density function $f(x)$ as $F(x) = \int_{-\infty}^{x} f(x)dx$. The complementary CDF (CCDF), on the other hand, is defined as $P(X > x) = 1 - F_X(x)$.

9 Since the OFDM signal has a high PAPR, it may be clipped in the transmitter power amplifier, because of its limited dynamic range or nonlinearity. Higher output back-off is required to prevent BER degradation and intermodulation products spilling into adjacent channels. However, clipping of an OFDM signal has similar effect as impulse interference against which an OFDM system is inherently robust. Computer simulations show that for a coded OFDM system, clipping of 0.5% of the time results in a BER degradation of 0.2 dB. At 0.1% clipping, the degradation is less than 0.1 dB.

10 In tone reservation method, the transmitter and the receiver reserve a subset of tones or subcarriers for generating PAPR reduction signals. Those reserved tones are not used for data transmission.

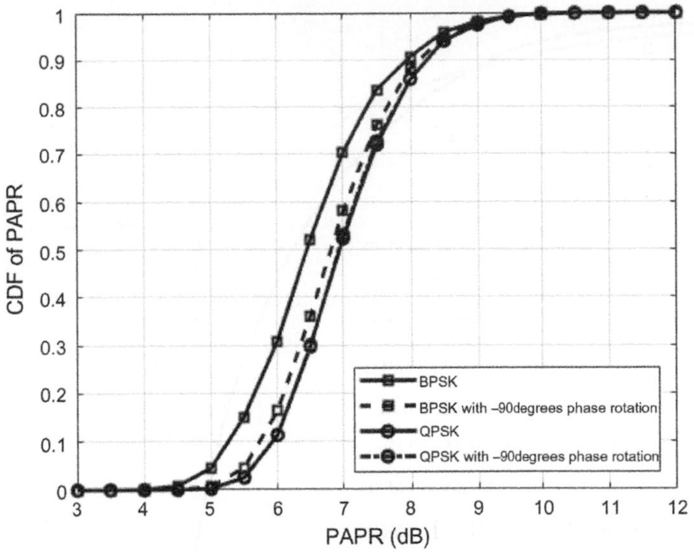

Figure 3.12
CDF of OFDM PAPR with BPSK/QPSK modulation and $N_{FFT} = 128$ [30].

Fig. 3.12 shows the CDF of OFDM PAPR for BPSK and QPSK modulation assuming a 40 MHz channel bandwidth and $N_{FFT} = 128$. We further applied 90 degrees phase rotation to the subcarriers in the upper 20 MHz of the channel and investigated the effect on the PAPR reduction [47]. It is shown that large PAPR values are less likely to occur with large FFT sizes as suggested earlier.

The PAs have generally a nonlinear amplitude response, where the output power is saturated for large input signals. Most applications require operation in the linear region of the PA where the output power is a linear function of the input. The larger the linear operation region or alternatively the higher saturation point, the more expensive the PA. Therefore, it is imperative to reduce the PAPR of the OFDM signal before processing through the PA. The wider bandwidth of NR compared to LTE increases the PAPR of the transmitted signal and makes it harder to achieve the same power efficiency as an LTE frontend. Current estimates suggest that for a single PA supplying an average transmit power of 23 dBm at the antenna, the power from the battery will be around 2.5 W, compared to around 1.8 W for current LTE UEs. The PAPR of the 5G NR signal is 3 dB higher than an equivalent LTE waveform [48], resulting in larger back-off or higher average transmit power. Another interesting observation is that for 5G CP-OFDM waveform using different modulations, there is no significant difference in the CCDF function- meaning higher order modulation has minimal impact on maximum power reduction (MPR) and PA back-off. Fig. 3.13 shows the

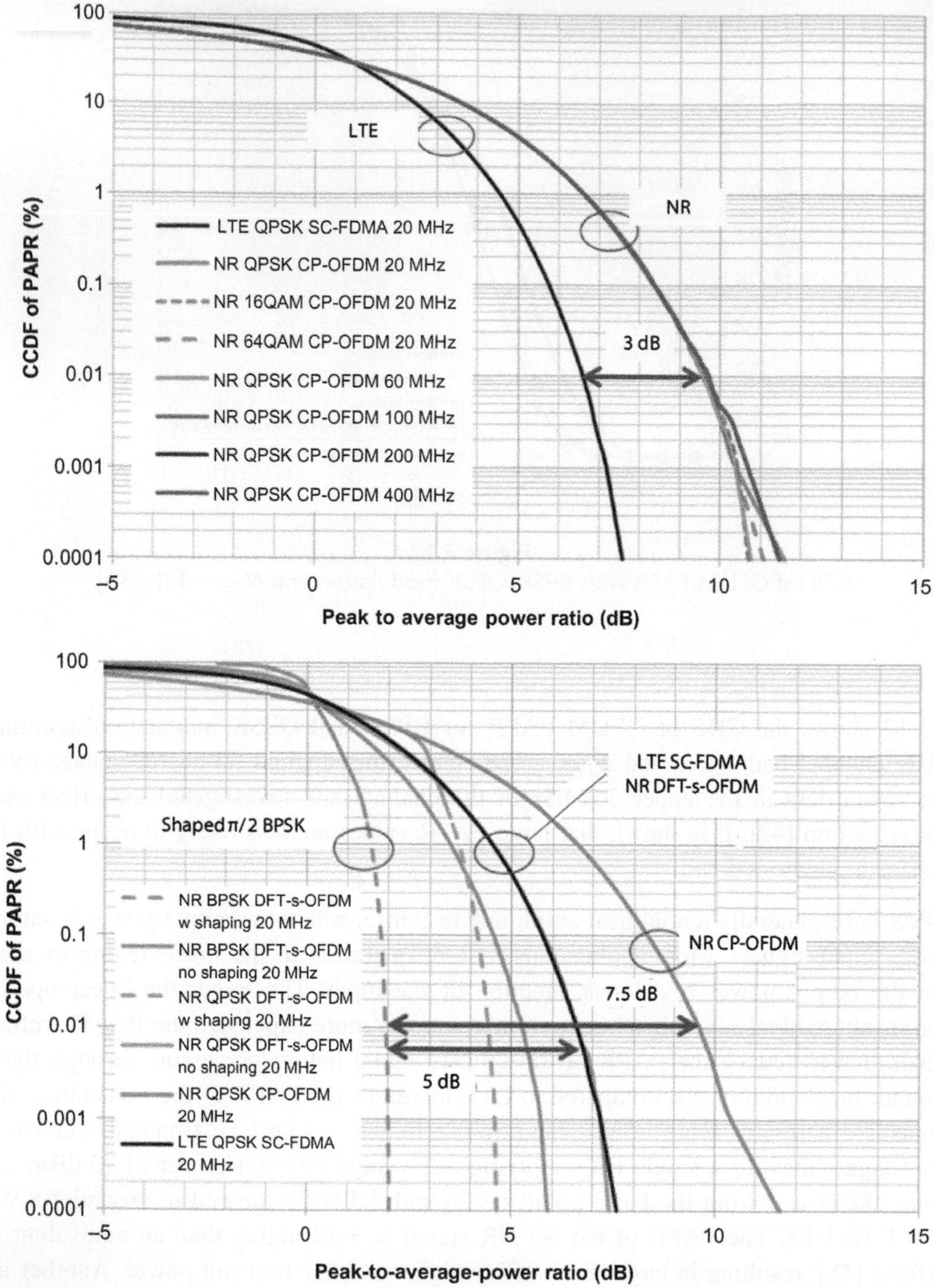

Figure 3.13

Comparison of CCDF of CP-OFDM and SC-FDMA PAPRs [48].

CCDF curves of some lower PAPR options, which can be used in cell-edge areas as well as mmWave frequency bands. We can observe that DFT-spread OFDM (DFT-S-OFDM) QPSK waveform in uplink exhibits very similar PAPR as the existing LTE single-carrier frequency division multiple access (SC-FDMA) used in the uplink; however, spectrally shaped $\pi/2-$ BPSK modulation can provide up to 7.5 dB PAPR reduction which can be used in sub-6 GHz and further in mmWave bands to improve uplink link budget and coverage [48].

3.2.1.4 Error Vector Magnitude

The modulation accuracy or the permissible signal constellation fuzziness is often measured in terms of error vector magnitude (EVM) metric. In general, the EVM is defined as the square root of the ratio of the mean error vector power to the mean reference-signal power expressed as a percentage. In other words, the EVM defines the average constellation error with respect to the farthest constellation point (i.e., the distance between the reference-signal and measured signal points in I–Q plane).

In NR, the EVM measurement is conducted for all bandwidths and each NR carrier over all allocated resource blocks and downlink subframes within 10 ms measurement period. The boundaries of the EVM measurement periods are not necessarily aligned with radio frame boundaries. 3GPP defines the reference points at which the [transmitter] EVM is measured at the receiver based on which the EVM must be measured after the FFT and a zero-forcing equalizer (per subcarrier amplitude/phase correction) in the receiver [9]. The basic unit of EVM measurement is defined over one subframe in the time domain and N_{BW}^{RB} subcarriers (180 kHz) in the frequency domain as follows [9]:

$$\mathrm{EVM}=\sqrt{\frac{\sum_{t\in T}\sum_{f\in F(t)}\left|Z'(t,f)-I(t,f)\right|^2}{\sum_{t\in T}\sum_{f\in F(t)}\left|I(t,f)\right|^2}}$$

where T is the set of symbols with the considered modulation scheme being active within the subframe, $F(t)$ is the set of subcarriers within the N_{BW}^{RB} subcarriers with the considered modulation scheme being active in symbol t, $I(t,f)$ is the ideal signal reconstructed by the measurement equipment according to the relevant transmitter model, and $Z'(t,f)$ is the modified signal under test defined as follows [9]:

$$Z'(t,f)=\frac{\mathrm{FFT}\left\{z(v-\Delta\tilde{t})e^{-j2\pi\Delta\tilde{f}v}\right\}e^{j2\pi f\Delta\tilde{t}}}{\tilde{a}(f)e^{j\tilde{\phi}(f)}}$$

where $z(v)$ is the time-domain samples of the signal under test, $\Delta\tilde{t}$ is the sample timing difference between the FFT processing window relative to the nominal timing of the ideal signal, $\Delta\tilde{f}$ is the RF frequency offset, $\tilde{\phi}(f)$ is the phase response of the transmitter chain, and

$\tilde{a}(f)$ is the amplitude response of the transmitter chain. In the above equations, the basic unit of measurement is one subframe and the equalizer is calculated over 10 subframes to reduce the impact of noise on the reference symbols. The boundaries of the 10 subframes measurement periods are not necessarily aligned with radio frame boundaries.

The EVM is averaged over all allocated downlink resource blocks with the considered modulation scheme in the frequency domain, and a minimum of 10 downlink subframes. For FDD systems, the averaging in the time domain equals the 10 subframe duration of the 10 subframes measurement period from the equalizer estimation step, whereas for TDD systems, the averaging in the time domain can be calculated from subframes of different frames and should have a minimum of 10 subframes averaging length.

$$\overline{\mathrm{EVM}_{frame}} = \left[\frac{\sum_{i=1}^{N_{dl}}\sum_{j=1}^{N_i}\mathrm{EVM}_{i,j}^2}{\sum_{i=1}^{N_{dl}}N_i}\right]^{1/2}$$

where N_i is the number of resource blocks with the considered modulation scheme in subframe i and N_{dl} is the number of allocated downlink subframes in one frame. While the above expressions for calculation of the EVM are the same for FR1 and FR2, the parameters are differently defined for the two frequency ranges [9].

The permissible EVM value can be estimated from the transmitter implementation margin, if the error vector is considered noise, which is added to the channel noise. The implementation margin is the excess power needed to maintain the carrier to noise ratio intact, when going from an ideal to a realistic transmitter design. The EVM cannot be measured at the antenna connector but should be measured by an ideal receiver with certain carrier recovery loop bandwidth specified in percent of the symbol rate. The measured EVM includes the effects of the transmitter filter accuracy, DAC, modulator imbalances, untracked phase noise, and PA nonlinearity. As mentioned earlier, the error vector magnitude is a measure of the difference between the reference waveform and the measured [transmitted] waveform. In practice, before calculating the EVM, the measured waveform is corrected by the sample timing offset and RF frequency offset, then the IQ origin offset is removed from the measured waveform. The measured waveform is further modified by selecting the absolute phase and absolute amplitude of the transmitter chain.

3.2.1.5 Carrier Frequency Offset

An OFDM system transmits information as a series of OFDM symbols. The time-domain samples $x_m(n)$ of the mth OFDM symbol are generated by performing IDFT on the information symbols $s_m(k)|_{k=0,1,\ldots,N_{FFT}-1}$, as follows [30]:

$$x_m(n) = \frac{1}{N_{FFT}}\sum_{k=0}^{N_{FFT}-1} s_m(k)e^{j2\pi k(n-N_{CP})/N_{FFT}}; \quad \forall 0 \le n \le N_{FFT}+N_{CP}-1$$

where N_{FFT} and N_{CP} denote the number of data samples and CP samples, respectively. The OFDM symbol $x_m(n)$ is transmitted through a channel $h_m(n)$ and is perturbed by a Gaussian noise $z_m(n)$. The channel $h_m(n)$ is assumed to be block-stationary, that is, it is time-invariant over each OFDM symbol. With this assumption, the output $y_m(n)$ of the channel can be represented as $y_m(n) = h_m(n)^* x_m(n) + z_m(n)$, where $h_m(n)^* x_m(n) = \sum_{k=-\infty}^{+\infty} h_m(k) x_m(k-n)$ and $z_m(n)$ is a zero-mean additive white Gaussian noise (AWGN) with variance σ_z^2. Since the channel impulse response $h_m(n)$ is assumed to be block-stationary, the channel response does not change over each OFDM symbol; however, the channel response $h_m(n)$ may vary across different OFDM symbols; thus it is a function of the OFDM symbol index m.

When the receiver oscillator is not perfectly synchronized to the transmitter oscillator, there can be a carrier frequency offset $\Delta f_{CFO} = f_{TX} - f_{RX}$ between the transmitter carrier frequency f_{TX} and the receiver carrier frequency f_{RX}. Furthermore, there may be a phase offset θ_0 between the transmitter carrier and the receiver carrier. The mth received symbol $y_m(n)$ can be represented as $y_m(n) = [h_m(n)^* x_m(n)]\exp(j\{2\pi\Delta f_{CFO}[n + m(N_{FFT} + N_{CP})]T_s + \theta_0\}) + z_m(n)$ where T_s is the sampling period. The carrier frequency offset Δf_{CFO} can be represented relative to the subcarrier bandwidth $1/(N_{FFT}T_s)$ by defining the relative frequency offset $\delta_{CFO} = \Delta f_{CFO} N_{FFT} T_s$. The carrier frequency offset attenuates the desired signal and introduces ICI, thus decreasing the SNR. The SNR of the kth subcarrier can be expressed as $\mathrm{SNR}_k(\delta_{CFO}) = \phi_{N_{FFT}}^2(\delta_{CFO}) P_h \sigma_x^2 / \left([1 - \phi_{N_{FFT}}^2(\delta_{CFO})] P_h \sigma_x^2 + \sigma_z^2\right)$ where $\phi_{N_{FFT}}(\delta_{CFO}) = \sin(\pi\delta_{CFO}) / \left[N_{FFT} sin(\pi\delta_{CFO}/N_{FFT})\right]$ in order to demonstrate the dependence of the SNR on the frequency offset. In the latter equation, P_h, σ_x^2, and σ_z^2 denote the total average power of channel impulse response, variance of the signal, and variance of the additive noise, respectively. The subcarrier index k is dropped since the SNR is the same for all subcarriers. From this SNR expression, it is clearly seen that the effect of the frequency offset is to decrease the signal power by $\phi_{N_{FFT}}^2(\delta_{CFO})$ and to convert the decreased power to interference power. The SNR depends not only on the frequency offset δ_{CFO}, but also on the number of subcarriers; however, as N_{FFT} increases, $\phi_{N_{FFT}}^2(\delta_{CFO})$ converges to $\mathrm{sinc}^2(\delta_{CFO})$. Therefore, the SNR converges to $\mathrm{SNR}(\delta_{CFO}) = \mathrm{sinc}^2(\delta_{CFO}) P_h \sigma_x^2 / ([1 - \mathrm{sinc}^2(\delta_{CFO})] P_h \sigma_x^2 + \sigma_z^2)$ as N_{FFT} becomes increasingly large. In the above equations, the power of inter-carrier interference as a function of relative carrier frequency offset is defined as $P_{ICI}(\delta_{CFO}) = [1 - \mathrm{sinc}^2(\delta_{CFO})] P_h \sigma_x^2$. In practice, the subcarrier spacing is not the same among different subcarriers due to mismatched oscillators (i.e., frequency offset), Doppler shift, and timing synchronization errors, resulting in inter-carrier interference and loss of orthogonality. It can be seen that the ICI increases with the increase of the OFDM symbol duration (or alternatively decrease of subcarrier spacing) and the frequency offset. The effects of timing offset are typically less than that of the frequency offset, provided that the CP is sufficiently large. It can be shown that the ICI power can be calculated as a function of generic Doppler power spectrum $\Lambda(\nu)$ as follows [30]:

$$P_{ICI} = \int_{-\nu_{\max}}^{\nu_{\max}} \Lambda(\nu)\left[1 - \operatorname{sinc}^2(T_s\nu)\right]d\nu$$

where $\nu_{\max}$ denotes the maximum Doppler frequency. We further assume that the transmitted signal power is normalized. It can be noted that the ICI generated as a result of carrier frequency offset is a special case of the above equation when $\Lambda(f) = \delta(f - f_{CFO})$ in which $\delta(f)$ represents the Dirac delta function. Using classic Jakes' model of Doppler spread where the spaced-time correlation function is defined as $\Phi(t) = J_0(2\pi\nu t)$ in which $J_0(x)$ denotes the zeroth-order Bessel function of the first kind, the ICI power can be written as follows:

$$P_{ICI_{Jakes}} = 1 - 2\int_0^1 (1 - f)J_0(2\pi\nu_{\max}T_s f)df$$

which approximately gives an upper bound on the ICI power due to Doppler spread. Comparison of the power of ICI generated by carrier frequency offset and Doppler spread suggests that the ICI impairment due to the former is higher than the latter.

3.2.1.6 Phase Noise

Oscillators are used in typical radio circuits to drive the mixer used for the up-conversion/down-conversion of the band-pass signal transmission. Ideally, the spectrum of the oscillator is expected to have an impulse at the frequency of oscillation with no frequency components elsewhere. However, the spectrum of a practical oscillator's output does have random variation around the oscillation frequency due to phase noise. The impact of local oscillator phase noise on the performance of an OFDM system has been extensively studied in the literature [30]. It has been shown that phase noise may have significant effects on OFDM signals with small subcarrier spacing (i.e., large OFDM symbol duration in time). Long symbol duration is required for implementing a long guard interval that can mitigate long multipath delay in single-frequency operation without excessive reduction of data throughput. The studies suggest that phase noise in OFDM systems can result in two effects: a common subcarrier phase rotation on all the subcarriers and a thermal-noise-like subcarrier de-orthogonality.

The common phase error, that is, constellation rotation, on all the demodulated subcarriers, is caused by the phase noise spectrum from DC (zero frequency) up to the frequency of subcarrier spacing. This low-pass effect is due to the long integration time of the OFDM symbol duration. This phase error can in principle be corrected by using pilots within the same symbol (in-band pilots). The phase error causes subcarrier constellation blurring rather than rotation. It results from the phase noise spectrum contained within the system bandwidth. This part of the phase noise is more crucial, since it cannot be easily corrected. The SNR degradation caused by the common phase error can be quantified as $\text{SNR}_{phase-rotation} = [I(\alpha)\beta\Delta f]^{-1}$ where Δf is the subcarrier spacing, β denotes the upper bound of the phase noise spectral mask, α is the ratio of the equivalent spectrum mask noise bandwidth and the subcarrier spacing, and $I(\alpha) = \int_{-\alpha}^{+\alpha} \text{sinc}(\pi x)dx$ with $I(0.5) = 0.774, I(1) = 0.903$, and $I(\infty) = 0.774$. It can be seen that, when $\alpha > 1$, the common phase error decreases as the subcarrier spacing decreases.

To mathematically model the effect of the phase noise, let's consider the noisy output of an oscillator which contains phase noise $\varphi(t)$ as follows

$$v(t) = A_c\cos(\omega_c t + \varphi(t))$$

Let's further assume that the stochastic variation of the phase can be modeled as the output of a system with a step function impulse response and input perturbation $n(t)$ as follows:

$$\varphi(t) = \int_{-\infty}^{t} n(\tau)d\tau$$

Based on the above assumption, the single-sided power spectral density (PSD) of the phase can be written as $S_\varphi(f) = S_n(f)/(2\pi f)^2$ in which $S_n(f)$ denotes the noise PSD function. As an example, if $S_n(f)$ is modeled as white noise, then $S_\varphi(f) \approx f^{-2}$ and if $S_n(f)$ is modeled as flicker noise, then $S_\varphi(f) \approx f^{-3}$. Considering that the PSD of the phase is difficult to observe, one may alternatively look at the PSD of the oscillator's noisy output $v(t)$. It can be shown that the PSD of $v(t)$ can be calculated as follows [30]:

$$S_v(f) = \sum_{k=-\infty}^{k=+\infty} \left(\frac{a_k^2 + b_k^2}{2}\right) \frac{\beta k^2 f_c^2}{\beta^2 \pi^2 k^4 f_c^4 + (f - kf_c)^2}$$

where $\{a_k\}$ and $\{b_k\}$ denote the Fourier series coefficients of $v(t)$ and β is a constant. Given that we are only interested in evaluating $S_v(f)$ at f_c, the above equation can be simplified as follows:

$$S_v(f) = \left(\frac{a_1^2 + b_1^2}{2}\right) \frac{\beta f_c^2}{\beta^2 \pi^2 f_c^4 + (f - f_c)^2}$$

The above function is a Lorentz distribution.[11] We now define function $\Omega(f)$ as the ratio of noise power in 1 Hz bandwidth at offset f from center frequency to carrier power which is expressed in dBc/Hz. As theoretically expected, having a higher phase noise in the signal

[11] The Cauchy distribution is a continuous probability distribution which is also known as Lorentz distribution or Cauchy–Lorentz distribution, or Lorentzian function. It describes the distribution of a random variable that is the ratio of two independent standard normal random variables, with the probability density function $f(x; 0, 1) = [\pi(1 + x^2)]^{-1}$.

does not increase the total power. A signal with higher phase noise will have smaller power near f_c and will have a broader spectrum around the center frequency. Conversely, a signal with lower phase noise has a sharper peak at the center frequency with less deviation. Therefore $\Omega(f)$ can be expressed as follows:

$$\Omega(f) = \frac{1}{\pi} \frac{\gamma}{\gamma^2 + (f - f_c)^2}, \quad \gamma = \beta \pi f_c^2$$

It can be shown that $\int_{-\infty}^{+\infty} \Omega(f)df = 1$. For higher values of β, the spectrum becomes wider with smaller magnitude of the main lobe of the spectrum. Note that a wider main lobe does not increase the total power of the carrier.

3.2.2 DFT-S-OFDM Basics and Transmission Characteristics

The LTE uplink uses SC-FDMA with CP in order to achieve inter-user orthogonality and to enable efficient frequency-domain equalization at the receiver. The DFT-spread-OFDM (DFT-S-OFDM) is a form of the single-carrier transmission technique, where the signal is generated in frequency domain, similar to OFDM as illustrated in Fig. 3.14, where the common processing blocks in OFDM and DFT-S-OFDM are distinguished from those that are specific to DFT-S-OFDM. This allows for a relatively high degree of commonality with the downlink OFDM baseband processing using the same parameters, for example, clock frequency, subcarrier spacing, FFT/IFFT size. The use of DFT-S-OFDM in the LTE uplink

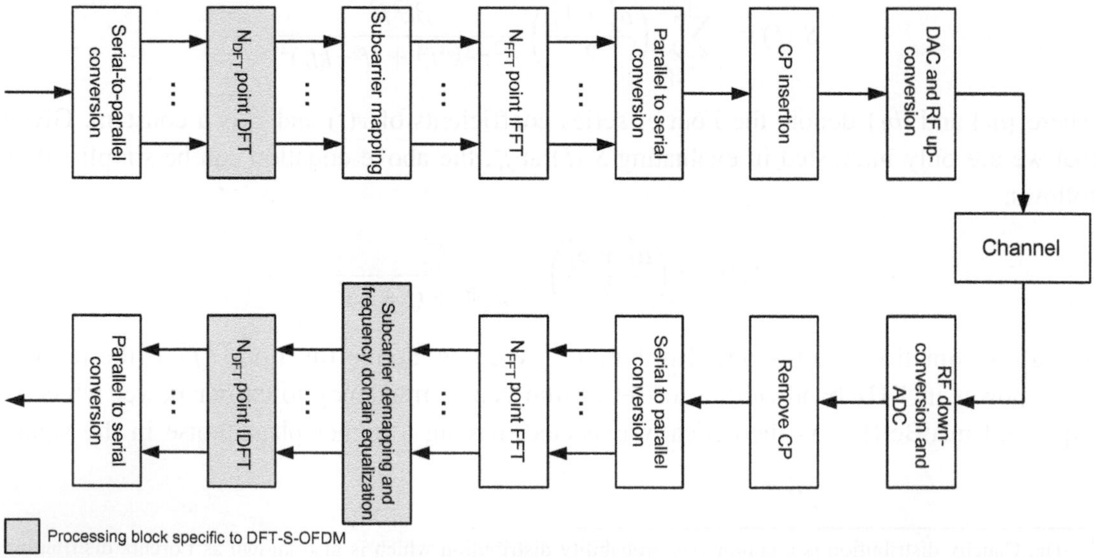

Figure 3.14
Transmitter structure for DFT-S-OFDM with localized subcarrier mapping schemes (note that $N_{DFT} < N_{FFT}$) [47].

was mainly due to relatively inferior PAPR properties of OFDM that resulted in worse uplink coverage compared to DFT-S-OFDM. The PAPR characteristics are important for cost-effective design of UE's PAs.

The principles of DFT-S-OFDM signal processing can be explained as follows. The ith transmitted symbol in a DFT-S-OFDM system without CP in single transmit/receive antenna case can be expressed as a vector of length N_{FFT} samples defined by $\mathbf{y} = \mathbf{F\Theta Dx}$ where $\mathbf{x} = (x_1, x_2, \ldots, x_M)^T$ is an $N_{DFT} \times 1$ vector with N_{DFT} QAM-modulated symbols (the superscript "T" denotes matrix transpose operation), $\mathbf{D}$ is an $N_{DFT} \times N_{DFT}$ matrix which performs N_{DFT}-point DFT operation, $\mathbf{\Theta}$ is the $N_{FFT} \times N_{DFT}$ mapping matrix for subcarrier assignment, and $\mathbf{F}$ performs N_{FFT}-point IFFT operation. After the propagation through the multipath fading channel and addition of the AWGN and removing the CP and going through the N_{FFT}-point FFT module, the received signal vector in the frequency domain can be expressed as $\mathbf{z} = \mathbf{F}^{-1}\mathbf{HF\Theta Dx} + \mathbf{w}$ where $\mathbf{H}$ is the diagonal matrix of channel response and $\mathbf{w}$ is the noise vector. Note that the maximum excess delay of the channel is assumed to be shorter than the CP; therefore, the ISI can be mitigated by the CP. The amplitude and phase distortion in the received signal due to the multipath channel is compensated by a frequency-domain equalizer (FDE) and the signal at the FDE output can be described as $\mathbf{v} = \mathbf{Cz}$ where $\mathbf{C} = \mathrm{diag}(c_1, c_2, \ldots, c_{N_{FFT}})$ is the diagonal matrix of FDE coefficients.

The FDE complex coefficients can be derived using minimum mean square error (MMSE) criterion as $c_k = H_k^* / \left(|H_k|^2 + \sigma_n^2/\sigma_s^2\right)$ where k denotes the subcarrier index, σ_n^2 denotes the variance of the additive noise, and σ_s^2 is the variance of the transmitted pilot symbol. Following the subcarrier demapping function and IDFT despreading, an $N_{DFT} \times 1$ vector $\hat{\mathbf{x}}$ containing N_{DFT} QAM-modulated symbols as an estimate to the input vector $\mathbf{x}$ is obtained at the receiver. The IDFT despreading block in the receiver averages the noise over each subcarrier. A particular subcarrier may experience deep fading in a frequency-selective fading channel. The IDFT despreading averages and spreads the fading effect, which results in a noise enhancement to all the QAM symbols. Therefore the IDFT despreading makes DFT-S-OFDM more sensitive to the noise.

As shown in Fig. 3.14, the modulation symbols in blocks of N_{DFT} symbols are processed through an N_{DFT}-point DFT processor, where N_{DFT} denotes the number of subcarriers assigned to the transmission of the data/control block. The rationale for the use of DFT precoding is to reduce the cubic metric of the transmitted signal. From an implementation point of view, the DFT size should ideally be a power of 2. However, such a constraint would limit the scheduler flexibility in terms of the amount of resources that can be assigned for an uplink transmission. In LTE, the DFT size and the size of the resource allocation is limited to products of the integers 2, 3, and 5. For example, the DFT sizes of 60, 72, and 96 are allowed, but a DFT size of 42 is not allowed [30]. Therefore, the DFT can be implemented as a combination of relatively low-complexity radix-2, radix-3, and radix-5 FFT processing blocks. The subcarrier mapping in DFT-S-OFDM determines which part of the spectrum is used for transmission by inserting a number of zeros (i.e., null subcarriers inserted between or around the data subcarriers) in the upper and/or lower end of the frequency region.

The goal of equalization is to compensate the effects of channel distortion due to frequency selectivity and to restore the original signal. One approach to signal equalization is in the time domain using a linear equalizer, which consists of a linear filter with an impulse response $w(t)$ operating on the received signal. By selecting different filter impulse responses, different receiver/equalizer strategies can be implemented. For example, the receiver filter can be selected to compensate the radio channel frequency selectivity. This can be achieved by configuring the receiver filter impulse response to satisfy $w(t)*h(t) = 1$ where the operator "*" denotes linear convolution. This method of filtering is known as zero-forcing equalization, which compensates the channel frequency selectivity. However, the ZF equalization may lead to significant increase in the noise level after equalization, degrading the overall link performance. This will be the case especially when the channel has large variations in its frequency response. Another alternative is to select a filter which provides a tradeoff between signal distortion due to channel frequency selectivity and the corruption due to noise/interference, resulting in a filter impulse response that minimizes the mean squared error between the equalizer output and the transmitted signal. The linear equalizers are typically implemented as a discrete-time FIR-filter with certain number of taps. In general, the complexity of such a discrete-time equalizer increases with increasing bandwidth of the signal [30,46].

An alternative to time-domain equalization is frequency-domain equalization which can significantly reduce the complexity of linear equalization. In this method, the equalization is performed on a block of data. The received signal is transformed to frequency domain using a DFT operation. The equalization is done as a frequency-domain filtering operation, where the frequency-domain filter $W(k)$ is the DFT of the corresponding time-domain impulse response $w(n)$. The equalized frequency-domain signal is then transformed to the time domain using an inverse-DFT operator. For processing of each signal block of size $N = 2^m$, the frequency-domain equalization would include two N-point DFT/IDFT operations and N complex multiplications.

With the introduction of a CP, the channel would appear as a circular convolution over a receiver processing block of size N. Therefore, there would be no need for overlap-and-discard in the receiver processing. Furthermore, the frequency-domain filter taps can now be calculated directly from an estimate of the sampled channel frequency response. Similar to the OFDM case, the drawback of using CP in conjunction with single-carrier transmission is the overhead in terms of extra power consumption and bandwidth. One method to reduce the relative CP overhead is to increase the block size N of the FDE. However, the accuracy of block equalization requires that the channel to be approximately constant over a period of time corresponding to the size of the processing block.

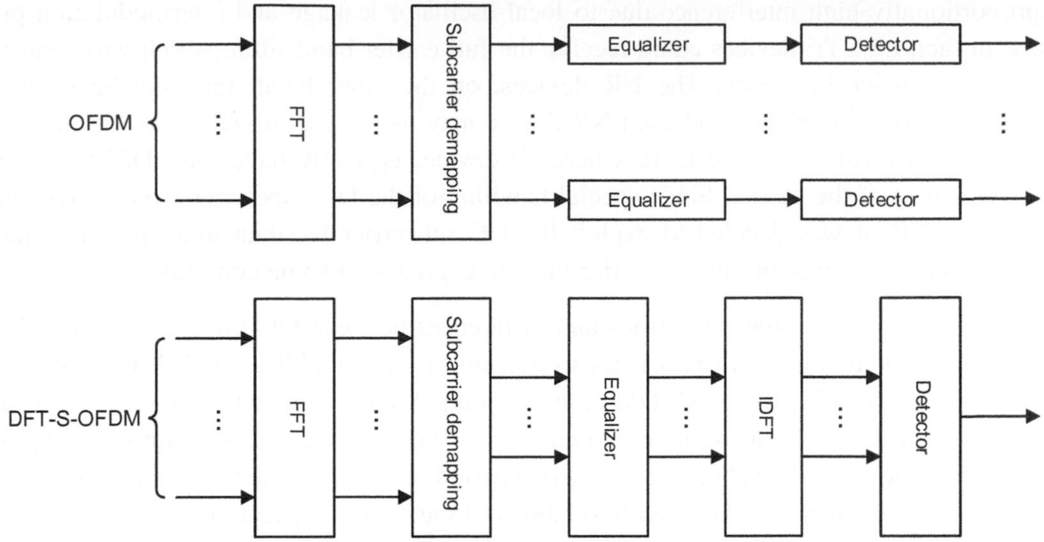

Figure 3.15
Illustration of different equalization/detection aspects of DFT-S-OFDM and OFDM [30].

The detection procedure for a DFT-S-OFDM signal is illustrated in Fig. 3.15 and is compared with that of an OFDM waveform. The transmission through a time-dispersive or equivalently a frequency-selective channel will distort the DFT-S-OFDM signal and an equalizer is needed to compensate for the effects of channel frequency selectivity. However, as shown in Fig. 3.15, a simple one-tap equalizer can be applied to each subcarrier in OFDM, whereas in the case of DFT-S-OFDM, the frequency-domain equalization function is applied to the complex-valued symbols at the output of the subcarrier demapping and prior to the IDFT operation.

It must be noted that in OFDM downlink parameterization, the DC subcarrier is unused in order to support direct conversion receiver architectures. In contrast, nulling DC subcarrier is not possible in DFT-S-OFDM since it affects the low cubic metric (CM)/PAPR property of the transmit signal. Direct conversion transmitters and receivers can introduce distortion at the carrier frequency (zero frequency or DC subcarrier in baseband) due to local oscillator leakage. In LTE downlink, this issue is avoided by inclusion of an unused DC subcarrier. However, for the uplink when using the DFT-S-OFDM waveform, the same solution may adversely impact the low CM property of the transmitted signal. In order to minimize the impact of such distortion on the packet error rate and the CM/PAPR, in LTE, the DC subcarrier of the DFT-S-OFDM signal is modulated in the same way as all the other subcarriers but the subcarriers are all frequency-shifted by half a subcarrier spacing $\Delta f/2$, resulting in an offset of 7.5 kHz relative to the DC subcarrier. Therefore two subcarriers straddle the DC location; hence, the amount of distortion affecting any individual resource block is reduced by half. In LTE, the DC subcarrier was not used because it might be subject to disproportionally high interference due to local-oscillator leakage and intermodulation products. In fact, all LTE devices could receive the full carrier bandwidth, which was centered around the carrier frequency. The NR devices, on the other hand, may not be centered around the carrier frequency and each NR device may have its own DC located at different locations in the carrier, unlike LTE, where all devices typically have their DC coinciding with the center of the carrier. Since special handling of the DC subcarrier would have been difficult in NR, it was decided to exploit the DC subcarrier for data transmission, understanding that the quality of this subcarrier may be degraded in some conditions.

In order to demonstrate the similarities and differences between OFDM and DFT-S-OFDM processing, let's assume that one wishes to transmit a sequence of eight QPSK symbols as shown in Fig. 3.16 [30]. In the OFDM case assuming $N_{DFT} = 4$, four QPSK symbols would be processed in parallel, each of them modulating its own subcarrier at the appropriate QPSK phase. After one OFDM symbol period, a guard period or CP is inserted to mitigate the multipath effects. For DFT-S-OFDM, each symbol is transmitted sequentially. With $N_{DFT} = 4$, there are four data symbols transmitted in one DFT-S-OFDM symbol period. The higher rate data symbols require four times the bandwidth and so each data symbol occupies $N_{DFT} \times \Delta f$ Hz of spectrum assuming a subcarrier spacing of Δf Hz. After four data symbols, the CP is inserted. Note the OFDM and DFT-S-OFDM symbol periods are the same [30].

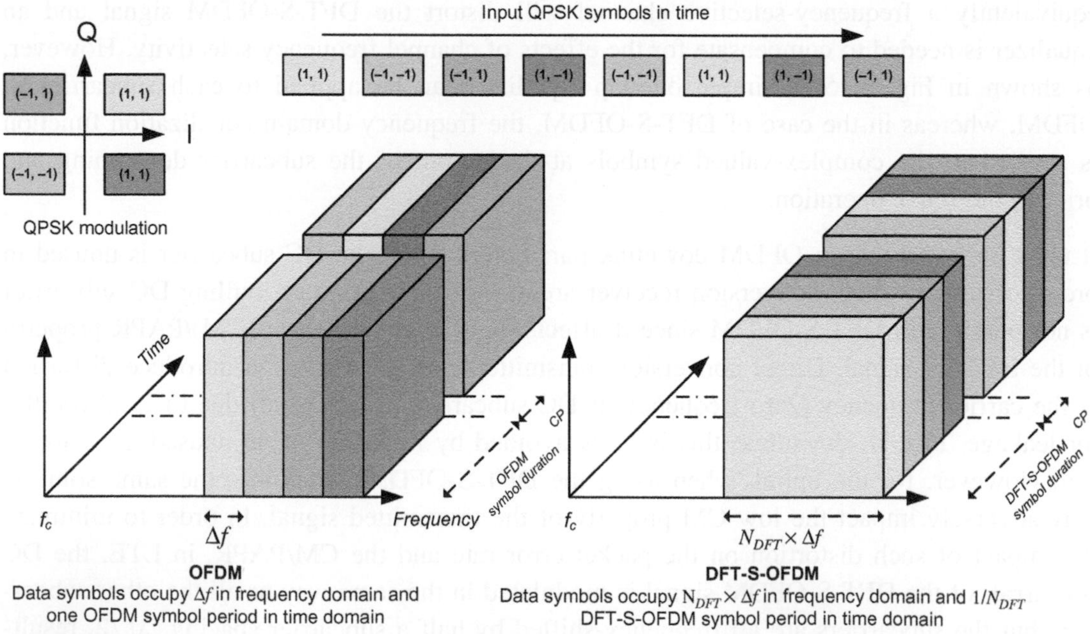

Figure 3.16
Comparison of OFDM and DFT-S-OFDM using QPSK modulation with $N_{DFT} = 4$ [30].

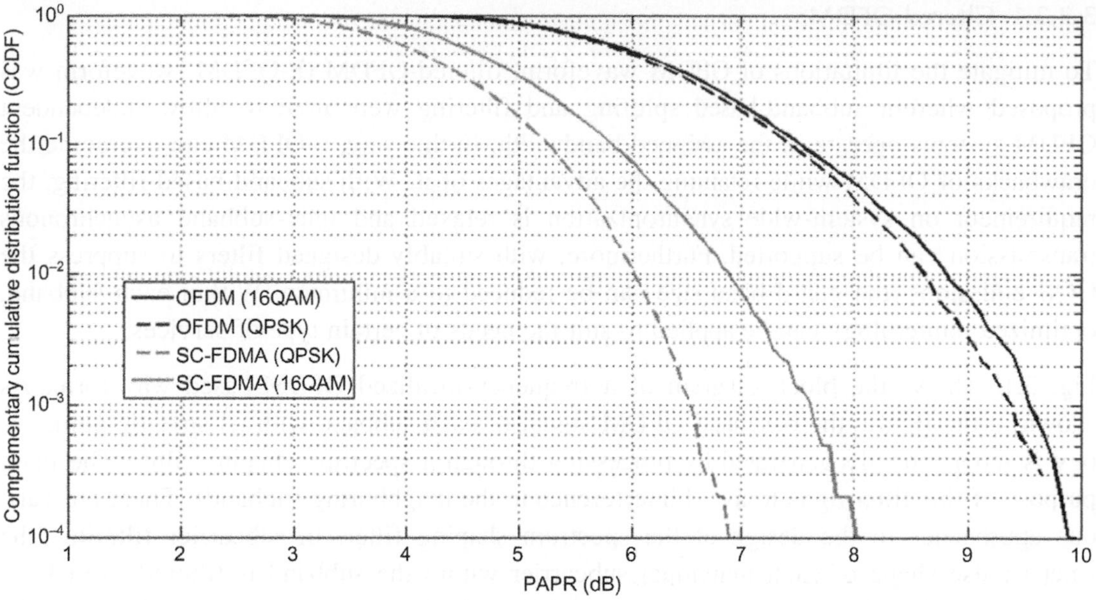

Figure 3.17
Comparison of OFDM and DFT-S-OFDM PAPRs (5 MHz bandwidth) [30,46].

As mentioned earlier, the PAPR of OFDM intrinsically is inferior to DFT-S-OFDM. Fig. 3.17 shows the comparison of complementary CDF (CCDF) of OFDM and DFT-S-OFDM PAPRs. It can be seen that the PAPR of DFT-S-OFDM is approximately 3 dB better than that of OFDM with probability of 0.99. In the case of 16QAM modulation, the PAPR of DFT-S-OFDM increases relative to that of DFT-S-OFDM with QPSK modulation, whereas in the case of OFDM, the PAPR distribution is independent of the modulation scheme because the OFDM signal is the sum of a large number of independently modulated subcarriers; thus the instantaneous power has an approximately exponential distribution, regardless of the modulation scheme applied to different subcarriers [30,46].

3.2.3 Other Waveform Candidates

In the study phase of 3GPP NR, a number of waveforms promising to improve upon OFDM waveform and to overcome the limitation of the latter were proposed and evaluated. However, when the practical aspects of implementation complexity and analog RF processing were considered, many of those candidate waveforms fail to provide any significant improvement over the status-quo, and thus 3GPP agreed to specify the OFDM as the baseline waveform for the new radio. In the following sections, we briefly describe those candidate waveforms and their respective advantages and disadvantages over OFDM.

3.2.3.1 Filtered-OFDM

To mitigate the limitations of OFDM waveform, filtered-OFDM (F-OFDM) waveform was proposed wherein subband-based splitting and filtering were used to allow independent OFDM systems operate in the assigned bandwidth. In this way, F-OFDM can overcome the drawbacks of OFDM while retaining the advantages of it. With subband-based filtering, the requirement on system-wide synchronization is relaxed and inter-subband asynchronous transmission can be supported. Furthermore, with suitably designed filters to suppress the OOB emissions, the guard band size can be reduced to a minimum. Within each subband, optimized numerology can be applied to suit the needs of certain type of services.

Fig. 3.18 shows the block diagram of a frequency-localized OFDM-based waveform. As shown in the figure, the baseband OFDM signal of each subband with its specific numerology is independently generated by processing through a spectrum shaping filter. The main purpose of this filtering is to avoid interference to the neighboring subbands. There are various approaches to the design of the spectrum shaping filter. In subcarrier filtering, the sinc(.) pulse shape of each individual subcarrier within the subband is filtered to make it more localized in frequency. An example of this method is the windowed OFDM where the subcarrier filtering is performed in the time domain by modifying the rectangular pulse shape of CP-OFDM to have smoother transitions in time at both ends. In an alternative approach known as subband filtering, the PSD of the entire subband is made well-localized without changing the CP-OFDM symbol's rectangular pulse. For this purpose, the subband CP-OFDM signal is passed through a frequency-localized filter whose bandwidth is close to the size of the subband. As a result, only a few subcarriers close to edges of the subband in frequency domain are affected by the filter, as the filter suppresses their out-of-subband side-lobes. This leads to F-OFDM signal generation. A key property of this approach is that the filter length can exceed the CP length, which allows better frequency localization than the subcarrier-based approach without causing any ISI. Although the subcarrier-based approach provides a lower complexity, it cannot achieve the frequency localization

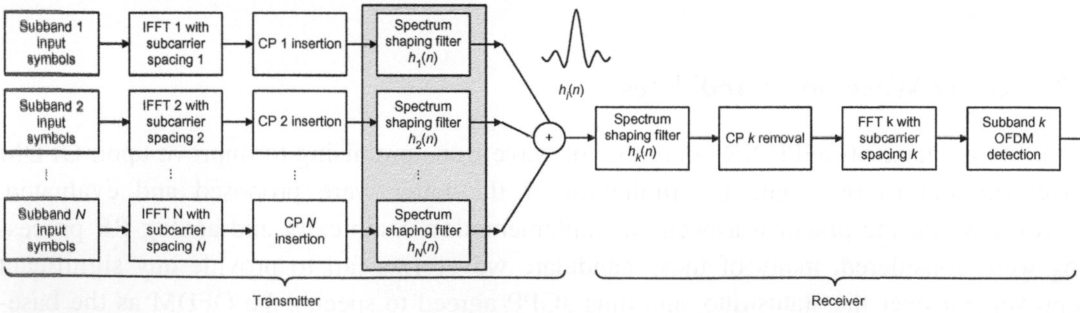

Figure 3.18
Illustration of transmit/receive processing for frequency-localized OFDM-based waveforms [49].

performance of the subband-based approach, and further causes ISI in multipath channels with large delay spread. A combination of the two approaches can provide a tradeoff between complexity and frequency localization. In particular, in the latter approach, the CP-OFDM signal of each subband is first subcarrier-filtered with an excess window length smaller than that of the original subcarrier-based approach. Then, the windowed signal is subband filtered with a filter length smaller than that of the original subband-based approach.

In the receiver side, in order to filter out the signals of the neighboring subbands, the received signal in baseband is first passed through the receiver spectrum shaping filter. Subcarrier-based, subband-based, or a composite approach can be employed by the receiver, independent of the approach employed by the transmitter. After the subband spectrum shaping, the resulting signal is processed by the regular OFDM processing within that subband.

The filters used in F-OFDM processing must satisfy a number of criteria. The passband of the filter should be as flat as possible over the subcarriers contained in the subband. This ensures that the distortion caused by the filter in the data subcarriers, especially the subband edge subcarriers, is minimal. The frequency roll-off of filter should start from the edges of the passband and the transition band of the filter should be sufficiently steep. This ensures that the system bandwidth is utilized as efficient as possible and the guard band overhead is minimized. Also, the neighboring subband signals with different numerologies can be placed next to each other in frequency with minimal number of guard subcarriers. The filter should further have sufficient stop-band attenuation to ensure that the leakage into the neighboring subbands is negligible.

The F-OFDM waveform processing introduces negligible delay at the receiver side. Signal processing delay of F-OFDM depends on the receiver processing capabilities. It should be noted that the only extra signal processing block in F-OFDM receiver compared to CP-OFDM is the receiver subband spectrum shaping filter. The delay due to spectrum shaping filter is implementation-specific and depends on the receiver processing capabilities. The rest of receiver processing blocks in F-OFDM, for example, FFT block size, channel estimation/equalization are the same as those in CP-OFDM [49]. The F-OFDM concept can be used in the asynchronous access of multiple UEs in the uplink as shown in Fig. 3.19.

3.2.3.2 Filter Bank Multicarrier

Filter bank multicarrier (FBMC) is an OFDM-based waveform wherein subcarriers are individually processed through filters that suppress their side-lobes, making them strictly band-limited. The transmitter and receiver may still be implemented through FFT/IFFT blocks or polyphase filter structures and band-limitedness may offer larger spectral efficiency than OFDM. During the study of waveforms, FBMC was found promising mainly due to signal band-limitedness in order to relax synchronization requirements in the uplink and/or in the

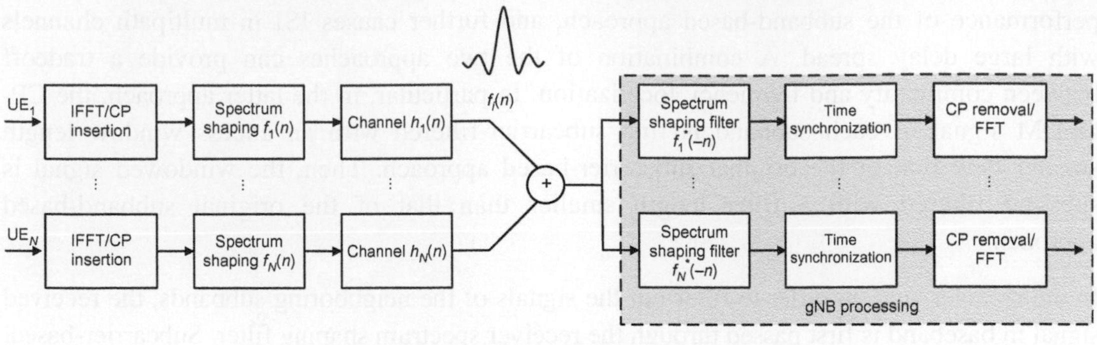

Figure 3.19
F-OFDM uplink asynchronous communication.

downlink with coordinated transmission, its greater robustness to frequency mis-alignments among users when compared to OFDM, and its more flexible exploitation of frequency white spaces in cognitive radio networks. The rectangular impulse adopted in OFDM systems is not well-localized in time and frequency, making it sensitive to timing and frequency offsets (e.g., introduced by channel, or local oscillator mismatch). As we discussed earlier, ideal time and frequency well-localized pulse does not exist in practice for the conventional OFDM according to Balian–Low theorem.[12] However, if pulse amplitude modulation (PAM) symbols instead of QAM symbols are considered, time and frequency well-localized pulse can be achieved in a multicarrier system called FBMC. The transmit signal can be expressed as

$$x(t) = \sum_{k=-\infty}^{\infty} \sum_{n=0}^{N_{\text{sub}}-1} j^{n+k} s_{k,n} g\left(t - kT/2\right) e^{j(2\pi/T)nt}$$

where $g(t)$ is a square-integrable function on real domain (Gabor set), which is manifested as the rectangular pulse in OFDM, and $s_{k,n}$ denotes real-valued data symbols.

In FBMC, the pulse $g(t)$ can be designed to achieve better time and frequency localization properties using filter design methods. Usually, the prototype filter $g(t)$ spans an integer K (overlapping factor) multiple length of symbol period $T_F = KT$. It must be noted that real

[12] In mathematics, the Balian–Low theorem in Fourier analysis states that there is no well-localized window function or Gabor function either in time or frequency domain for an exact Gabor frame. Let g denote a square-integrable function on the set of real numbers, and consider the so-called Gabor system $g_{m,n}(x) = g(x - na)e^{2\pi jmbx}$ for integers m and n, and $a, b > 0$ satisfying $ab = 1$. The Balian–Low theorem states that if $\{g_{m,n}: m, n \in \mathbf{Z}\}$ is an orthonormal basis for the Hilbert space $L^2(\mathbb{R})$, then either $\int_{-\infty}^{\infty} x^2 |g(x)|^2 dx = \infty$ or $\int_{-\infty}^{\infty} \phi^2 \left|\widehat{g}(\phi)\right|^2 d\phi = \infty$[Note : $\widehat{g}(\phi)$ is the Fourier transform of $g(x)$]. The Balian–Low theorem has been extended to exact Gabor frames.

and imaginary data values alternate on subcarriers and symbols, which is called offset QAM (OQAM). Since PAM symbols convey only one half of information content compared to QAM symbols, a data rate loss factor 2 is implicit. Nevertheless, the symbol period in FBMC is also halved to $T/2$ in order to compensate for the efficiency loss of OQAM modulation. Furthermore, CP is not essential anymore in FBMC due to the well-localized pulse shape. Filtering can use different overlap factors (i.e., K factor) to provide varying levels of OOB rejection. As K factor is reduced, the OOB characteristics have a spectrum-rejection profile similar to that of OFDM. The critical step for FBMC design is to implement filters for each subcarrier and to align multiple filters into a filter bank. One way to implement the filter bank is to design a prototype filter. Once the prototype filter is designed, the next step is to make a copy of the prototype filter and shift it to neighboring subcarriers as illustrated in Fig. 3.20.

The comparison of the PSDs of the OFDM and FBMC signals is shown in Fig. 3.21. The FBMC signal was processed with overlapping factors $K = 2$ and 3. It is shown that with the increase of the overlapping factor, the OOB emissions of the FBMC signal significantly decreases; however, the signal processing complexity and latency prohibitively increases relative to that of the OFDM processing.

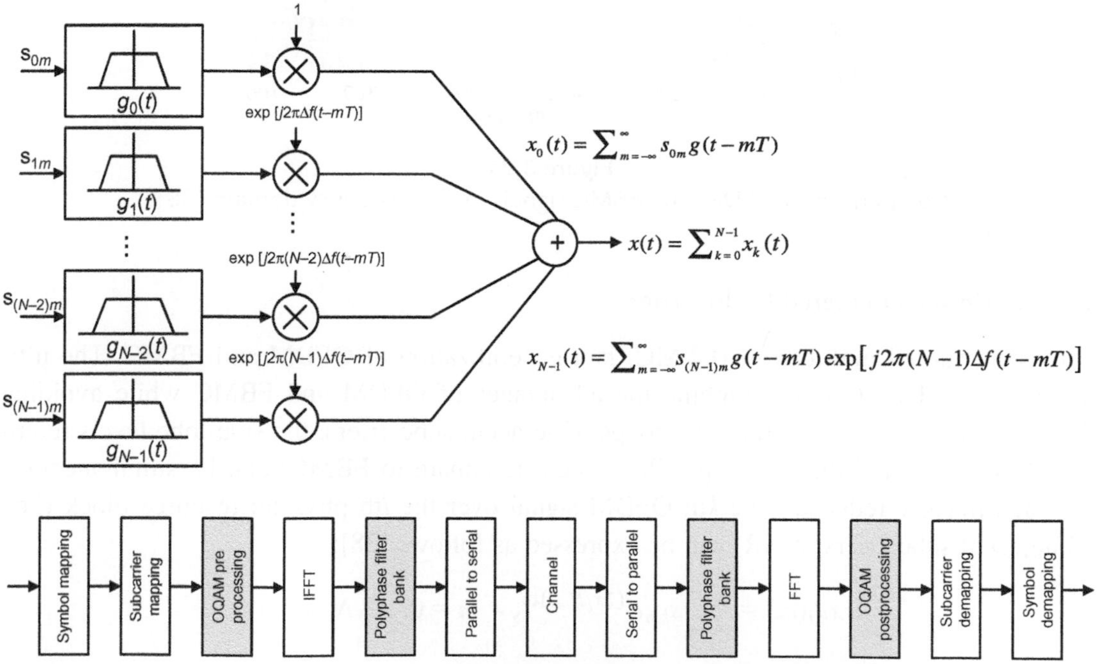

Figure 3.20
Illustration of FBMC concept and transmitter/receiver architecture [47].

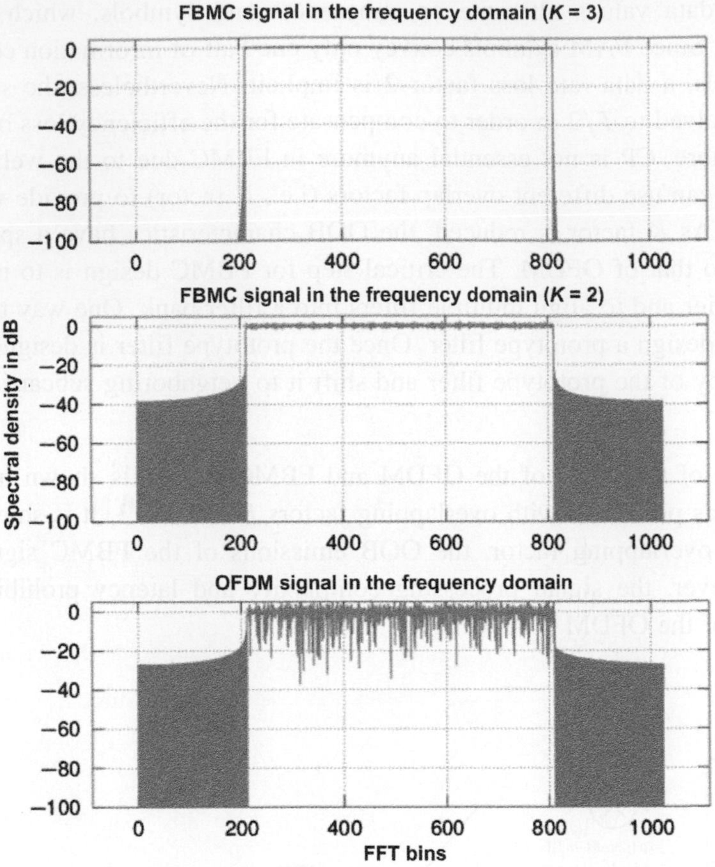

Figure 3.21
Comparison of OFDM and FBMC signals in the frequency domain [38].

3.2.3.3 Universal Filtered Multicarrier

Universal filtered multicarrier (UFMC) is a generalization of OFDM and FBMC. The ultimate goal of UFMC is to combine the advantages of OFDM and FBMC while avoiding their main drawbacks. By filtering groups of adjacent subcarriers, the side-lobe levels (compare to OFDM) and the prototype filter length (compare to FBMC) can be simultaneously and significantly reduced. The kth OFDM signal over the ith physical resource block (i.e., 12 adjacent subcarriers in NR) can be expressed as follows [38]:

$$x_{k,i}(m) = \sum_{n \in S_i} s_{k,n} e^{j(2\pi/N_{\text{sub}})kn}, \quad m = 0, \ldots, N_{\text{sub}} - 1$$

where S_i is a set which contains consecutive subcarrier indices that are assigned to the ith physical resource block. This signal is then filtered by an FIR-filter $f_i(n)$ with the length of

L_F. The UFMC scheme applies filtering on a per subband basis, reducing complexity of the baseband processing algorithms. Thus, the kth transmit symbol can be written as

$$\tilde{x}_k(m) = \sum_i \sum_{l=0}^{L_F-1} f_i(l) x_{k,i}(m-l), \quad m = 0, \ldots, N_{\text{sub}} + L_F - 2$$

The FIR-filter can be differently designed for each physical resource block. Let's assume that we use an identical Chebyshev filter with variable side-lobe attenuation for all physical resource blocks and the filter is shifted to the center frequencies of the physical resource blocks. The filter ramp-up and -down regions at the beginning and the end of individual UFMC symbols provide somewhat ISI protection, in the presence of channel delay spreads and timing offsets. With very high delay spreads, sophisticated multi-tap equalizers must be applied.

There is no time overlap between subsequent UFMC symbols. The symbol duration is $N_{sub} + L_F - 1$ with N_{sub} being the FFT size of the IFFT spreaders (the size of the subbands) and L_F the length of the filter. Similar to FBMC, in UFMC typically the FFT window size is increased, resulting in a higher implementation complexity. Also in UFMC the insertion of a guard interval or CP is optional. Another feature of the unified frame structure is the usage of multiple signal layers. The users can be separated based on their interleavers as it is done in interleave division multiple access scheme. This will introduce an additional degree of freedom for the system, improve robustness against cross-talk, and help to exploit the capacity of the multiple access channel (uplink) [45].

The comparison of the PSDs of the OFDM and UFMC signals is shown in Fig. 3.22. In the processing of the UFDM signal, a Chebyshev filter with $L_F = 74$ and side-lobe level attenuation of 40 dB has been used. Furthermore, we assume $N_{FFT} = 1024$ and $N_g = 0$. It is shown that while the relative complexity of UFMC is more manageable than FBMC, the OOB components are significantly more suppressed compared to that of the OFDM signal.

An alternative mathematical representation of the UFDM signal generation and processing can be given as follows:

$$\underbrace{\mathbf{x}[k]}_{[(N+L_F-1),1]} = \sum_{n=1}^{N_{SB}} \underbrace{\mathbf{F}_{n,k}}_{[N+L_F-1),1]} \underbrace{\mathbf{V}_{n,k}}_{[N,N_n]} \underbrace{\mathbf{S}_{n,k}}_{[N_n,1]}$$

where N, L_F, N_{SB}, and N_n denote the FFT size, the filter length, the number of subbands, and the number of complex QAM symbols, respectively; $[n, k]$ represents the subband index and user number; $\mathbf{F}_{n,k}, \mathbf{V}_{n,k}$, and $\mathbf{S}_{n,k}$ denote a Toeplitz matrix comprising the filter impulse response, an IDFT matrix corresponding to the subband location, and a symbol matrix, respectively. The above mathematical model is illustrated in Fig. 3.23.

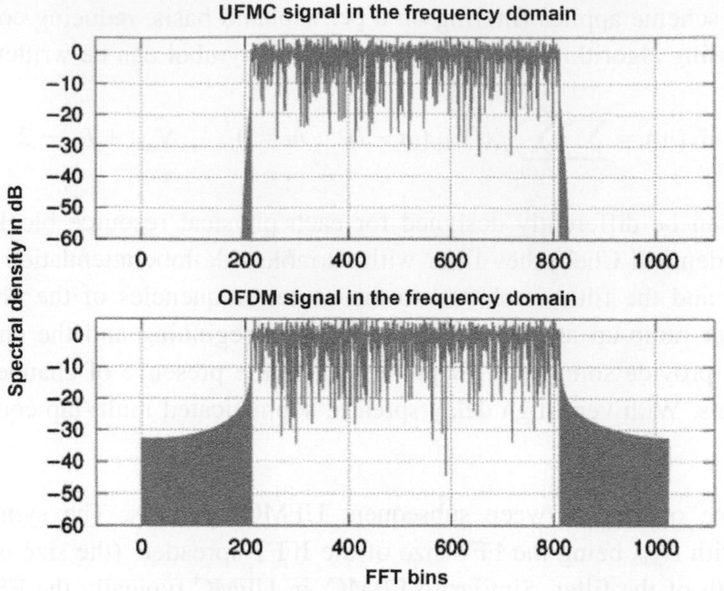

Figure 3.22
Comparison of OFDM and UFMC signals in the frequency domain [38].

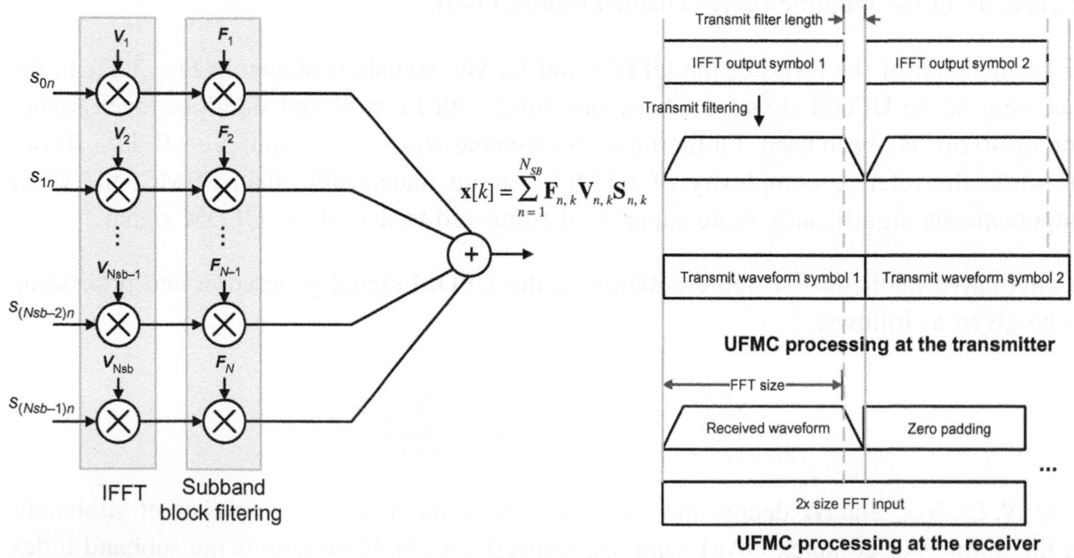

Figure 3.23
UFDM signal processing [42,47].

3.2.3.4 Generalized Frequency Division Multiplexing

Generalized frequency division multiplexing (GFDM) is a flexible multicarrier modulation scheme. The process is performed block-by-block, where each GFDM block consists of a number of K subcarriers and M sub-symbols. By setting the number of subcarriers and the number of sub-symbols to one, GFDM reduces to single-carrier frequency domain equalization and CP-OFDM as its special cases. Furthermore, pulse shaping with a prototype filter $g_{0,0}(n)$ is another flexibility in GFDM to reduce OOB emissions. In contrast to linear convolution used in FBMC, GFDM uses circular convolution. Let $g_{k,m}(n)$ denote the pulse-shaping filter corresponding to the data symbol $s_{k,m}$ that is transmitted at subcarrier m and time k. It can be shown that

$$g_{k,n}(n) = g_{0,0}[(n - mK)\mathrm{mod}N]e^{j2\pi kn/K}, \quad N = M \times K$$

In the above equation N denotes the number of symbols within a GFDM block (GFDM block size). Thus, the time-domain signal $x(n)$ of a GFDM block is expressed as

$$x(n) = \sum_{m=0}^{M-1}\sum_{k=0}^{K-1} s_{k,m}g_{k,m}(n), \quad n = 0, 1, \ldots, N-1$$

A CP and a cyclic suffix can be optionally added in the GFDM data block. Furthermore, a raised-cosine filter with configurable roll-off factor β is used for filtering. The GFDM signal processing stages are illustrated in Fig. 3.24.

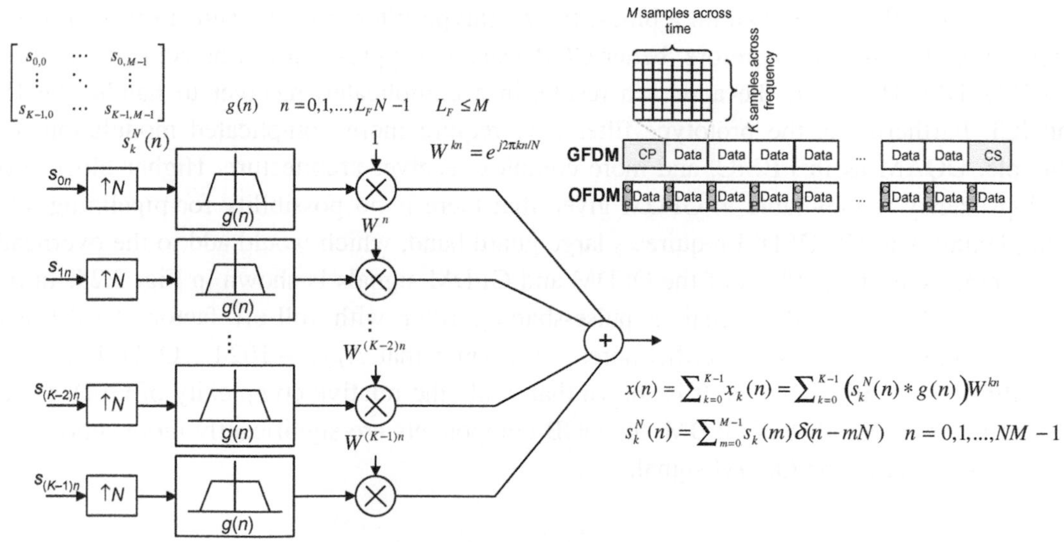

Figure 3.24
Illustration of GFDM signal processing [45,47].

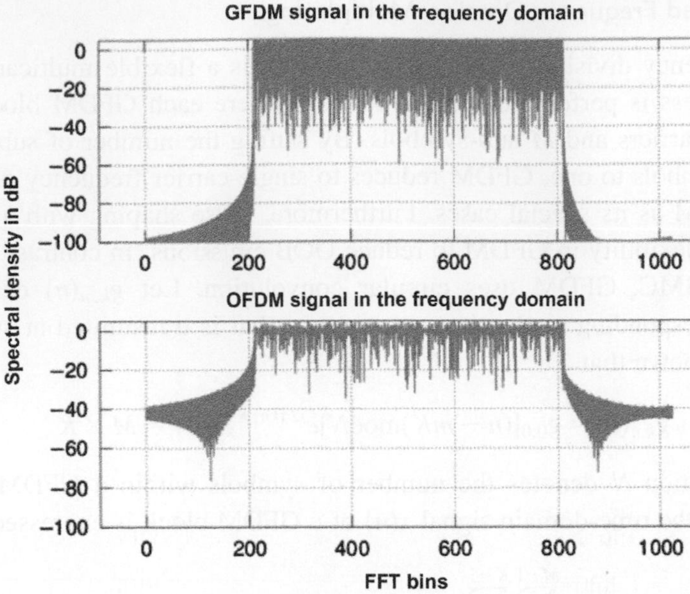

Figure 3.25
Comparison of OFDM and GFDM signals in the frequency domain [38].

In one aspect, GFDM is similar to FBMC where a prototype filter is used to suppress OOB emissions. However, for GFDM, multiple OFDM symbols are grouped into a block and a CP is added to the block. Within a block, the prototype filter is cyclic-shift in time, for different OFDM symbols. Therefore, better OOB leakage suppression can be achieved relative to CP-OFDM. However, the approach results in a complicated receiver to handle the ISI and ICI. Furthermore, the prototype filter may require more complicated modulation, for example, OQAM as in FBMC, and more complex receiver architecture. Higher block processing latency is inevitable in GFDM given that there is no possibility for pipelining, and multiplexing with CP-OFDM requires a large guard band, which would add to the overhead. The comparison of the PSDs of the OFDM and GFDM signals is shown in Fig. 3.25. In the processing of the GFDM signal, a pulse-shaping filter with roll-off factor $\beta = 0.1$ and $N = 10$ symbols were used. Furthermore, we assume that $N_{FFT} = 1024$, OFDM $N_g = 10$ and GFDM $N_g = 100$ samples. It is shown that while the relative complexity of GFDM processing is more than that of OFDM, the OOB components are significantly more suppressed compared to that of the OFDM signal.

3.2.3.5 Faster Than Nyquist Signal Processing

Faster than Nyquist (FTN) is a non-orthogonal transmission scheme which was one of the approaches initially considered for 5G systems that was expected to improve the spectrum efficiency by increasing the data rate. It was observed a few decades ago that the binary sinc pulse can be transmitted faster than what the Nyquist theorem states without increasing the bit error rate and despite of ISI. The idea was extended to frequency domain to reduce subcarrier spacing. The transmit signal of FTN can be expressed as follows [4]:

$$x(t) = \sum_{k=-\infty}^{\infty} \sum_{n=0}^{N_{\text{sub}}-1} s_{k,n} g(t - k\Delta_T T) e^{j2\pi\Delta_F nt/T}$$

where $\Delta T \leq 1$ is the time compression factor, which means that the pulses are transmitted faster a factor of $1/\Delta T$ and $\Delta F \leq 1$ is the frequency compression factor, which means the spectral efficiency is increased by a factor of $1/\Delta F$. The FTN transmitter structure is depicted in Fig. 3.26. Since the time and frequency spacing varies for different FTN systems, direct implementation cannot provide sufficient implementation flexibility. An FTN mapper based on projection scheme has been designed and used in FTN signaling systems to provide flexibility. The FTN mapper is shown in Fig. 3.26. A cyclic extension is needed in the modulation block, with which the system can switch easily between FTN and Nyquist modes. As mentioned before, FTN signaling inevitably introduces ISI in the time domain and/or ICI in the frequency domain when its baud rate is over the Nyquist one (see Fig. 3.27). Therefore a very important issue is how to design a receiver with ISI and/or ICI suppression capability to recover the original transmitted data.

3.2.3.6 Comparison of the Candidate Waveforms

A number of candidate waveforms were studied by 3GPP for NR, before CP-OFDM was selected as the default waveform for the downlink and uplink, which were based on FFT/IFFT processing, additional filtering, windowing, or precoding, in order to achieve higher time-frequency localization and lower OOB spectral leakage, and higher throughput.

Filtering is a straightforward way to suppress OOB emissions by applying a digital filter with prespecified frequency response. Some waveforms like F-OFDM and UFDM belong to this category. However, the delay spread of the equivalent composite channel may exceed the CP size and guard period in TDD systems, which results in ISI and imposes restrictions on downlink-to-uplink switching time. Furthermore, the promised OOB emission performance may diminish significantly when PA nonlinearity and other non-ideal RF processing effects are taken into account. At the cost of increased PAPR, filtering techniques are generally known to be unfriendly to communication at high carrier frequencies.

Windowing is used to prevent steep changes across two consecutive OFDM symbols in order to reduce the OOB emissions. Multiplying the time-domain samples located in the extended symbol edges by raised-cosine window coefficients is a widely used realization as chosen by windowed OFDM and weighted overlap-and-add OFDM waveforms. This

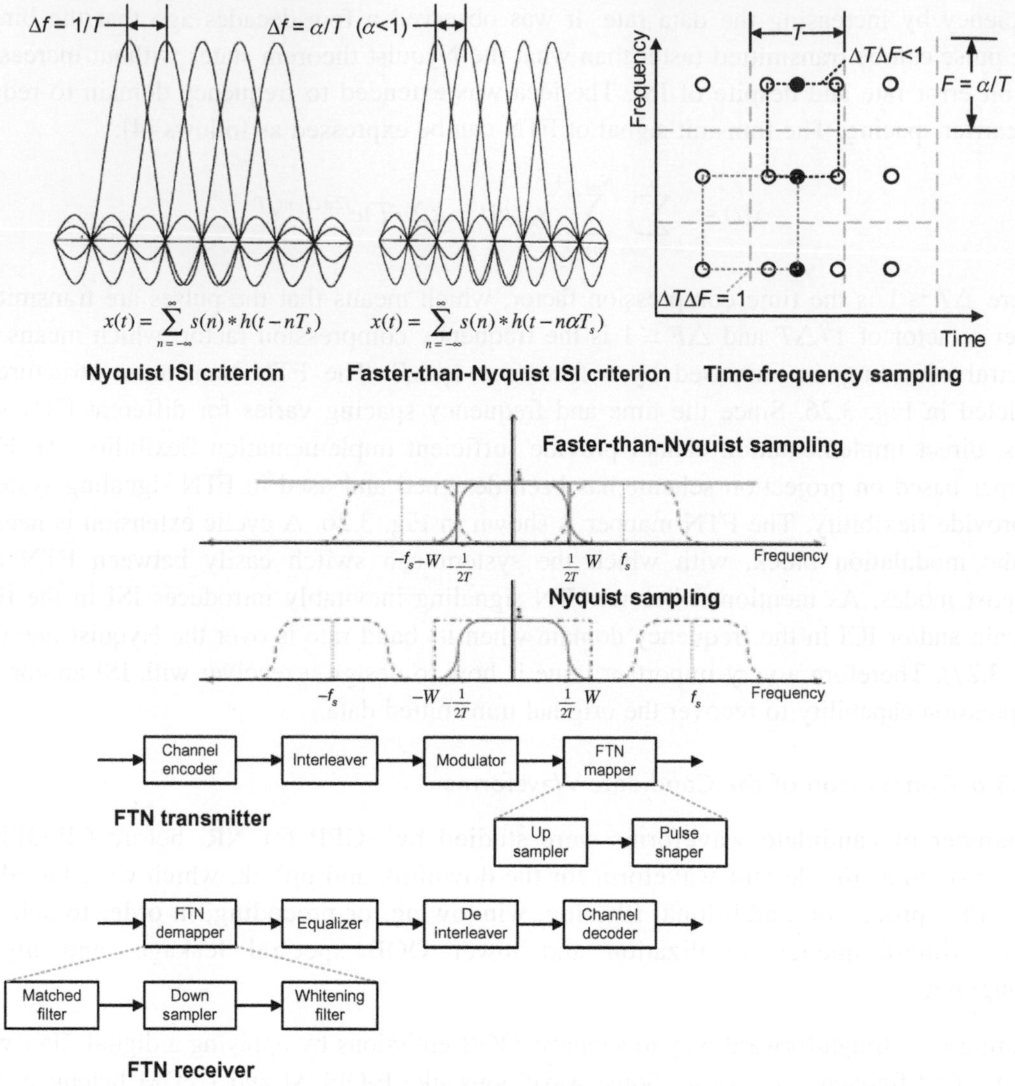

Figure 3.26
Illustration of FTN concept and transmitter and receiver architecture [44].

technique generally has little or no effect on PAPR increase and has lower complexity compared to that of filtering techniques. Nevertheless, the detection performance might be degraded because of ISI caused by symbol extension.

The linear processing of input data prior to IFFT is usually known as precoding, and may be helpful to improve OOB emissions and PAPR reduction. One example is DFT-S-OFDM

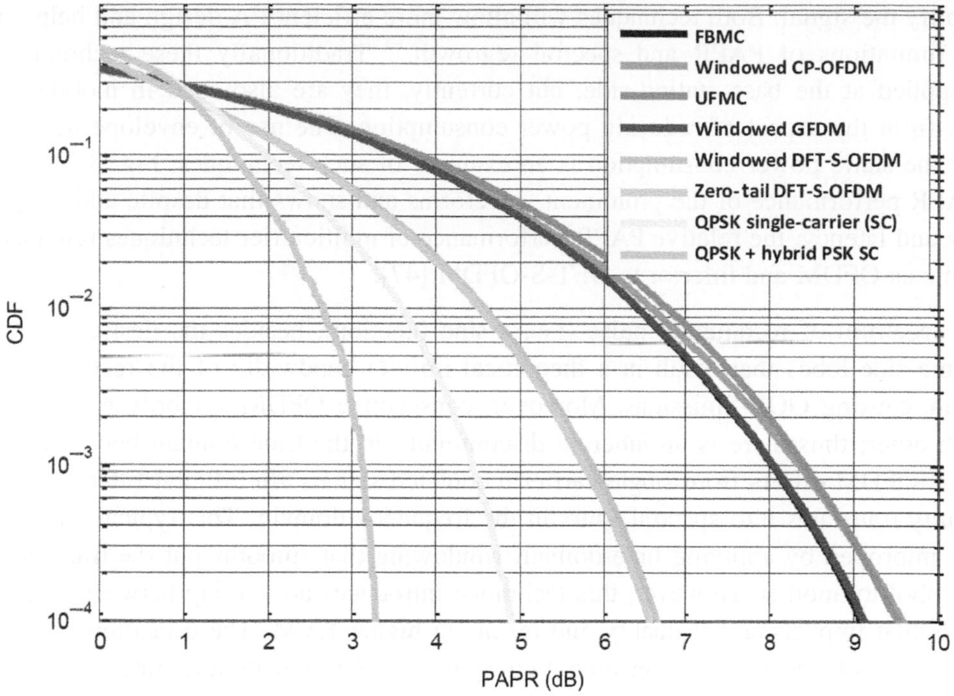

Figure 3.27
Comparison of PAPR performance of some prominent variants [47].

waveform that was adopted in LTE uplink transmission because of its low PAPR properties relative to conventional OFDM. Some variants of DFT-S-OFDM were proposed for NR such as zero-tail (ZT) DFT-S-OFDM by omitting the CP and letting the tail samples taper to zero. The DFT-S-OFDM-based waveforms, in contrast to filter-based waveforms, usually make it easier to maintain PA linear operation with less deterioration from lowering OOB emissions. Moreover, an appropriate modification of modulation schemes, such as $\pi/2$-BPSK can greatly assist such waveforms in achieving an extremely low PAPR. Note that in the absence of redundant intervals, ISI can still occur. Among DFT-based precoding techniques, other types of precoding matrices often have undesirable complexity and compatibility issues.

PAPR is one of the often-mentioned disadvantages of OFDM waveform. In practice, the crest factor reduction (CFR) techniques are applied to reduce the PAPR and digital predistortion algorithms will then correct for any distortion caused by the analog hardware used to amplify the signal. Both techniques will allow more efficient PA design and help mitigate major limitations of PAPR and spectral regrowth.[13] Traditionally these techniques were only applied at the base station side, but currently, they are also used in mobile devices, mainly from the aspect of reducing power consumption. The use of envelope tracking[14] to reduce the static power consumption is an example of such techniques. Fig. 3.27 compares the PAPR performance of the prominent waveforms and shows that despite additional complexity and latency, the relative PAPR performance of multicarrier techniques remains about the same as OFDM and inferior to DFT-S-OFDM [47].

In the spectrum of rectangular pulse (i.e., a sinc function), besides the desired peak, there are some side-lobes that result in a theoretical infinite bandwidth of the rectangular pulse function, causing OOB emissions. Moreover, consecutive OFDM symbols are independent of each other; thus there is an inherent discontinuity in the time domain between them. In this way OFDM differs from single-carrier modulated signals after digital filtering. This discontinuity translates into spectral spurs in the frequency domain. This typical characteristic can be improved by applying time-domain windowing that smooths out the transition from one symbol to another. However, this technique introduces an overlap between consecutive symbols that impacts signal quality and results in higher EVM. The transition time defines the duration of the overlap between two symbols. For a sampling rate of 30.72 MHz (20 MHz LTE signal), a transition time of 1 μs translates into 30 samples overlap.

Fig. 3.28 shows the ideal case by means of connecting the signal generator directly to a spectrum analyzer. The idea is to demonstrate the impact of a nonlinear device on the highlighted advantage of 5G waveform candidates, where any nonlinearity will result in a spectral regrowth, and there is a risk that this spectral regrowth may undermine any optimization due to the waveform design. The improved spectral characteristics of the candidate waveforms are clearly visible in the top snapshot. In a second step, a nonlinear amplifier is introduced into the signal path. The top and bottom snapshots in Fig. 3.28 compare the LTE OFDM signal with FBMC, UFMC, and GFDM waveforms under two different conditions. A generic PA, which supports a frequency range of 50 MHz to 4 GHz, is used to demonstrate the effects of nonlinearity on the waveforms. The maximum input power for the PA is 0 dBm, and it has a typical

[13] Intermodulation products or spurs can develop within the analog and digital transmitters in combined systems using high-level injection. In some cases spurs can result in suboptimal signal quality or even cause stations to interfere with each other's signals. The term spectral regrowth is used to describe intermodulation products generated when a digital transmitter is added to an analog transmission system.

[14] Envelope tracking is an approach to RF amplifier design in which the power supply voltage applied to the RF power amplifier is continuously adjusted to ensure that the amplifier is operating at peak efficiency for power required at each instant of transmission. A conventional RF amplifier is designed with a fixed supply voltage and operates most efficiently only when operating in compression region. Amplifiers operating with a constant supply voltage become less efficient as the crest factor of the signal increases, because the amplifier spends more time operating below peak power and, therefore, spends more time operating below its maximum efficiency.

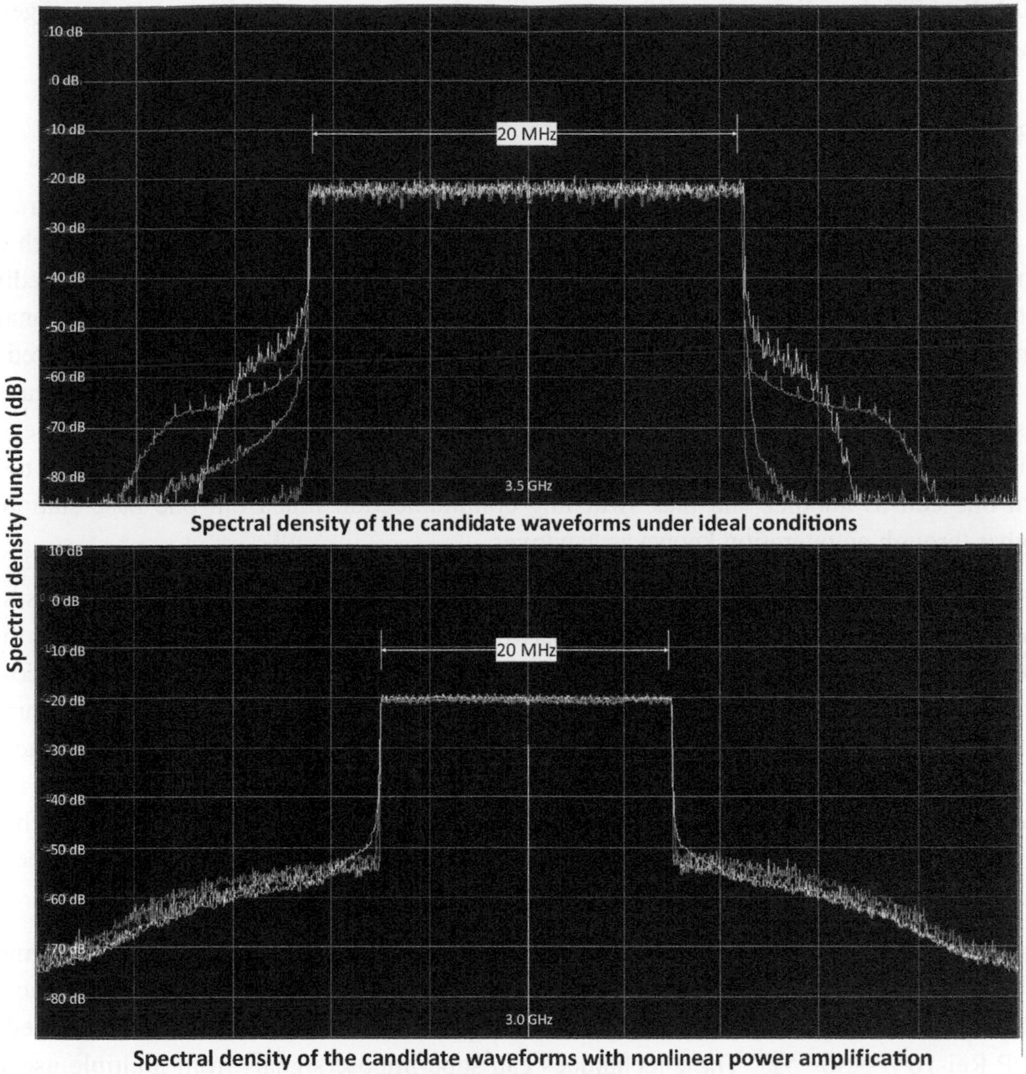

Figure 3.28

Comparison of a 20 MHz LTE downlink OFDM signal (yellow) with FBMC (blue), UFMC (green), and GFDM (orange) signals under ideal (top) and non-ideal conditions [45]. For color interpretation of this figure, please refer web version.

gain of 20 dB. Maximum achievable output power for the PA is +20 dBm. At 0 dBm input power, the PA starts to enter the saturation region. Higher input power would mean that the PA would be operating in the compression region. Fig. 3.28 further shows the result of a measurement with the same signal configuration for an input power of −2 dBm to the amplifier. The spectral advantages of the candidate waveforms seem to have almost disappeared compared to a 20 MHz LTE downlink signal. When using typical input power of 0 dBm, the advantages of non-OFDM waveforms will completely vanish [45].

3.3 Multiple-Access Schemes

A cellular network consists of a number of fixed base stations distributed across a geographical area. The coverage area is divided into cells and a mobile station communicates with one or more base stations in its proximity. There are two main issues in the physical and medium access layers of cellular communication schemes: multiple access and interference management. The first issue addresses how the overall radio resources of the system are shared by the users in the same cell (intra-cell) and the second issue addresses the interference caused by simultaneous signal transmissions in different cells (inter-cell). At the network layer, an important issue is to provide and maintain seamless connectivity to the users as they move from one cell to another and thus switching communication link from one base station to another through an operation known as handover.

There are various multiple-access schemes that have been studied and used in wireless systems in the past decades, which allow the network to share the available radio resources (i.e., time, frequency, code, space, power) among a number of active users in the cell in the downlink and uplink. Fig. 3.29 illustrates the concept of resource sharing in some prominent multiple-access schemes. As mentioned earlier, orthogonal frequency division multiple access (OFDMA) has been a promising MA scheme that has been used in mobile broadband radio access technologies such as NR and LTE. The new radio uses a symmetric OFDMA scheme in the downlink and uplink, whereas LTE uses OFDMA and SC-FDMA as the MA schemes in the downlink and uplink, respectively.

In addition, the non-orthogonal concept can be applied to MA scenarios. Sparse code multiple access (SCMA), non-orthogonal multiple access (NOMA), and mult-iuser shared access (MUSA) are examples of non-orthogonal multiple access schemes that were studied in 3GPP Rel-16 [23,50–52]. These techniques can superimpose signals from multiple users in the code domain or the power domain to enhance the system-access performance and potentially allow asynchronous access in the uplink.

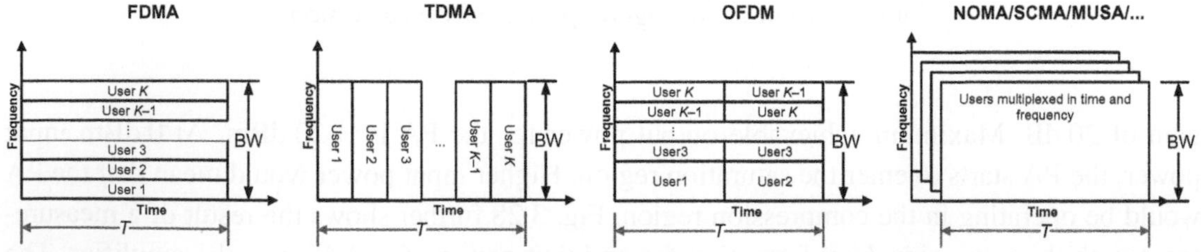

Figure 3.29
Illustration of various multiple access concepts [47].

3.3.1 Orthogonal Frequency Division Multiple Access

OFDMA is the multi-user variant of the OFDM scheme where multiple access is achieved by assigning subsets of time-frequency resources to different users, allowing simultaneous data transmission from several users. In OFDMA, the radio resources are 2D regions over time (an integer number of OFDM symbols) and frequency (a number of contiguous or non-contiguous subcarriers). Similar to OFDM, OFDMA employs multiple closely spaced subcarriers that are divided into groups of subcarriers where each group is called a resource block. The grouping of subcarriers into groups of resource blocks is referred to as sub-channelization. The subcarriers that form a resource block do not need to be physically adjacent. In the downlink, a resource block may be allocated to different users. In the uplink, a user may be assigned to one or more resource blocks. Sub-channelization defines subchannels that can be allocated to mobile stations depending on their channel conditions and service requirements. Using sub-channelization, within the same time slot (i.e., an integer number of OFDM symbols) an OFDMA system can allocate more transmit power to user devices with lower SNR and less power to user devices with higher SNR. Sub-channelization also enables the base station to allocate higher power to sub-channels assigned to indoor mobile terminals resulting in better indoor coverage. In OFDMA, an OFDM symbol is constructed of subcarriers, the number of which is determined by the FFT size. There are several subcarrier types: (1) data subcarriers are used for data transmission, (2) pilot or reference-signal subcarriers are utilized for channel estimation and coherent detection, and (3) null subcarriers that are not used for pilot/data transmission. The null subcarriers including the DC subcarrier (if it exists) are used for guard bands. The number of used (or occupied) subcarriers is always less than the FFT size. The guard bands are used to allow spectrum sharing and to reduce the adjacent channel interference and OOB emissions. The sampling frequency is selected to be greater than or equal the channel bandwidth. The number of time samples in a radio frame is always an integer and to further simplify the design of analog transmit filter, the sampling frequency is scaled by a factor greater than one (e.g., in LTE, the sampling frequency for 20 MHz bandwidth is 30.72 MHz).

In order to explain the signal processing concepts involved in an OFDMA transmission system, we use the generic transmitter model that is illustrated in Fig. 3.30, which shows the baseband structure of a general multicarrier transmitter that is applicable to a variety of multicarrier MA schemes such as OFDMA and SC-FDMA. Blocks of data represented by vector $\mathbf{s}$ of size $M \times 1$ are precoded with an $M \times M$ precoding matrix $\mathbf{P}$. The $M \times 1$ output vector is then mapped to M out of N inputs of the inverse-DFT block according to the subcarrier mapping $N \times M$ transform matrix $\mathbf{\Omega}$. To overcome the effects of frequency-selective channel fading, a CP of length N_{CP} is appended to beginning of each $N \times 1$ block output by the inverse-DFT function. Transmission with different rates among users is available according to each user's requirement, as a different number of

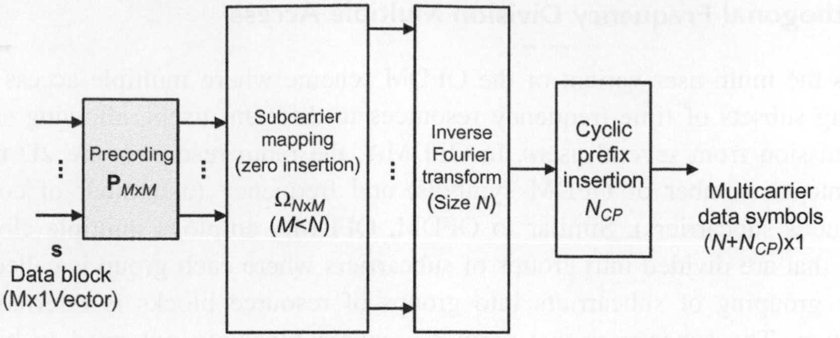

Figure 3.30
General multicarrier transmission scheme [47].

subcarriers and a different modulation and coding schemes can be applied to each user. Let $\mathbf{s}(n)$ denote the information symbols which are parsed into data blocks of size M. The ith data block $\mathbf{s}_i$ can be written as $\mathbf{s}_i = [s(iM), \ldots, s(iM+M-1)]^T$. Let's further denote by $\otimes$ the Kronecker product, by O the all-zero matrix of size $M \times N$ and by $\mathbf{I}$ the $M \times M$ identity matrix. We assume that the size of the inverse DFT is a multiple of the block size $N = MK$.

The special case of $\mathbf{P} = \mathbf{I}$ results in OFDMA where the user-specific data blocks are mapped to a subset of $M < N$ subcarriers, which are selected by the user-specific subcarrier mapping matrix $\mathbf{\Omega}$. The vector $\mathbf{\Omega s}$ is fed to the inverse-DFT function. The form of the matrix $\mathbf{\Omega}$ might lead to either a localized or a distributed subcarrier mapping as follows:

$$\mathbf{\Omega}_{N \times M} = \begin{pmatrix} \mathrm{O}_{L \times M} \\ \mathbf{I}_{M \times M} \\ \mathrm{O}_{(N-L-M) \times M} \end{pmatrix} \quad \text{or} \quad \mathbf{\Omega}_{N \times M} = \mathbf{I}_{M \times M} \otimes \begin{pmatrix} \mathrm{O}_{n \times 1} \\ 1 \\ \mathrm{O}_{(K-n-1) \times 1} \end{pmatrix}$$

By assigning different groups of subcarriers to different users, each user's transmit power can be concentrated in a restricted part of the channel bandwidth, resulting in significant coverage enhancement. Different user signals remain orthogonal only if time/frequency synchronization is maintained and an appropriate CP is appended to compensate for timing misalignments at the receiver. In order to maintain good performance in frequency-selective fading channels, robust forward error correction schemes must be employed.

Precoded OFDMA is a variant of OFDMA in which a precoding matrix $\mathbf{P}$ is used that spreads the energy of symbols over the subcarriers allocated to the user. Uniform energy distribution is desirable in practice. One of the most well-known precoding matrices is the Walsh–Hadamard matrix $\mathbf{P} = \begin{pmatrix} \mathbf{p}_0 & \mathbf{p}_1 & \cdots & \mathbf{p}_{M-1} \end{pmatrix}^T$ where the row vectors $\mathbf{p}_i$ are orthogonal Walsh–Hadamard sequences of length M [47].

3.3.2 Single-Carrier Frequency Division Multiple Access

SC-FDMA has been used as the uplink multiple-access scheme in LTE systems. The use of SC-FDMA was motivated by the fact that a single-carrier system with an OFDMA-type multiple-access would combine the advantages of the two techniques, that is, low PAPR and large coverage. The first SC-FDMA concept was interleaved FDMA (IFDMA), which was based on compression and block repetition of the modulated signal in the time domain. It can be theoretically shown that the spectrum of the compressed and K times repeated signal has the same shape as the original signal, with the difference that it presents exactly $K-1$ zeros between two data subcarriers. This feature enables us to easily interleave different users in the frequency domain by applying to each user a specific frequency shift, or equivalently, by multiplying the time-domain sequence by a specific phase ramp. In addition, similar to OFDMA, robustness to inter-cell interference can be achieved by coordinating resource allocation between adjacent cells. The same waveform can be obtained in the frequency domain, if discrete Fourier transform matrix $\mathbf{P} = [p_{k,n}]; p_{k,n} = \exp(j2\pi kn/M)$ is used as the precoding matrix in Fig. 3.30, resulting in DFT-precoded OFDMA, which is mathematically identical to IFDMA in a distributed scenario. The precoding operation $\mathbf{Ps}$ is equivalent to an $M-$ point DFT operation. With a subcarrier mapping matrix $\boldsymbol{\Omega}$ as given in Section 3.3.1, the spectrum of the distributed DFT-precoded OFDMA signal is identical to the IFDMA signal spectrum, thus it corresponds to the same waveform. This is also called DFT-spread OFDM. The two techniques are different implementations of SC-FDMA. The advantage of DFT-precoded OFDMA is in its more flexible structure. While IFDMA imposes a distributed signal structure, DFT-precoded OFDMA allows the use of an appropriate subcarrier mapping matrix $\boldsymbol{\Omega}$. Localized variants of implementation or channel-dependent mappings are also possible. A pulse-shaping filter can be further applied in the frequency domain, with a lower complexity than the time-domain filtering. Note that in frequency-selective channel scenarios, interference may occur among the M elements of each data block. This degradation, which is more important in a distributed subcarrier mapping, also impacts Walsh–Hadamard precoded OFDMA [47]

3.3.3 Non-orthogonal Multiple-Access Schemes

Previous generations of cellular standards relied on orthogonal MA, where each time/frequency resource block was exclusively assigned to one of the active users to ensure no inter-user interference would occur. In 3GPP NR, synchronous/scheduling-based orthogonal MA continues to play an important role in uplink/downlink transmissions. Non-orthogonal multiple access transmission, which allows multiple users to share the same time/frequency resource, was recently proposed to enhance the system capacity and to accommodate massive connectivity through asynchronous uplink access. Unlike orthogonal MA, multiple non-orthogonal multiple access users' signals are multiplexed using different power allocation coefficients or different signatures such as codebooks/codewords, sequences, interleavers, and preambles. The fundamental theory of non-orthogonal multiple access has been extensively studied in network information theory. The uplink and downlink non-orthogonal multiple access can be theoretically modeled as a multiple-access channel and a broadcast channel, respectively. The capacity region of the Gaussian broadcast channel can be achieved by power-domain superstition coding with a successive interference cancellation (SIC) receiver. Meanwhile, the capacity region of Gaussian multiple-access channel corresponds to CDMA, where different codes are used for the different transmitters, and the receiver decodes them in an SIC manner. In general, a user with poor-channel condition tends to allocate more transmission power, so this user would decode its own messages by treating the co-scheduled user's signal as noise. On the other hand, a user with good channel condition applies the SIC strategy by first decoding the information of the poor-channel user and then decoding its own, removing the other users' information. The results of studies in 3GPP suggest that using a non-SIC receiver results in negligible performance degradation in many cases [23]. Relaxing the need for an SIC receiver significantly would reduce the decoding complexity for the downlink case as the others' codebooks are no longer required.

In addition to the orthogonal MA scheme, the 3GPP NR may support an uplink non-orthogonal transmission (*Note: at the time of publication of this book, 3GPP has decided not to specify non-orthogonal multiple access in Rel-16 and instead only specify a two-step RACH*) to provide the massive connectivity that is desperately required for applications in mMTC as well as other scenarios. 3GPP has further studied grant-free uplink multiple-access schemes for mMTC scenarios. Since there is no need for a dynamic and explicit scheduling grant from gNB, latency reduction and control signaling minimization could be expected. For uplink non-orthogonal multiple access, network information theory suggests that CDMA with a SIC receiver provides a capacity achieving scheme. However, securing uplink non-orthogonal multiple access gain requires further system design enhancement. As the number of co-scheduled users increases, the decoding complexity of the SIC receiver increases. The message passing algorithm (MPA), as a less-complex decoding algorithm, as well as other low-complexity receiver designs have recently drawn attention. Several code-spreading-based techniques, including SCMA, MUSA, and PDMA and several others were the candidates under consideration in 3GPP Rel-16 NOMA study item. It has been shown that one can potentially achieve higher spectral efficiency, larger connectivity, and better user fairness with non-orthogonal multiple access relative to orthogonal MA schemes [50–52].

While the interference-free condition between orthogonally multiplexed users might facilitate multi-user detection at the receivers, it is widely known that orthogonal MA cannot achieve the sum capacity of a wireless system. The orthogonal MA also has limited granularity of resource scheduling, so it struggles to handle a large number of active connections. Non-orthogonal multi-user transmission/access has been recently investigated in a

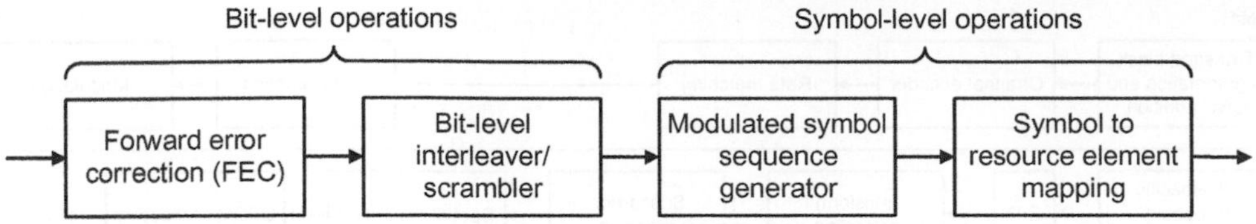

Figure 3.31
High-level block diagram for uplink non-orthogonal multiple-access schemes [22,23].

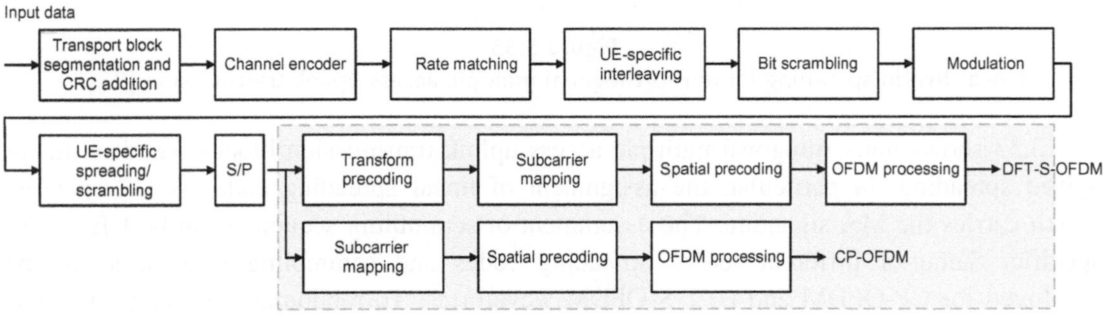

Figure 3.32
General framework for non-orthogonal multiple access uplink transmission [23].

systematic manner to deal with the above problems. Interference can be controlled by non-orthogonal resource allocation at the cost of increased receiver complexity.

The non-orthogonal multiple access schemes that were studied for the uplink transmission in 3GPP have the following features in common: (1) they use MA signature(s) at the transmitter side and (2) they allow multi-user detection at the receiver side. The MA signatures are typically used to differentiate the users. Thus all proposed non-orthogonal MA schemes for the uplink transmission at a high level can be described with the basic diagram shown in Fig. 3.31.

As we stated earlier, in non-orthogonal multiple access uplink transmission, multiple UEs share the same time/frequency resources via non-orthogonal resource allocation. There are various non-orthogonal multiple access schemes that can be derived from the general concept shown in Fig. 3.31. Fig. 3.32 shows a unified framework for non-orthogonal multiple access based on UE-specific spreading/scrambling/interleaving for the uplink data transmission. For data transmission based on UE-specific spreading, the existing solutions can be classified into two categories: linear spreading and nonlinear spreading. The category of linear spreading includes solutions such as resource spread multiple access (RSMA), MUSA, WSMA, and NCMA, while the category of nonlinear spreading includes SCMA. Linear spreading can be used in conjunction with DFT-S-OFDM and CP-OFDM waveforms. Although receiver implementation relatively less-complex and straightforward in orthogonal MA systems, the successful deployment of non-orthogonal multiple access depends on advanced receivers with inter-UE interference cancellation capabilities [23].

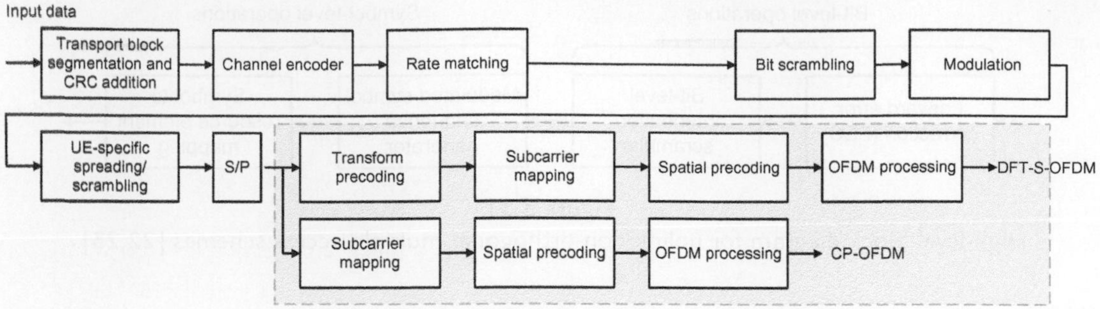

Figure 3.33
Linear hybrid spreading for non-orthogonal multiple access uplink transmission [23].

Fig. 3.33 shows non-orthogonal multiple access uplink transmission blocks based on linear hybrid spreading. In particular, the assignment of linear spreading codes is UE specific, which carries the MA signature. The assignment of scrambling sequence can be UE or cell specific. Same or different set of spreading codes and scrambling sequences can be employed for CP-OFDM and DFT-S-OFDM waveforms. To randomize the inter-UE interference and maximize the reuse of non-orthogonal multiple access resources, the mapping of spreading code and scrambling sequence can be made symbol dependent.

Compared to nonlinear spreading schemes such as SCMA, the studies indicate that solutions based on linear hybrid spreading exhibit similar BLER and significantly better performance in terms of scalability, complexity, PAPR, and flexibility. The use of long spreading codes can obtain a large codebook with good autocorrelation and low cross-correlation properties, which is suitable for grant-free non-orthogonal multiple access and is further robust against transceiver synchronization errors and timing inaccuracies associated with asynchronous transmission. The large processing gain of long spreading codes is also beneficial for inter-UE interference suppression. Therefore, it is an obvious candidate for grant-free, asynchronous transmission, relaxing the timing advance requirements. To improve the spectral efficiency, multi-layer transmission can be used, either in the form of standalone or in combination with spatial multiplexing schemes (in the presence of multiple transmit antennas). Compared to long spreading codes, short spreading codes need smaller spreading factor and higher spectral efficiency. The short spreading codes can be optimized to achieve the Welch bound[15] on cross-correlation, which can be leveraged for multi-user detection and inter-UE

[15] In mathematics, Welch bounds are a family of inequalities corresponding to the problem of evenly spreading a set of unit vectors in a vector space. If $\{\mathbf{x}_1, \ldots, \mathbf{x}_m\}$ are unit vectors in $\mathbb{C}^n$. We define $c_{\max} = \max_{i \neq j} |\langle \mathbf{x}_i, \mathbf{x}_j \rangle|$ where the inner product is defined in $\mathbb{C}^n$. Then, the following inequalities are given $(c_{\max})^{2k} \geq 1/(m-1)\left[m/\left(\binom{n+k-1}{k}-1\right)\right], \forall k = 1, 2, \ldots$. If $m \leq n$, then the vectors $\{\mathbf{x}_n\}$ can form an orthonormal set in $\mathbb{C}^n$. In this case, $c_{\max} = 0$ and the bounds are void. Consequently, interpretation of the bounds is only meaningful if $m > n$.

interference cancellation for synchronized reception. Moreover, it can be easily combined with spatial precoding to further mitigate the cross-correlation and enhance the non-orthogonal multiple access capacity.

A large number of the existing non-orthogonal multiple access schemes are based on linear spreading. For this type of spreading-based NOMA schemes, the same type of NOMA receivers can be utilized. The NOMA receiver consists of three parts. The first part is multi-user detector where the superposed received signal is processed jointly across the UEs to derive the LLR for each UE. The second part is the channel decoder which receives the LLR from multi-user detector and decodes the transmitted codeword. The output from the channel decoder can be either a decoded codeword when the decoding is successful or an intermediate LLR for each bit refined through the message passing decoding process. The third part is the iteration between the multi-user detector and the channel decoder, where they can exchange both soft-LLR information and hard decision information. We will focus on multi-user detector here and in particular, we will focus on elementary signal estimator (ESE) and linear MMSE (LMMSE) estimator. Without loss of generality, we consider single symbol processing. Let's assume that there are J users and K resources (spreading factor). The received signal at resource k can be written as $y_k = \sum_{j=1}^{J} h_{kj}x_{kj} + n_k$ where h_{kj} is the channel coefficient corresponding to resource k from user j, x_{kj} is the transmitted signal by user j on resource k, and $n \sim \mathbb{C}(0, N_0)$. For linear spreading codes, each user is assigned a spreading code sequence. Let c_{kj} be the kth coefficient of the spreading code for user j. We assume that all users share the same modulation alphabet $W = \{w_1, w_2, \ldots, w_M\}$. Then, $x_{kj} = c_{kj}s_j \forall k = 1, 2, \ldots, K$ where $s_j \in W$ is the transmitted symbol by user j. The received signal can be written as $\mathbf{y} = \mathbf{Hs} + \mathbf{n}$ where $\mathbf{s} = [s_1 \ldots s_J]^T$, $\mathbf{n} = [n_1 \ldots n_J]^T$, and $\mathbf{H} = [\tilde{h}_{kj}]_{k,j}$ is a $K \times J$ matrix with entries $\tilde{h}_{kj} = h_{kj}c_{kj}$. Let $\tilde{h}_j$ denote the jth column of matrix $\mathbf{H}$. The multi-user detector estimates the LLRs for $\mathbf{s}$ based on $\mathbf{y}$ [23].

Matched filter and ESE, which is a generalization of the matched filter that can accommodate soft interference cancellation, can be used as multi-user detectors. The ESE multi-user detector first compresses the received signals to scalar values for each UE by matched filtering. The output of the matched filter can be written as $\hat{\mathbf{y}} = [\hat{y}_1 \ldots \hat{y}_J]^T = \mathbf{H}^H\mathbf{y}$. In order to take advantage of soft information computed at the channel decoder, we can apply the ESE to $\hat{\mathbf{y}}$, which approximates signal and interference as Gaussian random variables as follows $\hat{y}_j = \left|\hat{\mathbf{h}}_j\right|^2 s_j + \sum_{i \neq j} \hat{\mathbf{h}}_j^H \hat{\mathbf{h}}_i s_i + \hat{\mathbf{h}}_j^H \mathbf{n} = \left|\hat{\mathbf{h}}_j\right|^2 s_j + \varepsilon_j$ where ε_j is residual interference plus noise. The residual interference ε_j is approximated as a Gaussian random variable which can be described by its mean and variance as follows:

$$\mu(\varepsilon_j) = \sum_{i \neq j} \hat{\mathbf{h}}_j^H \hat{\mathbf{h}}_i \mu(s_i), \quad \sigma^2(\varepsilon_j) = \sum_{i \neq j} \left|\hat{\mathbf{h}}_j^H \hat{\mathbf{h}}_i\right|^2 \sigma^2(s_i) + \left|\hat{\mathbf{h}}_j\right|^2 N_0$$

where $\mu(s_i)$ and $\sigma^2(s_i)$ are the a priori mean and variance of the symbol transmitted from the ith user, which can be computed using a priori bit LLRs. The LMMSE estimation can be used to estimate s_j from $\hat{y}_j$. From the estimation, LLR for each bit can be derived from conventional marginalization.[16] It can be noted that the ESE multi-user detector without matched filter and symbol spreading can also be used. For random symbol interleaver cases, the ESE multi-user detector can be applicable assuming that $p(\mathbf{y}|s_i = s) = \prod_{k=1}^{K} p(y_k|s_i = s)$. The ESE multi-user detector is also applicable to bit-level interleaving scenarios. It can be shown that the computational complexity of ESE multi-user detector scales as $O(K^2)$ for J UEs and spreading factor K [23].

Unlike ESE multi-user detector, the LMMSE estimator treats the received signal as a vector and applies LMMSE estimation matrix to the transmitted signal estimation of each UE. Let $\boldsymbol{\mu}_p$and$\mathbf{v}_p$ vectors denote the priori mean and variance of each UE transmitted signal derived from LLRs. The receiver applies the following LMMSE filter to the received vector where the output of the LMMSE filter is the mean and variance vectors.

$$\boldsymbol{\mu} = [\,\mu_1 \quad \ldots \quad m_J\,]^T = \boldsymbol{\mu}_p + \mathbf{v}_p\mathbf{H}^H(N_0\mathbf{I}+\mathbf{H}\mathbf{v}_p\mathbf{H}^H)^{-1}(\mathbf{y}-\mathbf{H}\boldsymbol{\mu}_p)$$

$$\mathbf{v} = [\,v_1 \quad \ldots \quad v_J\,]^2$$

$$v_i = v_{ip} - v_{ip}^2\hat{\mathbf{h}}_j^H(N_0\mathbf{I}+\mathbf{H}\mathbf{v}_p\mathbf{H}^H)^{-1}\hat{\mathbf{h}}_j$$

As mentioned earlier, $\boldsymbol{\mu}_p$ is the a priori mean vector and $\mathbf{v}_p$ is a diagonal matrix whose diagonal entries are a priori variance values for the corresponding transmitted symbols. A priori mean and variance values can be computed using the bit LLRs computed at the channel decoder. Based on the LMMSE output, the receiver generates extrinsic bit LLR values for channel decoder by marginalization. It can be shown that the computational complexity of LMMSE multi-user detector scales as $O(K^3 + K^2J + KJ^2)$ for J UEs and spreading factor of K [23].

Table 3.6 summarizes the use cases and operation modes of Rel-16 NOMA candidates. The features in the characteristics column reflect the potential benefits of NOMA over Rel-15 NR MA scheme [6], which are considered in the design, evaluation and comparison of NOMA transmitter and receiver schemes. The studies conducted in 3GPP suggest that when

16 In probability theory and statistics, the marginal distribution of a subset of a collection of random variables is the probability distribution of the variables contained in the subset. It gives the probabilities of various values of the variables in the subset without reference to the values of the other variables. This contrasts with a conditional distribution, which gives the probabilities contingent upon the values of the other variables. Marginal variables are those variables in the subset of variables being retained. The distribution of the marginal variables (the marginal distribution) is obtained by marginalizing; that is, focusing on the sums in the margin, over the distribution of the variables being discarded, and the discarded variables are said to have been marginalized.

Table 3.6: Rel-16 non-orthogonal multiple-access use cases and features supported by different operation modes [23].

Operation Mode		Dynamic MCS Support	Characteristics	Use Case
RRC_INACTIVE, grant-free with contention, tracking area-free (asynchronized)		No	Reduction in system overhead, latency, and power consumption	mMTC and eMBB
RRC_CONNECTED (Synchronized)	Grant-free with contention	No	Reduction in system overhead and latency	mMTC, URLLC, and eMBB
	Grant-based with overloading	Yes	Limited downlink overhead reduction	eMBB

the network operates in grant-based mode, transmission schemes proposed for NOMA can be applied to MU-MIMO; however, the relative spectral efficiency advantage of NOMA over MU-MIMO is not clear under underloaded scenarios. When the network operates in grant-free mode and the uplink access is contention-free, the relative gain of NOMA over MU-MIMO in terms of spectral efficiency is not proven. The most significant gain of NOMA over MU-MIMO may be attributed to contention-based, grant-free transmission and small data transmission in RRC_INACTIVE state scenarios.

3.3.3.1 Sparse Code Multiple Access

SCMA is a frequency-domain non-orthogonal multiple-access scheme, which can improve the spectral efficiency of wireless radio access. In SCMA, different incoming data streams are directly mapped to codewords of different multi-dimensional cookbooks, where each codeword represents a spread transmission layer. Each layer or user has its own dedicated codebook. Multiple SCMA layers share the same time-frequency resources of OFDMA. The sparsity of codewords makes the near-optimal detection feasible through iterative MPA.[17] Such low complexity of multi-layer detection allows excessive codeword overloading in which the dimension of multiplexed layers exceeds the dimension of codewords.

[17] Belief propagation, also known as sum-product message passing, is a message-passing algorithm for performing inference on graphical models such as Bayesian networks and Markov random fields. It calculates the marginal distribution for each unobserved node conditional on any observed nodes. Belief propagation is commonly used in artificial intelligence and information theory and has demonstrated empirical success in numerous applications including low-density parity-check codes, turbo codes, free energy approximation, and satisfiability. The algorithm was formulated as an exact inference algorithm on trees, which was later extended to polytrees. While it is not accurate for general graphs, it has been shown to be a useful approximation algorithm. If $X = \{X_i\}$ is a set of discrete random variables with a joint mass function p, the marginal distribution of a single X_i is the summation of p over all other variables $p_{X_i}(x_i) = \sum_{\mathbf{x}': x'_i \neq x_i} p(\mathbf{x}')$. However, this would become computationally prohibitive, whereas by exploiting the polytree structure, belief propagation would allow the marginals to be computed more efficiently [50–52].

Optimization of overloading factor along with modulation/coding levels of layers provides a more flexible and efficient link-adaptation mechanism. On the other hand, the signal spreading feature of SCMA can improve link-adaptation as a result of less colored interference. In SCMA, incoming bits are directly mapped to multi-dimensional complex codewords selected from predefined codebook sets. The co-transmitted spread data are carried over super-imposed layers. Since layers are not fully separated in a NOMA system, a nonlinear receiver is required to detect the intended layer of each user. Therefore, additional detection complexity is the cost of the nonorthogonal multiple-access especially when the system is heavily overloaded with a large number of multiplexed layers. Low-density spreading (LDS) is a special form of SCMA. In LDS, codewords are built by spreading of modulated symbols using LDS signatures with few non-zero elements within a large signature length. Despite the moderate complexity of detection, LDS suffers from poor performance especially for large constellation sizes beyond QPSK. All CDMA schemes, and in particular LDS, can be considered as different types of repetition coding in which different variations of a QAM symbol are generated by a spreading signature. Repetition coding is not able to provide desirable spectral efficiency for a wide range of SNR. In SCMA the QAM mapper and linear operation of sparse spreading are merged to directly map incoming bits to a complex sparse vector called a codeword. Both LDS and SCMA are based on the idea that one-user information is spread over multiple subcarriers. However, the number of subcarriers assigned to each user is smaller than the total number of subcarriers, and this low spreading (sparse) feature ensures that the number of users utilizing the same subcarrier is not too large, such that the system complexity remains manageable [50–52].

In LDS, each user spreads its data on a small set of subcarriers. There is no exclusivity in the subcarrier allocation and more than one user can share each subcarrier. The interference pattern at the receiver will generate a low-density graph, and graph theory-based techniques can be utilized. The main features of the LDS scheme can be summarized as follows. At each subcarrier, a user will have relativity small number of interferers comparing to the total number of users. Consequently, the search space will be smaller and more complex multi-user detection techniques can be implemented. Higher SINR can be achieved at each subcarrier, which results in reliable detection process. Each user will experience interference from different users at different subcarriers, which results in interference diversity by avoiding strong interferers to destroy the signal of a user on all the subcarriers. Belief propagation based multi-user detection can be implemented with linear complexity in the number of subcarriers [50–52].

The LDS and SCMA share the same concept which is to use a low-density or sparse non-zero element sequence to reduce the complexity of MPA processing at the receiver. However, in SCMA, bit streams are directly mapped to different sparse codewords. An example is illustrated in Fig. 3.34, where each user has a codebook and there are six users. All codewords in the same codebook contain zeros in the same two dimensions, and the positions

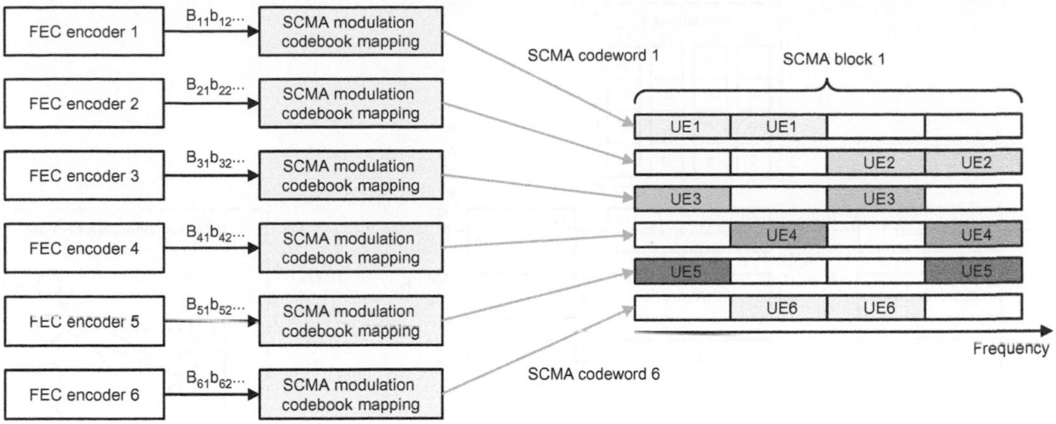

Figure 3.34
Illustration of SCMA concept [52].

of the zeros in the different codebooks are distinct to facilitate the collision avoidance of any two users. For each user, two bits are mapped to a complex codeword. Codewords for all users are multiplexed over four shared orthogonal resources. The key difference between LDS and SCMA is that a multi-dimensional constellation for SCMA is designed to generate codebooks, which provides a shaping gain that is not possible with LDS. In order to simplify the design of the multi-dimensional constellation, a baseline constellation can be generated by minimizing the average alphabet energy for a given minimum Euclidian distance between constellation points, and also taking into account the codebook-specific operations such as phase rotation, complex conjugate, and dimensional permutation [50–52].

We use the uplink multiple-access system shown in Fig. 3.35 to explain the SCMA processing stages. Let's assume that the system comprises K users whose information bits are spread over N resource elements. In an orthogonal scenario, $K \leq N$ to ensure that each user is assigned to an orthogonal resource element, while in the non-orthogonal scenarios, $K > N$ and the ratio K/N is defined as the overloading factor. The transceiver structure of SCMA can be mathematically modeled as follows. Let $\mathbf{b} = \begin{bmatrix} b_1 & b_2 & \dots & b_K \end{bmatrix}^T$ denote the information bits transmitted by K uplink users and $\mathbf{x}_k = [x_k^1 \ x_k^2 \ \dots \ x_k^N]^T$ represent the transmitted symbols by the kth user. An SCMA encoder at the kth user is defined to be a one-to-one mapping $f_k: B_k \rightarrow X_k$ with $b_k \in B_k$ and $\mathbf{x}_k \in X_k$, where the cardinality of B_k and X_k is 2^{N_B} where N_B denotes the number of information bits in b_k. Note that due to the sparse nature of SCMA scheme, $\mathbf{x}_k$ may contain zero symbols. The received signal at the base station $\mathbf{y}$, after passing through a block fading multiple-access (uplink) channel, can be expressed as $\mathbf{y} = \sum_{k=1}^{K} \mathbf{H}_k \mathbf{x}_k + \mathbf{z}$ where $\mathbf{H}_k = \mathrm{diag}(\mathbf{h}_k^1 \ \ \mathbf{h}_k^2 \ \ \dots \ \ \mathbf{h}_k^N)$ represents the channel between the base station and the kth user, $\mathbf{z} = \begin{bmatrix} z_1 & z_2 & \dots & z_N \end{bmatrix}$ is the additive white Gaussian

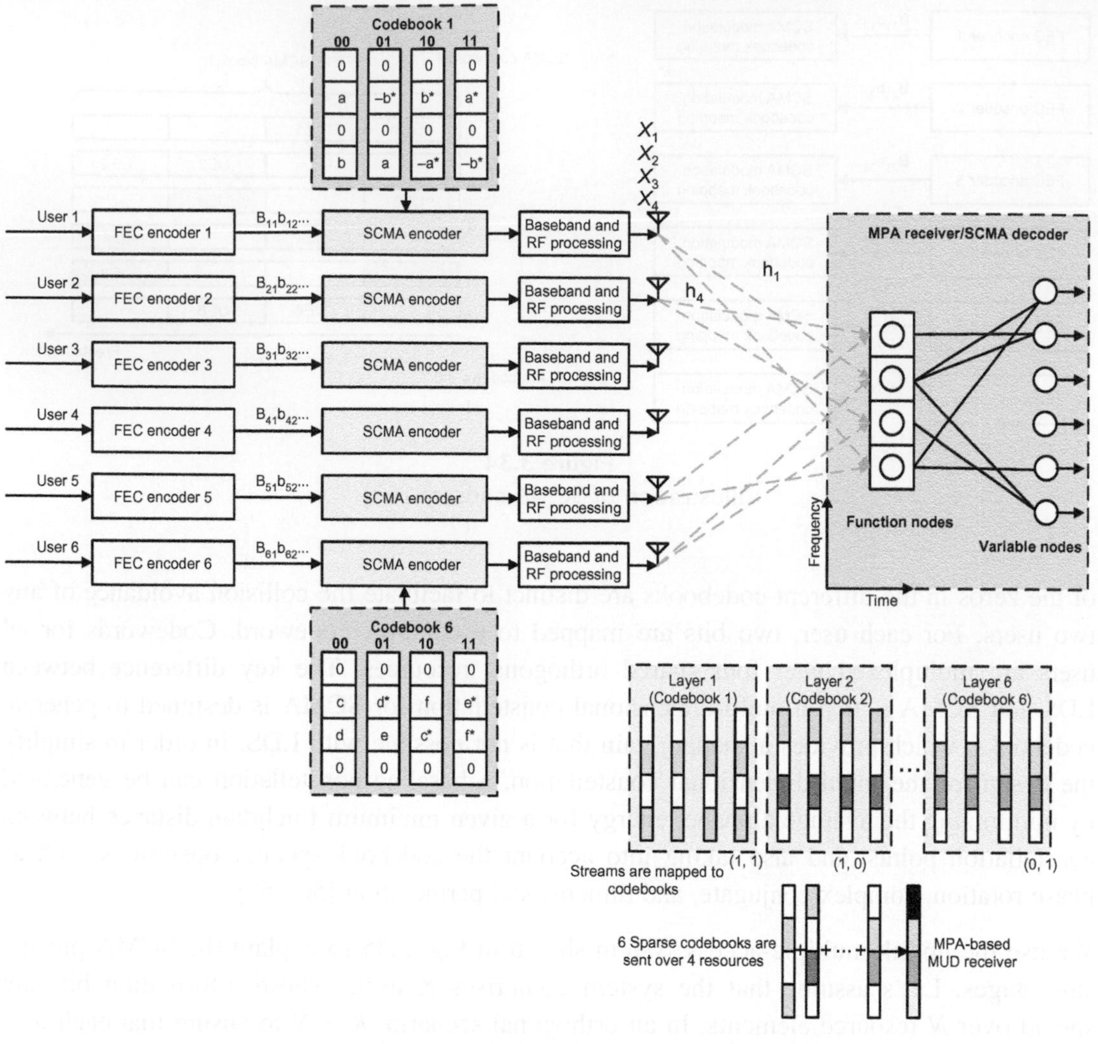

Figure 3.35
Example SCMA processing with $K = 6, N = 4$ and 150% overloading factor [52].

noise with zero mean and unity variance. Given the received signal $\mathbf{y} = \begin{bmatrix} y_1 & y_2 & \ldots & y_N \end{bmatrix}$ and the channel knowledge $\mathbf{H} = \{\mathbf{H}_k | k = 1, 2, \ldots, K\}$, the joint maximum-A-posteriori detection[18] of $\mathbf{X} = \begin{bmatrix} \mathbf{x}_1 & \mathbf{x}_2 & \ldots & \mathbf{x}_K \end{bmatrix}$ can be written as $\hat{\mathbf{X}} = \arg \max_{\mathbf{X} \in X_1 \times X_2 \times \cdots \times X_K} p(\mathbf{X}|\mathbf{y})$. In general, the solution to the above problem requires a global search over the joint space of K

[18] Maximum-a-posteriori (MAP) estimate of random variable X given that we have observed $Y = y$, is given by the value of x that maximizes $f_{X|Y}(x|y)$ if X is a continuous random variable, and $p_{X|Y}(x|y)$ if X is a discrete random variable. The MAP estimate is shown by $\hat{x}_{MAP} = \arg \max_x f_{X|Y}(x|y)$.

uplink users $X_1 \times X_2 \times \cdots \times X_K$. Due to the sparse nature of SCMA transmission scheme, the MPA detector can be applied to reduce the decoding complexity, which iteratively updates the belief associated with the underlying factor graph. Once $\hat{\mathbf{X}}$ has been estimated, we can use the inverse mapping function $(f_k)^{-1}$ to recover the original user information bits B_k [50–52].

3.3.3.2 Power-Domain Non-orthogonal Multiple Access

Power-domain NOMA can serve multiple users in the same time slot, OFDMA subcarrier, or spreading code, and multiple-access is realized by allocating different power levels to different users depending on their relative position to the base station.

Fig. 3.36 illustrates the concept of power-domain NOMA in the downlink with two UEs that utilize SIC receiver. For simplicity, we assume in this section the case of single transmit/receive antennas. The overall system transmission bandwidth is assumed to be 1 Hz. The base station transmits signal x_i to the ith UE with transmit power P_i where $E\{|x_i|^2\} = 1$ assuming $\sum_i P_i = P$. In power-domain NOMA, x_1 and x_2 are superposed as $x = \sqrt{P_1}x_1 + \sqrt{P_2}x_2$; thus the received signal at the ith UE can be written as $y_i = h_i x + w_i$, in which h_i is the complex-valued channel coefficient between the ith UE and the base station, and w_i denotes a zero-mean AWGN plus inter-cell interference. The PSD of w_i is N_{0i}. In the downlink NOMA, the SIC receiver is implemented at the UE receiver. The optimal order for decoding is in the order of decreasing channel gain normalized by noise and

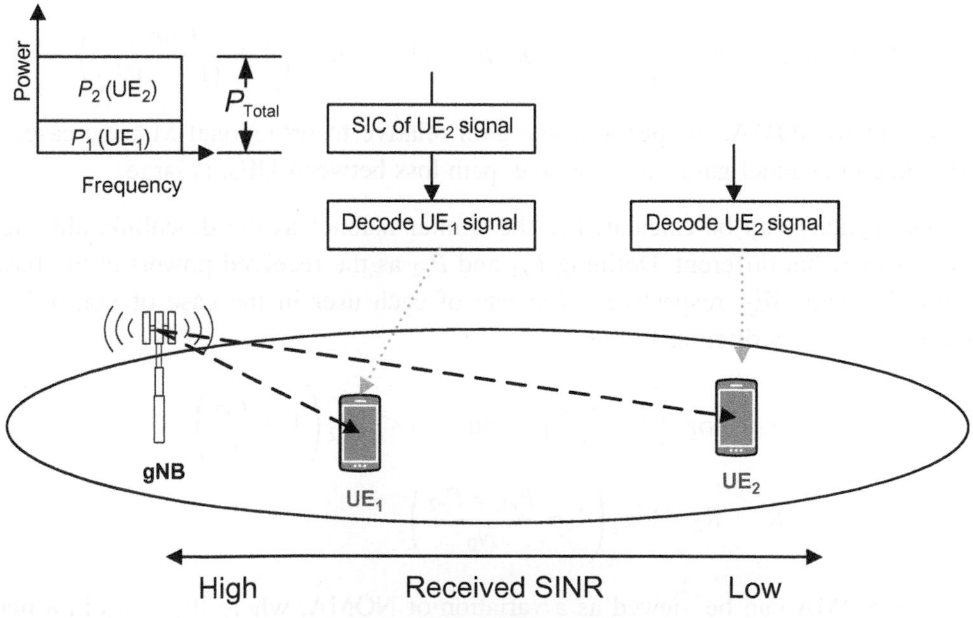

Figure 3.36
Illustration of the principle of downlink power-domain NOMA [52].

inter-cell interference power, that is, $|h_i|^2/N_{0i}$. Based on this order, we assume that any user can correctly decode the signals of other users whose decoding order comes before the corresponding user. Therefore, the ith UE can remove the inter-user interference from the jth user whose channel gain $|h_j|^2/N_{0j}$ is less than $|h_i|^2/N_{0i}$. In the example with two UEs, assuming that $|h_1|^2/N_{01} > |h_2|^2/N_{02}$, UE_2 does not have to perform interference cancellation since it comes first in the decoding order. UE_1, on the other hand, has to first decode x_2 and subtract that component from received signal y_1, then to decode x_1 without interference from x_2. Assuming successful decoding and no error propagation, the throughput of the ith UE, R_i, is given as follows [52]:

$$R_1 = \log_2\left(1 + \frac{P_1|h_1|^2}{N_{01}}\right) \quad \text{and} \quad R_2 = \log_2\left(1 + \frac{P_2|h_2|^2}{P_1|h_2|^2 + N_{02}}\right)$$

It can be seen that power allocation for each UE greatly affects the user throughput and thereby the modulation and coding scheme used for data transmission of each UE. By adjusting the power allocation ratio P_1/P_2, the base station can effectively control the throughput of each UE. The overall cell throughput, cell-edge throughput, and user fairness are closely related to the adopted power allocation scheme [52].

In a system that uses orthogonal MA scheme and hypothetically serves two UEs, if normalized bandwidth $0 < \beta < 1$ is assigned to the first UE, the remaining bandwidth $1 - \beta$ will be assigned to the second UE to maintain orthogonality between the users. The throughput of the ith UE R_i is given as follows:

$$R_1 = \beta\log_2\left(1 + \frac{P_1|h_1|^2}{\beta N_{01}}\right) \quad \text{and} \quad R_2 = (1-\beta)\log_2\left(1 + \frac{P_2|h_2|^2}{(1-\beta)N_{02}}\right)$$

In power-domain NOMA, the performance gain relative to orthogonal MA increases when the difference in channel gains, for example, path loss between UEs, is large.

The uplink capacity can be calculated in the similar manner as the downlink, although the formula is somewhat different. Defining P_{r1} and P_{r2} as the received powers at the base station from UE_1 and UE_2, respectively, the rate of each user in the case of non-orthogonal uplink access can be written as follows:

$$R_1 < \log_2\left(1 + \frac{P_{r1}}{N_0}\right) \quad \text{and} \quad R_2 < \log_2\left(1 + \frac{P_{r2}}{N_0}\right)$$

$$R_1 + R_2 < \log_2\left(1 + \frac{P_{r1} + P_{r2}}{N_0}\right)$$

Multicarrier NOMA can be viewed as a variation of NOMA, where the users in a network are divided into multiple groups. The users in each group are served in the same orthogonal resource block following the NOMA principle, and different groups are allocated to different orthogonal resource blocks. The motivation for employing hybrid NOMA is to reduce the system complexity. For example, assigning all the users in the network to a single group for the implementation of NOMA in one orthogonal resource block is problematic, since the user having the best channel conditions will have to decode all the other users' messages before decoding its own message, which results in high-complexity and high-decoding delay. Hybrid NOMA is an effective approach to make a tradeoff between system performance and complexity. Let's consider multicarrier NOMA as an example. The users in the cell are divided into multiple groups which are not necessarily mutually exclusive. The users within one group are assigned to the same subcarrier, and intra-group interference is mitigated using the NOMA principle. Different groups of users are assigned to different subcarriers, which effectively avoids inter-group interference. As a result, overloading the system, which is necessary in order to support more users than the number of available subcarriers and is required to enable massive connectivity, can be realized by the hybrid NOMA scheme. It is noted that, with hybrid NOMA, overloading is realized at reduced complexity since the number of users assigned to each subcarrier is limited [50–52].

The base station scheduler in power-domain NOMA searches and pairs multiple users for simultaneous transmission at each subband. To determine the set of paired users and the allocated power set at each subband, a multi-user proportional fairness (PF) scheduler may be used. The PF scheduling metric attempts to find the candidate user sets U and power sets P_s that maximize the following expression over each subband s:

$$Q(U,P_s) = \sum_{k \in U,P_s} \left(\frac{\eta(k,U,P_s,t)}{L(k,t)} \right)$$

where $(U_{\max}, P_{s_{\max}}) = \max\limits_{U,P_s} Q(U,P_s)$ denotes the maximum argument of PF scheduling metric $Q(U,P_s)$ for candidate user set U and allocated power set P_s over all users in the user set, $\eta(k,U,P_s,t)$ is the instantaneous throughput of user k in subband s at time instance t (the time index of a subframe), whereas $L(k,t)$ is the average throughput of user k. For power-domain NOMA, if we assume the possibility of dynamic switching between NOMA and orthogonal MA, then NOMA can be used only when it provides performance gains. Moreover, the number of users to be multiplexed over each subband is decided by searching all possible candidate user sets of different sizes up to m. The number of candidate user sets to be searched is given by [50–52]:

$$u = \binom{K}{1} + \binom{K}{2} + \cdots + \binom{K}{m}$$

In orthogonal MA schemes, the same MCS is selected over all subbands allocated to a single user. Therefore, the average signal-to-interference plus noise power ratio over all allocated subbands is used for MCS selection. However, when power-domain NOMA is utilized over each subband, user pairing and power allocation are performed over each subband. With such a mismatch between wideband MCS selection and subband power allocation granularities, the full-scale NOMA gains would not be realized. Furthermore, the higher the power allocation granularity, the more signaling overhead and thus performance degradation.

Power control of uplink NOMA is different from that of downlink in two aspects. In the downlink direction, the transmission power is limited by the maximum transmission power of the base station; however, in the uplink, the transmission power optimization is constrained by the maximum transmission power of individual UEs. In addition, there is a different approach to transmit power control in the uplink. In the downlink, the superposed signal received at a UE experiences the same channel, that is, the signals of different UEs have the same channel gain at each UE receiver. Therefore, the design of downlink power control tries to create sufficiently large difference among the signals of different UEs in the power domain in order to enable signal separation at the (SIC) receiver. For the uplink, due to different channels experienced by signals transmitted via different UEs, the received signal powers of different UEs already have differences in the power domain. On the other hand, when NOMA is applied in the uplink, the ICI greatly increases since multiple UEs are allowed to simultaneously transmit, whereas in the downlink, ICI does not increase when NOMA is applied because generally the base station has fixed transmission power regardless of the number of multiplexed UEs [23].

3.3.3.3 Scrambling-Based and Spreading-Based NOMA Schemes

In addition to the NOMA schemes discussed in the previous sections, there are other schemes proposed as part of 3GPP Rel-16 study item on NOMA. Scrambling-based NOMA schemes use different scrambling signatures for each user and utilize a low-rate forward error correction code or code repetition for multi-user decoding. The scrambling operation is carried out after the modulation. MMSE with SIC (MMSE-SIC) and ESE are used for the multi-user detection. RSMA is one of the scrambling-based schemes under consideration which utilizes low cross-correlation properties of long pseudo-random scrambling codes. Long scrambling sequences are used in RSMA. However, a long user signature causes high-decoding complexity and latency. Following the descrambling step, the ratio of signal-to-interference power is directly proportional to the scrambling code length. It must be noted that each user can transmit signal at any time using the asynchronous RSMA. Depending on the application scenarios, single-carrier RSMA or multicarrier RSMA can be utilized. Single-carrier RSMA can be used in the uplink access to reduce the PAPR of the UE. Multicarrier RSMA can be utilized in the downlink access to simplify the receiver complexity in the frequency-selective wireless fading channels. RSMA can be extended to multiple layers. Treating layers as virtual users, the data is split into multiple parallel layers for each user. The complexity of multi-layer RSMA is higher than that of single-layer RSMA. The RSMA uses hybrid short-code spreading and long-code scrambling as the MA signatures. The generation of scrambling sequences can be UE-group and/or cell specific, wherein the sequence ID of scrambling code is a function of the cell ID and UE-group ID. One or multiple UE groups can be configured in a cell. The sequences used for scrambling code can be Gold sequences, Zadoff–Chu sequences, or a combination of the two [12,23]. In Welch bound equality (WBE)-based spreading schemes such as RSMA, the design metric for the signature vectors is the total squared cross-correlation $\xi_c \triangleq \sum_{i,j} |\mathbf{s}_i{}^H \mathbf{s}_i|^2$. The lower bound on the total squared cross-correlation of any set of K vectors of length N is $K^2/N \leq \xi_c$. The WBE sequences are designed to meet the bound on the total squared cross-correlations of the vector set with equality $\xi_{Welch} \triangleq K^2/N$ [23].

The spreading-based NOMA schemes use non-orthogonal short spreading sequences with relatively low cross-correlation, for distinguishing multiple users, and the spreading sequences are non-sparse. The spreading sequences and the decoding algorithm are different for this category of NOMA schemes. In MUSA, modulation symbols of multiple users are spread by specially designed short sequences. All spreading symbols are transmitted over the same time-frequency resources. Multiple spreading sequences constitute a pool from which each user can randomly select. The spreading sequences of MUSA are complex-valued, in which the real part and imaginary part are both drawn from a real-valued multi-level set with uniform distribution. At the receiver, the codeword-level SIC detection is used to separate the target UE signal from the overlapped signals [23].

The average mutual information[19] can be used as a performance metric to compare the spectral efficiencies of various NOMA schemes with OFDMA. This performance metric provides the maximum information rate that can be reliably transmitted for a given channel state information. In a single-user case, the average mutual information is calculated for the signal after constellation mapping and before the soft demapping. This analysis can be extended to the multiuser case for evaluating the achievable sum-rate of the NOMA schemes. Let $\mathbf{x} = \begin{bmatrix} x_1 & x_2 & \ldots & x_J \end{bmatrix}$ denote the multi-user modulation symbol vector before the NOMA signature pattern, and $\mathbf{y} = \begin{bmatrix} y_1 & y_2 & \ldots & y_{KN_{rx}} \end{bmatrix}$ denote the channel output symbol vector at the receiver, in which $J, K, M, \text{and} N_{rx}$ represent the number of users, the NOMA signature length, the order of modulation, and the number of receive antennas, respectively. Assuming equi-probable input constellation points and the high-dimensional

---

[19] The average mutual information is defined as the weighted sum of the mutual information between each pair of the input and output events x_i and y_j. The average mutual information is a measure of independence between the two random variables X and Y. In mathematical terms $I(X;Y) = E[I(x_i;y_i)] = \sum_i \sum_j p(x_i, y_j) I(x_i; y_j)$ and $I(x_i;y_i) = \log_2 P(x_i, y_i)/P(x_i)P(y_i)$.

constellation set given by $\Omega = \{\omega_1, \omega_2, \ldots, \omega_{2^{MJ}-1}\}$, the average mutual information in the multi-user case can be written as [51]:

$$I(\mathbf{x};\mathbf{y}) = \frac{I(\mathbf{x};\mathbf{y}|\mathbf{H})}{K} = \frac{J\log_2 M}{K} - \frac{1}{K} E_{\mathbf{x},\mathbf{y},\mathbf{H}}\left\{ \log_2 \left[\frac{\sum_{\omega \in \Omega} P(\mathbf{y}|\mathbf{x}=\omega,\mathbf{H})}{P(\mathbf{y}|\mathbf{x},\mathbf{H})} \right] \right\}$$

A Monte Carlo simulation can be used to calculate the expectation function in the above equation. Assuming a six-user uplink multiple access channel, a tapped delay line TDL-A-30 ns channel **H** and ideal channel estimation, multi-user average mutual information of the NOMA schemes under consideration in 3GPP for Rel-16 and OFDMA have been calculated and compared in Fig. 3.37. In this analysis, the overloading factor is 150%, spectral efficiency per user is 0.25 bps/Hz, and the sum-rate is 1.5 bps/Hz. The SNR is defined as the ratio of average total received multi-users' power to the noise power at each receive antenna for given bandwidth. Multiple UEs are assumed to share the same six physical resource blocks. The number of users, SE per user, and transmission bandwidth are identical for both NOMA and OFDMA schemes to ensure fairness.

Fig. 3.37 further compares the BLER performance of the NOMA and OFDMA schemes for six and eight UEs, respectively. LTE turbo code is used for the channel coding and QPSK 1/2 for NOMA QPSK 3/4 is used for OFDMA, the overloading factor is 150%, spectral efficiency per user is 0.25 bps/Hz, and the sum-rate is 1.5 bps/Hz. In theory, the multi-user average mutual information analysis suggests that NOMA schemes provide higher capacity relative to OFDMA for given achievable sum-rate. In addition the, coding-based NOMA schemes (SCMA) have some performance advantage over other schemes. When the number of UEs increases the OFDMA system needs to use higher order modulation which would

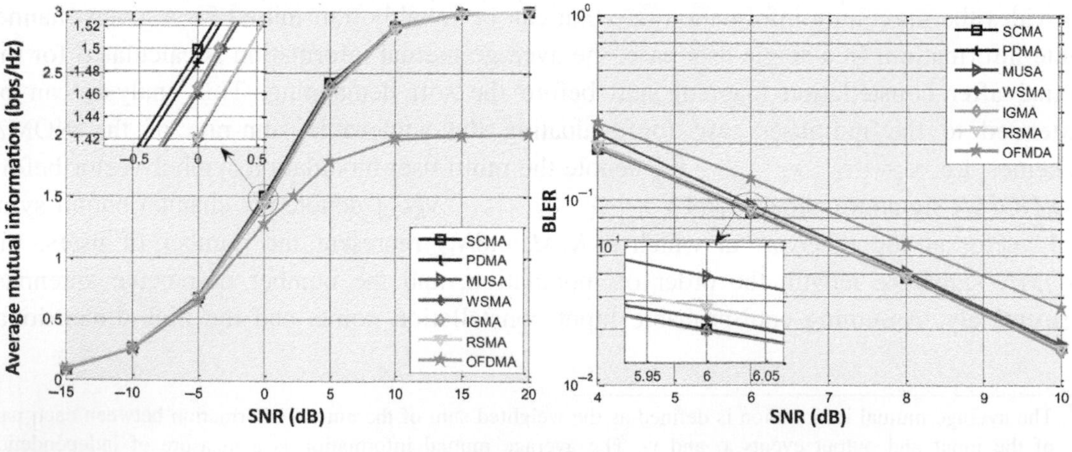

Figure 3.37
Comparison of average mutual information and BLER of NOMA and OFDMA schemes [51].

suffer from performance loss, while NOMA schemes with low order modulation can take advantage of superposition coding for higher overloading factor with slight performance degradation. The BLER performance advantage of NOMA schemes over OFDMA grows with the increase of the number of UEs [51].

3.4 Duplex Schemes

One of the key elements of any radio communication system is the way in which radio communications are maintained in the downlink and uplink. For cellular systems, it is necessary to enable simultaneous transmission of data in both directions, which creates a number of constraints on the schemes that may be used to control over the air transmission. As a result, the choice of duplex scheme becomes the basic part of the overall specification for the cellular or any radio communications system. The term duplex refers to the bidirectional communications between two devices. When unpaired spectrum or alternatively the same RF carrier is used for downlink and uplink communications, the transmit/receive functions are time-multiplexed. When paired spectrum or alternatively two RF carriers are used for downlink and uplink communications, the transmit/receive functions are frequency-multiplexed.

3.4.1 Frequency and Time Division Duplex Schemes

The Frequency Division Duplex is a duplex scheme in which uplink and downlink transmissions occur simultaneously using different frequencies. The downlink and uplink frequencies are separated by sufficiently large frequency offset. For the FDD scheme to properly operate, it is necessary that the frequency separation, that is, channel separation between the transmission and reception frequencies, to be sufficient in order to prevent the receiver blocking due to high-power transmitter signal. The receiver blocking is an important issue in FDD schemes and often highly selective filters may be required. For cellular systems using FDD, filters are required in the base station and the user terminal to ensure sufficient isolation of the transmitter signal without desensitizing the receiver. While implementation cost is not a significant constraint for the base stations, placing a filter in the user terminal involves higher complexity and cost. The use of an FDD system does enable simultaneous transmission and reception of signals. However, two RF channels are required, which in some cases may not be the efficient use of the available spectrum. The spectrum used for FDD systems is allocated by the regulatory bodies. Since there is a frequency separation between the uplink and downlink directions, it is not typically possible to reallocate spectrum to change the balance between the capacity of the uplink and downlink directions, if the capacity requirements for each direction vary over time.

The Time Division Duplex is a duplex scheme where uplink and downlink transmissions occur at different times but may share the same frequency. In other words, the downlink and uplink transmissions are multiplexed in time and are not concurrent. While FDD transmissions require a large frequency separation between the transmitter and receiver frequencies, TDD schemes require a guard time or guard interval between transmission and reception. This gap must be sufficient to allow the signals traveling from the remote transmitter to arrive before a transmission is started and the receiver is shut down. Although this delay is relatively short, switching between transmission and reception several times in a second, even a small guard time can reduce the spectral efficiency of the system since a percentage of time must be used for the guard interval. For small-sized cells, for example, up to one mile, the guard interval is typically small and acceptable. However, for large cell sizes, it may become an issue and may introduce significant overhead.

FDD has been the dominating duplex scheme since the beginning of the mobile communication era. In the 5G era, FDD will remain the main duplex scheme for lower frequency bands; however, for higher frequency bands, especially above 10 GHz and targeting very dense deployments, TDD will play more important role. In very dense deployments with low-power nodes, the TDD-specific interference scenarios (direct base station-to-base-station and device-to-device interference) will be similar to the base-station-to-device and device-to-base-station interference that also occurs in FDD schemes. Furthermore, for the dynamic traffic variations expected in very dense deployments, the ability to dynamically assign transmission resources (e.g., time slots) to different transmission directions may allow more efficient utilization of the available spectrum. To reach its full potential, 5G will allow for very flexible and dynamic assignment of TDD transmission resources. This contrasts with current TDD-based mobile technologies, including TD-LTE, for which there are restrictions on the downlink/uplink configurations; thus there typically exist assumptions about using the same configuration for neighboring cells and between neighboring operators.

The guard interval required for TDD will comprise two main elements: (1) A time-allowance for the propagation delay for any transmission from a remote transmitter to arrive at the receiver. This will depend upon the distances involved (i.e., cell radius) and (2) a time-allowance for the transceiver to switch from receive-to-transmit mode. The switching times can vary considerably depending on the implementation but can take a few microseconds. As a result, TDD is not normally suitable for use over very large cell sizes as the guard time increases and the spectral efficiency decreases.

It is often found that traffic in both directions is not balanced. Typically, there is more data transmitted in the downlink direction of a cellular system. This means that the capacity should be ideally greater in the downlink direction. Using a TDD system, it is possible to change the capacity in either direction simply by changing the number of time slots allocated to each direction. This is often dynamically configurable so it can be adapted to meet the demand. A further aspect to be noted with TDD transmission is the latency. Since data may not be able to be routed immediately to the transmission chain as a result of the time multiplexing between transmit and receive circuitry, there will be a small delay between the data being generated and being actually transmitted. Typically, this may be a few milliseconds depending on the frame timing. Both TDD and FDD have their advantages and each can be used in different deployment scenarios. Before deciding on a particular type of duplex scheme, it is necessary to analyze the advantages and disadvantages of each duplex mode. Table 3.7 summarizes and compares the relative attributes of TDD and FDD systems.

The new radio can operate in paired and unpaired spectrum using a common frame structure unlike LTE where two different frame structures were used, which was later expanded to three for support of unlicensed operation introduced in 3GPP Rel-13. The basic NR frame structure is designed such that it can support both half-duplex and full-duplex operation. In half duplex the device cannot transmit and receive at the same time.

A major issue in limiting the capacity of non-cooperative cellular massive MIMO networks operating in TDD mode is the pilot contamination. The rise of asymmetric uplink/downlink traffic patterns necessitates that the ratio between downlink/uplink traffic changes over the time; thus the static paired spectrum for downlink/uplink is not efficient for supporting such dynamic asymmetric traffic, particularly in UDNs. Flexible duplex can better adapt to dynamic asymmetric traffic. With flexible duplex, the uplink spectrum defined in FDD systems can be reallocated for downlink transmission with high flexibility. Considering the potential cross-link interference, that is, downlink to uplink and uplink to downlink, the transmission power for downlink transmission on the uplink spectrum should be constrained to a relatively low. Flexible duplex can be applicable to small cells with low transmission power and relay base stations.

3.4.2 Half-Duplex and Flexible-Duplex Schemes

The LTE and NR support TDD and FDD schemes with a great extent of commonality in the baseband processing. In order to reduce the implementation complexity and cost of FDD terminals and further to increase the reuse of baseband functional elements, a half-duplex FDD (H-FDD) operation is supported where the downlink and uplink transmissions are not simultaneous, but occur in two different frequencies. A classic H-FDD operation does not efficiently utilize the radio resources on the downlink and uplink RF carriers. The complementary grouping and scheduling of users would allow efficient use of downlink and uplink resources in an H-FDD operation. For H-FDD operation, a guard period is virtually

Table 3.7: Comparison of time division duplex (TDD) and frequency division duplex (FDD) attributes.

Attribute	TDD	FDD
Paired spectrum	Does not require paired spectrum as both transmit and receive occur on the same channel	Requires paired spectrum with sufficient frequency separation to allow simultaneous transmission and reception
Hardware cost	Lower cost as no diplexer[a] is needed to isolate the transmitter and receiver	Diplexer is needed and the implementation cost is higher
Channel reciprocity	Channel propagation is the same in both directions which enables estimation of the downlink channel from the uplink channel	Channel characteristics are different in both directions as a result of the use of different frequencies
UL/DL asymmetry	It is possible to dynamically change the UL and DL ratio based on the traffic volume in each direction	UL/DL bandwidths are determined by frequency allocation designated by the regulatory authorities. It is therefore not possible to dynamically adapt to the traffic volume
Guard period/ frequency separation	Guard period required to ensure uplink and downlink transmissions do not interfere. Large guard period will limit the capacity. Larger guard period normally required if distances are increased to accommodate larger propagation delays. Note that a guard band in frequency domain is required to suppress interference to adjacent bands.	Frequency separation is required to provide sufficient isolation between uplink and downlink. However, large frequency separation does not impact the capacity. Note that a guard band in frequency domain is required to suppress interference to adjacent bands.
Discontinuous transmission	Discontinuous transmission is required to allow both uplink and downlink transmissions. This can degrade the performance of the RF power amplifier in the transmitter	Continuous transmission is possible
Switching point synchronization	Base stations are required to be synchronized with respect to the uplink and downlink transmission times. If neighboring base stations use different uplink and downlink assignments and share the same channel, then interference may occur between cells.	Not applicable

[a]A diplexer is a passive device that implements frequency domain multiplexing. Two ports are multiplexed onto a third port. The signals on each input ports occupy nonoverlapping frequency bands. Consequently, the signals on the input ports can coexist on the output port without interfering with each other. On the other hand, a duplexer is a device that allows bidirectional communication over a single channel. In radar and radio communications systems the duplexer isolates the receiver from the transmitter while permitting them to share a common antenna. Most radio repeater systems include a duplexer. Duplexers are designed for operation in the frequency band used by the receiver and transmitter and must be capable of handling the output power of the transmitter. They must provide sufficient isolation between transmitter and receiver to prevent receiver desensitization [30].

created by the UE by not receiving the last OFDM symbol(s) of a downlink subframe immediately preceding an uplink subframe in which the UE is active. The length of the guard period is the sum of the maximum round-trip propagation delay in the cell, transmit-to-receive and receive-to-transmit switching delay at the UE.

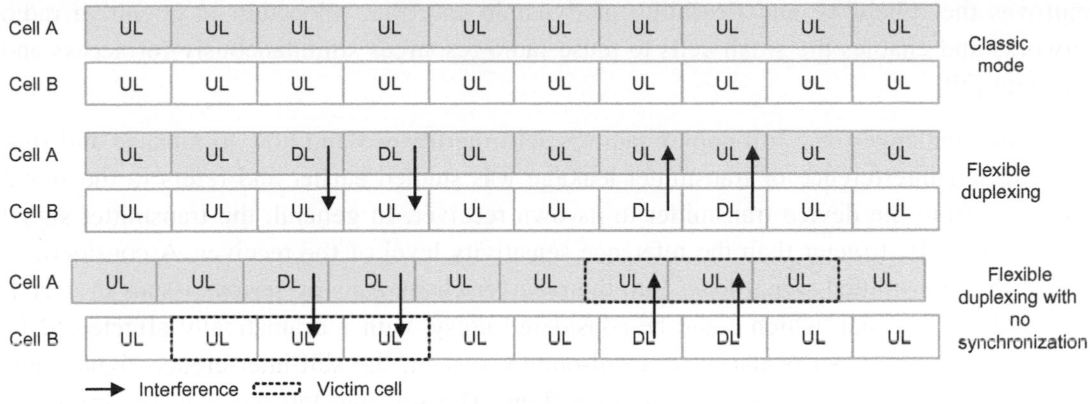

Figure 3.38
Examples of flexible duplexing schemes in TDD mode.

The choice of duplex scheme is typically determined by the spectrum allocation. For lower frequency bands, allocations are often paired, implying FDD mode is dominant. However, in higher frequency bands, unpaired spectrum allocations are increasingly common; thus TDD mode is used. 3GPP NR supports both duplexing methods. However, unlike LTE where the TDD uplink-downlink allocation does not change over time, NR supports dynamic TDD as a key technology component. In dynamic TDD, parts of slot can be dynamically allocated to either uplink or downlink based on the scheduler decision. This enables network adaptation to traffic variation, which is particularly common in dense deployments with a relatively small number of users per base station. The device follows any scheduling decisions, and it is up to the network scheduler, if necessary, to coordinate the scheduling decisions between neighboring sites to avoid inter-cell interference. There is also a possibility to semi-statically configure the transmission direction of some of the slots, a feature that can allow reduced device energy consumption as it is not necessary to monitor downlink control channels in slots that are a priori known to be reserved for uplink usage (Fig. 3.38).

3.4.3 Full-Duplex Schemes

The term full-duplex was traditionally used to describe a simultaneous bidirectional communication, in contrast to half-duplex, which assumed time division duplexing. However, in recent years, the term has carried a new concept and that is when a device can transmit and receive at the same time and over the same carrier frequency. Some authors refer to the latter scheme as in-band full duplex in the literature to distinguish this new concept from its traditional usage. It is intuitive that enabling wireless devices to operate in full-duplex mode offers the potential to double the spectral efficiency, considering that traditional approaches for increasing spectral efficiency such as adaptive coding and modulation, MIMO and smart antennas have almost reached their maximum limits. In addition, full-duplex scheme improves the reliability and flexibility of dynamic spectrum allocation in cognitive radio networks and enables the small cells to reuse radio resources simultaneously for access and backhaul [29].

The main challenge for a full-duplex radio is self-interference and how to manage and suppress it. Self-interference or transmitter leakage was studied earlier and refers to the signal that leaks from the device transmitter to its own receiver. In general, the transmitter signal is about 100 dB stronger than the reference sensitivity level of the receiver. A considerable part of this transmitted signal leaks into the receiver chain, causing serious issues in decoding the desired signal, which could be considered noisy, with a dramatically affected SNR. To achieve the best performance of a full-duplex system, the self-interference signal must be suppressed to reach the receiver's noise floor. The self-interference may be originated from the linear components and the RF carrier itself, which is attenuated and reflected from the environment. The received distortion can be modeled as a linear combination of different delayed copies of the original carrier. It can be further generated by nonlinear components because the imperfect radio circuits typically create third and fifth intermodulation products of the transmit signal. These higher order intermodulation terms have significant frequency content at frequencies close to the transmitted frequencies, which directly correspond to other harmonics. The self-interference can also be caused by the transmitter noise, which appears as an increase of about 50 dB over receiver noise floor. Due to random nature of this component, it can only be canceled by subtracting the appropriately weighted transmitter signal sampled in the analog domain from the received signal. For narrowband systems, the self-interference channel can be modeled as gain and delay functions, whereas wideband systems require a more complex model, because the reflected-path self-interference channel is often frequency-selective as a result of multipath propagation.

In general, one can model the process in the digital domain that is valid for both the narrowband and wideband scenarios as $r(n) = r_d(n) + w_r(n) + r_{DSI}(n) + r_{SSI}(n)$ where $r(n)$ is the total received complex-valued baseband samples, $r_d(n)$ is the desired signal from the remote node, $r_{DSI}(n)$ denotes the complex-valued samples caused by the direct self-interference component signal between the transmit and receive antennas in case of two antennas, or leaked signal in the circulator in case of one antenna, $r_{SSI}(n)$ represents the complex-valued samples caused by scattered self-interference components, and $w_r(n)$ denotes additive white Gaussian noise. A circulator is a passive three-port device in which an RF signal entering any port is only transmitted to the next port in rotation. Both direct and scattered self-interference can be represented as a combination of linear and nonlinear components. The suppression in the propagation domain can mitigate both linear and nonlinear self-interference at the same time and with the same isolation value. Meanwhile, cancellation techniques in the analog and digital domains have different performances for the two components. The self-interference cancellation is implemented in three domains: propagation, analog, and digital. Since none of these domains can meet the required cancellation requirements, hybrid solutions have been proposed in the literature [40,41]. The primary role of self-interference cancellation in the propagation and analog domains is to avoid the saturation of the receiver due to the high power of the self-interference signal as this power exceeds the ADC dynamic range and limits its precision after conversion because the desired signal is much weaker than the self-interference.

Self-interference comprises several components with different characteristics depending on the specifics of the full-duplex system implementation, such as the number of antennas, the characteristics of the RF components and the environment. The constituents of the self-interference can be classified by linearity. Linear components involve multipath propagation between the transmit/receive antennas. For a single-antenna system, linear components include the leakage of the circulator or the reflections from impedance mismatch. Components of RF circuits, such as attenuators and delay lines, are also modeled as linear systems for analog self-interference cancelation. The linear components of self-interference can be removed by the existing channel estimation methods, as in most conventional wireless communication systems. Nonlinear components are usually created by PAs in transmitters and low-noise amplifiers in receivers. The nonlinearity of the PA is generated because the power of the output signal is saturated for the high-power input signal, which worsens for modulation schemes with high PAPR such as OFDM and wideband CDMA. Intermodulation distortion, caused by the nonlinearity, interferes with the linear model of self-interference. The intermodulation can be theoretically calculated by a Volterra series[20] or approximated by a Taylor series. Since the even-ordered terms are out-of-band, the Taylor series would include only odd-ordered terms. Other RF imperfections/impairments, for example, I/Q imbalance, phase noise and transmitter noise can occur in the transmitter side. The I/Q imbalance occurs when there is mismatch between the gain and phase of the two sinusoidal signals, which deteriorates the baseband transmitter signal. The imperfection of the local oscillator can also degrade the linearity of the transmitter signal. In general, most of the impacts of the oscillator impairment is noticeable in random deviations in the output frequency, which can be modeled as phase noise. Transmitter noise also includes

[20] Volterra series is a mathematical model for nonlinear behavior of systems in which the output of the nonlinear system always depends on the input. This provides the ability to capture the memory effect of devices. In mathematics, Volterra series denotes functional expansion of a dynamic, nonlinear, time-invariant function. The Volterra series, which is used to prove the Volterra theorem, is an infinite sum of multidimensional convolutional integrals. A continuous time-invariant system with $x(t)$ as input and $y(t)$ as output can be expanded in Volterra series as $y(t) = h_0 + \sum_{n=1}^{N} \int_a^b \cdots \int_a^b h_n(\tau_1, \ldots, \tau_n) \prod_{m=1}^{n} x(t - \tau_m) d\tau_m$ where the constant term h_0 on the right-hand side is often set to zero by suitable choice of output level y. The function $h_n(\tau_1, \ldots, \tau_n)$ is called the nth order Volterra kernel. It can be regarded as a higher-order impulse response of the system. If N is finite, the series can be truncated. If $a, b,$ and$N$ are finite, the series is called doubly finite. Since in any physically realizable system, the output can only depend on previous values of the input, the kernels $h_n(t_1, \ldots, t_n)$ will be zero, if any of the variables $t_1, t_2, \ldots, t_n$ are negative. The integrals may then be written over the half range from zero to infinity. Therefore, if the operator is causal $a \geq 0$.

thermal noise which is typically generated by RF components. Using the estimated linear wireless channel, this method can mitigate transmitter impairments. Unlike other SIC methods, the distortion of the transmitter signal by PA nonlinearity or by I/Q imbalance is obtained directly by the auxiliary receive chain.

The specifications of a full-duplex system can be classified into three categories: main specifications, ADC specifications, and self-interference specifications. While the residual self-interference power P_{RSI} is higher than the receiver noise floor level, the signal to self-interference plus noise ratio in a full-duplex system is lower than the SNR of a half-duplex system receiver. This means that the maximum efficiency of full-duplex cannot be achieved. In general, self-interference cancellation solutions are a combination of several techniques to help meet the system requirements. Fig. 3.39 provides an example, showing the average performance value achieved in each domain.

In case of a shared transmit/receive antenna system, the suppression is performed using a three-port RF circulator. The ferrite within the device can be considered as a propagation domain. Achievable isolation by the circulator is between 15 and 30 dB, and in the case of wideband operation, the maximum value would decrease. In a separate-antenna system, several self-interference cancellation techniques can be used. The two transmit antennas can be placed at distances d and $d + \lambda/2$ away from the receive antenna. Separating the two transmitters by half a wavelength causes their signals to cancel one another. For narrowband signals, this technique is experimentally proved to be sufficiently robust; however, the suppression drastically decreases in case of wideband signals. The antenna directionality isolates the receiving antenna from the interfering signals of the transmitting antenna;

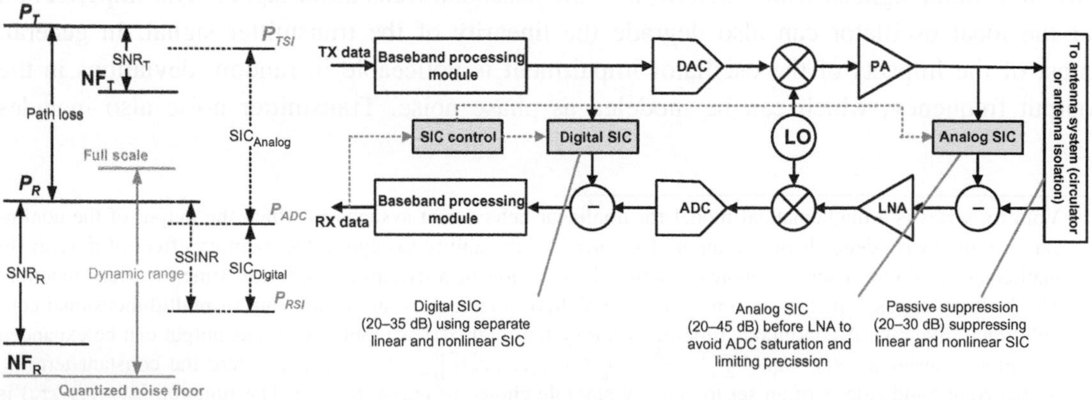

Figure 3.39

Example Wi-Fi or LTE signals, which are typically transmitted at an average power of 23 dBm (200 mW). The thermal noise level at 20 MHz bandwidth is about −101 dBm (−174 dBm/Hz + 73 dB) [40,41].

however, such an approach would not work for point-to-point full-duplex scenarios. Electromagnetic shielding can enhance the isolation between transmit and receive antennas. Nevertheless, one disadvantage is that the shielding affects the far-field antenna patterns because it prevents the antenna from transmitting to/receiving from the shielding direction; thus it is only relevant to the case of directional antennas. Self-interference can be mitigated using orthogonal polarization between the transmit/receive antennas, achieving about 10–20 dB isolation in an anechoic chamber and 6–9 dB in a reflective room at 2.4 GHz. The dual-polarized antenna can suppress self-interference by transmitting and receiving through orthogonal polarization. The isolation characteristic of a dual-polarized antenna can be expressed by its XPD factor. In transmit mode, XPD is the proportion of the signal that is transmitted in the orthogonal polarization to the desired direction. In receive mode, it is the antenna's ability to maintain the incident signal's polarization purity. For example, if a perfectly vertically polarized signal (containing no horizontal component) was incident upon a single polarized receive antenna, the electrical and mechanical imperfections would introduce a small amount of ellipticity to the polarization of the signal. The signal can be thought of as having both vertical and horizontal components. The ratio of the resulting horizontal to vertical components is the defined as XPD.

Active self-interference cancellation techniques in the analog domain generate a replica of the transmit signal and then adjust it to match the self-interference channel, making the replica similar to the self-interference signal in order to subtract it from the total received signal. This replica can be generated either in the analog domain or in the digital domain before the DAC. The self-interference cancellation signal is performed in the same domain from which it was created; thus no additional ADC/DAC is required. Replication of the transmission signal in the analog domain can be achieved by tapping the transmitter chain, using a power splitter, or using a balanced-to-unbalanced circuit in the case of two separate antennas. Experiments show the practical benefits of the latter approach relative to phase shifter, notably the flatter response within a wide frequency band. After creating an exact negative replication of the RF reference signal, the replica is adjusted by delay and attenuation elements to match the self-interference [40,41] (Fig. 3.40).

It can be observed in the experiments that the nonlinear components of self-interference can be 80 dB higher than the receiver noise floor. A large fraction of this component may be eliminated along with the linear self-interference by self-interference cancellation techniques in the analog and propagation domains, but the residual nonlinear self-interference in the digital domain, which amounts to 10–20 dB, needs to be canceled. In general, nonlinear self-interference cancellation methods are added to the linear methods to achieve optimal performance. A generic model for approximation of the nonlinear function is based on a Taylor series, where the output transmitted signal is represented as $y(n) = \sum_m a_m x_m^p(n)$ where $x_m^p(n)$ is the ideal passband analog signal for the digital representation of $x(t)$. It can

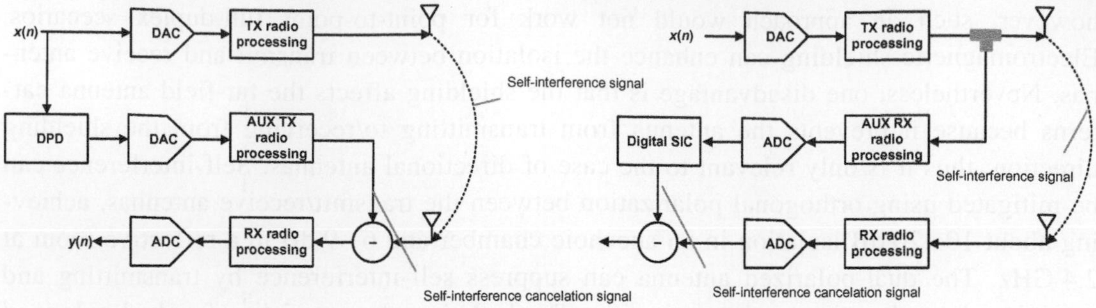

Figure 3.40
Auxiliary transmit/receive chain in full-duplex systems [40,41].

be shown that for practical wireless systems, only the odd orders of the polynomial contribute to the in-band distortion. Furthermore, only a limited number of odd orders contribute to the distortion, and higher orders can be neglected. The nonlinearity is typically characterized by the third-order intercept point, which is defined as the point at which the power of the third harmonic is equal to the power of the first harmonic. Another source of impairment is IQ imbalance which is caused by the gain and phase mismatches between I and Q branches of the transmitter and receiver chains. This imbalance results in the complex conjugate of the ideal signal to be superimposed with some level of attenuation. Thus the output of an imperfect IQ mixer is a transformation of an input signal $x(t)$ where both direct and conjugated signals are filtered and then summed together. The IQ imbalance can be modeled and compensated using widely linear filters.[21]

The results of studies suggest that the oscillator phase noise is one of the main self-interference cancellation challenges that limit the performance of full-duplex systems. It was assumed that when transmitter and receiver use a common local oscillator, the level of phase noise would remain at a tolerable level. However, this consideration is not always valid, especially in the case of OFDM systems. The studies show that with a phase noise variance between 0.4 and 1.0 degrees, the reduction in self-interference cancellation performance is about 20–25 dB for OFDM systems. This can be explained by the phase noise causing two effects: common phase error and inter-carrier interference. The former may have acceptable levels as previously assumed, but the latter stimulates an enhancement in

[21] Widely linear filters augment the data vector with its conjugate, thus providing the complete second-order statistical information when computed using the minimum mean square error cost function. Widely linear filters have been proposed for applications such as interference cancelation, demodulation, and equalization for direct sequence code-division-multiple-access systems, and array receivers. A widely linear filter forms the estimate of a desired sequence $d(n)$ through the inner product $y_{WL}(n) = \mathbf{w}^H \overline{\mathbf{x}}(n)$ where the weight vector $\mathbf{w} = [w_0 \; w_1 \; \ldots \; w_{2N-1}]^T$ has double dimension compared to the linear filter and $\overline{\mathbf{x}}(n) = [\mathbf{x}(n)\mathbf{x}^*(n)]^T$ with the definition $\mathbf{x}(n) = \left[x(n) \quad x(n-1) \quad \cdots \quad x(n-N+1)\right]^T$.

self-interference cancellation performance, which is achieved by consecutively estimating and suppressing the ICI.

The goal of digital self-interference cancellation is to remove the residual self-interference after analog self-interference cancellation especially that originated from NLoS reflections. Various signal processing and calculations performed after ADC for the purpose of self-interference cancellation are classified as digital self-interference cancellation. Since they operate in the digital domain, the baseband equivalent models of self-interference should be determined before calculation. The models include linear and nonlinear self-interference models. Once the self-interference signal is modeled as a function of the transmitted signal, it can be estimated from the transmitted and received signals. The residual self-interference is then reconstructed as the output of the function, and subtracted from the received signal [40,41].

3.5 Operating Frequency Bands

The new radio has been developed to operate in two distinct frequency regions: sub-6 GHz referred to as FR1, and mmWave referred to as FR2 in 3GPP radio specifications [9]. In each band, the new radio may operate in either FDD or TDD duplex modes. A band may be a supplementary downlink (SDL) or supplementary uplink (SUL) band used to provide additional capacity in the respective direction. Given the wide range of frequencies which 5G NR is required to support, three subcarrier-spacings (15, 30, and 60 kHz) are identified for operation in FR1 for sub-6 GHz bands; and two subcarrier-spacings (60 and 120 kHz) are designated for operation in FR2 for mmWave bands. Channel bandwidths vary based on SCS, with many supporting 5 MHz channel width with an SCS of 15 kHz. The FR1 covers a frequency range from 450 MHz to 6 GHz (this limit may be extended to 7.125 GHz in Rel-16) whereas FR2 spans a frequency range from 24.25 to 52.6 GHz. Note that frequency bands beyond 52.6 GHz will be supported in future releases of 3GPP once ratified by WRC-19 in late 2019. The NR operating bands have been shown in Table 3.8.

The gNB channel bandwidth supports a single NR RF carrier in the uplink or downlink. Different UE channel bandwidths may be supported within the same spectrum for transmitting to and receiving from UEs connected to the gNB. The assignment of the UE channel bandwidth is flexible and within the gNB channel bandwidth. The NR base station is able to transmit to and/or receive from one or more UE bandwidth parts (BWPs) that are smaller than or equal to the number of carrier resource blocks on the RF carrier [9]. The number of physical resource blocks configured in any channel bandwidth ensures that the required minimum guard band between neighboring channels is satisfied. The relationship between channel bandwidth, guard band and transmission bandwidth is shown in Fig. 3.41 for a single NR channel. The maximum number of resource blocks for each channel bandwidth and each frequency range is given in Table 3.9. Since the physical resource blocks always

Table 3.8: NR operating bands in frequency range (FR) 1 and FR2 [9].

NR Frequency Range	NR Operating Band	Uplink Operating Bands $F_{UL\_Low}-F_{UL\_High}$ (MHz)	Uplink Operating Bands Total Bandwidth (MHz)	Downlink Operating Bands $F_{DL\_Low}-F_{DL\_High}$ (MHz)	Downlink Operating Bands Total Bandwidth (MHz)	Duplex Mode
FR1	n1	1920–1980	60	2110–2170	60	FDD
	n2	1850–1910	60	1930–1990	60	FDD
	n3	1710 -1785	75	1805–1880	75	FDD
	n5	824–849	25	869–894	25	FDD
	n7	2500–2570	70	2620–2690	70	FDD
	n8	880–915	35	925–960	35	FDD
	n20	832–862	30	791–821	30	FDD
	n28	703–748	45	758–803	45	FDD
	n38	2570–2620	50	2570–2620	50	TDD
	n41	2496–2690	194	2496–2690	194	TDD
	n50	1432–1517	85	1432–1517	85	TDD
	n51	1427–1432	5	1427–1432	5	TDD
	n66	1710–1780	70	2110–2200	90	FDD
	n70	1695–1710	15	1995–2020	25	FDD
	n71	663–698	35	617–652	35	FDD
	n74	1427–1470	43	1475–1518	43	FDD
	n75	N/A		1432–1517	85	SDL
	n76	N/A		1427–1432	5	SDL
	n78	3300–3800	500	3300–3800	500	TDD
	n77	3300–4200	900	3300–4200	900	TDD
	n79	4400–5000	600	4400–5000	600	TDD
	n80	1710–1785	75	N/A		SUL
	n81	880–915	35	N/A		SUL
	n82	832–862	30	N/A		SUL
	n83	703–748	45	N/A		SUL
	n84	1920–1980	60	N/A		SUL
FR2	n257	26,500–29,500	3000	26,500–29,500	3000	TDD
	n258	24,250–27,500	3260	24,250–27,500	3260	TDD
	n260	37,000–40,000	3000	37,000–40,000	3000	TDD

consists of 12 subcarriers, the maximum number of resource blocks for each channel bandwidth depends on the subcarrier spacing.

The spacing between carriers will depend on the deployment scenario, the size of the frequency block available and the channel bandwidths. The nominal channel spacing df_s between two adjacent NR carriers with 100 kHz channel raster is defined as $df_s = (BW_{ch1} + BW_{ch2})/2$ where BW_{ch1} and BW_{ch2} denote the channel bandwidths of the two respective NR carriers. The channel spacing can be adjusted depending on the channel raster to optimize performance in a particular deployment scenario. For NR carriers in FR1

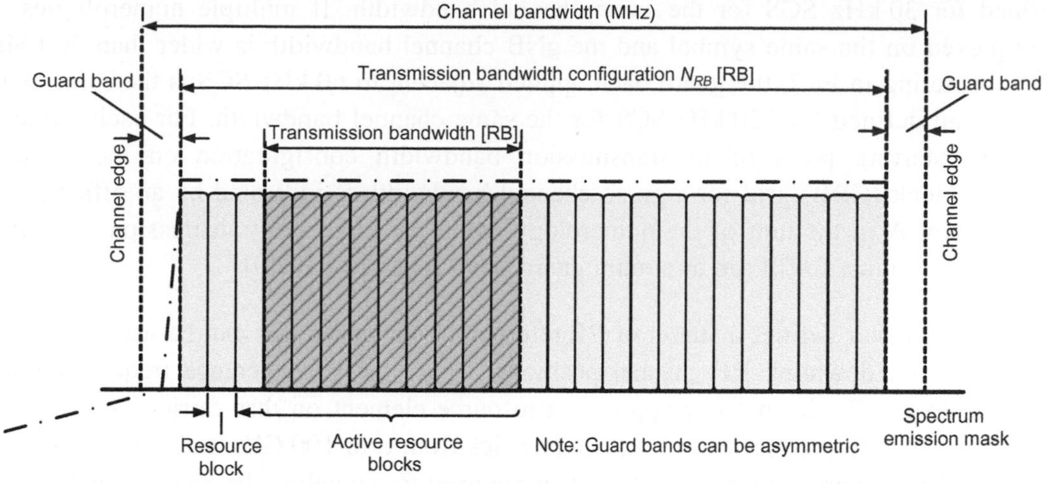

Figure 3.41
Definition of channel and transmission bandwidth configuration for one NR channel [9].

Table 3.9: Maximum transmission bandwidth configuration in N_{RB} for frequency ranges 1 and 2 [9].

		FR1													FR2			
		Bandwidth (MHz)													Bandwidth (MHz)			
Subcarrier Spacing (SCS) (kHz)		5	10	15	20	25	30	40	50	60	70	80	90	100	50	100	200	400
15	Number	25	52	79	106	133	160	216	270	N/A	N/A	N/A	N/A	N/A	N/A	N/A	N/A	N/A
30	of	11	24	38	51	65	78	106	133	162	189	217	245	273	N/A	N/A	N/A	N/A
60	resource	N/A	11	18	24	31	38	51	65	79	93	107	121	135	66	132	264	N/A
120	blocks N_{RB}	N/A	N/A	N/A	N/A	N/A	N/A	N/A	N/A	N/A	N/A	N/A	N/A	N/A	32	66	132	264

operating bands with 15 kHz channel raster, the nominal channel spacing is defined as $df_s = (\mathrm{BW}_{ch1} + \mathrm{BW}_{ch2})/2 \pm [5\mathrm{kHz}, 0]$. Furthermore, for NR carriers in FR2 operating bands with 60 kHz channel raster, the nominal channel spacing is defined as $df_s = (\mathrm{BW}_{ch1} + \mathrm{BW}_{ch2})/2 \pm [20\mathrm{kHz}, 0]$ [9].

In the case that multiple numerologies are multiplexed over the same symbol, the minimum guard band on each side of the carrier is the guard band applied at the configured gNB channel bandwidth for the numerology that is transmitted/received immediately adjacent to the guard band. Nevertheless, if multiple numerologies are multiplexed on the same symbol and the gNB channel bandwidth is wider than 50 MHz when operating in FR1, the guard band applied adjacent to 15 kHz SCS is the same as the guard band defined for 30 kHz SCS for the same channel bandwidth. If multiple numerologies are multiplexed on the same symbol and the gNB channel bandwidth is wider than 200 MHz when operating in FR2, the guard band applied adjacent to 60 kHz SCS is the same as the guard band defined for 120 kHz SCS for the same channel bandwidth. For each numerology, the starting point of its transmission bandwidth configuration on the common resource block (CRB) grid for a given channel bandwidth is indicated by an offset to reference point A in the unit of the numerology while the indicated transmission bandwidth configuration must fulfill the minimum guard band requirements [9].

The channel raster defines a subset of RF reference frequencies that can be used to identify the uplink and downlink RF channel positions, where the RF reference frequency corresponding to an RF channel is mapped to a resource element on that carrier. A global frequency raster is further defined for all frequencies from 0 to 100 GHz and is used to define the set of RF reference frequencies F_{REF} that are used for signaling the location of RF channels and synchronization signal blocks. The granularity of the global frequency raster is defined as ΔF_{global}. For each operating band, a subset of frequencies from the global frequency raster are applicable for that band and form a channel raster for that band with a granularity $\Delta F_{raster} \geq \Delta F_{global}$. The channel raster is mapped to physical resource block $n_{PRB} = \lfloor N_{RB}/2 \rfloor$ with resource element index $k = 0$ or 6 depending on whether $N_{RB} \bmod 2 = 0$ or 1, respectively [9].

The RF reference frequency in the uplink and downlink is identified by the NR absolute radio frequency channel number (NR-ARFCN) in the range [0...3279167] on the global frequency raster. The relationship between the NR-ARFCN and the RF reference frequency F_{REF} in MHz for the downlink and uplink is given as $F_{REF} = F_{REF-offset} + \Delta F_{raster}(N_{REF} - N_{REF-offset})$, where $F_{REF-offset}$ and $N_{REF-offset}$ are given in Table 3.10 and N_{REF} represents the NR-ARFCN [9].

For the supplementary bands and bands $(n1, n2, n3, n5, n7, n8, n20, n28, n66, n71)$ (see Table 3.8), the reference frequency is defined as $F_{REF\text{-}shift} = F_{REF} + \Delta_{shift}$ where $\Delta_{shift} = 0$ or 7.5kHz is signaled by the network.

A channel raster of 100 kHz is used for some NR operating bands, in which $\Delta F_{raster} = 20\Delta F_{global}$; thus every 20th NR-ARFCN within the operating band can be used for the channel raster with the step size of 20. For NR operating bands below 3 GHz with 15 kHz channel raster $\Delta F_{raster} = n_{step}\Delta F_{global}$. In this case, every $n_{step} \in \{3, 6\}$ NR-ARFCN within the operating band is a candidate channel raster. There are NR operating bands above 3 GHz where the channel raster is either 15 or 60 kHz. In that case, $\Delta F_{raster} = n_{step}\Delta F_{global}$ and every $n_{step} \in \{1, 2\}$ NR-ARFCN within the operating band is a candidate channel raster.

Table 3.10: NR absolute radio frequency channel number parameters for the global frequency raster [9].

Frequency Range (MHz)	ΔF_{global} (kHz)	$F_{REF-offset}$ (MHz)	$F_{REF-offset}$	N_{REF} Range
0–3000	5	0	0	0–599,999
3000–24,250	15	3000	600,000	600,000–2,016,666
24,250–100,000	60	24,250	2,016,667	2,016,667–3,279,167

Table 3.11: Global synchronization channel number (GSCN) parameters for the global frequency raster [9].

Frequency (MHz)	Synchronization Block Frequency Position SS_{REF}	GSCN	Range of GSCN
0–3000	1200 N kHz + 50 M kHz; N = 1:2499, $M\epsilon\{1,3,5\}$	$3N + (M-3)/2$	2–7498
3000–24,250	3000 MHz + 1.44 N MHz; N = 0:14,756	$7499 + N$	7499–22,255
24,250–100,000	24250.08 MHz + 17.28N MHz; N = 0:4383	$22,256 + N$	22,256–26,639

The synchronization raster signifies the frequency positions of the synchronization block that can be used by a UE for frequency acquisition when the UE has not received an explicit signaling indicating the synchronization block position. A global synchronization raster is defined for all frequencies. The frequency position of the synchronization block is defined by parameter SS_{REF} with a corresponding global synchronization channel number (GSCN). The GSCNs are numbered in increasing frequency order as shown in Table 3.11. The physical resource block number is $n_{PRB} = 10$ for the synchronization raster mapping to the synchronization block resource elements [9].

3.6 Frame Structure and Numerology

3GPP NR is designed to operate from 450 MHz to 100 GHz with a wide range of deployment scenarios, while supporting a variety of services. Given OFDMA was chosen as the multiple-access scheme for the downlink and uplink, it is not possible for a single numerology to satisfy the requirements of various use cases. Therefore, NR defines a family of OFDM numerologies for various frequency bands and deployment scenarios. The advantage of 3GPP NR relative to LTE is that it defines multiple numerologies which can be mixed and used simultaneously. A numerology is defined by a subcarrier spacing and a CP. The requirements for the OFDM subcarrier spacing is determined based on the carrier frequency, phase noise, delay spread, and Doppler spread. The use of smaller subcarrier spacing would result in either large EVM due to phase noise or more stringent requirements on the local oscillator. The small subcarrier spacing further leads to performance degradation in high Doppler scenarios. The required CP overhead and thus anticipated delay spread sets an

Table 3.12: Supported OFDM parameters in NR [12].

μ	Subcarrier Spacing $\Delta f = 2^{\mu} \times 15$ kHz	Cyclic Prefix Type	Supported for Data (PDSCH, PUSCH, etc.)	Supported for Synchronization Blocks (PSS, SSS, PBCH)	OFDM Useful Symbol Length μs	Cyclic Prefix Length μs	OFDM Symbol Length μs	N_{slot}^{symbol}	$N_{subframe}^{slot}$
0	15	Normal	Yes	Yes	66.67	4.69	71.35	14	1
1	30	Normal	Yes	Yes	33.33	2.34	35.68	14	2
2	60	Normal/ extended	Yes	No	16.67	1.17	17.84	14	4
3	120	Normal	Yes	Yes	8.33	0.57	8.92	14	8
4	240	Normal	No	Yes	4.17	0.29	4.46	14	16

upper limit for the subcarrier spacing. A large subcarrier spacing would result in unwanted overhead due to CP. The maximum FFT size of the OFDM modulation along with subcarrier spacing determines the channel bandwidth. Based on these observations, the subcarrier spacing should be as small as possible, while the system is still robust against phase noise and Doppler spread and supports the desired channel bandwidth. As shown in Table 3.12, in NR, the Primary and Secondary Synchronization Signals (PSS/SSS) and the Physical Broadcast Channel (PBCH), which are collectively known as SS block, will use 15/30 kHz SCS for sub-6 GHz and 120/240 kHz for above 6 GHz frequency bands. In the case of Physical Random-Access Channel (PRACH), the long preamble sequence utilizes 1.25/5 kHz SCS, in addition to the short preamble sequence using 15/30/60/120 kHz SCS. In other words, NR exploits a scalable OFDM subcarrier spacing (powers of 2) to support various frequency bands and deployment scenarios, where $\Delta f = 2^{\mu} \times 15\text{kHz}$ with $\mu = \{0, 1, 3, 4\}$ considered for PSS, SSS, and PBCH, and $\mu = \{0, 1, 2, 3\}$ designated for other physical channels. The normal CP is supported for all subcarrier spacing values, whereas extended CP is only supported for $\mu = 2$. From the network perspective, multiplexing of different numerologies over the same NR carrier bandwidth is possible in TDM and/or FDM manner in the downlink and uplink. From the UE perspective, multiplexing different numerologies is performed in TDM and/or FDM manner within or across a subframe. Regardless of the numerology used, the lengths of radio frame and subframe are always 10 and 1 ms, respectively. Different numerologies will then translate into the number of slots per subframe. The higher the subcarrier spacing, the more slots that can be accommodated per subframe [12]. Fig. 3.42 illustrates the NR frame, subframe, slot and mini-slot structure.

Different numerologies can be used in diverse deployment scenarios with their corresponding performance requirements. For example, the lower the subcarrier spacing, the larger the cell size, which will be suitable for the lower frequency deployments. At the same time, larger subcarrier spacing will allow shorter latency since the symbol duration will be shorter

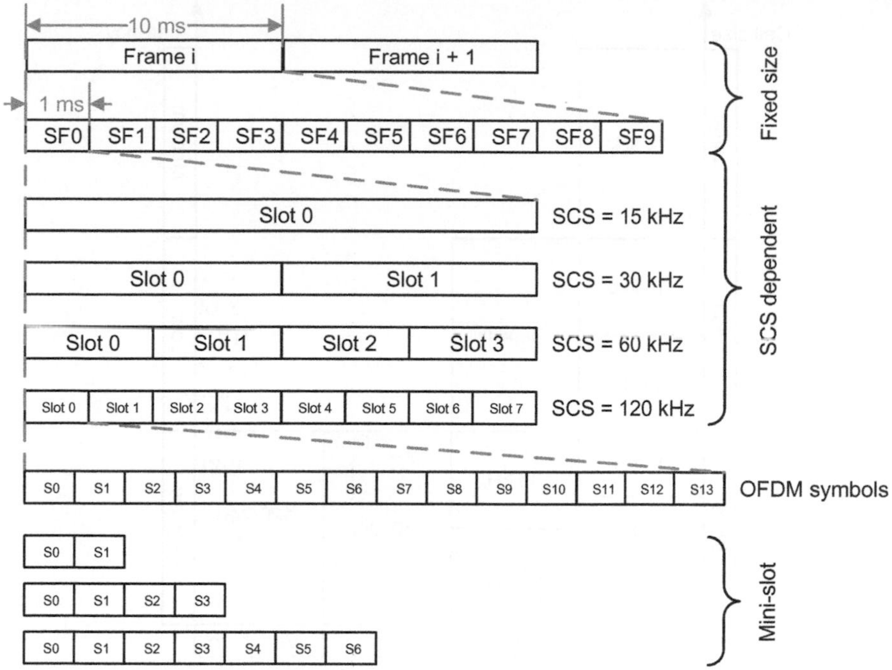

Figure 3.42
3GPP NR frame structure [12].

(see Fig. 3.42). Fig. 3.43 shows the relationship of the numerology, cell size, latency, and the carrier frequency [53].

The frame structure provides the basis for the timing of physical signal transmission. The timing scale is different for data, control, and synchronization physical channels. The sampling time in NR is defined as $T_c = 1/(\Delta f_{\max} N_{FFT})$ where $\Delta f_{\max} = 480\text{kHz}$ and $N_{FFT} = 4096$. In order to support multiple OFDM numerologies, the parameter μ and the corresponding CP for a BWP are signaled (configured) by the RRC parameters *DL-BWP-mu* and *DL-BWP-cp* for the downlink and *UL-BWP-mu* and *UL-BWP-cp* for the uplink [12,21]. The downlink and uplink transmissions are structured in the form of radio frames in the time domain with frame duration $T_{frame} = \left(\Delta f_{\max} N_{FFT}/100\right) T_c = 10$ ms, where each frame consists of ten subframes of $T_{subframe} = \left(\Delta f_{\max} N_f/1000\right) T_c = 1$ ms duration. The number of consecutive OFDM symbols in each subframe is defined as $N_{subframe}^{symbols}(\mu) = N_{slot}^{symbols} N_{subframe}^{slot}(\mu)$. In FDD mode, there is one set of frames in the uplink and one set of frames in the downlink on a carrier. As shown in Fig. 3.44, the *i*th uplink frame number starts at $T_{TA} = \left(N_{TA} + N_{TA\text{-}offset}\right) T_c$ before the start of the corresponding downlink frame at the UE in order to ensure uplink frame synchronization where $N_{TA\text{-}offset}$ depends on the frequency band of operation [18]. It can be seen that for the 15-kHz subcarrier spacing,

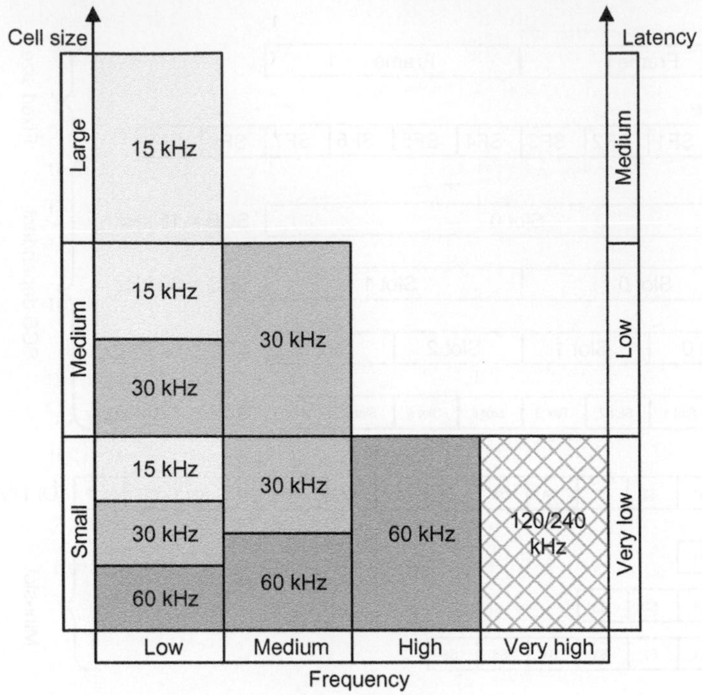

Figure 3.43
Relationship of numerology, carrier frequency, latency, and cell size [53].

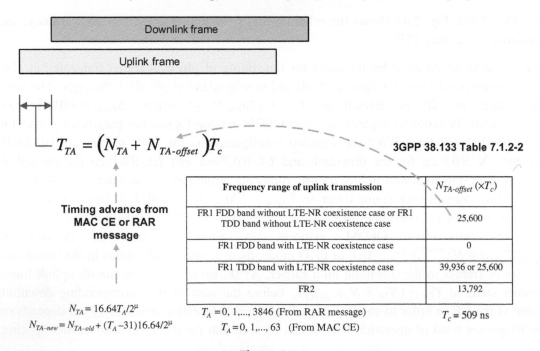

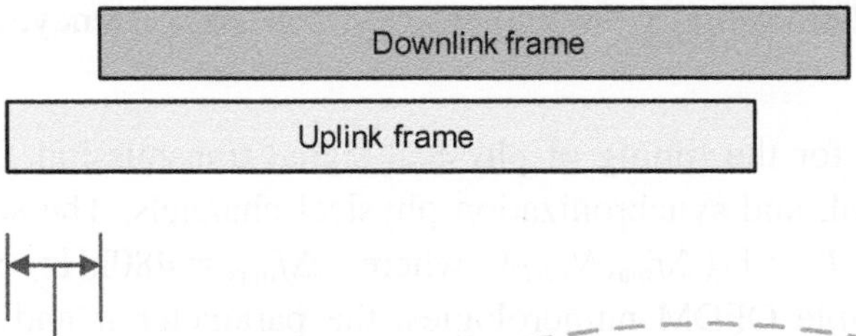

$$T_{TA} = (N_{TA} + N_{TA\text{-}offset})T_c$$

Timing advance from MAC CE or RAR message

$N_{TA} = 16.64T_A/2^{\mu}$

$N_{TA\text{-}new} = N_{TA\text{-}old} + (T_A - 31)16.64/2^{\mu}$

3GPP 38.133 Table 7.1.2-2

Frequency range of uplink transmission	$N_{TA\text{-}offset}$ ($\times T_c$)
FR1 FDD band without LTE-NR coexistence case or FR1 TDD band without LTE-NR coexistence case	25,600
FR1 FDD band with LTE-NR coexistence case	0
FR1 TDD band with LTE-NR coexistence case	39,936 or 25,600
FR2	13,792

$T_A = 0, 1, \ldots, 3846$ (From RAR message)

$T_A = 0, 1, \ldots, 63$ (From MAC CE)

$T_c = 509$ ns

Figure 3.44
Illustration of uplink timing calculation in NR [12,39].

an NR slot has the same structure as the LTE subframe, which is important for supporting LTE/NR coexistence scenarios. In the case of co-located deployment, slot and frame structures may be aligned to simplify cell search and inter-frequency measurements [9,16,17,19]. Coordination of control signals and channels in time domain will also be possible to avoid interference between LTE and NR. Given that a slot is defined as a fixed number of OFDM symbols, a larger subcarrier spacing results in a shorter slot duration, which can be used to support low-latency applications.

3GPP NR further supports a more efficient approach to low-latency transmissions by allowing scheduling shorter slot sizes known as mini-slots (see Fig. 3.42). The mini-slot-based transmissions can also preempt an already ongoing slot-based transmission to another device, allowing immediate transmission of application data requiring very low latency. Mini-slots can be used for low-latency applications such as URLLC and operation in unlicensed bands, for example, to start transmission directly after a successful listen-before-talk procedure without waiting for the slot boundary. Mini-slots can consist of two, four, or seven OFDM symbols, where the first symbol includes (uplink or downlink) control information (see Fig. 3.42). For low-latency applications, the HARQ protocol can be configured either on a slot or a mini-slot basis. For the regular frame structure used by delay-tolerant applications, slot bundling as in LTE is also possible. Mini-slots may also be used for fast flexible scheduling of services (preemption of URLLC over eMBB). However, mini-slots are likely to be supported by few UE categories. A major difference between LTE and NR in terms of scheduling granularity is that LTE transmission time interval is fixed at 1 ms whereas NR transmission interval is a slot or a fraction of slot whose length is a function of the subcarrier spacing.

For a given subcarrier spacing parameter μ, the slots are numbered as $n_s = \left\{0, 1, \ldots, N_{subframe}^{slot} - 1\right\}$ in ascending order within a subframe. There are N_{slot}^{symbol} consecutive OFDM symbols in a slot where N_{slot}^{symbol} depends on the CP. The start of slot n_{slot} in a subframe is aligned in time with the start of OFDM symbol $n_{slot}N_{slot}^{symbol}$ in the same subframe [12]. In the TDD mode, the OFDM symbols in a slot can be classified as downlink, flexible, or uplink. Table 3.13 shows possible slot formats. In a downlink frame slot, the UE assumes that downlink transmissions only occur in downlink or flexible symbols, whereas in an uplink frame slot, the UE only transmits in uplink or flexible symbols.

In LTE TDD, there are a number of predefined patterns for uplink/downlink OFDM symbol allocation in a radio frame, while NR does not define any preset uplink/downlink patterns (see Fig. 3.45). A slot format indication (SFI) parameter informs the UE whether an OFDM symbol is downlink, uplink or flexible. The SFI can indicate link direction over one or many slots when configured through RRC. The SFI carries an index to Table 3.13 (pre-configured UE-specific slot configuration table) configured through RRC. The SFI can be either dynamically configured through a DCI or statically or semi-statically configured through RRC. The UE assumes there is no conflict between dynamic SFI and downlink control

Table 3.13: Slot formats for normal cyclic prefix ("D/U" denotes flexible downlink/uplink symbols) [12].

	Symbol Number in a Slot													
Slot Format	0	1	2	3	4	5	6	7	8	9	10	11	12	13
0	DL	DL	DL	DL	DL	DL	DL	DL	DL	DL	DL	DL	DL	DL
1	UL	UL	UL	UL	UL	UL	UL	UL	UL	UL	UL	UL	UL	UL
2	D/U	D/U	D/U	D/U	D/U	D/U	D/U	D/U	D/U	D/U	D/U	D/U	D/U	D/U
3	DL	DL	DL	DL	DL	DL	DL	DL	DL	DL	DL	DL	DL	D/U
4	DL	DL	DL	DL	DL	DL	DL	DL	DL	DL	DL	DL	D/U	D/U
5	DL	DL	DL	DL	DL	DL	DL	DL	DL	DL	DL	D/U	D/U	D/U
6	DL	DL	DL	DL	DL	DL	DL	DL	DL	DL	D/U	D/U	D/U	D/U
7	DL	DL	DL	DL	DL	DL	DL	DL	DL	D/U	D/U	D/U	D/U	D/U
8	D/U	D/U	D/U	D/U	D/U	D/U	D/U	D/U	D/U	D/U	D/U	D/U	D/U	UL
9	D/U	D/U	D/U	D/U	D/U	D/U	D/U	D/U	D/U	D/U	D/U	D/U	UL	UL
10	D/U	UL	UL	UL	UL	UL	UL	UL	UL	UL	UL	UL	UL	UL
11	D/U	D/U	UL	UL	UL	UL	UL	UL	UL	UL	UL	UL	UL	UL
12	D/U	D/U	D/U	UL	UL	UL	UL	UL	UL	UL	UL	UL	UL	UL
13	D/U	D/U	D/U	D/U	UL	UL	UL	UL	UL	UL	UL	UL	UL	UL
14	D/U	D/U	D/U	D/U	D/U	UL	UL	UL	UL	UL	UL	UL	UL	UL
15	D/U	D/U	D/U	D/U	D/U	D/U	UL	UL	UL	UL	UL	UL	UL	UL
16	DL	D/U	D/U	D/U	D/U	D/U	D/U	D/U	D/U	D/U	D/U	D/U	D/U	D/U
17	DL	DL	D/U	D/U	D/U	D/U	D/U	D/U	D/U	D/U	D/U	D/U	D/U	D/U
18	DL	DL	DL	D/U	D/U	D/U	D/U	D/U	D/U	D/U	D/U	D/U	D/U	D/U
19	DL	D/U	D/U	D/U	D/U	D/U	D/U	D/U	D/U	D/U	D/U	D/U	D/U	UL
20	DL	DL	D/U	D/U	D/U	D/U	D/U	D/U	D/U	D/U	D/U	D/U	D/U	UL
21	DL	DL	DL	D/U	D/U	D/U	D/U	D/U	D/U	D/U	D/U	D/U	D/U	UL
22	DL	D/U	D/U	D/U	D/U	D/U	D/U	D/U	D/U	D/U	D/U	D/U	UL	UL
23	DL	DL	D/U	D/U	D/U	D/U	D/U	D/U	D/U	D/U	D/U	D/U	UL	UL
24	DL	DL	DL	D/U	D/U	D/U	D/U	D/U	D/U	D/U	D/U	D/U	UL	UL
25	DL	D/U	D/U	D/U	D/U	D/U	D/U	D/U	D/U	D/U	D/U	UL	UL	UL
26	DL	DL	D/U	D/U	D/U	D/U	D/U	D/U	D/U	D/U	D/U	UL	UL	UL
27	DL	DL	DL	D/U	D/U	D/U	D/U	D/U	D/U	D/U	D/U	UL	UL	UL
28	DL	DL	DL	DL	DL	DL	DL	DL	DL	DL	DL	DL	D/U	UL
29	DL	DL	DL	DL	DL	DL	DL	DL	DL	DL	DL	D/U	D/U	UL
30	DL	DL	DL	DL	DL	DL	DL	DL	DL	DL	D/U	D/U	D/U	UL
31	DL	DL	DL	DL	DL	DL	DL	DL	DL	DL	DL	D/U	UL	UL

(Continued)

Table 3.13: (Continued)

Slot Format	Symbol Number in a Slot													
	0	1	2	3	4	5	6	7	8	9	10	11	12	13
32	DL	DL	DL	DL	DL	DL	DL	DL	DL	DL	D/U	D/U	UL	UL
33	DL	DL	DL	DL	DL	DL	DL	DL	DL	D/U	D/U	D/U	UL	UL
34	DL	D/U	UL	UL	UL	UL	UL	UL	UL	UL	UL	UL	UL	UL
35	DL	DL	D/U	UL	UL	UL	UL	UL	UL	UL	UL	UL	UL	UL
36	DL	DL	DL	D/U	UL	UL	UL	UL	UL	UL	UL	UL	UL	UL
37	DL	D/U	D/U	UL	UL	UL	UL	UL	UL	UL	UL	UL	UL	UL
38	DL	DL	D/U	D/U	UL	UL	UL	UL	UL	UL	UL	UL	UL	UL
39	DL	DL	DL	D/U	D/U	UL	UL	UL	UL	UL	UL	UL	UL	UL
40	DL	D/U	D/U	D/U	UL	UL	UL	UL	UL	UL	UL	UL	UL	UL
41	DL	DL	D/U	D/U	D/U	UL	UL	UL	UL	UL	UL	UL	UL	UL
42	DL	DL	DL	D/U	D/U	D/U	UL	UL	UL	UL	UL	UL	UL	UL
43	DL	DL	DL	DL	DL	DL	DL	DL	DL	D/U	D/U	D/U	D/U	UL
44	DL	DL	DL	DL	DL	DL	D/U	D/U	D/U	D/U	D/U	D/U	UL	UL
45	DL	DL	DL	DL	DL	DL	D/U	D/U	UL	UL	UL	UL	UL	UL
46	DL	DL	DL	DL	DL	D/U	UL	DL	DL	DL	DL	DL	D/U	UL
47	DL	DL	D/U	UL	UL	UL	UL	DL	DL	D/U	UL	UL	UL	UL
48	DL	D/U	UL	UL	UL	UL	UL	DL	D/U	UL	UL	UL	UL	UL
49	DL	DL	DL	DL	D/U	D/U	UL	DL	DL	DL	DL	D/U	D/U	UL
50	DL	DL	D/U	D/U	UL	UL	UL	DL	DL	D/U	D/U	UL	UL	UL
51	DL	D/U	D/U	UL	UL	UL	UL	DL	D/U	D/U	UL	UL	UL	UL
52	DL	D/U	D/U	D/U	D/U	D/U	UL	DL	D/U	D/U	D/U	D/U	D/U	UL
53	DL	DL	D/U	D/U	D/U	D/U	UL	DL	DL	D/U	D/U	D/U	D/U	UL
54	D/U	D/U	D/U	D/U	D/U	D/U	D/U	DL	DL	DL	DL	DL	DL	DL
55	DL	DL	D/U	D/U	D/U	UL	UL	UL	DL	DL	DL	DL	DL	DL
56–255	Reserved													

information (DCI) DL/UL assignments, thus when operating in NR TDD mode, one has to clearly define how available time slots are allocated to downlink and uplink transmissions. The NR defines those patterns in more flexible manner using the following parameters (see Fig. 3.46) [21]:

- *dl-UL-TransmissionPeriodicity*: Periodicity of the DL–UL pattern
- *nrofDownlinkSlots*: Number of consecutive full DL slots at the beginning of each DL–UL pattern

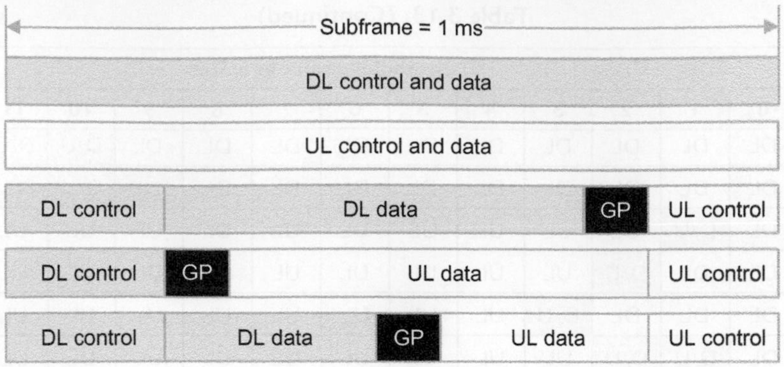

Figure 3.45
Different NR TDD-based subframe structures [53].

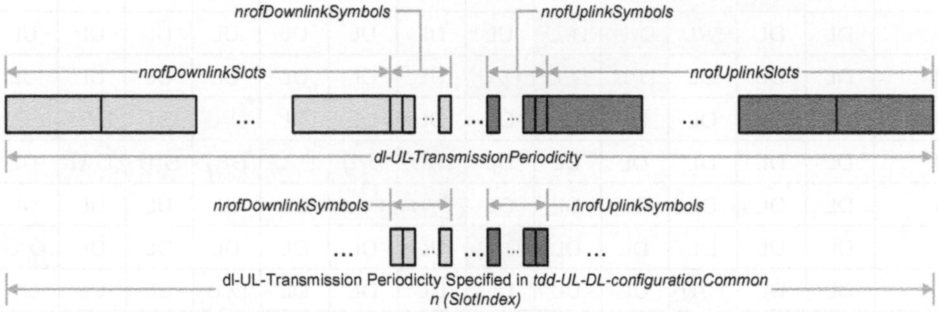

Figure 3.46
TDD UL/DL common and dedicated configurations [15,39].

- *nrofDownlinkSymbols*: Number of consecutive DL symbols in the beginning of the slot following the last full DL slot
- *nrofUplinkSlots*: Number of consecutive full UL slots at the end of each DL–UL pattern
- *nrofUplinkSymbols*: Number of consecutive UL symbols at the end of the slot preceding the first full UL slot

As shown in Table 3.13, a slot format includes a specific downlink, uplink, and flexible symbol configuration. In TDD mode, a slot can be all downlink, all uplink, or a combination of downlink and uplink segments. Data transmission can be scheduled to span one or multiple slots when slot aggregation is supported. For each serving cell, if a UE receives the RRC parameter *UL-DL-configuration-common*, it must set the slot format per slot over the number of slots indicated by this parameter. If the UE is additionally provided with RRC parameter *UL-DL-configuration-dedicated* for the slot format per slot over a number of slots, the latter parameter overrides only flexible symbols per slot over the number of slots indicated by *UL-DL-configuration-common* parameter. The UE determines the duration

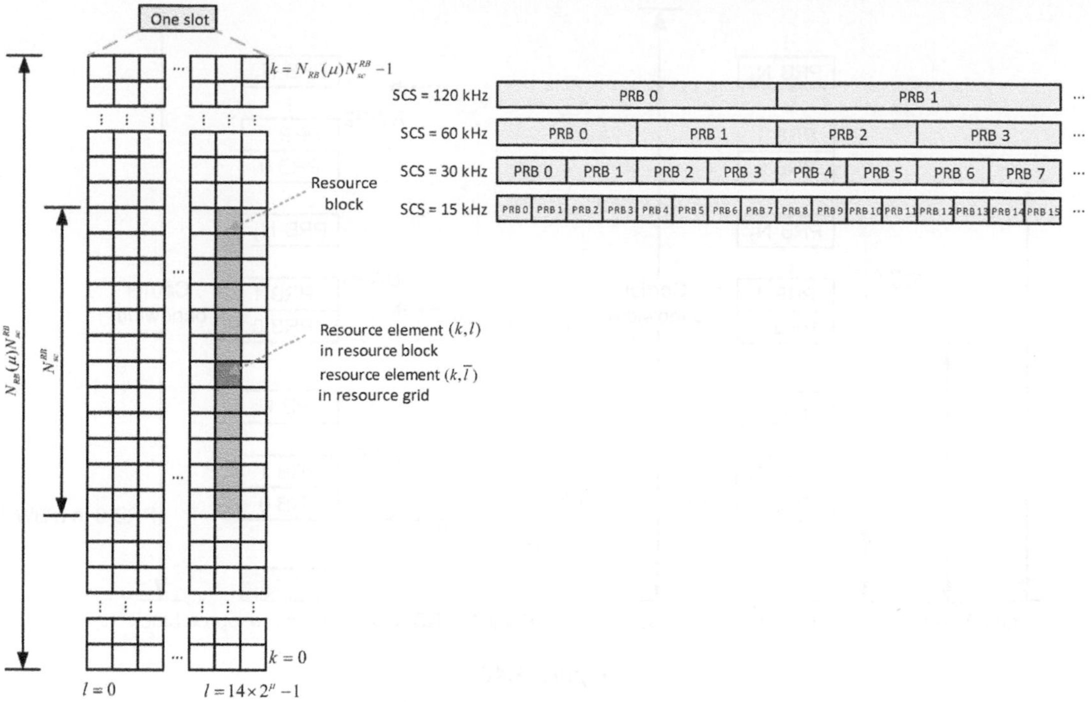

Figure 3.47
Illustration of NR resource grids and PRBs [12].

of each slot in the number of slots, in each configured BWP, based on the subcarrier spacing value provided by higher layer parameter *ref-scs*. The UE considers symbols in a slot indicated as downlink by higher layer parameter *UL-DL-configuration-common* or by higher layer parameter *UL-DL-configuration-dedicated* as available for receiving control/traffic. The UE further considers symbols in a slot as uplink indicated by higher layer parameter *UL-DL-configuration-common* or by higher layer parameter *UL-DL-configuration-dedicated* as available for transmission of control/traffic [14,15].

3.7 Time-Frequency Resources

3.7.1 Physical Resource Blocks

The basic scheduling unit in NR is a physical resource block (PRB) comprising 12 subcarriers in the frequency domain over one OFDM symbol. All subcarriers within a PRB have the same subcarrier spacing and CP length. When an NR system supports multiple numerologies, the corresponding PRBs are multiplexed in the time domain such that the boundaries of PRBs are aligned. For this purpose, multiple PRBs of the same bandwidth form a PRB grid, as illustrated in Fig. 3.47. A PRB grid formed by subcarriers spaced apart by

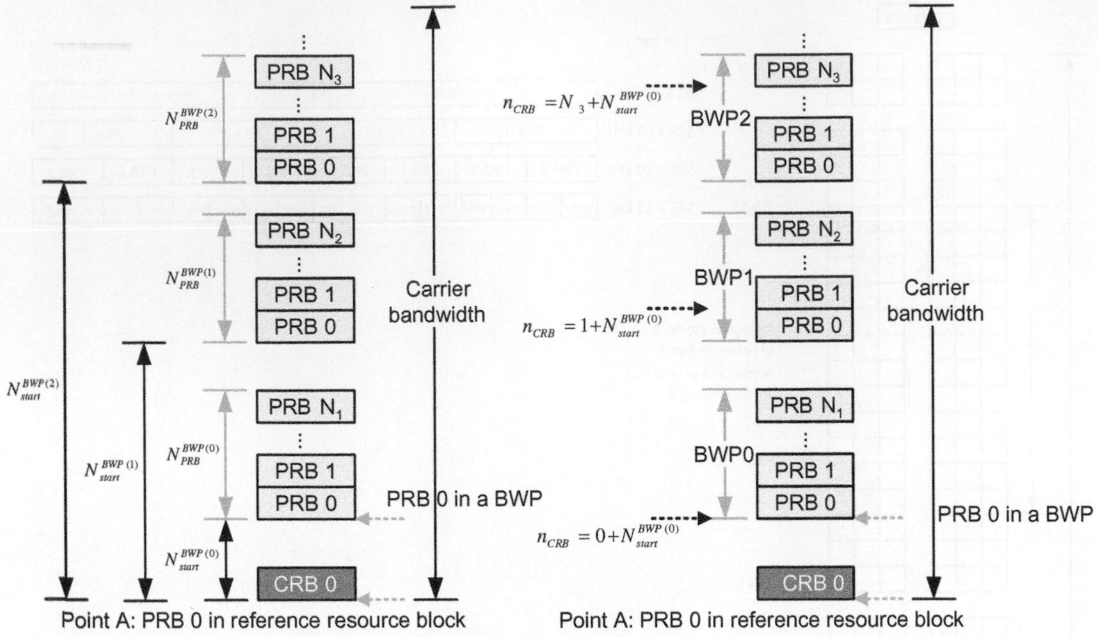

Figure 3.48
Mapping between n_{CRB} and n_{PRB} [12,39].

$\Delta f = 2^{\mu} \times 15$ kHz, where $\mu = 0, 1, \ldots, 4$ is a non-negative integer, is a superset of PRB grids with subcarrier spacing 15 kHz. For each numerology and carrier frequency, a resource grid of $N_{grid}(\mu)N_{sc}^{RB}$ subcarriers and $N_{symbol}(\mu)$ OFDM symbols is defined, starting at a CRB $N_{start}(\mu)$, whose value is signaled via RRC signaling. There is one set of resource grids per link direction (uplink or downlink). There is a single resource grid for a given antenna port p, numerology parameter μ and link direction [12,28].

An antenna port is a logical entity which is distinct from a physical antenna. Each antenna port is associated with a specific set of reference signals such that the channel over which a symbol is transmitted on that antenna port can be distinguished from the channel over which another symbol is conveyed on the same antenna port. Two antenna ports are said to be quasi-co-located, if the large-scale properties of the channel over which a symbol is conveyed on one antenna port can be inferred from the channel over which a symbol is conveyed on another antenna port. The large-scale properties include delay spread, Doppler spread, Doppler shift, average gain, average delay, and other spatial parameters [12]. In other words, a UE receiver can assume that the radio channels corresponding to two different antenna ports have the same large-scale properties (e.g., average delay spread, Doppler spread/shift, average delay, average gain, and spatial receive parameters), if the antenna ports are specified as being quasi-co-located. The UE can assume that two antenna ports are quasi-co-located with respect to certain channel properties either by NR specification or explicitly informed by the network via signaling.

Table 3.14: Minimum and maximum number of resource blocks/transmission bandwidths [12].

μ	min(N_{PRB})	max(N_{PRB})	Subcarrier Spacing (kHz)	Minimum Bandwidth (MHz)	Maximum Bandwidth (MHz)
0	24	275	15	4.32	49.5
1	24	275	30	8.64	99
2	24	275	60	17.28	198
3	24	275	120	34.56	396
4	24	138	240	69.12	397.44

Each element in the resource grid for antenna port p and numerology parameter μ is called a resource element and is uniquely identified by pair (k, l) where k is the index in the frequency domain (k is defined relative to point A such that $k = 0$ corresponds to the subcarrier centered around point A) and l refers to the symbol position in the time domain. A resource block is defined as $N_{sc}^{RB} = 12$ consecutive subcarriers in the frequency domain. Point A is a common reference point for resource block grids which is derived from the higher layer parameters [12].

As shown in Fig. 3.48, the resource blocks for each subcarrier spacing configuration are numbered from 0 to $N_{RB}(\mu)N_{sc}^{RB} - 1$ in upward direction in the frequency domain. Table 3.14 provides the minimum and maximum values of $N_{RB}(\mu)$ and their corresponding transmission bandwidths. The relationship between the CRB number n_{CRB} in the frequency domain and resource elements (k, l) for each subcarrier spacing configuration is given by $n_{CRB} = \lfloor k/N_{sc}^{RB} \rfloor$ where index k is defined relative to Point A such that $k = 0$ corresponds to the subcarrier centered around that point A. Physical resource blocks are defined within BWPs and are numbered from 0 to $N_{PRB}^{BWP(i)} - 1$. The relationship between the physical resource block n_{PRB} in the ith BWP and the CRB n_{CRB} is given by $n_{CRB} = n_{PRB} + N_{start}^{BWP(i)}$ where $N_{start}^{BWP(i)}$ is the CRB where BWP starts relative to CRB 0 as shown in Fig. 3.48. Similar to LTE, virtual resource blocks are defined within a BWP and are enumerated from 0 to $N_{VRB}^{BWP(i)} - 1$. The virtual resource blocks are resource blocks that are permuted across frequency dimension to take advantage of frequency diversity [12,28]. An interleaved mapping maps virtual resource blocks to physical resource blocks using an interleaver that spans the entire BWP and operates on pairs or quadruplets of resource blocks. A block interleaver with two rows is used, with pairs/quadruplets of resource blocks written in columns and read in rows. Whether to use pairs or quadruplets of resource blocks in the interleaving operation is configurable by higher layer signaling. The interleaved resource-block mapping in the frequency domain provides frequency diversity [13–15].

3.7.2 Bandwidth Part

3GPP NR supports very large operating bandwidths relative to the previous generations of 3GPP standards. Since the UEs in a cell may have different bandwidth capabilities, the use

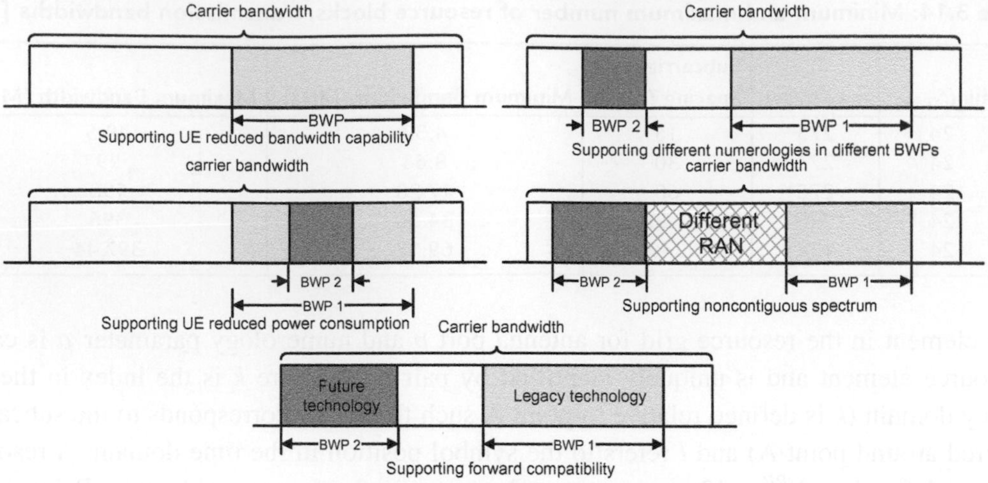

Figure 3.49
Bandwidth part use cases [42].

of wide bandwidth may cause more power consumption and may increase RF and baseband implementation complexity. Therefore, NR introduces the concept of BWP and allows the UEs with different bandwidth capabilities to operate in the cell with (configurable) smaller instantaneous bandwidth relative to the configured cell bandwidth, making NR more energy efficient solution despite its support of wideband channels (see Fig. 3.49). Alternatively, one may consider scheduling a UE such that it only transmits or receives within a certain frequency band. However, the difference of the latter approach with BWP is that the UE is not required to transmit or receive outside of the configured frequency band of the active BWP. The granularity of bandwidth allocation in NR is one PRB. For each serving cell, up to 4 downlink/uplink BWPs can be configured separately and independently for paired spectrum; nevertheless, only one BWP can be active at a given time and the UE is not expected to receive downlink/uplink physical signals/channels outside of an active BWP. For paired spectrum, a downlink BWP and an uplink BWP are jointly configured as a pair and up to four pairs can be configured. One can configure up to four BWPs on a supplemental uplink (SUL) carrier. Different use cases of the BWP are illustrated in Fig. 3.49 [28].

In other words, a BWP is a subset of contiguous CRBs for a given numerology (note that different numerologies may be used in different BWPs) on a given RF carrier. The starting position $N_{start}^{BWP(i)}$ and the number of resource blocks $N_{PRB}^{BWP(i)}$ in a BWP satisfy $N_{start}^{grid}(\mu) \leq N_{start}^{BWP(i)} < N_{start}^{grid}(\mu) + N_{size}^{grid}(\mu)$ and $N_{start}^{grid}(\mu) < N_{PRB}^{BWP(i)} + N_{start}^{BWP(i)} \leq N_{start}^{grid}(\mu) + N_{size}^{grid}(\mu)$, respectively. If a UE is configured with a SUL, then it can additionally be configured with up to four BWPs on the SUL with a single SUL BWP being active at a given time.

As we mentioned earlier, the transmit/receive bandwidth of a UE does not need to be as large as the bandwidth of the cell and it can be adaptively adjusted according to UE

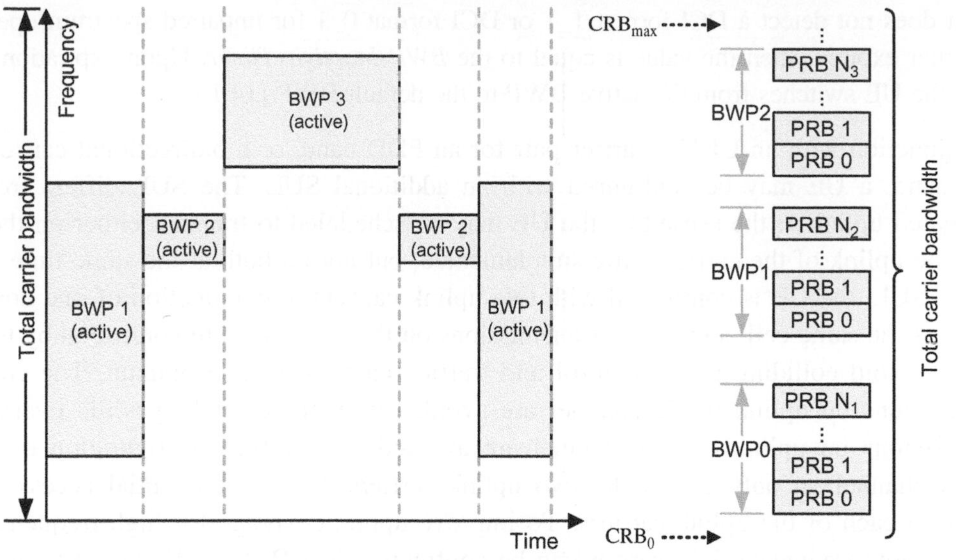

Figure 3.50
Illustration of the active bandwidth part adaptation concept [12,19,28].

operational conditions. With Bandwidth Adaptation (BA), a UE's bandwidth can be resized (e.g., reduced during a period of low activity for power saving); its location can be moved in the frequency domain (e.g., to increase scheduling flexibility); and the subcarrier spacing can be changed (e.g., to allow different services). A subset of the total cell bandwidth is called a BWP and the BA is achieved by configuring the UE with different BWP(s) and notifying the UE of the instantaneous active one. Fig. 3.50 shows a scenario where three different BWPs are configured: $BWP_1 = 50$ MHz and subcarrier spacing of 15 kHz; $BWP_2 = 10$ MHz and subcarrier spacing of 15 kHz; and $BWP_3 = 25$ MHz and subcarrier spacing of 60 kHz [19]. Note that at each time, only one BWP is active.

An initial active downlink BWP is defined by its location, number of contiguous PRBs, subcarrier spacing, and CP, for the control resource set corresponding to Type0-PDCCH common search space. For operation on the primary cell, a UE is provided by higher layer parameter *initial-UL-BWP* an initial uplink BWP to perform random-access procedure. If the UE is configured with a secondary carrier on the primary cell, it can also be configured with an initial BWP for random-access procedure on the secondary carrier [14,15]. A UE can be provided with a timer value by RRC parameter *BWP-InactivityTimer* for the primary cell. The UE subsequently starts the timer each time that it detects a DCI format 1_1 indicating an active downlink BWP, other than the default downlink BWP, for paired spectrum operation or each time the UE detects DCI format 1_1 or DCI format 0_1 indicating an active downlink or uplink BWP, other than the default downlink/uplink BWP, for unpaired spectrum operation. The UE increments the timer every 1 ms for sub-6 GHz carrier frequencies or every 0.5 ms for carrier frequencies above 6 GHz, if it does not detect a DCI format 1_1 for paired spectrum operation or if it does not detect a DCI format 1_1 or DCI format 0_1 for unpaired spectrum operation. The timer expires when the value is equal to the *BWP-InactivityTimer*. Upon expiration of the timer, the UE switches from the active BWP to the default BWP [14,15].

In conjunction with an UL/DL carrier pair for an FDD band, or a bidirectional carrier for a TDD band, a UE may be configured with an additional SUL. The SUL differs from the aggregated uplink in the sense that the UE may be scheduled to transmit either on the SUL or on the uplink of the carrier being supplemented, but not on both at the same time. In the case of SUL, the UE is configured with two uplink carriers in conjunction of one downlink carrier of the same cell, and uplink transmissions on those carriers are controlled by the network to avoid colliding uplink control and traffic channels in time domain. The colliding transmissions on uplink traffic channel are avoided through scheduling while overlapping transmissions on uplink control channel are avoided via limiting configuration of uplink control channel on only one of the two uplink carriers. In addition, initial access is supported on each of the uplink carriers. To improve uplink coverage for high-frequency scenarios, a low-frequency SUL carrier can be configured. In NR, the UE can take advantage of the bandwidth adaptation feature to save power while satisfying the requirements of various services/applications. The network can configure up to four BWPs for each UE, and dynamically send change indications to the UEs as required.

The BWP switching for a serving cell is used to activate an inactive BWP or to deactivate an active BWP at any given time, as shown in Fig. 3.51. The BWP switching is controlled by the PDCCH indicating a downlink assignment or an uplink grant, by the *bandwidthPartInactivityTimer*, or by the MAC entity itself upon initiation of random-access procedure. Upon addition of a Special Cell (SpCell)[22] or activation of an SCell, one BWP is initially active without receiving PDCCH indicating a downlink assignment or an uplink grant. The active BWP for a serving cell is indicated by either RRC or PDCCH. For paired spectrum, a downlink BWP is paired with an uplink BWP, and BWP switching applies to both uplink and downlink [20]. The BWP switching options are illustrated and compared in Fig. 3.52.

3.7.3 Resource Allocation

One of the main design objectives for signaling the resource allocation information, in the form of a set of resource blocks in each slot, to the active UEs in the cell is to find a balanced tradeoff between flexibility and signaling overhead. Indications of localized/distributed resource allocations to different UEs are transmitted via PDCCH. The resource allocation field in PDCCH is interpreted by the UE depending on the PDCCH DCI format.

[22] In the context of dual connectivity, the Special Cell refers to the primary cell of the MCG or the primary SCell of the SCG depending on whether the MAC entity is associated with the MCG or the SCG. Otherwise, the Special Cell refers to the PCell. A Special Cell supports PUCCH transmission and contention-based random-access procedure and is always activated [20].

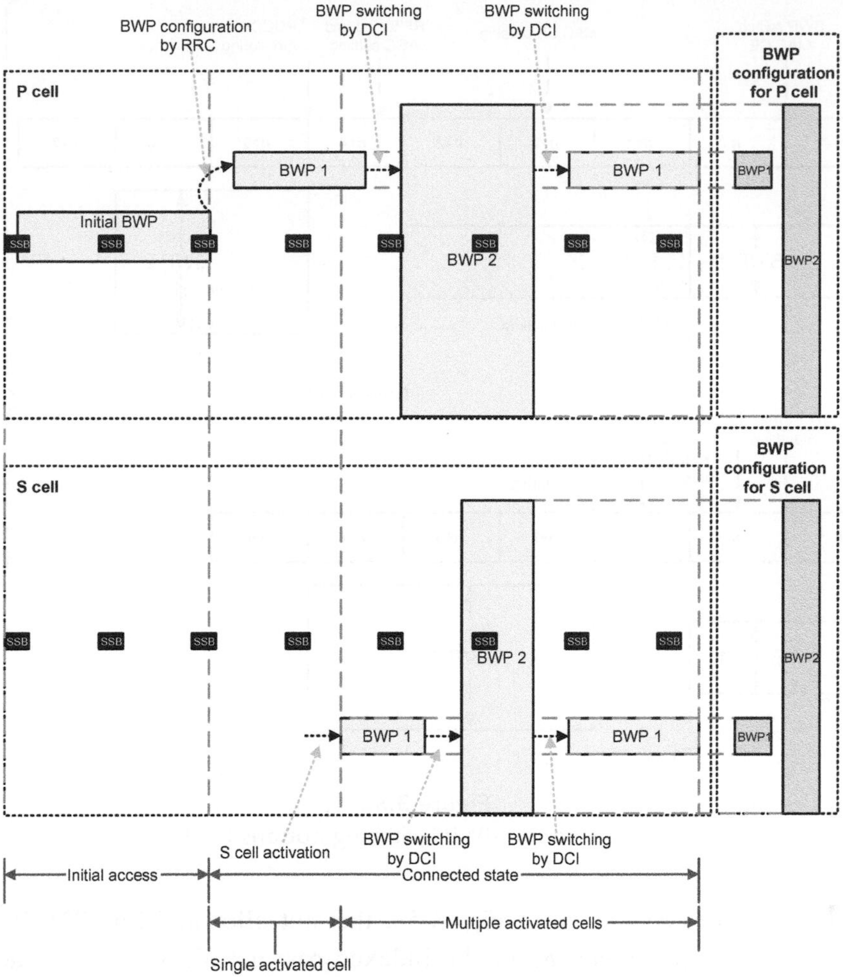

Figure 3.51
Illustration of BWP adaptation, activation, and switching [42].

The resource allocation in NR is defined both in time domain and frequency domain. Unlike LTE where the resource allocation in time domain was determined based on a fixed/predefined rule, in NR the resource allocation in time domain is more flexible, whereas the resource allocation in frequency domain is relatively similar to that of LTE [30]. Resource allocation type specifies the way in which the scheduler allocates physical resource blocks in frequency domain to each user for transmission in the downlink or uplink.

3.7.3.1 Resource Allocation in Time Domain

In the downlink, when the UE is scheduled to receive PDSCH via a DCI, that is, the *time domain resource assignment* field of DCI provides a row index to an allocation table, where the indexed row defines the slot offset K_0, the start and length indicator *SLIV* and the

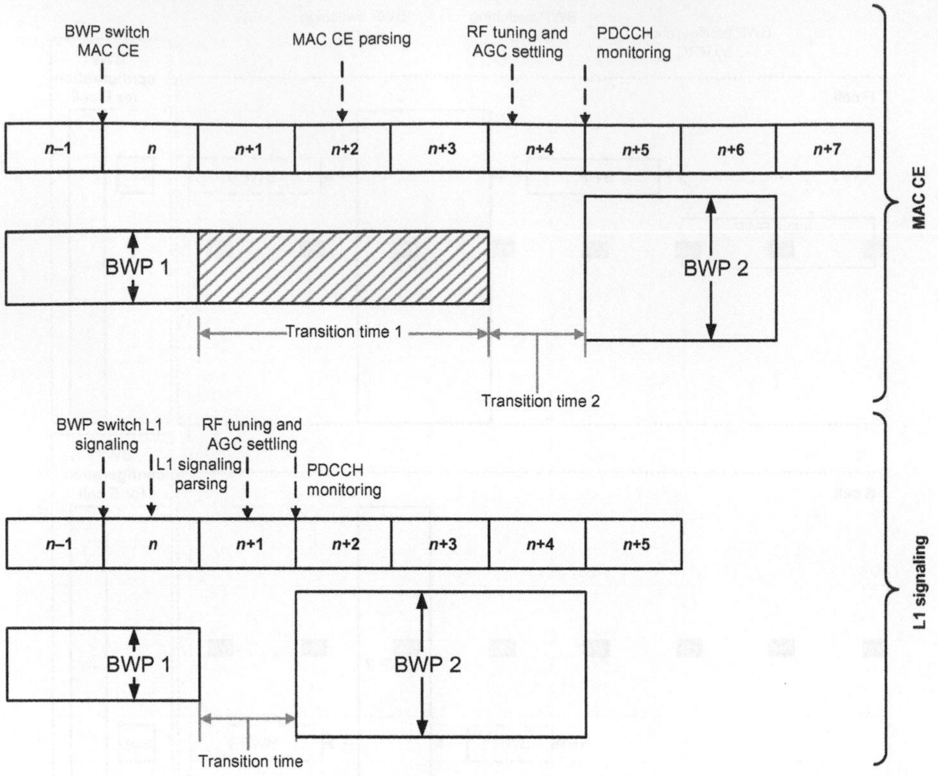

Figure 3.52
Comparison of BWP switching options [39].

PDSCH mapping type. As shown in Fig. 3.53, the slot allocated for PDSCH transmission is determined by parameter K_0 of the indexed row $n + K_0$ where n is the slot with the scheduling DCI, K_0 is based on the numerology of PDSCH. The starting symbol S relative to (the start of the slot), and the number of consecutive symbols L counting from the symbol S allocated for the PDSCH are determined from the start and length indicator $SLIV$ [14,15]. The slot allocated for the PDSCH is defined as $\lfloor (2^{\mu_{PDSCH}}/2^{\mu_{PDCCH}})n \rfloor + K_0$, where μ_{PDSCH} and μ_{PDCCH} are the subcarrier spacing configurations for PDSCH and PDCCH, respectively. If $(L-1) \leq 7$, then $SLIV = 14(L-1) + S$; otherwise $SLIV = 14(14 - L + 1) + (14 - S - 1)$, where $0 < L \leq 14 - S$ and the PDSCH mapping type is set to Type A or Type B [15]. The permissible S and L combinations corresponding to PDSCH allocations are shown in Table 3.15.

The PDSCH mapping type is related to the relative location of the demodulation reference signal (DM-RS) and the slot boundary as well as the size of the data. The mapping type A is used when the first DM-RS is located in the second or third OFDM symbol of

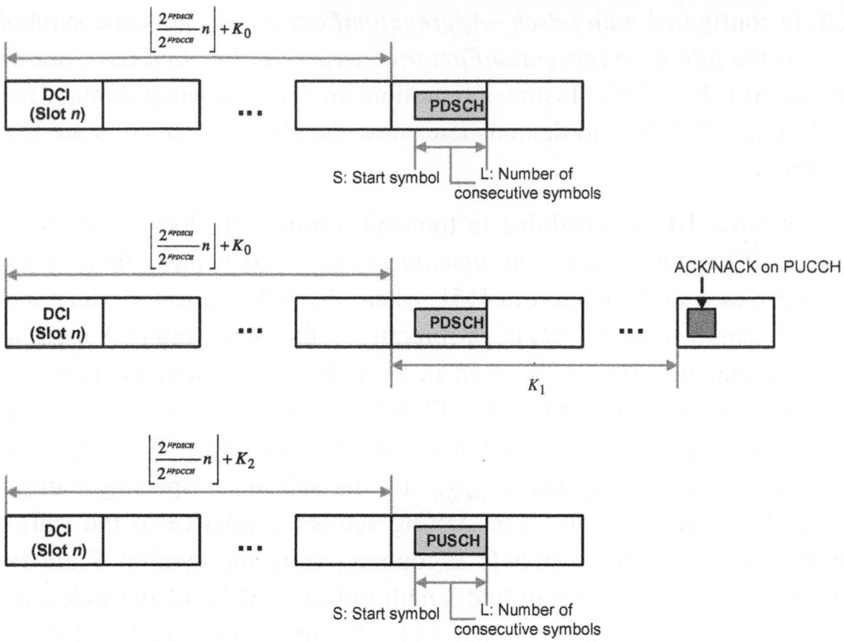

Figure 3.53
Illustration of downlink/uplink time-domain resource allocation [12,15,39].

Table 3.15: Permissible S and L values [15].

Mapping Type		Normal Cyclic Prefix			Extended Cyclic Prefix		
		S	L	$S+L$	S	L	$S+L$
PDSCH	Type A	{0,1,2,3}	{3,...,14}	{3,...,14}	{0,1,2,3}	{3,...,12}	{3,...,12}
	Type B	{0,...,12}	{2,4,7}	{2,...,14}	{0,...,10}	{2,4,6}	{2,...,12}
PUSCH	Type A	0	{4,...,14}	{4,...,14}	0	{4,...,12}	{4,...,12}
	Type B	{0,...,13}	{1,...,14}	{1,...,14}	{0,...,12}	{1,...,12}	{1,...,12}

the slot following a CORESET (i.e., a control region) at the beginning of a slot. The DM-RS is mapped relative to the start of the slot boundary regardless of the start of data transmission in the slot. This mapping type is primarily intended for the cases where the data occupies most of the slot. The mapping type B is used when the first DM-RS is located in the first symbol of the data allocation, that is, the DM-RS location is not given relative to the slot boundary, rather relative to where the data is located. This mapping is intended for transmissions over a small fraction of the slot to support very low latency and other transmissions that cannot wait until a slot boundary starts regardless of the transmission duration.

When the UE is configured with $pdsch{-}AggregationFactor > 1$, the same symbol allocation is applied across the $pdsch{-}AggregationFactor$ consecutive slots that have not been defined as uplink by the SFI. Fig. 3.53 illustrates time-domain and frequency-domain resource allocation procedure for PDSCH and demonstrates how the above parameters are used to locate the user allocation.

In the uplink, when a UE is scheduled to transmit a transport block on PUSCH by a DCI with or without CSI report(s), the *time domain resource assignment* field of the DCI provides a row index to a table defined in [15], where the indexed row defines the slot offset K_2, the start and length indicator $SLIV$, or directly by the start symbol S and the allocation length L, and the mapping type to be used in PUSCH transmission as shown in Fig. 3.53. The slot where the UE transmits the PUSCH is determined by parameter K_2 as $\lfloor (2^{\mu_{PUSCH}}/2^{\mu_{PDCCH}})n \rfloor + K_2$ where n is the slot with the scheduling DCI, K_2 is based on the numerology of PUSCH, μ_{PUSCH} and μ_{PDCCH} are the subcarrier spacing configurations for PUSCH and PDCCH, respectively. The starting symbol S relative to the start of the slot, and the number of consecutive symbols L counting from the symbol S allocated for the PUSCH are determined from the start and length indicator $SLIV$ of the indexed row as follows. If $(L-1) \leq 7$ then $SLIV = 14(L-1) + S$; otherwise $SLIV = 14(14 - L + 1) + (14 - S - 1)$, where $0 < L \leq 14 - S$ and the PUSCH mapping type is set to Type A or Type B [15]. The permissible S and L combinations corresponding to PDSCH allocations are shown in Table 3.15. When the UE is configured with $pusch - AggregationFactor > 1$, the same symbol allocation is applied across the $pusch - AggregationFactor$ consecutive slots and the PUSCH is limited to a single transmission layer. The UE repeats the transport block across $pusch - AggregationFactor$ consecutive slots applying the same symbol allocation in each slot [15].

3.7.3.2 Resource Allocation in Frequency Domain

A UE determines the frequency-domain resources on which it transmits or receives data by examining the resource-block allocation and BWP indicator fields in a DCI. The resource allocation fields determine the resources blocks in the active BWP on which data is transmitted. The gNB can signal the allocated resources to a UE using resource allocation type 0 or type 1, which are conceptually similar to LTE resource allocation type 0 and type 2 with the difference that in LTE, the resource allocation signales the allocations across the carrier bandwidth, whereas in NR, the indication is relevant only for the active BWP. Resource allocation type 0 is a bitmap-based allocation scheme, indicating the set of resource blocks that the UE is supposed to receive in the downlink transmission where the size of the bitmap is equal to the number of resource blocks in the BWP. This would allow an arbitrary combination of resource blocks to be scheduled for the UE at the expense of a large control/signaling overhead and some downlink coverage issues for larger BWP sizes due to limited capacity of a single OFDM symbol. Consequently, there

Table 3.16: Nominal resource block group size P [15].

Bandwidth Part Size $N_{PRB}^{BWP(i)}$	Configuration 1P	Configuration 2P
1−36	2	4
37−72	4	8
73−144	8	16
145−275	16	16

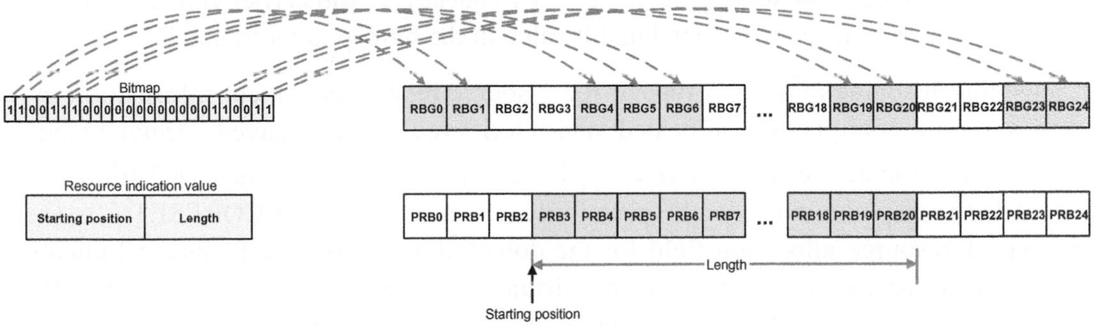

Figure 3.54
Illustration of frequency-domain type 0 and type 1 resource allocations [15,39].

is a need to reduce the bitmap size while maintaining the allocation flexibility. This can be achieved by addressing a group of contiguous resource blocks as opposed to individual PRBs. The size of the resource block group (RBG) is determined by the size of the BWP (see Table 3.16). Resource allocation type 1 indicates the allocated resources to the UE by means of a starting position and the length of the resource block allocation, thus only supporting contiguous allocations in frequency domain. In order to further reduce the signaling overhead, resource allocation type 1 combines the starting position and the length of resource allocation values into a single value referred to as resource indication value [15]. The two resource allocation types are illustrated in Fig. 3.54 (example).

The resource allocation scheme is configured using a bit in the DCI. Both resource allocation types refer to virtual resource blocks. For resource allocation type 0, a non-interleaved mapping from virtual to physical resource blocks is used, thus the virtual resource blocks are directly mapped to the corresponding physical resource blocks. For resource allocation type 1, both interleaved and non-interleaved mapping is supported. The VRB-to-PRB mapping bit, if present, indicates whether the allocation is based on interleaved or non-interleaved mapping.

In downlink resource allocation type 0, the resource block assignment information includes a bitmap representing the RBGs that are allocated to the scheduled UE where RBG is a set of consecutive physical resource blocks defined by a higher layer parameter and the size of the carrier BWP. The total number of RBGs N_{RBG} defined for a downlink carrier BWP of size $N_{PRB}^{BWP(i)}$ is $N_{RBG} = \left\lceil \left[N_{PRB}^{BWP(i)} + \left(N_{PRB}^{BWP(i)} \text{mod} P \right) \right] / P \right\rceil$ where the size of the first RBG is $RBG_0 = P - N_{start}^{BWP(i)} \text{mod} P$, the size of last RBG is $RBG_{last} = \left(N_{start}^{BWP(i)} + N_{PRB}^{BWP(i)} \right) \text{mod} P$, if $\left(N_{start}^{BWP(i)} + N_{PRB}^{BWP(i)} \right) \text{mod} P > 0$ and P otherwise. The size of all other RBGs is P [15]. The nominal values of P are given in Table 3.16. The bitmap is of size N_{RBG} bits with one bit per RBG such that each RBG is directly addressable. The RBGs is indexed in the order of increasing frequency and starting at the lowest frequency of the carrier BWP. An RBG is allocated to the UE, if the corresponding bit value in the bitmap is set to one [15].

In downlink resource allocation type 1, the resource block assignment information indicates a set of contiguously allocated non-interleaved or interleaved virtual resource blocks within the active BWP of size $N_{PRB}^{BWP(i)}$ except for the case when DCI format 1_0 is decoded in any common search space in which case the size of CORESET 0 is used. The type 1 resource allocation field for the downlink consists of a Resource Indication Value (RIV) corresponding to a starting virtual resource block RB_{start} and a length in terms of contiguously allocated resource blocks L_{RB}. In that case, if $(L_{RB} - 1) \leq \left\lfloor N_{PRB}^{BWP(i)} / 2 \right\rfloor$ then $\text{RIV} = N_{PRB}^{BWP(i)}(L_{RBs} - 1) + \text{RB}_{start}$; otherwise $\text{RIV} = N_{PRB}^{BWP(i)}(N_{PRB}^{BWP(i)} - L_{RB} + 1) + (N_{PRB}^{BWP(i)} - 1 - \text{RB}_{start})$, where $1 \leq L_{RB} < N_{PRB}^{BWP(i)} - RB_{start}$ [15]. However, when the DCI size for DCI format 1_0 in UE-specific search space is derived from the size of CORESET 0 and is applied to another active BWP with the size of N_{BWP}^{active}, a downlink type 1 resource block assignment field consists of a resource indication value corresponding to a starting resource block $\text{RB}_{start} = 0, K, 2K, \ldots, (N_{BWP}^{initial} - 1)K$ and a length in terms of virtually contiguously allocated resource blocks $L_{RBs} = K, 2K, \ldots, N_{BWP}^{initial}K$.

In the uplink, the UE determines the resource block assignment in frequency domain using the resource allocation field of DCI except for Msg.3 PUSCH initial transmission [15]. Two uplink resource allocations type 0 and type 1 are defined where resource allocation type 0 is used for PUSCH transmission when transform precoding is disabled. The uplink resource allocation type 1 is used for PUSCH transmission regardless of whether transform precoding is enabled or disabled. The UE can assume that when the scheduling PDCCH is received with DCI format 0_0, then uplink resource allocation type 1 is used. If a BWP indicator field is not configured in the scheduling DCI, the RB indexing for uplink type 0 and type 1 resource allocation is determined within the UE's active BWP. If a BWP indicator field is configured in the scheduling DCI, then the RB indexing for uplink type 0 and type 1 resource allocation is determined within the UE's BWP indicated by BWP indicator field value in the DCI. Upon detection of PDCCH intended for the UE, the UE must first determine the uplink BWP and then the resource allocation within the BWP [15].

The uplink resource allocation type 0 is similar to the downlink counterpart, where the resource block assignment information includes a bitmap indicating the RBGs that are allocated to the scheduled UE. The size of RBGs is given in Table 3.16. The uplink resource allocation type 1 is also similar to the downlink part described earlier, where the resource block assignment information informs the UE of a set of contiguously allocated noninterleaved virtual resource blocks within the active carrier BWP of size N_{PRB}^{BWP} except for the case when DCI format 0_0 is decoded in any common search space in which case the size of the initial BWP $N_{PRB}^{BWP(0)}$ is used [15].

3.7.3.3 Physical Resource Block Bundling

A UE cannot make any assumption about correlation of reference signals between different PDSCH scheduling occasions in the time domain. This is necessary to allow more flexibility in precoder-based beamforming and spatial signal processing. In the frequency domain, however, the UE can assume that there is correlation between reference signals within a precoding resource block group (PRG). Over the frequency span of one PRG, the UE may assume that the downlink precoder remains the same, and exploit this in the channel estimation process. The correlation assumption does hold between PRGs. It can be concluded that there is a tradeoff between the precoding flexibility and the channel estimation performance, that is, a large PRG size can improve the channel estimation accuracy at the cost of precoding flexibility. Therefore, the gNB may indicate the PRG size to the device where the possible PRG sizes are two, four or the total scheduled bandwidth in terms of PRBs. It is also possible to dynamically indicate the PRG size through the DCI. In addition, the UE can be configured to assume that the PRG size is equal to the scheduled bandwidth when the scheduled bandwidth is larger than half of the active BWP.

A UE may assume that the precoding granularity is $P'_{BWP(i)}$ consecutive resource blocks in the frequency domain, where $P'_{BWP(i)}$ values are taken from the limited set of {2, 4, wideband}. If $P'_{BWP(i)}$ is set to "wideband", the UE can expect to be scheduled with contiguous PRBs with the PRG and the use of the same precoding across the allocated resources by the gNB. If $P'_{BWP(i)}$ is set to 2 or 4, the ith BWP is partitioned into $P'_{BWP(i)}$ consecutive PRBs to form the PRGs. In practice, the number of consecutive PRBs in each PRG can be one or more. The first PRG size is given by $P'_{BWP(i)} - N_{start}^{BWP(i)} \bmod P'_{BWP(i)}$ and the last PRG size given by $(N_{start}^{BWP(i)} + N_{PRB}^{BWP(i)}) \mathrm{mod} P'_{BWP(i)}$, if $(N_{start}^{BWP(i)} + N_{PRB}^{BWP(i)}) \bmod P'_{BWP(i)} \neq 0$; otherwise, the last PRG size is $P'_{BWP(i)}$ [15]. The UE may assume the same precoding is applied for any downlink contiguous allocation of PRBs within a PRG. If a UE is scheduled a PDSCH with DCI format 1_0, it can assume that $P'_{BWP(i)}$ is equal to 2 PRBs [15].

3.7.4 Resource Allocation for Grant-Free/Semi-persistent Scheduling

In grant-free uplink transmissions, the UEs can transmit within a set of predetermined resource blocks without any explicit scheduling grants from the base station, resulting in lower control/signaling overhead and latency. However, uplink transmissions require the

UEs to transmit within a given set of resources that are pre-allocated by the base station with a certain periodicity. To avoid resource utilization inefficiency, multiple UEs might be allowed to share the same resources and to operate in a contention-based manner. Hence, collisions are inevitable, affecting connection reliability and latency, which is worsen as the UE traffic increases and retransmissions become necessary. In other words, grant-free transmission may not be scalable as the UE density increases in a network.

Fig. 3.55 shows a comparison between the contention-based (grant-free) and grant-based performance, where we observe that the efficiency of contention-based transmission degrades with increasing packet size and traffic load, while efficiency of grant-based transmission improves by increasing packet size and is stable over different load factors. The markings in the figure indicate that for packet sizes of 20, 30, and 40 bytes, there is a loading factor threshold below which contention-based transmission is more efficient than grant-based transmission. When the packet size is sufficiently small, for example, 10 bytes, contention-based transmission is always optimal within certain load factor range. Based on the analysis in [43], keep-alive messages of mobile Internet traffic are better suited to use

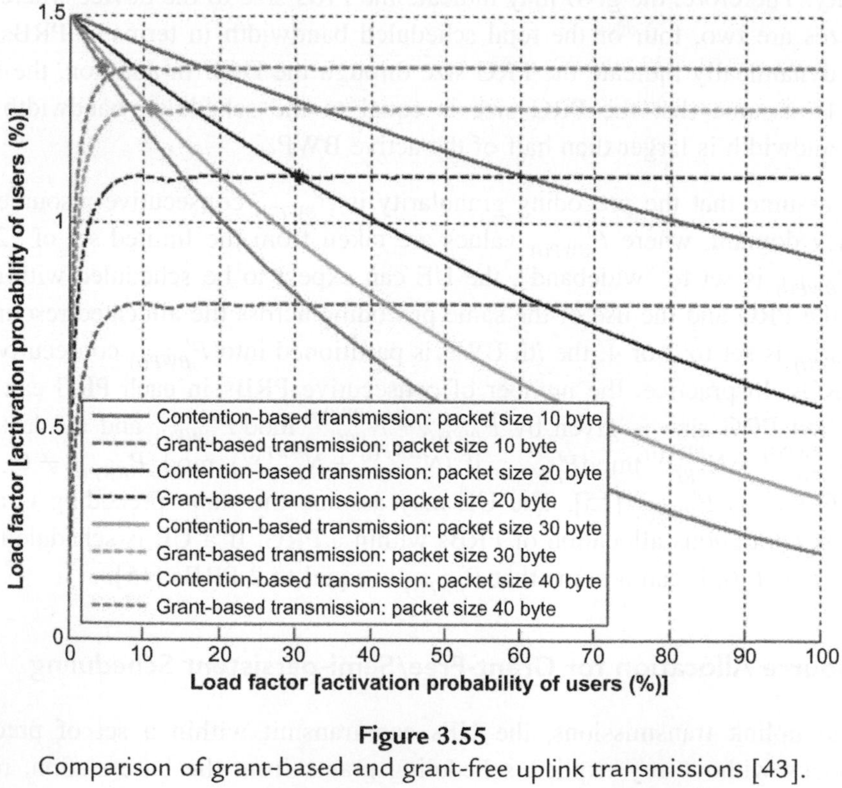

Figure 3.55
Comparison of grant-based and grant-free uplink transmissions [43].

contention-based transmission, while the rest of the mobile Internet packet types (e.g., video or voice) need to be transmitted by grant-based transmission, if not taking into account the latency constraints. This is because the former is of relatively small packet size and small loading, while packet size of the latter is large to be transmitted by contention (Fig. 3.55).

The semi-persistent scheduling (SPS)–based resource allocation refers to a transmission mode in which the serving base station allocates at least a part of resources and transport formats to the UE semi-statically over a certain time interval consisting of a number of TTIs. In the downlink, the semi-persistent scheduling is configured by RRC per serving cell and per BWP. Multiple configurations can be active simultaneously on different serving cells. Activation and deactivation of the downlink semi-persistent scheduling transmissions are independent among the serving cells [20].

In downlink semi-persistent scheduling, a downlink assignment is provided by PDCCH, and stored or cleared based on layer-1 signaling indicating semi-persistent scheduling activation or deactivation. A configured scheduling RNTI (CS-RNTI), which is similar to SPS C-RNTI in LTE, is configured by RRC for activation, deactivation, and retransmission. Furthermore, the number of configured HARQ processes and the periodicity of semi-persistent scheduling are signaled by RRC. When SPS resources are released by upper layers, all the corresponding configurations are subsequently released. In other words, the gNB can allocate downlink resources for the initial HARQ transmissions to UEs. The RRC defines the periodicity of the configured downlink assignments while PDCCH addressed with CS-RNTI can either signal and activate the configured downlink assignment, or deactivate it, that is, a PDCCH scrambled with CS-RNTI indicates that the downlink assignment can be implicitly reused according to the periodicity defined by RRC, until deactivated. After a downlink assignment is configured for semi-persistent scheduling, the MAC entity considers that the Nth downlink assignment will occur in a slot whose number meets the following criteria:

$$\left(N_{frame}^{slot} SFN + n_{slot}\right) = \left[\left(N_{frame}^{slot} SFN_{start-time} + T_{start-time}^{slot}\right) + N T_{SPS} N_{frame}^{slot}/10\right] \mathrm{mod}\left(1024 N_{frame}^{slot}\right)$$

where N_{frame}^{slot} denotes the number of slots per frame n_{slot} is the slot number in the frame, T_{SPS} denotes the SPS period, $SFN_{start-time}$ and $T_{start-time}^{slot}$ represent the SFN and slot of the first transmission of PDSCH where the configured downlink assignment was initialized, respectively [20].

In the uplink direction, there are two types of transmission without dynamic grant known as configured grant Type 1, where an uplink grant is provided by RRC, and stored as configured uplink grant; and configured grant Type 2, where an uplink grant is provided by PDCCH and stored or cleared as configured uplink grant based on physical layer signaling indicating configured grant activation or deactivation. Both Type 1 and Type 2 grants are configured by RRC per serving cell and per BWP. Multiple configurations can be active simultaneously on different serving cells. For Type 2 grant, activation and deactivation are independent among the serving cells. When the configured grant Type 1 is used, the RRC configures the following parameters: a CS-RNTI for retransmission; periodicity of the configured grant Type 1; the offset of a resource with respect to $SFN = 0$ in time domain; time-domain parameters which include the start symbol and the length of the assignment; as well as the number of HARQ processes [20]. Alternatively, when the configured grant Type 2 is going to be used, the RRC configures the following parameters: a CS-RNTI for activation, deactivation, and retransmission; the periodicity of the configured grant Type 2; and the number of HARQ processes. Once a configured grant Type 1 semi-persistent allocation is set up in a serving cell by upper layers, the corresponding MAC entity stores the uplink grant provided by upper layers and initializes the configured uplink grant to start in the symbol according to the provided parameters. After an uplink grant is set up for a configured grant Type 1 uplink transmission, the MAC entity considers the *N*th uplink transmit opportunity to occur in the symbol number which satisfies the following equation [20]:

$$\left[\left(N_{frame}^{slot} N_{slot}^{symbol} SFN\right) + \left(n_{slot} N_{slot}^{symbol}\right) + m_s\right] = \left(t_{offset} N_{slot}^{symbol} + S + NT_{SPS}\right) \bmod \left(1024 N_{frame}^{slot} N_{slot}^{symbol}\right), \quad \forall N \geq 0$$

where N_{slot}^{symbol}, m_s, and t_{offset} denote the number of symbols per slot, symbol number in the slot, and the time domain offset, respectively. Similarly, subsequent to an uplink grant set up for a configured grant Type 2, the MAC entity considers the time-domain location of the *N*th uplink grant-free transmission at the symbol for which the following criterion is satisfied [20]:

$$N_{frame}^{slot} N_{slot}^{symbol} SFN + n_{slot} N_{slot}^{symbol} + m_s = \left(SFN_{start-time} N_{frame}^{slot} N_{slot}^{symbol} + T_{start-time}^{slot} N_{slot}^{symbol} + T_{start-time}^{symbol} + NT_{SPS}\right) \bmod \left(1024 N_{frame}^{slot} N_{slot}^{symbol}\right), \quad \forall N \geq 0$$

where $SFN_{start-time}$, $T_{start-time}^{slot}$, and $T_{start-time}^{symbol}$ represent the SFN, slot, and symbol where the first transmission of PUSCH with the configured uplink grant was initialized, respectively. When a configured grant is released by upper layers, all corresponding configurations are cleared. Retransmissions except for repetition of the configured grants use uplink grants with CS-RNTI.

In summary, in the downlink, the gNB can allocate downlink resources for the initial HARQ transmissions to UEs with semi-persistent scheduling. The RRC signaling defines the periodicity of the configured downlink assignments while PDCCH addressed with

CS-RNTI can either signal and activate the configured downlink assignment, or deactivate it, that is, a PDCCH scrambled with CS-RNTI indicates that the downlink assignment can be implicitly reused according to the periodicity defined by RRC, until deactivated. In the uplink, the gNB can allocate uplink resources for the initial HARQ transmissions to UEs with configured uplink grants. Two types of configured uplink grants are defined: Type 1, where RRC directly provides the configured uplink grant (including the periodicity) and Type 2, where RRC defines the periodicity of the configured uplink grant while PDCCH addressed to CS-RNTI can either signal and activate the configured uplink grant, or deactivate it, that is, a PDCCH addressed with CS-RNTI indicates that the uplink grant can be implicitly reused according to the periodicity defined by RRC, until deactivated [19].

When carrier aggregation is configured, one configured uplink grant can be signaled per serving cell. Thus each serving cell can have one configured uplink grant active at any time. Similarly, when bandwidth adaptation is configured, one configured uplink grant can be signaled per BWP. A configured uplink grant for one serving cell can either be of Type 1 or Type 2. For Type 2, activation and deactivation of configured uplink grants are independent among the serving cells. When SUL is configured, a configured uplink grant can only be signaled for one of the two uplink carriers of the cell.

References

ITU-R Specifications[23]

[1] Report ITU-R P.2406-0, Studies for Short-Path Propagation Data and Models for Terrestrial Radiocommunication Systems in the Frequency Range 6 GHz to 100 GHz, September 2017.

[2] Recommendation ITU-R P.1238-9, Propagation Data and Prediction Methods for the Planning of Indoor Radiocommunication Systems and Radio Local Area Networks in the Frequency Range 300 MHz to 100 GHz, June 2017.

[3] Report ITU-R M.2376-0, Technical Feasibility of IMT in Bands Above 6 GHz, July 2015.

[4] Report ITU-R M.2412-0, Guidelines for Evaluation of Radio Interface Technologies for IMT-2020, October 2017.

[5] Report ITU-R M.2411-0, Requirements, Evaluation Criteria and Submission Templates for the Development of IMT-2020, November 2017.

3GPP Specifications[24]

[6] 3GPP TR 21.915, Summary of Rel-15 Work Items (Release 15), March 2019.

[7] 3GPP TR 36.873, Study on 3D Channel Model for LTE (Release 12), December 2017.

[8] 3GPP TR 37.910, Study on Self Evaluation towards IMT-2020 Submission (Release 15), December 2018.

[9] 3GPP TS 38.104, NR, Base Station (BS) Radio Transmission and Reception (Release 15), December 2018.

[10] 3GPP TS 38.201, NR, Physical Layer – General Description (Release 15), December 2018.

[11] 3GPP TS 38.202, NR, Services Provided by the Physical Layer (Release 15), December 2018.

---

23 ITU-R specifications can be accessed at the following URL: http://www.itu.int/en/ITU-R/study-groups/rsg5/rwp5d/imt-2020/.

24 3GPP specifications can be accessed at the following URL: http://www.3gpp.org/ftp/Specs/archive/.

[12] 3GPP TS 38.211, NR, Physical Channels and Modulation (Release 15), December 2018.
[13] 3GPP TS 38.212, NR, Multiplexing and Channel Coding (Release 15), December 2018.
[14] 3GPP TS 38.213, NR, Physical Layer Procedures for Control (Release 15), December 2018.
[15] 3GPP TS 38.214, NR, Physical Layer Procedures for Data (Release 15), December 2018.
[16] 3GPP TS 38.215, NR, Physical Layer Measurements (Release 15), March 2018.
[17] 3GPP TS 36.104, E-UTRA, Base Station (BS) Radio Transmission and Reception (Release 15), June 2018.
[18] 3GPP TS 38.133, NR, Requirements for Support of Radio Resource Management (Release 15), December 2018.
[19] 3GPP TS 38.300, NR, NR and NG-RAN Overall Description, Stage 2 (Release 15), December 2018.
[20] 3GPP TS 38.321, NR, Medium Access Control (MAC) Protocol Specification, (Release 15), December 2018.
[21] 3GPP TS 38.331, NR, Radio Resource Control (RRC), Protocol Specification (Release 15), December 2018.
[22] 3GPP TR 38.802, Study on New Radio Access Technology Physical Layer Aspects (Release 14), March 2017.
[23] 3GPP TR 38.812, Study on Non-Orthogonal Multiple Access (NOMA) for NR; (Release 15) October 2018.
[24] 3GPP TR 38.900, Study on Channel Model for Frequency Spectrum Above 6 GHz (Release 15), June 2018.
[25] 3GPP TR 38.901, Study on Channel Model for Frequencies From 0.5 to 100 GHz (Release 15), June 2018.

Articles, Books, White Papers, and Application Notes

[26] P. Marsch, Ö. Bulakci, 5G System Design: Architectural and Functional Considerations and Long-Term Research, Wiley, 2018.
[27] A. Zaidi, F. Athle, 5G Physical Layer: Principles, Models and Technology Components, Academic Press, 2018.
[28] E. Dahlman, S. Parkvall, 5G NR: The Next Generation Wireless Access Technology, Academic Press, 2018.
[29] F.-L. Luo, C.J. Zhang, Signal Processing for 5G: Algorithms and Implementations, Wiley-IEEE Press, 2016.
[30] S. Ahmadi, *LTE-Advanced: A Practical Systems Approach to Understanding 3GPP LTE Releases 10 and 11 Radio Access Technologies*, 2013, Academic Press.
[31] T.L. Marzetta, et al., Fundamentals of Massive MIMO, Cambridge University Press, 2016.
[32] T.S. Rappaport, R.W. Heath Jr., Millimeter Wave Wireless Communications, Prentice Hall, 2014.
[33] B. Sklar, Digital Communications: Fundamentals and Applications, Prentice Hall, 1987.
[34] F. Ademaj, et al., 3GPP 3D MIMO channel model: a holistic implementation guideline for open source simulation tools, EURASIP J. Wireless Commun. Netw. (2016) 55.
[35] B. Mondal, et al., 3D channel model in 3GPP, IEEE Commun. Mag. 53 (Issue 3) (2015).
[36] T.S. Rappaport et al., Overview of millimeter wave communications for fifth-generation (5G) wireless networks—with a focus on propagation models, IEEE Trans. Antennas Propag. (Special Issue on 5G) (2017).
[37] M. Zhang, et al., ns-3 Implementation of the 3GPP MIMO Channel Model for Frequency Spectrum above 6 GHz, ePrint of Cornell University Library, February 2017.
[38] X. Wang et al., Multicarrier Waveforms for 5G, Institut für Nachrichtenübertragung, WebDemos. <http://www.inue.uni-stuttgart.de/lehre/demo.html>.
[39] 5G New Radio, ShareTechNote. Available from: <http://www.sharetechnote.com>.
[40] D. Bharadia et al., Full duplex radios, in: Proceedings of ACM SIGCOMM, 2013.

[41] A. Sabharwal, et al., In-band full-duplex wireless: challenges and opportunities, IEEE J. Sel. Areas Commun. 32 (9) (2014).

[42] J. Campos, Understanding the 5G NR Physical Layer, Keysight Technologies, 2017.

[43] I. Chih-Lin, Seven fundamental rethinking for next-generation wireless communications, SIP 6 (2017).

[44] J. Fan, et al., Faster-than-Nyquist signaling: an overview, IEEE Access 5 (2017).

[45] A. Roessler, 5G Waveform Candidates, Application Note, Rohde & Schwarz, June 2016.

[46] B. Farhang-Boroujeny, OFDM vs. filter bank multicarrier, IEEE Signal Process Mag. (2011).

[47] White Paper, 5G Waveform & Multiple Access Techniques, Qualcomm Technologies, 2015.

[48] GTI 5G Device RF Component Research Report, 2018. Available from: <http://www.gtigroup.org>.

[49] 3GPP TSG RAN WG1, R1-165425, f-OFDM Scheme and Filter Design, 2016.

[50] Z. Ding, et al., A survey on non-orthogonal multiple access for 5G networks: research challenges and future trends, IEEE J. Sel. Areas Commun. 35 (10) (2017).

[51] Z. Wu et al., Comprehensive study and comparison on 5G NOMA schemes, IEEE Access, (2018).

[52] Y. Yuan, et al., Non-orthogonal transmission technology in LTE evolution, IEEE Commun. Mag. (2016).

[53] MediaTek, A New Era for Enhanced Mobile Broadband, White Paper, March 2018.

CHAPTER 4

New Radio Access Physical Layer Aspects (Part 2)

In this chapter, we discuss the theoretical and practical aspects of the downlink and uplink physical layer signal processing in NR and highlight the functional and procedural similarities and differences with the LTE physical layer processing. The chapter will describe generation, configuration, and beamformed transmission of various physical signals and physical channels as well as the HARQ protocols and power control schemes. Unlike LTE where the downlink and uplink waveforms and multiple access schemes are different, the NR uses OFDM waveform as the basis for both downlink and uplink transmission (except in certain cases where DFT precoding is used in the uplink), resulting in many similarities in functional blocks and their respective operation in the downlink and uplink. The physical channel processing in NR utilizes polar codes for robust coding of the control channels and low-density parity check (LDPC) codes for the data channels, deviating from channel coding schemes that are used in LTE.

Massive MIMO is one of the main enabling technologies in 5G wireless communications. A large number of antenna elements at the base station bring extra degrees of freedom for increasing the throughput and considerable beamforming gains for improving the coverage. In practice, a large number of antenna elements can be assembled into multiple antenna panels for the purpose of cost reduction and power saving. Multi-panel MIMO is expected to become promising for mmWave massive MIMO systems. The NR enables multi-panel antenna array operation through introduction of new reference signals, measurement, and reporting procedures. In this chapter, two of the unique NR MIMO features, that is, modular and high-resolution channel state information acquisition and beam management that distinguish NR from LTE, are described. The modular framework is composed of three components, namely resource setting, CSI reporting setting, and measurement setting, which associates a resource setting with a reporting setting. These settings serve as building blocks that allow the network to customize the CSI measurement and reporting for a UE. To improve user throughput, a high-resolution dual-stage precoding referred to as Type II CSI is supported to allow more accurate estimation of the channel, thereby improving the efficiency of the NR MU-MIMO schemes. The associated codebook features a frequency non-selective basis subset selection coupled with a frequency-selective linear combination of amplitude and phase of the precoding vectors within the basis subset. As a result, the

5G NR. DOI: https://doi.org/10.1016/B978-0-08-102267-2.00021-X

precoding matrix indicator (PMI) for Type II CSI consists of several components, each with different frequency resolution. Since the NR is primarily geared toward MU-MIMO operation, Type II CSI is complemented by Type I CSI designed for scenarios that do not require high spatial resolution, for example, SU-MIMO transmission.

In order to establish and sustain a link for data transmission and reception, beam management enables the network to perform beam switching using physical-layer measurement and link quality reporting. Beam management is especially relevant for above-6 GHz frequency planning where both gNB and UE employ narrow beams for data transmission and reception. The beam management can further be used for sub-6 GHz multi-TRP scenarios. When used in conjunction with CSI acquisition, the beam management allows the network to establish a seamless and low-latency link with the UE for data transmission. This is specifically important for over-6 GHz where a large number of narrow analog beams are used for data transmission, which in some scenarios requires frequent beam switching. Once the link is established via beam management, CSI acquisition can assist the network in link adaptation.

4.1 Downlink Physical Layer Functions and Procedures

4.1.1 Overall Description of Downlink Physical Layer

The NR downlink physical layer consists of higher layer configurable functional blocks and protocols that are configured according to the downlink physical channel characteristics, use case, deployment scenario, etc. As shown in Fig. 4.1, the downlink physical layer processing generally includes receiving higher layer data [e.g., MAC PDUs in the case of downlink shared channel or master information block (MIB) in the case of physical broadcast channel (PBCH)]; cyclic redundancy check (CRC) calculation and attachment; channel encoding and rate matching; modulation; mapping to physical resources and antennas; multi-antenna processing; and support of layer-1 control and HARQ-related signaling.

It was mentioned in Chapter 3 that OFDM was chosen as the default waveform in NR for both downlink and uplink directions due to its robustness to multipath delay spread and frequency-selectivity of wireless channels as well as scheduling flexibility for transmission of different channels and signals. Unlike LTE, the DFT-precoded OFDM is an optional transmission scheme in NR uplink that is used in link-budget–limited use cases. While the use of DFT-precoded OFDM in the uplink has certain advantages in reducing the PAPR (and alternatively the cubic metric) and achieving higher power–amplifier efficiency, it has several drawbacks including limitation in the use of spatial multiplexing, asymmetric downlink/uplink transmissions which would limit the sidelink operation, and scheduling complexity.

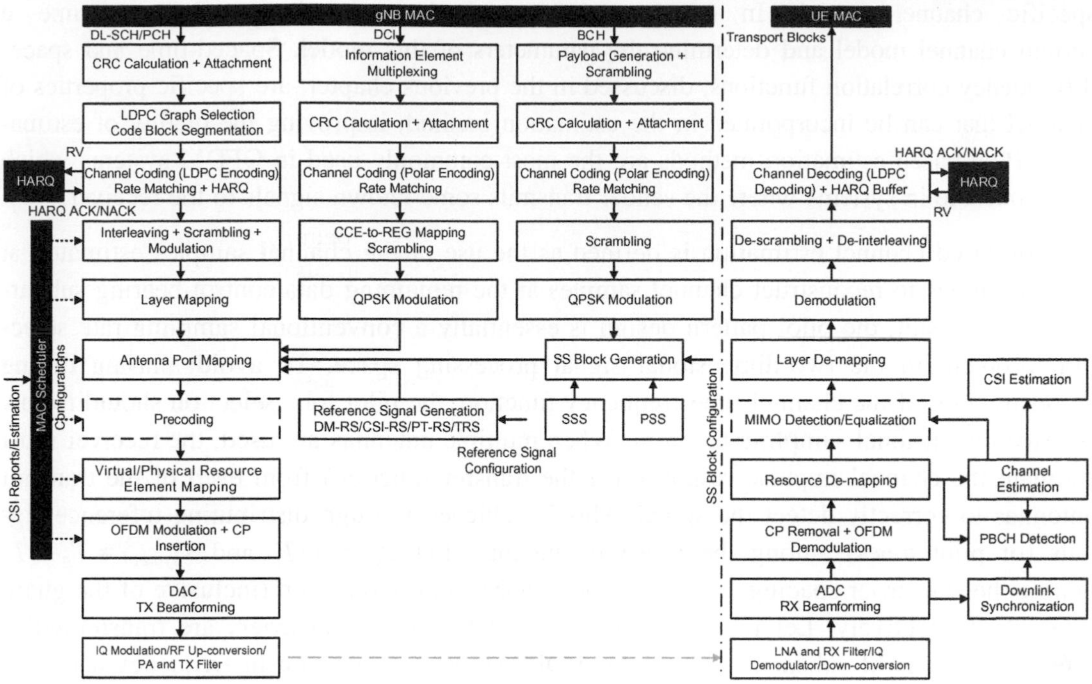

Figure 4.1
Overall downlink physical layer processing [5].

4.1.2 Reference Signals

To facilitate estimation of the multipath communication channel and reliable, coherent detection of traffic/control channels, an OFDM system makes use of reference signals (or pilot subcarriers). The pilot subcarriers provide estimate of the channel frequency response at the pilot locations over the time-frequency resource grid. It is possible to estimate the channel at other time-frequency locations using interpolation techniques. Using predefined pilot subcarriers to estimate the channel matrix, it is possible to equalize the effects of the channel and to reduce noise and interference effects on the received resource blocks. The NR specifications include several types of reference signals that are configured and transmitted in different manners, which are used for different purposes by a receiving device.

While perfect knowledge of the radio channel can be used to find an upper bound for system performance, such knowledge is not available in practice and the channel needs to be frequently estimated. Channel estimation can be performed in various ways including the use of frequency and/or time correlation properties of the wireless channel, blind or pilot-based channel estimation, and adaptive or non-adaptive channel estimation. Non-parametric methods attempt to estimate the frequency response without relying on a specific channel model. In contrast, the parametric estimation methods assume a certain channel model and determine the parameters of this model. Spaced-time and spaced-frequency correlation functions, discussed in the previous chapter, are specific properties of channel that can be incorporated in the estimation method, improving the quality of estimations. Pilot-based estimation methods are the most commonly used in OFDM systems which are applicable in systems where the sender transmits some known signals to the receiver.

A pilot-based channel estimation is defined as the use of the channel samples estimated at the pilot tones to reconstruct channel samples at the remaining data/control-bearing subcarriers. As a result, the pilot pattern design is essentially a conventional sampling rate selection problem in the two-dimensional signal processing space. To avoid aliasing during reconstruction of the channel time-frequency function, the pilot tone selection should follow the two-dimensional sampling theorem. When multiple antennas are used, the receiver must estimate the channel impulse response (or the transfer function) from each of the transmit antennas to correctly detect the signal. This is achieved through distributing reference signals (or pilot tones) among the transmit antennas. Let $\Delta f = 1/T_u$ and $T_{symbol} = T_g + T_u$ denote the subcarrier spacing (SCS) and the OFDM symbol duration (inclusive of the guard interval), respectively. Let us further assume that the pilot subcarriers are transmitted at integer multiples of subcarrier spacing and OFDM symbol duration in frequency and time directions, respectively (i.e., $f_p = m/T_u$ and $T_p = nT_{symbol}$ where m and n are integers). The (m, n) pair represents the pilots' separation in terms of subcarrier spacing and OFDM symbol duration. From the sampling theorem point of view, the channel's two-dimensional delay-Doppler response $h(\tau, \nu)$ can be fully reconstructed, if the two-dimensional transform function $H(t,f)$ is sampled greater than or equal to the Nyquist rate across time and frequency dimensions. Hence, the time-domain sampling rate must be greater than or equal to the channel's maximum Doppler spread, that is, $T_p \leq 1/\nu_{\max}$ (the sampling rate in time must be less than the coherence time) and the frequency-domain sampling rate must be greater than or equal to the channel's maximum delay spread, that is, $f_p \leq 1/\tau_{\max}$ (the sampling rate in frequency must be less than coherence bandwidth). Assuming a wide-sense stationary uncorrelated scattering channel model and further assuming the channel to be constant over one OFDM symbol, the frequency response $H(t,f)$ of an L-path channel is given $H(t,f) = \sum_{l=1}^{L} \exp[j(\psi_l + 2\pi\nu_l t - 2\pi f \tau_l)]/\sqrt{L}$ where ψ_l, ν_l, and τ_l denote the phase, Doppler frequency, and delay of the lth path, respectively. All these parameters are independent random variables. In general, the pilot signals are oversampled to ensure a good trade-off between performance and overhead. Therefore the choice of (m, n) depends on the channel's maximum delay spread and maximum Doppler spread and must satisfy the following equation according to two-dimensional sampling theorem:

$$n \leq \frac{1}{2T_{symbol}\nu_{\max}}, \quad m \leq \frac{T_u}{2\tau_{\max}}$$

It should be noted that the pilot density for a regular pattern can be calculated using the preceding equation. For large values of $\nu_{\max}$ (i.e., large Doppler spread, or alternatively small channel coherence time means that channel is time-varying), n should be small to appropriately track channel time variations. On the other hand, for large values of $\tau_{\max}$ (i.e., large delay spread, or alternatively small coherence bandwidth means that channel is frequency-selective), m should be small to closely follow channel frequency variation. In a regularly spaced pilot pattern, the pilot symbols are evenly spaced in frequency and in time.

Cell-specific reference signals (CRS) were originally defined in 3GPP LTE Rel-8 and have continued to be used as downlink wideband always-on power-boosted pilot subcarriers which are scaled with the number of transmit antenna ports that are essential in coherent decoding/detection of the LTE downlink control channels, mobility measurements, etc. The always-on and power-boosted properties of CRS have been the potential cause of inter-cell interference in LTE networks even in the absence of user traffic. The reference signals in NR are different from LTE in the sense that they are only present when the UE has data allocation; thus they are UE specific and confined to the time-frequency region where user data are allocated. Unlike LTE, NR does not utilize cell-specific reference signals, rather exclusively relies on user-specific demodulation reference signals (DM-RSs) for channel estimation, enabling efficient beamforming and other multi-antenna schemes. In NR, the reference signals are not transmitted unless there are data to transmit, thereby improving network energy efficiency and reducing inter-cell interference. Support for low-latency transmission is an important part of NR design. In a front-loaded DM-RS structure, the reference signals and downlink control channel carrying scheduling information are located at the beginning of the slot; thus a device can start processing the received data immediately without prior buffering and time-domain interleaving across OFDM symbols, thereby minimizing the decoding delay. There are other types of reference signals such as phase tracking reference signals (PT-RS) which are used to counter phase noise at higher frequencies. A question may arise that if there are no wideband cell-specific reference signals, how the UEs will measure the reference signal received power (RSRP) during initial access or cell selection (mobility measurements) as they will not have any allocation at that time? The answer lies in the fact that NR uses a primary synchronization signal (PSS)/secondary synchronization signal (SSS)/PBCH block structure, where the PBCH has its own set of reference signals which will always be present in the PBCH so the UE while detecting the PSS/SSS/PBCH block should be able to measure an RSRP value from the PBCH DM-RS.

To support channel tracking, different types of reference signals are transmitted in downlink and uplink. The reference signals in the downlink include the following [6]:

- UE-specific DM-RS for physical downlink control channel (PDCCH) can be used for downlink channel estimation and coherent demodulation of PDCCH. The DM-RS for PDCCH is transmitted together with the PDCCH and is present only in the resource blocks that are used for PDCCH transmission.

- UE-specific DM-RS for physical downlink shared channel (PDSCH) can be used for downlink channel estimation for coherent demodulation of PDSCH. The DM-RS for PDSCH is transmitted together with the PDSCH and is present only in the resource blocks that are allocated for PDSCH transmission.
- UE-specific PT-RS can be used in addition to the DM-RS for PDSCH for correcting common phase error (CPE) between PDSCH symbols not containing DM-RS. It may also be used for Doppler and time-varying channel tracking. The phase noise of the transmitters increases as the frequency of operation increases. The PT-RS plays a crucial role especially in mmWave frequencies to minimize the effect of the oscillator phase noise on system performance. The phase noise appears as a common phase rotation of all the subcarriers, known as CPE in an OFDM system. The NR system typically maps the PT-RS information to a few subcarriers per symbol because the phase rotation equally affects all subcarriers over an OFDM symbol but exhibits low correlation from symbol to symbol. The system configures the PT-RS depending on the quality of the oscillators, carrier frequency, SCS, and modulation and coding schemes that are used for the transmission. The PT-RS for PDSCH is transmitted together with the PDSCH on need basis. The PT-RS is denser in time domain but sparser in frequency domain compared to the DM-RS, and if configured, occurs only in conjunction with the DM-RS.
- UE-specific CSI-RS can be used for estimation of CSI to allow CSI measurement and reporting which assists the gNB in modulation coding scheme (MCS) selection, resource allocation, beamforming, and MIMO rank selection. The CSI-RS can be configured for periodic, aperiodic, or semi-persistent transmission with a configurable density by the gNB. The CSI-RS also can be used for interference measurement (IM) and fine frequency/time tracking purposes. Specific instances of CSI-RS can be configured for time/frequency tracking and mobility measurements. In the absence of the cell-specific reference signals in NR the CSI-RS can be used for radio resource management, measurements and mobility management purposes in connected mode.
- Tracking reference signals (TRS) are sparse set of reference signals, which are intended to assist the device in time and frequency tracking. The TRS does not exist independently, and a specific CSI-RS configuration is used as TRS. In addition to time and frequency tracking, the TRS is used for estimation of delay spread and Doppler spread at the UE side. It is transmitted with a limited bandwidth for a configurable period of time, controlled by the upper layer parameters.

Table 4.1 provides the L1 overhead associated with various NR downlink reference signals. In NR, the overhead due to the L1/L2 control signaling depends on the size and periodicity of the configured control resource set (CORESET) in the cell which includes the overhead from the PDCCH DM-RSs. If the CORESET is transmitted in every slot, maximum control channel overhead is 21% assuming three symbols and the entire carrier bandwidth used for CORESET, while a more typical overhead is 7% when one-third of the time and frequency

Table 4.1: Various downlink reference signals and their corresponding overhead [73,74].

Reference Signal Type	Description	Overhead
PDSCH DM-RS	The DM-RS can occupy 1/3, 1/2 or one full OFDM symbol. 1, 2, 3 or 4 symbols per slot can be configured to carry DM-RS.	2.4%–29%
PDSCH PT-RS	One resource element in frequency domain every second or fourth resource block. PT-RS is mainly intended for FR2.	0.2%-0.5%
CSI-RS	One resource element per resource block per antenna port per CSI-RS periodicity	0.25% for eight antenna ports transmitted every 20 ms with 15 kHz subcarrier spacing
TRS	Two slots with two symbols in each with comb-4 configuration	0.36% or 0.18% for 20 ms and 40 ms periodicity, respectively and 15 kHz subcarrier spacing

resources in the first three symbols of a slot are allocated to PDCCH. The overhead due to the SS/PBCH block is given by the number of SS/PBCH blocks transmitted within the SS/PBCH block period, the SS/PBCH block periodicity and the subcarrier spacing. Assuming 100 resource blocks across the carrier, the overhead for 20 ms periodicity is in the range of 0.6%–2.3% if the maximum number of SS/PBCH blocks is transmitted [73,74].

4.1.2.1 Demodulation Reference Signals

The main application of DM-RS in NR is to estimate the channel coefficients for coherent detection of the physical channels. In the downlink, the DM-RS is used for channel estimation and is subject to the same precoding as PDSCH; thus the (transmit-side) precoding is transparent to the receiver and is viewed as part of the overall channel. There is a trade-off between the channel estimation accuracy and DM-RS density/overhead. If the channel exhibits severe frequency-selectivity (i.e., narrower channel coherence bandwidth), the DM-RS density in the frequency-domain should be increased. Similarly, if the channel varies faster in time-domain (i.e., shorter channel coherence time), denser DM-RS allocation across time is required. After determining frequency/time-domain DM-RS densities, the DM-RS locations in the time-frequency resource grid should be considered. Assuming stationary channel conditions, uniform DM-RS allocation in both frequency and time-domain is preferred for minimizing interpolation error and reducing implementation complexity. Since no user data is transmitted by DM-RS per se, allocating DM-RS with a proper density is required to maximize the throughput.

In NR, a front-loaded DM-RS structure is used as a baseline to achieve low-latency decoding (see Fig. 4.2). In the time-frequency resource grid, the front-loaded DM-RS can be located just after the control region, followed by data region. As soon as channel is

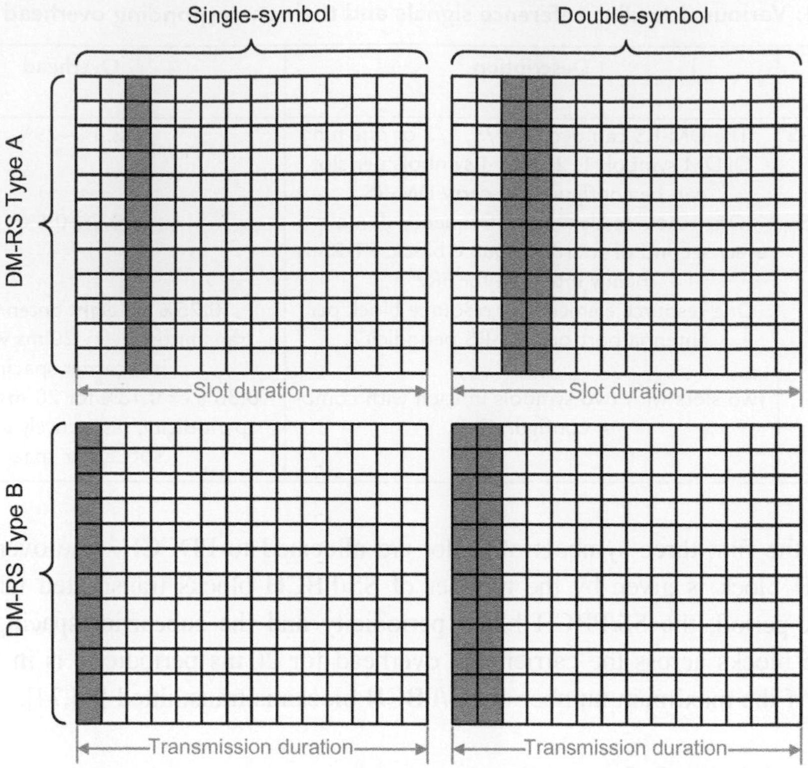

Figure 4.2
Comparison of DM-RS Type A and Type B mappings [6].

estimated based on the front-loaded DM-RS, the receiver can coherently demodulate data in the data region. The front-loaded DM-RS structure is particularly advantageous in decoding-latency reduction for low-mobility scenarios where channel coherence time is longer than the duration of the front-loaded DM-RS. However, allocating only the front-loaded DM-RS can degrade the link performance at higher UE speeds (i.e., channel coherence time becomes shorter). Although the channel information in the data region can be obtained by interpolation, the channel information accuracy diminishes with higher mobility. Therefore, we consider the front-loaded DM-RS patterns with 2 × and 4 × time-domain densities as shown in Fig. 4.3 [70]. To support high-speed scenarios, it is possible to configure up to three additional DM-RS occasions in a slot. The channel estimation in the receiver side can use these additional reference signals for more accurate channel estimation, for example, to perform interpolation between the DM-RS occasions within a slot. However, unlike LTE, it is not possible to interpolate channel estimations between slots, or in general different transmission occasions, since different slots may be transmitted to different devices and/or in different beam directions [14].

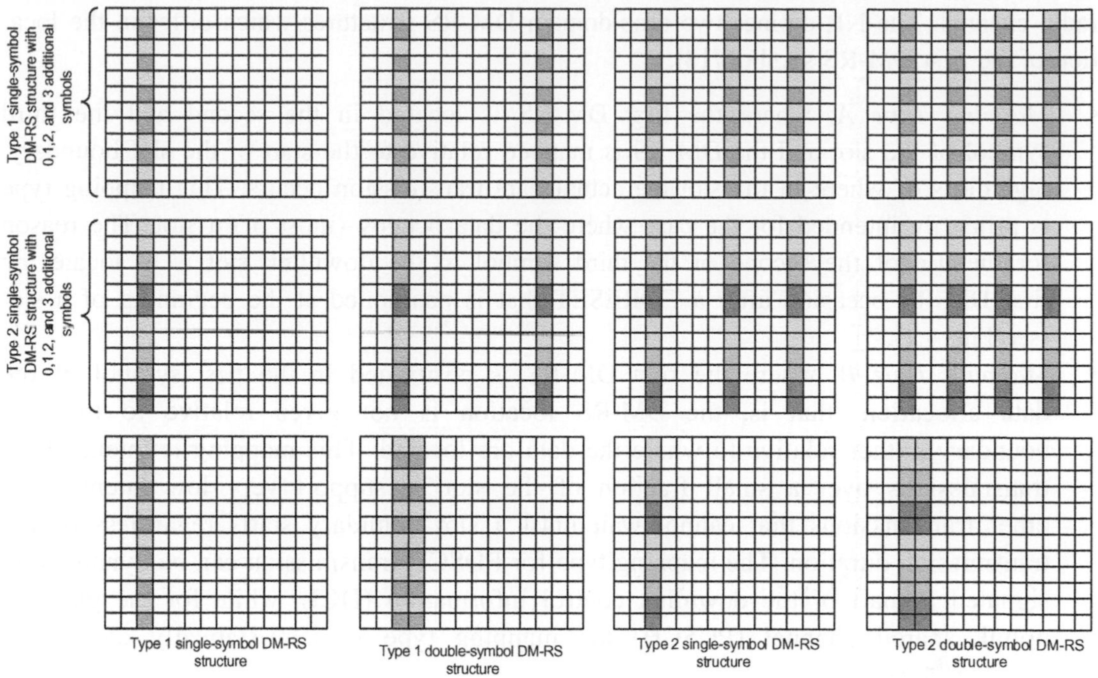

Figure 4.3
Various NR PDSCH DM-RS Type A time-frequency patterns [6].

In LoS-dominant channel conditions, the delay spread is expected to be shorter (or equivalently the channel coherence bandwidth becomes larger); thus one can consider reducing frequency-domain density of the DM-RS without significant degradation of channel estimation precision. By doing so, the overhead due to the DM-RS can be reduced. One example of such low-density DM-RS patterns in frequency-domain is shown in Fig. 4.3. For MIMO transmission up to two frequency-domain orthogonal DM-RS ports are supported. The DM-RS is UE-specific, can be beamformed, is confined in the UE scheduled resources, and is transmitted only when necessary, both in the downlink and uplink and is used to estimate the communication channel prior to coherent demodulation. To support multi-layer MIMO transmission, multiple orthogonal DM-RS ports can be scheduled, one for each layer. Orthogonality is achieved by means of FDM (comb structure), TDM, and/or CDM (with cyclic shift of the base sequence or orthogonal cover codes) methods. The basic DM-RS pattern is front loaded, as the DM-RS design considers the early decoding requirement to support low-latency applications. For low-speed scenarios, DM-RS uses low density in the time domain. However, for high-speed scenarios, the time density of DM-RS is increased to track fast changes in the radio channel. The NR defines two time-domain DM-RS structures which differ in the location of the first DM-RS symbol [14]:

- *Mapping Type A*, where the first DM-RS is located in the second and the third symbol of the slot and the DM-RS is mapped relative to the start of the slot boundary, regardless of where in the slot the actual data transmission occurs. This mapping type is primarily intended for the case where the data occupy (most of) a slot. The reason for the use of the second or the third symbol in the downlink slot is to locate the first DM-RS occasion after a CORESET that is positioned at the beginning of a slot (see Fig. 4.2).
- *Mapping Type B*, where the first DM-RS is positioned in the first symbol of the data allocation, that is, the DM-RS location is not given relative to the slot boundary, rather relative to where the data are located. This mapping is intended for transmissions over a small fraction of the slot to support very low latency and other transmissions that cannot wait until a slot boundary starts regardless of the transmission duration. The mapping type for PDSCH transmission can be dynamically signaled as part of the downlink control information (DCI), while for the physical uplink shared channel (PUSCH) the mapping type is semi-statically configured (see Fig. 4.2).

The different time-domain locations for (PDSCH) DM-RS mapping types are illustrated in Figs. 4.2 and 4.3, including both single-symbol and double-symbol DM-RS patterns. The purpose of the double-symbol DM-RS is primarily to provide a larger number of antenna ports than what is possible with a single-symbol structure as discussed later. Note that the time-domain location of the DM-RS depends on the scheduled data duration. Multiple orthogonal reference signals can be generated in each DM-RS occasion.

Different DM-RS patterns can be configured which are separated in time, frequency, and code domains. The DM-RS has two types: that is, Types 1 and 2, which are distinguished in frequency-domain mapping and the maximum number of orthogonal reference signals. Type 1 can provide up to four orthogonal signals using a single-symbol DM-RS and up to eight orthogonal reference signals using a double-symbol DM-RS, whereas Type 2 can provide 6 and 12 patterns depending on the number of symbols. The DM-RS Type 1 or 2 should not be confused with the mapping Type A or B, since different mapping types can be combined with different reference signal types. Reference signals should preferably have small power variations in the frequency domain to allow a similar channel-estimation quality for all frequencies spanned by the reference signal. Note that this is equivalent to a highly localized time-domain autocorrelation of the transmitted reference signal.

The PDSCH DM-RS sequence $r_{DM\text{-}RS}(n)$ is defined as $r_{DM\text{-}RS}(n) = \{[1 - 2c(2n)] + j[1 - 2c(2n+1)]\}/\sqrt{2}$, where $c(i)$ is a length-31 Gold sequence[1] generated by the pseudo-random sequence generator defined in [6] and initialized with $c_{init} = [2^{17}(N_{slot}^{symbol} n_{slot} + l + 1)(2N_{ID}^{n_{SCID}} + 1) + 2N_{ID}^{n_{SCID}} + n_{SCID}] \bmod 2^{31}$. In the latter expression, l denotes the OFDM symbol number within the slot, n_{slot} is the slot number within a frame, and $n_{SCID} = \{0, 1\}$ is given by the DM-RS sequence initialization field in the DCI associated with the PDSCH transmission, if DCI format 1_1 is used and $N_{ID}^{n_{SCID}} = \{0, 1, \ldots, 65535\}$ is the scrambling identifier when signaled by higher layer parameters *scramblingID0* and *scramblingID1*; otherwise $n_{SCID} = 0$ and $N_{ID}^{n_{SCID}} = N_{ID}^{cell}$. The PDSCH is scheduled by PDCCH using DCI format 1_1 (or DCI format 1_0) with CRC scrambled by C-RNTI, MCS-C-RNTI, or CS-RNTI [6].

The PDSCH DM-RS can be mapped to physical resources in two ways referred to as configuration Type 1 or 2, which is determined by RRC parameter *dmrs-Type*. Prior to resource mapping, the sequence $r_{DM\text{-}RS}(m)$ is multiplied by scaling factor β_{PDSCH}^{DMRS} to adjust the transmission power and is mapped to resource element (RE) (k, l) as $\alpha(k, l) = \beta_{PDSCH}^{DMRS} w_f(k') w_t(l') r_{DM\text{-}RS}(2n + k')$ where $w_f(k')$ and $w_t(l')$ are orthogonal cover codes or spreading functions across frequency and time that are given in [6]. In the latter expression, $n \in \mathbb{N}$, $l = \bar{l} + l'$, and $k' = 0, 1$ and $k = 4n + 2k' + \Delta$ for configuration Type 1 and $k = 6n + k' + \Delta$ for configuration Type 2.

In DM-RS Type 1, the underlying pseudo-random sequence is mapped to every other subcarrier in the frequency domain over the OFDM symbol used for reference signal transmission (see Fig. 4.3). Antenna ports 1000 and 1001 use even-numbered subcarriers in the frequency domain and are separated from each other by multiplying the underlying pseudo-random sequence with different length-2 orthogonal sequences in the frequency domain, resulting in transmission of two orthogonal reference signals for the two antenna ports. If the radio channel can be considered flat across four consecutive subcarriers, the two reference signals will maintain orthogonality at the receiver. Antenna ports 1000 and 1001 are said to belong to CDM group 0, since they use the same subcarriers but are separated in the code-domain using different orthogonal sequences. Reference signals for antenna ports 1002 and 1003 belong to CDM group 1 and are generated in the same way using odd-numbered

[1] A Gold sequence is a type of binary sequence that is often used in telecommunication and satellite navigation. Gold sequences have bounded small cross-correlations within a set. A set of Gold sequences consists of $2^n - 1$ sequences each with a period of $2^n - 1$. A set of Gold sequences can be generated by taking two maximum-length sequences of the same length $2^n - 1$ such that their absolute cross-correlation is less than or equal to $2^{(n+2)/2}$, where n is the size of the linear feedback shift register used to generate the maximum length sequence. The set of $2^n - 1$ (logical) exclusive OR of the two sequences in their various phases is a set of Gold codes. The highest absolute cross-correlation in this set of codes is $2^{(n+2)/2} + 1$ for even n, and $2^{(n+1)/2} + 1$ for odd n. The exclusive OR of two Gold codes from the same set is another Gold code with arbitrary phase.

subcarriers and are separated in the code domain within the CDM group and in the frequency domain between CDM groups. If more than four orthogonal antenna ports are needed, two consecutive OFDM symbols are used instead. The aforementioned structure is used over each of the OFDM symbols and a length-2 orthogonal sequence is used to extend the code-domain separation over time, resulting in up to eight orthogonal sequences.

The DM-RS Type 2 has a similar structure to Type 1, except some differences with to the number of antenna ports that are supported. Each CDM group for Type 2 consists of two neighboring subcarriers over which a length-2 orthogonal sequence is used to separate the two antenna ports sharing the same set of subcarriers. Four subcarriers are used in each resource block and in each CDM group. Since there are 12 subcarriers in a resource block, up to three CDM groups with two orthogonal reference signals can be created using one resource block over one OFDM symbol. If a second OFDM symbol is used along with a length-2 sequence in time-domain, up to 12 orthogonal reference signals can be generated [14].

The location of front-loaded DM-RS symbols, which can be either one or two symbols, is dependent on whether a slot based (DM-RS mapping Type A) or non-slot-based (DM-RS mapping Type B) scheduling is used. In the former type, fixed OFDM symbols regardless of the PDSCH assignment are used to map DM-RS (configurable via parameter $l_0 = \{2, 3\}$), whereas in latter type which corresponds to mini-slots, the first OFDM symbol assigned for PDSCH is used to map DM-RS. The reference point for l and the position l_0 of the first DM-RS symbol depends on the mapping type. Additional DM-RS symbols can be configured (e.g., for high-speed scenarios) as well as for broadcast/multicast PDSCH. In the preceding equations, the reference point for frequency index k depends on PDSCH payload. The reference point for frequency index k is subcarrier 0 of the lowest numbered resource block in CORESET 0, if the corresponding PDCCH is associated with CORESET 0 and Type0-PDCCH common search space and identified by system information (SI)-RNTI; otherwise, it is the subcarrier 0 in common resource block 0. Furthermore, the reference point for time index l and the reference position l_0 of the first DM-RS symbol depends on the mapping type, which for PDSCH mapping Type A, l is defined relative to the start of the slot, that is, $l_0 = 3$ if the RRC parameter *dmrs-TypeA-Position* equals 3; otherwise, $l_0 = 2$ and for PDSCH mapping Type B, l is defined relative to the start of the scheduled PDSCH resources, that is, $l_0 = 0$. The position of the DM-RS symbols is further dependent on parameter $\bar{l}$ where for PDSCH mapping Type A, the duration is between the first OFDM symbol of the slot, and the last OFDM symbol of the scheduled PDSCH resources in the slot; and for PDSCH mapping Type B, the duration is the number of OFDM symbols of the scheduled PDSCH resources given by the parameters specified in [6].

For PDSCH mapping Type B, if the PDSCH duration is 2, 4, or 7 OFDM symbols (i.e., mini-slot scheduling), and if the PDSCH allocation collides with resources reserved for a

CORESET, $\bar{l}$ is incremented such that the first DM-RS symbol is located immediately following the CORESET. If PDSCH duration is 2, 4, or 7 symbols, the UE would not expect to receive a DM-RS symbol beyond the second, third, and fourth symbol, respectively. If one additional single-symbol DM-RS is configured, the UE expects the additional DM-RS to be transmitted on the fifth or sixth symbol when the front-loaded DM-RS symbol is in the first or second symbol, respectively; otherwise, the UE should expect that the additional DM-RS is not transmitted. If PDSCH duration is two or four OFDM symbols, only a single-symbol DM-RS is supported. Furthermore, single-symbol or double-symbol DM-RS is used, if RRC parameter *maxLength* is equal to 1 or 2, respectively [8,9].

In the absence of CSI-RS configuration, the UE can assume PDSCH DM-RS and SS/PBCH block antenna ports are quasi-co-located with respect to Doppler shift, Doppler spread, average delay, delay spread, and spatial RX[2] parameters. The UE may assume that the PDSCH DM-RS within the same CDM group are quasi-co-located with respect to Doppler shift, Doppler spread, average delay, delay spread, and spatial RX parameters. Note that the spatial RX parameters are meant to describe angular/spatial channel properties at the UE to help the UE select and use one of the beams. The UE can use SS/PBCH block to obtain frequency offset, timing offset, Doppler spread, delay spread, and receive beam to process DM-RS. In other words, one can consider the spatial RX parameters as beam indication for the UE, where UE may use the acquired channel parameters from SS/PBCH to receive PDSCH.

4.1.2.2 Phase Tracking Reference Signals

The PT-RS was introduced in NR to enable compensation of oscillator phase noise in above-6 GHz frequency bands. Phase noise typically increases as a function of carrier frequency. Therefore, PT-RS can be utilized at high carrier frequencies (e.g., mmWave bands) to mitigate the phase noise effect. In the case of OFDM signals, the effect of phase noise is identical phase rotation of all the subcarriers, known as CPE. In NR, the PT-RS is designed

[2] To explain the spatial RX concept, let's consider receive antenna diversity where the transmitted signal is received by N antennas which are assumed to have sufficient spatial separation resulting in independent multipath channels between the transmitter and each receiver antenna. Denoting the channel vector, including the amplitude and phase coefficients representing the aggregated effects of multipath propagation, by $\mathbf{h} = (h_1, h_2, \ldots, h_N) \in \mathbb{C}^{N\times 1}$, the received baseband equivalent signal vector $\mathbf{r} \in \mathbb{C}^{N\times 1}$ can be expressed by $\mathbf{r} = \mathbf{h}x + \mathbf{n}$. Consequently, by combining the signals from the separate antenna branches with a proper weighting, we obtain the combiner output being equal to $y_{RX} = \mathbf{W}^H\mathbf{r} = \mathbf{W}^H\mathbf{h}x + \mathbf{W}^H\mathbf{n}$, where $\mathbf{W} \in \mathbb{C}^{N\times 1}$ denotes the weighting vector of the combiner. It is important to note that now the steering vector is replaced with a more generic channel response, including arbitrary amplitude and phase response for each antenna branch. Consequently, the combiner weights do not match with any particular physical direction anymore, rather they need to be adjusted according to the generic channel response and this form of multi-antenna-based signal combining is referred to as spatial RX processing. The actual weight selection and optimization, in turn, can be implemented by several methods, which differ in complexity and performance. The simplest method is selection combining where only the signal with the highest instantaneous SNR is used for detection.

so that it has low density in the frequency domain and high density in the time domain, because the phase rotation caused by CPE is identical for all subcarriers within an OFDM symbol; however, it has minimal correlation across OFDM symbols. The PT-RS is UE-specific, confined in a scheduled resource, and can be beamformed. The number of PT-RS ports can be lower than the total number of ports, and orthogonality between PT-RS ports is achieved by means of frequency-division multiplexing. The PT-RS is configurable depending on the quality of the oscillators, carrier frequency, OFDM subcarrier spacing, and modulation and coding schemes used for transmission.

The PT-RS introduced in NR is used for time and frequency tracking as well as estimation of delay spread, and Doppler spread at the UE side. They are transmitted in a confined bandwidth for a configurable time duration controlled by RRC parameters. The time-frequency structure of PT-RS depends on the waveform. For OFDM, the first reference symbol (prior to applying any orthogonal sequence) in a PDSCH/PUSCH allocation is repeated every $L_{PT\text{-}RS} \in \{1, 2, 4\}$ symbol, starting with the first OFDM symbol in the allocation. The repetition counter is reset at each DM-RS position since there is no need for PT-RS insertion immediately following a DM-RS occasion. In the frequency domain, PT-RS are transmitted in every second or fourth resource block, resulting in a sparse frequency-domain structure. The density in the frequency domain is dependent on the scheduled bandwidth in a sense that the higher the bandwidth, the lower the PT-RS density. For the smallest bandwidths, no PT-RS is transmitted. To reduce the risk of collision between PT-RS associated with different devices scheduled on overlapping frequency-domain resources, the subcarrier number and the resource blocks used for PT-RS transmission are determined by the C-RNTI of the device. The antenna port used for PT-RS transmission is given by the lowest numbered antenna port in the DM-RS antenna port group. An example time-frequency PT-RS structure is shown in Fig. 4.4.

The PT-RS for subcarrier k is given by $r_{PT\text{-}RS}(k) = r_{DM\text{-}RS}(2m + k')$ where $r_{DM\text{-}RS}(2m + k')$ is the DM-RS at time-domain position l_0 and subcarrier k. The PT-RS is present only in the resource blocks used for the PDSCH, and only if there is an explicit indication of their presence. In that case, the PT-RS is scaled by a factor of $\beta_{PT\text{-}RS}(i)$ to adjust the transmission power. The PT-RS is mapped to resource elements $a(k, l) = \beta_{PT\text{-}RS}(i) r_{PT\text{-}RS}(k)$, if l is located within the OFDM symbols allocated for the PDSCH transmission and the designated resource element is not used for DM-RS, CSI-RS, SS/PBCH block, PDCCH, or is declared as not available. The time indices l at which PT-RS are allocated are defined relative to the start of the PDSCH allocation l_0 and are given by $l = l_0 + iL_{PT\text{-}RS}$ where i is incremented as long as the PT-RS occasion falls inside the PDSCH allocation and the aforementioned conditions are met. For PT-RS resource mapping, the resource blocks allocated for PDSCH transmission are numbered from 0 to $N_{RB} - 1$ from the lowest scheduled resource block to the highest. The corresponding subcarriers in this set of resource blocks are numbered in increasing order starting from the lowest frequency to $N_{sc}^{RB} N_{RB} - 1$. The subcarrier indices that the PT-RS are

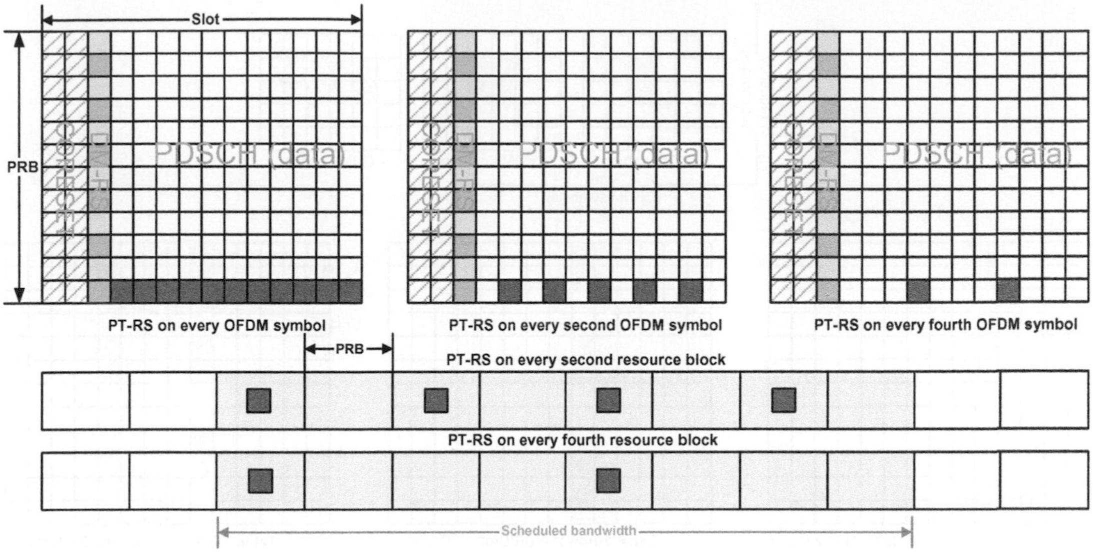

Figure 4.4
Illustration of PT-RS structures in time and frequency domain [6].

mapped to are given by $k = k_{ref}^{RE} + (iK_{PTRS} + k_{ref}^{RB})N_{sc}^{RB}; i \in \mathbb{N}$ where $k_{ref}^{RB} = n_{RNTI} \bmod K_{PT\text{-}RS}$ if $N_{RB} \bmod K_{PT\text{-}RS} = 0$; otherwise, $k_{ref}^{RB} = n_{RNTI} \bmod(N_{RB} \bmod K_{PT\text{-}RS})$ [6]. In the latter equation, n_{RNTI} denotes the RNTI associated with the DCI scheduling of the transmission; $K_{PT-RS} \in \{2, 4\}$, and k_{ref}^{RE} is the DM-RS port associated with the PT-RS port. The density of PT-RS in time and frequency domain is configurable to address different scenarios (e.g., different carrier frequency, modulation and coding scheme, and hardware quality). The PT-RS patterns in time and frequency domains are illustrated in Fig. 4.4.

4.1.2.3 Channel State Information Reference Signals

The CSI-RS in NR is used for downlink CSI estimation. It further supports RSRP measurements for mobility and beam management (including analog beamforming), time/frequency tracking for demodulation, and uplink reciprocity-based precoding. The CSI-RS is UE-specific; nevertheless, multiple users can share the same CSI-RS resource. The NR defines zero-power and non-zero-power CSI-RS. When a zero-power CSI-RS is configured, the resource elements (designated to CSI-RS) are not used for PDSCH transmission. In this case, the zero-power CSI-RS is used to mask certain resource elements, making them unavailable for PDSCH mapping. This masking not only supports transmission of UE-specific CSI-RS, but also the design allows introduction of new features while maintaining backward compatibility. The NR supports flexible CSI-RS configurations. A CSI resource can be configured with up to 32 antenna ports with configurable density. In the time domain, a CSI-RS resource may start at any OFDM symbol of a slot and span 1, 2, or 4

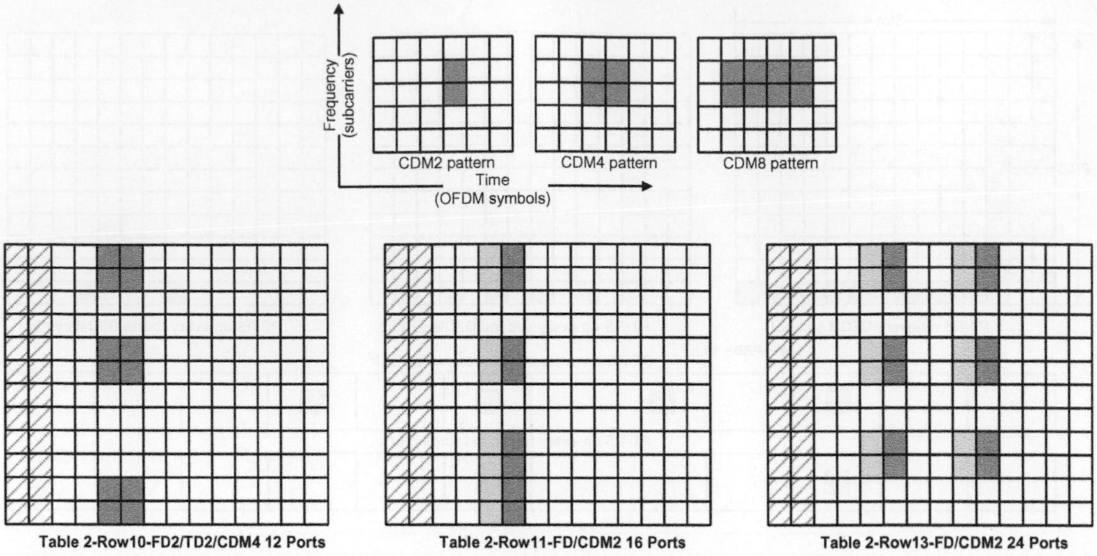

Figure 4.5
Example locations of CSI-RS in time and frequency [6].

OFDM symbols depending on the number of configured antenna ports. The CSI-RS can be periodic, semi-persistent, or aperiodic (DCI triggered). When used for time frequency tracking, the CSI-RS can be periodic or aperiodic. In this case, a single port is configured, and the signal is transmitted in the form of bursts over two or four symbols that are spread over one or two slots [71].

A configured CSI-RS resource may correspond to up to 32 different antenna ports. In NR, a CSI-RS is always configured on a per-device basis. It must be noted that the UE-specific configuration of CSI-RS does not necessarily mean that a transmitted CSI-RS can only be used by a single device, rather the same set of CSI-RS resources can be separately configured for multiple devices, which means that a single CSI-RS can be shared among multiple devices. As illustrated in Fig. 4.5, a single-port CSI-RS occupies a single resource element within a resource block in the frequency domain and one slot in the time domain. While the CSI-RS can be configured to occur anywhere within the resource block, in practice there are some restrictions on CSI-RS resource assignment to avoid collisions with other downlink physical channels and signals. The transmission of a configured CSI-RS is expected not to collide with a CORESET configured for the device; the DM-RS associated with PDSCH transmissions scheduled for the device; and the SS blocks transmissions.

A multiport CSI-RS can be viewed as multiple orthogonal per antenna-port CSI-RS sharing the set of resource elements assigned for transmission of the configured multiport CSI-RS. In general, the resource sharing is achieved through a combination of code-domain (i.e.,different per antenna-port CSI-RS are transmitted on the same set of resource elements with separation achieved by spreading the CSI-RS with different orthogonal codes), frequency-domain (i.e., different per antenna-port CSI-RS are transmitted on different subcarriers over an OFDM symbol), and time-domain (i.e., different per antenna-port CSI-RS are transmitted on different OFDM symbols within a slot) multiplexing schemes. As illustrated in Fig. 4.5, code-division multiplexing between different (per antenna-port) CSI-RS can be performed across frequency by spreading over two adjacent subcarriers (CDM2) to support two antenna ports; across frequency and time by spreading over two adjacent subcarriers and two adjacent OFDM symbols (CDM4) to enable four antenna ports; across frequency and time by spreading over two adjacent subcarriers and four adjacent OFDM symbols (CDM8) to support up to eight antenna port transmission. Combination of code/frequency and time-division multiplexing can be used to configure different multiport CSI-RS structures where, in general, an N-port CSI-RS occupies N resource elements within a resource block or a slot. When CSI-RS supports more than two antenna ports, there are multiple CSI-RS patterns/structures based on different combinations of CDM, TDM, and FDM that can be utilized as shown in Table 4.2.

The non-zero-power CSI-RS is mathematically represented by sequence $r_{CSI\text{-}RS}(m)$ which is defined as $r_{CSI\text{-}RS}(m) = \{[1-2c(2m)] + j[1-2c(2m+1)]\}/\sqrt{2}$, where $c(i)$ is a length-31 Gold sequence generated by the pseudo-random sequence generator defined in [6] and initialized with $c_{init} = [2^{10}(N_{slot}^{symbol} n_{slot} + l + 1)(2n_{ID} + 1) + n_{ID}] \bmod 2^{31}$ at the start of each OFDM symbol. In the latter equation, n_{slot} is the slot number within a radio frame, l is the OFDM symbol number within a slot, and n_{ID} is set by the RRC parameter *scramblingID* or *sequenceGenerationConfig* [6,13].

If CSI-RS is configured, the CSI-RS sequence $r_{CSI\text{-}RS}(m)$ is mapped to resources elements (k,l) such that $a(k,l) = \beta_{CSI\text{-}RS} w_f(k') w_t(l') r_{CSI\text{-}RS}(m')$ where $m' = \lfloor n\alpha \rfloor + k' + \lfloor k\rho/N_{sc}^{RB} \rfloor \forall n \in \mathbb{N}, k = nN_{sc}^{RB} + k'$ and $l = \bar{l} + l'$, $\alpha = \rho$ or 2ρ when $N_p = 1$ or $N_p > 1$, respectively, provided that the resource element (k,l) is within the resource blocks designated to the CSI-RS resource for which the UE is configured. The reference point for $k = 0$ is subcarrier 0 in common resource block 0. The value of density ρ and the number of antenna ports are given by RRC parameters [6,13]. The scaling factor $\beta_{CSI\text{-}RS}$ has a non-zero value for a non-zero-power CSI-RS to ensure that the power offset specified by the RRC parameter *powerControlOffsetSS* is satisfied. Other parameters $k', l', w_f(k')$, and $w_t(l')$ are given in Table 4.2 where each pair $(\bar{k}, \bar{l})$ corresponds to a CDM group of size 1 (no CDM) or size 2, 4, or 8. The indices k' and l' are used to index the resource elements within a CDM group [6,8]. The time-domain locations $l_0 = \{0, 1, \ldots, 13\}$ and $l_1 = \{2, 3, \ldots, 12\}$ are defined relative to the start of a slot with the starting positions of a CSI-RS in a slot configured by the RRC parameters provided in *CSI-RS-ResourceMapping* information element. The frequency-domain location of CSI-RS is determined by a bitmap signaled via the RRC parameter provided in

Table 4.2: Various CSI-RS patterns (locations) within a slot [6].

Row	Number of Ports N_p	CSI-RS Density ρ	CDM Type	$(\bar{k}, \bar{l})$	CDM Group Index j	k'	l'
1	1	3	No CDM	$(k_0, l_0), (k_0+4, l_0), (k_0+8, l_0)$	0,0,0	0	0
2	1	1,0.5	No CDM	(k_0, l_0)	0	0	0
3	2	1,0.5	FD-CDM2	(k_0, l_0)	0	0,1	0
4	4	1	FD-CDM2	$(k_0, l_0), (k_0+2, l_0)$	0,1	0,1	0
5	4	1	FD-CDM2	$(k_0, l_0), (k_0, l_0+1)$	0,1	0,1	0
6	8	1	FD-CDM2	$(k_0, l_0), (k_1, l_0), (k_2, l_0), (k_3, l_0)$	0,1,2,3	0,1	0
7	8	1	FD-CDM2	$(k_0, l_0), (k_1, l_0), (k_0, l_0+1), (k_1, l_0+1)$	0,1,2,3	0,1	0
8	8	1	CDM4 (FD2,TD2)	$(k_0, l_0), (k_1, l_0)$	0,1	0,1	0,1
9	12	1	FD-CDM2	$(k_0, l_0), (k_1, l_0), (k_2, l_0), (k_3, l_0), (k_4, l_0), (k_5, l_0)$	0,1,2,3,4,5	0,1	0
10	12	1	CDM4 (FD2,TD2)	$(k_0, l_0), (k_1, l_0), (k_2, l_0)$	0,1,2	0,1	0,1
11	16	1,0.5	FD-CDM2	$(k_0, l_0), (k_1, l_0), (k_2, l_0), (k_3, l_0)$ $(k_0, l_0+1), (k_1, l_0+1), (k_2, l_0+1), (k_3, l_0+1)$	0,1,2,3,4,5,6,7	0,1	0
12	16	1,0.5	CDM4 (FD2,TD2)	$(k_0, l_0), (k_1, l_0), (k_2, l_0), (k_3, l_0)$	0,1,2,3	0,1	0,1
13	24	1,0.5	FD-CDM2	$(k_0, l_0), (k_1, l_0), (k_2, l_0), (k_0, l_0+1), (k_1, l_0+1), (k_2, l_0+1)$ $(k_0, l_1), (k_1, l_1), (k_2, l_1), (k_0, l_1+1), (k_1, l_1+1), (k_2, l_1+1)$	0,1,2,3,4,5,6,7,8,9,10,11	0,1	0
14	24	1,0.5	CDM4 (FD2,TD2)	$(k_0, l_0), (k_1, l_0), (k_2, l_0), (k_0, l_1), (k_1, l_1), (k_2, l_1)$	0,1,2,3,4,5	0,1	0,1
15	24	1,0.5	CDM8 (FD2,TD4)	$(k_0, l_0), (k_1, l_0), (k_2, l_0)$	0,1,2	0,1	0,1,2,3
16	32	1,0.5	FD-CDM2	$(k_0, l_0), (k_1, l_0), (k_2, l_0), (k_3, l_0),$ $(k_0, l_0+1), (k_1, l_0+1), (k_2, l_0+1), (k_3, l_0+1)$ $(k_0, l_1), (k_1, l_1), (k_2, l_1), (k_3, l_1),$ $(k_0, l_1+1), (k_1, l_1+1), (k_2, l_1+1), (k_3, l_1+1)$	0,1,2,3, 4,5,6,7, 8,9,10,11, 12,13,14,15	0,1	0
17	32	1,0.5	CDM4 (FD2,TD2)	$(k_0, l_0), (k_1, l_0), (k_2, l_0), (k_3, l_0),$ $(k_0, l_1), (k_1, l_1), (k_2, l_1), (k_3, l_1)$	0,1,2,3,4,5,6,7	0,1	0,1
18	32	1,0.5	CDM8 (FD2,TD4)	$(k_0, l_0), (k_1, l_0), (k_2, l_0), (k_3, l_0)$	0,1,2,3	0,1	0,1,2,3

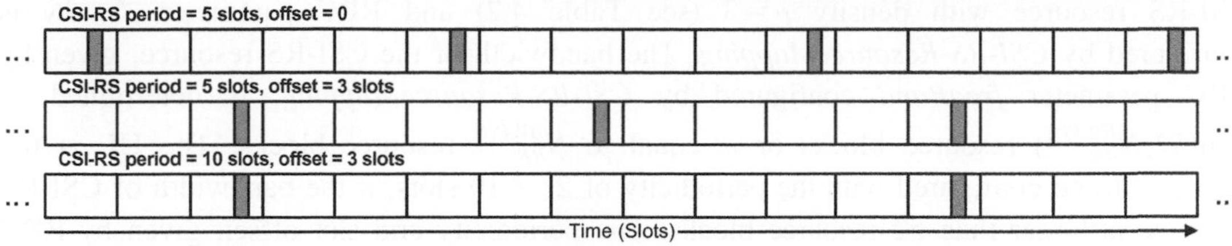

Figure 4.6
Example CSI-RS periodicity and offset [6].

CSI-RS-ResourceMapping information element [6,13]. The starting position and number of the resource blocks in which the CSI-RS is transmitted are provided via RRC signaling.

The CSI-RS is transmitted on antenna ports $p = 3000 + s + jL_{CDM}$ where $j = 0, 1, \ldots, N_p/L_{CDM} - 1$ and $s = 0, 1, \ldots, L_{CDM} - 1$. In the latter expression, $L_{CDM} \in \{1, 2, 4, 8\}$ is the CDM group size and N_p is the number of CSI-RS antenna ports. The CDM groups are numbered in order of increasing frequency-domain allocation first and then increasing time domain-allocation. For a CSI-RS resource configured as periodic or semi-persistent by the RRC parameter *resourceType*, the CSI-RS is transmitted in slot numbers satisfying $(N_{frame}^{slot} n_{slot} + n_{slot} - T_{offset}) \bmod T_{CSI\text{-}RS} = 0$ where the CSI-RS periodicity $T_{CSI\text{-}RS}$ (in number of slots) and slot offset T_{offset} are signaled by the RRC parameter *CSI-ResourcePeriodicityAndOffset* or *slotconfig* (see Fig. 4.6). The CSI-RS is transmitted in a slot only if, all OFDM symbols of that slot corresponding to the configured CSI-RS resource are designated for downlink transmission. The antenna ports within a CSI-RS resource are quasi-co-located with quasi-co-location (QCL) Type A, Type D (when applicable). In summary, the NR supports periodic, aperiodic, semi-persistent CSI-RS transmission. The NR CSI-RS patterns can be mapped to 1, 2, or 4 OFDM symbols and support CDM2, CDM4, CDM8 spreading functions, for example, CDM8 means that there are eight spreading functions $w_f(k')$ and $w_t(l')$. For CSI acquisition, the NR supports CSI-RS density $\rho = 0.5$ and 1 RE/RB/port and a PRB-level comb-type transmission. The number of antenna ports can be independently configured for periodic, aperiodic, semi-persistent CSI reporting. A CSI-RS resource configuration up to 32 ports is supported in NR. The UE-specific CSI-RS may be configured to support wideband CSI-RS and partial band CSI-RS. In order to reduce beam management overhead and latency, the NR supports subtime units of less than one OFDM symbol in a reference numerology [73]. Each CSI-RS resource is configured by the RRC parameter *NZP-CSI-RS-Resource*. The time-domain locations of the two periodic CSI-RS resources in a slot or four periodic CSI-RS resources in two consecutive slots are given by $l \in \{4, 8\}, l \in \{5, 9\}, \text{or } l \in \{6, 10\}$ for FR1 and FR2; or $l \in \{0, 4\}, l \in \{1, 5\}, l \in \{2, 6\}, l \in \{3, 7\}, l \in \{7, 11\}, l \in \{8, 12\}$ or $l \in \{9, 13\}$ for FR2. A single-port

CSI-RS resource with density $\rho = 3$ (see Table 4.2) and RRC parameter *density* is configured by *CSI-RS-ResourceMapping*. The bandwidth of the CSI-RS resource, given by RRC parameter *freqBand* configured by *CSI-RS-ResourceMapping*, is determined as $\min(52, N_{RB}^{BWP(i)})$ resource blocks or is equal to $N_{RB}^{BWP(i)}$ resource blocks. The UE is not expected to be configured with the periodicity of $2^{\mu} \times 10$ slots, if the bandwidth of CSI-RS resource is larger than 52 resource blocks. The periodicity and slot offset, given by RRC parameter *periodicityAndOffset* configured by *NZP-CSI-RS-Resource* parameter, is one of $2^{\mu}X_p$ slots where $X_p = 10, 20, 40,$ or 80 [9]. It should be noted that the property of periodic, semi-persistent, or aperiodic is not a property of the CSI-RS per se, rather the property of a CSI-RS resource set. As a result, activation/deactivation and triggering of semi-persistent and aperiodic CSI-RS must be done for a set of CSI-RS within a resource set. In the case of periodic CSI-RS transmission, the UE can assume that a configured CSI-RS transmission occurs every *N*th slot, where *N* ranges from 4 to 640. In addition to the periodicity, the device is also configured with a specific slot offset for the CSI-RS transmission. In the case of semi-persistent CSI-RS transmission, certain CSI-RS periodicity and slot offset are configured similar to periodic CSI-RS transmission. However, the CSI-RS transmission can be activated or deactivated via MAC control elements. Once the CSI-RS transmission has been activated, the device can assume that the CSI-RS transmission will continue according to the configured periodicity until it is deactivated. Similarly, once the CSI-RS transmission has been deactivated, the device can assume that there will be no CSI-RS transmission according to the configuration until it is reactivated. In the case of aperiodic CSI-RS, no periodicity is configured, rather the UE is triggered via signaling in the DCI [9].

The CSI-RS may be further used for RSRP measurements[3] and mobility management since NR does not include the cell-specific reference signals that were used in LTE for mobility management. The set of CSI-RS corresponding to a set of beams on which measurements are conducted should be included in the non-zero power (NZP)-CSI-RS resource set associated with the report configuration. Such a resource set may either include a set of configured CSI-RS or a set of SS blocks. Measurements for beam management can be carried out on either CSI-RS or SS block. In the case of L1-RSRP measurements based on CSI-RS, the CSI-RS should be limited to single-port or dual-port CSI-RS. In the latter case, the reported L1-RSRP should be a linear average of the L1-RSRP measured on each port. The device can report measurements corresponding to up to four reference signals (CSI-RS or SS blocks), that is, up to four beams, in a single reporting instance. Each report is related to up to four reference signals or beams and includes the measured L1-RSRP for the strongest

[3] CSI-RSRP is defined as the linear average over the power contributions (in Watts) of the resource elements that carry CSI-RS configured for RSRP measurements within the identified measurement frequency region in the configured CSI-RS occasions. The CSI reference signals are transmitted on specific antenna ports. This measurement is applicable for connected mode only for both intra- and inter-frequency measurements.

beam and the difference between other beams' L1-RSRP measurements and the measured L1-RSRP of the best beam [14].

If a UE is configured with an *NZP-CSI-RS-ResourceSet* via RRC parameter *repetition* set to "on," it may assume that the CSI-RS resources within the *NZP-CSI-RS-ResourceSet* are transmitted with the same downlink spatial domain transmission filter, where the CSI-RS resources in the *NZP-CSI-RS-ResourceSet* are transmitted on different OFDM symbols. If *repetition* is set to "off," the CSI-RS resources within the *NZP-CSI-RS-ResourceSet* are transmitted with the same downlink spatial domain transmission filter. If the UE is configured with a *CSI-ReportConfig* and parameter *reportQuantity* is set to "cri-RSRP," or "none" and if the *CSI-ResourceConfig* for channel measurement (RRC parameter *resourcesForChannelMeasurement*) contains a *NZP-CSI-RS-ResourceSet* that is configured with the higher layer parameter *repetition* and without the higher layer parameter *trs-Info*, the UE can only be configured with the same number (1 or 2) of ports with the higher layer parameter *nrofPorts* for all CSI-RS resources within the set. If the UE is configured with the CSI-RS resource on the same OFDM symbol(s) as an SS/PBCH block, the CSI-RS and the SS/PBCH block are quasi-co-located with QCL TypeD, if applicable. Furthermore, the UE will not be configured with the CSI-RS in PRBs that overlap with those of the SS/PBCH block, and the same subcarrier spacing is used for both the CSI-RS and the SS/PBCH block [9].

4.1.2.4 Tracking Reference Signals

A UE must track and compensate time and frequency variations of its local oscillator in order to successfully receive downlink transmissions. The problem is exacerbated in higher radio frequencies. To assist the device in this task, a tracking reference signal can be configured. The TRS is not a CSI-RS, rather a TRS is a resource set consisting of multiple periodic NZP-CSI-RS. More specifically, a TRS consists of four single-port, density-3 CSI-RS located within two consecutive slots as shown in Fig. 4.7. The CRS-RS within the resource

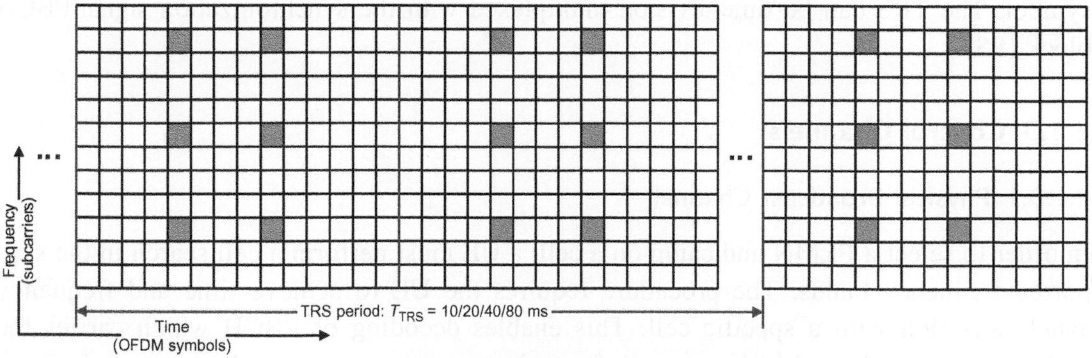

Figure 4.7
Example TRS structure (four single-port, Density-3 CSI-RS over two consecutive slots [9]).

set or the TRS can be configured with a periodicity of 10, 20, 40, or 80 ms. It must be noted that the exact set of time-frequency resource elements used for the TRS may vary; however, the two CSI-RS within a slot are always separated by four symbols in the time domain. This time-domain separation sets a limit for the maximum frequency error that can be compensated. Likewise, the frequency-domain separation of four subcarriers sets a limit for the maximum timing error that can be compensated. There is an alternative TRS structure with the same per-slot structure as the TRS structure shown in Fig. 4.7 with only two CSI-RS within a slot, compared to two consecutive slots for the TRS structure shown in the figure. In LTE, the CRS served the same purpose as the TRS; however, the TRS has relatively lower overhead, has one antenna port, and only present in two slots in every TRS period.

As we mentioned earlier, the CSI-RS for tracking or TRS is a special configuration of CSI-RS which is specifically configured for a UE. The TRS is used for fine time and frequency tracking as well as path delay spread and Doppler spread estimation. A UE in RRC_CONNECTED mode would receive information on a CSI-RS resource set which is configured specifically for the purpose of time/frequency tracking. In that case, the UE is configured with RRC parameter *trs-Info* and will assume that TRS is transmitted from the antenna port with the same port index of the configured NZP CSI-RS resources in the CSI-RS resource set. In frequency range 1, the UE may be configured with a CSI-RS resource set consisting of four periodic CSI-RS resources in two consecutive slots with two periodic CSI-RS resources in each slot, whereas in frequency range 2, the UE may be configured with a CSI-RS resource set of two periodic CSI-RS resources in one slot or with a CSI-RS resource set of four periodic CSI-RS resources in two consecutive slots with two periodic CSI-RS resources in each slot. The periodic CSI-RS resources in the CSI-RS resource set configured with RRC parameter *trs-Info* have the same periodicity, bandwidth and subcarrier location. The time-domain location of the TRS is determined by two periodic CSI-RS resources in a slot, or four periodic CSI-RS resources in two consecutive slots and via RRC signaling. The density of the TRS in one physical resource block is three REs per symbol. The TRS can be time-division multiplexed with the synchronization signal/PBCH block (SSB).

4.1.3 Control Channels

4.1.3.1 Physical Broadcast Channel

In order to select a PLMN and camp on a cell, a UE must perform a cell search in the supported frequency bands. The procedure requires the UE to achieve time and frequency synchronization with a specific cell. This enables decoding of PBCH which carries the MIB containing the critical system information necessary to decode transmissions on PDSCH. In NR, the SI is divided into minimum SI and other SI. The minimum SI is periodically broadcast and comprises basic information required for initial access and

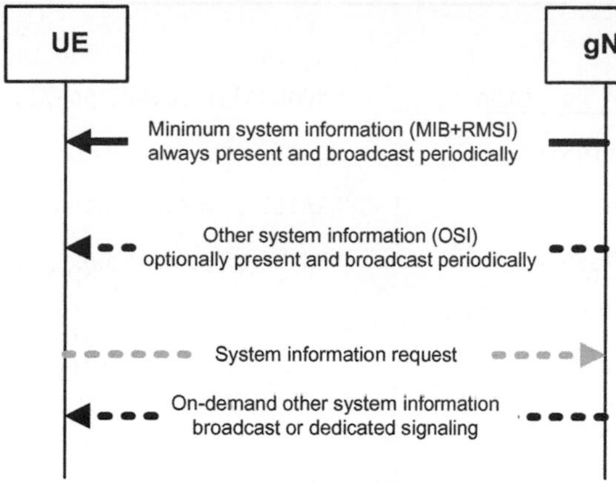

Figure 4.8
Transmission of system information in NR [11].

information for acquiring other SI broadcast periodically or scheduled on-demand. The other SI encompasses everything else not broadcast in the minimum SI message and may be either broadcast or individually transmitted to the UE. In the latter case, the on-demand transmission of other SI can be triggered by the network or based on a request from the UE. The change of SI can only occur at specific radio frames. Note that to ensure coverage and reliability, the SI may be transmitted a number of times with the same content within a modification period. For the minimum SI delivery, part of minimum SI is transmitted in PBCH. The remaining minimum SI (RMSI) is transmitted in the downlink shared channel. The initial BWP information is signaled by PBCH which contains the CORESET and PDSCH information for mapping the RMSI. Fig. 4.8 shows the process for transmission of various components of system information. Unlike LTE system where the minimum SI and a group of SI blocks were broadcast periodically, NR limits the amount of SI that is periodically broadcast and instead relies on less-frequent and on-demand transmission of the non-essential SI.

The MIB message (in ASN.1 format), which is carried in PBCH, consists of the following components [13]:

```
MIB ::= SEQUENCE {
        systemFrameNumber            BIT STRING (SIZE (6)),
        subCarrierSpacingCommon      ENUMERATED {scs15or60, scs30or120},
        ssb-SubcarrierOffset         INTEGER (0..15),
```

(*Continued*)

(Continued)

```
        dmrs-TypeA-Position            ENUMERATED {pos2, pos3},
        pdcch-ConfigSIB1
        cellBarred                     ENUMERATED {barred, notBarred},
        intraFreqReselection           ENUMERATED {allowed, notAllowed},
        spareBIT STRING (SIZE (1))
}
PDCCH-ConfigSIB1 :: = SEQUENCE {
controlResourceSetZero ControlResourceSetZero,
searchSpaceZero SearchSpaceZero
}
```

In the MIB message which is mapped to BCH logical channel, *subCarrierSpacingCommon* parameter indicates the subcarrier spacing for SIB1, Msg2/4 for the initial access and the SI messages where values 15 and 30 kHz are applicable to sub-6 GHz and values 60 and 120 kHz are applicable to carrier frequencies above 6 GHz; *ssb-subcarrierOffset* is the frequency-domain offset between SSB and the overall resource block grid in number of subcarriers; *dmrs-TypeA-Position* indicates the position of the first downlink DM-RS; and *pdcch-ConfigSIB1* determines the bandwidth of PDCCH/SIB1 or the size of the CORESET containing common search space for PDCCH. In other words, the first field of *pdcch-ConfigSIB1* determines the common CORESET corresponding to the initial downlink BWP and the second field identifies the common search space of initial downlink BWP [13].

The RMSI is transmitted via the PDSCH by downlink assignment in an RMSI CORESET. The concept of CORESET was introduced in NR to identify a set of time-frequency resources consisting of multiple resource blocks in the frequency domain and one to three OFDM symbols in the time domain. The NR enables UE to be configured with multiple CORESETs, and each CORESET is associated with a UE-specific configured resource mapping scheme.

The PBCH payload size is 56 bits including 24-bit CRC. In NR, PBCH uses a single-antenna port transmission scheme, using the same antenna port as PSS and SSS within the same SS block. The periodicity of PBCH is 80 ms. The MIB data arrive at the PBCH processing unit in the form of one transport block (TB) every 80 ms and goes through the following steps as shown in Fig. 4.9: payload generation, scrambling, TB CRC calculation and attachment, channel coding, and rate matching. The coded and modulated bits of PBCH are

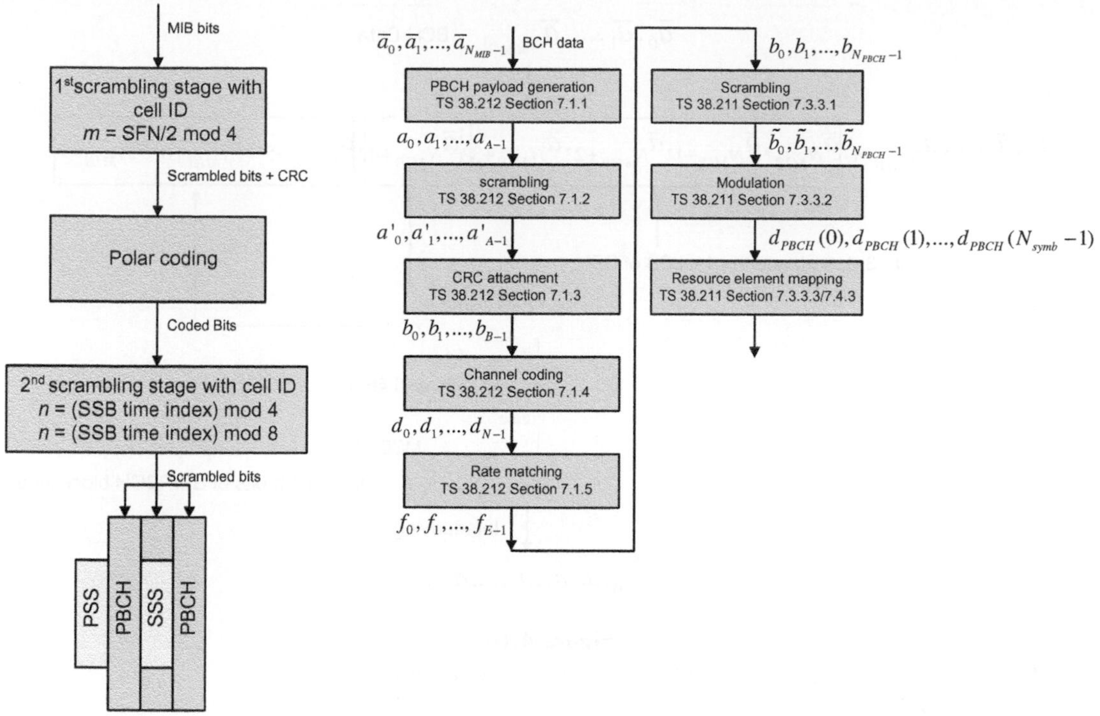

Figure 4.9
Physical layer processing of PBCH and mapping to time-frequency resources [5].

mapped onto the resource elements allocated for PBCH. As we will discuss later, the PBCH content is encoded using the polar code. Two scrambling operations are performed on PBCH which include one before CRC attachment and another one after the polar coding and rate matching. In the first scrambling stage, initialization based on Cell ID, the sequence is partitioned into four non-overlapping portions. The portion for transmission is selected based on the second and third least significant bits of the SFN. In the second scrambling stage, initialization based on Cell ID, the sequence is partitioned into four or eight non-overlapping portions. The portion for transmission is selected based on the second or third least significant bits of the SS block time index.

The physical layer processing of the PBCH is shown in Fig. 4.9. We denote the MIB bits in a TB delivered to the physical layer by $\overline{a}_0, \overline{a}_1, \ldots, \overline{a}_{N_{MIB-1}}$, where $N_{MIB} = 32$ bits is the payload size of the MIB. The lowest order information bit $\overline{a}_0$ is mapped to the most significant bit of the TB payload. Additional timing-related PBCH payload bits $\overline{a}_{N_{MIB}}, \overline{a}_{N_{MIB+1}}, \ldots, \overline{a}_{N_{MIB}+7}$ regenerated based on the least significant bits of the SFN, half frame, SS block index, and combined with the MIB payload (see Fig. 4.10). The $N_{MIB} + 8$ bits are interleaved according to an interleaving pattern specified in [7] prior to the first scrambling stage. In the first scrambling

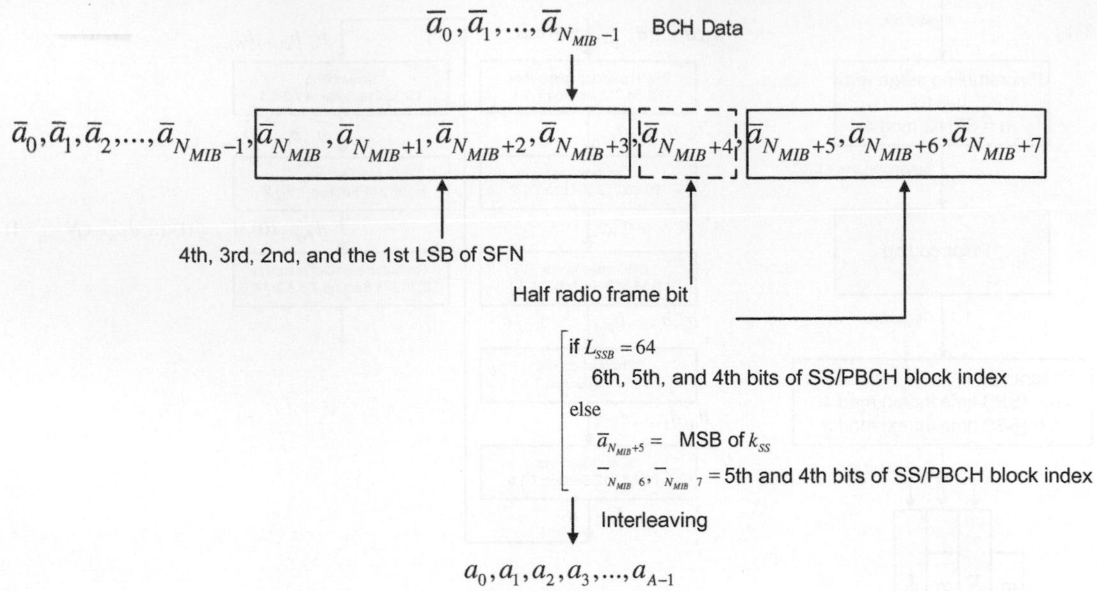

Figure 4.10
NR PBCH payload generation [7].

stage, the input bits to the scrambling unit a_i are scrambled according to $a'_i = (a_i + s_i)\bmod 2; \forall i = 0, 1, \ldots, N_{MIB} + 7$ where s_i is a sequence that is derived from a generic pseudo-random length-31 Gold sequence, which is initialized with $c_{init} = N_{ID}^{cell}$ at the start of each SFN whose value satisfies $SFN \bmod 8 = 0$. The s_i further depends on the half radio frame index, and the second and third least significant bits of the system frame number [7]. A CRC is calculated for the purpose of error detection on the entire BCH payload. The input bit sequence is denoted by $a'_0, a'_1, \ldots, a'_{A-1}$ and the parity bits by $p_0, p_1, \ldots, p_{L-1}$ where $A = N_{MIB} + 8$ is the payload size and $L = 24$ is the number of CRC parity bits. The parity bits are calculated and attached to the BCH payload using the generator polynomial $g_{CRC24C}(D) = D^{24} + D^{23} + D^{21} + D^{20} + D^{17} + D^{15} + D^{13} + D^{12} + D^8 + D^4 + D^2 + D + 1$, resulting in the sequence $b_0, b_1, \ldots, b_{B-1}$, where $B = A + L$. The information bits that are then delivered to the channel coding block are denoted by $c_0, c_1, \ldots, c_{K-1}$, where K is the number of input bits. They are encoded using a polar encoder by setting the parameters $n_{\max} = 9, I_{IL} = 1, n_{PC} = 0$, and $n_{PC}^{wm} = 0$. Detailed PBCH channel encoding and decoding block diagram is shown in Fig. 4.11. The output of polar encoder is denoted by $d_0, d_1, \ldots, d_{N-1}$ where N is the number of coded bits. The input sequence to the rate matching function is $d_0, d_1, \ldots, d_{N-1}$ and the output bit sequence after rate matching is denoted by $f_0, f_1, \ldots, f_{E-1}$ where the rate matching output sequence length is $E = 864$ and the rate matching is performed by setting the parameter I_{BIL} to zero. The preceding polar coding and rate matching parameters will be explained in Section 4.1.7.1 [7].

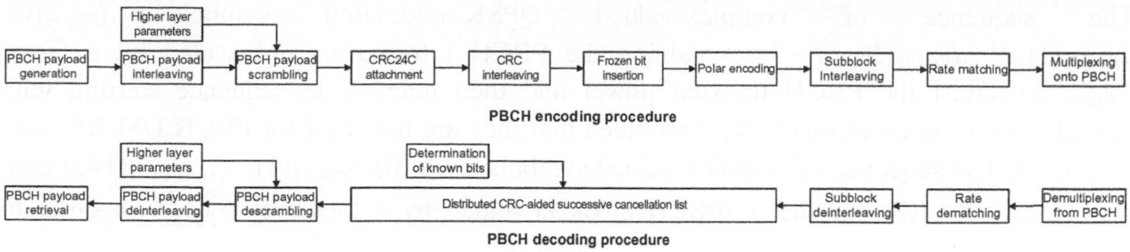

Figure 4.11
NR PBCH channel encoding/decoding block diagram [7,35].

The block of bits $b_0, b_1, \ldots, b_{N_{PBCH}-1}$, where N_{PBCH} denotes the number of bits transmitted on the PBCH, is scrambled prior to modulation, resulting in a block of scrambled bits $\tilde{b}_0, \tilde{b}_1, \ldots, \tilde{b}_{N_{PBCH}-1}$ in which $\tilde{b}_{(i)} = [b(i) + c(i + vN_{PBCH})] \bmod 2$ and $c(i)$ denotes a generic pseudo-random length-31 Gold sequence. The scrambling sequence is initialized with $c_{init} = N_{ID}^{cell}$ at the start of each SS/PBCH block. The parameter v is the two least significant bits of the SS/PBCH block index when $L_{\max} = 4$ and the three least significant bits of the SS/PBCH block index when $L_{\max} = 8$ or 64 where $L_{\max}$ denotes the maximum number of SS/PBCH blocks in an SS/PBCH period for a particular band (see Section 4.1.4.3). The block of scrambled bits $\tilde{b}_0, \tilde{b}_1, \ldots, \tilde{b}_{N_{PBCH}-1}$ is QPSK modulated, resulting in a block of complex-valued modulation symbols $d_{PBCH}(0), d_{PBCH}(1), \ldots, d_{PBCH}(N_{symb} - 1)$. The mapping of the modulated symbols to the physical resources is described in Section 4.1.4.3. The PBCH exploits a special type of DM-RS that is used for coherent detection and decoding of PBCH [6].

The total number of resource elements used for PBCH transmission per SS block is 576. Note that this number includes the resource elements for PBCH and the resource elements for the DM-RS needed for coherent demodulation of PBCH. Different numerologies can be used for SS/PBCH block transmission. However, to limit the need for devices to simultaneously search for SS/PBCH blocks of different numerologies, in many cases only a single SS block numerology is defined for a given frequency band. The DM-RS sequence $r_{PBCH}(m)$ for an SS/PBCH block is defined by $r_{PBCH}(m) = ([1 - 2c(2m)] + j[1 - 2c(2m + 1)]/\sqrt{2})$ where $c(n)$ is a length-31 Gold sequence generated by the pseudo-random sequence generator defined in [6] and initialized at the start of each SS/PBCH block with $c_{init} = 2^{11}(\bar{i}_{SSB} + 1)\left(\left\lfloor N_{ID}^{cell}/4 \right\rfloor + 1\right) + 2^6(\bar{i}_{SSB} + 1) + (N_{ID}^{cell} \bmod 4)$ [6]. If $L = 4$, $\bar{i}_{\mathrm{SSB}} = i_{\mathrm{SSB}} + 4n_{hf}$ where n_{hf} denotes the number of the half-frame in which PBCH is transmitted in a frame with $n_{hf} = 0$ for the first half-frame in the frame and $n_{hf} = 1$ for the second half-frame in the frame, and i_{SSB} is the two least significant bits of the SS/PBCH block index. In the case that $L = 8$ or $L = 64$, $n_{hf} = 0$, and $\bar{i}_{\mathrm{SSB}} = i_{\mathrm{SSB}}$ are the three least significant bits of the SS/PBCH block index. Note that L denotes the maximum number of SS/PBCH block beams in an SS/PBCH block period for a particular band [3].

The sequence of complex-valued QPSK-modulated symbols $d_{PBCH}(0)$, $d_{PBCH}(1), \ldots, d_{PBCH}(N_{symb}-1)$ containing the PBCH information are scaled by a factor β_{PBCH} to adjust the PBCH transmit power and then mapped in sequence starting with $d_{PBCH}(0)$ to resource elements (k, l) provided that they are not used for PBCH DM-RSs (see Fig. 4.12). The sequence of complex-valued symbols $r_{PBCH}(0), r_{PBCH}(1), \ldots, r_{PBCH}(143)$ containing the DM-RSs for the SS/PBCH block is scaled by a factor of $\beta_{PBCH}^{DM\text{-}RS}$ in order to

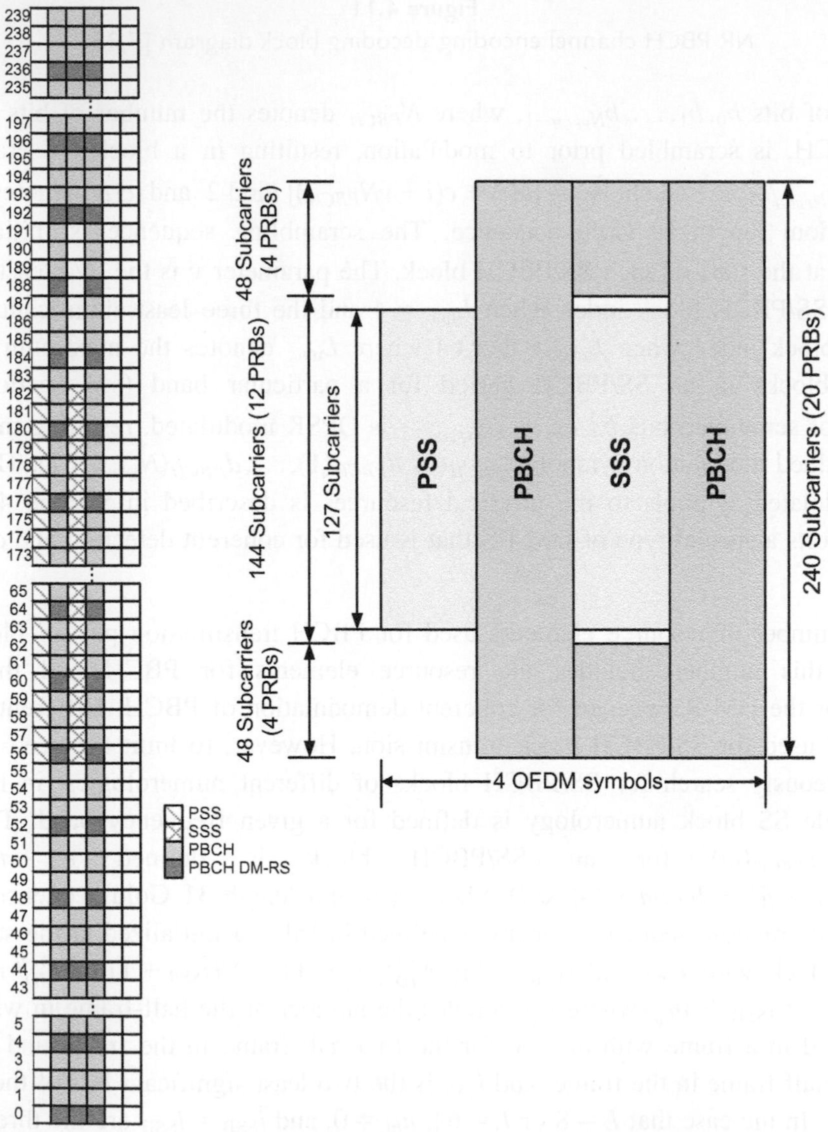

Figure 4.12
Structure of PBCH in time/frequency domain [30].

Table 4.3: Resource mapping within an SS/PBCH block for PSS, SSS, PBCH, and DM-RS for PBCH [6].

Physical Channel/Signal	OFDM Symbol Number l Relative to the Start of an SS/PBCH Block	Subcarrier Number k Relative to the Start of an SS/PBCH Block
PSS	0	56,57,...,182
SSS	2	56,57,...,182
Null	0	0,1,...,55,183,184,...,239
	2	48,49,...,55,183,184,...,191
PBCH	1,3	0,1,...,239
	2	0,1,...,47, 192,193,...,239
DM-RS for PBCH	1,3	$0+v, 4+v, 8+v, \ldots, 236+v$
	2	$0+v, 4+v, 8+v, \ldots, 44+v$ $192+v, 196+v, \ldots, 236+v$

adjust the PBCH DM-RS transmit power. They are then mapped to resource elements (k, l) in increasing order of first k (frequency index) and then l (time index), within one SS/PBCH block. As shown in Table 4.3, the location of PBCH DM-RS is dependent on parameter $v = N_{ID}^{cell} \bmod 4$ and is shifted in frequency with different N_{ID}^{cell} values (see Fig. 4.13 for an example).

4.1.3.2 Physical Downlink Control Channel

The data transmission in NR in downlink/uplink direction is generally controlled via MAC scheduling. Each device monitors a number of PDCCHs, typically once per slot, although it is possible to configure more frequent monitoring to support traffic requiring very low latency. Upon detection of a valid PDCCH, the device follows the scheduling decision and receives (or transmits) one unit of data, known as a transport block. The PDCCHs are transmitted in one or more CORESETs each of length one to three OFDM symbol(s). Unlike LTE, where control channels span the entire carrier bandwidth, the bandwidth of a CORESET can be configured. In NR, a flexible slot format can be configured for a UE by cell-specific and/or UE-specific higher layer signaling in a semi-static downlink/uplink assignment manner, or by dynamically signaling via DCI in group-common PDCCH (GC-PDCCH). When the dynamic signaling is configured, a UE should monitor GC-PDCCH which carries dynamic slot format indication (SFI). When a device enters the connected state, it has already obtained the information from PBCH about the CORESET where it can find the control channel used to schedule the RMSI. The CORESET configuration obtained from PBCH also defines and activates the initial bandwidth part in the downlink. The initial active uplink bandwidth part is obtained from the SI scheduled using the downlink PDCCH.

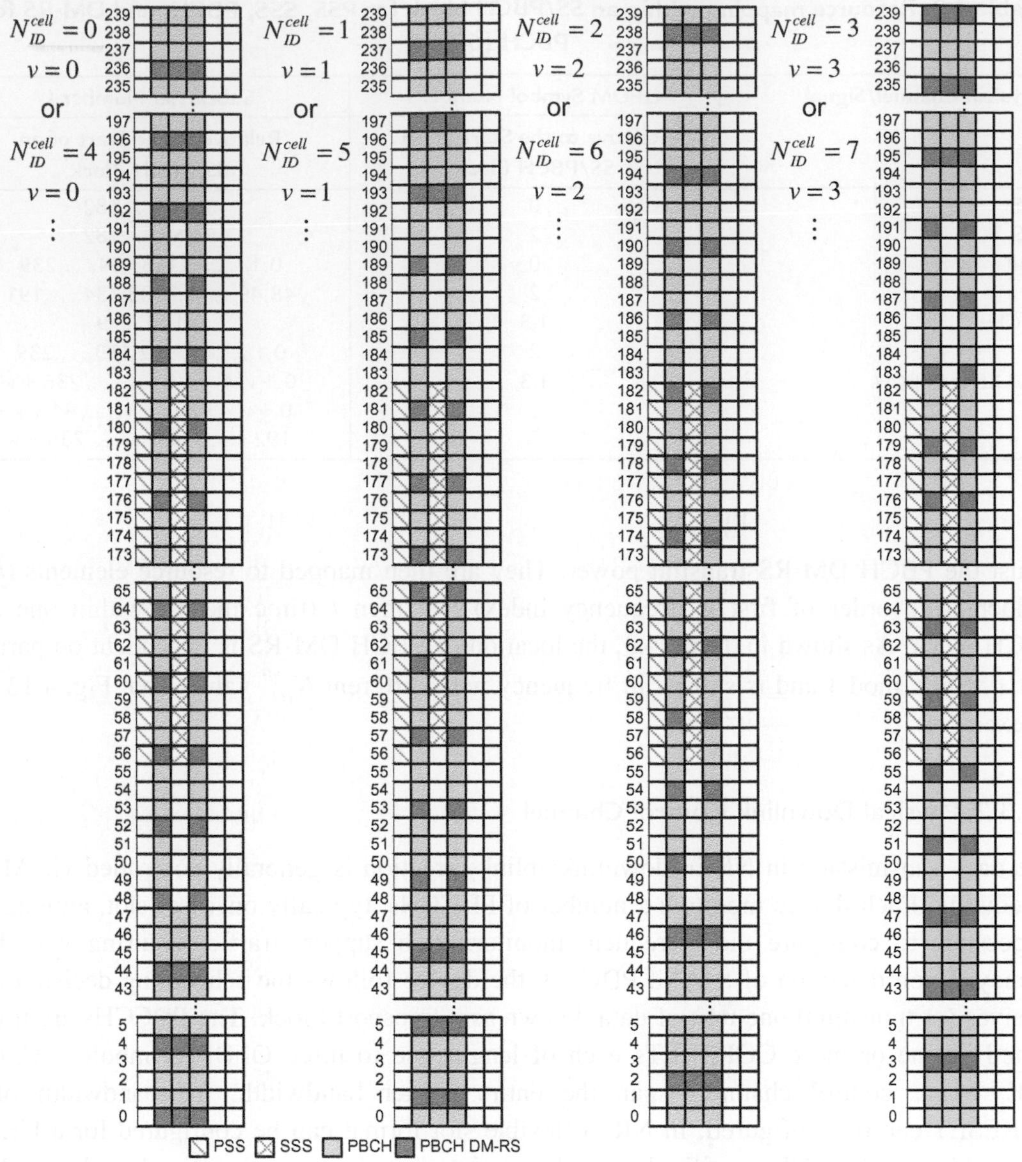

Figure 4.13
PBCH DM-RS location shift by physical cell ID [30].

In NR, the device typically attempts to blindly decode candidate PDCCHs transmitted from the network using one or more search spaces. However, there are some differences compared to LTE based on the different design targets for NR as well as the experience learned from LTE deployments. Unlike LTE PDCCH, the PDCCH in NR does not span the entire carrier bandwidth. This is due to the fact that NR devices may not be able to operate over the full carrier bandwidth of the gNB. The PDCCH in NR is designed to support UE-specific beamforming, which is the result of beam-centric physical-layer design of NR and a requirement when operating in mmWave bands with challenging link budgets.

4.1.3.2.1 Structure and Physical Layer Processing of PDCCH

A UE-specific PDCCH is used to schedule downlink and uplink transmissions on PDSCH and PUSCH, respectively. The DCI on PDCCH contains downlink assignments including modulation and coding format, resource allocation, and HARQ information related to DL-SCH; uplink scheduling grants including modulation and coding format, resource allocation, and HARQ information related to UL-SCH. The control channels are formed by aggregation of control channel elements (CCE), each control channel element consisting of a set of resource element groups (REG). Different code rates for the control channels are realized by aggregating different number of control channel elements. Polar code is used for PDCCH channel coding. Each resource element group carrying PDCCH includes its own DM-RS. QPSK modulation is used for PDCCH modulation.

A resource element is the smallest unit of the resource grid consisting of one subcarrier in frequency domain and one OFDM symbol in time domain. A PDCCH corresponds to a set of resource elements carrying DCI. Each NR control channel element consists of six REGs where a REG is equivalent to one resource block (12 resource elements in the frequency domain) over one OFDM symbol. The CCE size is designed such that at least one UE-specific DCI can be transmitted within one CCE with lower code rates. An NR REG bundle is further defined comprising 2, 3, or 6 REGs, which provides: (1) it determines the precoder cycling granularity (which affects the channel estimation performance) and (2) it is the interleaving unit for the distributed REG mapping. An NR PDCCH candidate consists of a set of CCEs, that is, 1, 2, 4, 8, or 16, corresponding to aggregation levels (ALs) 1, 2, 4, 8, 16, respectively. A control search space consists of a set of PDCCH candidates and is closely associated with the ALs, the number of decoding candidates for each AL, and the set of CCEs for each decoding candidate. A search spacein NR Rel-15 is associated with a single CORESET.

As shown in Fig. 4.14 CORESET is defined as a set of REGs with a given numerology. In the frequency domain, a CORESET is defined as a set of contiguous or distributed physical resource blocks configured using a six-PRB granularity, within which the UE attempts to blindly decode the DCI. There is no restriction on the maximum number of segments for a given CORESET. In the time domain, a CORESET spans 1, 2, or 3 contiguous OFDM symbol, and the exact duration is signaled to the UE via broadcast SI or UE-specific RRC signaling depending on whether it is a common CORESET or UE-specific CORESET. Compared to LTE PDCCH, the configurability of the CORESETs enable efficient resource sharing between downlink control and shared channels, thereby allowing efficient layer-1 signaling overhead management. One of the factors which could impact

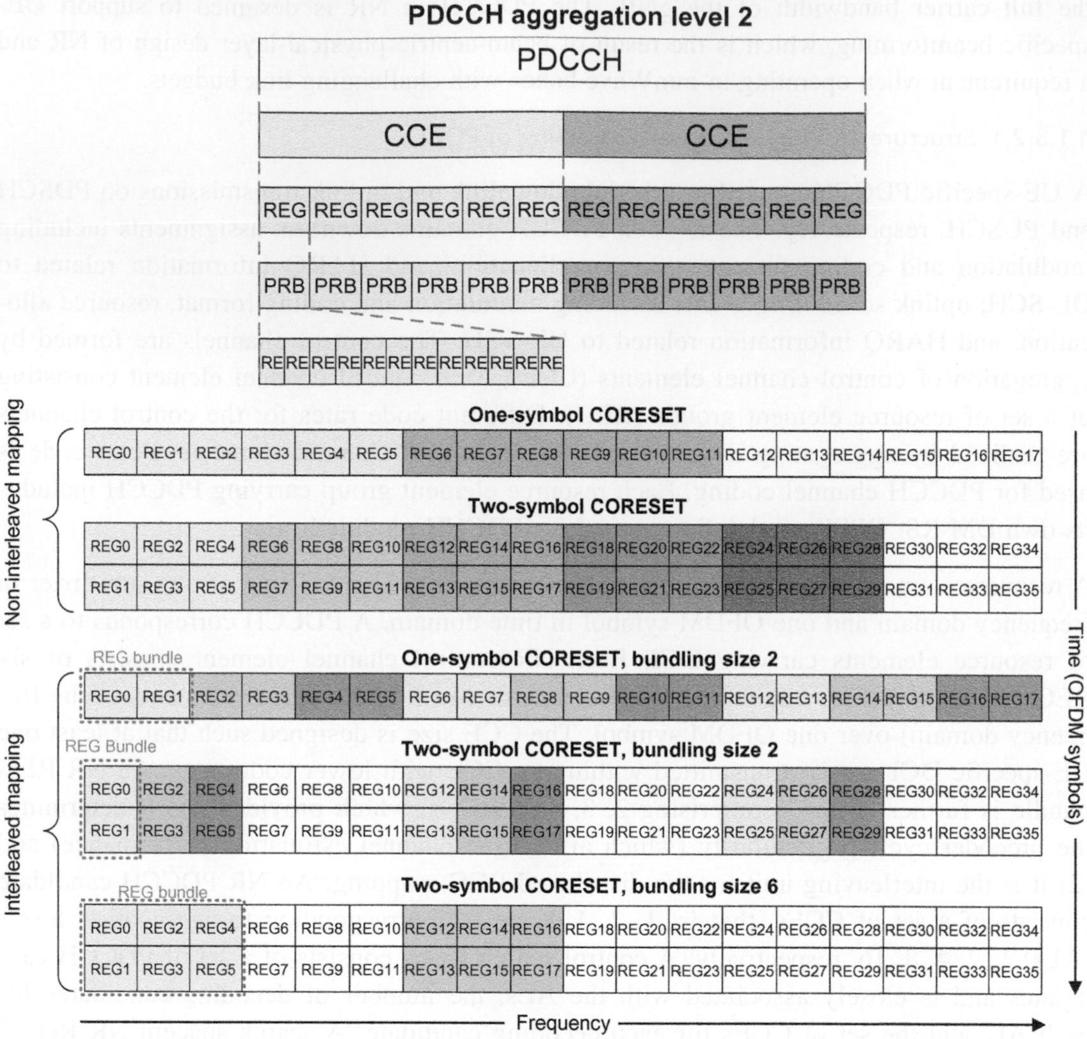

Figure 4.14

Illustration of RE, REG, CCE, REG bundle, and CORESET and example mappings of CCE to REG Bundles [14,68].

the time-domain duration of a CORESET is the bandwidth of the corresponding carrier, where more control symbols may be allowed for smaller bandwidths. For example, assuming a CORESET consists of 48 PRBs with 2 OFDM symbols, there are 16 CCEs that could accommodate up to 2 PDCCH candidates at AL-8 or a single candidate at AL-16. Furthermore, there can be multiple CORESETs inside the system bandwidth; thus the CORESET may not fully occupy the system bandwidth in the frequency domain. Downlink power adjustment can be applied to CORESETs that occupy narrower frequency regions within the carrier bandwidth, depending on desired coverage and link budget. In that case, one or two OFDM symbols may not be sufficient. One-symbol CORESET offers benefits from the perspective of latency and control overhead adjustments especially when there are few UEs in the cell or when the coverage target is limited (e.g., small cell deployments). The maximum CORESET duration that may be configured in a cell is implicitly signaled via PBCH. A UE may be configured with one or more CORESETs (using UE-specific or common higher layer signaling) with a maximum of three CORESETs per configured (downlink) BWP. Limiting the maximum number of CORESETs is beneficial for enabling more practical RRC signaling and better UE dimensioning. Note that the scheduling flexibility may not be impacted by limiting the maximum number of CORESETs since different monitoring occasions can be flexibly configured associated with the same CORESET. It is important to note that the concept of PDCCH monitoring periodicity is defined per search space set and is not configured at the CORESET level. Every configured search space with a certain monitoring periodicity (in terms of slots and starting symbols within the monitored slots) is associated with a CORESET. For a CORESET configured by UE-specific RRC signaling, some of the configured parameters include frequency-domain resources, starting OFDM symbol, CORESET duration, REG bundle size, transmission type (i.e., interleaved or non-interleaved), and precoding assumptions for channel estimation filtering [53].

As we mentioned earlier, a PDCCH consists of one or more CCEs. A CORESET consists of $N_{RB}^{CORESET}$ resource blocks in the frequency domain, determined by the RRC parameter *frequencyDomainResources* in *ControlResourceSet* information element, and $N_{symb}^{CORESET} \in \{1,2,3\}$ OFDM symbols in the time domain, defined by the RRC parameter *duration* in the *ControlResourceSet* information element, where $N_{symb}^{CORESET} = 3$ is supported, if the RRC parameter *dmrs-TypeA-Position* is set to 3 [6]. A control channel element consists of six REGs where a REG is equivalent to one resource block over one OFDM symbol. The REGs within a CORESET are numbered in increasing order in a time-first manner, starting with 0 for the first OFDM symbol and the lowest numbered resource block in the CORESET. A UE can be configured with multiple CORESETs. Each CORESET is associated with only one CCE-to-REG mapping (see Fig. 4.14). There is a direct correspondence between the number of CCEs and the AL, for example, for ALs 1, 2, 4, 8, and 16, there will be 1, 2, 4, 8, and 16 CCEs, respectively [6]. The time-frequency structure of REG, CCE, REG bundle, and CORESET as well as example mappings of CCE to REG bundles are illustrated Fig. 4.14.

The PDCCH processing steps are illustrated in Fig. 4.15. At a high level, the PDCCH processing in NR is similar to that of LTE ePDCCH than LTE PDCCH in the sense that each PDCCH is processed independently. As shown in the figure, the entire DCI bits are used to calculate the CRC parity bits. Let us assume that $a_0, a_1, \ldots, a_{N_{DCI}-1}$ denote the DCI input bits, and $p_0, p_1, \ldots, p_{L-1}$ represent the parity bits, where N_{DCI} and $L = 24$ are the payload size and the number of parity bits, respectively. Let us assume that $a'_0, a'_1, \ldots, a'_{N_{DCI}-1}$ is bit

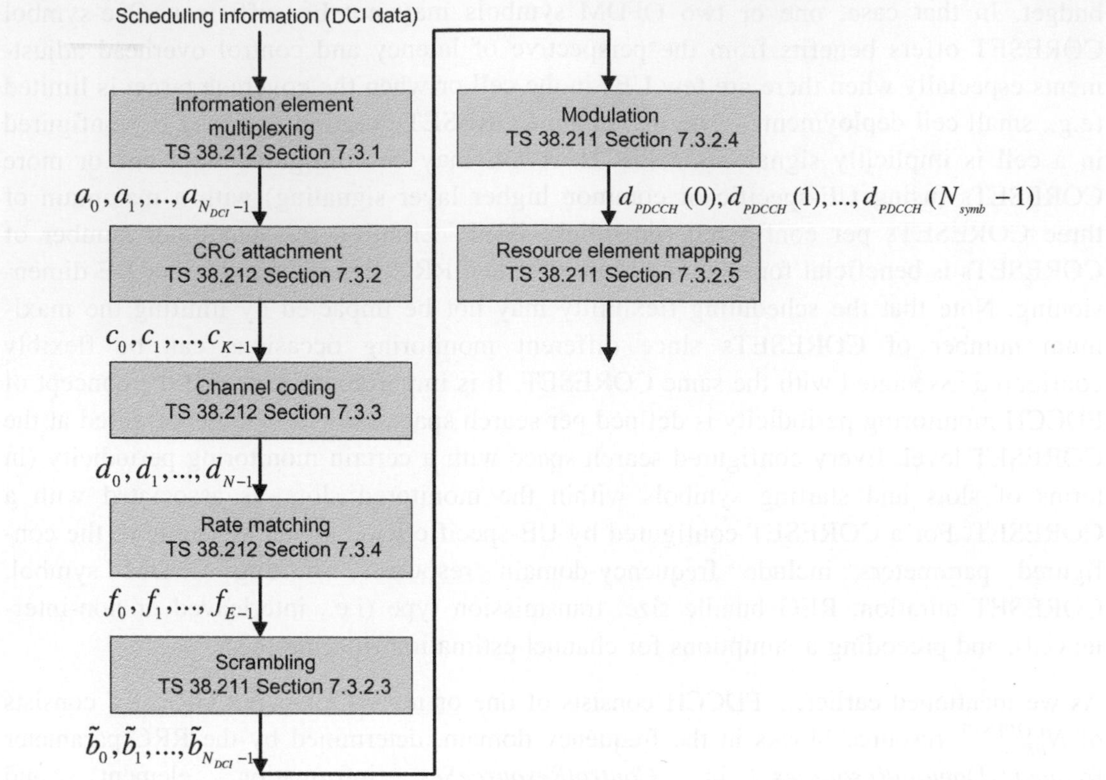

Figure 4.15
Physical layer processing of NR PDCCH [6,7].

sequence such that $a'_i = 1\ \forall i = 0, 1, \ldots, L-1$ and $a'_i = a'_{i-L}\ \forall i = L, L+1, \ldots, L+N_{DCI}-1$. The parity bits are computed with input bit sequence $a'_0, a'_1, \ldots, a'_{N_{DCI}+L-1}$ using the generator polynomial $g_{CRC24C}(D) = D^{24} + D^{23} + D^{21} + D^{20} + D^{17} + D^{15} + D^{13} + D^{12} + D^{8} + D^{4} + D^{2} + D + 1$. The output bit sequence is given as $b_0, b_1, \ldots, b_{B-1}$ where $b_k = a_k \forall k = 0, 1, \ldots, N_{DCI}-1$ and $b_k = p_{k-N_{DCI}}\ \forall k = N_{DCI}, N_{DCI+1}, \ldots, N_{DCI+L-1}$. Following the attachment of the CRC bits, the sequence is scrambled with the corresponding 16-bit RNTI $x_{RNTI}(0), x_{RNTI}(1), \ldots, x_{RNTI}(15)$, where $x_{RNTI}(0)$ corresponds to the MSB of the RNTI binary value, resulting in the sequence of bits $c_0, c_1, \ldots, c_{K-1}$ where $c_k = b_k \forall k = 0, 1, \ldots, N_{DCI}+7$ and $c_k = [b_k + x_{RNTI}(k - N_{DCI} - 8)] \bmod 2 \forall k = N_{DCI} + 8, N_{DCI}+9, \ldots, N_{DCI}+23$ [6,7].

The PDCCH encoding stages are shown in Fig. 4.15. The K scrambled information bits are delivered to the channel coding block and are polar coded, by setting the encoder parameters to the following $n_{\max} = 9, I_{IL} = 1, n_{PC} = 0$, and $n_{PC}^{wm} = 0$. The encoding process produces N bits which are denoted as $d_0, d_1, \ldots, d_{N-1}$. The rate matching for polar coded bits is performed on per coded block and consists of subblock interleaving, bit collection, and bit interleaving. Detailed PDCCH channel encoding and decoding block diagram is shown in Fig. 4.16.

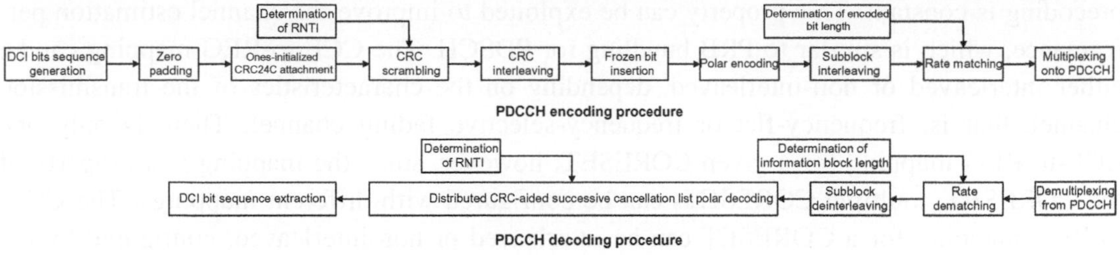

Figure 4.16
NR PDCCH channel encoding/decoding block diagram [7,35].

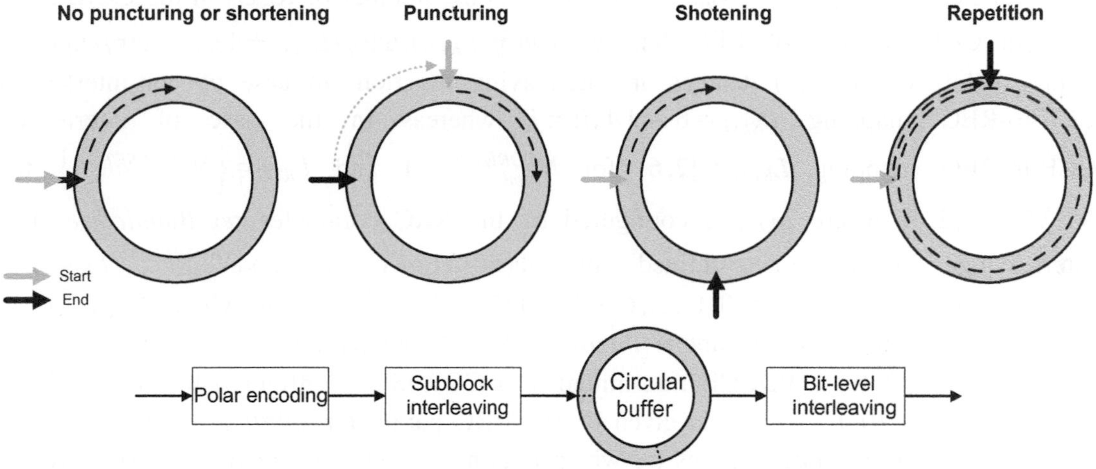

Figure 4.17
Encoding of NR PDCCH and rate matching variants [68].

The input bit sequence to rate matching function is denoted by $d_0, d_1, \ldots, d_{N-1}$ and the output is denoted by $f_0, f_1, \ldots, f_{E-1}$. The input bits to the subblock interleaver are divided into 32 subblocks and the output bits are generated according to $y_n = d_{J(n)}$ where $J(n) = NP(\lfloor 32n/N \rfloor)/32 + (n \bmod N/32)\,\forall n = 0, 1, \ldots, N-1$ and the subblock interleaver pattern is defined in [7]. The repetition, puncturing, or shortening of polar code is performed in the following manner: $N = 2^n$ coded bits at the output of polar encoder is written into a length-N circular buffer in an order that is predefined for a given value of N. As shown in Fig. 4.17, to obtain M coded bits for transmission, puncturing is realized by selecting bits from position $N - M$ to position $N - 1$ from the circular buffer, shortening is realized by selecting bits from position 0 to position $M - 1$ from the circular buffer, and repetition is realized by selecting all bits from the circular buffer, and additionally repeating $M - N$ consecutive bits from the circular buffer [7].

For each CORESET, there is an associated CCE-to-REG mapping based on the REG bundle (see Fig. 4.14). A REG bundle is a set of REGs across which the device can assume the precoding is constant. This property can be exploited to improve the channel estimation performance, which is similar to PRB bundling for PDSCH. The CCE-to-REG mapping can be either interleaved or non-interleaved, depending on the characteristics of the transmission channel, that is, frequency-flat or frequency-selective fading channel. There is only one CCE-to-REG mapping for a given CORESET; however, since the mapping is a property of the CORESET, multiple CORESETs can be configured with different mappings. The CCE-to-REG mapping for a CORESET can be interleaved or non-interleaved, configured by the RRC parameter *cce-REG-MappingType* in the *ControlResourceSet* information element and is described by REG bundles. The REG bundle i is defined as REGs $\{iL_{REG}, iL_{REG}+1,\ldots,iL_{REG}+L_{REG}-1\}\forall i=0,1,\ldots,N_{REG}^{CORESET}/L_{REG}-1$ where L_{REG} is the size of the REG bundle and $N_{REG}^{CORESET}=N_{RB}^{CORESET}N_{symb}^{CORESET}$ is the number of REGs in the CORESET. The jth CCE consists of REG bundles $\{\Phi(6j/L_{REG}),\Phi(6j/L_{REG}+1),\ldots,\Phi(6j/L_{REG}+6/L_{REG}-1)\}$, where $\Phi(\cdot)$ denotes an interleaving function. In case of non-interleaved CCE-to-REG mapping $L_{REG}=6$ and $\Phi(j)=j$, whereas in the case of interleaved CCE-to-REG mapping $L_{REG}=\{2,6\}$ for $N_{symb}^{CORESET}=1$ and $L_{REG}\in\left\{N_{symb}^{CORESET},6\right\}$ for $N_{symb}^{CORESET}\in\{2,3\}$ where L_{REG} is configured by the RRC parameter *reg-BundleSize*. The interleaving function is defined by $\Phi(j)=rC+c+n_{shift}\mathrm{mod}N_{REG}^{CORESET}/L_{REG};j=cR+r;r=0,1,\ldots,R-1;c=0,1,\ldots,C-1$ and $C=N_{REG}^{CORESET}/L_{REG}R$ where $R\in\{2,3,6\}$ is given by the higher layer parameter *interleaverSize*. Other parameters are defined as follows: $n_{shift}=N_{ID}^{cell}$ for a PDCCH transmitted in a CORESET configured by the PBCH or SIB1, and $n_{shift}\in\{0,1,\ldots,274\}$ is given by the RRC parameter *shiftIndex*. The UE is not expected to monitor configurations for which C is not an integer. For both interleaved and non-interleaved mappings, the same precoding is used within an REG bundle, if the higher layer parameter *precoderGranularity* equals L_{REG}. The same precoding is used across all REGs within the set of contiguous resource blocks in the CORESET, if the higher layer parameter *precoderGranularity* equals the size of the CORESET in the frequency domain. For a CORESET configured by PBCH, $L_{REG}=6, R=2$ and the same precoding is used within the REG bundle [6]. Unlike LTE, where the length of the control region can vary dynamically as indicated by PCFICH, a CORESET in NR is of fixed size. This is important from an implementation perspective, both for the UE and the network. From the UE perspective, a pipelined implementation is simpler, if the device can directly start to decode PDCCH without having to first decode another control channel. Various REG-to-CCE mapping options are shown in Fig. 4.18.

During the PDCCH detection and decoding process, the UE needs to estimate the channel using the reference signals associated with the PDCCH candidate being decoded. A single antenna port is used for PDCCH transmission which means any transmit diversity or multi-user MIMO scheme is handled in a device-transparent manner. The PDCCH has its own DM-RS, based on the same pseudo-random sequence that is used for PDSCH, that is, the pseudo-random sequence is generated across all the common resource blocks in the

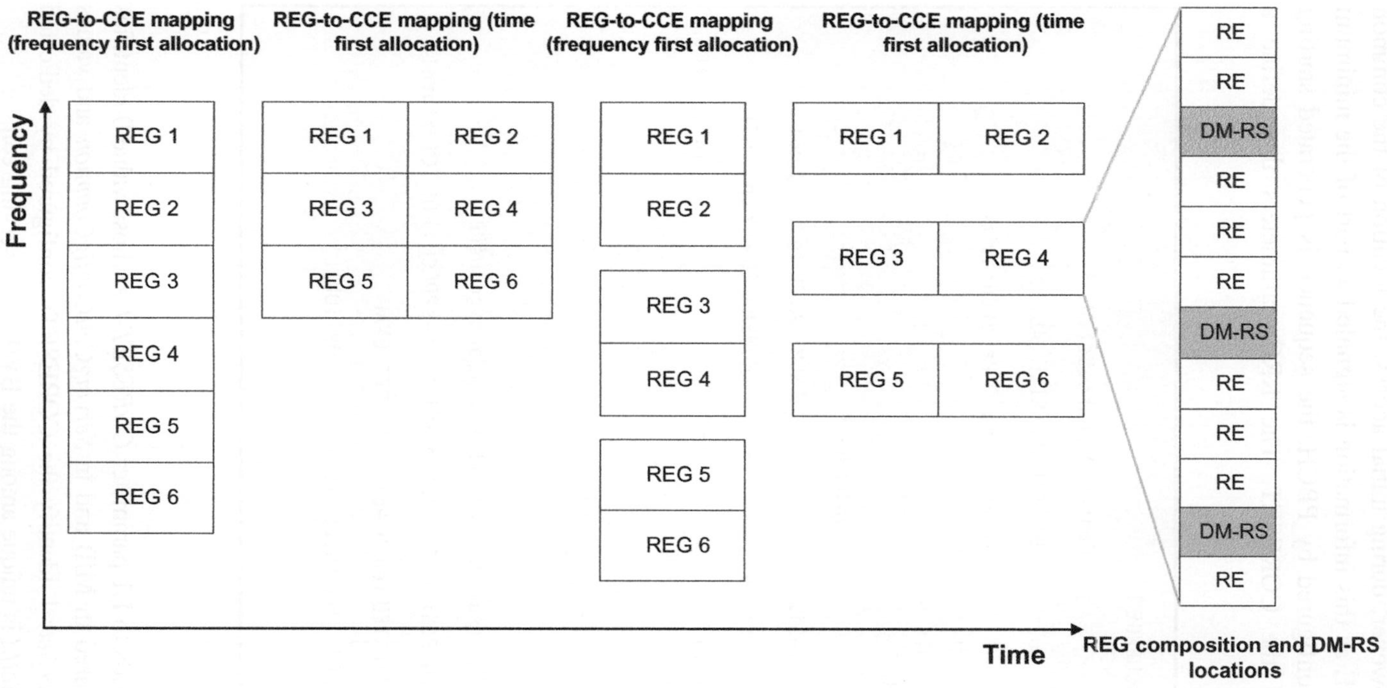

Figure 4.18
REG-to-CCE mapping options and REG composition [6].

frequency domain, but transmitted only in the resource blocks used for PDCCH (with one exception as discussed below). However, during initial access, the location of the common resource blocks is not known to UE as this information is signaled as part of the minimum SI. Therefore, for CORESET 0 configured by PBCH, the sequence is generated starting from the first resource block in the CORESET. The RRC parameters that define a CORESET are as follows [13]:

```
ControlResourceSet::=          SEQUENCE {
    controlResourceSetId                    ControlResourceSetId,
    frequencyDomainResources                BIT STRING (SIZE (45)),
    duration                               INTEGER (1..maxCoReSetDuration),
//maxCoReSetDuration = 3
    cce-REG-MappingType CHOICE {
            interleaved                     SEQUENCE {
                    reg-BundleSize                  ENUMERATED {n2, n3, n6},
                    interleaverSize                 ENUMERATED {n2, n3, n6},
                    shiftIndex                                      INTEGER(0..
maxNrofPhysicalResourceBlocks-1)
            },
            nonInterleaved          NULL
    },
    precoderGranularity ENUMERATED {sameAsREG-bundle, allContiguousRBs},
    tci-StatesPDCCH        SEQUENCE(SIZE (1..maxNrofTCI-StatesPDCCH)) OF TCI-StateId
    tci-PresentInDCI        ENUMERATED {enabled}        OPTIONAL
    pdcch-DMRS-ScramblingID   BIT STRING (SIZE (16))        OPTIONAL
}
```

In the preceding definition [6,9],

- *controlResourceSetId* corresponds to L1 parameter *CORESET-ID* whose value 0 identifies the common CORESET configured in MIB and in *ServingCellConfigCommon* and values 1, 2,..., *maxNrofControlResourceSets*-1 identify the CORESETs configured by dedicated signaling. The *controlResourceSetId* is unique among the BWPs of a serving cell.
- *frequencyDomainResources* corresponds to the L1 parameter *CORESET-freq-dom*. Each bit corresponds a group of six RBs, with the grouping start from PRB 0, which is fully

contained in the bandwidth part within which the CORESET is configured. The most significant bit corresponds to the group of the lowest frequency which is fully contained in the bandwidth part within which the CORESET is configured, each subsequent lower significant bit corresponds to the next lowest frequency group that are fully contained within the bandwidth part in which the CORESET is configured. The bits corresponding to a group not fully contained within the bandwidth part in which the CORESET is configured are set to zero.

- *duration* is the contiguous time duration of the CORESET in number of symbols.
- *cce-reg-MappingType* identifies the mapping method of CCE-to-REG.
- *reg-BundleSize* is the number of REGs within an REG bundle corresponding to L1 parameter *CORESET-REG-bundle-size.*
- interleaveSize corresponds to L1 parameter CORESET-interleaver-size.
- shiftIndex corresponds to CORESET-shift-index.
- *precoderGranularity* denotes the precoder granularity in frequency domain. It corresponds to L1 parameter *CORESET-precoder-granularity.*
- *tci-StatesPDCCH* is a reference to a configured transmission configuration indication (TCI)[4] state providing QCL configuration/indication for PDCCH.

[4] Downlink beamforming is typically transparent to the UE, that is, the device does not need to know which beam is used at the transmitter. However, NR also supports beam indication, which implies that the UE is informed of a certain PDSCH and/or PDCCH transmission using the same transmit beam as a configured reference signal (CSI-RS or SS block). The beam indication is based on the (downlink signaling of) transmission configuration indication (TCI) states. Each TCI state includes information about a reference signal, for example, a CSI-RS or an SS block. By associating a certain downlink transmission (PDCCH or PDSCH) with a certain TCI, the network informs the UE that it can assume the upcoming downlink transmission uses the same spatial filter as the reference signal associated with that TCI. A device can be configured with up to 64 candidate TCI states. The beam indication for PDCCH is done by assigning a subset of the M configured candidate states via RRC signaling to each configured CORESET. Using MAC signaling, the network can dynamically indicate a specific TCI state, within the per-CORESET-configured subset, to be valid. When monitoring PDCCH within a certain CORESET, the device can assume that the PDCCH transmission uses the same spatial filter as the reference signal associated with the MAC-indicated TCI. In other words, if the device has determined a suitable receiver-side beam direction for reception of the reference signal, it can assume that the same beam direction is suitable for reception of the PDCCH. For PDSCH beam indication, there are two alternatives depending on the scheduling offset, that is, depending on the transmission timing of PDSCH relative to the corresponding PDCCH carrying scheduling information for that PDSCH. If this scheduling offset is larger than N symbols, the DCI of the scheduling assignment may explicitly indicate the TCI state for the PDSCH transmission. To enable this, the device is initially configured with a set of up to eight TCI states from the originally configured set of candidate TCI states. A three-bit indicator within the DCI then indicates the exact TCI state which is valid for the scheduled PDSCH transmission. If the scheduling offset is smaller or equal to N symbols, the device should instead assume that the PDSCH transmission is QCL with the corresponding PDCCH transmission. In other words, the TCI state for the PDCCH state indicated by MAC signaling should be assumed to be valid for the corresponding scheduled PDSCH transmission. The reason for limiting the dynamic TCI selection based on DCI signaling to the scenarios where the scheduling offset is larger than a certain value is that for shorter scheduling offsets, there will not be sufficient time for the UE to successfully decode the TCI information within the DCI and to adjust the receive beam accordingly before the PDSCH is received [14].

- tci-PresentInDCI corresponds to L1 parameter CORESET-precoder-granularity.
- *pdcch-DMRS-ScramblingID* is the PDCCH DM-RS scrambling initialization.

The DM-RSs associated with a given PDCCH candidate are mapped to every fourth subcarrier in a REG, that is, the reference signal overhead is one-fourth. This is a denser reference signal pattern relative to LTE, which has a reference signal overhead of one-sixth; however, an LTE device can interpolate channel estimates across time and frequency as a result of cell-specific reference signals common to all devices and present regardless of control channel transmission. The use of a dedicated reference signal per PDCCH candidate is advantageous, despite the slightly higher overhead, since it allows different type of device-transparent beamforming schemes. By using a beamformed control channel, the coverage and performance can be enhanced compared to the non-beamformed control channels in LTE. This is an essential part of the beam-centric design of NR [14].

When attempting to decode a PDCCH candidate occupying certain number of CCEs, the device can compute the REG bundles that constitute the PDCCH candidate. Channel estimation must be performed per REG bundle as the network may change precoding across REG bundles. In general, this results in sufficiently accurate channel estimates for PDCCH detection. However, it is also possible to configure the device to assume the same precoding across contiguous resource blocks in a CORESET, thereby allowing the device to perform frequency-domain interpolation of the channel estimates. This also implies that the device may use reference signals referred to as wideband reference signals outside the PDCCH region that it is trying to detect (see Fig. 4.19). The QCL concept is also applicable to the reference signals. If the UE has a priori knowledge about the QCL of two reference signals, it can exploit this property to improve the channel estimation and to manage different receive beams at the device. If no QCL is configured for a CORESET, the UE assumes that PDCCH candidates are quasi-co-located with the SS/PBCH block with respect to delay spread, Doppler spread, Doppler shift, average delay, and spatial RX parameters. This is a reasonable assumption as the device has been able to receive and decode the PBCH in order to access the system.

The block of bits $b(0), b(1), \ldots, b(N_{PDCCH}-1)$, where N_{PDCCH} denotes the number of bits transmitted on PDCCH, is scrambled prior to modulation, resulting in a block of scrambled bits $\tilde{b}(0), \tilde{b}(1), \ldots, \tilde{b}(N_{PDCCH}-1)$ where $\tilde{b}(i) = [b(i) + c(i)] \bmod 2$, in which $c(i)$ is a length-31 Gold sequence generated by the pseudo-random sequence generator defined in [6] and initialized with $c_{init} = (n_{RNTI} 2^{16} + n_{ID}) \bmod 2^{31}$. For a UE-specific search space, $n_{ID} \in \{0, 1, \ldots, 65535\}$ is set by the RRC parameter *pdcch-DMRS-ScramblingID;* otherwise, $N_{ID} = N_{ID}^{cell}$ and n_{RNTI} is determined by the C-RNTI for a PDCCH in a UE-specific search space, when the RRC parameter *pdcch-DMRS-ScramblingID* is configured; otherwise $n_{RNTI} = 0$. The block of scrambled bits $\tilde{b}(0), \tilde{b}(1), \ldots, \tilde{b}(N_{PDCCH}-1)$ is then QPSK modulated, resulting in a block of complex-valued modulation symbols

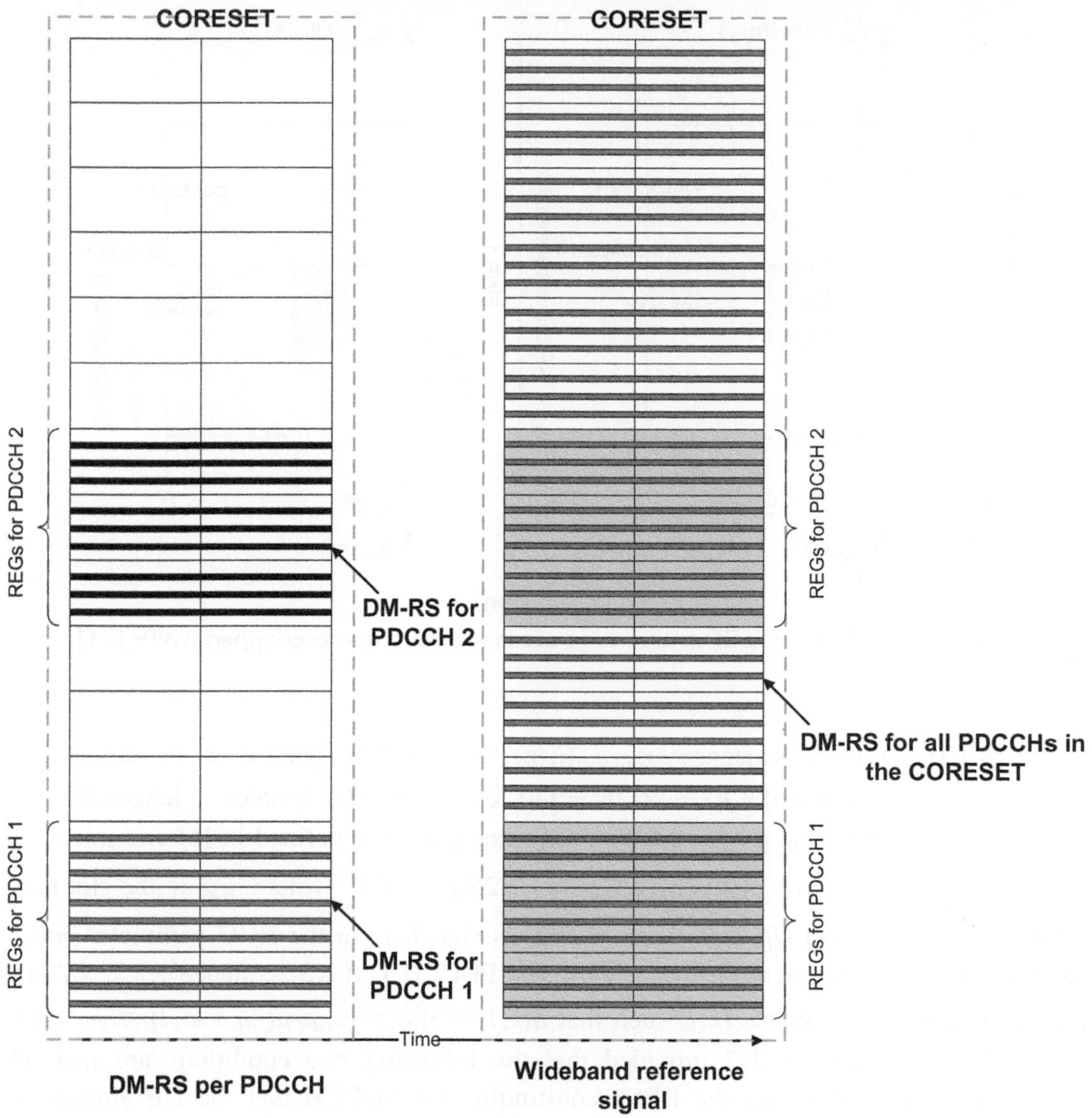

Figure 4.19
Illustration of the regular and wideband reference signals for PDCCH [68].

$d(0), d(1), \ldots, d(N_{symb} - 1)$. The block of complex-valued symbols $d(0), d(1), \ldots,$ $d(N_{symb} - 1)$ is scaled by a factor of β_{PDCCH} and is mapped to resource elements (k, l), which are designated to PDCCH to be monitored by the UE and are not used for the associated PDCCH DM-RS, in increasing order of first k (frequency index) and l (time index) [6]. Fig. 4.20 illustrates the CORESET structures in overlapped and non-overlapped BWPs.

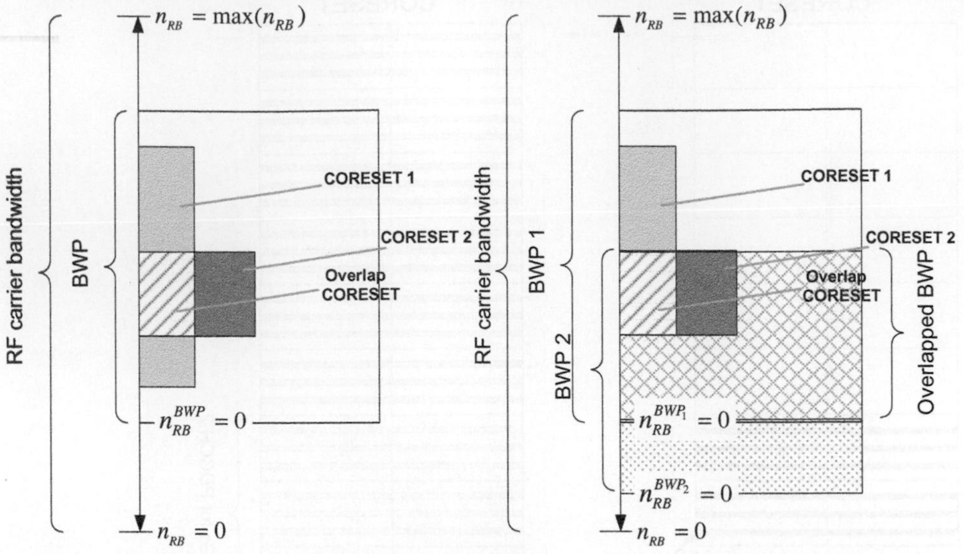

Figure 4.20
Illustration of the CORESET structures in overlapped and non-overlapped BWPs [71].

The PDCCH DM-RS sequence $r_{PDCCH}(l,m)$ on OFDM symbol l is defined by $r_{PDCCH}(l,m) = ([1-2c(2m)] + j[1-2c(2m+1)])/\sqrt{2}$ where $c(i)$ denotes a length-31 Gold sequence generated by the pseudo-random sequence generator defined in [6] and initialized with $c_{init} = \left[2^{17}\left(N_{slot}^{symbol} n_{slot} + l + 1\right)(2N_{ID}+1) + 2N_{ID}\right] \bmod 2^{31}$ where n_{slot} is the slot number within a frame, and $N_{ID} = \{0, 1, \ldots, 65535\}$ is signaled via the RRC parameter *pdcch-DMRS-ScramblingID*; otherwise $N_{ID} = N_{ID}^{cell}$. The PDCCH DM-RS sequence $r_{PDCCH}(l,m)$ is mapped to resource elements (k,l) such that $a(k,l) = \beta_{DMRS}^{PDCCH} r_{PDCCH}(3n + k', l); \forall n \in \mathbb{N}; k = nN_{sc}^{RB} + 4k' + 1$; and $k' = 0, 1, 2$ provided that the following two conditions are met: (1) resource elements are within the REGs constituting the PDCCH that the UE attempts to decode, if the RRC parameter *precoderGranularity* is equal *sameAsREG-bundle*; and (2) all REGs are within the set of contiguous resource blocks in the CORESET where the UE attempts to decode PDCCH, if the RRC parameter *precoderGranularity* is equal to the size of the CORESET in the frequency domain [6]. The reference point for k (frequency index) is subcarrier 0 of the lowest numbered common resource block in the CORESET, if the CORESET is configured by the PBCH or by the *controlResourceSetZero* field in the *PDCCH-ConfigCommon* information element; otherwise, subcarrier 0 in common resource block 0. The parameter l is the OFDM symbol number within the slot. In the absence of CSI-RS configuration, the PDCCH DM-RS and SS/PBCH blocks are quasi-co-located with respect to Doppler shift, Doppler spread, average delay, delay spread, and spatial RX parameters.

The notion of wideband DM-RS has been introduced to assist the NR UE in channel estimation for the control channel detection (see Fig. 4.19). For each CORESET, the precoder granularity in the frequency domain is configurable between REG bundle size and the number of contiguous RBs in the frequency domain within the CORESET. The UE may assume DM-RS is present in all REGs within the set of contiguous RBs of the CORESET where and when at least one REG of a candidate is mapped [6].

4.1.3.2.2 UE Group-Common Signaling

An NR gNB can simultaneously support diverse service categories and thus mitigation of the performance degradation of interrupted services is an important issue in the physical layer design. While the flexible frame structure may alleviate this problem, due to the implementation complexity and unpredictability of URLLC packet arrival times, a more elaborate solution is needed in real-world deployment scenarios. 3GPP has considered a number of solutions in the course of NR development. For infrequent URLLC transmissions, one can give priority to URLLC transmissions while ensuring the reliability of the other transmissions interrupted by URLLC traffic. A preemption indicator transmitted by the base station indicates which resources are used for the URLLC transmission. If the URLLC packet is stretched in the frequency domain, the URLLC transmission will interrupt the entire system bandwidth and thus degrade all data channels in use. To notify scheduled users of this event, the base station broadcasts a preemption indicator consisting of time and/or frequency information of the interrupted resources. This indicator helps users identify the reason for packet errors and what part of the packet is unaffected from the interruption. Retransmission of selected code blocks when the ongoing service is interrupted by the URLLC transmission is another solution, where part of the code block that has been affected by URLLC transmission is retransmitted. By transmitting a combining indicator or flush-out indicator, the receiver can perform the soft symbol combining of the transmitted and retransmitted code blocks. One can further achieve better coding gain by lowering the code rate of the retransmitted code block.

An efficient scheduling scheme in terms of resource allocation and latency is to multiplex data with different transmission time lengths (i.e., mini-slots and slots) and in case of resource limitations to let the high-priority service use resources from lower priority service. This type of multiplexing is also referred to as preemption. For example, in NR downlink, a mini-slot carrying high-priority or delay-sensitive data can preempt an already ongoing slot-based transmission on the first available OFDM symbols without waiting for the next free resource. This operation enables ultra-low latency for mini-slot-based transmission, especially in the scenario where a long slot-based transmission has already been scheduled. A similar concept is also considered for the uplink and in general for LTE. At the cost of degrading the longer transmission, no additional resources need to be reserved in advance for the URLLC service. The impacted longer transmission is then promptly repaired with a transmission containing a subset of the code block groups in a later transmission time, after providing the essential information to clean the corrupted soft values in the receive buffer from the preempted data. If the URLLC transmission occurs frequently, the efficiency of the above approach will be reduced due to the frequent retransmissions. To ensure the reliability of the ongoing services while supporting the URLLC transmission, robustness improvement and service sharing strategies may be adopted [68].

An NR UE can be configured to monitor group-common signaling via DCI format 2_1 which carries preemption indication related to multiplexing eMBB and URLLC traffic with different transmission durations in the downlink. Upon reception of preemption indication, a UE should apply an appropriate HARQ combining mechanism to retrieve the data despite the missing portion due to preemption. Group-common PDCCH is designed for the purpose of signaling to a group of UEs. It carries dynamic SFI, that is, via DCI format 2_0, to indicate which symbols in a slot are designated as downlink, uplink, or flexible symbols. The SFI carries an index to a UE-specific table containing permissible slot configurations [6]. The downlink preemption indicator is signaled via DCI format 2_1; however, whether a UE needs to monitor preemption indication is configured through RRC signaling. The UE is additionally configured with a set of serving cells; a mapping for each serving cell in the set of serving cells to the corresponding fields in DCI format 2_1; an information payload size for DCI format 2_1; and a bitmap for identification of punctured time-frequency resources via higher layer signaling. If the UE detects a DCI format 2_1 for a serving cell from the configured set of serving cells, the UE may assume that there is no transmission assigned to the UE in PRBs and symbols within the active downlink BWP, from a set of PRBs and a set of symbols of the last monitoring period, that are indicated by DCI format 2_1. Note that DCI format 2_1 indication is not applicable to the reception of SS/PBCH blocks [8].

A UE needs to monitor preemption indication carried by DCI format 2_1, if it is provided with RRC parameter *DownlinkPreemption* and it is configured with an INT-RNTI provided by RRC parameter *int-RNTI*. In that case, if the UE detects DCI format 2_1, the set of symbols indicated by a field in DCI format 2_1 includes the last $N_{slot}^{symb} T_{INT} 2^{\mu-\mu_{INT}}$ symbols prior to the first symbol of the CORESET in the slot. The parameter T_{INT} is the PDCCH monitoring periodicity provided by a higher layer parameter, N_{slot}^{symb} is the number of symbols per slot, μ is the subcarrier spacing configuration for a serving cell with mapping to a respective field in the DCI format 2_1, and μ_{INT} is the subcarrier spacing configuration of the downlink BWP where the UE receives the PDCCH conveying the DCI format 2_1. If the UE is configured with RRC parameter *TDD-UL-DL-ConfigurationCommon*, the symbols designated as uplink by the latter parameter are excluded from the last $N_{slot}^{symb} T_{INT} 2^{\mu-\mu_{INT}}$ symbols prior to the first symbol of the CORESET in the slot. The resulting set of symbols includes a number of symbols that is denoted as N_{INT} [8].

The UE is further provided with the (preemption) indication granularity for the set of PRBs and OFDM symbols (within a slot) that might be preempted through RRC parameter *timeFrequencySet*. If the value of latter parameter is zero (time-domain preemption region boundary), the 14-bit bitmap in DCI format 2_1 has a one-to-one correspondence with 14 groups of consecutive symbols from the set of symbols where each of the first $N_{INT} - 14\lfloor N_{INT}/14 \rfloor$ symbol groups includes $\lceil N_{INT}/14 \rceil$ symbols, each of the last $14 - N_{INT} + 14\lfloor N_{INT}/14 \rfloor$ symbol groups includes $\lfloor N_{INT}/14 \rfloor$ symbols, where in the bitmap, a bit value of "0" indicates transmission to the UE in the corresponding symbol group and a bit value of "1" indicates no transmission to the UE in the corresponding symbol group [8,13]. Interpretation of the bitmap is configurable such that each bit represents one OFDM symbol in the time domain and the full bandwidth part, or two OFDM symbols in the time domain and one half of the bandwidth part. Furthermore, the monitoring periodicity of the preemption indicator is configured in the device every *n*th slot. An example is shown in Fig. 4.21 where UE1 has been scheduled with a downlink transmission spanning one slot. During the transmission to UE1, delay-sensitive data for UE2 arrives at the gNB, which immediately schedules a transmission to UE2. Typically, if there are frequency resources available, the transmission to UE2 is scheduled using resources not overlapping with the ongoing transmission to UE1. However, in a heavy-loaded network, this may not be possible and there is no option but to use some of the resources originally allocated to UE1 for the delay-sensitive transmission to UE2. We refer to this case as the transmission to UE2 preempting the transmission to UE1, which would experience temporary performance degradation because some of the resources that UE1 assumes to contain its own data contain data for UE2.

If the value of the RRC parameter *timeFrequencySet* is one (frequency-domain preemption region boundary) then seven pairs of bits of a field in the DCI format 2_1 have a one-to-one mapping with seven groups of consecutive symbols where each of the first $N_{INT} - 7\lfloor N_{INT}/7 \rfloor$ symbol groups includes $\lfloor N_{INT}/7 \rfloor$ symbols, and each of the last $7 - N_{INT} + 7\lfloor N_{INT}/7 \rfloor$ symbol groups includes $\lfloor N_{INT}/7 \rfloor$ symbols (as shown in Fig. 4.21). The first bit in a pair of bits for a symbol group is applicable to the subset of first $\lfloor B_{INT}/2 \rfloor$ PRBs from the set of B_{INT} PRBs, and the second bit in the pair of bits for the symbol group is applicable to the subset of last $\lfloor B_{INT}/2 \rfloor$ PRBs from the set of B_{INT} PRBs, a bit value of "0" indicates transmission to the UE in the corresponding symbol group and subset of PRBs, and a bit value of "1" indicates no transmission to the UE in the corresponding symbol group and subset of PRBs [8,13].

In relation to the dynamic slot configuration, if a UE is configured by RRC parameter *SlotFormatIndicator*, it will be provided with SFI-RNTI via higher layer parameter *sfi-RNTI*, and the payload size of DCI format 2_0 via RRC parameter *dci-PayloadSize*. The UE is also provided, in one or more serving cells, with a configuration for search space set S and the corresponding CORESET P for monitoring the first $M_{P,S}^{L_{SFI}}$ PDCCH candidates

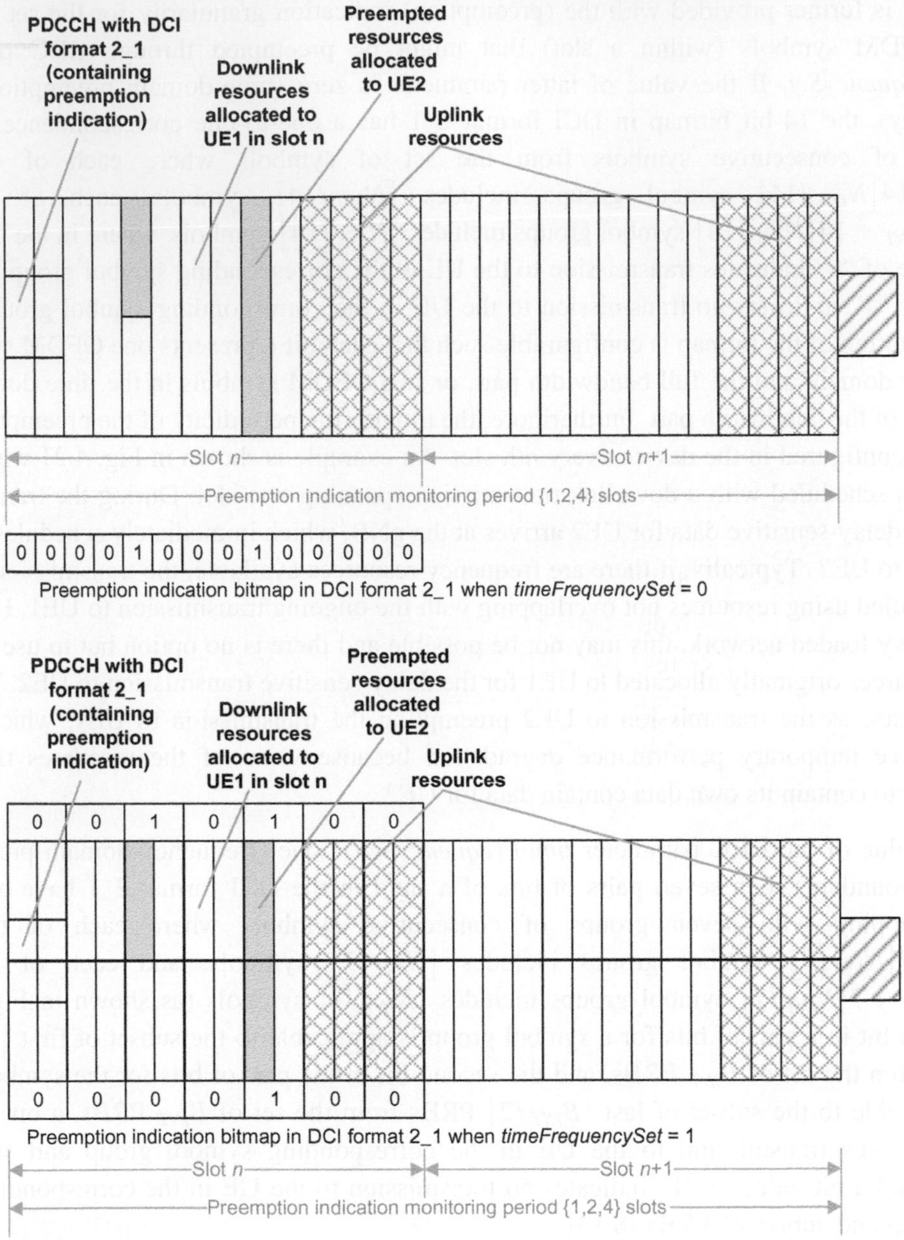

Figure 4.21
Illustration of downlink preemption indication (interrupted transmission indication) [8].

related to DCI format 2_0 at aggregation level of L_{SFI} CCEs to locate Type3-PDCCH common search space. The DCI format 2_0 indicates a slot format corresponding to each slot in the downlink or uplink BWPs of a serving cell. The indication is communicated by setting the value of the SFI index in DCI format 2_0 for the serving cell to a combination of slot formats for a number of slots. A slot format is identified by its corresponding index and the mapping between values of the SFI index and combinations of slot formats is signaled via RRC parameters [8].

In TDD operation mode, the UE is provided, via RRC signaling, with a reference subcarrier spacing configuration μ_{SFI} for each slot format in a set of slot formats indicated by the SFI index in DCI format 2_0. An active downlink and uplink BWP pair are associated with subcarrier spacing configuration $\mu \geq \mu_{SFI}$. Each slot format in the combination of slot formats is applicable to $2^{(\mu-\mu_{SFI})}$ consecutive slots in the active downlink and uplink BWP pair where the first slot starts at the same time as the first slot for the reference subcarrier spacing configuration and each downlink, uplink, or flexible symbol for the reference subcarrier spacing configuration corresponds to $2^{(\mu-\mu_{SFI})}$ consecutive downlink, uplink, or flexible symbols for the subcarrier spacing configuration μ [8]. In FDD operation mode, the SFI index field in DCI format 2_0 indicates an assortment of slot formats that include (separate) combinations of slot formats for reference downlink and uplink BWPs of the serving cell. The UE is provided via RRC signaling with the reference subcarrier spacing configurations μ_{SFI-DL} or μ_{SFI-UL} for the combination of slot formats indicated by the SFI index field value in DCI format 2_0 for the reference downlink and uplink BWPs of the serving cell, respectively.

4.1.3.2.3 Downlink Control Information Formats

The PDCCH payload is known as DCI to which a 24-bit CRC is attached to detect transmission errors and to aid the decoder in the UE receiver. Compared to LTE, the CRC size has been increased to reduce the risk of incorrectly received control information and to assist early termination of the decoding operation in the receiver. A DCI transports downlink and uplink scheduling information, requests for aperiodic channel quality indicator (CQI) reports, or uplink power control commands for one cell and one RNTI. Depending on the content and purpose of each DCI, different formats are defined. Each DCI payload is processed by information element multiplexing, CRC attachment, channel coding, and rate matching. The DCI formats defined in NR are shown in Table 4.4 along with their use cases. Similar to LTE, the UE identity modifies the CRC transmitted through a scrambling operation. When the DCI is received, the UE will calculate a scrambled CRC on the payload part using the same transmit-side procedure and compares it against the received CRC. If the CRC matches, the message is declared to be correctly received and intended for the UE. Therefore, the identity of the UE that is supposed to receive the DCI message is implicitly encoded in the CRC, which reduces the number of bits necessary to transmit on the PDCCH. Note that the RNTI which is further scrambled with the DCI CRC is not necessarily the identity of the device (in the case of C-RNTI), rather it can be different type of group or common RNTIs used to indicate paging or a random-access response.

The information fields in the DCI formats shown in Table 4.4 are mapped to the information bits a_0 to $a_{N_{DCI}-1}$ such that the first field is mapped to the lowest order information bit

Table 4.4: NR downlink control information (DCI) formats [7].

DCI Format	Purpose	Application
0_0	Uplink Scheduling	Scheduling of PUSCH in one cell
0_1		Scheduling of PUSCH in one cell
1_0	Downlink scheduling	Scheduling of PDSCH in one cell
		DCI format 1_0 with CRC scrambled by C-RNTI
		DCI format 1_0 with CRC scrambled by RA-RNTI
		DCI format 1_0 with CRC scrambled by TC-RNTI
		DCI format 1_0 with CRC scrambled by SI-RNTI
		DCI format 1_0 with CRC scrambled by P-RNTI
1_1		Scheduling of PDSCH in one cell
2_0	Other purposes	Notifying a group of UEs of the slot format
2_1		Notifying a group of UEs of the PRB(s) and OFDM symbol(s) where UE may assume that no transmission is intended for the UE
2_2		Transmission of TPC commands for PUCCH and PUSCH
2_3		Transmission of a group of TPC commands for SRS transmissions by one or more UEs

a_0, and each successive field is mapped to higher order information bits. If the number of information bits in a DCI format is less than 12 bits, zero padding is used until the payload size is equal to 12. The DCI size of the uplink DCI format 0_1 and downlink DCI format 1_1 are made equal with padding bits added to the smaller of the two in order to reduce the number of blind decoding. It may appear that parts of the DCI content are the same for the different formats; however, there are differences due to different capabilities supported by each DCI format. The content of various DCI formats is described in the following [7]. Note that the information fields and their interpretation may change according to the RNTI value that is used in conjunction with the DCI value.

DCI Format 0_0

- Identifier for DCI formats (1 bit): It is a bit to indicate whether the DCI is a downlink assignment or an uplink grant. The value of this bit is always set to 0, indicating an uplink DCI format.
- Frequency domain resource assignment: The number of bits for this field is determined by following formula $\lceil \log_2(N_{RB}^{UL,BWP}(N_{RB}^{UL,BWP}+1)/2) \rceil$. The meaning of $N_{RB}^{UL,BWP}$ varies depending on the search space where DCI format 0_0 is expected to be detected. When transmitted in common search space, it indicates the size of the initial bandwidth part, whereas in UE-specific search space, it would indicate the size of the active bandwidth part, if the following criteria are satisfied: the total number of different DCI sizes monitored per slot is less than 4 and the total number of different DCI sizes with C-RNTI

monitored per slot is less than 3. The value of this field is determined in two cases: If PUSCH hopping is enabled and the resource allocation is Type 1, N_{UL-hop} MSB bits are used to indicate the frequency offset. If PUSCH hopping is disabled and the resource allocation is type 1, the entire bits of this field would indicate PUSCH RIV.

- Time-domain resource assignment (4 bits): It carries the row index of the items in *pusch-TimedomainAllocationList* in RRC message for PUSCH configuration, where the indexed row defines the slot offset K_2, the start and length indicator $SLIV$, and the PUSCH mapping type to be applied in the PUSCH transmission.
- Frequency hopping flag (1 bit): It is used to handle frequency hopping for resource allocation Type 1.
- Modulation and coding scheme (5 bits): It is used to provide the device with information about the modulation scheme, the code rate, and the transport block size (TBS) (see Table 4.13).
- New data indicator (1 bit): It is used to indicate whether the grant relates to retransmission of a TB or transmission of a new TB.
- Redundancy version (RV) (2 bits): This field determines the value of the parameter $rv_{id} = 0, 1, 2, 3$ which is used to indicate the redundant information sent in a HARQ retransmission.
- HARQ process number (4 bits): It informs the device about the HARQ process to be used for soft combining.
- Transmit power control (TPC) command for scheduled PUSCH (2 bits): It is used to adjust the PUSCH transmission power.
- Uplink/Supplementary uplink (SUL) indicator (0 or 1 bit): One bit to indicate whether the grant relates to the SUL or the ordinary uplink, for UEs configured with SUL in the cell. It is only present if a SUL is configured as part of the SI.

DCI Format 0_1

- Identifier for DCI formats (1 bit): A bit to indicate whether the DCI is a downlink assignment or an uplink grant. The value of this bit is always set to 0, indicating an uplink DCI format.
- Carrier indicator (0 or 3 bits): This field is present if cross-carrier scheduling is configured and is used to indicate to which component carrier the DCI is related.
- Bandwidth part indicator (0, 1, 2 bits): It is used to activate one of up to four bandwidth parts configured by higher layer signaling. It is determined by the number of uplink BWPs n_{BWP} configured via RRC signaling, excluding the initial uplink bandwidth part. The size of this field is $\lceil \log_2(n_{BWP}) \rceil$ bits.
- Frequency domain resource assignment: This field indicates the resource blocks on one component carrier over which the device should transmit PUSCH. The number of bits is variable and dependent on the resource allocation type. The number of bits is equal to N_{RBG} bits, if resource allocation Type 0 is configured for the UE. For resource

allocation Type 1, the number of bits is equal to $\lceil \log_2(N_{RB}^{UL,BWP}(N_{RB}^{UL,BWP}+1)/2) \rceil$ bits or $\max\left(\lceil \log_2(N_{RB}^{UL,BWP}(N_{RB}^{UL,BWP}+1)/2)\rceil, N_{RBG}\right)+1$ bits if both resource allocation Type 0 and 1 are configured. Note that if both resource allocation Type 0 and 1 are configured, the MSB bit is used to distinguish resource allocation Type 0 from Type 1.

- Time domain resource assignment (0, 1, 2, 3, or 4 bits): This field indicates the resource allocation in the time domain. The number of bits is determined as $\lceil \log_2(I) \rceil$ bits, where I denotes the number of entries in RRC parameter *pusch-TimeDomainAllocationList*.
- VRB-to-PRB mapping (1 bit): It is used to indicate whether interleaved or non-interleaved VRB-to-PRB mapping should be used.
- Frequency hopping flag (1 bit): It is used to handle frequency hopping for resource allocation Type 1.
- Modulation and coding scheme (5 bits): It is used to provide the device with information about the modulation scheme, the code rate, and the TBS (see Table 4.13).
- New data indicator (1 bit): It is used to indicate whether the grant relates to retransmission of a TB or transmission of a new TB.
- Redundancy version (2 bits): This field determines the value of the parameter $rv_{id} = 0, 1, 2, 3$ which is used to indicate the redundant information sent in a HARQ retransmission.
- HARQ process number (4 bits): It informs the device about the HARQ process to be used for soft combining.
- First downlink assignment index (DAI) (1 or 2 bits): The DAI is used for handling of HARQ codebooks when UCI is transmitted on PUSCH. This field would be one bit for semi-static HARQ-ACK codebook and 2 bits for dynamic HARQ-ACK codebook.
- Second DAI (0 or 2 bits): This field would be 2 bits for dynamic HARQ-ACK codebook with two HARQ-ACK subcodebooks and zero bit otherwise.
- TPC command for scheduled PUSCH (2 bits): It is used to adjust the PUSCH transmission power.
- Sounding reference signals (SRS) resource indicator: The SRI is used to determine the antenna ports and uplink transmission beam to use for PUSCH transmission. The number of bits depends on the number of SRS groups configured and whether codebook-based or non-codebook-based precoding is used.
- Precoding information and number of layers (0, 2, 3, 4, 5, or 6 bits): This field is used to select the precoding matrix **W** and the number of layers for codebook-based precoding. The number of bits depends on the number of antenna ports and the maximum rank supported by the UE.
- Antenna ports (2, 3, 4, or 5 bits): This field indicates the antenna ports on which the data are transmitted as well as antenna ports that are scheduled for other users.
- SRS request (2 bits): This field is used to request transmission of a sounding RS.

- CSI request (0, 1, 2, 3, 4, 5, or 6 bits): This field is used to request transmission of a CSI report.
- Code block group (CBG) transmission information (0, 2, 4, 6, or 8 bits): This field indicates the code block groups for retransmission.
- PT-RS-DM-RS association (0 or 2 bits): This field is used to indicate the association between the DM-RS and PT-RS ports.
- *beta_offset* Indicator (0 or 2 bits): This information is used to control the amount of resources used by UCI on PUSCH in case dynamic beta offset signaling is configured for DCI format 0_1.
- DM-RS sequence initialization (0 or 1 bit): This information is used to select between two preconfigured initialization values for the DM-RS sequence. It would be zero bit, if the transform precoder is enabled and 1 bit, otherwise.
- Uplink/SUL indicator (0 or 1 bit): It would be zero bit for UEs that are not configured with SUL in the cell or UEs configured with SUL in the cell but only physical uplink control channel (PUCCH) carrier in the cell is configured for PUSCH transmission; otherwise, 0 bit for UEs configured with SUL in the cell.

DCI Format 1_0

- Identifier for DCI formats (1 bit): A bit to indicate whether the DCI is a downlink assignment or an uplink grant. The value of this field is always set to 1, indicating a downlink DCI format.
- Frequency-domain resource assignment: This field indicates the resource blocks on one component carrier on which the UE receives PDSCH. The size of this field depends on the size of the bandwidth part and on the resource allocation type, that is, Type 0 only, Type 1 only, or dynamic switching between the two types. The number of bits is $\lceil \log_2(N_{RB}^{DL,BWP}(N_{RB}^{DL,BWP}+1)/2) \rceil$ where the interpretation of $N_{RB}^{DL,BWP}$ depends on the search space where DCI format 1_0 is monitored. The parameter $N_{RB}^{DL,BWP}$ indicates the size of the active downlink bandwidth part, if DCI format 1_0 is monitored in the UE-specific search space and if the total number of different DCI sizes configured to monitor is less than 4 and the total number of different DCI sizes with C-RNTI configured to monitor is less than 3 for the cell; otherwise, $N_{RB}^{DL,BWP}$ would denote the size of CORESET 0.
- Time-domain resource assignment (4 bits): This field indicates the resource allocation in the time domain. When the UE is scheduled to receive PDSCH by a DCI, the time-domain resource assignment field of the DCI provides a row index of a higher layer-configured table *pdsch-symbolAllocation*, where the indexed row defines the slot offset K_0, the start and length indicator *SLIV*, and the PDSCH mapping type for PDSCH reception.
- VRB-to-PRB mapping (1 bit): It indicates whether interleaved or non-interleaved VRB-to-PRB mapping should be used and only presents for resource allocation Type 1.

- Modulation and coding scheme (5 bits): It is used to provide the UE with information about the modulation scheme, code rate, and TBS (see Table 4.13).
- New data indicator (1 bit): It is used to clear the UE soft buffer for initial transmissions.
- Redundancy version (2 bits): This field determines the value of the parameter $rv_{id} = 0, 1, 2, 3$ which is used to indicate the redundant information sent in a HARQ retransmission.
- HARQ process number (4 bits): This informs the device about the HARQ process to use for soft combining.
- Downlink assignment index (2 bits): The DAI only presents when a dynamic HARQ codebook is used. The DCI format 1_1 supports 0, 2, or 4 bits, while DCI format 1_0 uses 2 bits.
- TPC command for scheduled PUCCH (2 bits): It is used to adjust the PUCCH transmission power.
- PUCCH resource indicator (3 bits): It is used to select PUCCH resources from a set of configured resources.
- PDSCH-to-HARQ feedback timing indicator (3 bits): It provides information on when the HARQ acknowledgment should be transmitted relative to the PDSCH transmission.

DCI Format 1_1

- Identifier for DCI formats (1 bit): The value of this bit is always set to 1, indicating a downlink DCI format.
- Carrier indicator (0 or 3 bits): This field is present if cross-carrier scheduling is configured and is used to indicate the component carrier that the DCI corresponds to.
- Bandwidth part indicator (0, 1, or 2 bits): The number of bits is determined by the number of downlink BWPs $n_{BWP,RRC}$ configured by higher layers, excluding the initial downlink bandwidth part. The size of this field is equal to $\lceil \log_2(n_{BWP}) \rceil$ bits, where $n_{BWP} = n_{BWP,RRC} + 1$, if $n_{BWP,RRC} \leq 3$ in which case the bandwidth part indicator is equivalent to the higher layer parameter *BWP-Id*; otherwise $n_{BWP} = n_{BWP,RRC}$.
- Frequency-domain resource assignment: This field indicates the resource blocks on one component carrier on which the device should receive PDSCH. The number of bits is variable and dependent on the resource allocation type. The number of bits is equal to N_{RBG} bits, if resource allocation Type 0 is configured for the UE. For resource allocation Type 1, the number of bits is equal to $\lceil \log_2(N_{RB}^{UL,BWP}(N_{RB}^{UL,BWP} + 1)/2) \rceil$ bits or $\max\left(\lceil \log_2(N_{RB}^{DL,BWP}(N_{RB}^{DL,BWP} + 1)/2) \rceil, N_{RBG}\right) + 1$ bits, if both resource allocation Types 0 and 1 are configured. Note that if both resource allocation Types 0 and 1 are configured, the MSB bit is used to distinguish resource allocation Type 0 from Type 1.
- Time-domain resource assignment (0, 1, 2, 3, or 4 bits): The size of this field is determined as $\lceil \log_2(I) \rceil$ bits, where I is the number of entries in the higher layer parameter *pdsch-TimeDomainAllocationList*. When the UE is scheduled to receive PDSCH by a

DCI, the time-domain resource assignment field of the DCI provides a row index of a higher layer-configured table *pdsch-symbolAllocation*, where the indexed row defines the slot offset K_0, the start and length indicator *SLIV*, and the PDSCH mapping type for PDSCH reception.

- VRB-to-PRB mapping (0 or 1 bit): It indicates whether interleaved or non-interleaved VRB-to-PRB mapping should be used and only presents for resource allocation Type 1.
- PRB bundling size indicator (0 or 1 bit): It is used to indicate the PDSCH bundling size. It is zero bit if the RRC parameter *prb-BundlingType* is not configured or is set to "static." It is one bit, if the higher layer parameter *prb-BundlingType* is set to "dynamic."
- Rate matching indicator (0, 1, or 2 bits): The number of bits is determined according to RRC parameters *rateMatchPatternGroup1* and *rateMatchPatternGroup2*.
- Modulation and coding scheme (TB 1) (5 bits): It is used to provide the UE with information about the modulation scheme, code rate, and TBS (see Table 4.13) related to the first transport block.
- New data indicator (TB 1) (1 bit): It is used to clear the UE soft buffer for initial transmissions related to the first transport block.
- Redundancy version (TB 1) (2 bits): This field determines the value of the parameter which is used to indicate the redundant information sent in a HARQ retransmission related to the first transport block.
- Modulation and coding scheme (TB 2)[5] (5 bits): It is used to provide the UE with information about the modulation scheme, code rate, and TBS related to the second transport block (see Table 4.13).
- New data indicator (TB 2) (1 bit): It is used to clear the UE soft buffer for initial transmissions related to the second transport block.
- Redundancy version (TB 2) (2 bits): This field determines the value of the parameter which is used to indicate the redundant information sent in a HARQ retransmission related to the second transport block.
- HARQ process number (4 bits): This informs the device about the HARQ process to use for soft combining.
- Downlink assignment index (0, 2, or 4 bits): The number of bits is 4, if more than one serving cell is configured in the downlink and the RRC parameter *pdsch-HARQ-ACK-Codebook* = *dynamic*, where the two MSB bits are the counter DAI and the two LSB bits are the total DAI. The number of bits is 2, if only one serving cell is configured in the downlink and the RRC parameter *pdsch-HARQ-ACK-Codebook* = *dynamic*, where the 2 bits are the counter DAI; zero bits otherwise.

[5] If a second transport block is present (only if more than four layers of spatial multiplexing are supported in DCI format 1_1), the three fields above are repeated for the second transport block.

- TPC command for scheduled PUCCH (2 bits): It is used to adjust the PUCCH transmission power.
- PUCCH resource indicator (2 bits): It is used to select PUCCH resources from a set of configured resources.
- PDSCH-to-HARQ_Feedback timing indicator (3 bits): It provides information on when the HARQ acknowledgment should be transmitted relative to the PDSCH transmission.
- Antenna port(s) and number of layers (4, 5, or 6 bits): The antenna ports $\{p_0,\ldots,p_{N_A-1}\}$ are determined according to the ordering of DM-RS port(s). If a UE is configured with both *dmrs-DownlinkForPDSCH-MappingTypeA* and *dmrs-DownlinkForPDSCH-MappingTypeB*, the size of this field is equal to $\max(x_A, x_B)$, where x_A and x_B are the "antenna ports" bit sizes derived from *dmrs-DownlinkForPDSCH-MappingTypeA* and *dmrs-DownlinkForPDSCH-MappingTypeB*, respectively. A number of zeros are inserted in the $|x_A - x_B|$ MSB positions of this field, if the mapping type of the PDSCH corresponds to the smaller value of x_A or x_B.
- Transmission configuration indication (0 or 3 bits): The size of this field is zero bit, if RRC parameter *tci-PresentInDCI* is not enabled; otherwise it would carry 3 bits.
- SRS request (2 or 3 bits): It is used to request transmission of a sounding reference signals in the uplink. For UEs not configured with SUL in the cell, 2 bits are used, whereas for UEs that are configured with SUL in the cell, 3 bits are used where the first bit is the non-SUL/SUL indicator and the second and third bits are used to request periodic or aperiodic SRS transmission. This bit field may also indicate the associated CSI-RS.
- CBG transmission information (0, 2, 4, 6, or 8 bits): If CBG retransmissions are configured, this field indicates the code block groups that are retransmitted.
- CBG flushing out information (0 or 1): If CBG retransmissions are configured, the content of this field indicates the soft buffer flushing, which is determined by RRC parameter *codeBlockGroupFlushIndicator*.
- DM-RS sequence initialization (1 bit): This information is used to select between two preconfigured initialization values for the DM-RS sequence. It would be zero bit, if the transform precoder is enabled and 1 bit otherwise.

DCI Format 2_0 DCI format 2_0 is used for notifying the UE of slot format. The SFI is transmitted using regular PDCCH structure and SFI-RNTI, which is common to a group of UEs. To assist the device in the blind decoding, the device is configured with information about the up to two PDCCH candidates on which the SFI can be transmitted. DCI format 2_0 with CRC scrambled with SFI-RNTI carries *Slot format indicator* 1, *Slot format indicator* 2, ..., *Slot format indicator N*. The size of DCI format 2_0 is configurable by higher layers up to 128 bits.

DCI Format 2_1 DCI format 2_1 is used to signal the preemption indication to the device. It is transmitted using the regular PDCCH structure, using INT-RNTI which can be common to multiple devices. In other words, DCI format 2_1 is used for notifying the UEs of the PRB(s) and OFDM symbol(s) that are preempted and have no transmission intended for the UE. DCI format 2_1 with CRC scrambled by INT-RNTI carries *Preemption indication* 1, *Preemption indication* 2,..., *Preemption indication N*. The size of DCI format 2_1 is configurable by higher layers up to 126 bits where each preemption indication is 14 bits.

DCI Format 2_2 The main purpose of DCI format 2_2 is to support power control for semi-persistent scheduling (SPS) since there is no dynamic scheduling assignment or scheduling grant which can include the power control information for PUCCH and PUSCH in this case. The power-control message is addressed to a group of UEs using an RNTI specific for that group and each UE is configured with the power control bits in the message. DCI format 2_2 is further aligned with the size of DCI formats 0_0/1_0 to reduce the blind decoding complexity. DCI format 2_2 with CRC scrambled by TPC-PUSCH-RNTI or TPC-PUCCH-RNTI carries *block number* 1, *block number* 2,... *block number N*. The RRC parameters *tpc-PUSCH* or *tpc-PUCCH* determine the index to the block number for a cell uplink, with the following fields defined for each block: (1) closed-loop indicator (0 or 1 bit) and (2) TPC command (2 bits).

DCI Format 2_3 DCI format 2_3 is used for power control of uplink sounding reference signals for the UEs which have not linked the SRS power control to the PUSCH power control, either because independent control was desirable, or the UE was configured without PUCCH and PUSCH. DCI format 2_3 structure is similar to DCI format 2_2, with the possibility to individually configure 2 bits for SRS request in addition to the two power control bits. DCI format 2_3 is aligned with the size of DCI formats 0_0/1_0 to reduce the blind decoding complexity. DCI format 2_3 with CRC scrambled by TPC-SRS-RNTI carries *block number* 1, *block number* 2,..., *block number N* where the starting position of a block is determined by the parameter *startingBitOfFormat2-3* provided by the higher layers for the UE configured with the block. If the UE is configured with RRC parameter *srs-TPC-PDCCH-Group* = *typeA* for an uplink carrier without PUCCH and PUSCH or when the SRS power control is not linked to PUSCH power control, in DCI format 2_3 one block is configured for the UE containing 0 or 2 bits of SRS request. The TPC commands *TPC command number* 1, *TPC command number* 2,..., *TPC command number N* apply to the respective carriers. If the UE is configured with RRC parameter *srs-TPC-PDCCH-Group* = *typeB* for an uplink carrier without PUCCH and PUSCH or an uplink carrier on which the SRS power control is not tied to PUSCH power control, one or more blocks are configured for the UE by the higher layers. In that case, each block applies to an uplink carrier and DCI format 2_3 contains 0 or 2 bits of SRS request and 2 bits of TPC command.

4.1.3.2.4 Common and UE-Specific Search Spaces

A UE may be configured with one or more CORESETs (using UE-specific or common signaling) with a maximum of three CORESETs per configured downlink BWP. Note that the scheduling flexibility may not be impacted by limiting the maximum number of CORESETs since different monitoring occasions can be configured flexibly even in association with the same CORESET. It is important to further note that the concept of PDCCH monitoring periodicity is defined per search space set and is not configured at the CORESET-level. Every configured search space with a certain monitoring periodicity (in terms of slots and starting symbols within the monitored slots) is associated with a CORESET.

In LTE, the DCI format was closely coupled with the DCI size and monitoring for a certain DCI format in most cases implied monitoring for a new DCI size. In NR, the DCI formats and DCI sizes are decoupled. Different formats can have different DCI sizes, but several formats can share the same DCI size. This allows adding more formats in the future without increasing the number of blind decoding attempts. An NR device needs to monitor up to four different DCI sizes: one size used for the fallback DCI formats, one for downlink scheduling assignments, and unless the uplink downlink non-fallback formats are size-aligned, one for uplink scheduling grant. In addition, a device may need to monitor SFI and/or preemption indication DCIs using a fourth size, depending on the configuration.

An NR UE needs to monitor the PDCCH candidates at multiple aggregation levels for the detection and reception of PDCCH. Inside a configured CORESET, NR SS defines the PDCCH candidates of each AL [8]. In NR, PDCCH employs DM-RS-based transmission. Unlike LTE PDCCH where cell-specific reference signals were used for coherent demodulation, the NR channel estimation complexity scales with the number of CCEs being monitored. Thus it is important to balance the scheduling flexibility against UE implementation burden to facilitate cost-efficient UE implementation. For PDCCH DM-RS in a CORESET, the antenna port QCL configuration relating to the SS/PBCH block antenna port(s) or configured CSI-RS antenna port(s), is on a per-CORESET basis. This implies that in mmWave deployments, which rely on beam-sweeping operations, different CORESET and search space configurations corresponding to different received beams are necessary [8,53].

The CCE structure described in the previous section helps reduce the number of blind decoding attempts; however, it is required to have mechanisms to limit the number of PDCCH candidates that the device is expected to monitor. From a scheduling point of view, restrictions in the allowed aggregations are undesirable as they may reduce the scheduling flexibility and require additional processing at the transmitter side. At the same time, requiring the device to monitor all possible CCE aggregations in all configured CORESETs significantly increases device complexity and power consumption. A search space is a set of candidate control channels comprising a set of CCEs at a given aggregation level, which the device is supposed to monitor and decode. Due to multiple aggregation levels, a device can have multiple search spaces. There can be multiple SSs using the same CORESET or multiple CORESETs configured for a device. A device is not expected to monitor PDCCH outside its active bandwidth part. At a configured monitoring occasion for a search space, the devices will attempt to decode the candidate PDCCHs for that search space. Five different aggregation levels corresponding to 1, 2, 4, 8, and 16 CCEs can be configured. The highest aggregation level is meant to support extreme coverage requirements [6,8,14].

The number of PDCCH candidates can be configured per search space and per aggregation level. When the UE attempts to decode a candidate PDCCH, the content of the control channel is declared as valid, if the CRC checks and the device can successfully process the contained information, that is, scheduling assignment, and uplink grants. If the CRC does not pass, the information is either subject to uncorrectable transmission errors or intended for another UE and in either case the device ignores that PDCCH transmission. The gNB can only address a UE, if the corresponding control information is transmitted on a PDCCH formed by the CCEs in one of the UE's search spaces. Therefore, for efficient utilization of the CCEs in the system, the UE should be associated with different search spaces. Each device in the system can be configured with one or more UE-specific search spaces. Since a UE-specific search space is typically smaller than the number of PDCCHs that the network can transmit at the corresponding aggregation level, there must be a mechanism to determine a set of CCEs in UE-specific search space. One option is to allow the network to configure the UE-specific search space for each device, in the same way that CORESETs are configured. However, this would require explicit signaling exchange with each device and possibly reconfiguration at handover. Instead, the UE-specific search spaces for PDCCH are defined without explicit signaling and based on the device unique identity in the cell in the connected mode, that is, C-RNTI. Furthermore, the set of CCEs that the device should monitor at a certain aggregation level varies as a function of time to avoid two devices constantly blocking each other. If they collide at one time instant, they are not likely to collide at the next time instant. In each of these search spaces, the UE attempts to decode the PDCCHs using the UE-specific C-RNTI. There is also information intended for a group of UEs in the cell. These messages are scheduled with different predefined RNTIs, for example, SI-RNTI for scheduling system information, P-RNTI for transmission of a paging message, RA-RNTI for transmission of the random-access response, TPC-RNTI for uplink power control, INT-RNTI for preemption indication, and SFI-RNTI for slot format configuration. As part of random-access procedure, it is necessary to transmit information to a device before it is assigned a unique identity. These types of information cannot rely on a UE-specific search space as different devices would monitor different CCEs despite the message being intended for all of them. Thus common search spaces are defined, where a common search space is similar in structure to a UE-specific search space with the difference that the set of CCEs is predefined and known to all devices irrespective of their own identity [14].

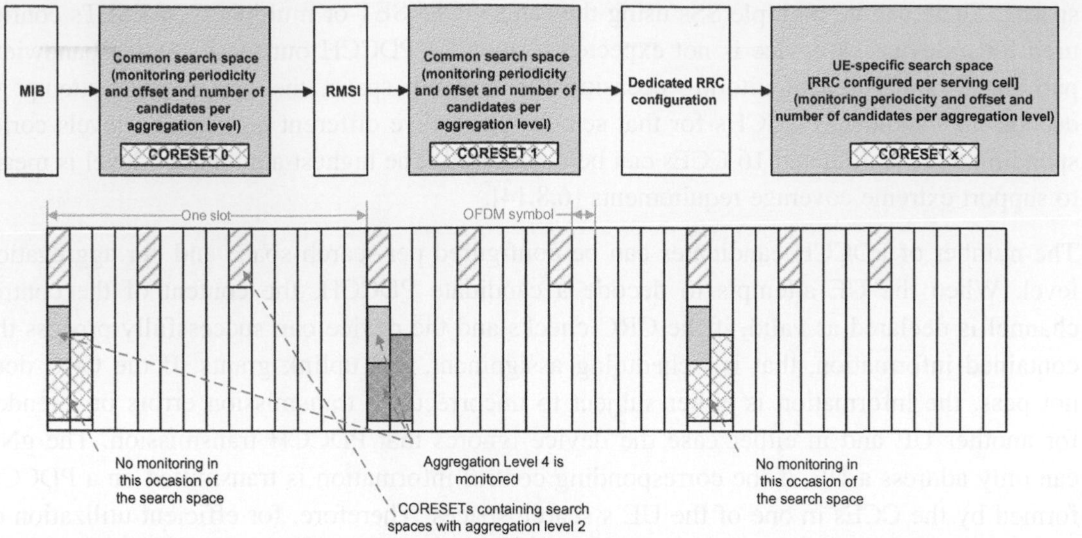

Figure 4.22
Procedure for PDCCH search space configuration and example search spaces [68].

The number of blind decoding attempts is proportional to subcarrier spacing and the slot duration. For 15/30/60/120 kHz subcarrier spacing, up to 44/36/22/20 blind decoding attempts per slot can be supported across all DCI payload sizes, respectively. It must be noted that the number of blind decoding attempts is not the only measure of UE complexity, channel estimation efforts also need to be taken in consideration. The number of channel estimations for subcarrier spacings of 15/30/60/120 kHz has been limited to 56/56/48/32 CCEs across all CORESETs in a slot [8]. Depending on the configuration, the number of PDCCH candidates may be limited either by the number of blind decoding attempts, or by the number of channel estimates. In order to minimize the device complexity, a device monitors a maximum of three different DCI sizes using C-RNTI and one DCI size using other RNTIs. In carrier aggregation scenarios, the general blind decoding operation described earlier is applied per component carrier. While the total number of channel estimates and blind decoding attempts increases compared to the single carrier case, there is no direct proportionality between the number of aggregated carriers and blind decoding attempts [14]. The procedure for PDCCH search space configuration and example search spaces are shown in Fig. 4.22.

As we stated earlier, a set of PDCCH candidates are defined for each UE to monitor, which are referred to as PDCCH search spaces. A search space can be categorized as common or UE-specific. In other words, a search space is defined by the PDCCH candidates that need to be monitored. These candidates are determined by a hashing function that operates within a set of CCEs in a particular CORESET and the monitoring periodicity and offsets that determine when the search space should be monitored (see Fig. 4.22). The UE is required to monitor PDCCH candidates in one or more of the following search spaces [8]:

- *Type0-PDCCH* common search space set configured by *pdcch-ConfigSIB1* in *MasterInformationBlock* or by *searchSpaceSIB1* in *PDCCH-ConfigCommon* or by *searchSpaceZero* in *PDCCH-ConfigCommon* for a DCI format, the CRC of which is scrambled with SI-RNTI in the primary cell.
- *Type0A-PDCCH* common search space set configured by *searchSpaceOtherSystemInformation* in *PDCCII-ConfigCommon* for a DCI format, the CRC of which is scrambled with SI-RNTI in the primary cell.
- *Type1-PDCCH* common search space set configured by *ra-SearchSpace* in *PDCCH-ConfigCommon* for a DCI format, the CRC of which is scrambled with RA-RNTI, TC-RNTI, or C-RNTI in the primary cell.
- *Type2-PDCCH* common search space set configured by *pagingSearchSpace* in *PDCCH-ConfigCommon* for a DCI format, the CRC of which is scrambled with P-RNTI in the primary cell.
- *Type3-PDCCH* common search space set configured by *SearchSpace* in *PDCCH-Config* with *searchSpaceType* = *common* for a DCI format, the CRC of which is scrambled with INT-RNTI, SFI-RNTI, TPC-PUSCH-RNTI, TPC-PUCCH-RNTI, TPC-SRS-RNTI, C-RNTI, CS-RNTI(s), or SP-CSI-RNTI.
- *UE-specific* search space set configured by *SearchSpace* in *PDCCH-Config* with *searchSpaceType* = *ue-Specific* for a DCI format, the CRC of which is scrambled with C-RNTI, CS-RNTI(s), or SP-CSI-RNTI.

An example search space configuration for two devices is shown in Fig. 4.23. The UE determines a CORESET and PDCCH monitoring occasions for Type0-PDCCH common search space set, if it is not provided with RRC parameter *searchSpace-SIB1*. The Type0-PDCCH common search space set is defined by the CCE aggregation levels and the number of PDCCH candidates per CCE aggregation level. The CORESET configured for this search space set has CORESET index 0 and search space set index 0. If the UE is not provided with a CORESET for any of Type0A-PDCCH/Type1-PDCCH/Type2-PDCCH common search spaces, the corresponding CORESET would be the same as the CORESET for Type0-PDCCH common search space. The CCE aggregation levels and the number of PDCCH candidates per CCE aggregation level for Type0-PDCCH, Type0A-PDCCH, and Type2-PDCCH common search space are given in Table 4.5 [8].

The DM-RS antenna port associated with PDCCH reception in the Type0-PDCCH/Type0A-PDCCH/Type2-PDCCH common search spaces and for the corresponding PDSCH receptions as well as the DM-RS antenna port associated with SS/PBCH block reception are quasi-co-located with respect to delay spread, Doppler spread, Doppler shift, average delay,

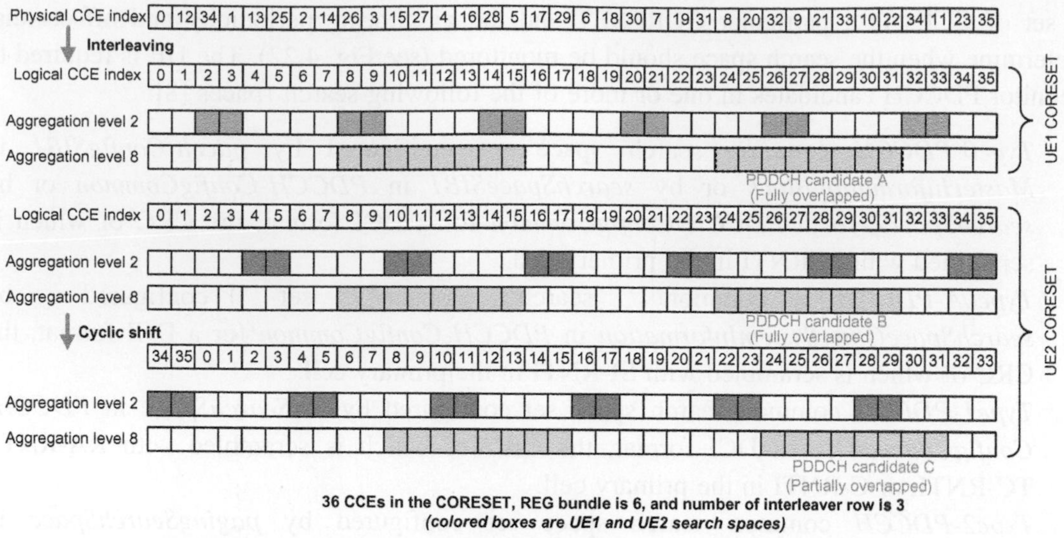

Figure 4.23
Example search space configuration for two devices [14].

Table 4.5: CCE aggregation levels and maximum number of PDCCH candidates per CCE aggregation level for Type0/Type0A/Type2-PDCCH common search space [8].

CCE Aggregation Level	Number of Candidates
4	4
8	2
16	1

and spatial RX parameters. The value for the DM-RS scrambling sequence initialization is the cell ID. The subcarrier spacing and the cyclic prefix length for PDCCH reception with Type0A-PDCCH/Type1-PDCCH/Type2-PDCCH common search spaces are the same as for PDCCH reception with Type0-PDCCH common search space. The DM-RS antenna port associated with PDCCH reception and the associated PDSCH reception in Type1-PDCCH common search space are quasi-co-located with the SS/PBCH block identified in initial access procedure or with a received CSI-RS with respect to delay spread, Doppler spread, Doppler shift, average delay, and spatial RX parameters [8].

For each downlink BWP configured for a UE in the serving cell, the UE can be provided with $N_{CORESET} \leq 3$ CORESETs. For each CORESET, the RRC signaling provides the UE with a CORESET index $0 \leq p < 12$; a DM-RS scrambling sequence initialization value; a precoder granularity for a number of REGs in the frequency domain where the UE can assume use of a same DM-RS precoder; a number of consecutive symbols; a set of resource blocks; CCE-to-REG mapping parameters; an antenna port QCL, from a set of antenna port QCLs, indicating QCL information of the DM-RS antenna port for PDCCH reception in a respective CORESET; and an indication for presence or absence of TCI field in DCI format 1_1 transmitted by PDCCH in the CORESET [8].

For each CORESET in a downlink BWP of a serving cell, the RRC parameter *frequencyDomainResources* provides a bitmap, whose bits have one-to-one correspondence with non-overlapping groups of six PRBs, in ascending order of the PRB index in the downlink BWP bandwidth of N_{RB}^{BWP} PRBs with starting position N_{start}^{BWP} where the first PRB of the first group of six PRBs is indexed as $6\lceil N_{start}^{BWP}/6 \rceil$. A group of six PRBs are allocated to a CORESET, if the corresponding bit value in the bitmap is set to one. If the UE receives the initial configuration of more than one TCI state through RRC parameter *TCI-States* but has not received a MAC CE activation command for at least one of the TCI states, the UE can assume that the DM-RS antenna port associated with PDCCH reception in the UE-specific search space is quasi-co-located with the SS/PBCH block that the UE has identified during the initial access procedure. If the UE has received a MAC CE activation command at least for one of the TCI states, it applies the activation command 3 ms after a slot where it transmits HARQ-ACK information for the PDSCH providing the activation command [8]. Table 4.6 provides the maximum number of PDCCH candidates $\max\left(M_{slot}^{PDCCH}(\mu)\right)$ across all CCE aggregation levels and across all DCI formats with different size in the same search space that the UE is expected to monitor per slot and per serving cell as a function of the subcarrier spacing. The table further provides the maximum number of non-overlapped CCEs that a UE is expected to monitor per slot and per serving cell as a function of the subcarrier spacing. The CCEs are considered non-overlapped, if they correspond to different CORESET indices or different first symbols for the reception of the respective PDCCH candidates [8].

Table 4.6: Maximum number of PDCCH candidates per slot and per serving cell as a function of subcarrier spacing [8].

μ	Maximum Number of Monitored PDCCH Candidates Per Slot and Serving Cell	Maximum Number of Non-overlapped CCEs Per Slot and Serving Cell
	$\max\left(M_{slot}^{PDCCH}(\mu)\right)$	$\max\left(C_{slot}^{PDCCH}(\mu)\right)$
0	44	56
1	36	56
2	22	48
3	20	32

For each downlink BWP that is configured for a UE in a serving cell, the UE is provided via RRC signaling with $S \leq 10$ search space sets. For each of those search space sets, the UE is provided with an search space set index $0 \leq s < 40$; an association between the search space set s and a CORESET p; a PDCCH monitoring periodicity of $k_{p,s}$ slots and a PDCCH monitoring offset of $\delta_{p,s}$ slots; a PDCCH monitoring pattern within a slot, indicating first symbol(s) of the CORESET within a slot for PDCCH monitoring; a number of PDCCH candidates $M_{p,s}^{L}$ per CCE aggregation level L; and an indication that search space set s is either a common or a UE-specific search space set via RRC signaling [8]. Alternative PDCCH mapping rules are illustrated in Fig. 4.24.

The UE can also be provided via RRC signaling with a time interval consisting of $T_{p,s} < k_{p,s}$ slots indicating a number of slots where the search space set s could exist. The information on the first symbol and the number of consecutive symbols for a CORESET, which results in a PDCCH candidate mapping to symbols of different slots, is not provided to the UE. The UE cannot assume that two PDCCH monitoring occasions, for the same search space set or for different search space sets, within the same CORESET are separated by a number of symbols that are less than the CORESET duration.

The UE determines the PDCCH monitoring occasion from the PDCCH monitoring periodicity, offset, and pattern within a slot. For search space set s in CORESET p, the UE determines the PDCCH monitoring occasion(s) in a slot with number n_{slot} in a frame with number n_{frame}, if $\left(n_{frame}N_{frame}^{slot} + n_{slot} - \delta_{p,s}\right) \bmod k_{p,s} = 0$. If the UE is informed in advance of the duration via RRC signaling, it would monitor PDCCH for search space set s in CORESET p for $T_{p,s}$ consecutive slots, starting from slot n_{slot} and would not monitor PDCCH for search space set s in CORESET p for the next $k_{p,s} - T_{p,s}$ consecutive slots [8].

A UE-specific search space at CCE aggregation level $L \in \{1, 2, 4, 8, 16\}$ is defined by a set of PDCCH candidates for CCE aggregation level L. For search space set s associated with CORESET p, the CCE indices for aggregation level L corresponding to PDCCH candidate $m_{s,n_{CI}}$ of the search space set in slot n_{slot} for an active BWP in the serving cell corresponding to carrier indicator field value n_{CI} are given as follows [8]:

$$\left\{Y_{p,n_{slot}} + \left\lfloor \frac{m_{s,n_{CI}} N_{CCE,p}}{L\max\left(M_{p,s}^{(L)}\right)} \right\rfloor + n_{CI} \bmod \lfloor N_{CCE,p}/L \rfloor \right\} L + i, \quad \forall i = 0, \ldots, L-1$$

For common search spaces $Y_{p,n_{slot}} = 0$, whereas for a UE-specific search spaces $Y_{p,n_{slot}} = \left(A_p Y_{p,n_{slot}-1}\right) \bmod D$, $Y_{p,-1} = n_{RNTI}$, $A_p = 39827,\ 39829,\ \text{or}\ 39839$ if $p \bmod 3 = 0, 1, \text{or}\ 2$, respectively. In the preceding expression, $D = 65537$; $N_{CCE,p}$ is the number of CCEs, numbered from 0 to $N_{CCE,p} - 1$, in CORESET p; and n_{CI} denotes the carrier indicator field value, if the UE is configured via RRC signaling with a carrier indicator field in the serving cell on which the PDCCH is monitored; otherwise, including for any common search space $n_{CI} = 0$. Furthermore, $m_{s,n_{CI}} = 0, \ldots, M_{p,s,n_{CI}}^{(L)} - 1$, where $M_{p,s,n_{CI}}^{(L)}$ is the

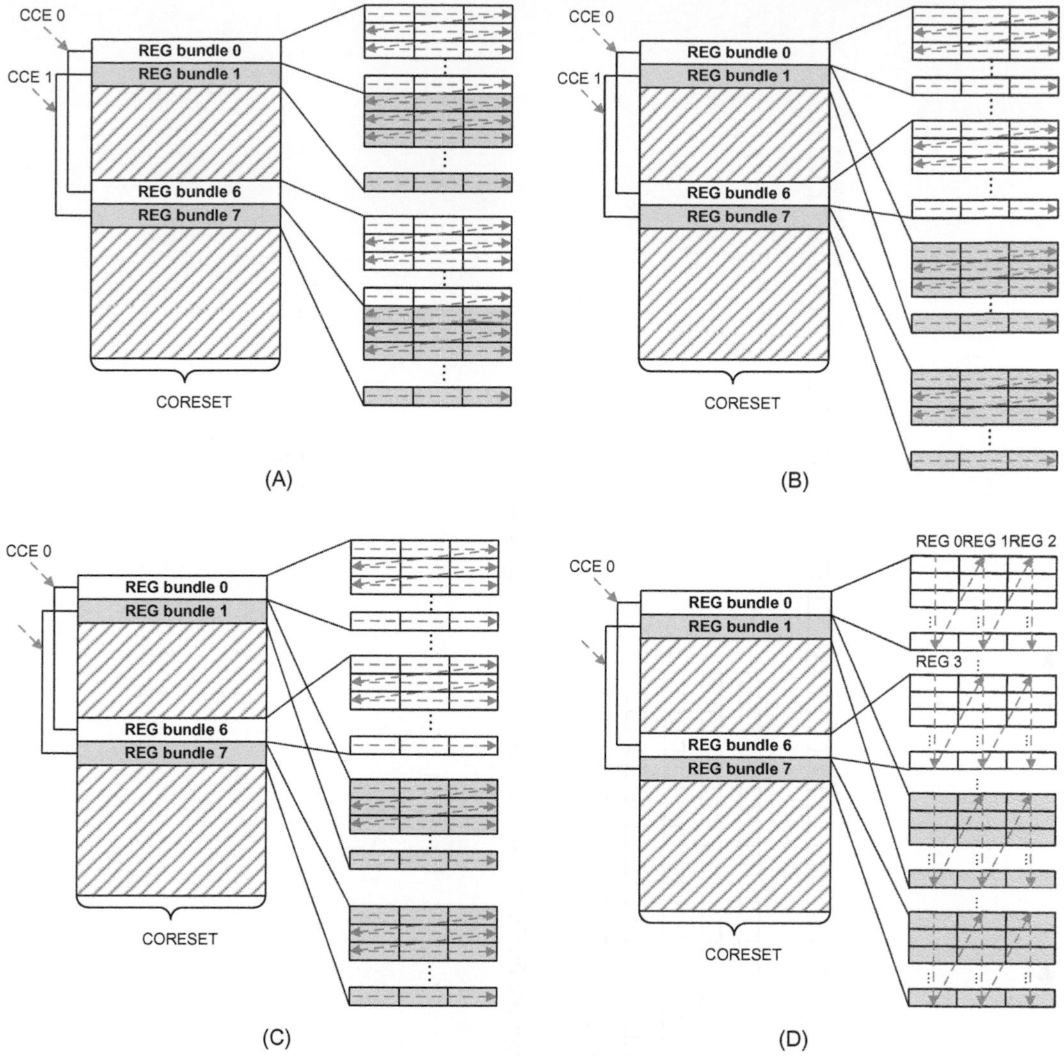

Figure 4.24
Illustration of alternative PDCCH mapping rules: (A) PDCCH candidate-level, (B) CCE-level, (C) REG bundle-level, and (D) REG-level.

number of PDCCH candidates that the UE is supposed to monitor at aggregation level L in a search space set s in a serving cell corresponding to n_{CI}. For any common search space, $\max(M^{(L)}_{p,s,n_{CI}}) = M^{(L)}_{p,s,0}$ whereas for a UE-specific search space, $\max(M^{(L)}_{p,s,n_{CI}})$ denotes the maximum of $M^{(L)}_{p,s,n_{CI}}$ over the configured values of n_{CI} for a CCE aggregation level L of search space set s in CORESET p; the RNTI value used for n_{RNTI} is the C-RNTI [8]. Example PDCCH search spaces at various aggregation levels are shown in Fig. 4.25.

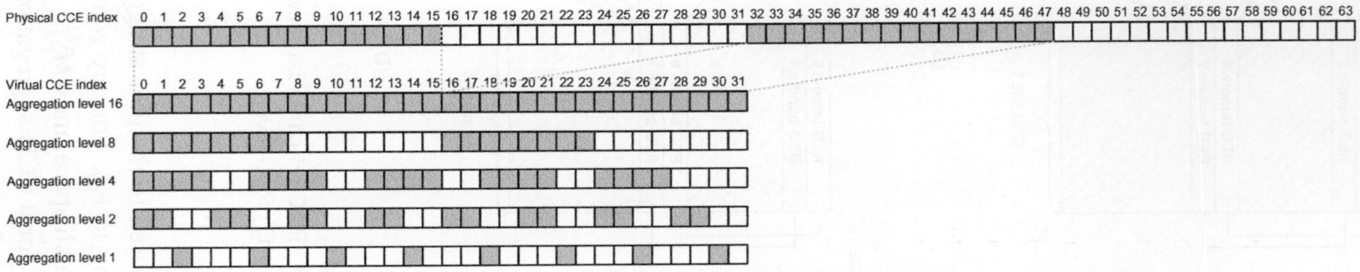

Figure 4.25
Example PDCCH search spaces at various aggregation levels.

When the total number of PDCCH candidates that a UE is configured to monitor in a slot exceeds the blind decoding limit of the UE, it must drop some of the candidates. While the number of blind decoding attempts can be controlled by the network through RRC configuration, the specification does not define a rule for dropping the PDCCH candidates, when the number of PDCCH candidates exceeds the blind decoding limit of the UE. As an example, the candidates can be prioritized according to the search space type and then according to search space set number within a search space type and finally according to aggregation level within a search space set. Once the blind decoding limit is reached, the remaining candidates can be discarded. If a UE has a limit of 56 CCEs and it is configured with a single CORESET in the slot that spans only one OFDM symbol then the CCE limit should not be an issue since the maximum number of PRBs on an RF carrier is 275, and this corresponds to less than 56 CCEs. However, if a UE is configured to monitor two-symbol or three-symbol CORESETs, or multiple CORESETs in a slot and the number of CCEs for all CORESETs is large, the CCE processing constraint can potentially limit the number of PDCCH candidates for which decoding can be attempted, considering that the CCE limit of 56 CCEs must be shared among many CORESETs in the slot.

4.1.3.2.5 Dynamic and Semi-persistent Scheduling

The MAC sublayer in a gNB includes dynamic resource schedulers that manage and allocate radio resources to active users in the downlink and uplink. Scheduling is performed in either dynamic or semi-static manner. Dynamic scheduling is the default mode-of-operation where the scheduler for each time interval, that is, a slot, determines which devices are going to transmit or receive and further configures the transmission parameters based on the measurement reports from the UEs. Since scheduling decisions are made frequently, it is possible to track fast variations of the user traffic as well as the channel quality, thus efficiently utilizing the available resources in order to maximize the network capacity. Semi-static scheduling implies that the transmission parameters are provided to the devices in advance and are not changed on a dynamic basis.

The scheduler operation takes into account the UE buffer status and the QoS requirements of each UE as well as the associated radio bearers when assigning radio resources among active UEs (see Fig. 4.26). The schedulers assign network resources to the UEs by considering the radio conditions as seen by the UEs, identified through measurements made at the gNB and/or reported by the UE. The schedulers assign radio resources in a unit of slot, for example, one mini-slot, one slot, or multiple slots and resource assignments consist of radio resources (time, frequency, code, space, power). The UEs identify the allocated resources by receiving a scheduling decision (resource assignment) through PDCCH. The UE periodically or on-demand basis conducts measurements to support scheduler operation. The uplink buffer status reports (measuring the data that is buffered in the logical channel queues in the UE) are used to provide support for QoS-aware packet scheduling. Power headroom reports

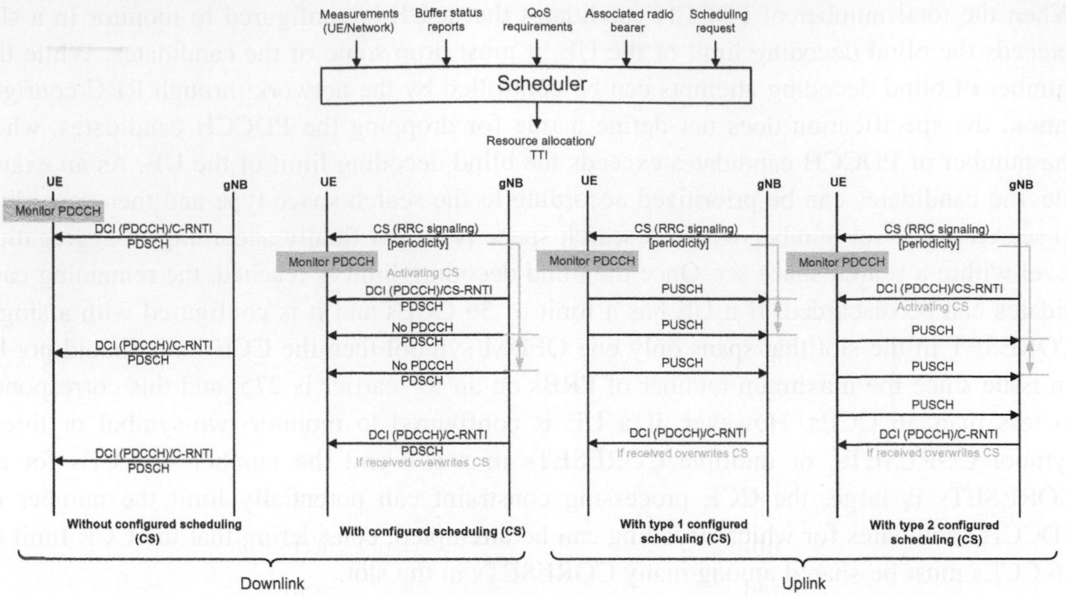

Figure 4.26
Illustration of downlink and uplink dynamic and configured scheduling [11].

(measuring the difference between the nominal UE maximum transmit power and the estimated power for uplink transmission) are used to provide support for power aware packet scheduling [11].

In the downlink, the gNB dynamically allocates radio resources to the active UEs using their respective C-RNTIs scrambled into their PDCCH(s) CRCs to uniquely identify the users. A UE always monitors certain PDCCH(s) candidates in order to find possible downlink/uplink assignments where the process is coordinated with the UE-specific DRX cycles, when configured. When carrier aggregation is configured for the UE, the same C-RNTI applies to all serving cells. In addition, with SPS, the gNB can allocate downlink resources for the initial HARQ transmissions to UEs. The RRC signaling defines the periodicity of the configured downlink assignments while PDCCH scrambled with CS-RNTI can either signal and activate the configured downlink assignment, or deactivate it, that is, a PDCCH addressed to a UE using its CS-RNTI indicates that the downlink assignment can be implicitly reused according to the periodicity defined by the RRC signaling, until deactivated. When a configured downlink assignment is active, if the UE cannot find its C-RNTI on the PDCCH(s), a downlink transmission according to the configured downlink assignment is assumed; otherwise, if the UE finds its C-RNTI on the PDCCH(s), the PDCCH allocation overrides the configured downlink assignment. When carrier aggregation is configured, one configured downlink assignment can be signaled per serving cell. When bandwidth adaptation is configured, one configured downlink assignment can be signaled per BWP. On each serving cell, there can be only one configured downlink assignment active at a time, and multiple configured downlink assignment can be simultaneously active on different serving cells. Activation and deactivation of configured downlink assignments are independent among the serving cells [11].

In the uplink, the gNB can dynamically allocate resources to the UEs by scrambling their respective C-RNTI with PDCCH(s) CRCs. A UE always monitors the PDCCH(s) in order to find possible grants for uplink transmission when its downlink reception is enabled where the activity is synchronized with the UE DRX cycles. When carrier aggregation is configured, the same C-RNTI applies to all serving cells. In addition, with configured grants, the gNB can allocate uplink resources for the initial HARQ transmissions to the UEs. Two types of configured uplink grants are defined in NR: Type 1, where the RRC signaling directly provides the configured uplink grant (including the periodicity); and Type 2, where the RRC signaling defines the periodicity of the configured uplink grant while PDCCH addressed to the UE using its CS-RNTI can either signal and activate the configured uplink grant, or deactivate it, that is, a PDCCH addressed to the UE using its CS-RNTI would indicate that the uplink grant can be implicitly reused according to the periodicity defined by RRC, until deactivated (see Fig. 4.26). When a configured uplink grant is active, if the UE cannot find its C-RNTI/CS-RNTI on the PDCCH(s), an uplink transmission according to the configured uplink grant can be attempted. Otherwise, if the UE finds its C-RNTI/CS-RNTI on the PDCCH(s), the PDCCH allocation overrides the configured uplink grant. Retransmissions other than repetitions are explicitly allocated via PDCCH(s). When carrier aggregation is configured, one configured uplink grant can be signaled per serving cell. Similarly, when bandwidth adaptation is configured, one configured uplink grant can be signaled per BWP. In each serving cell, there can be only one configured uplink grant active at a time. A configured uplink-grant for one serving cell can either be of Type 1 or Type 2. For Type 2, activation and deactivation of configured uplink grants are independent among the serving cells. When SUL is configured, a configured uplink grant can only be signaled for one of the two uplink carriers of the cell [11].

4.1.4 Synchronization Signals

In order to connect/attach to the network, a UE must perform initial cell search and downlink synchronization. The objective of initial cell search is to find a strong cell signal for connection establishment, to obtain an estimate of frame timing, to obtain cell identification, and to find the reference signals for coherent demodulation of PBCH and PDCCH. For this purpose, the PSS and SSS are used. The PSS and SSS are transmitted in SSBs together with PBCH. The blocks are transmitted per slot at a fixed slot location. During initial cell search, the UE correlates the received signals and the synchronization signal sequences by means

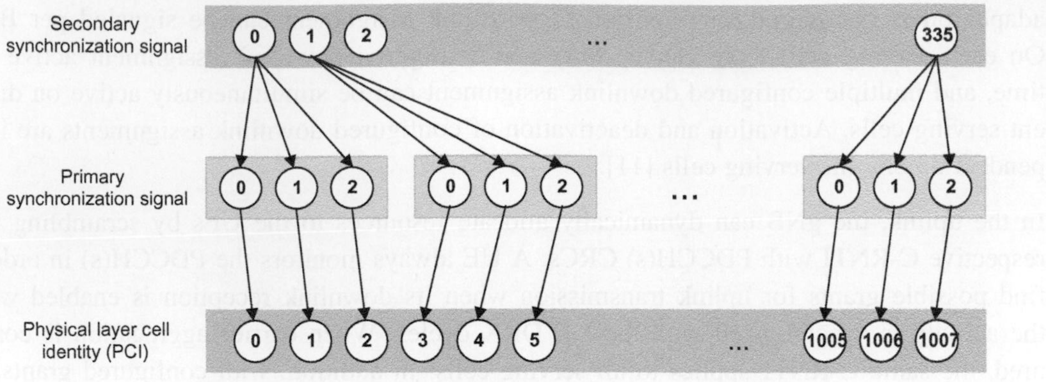

Figure 4.27
Derivation of PCI based on PSS and SSS sequences [6].

of matched filtering and attempts to locate the PSS in order to obtain symbol and half-frame timing. It then attempts to find the SSS in order to detect the cyclic prefix length as well as the duplexing scheme and to obtain the exact frame timing based on matched filter results for the PSS and SSS. It then proceeds to detect the cell identity from the reference signals sequence index and to decode the PBCH for the purpose of obtaining the minimum SI. The synchronization signal (SS) blocks are organized into SS bursts and SS bursts are organized into SS burst sets that are periodically transmitted in order to support beamforming operation.

In NR, there are 1008 unique physical-layer cell identities, that is, an increased number compared to 504 in LTE in order to provide sufficient deployment flexibility in dense network topologies. As shown in Fig. 4.27, the NR physical-layer cell identities are in 336 unique physical-layer cell-identity groups, each group containing three unique identities. Each NR cell ID can be jointly represented by a PSS/SSS combination. The PSS consists of three frequency-domain binary BPSK length-127 M-sequences,[6] and the SSS corresponds to 336 Gold sequences with length-127. Both of these signals are mapped into 127 contiguous subcarriers. A physical-layer cell identity is uniquely defined by a number in the range of 0−335, representing the physical-layer cell-identity group, and a number in the range of 0−2, representing the physical-layer identity within the physical-layer cell-identity group as

[6] Maximum length sequences are pseudo-random binary sequences that are generated using maximal linear feedback shift registers. The M-sequences are periodic and reproduce every binary sequence that can be reproduced by the shift registers (i.e., for length-m registers, they produce a sequence of length $2^m - 1$). An M-sequence is spectrally flat with the exception of a near-zero DC term. Since M-sequences are periodic and shift registers cycle through every possible binary value with the exception of the zero vectors, the registers can be initialized to any state with the exception of the zero vectors. A binary polynomial over GF(2) can be associated with the linear feedback shift register. The degree of polynomial is equal to the length of the shift register and the coefficients that are either 0 or 1 correspond to the taps of the register.

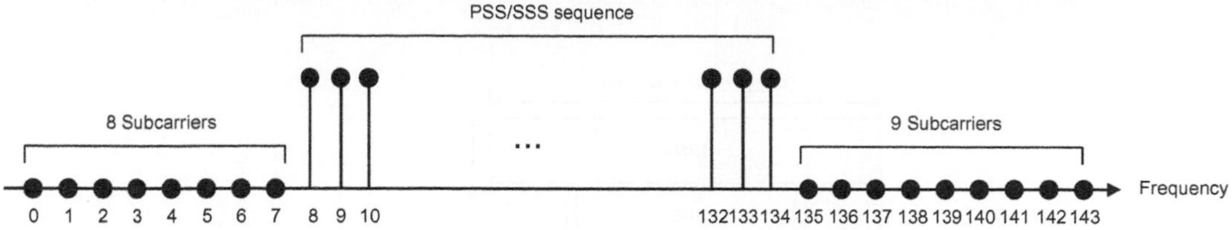

Figure 4.28
Illustration of PSS/SSS sequence mapping to the resource elements [6].

in the following formula $N_{ID}^{cell} = 3N_{ID}^{(1)} + N_{ID}^{(2)}$ where $N_{ID}^{(1)} \in \{0, 1, \ldots, 335\}$ and $N_{ID}^{(2)} = \{0, 1, 2\}$. The cell number information is carried in PSS, whereas the cell group number is carried in SSS. It is worthwhile mentioning that the physical beams associated with an SS blocks are transparent to the UE, since the latter only sees the equivalent synchronization signals and the PBCH after precoding and/or beamforming operations that are implementation specific.

4.1.4.1 Primary Synchronization Sequence

The PSS sequence $d_{PSS}(n)$ is defined by $d_{PSS}(n) = 1 - 2x(m)$ where $m = \left(n + 43N_{ID}^{(2)}\right) \mathrm{mod}\ 127$ and $0 \leq n < 127$. In the latter expression, $x(i+7) = [x(i+4) + x(i)] \mathrm{mod}\ 2$ and $\left[x(6)\ \ x(5)\ \ x(4)\ \ x(3)\ \ x(2)\ \ x(1)\ \ x(0)\right] = \left[1\ \ 1\ \ 1\ \ 0\ \ 1\ \ 1\ \ 0\right]$. The sequence of symbols $d_{PSS}(0), \ldots, d_{PSS}(126)$ containing the PSS is scaled by a factor of β_{PSS} in order to adjust its transmission power and is mapped to resource elements (k, l) in increasing order of k where k and l represent the frequency and time indices, respectively (see Table 4.3), within one SS/PBCH block. The PSS sequence mapping to resource elements in the frequency domain is illustrated in Fig. 4.28. Furthermore, the time-frequency structure and timing of the PSS transmission are depicted in Fig. 4.29.

4.1.4.2 Secondary Synchronization Sequence

The secondary synchronization signal sequence $d_{SSS}(n)$ is defined as $d_{SSS}(n) = [1 - 2x_0((n + m_0) \mathrm{mod}\ 127)][1 - 2x_1((n + m_1) \mathrm{mod}\ 127)]$ where $m_0 = 15\left\lfloor N_{ID}^{(1)}/112 \right\rfloor + 5N_{ID}^{(2)}$, $m_1 = N_{ID}^{(1)} \mathrm{mod}\ 112$ and $0 \leq n < 127$. In the latter expression, $x_0(i+7) = \left[x_0(i+4) + x_0(i)\right] \mathrm{mod}\ 2$ and $x_1(i+7) = \left[x_1(i+1) + x_1(i)\right] \mathrm{mod}\ 2$ where $\left[x_0(6)\ \ x_0(5)\ \ x_0(4)\ \ x_0(3)\ \ x_0(2)\ \ x_0(1)\ \ x_0(0)\right] = \left[0\ \ 0\ \ 0\ \ 0\ \ 0\ \ 0\ \ 1\right]$ and $\left[x_1(6)\ \ x_1(5)\ \ x_1(4)\ \ x_1(3)\ \ x_1(2)\ \ x_1(1)\ \ x_1(0)\right] = \left[0\ \ 0\ \ 0\ \ 0\ \ 0\ \ 0\ \ 1\right]$. The sequence of symbols $d_{SSS}(0), \ldots, d_{SSS}(126)$ containing the secondary synchronization signal is scaled by a factor of β_{SSS} and is mapped to resource elements (k, l) in increasing order of k where k and l represent the frequency and time indices, respectively (see Table 4.3),

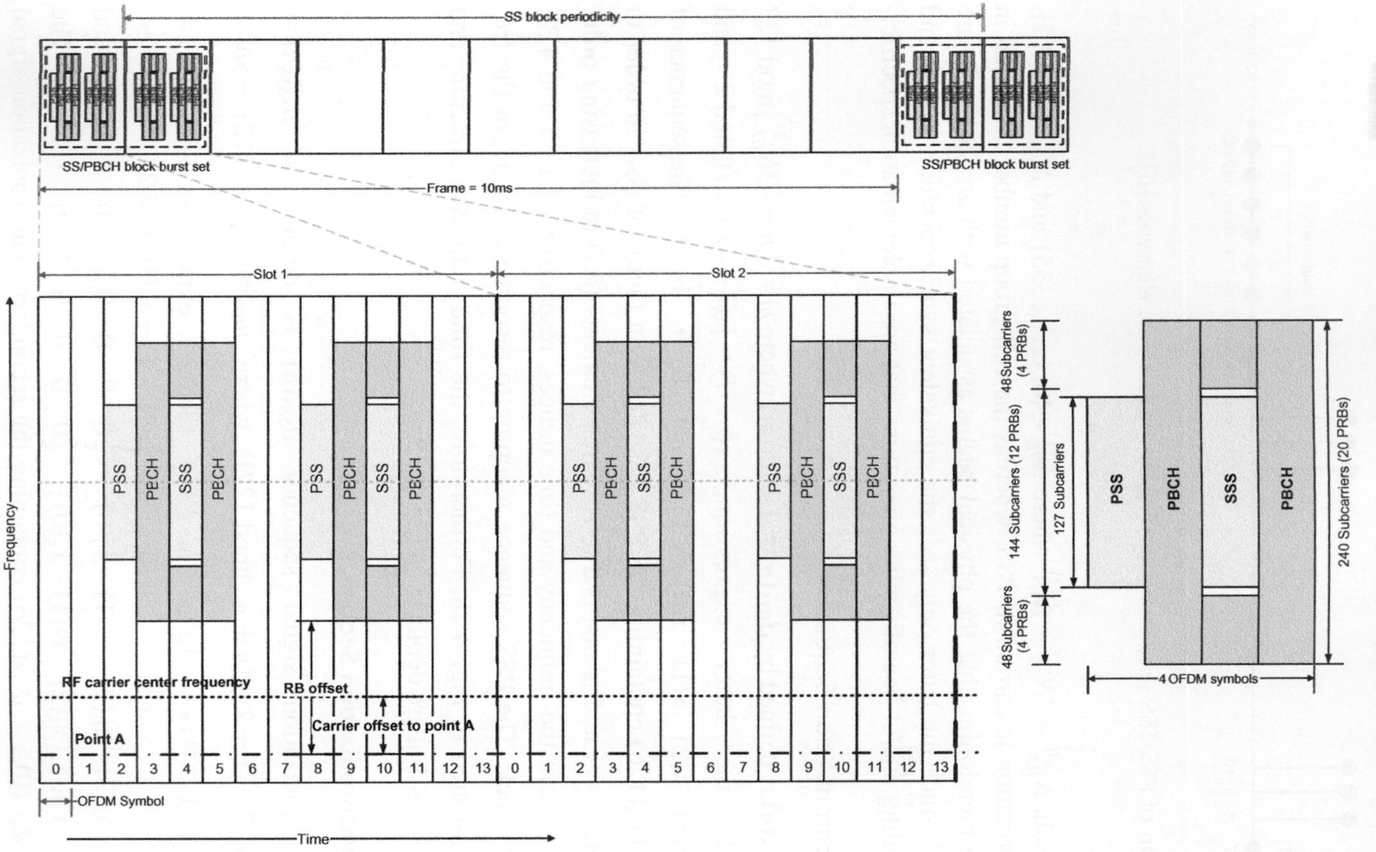

Figure 4.29
Time-frequency structure of the PSS within an SS block [6,56].

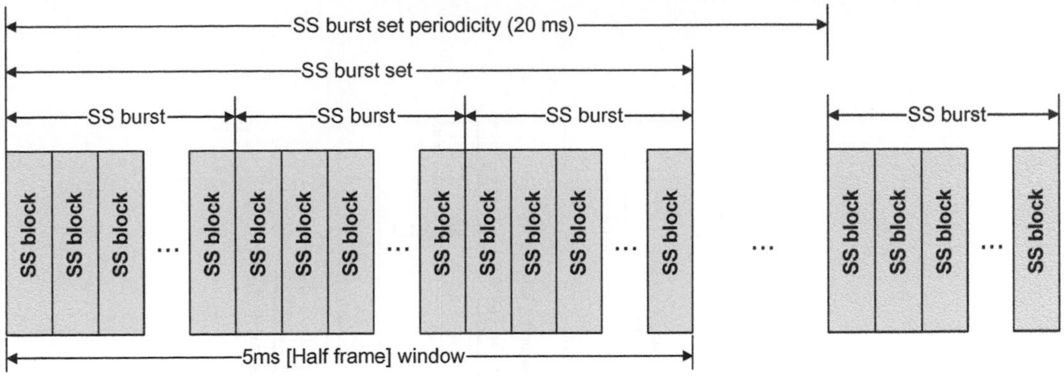

Figure 4.30
SS block structure and timing [6,8].

within one SS/PBCH block [6]. The SSS sequence mapping to resource elements in the frequency domain is illustrated in Fig. 4.28. Furthermore, the time-frequency structure and timing of the SSS transmission are depicted in Fig. 4.29.

4.1.4.3 Synchronization Signal Blocks

In NR, the primary and secondary synchronization signals are used by the UE for initial cell search and to obtain frame timing, Cell ID, and to find the reference signals for coherent demodulation of other channels. The PSS, SSS, and PBCH are time-multiplexed and transmitted in an SSB with the same numerology. One or more SS block(s) constitute an SS burst, and one or more SS bursts form an SS burst set as illustrated in Fig. 4.30. The SS burst sets are transmitted periodically. An SS block consists of four consecutive OFDM symbols. Regardless of the SS burst set composition, the transmission of SSBs within an SS burst set is confined to a 5 ms window to help the UEs reduce power consumption and complexity for radio resource management-related measurements. Fig. 4.31 compares the synchronization signals and broadcast channel transmission timings of LTE and NR.

The SS block is transmitted periodically with a period which may be configured between 5 and 160 ms. However, the UEs performing initial cell search or handover can assume that the SS block is repeated every 20 ms. Each SS burst set is always confined to a 5 ms window located either in the first or the second half of a 10 ms radio frame. This allows a UE that is searching for an SS block in the frequency domain to know the time duration it should pause at each frequency before retuning to the next frequency within the synchronization raster, concluding that there is no PSS/SSS present at that frequency. The 20 ms SS block periodicity is four times longer than the corresponding 5 ms periodicity of LTE PSS/SSS transmission (see Fig. 4.31). The longer SS block period was selected to improve the NR network energy efficiency and to reduce the layer-1 overhead. The disadvantage of a

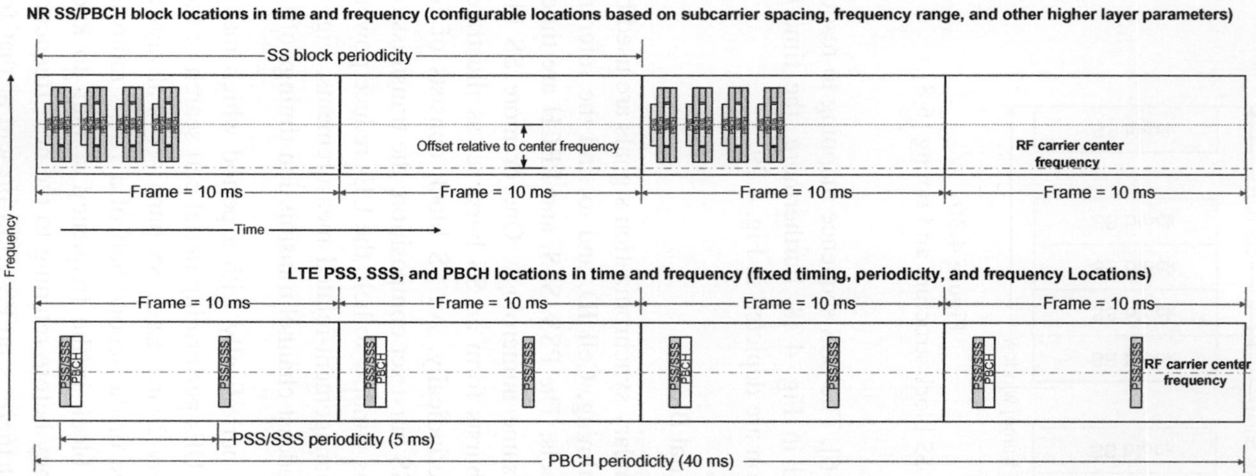

Figure 4.31
Comparison of LTE and NR synchronization and broadcast channel transmission timing.

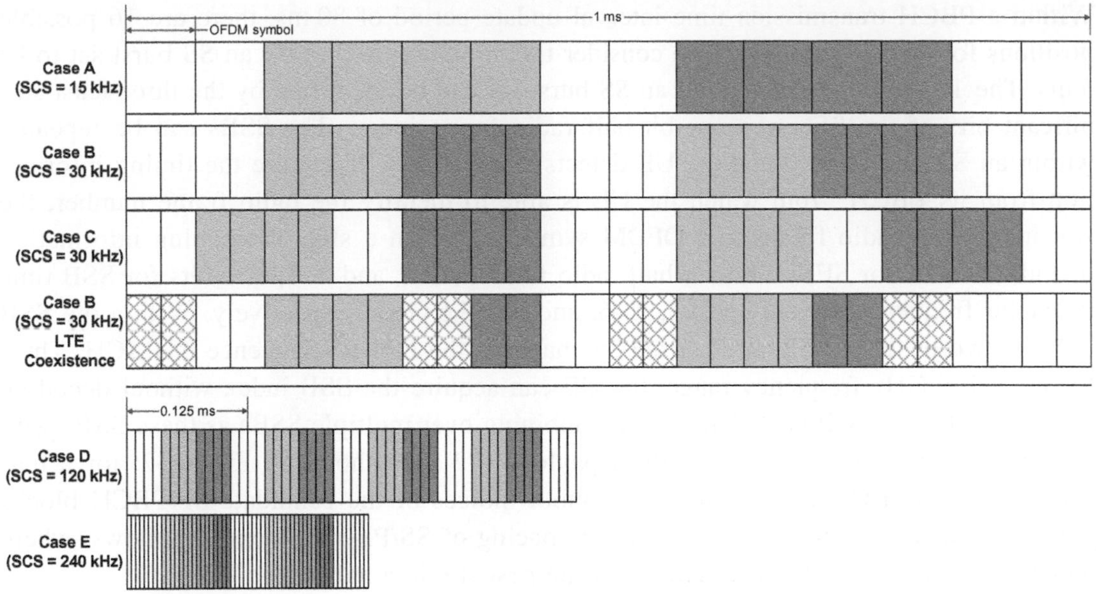

Figure 4.32
Structure and timing of SS/PBCH block transmission with various numerologies [6,31].

longer SS block period is that a device must pause at each frequency for a longer time in order to conclude that there is no PSS/SSS at the frequency. However, this is compensated by the sparse synchronization raster relative to LTE, which reduces the number of frequency-domain locations at which a device must search for an SS block.

The maximum number of SSBs within an SS burst set is 4 for frequency ranges up to 3 GHz, 8 for 3–6 GHz, and 64 for above 6 GHz in order to achieve a trade-off between coverage and layer-1 overhead. Furthermore, the number of actually transmitted SSBs could be less than the maximum number. The position(s) of actually transmitted SSBs can be signaled to the UEs in order to assist their RRC_CONNECTED or RRC_IDLE mode measurements and to help the UEs in RRC_CONNECTED and potentially RRC_IDLE mode to receive downlink data/control in unused SSBs. The structure and timing of SSB transmission with various numerologies is illustrated in Fig. 4.32, where a number of symbols are reserved for downlink control at the beginning of the slot, and some symbols are reserved for guard period and uplink control to allow UL/DL switching and fast uplink feedback. The SSB pattern corresponding to 15 and 30 kHz subcarrier spacing can provide more UL/DL switching opportunities in TDD mode. The SSB pattern for 30 kHz subcarrier spacing can be used to facilitate LTE-NR coexistence in the downlink in an FDD system, considering the locations of LTE PDCCH and cell-specific reference signals in the symbols with LTE default 15 kHz subcarrier spacing.

Within a PBCH transmission time interval update period of 80 ms, there are 16 possible positions for an SS burst set, if we consider the minimum period for an SS burst set to be 5 ms. The 16 possible positions of an SS burst set can be identified by the three least significant bits of the SFN and one-bit half radio frame index. The SSBs can be repeated within an SS burst set. When the UE detects an SSB, it will acquire the timing information from its PBCH, from which the UE is able to identify the radio frame number, the slot index in a radio frame, and OFDM symbol index in a slot. The timing information includes 10 bits for SFN, 1 bit for half radio frame index, and 2, 3, or 6 bits for SSB time index for frequency ranges up to 3, 3–6, and 6–52.6 GHz, respectively. Within the SSB indices, two or three LSBs are carried by changing the DM-RS sequence of PBCH. Thus, for the sub-6 GHz frequency range, the UE can acquire the SSB index without decoding the PBCH. It also facilitates PBCH soft combining over multiple SSBs as these SSBs with different indices carry the same PBCH payload [72]. As shown in Fig. 4.32, for a half frame with SS/PBCH blocks, the first symbol indices of the candidate SS/PBCH blocks are determined according to the subcarrier spacing of SS/PBCH blocks as follows, where index 0 corresponds to the first symbol of the first slot in a half-frame [8]:

- *Case A (SCS = 15 kHz):* The first symbols of the candidate SS/PBCH blocks have indices of $\{2,8\}+14n$ where $n=0,\ 1$ for carrier frequencies $f_c \leq 3$ GHz and $n=0,1,2,3$ for carrier frequencies 3 GHz $\leq f_c \leq 6$ GHz.
- *Case B (SCS = 30 kHz):* The first symbols of the candidate SS/PBCH blocks have indices $\{4,8,16,20\}+28n$ where $n=0$ for carrier frequencies $f_c \leq 3$ GHz, $n=0,\ 1$ for carrier frequencies 3 GHz $\leq f_c \leq 6$ GHz.
- *Case C (SCS = 30 kHz):* The first symbols of the candidate SS/PBCH blocks have indices $\{2,8\}+14n$ where $n=0,\ 1$ for paired spectrum operation and carrier frequencies $f_c \leq 3$GHz and $n=0,1,2,3$ for carrier frequencies 3 GHz $\leq f_c \leq 6$ GHz. For unpaired spectrum operation, $n=0,\ 1$ for carrier frequencies $f_c \leq 2.4$ GHz, $n=0,1,2,3$ for carrier frequencies 2.4 GHz $\leq f_c \leq 6$ GHz.
- *Case D (SCS = 120 kHz):* The first symbols of the candidate SS/PBCH blocks have indices $\{4,8,16,20\}+28n$ where $n=0,1,2,3,5,6,7,8,10,11,12,13,15,16,17,18$ for carrier frequencies $f_c \geq 6$ GHz.
- *Case E (SCS = 240 kHz):* The first symbols of the candidate SS/PBCH blocks have indices $\{8,12,16,20,32,36,40,44\}+56n$ where $n=0,1,2,3,5,6,7,8$ for carrier frequencies $f_c \geq 6$ GHz.

In order to support multi-beam operation, particularly in high-frequency scenarios, NR introduced the SSB, which comprises primary and secondary synchronization signals and PBCH. As illustrated in Fig. 4.33, a given SSB is repeated within an SS burst set, which is potentially used for gNB beam-sweeping transmission. The SS burst set is confined to a 5 ms window and transmitted periodically. For initial cell selection, the UE assumes a default SS burst set periodicity of 20 ms. The main advantage of SS burst set is that

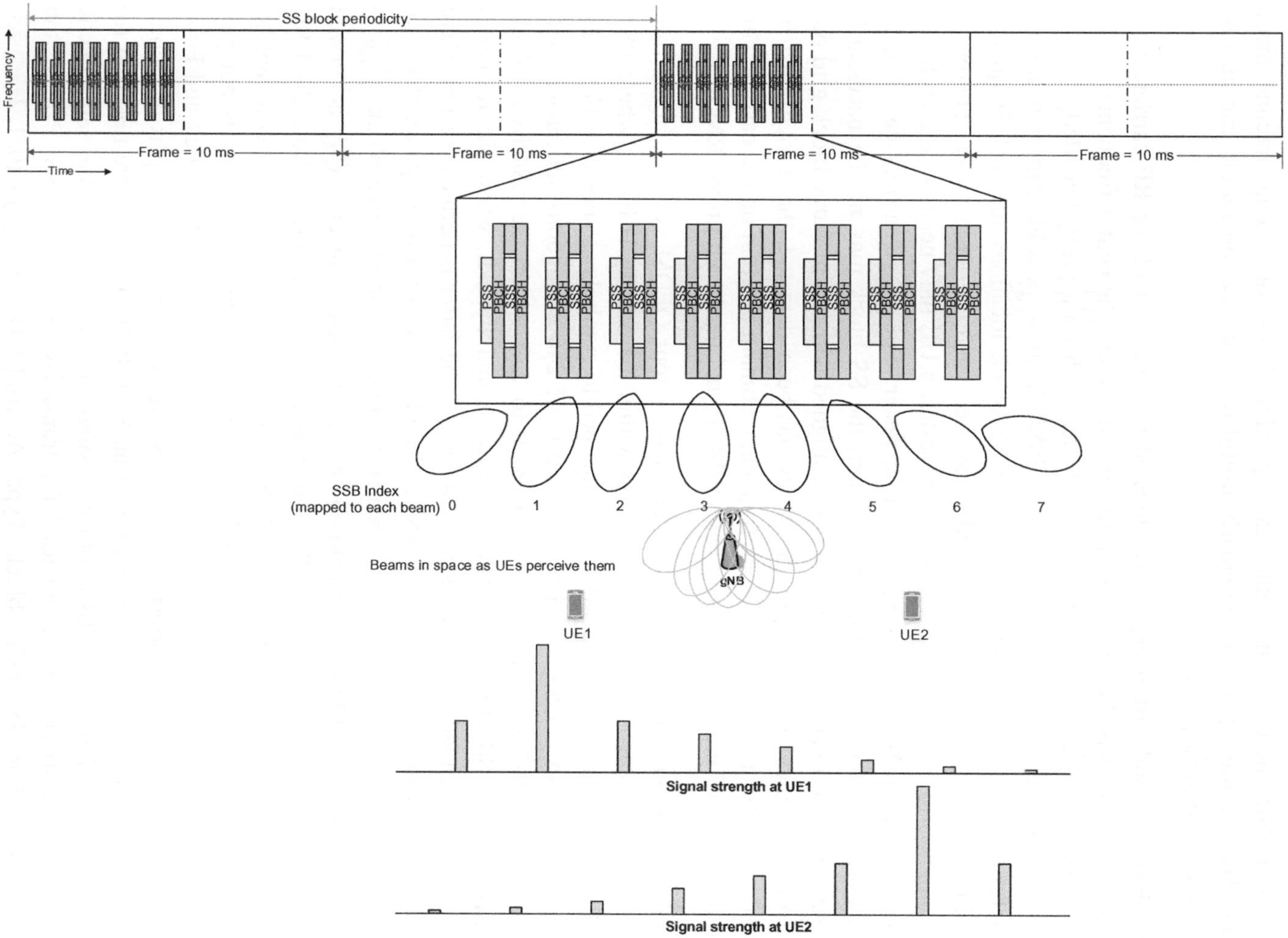

Figure 4.33
Example beam sweeping and correspondence to the transmission of SS burst set [30].

time-division multiplexing beam-sweeping allows for low-cost analog antenna array architectures. Frequency-division multiplexing is another approach that could have been potentially adopted in NR; however, it would have precluded use of analog antenna array architectures. This feature, although particularly useful for mmWave operation, can also be leveraged at lower frequency bands.

In LTE, the PSS/SSS and PBCH are always located at the center of the RF carrier. Thus once an LTE device detects the PSS/SSS, it has already found the center frequency of the carrier. The drawback of this approach is that a device with no a priori knowledge of the frequency-domain carrier position must search for PSS/SSS at all possible carrier positions. To allow a faster cell search in NR, the possible SS block locations for each frequency band are a limited set of frequencies referred to as the synchronization raster. Therefore, instead of searching for an SS block at each carrier raster, a UE only needs to search for an SS block within the sparse set of synchronization raster. Since NR carriers can still be located at an arbitrary position on the carrier raster, the SS block may not be necessarily located at the center of a carrier, and it may not be aligned with the resource block grid due to different numerologies. Thus once the SS block has been detected, the device must be explicitly informed about the exact SS block frequency-domain position on the carrier. This is achieved by means of information partly within PBCH and partly within the RMSI.

As we mentioned earlier, an SS/PBCH block consists of four OFDM symbols in the time domain, numbered in increasing order from 0 to 3 within the SS/PBCH block, where PSS, SSS, and PBCH with the associated DM-RS are mapped to symbols as shown in Fig. 4.29. In the frequency domain, an SS/PBCH block consists of 240 contiguous subcarriers with the subcarriers numbered in increasing order from 0 to 239 within the SS/PBCH block. There are two types of SS/PBCH block, that is, Type A and Type B, where the former is specified for operation in sub-6 GHz frequency range and the latter is defined for mmWave bands. The frequency-domain location of SS/PBCH block is defined by parameter k_{SSB} which provides the subcarrier offset from subcarrier 0 in common resource block N_{CRB}^{SSB} to subcarrier 0 of the SS/PBCH block. The common resource block parameter N_{CRB}^{SSB} is derived from the RRC parameter *offsetToPointA*. The four LSBs of k_{SSB} are derived from the RRC parameter *ssb-SubcarrierOffset* where, for SS/PBCH block Type A, the most significant bit of k_{SSB} is given by $a_{N_{MIB}+5}$ in the PBCH payload [6,7]. If *ssb-SubcarrierOffset* is not provided, k_{SSB} is derived from the frequency difference between the SS/PBCH block and Point A. The complex-valued symbols corresponding to resource elements that are part of a common resource block partially or fully overlap with an SS/PBCH block and are not used for SS/PBCH transmission. For an SS/PBCH block, a single-antenna port and the same cyclic prefix length and subcarrier spacing are used for transmission of PSS, SSS, PBCH, and DM-RS for PBCH. For SS/PBCH block Type A, $\mu \in \{0, 1\}$ and $k_{SSB} \in \{0, 1, 2, \ldots, 23\}$ with the quantities k_{SSB} and N_{CRB}^{SSB} expressed in terms of 15 kHz subcarrier spacing. For

SS/PBCH block Type B, $\mu \in \{3,4\}$ and $k_{SSB} \in \{0,1,2,\ldots,11\}$ where the quantity k_{SSB} expressed in terms of the subcarrier spacing provided by the RRC parameter *subCarrierSpacingCommon*, and N_{CRB}^{SSB} is defined in terms of 60 kHz subcarrier spacing. The center of subcarrier 0 of resource block N_{CRB}^{SSB} coincides with the center of subcarrier 0 of a common resource block with the subcarrier spacing provided by the RRC parameter *subCarrierSpacingCommon*. This common resource block overlaps with subcarrier 0 of the first resource block of the SS/PBCH block. The SS/PBCH blocks are transmitted with the same block index on the same center frequency location which are quasi-co-located with respect to Doppler spread, Doppler shift, average gain, average delay, delay spread, and spatial RX parameters (when applicable) [6].

4.1.5 Physical Downlink Shared Channel

The downlink physical layer processing of transport channels consists of several steps as shown in Fig. 4.34 including CRC calculation and attachment to the TBs where a 24-bit CRC for payloads larger than 3824 bits or otherwise 16-bit CRC is attached; code block segmentation and code block CRC attachment; channel coding based on LDPC codes; physical-layer HARQ processing and rate matching; bit-interleaving; modulation; layer mapping and precoding; and mapping to assigned resources and antenna ports. At least one symbol with DM-RSs is present on each layer in which PDSCH is transmitted to a UE. The number of DM-RS symbols and RE mapping is configured by the RRC parameters. The PT-RS may be transmitted on additional symbols to aid receiver phase tracking.

As shown in Fig. 4.34, in the first stage of PDSCH processing, the entire TB $a_0, a_1, \ldots, a_{N_{TB}-1}$ is used to calculate the CRC parity bits $p_0, p_1, \ldots, p_{L_{CRC}-1}$ where N_{TB} and L_{CRC} denote the (TB) payload size and the number of CRC parity bits, respectively. The number of parity bits is set to 24 and the CRC generator polynomial $g_{CRC24A}(D) = [D^{24} + D^{23} + D^{18} + D^{17} + D^{14} + D^{11} + D^{10} + D^7 + D^6 + D^5 + D^4 + D^3 + D + 1]$ is used, however, if $N_{TB} \geq 3824$, L_{CRC} is set to 16 bits and the generator polynomial $g_{CRC16}(D) = [D^{16} + D^{12} + D^5 + 1]$ is used. The output bits following the CRC attachment are denoted by $b_0, b_1, \ldots, b_{B-1}$ where $B = N_{TB} + L_{CRC}$. For initial transmission of a TB with coding rate R, which is determined by the MCS index contained in the DCI, and the retransmissions of the same TB, each code block of the TB is encoded with LDPC base graph 2 (see the section on channel coding), if $N_{TB} \leq 292$, if $N_{TB} \leq 3824$ and $R \leq 0.67$ or if $R \leq 0.25$; otherwise, LDPC base graph 1 is used as depicted in Fig. 4.35.

As shown in Fig. 4.35, the maximum size of a code block K_{cb} is 8448 bits for LDPC base graph 1, 3840 for LDPC base graph 2. The code blocks whose size exceeds these limits would be segmented and appended with an additional CRC of length $L_{CRC} = 24$ bits. The input bits to the code block segmentation denoted as $b_0, b_1, \ldots, b_{B-1}$, where B is the number of bits in the TB (including the CRC), are then processed by code block segmentation

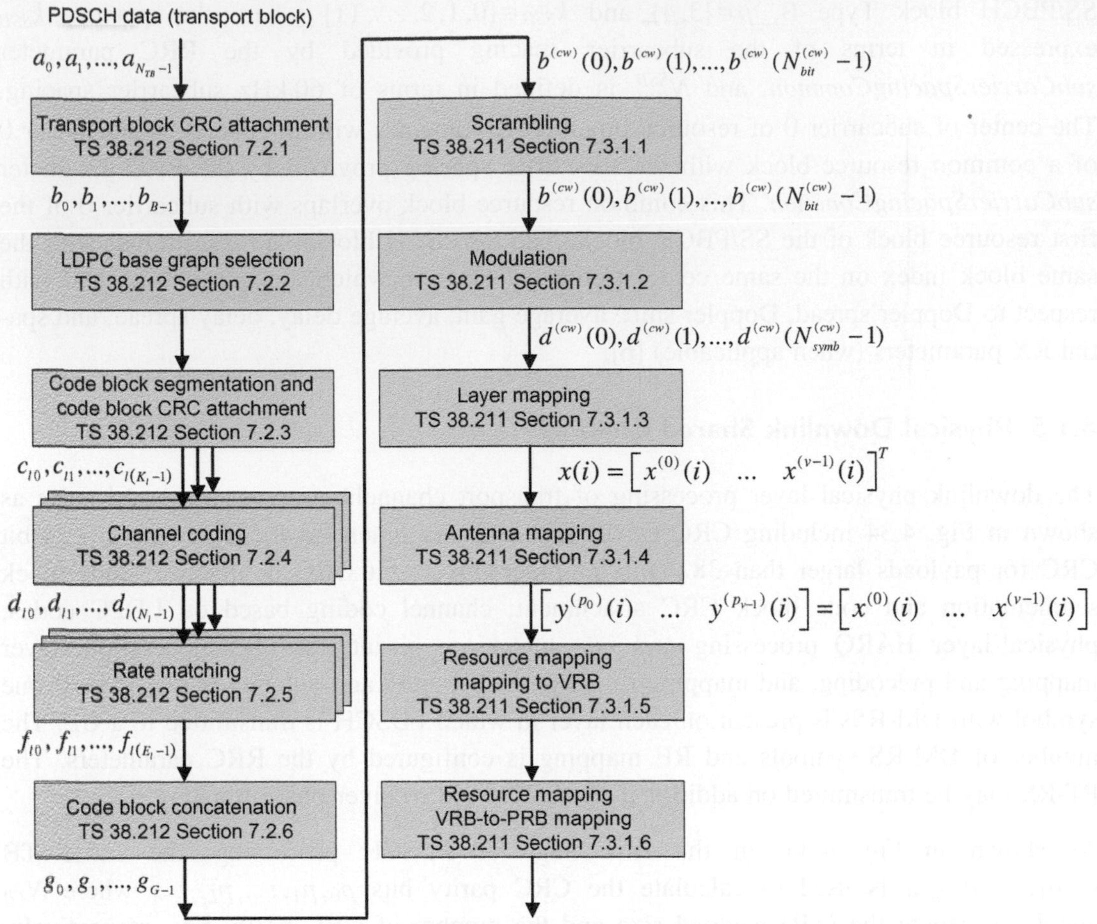

Figure 4.34
Physical layer processing of PDSCH [6,30].

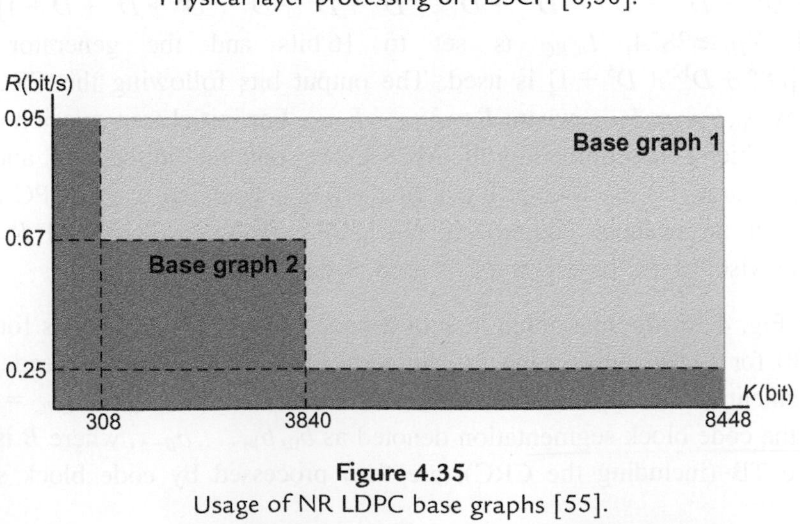

Figure 4.35
Usage of NR LDPC base graphs [55].

Table 4.7: NR low-density parity check code lifting factors Z_c [7].

Set Index i_{LS}	Set of Lifting Sizes Z_C
0	{2,4,8,16,32,64,128,256}
1	{3,6,12,24,48,96,192,384}
2	{5,10,20,40,80,160,320}
3	{7,14,28,56,112,224}
4	{9,18,36,72,144,288}
5	{11,22,44,88,176,352}
6	{13,26,52,104,208}
7	{15,30,60,120,240}

Table 4.8: NR low-density parity check base graphs parameters [7].

Parameter	Base Graph 1	Base Graph 2
Minimum code rate R_{min}	1/3	1/5
Base matrix size	46×68	42×52
Number of systematic columns K_b	22	10
Maximum information block size K_{cb}	$8448(=22 \times 384)$	$3840(=10 \times 384)$
Number of non-zero elements	316	197

followed by code block CRC attachment, resulting in the output bits $c_{l0}, c_{l1}, \ldots, c_{l(K_l-1)}$ where index $0 \le l < C$ represents the code block number and K_l denotes the number of bits for lth code block. The total number of code blocks C is determined by $C = \lceil B/(K_{cb} - L_{CRC}) \rceil$. The code blocks are then fed into the channel coding unit. The LDPC encoded bits are denoted by $d_{l0}, d_{l1}, \ldots, d_{l(N_l-1)}$ where the value of N_l is calculated as follows: If the bit sequence input for a given code block to channel coding is denoted by $c_0, c_1, \ldots, c_{K-1}$ where K is the number of bits to encode, and the LDPC encoded bits are denoted by $d_0, d_1, \ldots, d_{N-1}$ then $N = 66Z_c$ for LDPC base graph 1 and $N = 50Z_c$ for LDPC base graph 2, where the lifting factor Z_c is given in Table 4.7 [7]. The NR LDPC base graphs parameters are shown in Table 4.8.

The rate matching for LDPC code is performed on code block basis and consists of bit selection and bit-level interleaving. The input bit sequence to rate matching block is denoted as $d_0, d_1, \ldots, d_{N-1}$ which is written into a circular buffer of length N_{cb} for code block l, where code length N was defined earlier. Let us assume $N_{cb} = N$ for the lth code block, if $I_{LBRM} = 0$[7] and in other cases $N_{cb} = \min(N, N_{REF})$, in which $N_{REF} = \lfloor TBS_{LBRM}/(R_{LBRM}C) \rfloor$, C is the number of code blocks of the transport block, $R_{LBRM} = 2/3$ and TBS_{LBRM} for DL-SCH/PCH is obtained from Table 4.16, taking into consideration the maximum number of layers for one TB supported by the UE in the serving cell; the maximum modulation order

[7] Limited-buffer rate matching (LBRM) is a technique to process HARQ with reduced requirements for soft buffer sizes while maintaining the peak data rates. LBRM shortens the length of the virtual circular buffer of the code block segments for certain large sizes of transport blocks, thus sets a lower bound on the code rate.

configured for the serving cell, if configured by higher layers; otherwise, a maximum modulation order of $Q_m = 6$ is assumed for DL-SCH; and the maximum coding rate of 948/1024. Due to unequal amplitude of demodulated log likelihood ratios (LLRs) for 16QAM/64QAM/256QAM modulated symbols, it is necessary to consider a bit interleaving scheme for high-order modulations (see Fig. 4.37) in order to enhance the performance of the LDPC codes. The output bit sequence of the bit-interleaving function is the input to code block concatenation. The code block concatenation consists of sequentially concatenating the rate-matched outputs of different code blocks. The output bit sequence of code block concatenation is denoted by $g_0, g_1, \ldots, g_{G-1}$ where G is the total number of coded bits for transmission [7].

Rate matching is performed separately for each code block by puncturing a fraction of systematic bits. Depending on the code block size, the fraction of punctured systematic bits can be up to one-third of the systematic bits. The remaining coded bits are written into a circular buffer, starting with the non-punctured systematic bits and continuing with parity bits as shown in Fig. 4.36. The selection of the bits for transmission is based on reading the

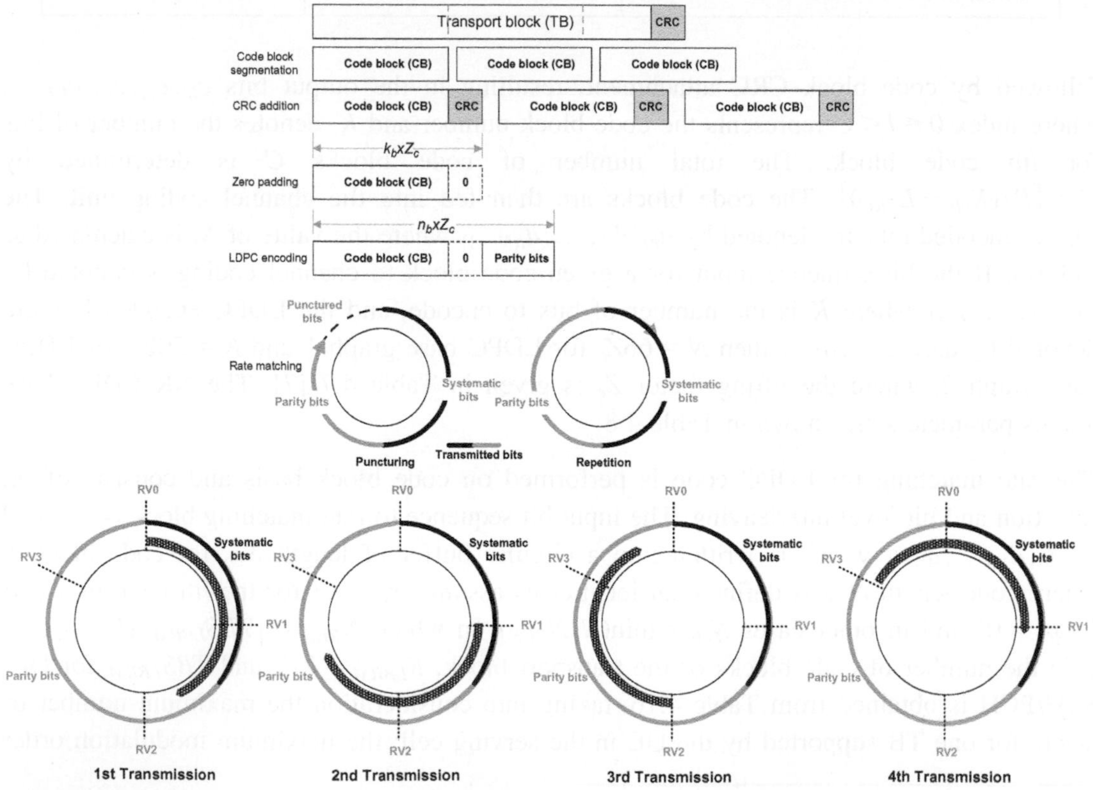

Figure 4.36
Example of rate-matching and code block concatenation processes [14].

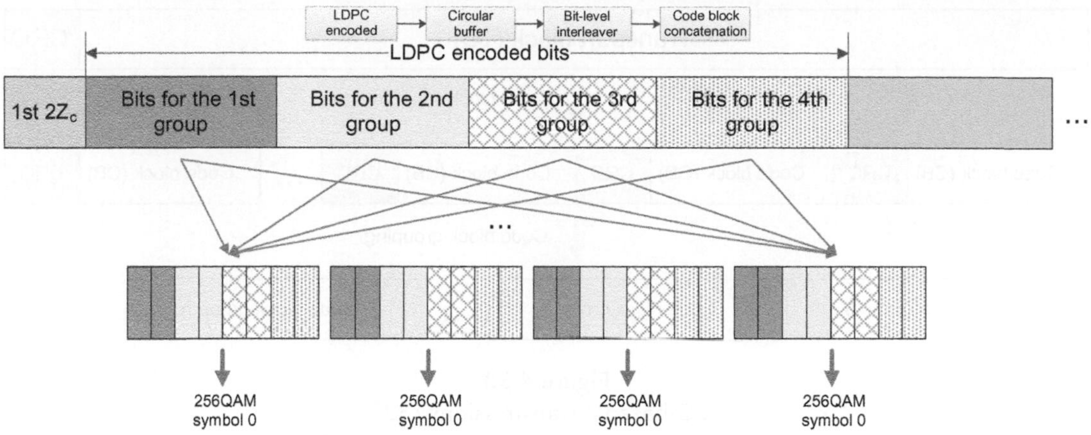

Figure 4.37
Example bit-level interleaving for 256QAM modulation.

Table 4.9: Starting position of different redundancy versions [7].

rv_{id}	k_0	
	LDPC Base Graph 1	**LDPC Base Graph 2**
0	0	0
1	$\left\lfloor \frac{17N_{cb}}{66Z_c} \right\rfloor Z_c$	$\left\lfloor \frac{13N_{cb}}{50Z_c} \right\rfloor Z_c$
2	$\left\lfloor \frac{33N_{cb}}{66Z_c} \right\rfloor Z_c$	$\left\lfloor \frac{25N_{cb}}{50Z_c} \right\rfloor Z_c$
3	$\left\lfloor \frac{56N_{cb}}{66Z_c} \right\rfloor Z_c$	$\left\lfloor \frac{43N_{cb}}{50Z_c} \right\rfloor Z_c$

required number of bits from the circular buffer where the exact set of bits to transmit depends on the RV corresponding to different starting positions in the circular buffer. Thus by selecting different RVs, different sets of coded bits representing the same set of information bits can be generated, which is used when implementing HARQ with incremental redundancy. The starting points in the circular buffer (RV0,RV1,RV2,RV3) are defined such that both RV0 and RV3 codes are self-decodable which means that they include the systematic bits under typical conditions. The RV index of the incremental redundancy HARQ in NR is derived differently compared to LTE. Unlike LTE that RV index positions are sequentially incremented in the circular buffer, in NR, if $rv_{id} = 0, 1, 2, 3$ denotes the RV number for the current transmission, the rate matching output bit sequence $\{e_k | k = 0, 1, \ldots, E-1\}$ is generated as $e_k = d_{(k_0+j)\bmod N_{cb}}$ where k_0 is given by Table 4.9 according to the value of rv_{id} and the LDPC base graph lifting factor [7,14].

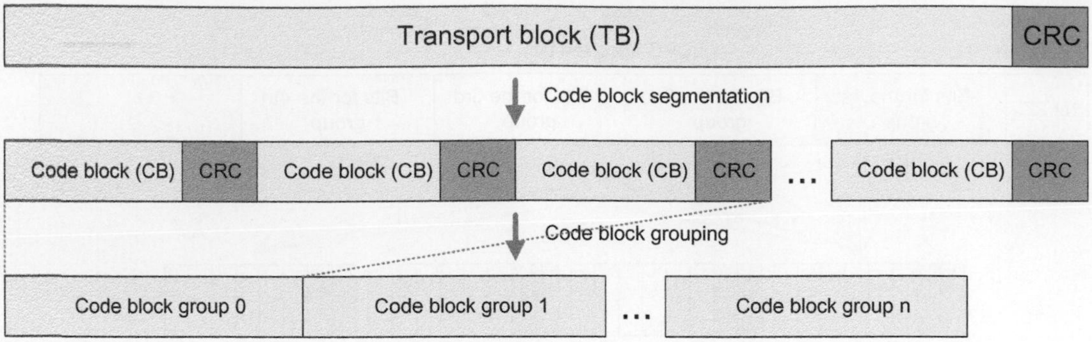

Figure 4.38
CBG-based retransmissions [32].

It is possible to perform HARQ retransmissions with a code block granularity. In that case the information included in the DCI would determine the code block group (CBG) which is (re)transmitted, and information about handling the CBGs for soft-buffer/HARQ combining purposes (see Fig. 4.38). The NR supports code block group-based transmission with single or multi-bit HARQ-ACK feedback. For the case of CBG-based retransmission, HARQ-ACK multiplexing is supported. The motivation for CBG-based retransmission is to improve the spectrum efficiency because if CBG-based retransmission is configured, the HARQ feedback is provided per CBG and only the erroneously received code block groups are retransmitted. This can consume less radio resources than retransmitting the entire TB. If the retransmission is caused due to low SNR then combining in the soft-buffer would improve the decoding quality during retransmissions; however, if the retransmitted code block was affected by preemption, the buffer content is not correct, and it is better to discard the content of the buffer and to request a fresh transmission.

If a UE is configured to receive CBG-based transmissions when the RRC parameter *codeBlockGroupTransmission* set for PDSCH, it determines the number of CBGs for a PDSCH transmission by calculating $M = \min(N_{CBG}, C)$ where N_{CBG} denotes the maximum number of CBGs per TB which is configured by RRC parameter *maxCodeBlockGroups PerTransportBlock* for PDSCH, and C is the number of code blocks. We define $M_1 = C \bmod M, K_1 = \lceil C/M \rceil$ and $K_2 = \lfloor C/M \rfloor$. If $M_1 > 0$, the mth CBG when $m = 0, 1, \ldots, M_1 - 1$ consists of code blocks with indices $mK_1 + k, \forall k = 0, 1, \ldots, K_1 - 1$. The mth CBG when $m = M_1, M_1 + 1, \ldots, M - 1$ consists of code blocks with indices $M_1K_1 + (m - M_1)K_2 + k, \forall k = 0, 1, \ldots, K_2 - 1$. If a UE is configured with CBG-based retransmissions, the scheduling assignment for the UE would contain the necessary HARQ-related control signaling including process number, new-data indicator, CBG transmit indicator (CBGTI), and the CBG flush indicator (CBGFI) as well as information to handle the transmission of the HARQ acknowledgment in the uplink such as timing and resource indication information. Upon receiving a scheduling assignment in the DCI, the receiver would attempt to decode the TB. Since transmissions and retransmissions are scheduled using the same framework, the UE needs to know whether this is a new transmission, in that case the soft buffer should be flushed, or a retransmission, where soft combining should be performed. Therefore, a single-bit new data indicator is included as part of the scheduling information. The new data indicator operates at TB level. However, if CBG-based retransmissions are configured, the device needs to know which CBGs are retransmitted and whether the corresponding soft buffer should be flushed. This is handled by additional information fields in the DCI when CBG-based retransmissions are configured, that is, CBGTI and CBGFI fields. The CBGTI is a bitmap indicating whether a certain CBG is present in the downlink transmission. The CBGFI is a single bit field, indicating whether the CBGs identified by CBGTI should be flushed or whether soft combining should be performed. The decoding operation results in either a positive acknowledgment in the case of a successful decoding or a negative acknowledgment in the case of unsuccessful decoding, and it is fed back to the gNB as part of the uplink control information. If CBG-based retransmissions are configured, a bitmap with one bit per CBG is fed back instead of a single bit representing the entire TB. The uplink uses the same asynchronous HARQ protocol as the downlink. The necessary HARQ-related information including process number, new-data indicator, and CBGTI (when configured) are included in the scheduling grant [9]. CBG-based retransmissions are transparent to the MAC sublayer and are handled in the physical layer despite being part of the HARQ mechanism. From the MAC perspective, the TB is not correctly received until all CBGs are correctly received and decoded. It is not possible to combine transmission of new CBGs associated with another TB with retransmissions of CBGs belonging to the incorrectly received TB in the same HARQ process [14].

The bit sequence $f_0, f_1, \ldots, f_{E-1}$ is generated by interleaving bit sequence $e_0, e_1, \ldots, e_{E-1}$ according to the value of the modulation order Q_m as follows $f_{i+jQ_m} = e_{iE/Q_m+j}$, $\forall j = 0, 1, \ldots, E/Q_m - 1; i = 0, 1, \ldots, Q_m - 1$ [7]. The new radio supports up to two codewords in the downlink transmission. For each codeword cw, the block of bits $b^{(cw)}(0), b^{(cw)}(1), \ldots, b^{(cw)}(N_{bit}^{(cw)} - 1)$, where $N_{bit}^{(cw)}$ denotes the number of bits in codeword cw transmitted on the physical shared channel, is scrambled prior to modulation, resulting in a block of scrambled bits $\tilde{b}^{(cw)}(0), \tilde{b}^{(cw)}(1), \ldots, \tilde{b}^{(cw)}(N_{bit}^{(cw)} - 1)$ such that $\tilde{b}^{(cw)}(i) = [b^{(cw)}(i) + c^{(cw)}(i)] \bmod 2$. The scrambling sequence $c^{(cw)}(i)$ is a generic pseudo-random length-31 Gold sequence that is initialized by setting $c_{init} = n_{RNTI}2^{15} + cw2^{14} + n_{ID}$ where $n_{ID} \in \{0, 1, \ldots, 1023\}$, if configured through RRC signaling; otherwise $n_{ID} = N_{ID}^{cell}$, n_{RNTI} corresponds to the RNTI associated with the PDSCH transmission [6].

The scrambled bit sequence $\tilde{b}^{(cw)}(0), \tilde{b}^{(cw)}(1), \ldots, \tilde{b}^{(cw)}(N_{bit}^{(cw)} - 1)$ is modulated using one of the modulation schemes, for example, QPSK, 16QAM, 64QAM, or 256QAM. The complex-valued modulation symbols for each of the codewords that are going to be transmitted are mapped to one or several layers for spatial multiplexing. As shown in Table 4.10, the

Table 4.10: Codeword to layer mapping for spatial multiplexing [6].

Number of Layers υ	Number of Codewords cw	Mapping	Parameters
1	1	$x^{(0)}(i)=d^{(0)}(i)$	$N_{layer}^{symb}=N_{symb}^{(0)}, i=0,1,\ldots,N_{layer}^{symb}-1$
2	1	$x^{(0)}(i)=d^{(0)}(2i)$ $x^{(1)}(i)=d^{(0)}(2i+1)$	$N_{layer}^{symb}=N_{symb}^{(0)}/2$
3	1	$x^{(0)}(i)=d^{(0)}(3i)$ $x^{(1)}(i)=d^{(0)}(3i+1)$ $x^{(2)}(i)=d^{(0)}(3i+2)$	$N_{layer}^{symb}=N_{symb}^{(0)}/3$
4	1	$x^{(0)}(i)=d^{(0)}(4i)$ $x^{(1)}(i)=d^{(0)}(4i+1)$ $x^{(2)}(i)=d^{(0)}(4i+2)$ $x^{(3)}(i)=d^{(0)}(4i+3)$	$N_{layer}^{symb}=N_{symb}^{(0)}/4$
5	2	$x^{(0)}(i)=d^{(0)}(2i)$ $x^{(1)}(i)=d^{(0)}(2i+1)$ $x^{(2)}(i)=d^{(1)}(3i)$ $x^{(3)}(i)=d^{(1)}(3i+1)$ $x^{(4)}(i)=d^{(1)}(3i+2)$	$N_{layer}^{symb}=N_{symb}^{(0)}/2=N_{symb}^{(1)}/3$
6	2	$x^{(0)}(i)=d^{(0)}(3i)$ $x^{(1)}(i)=d^{(0)}(3i+1)$ $x^{(2)}(i)=d^{(0)}(3i+2)$ $x^{(3)}(i)=d^{(1)}(3i)$ $x^{(4)}(i)=d^{(1)}(3i+1)$ $x^{(5)}(i)=d^{(1)}(3i+2)$	$N_{layer}^{symb}=N_{symb}^{(0)}/3=N_{symb}^{(1)}/3$
7	2	$x^{(0)}(i)=d^{(0)}(3i)$ $x^{(1)}(i)=d^{(0)}(3i+1)$ $x^{(2)}(i)=d^{(0)}(3i+2)$ $x^{(3)}(i)=d^{(1)}(4i)$ $x^{(4)}(i)=d^{(1)}(4i+1)$ $x^{(5)}(i)=d^{(1)}(4i+2)$ $x^{(6)}(i)=d^{(1)}(4i+3)$	$N_{layer}^{symb}=N_{symb}^{(0)}/3=N_{symb}^{(1)}/4$
8	2	$x^{(0)}(i)=d^{(0)}(4i)$ $x^{(1)}(i)=d^{(0)}(4i+1)$ $x^{(2)}(i)=d^{(0)}(4i+2)$ $x^{(3)}(i)=d^{(0)}(4i+3)$ $x^{(4)}(i)=d^{(1)}(4i)$ $x^{(5)}(i)=d^{(1)}(4i+1)$ $x^{(6)}(i)=d^{(1)}(4i+2)$ $x^{(7)}(i)=d^{(1)}(4i+3)$	$N_{layer}^{symb}=N_{symb}^{(0)}/4=N_{symb}^{(1)}/4$

complex-valued modulation symbols $d^{(cw)}(0), d^{(cw)}(1), \ldots, d^{(cw)}(N_{symb}^{(cw)} - 1)$ corresponding to codeword cw are mapped to layers $x(i) = \left[x^{(0)}(i) \quad \ldots \quad x^{(v-1)}(i)\right]^T, \forall i = 0, 1, \ldots, N_{layer}^{symb} - 1$ where v denotes the number of layers and N_{layer}^{symb} is the number of modulation symbols per layer [6].

The block of vectors $x(i) = [x^{(0)}(i) \ldots x^{(v-1)}(i)]^T, \forall i = 0, 1, \ldots, N_{layer}^{symb} - 1$ are mapped to antenna ports as follows:

$$[y^{(p_0)}(i) \ldots y^{(p_{v-1})}(i)] = [x^{(0)}(i) \ldots x^{(v-1)}(i)], \forall i = 0, 1, \ldots, N_{layer}^{symb} - 1$$

The set of antenna ports $\{p_0, p_1, \ldots, p_{v-1}\}$ are determined according to the procedure specified in [7].

For each antenna port that is used for transmission of the physical shared channel, the properly scaled block of complex-valued symbols $y^{(p)}(0), y^{(p)}(1), \ldots, y^{(p)}(N_{AP}^{symb} - 1)$ are sequentially mapped to the virtual resource elements (k', l) allocated for transmission of PDSCH that have not been designated for reference signals. Any common resource block partially or fully overlapping with an SS/PBCH block is considered occupied and is not used for transmission. The virtual resource elements are mapped in frequency-first manner as illustrated in Fig. 4.39. The virtual resource blocks are mapped to physical resource blocks in

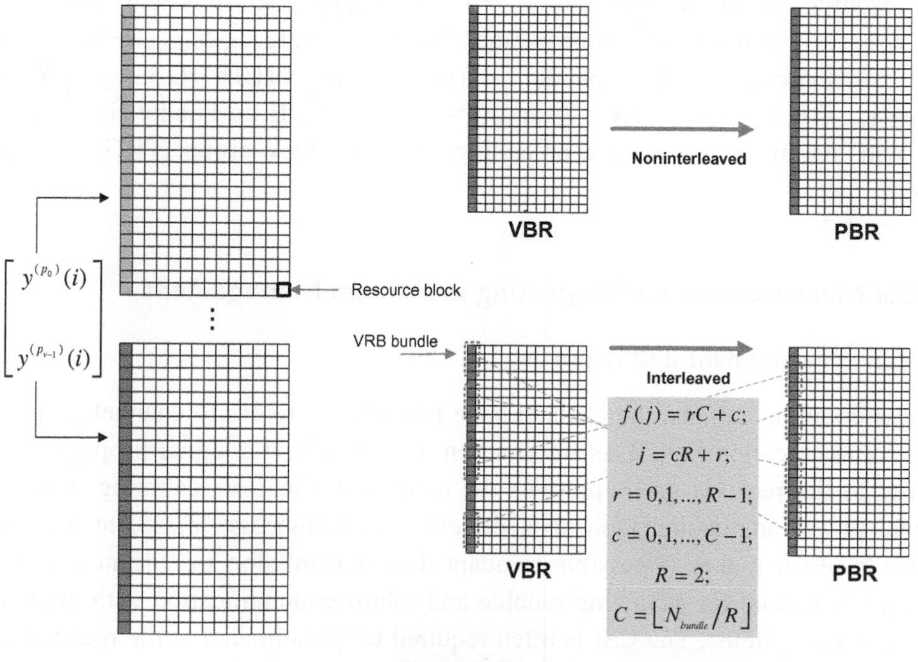

Figure 4.39
Mapping to VRBs and from VRBs to PRBs [6].

form of either non-interleaved or interleaved, wherein the non-interleaved is the default mapping scheme.

For non-interleaved VRB-to-PRB mapping, the virtual resource block n is mapped to physical resource block n, except for PDSCH transmissions scheduled with DCI format 1_0 in a common search space where virtual resource block n is mapped to physical resource block $n + N_{CORESET}^{start}$, where $N_{CORESET}^{start}$ is the lowest numbered physical resource block in the CORESET where the corresponding DCI was received [6].

In interleaved mapping scheme, the mapping process is defined in terms of resource block bundles. The set of $N_{BWP}^{size}(i)$ resource blocks in the ith bandwidth part with starting position $N_{BWP}^{start}(i)$ are divided into $N_{bundle} = \lceil [N_{BWP}^{size}(i) + (N_{BWP}^{start}(i) \bmod L_i)]/L_i \rceil$ resource-block bundles in increasing order of the resource-block number and bundle number where L_i is the bundle size for the ith bandwidth part defined by RRC parameter *vrb-ToPRB-Interleaver* and resource block bundle 0 consists of $L_i - (N_{BWP}^{start}(i) \bmod L_i)$ resource blocks, resource block bundle $N_{bundle} - 1$ consists of $(N_{BWP}^{start}(i) + (N_{BWP}^{size}(i)) \bmod L_i, \forall L_i > 0$ resource blocks; otherwise, all resource block bundles consists of L_i resource blocks (except for PDSCH transmissions scheduled with DCI format 1_0 with the CRC scrambled by SI-RNTI in Type0-PDCCH common search space in CORESET 0 and in any common search space other than Type0-PDCCH common search space). The virtual resource blocks in the region $j \in \{0, 1, \ldots, N_{bundle} - 2\}$ are mapped to the physical resource blocks such that virtual resource block bundle $N_{bundle} - 1$ is mapped to physical resource block bundle $N_{bundle} - 1$ and virtual resource block bundle $j \in \{0, 1, \ldots, N_{bundle} - 2\}$ is mapped to physical resource block bundle $f(j)$ where $f(j) = rC + c; j = cR + r; r = 0, 1, \ldots, R - 1; c = 0, 1, \ldots, C - 1; R = 2$; and $C = \lceil N_{bundle}/R \rceil$. If no bundle size is configured, the UE will assume $L_i = 2$ with a precoding resource block group (PRG) size of 4 (see Fig. 4.39).

4.1.6 CSI Measurement and Reporting and Beam Management

4.1.6.1 CSI Measurement and Reporting

In wireless communications, the channel state information refers to channel properties of a wireless communication link. This information describes how a signal propagates from the transmitter to the receiver and represents the combined effect of scattering, multipath fading, signal power attenuation with distance, etc. The knowledge of CSI at the transmitter and/or the receiver makes it possible to adapt data transmission to current channel conditions, which is crucial for achieving reliable and robust communication with high data rates in multi-antenna systems. The CSI is often required to be estimated at the receiver, and usually quantized and fed back to the transmitter. The downlink channel can be estimated from uplink reference signals in TDD systems under certain conditions due to reciprocity.

In general, the transmitter and receiver can observe different CSI. There are two types of CSI, that is, instantaneous CSI and statistical CSI. In instantaneous CSI or short-term CSI the current channel conditions are known, which can be interpreted as knowing the impulse response of a digital filter. This provides an opportunity to adapt the transmit signal to the impulse response and thereby to optimize the received signal for spatial multiplexing or to achieve low bit-error-rates. In statistical CSI or long-term CSI, the statistical characteristics or statistics of the channel are known. The latter information may include the type of fading distribution, the average channel gain, the line-of-sight component, and the spatial correlation. Similar to the instantaneous CSI, this information can be used for optimization of transmission parameters. The CSI estimation accuracy is practically limited by how fast the channel conditions are varying. In fast-fading channels where the channel conditions may vary rapidly during transmission of a single information symbol, only statistical CSI is reasonable. On the other hand, in slow-fading scenarios, the instantaneous CSI can be estimated with reasonable precision and used for transmission adaptation for a period of time before becoming obsolete. In practical scenarios, the available CSI is often manifested as instantaneous CSI with some estimation/quantization error combined with some statistical information.

To support diverse use cases, NR features a highly flexible and unified CSI framework, in which there is reduced coupling between CSI measurements, CSI reporting and the actual downlink transmission compared to LTE. The CSI framework can be seen as a toolbox, where different CSI reporting settings and CSI-RS resource settings for channel and interference measurements can be selected, so that they correspond to the antenna configuration and transmission scheme in use such that the CSI reports on different beams can be dynamically triggered. The framework also supports more advanced schemes such as multi-point transmission and coordination. The control and data transmissions follow a self-contained principle, where all information required to decode the transmission (such as accompanying DM-RS) is contained within the transmission itself. As a result, the network can seamlessly change the transmission point or beam as the UE moves in the network. The CSI-RS reference signals are used for CSI acquisition and beam management. The CSI-RS resources for a UE are configured by RRC information elements and can be dynamically activated/deactivated via MAC control elements or DCI [57].

The configuration and use of CSI-RS in NR can be defined via the CSI framework. As shown in Fig. 4.40, the basic units of CSI framework in NR are CSI reporting setting and CSI resource setting. The CSI reporting setting is linked to M resource settings for channel and interference measurements (CM and IM), where $M = 1$ indicates resource setting for channel measurement and beam management; $M = 2$ indicates resource settings for channel measurements and CSI-interference measurement (CSI-IM) or NZP CSI-RS for interference measurement; and $M = 3$ is an indication for resource settings for channel measurements and two resource settings for CSI-IM and NZP CSI-RS–based interference measurement.

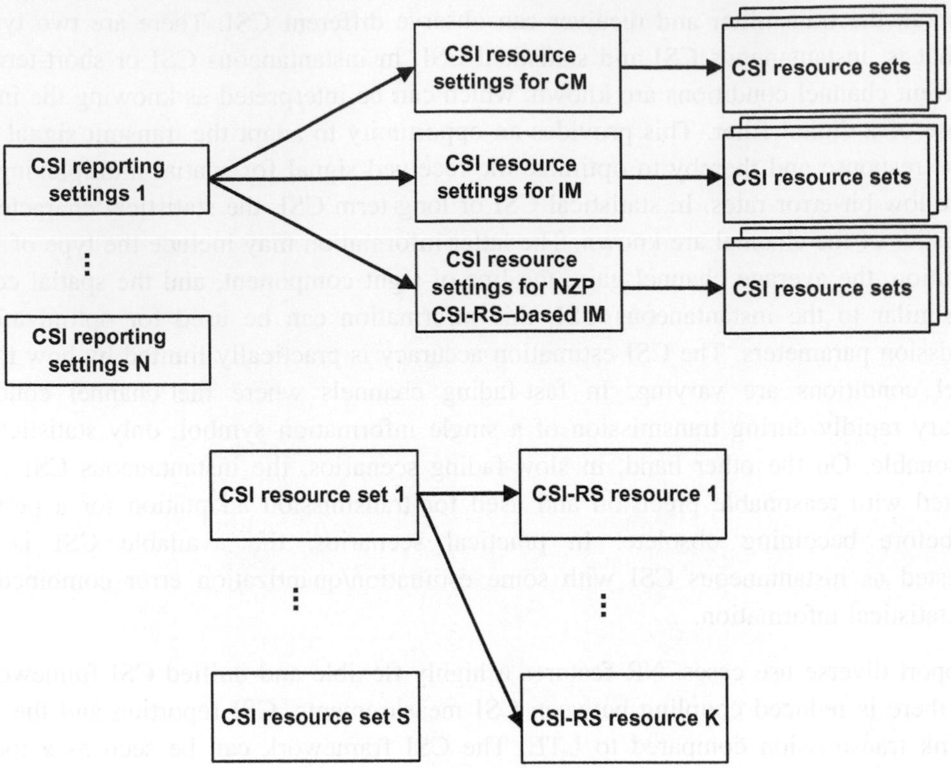

Figure 4.40
CSI framework in NR.

The above-mentioned resource settings are linked to *S* resource sets each resource set comprises SS/PBCH block resources for beam management and is linked to CSI-RS resources [9].

The time and frequency resources that can be used by the UE to report CSI are controlled by the gNB. The CSI may consist of CQI, PMI, CSI-RS resource indicator (CRI), SS block resource indicator, layer indication (LI), rank indicator (RI), and/or and L1-RSRP measurements. For CQI, PMI, CRI, LI, RI, L1-RSRP, the UE is configured via RRC signaling with more than one *CSI-ReportConfig* reporting settings, *CSI-ResourceConfig* resource settings, and one or two lists of trigger states, indicating the resource set IDs for channel and optionally for interference measurement. Each trigger state contains an associated *CSI-ReportConfig* [9].

Each reporting setting *CSI-ReportConfig* is associated with a single downlink BWP and contains the reported parameter(s) for one CSI reporting band including CSI Type-I or II, codebook configuration comprising codebook subset restriction, time-domain behavior, frequency granularity for CQI and PMI, measurement restriction configurations, LI, reported L1-RSRP parameter(s), CRI, and the SSB resource indicator. The time-domain behavior of the *CSI-ReportConfig* is determined by RRC signaling and can be set to aperiodic, semi-persistent, or periodic. For periodic and semi-persistent CSI reporting, the configured periodicity and slot offset applies in the numerology of the uplink BWP in which the CSI report is configured to be transmitted. The higher layer parameter *ReportQuantity* identifies the CSI-related or L1-RSRP-related quantities to report. Another RRC parameter indicates the reporting granularity in the frequency domain including the CSI reporting band and whether PMI/CQI reporting is wideband or subband. The *CSI-ReportConfig* can also contain *CodebookConfig*, which contains configuration parameters for Type-I or Type-II CSI including codebook subset restriction, and configurations of group-based reporting [9].

Each CSI resource setting contains a configuration of more than one CSI resource sets, each consisting of CSI-RS resources (either NZP CSI-RS or CSI-IM) and SS/PBCH block resources used for L1-RSRP computation. Each CSI resource setting located in the downlink BWP is defined by RRC signaling, and all CSI resource settings are linked to a CSI report setting within the same downlink BWP. The reporting configuration for CSI can be aperiodic (using PUSCH), periodic (using PUCCH), or semi-persistent (using PUCCH and DCI activated PUSCH). The CSI-RS resources can be periodic, semipersistent, or aperiodic. The supported combinations of CSI reporting configurations and CSI-RS resource in NR are shown in Tables 4.11 and 4.12. If interference measurement is performed using CSI-IM, each CSI-RS resource for CM is resource-wise associated with a CSI-IM resource based on the ordering of the CSI-RS resource and CSI-IM resource in the corresponding resource sets. The number of CSI-RS resources for channel measurement equals to the number of CSI-IM resources [9].

The CSI reports are used to provide the gNB with an estimate of the downlink communication channel observed by the UE in order to assist channel-dependent scheduling. The new

Table 4.11: Triggering/activation of CSI reporting for CSI-RS configurations [9].

CSI-RS Configuration	Periodic CSI Reporting	Semi-persistent CSI Reporting	Aperiodic CSI Reporting
Periodic CSI-RS	No dynamic triggering/ activation	For reporting on PUCCH, the UE receives an activation command and for reporting on PUSCH, the UE receives triggering on DCI	Triggered by DCI or activation command
Semi-persistent CSI-RS	Not supported	For reporting on PUCCH, the UE receives an activation command, whereas for reporting on PUSCH, the UE receives the trigger on DCI	Triggered by DCI or activation command
Aperiodic CSI-RS	Not supported	Not supported	Triggered by DCI or activation command

Table 4.12: Major components of NR CSI framework [9].

CSI Report Settings	CSI Resource Settings	CSI Trigger States
It defines what CSI to report and when to report it. • Quantities to report: CSI related or L1-RSRP related • Time-domain behavior: aperiodic, semi-persistent, periodic • Frequency-domain granularity: reporting band, wideband, subband • Time-domain restrictions: For channel and interference measurements • Codebook configuration parameters: Type-I and Type-II	It defines what signals to use to compute CSI. • A resource setting configures more than one CSI resource sets where each CSI resource set consists of CSI-RS resources (either NZP CSI-RS or CSI-IM); and SS/PBCH Block Resources that are used for L1-RSRP calculation • Time-domain behavior: aperiodic, semi-persistent, periodic as well as periodicity and slot offset *Note: The number of CSI-RS Resource Sets is limited to one, if CSI Resource Setting is periodic or semi-persistent*	It associates "what CSI to report and when to report it" with "what signals to use to compute CSI." • Links report settings with resource settings • Contains the list of associated *CSI-ReportConfig*

radio supports analog beamforming and high-resolution CSI feedback through beam management, where a UE measures a set of analog beams for each digital port and reports the beam quality. The gNB then assigns a number of analog beams to the UE. As the downlink channel experienced by the UE varies, the gNB can change this assignment when the link associated with an assigned beam deteriorates. While beam management is especially instrumental in above 6 GHz frequency bands, it can also be applied to sub-6 GHz bands. Furthermore, NR supports a modular and scalable CSI framework, where high-resolution spatial channel information is provided via two-stage precoding. The first stage involves the choice of a basis subset, and the second stage incorporates a set of coefficients for approximating the channel eigenvector with a linear combination of the basis subsets. It must be noted that while beam management and CSI acquisition can be independently operated, they can be used together to support mobile UEs.

For UEs in RRC_CONNECTED state, in addition to the SS block, UE-specific CSI-RS can be configured in order to improve the quality of UE measurements and to provide better user-centric mobility experience. For example, in high-frequency bands, narrow-beam CSI-RS can be configured for the UEs at the edge of the cell in order to achieve better signal-to-interference-plus-noise ratio (SINR) range and measurement accuracy. As shown in Fig. 4.41,

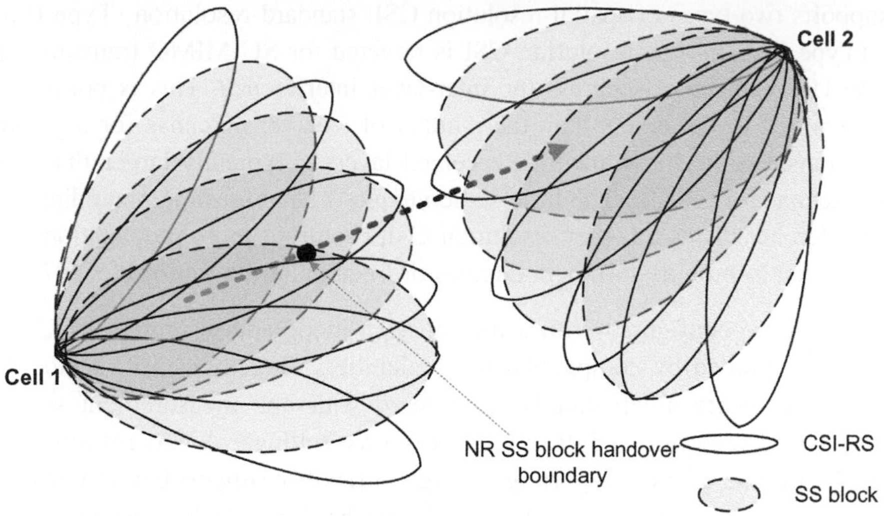

Figure 4.41
Configurable CSI-RS for downlink mobility measurements [72].

assuming the same energy per resource element (EPRE) is applied to CSI-RS and SSB resources for transmission, narrow-beam CSI-RS measurement can provide better SINR range which can improve RSRP measurement accuracy compared to wide-beam SSB measurements. The CSI-RS properties of the serving and neighboring cells for the mobility measurements can include NR cell ID, slot configuration used to obtain the slot offset for CSI-RS and the periodicity, for example, 5, 10, 20, 40 ms, configurable measurement bandwidth of CSI-RS, configurable parameter for CSI-RS scrambling sequence, configurable numerologies, and association between CSI-RS for mobility measurement and SSB, such as spatial QCL[8] information. The above CSI-RS properties are signaled to the UE via dedicated RRC signaling [72].

[8] Two antenna ports are said to be quasi co-located, if the properties of the channel over which a symbol on one antenna port is transmitted can be inferred from the channel over which a symbol on the other antenna port is transmitted. The QCL supports beam management (spatial parameter), frequency/timing offset estimation (Doppler/delay), and RRM measurements (average gain). The reference signal set contains a reference to either one or two downlink reference signals and an associated quasi co-location type (QCL-Type) for each one configured by an RRC parameter. The quasi co-location relationship is configured by the RRC parameter *qcl-Type1* for the first downlink reference signal, and *qcl-Type2* for the second downlink reference signal (if configured). The quasi co-location types corresponding to each reference signal are given by the RRC parameter *qcl-Type* in *QCL-Info* and may take one of the following values [9]:

- QCL-TypeA': {Doppler shift, Doppler spread, average delay, delay spread}
- QCL-TypeB': {Doppler shift, Doppler spread}
- QCL-TypeC': {average delay, Doppler shift}
- QCL-TypeD': {Spatial RX parameter}

The NR supports two types of spatial-resolution CSI: standard-resolution (Type I) and high--resolution (Type II). The low-resolution CSI is targeted for SU-MIMO transmission since it relies on the UE receiver to suppress the inter-layer interference. This is possible since the number of received layers is less than the number of receiver antennas for a given UE. For MU-MIMO transmission, the number of received layers is typically larger than the number of receive antennas for the UE. The base station exploits beamforming/precoding to suppress inter-UE interference. Thus a higher resolution CSI, capturing more propagation paths of the channel, is needed to provide sufficient degrees of freedom at the transmitter [57].

In LTE, the UEs are configured with a transmission mode and a number of CSI reporting modes which are limited by complexity and scalability, whereas in NR, a modular framework is specified where a UE can be configured with one measurement setting, which includes $N \geq 1$ CSI reporting and $M \geq 1$ CSI resource settings. A CSI resource setting can be associated with one or more reporting settings to flexibly support beam management and CSI acquisition, resulting in $L \geq 1$ links. A UE can be dynamically assigned one or more reporting settings or links to generate the desired CSI report, which may include CRI, which is used to indicate a preferred CSI-RS resource from a configured set since different CSI-RS resources in the set can be differently precoded, rank indicator (RI), CQI, and PMI. The CRI, RI, CQI, and PMI are associated with resource selection (when a UE measures multiple CSI-RS resources), the number of dominant downlink channel directions, sustained spectral efficiency or related SINR values, and the dominant channel directions chosen from a codebook of vectors or matrices. Since CSI requirements for different operational modes are different, the CSI reporting settings can include different CSI components for CSI acquisition (see Fig. 4.40 and Table 4.12).

The UE measures the spatial channel between itself and the serving base station using the CSI-RS transmitted from the gNB transmit antenna ports in order to generate a CSI report. The UE then calculates the CSI-related metrics and reports the CSI to the gNB. Using the reported CSIs from all UEs, the gNB performs link adaptation and scheduling. The goal of CSI measurement and reporting is to obtain an approximation of the CSI. This can be achieved when the reported PMI accurately represents the dominant channel eigenvector(s), thereby enabling accurate beamforming.

The standard-resolution (Type I) CSI utilizes a dual-stage codebook with precoding matrix $\mathbf{W} = \mathbf{W_1 W_2}$ incorporating a wideband $\mathbf{W_1}$ matrix that is common for all subbands, capturing long-term channel characteristics, and a subband $\mathbf{W_2}$ matrix representing fast fading properties of the channel. In this context a subband comprises multiple consecutive resource blocks. Type I codebooks support up to rank 8, that is, the rank indicator $RI \in \{1, 2, \ldots, 8\}$. Designed for N_{panel} panels of dual-polarized arrays, the $\mathbf{W_1}$ matrix factor is constructed from $2N_{panel}$ blocks of two-dimensional DFT matrices. The $\mathbf{W_2}$ matrix selects a subset of DFT vectors from $\mathbf{W_1}$ and applies phase shifts (taken from the phase shift keying alphabet) across panels and polarizations. For Type I, when $RI = 1$, the selected subset includes only one DFT vector, and when $RI > 1$, the subset includes multiple DFT vectors to generate orthogonal DFT beams. The phase shifts are introduced across two polarizations since the associated channels tend to be uncorrelated. This is also utilized for multi-panel arrays since the spacing between the last element of a panel and the first element of the next panel is different from the inter-element spacing within a panel. In addition, the panels may not be sufficiently phase- and/or timing-calibrated [57].

For high-spatial-resolution (Type II) CSI, feedback of up to two layers, that is, $RI \in \{1, 2\}$ is supported with a linear combination codebook. The codebook resolution is sufficiently high to facilitate sufficiently accurate approximation of the downlink channel. In this scheme the UE reports a PMI that represents a linear combination of multiple beams, as shown in Fig. 4.42. Similar to Type I, Type II employs a dual-stage $\mathbf{W} = \mathbf{W}_1\mathbf{W}_2$ codebook wherein $\mathbf{W}_1$ is a wideband and $\mathbf{W}_2$ is a subband precoder. However, unlike Type I, the recommended precoding matrix from Type II CSI is non-constant modulus since different precoder elements can have different magnitudes [57].

For Type II CSI and $RI \in \{1, 2\}$, the $\mathbf{W}_1$ matrix selects a subset of DFT vectors of size $L \in \{2, 3, 4\}$ which serves as a basis set for linear combination performed by $\mathbf{W}_2$. This subset selection is common across two polarizations and, for $RI = 2$, two transmission layers. The linear combination is performed per subband as well as independently across polarizations and layers. To reduce the feedback overhead for $\mathbf{W}_2$, some partial information pertaining to linear combination such as the strongest of $2L$ linear combination coefficients and $2L - 1$ wideband reference amplitudes for subband differential encoding of the linear combination coefficients in $\mathbf{W}_2$ is also included in $\mathbf{W}_1$. Therefore, the amplitude component of the linear combination coefficients comprises wideband and subband components. The phase component is per subband and configurable as QPSK or 8-PSK. Due to large degrees of freedom offered by Type II CSI, the number of precoder hypotheses is large. However, exhaustive codebook search, which is prohibitively complex, is not needed. Due to high spatial resolution, the precoder can be efficiently determined by performing scalar quantization of each of the channel eigenvector coefficients. Since Type II CSI is configurable in terms of its basis set size $L \in \{2, 3, 4\}$, the amplitude frequency granularity, that is, wideband-only or wideband + subband, and phase shift (QPSK or 8-PSK), a range of performance-overhead trade-offs would be possible [57].

There are two subtypes of Type I CSI that are referred to as Type I single-panel CSI and Type I multi-panel CSI, corresponding to different codebooks. These codebooks are designed assuming different antenna configurations on the gNB transmitter. The codebooks for Type I single-panel CSI are designed assuming a single antenna panel with $N_H \times N_V$ cross-polarized antenna elements. In general, the precoder matrix $\mathbf{W}$ for Type I single-panel CSI can be constructed as the product of two matrices $\mathbf{W}_1$ and $\mathbf{W}_2$ where the information

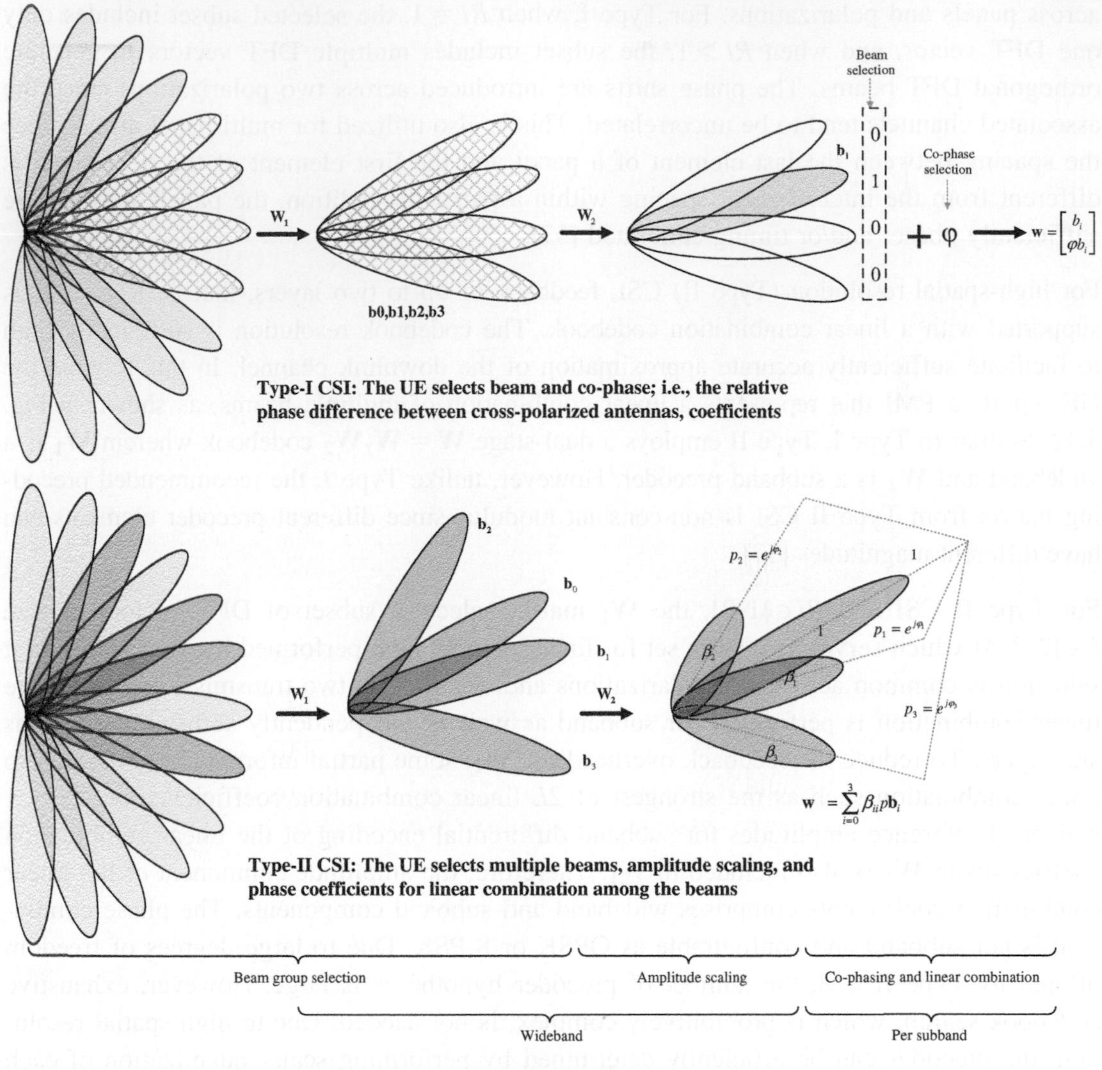

Figure 4.42
Illustration of Type-I and Type-II CSI in NR [57,69].

about the selected $\mathbf{W_1}$ and $\mathbf{W_2}$ is reported separately in different parts of the PMI. The matrix $\mathbf{W_1}$ captures long-term frequency-independent characteristic of the channel. A single $\mathbf{W_1}$ is selected and reported for the entire reporting bandwidth (wideband feedback). In contrast, the matrix $\mathbf{W_2}$ encompasses short-term frequency-dependent characteristic of the channel. Thus the precoder matrix can be selected and reported on a subband basis where a subband covers a fraction of the overall reporting bandwidth. Alternatively, the device may decide not to report $\mathbf{W_2}$ when subsequently selecting CQI. In that case, it should assume that the network randomly selects $\mathbf{W_2}$ on a per physical resource block group basis.

Note that this does not impose any restrictions on the actual precoding applied at the gNB side, rather it is only about the assumptions that the device would make when selecting CQI. The matrix $\mathbf{W_1}$ can be considered as defining a beam or a group of beams pointing toward a specific direction. More specifically, the matrix $\mathbf{W_1}$ can be written as $\mathbf{W_1} = \begin{bmatrix} \mathbf{b} & \mathbf{0} \\ \mathbf{0} & \mathbf{b} \end{bmatrix}$ where each column of the matrix $\mathbf{b}$ defines a beam. The 2×2 structure of the matrix corresponds to two polarizations. Note that, as the matrix $\mathbf{W_1}$ is assumed to represent long-term frequency-independent channel characteristics, the same beam direction can be assumed to fit both polarization directions. Selecting matrix $\mathbf{W_1}$ or equivalently $\mathbf{b}$ can be seen as selecting a specific beam direction from a large set of possible beam directions defined by the full set of $\mathbf{W_1}$ matrices within the codebook. In the case of rank 1 or rank 2 transmissions, either a single beam or four adjacent beams are defined by the matrix $\mathbf{W_1}$ (see Fig. 4.42). In the case of four adjacent beams corresponding to four columns in matrix $\mathbf{b}$, matrix $\mathbf{W_2}$ would select the exact beam to be used for the transmission. Since $\mathbf{W_2}$ can be reported on a subband basis, it is possible to adjust the beam direction per subband. In addition, $\mathbf{W_2}$ provides co-phasing between the two polarizations. In the case where $\mathbf{W_1}$ only defines a single beam corresponding to $\mathbf{b}$ being a vector, matrix $\mathbf{W_2}$ would only provide co-phasing between the two polarizations. For transmission ranks $R > 2$ the matrix $\mathbf{W_1}$ defines N_{beams} orthogonal beams where $N_{beams} = \lceil R/2 \rceil$. The N_{beams} beams, together with the two polarization directions in each beam, are then used for transmission of the R layers, with the matrix $\mathbf{W_2}$ only providing co-phasing between the two polarizations. The NR supports transmission of up to eight layers to the same device [14].

In contrast to single-panel CSI, codebooks for Type I multi-panel CSI are designed assuming the joint use of multiple antenna panels at the network side considering that it may be difficult to ensure coherence between transmissions from different panels. More specifically, the design of the multi-panel codebooks assumes an antenna configuration with two or four two-dimensional panels, each with $N_H \times N_V$ cross-polarized antenna elements. The operation principles of Type I multi-panel and single-panel CSI are similar, except that the matrix $\mathbf{W_1}$ defines one beam per polarization and panel, whereas matrix $\mathbf{W_2}$ provides per-subband co-phasing between polarizations as well as panels. The Type I multi-panel CSI supports spatial multiplexing with up to four layers.

Type II CSI provides channel information with significantly higher spatial granularity compared to Type I CSI. Similar to Type I CSI, Type II CSI is based on wideband selection and reporting of beams from a large set of beams. However, while Type I CSI selects and reports a single beam, Type II CSI may select and report up to four orthogonal beams. For each selected beam and each of the two polarizations, the reported PMI then provides an amplitude value (partly wideband and partly subband) and a phase value (subband). This allows constructing a more detailed model of the channel, capturing the main rays and their respective amplitudes and phases. At the network side, the PMI received from multiple devices can be used to identify a set of devices with which transmission can be done simultaneously on a set of time/frequency resources, that is, MU-MIMO, and what precoder to use for each transmission. Since Type II CSI targets MU-MIMO scenarios, transmission is limited to a maximum of two layers per device [14].

In order to reduce the overhead due to transmission of reference signals, 3GPP NR has moved away from the notion of continuously transmitted wideband cell-specific reference signals and instead has defined more flexible and configurable UE-specific reference signals transmitted on-demand. This transition has resulted in the definition of new UE measurement procedures based on the CSI-RSs. For this purpose, new metrics have been defined that will be explained in detail in the following:

- *Synchronization Signal-Reference Signal Received Power (SS-RSRP)* is defined as the linear average over the power contributions [in (Watts)] of the resource elements that carry secondary synchronization signals. The measurement time resource(s) for SS-RSRP are confined within SS/PBCH block measurement time configuration (SMTC)[9] window size. If SS-RSRP is used for L1-RSRP as configured by reporting configurations, the measurement time resource(s) restriction by SMTC window size is not applicable. For SS-RSRP calculation, DM-RSs for PBCH and, if indicated by higher layers, CSI-RSs in addition to the secondary synchronization signals may be used. The SS-RSRP using DM-RS for PBCH or CSI-RS are measured by linear averaging over the power contributions of the resource elements that carry corresponding reference signals by considering the power scaling for the reference signals. If SS-RSRP is not used for L1-RSRP, the additional use of CSI-RSs for SS-RSRP determination is not applicable. The SS-RSRP is measured only over the reference signals corresponding to SS/PBCH blocks with the same SS/PBCH block index and the same physical-layer cell identity. If SS-RSRP is not used for L1-RSRP and higher layers indicate certain SS/PBCH blocks for performing SS-RSRP measurements, then SS-RSRP is measured only from the indicated set of SS/PBCH block(s). For frequency range 1, the reference point for the SS-RSRP is the antenna connector of the UE. For frequency range 2; however, the SS-RSRP is measured based on the combined signal from antenna elements corresponding to a given receiver branch. For frequency ranges 1 and 2 if receiver diversity is used by the UE, the reported SS-RSRP value must not be lower than the corresponding SS-RSRP of any of the individual receiver branches [10].

[9] SS block-based RRM measurement timing configuration or SMTC is the measurement window periodicity/duration/offset information for UE RRM measurement per carrier frequency. For intra-frequency connected mode measurement, up to two measurement window periodicities can be configured. For the idle mode measurements, a single SMTC is configured per carrier frequency. For inter-frequency connected mode measurements, a single SMTC is configured per carrier frequency.

Table 4.13: NR carrier received signal strength indicator measurement symbols [10].

OFDM Signal Indication *SS-RSSI-MeasurementSymbolConfig*	OFDM Symbol Indices
0	{0,1}
1	{0,1,2,...,10,11}
2	{0,1,2,...,5}
3	{0,1,2,...,7}

- *Secondary Synchronization Signal-Reference Signal Received Quality (SS-RSRQ)* is defined as the ratio of $N_{RB} \times SS\_RSRP/NR\_carrier\_RSSI$, where N_{RB} is the number of resource blocks in the received signal strength indicator (RSSI) measurement bandwidth of the NR carrier. The latter measurements are conducted over the same set of resource blocks. The NR carrier RSSI, *NR_carrier_RSSI*, comprises the linear average of the total received power [in (Watts)] observed over certain OFDM symbols within measurement time resource(s), in the measurement bandwidth, over N_{RB} number of resource blocks from all sources, including co-channel serving and non-serving cells, adjacent channel interference, and thermal noise. The measurement time resource(s) for *NR_carrier_RSSI* are confined within SMTC window duration. If indicated by higher layers, for a half-frame with SS/PBCH blocks, *NR_carrier_RSSI* is measured over OFDM symbols of the indicated slots shown in Table 4.13; otherwise, if measurement gap is not used, *NR_carrier_RSSI* is measured over OFDM symbols within SMTC window duration and, if measurement gap is used, *NR_carrier_RSSI* is measured over OFDM symbols corresponding to overlapped time span between SMTC window duration and minimum measurement time within the measurement gap [10]. If higher layers indicate certain SS/PBCH blocks for performing SS-RSRQ measurements, then SS-RSRP is measured only over the indicated set of SS/PBCH block(s). For frequency range 1, the reference point for the SS-RSRQ is the antenna connector of the UE. For frequency range 2, *NR_carrier_RSSI* is measured based on the combined signal from antenna elements corresponding to a given receiver branch, where the combining function for *NR_carrier_RSSI* is the same as the one used for SS-RSRP measurements. For frequency range 1 and 2, if receiver diversity is used by the UE, the reported SS-RSRQ value must not be lower than the corresponding SS-RSRQ of any of the individual receiver branches [10].
- *CSI-Reference Signal Received Power (CSI-RSRP)* is defined as the linear average over the power contributions (in Watts) of the resource elements that carry CSI-RSs configured for RSRP measurements within the measurement frequency region in the predefined CSI-RS occasions. For CSI-RSRP calculation, the CSI-RSs transmitted on antenna port 3000 or antenna ports 3000, 3001 are used. For frequency range 1, the reference point for the CSI-RSRP is the antenna connector of the UE. For frequency range 2;

however, the CSI-RSRP is measured based on the combined signal from antenna elements corresponding to a given receiver branch. For frequency ranges 1 and 2, if receive diversity is used by the UE, the reported CSI-RSRP value must not be lower than the corresponding CSI-RSRP of any of the individual receiver branches [10].

- *CSI-Reference Signal Received Quality (CSI-RSRQ)* is defined as the ratio of $N_{RB} \times CSI\_RSRP/CSI\_RSSI$, where N_{RB} denotes the number of resource blocks used in the CSI-RSSI measurement bandwidth. The latter measurements are conducted over the same set of resource blocks.
- *CSI-Received Signal Strength Indicator (CSI-RSSI)* is the linear average of the total received power (in Watts) observed over the OFDM symbols within measurement time resource(s), in the measurement bandwidth, over N_{RB} number of resource blocks from all sources, including co-channel serving and non-serving cells, adjacent channel interference, and thermal noise. The measurement time resource(s) for CSI-RSSI corresponds to OFDM symbols containing configured CSI-RS occasions. For CSI-RSRQ calculation, CSI-RSs transmitted on antenna port 3000 are used. For frequency range 1, the reference point for the CSI-RSRQ is the antenna connector of the UE, whereas for frequency range 2, CSI-RSSI is measured based on the combined signal from antenna elements corresponding to a given receiver branch, where the combining for CSI-RSSI is the same as the one used for CSI-RSRP measurements. For frequency ranges 1 and 2, if receive diversity is used by the UE, the reported CSI-RSRQ value must not be lower than the corresponding CSI-RSRQ of any of the individual receiver branches [10].
- *Synchronization Signal-Signal-to-Interference Plus Noise Ratio (SS-SINR)* is defined as the linear average over the power contribution (in Watts) of the resource elements carrying secondary synchronization signals divided by the linear average of the noise and interference power contribution (in Watts) over the resource elements carrying secondary synchronization signals within the same frequency region. The measurement time resource(s) for SS-SINR are confined within SMTC window duration. For SS-SINR calculation, the DM-RSs associated with PBCH in addition to the secondary synchronization signals may be used. If the RRC signaling identifies certain SS/PBCH blocks for conducting SS-SINR measurements, then SS-SINR is measured only over the set of SS/PBCH block(s) identified via signaling. For frequency range 1, the reference point for the SS-SINR is the antenna connector of the UE, whereas for frequency range 2, the SS-SINR is measured based on the combined signal from antenna elements corresponding to a given receiver branch. For frequency ranges 1 and 2, if receiver diversity is used by the UE, the reported SS-SINR value must not be lower than the corresponding SS-SINR of any of the individual receiver branches [10].
- *CSI-SINR* is defined as the linear average over the power contribution (in Watts) of the resource elements carrying CSI-RSs divided by the linear average of the noise plus interference power contributions (in Watts) over the resource elements carrying CSI-RSs within the same frequency region. For CSI-SINR calculation, the CSI-RSs

transmitted on antenna port 3000 are used. For frequency range 1, the reference point for the CSI-SINR is the antenna connector of the UE, whereas for frequency range 2, the CSI-SINR is measured based on the combined signal from antenna elements corresponding to a given receiver branch. For frequency ranges 1 and 2, if receive diversity is used by the UE, the reported CSI-SINR value must not be lower than the corresponding CSI-SINR of any of the individual receiver branches [10].

- *SRS-RSRP* is defined as linear average of the power contributions (in Watts) of the resource elements carrying the uplink SRS. The SRS-RSRP is measured over the configured resource elements within the measurement frequency region and over the time resources in the predefined measurement occasions. For frequency range 1, the reference point for the SRS-RSRP is the antenna connector of the UE. If receive diversity is used by the UE, the reported SRS-RSRP value must not be lower than the corresponding SRS-RSRP of any of the individual receiver branches [10].

The NR supports operating frequencies in a wide range from 450 MHz to 52.6 GHz (and higher in the future releases). The main challenge in NR is to overcome the higher propagation loss and sensitivity to blockage in the above 6 GHz frequency bands. To overcome this issue, efficient usage of highly directional beamformed transmission and reception using a larger number of antenna elements is crucial at the gNB and the UE. To achieve large beamforming gain with reasonable implementation complexity, hybrid beamforming was found to be a suitable solution. The analog beams on each panel/subarray are adapted through phase shifters and/or amplitude scaling. Digital beamforming is adapted by applying different digital precoders across panels/subarrays. At the gNB, downlink transmission with analog beamforming, that is, transmitter beam pointing toward a certain direction can only cover a limited area due to its relatively narrow beam-width. Therefore, the gNB needs to utilize multiple transmit beams to cover the entire cell. Similarly, in the uplink the gNB needs to utilize multiple receive beams to receive the uplink transmissions from the entire cell.

4.1.6.2 Beam Management

The new radio provides a set of mechanisms by which the UEs and the gNB can establish highly directional transmission links, typically using large-scale phased arrays, to benefit from the resulting beamforming gain and to sustain an acceptable communication quality. Directional links, nevertheless, require fine alignment of the transmitter and receiver beams that can be only achieved through a set of procedures known as beam management. The beam management procedures are essential to perform a variety of radio access network functions including the initial access for idle users, which allows a mobile UE to establish a physical connection with a gNB and beam tracking for connected users, which enables beam adaptation schemes, or handover, path selection and radio link failure (RLF) recovery procedures. In LTE, these control procedures are performed using omnidirectional transmission, and beamforming or other directional transmissions can only be performed after a physical link (user-plane) is established. In certain conditions such as operation in the mmWave bands, it may be necessary to exploit high antenna gains even during initial access and in general for improving the control channel coverage. However, directionality can significantly delay the access procedures and make the performance more sensitive to the beam alignment.

The Rel-15 NR is designed to support analog beamforming in addition to digital precoding/beamforming. In high frequencies, analog beamforming may be beneficial from an implementation viewpoint despite the fact that analog beamforming may constrain the transmit/receive beam to be formed in one direction at a given time and further requires beam sweeping, where the same signal is repeated in multiple OFDM symbols but on different transmit beams. Beam sweeping would ensure that the signal can be transmitted with high directionality in order to cover the entire cell area. The NR has specified control/signaling schemes to support beam management procedures including an indication to the device to assist the selection of an appropriate receive beam. For large number of antennas, beams are narrow and beam tracking may fail; therefore, beam-recovery procedures have been defined where a device can trigger the beam-recovery procedure. A cell may have multiple transmission points each with beams and the beam-management procedures to allow UE-transparent mobility for seamless handover between the beams of different transmission points. In addition, uplink-centric and reciprocity-based beam management is also possible by utilizing uplink signals [14]. In some cases, a suitable transmit/receive beam pair for the downlink transmission will also be a suitable beam pair for the uplink transmission direction and vice versa. The NR refers to this as downlink/uplink beam correspondence, where it is sufficient to determine a suitable beam pair in one direction and use the same beam pair in the opposite direction. Since beam management is not intended to track fast-varying and frequency-selective channels, beam correspondence does not require that downlink and uplink transmissions to take place on the same carrier frequency; thus the concept of beam correspondence is also applicable to FDD systems in paired spectrum.

The beam management is defined as the process of acquiring and maintaining a set of beams, which are originated at the gNB and/or the UE and can be used for downlink and uplink transmission and reception. The beam management process comprises the following functions as shown in Fig. 4.43 [11]:

- *Beam sweeping:* Covering a spatial area with a set of beams transmitted and received according to prespecified intervals and directions. The measurement process is carried out with an exhaustive search, that is, both UE and the base station have a predefined codebook of directions (each identified by a beamforming vector) that cover the entire angular space and are used sequentially to transmit/receive synchronization and reference signals.

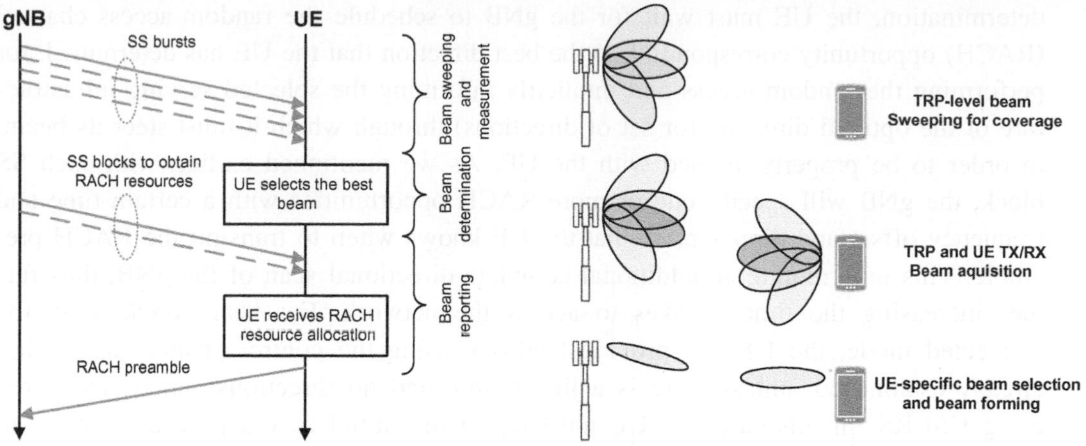

Figure 4.43
Signals and messages exchanged during downlink beam management procedure [69].

- *Beam measurement:* Evaluation of the quality of the received signal at the gNB or at the UE. Different metrics may be used for this purpose such as SNR, which is the average of the received power on synchronization signals divided by the noise power. The measurements for the initial access are based on the SS blocks. The tracking is done using the measurements conducted on the SS bursts and the CSI-RSs, which include a set of directions that may cover the entire set of available directions based on the UE requirements.
- *Beam determination:* Selection of the suitable beam or beams either at the gNB or at the UE, according to the measurements obtained via the beam measurement procedure. This process further allows the TRP(s) or the UEs to select their own transmit/receive beam(s). In beam determination, the gNB and the UE find a beam direction to ensure good radio link quality for the unicast control and data channel transmission. Once a link is established, the UE measures the link quality of multiple transmit and receive beam pairs and reports the measurement results to the gNB. Furthermore, the UE mobility, orientation, and channel blockage can change the radio link quality of the transmitted and received beam pairs. When the quality of the current beam pair degrades, the gNB and the UE can switch to another beam pair with better radio link quality. The gNB and the UE can monitor the quality of the current beam pair along with some other beam pairs and perform beam switching when necessary. When the gNB assigns a transmit beam to the UE via downlink control signaling, the beam indication procedure is used.
- *Beam reporting:* The procedure used by the UE to send beam quality and beam decision information to the gNB where the UE reports its observation of beamformed signal(s) based on beam measurement to the gNB. For the initial access, after beam

determination, the UE must wait for the gNB to schedule the random-access channel (RACH) opportunity corresponding to the best direction that the UE has determined, for performing the random access and implicitly informing the selected serving infrastructure of the optimal direction (or set of directions) through which it must steer its beam, in order to be properly aligned with the UE. As we mentioned earlier, with each SS block, the gNB will specify one or more RACH opportunities with a certain time and frequency offset and direction, so that the UE knows when to transmit the RACH preamble. This may require an additional complete directional scan of the gNB, thus further increasing the time it takes to access the network. For beam tracking in the connected mode, the UE can provide feedback using the control channel that it has already established, unless there is a link failure and no directions can be recovered using CSI-RS. In this case, the UE must repeat the initial access procedure or try to recover the link using the SS block bursts while the user experiences a service unavailability.

- *Beam switching and recovery:* Beam recovery involves a procedure when the link between the gNB and the UE can no longer be maintained and needs to be reestablished.

These procedures are periodically repeated to update the optimal transmitter and receiver beam pair over time. There are two network deployment scenarios for NR, that is, non-standalone and standalone, which can affect the way these procedures are performed. In non-standalone scenario, an NR gNB uses an LTE cell as an anchor for the control-plane management and mobile terminals exploit multi-connectivity to maintain multiple connections to different cells so that any link failure can be overcome by switching data paths. However, such an option may not be available in standalone deployments. The beam management procedure is further illustrated in Fig. 4.44.

The beam management operation in general is based on the control messages which are periodically exchanged between transmit and receive nodes. The reference signals used for beam management depend on the state of the UE. In idle mode, the PSS, SSS, and PBCH DM-RS are used, whereas in the connected mode, the CSI-RS and SRS are used in the downlink and uplink, respectively. A radio connection between a gNB with $N_{gNB-beam}$ analog beams and a UE with $N_{UE-beam}$ analog beams has a total of $N_{gNB-beam} \times N_{UE-beam}$ TX-RX beam pairs. Given that the number of TX/RX beams is typically large in mmWave bands to achieve sufficient coverage, efficient beam measurement and reporting procedures are important to ensure minimal overhead and UE complexity. A gNB capable of transmitting $N_{gNB-beam}$ analog beams can configure up to N_{RS} reference signals for beam measurement. Each RS is beamformed with its associated analog beam pointing in a particular direction. The analog beam associated with each reference signal may also be kept fixed over certain time intervals to allow the UE to test different RX beams for a given TX beam. In this manner, up to $N_{gNB-beam} \times N_{UE-beam}$ beam pairs can be measured.

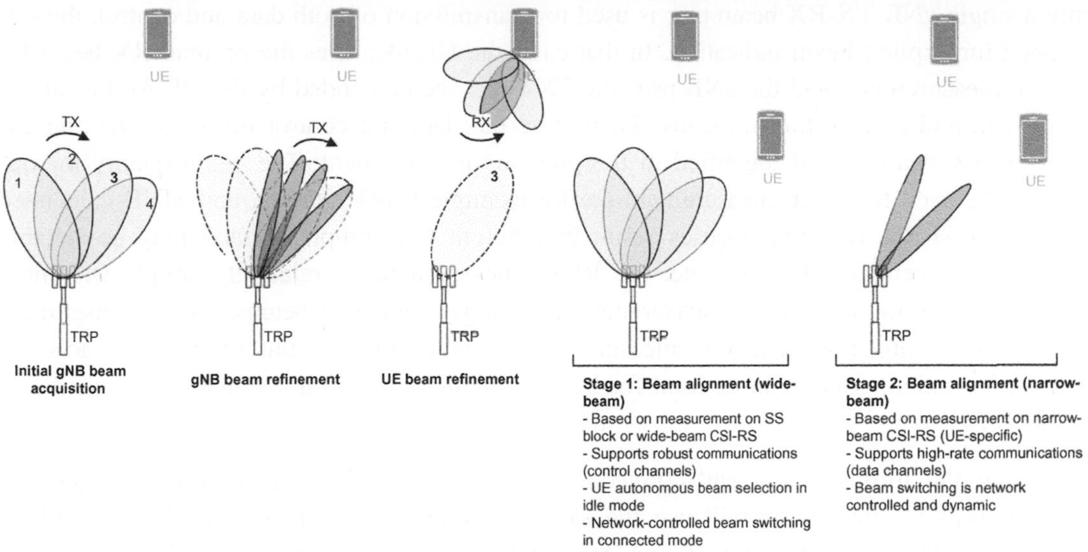

Figure 4.44
Illustration of beam management procedure [26].

The beamformed CSI-RS used for beam management can be transmitted either aperiodically or periodically. When the system is underloaded, beam sweeping across a small number of TX beams over a narrow angular area on an aperiodic basis for a given UE is sufficient. When the system loading increases, it would be more efficient to perform a periodic sweep over a wider angular area covering a larger number of UEs. Despite the fact that $N_{gNB-beam} \times N_{UE-beam}$ beam pairs correspond to $N_{gNB-beam} \times N_{UE-beam}$ beam qualities along with their pair indicators, not all $N_{gNB-beam} \times N_{UE-beam}$ beam qualities need to be reported, because for each TX beam, only the measurement quality of one beam pair, that is, the optimal RX beam for the given TX beam, needs to be reported. The optimal RX beam is known at the UE based on the beam measurement, which is not reported. In the subsequent data or control transmission to the UE, the gNB indicates the index of the selected TX beam to the UE. The UE can use the latest optimal RX beam of the indicated TX beam, which is stored in the UE local memory upon beam measurement, for the purpose of reception. Moreover, within the $N_{gNB-beam}$ candidate TX beams, the qualities and/or indices of $1 \leq n_{beam} \leq N_{gNB-beam}$ TX beams can be reported. If $n_{beam} = 1$, the UE may report the optimal TX beam index without the associated beam quality as a recommendation to the gNB for downlink beamforming. If $n_{beam} > 1$, the UE can report the indices of n_{beam} selected TX beams, for example, the best n_{beam} beams, along with their measured relative/absolute qualities to the gNB. The gNB compares the qualities of the reported TX beams and selects one TX beam for downlink transmission. The NR also supports lower complexity beam management procedures that do not require explicit indication of the selected TX beam to the UE. If the UE only reports the best TX beam index and only a single gNB TX-RX beam pair is used for transmission of both data and control, there is no need for explicit beam indication. In that case, the UE identifies the optimal RX beam for a given measurement, and the gNB uses the TX beam recommended by the UE for the subsequent data and control transmissions. To receive the data and control transmissions, the UE uses the RX beam that it identified in the previous measurement. The beam quality metrics that can be used for beam measurement are for example RSRP, RSRQ, and SINR calculated based on CSI-RS. Different metrics result in different UE complexity and may be suitable for different scenarios, for example, the RSRP measurement is relatively simple and more power efficient and allows fast measurement of a large number of beams, which is useful for initial beam acquisition. The CSI measurement is more complex but offers more accurate beam information, which can be used for beam refinement within a small group of candidate beams [57].

Following the initial access and connection set up, the device can assume that network transmissions to the device will use the same transmit beam that was used for SS block acquisition. Therefore, the device can assume that the receive beam which was used to acquire the SS block will also be a suitable beam for the reception of subsequent downlink transmissions. Similarly, subsequent uplink transmissions would use the same beam that was used for the random-access preamble transmission, implying that the network can assume that the uplink receive beam established at the initial access will remain valid.

In LTE, a UE continuously performs radio link monitoring of the channel quality of its serving cell to ensure sufficient coverage of control channels. If the link quality is considered poor, the UE declares RLF and triggers a higher layer connection reestablishment procedure, resulting in cell reselection. For a large number of antennas, the UE-specific beams are narrow, and beam tracking can fail, for example, when a moving object blocks the LoS path to the UE. This event is regarded as beam failure in NR. In this case declaring RLF and performing cell reselection is unnecessary since another beam from the same cell can be used to cover the UE. This physical layer procedure is an example of beam recovery, in which the UE continuously monitors a UE-specific periodic reference signal associated with the TX beam with which a control channel is transmitted. If the measured beam quality deteriorates, the UE declares beam failure and proceeds to identify an alternative candidate TX beam, selected from a set of UE-common TX beams used for the periodic beam sweeping of initial access signals. These beams are typically wider compared to UE-specific data beams. As a part of this TX beam determination, the UE also determines an appropriate RX beam to receive the indicated TX beam. When a new beam is detected, the UE transmits a beam recovery request message using a preconfigured uplink resource (which includes an identifier of the new beam) to the serving cell. The network then transmits a recovery response to the UE. If the response is successfully received by the UE, the beam recovery procedure is successful, and a new beam pair is established; otherwise, the UE may perform additional beam recovery attempts, which upon failure would force the UE to initiate the RLF procedure, which includes cell reselection [57].

The beamforming weights for data transmission are typically obtained from codebook-based CSI reporting. Once the codebook is specified, the beamforming capability in azimuth and elevation dimension is restricted by the codebook size, parameters, and structure. Alternatively, non-codebook-based beamforming can be extended to enable a gNB to reuse beam management procedures for acquiring the beamforming weights. In particular, a UE is configured with $K > 1$ CSI-RS resources, each associated with a TX beam. The beamforming can be performed in either digital or analog domain. However, in lower carrier frequencies, digital beamforming may be predominantly utilized. The UE measures the configured K CSI-RS resources and selects $N_{CSI} \leq K$ CSI-RS resources based on their respective qualities and reports the N_{CSI} CRIs. The actual CSI (e.g., CQI/PMI/RI) measured from the selected CSI-RS resources are also reported together with CRIs. As such, beam reporting and CSI reporting can be combined to acquire CSI. When a certain degree of channel reciprocity is achievable, beamforming weights can be derived at the gNB by measuring uplink signal(s) transmitted from the UE. The CSI reporting from the UE can also be utilized by the gNB to determine the beamforming weights [57].

To enable analog beamforming at the UE receiver, different reference signals within the resource set should be transmitted in different symbols, allowing the receiver-side beam to sweep over the set of reference signals. At the same time, the device can assume that different reference signals in the resource set are transmitted using the same spatial filter or alternatively the same transmit beam. In general, a configured resource set includes a repetition flag that indicates whether a device can assume that all reference signals within the resource set are transmitted using the same spatial filter. For a resource set to be used for downlink receiver-side beam adjustment, the repetition flag should be set [9].

4.1.7 Channel Coding and Modulation Schemes

Channel coding is one of the areas where NR is taking a completely different approach from LTE. In NR, LDPC coding has replaced the turbo coding that was previously used for LTE PDSCH/PUSCH coding and polar codes have replaced the tail biting convolutional codes used previously for LTE PDCCH/PUCCH/PBCH coding, except for very small block lengths where repetition/block coding may be used. Turbo codes generally have a low encoding complexity and high decoding complexity, whereas LDPC codes have more complex encoding and less complex decoding algorithms. Considering eMBB use cases with large code block sizes and the code rates up to 8/9, turbo codes may not meet the implementation complexity required for the decoder. The LDPC codes, on the other hand, have relatively simple and practical decoding algorithms. The decoding is performed by iterative belief propagation (BP). The accuracy of decoding will be improved in each iteration, and the number of iterations is decided based on the requirement of the application, providing a trade-off between the bit error performance, latency, and complexity. In terms of latency, LDPC codes are parallel in nature, while turbo codes are serial in nature, allowing the LDPC codes to better support low latency applications than turbo codes. Furthermore, the bit error rate (BER) of the turbo codes have higher error floor compared to that of LDPC, and the LDPC matrix can be extended to lower rates than LTE turbo codes, achieving higher coding gains for low-rate applications targeting high reliability. As for the polar codes, they were introduced in 2009 and they are among the capacity-achieving codes with low encoding and decoding complexity. They provide full flexibility with very good performance with any code length and code rate without error floor, that is, they do not suffer a decrease in the slope of BLER versus SNR.

4.1.7.1 Principles of Polar Coding

In order to describe the polar encoding/decoding concept, let us review a few prerequisites. The symmetric capacity is defined as the highest possible rate that can be achieved when all of the input symbols to the channel are equiprobable. The mutual information of a binary-input discrete memoryless channel (symmetric capacity) with input alphabet $\mathbb{X} = \{0, 1\}$ is defined as follows [29]:

$$I(W) = \sum_{y \in \mathbb{Y}} \sum_{x \in \mathbb{X}} \frac{1}{2} W(y|x) \log_2 \frac{2W(y|x)}{W(y|0) + W(y|1)}$$

The symmetric capacity is equal to the Shannon capacity when the channel W is a symmetric channel. The Bhattacharyya parameter $Z(W)$ is the upper bound on the probability of a maximum likelihood decision error when transmitting 0 or 1 over the channel W. Thus the Bhattacharyya parameter $Z(W)$ is a channel reliability measure. The Bhattacharyya parameter can be calculated as follows:

$$Z(W) = \sum_{y \in \mathbb{Y}} \sqrt{W(y|0)W(y|1)}$$

The relationship between $I(W)$ and $Z(W)$ for any binary-input, discrete, memoryless channel W can be described as follows:

$$I(W) \geq \log \frac{2}{1 + Z(W)}, \quad I(W) \leq \sqrt{1 - Z(W)^2}$$

which means $I(W) = 1$ or $I(W) = 0$, if and only if $Z(W) = 0$ or $Z(W) = 1$, respectively.

Polar codes are the first type of forward error correction codes achieving the symmetric capacity for arbitrary binary-input discrete memoryless channel under low-complexity encoding and low-complexity successive cancelation (SC) decoding with order of $\mathcal{O}(N \log N)$ for infinite length codes. Polar codes are founded based on several concepts including channel polarization, code construction, polar encoding, which is a special case of the normal encoding process (i.e., more structural) and its decoding concept [29].

Channel polarization is the first phase of polar coding where N distinct channels are synthesized such that each of these channels is either completely noisy or completely noiseless, that is, strictly valid for infinite code length N. The measure of how much a channel is noisy in the context of the polar codes was first determined by the symmetric capacity or the Bhattacharyya parameter of the channel; however, BER was used later as a common measure.

Code construction phase involves selecting channels in which the information bits are transmitted. In other words, constructing a polar code means using a vector of bit-channel indices that would be used to transmit information. The rest of the bit-channels would have no data and contain the frozen bits. Several code construction algorithms that vary in complexity, precision, and BER performance exist which include evolution of Z-parameters–based code construction algorithm; Monte-Carlo simulation–based construction algorithm; density evolution–based code construction algorithm; Gaussian approximation–based algorithm; and transition probability matrix–based algorithm [29].

Polar codes are a member of the coset[10] linear block code family, where the information bits are multiplied by a submatrix out of the traditional polar generator matrix, and the frozen bits are multiplied by another submatrix. Polar encoding is characterized by its structural manner, in the sense that all parameters are static, independent of the code rate. Different code rates correspond to different number of information bits, while using the same generator matrix. The systematic polar encoding is an extended version of the nonsystematic polar encoding, where the codeword is first non-systematically encoded, bits at frozen bit-channel positions are reset to the values of the frozen bits, and then non-systematically encoded. Systematic encoding provides better performance in terms of BER than non-systematic encoding. However, both have the same BLER performance. There are various polar decoding algorithms including SC, SC list (SCL), SCL with (SCL-CRC), and BP.

The SC decoder is based on the concept of successively decoding bits, where each stage of bit decoding is based on previously decoded bits. It suffers from inter-bit dependence due to its successive nature and thus error propagation. As a standalone decoder for polar codes, it is outperformed by most polar decoders in terms of BER performance. However, it enjoys a

[10] For a subgroup H of a group G and an element x of G, define xH to be the set $\{xh{:}h \in H\}$ and Hx to be the set $\{hx{:}h \in H\}$. A subset of G in the form of xH for some $x \in G$ is said to be a left coset of H and a subset of the form Hx is said to be a right coset of H. For any subgroup H, we can define an equivalence relation $\sim$ by $x \sim y$ if $x = yh$ for some $h \in H$. The equivalence classes of this equivalence relation are exactly the left cosets of H, and an element x of H is in the equivalence class xH. Thus the left cosets of H form a partition of G. It is also true that any two left cosets of H have the same cardinal number, and in particular, every coset of H has the same cardinal number as $eH = H$, where e is the identity element. Thus the cardinal number of any left coset of H has cardinal number the order of H. The same results are true of the right cosets of G and, in fact, one can prove that the set of left cosets of H has the same cardinal number as the set of right cosets of H.

potential for list decoding, because of its sequential hierarchical structure. It was proved that polar codes achieve Shannon capacity of any symmetric binary-input discrete memoryless channel under SC decoding [33,34].

SCL decoder was proposed as an extended version of SC decoder where instead of successively computing hard decisions for each bit, it branches one SC decoder into two parallel SC-decoders at each stage of decision where each branch has its path metric that is continuously updated for each path. It can be shown that a list of size 32 is enough to almost achieve the ML bound.

SCL with CRC decoder is an extension of SCL decoder, where a high-rate CRC code is appended to the polar code, so that the correct codeword is selected among the candidate codewords from the final list of paths. It was observed that whenever an SCL-decoder fails, the correct codeword exists in the list. Therefore, the CRC was proposed as a validity check for each candidate codeword in the list.

In BP decoder, unlike SC-based decoding techniques, there is no inter-bit dependence and thus no error propagation. It does not encounter any intermediate hard decisions. It updates the LLR values iteratively through right-to-left and left-to-right iterations using the same update functions that were used in LDPC domain. For finite length codes, BP decoder outperforms SC decoder in terms of BER performance.

Channel polarization is the concept upon which the polar codes are built. It is the process through which N distinct channels are generated $W_N^{(i)}:1 \leq i \leq N$ from N independent copies of a binary-input discrete memoryless channel. The N generated channels are polarized and have mutual information either close to 0 (i.e., noisy channels) or close to 1 (i.e., noiseless channels). The synthesized channels become perfectly noisy/noiseless as N approaches infinity. The process of channel polarization consists of two phases namely channel combining and channel splitting. In the former phase, N distinct channels are created in $n = \log_2 N$ steps, through recursively combining N copies of a binary-input discrete memoryless channel to form a vector channel $W_N:X^N \rightarrow Y^N$, where N must be an integer power of two. In the second phase the channel W_N is split into N binary-input channels $W_N^{(i)}:X \rightarrow Y^N \times X^{i-1}, 1 \leq i \leq N$. Uncoded information bits are transmitted over the reliable or noiseless channels with rate 1 and frozen bits are transmitted over the unreliable or noisy channels [33,34].

Polar code construction is the process of selecting the set of K good channels out of N channels over which uncoded information bits will be transmitted. The selection of the information set $\mathbb{A}$ is done in a channel dependent manner. For finite-length polar codes, the synthesized channels are not fully polarized. Bit errors over the quasi-polarized channels are inevitable. Thus the polar code construction phase is critical to obtain the best possible performance. To construct a polar code, the K reliable channels are chosen to minimize the sum of their Bhattacharyya parameter values $\sum_{i \in \mathbb{A}} Z\left(W_N^{(i)}\right)$ in order to minimize the upper bound on the block error probability of the constructed polar code. For the binary erasure channel, the Bhattacharyya parameter can be calculated using recursive formulas. Thus the polar code construction problem can be solved without a need for approximation. The Bhattacharyya parameter can be calculated using the following recursive formulas with a complexity of $\mathcal{O}(N)$ [29]:

$$Z(W_N^{(2j-1)}) = 2Z(W_{N/2}^{(j)}) - Z(W_{N/2}^{(j)})^2$$
$$Z(W_N^{(2j)}) = Z(W_{N/2}^{(j)})^2$$
$$Z(W_1^{(1)}) = \epsilon$$

For the AWGN channel no efficient algorithm for calculating the Bhattacharyya parameter per synthesized channel is known. Approximating the exact polar code construction is possible to reduce complexity by calculating an estimate of the Bhattacharyya parameter per synthesized channel. Several suboptimal construction methods were proposed in the literature with different computational complexities. The main difference between polar codes and Reed–Muller codes is the choice of the information set $\mathbb{A}$. In the case of Reed–Muller codes, the indices of the highest weight rows of the generator matrix **G** are selected to carry the information. A polar code of length $N = 2^n$ is generated using generator matrix **G** of size $N \times N$. A block of length N, consisting of $N - K$ frozen bits and K information bits, is multiplied by **G** to produce the polar codeword $\mathbf{x} = \mathbf{uG}$. The generator matrix **G** is based on a kernel that is used to construct the code, where

$$\mathbf{G} = \begin{bmatrix} 1 & 0 \\ 1 & 1 \end{bmatrix}^{\otimes \log_2 N}$$

in which $\otimes$ denotes the Kronecker product.[11] A polar encoding lattice, equivalent to **G**, can also be used as a polar encoder, as shown in Fig. 4.45. Note that successive graph representations have recursive relationships. More specifically, the graph representation for a polar encoding kernel operation having a kernel block size of $N = 2$ comprises a single stage, containing a single XOR. The first of the $N = 2$ kernel encoded bits is obtained as the XOR of the $N = 2$ kernel information bits, while the second kernel encoded bit is equal to the second kernel information bit. For greater kernel block sizes N, the graph representation may be considered to be a vertical concatenation of two graph representations for a kernel block size of $N = 2$, followed by an additional stage of XORs, as shown in Fig. 4.45. In analogy with the $N = 2$ kernel described above, the first $N = 2$ of the N kernel encoded

[11] Given a $m \times n$ matrix **A** and a $p \times q$ matrix **B**, the Kronecker product $\mathbf{C} = \mathbf{A} \otimes \mathbf{B}$ also called matrix direct product, is an $mp \times nq$ matrix with elements defined by $c_{\alpha\beta} = a_{ij} b_{kl}$ where $\alpha \equiv p(i-1) + k$ and $\beta = q(j-1) + l$. The matrix direct product provides the matrix of the linear transformation induced by the vector space tensor product of the original vector spaces. Assuming that operators $S{:}V_1 \to W_1$ and $T{:}V_2 \to W_2$ are given by $S(x) = Ax$ and $T(y) = By$, then $S \otimes T{:}V_1 \otimes V_2 \to W_1 \otimes W_2$ is determined by $S \otimes T(x \otimes y) = (Ax) \otimes (By) = (A \otimes B)(x \otimes y)$.

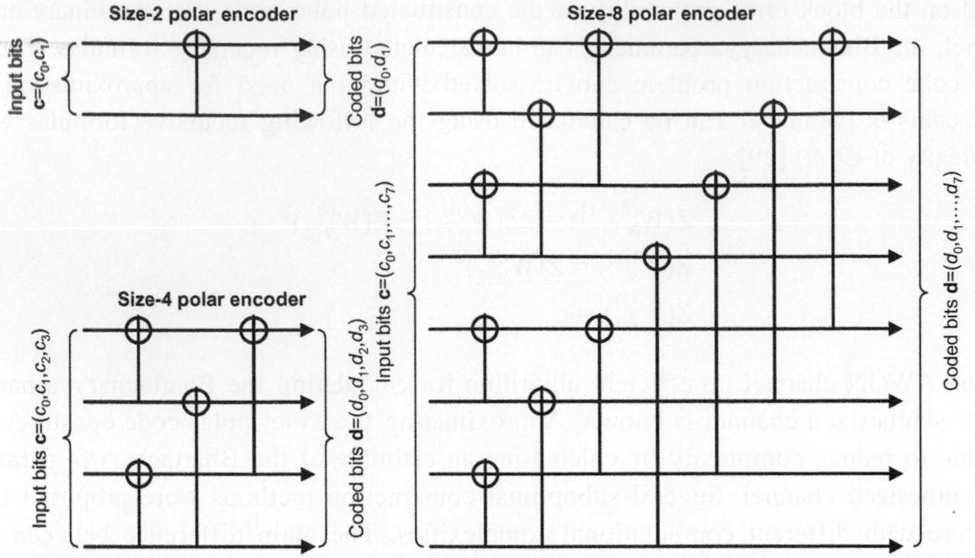

Figure 4.45
Polar encoder of different sizes [32].

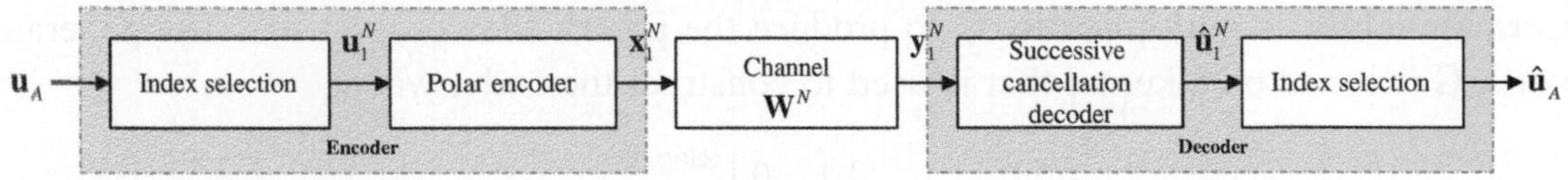

Figure 4.46
High-level polar code encoding and SC list decoding [40].

bits are obtained as XORs of corresponding bits from the outputs of the two $N = 2$ kernels, while the second $N = 2$ of the kernel encoded bits are equal to the output of the second $N = 2$ kernel [35].

Polar codes were introduced as non-systematic codes. Any linear code can be transformed from non-systematic to a systematic code. The systematic polar encoding can be performed using the standard non-systematic polar encoding apparatus in a three-phase operation as follows (see Fig. 4.46):

1. The vector $\mathbf{u} = (u_{\mathbb{A}}, u_{\mathbb{A}^c})$ is encoded in the standard non-systematic fashion producing the vector $\bar{\mathbf{u}}$.
2. The frozen bit positions $\mathbb{A}^c$ in the vector $\bar{\mathbf{u}}$ are set to zero $\bar{\mathbf{u}}_{\mathbb{A}^c} = 0$ (the frozen bits are always set to zero here).
3. The modified vector $\bar{\mathbf{u}}$ is encoded in the standard non-systematic fashion producing the codeword $\mathbf{x}$ which is a systematic polar codeword, in the sense that the information bits $\mathbf{u}_{\mathbb{A}}$ appear in the final codeword $\mathbf{x}$ in the information bits position $\mathbf{x}_{\mathbb{A}}$.

The main advantage of the systematic polar codes is that its BER performance is better than non-systematic polar codes. However, both systematic and non-systematic polar codes have the same BLER performance. It is observed that the systematic polar coding is more robust to error propagation using SC decoder when compared to that in the non-systematic polar coding. There are two main methods for polar decoding namely SC decoder and its variants and BP decoder. The SC decoder and its variants have serial decoding characteristics which cannot be parallelized; thus these decoders suffer from long decoding latency and low throughput, making them not suitable for high-speed applications. The BP decoder can use parallel processing and thus can enhance its throughput, making it suitable for high-speed applications.

In the receiver, the role of the demodulator is to recover information pertaining to the encoded block. However, the demodulator is typically unable to obtain absolute confidence about the value of the bits in the encoded block due to the random nature of the noise in the communication channel. The demodulator may express its confidence about the values of the bits in the encoded block by generating a soft encoded block, which comprises N number of encoded soft bits. Each soft bit may be represented in the form of an LLR as follows: $LLR = \log\,[p(bit = 0)] - \log\,[p(bit = 1)]$ where $p(bit = 0)$ and $p(bit = 1)$ are the probabilities that the corresponding bit has a value of "0" and "1", respectively. A positive LLR indicates that the demodulator has greater confidence that the corresponding bit has a value of "0", while a negative LLR indicates greater confidence in the bit value of "1". The magnitude of the LLR corresponds to the confidence level, where an infinite value corresponds to absolute confidence, while a magnitude of zero indicates that the demodulator has no information about the bit value [29].

The polar decoder may operate based on different algorithms including SC decoding and SCL decoding. The SC decoder was the first decoder that was used for polar codes which consists of N decision elements for the N bits of $\hat{\mathbf{u}}_1^N$ (see Fig. 4.46). Each of the decision elements computes the hard-decision output based on the observed channel output y_1^N and the previously decoded bits, for example, the kth decision element would compute $\hat{\mathbf{u}}_k$ using y_1^N and $\hat{\mathbf{u}}_1^{k-1}$. The decision element computes the likelihood ratio as follows [29]:

$$L_i \triangleq \frac{W_N^{(i)}\left(y_1^N, \hat{\mathbf{u}}_1^{i-1} | 0\right)}{W_N^{(i)}\left(y_1^N, \hat{\mathbf{u}}_1^{i-1} | 1\right)}$$

The hard decision per decision element is generated according to the following rule:

$$\hat{\mathbf{u}}_k = \begin{cases} 0, & L_k \geq 1 \\ 1, & L_k < 1 \end{cases}$$

The decision elements of indices that belong to the set $\mathbb{A}^c$ are set to zero $\hat{\mathbf{u}}_k = \mathbf{0}$, that is, the frozen bit positions and values can be considered as the decoder's prior knowledge.

In the SC decoding process, the value selected for each bit in the recovered information block depends on the sign of the corresponding LLR, which in turn depends on the values selected for all previous recovered information bits. If this approach results in the selection of the incorrect value for a particular bit, then this will often result in propagation of errors in all subsequent bits. The selection of an incorrect value for an information bit may be detected with consideration of the subsequent frozen bits, since the decoder knows that these bits should have zero values. More specifically, if the corresponding LLR has a sign that would imply a value of "1" for a frozen bit, then this suggests that an error may have occurred during the decoding of one of the preceding information bits. However, in the SC decoding process, there is no opportunity to consider alternative values for the preceding information bits. Once a value has been selected for an information bit, the SC decoding process is final. This motivates SCL decoding, which enables a list of alternative values for the information bits to be considered. As the decoding process continues, it considers both options for the value of each successive information bit. More specifically, an SCL decoder maintains a list of candidate kernel information blocks, where the list and the kernel information blocks are built up as the SCL decoding process proceeds. At the start of the process, the list comprises only a single-kernel information block having a length of zero bits. Whenever the decoding process reaches a frozen bit, a bit value of "0" is appended to the end of each candidate kernel information block in the list. However, whenever the decoding process reaches an information bit, two replicas of the list of candidate kernel information blocks is created. The bit value of "0" is appended to each block in the first replica and the bit value of "1" is appended to each block in the second replica. Following this, the two lists are merged to form a new list having a length which is double the length of the original list. This continues until the length of the list reaches a limit L, that is, the list size, which is typically a power of two. From this point onwards, each time the length of the list is doubled when considering an information bit, the worst L among the $2L$ candidate kernel information blocks are identified and pruned from the list. Thus the length of the list is maintained at L until the SCL decoding process is completed. In this process, the worst candidate kernel information blocks are identified by comparing and sorting appropriate metrics that are computed for each block, based on the LLRs obtained on the left-hand edge of the polar code graph [35,38].

There are several challenges associated with the hardware implementation of polar encoders and, in particular, polar decoders. For example, the complexity of a polar decoder is much greater than that of a polar encoder for three reasons: (1) while polar encoders operate on the basis of bits, polar decoders operate on the basis of the probabilities of bits, which require more memory to store and more complex computations; (2) while polar encoders only have to consider the particular permutation of the information block that they are presented, polar decoders must consider all possible permutations of the information block and must select the one which is the most likely; and (3) while polar encoders only process each information block once, an SCL polar decoder must process each information block L times in order to achieve sufficiently strong error correction. For these reasons, the latency, hardware resource usage, and power consumption of polar decoders are typically much greater than those of polar encoders. Another challenge in the implementation of the SCL decoding process is imposed by metric sorting. As described earlier, the sorting is required in order to identify and prune the worst L candidate kernel information blocks, among the merged list of $2L$ candidates. One option is to employ a large amount of hardware to simultaneously compare each of the $2L$ candidates with every other candidate, so that the sorting can be completed within a short time. Alternatively, the hardware resource requirement can be reduced by structuring successive comparisons to efficiently reuse intermediate results at the cost of increasing the latency required to rank the $2L$ candidates. The CRC bits are employed by the NR polar code in order to facilitate error detection and to improve the error correction capability of the polar decoder. However, there is a trade-off between the error detection capability and the error correction capability. In order to meet the BLER requirements of NR for the control channels, the CRC bits must be handled very carefully, in a manner which is not captured in the NR specifications. In particular, the CRC (and parity check or PC) bits must be decoded as an integral part of the polar decoding process, using an unconventional decoding technique [35].

4.1.7.2 NR Polar Coding

3GPP NR uses a variant of the polar code called distributed CRC (D-CRC) polar code, that is, a combination of CRC-assisted and PC polar codes, which interleaves a CRC-concatenated block and relocates some of the PC bits into the middle positions of this block prior to performing the conventional polar encoding described earlier. This allows a decoder to early terminate the decoding process as soon as any parity check is not successful. The D-CRC scheme is important for early termination of decoding process, because the post-CRC interleaver can distribute information and CRC bits such that partial CRC checks can be performed during list decoding and paths failing partial CRC check can be pruned, leading to early termination of decoding. The post-CRC interleaver design is closely tied to the CRC generator polynomial, thus by appropriately selecting the CRC polynomial, one can achieve better early termination gains and maintain acceptable false alarm rate. The signal flow graph of D-CRC polar encoding and decoding is shown in Fig. 4.47.

In NR, the polar code is used to encode broadcast channel as well as DCI and uplink control information (UCI). Let us denote by N_c the number of control bits that must be transmitted using a code of length E bits. We add L_{CRC} CRC bits to the information bits, resulting in K bits that will be encoded by an (N, K) polar encode with $N = 2^n$. Rate matching is performed to obtain a code of length E and effective rate $R = N_c/E$. To each vector $\mathbf{a} = (a_0, a_1, \ldots, a_{N_c-1})$, containing N_c control information bits to be transmitted, L_{CRC}-bit CRC is attached. The resulting vector $\mathbf{c} = (c_0, c_1, \ldots, c_{K-1})$ comprising $K = N_c + L_{CRC}$ bits

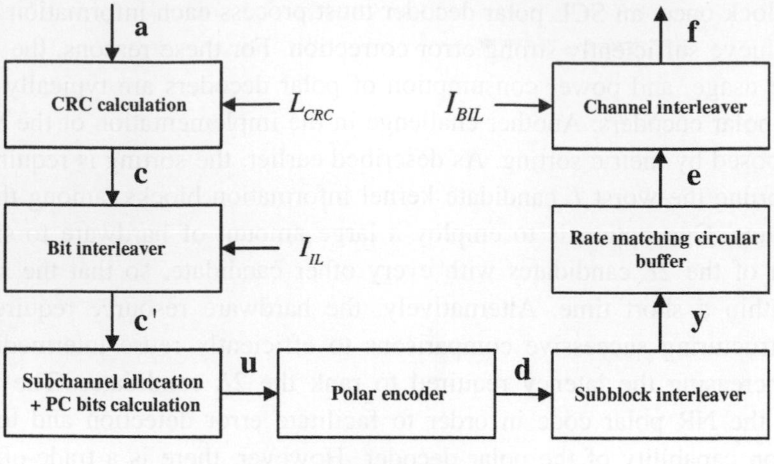

Figure 4.47
3GPP NR polar encoding flow graph [36].

is passed through an interleaver. Based on the desired code rate R and codeword length E, a polar code of length N is utilized along with the relative bit channel reliability sequence and the frozen set. The interleaved vector $\mathbf{c}'$ is assigned to the information set along with the PC bits, while the remaining bits in the N-bit vector $\mathbf{u}$ are frozen. Vector $\mathbf{u}$ is encoded with $\mathbf{d} = \mathbf{uG}$, where the generator matrix $\mathbf{G}$ was defined earlier. After encoding, a subblock interleaver divides $\mathbf{d}$ into 32 equal-length blocks, scrambling them and creating vector $\mathbf{y}$ that is fed into the circular buffer as illustrated in Fig. 4.47. For rate matching, puncturing, shortening, or repetition are applied to change the N-bit vector $\mathbf{y}$ into the E-bit vector $\mathbf{e}$. A channel interleaver is finally applied to compute the vector $\mathbf{f}$ that is now ready to be modulated and transmitted [36].

The NR polar encoder relies on several parameters that depend on the amount and type of information to be transmitted and the physical channel used. The first parameter that needs to be identified is the code length of the polar code, $N = 2^n$. The number n is calculated as $n = \max\{\min\{n_1, n_2, n_{\max}\}, n_{\min}\}$, where $n_{\min}$ and $n_{\max}$ provide a lower and an upper bound on the code length, respectively. In particular, $n_{\min} = 5$ and $n_{\max} = 9$ for the downlink control channel, whereas $n_{\max} = 10$ for the uplink control channel. The parameter $n_2 = \lceil \log_2(K/R_{\min}) \rceil$ gives an upper bound on the code rate based on the minimum code rate admitted by the encoder, that is, 1/8. The value of parameter n_1 is dependent on the rate-matching scheme. It is usually calculated as $n_1 = \lceil \log_2 E \rceil$ so that 2^{n_1} is the smallest power of two larger than E. However, a correction factor is introduced to avoid too severe rate matching: if $\log_2 E < 0.17$, that is, if the smallest power of two larger than E is too far from E, the parameter is set to $n_1 = \lfloor \log_2 E \rfloor$. In this case, an additional constraint on the code dimension is added by imposing $K/E < 9/16$ to ensure that $K < N$. If a code length

Table 4.14: 3GPP NR polar encoding parameters [7].

<table>
<tr><th rowspan="3"></th><th colspan="3">PUCCH/PUSCH</th><th rowspan="3">PDCCH/PBCH</th></tr>
<tr><th rowspan="2">$N_c \geq 20$</th><th colspan="2">$12 \leq N_c \leq 19$</th></tr>
<tr><th>$E - N_c \leq 175$</th><th>$E - N_c > 175$</th></tr>
<tr><td>n_{max}</td><td colspan="3">10</td><td>9</td></tr>
<tr><td>I_{IL}</td><td colspan="3">0</td><td>1</td></tr>
<tr><td>I_{BIL}</td><td colspan="3">1</td><td>0</td></tr>
<tr><td>L_{CRC}</td><td>11</td><td colspan="2">6</td><td>24</td></tr>
<tr><td>n_{PC}</td><td>0</td><td colspan="2">3</td><td>0</td></tr>
<tr><td>n_{PC}^{wm}</td><td>0</td><td>0</td><td>1</td><td>0</td></tr>
</table>

$N > E$ is selected, the polar code will be punctured or shortened, depending on the code rate before the transmission. In particular, if $K/E \leq 7/16$, the code will be punctured; otherwise it will be shortened. If $N < E$, repetition is used, and some encoded bits will be transmitted twice. In this case, the code construction ensures that $K < N$.

As shown in Table 4.14, a set of parameters are defined to differentiate between different type of control information. The parameters I_{IL} and I_{BIL} refer to the activation of the input bits interleaver and the channel interleaver, respectively. The value of the two types of assistant PC bits are given by n_{PC} and n_{PC}^{wm}. The length of the control information vector N_c and the length of the transmitted codeword E are dependent on the type, content, and number of consecutive transmissions and are reliant on the decisions taken in the higher layers [36].

The K bit output of the CRC encoder is interleaved before being fed to the polar encoder. The interleaver is activated through the I_{IL} flag. In particular, the input bit interleaver is activated for PBCH and PDCCH payloads, while it is set to zero in the case of PUCCH and PUSCH control information. The input bit interleaver interleaves up to $K_{IL}^{\max} = 164$ input bits, where the interleaving pattern is calculated based on the sequence $\Pi_{IL}^{\max}(m)$ given in [7]. The maximum number of input bits $K_{IL}^{\max}$ is set to 164 suggesting that the maximum number of control information bits without CRC is limited to 140. In more detail, $\left(K_{IL}^{\max} - K\right)$ is subtracted from all the entries of $\Pi(k)$, such that $\Pi(k)$ contains the integers smaller than K in permuted order. This scrambling sequence has been proposed to facilitate early termination, both during normal decoding and in DCI blind detection. This is made possible by the fact that after interleaving, every CRC remainder bit is placed after its relevant information bits. The interleaving function is applied to vector $\mathbf{c}$ to obtain the $K-$ bit vector $\mathbf{c}' = (c_{\Pi(0)}, c_{\Pi(1)}, \ldots, c_{\Pi(K-1)})$ [7].

In the subchannel allocation process prior to polar encoding, the vector $\mathbf{c}'$ is expanded into the N-bit vector $\mathbf{u}$ with the addition of assistant bits and frozen bits. As a first step, n_{PC} PC bits are inserted among the K information and CRC bits. Thus the polar encoder represents a $(N, K+n_{PC})$ code. To create the input vector $\mathbf{u}$ to be encoded, the frozen set of subchannels needs to be identified. The number and position of frozen bits depend on N, E and the selected rate-matching scheme. To begin with, the frozen set $\overline{\mathbf{Q}}_F^N$ and the complementary information set $\overline{\mathbf{Q}}_I^N$ are computed based on the polar reliability sequence $\mathbf{Q}_0^{N_{\max}-1}=\left\{Q_0^{N_{\max}}, Q_1^{N_{\max}}, \ldots, Q_{N_{\max}-1}^{N_{\max}}\right\}$ and the rate matching scheme. The information bits are subsequently assigned to vector $\mathbf{u}$ according to the information set. The assistant PC bits are calculated and stored in $\mathbf{u}$, if necessary [7,36]. The first bits identified in the frozen set correspond to the indices of the $N-E$ bits that are not transmitted, that is, the bits that are punctured from the codeword by the rate matching. These indices correspond to the first $N-E$ or the last $N-E$ codeword bits in the case of puncturing and shortening, respectively. Due to the presence of an interleaver between the encoding and the rate matching, the actual indices to be added to the frozen set correspond to the first or the last bits after the interleaving process. If $K/E \leq 7/16$ and henceforth the polar code must be punctured, additional indices are included in the frozen set to prevent bits in the information set to become ineffective due to puncturing. Furthermore, new indices are added to the frozen set from the reliability sequence starting from the least reliable bits. The polar reliability sequence $\mathbf{Q}_0^{N_{\max}-1}=\left\{Q_0^{N_{\max}}, Q_1^{N_{\max}}, \ldots, Q_{N_{\max}-1}^{N_{\max}}\right\}$ is a list of integers smaller than 1024 sorted in reliability order, from the least reliable to the most reliable; indices larger than N are skipped during the creation of $\overline{\mathbf{Q}}_F^N$.

Unlike the conventional polar encoding process where the Bhattacharyya parameters or in general the reliability factors are calculated prior to encoding process, in 3GPP NR, those reliability factors are tabulated in the standard specification. Prior to encoding, the polar sequence $\mathbf{Q}_0^{N_{\max}-1}=\left\{Q_0^{N_{\max}}, Q_1^{N_{\max}}, \ldots, Q_{N_{\max}-1}^{N_{\max}}\right\}$, in which $0 \leq Q_i^{N_{\max}} \leq N_{\max}-1$ for $i=0,1,\ldots,N_{\max}-1$ denotes a bit index, is sorted in ascending order of reliability factors $W\left(Q_0^{N_{\max}}\right)<W\left(Q_1^{N_{\max}}\right)<\cdots<W\left(Q_{N_{\max}-1}^{N_{\max}}\right)$, where $W\left(Q_i^{N_{\max}}\right)$ denotes the reliability of bit index $Q_i^{N_{\max}}$. For any code block of length N bits, the same polar sequence $\mathbf{Q}_0^{N-1}=\{Q_0^N, Q_1^N, Q_2^N, \ldots, Q_{N-1}^N\}$ is utilized. The polar sequence $\mathbf{Q}_0^{N-1}$ is a subset of polar sequence $\mathbf{Q}_0^{N_{\max}-1}$ with all elements $Q_i^{N_{\max}}$ of values less than N, ordered in ascending order of reliability factors $W(Q_0^N)<W(Q_1^N)<W(Q_2^N)<\cdots<W(Q_{N-1}^N)$. In the preceding expressions, sequence $\overline{\mathbf{Q}}_I^N$ denotes the set of information bit indices in the polar sequence $\mathbf{Q}_0^{N-1}$, and $\overline{\mathbf{Q}}_F^N$ is the set of other bit indices in polar sequence $\mathbf{Q}_0^{N-1}$, where $\overline{\mathbf{Q}}_I^N$ and $\overline{\mathbf{Q}}_F^N$ are derived through subblock interleaving, $\left|\overline{\mathbf{Q}}_I^N\right|=K+n_{PC}$ (i.e., cardinality of the set or the length of the sequence), $\left|\overline{\mathbf{Q}}_F^N\right|=N-\left|\overline{\mathbf{Q}}_I^N\right|$ and n_{PC} is the number of PC bits (see Table 4.15).

Table 4.15: Polar sequence $Q_0^{N_{max}-1}$ and the associated reliability factor $W(Q_i^{N_{max}})$ [7].

$W(Q_i^{N_{max}})$	$Q_i^{N_{max}}$	$W(Q_i^{N_{max}})$	$Q_i^{N_{max}}$	$W(Q_i^{N_{max}})$	$Q_i^{N_{max}}$	$W(Q_i^{N_{max}})$	$Q_i^{N_{max}}$	$W(Q_i^{N_{max}})$	$Q_i^{N_{max}}$	$W(Q_i^{N_{max}})$	$Q_i^{N_{max}}$	$W(Q_i^{N_{max}})$	$Q_i^{N_{max}}$	$W(Q_i^{N_{max}})$	$Q_i^{N_{max}}$
0	0	128	518	256	94	384	214	512	364	640	414	768	819	896	966
1	1	129	54	257	204	385	309	513	654	641	223	769	814	897	755
2	2	130	83	258	298	386	188	514	659	642	663	770	439	898	859
3	4	131	57	259	400	387	449	515	335	643	692	771	929	899	940
4	8	132	521	260	608	388	217	516	480	644	835	772	490	900	830
5	16	133	112	261	352	389	408	517	315	645	619	773	623	901	911
6	32	134	135	262	325	390	609	518	221	646	472	774	671	902	871
...	...	...	...	...	...	...	...	...	...	...	...	...	...	...	...
...	...	...	...	...	...	...	...	...	...	...	...	...	...	...	...
125	768	253	209	381	539	509	248	637	806	765	914	893	506	1021	1021
126	268	254	284	382	111	510	369	638	427	766	752	894	749	1022	1022
127	274	255	648	383	331	511	190	639	904	767	868	895	945	1023	1023

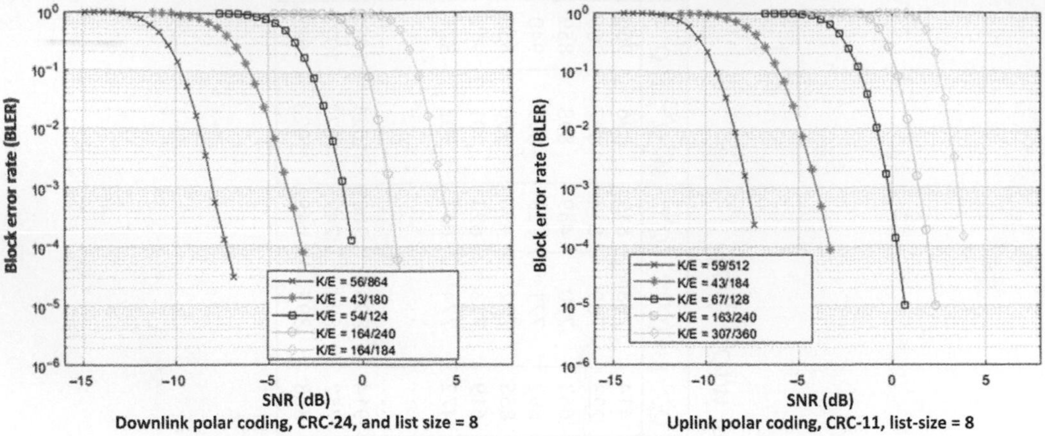

Figure 4.48
Downlink/uplink polar coding with various K/E ratios [67].

Once the input bits are reordered and moved to the reliable positions according to the above reliable bit positions determination procedure, the reordered bits are passed through the polar encoder [7].

As we defined earlier, the polar code generator matrix $\mathbf{G}_N = (\mathbf{G}_2)^{\otimes n}$ is constructed as the nth Kronecker power of matrix $\mathbf{G}_2 = \begin{bmatrix} 1 & 0 \\ 1 & 1 \end{bmatrix}$. For a bit index $j = 0, 1, \ldots, N-1$, let $\mathbf{g}_j$ denote the jth row of $\mathbf{G}_N$ and $w(\mathbf{g}_j)$ the row weight of $\mathbf{g}_j$, where $w(\mathbf{g}_j)$ is the number of ones in $\mathbf{g}_j$. Let us further assume that $\mathbf{Q}_{PC}^N$ is the set of bit indices for PC bits, where the cardinality of the set is $|\mathbf{Q}_{PC}^N| = n_{PC}$. A number of PC bits are placed in the $(n_{PC} - n_{PC}^{wm})$ least reliable bit indices in $\overline{\mathbf{Q}}_I^N$. Other n_{PC}^{wm} PC bits are placed in the bit indices of minimum row weight in $\tilde{\mathbf{Q}}_I^N$, where $\tilde{\mathbf{Q}}_I^N$ denotes the $\left(\left|\overline{\mathbf{Q}}_I^N\right| - n_{PC}\right)$ most reliable bit indices in $\overline{\mathbf{Q}}_I^N$. If there are more than n_{PC}^{wm} bit indices of the same minimum row weight in $\tilde{\mathbf{Q}}_I^N$, the other n_{PC}^{wm} PC bits are placed in the n_{PC}^{wm} bit indices of the highest reliability and the minimum row weight in $\tilde{\mathbf{Q}}_I^N$. The output bit sequence following polar encoding $\mathbf{d} = [d_0, d_1, \ldots, d_{N-1}]$ is obtained as $\mathbf{d} = \mathbf{u}\mathbf{G}_N$ where vector $\mathbf{u} = [u_0, u_1, \ldots, u_{N-1}]$ is derived from interleaved sequence $c'_0, c'_1, \ldots, c'_{K-1}$ following the above bit reordering process. The encoding is performed in $GF(2)$[12].

The performance of the polar codes with different K/E ratios is shown in Fig. 4.48.

[12] $GF(2)$ is the Galois field comprising two elements and the smallest finite field. One may also define $GF(2)$ as the quotient ring of the ring of integers $\mathbb{Z}$ by the ideal $2\mathbb{Z}$ of all even numbers $GF(2) = \mathbb{Z}/2\mathbb{Z}$.

4.1.7.3 Principles of Low Density Parity Check Coding

LDPC codes belong to the class of forward error correction codes which are used for sending a message over noisy transmission channel. These codes can be described by a parity check matrix which contains mostly zeros and a relatively small number of ones. Thus the decoding complexity is small when compared to other code constructions. A very efficient iterative decoding algorithm known as belief propagation is used in the decoder. The LDPC codes can be divided into two groups: regular LDPC codes when the column weight and the row weight of the PC matrix are constant and equal and irregular LDPC codes when the column weight and the row weight are not constant and equal, meaning that the number of ones per row and column is different.

The LDPC codes are represented in different ways. Similar to all linear block codes, a matrix representation by the corresponding generator matrix **G** or the PC matrix **H** is possible. Thus if the number of input information bits is K and the number of output bits is N, the PC matrix **H** is expressed as an $M \times N$ matrix, where $M = N - K$. The resultant code rate K/N defines the size of the PC matrix. The LDPC codes can be further graphically represented with a Tanner graph, which is one of the most common graphical representations for the LDPC codes. It provides the complete representation of the code and helps to describe the decoding algorithm. Tanner graphs are bipartite graphs, that is, there are two disjunct sets of nodes. The two types of nodes are variable nodes (VND) and check nodes (CND). The VNDs represent the code bits, thus, each of the N columns of matrix **H** is represented by one VND. The CNDs represent the code constraints; thus each of M rows of matrix **H** is represented by one CND. Each VND v_i is connected to a CND c_j, if $h_{ij} = 1$. For the following example PC matrix, the Tanner graph is as shown in Fig. 4.49.

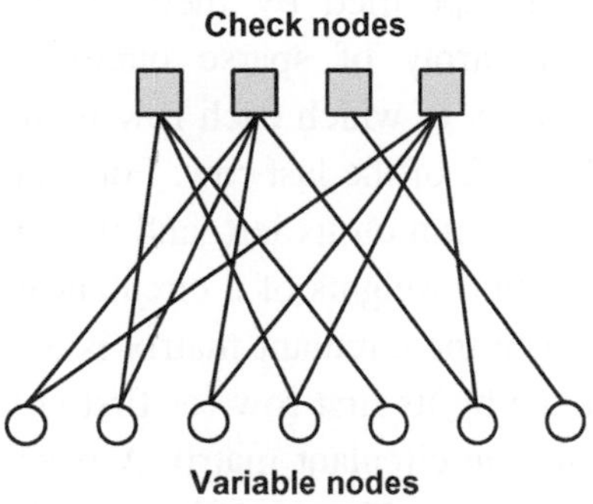

Figure 4.49
Example Tanner graph [38].

$$\mathbf{H}=\begin{pmatrix}0&1&0&1&1&0&0\\1&1&1&0&0&1&0\\0&0&0&0&0&0&1\\1&0&1&1&0&1&0\end{pmatrix}$$

Quasi-cyclic (QC) LDPC codes belong to the class of structured codes that are relatively easier to implement without significantly compromising the performance of the code. The QC-LDPC codes can be implemented using simple shift registers with linear complexity based on their generator matrices. Well-designed QC-LDPC codes have been shown to outperform computer-generated random LDPC codes, in terms of bit-error rate and block-error rate performance and the error floor. These codes also have advantages in decoder hardware implementation due to their cyclic symmetry, which results in simple regular interconnection and modular structure. In most of the wireless communication standards including 3GPP NR, a base graph $\mathbf{u}$ is used to define the LDPC code. However, $\mathbf{u}$ needs to be transformed into a PC matrix $\mathbf{H}$ using a lifting factor Z. Lifting means that each (integer) entry of base graph $\mathbf{u}$ is replaced by a permuted $Z \times Z$ identity matrix. We start with an identity matrix $\mathbf{I}$ and circularly shift the entries of this matrix according to the base graph entry u_{ij} to obtain the desired matrix $\mathbf{H}$. As an example, suppose a 2×2 base graph matrix $\mathbf{u}$ and lifting factor $Z=3$ are given. The transformation from $\mathbf{u}$ to $\mathbf{H}$ can be performed as follows:

$$\mathbf{u}=\begin{bmatrix}2&3\\0&1\end{bmatrix},\quad \mathbf{I}=\begin{bmatrix}1&0&0\\0&1&0\\0&0&1\end{bmatrix}\rightarrow\mathbf{H}=\begin{bmatrix}\begin{bmatrix}0&0&1\\1&0&0\\0&1&0\end{bmatrix}&\begin{bmatrix}1&0&0\\0&1&0\\0&0&1\end{bmatrix}\\\begin{bmatrix}1&0&0\\0&1&0\\0&0&1\end{bmatrix}&\begin{bmatrix}0&1&0\\0&0&1\\1&0&0\end{bmatrix}\end{bmatrix}\rightarrow\mathbf{H}=\begin{bmatrix}0&0&1&1&0&0\\1&0&0&0&1&0\\0&1&0&0&0&1\\1&0&0&0&1&0\\0&1&0&0&0&1\\0&0&1&1&0&0\end{bmatrix}$$

The LDPC codes are universally specified by their PC matrices. The PC matrix of a QC-LDPC code is given as an array of sparse circulant matrices of the same size. A circulant matrix is a square matrix in which each row is the cyclic shift of the row above it, and the first row is the cyclic shift of the last row. For a circulant matrix, each column is the downward cyclic shift of the column on its left, and the first column is the cyclic shift of the last column. The row and column weights of a circulant matrix are the same. If the row/column weight is equal to one then the circulant matrix is also a permutation matrix. A circulant matrix is fully characterized by its first row (or first column), which is called the generator of the circulant. For an $m \times m$ circulant matrix $\mathbf{A}$ over $GF(2)$, if its rank $r=m$, then all of its rows are linearly independent. However, if its rank $r<m$ then any consecutive r rows (or columns) of $\mathbf{A}$ may be regarded as linearly independent, and the other $m-r$ rows (or columns) are linearly dependent. This is due to the cyclic structure of $\mathbf{A}$. A QC-LDPC code is given by the null space of an array of sparse circulant matrices of the same size.

An LDPC code can be defined by PC matrix $\mathbf{H}$. For each codeword $\mathbf{v}$, it can be shown that $\mathbf{H}\mathbf{v}^T = \mathbf{0}$. A non-codeword (corrupted codeword) on the other hand, will generate a non-zero vector, which is called syndrome. Mathematically, an LDPC code is the null-space of the PC matrix $\mathbf{H}$. A regular LDPC block code (d_v, d_c) has VND degree d_v and CND degree d_c, which is equal to the column (row)-weight of $\mathbf{H}$. The PC matrix represents a system of linear equations where each row can be represented by a linear combination of the CNDs. Any set of vectors that span the row-space generated by $\mathbf{H}$ can serve as the rows of a PC matrix. The degree of a node is the number of edges (lines) connected to it in a Tanner graph.

The decoding algorithms for LDPC codes were discovered independently, and they come under different names. The most common ones are the BP algorithm, the message passing algorithm (MPA), and the sum-product algorithm. In a BP algorithm, the probabilistic messages are iteratively exchanged between variable and check nodes until either a valid codeword is found or the maximum number of iterations is exceeded. The LDPC codes can be decoded using message passing or BP on the bipartite Tanner graph where, the CNDs and VNDs communicate with each other, successively passing revised estimates of the associated LLR in each decoding iteration. The bit reliability metric is defined as $LLR(b_i) = \log p(b_i = 0) - \log(b_i = 1)$ where b_i denotes the ith bit in the received codeword. If $LLR > 0$, it implies that $b_i = 0$ is more likely, while $LLR < 0$ implies that $b_i = 1$ is more probable. As an example, assume that codeword $(0, 0, 0, 0, 0, 0, 0, 0, 0, 0, 0)$ is transmitted, and $(0, 0, 0, 0, 1, 0, 0, 0, 0, 0, 0)$ is received by decoder. Each valid 11-bit codeword $\mathbf{c} = (c_0, c_1, \ldots, c_{10})$ has the sum (modulo 2) of all bits equal to zero. The received vector does not satisfy this code constraint, indicating that there are errors present in the received codeword. Furthermore, assume that the decoder is provided with bit-level reliability metric in the form of probability (confidence in the received values) of being correct as $(0.9, 0.8, 0.86, 0.7, 0.55, 1, 1, 0.8, 0.98, 0.68, 0.99)$. From the soft information, it follows that bit c_4 is the least reliable and should be flipped to bring the received codeword in compliance with code constraint. Using LLRs as messages, the hardware implementation has become much easier when compared to the message passing algorithm. The implementation complexity is further reduced by simplifying the process for updating the CNDs, which is the most complex part of the message passing algorithm. This algorithm is known as the min-sum algorithm. An LDPC decoder can be implemented using serial, parallel, or partially parallel architectures. The performance of the LDPC decoder depends on various factors such as decoder algorithm and architecture, quantization of LLRs and the maximum number of decoding iterations. The maximum number of decoding iterations used for the decoding process determines the data rate and latency of the LDPC decoder. After performing maximum number of decoding iterations, the codeword is then estimated. In order to save decoder power consumption and to decrease the latency, a decoder design that verifies the codeword after each iteration and stops the decoding process when the estimated codeword is correct, is needed. If parity check is satisfied then the codeword is estimated at the beginning of the next iteration and the decoding process is stopped.

4.1.7.4 NR Low Density Parity Check Coding

3GPP NR has taken a different approach to LDPC coding for the downlink and uplink traffic channels. In order to ensure support of a wide range of code rates with sufficient granularity and HARQ-IR, two base graphs with the structures that are explained later in this section have been adopted. Code extension of a PC matrix (lower triangular extension, which includes diagonal-extension as a special case) is used to support HARQ-IR and rate-matching. The 3GPP NR LDPC base graphs consist of five submatrices **A**,**B**,**C**,**D**,**E** as shown in Fig. 4.50. As depicted in the figure, matrix **A** corresponds to the systematic bits; **B** is a square matrix and corresponds to the parity bits. The first or last column of matrix **B** has a weight equal to one. The last row of **B** has a non-zero value and a weight equal to one. If there is a column with weight of one then the remaining columns contain a square matrix such that the first column has weight three. The columns after the weight three column have a dual diagonal structure (i.e., main diagonal and off diagonal elements). If there is no column with weight one, **B** consists of only a square matrix such that the first column has weight three, **C** is a zero matrix, **D** corresponds to single parity check rows, and **E** is an identity matrix for the base graph [55].

The rate matching for the LDPC code uses a circular buffer similar to LTE. The circular buffer is filled with an ordered sequence of systematic bits and parity bits. For HARQ-IR, each RV RV_i is assigned a starting bit location s_i in the circular buffer. For HARQ-IR retransmission of RV_i, the coded bits are read out sequentially from the circular buffer, starting with the bit location s_i. Limited buffer rate matching is further supported. Before code block segmentation, $L_{CRC} = 24$ TB-level CRC bits are attached to the end of the transport block. The value of L_{CRC} was determined to satisfy the probability of misdetection of the TB with $BLER < 10^{-6}$ as well as the inherent error detection of LDPC codes.

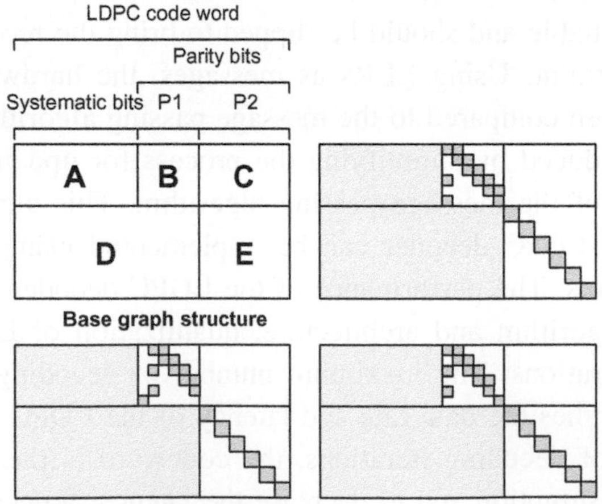

Figure 4.50
Structure of 3GPP NR base graphs with dual diagonal property [55].

The NR LDPC coding chain includes code block segmentation, CRC attachment, LDPC encoding, rate matching, and systematic-bit-priority interleaving. More specifically, code block segmentation allows very large transport blocks (MAC PDUs) to be divided into multiple smaller sized code blocks that can be efficiently processed by the LDPC encoder/decoder. The CRC bits are then attached for error detection purposes. When combined with the inherent error detection of the LDPC codes through the parity check equations, very low probability of undetected errors can be achieved. The rectangular interleaver with number of rows equal to the QAM order improves the performance by making systematic bits more reliable than parity bits for the initial transmission of the code blocks. The NR LDPC codes use a QC structure, where the PC matrix is defined by a smaller base graph. Each entry of the base graph represents either a $Z \times Z$ zero matrix or a shifted $Z \times Z$ identity matrix, where a cyclic shift to the right of each row is applied. Unlike the LDPC codes specified in other wireless technologies, the NR LDPC codes have a rate-compatible structure, which means codewords with different rates can be generated by including a different number of parity bits, or equivalently by using a smaller subset of the full PC matrix. This is especially useful for communication systems employing HARQ-IR for retransmissions. Another advantage of this structure is that for higher rates, the PC matrix and the decoding complexity and latency are smaller. This is in contrast with the LTE turbo codes, which have constant decoding complexity and latency irrespective of the code rate [55].

The NR data channel supports two base graphs to ensure good performance and decoding latency can be achieved for the full range of code rates and information block sizes. The base graph 1 is optimized for large information block sizes and high code rates. It is designed for maximum code rate of 8/9 and may be used for code rates up to 0.95. The base graph 2 is optimized for small information block sizes and lower code rates. The lowest code rate for base graph 2 without using repetition is 1/5. This is significantly lower than that of the LTE turbo codes, which rely on repetition for code rates below 1/3. The NR LDPC codes can also achieve an additional coding gain at low-code rates, which makes them suitable for high reliability scenarios. From decoding complexity perspective, for a given number of input bits, it is beneficial to use base graph 2, since it is more compact and utilizes a larger lifting factor, that is, more parallelism, relative to base graph 1. The decoding latency is typically proportional to the number of non-zero elements in the base graph. Since base graph 2 has much fewer non-zero elements compared to base graph 1 for a given code rate, its decoding latency is significantly lower.

In 3GPP NR, the input bit sequence is represented as $\mathbf{c} = [c_0, c_1, \ldots, c_{K-1}]^T$, where K is the number of information bits to encode. The output LDPC-coded bits are denoted by $d_0, d_1, \ldots, d_{N-1}$ where $N = 66Z_c$ for base graph 1 and $N = 50Z_c$ for base graph 2, where the value of lifting factor Z_c is given in Table 4.7. A code block is encoded by the LDPC encoder based on the following procedure [7]:

1. Find the set with index i_{LS} in Table 4.7 which contains Z_c.
2. Set $d_{k-2Z_c} = c_k, \forall k = 2Z_c, \ldots, K-1$.
3. Generate $N + 2Z_c - K$ parity bits $\mathbf{w} = [w_0, w_1, \ldots, w_{N+2Z_c-K+1}]^T$ such that $\mathbf{H}[\mathbf{c} \quad \mathbf{w}]^T = \mathbf{0}$.
4. The encoding is performed in $GF(2)$. For base graph 1, matrix $\mathbf{H_{BG}}$ (representing the base graph) has 46 rows and 68 columns. For base graph 2, matrix $\mathbf{H_{BG}}$ has 42 rows and 52 columns. The elements of $\mathbf{H_{BG}}$ matrices are given in [7]. The PC matrix $\mathbf{H}$ is obtained by replacing each element of $\mathbf{H_{BG}}$ with a $Z_c \times Z_c$ matrix such that each element of value 0 in $\mathbf{H_{BG}}$ is replaced by an all zero matrix $\mathbf{0}$ of size $Z_c \times Z_c$; and each element of value 1 in $\mathbf{H_{BG}}$ is replaced by circular permutation matrix $\mathbf{I}(P_{i,j})$ of size $Z_c \times Z_c$, where i and j are the row and column indices of the element, and $\mathbf{I}(P_{i,j})$ is obtained by circularly shifting the identity matrix $\mathbf{I}$ of size $Z_c \times Z_c$ to the right $P_{i,j}$ times. The value of $P_{i,j}$ is given by $P_{i,j} = V_{i,j} \bmod Z_c$ and the value of $V_{i,j}$ is given in [7].
5. Set $d_{k-2Z_c} = w_{k-K}, \forall k = K, \ldots, N + 2Z_c - 1$.

The performance of the NR LDPC codes over an AWGN channel was evaluated using a normalized min-sum decoder, layered scheduling, and a maximum of 20 decoder iterations. Fig. 4.51 shows the required SNR to achieve certain BLER targets as a function of information block size K for code rate 1/2 and QPSK modulation. The results show that the NR LDPC codes provide consistently good performance over a large range of block sizes. For this code rate, base graph 2 is used for all K for which it is defined, that is, for $K \leq 3840$, while base graph 1 is used for larger values of K (note the discontinuity point in the curves at $K = 3840$). As shown in the figure, there is a small gap in performance at the block size where the transition occurs between base graph 1 and base graph 2 [55].

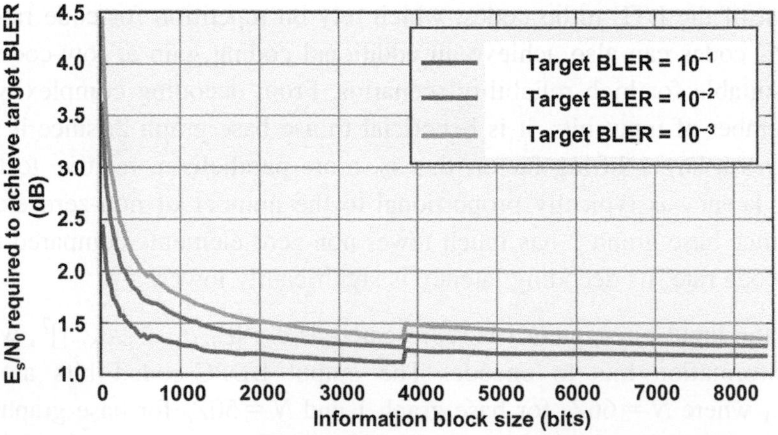

Figure 4.51
Performance of NR LDPC codes at code rate 1/2, QPSK modulation, and 20 iterations [55].

4.1.7.5 Modulation Schemes and MCS Determination

3GPP NR supports various modulation schemes (QPSK, 16QAM, 64QAM, and 256QAM) for CP-OFDM in both downlink and uplink. For the DFT-s-OFDM, however, NR uses an additional modulation $\pi/2$-BPSK in the uplink to achieve better efficiency for power amplifiers and lower PAPR in very low data rate cases. While 1024QAM can theoretically provide 25% throughput gain relative to 256QAM, due to real-world implementation complexity and the need for a very high SINR levels to achieve acceptable BLER targets, it was not included in the NR modulation schemes. A constellation diagram is a graphical representation of the complex envelope of each possible symbol state. The power efficiency is related to the minimum distance between the points in the constellation. The bandwidth efficiency is related to the number of points in the constellation. The gray coding is used to assign groups of bits to each constellation point. In gray coding, adjacent constellation points differ by a single bit. The modulation mapping function takes input bit sequence $b(i)b(i+1)\cdots b(i+Q_m-1)$ and generates the corresponding complex-valued modulation symbol $x=\gamma(I+jQ)$ in the output with the value of γ is chosen to achieve equal average power. More specifically, the NR supports the following modulation schemes depending on the channel conditions experienced by the users [6]:

- *$\pi/2$-BPSK:* In this case, bit $b(i)$ is mapped to complex-valued modulation symbol $d(i)$ according to $d(i)=1/\sqrt{2}[(1-2b(i))+j(1-2b(i))]\exp\left(j(\pi/2)(i \bmod 2)\right)$.
- *BPSK:* In this case, bit $b(i)$ is mapped to complex-valued modulation symbol $d(i)$ according to $d(i)=1/\sqrt{2}[(1-2b(i))+j(1-2b(i))]$.
- *QPSK:* In this case, a pair of bits $b(2i), b(2i+1)$ is mapped to complex-valued modulation symbol $d(i)$ according to $d(i)=1/\sqrt{2}[(1-2b(2i))+j(1-2b(2i+1))]$.
- *16QAM:* In this case, bit quadruplet $b(4i), b(4i+1), b(4i+2), b(4i+3)$ are mapped to complex-valued modulation symbol $d(i)$ according to $d(i)=1/\sqrt{10}\left\{(1-2b(4i))[2-(1-2b(4i+2))]+j(1-2b(4i+1))[2-(1-2b(4i+3))]\right\}$.
- *64QAM:* In this case, bit sextuplet $b(6i), b(6i+1), b(6i+2), b(6i+3), b(6i+4), b(6i+5)$ are mapped to complex-valued modulation symbol $d(i)$ according to $d(i)=1/\sqrt{42}\left\{(1-2b(6i))[4-(1-2b(6i+2))[2-(1-2b(6i+4))]]+j(1-2b(6i+1))[4-(1-2b(6i+3))[2-(1-2b(6i+5))]]\right\}$.
- *256QAM:* In this case, bit octuplet $b(8i), b(8i+1), b(8i+2), b(8i+3), b(8i+4), b(8i+5), b(8i+6), b(8i+7)$ are mapped to complex-valued modulation symbol $d(i)$ according to $d(i)=1/\sqrt{170}\left\{1-2b(8i)[8-1-2b(8i+2)[4-1-2b(8i+4)[2-1-2b(8i+6)]]]+j1-2b(8i+1)[8-1-2b(8i+3)[4-1-2b(8i+5)[2-1-2b(8i+7)]]]\right\}$.

To determine the modulation order, target code rate, and TBS in the PDSCH, the UE needs to read the 5-bit modulation and coding scheme field I_{MCS} in the DCI to determine the modulation order Q_m and target code rate R based on the procedure that we defined in the

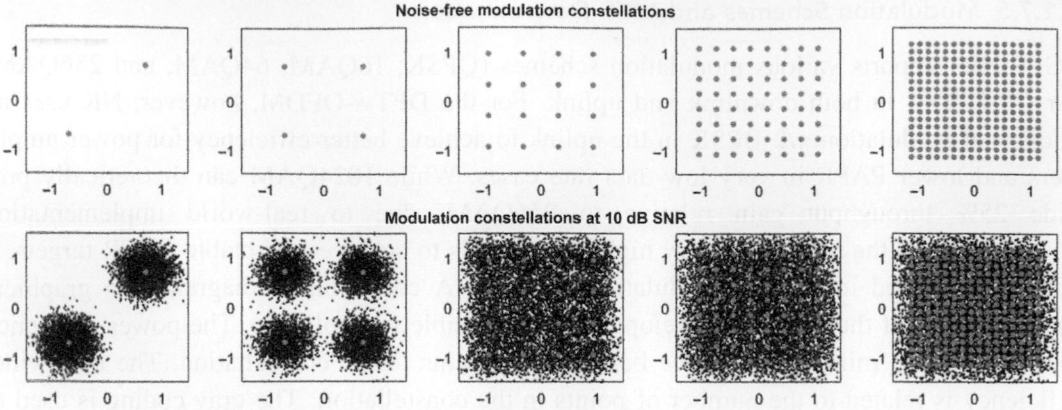

Figure 4.52
Modulation constellations at noise-free and SNR = 10 dB conditions.

previous section. The UE then use the number of layers v and the total number of allocated PRBs before rate matching n_{PRB} to determine to the TBS. The UE may skip decoding of a TB in an initial transmission, if the effective channel code rate is higher than 0.95. The effective channel code rate is defined as the number of downlink information bits (including CRC bits) divided by the number of physical channel bits transmitted on PDSCH [9]. Fig. 4.52 shows the constellations of NR modulation schemes in noise-free and SNR = 10 dB conditions, which demonstrate the effect of achievable SNR at the receiver detector and the choice of modulation order for transmission.

The concepts of MCS, code rate, TB, and TBS in 3GPP NR are similar to those of 3GPP LTE. In NR, the DL-SCH and UL-SCH MCS and code rate for transmission are determined by predefined tables given in [9]. However, TBS determination in NR is more complicated than that of LTE. Unlike LTE where all possible TBSs were precalculated and listed in the MCS table, in NR, the TBS determination process is described as a procedure that has been illustrated in Fig. 4.53. As shown in the flow chart, the initial input to this algorithm is N_{info}; however, to determine this value, the following calculations are necessary [9]:

$$\underbrace{N'_{RE}}_{\text{The number of REs allocated for PDSCH within a PRB}} = 12\underbrace{N_{slot}^{symb}}_{\text{The number of symbols of the PDSCH allocation within the slot}} - \underbrace{N_{DM-RS}^{PRB}}_{\text{The number of REs for DM-RS per PRB in the scheduled duration including the overhead of the DM-RS CDM groups without data}} - \underbrace{N_{overhead}^{PRB}}_{\text{The overhead configured by higher-layer parameter xOverhead in PDSCH-ServingCellConfig}}$$

$$\underbrace{N_{RE}}_{\text{The total number of REs allocated for PDSCH}} = \min(156, N'_{RE})\underbrace{n_{PRB}}_{\text{The total number of allocated PRBs for the UE}}$$

$$\underbrace{N_{info}}_{\text{Intermediate number of information bits}} = N_{RE}\underbrace{R}_{\text{Target code rate}}\underbrace{Q_m}_{\text{Modulation order}}\underbrace{v}_{\text{Number of layers}}$$

For downlink shared channel, the supported modulation schemes include QPSK, 16QAM, 64QAM, and 256QAM. After detecting the CSI-RS and estimating the channel quality, UE

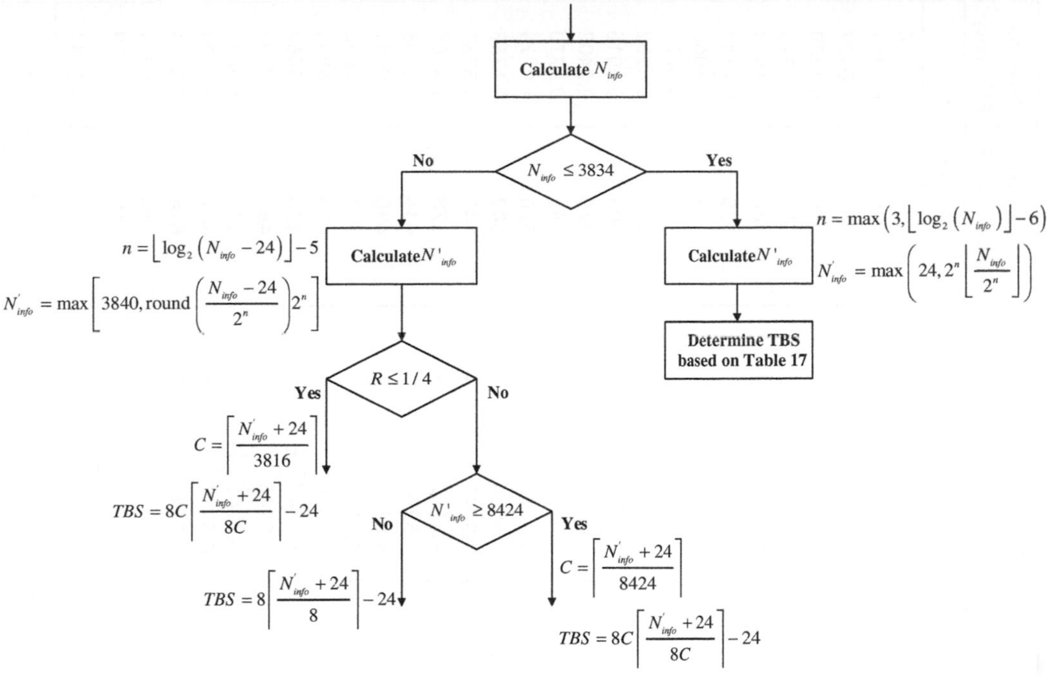

Figure 4.53
TBS determination procedure in NR [30].

reports the CQI to the gNB, which includes the information such as modulation scheme and coding rate. To balance the overhead and the granularity of CQI indication, two CQI/MCS tables are defined for eMBB, where the maximum order of modulation in one CQI/MCS table is 64QAM and in another table is 256QAM (see MCS Tables I and II in Table 4.16). The network will instruct the UE to select CQI/MCS table through RRC signaling. The third MCS table is meant for URLLC use cases where the target BLER is 10^{-5}, which is signaled to UE when the CRC of the PDCCH is scrambled with MCS-C-RNTI. This MCS table was designed to allow single transmissions to the UEs with delay sensitive applications to ensure maximum likelihood of correct reception.

Given the modulation order, the number of resource blocks scheduled, and the scheduled transmission duration, the number of available resource elements can be computed. From this number, the resource elements used for DM-RS are subtracted. A constant, configured by higher layers, modeling the overhead of other signals such as CSI-RS or SRS, is also subtracted. The resulting estimate of resource elements available for data allocation is then, together with the number of transmission layers, the modulation order, and the code rate obtained from the MCS table, are used to calculate an intermediate number of information bits. This intermediate number is then quantized to obtain the final transport block size,

Table 4.16: MCS index tables 1/2/3 for physical downlink shared channel [9].

MCS Index I_{MCS}	Modulation Order Q_m	Target Code Rate $1024 \times R$	Spectral Efficiency	Modulation Order Q_m	Target Code Rate $1024 \times R$	Spectral Efficiency	Modulation Order Q_m	Target Code Rate $1024 \times R$	Spectral Efficiency
	MCS Table I			MCS Table II			MCS Table III		
0	2	120	0.2344	2	120	0.2344	2	30	0.0586
1	2	157	0.3066	2	193	0.3770	2	40	0.0781
2	2	193	0.3770	2	308	0.6016	2	50	0.0977
3	2	251	0.4902	2	449	0.8770	2	64	0.1250
4	2	308	0.6016	2	602	1.1758	2	78	0.1523
5	2	379	0.7402	4	378	1.4766	2	99	0.1934
6	2	449	0.8770	4	434	1.6953	2	120	0.2344
7	2	526	1.0273	4	490	1.9141	2	157	0.3066
8	2	602	1.1758	4	553	2.1602	2	193	0.3770
9	2	679	1.3262	4	616	2.4063	2	251	0.4902
10	4	340	1.3281	4	658	2.5703	2	308	0.6016
11	4	378	1.4766	6	466	2.7305	2	379	0.7402
12	4	434	1.6953	6	517	3.0293	2	449	0.8770
13	4	490	1.9141	6	567	3.3223	2	526	1.0273
14	4	553	2.1602	6	616	3.6094	2	602	1.1758
15	4	616	2.4063	6	666	3.9023	4	340	1.3281
16	4	658	2.5703	6	719	4.2129	4	378	1.4766
17	6	438	2.5664	6	772	4.5234	4	434	1.6953
18	6	466	2.7305	6	822	4.8164	4	490	1.9141
19	6	517	3.0293	6	873	5.1152	4	553	2.1602
20	6	567	3.3223	8	682.5	5.3320	4	616	2.4063
21	6	616	3.6094	8	711	5.5547	6	438	2.5664
22	6	666	3.9023	8	754	5.8906	6	466	2.7305
23	6	719	4.2129	8	797	6.2266	6	517	3.0293
24	6	772	4.5234	8	841	6.5703	6	567	3.3223
25	6	822	4.8164	8	885	6.9141	6	616	3.6094
26	6	873	5.1152	8	916.5	7.1602	6	666	3.9023
27	6	910	5.3320	8	948	7.4063	6	719	4.2129
28	6	948	5.5547	2	Reserved		6	772	4.5234
29	2	Reserved		4	Reserved		2	Reserved	
30	4	Reserved		6	Reserved		4	Reserved	
31	6	Reserved		8	Reserved		6	Reserved	

Table 4.17: NR transport-block size for $N_{info} \leq 3824$ [9].

Index	TBS (bits)	Index	TBS (bits)	Index	TBS (bits)	Index	TBS (bits)	Index	TBS (bits)	Index	TBS (bits)	Index	TBS (bits)
1	24	16	144	31	336	46	704	61	1288	76	2216	91	3624
2	32	17	152	32	352	47	736	62	1320	77	2280	92	3752
3	40	18	160	33	368	48	768	63	1352	78	2408	93	3824
4	48	19	168	34	384	49	808	64	1416	79	2472		
5	56	20	176	35	408	50	848	65	1480	80	2536		
6	64	21	184	36	432	51	888	66	1544	81	2600		
7	72	22	192	37	456	52	928	67	1608	82	2664		
8	80	23	208	38	480	53	984	68	1672	83	2728		
9	88	24	224	39	504	54	1032	69	1736	84	2792		
10	96	25	240	40	528	55	1064	70	1800	85	2856		
11	104	26	256	41	552	56	1128	71	1864	86	2976		
12	112	27	272	42	576	57	1160	72	1928	87	3104		
13	120	28	288	43	608	58	1192	73	2024	88	3240		
14	128	29	304	44	640	59	1224	74	2088	89	3368		
15	136	30	320	45	672	60	1256	75	2152	90	3496		

while at the same time ensuring byte-aligned code blocks, and that no padding bits are needed in the LDPC coding. The quantization also results in the same transport block size being obtained, even if there are small variations in the amount of resources allocated, a property that is useful when scheduling retransmissions on a different set of resources than the initial transmission (see Table 4.17). In the case of a retransmission, the transport block size by definition, is unchanged and there is no need to signal this information. Instead, the reserved entries represent the modulation scheme (QPSK, 16QAM, 64QAM, or if configured 256QAM), which allows the scheduler to use an (almost) arbitrary combination of resource blocks for the retransmission. The use of the reserved entries assumes that the UE properly received the control signaling for the initial transmission, if this is not the case, the retransmission should explicitly indicate the transport block size [14].

4.1.8 HARQ Operation and Protocols

4.1.8.1 HARQ Principles

While ARQ error control mechanism is simple and provides high transmission reliability, the throughput of ARQ schemes drop rapidly with increasing channel error rates, and the latency, due to retransmissions, could be excessively high and intolerable for some delay--sensitive applications. Systems using forward error correction (FEC), on the other hand, can maintain constant throughput regardless of channel error rate. However, FEC schemes have some drawbacks. High reliability is hard to achieve with FEC and requires the use of long and powerful error correction codes that increase the complexity of implementation.

The drawbacks of ARQ and FEC can be overcome, if the two error control schemes are properly combined. In order to achieve increased throughput and lower latency in packet transmission, hybrid ARQ (HARQ) scheme was designed to combine ARQ error-control mechanism and FEC coding. A HARQ system consists of a FEC subsystem contained in an ARQ system. In this approach, the average number of retransmissions is reduced by using FEC through correction of the error patterns that occur more frequently; however, when the less frequent error patterns are detected, the receiver requests a retransmission where each retransmission carries the same or some redundant information to help the packet detection. The HARQ uses FEC to correct a subset of errors at the receiver and rely on error detection to detect the remaining errors. Most practical HARQ schemes utilize CRC codes for error detection and some form of FEC for correcting the transmission errors. The HARQ schemes are typically classified into two groups depending on the content of subsequent retransmissions, as follows [15]:

1. *HARQ with chase combining:* In this HARQ scheme, the same data packet is transmitted in all retransmissions. Soft combining may be used to improve the reliability. The blocks of data along with the CRC code are encoded using FEC encoder before transmission. If the receiver is unable to correctly decode the data block, a retransmission is requested. When a retransmitted coded block is received, it is combined with the previously received block corresponding to the same information bits (using for example maximum ratio combining method) and fed to the decoder. Since each retransmission is an identical replica of the original transmission, the received E_b/N_0, that is, the energy per information bit divided by the noise spectral power density, increases per each retransmission, improving the likelihood of correct decoding. In chase combining HARQ, the redundancy version of the encoded bits is not changed from one transmission to the next; therefore, the puncturing pattern remains the same. The receiver uses the current and all previous HARQ transmissions of the code block in order to decode the information bits. The process continues until either the information bits are correctly decoded and pass the CRC test or the maximum number of HARQ retransmissions is reached. When the maximum number of retransmissions is reached, the MAC sublayer resets the process and continues with fresh transmission of the same code block. A number of parallel channels for HARQ can help improve the throughput as one process is awaiting an acknowledgment; another process can utilize the channel and transmit subpackets. Fig. 4.54 illustrates the operation of chase combining HARQ (HARQ-CC) scheme and how the retransmission of the same coded bits changes the combined energy per bit E_b while maintaining the effective code rate intact.
2. *HARQ with incremental redundancy:* In this HARQ scheme, additional parity bits are sent in subsequent retransmissions. Therefore, after each retransmission, a richer set of parity bits is available at the receiver, improving the probability of reliable decoding. In incremental redundancy schemes, however, information cannot be recovered from parity

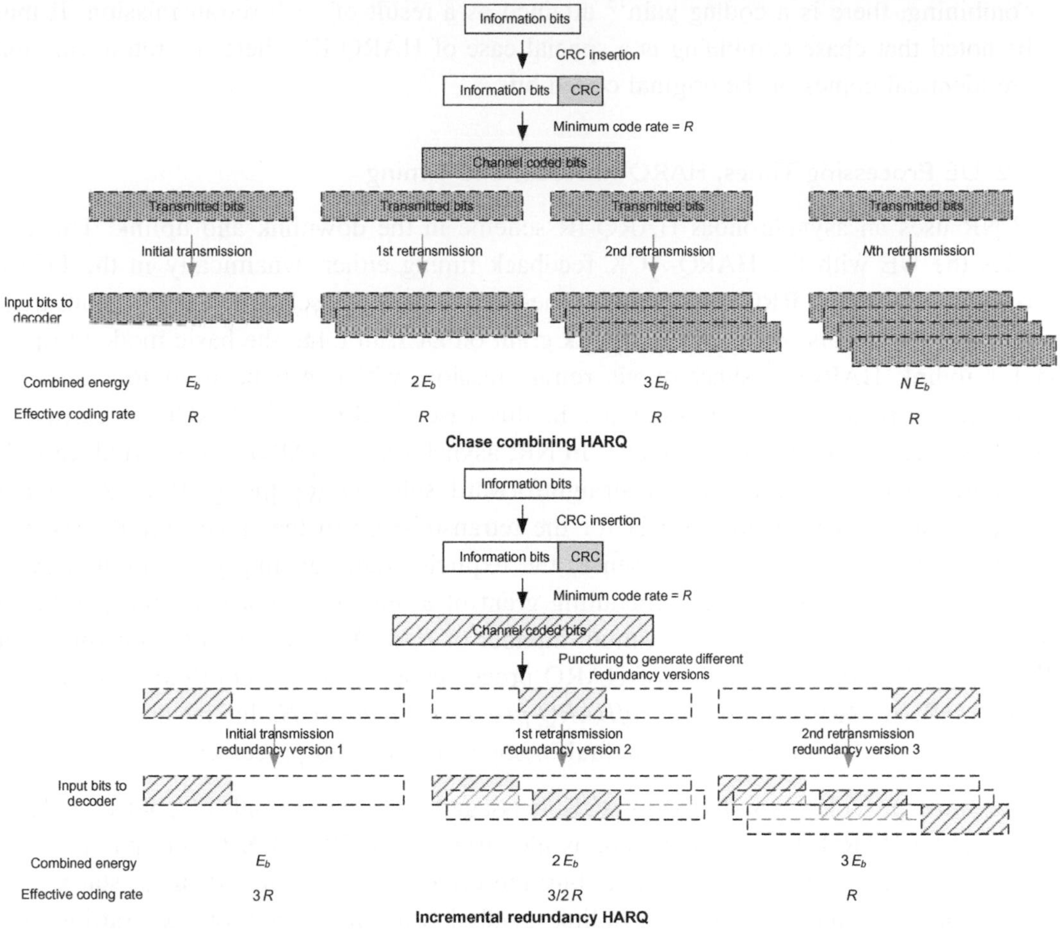

Figure 4.54

Illustration of the chase combining and incremental redundancy HARQ schemes [15].

bits alone. In incremental redundancy HARQ (HARQ-IR) scheme, a number of coded bits with increasing redundancy are generated and transmitted to the receiver when a retransmission is requested to assist the receiver with the decoding of the information bits. The receiver combines each retransmission with the previously received bits belonging to the same packet. Since each retransmission carries additional parity bits, the effective code rate is lowered by each retransmission as shown in Fig. 4.54. The IR is based on low-rate code and the different redundancy versions are generated by puncturing the channel coder output. In the example shown in Fig. 4.54, the basic code rate is R, and one-third of the coded bits are transmitted in each retransmission. Aside from increasing the received signal to noise ratio E_b/N_0 by each retransmission due to combining, there is a coding gain[13] attained as a result of each retransmission. It must be noted that chase combining is a special case of HARQ-IR where the retransmissions are identical copies of the original coded bits.

4.1.8.2 UE Processing Times, HARQ Protocol and Timing

3GPP NR uses an asynchronous HARQ-IR scheme in the downlink and uplink. The gNB provides the UE with the HARQ-ACK feedback timing either dynamically in the DCI or semi-statically through RRC configuration messages. The gNB schedules each uplink transmission and retransmission using the uplink grant on DCI. In LTE, the basic mode of operation for uplink HARQ is synchronous retransmission, which can be used to reduce the scheduling overhead for retransmissions. In this case, HARQ ACK/NACK is carried on PHICH as a short and efficient message. In NR, asynchronous HARQ is supported. In order to support asynchronous HARQ, a straightforward solution for the gNB is to send an explicit uplink grant through PDCCH for the retransmission in the same way that is done for transmissions in LTE. In some sense, the explicit grant can imply an implicit ACK/NACK. For example, an explicit scheduling grant of a retransmission may imply a NACK for the initial transmission. The maximum number of HARQ processes in the downlink and uplink per cell is 16. The number of HARQ processes is separately configured for the UE for each cell by RRC parameter *nrofHARQ-processesForPDSCH*. In the absence of any configuration, the UE may assume a default number of 8 HARQ processes.

The UE must provide a valid HARQ-ACK message, if the first uplink symbol of PUCCH conveying the HARQ-ACK information, as identified by HARQ-ACK timing parameter K_1 and the assigned PUCCH resource including the effect of the timing advance, starts on or after symbol L_1, that is, the next uplink symbol with its cyclic prefix starting after $T_{proc} = 2192(N_1 + d_x)\kappa 2^{-\mu} T_c$ (processing time) following the end of the last symbol of the PDSCH carrying the transport block being acknowledged. As shown in Fig. 4.55, parameter N_1 is based on the numerology and corresponds to $(\mu_{PDCCH}, \mu_{PDSCH}, \mu_{UL})$ where $\mu_{PDCCH}, \mu_{PDSCH},$ and μ_{UL} correspond to the subcarrier spacing of the PDCCH scheduling, PDSCH transmission, and the uplink channel on which the HARQ-ACK is transmitted, respectively. As shown in Table 4.18, the value of parameter N_1 further depends on the PDSCH DM-RS pattern and whether additional DM-RS is used as well as UE PDSCH processing capability. The value of parameter d_x is dependent on the PDSCH mapping type (A or B), UE PDSCH processing capability, and the number of PDSCH symbols [9]. The timing relationship between HARQ-ACK and PDSCH data transmission depends on the value of the above parameters and is depicted in Fig. 4.55.

[13] The coding gain is defined as the difference between E_b/N_0 required to achieve a given bit error rate in a coded system and the E_b/N_0 required to achieve the same BER in an uncoded system.

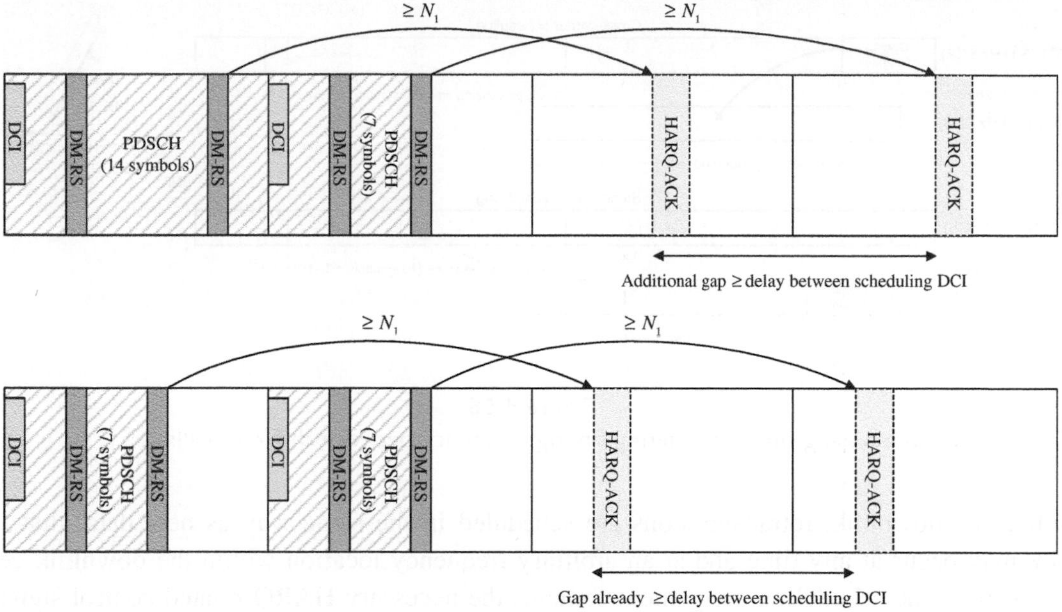

Figure 4.55
Timing relationship between PDSCH and HARQ-ACK transmission [9].

Table 4.18: Physical downlink shared channel processing times [9].

μ	PDSCH Decoding Time N_1 (OFDM Symbols)		PDSCH Decoding Time N_1 (OFDM Symbols)
	PDSCH Processing Capability 1		PDSCH Processing Capability 2
	No Additional PDSCH	Additional PDSCH	No Additional PDSCH
	DM-RS Configured (Front-Loaded DM-RS)	DM-RS Configured	DM-RS Configured (Front-Loaded DM-RS)
0	8	13	3
1	10	13	4.5
2	17	20	9 (in Frequency Range 1)
3	20	24	–

In the case of carrier aggregation, the multi-carrier nature of the physical layer is only exposed to the MAC sublayer for which one HARQ entity is required per serving cell. In both uplink and downlink, there is one independent HARQ entity per serving cell and one TB is generated per TTI in the absence of spatial multiplexing. Each TB and its associated HARQ retransmissions are mapped to a single serving cell [11]. Fig. 4.56 depicts HARQ-ACK timing requirements in a cross-carrier scheduling scenario when a UE receives and transmits to from/to two gNBs.

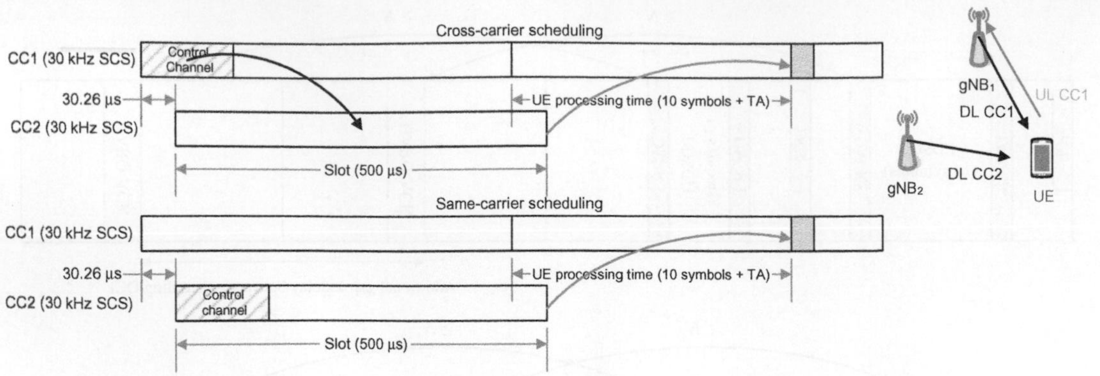

Figure 4.56
UE processing time considering timing difference between different cells [65].

In the NR downlink, retransmissions are scheduled in the same way as new data, that is, they may occur at any time and at an arbitrary frequency location within the downlink cell bandwidth. The scheduling assignment contains the necessary HARQ-related control signaling such as process number, new-data indicator, CBGTI, and CBGFI in case of CBG-based retransmission as well as information to handle the transmission of the acknowledgment in the uplink such as timing and resource indication. Upon receiving a scheduling assignment in the DCI, the receiver would attempt to decode the TB after soft combining with previous retransmissions. Since transmissions and retransmissions are scheduled using the same framework, the UE needs to know whether the transmission is a new transmission, in which case the soft buffer should be flushed, or a retransmission, in which case soft combining should be performed. Therefore, an explicit new-data indicator is included for the scheduled TB as part of the scheduling information transmitted in the downlink. The new-data indicator is toggled for a new TB. Upon reception of a downlink scheduling assignment, the UE checks the new-data indicator to determine whether the current transmission should be soft combined with the received data currently in the soft buffer for the HARQ process or the soft buffer should be cleared [9,11].

The NR uplink uses asynchronous HARQ protocol in the same way as downlink. The HARQ-related information including process number, new-data indicator, CBG-based retransmission (if configured), and the CBGTI are included in the scheduling grant. The uplink CBGTI is used in the same way as in the downlink, that is, to indicate the CBGs that need to be retransmitted in the case of CBG-based retransmission. Note that no CBGFI is needed in the uplink as the soft buffer is located in the gNB which can decide whether to flush the buffer or not based on the scheduling decisions.

The use of HARQ in NR allows reliable delivery of layer-1 packets between peer entities. Each HARQ process supports one TB when the physical layer is not configured for

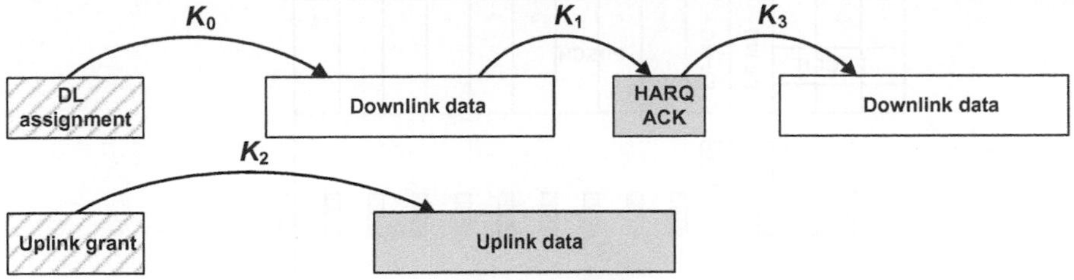

Figure 4.57
NR HARQ protocol timing [9].

downlink/uplink spatial multiplexing; otherwise, each HARQ process supports one or multiple TBs. The NR HARQ operation and timing are illustrated in Fig. 4.57, where the parameters used in the figure are defined as follows [8,9]:

- K_0 denotes the delay between downlink grant and corresponding PDSCH data reception.
- K_1 is the delay between PDSCH data reception and the corresponding ACK/NACK transmission on the uplink.
- K_2 denotes the delay between uplink grant reception in downlink and the uplink data transmission on PUSCH.
- K_3 is the delay between ACK/NACK reception in the uplink and the corresponding retransmission of downlink data on PDSCH.
- The parameters K_0, K_1, and K_2 are signaled via DCI and if $K_1 = 0$, a self-contained subframe/slot is configured.

The choice of N_1 value has a significant impact on the UE processing time, where N_1 is defined as the number of OFDM symbols from the end of PDSCH reception to the start of the corresponding ACK/NACK transmission from UE (as shown Fig. 4.58). Depending on the frame structure, some UE data processing can be done in parallel with PDSCH reception in order to allow faster HARQ ACK/NACK transmission. For example, for front-loaded DM-RS pattern and slot-based scheduling with subcarrier spacing of 15 kHz, it can be shown that PDCCH processing, the demodulation/detection of the symbols other than the last symbol of PDSCH can be performed in $T_1 = T_{FFT} + T_{demodulation} + T_{decode} + T_{UL\text{-}HARQ} + T_{other}$ where T_{FFT}, $T_{demodulation}$, T_{decode}, $T_{UL\text{-}HARQ}$, and T_{other} denote the processing time for FFT/IFFT per symbol, the demodulation time for one symbol, the decoding time for one symbol code blocks, the processing time for uplink ACK/NACK, and the other implementation-specific processing times, respectively.

For the non-slot-based scheduling (e.g., two-symbol mini-slot) and subcarrier spacing 15 kHz, it can be shown that the UE processing requirement $T_2 = T_{PDCCH} + 2T_{demodulation} + T_{decode} + T_{UL\text{-}HARQ} + T_{other}$ where T_{PDCCH} denotes the

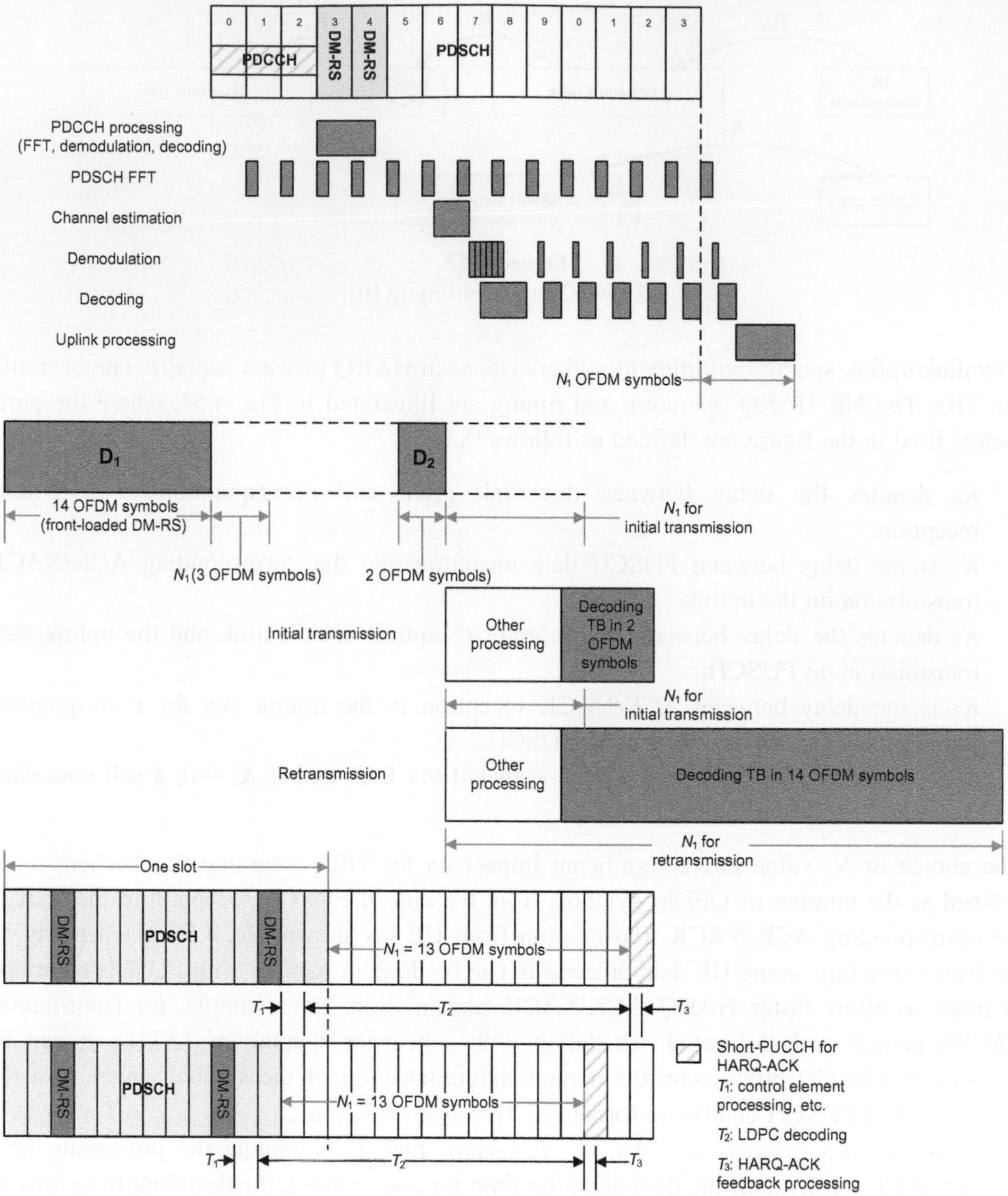

Figure 4.58

UE processing time components (front-loaded DM-RS, slot-based scheduling, and 15 kHz SCS).

processing time for PDCCH including decoding, demodulation, and parsing. The above processing time calculation was done under certain conditions which may vary in different scenarios. There is no limit for scheduling transmission/retransmission; however, the initial transmission is slot-based, and the retransmissions may be non-slot based as shown in Fig. 4.58. In the following example, we assume subcarrier spacing is 15 kHz, D_1 is a slot-based downlink scheduling period with front-loaded DM-RS and D_2 is a non-slot-based scheduling period. If D_2 is the initial transmission then the processing time is calculated as T_2; nevertheless, if D_2 is the retransmission of D_1 then we need to perform TB decoding with 14 OFDM symbols; thus the processing time for retransmission is shown to be $T_3 = T_{PDCCH} + 2T_{demodulation} + 13T_{decode} + T_{UL\text{-}HARQ} + T_{other}$, meaning that for the same conditions, the processing time for initial transmission and retransmission are different [65].

As we mentioned earlier, the maximum number of HARQ processes per carrier supported in NR is 8 or 16. For continuous downlink transmission at the peak data rate, the minimum number of HARQ processes is $\min(N_{DL-HARQ}) = K_1 + K_3 + \lfloor 2T_d/TTI_{DL} \rfloor$, in which T_d denotes the transmission delay. The required number of HARQ processes may vary depending on UE HARQ processing capability, numerology, and network configurations. The determination of the number of HARQ processes is up to gNB scheduler and thus signaled via the DCI. To reduce the overhead, the gNB can semi-statically configure a UE with a smaller number of HARQ processes than 16 per bandwidth part. In order to reduce the latency due to HARQ retransmissions and to avoid retransmission of the entire TB and the performance degradation of HARQ due to large transport block size, NR defined CBG-based transmission and HARQ operation which supports single-/multi-bit HARQ-ACK feedback in Rel-15. The CBG-based (re)transmissions are only allowed for the same TB of a HARQ process. The CBG can include all code blocks of a TB regardless of the size of the TB. In such conditions, the UE reports single HARQ-ACK bits for the TB. The CBG can include one code block and its granularity is configurable. The UE is semi-statically configured by RRC signaling to enable CBG-based retransmission.

When the *CSI request* field in a DCI triggers a CSI report(s) on PUSCH, the UE is required to provide a valid CSI report(s), if the first uplink symbol to carry the corresponding CSI report(s), including the effect of the timing advance, starts no earlier than at symbol l_{CSI}, and if the first uplink symbol to carry the corresponding CSI report, including the effect of the timing advance, starts no earlier than at symbol l'_{CSI}. The reference symbol l_{CSI} is defined as the next uplink symbol with starting $T_{proc\text{-}CSI} = 2192 l_{CSI} \kappa 2^{-\mu} T_C$ after the end of the last symbol of the PDCCH triggering the CSI report. The reference symbol l'_{CSI} is defined as the next uplink symbol starting $T'_{proc\text{-}CSI} = 2192 l'_{CSI} \kappa 2^{-\mu} T_C$ after the end of the last symbol of the latest aperiodic CSI-RS resource for channel measurements, aperiodic CSI-IM used for interference measurements, and aperiodic NZP CSI-RS for interference measurement, when aperiodic CSI-RS is used for channel measurement for the triggered nth CSI report [9].

Table 4.19: NR channel state information computation delay requirements 1 and 2 [9].

μ	Delay Requirement 1		Delay Requirement 2			
	Z_1 (Symbols)		Z_1 (Symbols)		Z_2 (Symbols)	
	Z_1	Z'_1	Z_1	Z'_1	Z_2	Z'_2
0	10	8	22	16	40	37
1	13	11	33	30	72	69
2	25	21	44	42	141	140
3	43	36	97	85	152	140

As a result, l_{CSI} and l'_{CSI} are defined as $l_{CSI} = \max_{m=0,1,\ldots,N_{CSI}-1} l_{CSI}(m)$ and $l'_{CSI} = \max_{m=0,1,\ldots,N_{CSI}-1} l'_{CSI}(m)$, where N_{CSI} is the number of updated CSI report(s), $l_{CSI}(m)$ and $l'_{CSI}(m)$ corresponds to the mth updated CSI report. The values of $l_{CSI}(m)$ and $l'_{CSI}(m)$ are set to l_1, l'_1, l_2, or l'_2 depending on the CSI computation delay requirements as shown in Table 4.19, μ corresponds to the $\min(\mu_{PDCCH}, \mu_{CSI\text{-}RS}, \mu_{UL})$ where the μ_{PDCCH} corresponds to the subcarrier spacing of the PDCCH in which the DCI was transmitted, and μ_{UL} corresponds to the subcarrier spacing of the PUSCH in which the CSI report is transmitted, and $\mu_{CSI\text{-}RS}$ corresponds to the minimum subcarrier spacing of the aperiodic CSI-RS triggered by the DCI.

4.1.8.3 Semi-static/Dynamic Codebook HARQ-ACK Multiplexing

The NR supports multiplexing of HARQ acknowledgments for multiple transport blocks into an acknowledgment bitmap, when multiple TBs need to be acknowledged at the same time or alternatively multiple acknowledgments need to be transmitted in the uplink at the same time in carrier aggregation and CBG-based retransmission scenarios. This bitmap can be signaled either via a semi-static codebook or a dynamic codebook both configured through RRC signaling. The semi-static codebook can be viewed as a matrix consisting of a time-domain dimension and a component-carrier, CBG, or MIMO layer dimension, both of which are semi-statically configured. The size in the time domain is given by the maximum and minimum HARQ acknowledgment timings and the size in the carrier domain is given by the number of simultaneous transport blocks or CBGs across all component carriers. An example is shown in Fig. 4.59, where the acknowledgment timings are one, two, three, and four slots, respectively, and three carriers, one with two TBs, one with one TB, and one with four CBGs, are configured. Since the codebook size is fixed, the number of bits to transmit in a HARQ-ACK is known and an appropriate format for the uplink control signaling can be selected. Each entry in the matrix represents successful/unsuccessful outcome of the decoding of the corresponding downlink transmission. A NACK is sent in position of unused transmission opportunities in the codebook, resulting in improved robustness in the case of missed downlink assignment where the gNB can retransmit the missing TB or the CBG [14].

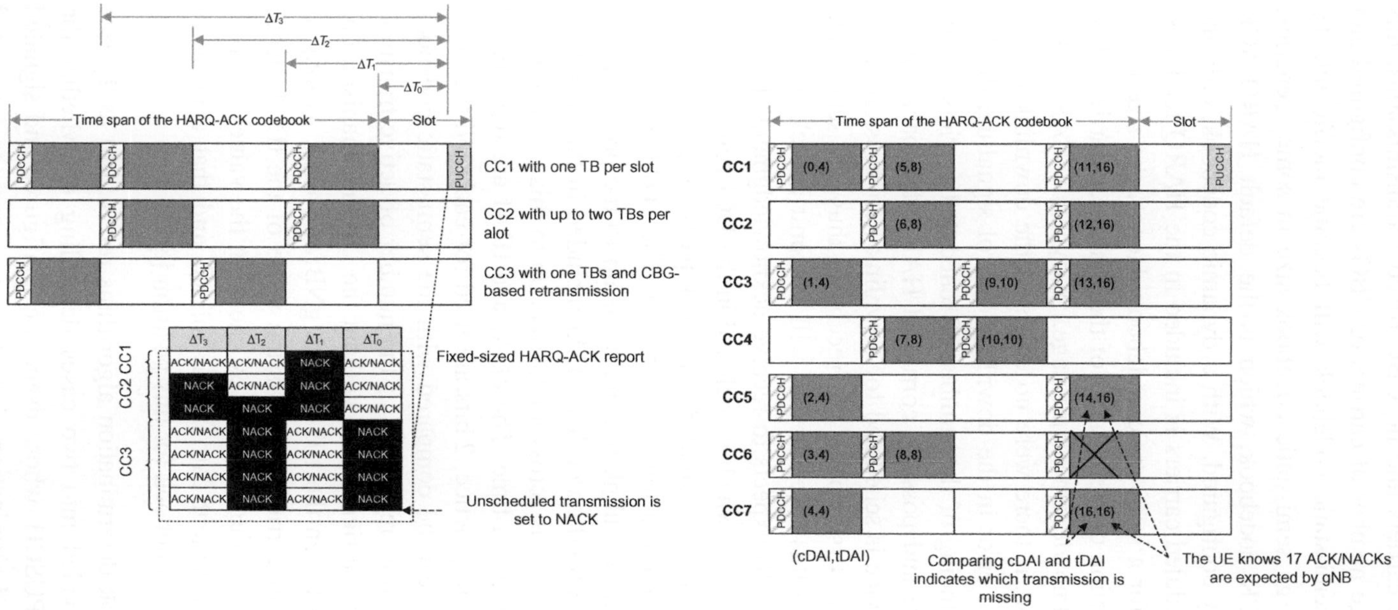

Figure 4.59
Illustration of the semi-static and dynamic codebooks (example) [14].

One drawback of the semi-static codebook is the potentially large size of the HARQ feedback. For a small number of component carriers and no CBG-based retransmissions, this may not be a problem; however, if a large number of carriers or CBGs are configured and only a fraction of them are used, the semi-static codebook will become inefficient. To address the drawback of a potentially large semi-static codebook size in some scenarios, NR also supports a dynamic HARQ-ACK codebook, which is the default HARQ-ACK codebook, unless the system is differently configured. With a dynamic codebook, only the acknowledgment information for the scheduled carriers is included in the HARQ feedback, as opposed to all carriers as is the case for a semi-static codebook. Hence, the size of the codebook may dynamically vary as a function of the number of the scheduled carriers. This would reduce the size of the HARQ acknowledgment message. A dynamic HARQ-ACK codebook would be straightforward option, if there were no errors in the downlink control signaling. However, in the presence of an error in the downlink control signaling, the UE and the gNB may have different understanding of the number of scheduled carriers, which would lead to an incorrect codebook size and possibly corrupted HARQ feedback for all carriers. As an example, assume that a device is scheduled for downlink transmission in two consecutive slots, but the PDCCH of the first slot is not received; thus the scheduling assignment for the first slot is missing. In this case, the UE will transmit an acknowledgment for the second slot only, while the gNB expects to receive acknowledgments for two slots. To mitigate such cases, NR supports DAI which is included in the DCI containing the downlink assignment. The DAI field is further divided into two parts, a counter DAI (cDAI) and, in the case of carrier aggregation, a total DAI (tDAI). The cDAI included in the DCI indicates the number of scheduled downlink transmissions up to the point that the DCI was received in a carrier first, time second order. The tDAI included in the DCI indicates the total number of downlink transmissions across all carriers up to this point in time, that is, the highest cDAI at the current point in time. The cDAI and tDAI are represented with decimal numbers with no limitation. In practice, 2 bits are used for each and the numbering is calculated in modulo four. This can be compared with the semi-static codebook which would require a fixed number of entries regardless of the number of active transmissions. If one transmission on a component carrier is lost, without the DAI mechanism, this would result in mismatched codebooks between the UE and the gNB; however, as long as the device receives at least one component carrier, it knows the value of the tDAI, hence the size of the codebook at this point in time. Furthermore, by checking the values received for the cDAI, it can conclude which component carrier was missed and that a negative acknowledgment should be assumed in the codebook for this position [8].

There are two different types of codebook determination algorithms referred to as Type 1 and Type 2. Each of these types is divided into two cases depending on whether the HARQ-ACK is reported on PUCCH or PUSCH, whose usage are configured and signaled through RRC parameters. The codebook determination algorithm types and the associated RRC parameters are summarized in Table 4.20.

Table 4.20: HARQ-ACK codebook determination [8].

Codebook Determination Type	Condition
CBG-based HARQ-ACK codebook determination	*PDSCH-CodeBlockGroupTransmission* = ON
Type-1 HARQ-ACK codebook determination	*pdsch-HARQ-ACK-Codebook* = *semistatic*
Type-1 HARQ-ACK codebook in PUCCH	–
Type-1 HARQ-ACK codebook in PUSCH	–
Type-2 HARQ-ACK codebook determination	*pdsch-HARQ-ACK-Codebook* = *dynamic*
Type-2 HARQ-ACK codebook in PUCCH	*PDSCH-CodeBlockGroupTransmission* = OFF
Type-2 HARQ-ACK codebook in PUSCH	–

4.1.9 Downlink MIMO Schemes

The new radio downlink control and traffic channels rely on DM-RSs to facilitate coherent detection where a UE can assume that the DM-RSs are jointly precoded with the data. In NR Rel-15, downlink multi-antenna precoding is transparent to the UE and the network can apply any transmitter-side precoding without informing the device about the selected precoder. The specification impact of downlink multi-antenna precoding is therefore mainly related to the measurements and reporting mechanisms conducted by the device to support network selection of precoder for downlink transmissions. These precoder-related measurements and reporting are part of the more general CSI reporting framework based on report configurations which was described in previous sections.

A CSI report may consist of rank indicator (RI), precoder-matrix indicator (PMI), and CQI indicating a suitable transmission rank, precoding matrix given the selected rank, coding and modulation scheme given the selected precoder matrix, respectively, from the device perspective. As we mentioned earlier, the reported PMI is a suitable precoder matrix from the UE perspective to be used for downlink transmissions. Each PMI value corresponds to a specific precoder matrix or codebook. Note that the device selects PMI based on a certain number of antenna ports, given by the number of antenna ports of the configured CSI-RS associated with the report configuration, and the selected rank. There is at least one codebook for each valid combination of antenna ports and rank. It must be understood that the suggested codebooks by the UE do not compel the gNB, in practice, not to use another precoding matrix for downlink transmission to the device.

The MU-MIMO schemes typically require detailed knowledge of the channel experienced by each device at the gNB compared to SU-MIMO precoding and transmission to a single device. Therefore, NR defines two types of CSI that differ in the structure and size of the precoding matrices (codebooks), that is, Type I CSI and Type II CSI. Type I CSI primarily targets scenarios where a single user is scheduled within a given time/frequency resources potentially with transmission of a relatively large number of layers in parallel (high-order spatial multiplexing) and Type II CSI mainly targets MU-MIMO scenarios with multiple devices being scheduled simultaneously within the same time/frequency resources but with only a limited number of spatial layers (maximum of two layers) per scheduled device.

The codebooks for Type I CSI are relatively simple and used to focus the transmit energy at the target receiver. Inter-layer interference is assumed to be primarily handled by utilizing multiple receive antennas and advanced receiver architectures. In contrast, the codebooks for Type II CSI are significantly more extensive, allowing the PMI to provide channel information with much higher spatial granularity. The more extensive channel information allows the network to select a downlink precoder that not only focuses the transmitted energy at the target device but also limits the interference to other devices simultaneously scheduled on the same time/frequency resources. The higher spatial granularity of the PMI feedback is made possible with significantly higher signaling overhead. While a PMI report for Type I CSI may consist of a few tens of bits, the PMI report for Type II CSI may consist of several hundred bits. Therefore, Type II CSI is mainly applicable for low-mobility scenarios where the feedback periodicity in time can be reduced.

4.1.9.1 Capacity of MIMO Channels

A generic MIMO system consists of a MIMO transmitter with N_{TX} transmit antennas, a MIMO receiver with N_{RX} receive antennas, and $N_{RX} \times N_{TX}$ paths or channels between transmit and receive antennas. Let $x_k(t)$ denote the transmitted signal from the kth transmit antenna at time t then the received signal at the lth antenna can be expressed as $y_l(t) = \sum_{k=0}^{N_{TX}-1} h_{lk}(t)^* x_k(t) + n_l(t)$ where $h_{lk}(t)$ and $n_l(t)$ are the channel impulse response between the kth transmit antenna and lth receive antenna and the additive noise at the lth receive antenna port, respectively. The preceding equation can be written in the frequency-domain as $Y_l(\omega) = H_{lk}(\omega) X_k(\omega) + N_l(\omega)$. If $\mathbf{x}(\omega) = \left[X_1(\omega), X_2(\omega), \ldots, X_{N_{TX}}(\omega)\right]^T$, $\mathbf{y}(\omega) = \left[Y_1(\omega), Y_2(\omega), \ldots, Y_{N_{RX}}(\omega)\right]^T$, and $\mathbf{n}(\omega) = \left[N_1(\omega), N_2(\omega), \ldots, N_{N_{RX}}(\omega)\right]^T$ denote the Fourier transform vectors of $\mathbf{x}_k(t)$, $\mathbf{y}_l(t)$, and $\mathbf{n}_l(t)$, respectively then $\mathbf{y}(\omega) = \mathbf{H}(\omega)\mathbf{x}(\omega) + \mathbf{n}(\omega)$ where $\mathbf{H}(\omega)$ is an $N_{RX} \times N_{TX}$ channel matrix with $H_{lk}(\omega)|_{k=1,2,\ldots,N_{TX}; l=1,2,\ldots,N_{RX}}$ entries. Assuming a linear time-invariant MIMO channel, the channel input–output relationship can be further described in the discrete time-domain as follows:

$$y_l(nT) = \sum_{k=1}^{N_{TX}} \sum_{m=0}^{M-1} h_{lk}(mT) x_k[(n-m)T] + n_l(nT), \quad 0 \le n \le M-1; \quad 1 \le l \le N_{RX}$$

where $x_k(n)|_{k=1,2,\ldots,N_{TX}}$ and $y_l(n)|_{l=1,2,\ldots,N_{RX}}$ represent the channel input and output time-domain signals, respectively. In the case of time-varying channels, the preceding equation can be written as $y_l(t) = \sum_{k=0}^{N_{TX}-1} h_{lk}(t,\tau) x_k(\tau) + n_l(t)$ where $h_{lk}(t,\tau)$ denotes the time-varying impulse response of the lkth channel.

The matrix form of MIMO channel output in frequency-domain sampled at single frequency ω_m can be written as $\mathbf{y}(\omega_m) = \mathbf{H}(\omega_m)\mathbf{x}(\omega_m) + \mathbf{n}(\omega_m)$. In an OFDM system, the signal processing is inherently performed in frequency domain. The OFDM transforms a frequency-selective fading channel to a flat-fading channel when considering narrowband orthogonal subcarriers. In such system, the MIMO signal processing can be performed at each subcarrier. This is the main reason for suitability of MIMO extension to an OFDM system. When MIMO processing is performed at each subcarrier, the MIMO channel input–output relationship can be demonstrated as $\mathbf{y} = \mathbf{Hx} + \mathbf{n}$, where the channel between the transmitter and the receiver is typically modeled as a finite impulse response (FIR) filter. In this case, each tap is typically a complex-valued Gaussian random variable with exponentially decaying magnitudes. The tap delays correspond to the RMS delay spread and the channel type (e.g., low-delay spread or flat fading, high-delay spread or frequency-selective fading). There is a new realization of the channel at every transmitted packet, if the channel remains invariant for the duration of the packet; otherwise, the variation of the channel is explicitly modeled in the signal detection. As mentioned earlier, there are $N_{RX} \times N_{TX}$ paths between the transmitter and the receiver where each channel is the sum of several FIR filters with different delay spreads. The channels may or may not be correlated. The MIMO schemes can be used with non-OFDM systems when the channel is modeled as flat fading such that $y_l(nT) = \sum_{k=1}^{N_{TX}} h_{lk} x_k(nT) + n(nT)$. An important question is to what extent MIMO techniques can increase the throughput and improve the reliability of the wireless communication systems. This question can be answered by calculating the information theoretic capacity of a single-input–single-output (SISO) channel and comparing it with the capacity of single-input–multiple-output (SIMO), multiple-input–single-output (MISO), and MIMO channels.

For a memoryless SISO channel (i.e., one transmit and one receive antenna), the channel capacity is given by $C_{SISO} = \log_2(1 + \gamma |h|^2)$ where h is the normalized complex-valued gain/attenuation of a fixed wireless channel or that of a particular realization of a random channel and $\gamma = E_s/N_0$ denotes the SNR at the receive antenna port.

As the number of receive antennas increases, the statistics of channel capacity improve. Using N_{RX} receive antennas and one transmit antenna, a SIMO system is formed with a capacity given by (when the channel is unknown to the transmitter) $C_{SIMO} = \log_2(1 + \gamma \sum_{i=1}^{N_{RX}} |h_i|^2)$ where h_i is the gain of the ith channel corresponding to the ith receive-antenna. Note that increasing the value of N_{RX} results in a logarithmic increase in average channel capacity. It can be shown that knowledge of the channel at the transmitter in SIMO cases does not provide any capacity benefit.

In the case of MISO or transmit diversity, where the transmitter does not typically have knowledge of the channel, the capacity is given by $C_{MISO} = \log_2(1 + \gamma / N_{TX} \sum_{i=1}^{N_{TX}} |h_i|^2)$. It is noted from the latter equation that when the channel information is not available at the transmitter $C_{MISO} < C_{SIMO}$. The power normalization factor N_{TX} ensures that the total transmit power is uniformly distributed among the transmit antennas. Furthermore, one notes the absence of an array gain in this case (MISO) compared to that of the receive diversity scenario where the energy of the multipath channels can be coherently combined. In addition, the MISO capacity has a logarithmic relationship with N_{TX} similar to that of a SIMO scheme. When the channel information is available at the transmitter, the capacity of a MISO system can approach to that of a SIMO system.

The use of diversity at both transmitter and receiver sides creates a MIMO system. The capacity of a MIMO system with N_{TX} transmit-antennas and N_{RX} receive antennas is expressed as $C_{MIMO} = \log_2\left(\det\left[\mathbf{I} + \gamma/N_{TX}\mathbf{H}\mathbf{H}^H\right]\right)$ where $\mathbf{I}$ is an $N_{RX} \times N_{RX}$ identity matrix and $\mathbf{H}$ is the $N_{RX} \times N_{TX}$ channel matrix. Note that both MISO and MIMO channel capacities are based on equal power and uncorrelated sources. It is demonstrated in the literature that the capacity of the MIMO channel increases linearly with $\min(N_{RX}, N_{TX})$ rather than logarithmically as in the case of MISO or SIMO channel capacity since the determinant operator yields the product of non-zero eigenvalues of its channel-dependent matrix argument, each eigenvalue characterizing the SNR over a SISO eigen-channel. It will be shown later that the overall MIMO channel capacity is the sum of capacities of each of these SISO eigen-channels. The increase in capacity is dependent on properties of the channel eigenvalues. If the channel eigenvalues decay rapidly then linear growth in capacity will not occur. However, the eigenvalues have a known limiting distribution and tend to be spaced out along the range of this distribution. Thus it is unlikely that most eigenvalues are very small, and the linear growth is indeed achieved.

The capacity of the MIMO channel can be calculated under various conditions and different assumptions. Depending on whether the receiver has perfect channel knowledge or whether the channel exhibits a flat fading or frequency-selective fading behavior, different expressions for the channel capacity can be obtained. The MIMO channel capacity can be analyzed under two different assumptions: (1) transmitter has no channel knowledge and (2) the transmitter has perfect channel knowledge through feedback from the receiver or reciprocity of the downlink and uplink channels. Let us denote by Ξ the $N_{TX} \times N_{TX}$ covariance matrix of the channel input vector $\mathbf{x}$ and let further assume that the channel is unknown to the transmitter then it can be shown that the MIMO channel capacity can be written as $C_{MIMO} = \log_2\left(\det\left[\mathbf{I} + \mathbf{H}\Xi\mathbf{H}^H\right]\right)$ where $\text{tr}(\Xi) \leq \gamma$ ensures the total signal-power does not exceed a certain limit. It can be shown that for equal transmit power and uncorrelated sources $\Xi = (\gamma/N_{TX})\mathbf{I}$. This is true when the channel matrix is unknown to the transmitter and the input signal is Gaussian-distributed, maximizing the mutual information. If the receiver measures and sends channel quality feedback or CSI to the transmitter, the covariance matrix Ξ is not proportional to the identity matrix, rather it is constructed from a water-filling algorithm. If one compares the capacity achieved assuming equal transmit-power and unknown channel with that of perfect channel estimation through feedback then the capacity-gain due to use of feedback is obtained.

For the independent identically distributed Rayleigh fading scenario, the linear capacity growth discussed earlier will be observed. It is shown that MIMO channel capacity can be written as $C_{MIMO} = \sum_{i=1}^{\min(N_{TX},N_{RX})} \log_2(1+\gamma\lambda_i^2/N_{TX})$ where $\lambda_i, i=1,2,\ldots, \min(N_{TX},N_{RX})$ are the non-zero eigenvalues of $\mathbf{HH}^H$. We can decompose the MIMO channel into $K \le \min(N_{TX},N_{RX})$ equivalent parallel SISO channels using the singular value decomposition (SVD) theorem.[14] Let $\mathbf{y}=\mathbf{Hx}+\mathbf{n}$ describe the input–output relationship of the MIMO channel where $\mathbf{y}$ is the output vector with N_{RX} components, $\mathbf{x}$ is the input vector with N_{TX} components, $\mathbf{n}$ is the additive noise vector with N_{RX} components, and $\mathbf{H}$ is the $N_{RX}\times N_{TX}$ channel matrix. Using the SVD theorem, we can show $\mathbf{H}=\mathbf{U\Sigma V}^H$. Let $\widehat{\mathbf{x}}=\mathbf{V}^H\mathbf{x}, \widehat{\mathbf{y}}=\mathbf{U}^H\mathbf{y}$, and $\widehat{\mathbf{n}}=\mathbf{U}^H\mathbf{n}$ denote the unitary transformation of the channel input and output and noise vectors, it can be shown that $\widehat{\mathbf{y}}=\mathbf{\Sigma}\widehat{\mathbf{x}}+\widehat{\mathbf{n}}$. Since $\mathbf{U}$ and $\mathbf{V}$ are unitary matrices and $\mathbf{\Sigma}=\mathbf{diag}(\lambda_1,\lambda_2,\ldots,\lambda_{\min(N_{RX},N_{TX})},0,0,\ldots,0)$, it is clear that the capacity of this model is the same as the capacity of the model $\mathbf{y}=\mathbf{Hx}+\mathbf{n}$. However, $\mathbf{\Sigma}$ is a diagonal matrix with K non-zero elements on the main diagonal, thus $\widehat{y}_1=\lambda_1\widehat{x}_1+\widehat{n}_1,\ldots,\widehat{y}_K=\lambda_K\widehat{x}_K+\widehat{n}_K,\widehat{y}_{K+1}=\widehat{n}_{K+1}$. The latter equations are conceptually equivalent to K parallel SISO eigenchannels, each with signal power of $\lambda_i^2, i=1,2,\ldots, \min(N_{TX},N_{RX})$. As a result, the MIMO channel capacity can be rewritten in terms of the eigenvalues of the input signal covariance matrix Ξ.

When the channel knowledge is available at the transmitter and receiver then $\mathbf{H}$ is known, and we can optimize the capacity over Ξ subject to the power constraint $\mathrm{tr}(\Xi)\le\gamma$. It is shown in the literature that the optimal Ξ in this case exists and is known as water-filling solution. The channel capacity in this case is given by

$$C=\sum_{k=1}^{K}\log_2\left(\eta\lambda_i^2\right)^+, \quad \gamma=\sum_{k=1}^{K}\left(\eta-\lambda_i^{-2}\right)^+$$

where $(x)^+=x\forall x\ge 0, (x)^+=0\forall x<0$ and η is a nonlinear function of eigenvalues of the channel input covariance matrix. The effect of various channel conditions on the channel capacity has been extensively studied in the literature. For example, increasing the LoS signal strength at fixed SNR reduces capacity in Rician channels [22,24]. This can be explained in terms of the channel matrix rank or through various eigenvalue properties. The issue of correlated fading is of considerable importance for implementations where the

[14] The concept of decomposition of an $N\times N$ Hermitian matrix in terms of quadratic product of $N\times N$ unitary matrix composed of eigenvectors and $N\times N$ diagonal matrix of eigenvalues can be generalized to $M\times N$ complex-valued matrices of rank K. If $\mathbf{A}$ is a $M\times N$where$(M>N)$ complex-valued matrix of rank K then $\mathbf{A}=\mathbf{U\Sigma V}^H$ denotes the singular value decomposition of $\mathbf{A}$. The $M\times M$ unitary matrix $\mathbf{U}$ is composed of the eigenvectors of $\mathbf{AA}^H$, that is, $\mathbf{U}=(\mathbf{u_1},\mathbf{u_2},\ldots,\mathbf{u_m})$, where $\mathbf{AA}^H\mathbf{u}_i=\sigma_i^2\mathbf{u}_i$. The $N\times N$ unitary matrix $\mathbf{V}$ is composed of eigenvectors of $\mathbf{A}^H\mathbf{A}$, that is, $\mathbf{V}=(\mathbf{v_1},\mathbf{v_2},\ldots,\mathbf{v_n})$, where $\mathbf{A}^H\mathbf{v}_i=\sigma_i^2\mathbf{v}_i$ The elements of the $M\times N$ are the square roots of the eigenvalues of matrix $\mathbf{A}^H\mathbf{A}$o as the singular values of matrix $\mathbf{A}$ may be written as $\mathbf{A}=\sum_{i=1}^{K}\sigma_i\mathbf{u}_i\mathbf{v}_i^H$singular values of matrix $\mathbf{A}$ is the rank of $\mathbf{A}$. The singular values of matrix $\mathbf{A}$ are positive real numbers which satisfy $\sigma_1\ge\sigma_2\ge\cdots\ge\sigma_K>0$ [22].

antennas are required to be closely spaced. The optimal water-filling allocation strategy is obtained when the power allocated to each spatial subchannel is non-negative.

In the design of wireless communication systems, the main objective is to exploit the transmission schemes whose performance can approach the channel capacity as much as possible. Therefore, it is important to understand the underlying concepts and various information theoretic definitions of channel capacity and what can be pragmatically achieved under realistic channel conditions and transceiver implementations. Let us begin our concise study with the most generic definition of channel capacity. We denote the input and output of a memoryless SISO wireless channel with the random variables X and Y, respectively, the channel capacity is defined as $C = \max_{p(x)} I(X;Y)$ where $I(X;Y)$ represents the mutual information between X and Y. Shannon's theorem [19] provides an operational meaning to the definition of the instantaneous capacity as the number of bits that can be transmitted reliably over the channel with vanishing probability of error. The mutual information is maximized with respect to all possible transmit signal statistical distributions $p(x)$. Mutual information is a measure of the amount of information that one random variable contains about another variable. The mutual information between X and Y can also be written as $I(X;Y) = H(Y) - H(Y|X)$ where $H(Y|X)$ represents the conditional entropy between the random variables X and Y. The entropy of a random variable can be described as the measure of uncertainty in the random variable or the amount of information required on the average to describe the random variable. Thus the mutual information representation of channel capacity can be described as the reduction in the uncertainty of one random variable due to the knowledge of the other. Note that the mutual information between X and Y depends on the properties of the channel through a channel matrix $\mathbf{H}$ and the properties of X through the probability distribution of X [22].

Throughout this section, it is assumed that the channel matrix $\mathbf{H}$ is random and that the receiver has perfect channel knowledge. It is also assumed that the channel is memoryless, that is, for each use of the channel an independent realization of $\mathbf{H}$ is drawn. This means that the capacity can be computed as the maximum of the mutual information as defined earlier. The results are also valid when $\mathbf{H}$ is generated by an ergodic process because as long as the receiver observes the $\mathbf{H}$ process, only the first order statistics are needed to determine the channel capacity.

The ergodic (mean) capacity of a random channel with $N_{RX} = N_{TX} = 1$ and an average transmit power constraint P_T can be expressed as $C_{ergodic} = E_{\mathbf{H}}\left\{ \max_{p(x):P \leq P_T} I(X;Y) \right\}$ where P is the average power of a single codeword, transmitted over the channel, and $E_{\mathbf{H}}\{.\}$ denotes the expectation over all channel realizations. Compared to the generic definition, the capacity of the channel is now defined as the maximum of the mutual information between the input and the output over all statistical distributions on the input that satisfy the power constraint. In general, the capacity of a random MIMO channel with power constraint P_T can be expressed as $C_{ergodic} = E_{\mathbf{H}}\left\{ \max_{p(x):\mathrm{tr}(\Phi) \le P_T} I(X;Y) \right\}$ where $\mathbf{\Phi} = E\left[\mathbf{x}\mathbf{x}^H\right]$ is the covariance matrix of the transmit signal vector $\mathbf{x}$. The total transmit power is limited to P_T irrespective of the number of transmit antennas. For a fading channel the channel matrix $\mathbf{H}$ is a stochastic process; thus the associated channel capacity $C(\mathbf{H})$ is a random variable. In this case, the ergodic channel capacity is defined as the average of instantaneous channel capacity over the distribution of $\mathbf{H}$. The ergodic channel capacity of the MIMO transmission scheme is given by

$$C_{ergodic} = E\left\{ \max_{\mathrm{tr}(\Phi) = N_{TX}} \log_2\left(\det\left[\mathbf{I} + \gamma/N_{TX}\mathbf{H}\mathbf{\Phi}\mathbf{H}^H\right]\right) \right\}$$

where $\mathbf{\Phi}$ denotes the $N_{TX} \times N_{TX}$ covariance matrix of the channel input vector $\mathbf{x}$. According to information theoretic concepts, this capacity cannot be achieved unless channel coding is employed across an infinite number of independently fading blocks. Let us focus on the case of perfect CSI at the receiver side and no CSI at the transmitter side, which implies that the maximization of latter equation is now more restricted than in the previous case. Nevertheless, it has been shown in the literature that the optimal signal covariance matrix must be chosen according to $\mathbf{\Phi} = \mathbf{I}$. This means that the antennas should transmit uncorrelated streams with the same average power. With this result, the ergodic MIMO channel capacity reduces to [15]

$$C_{ergodic} = E\left\{\log_2\left(\det\left[\mathbf{I} + \gamma/N_{TX}\mathbf{H}\mathbf{H}^H\right]\right)\right\}$$

It is obvious that this is not the Shannon capacity in a true sense, since as mentioned earlier one with perfect channel knowledge at the transmitter can choose a signal covariance matrix that outperforms $\mathbf{\Phi} = \mathbf{I}$ case. Nevertheless, we refer to the preceding expression as the ergodic channel capacity with CSI at the receiver and no CSI at the transmitter.

The capacity under channel ergodicity is defined as the average of the maximum value of the mutual information between the transmitted and the received signals, where the maximization is carried out with respect to all possible transmit signal statistical distributions. Another measure of channel capacity that is frequently used is outage capacity. With outage capacity, the channel capacity is associated to an outage probability. Capacity is treated as a random variable which depends on the channel instantaneous response and remains constant during the transmission of a finite-length coded block of information. If the channel capacity falls below the outage capacity, there is no possibility that the transmitted block of information can be decoded with no errors, no matter which coding scheme is employed. The probability that the capacity is less than the outage capacity denoted by C_{outage} is ρ. This can be expressed in mathematical terms by $p(C < C_{outage}) = \rho$. In this case the latter expression represents an upper bound since there is a finite probability ρ that the channel capacity is less than the outage capacity. It can also be written as a lower bound, representing the case where there is a finite probability $(1-\rho)$ that the channel capacity is higher than

C_{outage}, which means $p(C > C_{outage}) = 1 - \rho$. In other words, since the MIMO instantaneous channel capacity is a random variable, it is meaningful to consider its statistical distribution, thus a useful measure of its statistical behavior is the outage capacity. Outage analysis quantifies the level of performance (in this case capacity) that is guaranteed with a certain level of reliability. The ρ% outage capacity $C_{outage}(\rho)$ is defined as the information rate that is guaranteed for $(100 - \rho)$% of the channel realizations ($p(C < C_{outage}) = \rho$). The outage capacity is often a more relevant measure than the ergodic channel capacity, because it describes in some way the quality of the channel. This is due to the fact that the outage capacity measures the probabilistic distribution of the instantaneous rate supported by the channel. Thus, if the rate supported by the channel is spread over a wide range, the outage capacity for a fixed probability level may become small, whereas the ergodic channel capacity may be high [15].

4.1.9.2 Single-User and Multi-user MIMO

Single-user MIMO (SU-MIMO) techniques are point-to-point schemes that improve channel capacity and reliability through the use of space-time/space-frequency codes (transmit/receive diversity) in conjunction with spatial multiplexing schemes. In an SU-MIMO transmission, the advantage of MIMO processing is obtained from the coordination of processing among all the transmitters or receivers. In the multi-user channel, on the other hand, it is usually assumed that there is no coordination among the users. As a result of the lack of coordination among users, uplink and downlink multi-user MIMO channels are different. In the uplink scenario, users transmit to the base station over the same channel. The challenge for the base station is to separate the signals transmitted by the users, using array processing or multi-user detection methods. Since the users are not able to coordinate with each other, there is not much that can be done to optimize the transmitted signals with respect to each other. If some channel feedback is allowed from the transmitter back to the users, some coordination may be possible, but it may require that each user know all the other users' channels rather than only its own. Otherwise, the challenge in the uplink is mainly in the processing done by the base station to separate the users. In the downlink channel, where the base station simultaneously transmits to a group of users over the same channel, there is some inter-user interference for each user which is generated by the signals transmitted to other users. Using multi-user detection techniques, it may be possible for a given user to overcome the multiple access interference, but such techniques are often extremely complicated for use at the receivers. Ideally, one would like to mitigate the interference at the transmitter by carefully designing the transmit signal. If CSI is available at the transmitter, it is aware of what interference is caused for each user by the signal it is transmitted to other users. The inter-user interference can be mitigated by beamforming or the use of dirty paper codes. In general, single-user and multi-user MIMO schemes are compared as follows [15]:

- SU-MIMO is a point-to-point link with predictable link capacity, whereas MU-MIMO channel is a broadcast channel (BC) in the downlink direction and a multiple access channel (MAC) in the uplink direction whose link-level data rates are characterized in terms of capacity regions.
- Multi-layer SU-MIMO schemes offer layer/stream diversity in the sense that if one stream has a poor SNR, the system will not necessarily experience an outage, whereas in the same situation, a MU-MIMO system will be in outage. This is because in MU-MIMO schemes, users have typically an equal target data rate and symbol error rate on their respective links, while in SU-MIMO systems, only the sum rate of the overall link is considered since all streams are delivered to same user.
- The MU-MIMO schemes suffer from near-far problem due to significant difference between the path losses experienced by each user, resulting in large deviation in the SINR of the corresponding user links. This would benefit the users with better channel conditions, while there is no near-far problem in SU-MIMO systems. The near-far problem in MU-MIMO systems may be alleviated via appropriate grouping of the users with similar channel conditions.
- The use of cooperative collocated transmit antennas in SU-MIMO schemes can facilitate the encoding at the transmitter and decoding at the receiver. In contrast, the users in a MU-MIMO scheme can cooperate in encoding at the base station in the downlink and decoding in the uplink; however, the users cannot cooperate in decoding in the downlink or encoding in the uplink directions.
- The capacity of the downlink and uplink is theoretically identical in the SU-MIMO systems (given the same transmit power and the perfect channel knowledge in the transmitter and the receiver); however, the capacities of the MU-MIMO BC and MAC are not identical.
- The capacity of the SU-MIMO schemes is less impacted by lack of CSI at the transmitter, whereas the capacity of the MU-MIMO BC significantly suffers from lack of CSI at the transmitter.
- SU-MIMO suffers from limited exploitation of multi-user diversity. The number of spatial dimensions is limited by number of antennas at the UE. There is a potential that spatial dimensions are wasted, if the UEs have a smaller number of antennas compared to the base station.
- MU-MIMO more efficiently exploits the multi-user diversity since all spatial dimensions which are supported by the base station can be exploited. It will achieve capacity gain, if UEs have a smaller number of antennas relative to the base station. Stronger spatial dimensions are exploited, particularly in the case of low-rank channel. The utilized spatial dimensions may be weak in the case of low-rank channel due to spatial correlation.

The advantages/disadvantages of SU-MIMO and MU-MIMO schemes are summarized in Table 4.21.

Table 4.21: Comparison of SU-MIMO and MU-MIMO schemes.

	SU-MIMO	MU-MIMO
Advantages	High user throughput High peak data rates	High system capacity Full exploitation of multiuser diversity
Disadvantages	Multiple transmit antennas at the base station are not fully exploited Multiuser diversity is not fully exploited	Degradation of peak data rates due to interuser interference

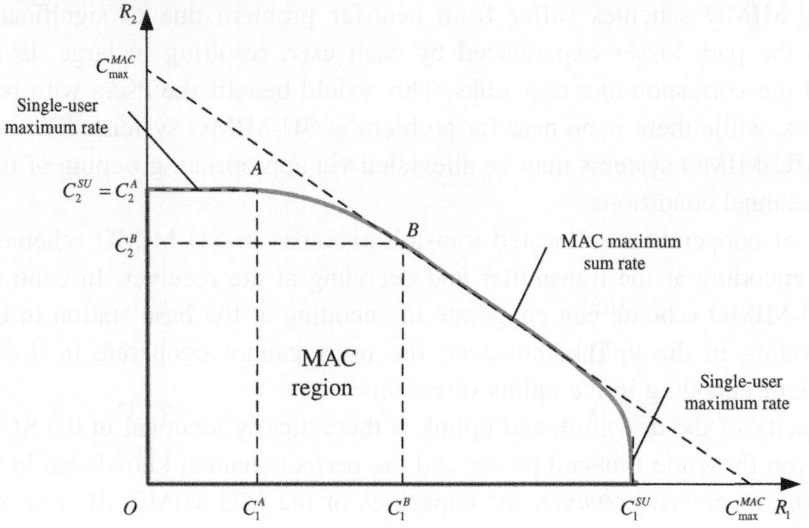

Figure 4.60
Capacity region of MU-MIMO BC with two users compared to SU-MIMO.

An important metric for measuring the performance of any communication channel is the information theoretic capacity. In an SU-MIMO channel, the capacity is the maximum amount of information that can be transmitted as a function of available bandwidth given a constraint on transmitted power. In SU-MIMO channels, it is common to assume that the total power distributed among all transmit antennas is limited. For the multi-user MIMO channel, the problem is somewhat more complex. Given a constraint on the total transmit power, it is possible to allocate varying fractions of that power to different users in the network; thus for any value of total power, different information rates are obtained. The result is a capacity region shown in Fig. 4.60 for two-user MU-MIMO channel. The maximum capacity for user 1 is achieved when 100% of the power is allocated to user 1, and for user 2, the maximum capacity is also obtained when it is allocated the full power. For every possible power distribution, there is an achievable information rate, which results in the capacity regions depicted in the figure. Two regions are shown in Fig. 4.60, the larger one for the case where both users have roughly the same maximum capacity (similar channel conditions), and the other region for a case where one of the users has much better channel condition than the other user. For N_{user} users, the capacity region is characterized by an N_{user}-dimensional hyper-region.

Let us use a simple MU-MIMO system model to demonstrate how the sum rate of the system is calculated. As shown in Fig. 4.61, the transmit vector $\mathbf{x}$ can be expressed as the weighted sum (precoded) of the input data symbols $s_k|_{k=1,2,\ldots,N_{user}}$ as follows $\mathbf{x} = \sum_k \mathbf{p}_k s_k = \mathbf{Ps}$ where $N_{user} \times 1$ vector $\mathbf{s} = \left(s_1, s_2, \ldots, s_{N_{user}}\right)^T$ denotes the data symbols from N_{user} users and $\mathbf{P} = \left(\mathbf{p}_1, \mathbf{p}_2, \ldots, \mathbf{p}_{N_{user}}\right)$ is the precoding matrix comprising N_{user} precoding vectors. It is assumed that the finite transmit power at the transmitter can be calculated as follows $P_{TX} = E\{\mathbf{x}^H \mathbf{x}\}$. The kth complex-valued output of the system can be written as $\mathbf{y}_k = \mathbf{H}_k \mathbf{x}_k + \mathbf{n}_k \in \mathbb{C}^N$ where N denotes the dimension of vector $\mathbf{y}$. The kth branch user data can be detected using a linear minimum mean squared error (MMSE) receiver as follows $\hat{s}_k = \mathbf{w}_k^T \mathbf{x}_k \in \mathbb{C}$ in which the MMSE weighting matrix is given by

$$\mathbf{w}_k = \mathbf{H}_k^* \mathbf{P}^* \mathbf{P}^T \mathbf{H}_k^T + \frac{N_{user}}{P_{TX}} \mathbf{I}_N{}^{-1} \mathbf{H}_k^* \mathbf{p}_k^*$$

It can be further shown that the SINR at the kth output is given by

$$\gamma_k = \frac{\left|\mathbf{w}_k^T \mathbf{H}_k \mathbf{p}_k\right|^2}{\left\|\mathbf{w}_k\right\|_2^2 N_{user}/P_{TX} + \sum_{i(i \neq k)} \left|\mathbf{w}_k^T \mathbf{H}_k \mathbf{p}_i\right|^2}$$

The sum rate of the system is given as $R_{sum} = \sum_k \log_2(1 + \gamma_k)$.

In the uplink of a multi-user MIMO system, the received signal at the gNB can be written as $\mathbf{y} = \sum_{k=1}^{N_{user}} \mathbf{H}_k^H \mathbf{x}_k + \mathbf{n}$ where $\mathbf{x}_k$ is the $N_{TX_k} \times 1$ transmitted signal vector of the kth UE with N_{TX_k} transmit antennas, $\mathbf{H}_k \in \mathbb{C}^{N_{TX_k} \times N_{RX}}$ denotes the flat-fading channel matrix from the kth user to the gNB and $\mathbf{n} = (n_1, n_2, \ldots, n_{N_{RX}}), n_k \sim N(0, 1)$ is an independent and identically distributed additive white Gaussian noise vector at the gNB. We assume that the receiver k has perfect and instantaneous knowledge of the channel matrix $\mathbf{H}_k$. Note that the gNB is equipped with N_{TX} transmit and N_{RX} receive antennas.

In the downlink, the received signal at the kth receiver can be written as $\mathbf{y}_k = \mathbf{H}_k \mathbf{x} + \mathbf{n}_k \forall k = 1, 2, \ldots, N_{user}$ where $\mathbf{H}_k \in \mathbb{C}^{N_{RX_k} \times N_{TX}}$ is the downlink channel, and $\mathbf{n}_k \in \mathbb{C}^{N_{RX_k} \times N_{TX}}$ is the complex-valued additive Gaussian noise at the kth receiver. We assume that each receiver also has perfect and instantaneous knowledge of its own channel matrix $\mathbf{H}_k$. The transmitted signal $\mathbf{x}$ is a function of the multiple users' information data, that is, $\mathbf{x} = \sum_{k=1}^{N_{user}} \mathbf{x}_k$ where $\mathbf{x}_k$ is the signal carrying kth user's message with covariance matrix $\boldsymbol{\Omega}_k = E\{\mathbf{x}_k \mathbf{x}_k^H\}$. The power allocated to the kth user is given by $\rho_k = \mathrm{tr}\{\boldsymbol{\Omega}_k\}$. Under a

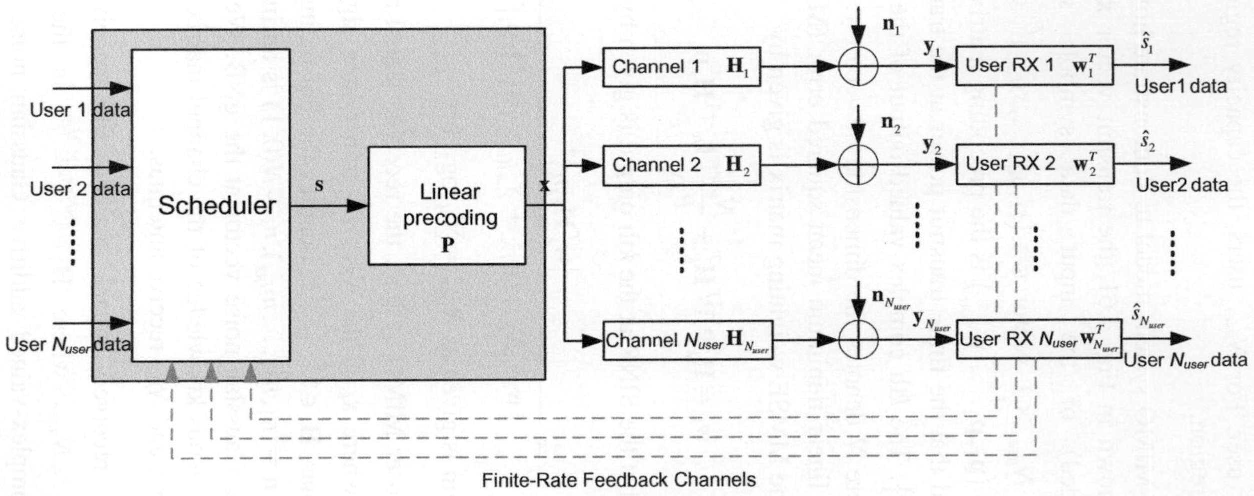

Figure 4.61
MU-MIMO BC model [15].

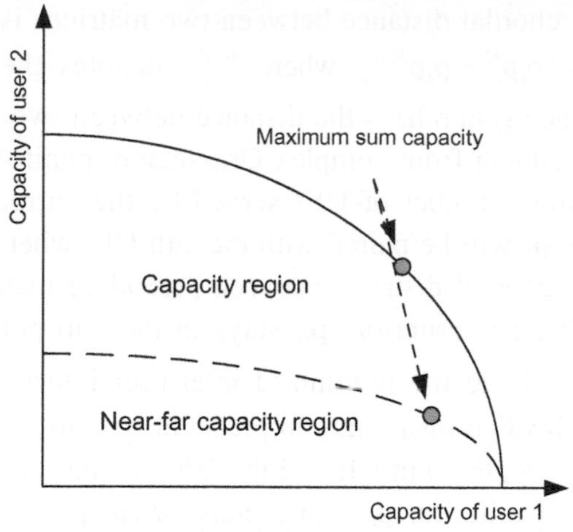

Figure 4.62
An example illustration of capacity region.

sum power constraint at the gNB, the power allocation needs to maintain $\sum_{k=1}^{N_{user}} \rho_k \leq P_{total}$. Assuming a unit variance for the noise, it can be shown that the capacity region for a given matrix channel realization can be written as [15]

$$C_{DL} = \bigcup_{\left(\rho_1,\rho_2,\ldots,\rho_{N_{user}} \mid \sum \rho_k \leq P_{total}\right)} \left\{ \left(R_1,R_2,\ldots,R_{N_{user}}\right) \in \mathbb{R}^{+N_{user}}, R_l \leq \log_2 \frac{\det\left[\mathbf{I}+\mathbf{H}_i\left(\sum_{j\geq i}\boldsymbol{\Omega}_j\right)\mathbf{H}_i^H\right]}{\det\left[\mathbf{I}+\mathbf{H}_i\left(\sum_{j>i}\boldsymbol{\Omega}_j\right)\mathbf{H}_i^H\right]} \right\}$$

where $\mathbb{R}^{+N_{user}}$ is the N_{user}-dimensional set of positive real numbers. The preceding equation may be optimized over each possible user ordering. Although difficult to realize in practice, the computation of the capacity region can be simplified using the assumption that the downlink capacity region can be calculated through the union of regions of the dual MAC with all uplink power allocation vectors meeting the sum power constraint. The fundamental effect of the use of multiple antennas at either the gNB or the user terminals in increasing the channel capacity is best realized by examining how the sum capacity, that is, the point obtained by the maximum $\sum_{k=1}^{N_{user}} R_k$ in the capacity region, scales with the number of active users (see Fig. 4.62).

An efficient UE pairing scheme is required at the gNB to choose the correct pair of UEs for transmission in MU-MIMO systems. This pairing scheme is required to maintain minimal interference among scheduled UEs in MU-MIMO transmission. A proper pairing scheme can be designed by maximizing the chordal distance[15] between the feedback precoding

[15] The asymptotic performance of a coding scheme is dominated by the shortest distance between any pair of codewords. The relevant distance measure between two codewords $\mathbf{x}_1$ and $\mathbf{x}_2$ of an orthogonal code for a non-coherent MIMO system is the chordal distance defined as $d^2(\mathbf{x}_1, \mathbf{x}_2) = M - \|\mathbf{x}_1\mathbf{x}_2^H\|_F^2$.

matrices of the UEs. The chordal distance between two matrices is given in [15] and represented by $d_c(\mathbf{p}_i, \mathbf{p}_j) = \frac{1}{\sqrt{2}} \left\| \mathbf{p}_i \mathbf{p}_i^H - \mathbf{p}_j \mathbf{p}_j^H \right\|_F$ where $\|.\|_F$ denotes the Frobenius norm of the matrix. The chordal distance generalizes the distance between two points on the unit sphere through an isometric embedding from complex Grassmann manifold $Gr(N_{TX}, N_l)$ to the unit sphere. Assuming an infinite number of UEs served by the current gNB, the kth UE with reported precoding matrix $\mathbf{p}_k$ will be paired with the mth UE, where the mth UE reports precoding matrix $\mathbf{p}_m$ and the chordal distance between precoding matrices is maximized. With the maximized chordal distance criterion, $\mathbf{p}_m$ stays in the anti-polar position of $\mathbf{p}_k$; hence, $\left\| \mathbf{H}_k \mathbf{p}_m \right\|^2$ is minimized, yielding the minimized inter-user interference. Therefore, the UE pairing scheme for MU-MIMO transmission in practical systems is designed to find the best match between two UEs (e.g., the mth UE and the kth UE) based on the reported precoding matrices and the following criterion $\mathbf{p}_m = \arg \max_{\forall \mathbf{p}_i \in \mathcal{P}_{UE}} d_c(\mathbf{p}_i, \mathbf{p}_k)$ with $\mathcal{P}_{UE}$ representing the pool containing all reported precoding matrices at a certain gNB.

4.1.9.3 Analog, Digital, and Hybrid Beamforming

Large antenna arrays and beamforming play an important role in 5G implementations since both base stations and devices can accommodate a larger number of antenna elements at mmWave frequencies. Aside from a higher directional gain, these antenna types offer complex beamforming capabilities. This allows increasing the capacity of cellular networks by improving the signal-to-interference ratio through direct targeting of user groups. The narrow transmit-beams simultaneously lower the amount of interference in the radio propagation environment and make it possible to maintain sufficient signal power at the receiver terminal at larger distances in rural areas. An important prerequisite for any beamforming architecture is a phase coherent signal, which means that there is a defined and stable phase relationship between all RF carriers. A fixed phase offset between the carriers can be used to steer the main antenna lobe to a desired direction. A major difference between NR and LTE is the support for beamformed control channels, which resulted in different reference signal design for each control channel. The NR physical channels and signals, including those used for control and synchronization, have all been designed to support beamforming. The CSI for operation with a large number of antennas can be obtained via CSI reports from the devices based on transmission of CSI-RSs in the downlink, as well as using uplink measurements exploiting channel reciprocity.

4.1.9.3.1 Analog Beamforming

Analog beamforming typically relies on conditioning the amplitude and phase of the signals that feed the antenna array. The combination of these two factors is used to improve sidelobe suppression or steering nulls. Phase and amplitude for each antenna element are combined by applying a complex-valued weighting factor to the signal that is fed to the

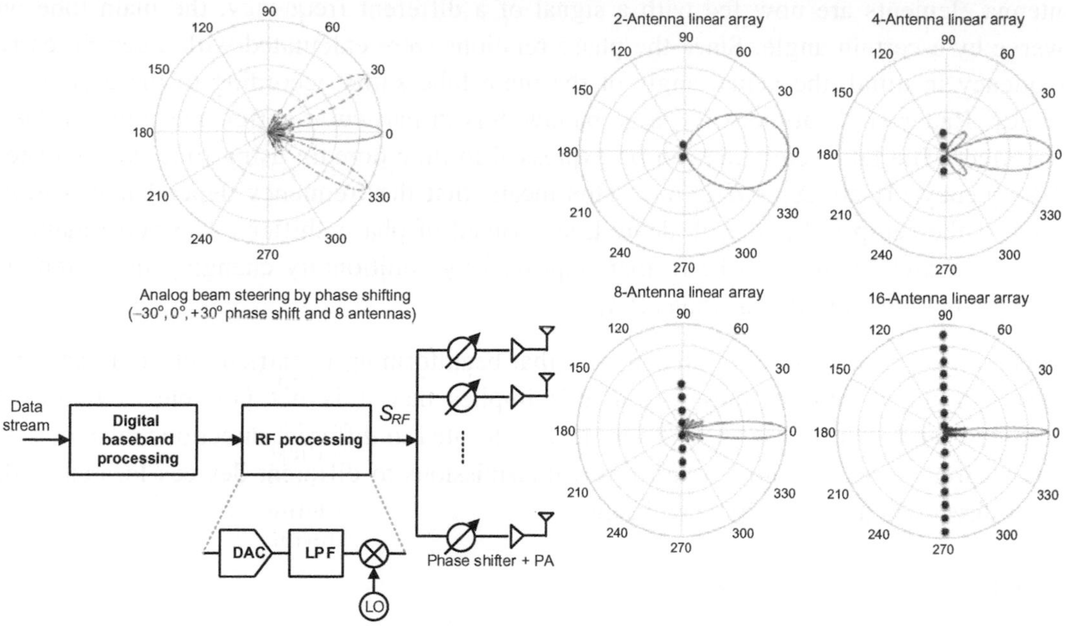

Figure 4.63
Analog beamforming transmitter architecture.

corresponding antenna. Fig. 4.63 shows a basic implementation of an analog beamforming transmitter architecture. This architecture consists of only one RF chain and multiple phase shifters that feed an antenna array. The phased arrays have been used in practical systems (e.g., radar systems) for the past several decades. Beam steering was often carried out with a selective RF switch and fixed phase shifters. This concept is still used in modern communication systems using advanced hardware and improved precoding techniques. These enhancements enable separate control of the phase of each element. Unlike traditional passive architectures, the beam can be steered not only to discrete but virtually any angle using active beamforming antennas. Analog beamforming is performed in the analog domain at RF frequencies or an intermediate frequency. However, implementing multi-stream transmission with analog beamforming is a highly complex task. In order to calculate the phase shifts, a uniformly spaced linear array with element spacing is assumed. Considering the receive scenario, the antenna array must be in the far field of the incoming signal, so that the arriving wave front is approximately planar. If the signal arrives at an angle θ relative to the antenna boresight, the wave must travel an additional distance $d \sin \theta$ to arrive at each successive element. This translates to an element specific delay which can be converted to a frequency-dependent phase shift of the signal as $\Delta\varphi = 2\pi d(\sin \theta)/\lambda$. The frequency dependency translates into an effect called beam unevenness. The main lobe of an antenna array at a defined frequency can be steered to a certain angle using phase offsets calculated by the latter equation (see Fig. 4.63). If the antenna elements are now fed with a signal of a different frequency, the main lobe will swerve by a certain angle. Since the phase relations were calculated with a certain carrier frequency in mind, the actual angle of the main lobe shifts according to the current frequency. Radar applications with large bandwidths in particular suffer inaccuracies due to this effect. The latter equation can be expressed in time domain using time delays instead of frequency offsets $\Delta\tau = d/c \sin\theta$. This means that the frequency dependency is eliminated, if the setup is fitted with delay lines instead of phase shifters. The performance of the analog architecture can be further improved by additionally changing the magnitude of the signals feeding the antennas [56].

Analog signal processing typically implies that beamforming is carried out on a per-carrier basis. For the downlink transmission, this implies that it is not possible to frequency-multiplex beamformed transmissions to devices located in different directions relative to the base station. In other words, beamformed transmissions to different devices located in different directions must be separated in time.

4.1.9.3.2 Digital Beamforming

While analog beamforming is generally restricted to one RF chain even when using large number of antenna arrays, digital beamforming in theory supports as many RF chains as there are antenna elements. If suitable precoding is performed in the digital-domain baseband, this would yield higher flexibility regarding the transmission and reception. The additional degree of freedom can be leveraged to perform advanced techniques such as multi-beam MIMO. These advantages result in the highest theoretical performance possible compared to other beamforming architectures. Fig. 4.64 illustrates the high-level digital beamforming transmitter architecture with multiple RF chains. Uneven beam is a problem

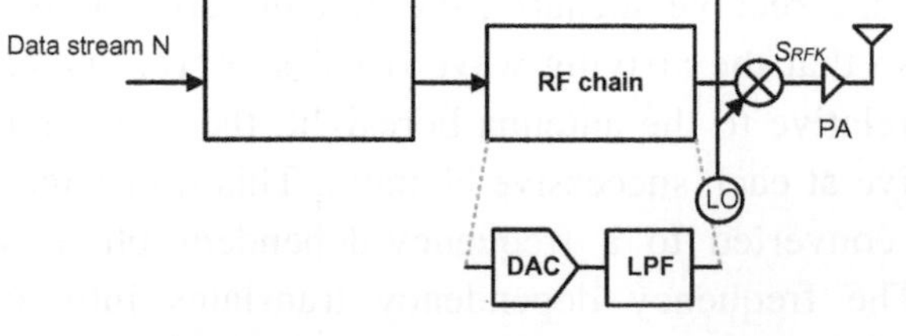

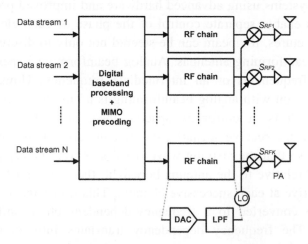

Figure 4.64
Digital beamforming transmitter architecture.

of analog beamforming architectures using phase shifters. This is a drawback considering 5G plans to make use of large bandwidths in the mmWave bands. Digital control of the RF chain enables optimization of the phases according to the frequency over a large band. Nonetheless, digital beamforming may not always be ideally suited for practical implementations of 5G applications. The very high complexity and requirements regarding the hardware may significantly increase cost, energy consumption, and complicate integration in mobile devices. Digital beamforming is better suited for use in base stations, since performance outweighs mobility in this case. Digital beamforming can accommodate multi-stream transmission and serve multiple users simultaneously, which is a key driver of the technology [56].

Multiple antennas at the transmitter and receiver can be used to achieve array and diversity gain instead of capacity gain. In this case, the same symbol weighted by a complex-valued scale factor is sent from each transmit antenna, so that the input covariance matrix has unit rank. This scheme is referred to as beamforming. It must be noted that there are two conceptually and practically different classes of beamforming: (1) direction-of-arrival beamforming (i.e., adjustment of transmit or receive antenna directivity) and (2) eigen-beamforming (i.e., a mathematical approach to maximize signal power at the receive antenna based on certain criterion). In this section, we only consider eigen-beamforming schemes.

A classic eigen-beamforming scheme usually performs linear, single-layer, complex-valued weighting on the transmit symbols such that the same signal is transmitted from each transmit antenna using appropriate weighting factors. In this scheme, the objective is to maximize the signal power at the receiver output. When the receiver has multiple antennas, the single-layer beamforming cannot simultaneously maximize the signal power at the receive antennas; hence, precoding is used for multi-layer beamforming in order to maximize the throughput of a multi-antenna system. Precoding is a generalized beamforming scheme to support multi-layer transmission in a MIMO system. Using precoding, multiple streams are transmitted from the transmit antennas with independent and appropriate weighting per each antenna such that the throughput is maximized at the receiver output.

Let us begin our concise study of eigen-beamforming and MIMO precoding using a simplified model, where we have two transmit antennas at the base station and a UE with a single receive antenna. The goal is to find the complex-valued precoding weights such that the SNR at the receiver is maximized. The channel in this example is a vector $\mathbf{h} = (h_1, h_2)$ where h_1 and h_2 are channel coefficients. It can be shown that the complex-valued weighting factors p_1 and p_2 can be calculated as shown in Fig. 4.65. This example illustrates the concept of digital precoding and how in theory the weighting vectors/matrices are calculated to maximize the SNR at the receiver.

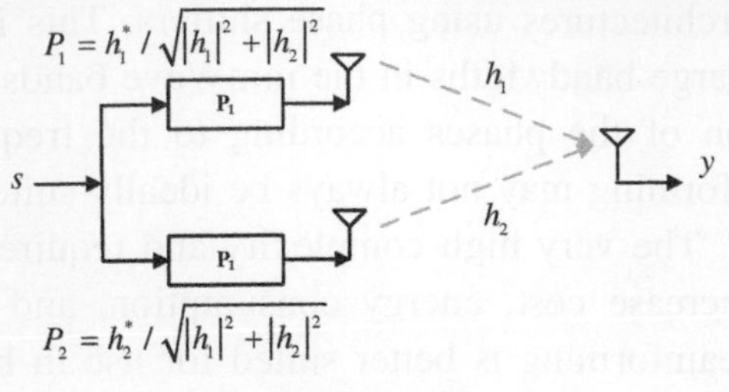

Figure 4.65
Concept of the digital precoding.

In an SU-MIMO system, the identity matrix precoding (for open-loop) and SVD precoding (for closed-loop) can be used to achieve link-level MIMO channel capacity. In addition, random unitary precoding can achieve the open-loop MIMO channel capacity with no signaling overhead in the uplink. The SVD precoding, on the other hand, has been shown to achieve the MIMO channel capacity when CSI is known at the transmitter. In a precoded SU-MIMO system with N_{TX} transmit antennas and N_{RX} receive antennas, the input–output relationship can be described as $\mathbf{y} = \mathbf{HWs} + \mathbf{n}$ where $\mathbf{s} = (s_1, s_2 \ldots, s_M)^T$ is an $M \times 1$ vector of normalized complex-valued modulated symbols, $\mathbf{y} = (y_1, y_2, \ldots, y_{N_{RX}})^T$ and $\mathbf{n} = (n_1, n_2, \ldots, n_{N_{RX}})^T$ are the $N_{RX} \times 1$ vectors of received signal and noise, respectively, $\mathbf{H}$ is the $N_{RX} \times N_{TX}$ complex-valued channel matrix, and $\mathbf{W}$ is the $N_{TX} \times M$ linear *precoding* matrix.

In the receiver, a hard-decoded symbol vector $\hat{\mathbf{s}}$ is obtained by decoding the received vector $\mathbf{y}$ using a vector decoder, assuming perfect knowledge of the channel and the optimal selection of precoding matrices. We assume that the entries of $\mathbf{H}$ are independent and identically distributed according to $\mathbb{C}(0, 1)$ (complex-valued normal distribution) and the entries of noise vector $\mathbf{n}$ are independent and identically distributed according to $\mathbb{C}(0, N_0)$. The input vector $\mathbf{s}$ is assumed to be normalized, thus $E[\mathbf{ss}^H] = \mathbf{I}$ where $\mathbf{I}$ is an identity matrix. Let us further assume that precoding matrix $\mathbf{W}$ is unitary, thus $\mathbf{WW}^H = \mathbf{I}$. The receiver selects a precoding matrix $\mathbf{W}_i, i = 1, 2, \ldots, N_{codebook}$ from a finite set of quantized precoding matrices $\boldsymbol{\Omega} = \{\mathbf{W}_1, \mathbf{W}_2, \ldots, \mathbf{W}_{N_{codebook}}\}$ and sends the index of the chosen precoding matrix back to the transmitter over a low-delay feedback channel. There are two important issues concerning the above precoding scheme: (1) optimal selection criterion for choosing a precoding matrix from set $\boldsymbol{\Omega}$ and (2) design of codebook $\boldsymbol{\Omega}$. The matrix $\mathbf{W}_i, i = 1, 2, \ldots, N_{codebook}$ can be selected from $\boldsymbol{\Omega}$ by using either of the following optimization criteria [75]: (1) minimizing the trace of the mean squared error (MMSE-trace selection), (2) minimizing the determinant of the mean squared error (MMSE-determinant selection), (3) maximizing the minimum singular value of $\mathbf{HW}$ (singular value selection), (4) maximizing the instantaneous capacity (capacity selection), or (5) maximizing the minimum received symbol vector distance (minimum distance selection). The above selection criteria may be evaluated at the receiver using a full search over all matrices in $\boldsymbol{\Omega}$. Using distortion functions based on the selection criteria, it can be shown that the codebook $\boldsymbol{\Omega}$ is designed using Grassmannian subspace packing [75]. If MMSE-trace, singular value, or minimum distance selection is used, the codebook is designed such that $\varepsilon = \min_{\mathbf{W}_i \neq \mathbf{W}_j} \|\mathbf{W}_i \mathbf{W}_i^H - \mathbf{W}_j \mathbf{W}_j^H\|_2$ [16] is maximized. If MMSE-determinant or capacity selection optimization method is used, the codebook is designed such that $\varepsilon = \min_{\mathbf{W}_i \neq \mathbf{W}_j} \arccos|\det(\mathbf{W}_i^H \mathbf{W}_j)|$ is maximized.

The precoding matrices (MIMO codebooks) are designed based on a trade-off between performance and complexity. The following are some desirable properties of the codebooks:

1. Low-complexity codebooks can be designed by choosing the elements of each constituent matrix or vector from a small binary set, for example, a four alphabet ($\pm 1, \pm j$) set, which eliminates the need for matrix or vector multiplication. In addition, nested property of the codebooks can further reduce the complexity of CQI calculation when rank adaptation is performed.
2. Base station may perform rank overriding which results in significant CQI mismatch, if the codebook structure cannot adapt to it. A nested property with respect to rank overriding can be exploited to mitigate the mismatch effects.
3. Power amplifier balance is taken into consideration when designing codebooks with constant modulus property, which may eliminate unnecessary increase in PAPR.
4. Good performance for a wide range of propagation scenarios, for example, uncorrelated, correlated, and dual-polarized channels, is expected from the codebook design algorithms. A DFT-based codebook is optimal for linear array with small antenna spacing since the vectors match with the structure of the transmit array response. In addition, with an optimal selection of the matrices and the entries of the codebook (rotated block diagonal structure), significant gains can be obtained in dual-polarized scenarios.
5. Low feedback and signaling overhead are desirable from operation and performance perspective.
6. Low memory requirement is another design consideration for the MIMO codebooks.

Let us consider multi-user MIMO systems and briefly study how precoding is applied in those scenarios. In the downlink direction of a precoded MU-MIMO system (alternatively known as BC in the literature) with N_{TX} transmit antennas at the base station and one receive antenna at the kth mobile station the input–output relationship can be written as $y_k = \mathbf{h}_k^H \mathbf{x} + n_k, k = 1, 2, \ldots, N_{user}$, where $\mathbf{x} = \sum_{i=1}^{N_{user}} s_i \mathbf{w}_i$ is the $N_{TX} \times 1$ vector of weighted transmitted symbols s_i, y_k and n_k are the received signal and noise, respectively, $\mathbf{h}_k$ is the kth $N_{TX} \times 1$ channel vector, where matrix $\mathbf{H} = (\mathbf{h}_1, \mathbf{h}_2, \ldots, \mathbf{h}_{N_{user}})^T$ is the $N_{user} \times N_{TX}$

[16] The Euclidean norm of square matrix $\mathbf{A}$ is defined as $\|\mathbf{A}\|_2 = \sup_{\|\mathbf{x}\|=1} \|\mathbf{A}\mathbf{x}\|_2$. The spectral norm of matrix $\mathbf{A}$ is the largest singular value of $\mathbf{A}$ or the square root of the largest eigenvalue of the positive-semi-definite matrix $\mathbf{A}^H\mathbf{A}$, that is, $\|\mathbf{A}\|_2 = \sqrt{\lambda_{\max}(\mathbf{A}^H\mathbf{A})}$ where $\mathbf{A}^H$ denotes the conjugate transpose of $\mathbf{A}$ [22].

complex-valued downlink channel matrix and $\mathbf{w}_k$ is the kth $N_{TX} \times 1$ normalized linear precoding vector.

The mathematical relationship for the input and output of a precoded MU-MIMO system in the uplink (alternatively known as MAC in the literature) with N_{RX} receive antennas at the base station and one transmit antenna at each user terminal can be written as $\mathbf{y} = \sum_{k=1}^{N_{user}} s_k v_k \mathbf{h}_k + \mathbf{n}$ where $s_k v_k$ is the weighted complex-valued modulated symbol from the kth user, $\mathbf{y} = (y_1, y_2, \ldots, y_{N_{RX}})^T$ and $\mathbf{n} = (n_1, n_2, \ldots, n_{N_{RX}})^T$ are the $N_{RX} \times 1$ vectors of received signal and noise, respectively, $\mathbf{h}_k$ is the kth $N_{RX} \times 1$ channel vector, where matrix $\mathbf{H} = (\mathbf{h}_1, \mathbf{h}_2, \ldots, \mathbf{h}_{N_{user}})^T$ is the $N_{RX} \times N_{user}$ complex-valued uplink channel matrix. As mentioned earlier, perfect knowledge of the CSI is necessary at the transmitter in order to achieve the capacity of a multi-user MIMO channel. However, in practical systems, the receiver only provides partial CSI through uplink feedback channels to the transmitter, that is, the multi-user MIMO precoding with limited feedback. The received signal in the downlink of a MU-MIMO system with limited feedback precoding is mathematically expressed as $y_k = \mathbf{h}_k^H \sum_{i=1}^{N_{user}} s_i \hat{\mathbf{w}}_i + n_k, k = 1, 2, \ldots, N_{user}$. The transmit vector for limited feedback precoding is modeled as $\hat{\mathbf{w}}_i = \mathbf{w}_i + \varepsilon_i$ where ε_i is the error vector generated as a result of the limited feedback and vector quantization, the expression for the received signal can be rewritten as $y_k = \mathbf{h}_k^H \sum_{i=1}^{N_{user}} s_i \mathbf{w}_i + \mathbf{h}_k^H \sum_{i=1}^{N_{user}} s_i \varepsilon_i + n_k, k = 1, 2, \ldots, N_{user}$ where $\mathbf{h}_k^H \sum_{i=1}^{N_{user}} s_i \varepsilon_i$ is the residual interference due to the limited-feedback precoding.

To reduce the residual interference term, one should use more accurate CSI feedback which results in the use of more uplink resources for the feedback. It is shown in the literature that the number of feedback bits per user $N_{feedback}$ must be increased linearly with the SNR γ_{dB} (in decibels) at the rate of $N_{feedback} = (N_{TX} - 1)\log_2 \gamma = \gamma_{dB}(N_{TX} - 1)/3$ in order to achieve the full multiplexing gain of N_{TX} antennas [76]. In addition, the scaling of $N_{feedback}$ guarantees that the throughput loss relative to zero-forcing (ZF) precoding with perfect CSI knowledge at the transmitter is upper bounded by N_{TX} bps/Hz, which corresponds to approximately 3 dB power offset. The throughput of a feedback-based ZF system is bounded, if the SNR approaches infinity and the number of feedback bits per user is fixed. Reducing the number of feedback bits according to $N_{feedback} = \alpha \log_2 \gamma$ for any $\alpha < N_{TX} - 1$ results in a strictly inferior multiplexing gain of $N_{TX}[\alpha/(N_{TX} - 1)]$ where N_{TX} is the number of transmit antennas and γ is the SNR of the downlink channel.

In order to calculate the amount of feedback required to maintain certain throughput, the difference between the feedback rates of ZF precoding with perfect feedback $R_{PF-ZF}(\gamma)$ and with limited feedback $R_{LF-ZF}(\gamma)$ is required to satisfy the following constraint $\Delta R(\gamma) = R_{PF-ZF}(\gamma) - R_{LF-ZF}(\gamma) \leq \log_2 b$. In order to maintain a rate offset less than $\log_2 b$ (per user) between ZF with perfect CSI and with finite-rate feedback (i.e., $\Delta R(\gamma) \leq \log_2 b, \forall \gamma$), it is sufficient to scale the number of feedback bits per user according to $N_{feedback} = \gamma_{dB}(N_{TX} - 1)/3 - (N_{TX} - 1)\log_2(b - 1)$. The rate offset of $\log_2 b$ (per user) is translated into a power offset, which is a more useful metric from the design perspective. Since a multiplexing gain of N_{TX} is achieved with ZF, the ZF curve has a slope of N_{TX} bps/Hz/3 dB at asymptotically high SNR. Therefore, a rate offset of $\log_2 b$ bps/Hz per user corresponds to a power offset of $3\log_2 b$ decibels. To feedback $N_{feedback}$ bits through uplink channel, the throughput of the uplink feedback channel should be larger than or equal to $N_{feedback}$, that is, $w_{FB}\log_2(1+\gamma_{FB}) \geq N_{feedback}$ where γ_{FB} denotes the SNR of the feedback channel. Thus the required feedback resource to satisfy the constraint $\Delta R(\gamma) \leq \log_2 b$ can be shown to be given as follows $w_{FB} \geq \left[\gamma_{dB}(N_{TX}-1)/3-(N_{TX}-1)\log_2(b-1)\right]/\log_2(1+\gamma_{FB})$, that is, the required feedback resource is a function of both downlink and uplink channel conditions [76].

We defined digital precoding as adaptive or non-adaptive weighting of the spatial streams prior to transmission from each antenna port (in a multi-antenna configuration) using a precoding matrix for the purpose of improving the reception or separation of the spatial streams at the receiver. Both feed-back and feed-forward precoder matrix selection schemes can be used in order to select the optimal weights. Feed-back precoding matrix selection techniques do not rely on channel reciprocity, rather they use feedback channels provided that the feedback latency is less than the channel coherence time. In feed-forward approaches, the necessary CSI can be theoretically obtained through direct feedback, where the CSI is explicitly signaled to the transmitter by the receiver or estimated using the SRS. The direct channel feedback methods preclude the channel reciprocity requirement, whereas channel sounding methods rely on channel reciprocity. Therefore, explicit control signaling is required for the PMI-based (feedback) schemes. However, in reciprocity-based schemes, the sounding signals in the uplink and precoded pilots in the downlink are used to assist the transmitter and receiver to appropriately select the precoding matrix. The reciprocity-based schemes have the additional advantage of not being constrained to a finite set of codebooks. Beamforming relies on long-term statistics of the radio channel and, unlike reciprocity-based techniques, does not require short-term correlation between the uplink and downlink in order to properly function.

4.1.9.3.3 Hybrid Beamforming

Hybrid beamforming has been proposed as a possible solution that is able to combine the advantages of both analog and digital beamforming architectures. The idea of hybrid beamforming is based on the concept of phased array antennas commonly used in radar applications. Due to the reduced power consumption, it is also seen as a possible solution for mmWave mobile broadband communication. If the phased array approach is combined with digital beamforming, the phased array approach might also be feasible for non-static or quasi-static scenarios. Considering the inefficiency of mmWave amplifiers and the high insertion loss of RF phase shifters, it is more desirable to perform the phase shifting in the baseband. The power consumption associated with both cases is comparable, as long as the

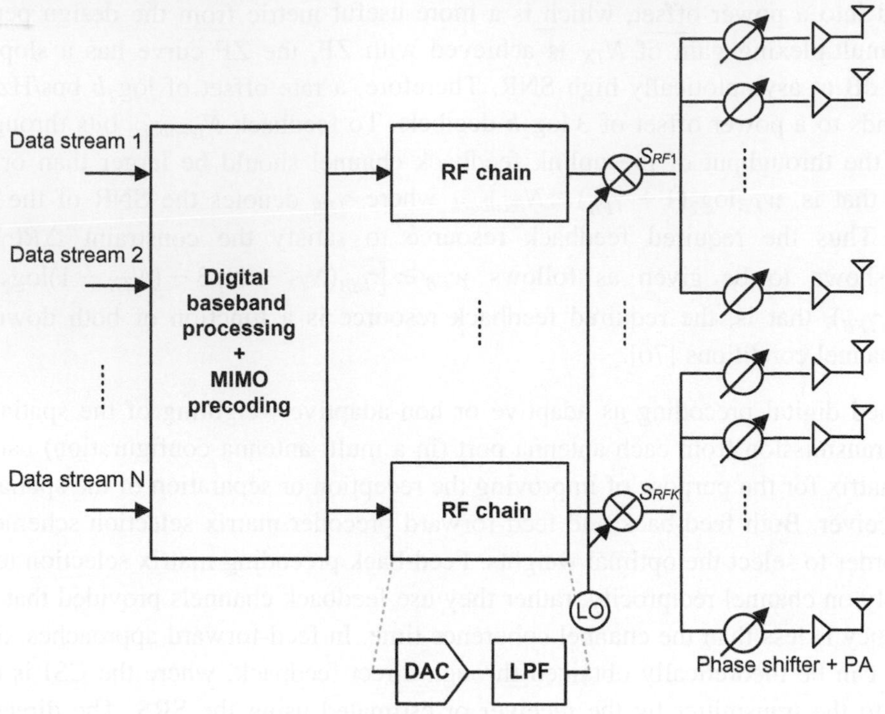

Figure 4.66
Hybrid beamforming architecture.

number of antennas per RF-chain remains relatively small. A significant cost reduction can be achieved by reducing the number of complete RF chains. This does also lead to lower the overall power consumption. Since the number of data converters is significantly lower than the number of antennas, there are less degrees of freedom for digital baseband processing. Thus the number of simultaneously supported streams is reduced compared to digital beamforming. The resulting performance gap is expected to be relatively low in mmWave bands, which this scheme is more suitable, due to the specific channel characteristics. The high-level block diagram of a hybrid beamforming transmitter is shown in Fig. 4.66. The precoding is divided between the analog and digital domains. In theory, it is possible to assume that each amplifier is interconnected to each radiating element.

In recent years, hybrid beamforming with low-resolution data conversion (digital-to-analog/analog-to-digital) has been studied including the energy efficiency/spectral efficiency trade-off of fully connected hybrid and digital beamforming with low-resolution data converters. One of the challenges of large antenna arrays is the increasing cost and complexity of the use of many analog-to-digital and digital-to-analog converters and other RF components to drive individual elements or subarrays. Thus the feasibility study of low-resolution and in the extreme case one-bit resolution data converters would be very important for practical implementation of massive MIMO systems [16].

In summary, there are three types of beamforming architectures used for antenna arrays [56]:

- Analog beamforming: The traditional way to form beams is to use attenuators and phase shifters as part of the analog RF circuit where a single data stream is divided into separate paths. The advantage of this method is that only one RF chain (PA, LNA, filters, switch/circulator) is required. The disadvantage is the loss from the cascaded phase shifters at high power.
- Digital beamforming: It assumes there is a separate RF chain for each antenna element. The beam is then formed by matrix-type operations in the baseband where the amplitude and phase weighting are applied. For frequencies lower than 6 GHz, this is the preferred method since the RF chain components are comparatively inexpensive and can combine MIMO and beamforming into a single array. For frequencies of 28 GHz and above, the PAs and ADC/DACs are very lossy for standard CMOS components. Gallium arsenide and gallium nitrate can be used in high frequencies to decrease losses at the expense of higher cost.
- Hybrid beamforming: It combines digital and analog beamforming in order to allow the flexibility of MIMO and beamforming while reducing the cost and losses of the beamforming unit. Each data stream has its own separate analog beamforming unit with a set of $N_{antennas}$ antennas. If there are $N_{streams}$ data streams, then there are $N_{streams} \times N_{antennas}$ antennas. The analog beamforming unit loss due to phase shifters can be mitigated by replacing the adaptive phase shifters with a selective beamformer such as a Butler matrix.[21] Some architectures use the digital beamforming unit to steer the direction of the main beam while the analog beamforming unit steers the beam within the digital envelop.

4.1.9.4 Full-Dimension MIMO

Beamforming is a signal processing method that generates directional antenna beam patterns using multiple antennas at the transmitter. It is possible to steer the transmitted signal toward a desired direction and, at the same time, avoid receiving the unwanted signal from an undesired direction. Traditional beamforming schemes controlled the beam pattern only in the horizontal (azimuth) plane. The three-dimensional beamforming adapts the radiation beam pattern in both elevation and azimuth planes to provide more degrees of freedom in supporting users. Higher average user throughput, less inter-cell and inter-sector interference, higher energy efficiency, improved coverage, and increased spectral efficiency are some of the advantages of 3D beamforming or full-dimensional MIMO. In order to exploit the vertical dimension, the antenna tilt can be considered in the vertical axis. The antenna tilt angle is defined as the angle between the horizontal plane and the boresight direction of the antenna pattern. Mechanical alignment of the antenna was traditionally used to adjust

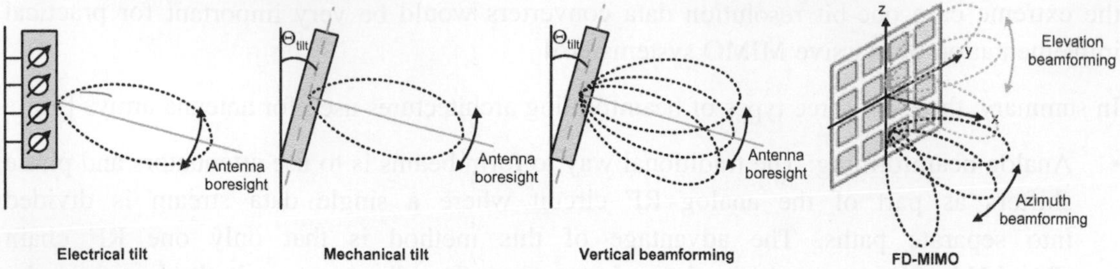

Figure 4.67

Illustration of mechanical/electrical tilting, vertical, and 3D beamforming [23].

the tilt angle of the antenna along the vertical axis. As depicted in Fig. 4.67, some adjustable brackets were used to mechanically change the tilting angle of the antenna.

It is possible to control the tilt angle electrically by applying an overall phase shift to all antenna elements in the array. An active antenna system (AAS) is a recent technology that allows more individual control on antenna elements, where each array element (or group of elements) is integrated with a separate RF transceiver unit that provides remote control to the elements. By employing AAS at the base station, the vertical radiation pattern can also be adjusted dynamically in each sector, and multiple elevation beams can be generated to support multiple users or cover multiple regions; thus full-dimension MIMO (FD-MIMO) is a combination of azimuth and elevation beamforming. Depending on the way that the antenna down tilt is changed, 3D beamforming can be classified into static and dynamic schemes. The static 3D beamforming refers to a system where the antenna tilt at the base station is set to a fixed value according to some statistical metrics, for example, the mean value of the vertical angles of users. This method cannot be adapted to the dynamic patterns of users' movements, that is, once the tilt angle is selected, it will remain unchanged. In contrast the dynamic 3D beamforming is a technique that steers the base station antenna tilt angle according to specific user locations. As mentioned earlier, the antennas at the base station are usually configured as a linear array of a limited number of antennas in the azimuth plane. However, these geometries can shape the radiation pattern only in the horizontal plane; hence, to change the beam in the elevation plane for 3D beamforming, more general 2D or 3D array topologies are necessary. Those arrays are active antenna systems that are spaced in both azimuth and vertical planes with different configurations such as planar, circular, spherical, or cylindrical structures. In addition, the array may include copolarized or cross-polarized antenna elements [23].

In general, adding more antenna elements to the array provides more flexibility in beam steering designs and increases the number of radiation beams of the array. For vertical sectorization in which the number of vertical sectors is usually small (e.g., two or three), only a small number of antennas are required in the vertical plane. However, in 3D beamforming with per-user beam pattern adaptation (i.e., user tracking), a large number of antennas are needed. One of the challenges of the 3D beamforming is physical constraints and placement of a large number of antennas at the base station (see Fig. 4.68). This problem may be alleviated in higher frequencies that are used for 5G networks [45].

The study of elevation beamforming and FD-MIMO began in 3GPP LTE Rel-12. In an FD-MIMO system, a base station with two-dimensional active antenna array supports multi-user joint elevation and azimuth or 3D beamforming, which results in much higher cell capacity compared to conventional systems. In an FD-MIMO architecture using 2D AASs with $N_{column} \times N_{row}$ physical antennas, the precoding of a data stream is performed in two stages: (1) antenna-port virtualization that is a stream on an antenna port is precoded on N_{TXRU} transceiver units; and (2) transceiver-unit virtualization where a signal is precoded on $N_{antenna}$ antenna elements. It is noted that in traditional transceiver architecture modeling, a fixed one--to-one mapping is assumed between antenna ports and transceiver units, and TXRU virtualization effect is combined into a fixed antenna pattern which captures the effects of both TXRU virtualization and antenna element pattern. Antenna-port virtualization is an operation in the digital domain, and it refers to digital precoding that can be performed in frequency-selective manner. An antenna port is typically defined in conjunction with a reference signal. For example, for precoded data transmission on an antenna port, a DM-RS is transmitted on the same bandwidth as the data and both are precoded with the same digital precoder. For CSI estimation, on the other hand, CSI-RSs are transmitted on multiple antenna ports. For CSI-RS transmissions, the precoder characterizing the mapping between CSI-RS ports to TXRUs can be designed as an identity matrix, to facilitate device's estimation of TXRU virtualization precoding matrix for data precoding vectors. The TXRU virtualization is an analog operation; thus it refers to time-domain analog precoding. The TXRU virtualization can be made time-adaptive. When TXRU virtualization is semi-static (or rate of change in TXRU virtualization is slow), the TXRU virtualization weights of a serving cell can be chosen to provide good coverage to its serving mobiles and to reduce interference to other cells. There will be more challenges, if TXRU virtualization is dynamic in terms of hardware implementation and protocol design [46].

In 1D TXRU virtualization, N_{TXRU} TXRUs are associated with N_{column} antennas comprising a column antenna array with the same polarization. In 2D antenna arrays with dual-polarized configuration, that is, $P = 2$, and the addition of N_{row} rows, the total number of TXRUs would be $Q = N_{TXRU}N_{row}P$. In 2D TXRU virtualization, Q TXRUs can be associated with any of $N_{column}N_{row}P$ antenna elements. These two different TXRU architectures have different trade-offs in terms of hardware complexity, power efficiency, cost and performance. For each method, subarray partition and full-connection architectures are considered. In subarray partition the antenna elements are partitioned into multiple groups with the same number of elements. In 1D subarray partition, N_{column} antenna elements comprising a column are partitioned into groups of K elements. In 2D subarray partition the total number of antenna elements

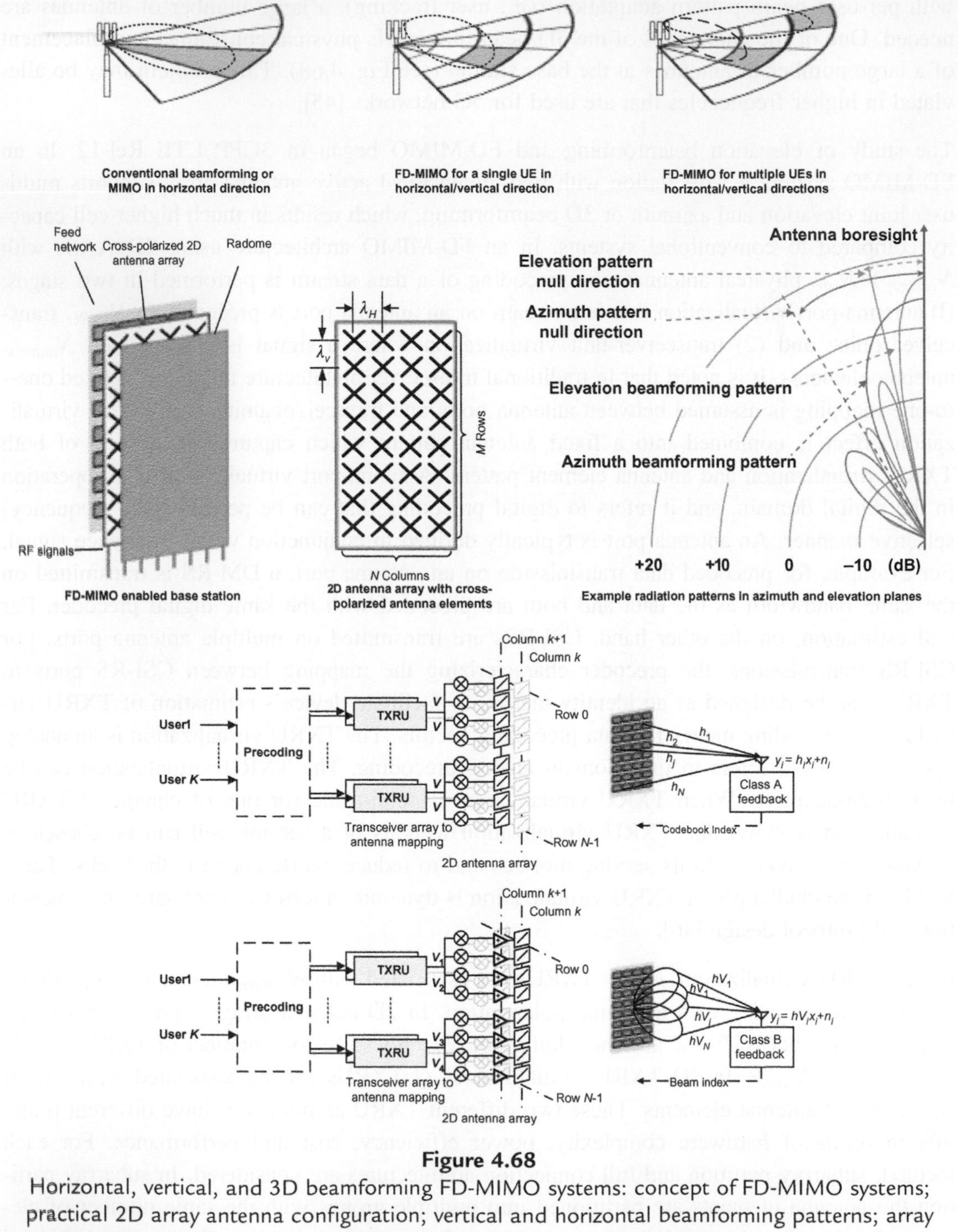

Figure 4.68

Horizontal, vertical, and 3D beamforming FD-MIMO systems: concept of FD-MIMO systems; practical 2D array antenna configuration; vertical and horizontal beamforming patterns; array partitioning architecture with the conventional CSI-RS transmission; and array connected architecture with beamformed CSI-RS transmission [46].

$N_{column}N_{row}P$ is partitioned into rectangular arrays of $K_1 \times K_2$ elements. On the other hand, in 1D full-connection, the output signal of each TXRU associated with a column antenna array with a same polarization is split into N_{column} signals, and those signals are precoded by a group of N_{column} phase shifters or variable gain amplifiers. Then N_{TXRU} weighted signals are combined at each antenna element. In 2D full-connection the output signal of each TXRU is split into $N_{column}N_{row}P$ signals, and those signals are precoded by a group of $N_{column}N_{row}P$ phase shifters or variable gain amplifiers. Then, Q weighted signals are combined at each antenna element. An illustration of 1D subarray partition and full-connection as well as general FD-MIMO architectures are shown in Fig. 4.69 [47].

Multi-antenna systems with a large number of base station antennas, often called massive MIMO, have received much attention in academia and industry as a means to improve the spectral efficiency, energy efficiency, and processing complexity of the cellular systems.

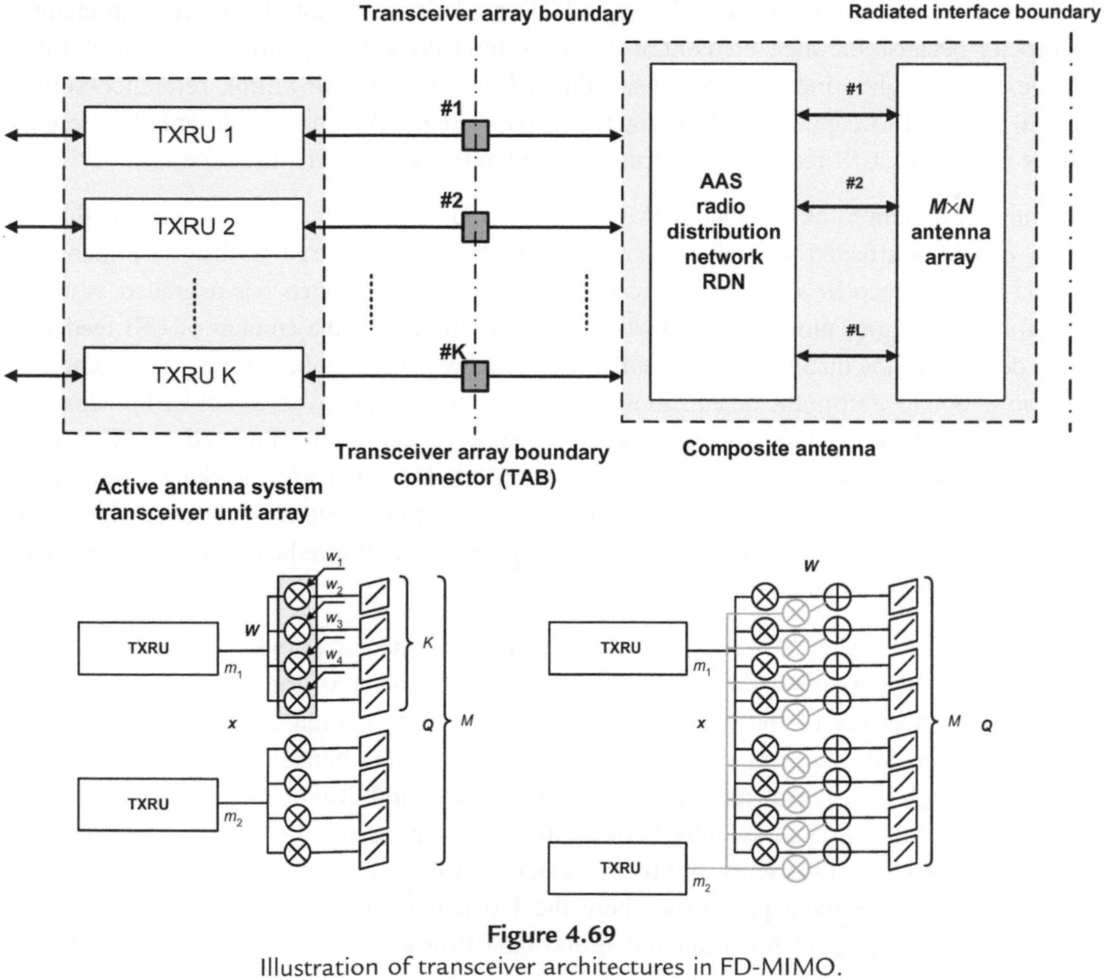

Figure 4.69
Illustration of transceiver architectures in FD-MIMO.

While massive MIMO is a promising technology, there are many practical and technical challenges on the path to its successful commercialization, including design and implementation of low-cost and low-power base station with large antenna arrays, capacity improvement of fronthaul links between remote radio heads and baseband units, measurement and reporting of high dimensional/resolution CSI, etc.

One of the main features of FD-MIMO systems is the potential to use a large number of antennas at the base station. Theoretically, as the number of base station antennas increases, the cross-correlation of two random channel realizations approaches zero; thus inter-user interference in the downlink can be controlled via a simple linear precoder. However, such a benefit can be realized only when the perfect CSI is available at the base station. While the CSI acquisition in TDD systems is relatively simple due to the channel reciprocity, that is not the case for FDD systems, because the time variation and frequency response of the channel in FDD systems are measured via the downlink reference signals and fed back to the base station after quantization. Even in TDD mode, one cannot always rely on channel reciprocity because the measurement at the transmitter does not capture the downlink interference from neighboring cells or co-scheduled UEs. As such, downlink reference signals are still required to capture the CQI for the TDD systems. As a result, downlink reference signals and uplink CSI feedback are crucial for operation of both duplex schemes.

A common problem in closed-loop MIMO systems, and in particular FDD systems, is that the quality of CSI is affected by limitation of the feedback resources. As CSI distortion increases, the MU-MIMO precoder's capability to control the inter-user interference is degraded, resulting in performance degradation of the FD-MIMO system. In general the amount of CSI feedback, which determines the quality of CSI, needs to be scaled with the number of transmit antennas of the base station to control the quantization error, while limiting the overhead of CSI feedback to avoid adverse impact on the system performance. An important problem related to CSI acquisition at the base station is the reference signal overhead. The UE performs channel estimation using the reference signals transmitted from the base station. Since the reference signals are typically distinguished through their orthogonal signatures, their overhead grows linearly with the number of transmit antennas.

As we mentioned earlier, FD-MIMO systems employ 2D planar arrays; thus propagation in both vertical and horizontal directions as well as the geometry of the transmitter array and the propagation effect of the 3D objects between the base station and the mobile station should be taken into account in channel modeling. The 3D channel propagation behavior obtained through measurements show the effect of height and distance-dependent LoS channel and the fact that LoS probability between the base station and the UE increases with the UE's height and increases when the distance between them decreases. Further it shows the effect of height-dependent path loss where the UE experiences less path loss on a higher floor (e.g., 0.6 dB/m gain for a macrocell and 0.3 dB/m gain for a micro cell). The height and distance-dependent elevation spread of departure (ESD) angles effect is exhibited when the location of the base station is higher than the UE, ESD decreases with the height of the UE and as the UE moves away from the base station [46,47].

FD-MIMO systems make use of a beamformed reference signals for CSI acquisition. Beamformed reference signal transmission is a channel training technique that uses multiple precoding weights in the spatial domain. In this scheme, the UE selects the best weight among those transmitted and then feeds back this index. This scheme provides many benefits compared to the case with non-precoded reference signals and especially when the number of transmit antennas is large. It can be shown that this scheme has less uplink feedback overhead relative to the case with perfect CSI, where the number of feedback bits used for channel vector quantization is linearly proportional to the number of transmit antennas, whereas the amount of feedback for the beamformed reference signals scales logarithmically with the number of reference signals, because the UE only feeds back the index of the best beamformed reference signal. It can be further shown that there is less downlink pilot overhead when the non-precoded reference signal is used. The non-beamformed reference signal overhead increases with the number of transmit antennas, resulting in substantial loss of sum capacity in the FD-MIMO, whereas the beamformed reference signal overhead is proportional to the number of reference signals and independent of the number of transmit antennas; therefore, the rate loss of the beamformed reference signals is marginal even when the number of transmit antennas increases [46,47].

As we mentioned earlier, an AAS transceiver contains integrated PA and LNA so that the gNB can control the gain and phase of individual antenna elements. A radio signal distribution/combining network between TXRUs and antenna elements was introduced (see Fig. 4.69) whose role is to deliver the transmit signal from the PA to the antenna array elements and the received signal from the antenna array to the LNA. Depending on the CSI-RS transmission and feedback mechanism, two architecture options, array partitioning and array connected, may be used. The former architecture is more suitable for the conventional codebook scheme, and the latter is for the beamforming scheme. In the array partitioning architecture, antenna elements are divided into multiple groups, and each TXRU is connected to one of them, whereas in the array connected architecture, the radio distribution network is designed such that the RF signals of multiple TXRUs are delivered to the single-antenna element. To combine RF signals from multiple TXRUs, additional RF combining circuitry is needed. In the array partitioning architecture, the total number of antenna elements L is partitioned into several groups of TXRUs, and an orthogonal CSI-RS is assigned for each group. Each TXRU transmits its own CSI-RS so that the UE can measure channel $\mathbf{h}$ from the CSI-RS observation. In the array connected architecture, each antenna element is connected to $L' < L$ TXRUs and an orthogonal CSI-RS is assigned for each TXRU. Denoting $\mathbf{h} \in \mathbb{C}^{1 \times N}$ as the channel vector and $\mathbf{v} \in \mathbb{C}^{N \times 1}$ as the precoding vector for each beamformed CSI-RS, the beamformed CSI-RS observation can be expressed as $y = \mathbf{h}\mathbf{v}x + n$ and the UE measures the precoded channel $\mathbf{hv}$. Due to the narrow and directional

CSI-RS beam transmission with a linear array, the SNR of the precoded channel is maximized at the target direction, that is, $SNR = |\mathbf{hv}(\varphi)|^2/\sigma^2$ where φ is the beam direction and σ^2 is the noise variance. In non-beamformed scenario, the UE selects and sends a precoder index which maximizes certain performance criterion to the gNB and adapts to the channel variation. In the beamformed scenario, the gNB transmits multiple beamformed CSI-RSs using the connected array architecture and the UE selects the preferred beam and then feeds back its index. When the gNB receives the beam index, the weight corresponding to the selected beam is used for data transmission to the UE [48].

Let us consider a cellular system consisting of N_{cell} cells each with one base station and N_{UE} terminals in each cell, as shown in Fig. 4.70. Each gNB is equipped with a 2D antenna array of $N_V \times N_H$ vertical and horizontal antennas, and each UE has a single antenna. We assume that all gNBs and UEs are synchronized and operate in TDD mode with universal frequency reuse. In the downlink, the lth base station applies a $N_V N_H \times N_{UE}$ precoder $\mathbf{F}_n, n = 1, 2, \ldots, N_{cell}$ to transmit a symbol for each user, with a power constraint $\|[\mathbf{F}_n]_{:,k}\|^2 = 1, k = 1, 2, \ldots, N_{UE}$. Uplink and downlink channels are assumed to be reciprocal. If $\mathbf{h}_{nck}$ denotes the $N_V N_H \times 1$ uplink channel from user k in cell c to the nth base station, then the received signal by this user in the downlink can be

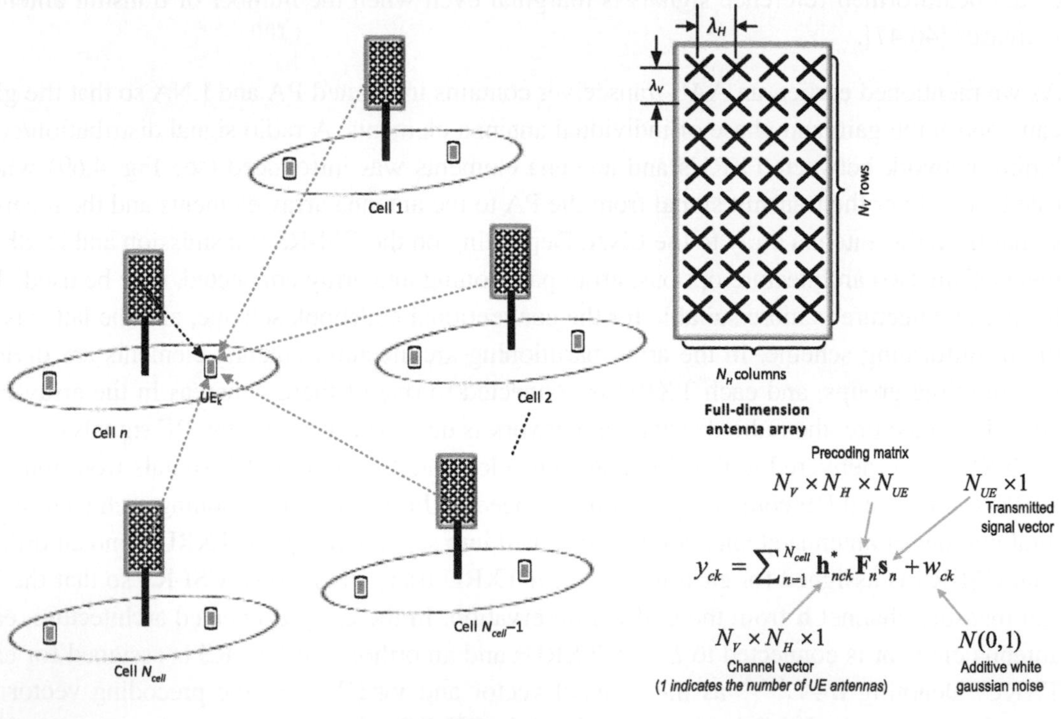

Figure 4.70
Illustration of full-dimension MU-MIMO concept [48].

written as $y_{ck} = \sum_{n=1}^{N_{cell}} \mathbf{h}_{nck}^{*}\mathbf{F}_n\mathbf{s}_n + w_{ck}$ where $\mathbf{s}_n$ represents the $N_{UE} \times 1$ vector of transmitted symbols from the nth base station. We further assume $E\{\mathbf{s}_n\mathbf{s}_n^{*}\} = P/N_{UE}\mathbf{I}$ with P denoting the total transmit power and $w_{ck} \sim N(0, \sigma^2)$ is the additive white Gaussian noise at kth user receiver in cell c. Given the 2D antenna arrays deployed at the gNBs, the channels from the base stations to each UE have a 3D structure. Using the Kronecker product correlation model, which has been shown to provide a good approximation to 3D covariance matrices, the covariance of the 3D channel $\mathbf{h}_{nck}$, which is defined as $\mathbf{R}_{nck} = E\{\mathbf{h}_{nck}\mathbf{h}_{nck}^{*}\}$, is approximated by $\mathbf{R}_{nck} = \mathbf{R}_{nck}^{\mathbf{A}} \otimes \mathbf{R}_{nck}^{\mathbf{E}}$ where $\mathbf{R}_{nck}^{\mathbf{A}}$ and $\mathbf{R}_{nck}^{\mathbf{E}}$ represent the covariance matrices in the azimuth and elevation directions, respectively. If $\mathbf{R}_{nck}^{\mathbf{A}} = \mathbf{U}_{nck}^{\mathbf{A}}\mathbf{\Lambda}_{nck}^{\mathbf{A}}\mathbf{U}_{nck}^{\mathbf{A}^*}$ and $\mathbf{R}_{nck}^{\mathbf{E}} = \mathbf{U}_{nck}^{\mathbf{E}}\mathbf{\Lambda}_{nck}^{\mathbf{E}}\mathbf{U}_{nck}^{\mathbf{E}^*}$ are the SVDs of $\mathbf{R}_{nck}^{\mathbf{A}}$ and $\mathbf{R}_{nck}^{\mathbf{E}}$ then using Karhunen–Loève transformation[17], the channel $\mathbf{h}_{nck}$ can be expressed as

$\mathbf{h}_{nck} = [\mathbf{U}_{nck}^{\mathbf{A}}\mathbf{\Lambda}_{nck}^{\mathbf{A}\ 1/2} \otimes \mathbf{U}_{nck}^{\mathbf{E}}\mathbf{\Lambda}_{nck}^{\mathbf{E}\ 1/2}]\mathbf{w}_{nck}$ where $\mathbf{w}_{nck} \sim N(0,\mathbf{I})$ is a $\text{rank}(\mathbf{R}_{nck}^{\mathbf{A}})\text{rank}(\mathbf{R}_{nck}^{\mathbf{E}}) \times 1$ vector, with rank($\mathbf{A}$) representing the rank of the matrix $\mathbf{A}$ [48]. The SINR at the kth user receiver in cell c can be shown to be as follows:

$$\text{SINR}_{ck} = \frac{\overbrace{(P/N_{UE})\left|\mathbf{h}_{cck}^{*}[\mathbf{F}_c]_{:,k}\right|^2}^{\text{Received signal power of the } k\text{th UE}}}{\underbrace{(P/N_{UE})\sum_{m\neq k}\left|\mathbf{h}_{cck}^{*}[\mathbf{F}_c]_{:,m}\right|^2}_{\substack{\text{Sum of interference from the signal}\\ \text{transmited to all other UEs in cell c}\\ \text{(interference)}}} + \underbrace{(P/N_{UE})\sum_{n\neq c}\left\|\mathbf{h}_{nck}^{*}\mathbf{F}_n\right\|^2}_{\substack{\text{Sum of interference from the signal}\\ \text{transmitted by all neighboring cells}\\ \text{(interference)}}} + \underbrace{\sigma^2}_{\text{Receiver noise}}}$$

The objective is to design the precoding matrices $\mathbf{F}_n, n = 1, 2, \ldots, N_{cell}$ such that they minimize inter-cell interference with minimal requirements on the channel knowledge, and they

[17] In the theory of stochastic processes, the Karhunen–Loève theorem is a representation of a stochastic process as an infinite linear combination of orthogonal functions. The transformation is also known as Hotelling transform and eigenvector transform, and it is closely related to principal component analysis (PCA). In contrast to a Fourier series where the coefficients are fixed numbers and the basis functions are sinusoidal the coefficients in the Karhunen–Loève theorem are random variables and the basis functions depend on the process. In fact the orthogonal basis functions used in this representation are determined by the covariance matrix of the process. Therefore, the Karhunen–Loève transform adapts to the process in order to produce the optimal basis for its expansion. In the case of a centered stochastic process $\{X(t)|t\in[a,b]\}$, that is, $E[X(t)] = 0\forall t\in[a,b]$, satisfying a continuity condition, it can be shown that $X(t)$ can be expanded as $X(t) = \sum_{k=1}^{\infty} Z_k e_k(t)$ where Z_k's are pairwise uncorrelated random variables and the functions $e_k(t)$ are continuous real-valued functions on $[a,b]$ that are pairwise orthogonal in $L^2[a,b]$. It is therefore sometimes said that the expansion is bi-orthogonal since the random coefficients Z_k are orthogonal in the probability space while the deterministic functions $e_k(t)$ are orthogonal in the time domain. The general case of a process $X(t)$ that is not centered can be converted into a centered process by considering $X(t) - E[X(t)]$, which is a centered process. If the process is Gaussian then the random variables Z_k are Gaussian and stochastically independent. This result generalizes the Karhunen–Loève transform. An important example of a centered real stochastic process on [0, 1] is the Wiener process, where the Karhunen–Loève theorem can be used to provide a canonical orthogonal representation for it. In this case, the expansion consists of sinusoidal functions [22].

can be implemented using low-complexity hybrid analog/digital architectures, that is, with a small number of RF chains. The main idea of multi-layer precoding is to design the precoder matrix as a product of three precoding matrices (layers) where each layer is designed to achieve only one precoding objective, for example, maximizing desired signal power, minimizing inter-cell interference, or minimizing multi-user interference [48].

$$\mathbf{F}_c = \underbrace{\mathbf{F}_c^{(1)}}_{\substack{\text{Inter – cell interference management} \\ \downarrow \\ \text{Minimize inter – cell interference}}} \underbrace{\mathbf{F}_c^{(2)}}_{\substack{\text{Desired signal} \\ \text{beamforming}}} \underbrace{\mathbf{F}_c^{(3)}}_{\substack{\text{Multi – user interference management} \\ \downarrow \\ \text{Minimizing multi – user interference}}}$$

4.1.9.5 Large-Scale (Massive) MIMO Systems

Massive MIMO is the generalization of a multi-user MIMO system that serves multiple users through spatial multiplexing over a channel with favorable propagation[18] conditions using time-division duplex scheme and relying on channel reciprocity and uplink reference signals to obtain CSI of each user. The base station is equipped with $N_{antennas}$ antennas to communicate with N_{user} (typically modeled with single-antenna) UEs on each time/frequency resource, where $N_{user} \ll N_{antennas}$. Each base station in the network operates individually and processes its signals using linear transmit precoding and linear receive combining [16,52]. By coherent processing of the signals over the array, transmit precoding can be used in the downlink to focus each signal at its target user, and receive combining can be used in the uplink to distinguish between signals received from different user terminals, thus the larger the number of antennas, the finer the spatial precision. A generic massive MIMO system operates in TDD mode, where the uplink and downlink transmissions take place on the same frequency resource but are separated in time. The physical propagation channels are reciprocal, meaning that the channel responses are theoretically the same in both directions, which can be utilized in TDD operation. In practice, the transceiver hardware is not reciprocal, thus transceiver calibration is required to exploit the channel reciprocity. Since uplink–downlink hardware mismatches only slowly and slightly change over time, they can be mitigated by simple calibration methods even without extra reference transceivers by relying on mutual coupling between antennas in the array. There are several reasons for the suitability of the TDD mode for massive MIMO which include the following [16,52]:

- The base station needs to know the CSI to process the antennas coherently.

[18] Favorable propagation means that the channel matrix between the base station antenna array and the users is well-conditioned. In a massive MIMO system, under some conditions, the favorable propagation property holds due to the law of large numbers. In other words, the propagation is said to be favorable when users are mutually orthogonal in some practical sense.

- The uplink channel estimation overhead is proportional to the number of terminals and independent of the number of antennas, making the scheme scalable with respect to the number of antennas. Furthermore, basic estimation theory indicates that the estimation quality (per antenna) cannot be reduced by adding more antennas at the base station. In fact the estimation quality improves with the number of antennas, if there is a known correlation structure between the channel responses over the array.

The data transmission in massive MIMO is based on linear processing at the gNB. In the uplink, the gNB has N_{RX} observations of the multiple access channel from the N_{user} terminals. The gNB applies maximal ratio combining to separate the signal transmitted by each terminal from the interfering signals, using the channel estimate of a terminal to maximize the signal power of that terminal by coherently adding the signal components. This results in a signal amplification proportional to N_{RX}, which is known as the array gain. Alternatively, ZF combining can be used, which suppresses inter-cell interference at the cost of reducing the array gain to $N_{RX} - N_{user} + 1$, or MMSE combining can be utilized that balances between amplifying signals and suppressing interference. Receive combining creates one effective scalar channel per terminal where the intended signal is amplified, and/or the interference is suppressed. The performance of the received combining methods will be improved by adding more gNB antennas, since there are more channel observations to utilize. The remaining interference is typically treated as additive noise; thus, conventional single-user detection algorithms can be applied. Another benefit of the combining is that small-scale fading averages over the array, in the sense that its variance decreases with N_{RX}. This is known as channel hardening and is a consequence of the law of large numbers. Since the uplink and downlink channels are ideally reciprocal in TDD systems, there is a strong connection between receive combining in the uplink and transmit precoding in the downlink. This is known as uplink–downlink duality. Linear precoding based on MRC, ZF, or MMSE principles can be applied to focus each signal on its target user and possibly to minimize interference toward other users [16,52].

It can be shown that the achievable spectral efficiency per cell of massive MIMO systems under ideal conditions and independent identically distributed Rayleigh fading can be expressed in the following form [16,52]:

$$\eta = N_{user}\left(1 - \frac{N_{user}}{\tau}\right)\log_2\left(1 + \frac{\varepsilon_{CSI}\gamma N_{TX}}{N_{user}\gamma + 1}\right) \quad \text{bps/Hz/Cell}, \quad \varepsilon_{CSI} = \left(1 + \frac{1}{N_{user}\gamma_{uplink}}\right)^{-1}$$

where $\left(1 - N_{user}/\tau\right)$ is the loss due to pilot transmission, γ is the downlink/uplink SNR, and ε_{CSI} is the quality of the estimated CSI, proportional to the mean-squared power of the MMSE channel estimate, where $\varepsilon_{CSI} = 1$ represents perfect CSI. Note that the numerator of the logarithm argument increases proportionally with respect to N_{TX} due to the array gain and that the denominator represents the interference plus noise. While the generic theory of massive MIMO systems assumed single-antenna terminals, the technology can support terminals with N'_{RX} antennas. In this case, N_{user} denotes the number of simultaneous data streams and the preceding equation describes the spectral efficiency per stream. These streams can be divided over anything from N_{user}/N'_{RX} to N_{user} terminals [16,52].

We discussed the capacity of MIMO systems earlier in this chapter, which can be written as follows:

$$C = \log_2 \det\left(\mathbf{I} + \frac{\gamma}{N_{TX}}\mathbf{H}\mathbf{H}^H\right) = \sum_{i=1}^{\min(N_{TX},N_{RX})} \log_2\left(1 + \frac{\gamma\sigma_i^2}{N_{TX}}\right)$$

In the preceding equation, it is assumed that transmitter has the full knowledge of CSI and that the channel matrix $\mathbf{H}$ can be decomposed using SVD method, where σ_i^2's are the eigenvalues of $\mathbf{H}\mathbf{H}^H$. In the preceding equation, if the SNR γ is extremely small, the capacity asymptotically approaches $C_{\gamma\to 0} \approx \gamma N_{RX}/\ln 2$, which is independent of N_{TX}, thus, even under the most favorable propagation conditions, the multiplexing gains are lost, and from the perspective of achievable rate, multiple transmit antennas are of no value. Next, let the number of transmit antennas grow large while keeping the number of receive antennas constant. We further assume that the row vectors of the channel matrix are asymptotically orthogonal, $C_{N_{TX}\gg N_{RX}} \approx N_{RX}\log_2(1+\gamma)$. Then, let the number of receive antennas increase while keeping the number of transmit antennas constant. We also assume that the column vectors of the channel matrix are asymptotically orthogonal, $C_{N_{RX}\gg N_{TX}} \approx N_{TX}\log_2(1+\gamma N_{RX}/N_{TX})$. Therefore, an excess number of transmit or receive antennas, combined with asymptotic orthogonality of the propagation vectors, constitutes a highly desirable scenario. Additional receive antennas continue to improve the effective SNR and could in theory compensate for a low SNR and restore multiplexing gains that would otherwise be lost. Furthermore, orthogonality of the propagation vectors implies that independent and identically distributed complex-Gaussian inputs are optimal so that the achievable rates are in fact the true channel capacities [61].

The studies on massive MIMO have been mainly focused on frequencies below 6 GHz, where the transceiver hardware technology is very mature. The same concept can be applied in mmWave bands, where many antennas might be required since the effective aperture of the antenna is much smaller. However, the hardware implementation will be more challenging. The support of mobility will be more difficult because the coherence time will be an order of magnitude shorter due to higher Doppler spread, which reduces the spatial multiplexing capability.

The channel impulse response between a user terminal and a base station can be represented by an $N_{antennas}$-dimensional vector. Since the N_{user} channel vectors are mutually nonorthogonal in general, advanced interference cancellation receivers are needed to suppress interference and achieve the sum capacity of the multi-user channel. As we mentioned earlier, the favorable propagation is an environment where the N_{user} users' channel vectors are mutually orthogonal (i.e., their inner products are zero). The favorable propagation channels are ideal for multi-user transmission since the interference is removed by simple linear processing (i.e., MRC and ZF) that utilizes the channel orthogonality. The question is whether there are any favorable propagation channels in practice. An approximate form of favorable propagation is achieved in non-LOS environments with rich scattering, where each channel vector has independent stochastic entries with zero mean and identical distribution. Under these conditions, the inner products (normalized by $N_{antennas}$) approach to zero as the number of antennas increases; meaning that the channel vectors tend to be orthogonal as $N_{antennas}$ increases. The sufficient condition above is satisfied for Rayleigh fading channels, which are considered in the studies on massive MIMO, but approximate favorable propagation can also be obtained in other conditions [16,52].

The conventional open-loop beamforming provides meaningful array gains for small arrays in LoS propagation environments; however, this scheme is not scalable and not able to handle isotropic fading (*isotropic fading encompasses a broad range of fading channels with the common property that the transmitter is unable to track the directions of the users' time-varying channel vectors. Thus, from the transmitter standpoint all directions are statistically equivalent*). In practice, the channel of a particular user terminal might not be isotropically distributed, rather it might have distinct statistical spatial properties. The codebook in open-loop beamforming cannot be tailored to a specific terminal, rather needs to explore all channel directions that are possible for the array. For large arrays with arbitrary propagation properties, the channels must be measured by reference signals as is done in the massive MIMO [16,52].

The studies on massive MIMO were mainly focused on the asymptotic regime where the number of service antennas $N_{antennas} \to \infty$. Recent studies have derived closed-form achievable spectral efficiency expressions that are valid for any number of antennas and user terminals, SNR, and choice of reference signals. Those expressions do not rely on idealized assumptions such as perfect CSI, rather on worst case assumptions regarding the channel acquisition and signal processing. Although the total spectral efficiency per cell is greatly improved with massive MIMO, the anticipated performance per user lies in the conventional range of 1–4 bps/Hz. This is part of the range where conventional channel codes perform close to the Shannon limits.

There are no strict requirements on the relation between $N_{antennas}$ and N_{user} in massive MIMO systems. A simple definition of massive MIMO would be a system with many active antenna elements that can serve a large number of user terminals. One should avoid specifying a certain ratio $N_{antennas}/N_{user}$, since it depends on a variety of conditions such as the system performance metric, propagation environment, and coherence block length.

The massive MIMO gains do not require high-precision hardware; in fact, lower hardware precision can be handled compared to other systems since additive distortions are suppressed in the processing. Another reason for the robustness is that massive MIMO can achieve high spectral efficiencies by transmitting low-order modulations to a multitude of terminals, while contemporary systems require high-precision hardware to support transmission of high-order modulations to a few terminals [16,52].

In an OFDM system, resource allocation means that the time-frequency resources are divided between the terminals to satisfy user-specific performance constraints, finding the best subcarriers for each terminal, and overcoming the small-scale fading effects by power control. Frequency-selective resource allocation can provide significant improvements when there are large variations in channel quality over the subcarriers, but it is also demanding in terms of channel estimation and computational overhead since the decisions depend on the small-scale fading, which varies in time. If the same resource allocation concepts were applied to massive MIMO systems, with tens of terminals at each of the thousands of subcarriers, the system complexity would have been excessively prohibitive. However, the channel hardening effect in massive MIMO means that the channel variations are negligible over the frequency-domain and mainly depend on large-scale fading in the time domain, which typically varies much slower than small-scale fading, making the conventional resource allocation concepts unnecessary for massive MIMO. The available bandwidth can be simultaneously allocated to each active terminal, and the power control decisions are made jointly for all subcarriers based only on the large-scale fading characteristics [16,52]. Thus, the resource allocation can be greatly simplified in massive MIMO systems.

4.1.9.6 NR Multi-antenna Transmission Schemes

Multi-antenna transmission and beamforming of control and traffic channels are the distinct features of the new radio relative to its predecessors. In above 6 GHz frequency bands, the large number of antenna elements is primarily used for beamforming to achieve enhanced coverage, while at lower frequency bands, they enable FD-MIMO and interference avoidance by spatial filtering. The NR physical channels and signals, including those used for control and synchronization, have all been designed for beamforming. Unlike LTE whose downlink control channels used transmit diversity to ensure sufficient control channel link budget, NR control channels rely on a single-antenna port and beamforming to achieve coverage requirements. The CSI for operation of massive MIMO schemes can be obtained by feedback of CSI reports based on transmission of CSI reference signals in the downlink, either per antenna element or per beam, as well as using uplink measurements exploiting channel reciprocity. In order to simplify the implementation, the new radio supports analog beamforming in addition to digital precoding and beamforming. The support of analog beamforming, where the beam is shaped after digital-to-analog conversion, is necessary in high frequencies. Analog beamforming requires that the receive or transmit beams to be formed in one direction at a given time and further requires beam-sweeping, in which the same signal is repeated over multiple OFDM symbols and in different transmit beams. This is to ensure that control/traffic signals can be transmitted with high gain to sufficiently cover the service area of the base station. Beam management procedures and signaling are further specified in NR including indication to the device to assist selection of the receive beam (in case of analog receive beamforming) during data and control reception [11].

The beams corresponding to the large antenna arrays are narrower and beam tracking may fail; therefore, beam recovery procedures have been specified in NR where a device can trigger a beam recovery procedure. Furthermore, a cell may consist of multiple transmission points, each transmitting its own beams; in that case, beam management procedures would allow device-transparent mobility for seamless handover between the beams of different transmission points. In addition, uplink-centric and reciprocity-based beam management is possible by utilizing uplink reference signals. The possibility of spatially separating users increases when using a large number of antenna elements in lower frequency bands in both uplink and downlink; however, this requires the knowledge of channel at the transmitter. In NR, extended support for such multi-user spatial multiplexing was introduced, either through high-resolution CSI feedback with a linear combination of DFT vectors, or uplink SRS improvements based on channel reciprocity. Moreover, support for distributed MIMO has been introduced, where the device can receive multiple PDCCHs and PDSCHs per slot to enable simultaneous data transmission from multiple transmission points to the same user [69].

The support of hybrid beamforming and high-resolution CSI feedback have been two important design principles in NR MIMO. The first goal is addressed by beam management, where a UE measures a set of analog beams for each digital port and reports the beam quality. The gNB then assigns one or a small number of analog beams to the UE. As the downlink channel experienced by the UE changes, the gNB can modify this assignment, particularly when the link associated with the assigned beam fails. While beam management is instrumental in above 6 GHz frequency bands, it is also applicable to sub-6 GHz bands. For instance, in multipoint transmission scenario, where multiple transmission-reception points are associated with a UE, each link corresponds to a beam. The second goal is addressed by designing a modular and scalable CSI framework. High-resolution spatial channel information is provided via two-stage high-resolution precoding. The first stage involves the choice of basis subset, and the second stage comprises a set of coefficients for approximating a channel eigenvector with a linear combination of the basis subset. Note that while beam management and CSI acquisition can be operated independently, they can be used together for mobile UEs.

In NR, the DM-RS and the TRS are used to estimate the timing offset and frequency errors. To guarantee that the timing-offset and frequency-error of the antenna panels can be estimated independently, the DM-RS and TRS are grouped, and the grouping information is signaled to the UE. The UE distinguishes the time-frequency resource of reference signals allocated for different data streams. In this way, timing-offset and frequency-errors can be independently estimated for different MIMO layers in non-coherent MIMO transmission to improve the performance.

For downlink transmission with multiple antenna panels, the accuracy of CSI acquisition is critical for the system performance. To characterize the channel directional information, the codebook and the related feedback mechanism are typically designed for CSI acquisition, especially for FDD systems. For users in the cell edge of TDD systems, reciprocal channel estimation based on uplink reference signals is also inaccurate, and codebook-based CSI feedback can help improve the performance. Codebook design for massive MIMO should be flexible for different antenna array structures and applicable to various numbers of antennas. For a uniform panel array, the codebook is designed similar to the one used for FD-MIMO. While a single DFT vector can only characterize one spatial channel path, an advanced codebook taking the combination of two or more DFT vectors as the precoder can capture the characteristics of multipath channels. For non-uniform panel array, the antenna elements in the horizontal/vertical direction cannot be viewed as a uniform array. Thus array response vector is not in DFT form, and additional phase difference exists between the panels. In other words, the DFT vector cannot capture the actual channel response but can cause beam distortion and beam gain reduction. Moreover, panels at one transmit/receive point are not easily calibrated in a practical implementation, where a fixed or random phase may exist among different panels. A good codebook design should use phase/amplitude factors among panels to combine the beamforming vectors of the panels to match the array response [59].

In multi-panel MIMO codebook design, 2D-DFT vectors may be used as the per-panel beamforming vectors representing the spatial propagation properties for each panel. Furthermore, an inter-panel co-phasing factor is added across the DFT-based codeword to reduce performance loss. Let us use the two-panel case as an example, one can design the multi-panel codebook $\mathbf{W}$ for dual-polarized antenna array as $\mathbf{W} = \left[\mathbf{W}_1^1\mathbf{W}_2^1 \quad \mathbf{W}_1^2\mathbf{W}_2^2\varphi \right]^T$ where $\mathbf{W}_1^1$ and $\mathbf{W}_2^1$ consist of DFT vectors reflecting long-term channel characteristics of antenna panel 1 and panel 2, respectively; $\mathbf{W}_1^2$ and $\mathbf{W}_2^2$ consist of phase factors reflecting the short-term and frequency-selective channel elements of panels 1 and 2, respectively, and φ is the inter-panel co-phasing factor, which is designed for feedback in a wideband manner or subband manner according to the use cases. The precoding vectors for both panels are restricted to be the same in order to reduce the feedback overhead [59].

For a uniform or non-uniform planer array with ideal synchronization or small phase offset, coherent transmission can be used to achieve spatial multiplexing/diversity gain. However, the performance of coherent transmission may be degraded in practice due to different reflection and refraction propagation paths which cause different average channel delay and different average channel gain for a given direction of arrival and departure. If antenna ports from different panels experience different large-scale fading, it causes the eigenvectors of the channel matrix to become ill-conditioned (rank deficiency), resulting in inaccurate CSI for coherent MIMO transmission. Moreover, time-varying amplitude or phase calibration error among the panels may occur in practice, especially when the antenna panels have independent clocks and different operating temperatures. In addition, the frequency offset for different panels observed at the receiver may be different when there is a time-varying relative phase between different antenna panels.

The reference signals transmitted from geographically separated panels may experience different Doppler shift and Doppler spread, because the UE may move in a different directions relative to the panels. If the UE assumes the panels have the same Doppler shift and Doppler spread, the frequency offset estimation would be inaccurate, which results in performance degradation. The relative frequency offset between panels can be from 0 to 300 Hz, due to the following factors [59]:

- Doppler shift: For a UE moving with the speed of 30 km/h at the carrier frequency of 2 GHz, the maximum difference between the experienced Doppler shifts is 111 Hz.
- Frequency error: In LTE, the maximum tolerable oscillator inaccuracy is ± 0.05 ppm for the wide area base station classes; thus the maximum frequency error between two non-calibrated panels with independent oscillators at the carrier frequency of 2 GHz is about 200 Hz. Thus the total amount of relative frequency offset is around 300 Hz. As the carrier frequency increases, the frequency offset problem would be more severe.

The same timing offset assumption for different panels may impact the accuracy of channel estimation. With a timing offset, a random linear phase factor will be added to the channel coefficient on adjacent subcarriers, which is difficult to accurately estimate. The inter-symbol interference caused by the timing offset impacts the channel estimation accuracy in both CSI measurement and data demodulation, resulting in significant MIMO performance degradation. It can be shown that the performance loss due to timing offset is not negligible, especially in the case of negative timing offset. For positive timing offset, when the aggregated signals are received with the timing difference shorter than the cyclic prefix length, less ISI is incurred. For the negative timing offset, since the incurred ISI cannot be mitigated by removing the cyclic prefix, severe performance loss is caused. The capability of interference suppression in the receiver is critical for the performance of non-coherent MIMO transmission. For coherent MIMO transmission, a traditional linear receiver such as MRC/MMSE can be used to achieve acceptable performance, while for non-coherent MIMO transmission, an advanced receiver is required. If the data streams from different directions (especially in the case of widely spaced panels) arrive at the UE in different angles, a simple linear MMSE receiver may be sufficient. However, the performance gain depends on the remaining interference after the space-domain interference rejection by the MMSE processing. If different streams arrive at the UE from the same or adjacent directions (e.g., different antenna panels located in a centralized manner),

Downlink MIMO operation in sub-6 GHz

Single CSI-RS	Multiple CSI-RS	SRS-based
▪ CSI-RS may be beamformed ▪ Allows codebook feedback ▪ Similar to LTE class-A CSI feedback (gNB transmit CSI-RS; UE computes RI/PMI/CQI) ▪ Maximum of 32 ports in the CSI-RS (codebooks are defined for up to 32 ports) ▪ Typically intended for arrays having 32 TXRUs or less with no beam selection (no CRI)	▪ Combines beam selection with codebook feedback (multiple beamformed CSI-RS with CRI feedback) ▪ Similar to LTE Class-B CSI Feedback (gNB transmits one or more CSI-RS, each in different directions; UE computes CRI/PMI/CQI) ▪ Supports arrays with arbitrary number of TXRUs ▪ Maximum of 32 ports per CSI-RS	▪ Exploits TDD reciprocity feature ▪ Similar to SRS-based operation in LTE ▪ Supports arrays with arbitrary number of TXRUs ▪ Procedure (UE transmits SRS, gNB computes precoding weights)

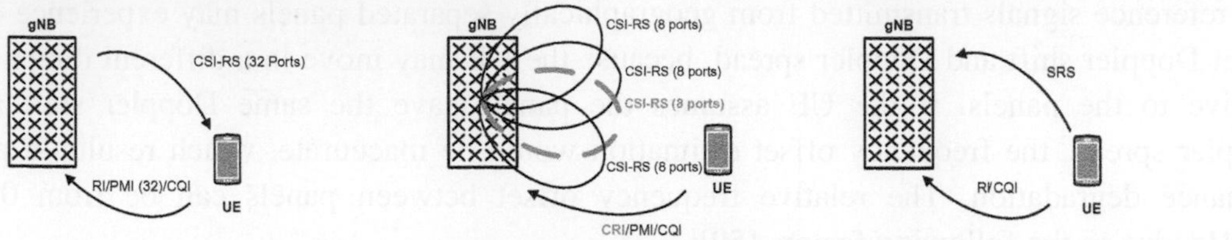

Downlink MIMO operation in above 6 GHz

Single-panel array	Multipanel array
▪ Combination of analog beamforming and digital precoding at baseband ▪ Analog beamforming is typically one RF beamforming weight-vector per polarization (a single cross-polarized beam) ▪ Supports two TXRUs and single-user MIMO only ▪ Digital precoding options: None (rank-2 all the time); CSI-RS based (RI/PMI/CQI); and SRS-based (RI/CQI)	▪ Combination of analog beamforming and digital precoding at baseband ▪ Analog beamforming is typically one RF beamforming weight-vector per polarization per panel ▪ One cross-polarized beam per subpanel ▪ Number of TXRUs = 2 x number of panels ▪ Digital precoding options: CSI-RS based (RI/PMI/CQI); SRS-based (RI/CQI); and SU-MIMO and MU-MIMO (typically one UE per cross-polarized beam)

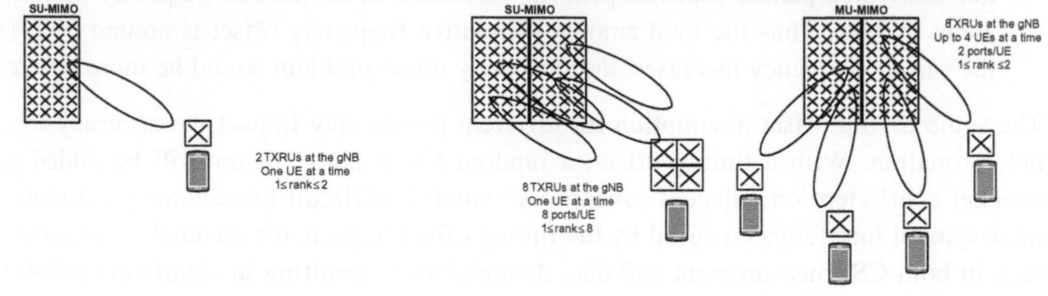

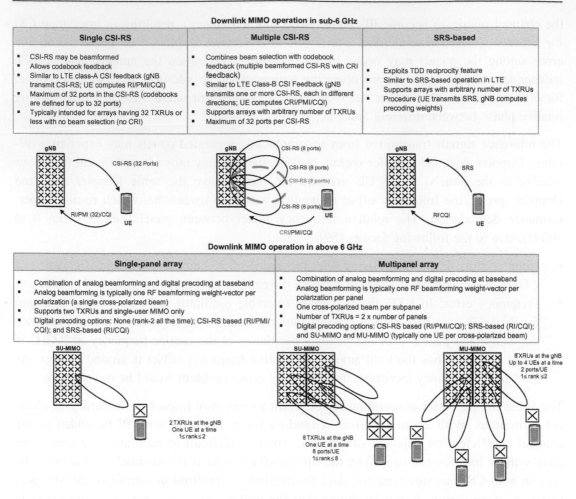

Figure 4.71
Summary of NR downlink multi-antenna operation [26].

inter-stream interference would be inevitable, and it would be the dominant factor to degrade the demodulation performance. In this case, a nonlinear receiver may be needed. When the received signal power difference among the streams is larger than 3 dB, the codeword–level successive interference cancellation or SIC receiver can satisfactorily perform. However, when the received signal power is almost the same for different data streams, the performance of SIC will deteriorate. In that case, a parallel interference cancellation receiver would be a more suitable choice. In NR, the improved DM-RSs and tracking reference signals can be used to estimate the timing offset and frequency errors in above 6 GHz frequencies [59]. The NR downlink multi-antenna operation has been discussed in different sections of this chapter and is summarized in Fig. 4.71.

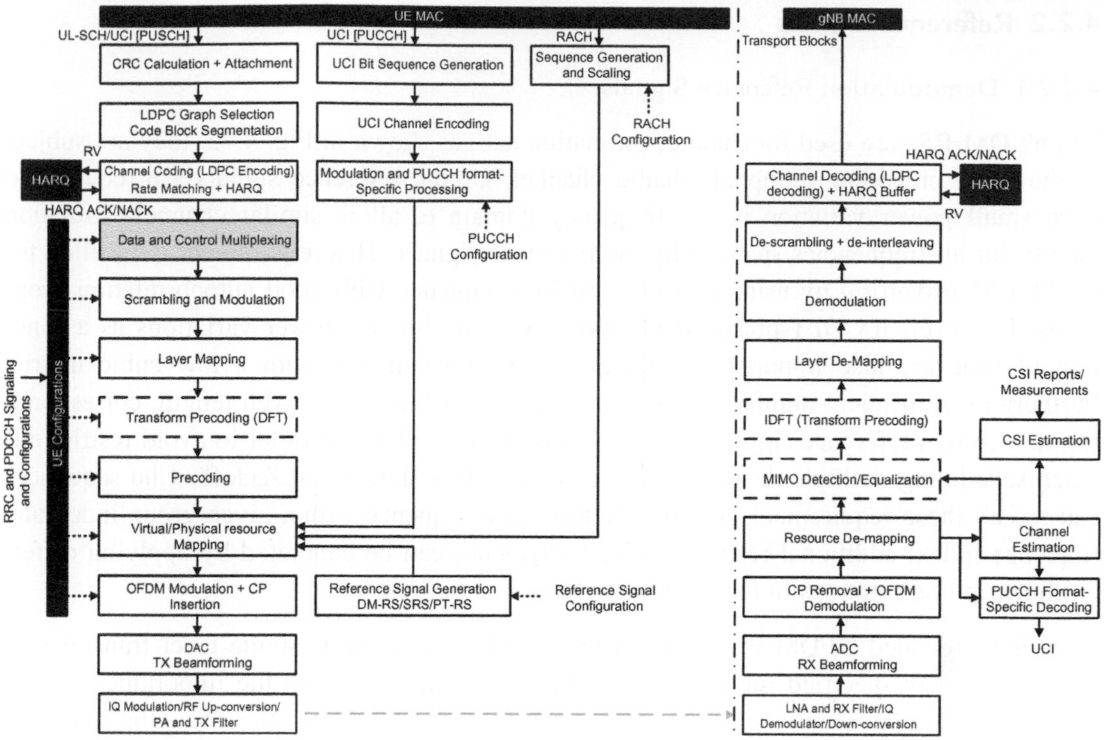

Figure 4.72
Overall uplink physical layer processing [5].

4.2 Uplink Physical Layer Functions and Procedures

4.2.1 Overall Description of Uplink Physical Layer

The NR uplink physical-layer consists of higher layer configurable functional blocks that are configured according to the uplink physical channel characteristics, use case, deployment scenario, etc. As shown in Fig. 4.72, the physical-layer processing generally includes receiving higher layer data (e.g., MAC PDUs in the case of uplink shared channel); CRC calculation and attachment; channel encoding and rate matching; modulation; mapping to physical resources and antennas; multi-antenna processing; and support of layer-1 control and HARQ-related signaling. The physical-layer model for RACH transmission is characterized by a physical RACH (PRACH) preamble format that consists of a cyclic prefix, a preamble, and a guard time during which no signal is transmitted.

4.2.2 Reference Signals

4.2.2.1 Demodulation Reference Signals

Uplink DM-RSs are used for channel estimation and, as shown in Fig. 4.73, they are subject to the same precoding as uplink shared channel. Uplink reference signals are required to have small power variation in the frequency domain to allow similar channel-estimation quality for all frequencies spanned by the reference signals. This requirement is fulfilled for CP-OFDM waveform by using a pseudo-random sequence with good autocorrelation properties. However, for DFT-precoded OFDM waveform, limited power variations as a function of time are also important to achieve signal transmission with a low cubic metric. Furthermore, sufficient number of reference signal sequences of a given length, corresponding to a certain reference signal bandwidth, should be available in order to avoid restrictions when scheduling multiple devices in different cells. It is shown that Zadoff–Chu sequences can satisfy these requirements. From a Zadoff–Chu sequence with a given group index and sequence index, additional reference signal sequences can be generated by applying different linear phase rotations in the frequency domain.

The DFT-precoded OFDM waveform in the uplink only supports single-layer transmission and is primarily specified for limited link-budget scenarios. Due to the importance of low cubic metric property and the corresponding high-power–amplifier efficiency, the reference signal structure is somewhat different compared to the CP-OFDM uplink case. In general, transmitting reference signals that are frequency-multiplexed with other uplink transmissions from the same device is not suitable for the uplink as it would negatively impact the device power-amplifier efficiency due to increased cubic metric. Instead, certain OFDM

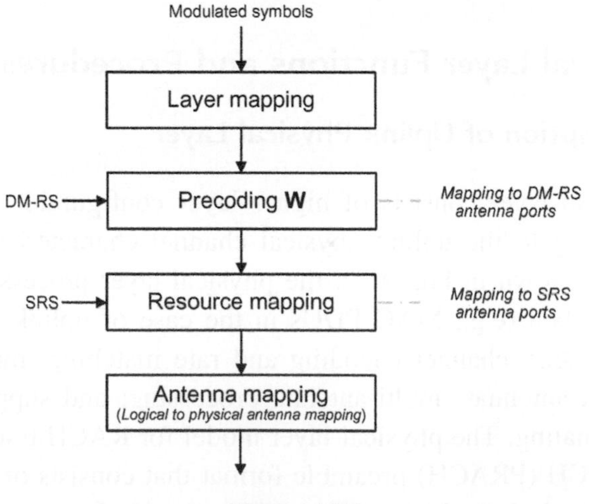

Figure 4.73
Processing of uplink DM-RS with PUSCH.

symbols within a slot are used exclusively for DM-RS transmission, that is, the reference signals are time-multiplexed with the data transmitted on PUSCH from the same device. The NR uses the same DM-RS structure for downlink and uplink in the case of CP-OFDM waveform. For DFT-spread OFDM waveform in the uplink, the DM-RS is based on Zadoff–Chu sequences and supports contiguous allocations and single-layer transmission, similar to LTE, in order to improve the power-amplifier efficiency. Multiple orthogonal reference signals can be generated in each DM-RS occasion where different reference signals are separated in the time, frequency, and code domains. Two different types of DM-RSs can be configured, namely, DM-RS Type 1 and Type 2, which differ in the maximum number of orthogonal reference signals and the mapping to the resource elements in the frequency domain. Type 1 provides up to four orthogonal reference signals using a single-symbol DM-RS and up to eight orthogonal reference signals using a double-symbol DM-RS, whereas Type 2 provides 6 and 12 patterns for single and double-symbol DM-RS, respectively. The DM-RS type 1 or 2 should not be confused with the mapping type A or B (see Section 4.1.2.1), as different mapping types can be combined with different reference signal types. In the following sections, we describe the DM-RSs corresponding to PUSCH and various PUCCH formats for the CP-OFDM uplink waveform.

4.2.2.1.1 PUSCH DM-RS

The DM-RS sequence for PUSCH $r_{DM-RS}(m)$ is generated according to $r_{DM\text{-}RS}(m) = [(1-2c(2m)) + j(1-2c(2m+1))]/\sqrt{2}$ where $c(i)$ is a length-31 Gold sequence generated by the pseudo-random sequence generator defined in [6] and initialized with $c_{init} = [2^{17}(N_{slot}^{symbol} n_{slot} + l + 1)(2N_{ID}^{n_{SCID}} + 1) + 2N_{ID}^{n_{SCID}} + n_{SCID}] \bmod 2^{31}$, l is the OFDM symbol number within the slot, n_{slot} is the slot number within a frame, and $N_{ID}^0, N_{ID}^1 \in \{0, 1, \ldots, 65535\}$ are given by the higher layer parameters *scramblingID0* and *scramblingID1* in the *DMRS-UplinkConfig* information element, respectively. The parameter $n_{SCID} \in \{0, 1\}$ may be signaled through the DM-RS initialization field of DCI format 0_1, otherwise, $n_{SCID} = 0$ [6].

The sequence $r_{DM\text{-}RS}(n)$ is mapped to an intermediate quantity $b_{k,l}^{(p,\mu)}$ according to $b_{k,l}^{(p,\mu)} = w_f(k')w_t(l')r_{DM\text{-}RS}(2n+k')$ where $k = 4n + 2k' + \Delta$ for DMR-RS Type 1 and $k = 6n + k' + \Delta$ for DMR-RS Type 2, and $k' = 0, 1$; $l = \bar{l} + l'$; $n \in \mathbb{N}$; $p = 0, 1, \ldots, N_{AP} - 1$ [6]. The spreading sequences $w_f(k')$, $w_t(l')$, and the offset Δ are given in Table 4.22 [6].

The intermediate quantity $b_{k,l}^{(p,\mu)}$ is precoded and multiplied with the amplitude scaling factor $\beta_{DM\text{-}RS}^{PUSCH}$ in order to adjust the transmit power of the reference signals and are then mapped to the physical resources according to $\left[a_{k,l}^{(p_0,\mu)} \ldots a_{k,l}^{(p_{\rho-1},\mu)}\right]^T = \beta_{DM\text{-}RS}^{PUSCH} \mathbf{W} \left[b_{k,l}^{(0,\mu)}(m) \ldots b_{k,l}^{(N_{AP}-1,\mu)}(m)\right]^T$. In the latter equation, $\mathbf{W}$ is the precoding matrix, and

Table 4.22: Parameters for PUSCH DM-RS Types 1 and 2 [6].

p	CDM Group	Δ	$w_f(k')$		$w_t(l')$	
			$k'=0$	$k'=1$	$l'=0$	$l'=1$
DM-RS Type 1						
0	0	0	+1	+1	+1	+1
1	0	0	+1	−1	+1	+1
2	1	1	+1	+1	+1	+1
3	1	1	+1	−1	+1	+1
4	0	0	+1	+1	+1	−1
5	0	0	+1	−1	+1	−1
6	1	1	+1	+1	+1	−1
7	1	1	+1	−1	+1	−1
DM-RS Type 2						
0	0	0	+1	+1	+1	+1
1	0	0	+1	−1	+1	+1
2	1	2	+1	+1	+1	+1
3	1	2	+1	−1	+1	+1
4	2	4	+1	+1	+1	+1
5	2	4	+1	−1	+1	+1
6	0	0	+1	+1	+1	−1
7	0	0	+1	−1	+1	−1
8	1	2	+1	+1	+1	−1
9	1	2	+1	−1	+1	−1
10	2	4	+1	+1	+1	−1
11	2	4	+1	−1	+1	−1

it is assumed that the resource elements $b_{k,l}^{(p,\mu)}$ are within the common resource blocks allocated for PUSCH transmission. It is further assumed that the reference point for the frequency index k is subcarrier 0 in common resource block 0; the reference point for time index l and the position l_0 of the first DM-RS symbol depends on the mapping type, that is, for PUSCH mapping type A, l is defined relative to the start of the slot, if frequency hopping is disabled; otherwise, it is measured relative to the start of each hop, and l_0 is given by the higher layer parameter *dmrs-TypeA-Position*. For PUSCH mapping type B, l is defined relative to the start of the scheduled PUSCH resources, if frequency hopping is disabled and relative to the start of each hop otherwise and $l_0 = 0$ [6].

In other words, the pseudo-random sequence corresponding to DM-RS Type 1 is mapped to every second subcarrier in the frequency domain over the OFDM symbol assigned to DM-RS transmission. As shown in Table 4.22, antenna ports 0 and 1 are associated with CDM group 0 and mapped to even-numbered subcarriers and are separated in the code-domain using different orthogonal sequences. The DM-RS corresponding to PUSCH for antenna ports 2 and 3 belong to CDM group 1 and are generated in the same way using odd-numbered subcarriers which are separated in the code domain. If more than four orthogonal antenna ports are needed, two consecutive OFDM symbols are used. The above DM-RS structure is used over each OFDM symbol and a length-2 orthogonal sequence is applied across time, resulting in up to eight orthogonal sequences.

DM-RS Type 2 has a similar structure as Type 1; however, there are some differences, most notably the number of antenna ports supported. As shown in Fig. 4.74, each CDM group for DM-RS Type 2 consists of two neighboring subcarriers over which a length-2 orthogonal sequence is applied to separate the two antenna ports sharing the same set of subcarriers. Two such pairs of subcarriers are used in each resource block for one CDM group. Since there are 12 subcarriers in a resource block, up to three CDM groups each with two orthogonal reference signals can be created using one resource block over one

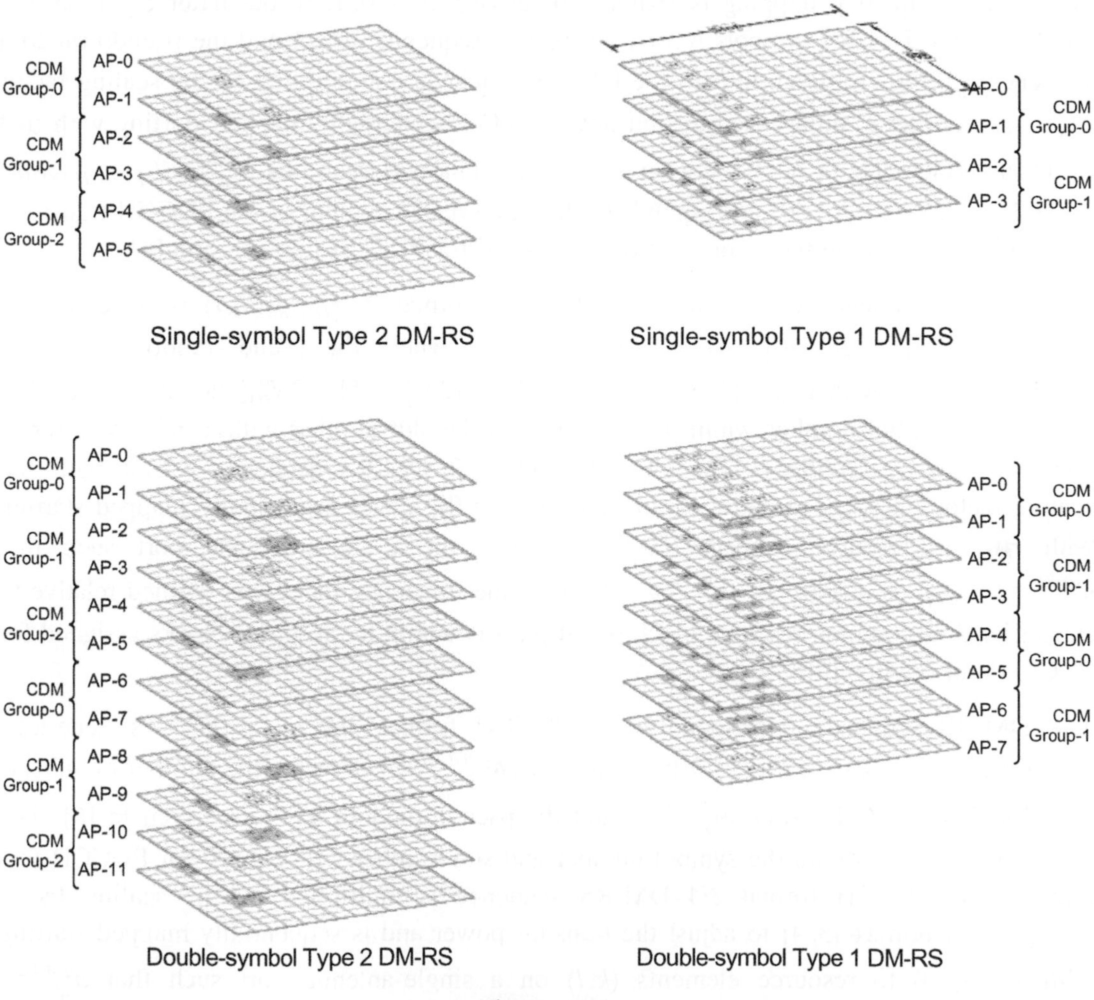

Figure 4.74
Illustration of DM-RS Type-1 and Type-2 time-frequency-code structures [14].

OFDM symbol. The number of Type 2 DM-RSs can be increased up to 12 by applying a length-2 orthogonal sequence across time domain. While the basic structures of DM-RS Type 1 and Type 2 are similar, the frequency-domain density of DM-RS Type 1 is higher than that of Type 2; however, Type 2 provides a larger number of orthogonal patterns, which useful for MU-MIMO use cases. The type of the reference signal structure is determined by dynamic scheduling and higher layer configuration.

4.2.2.1.2 PUCCH DM-RS

The DM-RS sequence for PUCCH format 1 is defined by $z(m'N_{sc}^{RB}N_{SF,0}^{PUCCH,1} + mN_{sc}^{RB} + n) = w_i(m)r_{u,v}^{(\alpha,\delta)}(n)$ where $n = 0, 1, \ldots, N_{sc}^{RB} - 1$, $m = 0, 1, \ldots, N_{SF,m'}^{PUCCH,1} - 1$, and $m' = 0$ when intra – slot frequency hopping is enabled; otherwise, $m' = 0, 1$. In the latter equation, the number of the DM-RS symbols $N_{SF,m'}$, orthogonal sequence $w_i(m)$, and the pseudo-random sequence $r_{u,v}^{(\alpha,\delta)}(n)$ are given in [6]. The DM-RS sequence is multiplied by the scaling factor $\beta_{PUCCH,1}$ in order to adjust the transmit power and is sequentially mapped starting with $z(0)$ to resource elements (k, l) in a slot on a single-antenna port such that $a_{k,l}^{(p,\mu)} = \beta_{PUCCH,1}z(m)$; $l = 0, 2, 4, \ldots$ wherein $l = 0$ corresponds to the first OFDM symbol of the PUCCH transmission and (k, l) is within the resource blocks assigned for PUCCH transmission [8].

The DM-RS sequence corresponding to PUCCH format 2, $r_{DM\text{-}RS}(m, l)$ is generated as $r_{DM\text{-}RS}(m, l) = \{[1 - 2c(2m)] + j[1 - 2c(2m + 1)]\}/\sqrt{2}$ where the pseudo-random sequence $c(i)$ is initialized with $c_{init} = [2^{17}(N_{slot}^{symbol} n_{slot} + l + 1)(2N_{ID}^0 + 1) + 2N_{ID}^0] \bmod 2^{31}$. Index l is the OFDM symbol number within the slot, n_{slot} is the slot number within the radio frame, and $N_{ID}^0 = \{0, 1, \ldots, 65535\}$. The PUCCH format 2 DM-RS sequence scaled with the scaling factor $\beta_{PUCCH,2}$ to adjust the transmit power [8] and is sequentially mapped starting with $r(0)$ to resource elements (k, l) in a slot on a single-antenna port such that $a_{k,l}^{(p,\mu)} = \beta_{PUCCH,2}r_{DM-RS}(m, l)$; $k = 3m + 1$ where the frequency index k is defined relative to subcarrier 0 of common resource block 0 and (k, l) is within the resource blocks assigned for PUCCH transmission [6].

The DM-RS sequence corresponding to PUCCH format 3/4 $r_{DM\text{-}RS}(m, l)$ is generated according to $r_{DM\text{-}RS}(m, l) = r_{u,v}^{(\alpha,\delta)}(m); m = 0, 1, \ldots, M_{sc}^{PUCCH,s} - 1$ where the number of subcarriers for this PUCCH format $M_{sc}^{PUCCH,s}$ and the pseudo-random $a_{k,l}^{(p,\mu)}$ are given in [6]. The cyclic shift α varies with the symbol number and slot number with $m_0 = 0$ for PUCCH format 3. The PUCCH format 3/4 DM-RS sequence is multiplied by the scaling factor $\beta_{PUCCH,s}$ wherein $s \in \{3, 4\}$ to adjust the transmit power and is sequentially mapped starting with $r_{DM\text{-}RS}(0)$ to resource elements (k, l) on a single-antenna port such that $a_{k,l}^{(p,\mu)} = \beta_{PUCCH,s}r_{DM\text{-}RS}(m, l)$; $m = 0, 1, \ldots, M_{sc}^{PUCCH,s} - 1$. The frequency index k is defined relative to subcarrier 0 of the lowest numbered resource block assigned for PUCCH transmission,

Table 4.23: DM-RS positions for PUCCH format 3 and 4 [6].

PUCCH Length	DM-RS Position / within PUCCH Allocation			
	No Additional DM-RS		Additional DM-RS	
	No Hopping	Hopping	No Hopping	Hopping
4	1	0,2	1	0,2
5	0,3		0,3	
6	1,4		1,4	
7	1,4		1,4	
8	1,5		1,5	
9	1,6		1,6	
10	2,7		1,3,6,8	
11	2,7		1,3,6,9	
12	2,8		1,4,7,10	
13	2,9		1,4,7,11	
14	3,10		1,5,8,12	

and time index l is given in Table 4.23 with and without intraslot frequency hopping as well as with and without additional DM-RS, where $l = 0$ corresponds to the first OFDM symbol of the PUCCH transmission [6].

4.2.2.2 Phase-Tracking Reference Signals

There is typically a mismatch between the oscillator frequencies of the transmitter and receiver in a communication system, resulting in a shift of the received signal spectrum at the baseband. In OFDM, this effect creates a misalignment between the FFT bins and the peaks of the sinc(.) pulses of the received signal, which would compromise the orthogonality between the subcarriers and would result in a spectral leakage. Each subcarrier interferes with other subcarriers, although the effect is dominant between adjacent subcarriers. Since there are many subcarriers, this is a random process is equivalent to a noise with Gaussian distribution. This random frequency offset degrades the SINR of the receiver. Therefore, an OFDM receiver will need to track and compensate the phase noise. The PT-RS is mainly introduced to compensate the CPE; however, PT-RS can also be used for ICI mitigation in higher frequency bands and potentially for CFO and Doppler estimation [54].

There is a trade-off between phase tracking accuracy and signaling overhead. If the density of PT-RS is high, phase tracking accuracy is high, and the CPE can be better compensated to achieve better performance. However, higher PT-RS density also means larger signaling overhead, which might lead to lower spectral efficiency or effective transmission rate. The studies show that the reduction of PT-RS density in time domain will degrade the BLER performance regardless of modulation order. However, the performance degradation is particularly significant when time density is reduced from 1 to 2, that is, from PT-RS for each OFDM symbol to PT-RS for every other OFDM symbol in the case of 256 QAM [54]. In that case, although signaling overhead is halved for time density 2, that is, more information bits can be transmitted in each RB, the effective data transmission rate suffers performance loss due to degraded BLER. In contrast, for 64QAM the degradation due to time density reduction is much less. Therefore, the time density of PT-RS can be a function of modulation order and it should increase with higher modulation order.

Since the receiver only needs to track the phase difference using the PT-RS, it does not need to know the amplitude of the PT-RS. Therefore, unlike other reference signals, for example, DM-RS, where the reference signals are formed by a pseudo-random sequence of symbols, PT-RS reference signals can use the exact same symbol. In MU-MIMO, PT-RS can be configured for each user and it is possible that the same subcarrier is used for multiple users, thus PT-RS collisions may happen, which would degrade the CPE compensation for two reasons: (1) the interference pattern is not completely random since the same symbol is used for PT-RS and (2) the interference level is higher when the power of PT-RS is boosted for more accurate CPE compensation. In such cases, it would be better to avoid PT-RS collision so that the interference is randomized without power boosting. This can be avoided by introducing an RB-level offset when configuring PT-RS for each user [54].

In NR, both DFT-S-OFDM and CP-OFDM waveforms are supported for the uplink transmissions and PT-RS signals are necessary for both waveforms. The PT-RS insertion follows a common framework for both downlink and uplink in case of CP-OFDM waveforms. In the case of DFT-S-OFDM, two types of insertion mechanisms for PT-RSs were studied, that is, pre-DFT and post-DFT insertion. In the former mechanism, PT-RS signals are inserted in the frequency domain before DFT precoding so that the resulting waveform still maintains the single-carrier properties. In the latter mechanism, the PT-RS are inserted after DFT precoding of the data symbols via various mechanisms such as puncturing. The PAPR of such a mechanism can however be controlled by using some signal processing techniques. In the time domain, the PT-RS locations can be configured to be either present in every symbol or every other symbol. The NR supports pre-DFT PT-RS insertion to conserve the single carrier property.

In the frequency domain, PT-RSs are transmitted in every second or fourth resource block, resulting in a sparse frequency-domain structure. The density in the frequency domain corresponds to the scheduled transmission bandwidth such that the higher the bandwidth, the lower the PT-RS density. For the smallest bandwidths, no PT-RS is transmitted. To reduce the risk of collisions between PT-RSs associated with different devices scheduled on overlapping time-frequency resources, the subcarrier number and the resource blocks used for PT-RS transmission are determined by the C-RNTI of the UE. The antenna port used for PT-RS transmission is given by the lowest numbered antenna port in the DM-RS antenna port group [14].

In a multi-TRP deployment, a UE can be supported by multiple co-located or non-co-located transmission points belonging to the same or different gNBs. The NR further supports large number of antenna elements at the gNB. These antenna elements are typically grouped as panels, where the signals feeding the antenna panels are generated by separate oscillators, which require individual phase noise compensation. In designing PT-RS for multi-TRP deployments, the orthogonal time-frequency allocation of PT-RS is crucial to minimize interference. The orthogonality of the PT-RS can be ensured in the frequency domain via frequency-division multiplexing of the PT-RS bearing resource elements. The increase in signaling overhead is a valid concern when there is a higher number of gNBs or many transmit panels. The typical CPE caused by the phase noise rotates the constellations by a limited margin, so only the higher order modulation schemes are impacted by the CPE. The users with higher MCS receive good SNR levels and are usually located closer to the gNB. When the MU-MIMO users are grouped, there will be higher and lower MCS users in these groups. If the PT-RS is transmitted without power boosting and with a wider beam than the narrow-beam data transmissions, the received EIRP for the PT-RS will be lower relative to the data transmissions. The lower MCS users that are generally further away from the cell center will receive PT-RS with a much lower effective power and will be able to discard PT-RS as interference, that is, they will not need CPE correction. They will be able to request the gNB to allocate data within these resource elements transmitted through narrow-beams. The same resource elements are used for the PT-RS in the wider beam transmissions for the benefit of higher MCS users, for whom the same resource elements will not be utilized in the narrow-beam data transmissions. With this effective power discrimination, non-orthogonal multiplexing of PT-RS and data is possible for the MU-MIMO configurations, which effectively increases the system spectral efficiency [54].

In CP-OFDM uplink, the precoded phase-tracking reference signal for subcarrier k on layer j is given by $r_{PT\text{-}RS}(p,m)=[(1-2c(2m))+j(1-2c(2m+1))]/\sqrt{2}$ if $p=p'$ or $p=p''$ where antenna ports p' or (p',p'') are associated with PT-RS transmission. The pseudo-random sequence $c(i)$ is initialized with $c_{init}=[2^{17}(N_{slot}^{symbol}n_{slot}+l+1)(2N_{ID}^{n_{SCID}}+1)+2N_{ID}^{n_{SCID}}+n_{SCID}] \bmod 2^{31}$, in which l is the OFDM symbol number within the slot, n_{slot} is the slot number within a frame, and $N_{ID}^{0}, N_{ID}^{1} \in \{0,1,\ldots,65535\}$ are given by the higher layer parameters *scramblingID0* and *scramblingID1*, respectively. The parameter $n_{SCID} \in \{0,1\}$ may be signaled through the DM-RS initialization field of DCI format 0_1; otherwise, $n_{SCID}=0$ [6].

The UE transmits PT-RSs (if configured) only in the resource blocks designated to PUSCH transmission. The PT-RS is mapped to resource elements according to $[a_{k,l}^{(p_0,\mu)} \cdots a_{k,l}^{(p_{\rho-1},\mu)}]^T = \beta_{PT\text{-}RS}\mathbf{W}[r_{PT\text{-}RS}(p_0,2n+k') \cdots r_{PT\text{-}RS}(p_{v-1},2n+k')]^T$ where $k=4n+2k'+\Delta$ for configuration Type 1 or $k=6n+k'+\Delta$ for configuration Type 2, if l is within the OFDM symbols allocated for the PUSCH transmission, the resource element (k,l) is not used for DM-RS. The parameters k' and Δ as well as the precoding matrix $\mathbf{W}$ are given in [6]. The configuration type is provided by the higher layer parameter *DMRS-UplinkConfig*. In the preceding expression, scaling factor $\beta_{PT\text{-}RS}$ is used to adjust the transmit power. The set of time indices l is defined relative to the start of the PUSCH allocation is defined by $\max(l_{ref}+(i-1)L_{PT\text{-}RS}+1, l_{ref}), \ldots, l_{ref}+iL_{PT\text{-}RS} \forall L_{PT\text{-}RS} \in \{1,2,4\}$ where any symbol in this interval which overlaps with a DM-RS symbol is skipped.

Table 4.24: Time-domain/frequency-domain density of phase tracking reference signal as a function of scheduled modulation coding scheme/ bandwidth [9].

Scheduled MCS	Time-Domain Density $L_{PT\text{-}RS}$
$0 \le MCS < MCS_1$	No PT-RS
$MCS_1 \le MCS < MCS_2$	Every OFDM symbol
$MCS_2 \le MCS < MCS_3$	Every second OFDM symbol
$MCS_3 \le MCS < MCS_4$	Every fourth OFDM symbol
Scheduled Bandwidth	**Frequency Domain Density $K_{PT\text{-}RS}$**
$0 \le N_{RB} < N_{RB_1}$	No PT-RS
$N_{RB_1} \le N_{RB} < N_{RB_2}$	Every second RB
$N_{RB_2} \le N_{RB}$	Every fourth RB

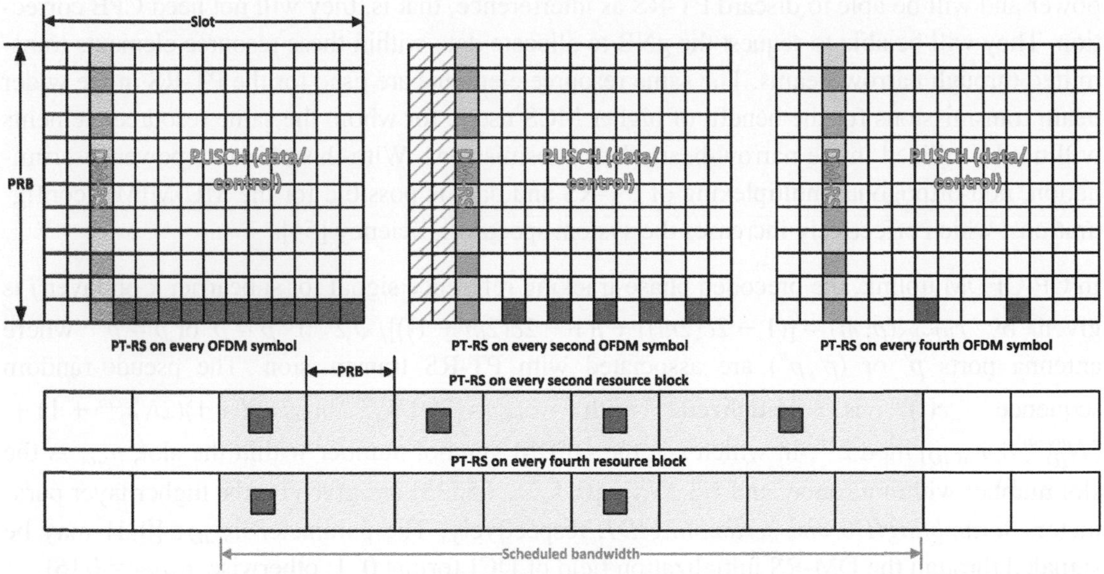

Figure 4.75
Example CP-OFDM uplink time-frequency resource mappings.

The resource blocks allocated for PUSCH transmission are numbered from 0 to N_{RB-1} relative to the lowest scheduled resource block for the purpose of PT-RS transmission. The subcarriers associated with these resource blocks are numbered in increasing order from 0 to $N_{sc}^{RB}N_{RB} - 1$ starting with the lowest frequency. The PT-RS is mapped to subcarriers $k = k_{ref}^{RE} + (iK_{PT\text{-}RS} + k_{ref}^{RB})N_{sc}^{RB}$ where $K_{PT\text{-}RS} \in \{2, 4\}$ (see Table 4.24). The parameter $k_{ref}^{RB} = n_{RNTI} \bmod K_{PT\text{-}RS}$, if $N_{RB} \bmod K_{PT\text{-}RS} = 0$; otherwise, $k_{ref}^{RB} = n_{RNTI} \bmod (N_{RB} \bmod K_{PT\text{-}RS})$. The parameter n_{RNTI} is the RNTI associated with the DCI that is used to schedule PUSCH [6]. Example time-frequency mapping of CP-OFDM uplink PT-RS is depicted in Fig. 4.75.

4.2.2.3 Sounding Reference Signal

A UE can be configured to transmit SRS in order to enable the gNB to estimate the uplink channel. Similar to the downlink CSI-RS, the SRS can serve as QCL reference for other physical channels such that they can be configured and transmitted quasi-co-located with SRS. As a result, if the knowledge of a suitable receive beam for the SRS is available, the receiver would know that the same receive beam should be suitable for other physical channels. The SRS supports up to four antenna ports, and it is designed to have low cubic metric, enabling efficient operation of the high-power amplifier. In general, the SRS can span one, two, or four consecutive OFDM symbols and is located within the last six symbols of a slot. In the frequency domain, an SRS occasion has a comb structure where the SRS is transmitted on every Nth subcarrier where $N = 2$ or 4, referred to as comb-2 or comb-4. The SRS transmissions from different devices can be frequency-multiplexed within the same frequency range using different comb patterns corresponding to different frequency offsets. For comb-2, that is, transmitting SRS on every other subcarrier, two SRSs can be frequency multiplexed, whereas for comb-4, up to four SRSs can be frequency multiplexed. Fig. 4.76 illustrates example SRS multiplexing assuming a comb-2 structure spanning two OFDM symbols.

The sequences used to represent a set of SRS are based on Zadoff–Chu sequences.[19]

---

[19] A Zadoff–Chu sequence is a complex-valued sequence with constant amplitude property whose cyclically shifted versions exhibit low cross-correlations. Thus under certain conditions, the cyclically shifted versions of each sequence remain orthogonal to one another. A Zadoff–Chu sequence that has not been shifted is referred to as a root sequence. The uth root Zadoff–Chu sequence of prime length N is defined as follows:

$$xu(n) \triangleq \begin{cases} e^{-j[\pi un(n+1)]/N} & 0 \leq n < N-1 \quad (N \text{ is an odd integer}) \\ e^{-j\pi un^2/N} & 0 \leq n < N-1 \quad (N \text{ is an even integer}) \end{cases}$$

where N is an integer, denoting the length of the Zadoff–Chu sequence. One can verify that $x_u(n)$ is periodic with period N, that is, $x_u(n) = x_u(n+N), \forall n$. In other words, the sequence index $u$is a prime relative to N. For a fixed value of u the Zadoff–Chu sequence has an ideal periodic autocorrelation property (i.e., the periodic autocorrelation is zero for all time shifts other than zero). For different values of index u the Zadoff–Chu sequences are not orthogonal, rather exhibit low cross-correlation. If the sequence length N is selected as a prime number, there are different sequences with periodic cross-correlation of $1/\sqrt{N}$ between any two sequences regardless of time shift. The Zadoff–Chu sequences are a subset of constant amplitude zero autocorrelation sequences. The properties of Zadoff–Chu sequences can be summarized as follows:

- They are periodic with period N, if N is a prime number, that is, $x_u(n+N) = x_u(n)$.
- Given N is a prime number, the DFT of a Zadoff–Chu sequence is another Zadoff–Chu sequence conjugated and time-scaled multiplied by a constant factor, that is, $X_u[k] = x_u^*(vk)X_u[0]$ where v is the multiplicative inverse of u modulo N. It can be shown that $x_u^*(vk) = x_v^*(k)e^{j\pi(1-v)k/N}$.
- The autocorrelation of a prime-length Zadoff–Chu sequence with a cyclically shifted version of itself also yields zero autocorrelation sequence, that is, it is non-zero only at one instant which corresponds to the cyclic shift zero.
- The cross-correlation between two prime-length Zadoff–Chu sequences, that is, different u, is constant and equal to $1/\sqrt{N}$.
- The Zadoff–Chu sequences have low PAPR.

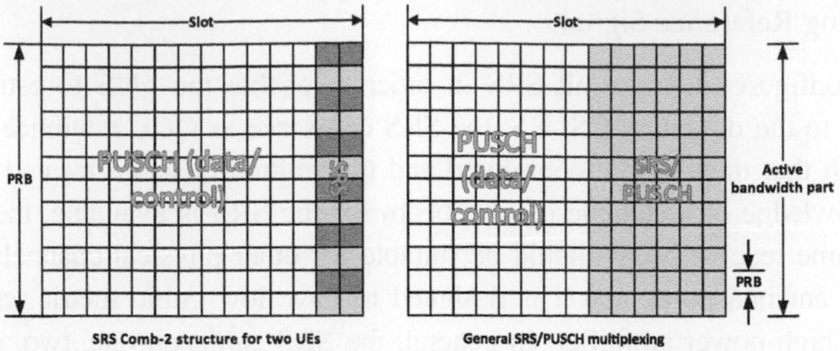

Figure 4.76
Example illustration of NR SRS structure in time and frequency domains [6].

Although Zadoff–Chu sequences of prime length are preferred in order to maximize the number of available sequences, the SRS sequences are not of prime length. The SRS sequences are extended Zadoff–Chu sequences based on the longest prime-length Zadoff–Chu sequence with a length N less than or equal to the desired SRS sequence length. The sequence is then cyclically extended in the frequency domain up to the desired SRS sequence length. As the extension is performed in the frequency domain, the extended sequence has a constant spectrum and a perfect cyclic autocorrelation, but the time-domain amplitude is not constant and slightly varies. The extended Zadoff–Chu sequences are used as SRS sequences for sequence lengths of 36 or larger, corresponding to an SRS extending over 6 and 12 resource blocks in the cases of comb-2 and comb-4, respectively. For shorter sequence lengths, special flat-spectrum sequences with good time-domain envelope properties are found through computer search since there would not be sufficient number of Zadoff–Chu sequences available.

An SRS resource is configured by the *SRS-Resource* information element and consists of 1, 2, or 4 antenna ports, where the number of antenna ports is set by the higher layer parameter *nrofSRS-Ports*, and $N_{symb}^{SRS} \in \{1, 2, 4\}$ consecutive OFDM symbols provided via parameter *nrofSymbols* contained in the higher layer parameter *resourceMapping*. The starting position l_0 in the time domain given by $l_0 = N_{slot}^{symb} - 1 - l_{offset}$ where the offset parameter $l_{offset} \geq N_{symb}^{SRS} - 1, l_{offset} \in \{0, 1, \ldots, 5\}$ counts symbols backwards from the end of the slot and is given by the field *startPosition* contained in the higher layer parameter *resourceMapping*. The SRS sequence for an SRS resource is generated as $r^{(p_i)}(n, l') = r_{u,v}^{(\alpha_i, \delta)}(n) \forall 0 \leq n \leq M_{sc,b}^{RS} - 1, l' \in \{0, 1, \ldots, N_{symb}^{SRS} - 1\}$ where the length of the SRS sequence is given by $M_{sc,b}^{RS} = m_{SRS,b} N_{sc}^{RB} / K_{TC}$, $r_{u,v}^{(\alpha_i, \delta)}(n)$ is a Zadoff–Chu sequence, $\delta = \log_2(K_{TC})$ in which K_{TC} denotes the transmission comb number given by the higher layer parameter *transmissionComb*. The cyclic shift α_i for antenna port p_i is given as $\alpha_i = 2\pi n_{SRS}^{cs,i}/n_{SRS}^{cs,\max}$ where $n_{SRS}^{cs,i} = (n_{SRS}^{cs} + n_{SRS}^{cs,\max}(p_i - 1000)/N_{ap}^{SRS}) \bmod n_{SRS}^{cs,\max}$ and $n_{SRS}^{cs} \in \{0,1,\ldots,n_{SRS}^{cs,\max} - 1\}$ is provided by the higher layer parameter *transmissionComb*. The maximum number of cyclic shifts is $n_{SRS}^{cs,\max} = 12$ when $K_{TC} = 4$ and $n_{SRS}^{cs,\max} = 8$ when $K_{TC} = 2$ [6]. The sequence group is defined as $u = [f_{gh}(n_{slot}, l') + n_{ID}^{SRS}] \bmod 30$ and the sequence number v depends on the higher layer parameter *groupOrSequenceHopping* in the *SRS-Config* information element. Furthermore, the SRS sequence identity n_{ID}^{SRS} is given by the higher layer parameter *sequenceId* in the *SRS-Config* information element. If *groupOrSequenceHopping* parameter indicates neither group nor sequence hopping, $f_{gh}(n_{slot}, l') = 0$ and $v = 0$ are used; otherwise, if parameter *groupOrSequenceHopping* indicates *groupHopping*, group hopping is used and $f_{gh}(n_{slot}, l') = \left(\sum_{m=0}^{7} c\left[8(n_{slot}N_{slot}^{symb} + l_0 + l') + m\right]2^m\right) \bmod 30$ and $v = 0$ where the pseudo-random sequence $c(i)$ is initialized with $c_{init} = n_{ID}^{SRS}$ at the beginning of each radio frame. If parameter *groupOrSequenceHopping* indicates *sequenceHopping*, sequence hopping is used and $f_{gh}(n_{slot}, l') = 0$ and $v = c(n_{slot}N_{slot}^{symb} + l_0 + l')$ for $M_{sc,b}^{SRS} \geq 6N_{sc}^{RB}$; otherwise, $v = 0$. The pseudo-random sequence $c(i)$ is initialized similar to the group hopping case [6].

Each SRS is transmitted on a designated SRS resource, and the SRS sequence $r^{(p_i)}(n, l')$ for each OFDM symbol l' and antenna port p_i is multiplied by a scaling factor β_{SRS} to adjust its transmit power. The scaled sequence is then sequentially mapped to resource elements (k, l) starting with $r^{(p_i)}(0, l')$ in a slot for each antenna port such that $a_{K_{TC}k' + k_0^{(p_i)}, l' + l_0}^{(p_i)} = \beta_{SRS} r^{(p_i)}(k', l')/\sqrt{N_{ap}} \forall k' = 0, 1, \ldots, M_{sc,b}^{RS} - 1$ and $l' = 0, 1, \ldots, N_{symb}^{SRS} - 1$. The length of the SRS sequence is given by $M_{sc,b}^{RS} = m_{SRS,b}N_{sc}^{RB}/K_{TC}$ where $m_{SRS,b}$ is given in [6]. The frequency-domain starting position $k_0^{(p_i)}$ is defined by $k_0^{(p_i)} = \bar{k}_0^{(p_i)} + \sum_{b=0}^{B_{SRS}} K_{TC} M_{sc,b}^{SRS} n_b$ where B_{SRS} denotes the SRS bandwidth, $\bar{k}_0^{(p_i)} = n_{shift}N_{sc}^{RB} + k_{TC}^{(p_i)}$, $k_{TC}^{(p_i)} = (\bar{k}_{TC} + K_{TC}/2) \bmod K_{TC}$ if $n_{SRS}^{cs} \in \{n_{SRS}^{cs,\max}/2, \ldots, n_{SRS}^{cs,\max} - 1\}$ and $N_{ap}^{SRS} = 4$; otherwise, $k_{TC}^{(p_i)} = \bar{k}_{TC}$ [6]. The frequency domain shift value n_{shift} adjusts the SRS allocation with respect to the common resource block grid. The transmission comb offset $\bar{k}_{TC} \in \{0, 1, K_{TC} - 1\}$ is contained in the higher layer parameter *transmissionComb* in the *SRS-Config* information element and n_b is a frequency position index. Frequency hopping of the SRS is configured by the parameter $b_{hop} \in \{0, 1, 2, 3\}$. If $b_{hop} \geq B_{SRS}$, frequency hopping is disabled and the frequency position index n_b is set to a constant as follows $n_b = \lfloor 4n_{RRC}/m_{SRS,b} \rfloor \bmod N_b$ for all N_{symb}^{SRS} OFDM symbols of the SRS resource. The value of the parameter n_{RRC} is given by the higher layer parameter *freqDomainPosition* and the values of $m_{SRS,b}$ and N_b for $b = B_{SRS}$ are given in [6]. If $b_{hop} < B_{SRS}$, the frequency hopping is enabled and the frequency position indices n_b are defined as $n_b = \lfloor 4n_{RRC}/m_{SRS,b} \rfloor \bmod N_b$ if $b \leq b_{hop}$; otherwise, $n_b = \{F_b(n_{SRS}) + \lfloor 4n_{RRC}/m_{SRS,b} \rfloor\} \bmod N_b$ where N_b and $F_b(n_{SRS})$ are defined in [6]. The quantity n_{SRS} counts the number of SRS transmissions. For the case of an SRS resource configured as aperiodic by the higher layer parameter *resourceType*, $n_{SRS} = \lfloor l'/R \rfloor$ within the slot where N_{symb}^{SRS} symbol SRS resource is transmitted. The quantity $R \leq N_{symb}^{SRS}$ is a repetition factor configured via higher layer signaling.

For the case of an SRS resource configured as periodic or semi-persistent by the higher layer parameter *resourceType*, the value of the SRS counter is given by $n_{SRS} = \left(N_{frame}^{slot} n_{frame} + \quad n_{slot} - T_{offset}\right)\left(N_{symb}^{SRS}/RT_{SRS}\right) + \lfloor l'/R \rfloor$ for slots that satisfy $\left(N_{frame}^{slot} n_{frame} + n_{slot} - T_{offset}\right) \bmod T_{SRS} = 0$, where T_{SRS} is SRS periodicity defined in the number of slots and T_{offset} is the slot offset [6]. Note that when supporting more than one SRS antenna port, different antenna ports share the same set of resource elements and the same baseline SRS sequence; nevertheless, different phase rotations are applied to separate them. Applying a phase rotation in the frequency domain is equivalent to cyclic shift in the time domain.

As we mentioned earlier, the SRS can be configured as periodic, semi-persistent, or aperiodic transmission A periodic SRS is transmitted with a certain configured periodicity and a certain configured slot offset within that period. A semi-persistent SRS has a configured periodicity and slot offset in the same way as a periodic SRS; however, the SRS transmission is performed according to the configured periodicity and slot offset that is activated or deactivated via MAC control element signaling. An aperiodic SRS is only transmitted when explicitly triggered by means of a DCI. It should be noted that SRS activation/deactivation or triggering for semi-persistent and aperiodic cases is not done for a specific SRS, rather for an SRS resource set which may include multiple SRSs [14].

A UE can be configured with one or several SRS resource sets, where each resource set includes one or more configured SRSs. All SRS occasions included within a configured SRS resource set are of the same type. In other words, periodic, semi-persistent, or aperiodic transmission is a property of an SRS resource set. A UE can be configured with multiple SRS resource sets that can be used for different purposes, including both downlink and uplink multi-antenna precoding and/or downlink and uplink beam management. The transmission of the set of configured SRS included in an aperiodic SRS resource set is triggered by a DCI. More specifically, DCI format 0_1 containing uplink grant and DCI format 1_1 containing downlink scheduling assignment include a 2-bit SRS-request that can trigger the transmission of one of the three different aperiodic SRS resource sets configured for the UE and the fourth bit combination corresponds to no trigger.

The SRS antenna ports are typically not mapped directly to the UE's physical antennas, rather via some antenna mapping scheme. In order to provide connectivity regardless of the rotational direction of the device, the NR devices supporting high-frequency operation will typically include multiple antenna panels pointing in different directions. The SRS may be mapped to one of those panels and transmission from different panels will then correspond to different antenna mapping schemes. The antenna mapping scheme has a real impact despite the fact that it is transparent to the gNB receiver; thus it is seen as an integral part of the overall channel from the UE to the gNB. The gNB may estimate the channel based on SRS transmission from a UE and subsequently select a precoding matrix that the device should use for uplink transmission. The device is then assumed to use that precoding matrix in combination with the antenna mapping scheme that is applied to the SRS. In other cases, a device may be explicitly scheduled for data transmission using the antenna ports defined for SRS transmission. In practice, this implies that the device will transmit the data using the same antenna mapping scheme that was used for SRS transmission, meaning that the UE should use the same beam or panel that was used for SRS transmission [14].

4.2.3 Control Channels

4.2.3.1 Physical Uplink Control Channel

Physical uplink control channel carries the uplink control information (UCI) from the UEs to the gNB. The new radio has specified five PUCCH formats (as shown in Table 4.25) that are identified depending on the duration of PUCCH and the UCI payload size. The NR further supports simultaneous transmission of data and control information on PUSCH. Thus if the device is transmitting on PUSCH, the UCI is multiplexed with data on the allocated resources instead of being transmitted on PUCCH. It must be noted that simultaneous transmission of PUSCH and PUCCH is not supported in NR Rel-15. The NR PUCCH can be beamformed by configuring one or more spatial correspondence between PUCCH and downlink physical

Table 4.25: PUCCH formats [6].

Type	PUCCH Format	Number of OFDM Symbols N_{symb}^{PUCCH}	Number of Bits	Description
Short PUCCH	0	1–2	≤2	Short PUCCH of one or two symbols with small UCI payloads of up to 2 bits with UE multiplexing in the same PRB
	2	1–2	>2	Short PUCCH of one or two symbols with large UCI payloads of more than 2 bits with no multiplexing in the same PRB
Long PUCCH	1	4–14	≤2	Long PUCCH of 4–14 symbols with small UCI payloads of up to 2 bits with multiplexing in the same PRB
	3	4–14	>2	Long PUCCH of 4–14 symbols with medium UCI payloads with some multiplexing capacity in the same PRB
	4	4–14	>2	Long PUCCH of 4–14 symbols with large UCI payloads with no multiplexing capacity over the same PRB

signals such as CSI-RS or SS block. As a result, the device can transmit PUCCH using the same beam as it used for receiving the corresponding downlink signal. For example, if the spatial relation between PUCCH and SS block is configured, the device will transmit PUCCH using the same beam as it used for receiving the SS block. Multiple spatial relations can be configured and selected via MAC control elements [71].

It will be shown later in this section that the short PUCCH format of up to two UCI bits is based on sequence selection, while the short PUCCH format of more than two UCI bits frequency multiplexes UCI and DM-RS. The long PUCCH formats time-multiplex the UCI and DM-RS. Frequency hopping is supported for long PUCCH formats and for short PUCCH formats of duration two symbols. Long PUCCH formats can be repeated over multiple slots. The UCI multiplexing in PUSCH is further supported when UCI and PUSCH transmissions coincide in the same slot, that is, UCI carrying HARQ-ACK feedback with 1 or 2 bits is multiplexed by puncturing and rate-matching PUSCH. In all other cases, the UCI is multiplexed by rate matching PUSCH. The UCI may carry CSI, HARQ ACK/NACK, or scheduling request (SR). The QPSK modulation is used for long PUCCH with 2 or more bits of information, and short PUCCH with more than 2 bits of information. The BPSK modulation is used for long PUCCH with a single information bit. Transform precoding is applied to long PUCCH [11].

In deployment scenarios where a SUL is used the UE is configured with two uplink carriers and one downlink carrier in the same cell. The transmissions on those uplink carriers are controlled by the network to avoid overlapping PUSCH/PUCCH transmissions in the time domain. The overlapping transmissions on PUSCH are avoided with properly scheduling the uplink transmissions, while overlapping transmissions on PUCCH are circumvented by proper configuration, that is, PUCCH can only be configured for one of the two uplink carriers of the cell. In addition, initial access is supported on each of those uplink carriers [11].

A UE is semi-statically configured via RRC signaling to perform periodic CSI reporting using PUCCH and can be configured for multiple periodic CSI reports corresponding to one or more CSI reporting setting indications, where the associated CSI measurement links and CSI Resource Settings are also configurable. Periodic CSI reporting on PUCCH formats 2, 3, 4 supports Type I CSI with wideband granularity. A UE performs semi-persistent CSI reporting on PUCCH after successfully decoding a selection command, which contains one or more reporting setting indications where the associated CSI measurement links and CSI resource settings are configured. Semi-persistent CSI reporting on PUCCH supports Type I CSI. The semi-persistent CSI reporting on PUCCH format 2 supports Type I CSI with wideband granularity, whereas semi-persistent CSI reporting on PUCCH format 3 or 4 supports Type I subband CSI and Type II CSI with wideband frequency granularity. When PUCCH carry Type I CSI with wideband frequency granularity, the CSI payloads carried by PUCCH format 2, and PUCCH format 3 or 4 are identical irrespective of RI and CRI. For

Type I CSI subband reporting on PUCCH format 3 or 4, the payload is split into two parts. The first part may contain RI, CRI, and/or CQI for the first codeword. The second part contains PMI and the CQI for the second codeword when $RI > 4$. A semi-persistent report carried on PUCCH format 3 or 4 supports only part 1 of Type II CSI feedback. Supporting Type II CSI reporting on PUCCH format 3 or 4 is considered a UE capability. A Type II CSI report (part 1 only) is carried on PUCCH format 3 or 4 and is calculated independent of any Type II CSI reports carried on PUSCH. When the UE is configured with CSI reporting on PUCCH format 2, 3, or 4, each PUCCH resource is configured for each candidate uplink BWP. The UE will never report CSI with a payload size larger than 115 bits when configured with PUCCH format 4 [8].

The NR physical uplink control channel is used for transmission of HARQ ACK/NACK for received downlink data; CSI feedback related to the downlink channel conditions to assist dynamic scheduling, and SR indicates that a UE needs uplink resources for data transmission. The NR PUCCH supports two transmission modes (see Fig. 4.77):

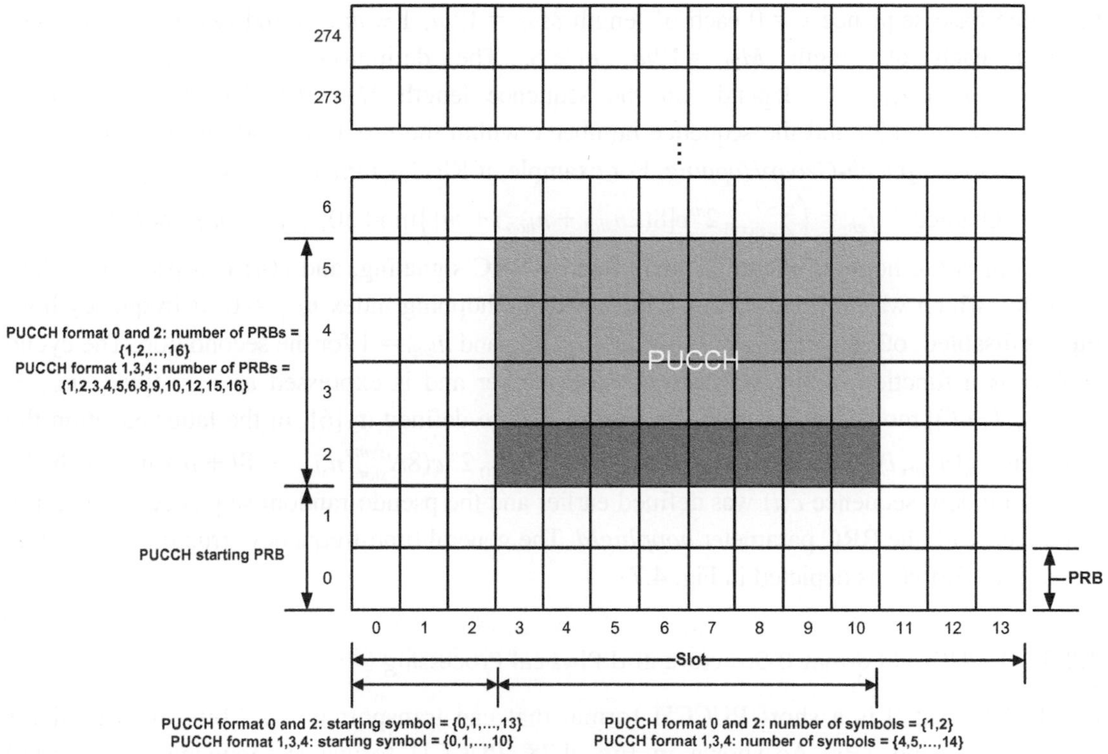

Figure 4.77
General time-frequency structure of PUCCH for various formats [6].

- *Short-PUCCH*: NR PUCCH can be transmitted in short duration around the last uplink symbol(s) of a slot. The transmission can span one or two OFDM symbols.
- *Long-PUCCH*: NR PUCCH can be transmitted in long duration at least over four uplink symbols to improve coverage. It is also considered to support a long-PUCCH transmission over multiple slots.

One or multiple PRBs can be allocated as the minimum resource unit size in frequency domain for short-PUCCH and long-PUCCH. Intra-slot frequency hopping can be configured for PUCCH format 1, 3, or 4, that is, the long-PUCCH, where the number of symbols in the first hop is given by $\lfloor N_{symb}^{PUCCH}/2 \rfloor$ in which N_{symb}^{PUCCH} is the length of PUCCH transmission in the number of OFDM symbols. PUCCH formats 0, 1, 3, and 4 use the low-PAPR Zadoff–Chu sequences $r_{u,v}^{(\alpha,\delta)}(n)|_{\delta=0}$ where α is a cyclic shift of a base sequence $\bar{r}_{u,v}(n)$ such that $r_{u,v}^{(\alpha,0)}(n) = e^{j\alpha n}\bar{r}_{u,v}(n),\ 0 \le n < M_{ZC}$ in which $M_{ZC} = 12m$ is the length of the sequence (assuming $\delta = 0$). Multiple sequences can be generated from a single base sequence based on different values of α (cyclic shift). Base sequences $\bar{r}_{u,v}(n)$ are divided into groups, where $u \in \{0, 1, \ldots, 29\}$ is the group number and v is the base sequence number within the group, such that each group contains one base sequence $v = 0$ each of length $M_{ZC} = 12m,\ 1 \le m \le 5$ and two base sequences $v = 0, 1$ each of length $M_{ZC} = 12m,\ m \ge 6$. The definition of the base sequence $\bar{r}_{u,v}(0), \ldots, \bar{r}_{u,v}(M_{ZC} - 1)$ depends on the sequence length M_{ZC} [6]. The sequence group $u = (f_{gh} + f_{ss}) \bmod 30$ and the sequence number v within the group depend on the value of the RRC parameter *pucch-GroupHopping*. For example, if RRC parameter *pucch-GroupHopping* is set to "enabled," $f_{gh} = \left(\sum_{m=0}^{7} 2^m c[8(2n_{slot} + n_{hop}) + m]\right) \bmod 30;\ f_{ss} = n_{ID} \bmod 30;\ v = 0$ where n_{ID} is the hopping identifier identified by RRC signaling, and $c(n)$ is a pseudo-random sequence which was defined earlier. The frequency hopping index $n_{hop} = 0$, if frequency hopping is disabled; otherwise, $n_{hop} = 0$ for the first hop and $n_{hop} = 1$ for the second hop. The cyclic shift α is a function of the symbol and slot number and is expressed as $\alpha_l = ([m_0 + m_{cs} + n_{cs}(n_{slot}, l + l')] \bmod 12)\pi/6$ where the parameters are defined in [6]. In the latter equation the function $n_{cs}(n_{slot}, l)$ is defined as $n_{cs}(n_{slot}, l) = \sum_{m=0}^{7} 2^m c(8N_{slot}^{symb} n_{slot} + 8l + m)$ in which the pseudo-random sequence $c(i)$ was defined earlier and the pseudo-random sequence generator is initialized with the RRC parameter *hoppingId*. The general time-frequency structure of PUCCH for various formats is depicted in Fig. 4.77.

4.2.3.1.1 PUCCH Format 0 Structure and Physical Processing

PUCCH format 0 is a short PUCCH format that can transport up to 2 bits. It is used for HARQ-ACK and SRs. As shown in Fig. 4.78, PUCCH format 0 sequence is generated $x(12l + n) = r_{u,v}^{(\alpha,\delta)}(n); n = 0, 1, \ldots, 11$ and $l = 0$ for single-symbol PUCCH transmission and $l = 0, 1$ for double-symbol PUCCH transmission. The sequence $r_{u,v}^{(\alpha,\delta)}(n)$ was defined in the previous section, in which the parameter m_{cs} depends on the UCI. The sequence $x(n)$ is

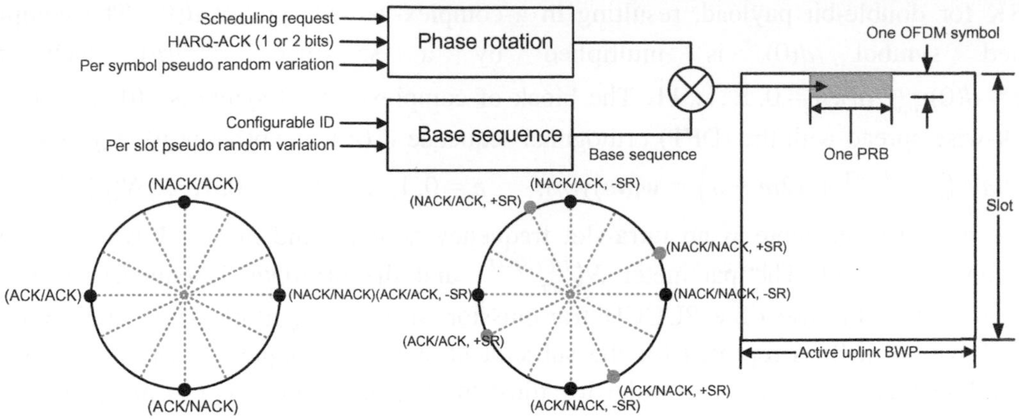

Figure 4.78
Time-frequency structure of PUCCH format 0 and signaling via phase rotation [6,14].

multiplied by amplitude scaling factor $\beta_{PUCCH-F0}$ in order to adjust the transmit power and is mapped sequentially to resource elements (k, l) assigned for PUCCH transmission in frequency-first manner on a single-antenna port [6].

As shown in Fig. 4.78, the phase rotations through parameter α represent different information bits that are separated with π and $\pi/2$ for 1- and 2-bit HARQ-ACK, respectively. In the case of a simultaneous SR, the phase rotation is increased by $\pi/4$ for 1-bit acknowledgments and by $\pi/6$ for 2-bit acknowledgments. The base sequences are configured per cell using an identity provided as part of the SI. Furthermore, a sequence hopping, where the base sequence varies on a slot-by-slot basis, can be used to randomize the interference between different cells. PUCCH format 0 is typically transmitted at the end of a slot. If two OFDM symbols are used for PUCCH format 0, the same information is transmitted on both OFDM symbols. However, the reference phase rotation as well as the frequency-domain resources may vary between the symbols, effectively creating a frequency-hopping effect [6,14].

4.2.3.1.2 PUCCH Format 1 Structure and Physical Processing

PUCCH format 1 is the long-format version of PUCCH format 0, which can carry up to 2 bits, using 4–14 OFDM symbols over one resource block per symbol in frequency. The OFDM symbols are split between symbols used for control information and symbols designated to the reference signals to enable coherent reception. The number of symbols used for control information and reference signals is a trade-off between channel estimation accuracy and energy in the information part. It was shown that a reasonable trade-off would be achieved if half of the symbols are used for reference signals. The block of bits $b(0), \ldots, b(N_{UCI} - 1) | N_{UCI} = 0, 1$ is modulated using BPSK for a single-bit payload and

QPSK for double-bit payload, resulting in a complex-valued symbol $d(0)$. The complex-valued symbol $d(0)$ is multiplied by a sequence $r_{u,v}^{(\alpha,\delta)}(n)$ such that $y(n) = d(0)r_{u,v}^{(\alpha,\delta)}(n); n = 0, 1, \ldots, 11$. The block of complex-valued symbols $y(0), \ldots, y(11)$ is block-wise spread with the (DFT) orthogonal sequence $w_i(m) = \exp(j2\pi\phi(m)/N_{SF})$ such that $z\left(12m'N_{SF,0}^{PUCCH-F1} + 12m + n\right) = w_i(m)y(n); \quad n = 0, 1, \ldots, 11; m = 0, 1, \ldots, N_{SF,m'}^{PUCCH-F1} - 1$ where $m' = 0$ when there is no intra-slot frequency hopping and $m' = 0, 1$ when intra-slot hopping is enabled. The parameter $N_{SF,m'}^{PUCCH-F1}$ and the orthogonal sequence $w_i(m)$ are defined in [6]. In case of a PUCCH transmission spanning multiple slots, the complex-valued symbol $d(0)$ is repeated for the subsequent slots. The sequence $z(n)$ is scaled with the scaling factor $\beta_{PUCCH-F1}$ in order to adjust the transmit power and is mapped sequentially to resource elements (k, l) if they are not used by DM-RS. The mapping to resource elements (k, l) designated to PUCCH transmission is in increasing order of frequency and then time over the assigned physical resource block on a single-antenna port. The time-frequency structure of PUCCH format 1 is illustrated in Fig. 4.79.

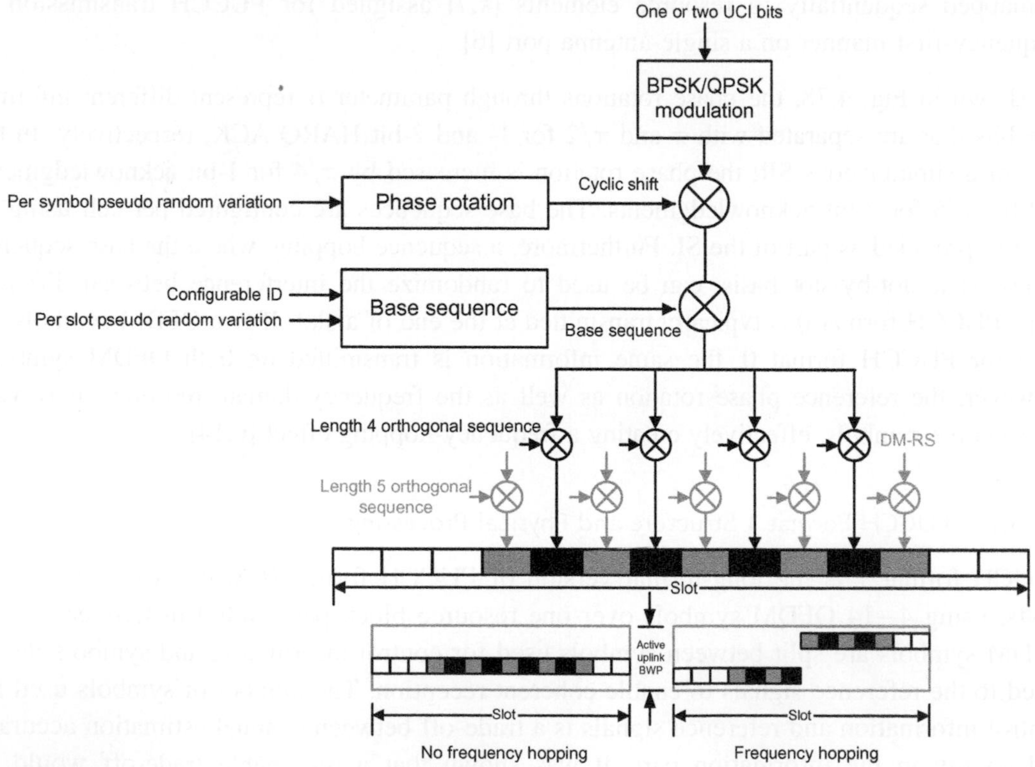

Figure 4.79
Time-frequency structure of PUCCH format 1 [6].

Unlike LTE, where PUCCH frequency-hopping was always done at the slot boundary, the NR provides additional flexibility by allowing variable PUCCH duration depending on the scheduling decisions and overall system configuration. Furthermore, since the devices are supposed to only transmit within their active bandwidth part, hopping is typically not between the edges of the overall carrier bandwidth as in LTE. Therefore, frequency hopping is configurable and determined as part of PUCCH resource configuration. The position of the hop is obtained from the length of PUCCH. If frequency hopping is enabled, one orthogonal block-spreading sequence is used per hop.

4.2.3.1.3 PUCCH Format 2 Structure and Physical Processing

PUCCH format 2 is a short PUCCH format which is used to carry more than two uplink control bits, for example, CSI report and HARQ acknowledgments, or HARQ acknowledgments per se. An SR can also be included in the bits and jointly encoded. If the payload to be transmitted by PUCCH format 2 is too large, the CSI reports are not transmitted in order to make room for more important HARQ acknowledgment bits. The overall PUCCH format 2 physical layer processing is depicted in Fig. 4.80. A CRC is added for large payload sizes

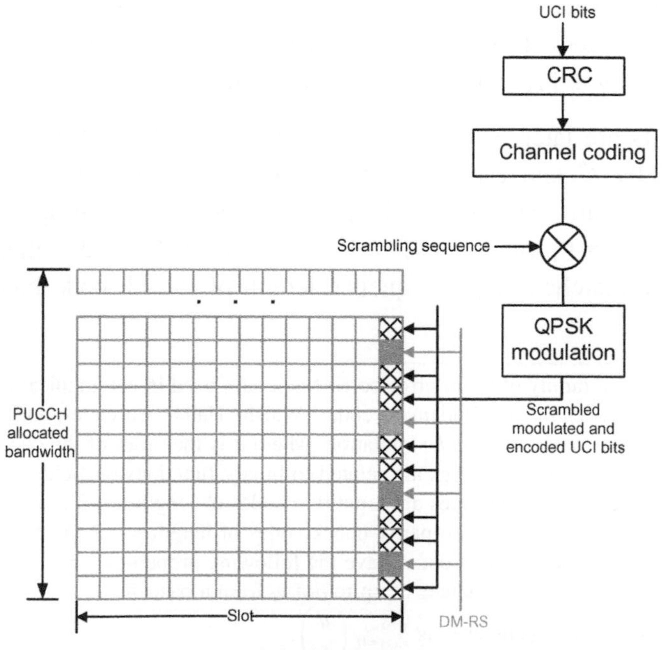

Figure 4.80
Processing and mapping of PUCCH format 2 [6].

and the resulting block of bits are encoded using Reed–Muller codes[20] for payload sizes up to 12 bits or polar codes for larger payloads. The encoded block is scrambled and QPSK modulated. The scrambling sequence is based on the device identity (C-RNTI) together with the physical-layer cell identity (or a configurable virtual cell identity), ensuring interference randomization across cells and devices using the same set of time-frequency resources. The QPSK-modulated symbols are then mapped to subcarriers across multiple resource blocks using one or two OFDM symbols. A pseudo-random QPSK sequence, mapped to every third subcarrier on each OFDM symbol, is used as a DM-RS to facilitate coherent detection at the base station [14].

The physical layer processing of UCI starts with CRC attachment and channel coding (see Fig. 4.80). Let us $a_0, a_1, \ldots, a_{N_{UCI}-1}$ denote the input UCI bit sequence, in which N_{UCI} is the payload size. If $N_{UCI} \geq 12$ the UCI bits are encoded with the polar codes and if $N_{UCI} \leq 11$, Reed–Muller codes, simplex code ($N_{UCI} = 2$), or repetition coding ($N_{UCI} = 1$) would be used to encode the UCI bits [7]. If the payload size $N_{UCI} \geq 12$, code block segmentation and CRC attachment may be performed prior to channel coding. If $N_{UCI} \geq 360, E \geq 1088$ bits or if $N_{UCI} \geq 1013, I_{seg} = 1$ segmentation is performed; otherwise, $I_{seg} = 0$ and no segmentation is done, where E is the rate-matched output sequence length. If $12 \leq N_{UCI} \leq 19$, the parity bits $p_{r0}, p_{r1}, \ldots, p_{r(N_{CRC}-1)}$ are computed by setting $N_{CRC} = 6$ bits using the generator polynomial $g_{CRC6}(D)$. The resulting sequence is $c_{l0}, c_{l1}, \ldots, c_{l(K_l-1)}$ where l is the code block number and K_l is the number of bits associated with the lth code block. If $N_{UCI} \geq 20$, the parity bits are computed by setting $N_{CRC} = 11$ bits using the generator polynomial $g_{CRC11}(D)$ [7]. Note that if the payload size $N_{UCI} \leq 11$, no CRC bits are attached.

The information bits are later encoded by channel coding block. The total number of code blocks is denoted by C and each code block is individually encoded. If $18 \leq K_l \leq 25$, the information bits are encoded with polar encoder by setting the parameters as follows: $n_{\max} = 10$, $I_{IL} = 0$, $n_{PC} = 3$, $n_{PC}^{wm} = 1$, if $E_l - K_l + 3 > 192$; otherwise, $n_{PC}^{wm} = 0$, in which E_l is the rate-matched output sequence length of the lth code block. If $K_l > 30$, the

[20] Reed–Muller codes are a family of linear error-correcting codes used in communication systems. The special cases of Reed–Muller codes include Hadamard codes, Walsh–Hadamard codes, and Reed–Solomon codes. Reed–Muller codes are denoted by $RM(d, r)$ notation, where d is the order of the code, and r determines the length of code $n = 2^r$. Reed–Muller codes are related to binary functions on field $GF(2^r)$ over the elements $\{0, 1\}$. It can be shown that $RM(0, r)$ codes are repetition codes of length $n = 2^r$, rate $R = 1/n$, and minimum distance $d_{\min} = n$, and $RM(1, r)$ codes are parity check codes of length $n = 2^r$, rate $R = (r+1)/n$, and minimum distance $d_{\min} = n/2$. Reed–Muller codes have the following properties [15]:

1. The set of all possible exterior products of up to $d$of v_i form a basis for $\mathbf{F}_2^n$.
2. The rank of $RM(d, r)$ code is defined as $\sum_{s=0}^{r} \binom{d}{s}$.
3. $RM(d, r) = RM(d, r-1)|RM(d-1, r-1)$, where $|'$ denotes the bar product of two codes.
4. $RM(d, r)$ has minimum Hamming weight 2^{d-r}.

Table 4.26: Rate-matched UCI output sequence length E_{UCI} [7].

UCI(s) for Transmission on a PUCCH	UCI for Encoding	Value of E_{UCI}
HARQ-ACK	HARQ-ACK	$E_{UCI}=E_T$
HARQ-ACK, SR	HARQ-ACK, SR	$E_{UCI}=E_T$
CSI (consisting of one part)	CSI	$E_{UCI}=E_T$
HARQ-ACK, CSI (consisting of one part)	HARQ-ACK, CSI	$E_{UCI}=E_T$
HARQ-ACK, SR, CSI (CSI not of two parts)	HARQ-ACK, SR, CSI	$E_{UCI}=E_T$
CSI (consisting of two parts)	CSI Part 1	$E_{UCI}=\min\left(E_T, \left\lceil (O^{CSI\text{-}part1}+N_{CRC})/R_{UCI}^{\max}/Q_m \right\rceil Q_m\right)$
	CSI Part 2	$E_{UCI}=E_T-\min\left(E_T, \left\lceil (O^{CSI\text{-}part1}+N_{CRC})/R_{UCI}^{\max}/Q_m \right\rceil Q_m\right)$
HARQ-ACK, CSI (consisting of two parts)	HARQ-ACK, CSI Part 1	$E_{UCI}=\min\left(E_T, \left\lceil (O^{ACK}+O^{CSI\text{-}part1}+N_{CRC})/R_{UCI}^{\max}/Q_m \right\rceil Q_m\right)$
	CSI Part 2	$E_{UCI}=E_T-\min\left(E_T, \left\lceil (O^{ACK}+O^{CSI\text{-}part1}+N_{CRC})/R_{UCI}^{\max}/Q_m \right\rceil Q_m\right)$
HARQ-ACK, SR, CSI (consisting of two parts)	HARQ-ACK, SR, CSI Part 1	$E_{UCI}=\min\left(E_T, \left\lceil (O^{ACK}+O^{SR}+O^{CSI\text{-}part1}+N_{CRC})/R_{UCI}^{\max}/Q_m \right\rceil Q_m\right)$
	CSI Part 2	$E_{UCI}=E_T-\min\left(E_T, \left\lceil (O^{ACK}+O^{SR}+O^{CSI\text{-}part1}+N_{CRC})/R_{UCI}^{\max}/Q_m \right\rceil Q_m\right)$

information bits are encoded with polar encoder by setting the parameters as follows: $n_{\max}=10$, $I_{IL}=0$, $n_{PC}=3$, $n_{PC}^{wm}=0$. The output bits following the encoding are denoted by $d_0, d_1, \ldots, d_{N-1}$, where N is the number of coded bits.

For PUCCH format 2/3/4, the total rate-matched output sequence length E_T is given by Table 4.26, where $N_{symb,UCI}^{PUCCH,2}$, $N_{symb,UCI}^{PUCCH,3}$, *and* $N_{symb,UCI}^{PUCCH,4}$ denote the number of symbols carrying UCI for PUCCH format 2/3/4; $N_{PRB}^{PUCCH,2}$, $N_{PRB}^{PUCCH,3}$ are the number of PRBs that are determined by the UE for PUCCH format 2/3 transmission; and $N_{SF}^{PUCCH,4}$ is the spreading factor for PUCCH format 4 [7,8]. The rate matching is performed by setting $I_{BIL}=1$ and the rate matching output sequence length to $E_l=\lfloor E_{UCI}/C_{UCI} \rfloor$, where C_{UCI} is the number of code blocks for UCI and the value of E_{UCI} is given by Table 4.26. In this table the following parameters have been used [7]:

- O^{ACK} is the number of bits for HARQ-ACK for transmission on the current PUCCH.
- O^{SR} denotes the number of bits for SR for transmission on the current PUCCH.
- $O^{CSI\text{-}part\ 1}$ is the number of bits for CSI part 1 for transmission on the current PUCCH.
- $O^{CSI\text{-}part\ 2}$ denotes the number of bits for CSI part 2 for transmission on the current PUCCH.
- $R_{UCI}^{\max}$ is the configured PUCCH maximum coding rate.

The output bit sequence after rate matching is denoted by $f_{l0}, f_{l1}, \ldots, f_{l(E_l-1)}$ where E_l is the length of rate-matched output sequence in lth code block number.

The encoded UCI payload $b(0),\ldots,b(N_E-1)$, where N_E denotes the number of encoded bits in the payload transmitted on the physical uplink control channel, is scrambled prior to modulation, resulting in a block of scrambled bits $\tilde{b}(0),\ldots,\tilde{b}(N_E-1)$ such that $\tilde{b}(i)=[b(i)+c(i)] \bmod 2$, in which the scrambling sequence $c(i)$ is a pseudo-random sequence that is initialized with the value $c_{init}=n_{RNTI}2^{15}+n_{ID}$ that is derived from RRC configured parameter $n_{ID}\in\{0,1,\ldots,1023\}$; otherwise, $n_{ID}=N_{ID}^{cell}$. The block of scrambled bits $\tilde{b}(0),\ldots,\tilde{b}(N_E-1)$ is QPSK modulated, resulting in a block of complex-valued modulation symbols $d(0),\ldots,d(N_E/2-1)$. The block of modulation symbols is scaled with the scaling factor $\beta_{PUCCH\text{-}F2}$ to adjust the transmit power and is sequentially mapped, starting with $d(0)$, to resource elements (k,l) which are reserved for PUCCH transmission and are not used by the associated DM-RS. The mapping to resource elements (k,l) is in increasing order of the frequency index k followed by time index l on a single-antenna port. PUCCH format 2 is typically transmitted at the end of a slot as illustrated in Fig. 4.80; however, it is also possible to transmit PUCCH format 2 in other positions within a slot [6].

4.2.3.1.4 PUCCH Formats 3 and 4 Structure and Physical Processing

PUCCH format 3 is the long PUCCH counterpart to PUCCH format 2, wherein more than two UCI bits can be transmitted over 4–14 OFDM symbols, where there can be multiple resource blocks on each symbol. As a result, it is the PUCCH format with the largest payload capacity. The OFDM symbols are used for carrying the UCI and the PUCCH DM-RS. The control information is encoded using Reed–Muller codes for payload sizes less than 11 bits and polar codes for larger payloads and then scrambled and modulated. The scrambling sequence is based on the UE identity (C-RNTI) together with the physical-layer cell identity (or a configurable virtual cell identity), ensuring interference randomization across cells and devices that use the same set of time-frequency resources. Prior to channel coding stage, a CRC is attached to the control information for large payloads. The encoded bits are QPSK modulated; however, there is an option to use $\pi/2$-BPSK modulation to lower the cubic metric at the expense of some loss in link performance. The complex-valued modulation symbols are distributed among the OFDM symbols and DFT precoding is performed to reduce the cubic metric and improve the power amplifier efficiency.

The structure of PUCCH format 4 is similar to that of PUCCH format 3 with the possibility to code-multiplex multiple devices over the same resources using one resource block in the frequency domain. Each OFDM symbol carries $12/N_{SF}$ unique modulation symbols. Prior to DFT-precoding, each modulation symbol is block-spread with an orthogonal sequence of length N_{SF}. The spreading factors of length two and four are supported, implying that the multiplexing capacity would be two or four devices on the same set of resource blocks.

As shown in Fig. 4.81, frequency hopping can be configured for PUCCH format 3/4 to exploit frequency diversity; however, these PUCCH formats can operate without frequency hopping. The location of the reference signal symbols depends on the frequency hopping and the length

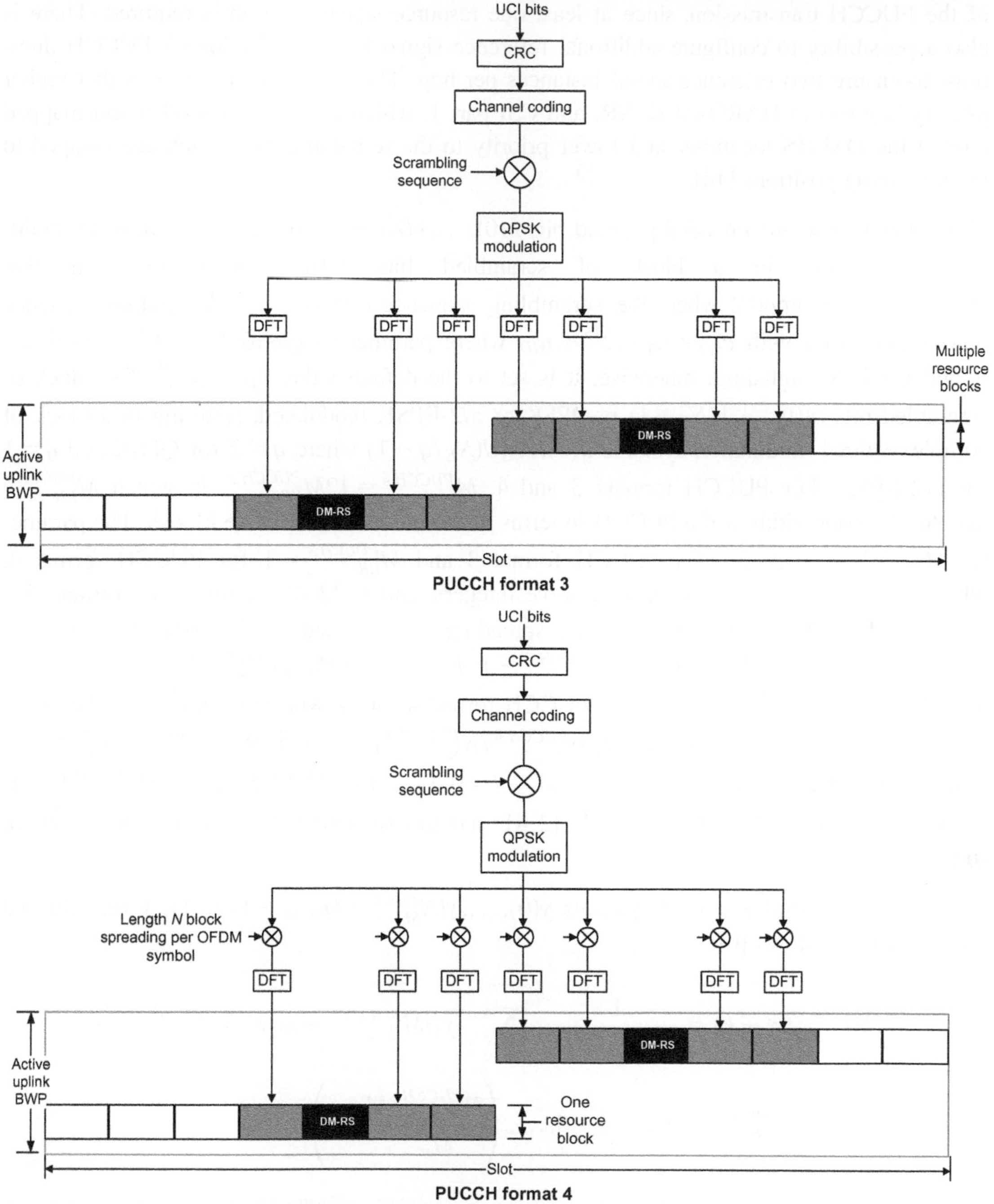

Figure 4.81
Short-PUCCH time-frequency structure [6,14].

of the PUCCH transmission, since at least one resource signal per hop is required. There is also a possibility to configure additional reference signal locations for longer PUCCH durations to ensure two reference signal instances per hop. The UCI mapping is such that higher priority is given to HARQ-ACK, SR, and CSI part 1, which are jointly encoded and mapped around the DM-RS locations, and lower priority to the remaining bits which are mapped to the remaining positions [14].

More specifically, the encoded payload bits $b(0),\ldots,b(N_E-1)$ are scrambled prior to modulation, resulting in a block of scrambled bits $\tilde{b}(0),\ldots,\tilde{b}(N_E-1)$ such that $\tilde{b}(i)=(b(i)+c(i))\text{mod}\,2$ where the scrambling sequence $c(i)$ is a pseudo-random sequence that is initialized with $c_{init}=n_{RNTI}2^{15}+n_{ID}$ where parameter $n_{ID}\in\{0,1,\ldots,1023\}$ is determined via RRC signaling; otherwise, it is set to the default value $n_{ID}=N_{ID}^{cell}$. The block of scrambled bits $\tilde{b}(0),\ldots,\tilde{b}(N_E-1)$ is QPSK or $\pi/2$-BPSK modulated, resulting in a block of complex-valued modulation symbols $d(0),\ldots,d(N_E/q-1)$ where $q=2$ for QPSK and $q=1$ for $\pi/2$-BPSK. For PUCCH formats 3 and 4, $M_{sc}^{PUCCH,s}=12M_{RB}^{PUCCH,s}$, in which $M_{RB}^{PUCCH,s}$ denotes the bandwidth of the PUCCH in terms of the number of resource blocks. The parameter $M_{RB}^{PUCCH,s}=2^{\alpha_2}3^{\alpha_3}5^{\alpha_5}$ for PUCCH format 3 and $M_{RB}^{PUCCH,s}=1$ for PUCCH format 4, wherein $\alpha_2,\alpha_3,\alpha_5$ is a set of non-negative integers, and $s\in\{3,4\}$ identifies the format. For PUCCH format 3, no block-wise spreading is applied and $y(lM_{sc}^{PUCCH,3}+k)=d(lm_{sc}^{PUCCH,3}+k)$; $k=0,1,\ldots,M_{sc}^{PUCCH,3}-1$; $l=0,1\ldots,(M_{symb}/M_{sc}^{PUCCH,3})-1$ where $M_{RB}^{PUCCH,3}\geq 1$ and $N_{SF}^{PUCCH,3}=1$. For PUCCH format 4, block-wise spreading is applied such that $y\left(lM_{sc}^{PUCCH,3}+k\right)=w_m(k)d\left(l\left(M_{sc}^{PUCCH,4}/N_{SF}^{PUCCH,4}\right)+k\bmod\left(M_{sc}^{PUCCH,4}/\ N_{SF}^{PUCCH,4}\right)\right)$ where $k=0,1,\ldots,M_{sc}^{PUCCH,4}-1$ and $l=0,1,\ldots,(N_{SF}^{PUCCH,4}M_{symb}/\ M_{sc}^{PUCCH,4})-1$. Furthermore, $M_{RB}^{PUCCH,4}=1$, $N_{SF}^{PUCCH,4}\in\{2,4\}$ and the spreading codes $w_m(k)$ are given in [6,8].

The block of complex-valued symbols $y(0),\ldots,y(N_{SF}^{PUCCH,s}M_{symb}-1)$ is DFT transformed (precoded) as follows [6]:

$$z\left(lM_{sc}^{PUCCH,s}+k\right)=\frac{1}{\sqrt{M_{sc}^{PUCCH,s}}}\sum_{m=0}^{M_{sc}^{PUCCH,s}-1}y\left(lM_{sc}^{PUCCH,s}+m\right)e^{-j\left(2\pi mk/M_{sc}^{PUCCH,s}\right)}$$

$$k=0,\ldots,M_{sc}^{PUCCH,s}-1;\quad l=0,\ldots,\left(\frac{N_{SF}^{PUCCH,s}M_{symb}}{M_{sc}^{PUCCH,s}}\right)-1$$

resulting in a block of complex-valued symbols $z(0),\ldots,z(M_{SF}^{PUCCH,s}M_{symb}-1)$, which are scaled with a scaling factor $\beta_{PUCCH,s}$ to properly adjust the transmit power and are subsequently mapped starting with $z(0)$ to resource elements (k,l) which are designated to PUCCH transmission and are not used by the associated PUCCH DM-RS. The mapping to resource elements is done in frequency-first manner on a single-antenna port. In case of intra-slot frequency hopping, $\lfloor N_{symb}^{PUCCH,s}/2 \rfloor$ OFDM symbols are transmitted in the first hop and the remaining $N_{symb}^{PUCCH,s} - \lfloor N_{symb}^{PUCCH,s}/2 \rfloor$ symbols in the second hop where $N_{symb}^{PUCCH,s}$ is the total number of OFDM symbols used in one slot for PUCCH transmission [6]. The physical processing and structure of PUCCH formats 3 and 4 are illustrated in Fig. 4.81.

4.2.3.2 Physical Random-Access Channel

The NR supports a four-step random-access procedure similar to LTE. However, beamforming and beam tracking aspects introduced in NR random-access procedure make the overall process different from that of LTE in frequencies above 6 GHz. The UEs need to detect and select the best beam for RACH process (beam selection process) prior to PRACH sequence selection and transmission. The PRACH design in NR relies on Zadoff–Chu sequences for the preamble construction. There are three PRACH preamble formats with long sequence length of 839, two of which, with subcarrier spacing of 1.25 kHz (same as LTE), are used for LTE refarming and large cells (up to 100 km); another format with subcarrier spacing of 5 kHz is defined for high-speed scenarios (up to 500 km/h) and cell radius up to 14 km (see Tables 4.27 and 4.28). Long sequences support

Table 4.27: PRACH preamble formats for $L_{RA} = 839$ and $\Delta f_{RA} \in \{1.25, 5\}$ kHz [6].

PRACH Format	L_{RA}	Subcarrier Spacing Δf_{RA} (kHz)	Bandwidth (MHz)	N_{SEQ}	T_{SEQ}	T_{CP}	T_{GP}	Support for Restricted Sets
0	839	1.25	1.08	1	$24576T_s$	$3168T_s$	$2976T_s$	Type A, Type B
1		1.25	1.08	2	$24576T_s$	$21024T_s$	$21984T_s$	
2		1.25	1.08	4	$24576T_s$	$4688T_s$	$29264T_s$	
3		5	4.32	1	$24576T_s$	$3168T_s$	$2976T_s$	

Table 4.28: Preamble formats for $L_{RA} = 139$ and $\Delta f_{RA} = 2^{\mu} \times 15$ kHz where $\mu \in \{0, 1, 2, 3\}$ and $\kappa = 64$ [6].

PRACH Format	L_{RA} Samples	Δf_{RA} (kHz)	T_{SEQ}	N_{SEQ}	N_u^{RA} Samples	N_{CP}^{RA} Samples
A1	139	$2^{\mu} \times 15$	$2048T_s$	2	$2 \times 2048\kappa 2^{-\mu}$	$288\kappa 2^{-\mu}$
A2				4	$4 \times 2048\kappa 2^{-\mu}$	$576\kappa 2^{-\mu}$
A3				6	$6 \times 2048\kappa 2^{-\mu}$	$864\kappa 2^{-\mu}$
B1				2	$2 \times 2048\kappa 2^{-\mu}$	$216\kappa 2^{-\mu}$
B2				4	$4 \times 2048\kappa 2^{-\mu}$	$360\kappa 2^{-\mu}$
B3				6	$6 \times 2048\kappa 2^{-\mu}$	$504\kappa 2^{-\mu}$
B4				12	$12 \times 2048\kappa 2^{-\mu}$	$936\kappa 2^{-\mu}$
C0				1	$1 \times 2048\kappa 2^{-\mu}$	$1240\kappa 2^{-\mu}$
C2				4	$4 \times 2048\kappa 2^{-\mu}$	$2048\kappa 2^{-\mu}$

unrestricted sets and restricted sets of Type A and Type B, while short sequences support unrestricted sets only. Considering network beam-sweeping reception within a RACH occasion, NR introduced a new set of PRACH preamble formats of shorter sequence length of 139 on 1, 2, 4, 6, and 12 OFDM symbols and subcarrier spacings of 15, 30, 60, and 120 kHz. The new formats are composed of single or consecutive repeated RACH sequences. The cyclic prefix is inserted at the beginning of the preambles, and the guard time is appended at the end of the preambles, while the cyclic prefix and gap between RACH sequences is omitted. For both short and long PRACH preamble sequences, the network can also conduct beam-sweeping reception between RACH occasions.

Multiple RACH preamble formats are defined with one or more PRACH symbols, and different cyclic prefix and guard time lengths. The PRACH preamble configuration is signaled to the UE in the SI. The UE calculates the PRACH transmit power for the retransmission of the preamble based on the most recent estimate of pathloss and power ramping counter. If the UE conducts beam switching, the counter of power ramping remains unchanged. The SI informs the UE of the association between the SS blocks and the RACH resources. The threshold of the SS block for RACH resource association is based on the RSRP and is network configurable.

Prior to initiation of the physical random-access procedure, the physical layer of the UE must receive a set of SS/PBCH block indices and provide the UE RRC sublayer with the corresponding set of RSRP measurements conducted on those SS/PBCH candidates. The information required for the UE physical layer prior to PRACH transmission includes PRACH preamble format, time resources, and frequency resources for PRACH transmission as well as the parameters for determining the root sequences and their cyclic shifts in the PRACH preamble sequence set including index to logical root sequence table, cyclic shift N_{CS}, and set type, that is, unrestricted, restricted set A, or restricted set B.

Physical random-access procedure is triggered following a request from the UE RRC sublayer or by a PDCCH command (control element) with the following parameters [8]: configuration for PRACH transmission, preamble index, preamble subcarrier spacing, transmit power P_{PRACH}^{target}, RA-RNTI, and a PRACH resource. The PRACH preamble is transmitted using the selected PRACH format with transmission power value over the designated PRACH resource.

The SS/PBCH block indices are mapped to PRACH occasions (see Fig. 4.82); first, in increasing order of preamble indices within a single PRACH occasion followed by, in increasing order of frequency resource indices of frequency-multiplexed PRACH occasions, then, in increasing order of time resource indices of time-multiplexed PRACH occasions within a PRACH slot and, finally, in increasing order of indices of PRACH slots. The association period, starting from frame 0, for the mapping of SS/PBCH blocks to PRACH occasions

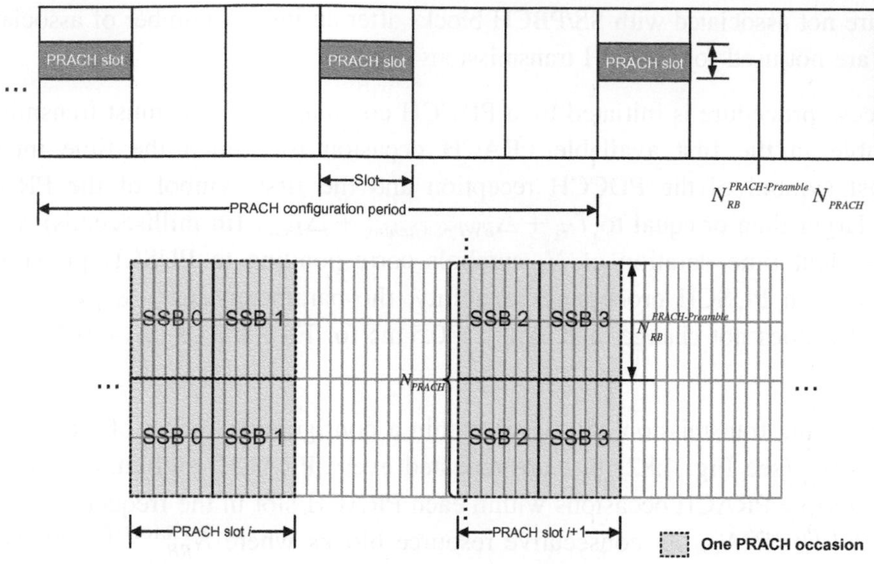

Figure 4.82
Structure of NR PRACH opportunities in time and frequency domain [14].

Table 4.29: Mapping between PRACH configuration period and SS/PBCH block to PRACH occasion association period [8].

PRACH Configuration Period (ms)	Association Period (Number of PRACH Configuration Periods)
10	{1,2,4,8,16}
20	{1,2,4,8}
40	{1,2,4}
80	{1,2}
160	{1}

is the smallest value in a set (see Table 4.29) determined by the PRACH configuration period such that N_{TX}^{SSB} SS/PBCH blocks are mapped at least once to the PRACH occasions within the association period. A UE obtains the parameter N_{TX}^{SSB} from *SystemInformationBlockType1*. If after an integer number of SS/PBCH blocks to PRACH occasions mapping cycles within the association period, there is a set of PRACH occasions that are not mapped to N_{TX}^{SSB} SS/PBCH blocks; no SS/PBCH blocks are mapped to the set of PRACH occasions. An association pattern period includes one or more association periods and is calculated such that a pattern between PRACH occasions and SS/PBCH blocks repeats at most every 160 ms. The PRACH occasions that are not associated with SS/PBCH blocks after an integer number of association periods, if any, are not used for PRACH transmissions [8].

If a random-access procedure is initiated by a PDCCH command, the UE must transmit the PRACH preamble in the first available PRACH occasion for which the time interval between the last symbol of the PDCCH reception and the first symbol of the PRACH transmission is larger than or equal to $T_{N_2} + \Delta_{BWP\text{-}Switching} + \Delta_{Delay}$ (in milliseconds), where T_{N_2} is the equivalent time duration of N_2 symbols corresponding to PUSCH preparation time assuming certain PUSCH processing capability, the parameter $\Delta_{BWP-Switching} = 0$, if uplink active BWP does not change, and $\Delta_{Delay} = 0.5$ ms for FR1 and $\Delta_{Delay} = 0.25$ ms for FR2 [8].

The PRACH preamble transmission can occur within a configurable subset of slots known as the PRACH slots (see Fig. 4.82) that are repeated every PRACH configuration period. There may be multiple PRACH occasions within each PRACH slot in the frequency-domain that cover $N_{RB}^{PRACH\text{-}Preamble} N_{PRACH}$ consecutive resource blocks where $N_{RB}^{PRACH\text{-}Preamble}$ is the preamble bandwidth measured in number of resource blocks and N_{PRACH} is the number of frequency-domain PRACH occasions. For a given preamble type, corresponding to a certain preamble bandwidth, the overall available time-frequency PRACH resources within a cell can be described by the following parameters: a configurable PRACH periodicity that can range from 10 to 160 ms; a configurable set of PRACH slots within the PRACH period; a configurable frequency-domain PRACH resource given by the index of the first resource block in the resource and the number of frequency-domain PRACH occasions [14].

In NR, the set of random-access preambles $x_{u,v}(n)$ is generated based on Zadoff–Chu sequences such that $x_{u,v}(n) = x_u[(n + C_v) \bmod L_{RA}]$ wherein $x_u(i) = \exp(-j\pi ui(i+1)/L_{RA})$ is a Zadoff–Chu sequence of length L_{RA} and root index u; and $i = 0, 1, \ldots, L_{RA} - 1$. The frequency-domain representation of the preamble sequence is obtained by taking L_{RA} − point DFT of the sequence $x_{u,v}(n)$, resulting in $y_{u,v}(n) = \sum_{m=0}^{L_{RA}-1} x_{u,v}(m) \exp(-j2\pi mn/L_{RA})$ where the sequence length $L_{RA} = 839$ or $L_{RA} = 139$ depend on the PRACH preamble format (see Tables 4.27 and 4.28). There are 64 preambles in each time--frequency PRACH occasion, numbered in increasing order of cyclic shift C_v of a logical root sequence and increasing order of the logical root sequence index, starting with the index obtained via RRC signaling. The sequence number u is obtained from the logical root sequence index [6]. The output of the DFT is then repeated N_{SEQ} times, after which a cyclic prefix is inserted. For the PRACH preamble, the cyclic prefix is not inserted per OFDM symbol, rather it is inserted only once for the block of N_{SEQ} repeated symbols (see Fig. 4.83).

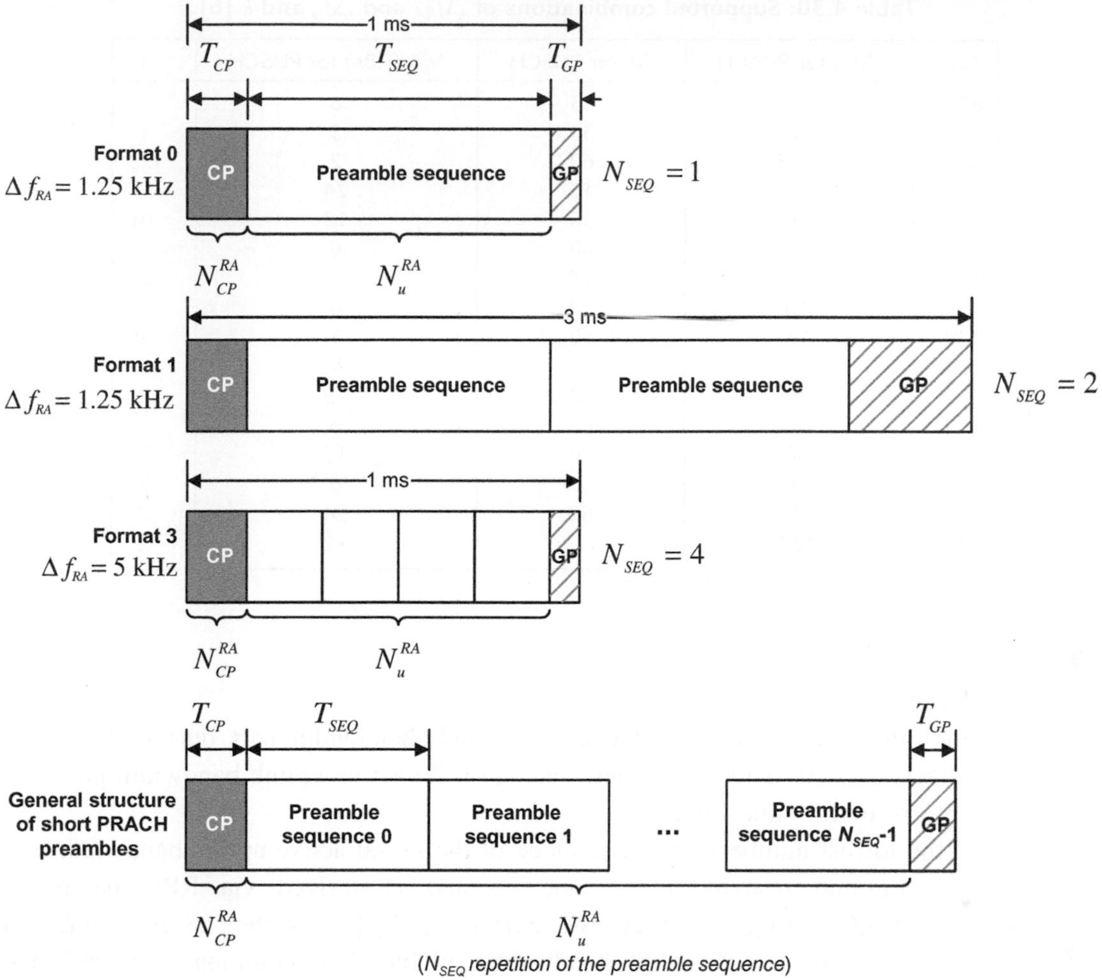

Figure 4.83
NR PRACH preamble structures [6].

The time-domain representation of PRACH signal $s_{PRACH}(t, l)$ on a single-antenna port is given as follows [6]:

$$s_{PRACH}(t,l) = \sum_{k=0}^{L_{RA}-1} a_k e^{j2\pi(k+Kk_1+\bar{k})\Delta f_{RA}(t-N_{CP,l}^{RA}T_c-t_{start}^{RA})}; \quad K = \frac{\Delta f}{\Delta fRA}$$

$$k_1 = k_o^{\mu} + N_{BWP,i}^{start}N_{sc}^{RB} + n_{RA}^{start}N_{sc}^{RB} + n_{RA}N_{RB}^{RA}N_{sc}^{RB} - \frac{N_{grid}^{size,\mu}N_{sc}^{RB}}{2}$$

$$t_{start}^{RA} \le t < t_{start}^{RA} + \left(N_u + N_{CP,l}^{RA}\right)T_c$$

$$k_0^{\mu} = \left(N_{grid}^{start}(\mu) + \frac{N_{grid}^{size}(\mu)}{2}\right)N_{sc}^{RB} - \left(N_{grid}^{start}(\mu_0) + \frac{N_{grid}^{size}(\mu_0)}{2}\right)N_{sc}^{RB}2^{\mu_0-\mu}$$

Table 4.30: Supported combinations of Δf_{RA} and Δf, and $\bar{k}$ [6].

L_{RA}	Δf_{RA} for PRACH	Δf for PUSCH	N_{RB}^{RA} (RBs) for PUSCH	$\bar{k}$
839	1.25	15	6	7
839	1.25	30	3	1
839	1.25	60	2	133
839	5	15	24	12
839	5	30	12	10
839	5	60	6	7
139	15	15	12	2
139	15	30	6	2
139	15	60	3	2
139	30	15	24	2
139	30	30	12	2
139	30	60	6	2
139	60	60	12	2
139	60	120	6	2
139	120	60	24	2
139	120	120	12	2

In the preceding expression

- $\bar{k}$ is given in Table 4.30 [6].
- Δf is the subcarrier spacing of the active uplink bandwidth part during the initial access; otherwise, Δf is the subcarrier spacing of the active uplink bandwidth part (see Table 4.30 for permissible values).
- $N_{BWP,i}^{start}$ is the lowest numbered resource block of the initial active uplink bandwidth part based on common resource block indexing and is derived via RRC parameter *initialUplinkBWP* during initial access; otherwise, $N_{BWP,i}^{start}$ is the lowest numbered resource block of the active uplink bandwidth part based on common resource block indexing and is derived by the higher layer parameter *BWP-Uplink*.
- n_{RA}^{start} is the frequency offset of the lowest PRACH transmission occasion in the frequency-domain relative to *PRB_0* of the initial active uplink bandwidth part given by the RRC parameter *msg1-FrequencyStart* during initial access associated with the initial active uplink bandwidth part; otherwise, n_{RA}^{start} is the frequency offset of lowest PRACH transmission occasion in frequency domain with respect to physical resource block 0 of the active uplink bandwidth part given by the RRC parameter *prach-frequency-start* associated with the active uplink bandwidth part.
- n_{RA} is the PRACH transmission occasion index in the frequency-domain for a given PRACH transmission occasion in time.
- N_{RB}^{RA} is the number of resource blocks that are occupied by PRACH preamble.

The starting position of PRACH preamble in a subframe t_{start}^{RA} when $\Delta f_{RA} \in \{1.25, 5, 15, 30\}$ kHz or in a slot with 60 kHz subcarrier spacing when

$\Delta f_{RA} \in \{60, 120\}$ kHz is defined as $t_{start}^{RA} = 0$ when $l = 0$; otherwise, $t_{start}^{RA} = t_{start,l-1}^{\mu} + \left(N_u^{\mu} + N_{CP,l-1}^{\mu}\right)T_c$. The subframe or 60 kHz slot is assumed to start at $t = 0$. The timing advance is assumed to be zero $N_{TA} = 0$. The numerology corresponding to $\Delta f_{RA} \in \{1.25, 5\}$ kHz is assumed to be $\mu = 0$; otherwise, it is given by $\Delta f_{RA} \in \{15, 30, 60, 120\}$ kHz, and the symbol position l is given by $l = l_0 + n_t^{RA} n_{duration}^{RA} + 14 n_{slot}^{RA}$ where l_0 is the starting symbol, n_t^{RA} is the PRACH transmission occasion within the PRACH slot, numbered in increasing order from 0 to $N_t^{RA,slot} - 1$ within a RACH slot, $N_{duration}^{RA}$ is given in [6], and the n_{slot}^{RA} depends on Δf_{RA}, that is, if $\Delta f_{RA} \in \{1.25, 5, 15, 60\}$ kHz then $n_{slot}^{RA} = 0$; otherwise if $\Delta f_{RA} \in \{30, 120\}$ kHz and either of the *number of PRACH slots within a subframe* or *number of PRACH slots within a 60 kHz slot* is equal to 1, then $n_{slot}^{RA} = 1$; otherwise, $n_{slot}^{RA} = 0, 1$. The quantities L_{RA} and N_u are the length of the PRACH sequence and the number of samples in a PRACH symbol, and $N_{CP,l}^{RA} = N_{CP}^{RA} + 16\kappa n$ wherein $n = 0$ for $\Delta f_{RA} \in \{1.25, 5\}$ kHz. For $\Delta f_{RA} \in \{15, 30, 60, 120\}$ kHz, n is the number of times the interval $[t_{start}^{RA}, t_{start}^{RA} + (N_u^{RA} + N_{CPt}^{RA})T_c]$ overlaps with either time instance zero or time instance $(\Delta f_{\max} N_f / 2000) T_c = 0.5$ ms in a subframe [6].

The parameters *ZeroCorrelationZoneConfig* and *prach-RootSequenceIndex* (defined in [6]) are used to generate the random-access signatures for each cell, which are required to be distinct across neighboring cells. There is a relationship between the preamble format and the cell radius, which means that the selection of *ZeroCorrelationZoneConfig* parameter is related to the cell radius. The parameters *ZeroCorrelationZoneConfig* and *prach-RootSequenceIndex* are derived from *SystemInformationBlockType1*. The random-access sequences are generated via selection of a Zadoff–Chu sequence (1 out of 839 or 139) given by *prach-RootSequenceIndex* and a cyclic shift that is used 64 times to generate the 64 random-access signatures from the Zadoff–Chu sequence selected. The cyclic shift is indirectly provided to the UE via the parameter *ZeroCorrelationZoneConfig*. The cyclic shift is also related to the cell size. The relationship between the cyclic shift and the cell size is given by $(N_{CS} - 1)(800\ \mu s/839) \geq \text{RTD} + \tau_{Delay\_Spread}$. If $\Delta f_{RA} = 1.25$ kHz, the PRACH symbol duration is 0.8 ms (0.133 ms in case of 139). The round-trip delay can be written as $\text{RTD} = 2R_{cell}/c$; therefore $R_{cell} \leq c[(N_{CS} - 1)(800\ \mu s/839) - \tau_{Delay\_Spread}]/2$. As an example, if we assume that *ZeroCorrelationZoneConfig* is 12 then from [6] and assuming $\Delta f_{RA} = 1.25$ kHz, $N_{CS} = 119$. Furthermore, if $\tau_{Delay\_Spread} = 6\ \mu s$ then the cell size will be approximately 15.97 km. Note that the smaller the cyclic shift, the smaller cell size. The delay spread in the preceding expression is derived empirically and the value of the delay spread is typically different for rural, suburban, urban and dense urban environments. In practice, the *ZeroCorrelationZoneConfig* parameter is a pointer to a table that provides a set of available cyclic shifts in the cell, where different tables indicated by this parameter have different distances between the cyclic shifts, thus providing larger or smaller zones or timing errors for which orthogonality or zero correlation can be maintained.

Table 4.31: Random-access configurations for TDD mode in FR1 [6].

PRACH Configuration Index	Preamble Format	n_{SFN} mod $x = y$		Subframe Number	Starting Symbol	Number of PRACH Slots Within a Subframe	$N_t^{PRACH,slot}$ Number of Time-Domain PRACH Occasions Within a RACH Slot	$N_{duration}^{RA}$ PRACH Duration
		x	y					
0	0	16	1	9	0	–	–	0
1	0	8	1	9	0	–	–	0
2	0	4	1	9	0	–	–	0
3	0	2	0	9	0	–	–	0
4	0	2	1	9	0	–	–	0
5	0	2	0	4	0	–	–	0
6	0	2	1	4	0	–	–	0
7	0	1	0	9	0	–	–	0
8	0	1	0	8	0	–	–	0
9	0	1	0	7	0	–	–	0
10	0	1	0	6	0	–	–	0
11	0	1	0	5	0	–	–	0
12	0	1	0	4	0	–	–	0
…	…	…	…	…	…	…	…	…
251	C2	1	0	3,4,8,9	2	2	2	6
252	C2	1	0	0,1,2,3,4,5,6,7,8,9	8	1	1	6
253	C2	1	0	1,3,5,7,9	2	1	2	6
254	C2	8	1	9	8	2	1	6
255	C2	4	1	9	8	1	1	6

The PRACH preamble sequence is mapped to physical resources such that $a_k^{(p,RA)} = \beta_{PRACH} y_{u,v}(k);\; k = 0, 1, \ldots, L_{RA} - 1$ where β_{PRACH} is a transmit power adjustment scaling factor, p is the antenna port from which the PRACH is transmitted. The PRACH preambles can only be transmitted in the time resources that are signaled via RRC parameter *prach-ConfigurationIndex* and further depend on frequency range FR1 or FR2 where the system is deployed and the spectrum type. The PRACH preambles can only be transmitted in the frequency resources specified by parameter *msg1-FrequencyStart*. The PRACH frequency resources $n_{RA} \in \{0, 1, \ldots, M-1\}$, in which the parameter M is derived from the RRC parameter *msg1-FDM*, are numbered in increasing order within the initial active uplink bandwidth part during initial access, starting from the lowest frequency. For the purpose of slot numbering, it is assumed that subcarrier spacing is 15 kHz for FR1 and 60 kHz for FR2. Table 4.31 provides random-access configurations for TDD mode in FR1 [6,8].

The transmission power for PRACH $P_{PRACH}^{klm}(i)$ on the kth uplink BWP of the lth carrier based on a certain SS/PBCH block determination for the mth serving cell in transmission period i is determined by the UE as $P_{PRACH}^{klm}(i) = \min\,[P_{CMAX}^{lm}(i), P_{PRACH-target}^{lm} + PL_{klm}]$ (dBm), wherein $P_{CMAX}^{lm}(i)$ is the configured UE transmission power for the lth carrier in the *m*th serving cell within transmission period *i*, $P^{lm}_{PRACH-target}$ is the PRACH preamble target reception power *PREAMBLE_RECEIVED_TARGET_POWER* signaled via RRC for the *k*th uplink BWP on *l*th carrier in the *m*th serving cell, and PL_{klm} is the calculated pathloss for the *k*th uplink BWP corresponding to the *l*th carrier for the current SS/PBCH block in the *m*th serving cell calculated by the UE in decibels. If within a random-access response window, the UE cannot receive a random-access response that contains a preamble identifier corresponding to the preamble sequence transmitted by the UE, the UE will typically ramp up (in steps) the transmission power up to a certain limit for the subsequent PRACH transmissions. If prior to PRACH retransmission, the UE changes the spatial domain transmission filter; the physical layer will notify the higher layers to suspend the power ramping counter [8].

The random-access preamble sequence can be generated at the system sampling rate, by means of a large IDFT unit. The cyclic shift can be implemented either in the time domain after the IDFT or in the frequency domain before the IDFT through a phase shift. For all possible system sampling rates, both cyclic prefix and sequence duration correspond to an integer number of samples. The method of Fig. 4.84 does not require any time-domain filtering in the baseband but requires large IDFT sizes (up to 24576 for a 20 MHz spectrum allocation), which are practically prohibitive. Therefore, another option for generating the PRACH preamble consists of using small-sized IDFT and shifting the preamble to the required frequency location through time-domain up-sampling and filtering (hybrid frequency/time-domain generation). Assuming that the preamble sequence length is 839, the smallest IFFT size that can be used is 1024. The sizes of the random-access cyclic prefix and preamble sequence duration have been chosen to provide an integer number of samples at the system sampling rate. The cyclic prefix can be inserted before the up-sampling and time-domain frequency shift, in order to minimize the intermediate storage requirements.

Assuming sampling rate of 30.72 MHz and considering that the random-access preamble spans 0.8 ms, it can be concluded that the number of samples in time is equal to 24576. Furthermore, if the PRACH subcarrier spacing is assumed to be 1.25 kHz and the subcarrier spacing for PUSCH and PUCCH is 15 kHz, in order to maintain the same sampling rate, a 12×2048 point DFT operation would be needed for the PRACH signal generation at the transmitter side, if the entire processing is done in the frequency domain. An alternative approach is to use time-domain signal generation and extraction which involves up-sampling and filtering operations at the transmitter. The drawback of time-domain implementation is that the up-sampling from 1.08 MHz to the system sampling-rate of 30.72 MHz is difficult to implement.

The implementation of the PRACH signal at the gNB receiver can take a frequency-domain or a hybrid time/frequency-domain approach. As illustrated in Fig. 4.84 as an example, the common parts to both approaches are the cyclic prefix removal, which always occurs at the

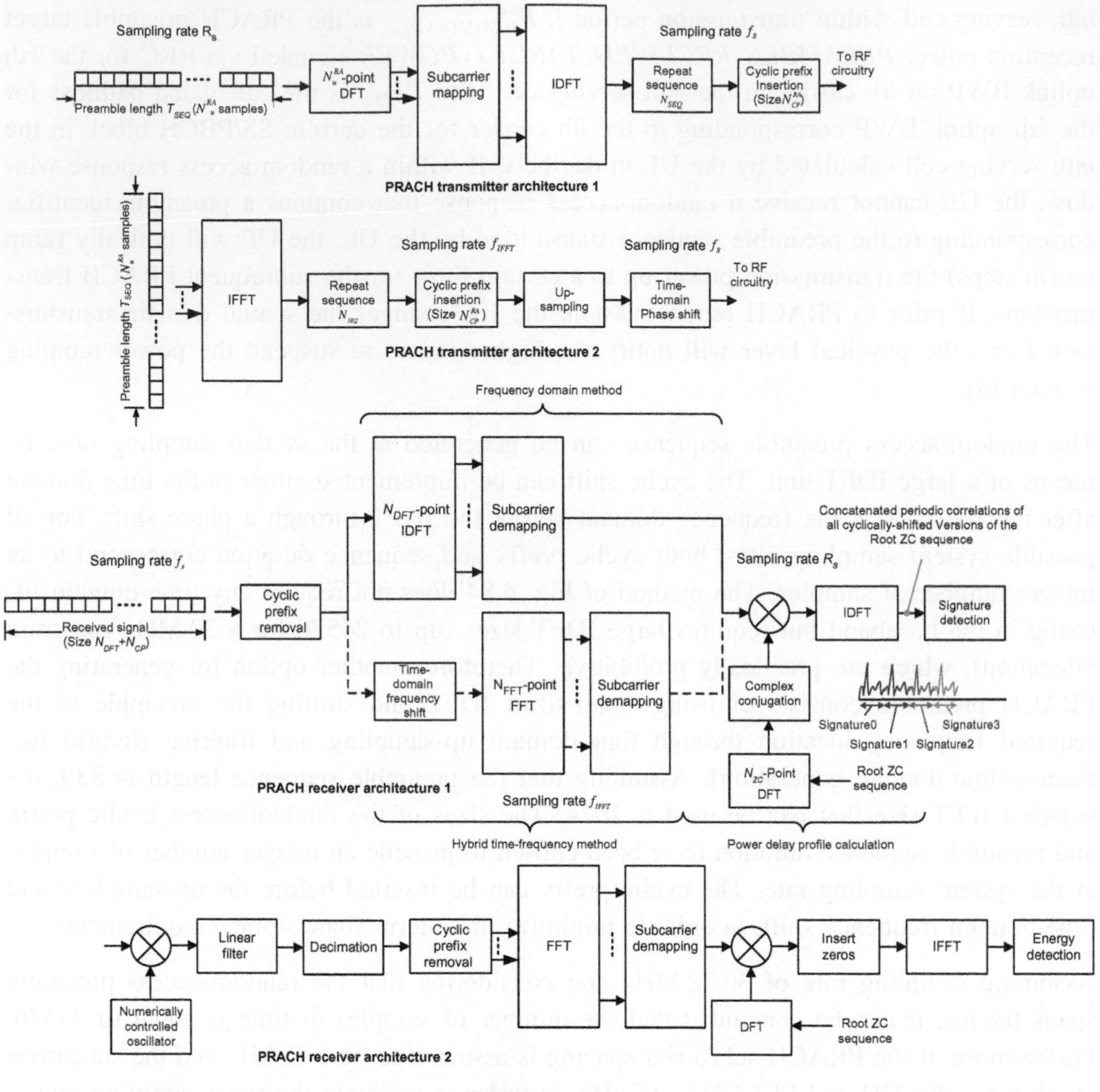

Figure 4.84
Example PRACH transmitter and receiver structure [15].

front-end at the system sampling rate, the power delay profile calculation, and signature detection. The two approaches differ only in the computation of the subcarriers carrying the PRACH signal(s). The frequency-domain method computes the full range of subcarriers used for uplink transmission over the system bandwidth from 0.8 ms-long received input samples. As a result, the PRACH subcarriers are directly extracted from the set of uplink subcarriers, which does not require any frequency shift or time-domain filtering but involves an extremely large DFT computation. Note that even though DFT size $N_{DFT} = n2^m$, and we can use fast and effcient DFT computation algorithms, the DFT computation cannot start until the complete sequence is stored in memory, which increases the processing delay.

4.2.3.2.1 Four-Step Random-Access Procedure

From the UE physical layer perspective, the RACH procedure consists of transmission of random-access preamble (Msg1) in a PRACH occasion, receiving random-access response message via PDCCH/PDSCH (Msg2), and transmission of Msg3 in PUSCH, and receiving PDSCH for contention resolution. If the random-access procedure is initiated by a PDCCH command, the PRACH preamble is transmitted with the same subcarrier spacing. The random-access procedure comprises four steps. However, before the UE can attempt to access the network, it must synchronize to the downlink and receive the SI via PBCH and PDCCH/PDSCH. Upon receiving the SI, the UE would have the knowledge of PRACH configuration and transmission parameters such as PRACH preamble format, time-frequency resources to transmit PRACH, the parameters for determining the root sequences and their cyclic shifts in the PRACH preamble sequence set, index to the logical root sequence table, cyclic shifts, and the associated set type, that is, unrestricted, restricted Type A, or restricted Type B [6,8,12]. More specifically, the RACH procedure consists of the following 4 steps [11,12]:

In the first step, the UE transmits a PRACH preamble associated with an RA-RNTI, if all conditions for PRACH transmission are met [12]. The gNB calculates the RA-RNTI associated with the PRACH occasion, in which the random-access preamble is transmitted, as follows $RA\text{-}RNTI = 1 + s_{id} + 14t_{id} + 14 \times 80f_{id} + 14 \times 80 \times 8ul_{carrier\_id}$ where $0 \leq s_{id} < 14$ is the index of the first OFDM symbol of the specified PRACH;$0 \leq t_{id} < 80$ denotes the index of the first slot symbol of the specified PRACH in a system frame;$0 \leq f_{id} < 8$ is the index of the specified PRACH in the frequency domain; and $ul_{carrier\_id}$ is the uplink carrier used for Msg1 transmission ($ul_{carrier\_id} = 0$: NR uplink carrier, $ul_{carrier\_id} = 1$: SUL carrier). The frequency-domain location (resource) for PRACH preamble is determined by the RRC parameter *msg1-FDM* and *msg1-FrequencyStart*. The time-domain location (resource) for PRACH preamble is determined by the RRC parameter *prach-ConfigurationIndex*.

In the second step, following the PRACH transmission, the UE awaits random-access response from the gNB which would be sent through a DCI scrambled with RA-RNTI value calculated as above. The UE attempts to detect a PDCCH with the corresponding RA-RNTI within the period of *ra-ResponseWindow*. The UE searches for the DCI in the Type 1 PDCCH common search space. The DCI format for scheduling RAR message on PDSCH is DCI format 1_0 scrambled with RA-RNTI. The resource allocation type for the Msg2 on PDSCH is resource allocation Type 1. The frequency-domain resource allocation for the PDSCH carrying RAR message is specified by DCI format 0_1. The time-domain resource allocation for the RAR message on PDSCH is specified by DCI format 1 and *PDSCH-ConfigCommon*. The RAR window is configured by *rar-WindowLength* information element in a SIB message. If the UE successfully detects the PDCCH, it can decode PDSCH carrying the RAR message. After decoding the RAR message, the UE checks, if the random-access preamble ID (RAPID) in the RAR message matches the RAPID assigned to the UE. The PDCCH and PDSCH associated with the process are expected to use the same subcarrier spacing and cyclic prefix as SIB1. Note that the gNB is not expecting any HARQ-ACK for the RAR message. The gNB may conclude that UE has successfully received and decoded the RAR message, if the UE does not retransmit PRACH, which would happen if the UE does not detect the DCI format 1_0 with CRC scrambled with the corresponding RA-RNTI within the RAR window, or if the UE does not correctly receive the transport block in the corresponding PDSCH within that window.

In the third step, the UE must determine whether it should apply transform precoding for Msg3 on PUSCH, based on the RRC parameter *msg3-transformPrecoder*. The UE determines the subcarrier spacing for Msg3 on PUSCH based on the RRC parameter *SubcarrierSpacing* in *BWP-UplinkCommon*. The UE then transmits Msg3 on PUSCH to the same serving cell to which it had sent the PRACH.

In the fourth step, Msg4 is transmitted to the UE for contention resolution. The UE starts *ra-ContentionResolutionTimer* and monitors PDCCH with TC-RNTI while *ra-ContentionResolutionTimer* is running. The UE looks for the DCI in Type 1 PDCCH common search space. If the PDCCH is successfully detected, the UE proceed to decode PDSCH carrying the MAC control element, and at the same time, it sets the value of the C-RNTI to TC-RNTI and discards *ra-ContentionResolutionTimer*. The UE considers the RACH procedure as successfully completed. Once the UE successfully decodes Msg4 (contention resolution), it sends HARQ-ACK for the data (PDSCH carrying Msg4). In response to the PDSCH reception with the UE contention resolution identity, the UE transmits HARQ-ACK information on a PUCCH. This procedure is illustrated in Fig. 4.85.

4.2.3.2.2 Two-Step Random-Access Procedure

As we mentioned in the previous section, the NR Rel-15 supports a four-step RACH procedure. A two-step RACH procedure can be utilized, wherein the UE combines Msg1 and Msg3 of the four-step RACH procedure into one message, for example, MsgA, and transmits it to the base station. The base station also combines Msg2 and Msg4 of the four-step RACH procedure and sends it as a response, for example, MsgB, to the UE. The combining of the messages provides a low-latency RACH procedure which is useful for low-latency applications and services. More specifically, in two-step RACH, MsgA is a combination of the PRACH preamble in Msg1 and the data contained in Msg3, while MsgB combines the random-access response in Msg2 and the contention resolution information of Msg4. Fig. 4.86 shows and compares the two-step and four-step contention-based random-access procedures.

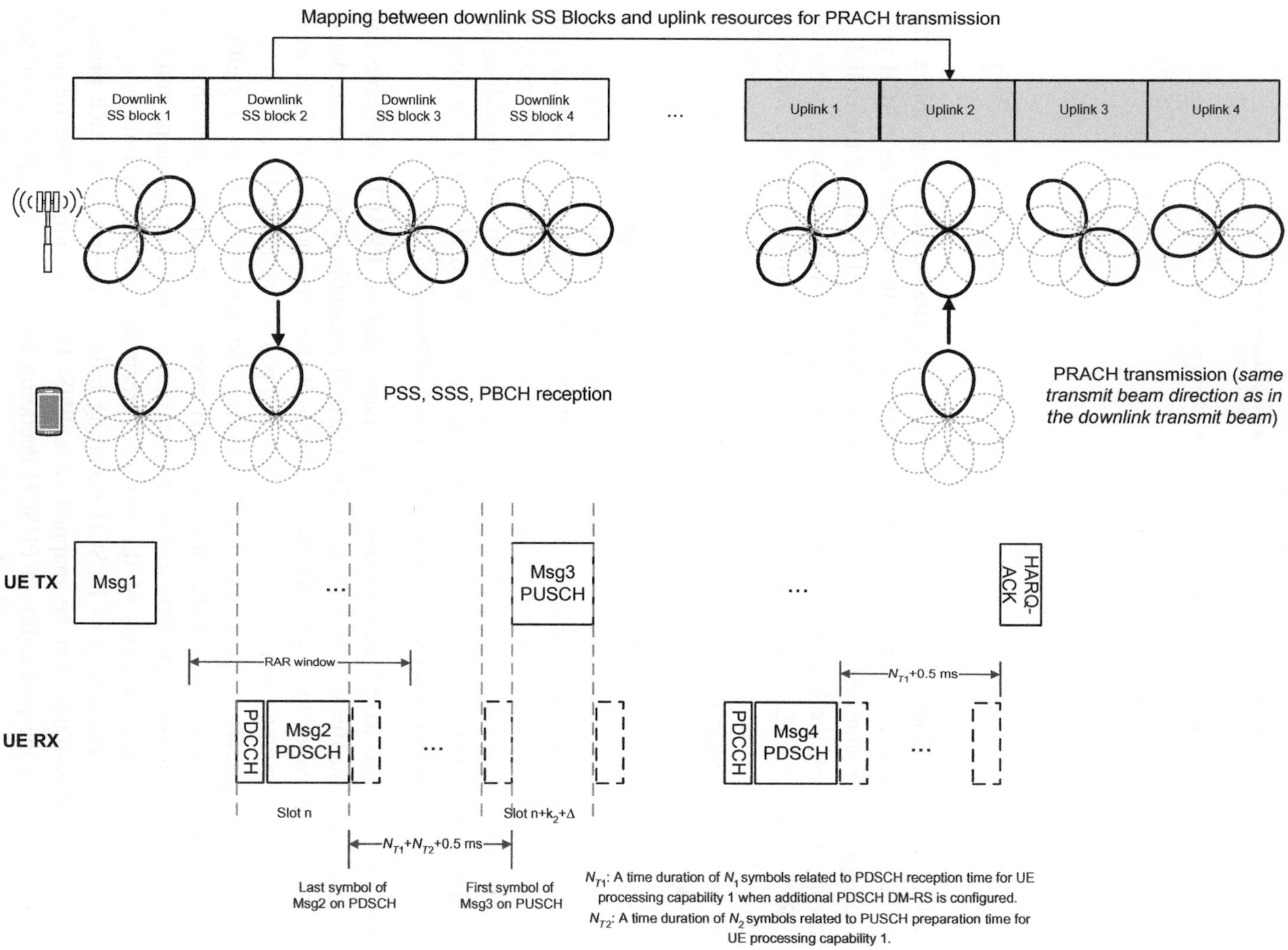

Figure 4.85
Random-access procedure [8,12].

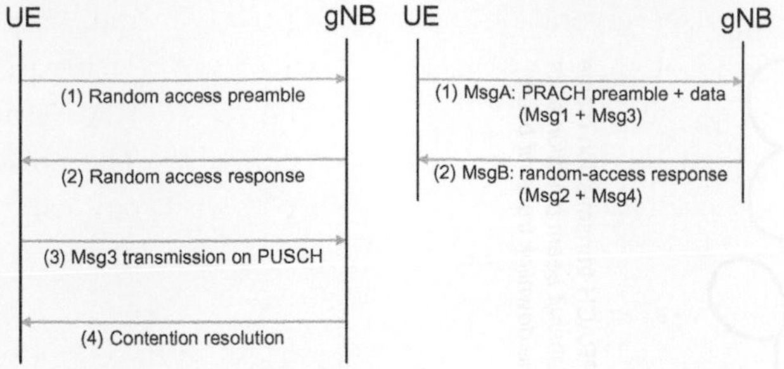

Figure 4.86
Comparison of two-step and four-step contention-based random-access procedures [11].

In NR Rel-15, the RACH procedure is triggered, when uplink data becomes available at the UE buffer and the UE is either in the RRC_IDLE/INACTIVE state, where the RACH procedure is triggered for state transition, or in the RRC_CONNECTED state, if the uplink is not synchronized, where the RACH procedure is used to reestablish uplink synchronization, or in the RRC_CONNECTED state, if the UE has no PUCCH resources available for SR or the SR procedure fails, where the RACH procedure serves as an SR. In addition, the RACH procedure is used for beam failure and recovery, on-demand SI request, or it can be explicitly triggered by the network with RRC for handover.

In NR Rel-15, the uplink data cannot be transmitted until the RACH procedure is successfully completed. It is observed that for small packet transmission, four-step RACH is not efficient in terms of latency and signaling overhead; thus the two-step RACH has been proposed to simplify the RACH procedure to achieve lower signaling overhead and latency. It is possible to allow Msg3 in four-step RACH procedure to carry data in order to reduce the latency and overhead. However, even in that case, the four-step RACH would still involve more signaling and latency relative to that of the two-step approach.

In two-step RACH, the MsgA may consist of two parts, that is, PRACH preamble and PUSCH which are time-division multiplexed. The PRACH preamble is used for UE detection, allowing the network to prepare for the reception of the corresponding PUSCH message. In NR Rel-15, up to 64 preamble signatures are mapped to one PRACH occasion. The preambles are orthogonal, or quasi-orthogonal, allowing the network to receive multiple preambles (from different UEs) in the same PRACH occasion. If all the preambles are mapped to a PUSCH in the same time-frequency occasion and more than one preamble is detected, the PUSCH transmissions of the detected preambles overlap in time and frequency, increasing the probability of PUSCH decoding failure. Alternatively, each preamble, or subset of preambles, can be mapped to a PUSCH in a unique time-frequency resource. This reduces the probability of PUSCH decoding failure due to collision but significantly increases the two-step RACH physical-layer overhead in the uplink. The MsgB in two-step procedure comprises several fields including the detected unique ID for contention resolution, where the size of the detected ID depends on the use case; a timing advance field; an uplink-grant for scheduling the data packets after the RACH procedure; and small user-plane/control-plane packets for downlink communication. The presence and the size of each field depend on the use case; thus the total size of MsgB may vary.

4.2.4 Physical Uplink Shared Channel

The physical uplink shared channel is used to transmit the user traffic and control information in the uplink. It supports two transmission modes namely codebook-based and non-codebook-based multi-antenna transmission. For codebook-based transmission, the gNB provides the UE with a transmit precoding matrix indication in DCI. The UE uses the indicator to select the PUSCH transmit precoder from a set of codebooks. For non-codebook-based transmission, the UE determines its PUSCH precoder based on (wideband) SRS resource indication (SRI) field from DCI. A closed-loop DM-RS-based spatial multiplexing is supported for PUSCH with up to four transmission layers for SU-MIMO with CP-OFDM waveform. Uplink SU-MIMO uses one codeword. Support of DFT-S-OFDM in the uplink is optional, and when transform precoding is used, only a single MIMO transmission layer is supported.

As shown in Fig. 4.87, the uplink physical layer processing of transport channels consists of the following stages:

- Transport block CRC attachment, where TBSs larger than 3824 use 24-bit CRC and other TBSs utilize 16-bit CRC, followed by LDPC base graph selection
- Code block segmentation and code block CRC attachment, which always uses 24-bit CRC
- Channel coding which makes use of LDPC coding (base graph 1 or 2)
- Rate matching and code block concatenation followed by data and control multiplexing
- Scrambling and modulation where any of the modulation schemes may be used, that is, $\pi/2$-BPSK (with transform precoding only), QPSK, 16QAM, 64QAM, or 256QAM
- Layer mapping
- Transform precoding (enabled/disabled by configuration) and precoding
- Mapping to assigned resources and antenna ports

The UE transmits at least one symbol with DM-RS on each layer in which PUSCH is transmitted. The number of DM-RS symbols and resource element mapping is configured via RRC signaling. The PT-RS may be transmitted on additional symbols to assist the gNB receiver with phase tracking [11].

Following the above summary, let us discuss the physical layer processing of the UL-SCH in more detail. The CRC is calculated over the entire transport block that is constructed

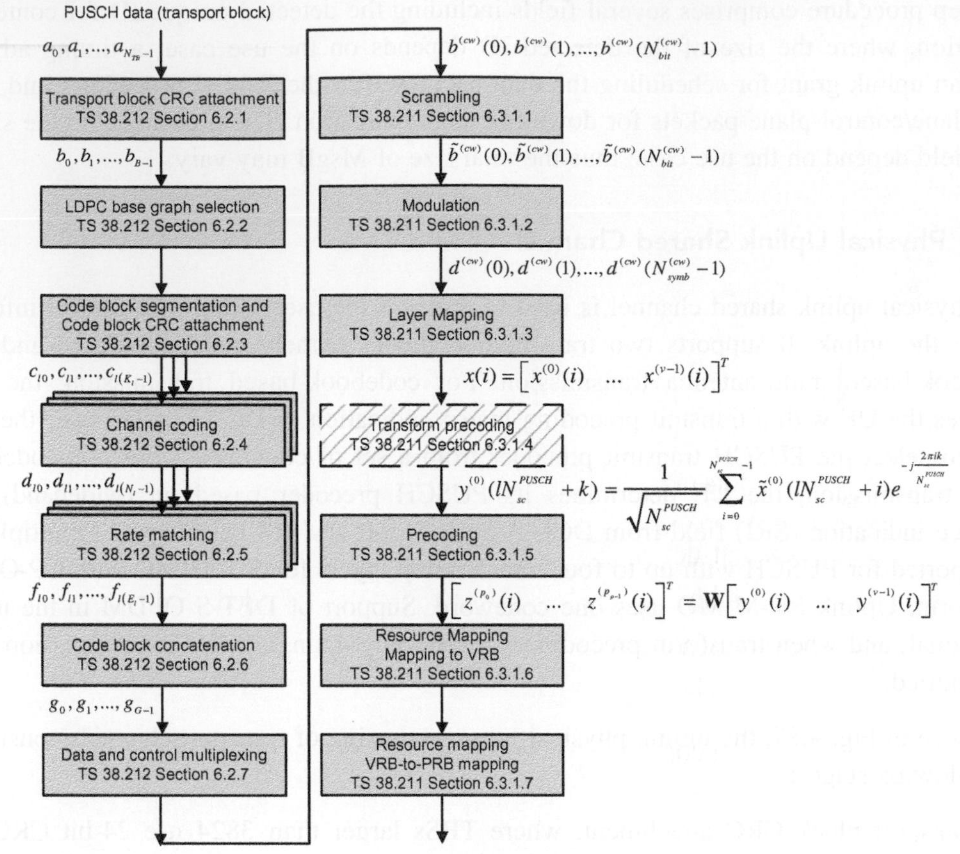

Figure 4.87
Physical processing of the uplink shared channel [30].

from the MAC PDU(s). We denote the TB bits by $a_0, a_1, \ldots a_{N_{PUSCH}-1}$ and the parity bits by $p_0, p_1, \ldots, p_{N_{CRC}-1}$, where N_{PUSCH} is the payload size and N_{CRC} is the number of parity bits. The number of parity bits depends on the PUSCH payload size. If $N_{PUSCH} > 3824$, $N_{CRC} = 24$ CRC bits are computed and attached to the TB using the generator polynomial $g_{CRC24A}(D) = D^{24} + D^{23} + D^{18} + D^{17} + D^{14} + D^{11} + D^{10} + D^7 + D^6 + D^5 + D^4 + D^3 + D + 1$; otherwise $N_{CRC} = 16$ CRC bits are calculated using the generator polynomial $g_{CRC16}(D) = D^{16} + D^{12} + D^5 + 1$. The code block bits after CRC attachment are denoted by $b_0, b_1, \ldots, b_{N_{PUSCH}+N_{CRC}-1}$ [7].

For initial transmission of a TB with coding rate R, determined by the MCS index and the subsequent retransmissions of the same TB, each code block of the TB is encoded with either LDPC base graph 1 or 2 depending on the values of N_{PUSCH} and R. If $N_{PUSCH} \leq 292$ or if $N_{PUSCH} \leq 3824$ and $R < 0.67$ or if $R < 0.25$, LDPC base graph 2 is used; otherwise, LDPC base graph 1 is used [7].

The input bit sequence to the code block segmentation is denoted by $b_0, b_1, \ldots, b_{N_{PUSCH}+N_{CRC}-1}$. If $B = N_{PUSCH} + N_{CRC}$ is larger than the maximum code block size K_{cb}, the bit sequence is segmented, and a 24-bit CRC is attached to each (segmented) code block. The value of K_{cb} for LDPC base graph 1 is $K_{cb} = 8448$ and for LDPC base graph 2, $K_{cb} = 3840$. The number of segmented code blocks is determined by $C = \lceil B/(K_{cb} - N_{CRC}) \rceil$. The output bits from code block segmentation are denoted by $c_{l0}, c_{l1}, \ldots, c_{l(K_l-1)}$ where $0 \le l \le C$ is the code block number, and $K_l = K$ is the number of bits in the lth code block.

The code blocks are then delivered to the channel coding unit where each code block is individually encoded with the LDPC encoder. The encoded bits are denoted by $d_{l0}, d_{l1}, \ldots, d_{l(N_l-1)}$, in which $N_l = 66Z_c$ for LDPC base graph 1 and $N_l = 50Z_c$ for LDPC base graph 2, where the value of the lifting factor Z_c is given in Table 4.7.

The encoded bits for each code block are processed through the rate matching function. The total number of code blocks is denoted by C, and each code block is individually rate matched by setting $I_{LBRM} = 1$, if RRC parameter *rateMatching* is set to *limitedBufferRM*; otherwise, by setting $I_{LBRM} = 0$. After the rate matching stage, the bits are denoted by $f_{l0}, f_{l1}, \ldots, f_{l(E_l-1)}$, where E_r is the number of rate matched bits in the lth code block.

The input bit sequence to the code block concatenation module are the sequences $\{f_{lk} | l = 0, 1, \ldots, C-1;\ k = 0, 1, \ldots, E_l - 1\}$ where E_l is the number of rate-matched bits in the lth code block. The output bit sequence from the code block concatenation function is the sequence $g_0, g_1, \ldots, g_{G-1}$ where G is the total number of coded bits for transmission. The code block concatenation function sequentially concatenate different rate-matched code blocks [7].

The NR supports UCI multiplexing on PUSCH when the UCI and PUSCH transmissions coincide in time, either due to transmission of an uplink TB or due to triggering of aperiodic CSI transmission without an uplink TB. The UCI carrying HARQ-ACK feedback with 1 or 2 bits is multiplexed by puncturing PUSCH. In all other cases, the UCI is multiplexed by rate matching PUSCH. In case of SUL, the UE is configured with two uplink carriers and one downlink carrier in the same cell. The uplink transmissions on the uplink carriers are controlled by the network to avoid overlapping PUSCH/PUCCH transmission in time-domain. Overlapping transmissions on PUSCH are avoided through scheduling, while overlapping transmissions on PUCCH are avoided through configuration since PUCCH can only be configured for one of the two uplink carriers in the cell. The initial access is supported on each uplink carrier. The mapping of the UCI to PUSCH resources is such that more (operationally) important bits (HARQ-ACK) are mapped to the first OFDM symbol after the first DM-RS, and less (operationally) important bits (CSI reports) are mapped to the subsequent symbols. Unlike the data part, which relies on rate adaptation to overcome the effects of radio propagation, the L1/L2 control signaling part cannot be rate-adapted. Power control may theoretically be used, but it would imply fast power variations in the time domain, which negatively impact the RF properties. Therefore, the transmission power is maintained constant over the PUSCH duration and the amount of resource elements allocated to L1/L2 control signaling is changed by changing the code rate of the control signaling. In addition to a semi-static value controlling the amount of PUSCH resources used for UCI, it is also possible to signal this information as part of a DCI.

The bit sequence $b^{(q)}(0), b^{(q)}(1), \ldots, b^{(q)}(N_{bit}^{(q)}-1)$ from the output of code block concatenation and multiplexing, where $N_{bit}^{(q)}$ is the number of bits in codeword q transmitted on the physical shared channel, are scrambled prior to modulation, resulting in a block of scrambled bits $\tilde{b}^{(q)}(0), \tilde{b}^{(q)}(1), \ldots, \tilde{b}^{(q)}(N_{bit}^{(q)}-1)$ such that the UL-SCH bits (except the UCI place-holder bits) are scrambled with pseudo-random sequence $c(n)$ that is initialized with the 16-bit RNTI bits as follows $\tilde{b}^{(q)}(i) = (b^{(q)}(i) + c^{(q)}(i)) \bmod 2$, where $c_{init} = n_{RNTI}2^{15} + n_{ID}$ and $n_{ID} \in \{0, 1, \ldots, 1023\}$ is set equal to the RRC parameter *dataScramblingIdentityPUSCH*; otherwise, $n_{ID} = N_{ID}^{cell}$. The parameter n_{RNTI} corresponds to the RNTI associated with the PUSCH transmission, if the input bits corresponding to UCI place-holder bits are set to one or the previous scrambled bit, depending on the value of the bits [6].

The block of scrambled bits $\tilde{b}^{(q)}(0), \tilde{b}^{(q)}(1), \ldots, \tilde{b}^{(q)}(N_{bit}^{(q)}-1)$ are modulated using $\pi/2$-BPSK (with transform precoding only), QPSK, 16QAM, 64QAM, or 256QAM modulation schemes, resulting in a block of complex-valued modulation symbols $d^{(q)}(0), d^{(q)}(1), \ldots, d^{(q)}(N_{symb}^{(q)}-1)$.

The complex-valued modulation symbols can be mapped to a maximum of four layers. More specifically, the complex-valued modulation symbols are mapped to $x(i) = [x^{(0)}(i) \ldots x^{(v-1)}(i)]^T$ layer, $i = 0, 1, \ldots, N_{layer}^{symb}$ where v is the number of layers and N_{layer}^{symb} is the number of modulation symbols per layer. If transform precoding is not enabled; for uplink CP-OFDM, $y^{(\lambda)}(i) = x^{(\lambda)}(i)$ for each layer $\lambda = 0, 1, \ldots, v-1$. However, for DFT-S-OFDM uplink, $v = 1$, and $\tilde{x}^{(0)}(i)$ depends on the configuration of phase-tracking reference signals. If phase-tracking reference signals are not configured, the block of complex-valued symbols $x^{(0)}(0), \ldots, x^{(0)}(N_{layer}^{symb}-1)$ for the single-layer $v = 1$ are divided into $N_{layer}^{symb}/N_{sc}^{PUSCH}$ sets, each corresponding to one OFDM symbol and $\tilde{x}^{(0)}(i) = x^{(0)}(i)$. In case phase-tracking reference signals are configured, the block of complex-valued symbols $x^{(0)}(0), \ldots, x^{(0)}(N_{layer}^{symb}-1)$ are divided into a number of groups, where each group corresponds to one OFDM symbol. The lth group contains $N_{sc}^{PUSCH} - \varepsilon_l N_{samp}^{group} N_{group}^{PT\text{-}RS}$ subcarriers and is mapped to the complex-valued symbols $\tilde{x}^{(0)}(lN_{sc}^{PUSCH} + i')$ corresponding to the lth OFDM symbol prior to transform precoding, wherein $i' = \{0, 1, \ldots, N_{sc}^{PUSCH} - 1\}$ and $i' \neq m$. The index m of PT-RS samples in the lth group, the number of samples per PT-RS group N_{samp}^{group}, and the number of PT-RS groups $N_{group}^{PT\text{-}RS}$ are defined in [6]. The quantity $\varepsilon_l = 1$, when the lth OFDM symbol contains one or more PT-RS samples, otherwise $\varepsilon_l = 0$.

The transform precoding is then performed, resulting in a block of complex-valued symbols $y^{(0)}(0),\ldots,y^{(0)}(N_{layer}^{symb}-1)$ as follows [6]:

$$y^{(0)}\left(lN_{sc}^{PUSCH}+k\right)=\frac{1}{\sqrt{N_{sc}^{PUSCH}}}\sum_{i=0}^{N_{sc}^{PUSCH}-1}\tilde{x}^{(0)}\left(lN_{sc}^{PUSCH}+i\right)e^{-j\left(2\pi ik/N_{sc}^{PUSCH}\right)};$$

$$k=0,\ldots,N_{sc}^{PUSCH}-1;\quad l=0,\ldots,\frac{N_{layer}^{symb}}{N_{sc}^{PUSCH}}-1$$

The parameter $N_{sc}^{PUSCH}=N_{RB}^{PUSCH}N_{sc}^{RB}$ where N_{RB}^{PUSCH} represents the bandwidth of the PUSCH in terms of resource blocks which must satisfy $N_{sc}^{PUSCH}=2^{\alpha_2}3^{\alpha_3}5^{\alpha_5}$ where $\alpha_2,\alpha_3,\alpha_5$ are non-negative integers. The DFT precoding is used to reduce the cubic metric of the uplink signal, thereby enabling higher power-amplifier efficiency. From implementation point of view, it is better to constrain the DFT size to a power of 2. However, such a constraint would limit the scheduler flexibility in terms of the amount of resources that can be assigned for an uplink transmission. In NR, the DFT precoding size, and thus the size of the resource allocation, is limited to products of the integers 2, 3, and 5 such that the DFT can be implemented as a combination of relatively less complex radix-2, radix-3, and radix-5 FFT processing.

The block of vectors $[y^{(0)}(i)\ldots y^{(v-1)}(i)]^T$, $i=0,1,\ldots,N_{layer}^{symb}-1$ corresponding to layers are precoded as follows:

$$\left[z^{(p_0)}(i)\ldots z^{(p_{\rho-1})}(i)\right]^T=\mathbf{W}\left[y^{(0)}(i)\ldots y^{(v-1)}(i)\right]^T$$

where $i=0,1,\ldots,N_{ap}^{symb}-1$, $N_{ap}^{symb}=M_{layer}^{symb}$, and p_i denotes the antenna port. For non-codebook-based transmission, the precoding matrix **W** is an identity matrix. However, for codebook-based transmission, the precoding matrix **W** is a scaler equal to one for single-layer transmission on a single-antenna port; otherwise, depending on value of the transmitted precoding matrix indicator (TPMI) index obtained from the DCI scheduling the uplink transmission, it will be chosen from a set of predefined matrices [6]. As an example, Table 4.32 provides the entries of the precoding vectors for single-layer transmission using two antenna ports.

Table 4.32: Precoding matrix W for single-layer transmission using two antenna ports [6].

	TPMI Index 0	TPMI Index 1	TPMI Index 2	TPMI Index 3	TPMI Index 4	TPMI Index 5
W	$\frac{1}{\sqrt{2}}\begin{bmatrix}1\\0\end{bmatrix}$	$\frac{1}{\sqrt{2}}\begin{bmatrix}0\\1\end{bmatrix}$	$\frac{1}{\sqrt{2}}\begin{bmatrix}1\\1\end{bmatrix}$	$\frac{1}{\sqrt{2}}\begin{bmatrix}1\\-1\end{bmatrix}$	$\frac{1}{\sqrt{2}}\begin{bmatrix}1\\j\end{bmatrix}$	$\frac{1}{\sqrt{2}}\begin{bmatrix}1\\-j\end{bmatrix}$

For each antenna port used for transmission of PUSCH, the block of complex-valued symbols $z^{(p)}(0), \ldots, z^{(p)}(M_{ap}^{symb} - 1)$ are scaled by a factor of β_{PUSCH} to adjust the transmit power and sequentially mapped, starting with $z^{(p)}(0)$, to virtual resource elements (k', l) in the virtual resource blocks assigned for PUSCH transmission. The physical resource blocks corresponding to the latter virtual resources must not be used for transmission of DM-RS, PT-RS, or DM-RS intended for other co-scheduled UEs. The mapping to virtual resource elements (k', l) is in increasing order of frequency index k' over the assigned virtual resource blocks, where $k' = 0$ is the first subcarrier in the lowest numbered virtual resource block followed by time index l. The virtual resource blocks are then mapped to physical resource blocks in a non-interleaved manner. For non-interleaved VRB-to-PRB mapping, virtual resource block n is mapped to physical resource block n.

While dynamic scheduling is the basic mode of operation in NR, the resources for uplink data transmission or downlink data reception can be configured in advance for the UE. Once the uplink data are available at UE's buffer, it can immediately start uplink transmission without going through the SR and grant cycle, thus reducing the latency. In other words, the NR PUSCH transmissions can be dynamically scheduled by an uplink grant provided by a DCI, or the transmission can correspond to a configured grant Type 1 or Type 2. As shown in Fig. 4.88, the configured grant Type 1 PUSCH transmission is semi-statically configured to operate upon the reception of RRC parameter *configuredGrantConfig* including *rrc-ConfiguredUplinkGrant* without the detection of an uplink grant in a DCI. The configured grant Type 2 PUSCH transmission is semi-persistently scheduled by an uplink grant in a valid activation DCI after the reception of RRC parameter *configurdGrantConfig* that does not include *rrc-ConfiguredUplinkGrant*. The UE transmits PUSCH upon detection of a PDCCH with DCI format 0_0 or 0_1, and it is not expected to be scheduled to transmit

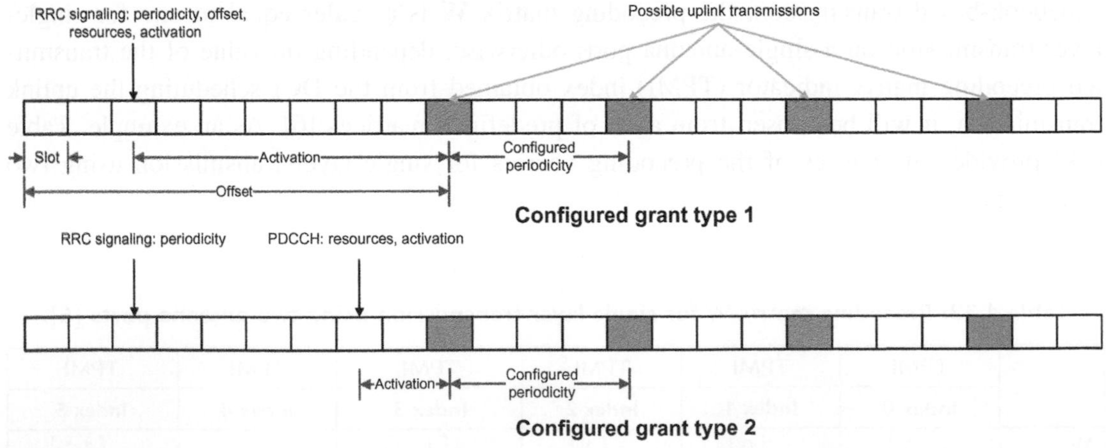

Figure 4.88
Illustration of uplink transmission with configured grants Type 1 and Type 2 [9].

another PUSCH by DCI format 0_0 or 0_1 scrambled by C-RNTI or MCS-C-RNTI for a given HARQ process until the end of the expected transmission of the last PUSCH for that HARQ process [9].

When the UE is scheduled to transmit a TB without a CSI report, or it is scheduled to transmit a TB with CSI report(s) on PUSCH by a DCI, the *time domain resource assignment* field value m of the DCI provides a row index $m+1$ to an allocated table. The indexed row defines the slot offset K_2, the start and length indicator *SLIV*, or directly the start symbol S and the allocation length L, and the PUSCH mapping type to be applied. Alternatively, when a UE is scheduled to transmit a CSI report(s) on PUSCH without a TB, the *time domain resource assignment* field value m of the DCI provides a row index $m+1$ to an allocated table which is defined by RRC parameter *pusch-TimeDomainAllocationList* in *pusch-Config*. The indexed row defines the start and length indicator *SLIV*, and the PUSCH mapping type. The parameter K_2 value is determined as $K_2 = \max_j Y_j(m+1)$, where $Y_j, j=0,\ldots,N_{repetition}-1$ are the corresponding list entries of the RRC parameter *reportSlotOffsetList* in *CSI-ReportConfig* for the $N_{repetition}$ triggered CSI Reporting Settings and $Y_j(m)$ is the mth entry of Y_j [9]. The slot where the UE is expected to transmit PUSCH is determined by $\lfloor (2^{\mu_{PUSCH}}/2^{\mu_{PDCCH}})n \rfloor + K^2$ where n is the slot with the scheduling DCI, K_2 is based on PUSCH numerology, and μ_{PUSCH} and μ_{PDCCH} are the subcarrier spacing for PUSCH and PDCCH, respectively. The starting symbol S relative to the start of the slot, and the number of consecutive symbols L counting from symbol S allocated for PUSCH are determined from the start and length indicator *SLIV* of the indexed row such that if $(L-1) \leq 7$ then $SLIV = 14(L-1)+S$; otherwise $SLIV = 14(15-L)+(13-S)$ where $0 < L \leq 14-S$. The PUSCH mapping type is set to Type A or Type B as given by the indexed row [9]. We have mentioned before that the time-domain reference point of the first PUSCH DM-RS symbol depends on the mapping type, where for PUSCH mapping type A, l is defined relative to the start of the slot, if frequency hopping is disabled and relative to the start of each hop in case frequency hopping is enabled, and the offset l_0 is given by the higher layer parameter *dmrs-TypeA-Position*. For PUSCH mapping type B, time-domain index l is defined relative to the start of the scheduled PUSCH resources, if frequency hopping is disabled and relative to the start of each hop when frequency hopping is enabled and the offset $l_0 = 0$ [6].

The UE determines the resource block assignment in frequency domain using the resource allocation field in the detected PDCCH DCI except for Msg3 PUSCH initial transmission. Two uplink resource allocation types are supported. The uplink resource allocation Type 0 is supported for PUSCH only when transform precoding is disabled, whereas the uplink resource allocation Type 1 is supported for PUSCH regardless of whether transform precoding is disabled. If the scheduling DCI is configured to indicate the uplink resource allocation type as part of the *Frequency domain resource* assignment field by setting RRC parameter *resourceAllocation* in *pusch-Config* to "dynamicswitch", the UE must use uplink resource allocation Type 0 or Type 1; otherwise, it uses the uplink frequency resource allocation type as defined by the RRC parameter *resourceAllocation* [9]. When the UE is scheduled with DCI format 0_0, the uplink resource allocation Type 1 is used. If a bandwidth part indicator field is not configured in the scheduling DCI, the RB indexing for uplink Type 0 and Type 1 resource allocation is determined within the UE's active bandwidth part. However, if a bandwidth part indicator field is configured in the scheduling DCI, the RB indexing for uplink Type 0 and Type 1 resource allocation is determined within the UE's bandwidth part indicated by bandwidth part indicator field value in the DCI. Upon detection of PDCCH, the UE first determines the uplink bandwidth part and then the resource allocation within the bandwidth part where RB numbering starts from the lowest RB in the determined uplink bandwidth part [9].

4.2.5 Uplink MIMO Schemes

The NR supports multi-antenna precoding with up to four layers for PUSCH transmission. However, when uplink DFT-based transform precoding is enabled, only single-layer (rank 1) transmission is supported. The UE can be configured in either codebook-based or non-codebook-based modes for PUSCH transmission. The selection between these two modes depends on the extent to which the uplink channel conditions can be estimated by the UE based on downlink measurements. PUSCH DM-RSs are precoded in the same way that PUSCH data subcarriers are precoded to allow coherent demodulation, making uplink precoding transparent to the gNB receiver. When codebook-based precoding is used in the uplink, the scheduling grant includes information about the precoder in the same way that the UE provides the network with PMI to assist downlink multi-antenna precoding. However, in contrast to the downlink, where the network may or may not use the precoder matrix indicated by the PMI, in the uplink direction, the UE is expected to use the precoder suggested by the network. In case of non-codebook-based transmission, the network can influence the selection of the uplink precoder. Another aspect that may impose constraints on the uplink multi-antenna transmission is to what extent one can assume coherence between different device antennas, that is, to what extent the relative phase between the signals transmitted on two antennas can be controlled. The phase coherence is necessary when antenna port−specific weight factors, including specific phase shifts, are applied to the signals transmitted on the different antenna ports. The NR specifications allow different UE capabilities concerning inter-antenna-port phase coherence, referred to as full coherence, partial coherence, and no coherence. In case of full coherence, it can be assumed that the device can control the relative phase between any of its antenna ports that are used for uplink transmission. In case of partial coherence, the device is capable of pairwise coherence, that is, the device can control the relative phase between antenna-port pairs. However, there is no guarantee that coherence can be achieved. In case of no coherence, there is no guarantee of phase coherence between any pair of the device antenna ports [6,9,14].

In codebook-based uplink shared-channel transmission, the network selects the transmission rank and the corresponding precoding matrix and informs the device through uplink scheduling grant. At the UE side, the precoding matrix is applied to the scheduled PUSCH transmission and the indicated number of layers is mapped to the antenna ports. To select a suitable rank and a corresponding precoding matrix, the gNB needs estimates of the channels between the device antenna ports and the corresponding network receive antennas. To enable this, a UE configured for codebook-based PUSCH transmission would typically be configured for transmission of at least one multi-port SRS. Based on measurements on the configured SRS, the network can estimate the channel and determine a suitable rank and precoding matrix. The network cannot select an arbitrary precoder, rather for a given combination of the antenna ports and transmission rank, the network can select the precoding matrix from a limited set of available precoders [6,9,14].

When selecting the precoding matrix, the network needs to consider the device capability in terms of antenna-port phase coherence. If the UE does not support antenna-port phase coherence, only the first two precoding matrices can be used with rank-1 transmission. It must be noted that restricting the codebook selection to these two matrices is equivalent to selecting either the first or the second antenna port for transmission. In this type of antenna selection, phase coherence between the antenna ports is not required. The selection of the remaining precoding vectors would imply linear combination of the signals of different antenna ports, which requires phase coherency among the antenna ports. A fundamental difference between NR codebook-based PUSCH transmission and LTE uplink is that a device can be configured to transmit multiple SRS from multiple antenna ports. In multi-SRS transmission, the network feedback includes SRI, identifying one of the configured SRSs. The UE should then use the precoder identified in the scheduling grant and map the output of the precoder to the antenna ports corresponding to the SRSs indicated in the SRI. The device should then transmit the precoded signal using the same antenna configuration and mapping that was used for the SRS transmission (indicated by the SRI). The use of multiple SRSs for codebook-based PUSCH transmission assumes that the UE transmits multi-port SRSs over separate and relatively wide beams. These beams may correspond to different UE antenna panels with different directions, where each panel includes a set of antenna elements corresponding to the antenna ports of each multi-port SRS (see Fig. 4.89). The SRI received from the network determines which beam should be used for the transmission, while the precoder information (number of layers and precoder) determines how the transmission should be done within the selected beam. Codebook-based precoding is typically used when uplink/downlink reciprocity cannot be achieved and when uplink measurements are needed in order to determine a suitable uplink precoding [6,9,14].

In contrast to codebook-based precoding, which is based on network measurements and selection of uplink precoder, non-codebook-based precoding is based on device

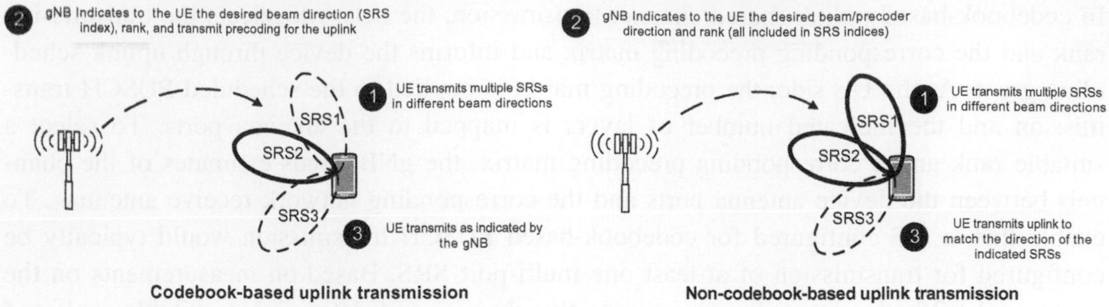

Figure 4.89
Codebook-based uplink transmission versus non-codebook-based uplink transmission [69].

measurements and precoder indications to the network. The concept of uplink non-codebook-based precoding is illustrated in Fig. 4.89. Based on downlink measurements conducted on configured CSI-RS resources, the UE selects a suitable uplink multi-layer precoder. Non-codebook-based precoding relies on channel reciprocity and assumes that the device can acquire accurate knowledge of the uplink channel based on downlink measurements. Note that there are no restrictions on the selection of precoder by the UE. Each column of a precoding matrix defines a digital beam for the corresponding layer. Therefore, selection of precoder for each layer can be perceived as selection of different beam directions, where each beam corresponds to one possible layer. It must be noted that the UE precoder selection is typically done based on downlink measurements, which may not be necessarily the best precoder from network point of view. As a result, the NR non-codebook-based precoding includes an additional step where the network can modify the device-selected precoder by removing some of the beams or equivalently some columns from the selected precoder [6,9,14].

As we mentioned earlier, codebook-based and non-codebook-based transmission modes are supported for PUSCH transmission. The UE will be configured with codebook-based transmission, when the RRC parameter *txConfig* in *pusch-Config* is set to "*codebook*", and it will be configured with non-codebook-based transmission, when the RRC parameter *txConfig* is set to "*nonCodebook*". If the RRC parameter *txConfig* is not provided, the PUSCH transmission will be based on a single-antenna port, triggered by DCI format 0_0. For codebook-based transmission, the UE determines the transmission precoder based on the information obtained from the SRI, TPMI, and the transmission rank, where the SRI, TPMI, and the transmission rank are given by the corresponding fields of the DCI. The TPMI is used to identify the preferred precoder over the SRS ports in the selected SRS resource by the SRI when single or multiple SRS resources are configured. Note that the indicated SRI in *n*th slot is associated with the most recent transmission of SRS resource identified by the SRI, where the SRS resource is prior to the PDCCH carrying the SRI. In codebook-based transmission mode, the UE determines its codebook subsets based on

TPMI and following reception of RRC parameter *codebookSubset* in *pusch-Config,* which may be configured with "*fullyAndPartialAndNonCoherent*", or "*partialAndNonCoherent*", or "*nonCoherent*" depending on the UE capability. The maximum transmission rank may be configured by the higher parameter *maxRank* in *pusch-Config.* Furthermore, for codebook-based transmissions, the UE may be configured with a single *SRS-ResourceSet,* and only one SRS resource can be indicated based on the SRI in the SRS resource set. The maximum number of configured SRS resources for codebook-based transmission is 2. If aperiodic SRS is configured for a UE, the SRS request field in DCI triggers the transmission of aperiodic SRS resources [9].

The codebook subset restriction concept was introduced in LTE [15]. It helps avoid CSI reporting for the undesired (spatial) directions. The LTE codebook subset restriction includes RI and PMI restriction, which provides sufficient flexibility to control PMI calculation and transmission from the UE. The content of CSI in NR is very similar to LTE, in the sense that NR supports CSI components such as RI and PMI.

Since the number of possible RI values is small, bitmap with one-to-one correspondence between each bit in bitmap, and RI value can be specified for the purpose of RI restriction. At the same time, the number of possible PMI values especially for larger number of antenna ports is very large to support one-to-one correspondence between PMIs and bits in the bitmap. Therefore, a solution with reduced signaling overhead should be considered. More specifically, similar to LTE, a DFT beam restriction is introduced, so that the PMI can be considered as restricted, if at least one beam is restricted by the corresponding DFT beam restriction bitmap. For co-phasing of the polarization, the bitmap should not be used as it does not affect the beamforming direction.

The NR Type-I single-panel codebook structure is similar to that of LTE FD-MIMO codebooks, except rank 3/4 codebooks for 16, 24, and 32 antenna ports at the gNB. Let us consider codebook subset restriction for less than 16 antenna ports at the gNB. In this case, the beamforming vector for PMIs of all ranks is represented by 2D DFT beam denoted as $\mathbf{b}_i$, which is represented as Kronecker product of two 1D DFT vectors $\mathbf{b}_i = \mathbf{u}_n \otimes \mathbf{q}_l$ wherein [66]

$$\mathbf{u}_n = \frac{1}{\sqrt{N_1}}\left[1 \quad e^{2\pi j\left(1/N_1O_1\right)n} \quad \ldots \quad e^{2\pi j\left((N_1-1)/N_1O_1\right)n}\right]^T$$

$$\mathbf{q}_l = \frac{1}{\sqrt{N_2}}\left[1 \quad e^{2\pi j\left(1/N_2O_2\right)l} \quad \ldots \quad e^{2\pi j\left((N_2-1)/N_2O_2\right)l}\right]^T$$

In this case, in order to restrict transmission in specific direction, a bitmap with size $N_1N_2O_1O_2$ can be specified, where each bit a_i corresponds to DFT beam $\mathbf{b}_i$. If at least one layer of the PMI consists of $\mathbf{b}_i$, the PMI is considered to be restricted and cannot be

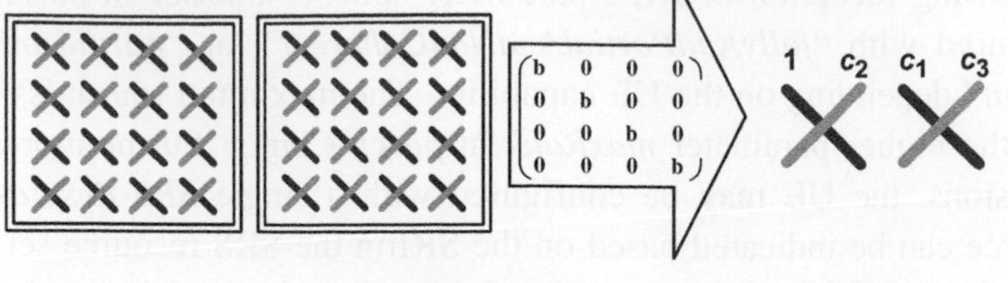

Figure 4.90
Illustration of precoding for multi-panel antenna array [66].

reported by the UE. Rank 3/4 PMIs for 16, 24, and 32 antenna ports have different structures compared to Type-I single-panel PMI. Therefore, it may not be possible to use the same approach for all ranks. The multi-panel codebook is constructed by DFT-based beamforming per each panel and co-phasing of polarization and panels, where the same DFT beam is applied for all panels and polarizations. An example is shown in Fig. 4.90, where precoder **p** for rank-1 multi-panel codebook and two-panel antenna can be computed as follows, in which c_1, c_2, c_3 coefficients are independently reported in accordance to mode 2 of multi-panel codebooks [66].

$$\mathbf{p} = \begin{pmatrix} \mathbf{b} & \mathbf{0} & \mathbf{0} & \mathbf{0} \\ \mathbf{0} & \mathbf{b} & \mathbf{0} & \mathbf{0} \\ \mathbf{0} & \mathbf{0} & \mathbf{b} & \mathbf{0} \\ \mathbf{0} & \mathbf{0} & \mathbf{0} & \mathbf{b} \end{pmatrix} \begin{pmatrix} 1 \\ c_1 \\ c_2 \\ c_3 \end{pmatrix} = \begin{pmatrix} \mathbf{b} \\ c_1\mathbf{b} \\ c_2\mathbf{b} \\ c_3\mathbf{b} \end{pmatrix}$$

In the preceding example, it is assumed that the antenna port indexing is performed in such way that $[\mathbf{b} \quad c_1\mathbf{b}]$ corresponds to beamforming vector of the first polarization and $[\mathbf{b} \quad c_3/c_2\mathbf{b}]$ to beamforming vector of the second polarization.

The direction of the transmission in the above PMI structure is determined by the DFT beam denoted by vector **b** and co-phasing coefficients c_2 and c_3/c_2. Therefore, codebook subset restriction for multi-panel codebook PMI restriction consider all possible combinations of DFT beams and co-phasing coefficients, which are determining the direction of the transmission. The resulting size of bitmap in that case equals to $N_1N_2O_1O_2 4^{(N_g-1)}$ [66].

The Type-II CSI was designed to enhance the performance of MU-MIMO transmission. The accuracy of spatial channel feedback in case of Type-II CSI allows interference suppression improvement through use of advanced precoding schemes such MMSE precoding. An accurate knowledge of the channel increases suppression capabilities of intra-cell interference. The beamforming vector in Type-II codebook is represented as linear combination of 2, 3, or 4 DFT beams as $\mathbf{w}_{rl} = \sum_{i=0}^{L-1} p_{rli}^{(WB)} p_{rli}^{(SB)} c_{rli} \mathbf{v}_{k_i}$, where $p_{rli}^{(WB)}$ denotes the wideband beam amplitude scaling factor; $p_{rli}^{(SB)}$ is the subband beam amplitude scaling factor, and c_{rli} is the beam combining coefficient (phase) for beam i, polarization r, and layer l [66].

For non-codebook-based uplink transmission, PUSCH transmission can be scheduled by DCI format 0_0, DCI format 0_1 or semi-statically configured. The UE can determine its PUSCH precoder and transmission rank based on the SRI when multiple SRS resources are configured, where the SRI is given by the SRI in DCI or the SRI is given by RRC parameter *srs-ResourceIndicator*. The UE must use one or multiple SRS resources for SRS transmission. In an SRS resource set, the maximum number of SRS resources which can be configured for the UE for simultaneous transmission on the same symbol and the maximum number of SRS resources depend on the UE capability. It must be noted that only one SRS port for each SRS resource can be configured and only one SRS resource set can be configured with higher layer parameter *usage* in *SRS-ResourceSet* set to "*nonCodebook*". The maximum number of SRS resources that can be configured for non-codebook-based uplink transmission is 4. The indicated SRI in the *n*th slot is associated with the most recent transmission of SRS resource(s) identified by the SRI, where the SRS transmission is prior to the PDCCH carrying the SRI [9].

For non-codebook-based uplink transmission, the UE can calculate the precoder used for the transmission of SRS based on measurement of an associated NZP CSI-RS resource. A UE can be configured with only one NZP CSI-RS resource for the SRS resource set with higher layer parameter *usage* in *SRS-ResourceSet* set to "*nonCodebook*". If aperiodic SRS resource set is configured, the associated NZP-CSI-RS is indicated via SRS request field in DCI formats 0_1 and 1_1. A UE is not expected to update the SRS precoding information, if the gap between the last symbol of the reception of the aperiodic NZP-CSI-RS resource and the first symbol of the aperiodic SRS transmission is less than 42 OFDM symbols. The CSI-RS is located in the same slot as the SRS request field. If the UE is configured with aperiodic SRS associated with aperiodic NZP CSI-RS resource, none of the TCI states configured on the scheduled component carrier are configured with "QCL-TypeD". The UE performs one-to-one mapping between the indicated SRI(s) and the indicated DM-RS port (s) and their corresponding PUSCH layers provided by DCI format 0_1 or by *configuredGrantConfig*. The UE transmits PUSCH using the same antenna ports as the SRS port(s) in the SRS resource(s) indicated by SRI(s) given by DCI format 0_1 or by *configuredGrantConfig*. For non-codebook-based uplink transmission, the UE can be scheduled with DCI format 0_1 when at least one SRS resource is configured in *SRS-ResourceSet* with *usage* set to "*nonCodebook*" [9].

4.2.6 Link Adaptation and Power Control

Power control is a mechanism where the transmit power of the downlink or uplink control or traffic channels are adjusted at the gNB or at UEs, based on instructions from the serving base station such that with minimal impact on the reliability of the downlink/uplink transmissions and throughput, the inter-user/inter-cell interference among users and base stations are reduced. Therefore, power control can be considered as a link adaptation mechanism that is utilized for interference mitigation in cellular systems.

While increasing the transmit power over a communication link has certain advantages such as higher SNR at the receiver, which reduces the BER and allows higher data rate and results in greater spectral efficiency as well as more protection against signal attenuation over fading channels, a higher transmit power; however, has several drawbacks including increased power consumption of the transmitting device, reducing the UE battery life, and increased interference to other users in the same or adjacent frequency bands. The following sections describe the power control algorithms that are incorporated in NR.

The NR provides uplink power control mechanisms to compensate the effects of path loss, shadowing, fast fading, and implementation loss. The uplink power control is further used to mitigate inter-cell and intra-cell interference, thereby enhancing the overall throughput and reducing the effective UE power consumption. The uplink power control includes open-loop and closed-loop power control. The base station transmits necessary power control information through transmission of power control messages. The parameters of the power control algorithm are optimized on a system-wide basis by the gNB and are broadcast periodically or trigged by events. The UE provides the necessary information through higher layer control messages to the serving gNB in order to enable uplink power control. The gNB can exchange necessary information with neighboring base stations through backhaul to support uplink power control to facilitate the handover process.

The power control scheme may not be effective in high mobility scenarios for compensating the effects of a fast fading channel due to variation of the channel impulse response. As a result, the power control is used to mitigate the distance-dependent path loss, shadowing, and implementation loss. The uplink power control takes into consideration the MIMO transmission mode and whether a single user or multiple users are supported on the same resource at the same time. The open-loop power control compensates the channel variations and implementation loss without requiring frequent interactions with the serving gNB. The UE can determine the transmit power based on the transmission parameters sent by the gNB, uplink channel quality, downlink CSI, or the interference knowledge obtained from downlink transmissions. The open-loop power control provides a coarse initial transmit power setting for the device before establishing connection with the base station. It is believed that rate control is more efficient than power control under certain conditions. Rate control in principle implies that the power amplifier is always transmitting at full power and therefore is efficiently utilized. On the other hand, power control often results in inefficient utilization of the power amplifier because the transmission power is often less than its maximum. In practice, the radio-link data rate is controlled by adjusting the modulation scheme and/or the channel coding rate. In good channel conditions, the value of E_b/N_0 at the receiver is high, and the main limitation of the data rate is the bandwidth of the radio link.

In such conditions, use of higher order modulation, for example, 16QAM 64QAM, or 256QAM together with a high coding rate, is more appropriate for link adaptation. Similarly, in the case of poor channel conditions, the use of QPSK and low-rate coding is preferred. Link adaptation by means of rate control is referred to as adaptive modulation and coding.

A power control mechanism takes into consideration the serving gNB target link SINR and/or interference level to other cells/sectors for mitigating inter-cell interference. In order to achieve the target SINR, the serving gNB path loss can be fully or partially compensated based on a trade-off between overall system throughput and cell-edge performance. The UE transmit power is adjusted in order to ensure the level of interference is less than the permissible interference level. The closed-loop power control, on the other hand, compensates channel variations through periodic power-control commands from the serving gNB. The base station measures the uplink CSI and interference level using uplink data and/or control channel transmissions and sends power control commands to the devices. Upon receiving the power control command from the gNB, the UE adjusts its uplink transmit power. The closed-loop power control is active during data and control channel transmissions. A UE is expected to maintain the transmit power density (i.e., total transmit power normalized by transmission bandwidth) for each data and control channel below a certain level that is determined by the maximum permissible power level for the UE, emission mask, and other regulatory constraints. In other words, when the number of active logical resource units assigned to a particular user is reduced, the total transmitted power must be reduced proportionally by the UE in the absence of any additional change of power control parameters.

When the number of resource blocks is increased, the total transmitted power must be proportionally increased such that the transmitted power level does not exceed the permissible power levels specified by 3GPP and the regulatory specifications [1–3]. For interference level control, the information about the current interference level of each gNB may be shared among the base stations via backhaul.

The (uplink) TPC in mobile communication systems is meant to balance the transmitted energy per bit in order to maintain the link quality corresponding to the minimum QoS requirements, to minimize interference to other users in the system, and to minimize the power consumption of the device. In achieving these goals, the power control has to adapt to the characteristics of the propagation channel, including path loss, shadowing, and fast fading, as well as overcoming interference from other users both within the same cell and in neighboring cells. The NR uplink power control is similar to LTE and is based on a combination of open-loop power control, including support for fractional path loss compensation, where the device estimates the uplink path loss based on downlink measurements and sets the transmit power accordingly, and closed-loop power control based on explicit power-control commands provided by the network. In practice, these power-control commands are determined based on prior measurements of the received uplink power. The main difference with the LTE control is the possibility of beam-based power control.

Uplink power control determines the power level for PUSCH, PUCCH, SRS, and PRACH transmissions. The ith PUSCH/PUCCH/SRS/PRACH transmission occasion is defined by slot index n_{slot} within a frame with system frame number SFN, the first symbol S within the slot, and the number of consecutive symbols L. For a PUSCH transmission on active uplink BWP k of carrier l of serving cell m and parameter set configuration with index j and PUSCH power control adjustment state with index u, a UE first calculates a linear value $\hat{P}^{klm}_{PUSCH}(i,j,q_d,u)$ of the transmit power $P^{klm}_{PUSCH}(i,j,q_d,u)$. If PUSCH transmission is scheduled by DCI format 0_1 and when *txConfig* in *PUSCH-Config* is set to "*codebook*", the UE scales the linear value by the ratio of the number of antenna ports with a non-zero PUSCH transmission power to the maximum number of SRS ports supported by the UE in one SRS resource. The UE divides the power equally among the antenna ports on which it transmits PUSCH with non-zero power. The PUSCH transmit power is calculated as follows [8]:

$$P^{klm}_{PUSCH}(i,j,q_d,u) = \min\left\{ P_{CMAX_{lm}}(i) P_{o\_PUSCH_{klm}}(j) + 10\log_{10}\left[2^{\mu}\mathrm{N}^{PUSCH}_{RB_{klm}}(i)\right] + \alpha_{klm}(j)PL_{klm}(q_d) + \Delta TF_{klm}(i) + f_{klm}(i,u) \right\} \text{ (dBm)}$$

where

- (i,j,q_d,u) denote transmission occasion, parameter set configuration index, reference signal index for the active downlink BWP, and PUSCH power control adjustment state index, respectively.
- $P_{CMAX_{lm}}(i)$ denotes the maximum permissible UE transmit power for carrier l, serving cell m, and in PUSCH transmission occasion i.
- $P_{O\_PUSCH_{klm}}(j) = P_{O\_NOMINAL\_PUSCH_{lm}}(j) + P_{O\_UE\_PUSCH_{klm}}(j);\ j \in \{0,1,\ldots,J-1\}$ is a parameter determined by RRC parameters *preambleReceivedTargetPower*, *msg3-DeltaPreamble*, *ConfiguredGrantConfig*, *p0-NominalWithGrant*, *p0-PUSCH-Alpha*, *P0-PUSCH-AlphaSet*, *SRI-PUSCH-PowerControl* as well as the SRI field in DCI format 0_0 and 0_1. The quantity P_O is provided as part of the power-control configuration and would typically depend on the target data rate but also on the noise and interference level experienced at the receiver.
- $N^{PUSCH}_{RB_{klm}}(i)$ denotes the bandwidth of PUSCH resource assignment expressed in number of resource blocks.
- $\alpha_{klm}(j)$ is a network-configurable parameter corresponding to fractional path loss compensation which is determined by RRC parameters *msg3-Alpha, ConfiguredGrantConfig, p0-PUSCH-Alpha, P0-PUSCH-AlphaSet, SRI-PUSCH-PowerControl* as well as the SRI field in DCI format 0_0 and 0_1. In the case of fractional path loss compensation ($\alpha < 1$), the path loss will not be fully compensated, and

the average received power will vary depending on the location of the device within the cell. In this case, the received power is lower for devices with higher path loss located at larger distances from the cell site. This must then be compensated by adjusting the uplink data rate. The advantage of fractional path loss compensation is reduced interference to neighboring cells, which is achieved at the expense of larger variation in the service quality, with reduced peak data-rate availability for devices closer to the cell edge.

- $PL_{klm}(q_d)$ is the downlink path loss estimate in dB calculated by the UE using reference signal index q_d for the active downlink BWP.
- $\Delta_{TF_{klm}}(i) = 10\log_{10}\left[(2^{BPRE \times K_s} - 1)\beta_{offset}^{PUSCH}\right]$ for $K_s = 1.25$ (provided by *deltaMCS*); otherwise $\Delta_{TF_{klm}}(i) = 0$ is related to the modulation scheme and channel coding rate used for the PUSCH transmission. The parameter bits per resource element (BPRE) is given by $BPRE = \sum_{r=0}^{C-1} K_r/N_{RE}$ for PUSCH with UL-SCH data and $BPRE = Q_m R/\beta_{offset}^{PUSCH}$ for CSI transmission in PUSCH without UL-SCH data wherein Q_m is the modulation order and R is the target code rate. This term models how the required received power varies when the number of information BPRE changes due to different modulation schemes and channel-coding rates.
- $f_{klm}(i,u)$ denotes PUSCH power control adjustment state which is given by $f_{klm}\left(i,u\right) = f_{klm}\left(i - i_0, u\right) + \sum_{v=0}^{C(D_i)-1} \delta_{PUSCH_{klm}}(v,u)$ wherein $\delta_{PUSCH}(.)$ is the power adjustment due to closed-loop power control. The power control commands can be sent to multiple devices by means of DCI format 2_2. Each power control command consists of 2 bits corresponding to four different update steps (−1, 0, +1, +3 dB). The reason for including 0 dB as an update step is that a power-control command is included in every scheduling grant, and it is desirable not to have to adjust the PUSCH transmit power for each grant.

The PUCCH power control follows the same principles as PUSCH power control with some minor differences. For PUCCH power control, there is no fractional path loss compensation ($\alpha = 1$). Furthermore, for PUCCH power control, the closed-loop power control commands are carried within DCI formats 1_0 and 1_1, which are used for downlink scheduling assignments rather than within uplink scheduling grants. This is partly due to the fact that PUCCH transmission is used to carry HARQ-ACKs in response to a downlink transmission and such downlink transmissions are typically associated with downlink scheduling assignments on PDCCH and the corresponding power control commands could be used to adjust the PUCCH transmit power prior to the transmission of HARQ-ACKs. Similar to PUSCH, power control commands can also be carried jointly to multiple devices by means of DCI format 2_2.

When the UE transmits PUCCH on active uplink BWP k of carrier l of serving cell m and PUCCH power control adjustment state with index u, it determines the PUCCH transmission power $P_{PUCCH_{klm}}(i, q_u, q_d, u)$ in the ith PUCCH transmission occasion as follows [8]:

$$P_{PUCCH_{klm}}(i, q_u, q_d, u) = \min\left\{P_{CMAX_{lm}}(i), P_{O\_PUCCH_{klm}}(q_u) + 10\log_{10}[2^{\mu} N_{RB_{klm}}^{PUCCH}(i)] + PL_{klm}(q_d) + \Delta_{F\_PUCCH}(F) + \Delta_{TF_{klm}}(i) + g_{klm}(i, l)\right\}$$

where

- $P_{CMAX_{lm}}(i)$ is the configured UE transmit power.
- $P_{O\_PUCCH_{klm}}(q_u) = P_{O\_NOMINAL\_PUCCH} + P_{O\_UE\_PUCCH}(q_u)$ whose parameters are provided by *p0-nominal* and *p0-PUCCH-Value*.
- $N_{RB_{klm}}^{PUCCH}(i)$ is the bandwidth of PUCCH resource assignment expressed in number of resource blocks.
- $PL_{klm}(q_d)$ is the downlink path loss estimate in dB calculated by the UE using reference signal resource index q_d.
- $\Delta_{F\_PUCCH}(F)$ is PUCCH format-dependent power adjustment parameter provided by *deltaF-PUCCH-f0* for PUCCH format 0, *deltaF-PUCCH-f1* for PUCCH format 1, *deltaF-PUCCH-f2* for PUCCH format 2, *deltaF-PUCCH-f3* for PUCCH format 3, and *deltaF-PUCCH-f4* for PUCCH format 4.
- $\Delta TF_{klm}(i)$ is the PUCCH transmission power adjustment component which is dependent on the PUCCH format and the corresponding transport parameters.
- $g_{klm}(i, u)$ denotes the PUCCH power control adjustment state.

The UE calculates the transmission power for PRACH as $P_{PRACH_{klm}}(i) = \min(P_{CMAX_{lm}}(i), P_{PRACH\text{-}Target_{lm}} + PL_{klm})$ where $P_{CMAX_{lm}}(i)$ is the configured maximum UE transmission power, $P_{PRACH\text{-}Target_{lm}}$ is the target PRACH reception power corresponding to *PREAMBLE_RECEIVED_TARGET_POWER* parameter provided via RRC signaling, and PL_{klm} is path loss estimated for the active uplink BWP k of carrier l of serving cell m. The path loss is estimated based on the downlink reference signal associated with the PRACH transmission on the active downlink BWP of the mth serving cell and it is calculated by the UE as [*reference signal power*] − [*higher layer filtered RSRP*] in (dBm). If the active downlink BWP is the initial downlink BWP and for the SS/PBCH block and CORESET multiplexing pattern 2 or 3, the UE determines PL_{klm} based on the SS/PBCH block associated with the PRACH transmission [8].

The UE can set its configured maximum output power $P_{CMAX_{lm}}$ for the lth carrier of the mth serving cell in each slot. The configured maximum output power $P_{CMAX_{lm}}$ is set within $P_{CMAX-Low_{lm}} \le P_{CMAX_{lm}} \le P_{CMAX-High_{lm}}$ where $P_{CMAX-Low_{lm}} = \min[P_{EMAX_m}\Delta T_{C_m}, (P_{PowerClass} - \Delta P_{PowerClass}) - \max(MPR_m + A\text{-}MPR_m + \Delta T_{IB_m} + \Delta T_{C_m} + \Delta T_{RX_{SRS}}, P\text{-}MPR_m)]$ and $P_{CMAX-High_{lm}} = \min(P_{EMAX_m}, P_{PowerClass} - \Delta P_{PowerClass})$. In the latter equation, P_{EMAX_m} is the value given by information element *P-Max* for the mth serving cell, and $P_{PowerClass}$ is the maximum UE power without taking into account the tolerance specified by 3GPP specifications [1,2].

The PRACH preamble transmission involves some uncertainty about the required transmit power. As a result, the PRACH preamble transmission includes a power-ramping mechanism where the preamble may be retransmitted with a transmit power that is increased in steps during each transmission attempt. The device selects the initial PRACH preamble transmit power based on estimates of the downlink path loss in combination with a target received preamble power configured by the network. The path loss should be estimated based on the received power of the SS/PBCH block that the device has acquired and has determined the RACH resources for preamble transmission. If no random-access response is received within a predetermined window, the device can assume that the preamble was not correctly received by the network. In that case, the device repeats the preamble transmission with an increased transmit power. This power ramping continues until a random-access response has been received or until a configurable maximum number of retransmissions are attempted. Under such condition the random-access transmission has failed [14].

When uplink beamforming is used, the uplink path loss estimate $PL_{klm}(q_d)$ is used to determine the transmit power. The latter path loss estimate includes the effect of beamforming gain of the uplink beam pair to be used for PUSCH transmission. Assuming beam correspondence, this can be achieved by estimating the path loss based on measurements on a downlink reference signal transmitted over the corresponding downlink beam pair. As the uplink beam used for the transmission pair may change between PUSCH transmissions, the device may have to perform multiple path loss estimates corresponding to different candidate beam pairs. During PUSCH transmission over a specific beam pair, the path loss estimate corresponding to that beam pair is used to determine PUSCH transmit power. This is enabled by the parameter q in the path loss estimate $PL_{klm}(q_d)$[14].

The gNB configures the UE with a set of downlink reference signals based on which the path loss is estimated. Each reference signal is associated with a specific value of q. To limit the number of path loss estimations, the UE is not required to perform more than four parallel path loss estimations corresponding to different directions. The network configures a mapping between possible SRI values, provided in the scheduling grant, and different values of q. When a PUSCH transmission is scheduled by a scheduling grant including SRI, the path loss estimate associated with that SRI is used for determining the transmit power for the scheduled PUSCH transmission [14].

In the preceding equation for PUSCH power control, the open-loop parameters $P_{0\_PUSCH_{klm}}(j)$ and $\alpha_{klm}(j)$ are associated with parameter j, suggesting that there are multiple open-loop parameter pairs that can be used for different types of PUSCH transmissions, for example, Msg3 on PUSCH, grant-free PUSCH transmission, and scheduled PUSCH transmission. However, there is also a possibility to have multiple pairs of open-loop parameter for scheduled PUSCH transmission, where the pair to use for a certain PUSCH transmission can be selected based on the SRI similar to the selection of path loss estimates. In practice, this implies that the open-loop parameters $P_{0\_PUSCH_{klm}}(j)$ and $\alpha_{klm}(j)$ will depend on the uplink beam [14].

For PUSCH transmissions, the device can be configured with different open-loop parameter pairs $P_{0\_PUSCH_{klm}}(j)$ and $\alpha_{klm}(j)$ corresponding to different values of parameter j, where $j=0$ is associated with Msg3 transmission and $j=1$ is used in the case of grant-free PUSCH transmission. Each possible value of the SRI that can be provided as part of the uplink scheduling grant is associated with one of the configured open-loop parameter pairs. When a PUSCH transmission is scheduled with a certain SRI included in the scheduling grant, the open-loop parameters associated with that SRI are used when determining the transmit power for the scheduled PUSCH transmission [8].

The other parameter in PUSCH power control equation is the power control adjustment state with index u which is related to the closed-loop mechanism. PUSCH power control allows two independent closed-loop processes associated with $u=0$ and $u=1$. The u value(s) are provided by *sri-PUSCH-ClosedLoopIndex* RRC parameter. If PUSCH transmission is scheduled by a DCI format 0_1 and if DCI format 0_1 includes an SRI field, the UE determines the u value that is mapped to the SRI field. This means that similar to the case for multiple path loss estimates and multiple open-loop parameter pairs, the selection of u indicates the selection of the closed-loop process which is associated to the SRI included in the scheduling grant [8].

As we mentioned earlier, the UE relies on downlink measurements to calculate the path loss and to determine the power control parameters. The gNB determines the downlink transmit EPRE. For SS-RSRP, SS-RSRQ, and SS-SINR measurements, the UE may assume that downlink EPRE is constant across the bandwidth. The UE may further assume that downlink EPRE is constant over SSS carried in different SS/PBCH blocks, and the ratio of SSS EPRE to PBCH DM-RS EPRE is 0 dB [8]. For CSI-RSRP, CSI-RSRQ, and CSI-SINR measurements, the UE may assume downlink EPRE of a port of CSI-RS resource configuration is constant across the configured downlink bandwidth and constant across all configured OFDM symbols. The downlink SS/PBCH SSS EPRE can be derived from the SS/PBCH downlink transmit power given by the parameter *SS-PBCH-BlockPower* provided by RRC signaling. The downlink SSS transmit power is defined as the linear average over the power contributions (in Watts) of all REs that carry the SSS within the operating system bandwidth. The downlink CSI-RS EPRE can be derived from the SS/PBCH block downlink transmit power given by the parameter *SS-PBCH-BlockPower* and CSI-RS power offset given by the parameter *powerControlOffsetSS* provided through RRC signaling. The downlink reference signal transmit power is defined as the linear average over the power contributions [in (W)] of the resource elements that carry the configured CSI-RS within the operating system bandwidth.

For the purpose of downlink power allocation, the ratio of PDSCH EPRE to DM-RS EPRE ($\rho_{DM\text{-}RS}$ dB]) is given in Table 4.33 for downlink DM-RS associated with PDSCH, which depends on the number of DM-RS CDM groups without data [9]. It can be shown that the

Table 4.33: Ratio of PDSCH EPRE to DM-RS EPRE [9].

Number of DM-RS CDM Groups without Data	DM-RS Configuration Type 1 (dB)	DM-RS Configuration Type 2 (dB)
1	0	0
2	− 3	− 3
3	−	− 4.77

Table 4.34: PT-RS EPRE to PDSCH EPRE per layer per resource element [9].

EPRE-Ratio $\rho_{PT\text{-}RS}$	Number of PDSCH Layers					
	1	2	3	4	5	6
0	0	3	4.77	6	7	7.78
1	0	0	0	0	0	0
2	Reserved					
3	Reserved					

DM-RS power scaling factor $\beta_{PDSCH}^{DM\text{-}RS}$ applied prior to DM-RS resource mapping is given by $\beta_{PDSCH}^{DM\text{-}RS} = 10^{(-\rho_{DM\text{-}RS}/20)}$.

When the UE is scheduled with a PT-RS port associated with the PDSCH, the ratio of PT-RS EPRE to PDSCH EPRE per layer per resource element for PT-RS port $\rho_{PT\text{-}RS}$ is given by Table 4.34. In that case, the PT-RS power scaling factor $\beta_{PT\text{-}RS}$ is given by $\beta_{PT\text{-}RS} = 10^{(\rho_{PT\text{-}RS}/20)}$; otherwise, it can be assumed that *epre-Ratio* is set to state "0" in Table 4.34.

References

3GPP *Specifications*[21]

[1] 3GPP TS 38.101-1, NR, User Equipment (UE) Radio Transmission and Reception; Part 1: Range 1 Standalone (Release 15), December 2018.

[2] 3GPP TS 38.101-2: NR, User Equipment (UE) Radio Transmission and Reception; Part 2: Range 2 Standalone (Release 15), December 2018.

[3] 3GPP TS 38.104, NR, Base Station (BS) Radio Transmission and Reception (Release 15), December 2018.

[4] 3GPP TS 38.133, NR, Requirements for Support of Radio Resource Management (Release 15), December 2018.

[5] 3GPP TS 38.202, NR, Services Provided by the Physical Layer (Release 15), December 2018.

[6] 3GPP TS 38.211, NR, Physical Channels and Modulation (Release 15), December 2018.

[21] 3GPP specifications can be accessed at the following URL: http://www.3gpp.org/ftp/Specs/archive/.

[7] 3GPP TS 38.212, NR, Multiplexing and Channel Coding (Release 15), December 2018.
[8] 3GPP TS 38.213, NR, Physical Layer Procedures for Control (Release 15), December 2018.
[9] 3GPP TS 38.214, NR, Physical Layer Procedures for Data (Release 15), December 2018.
[10] 3GPP TS 38.215, NR, Physical Layer Measurements (Release 15), December 2018.
[11] 3GPP TS 38.300, NR, NR and NG-RAN Overall Description, Stage 2 (Release 15), December 2018.
[12] 3GPP TS 38.321, NR, Medium Access Control (MAC) Protocol Specification (Release 15), December 2018.
[13] 3GPP TS 38.331, NR, Radio Resource Control (RRC), Protocol Specification (Release 15), December 2018.

Articles, Books, White Papers, and Application Notes

[14] E. Dahlman, S. Parkvall, 5G NR: The Next Generation Wireless Access Technology, Academic Press, August 2018.
[15] S. Ahmadi, LTE-Advanced: A Practical Systems Approach to Understanding 3GPP LTE Releases 10 and 11 Radio Access Technologies, Academic Press, November 2013.
[16] T.L. Marzetta, et al., Fundamentals of Massive MIMO, Cambridge University Press, December 2016.
[17] T.S. Rappaport, R.W. Heath Jr, Millimeter Wave Wireless Communications, Prentice Hall, September, 2014.
[18] S. Lin, D.J. Costello, Error Control Coding, second ed., Prentice Hall, June 2004.
[19] C.E. Shannon, W. Weaver, The Mathematical Theory of Communication, University of Illinois Press, September 1949.
[20] A. Papoulis, Probability, Random Variables and Stochastic Processes, fourth ed., McGraw Hill Higher Education, January 2002.
[21] T.M. Cover, J.A. Thomas, Elements of Information Theory, John Wiley & Sons, July 2006.
[22] J.G. Proakis, Digital Communications, fifth ed., McGraw-Hill, November 2007.
[23] F. Ademaj, et al., 3GPP 3D MIMO channel model: a holistic implementation guideline for open source simulation tools, EURASIP J. Wirel. Commun. Netw. 2016 (2016) 55.
[24] R.G. Gallager, Principles of Digital Communication, Cambridge University Press, March 2008.
[25] B. Mondal, et al., 3D channel model in 3GPP, IEEE Commun. Mag. 53 (3) (2015).
[26] F. Vook, 3GPP new radio at sub-6GHz: features and performance, in: IWPC Workshop on 5G NR Mobile Networks and User Equipment, January 2018.
[27] T.S. Rappaport, et al., Overview of millimeter wave communications for fifth-generation (5G) wireless networks—with a focus on propagation models, in: IEEE Transactions on Antennas and Propagation, Special Issue on 5G, November 2017.
[28] S. Cammerer, Spatially Coupled LDPC Codes, Institut für Nachrichtenübertragung. WebDemos. Available from: <http://www.inue.uni-stuttgart.de/lehre/demo.html>.
[29] A. Elkelesh, M. Ebada, Polar Codes, Institut für Nachrichtenübertragung. WebDemos. Available from: <http://www.inue.uni-stuttgart.de/lehre/demo.html>.
[30] 5G New Radio, ShareTechNote. Available from: <http://www.sharetechnote.com>.
[31] J. Campos, Understanding the 5G NR Physical Layer, Keysight Technologies, November 2017.
[32] MediaTek, 5G NR: A New Era for Enhanced Mobile Broadband, White paper, March 2018.
[33] E. Arikan, Channel polarization: a method for constructing capacity achieving codes for symmetric binary-input memoryless channels, IEEE Trans. Inf. Theory 55 (7) (2009) 3051−3073.
[34] E. Arikan, Systematic polar coding, IEEE Commun. Lett. 15 (8) (2011) 860−862.
[35] R.G. Maunder, The implementation challenges of polar codes, in: AccelerComm White Paper, February 2018.
[36] V. Bioglio, C. Condo, I. Land, Design of Polar Codes in 5G New Radio, Cornell University Library, 2019.
[37] F. Hamidi-Sepehr, et al., Analysis of 5G LDPC codes rate-matching design, in: IEEE Vehicular Technology Conference (VTC), June 2018.

[38] 3GPP TSG RAN WG1, R1-1608862, Polar Code Construction for NR, Huawei, HiSilicon, 2016.

[39] R.G. Gallager, MIT Press Classic Series Low-Density Parity-Check Codes, MIT Press, Cambridge, MA, 1963.

[40] F. Sabatier, Polar Coding Tutorial. Available from: <http://ipgdemos.epfl.ch/polarcodestutorial/index.html>.

[41] C. Berrou, A. Glavieux, P. Thitimajshima, Near Shannon limit error-correcting coding and decoding: turbo-codes, in: Proceedings of IEEE International Communications Conference (ICC), May 1993.

[42] M. Giordani, et al., A tutorial on beam management for 3GPP NR at mmWave frequencies, IEEE Commun. Surv. Tutorials (2018).

[43] H. Shariatmadari, et al., Fifth-generation control channel design, achieving ultra-reliable low-latency communications, IEEE Veh. Technol. Mag. 13 (2) (2018) 84–93.

[44] 5G New Radio: Introduction to the Physical Layer, White Paper, National Instruments, 2018.

[45] S.M. Razavizadeh, et al., Three-dimensional beamforming: a new enabling technology for 5G wireless networks, IEEE Signal Process Mag. 31 (6) (2014) 94–101.

[46] H. Ji, et al., Overview of full-dimension MIMO in LTE-Advanced Pro, IEEE Commun. Mag. 55 (2) (2017) 176–184.

[47] Y.-H. Nam, et al., Full dimension MIMO for LTE-advanced and 5G, in: Information Theory and Applications Workshop (ITA), February 2015.

[48] A. Alkhateeb, G. Leus, R.W. Heath, Multi-layer precoding: a potential solution for full-dimensional massive MIMO systems, IEEE Trans. Wireless Commun. 16 (9) (2017) 5810–5824.

[49] J. Jeon, NR wide bandwidth operations, IEEE Commun. Mag. 56 (3) (2018) 42–46.

[50] D.H.N. Nguyen, T. Le-Ngoc, MMSE precoding for multiuser MISO downlink transmission with non-homogeneous user SNR conditions, EURASIP J. Adv. Signal Process. 2014 (2014) 85.

[51] R.-A. Pitaval, et al., Overcoming 5G PRACH capacity shortfall by combining Zadoff-Chu and M-sequences, IEEE Int. Conf. Commun. (ICC) (2018).

[52] T.L. Marzetta, Massive MIMO: an introduction, Bell Labs Tech. J. 20 (2015) 11–22.

[53] F. Hamidi-Sepehr, et al., 5G NR PDCCH: design and performance, in: 2018 IEEE 5G World Forum (5GWF), July 2018.

[54] Y. Qi, et al., On the Phase Tracking Reference Signal (PT-RS) Design for 5G New Radio (NR), Cornell University Library, 2018.

[55] Y. Blankenship, et al., Channel coding in 5G new radio: a tutorial overview and performance comparison with 4G LTE, IEEE Veh. Technol. Mag. 13 (4) (2018) 60–69.

[56] M. Reil, G. Lloyd, Millimeter-wave beamforming: antenna arrays & characterization, White Paper, Rohde & Schwarz, 2016.

[57] E. Onggosanusi, et al., Modular and high-resolution channel state information and beam management for 5G new radio, IEEE Commun. Mag. 56 (3) (2018) 48–55.

[58] A.L. Swindlehurst, et al., Millimeter-wave massive MIMO: the next wireless revolution? IEEE Commun. Mag. 52 (9) (2014) 56–62.

[59] Y. Huang, et al., Multi-panel MIMO in 5G, IEEE Commun. Mag. 56 (3) (2018) 56–61.

[60] T.L. Marzetta, Noncooperative cellular wireless with unlimited numbers of base station antennas, IEEE Trans. Wireless Commun. 9 (11) (2010) 3590–3600.

[61] F. Rusek, et al., Scaling up MIMO: opportunities and challenges with very large arrays, IEEE Signal Process Mag. 30 (1) (2013) 40–60.

[62] E.G. Larsson, et al., Massive MIMO for next generation wireless systems, IEEE Commun. Mag. 52 (2) (2014) 186–195.

[63] L. Lu, et al., An overview of massive MIMO: benefits and challenges, IEEE J. Sel. Top. Signal Process. 8 (5) (2014) 742–758.

[64] W. Roh, et al., Millimeter-wave beamforming as an enabling technology for 5G cellular communications: theoretical feasibility and prototype results, IEEE Commun. Mag. 52 (2) (2014) 106–113.

[65] 3GPP TSG RAN WG1, R1-1800036, Summary of Remaining Issues on HARQ Management, Huawei, HiSilicon, 2018.
[66] 3GPP TSG RAN WG1, R1-1712546, Discussion on Codebook Subset Restriction for NR, Intel Corporation, 2017.
[67] 5G Toolbox, The MathWorks, Inc. Available from: <https://www.mathworks.com/help/5g/index.html>.
[68] H. Koorapaty, NR physical layer design: physical layer structure, numerology and frame structure, in: RWS-180007, Workshop on 3GPP Submission Towards IMT-2020, October 2018.
[69] Y. Kim, NR physical layer design: NR MIMO, in: RWS-180008, Workshop on 3GPP Submission Towards IMT-2020, October 2018.
[70] G. Noh, et al., DM-RS design and evaluation for 3GPP 5G new radio in a high speed train scenario, IEEE Global Commun. Conf, 2017.
[71] X. Lin, et al., 5G New Radio: Unveiling the Essentials of the Next Generation Wireless Access Technology, Cornell University Library, 2018.
[72] J. Liu, et al., Initial access, mobility, and user-centric multi-beam operation in 5G new radio, IEEE Commun. Mag. 56 (3) (2018) 35–41.
[73] Updated information on China's IMT-2020 Submission, IEEE ComSoc Technology Blog, October 2018 (https://techblog.comsoc.org/2018/10/19/updated-information-on-chinas-imt-2020-submission/).
[74] W.U. Yong, Self evaluation: enhanced mobile broadband (eMBB) evaluation results, in: RWS-180018, Workshop on 3GPP Submission Towards IMT-2020, October 2018.
[75] D.J. Love, R.W. Heath Jr., Grassmannian precoding for spatial multiplexing systems, in: Proceedings of the Allerton Conference on Communication Control and Computing, Monticello, 2003.
[76] N. Jindal, MIMO broadcast channels with finite-rate feedback, IEEE Trans. Inf. Theory 52 (11) (2006).

CHAPTER 5

New Radio Access RF and Transceiver Design Considerations

The NR RF transceiver characteristics are related to the frequency bands in which the 5G systems will be deployed. Due to wide range of the target frequency bands, spectrum flexibility was required for the new radio in order to operate in diverse spectrum allocations. While spectrum flexibility has been used in the previous generations of radio access technologies, it has become more important for NR development and deployment. Such spectrum flexibility is manifested as feasibility of deployment and resource allocations in frequency blocks of different sizes over an extremely wide range of contiguous or non-contiguous spectrum, both in the form of paired and unpaired frequency bands along with aggregation of different spectrum blocks within and across different bands. The NR has the capability to operate with mixed OFDM numerologies over the same or different RF carrier(s) and has relatively more flexibility compared to LTE in terms of frequency-domain scheduling and multiplexing of devices over the serving base station (BS) RF carrier(s). The use of OFDM waveform in NR provides the desired flexibility in terms of the size of the spectrum allocation and the instantaneous transmission bandwidth adaptation. The application of active antenna system (AAS) concept and multiple antennas in the base stations and the devices, which emerged during LTE development, has taken a giant leap in NR with the support of massive MIMO and control/data channel beamforming both in the existing LTE bands and in the new mmWave bands. Aside from physical layer design implications, the advent of the latter features significantly impact the analog/digital RF hardware system design/implementation including filters, amplifiers, data converters, antennas, etc.

In this chapter, we will discuss the new radio spectrum, RF characteristics, implementation considerations of the NR base stations and the devices as well as the hardware technologies that are used to implement various features of the new radio. In the course of this chapter, we will discuss various types of NR base stations and their external interfaces over which the RF requirements are defined. We further explore the conducted and over-the-air (OTA) RF requirements specified by 3GPP for testing and evaluating the performance of the NR base stations and devices.

5G NR. DOI: https://doi.org/10.1016/B978-0-08-102267-2.00005-1

5.1 NR Radio Parameters and Spectrum

The requirement for spectrum flexibility was a key driving factor for the adoption of OFDM-based technologies in 3GPP LTE which continues to be a major driver for the NR frequency planning and deployments. The need for diverse spectrum allocations in terms of spectral bands, operation bandwidths, duplex schemes (paired and unpaired spectrum) and multiple-access schemes emerged during the 3G and 4G deployments and has led to one of the most distinctive characteristics of 5G NR, which supports a large and diverse spectrum from 450 MHz to 52.6 GHz (and up to 100 GHz in future releases). The maximum frequency currently under study in ITU-R is 86 GHz (see Fig. 5.1). The NR supports operating bandwidths of 5 MHz to 3.2 GHz for both paired and unpaired spectrum as well as supplementary downlink (SDL) or supplementary uplink (SUL) carriers. The NR defines various frequency bands within the 5G spectrum. Although the boundaries of the NR frequency bands can vary in different countries and regions, it must be possible to efficiently allocate RF carriers at positions where the spectrum blocks are used with minimal spectrum wastage. This requires carefully defined channel raster for carrier allocation. There are also a number of other spectrum blocks being considered by regulatory bodies that were not initially considered by 3GPP specifications, which include 5925–7150 MHz in the United States and 5925–6425 MHz by CEPT[1] for unlicensed use as well as frequency range 64–86 GHz for extremely wideband applications [12]. Note that various parts of the 64–86 GHz range are allocated differently in different regions of the world. For instance, 64–71 GHz is set aside for unlicensed use in North America and is under consideration in CEPT (66–71 GHz); 66–76 and 81–86 GHz are under study in ITU-R as possible bands for IMT-2020 to be ratified during WRC-19 [12].

In general, the NR UEs do not receive or transmit using the full channel bandwidth of the gNBs, rather they can be assigned to what is referred to as bandwidth parts. While the concept does not have any direct RF implications, it is important to note that the gNB and the UE channel bandwidths are defined independently, and the device bandwidth capability does not have to match the gNB channel bandwidth. A unified frame structure is defined in NR that supports TDD, FDD, and half-duplex FDD operations. The duplex scheme is specifically defined for each operating band. Some bands are also designated as SDL or SUL bands to be used in FDD operation. Some of the frequency bands that have been identified for NR deployment are the existing ITU-R IMT[2] bands which may have already been accommodating 2G, 3G, and/or 4G [incumbent] deployments. In some regions, a number of

[1] The European Conference of Postal and Telecommunications Administrations (CEPT) was established in 1959 and is an organization where policy makers and regulators from 48 countries across Europe collaborate to harmonize telecommunication, radio spectrum, and postal regulations across Europe. The CEPT conducts its work through three autonomous business committees (ECC, Com-ITU and CERP).

[2] The term International Mobile Telecommunications (IMT) is the generic term used by the ITU community to designate broadband mobile systems. It encompasses IMT-2000, IMT-Advanced, and IMT-2020 collectively.

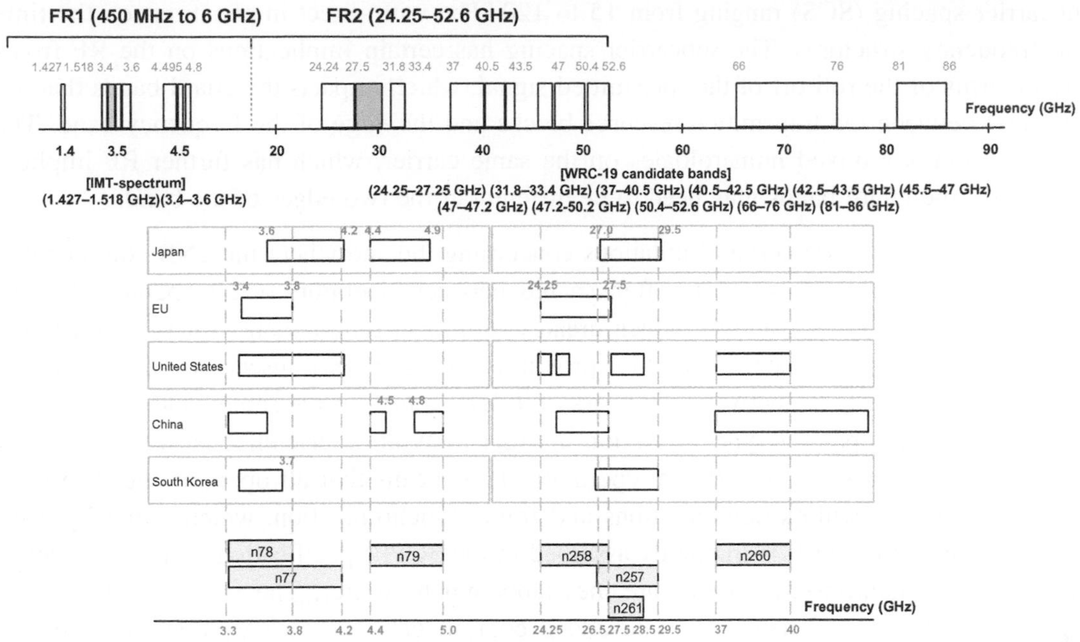

Figure 5.1
NR bands and 5G frequency bands under study in ITU-R WRC-19 [14,29].

frequency bands are designated and regulated as technology-neutral, which means that the coexistence between different technologies is a requirement for deployment. The capability to operate in this wide range of frequency bands for any cellular system, including NR, has direct implications on the RF requirements. The operators in the same band may deploy NR or other IMT technologies such as LTE [11]. Such coexistence requirements are typically specified within 3GPP (and/or ITU-R), but there may also be regional requirements defined by regulatory bodies that must be satisfied in order to be able to deploy the technology. Mobile operators have diverse spectrum holdings and allocations, which in many cases consist of a spectrum block that cannot accommodate one wideband carrier, or the allocation may be non-contiguous, consisting of multiple frequency blocks across multiple bands. In these cases, the NR specifications support carrier aggregation, where multiple carriers within a band, or in multiple bands, can be combined to create effectively wider transmission bandwidths. As shown in Fig. 5.1, the 5G spectrum in various countries is different which makes it difficult to harmonize spectrum and achieve global roaming.

The LTE and NR coexistence in the same spectrum, which is required for non-standalone and early deployments of 5G systems, makes it possible to deploy NR in the existing LTE frequency allocations. Since the co-channel NR and LTE carriers need to be aligned at subcarrier level, some restrictions are imposed on the NR channel raster in order to align the position of the NR and LTE carriers. The NR further supports multiple numerologies with subcarrier spacing (SCS) ranging from 15 to 120 kHz, with direct implications on the time and frequency structures. The subcarrier spacing has certain implications on the RF front-end in terms of the roll-off of the transmitted signal, which impacts the guard bands that are allocated between the transmitted resource blocks and the edge of the frequency band. The NR also supports mixed numerologies on the same carrier, which has further RF implications since the guard bands may need to be different at the two edges of the band.

In some cases, there are certain limitations concerning the site where the gNB equipment is deployed. The candidate sites are often shared between operators or an operator deploys multiple technologies at one site, which creates additional requirements for the gNB transmitters and receivers to operate in close proximity of other BS transceivers (e.g., blocking effects). The coexistence between operators of TDD systems in the same band is in general provided by inter-operator synchronization in order to avoid interference between downlink and uplink transmissions of different operators. This means that all operators need to have the same uplink/downlink configurations and frame synchronization, which is not by itself an RF requirement, but it is implicitly assumed in the 3GPP specifications. The RF requirements for unsynchronized systems are inevitably much stricter. The frequency bands are regionally defined, and new bands are added continuously for each generation of mobile systems, which means each new release of 3GPP specifications will incorporate additional bands. Using a release-independent principle, it is possible to design devices based on an early release of 3GPP specifications that support a frequency band added in a later release. The first set of NR bands is defined in 3GPP Rel-15, and additional bands will be added in a release-independent manner.

The NR can operate over a wide range of frequencies which include sub-6 GHz and above 6 GHz spectrum. The support of this extremely wide range of frequencies implies that the radio characteristics and operating parameters can significantly vary depending on the frequency band and the operating bandwidth; thus it is desirable to be able to configure the radio-related parameters such as subcarrier spacing, OFDM symbol length, cyclic prefix length, channel bandwidth according to the allocated frequency band. The NR specifies two frequency ranges, namely, FR1 and FR2 where each support certain OFDM numerologies. In FR1, which covers 450 MHz to 6 GHz, the subcarrier spacings that can be used for data transmission are 15, 30, and 60 kHz. The FR2 supports subcarrier spacings of 60 and 120 kHz. The UEs are required to support all subcarrier spacings except 60 kHz in FR1. The ratio of cyclic prefix length over OFDM symbol length is the same for all supported subcarrier spacings. Therefore, as the subcarrier spacing increases, the cyclic prefix length decreases, resulting in more susceptibility of NR signals to multipath delay distortion and coverage reduction. The use of larger subcarrier spacings further increases the NR signal tolerance to Doppler shift. It must be noted that the effect of oscillator phase noise is more pronounced in FR2 bands, resulting in inter-carrier interference. The use of larger subcarrier spacing helps mitigate the effects of phase noise. The subcarrier spacing is configured by the network using higher layer signaling. The most appropriate subcarrier spacing can be configured according to the deployment scenario. The FR2 signal propagation is characterized with less diffraction, higher penetration loss, and in general higher path loss. This can be compensated by incorporating more antenna elements at the transmit and receive sides, narrower beams with higher antenna gains, leading to massive MIMO systems.

LTE initially supported channel bandwidths up to 20 MHz, and later through the use of carrier aggregation, it was able to operate in up to 100 MHz bandwidths. The number of component carriers was later increased to 32 in 3GPP Rel-13, resulting in the maximum operating bandwidth of up to 640 MHz. The Rel-15 NR supports inter-band and intra-band contiguous and non-contiguous carrier aggregation as well as dual connectivity which allows simultaneous communication to NR and LTE base stations. Due to much wider channel bandwidths targeted in NR, the maximum channel bandwidth per component carrier has been increased to 100 MHz in FR1 (using SCSs of 30 and 60 kHz) and 400 MHz in FR2 (using 120 kHz SCS). Support of 400 MHz bandwidth in FR2 is optional; however, the NR devices are required to support up to 200 MHz in FR2. The NR new bands and the corresponding channel bandwidths and SCSs are shown in Table 5.1 [29].

The LTE spectrum utilization (defined as the ratio of the transmission bandwidth over the channel bandwidth) was 90% and guard bands were provisioned on both sides of the transmission bandwidth to protect communication in the adjacent channels by limiting the adjacent channel interference. In NR, the spectrum utilization has been increased to 98% (depending on transmission bandwidth), as a result of time and frequency-domain preprocessing of the OFDM signal (i.e., windowing and spectral shaping). While the size of the guard bands have been reduced in NR due to increased spectrum utilization, the out-of-band (OOB) emission and the permissible leakage power requirements have remained the same as LTE.

Table 5.1: NR new bands and channel bandwidths ("x" configurable and "*" optional) [5,29].

Frequency Range	Frequency Band		Subcarrier Spacing (kHz)	Component Carrier Bandwidth (MHz)														
	Band Number	Frequency Band (GHz)		5	10	15	20	25	30	40	50	60	70	80	90*	100	200	400*
FR1	n77/ n78	3.7	15		x	x	x			x	x						—	—
			30		x	x	x			x	x	x		x	x	x	—	—
			60		x	x	x			x	x	x		x	x	x	—	—
	n79	4.5	15							x	x						—	—
			30							x	x			x		x	—	—
			60							x	x			x		x	—	—
FR2	n257/ n258	24–40	60	—	—	—	—	—	—	—	x	—	—	—	—	x	x	
	n260/ n261		120	—	—	—	—	—	—	—	x	—	—	—	—	x	x	x

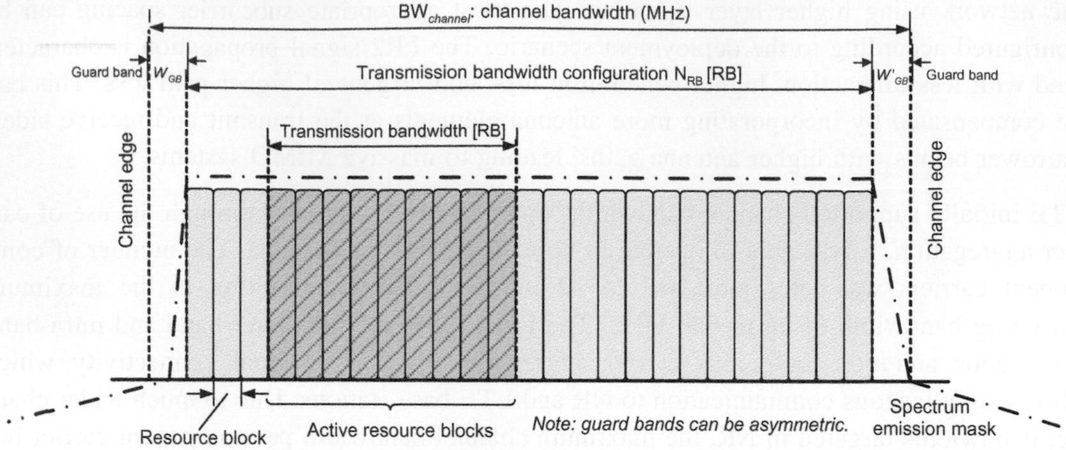

Figure 5.2
Illustration of channel bandwidth, transmission bandwidth, and guard bands [5].

The channel bandwidth of a gNB ($BW_{channel}$) supports transmission on a single (NR) uplink or downlink RF carrier. Different UE channel bandwidths may be supported within the same gNB channel bandwidth for bidirectional communication with the UE. The relative location of the UE channel bandwidth is flexible and within the gNB channel bandwidth. The gNB is able to transmit/receive to/from one or more UE bandwidth parts that are smaller than or equal to the total number of carrier resource blocks on the RF carrier. The relationship between the channel bandwidth, guard band, and the transmission bandwidth is shown in Fig. 5.2. The transmission bandwidth configuration in terms of the number of physical resource blocks (N_{RB}) for each *BS channel bandwidth* and SCS was given in Table 3.9, for FR1 and FR2. The minimum guard band for each *BS channel bandwidth* and SCS is shown in Table 5.2 for FR1 and FR2. The minimum guard band shown for SCS of 240 kHz is only applicable when an SS/PBCH block (SSB) with SCS of 240 kHz is placed adjacent to the edge of the *BS channel bandwidth* within which the SSB is located; otherwise, 240 kHz subcarrier spacing is not used for any other configurations. The number of RBs configured within any *BS channel bandwidth* will guarantee that the minimum guard band shown in Table 5.2 is satisfied.

If multiple numerologies are multiplexed over the same OFDM symbol, the minimum guard band on each side of the carrier is the guard band applied at the configured *BS channel bandwidth* for the numerology that is transmitted/received adjacent to the guard band. In FR1, if multiple numerologies are multiplexed over the same OFDM symbol and the *BS channel bandwidth* is greater than 50 MHz, then the guard band that is inserted adjacent to 15 kHz SCS is the same as the guard band defined for 30 kHz SCS for the same *BS channel bandwidth*. In FR2, if multiple numerologies are multiplexed over the same OFDM symbol

Table 5.2: Minimum guard band (kHz) for FR1 and FR2 [5].

Subcarrier Spacing (kHz)	Bandwidth (MHz)																
	FR1													FR2			
	5	10	15	20	25	30	40	50	60	70	80	90	100	50	100	200	400
15	242.5	312.5	382.5	452.5	522.5	592.5	552.5	692.5	N/A	N/A	N/A	N/A	N/A	N/A	N/A	N/A	N/A
30	505	665	645	805	785	945	905	1045	825	965	925	885	845	N/A	N/A	N/A	N/A
60	N/A	1010	990	1330	1310	1290	1610	1570	1530	1490	1450	1410	1370	1210	2450	4930	N/A
120	N/A	N/A	N/A	N/A	N/A	N/A	N/A	N/A	N/A	N/A	N/A	N/A	N/A	1900	2420	4900	9860
240	N/A	N/A	N/A	N/A	N/A	N/A	N/A	N/A	N/A	N/A	N/A	N/A	N/A	N/A	3800	7720	215,560

and the *BS channel bandwidth* is larger than 200 MHz, then the guard band inserted adjacent to 60 kHz SCS is the same as the guard band defined for 120 kHz SCS for the same *BS channel bandwidth* [5].

For each *BS channel bandwidth* and each numerology, the *BS transmission bandwidth configuration* is required to satisfy the minimum guard band requirement. The common resource blocks (see Section 3.7.1) are specified for each numerology and the starting point of the associated transmission bandwidth configuration on the common resource block grid for a given channel bandwidth is indicated by an offset to reference point A in the unit of the numerology. For each numerology, all *UE transmission bandwidth configurations* that are indicated to the UEs by the serving gNB through higher layer parameter *carrierBandwidth* must be located within the *BS transmission bandwidth configuration* [5].

In carrier aggregation scenarios, the transmission bandwidth configuration is defined per component carrier. An *aggregated BS channel bandwidth* and *guard bands* are defined for intra-band contiguous carrier aggregation. The *aggregated BS channel bandwidth* ($BW_{channel\_CA}$) is defined as $BW_{channel\_CA} = f_{edge\text{-}high} - f_{edge\text{-}low}$ (MHz). As shown in Fig. 5.3, the lower bandwidth edge $f_{edge\text{-}low}$ and the upper bandwidth edge $f_{edge\text{-}high}$ of the aggregated BS channel bandwidth are used as frequency reference points for transmitter and receiver requirements, which are defined as $f_{edge\text{-}low} = f_{C\text{-}low} - f_{offset-low}$ and

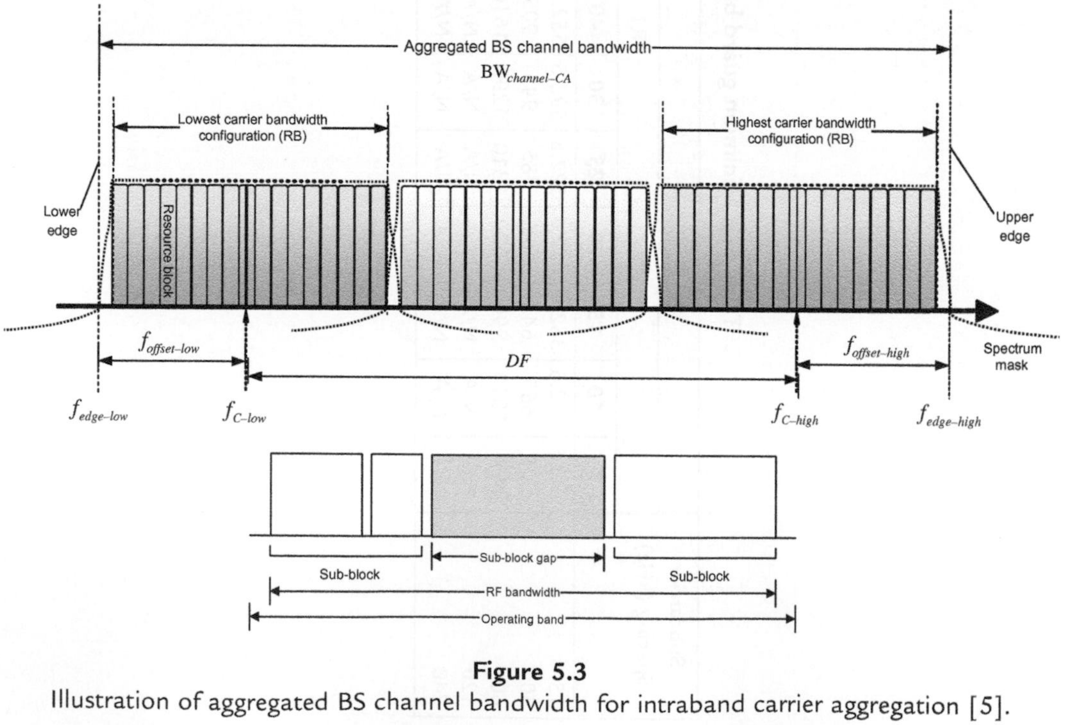

Figure 5.3
Illustration of aggregated BS channel bandwidth for intraband carrier aggregation [5].

$f_{edge\text{-}high} = f_{C\text{-}high} + f_{offset\text{-}high}$. The lower and upper frequency offsets depend on the transmission bandwidth configurations of the lowest and highest assigned edge component carrier and are defined as $f_{offset-low} = (12N_{RB-low} + 1)SCS_{low}/2 + W_{GB}$ (MHz) and $f_{offset\text{-}high} = (12N_{RB\text{-}high} - 1)\mathrm{SCS}_{high}/2 + W_{GB}$(MHz), where W_{GB} denotes the lower or upper minimum guard band defined for the lowest and highest assigned component carrier. In the latter equations, the parameters $N_{RB\text{-}low}$ and $N_{RB\text{-}high}$ are the transmission bandwidth configurations for the lowest and highest assigned component carrier, SCS_{low} and SCS_{high} denote the subcarrier spacing for the lowest and highest assigned component carrier, respectively [5]. It can be shown that the minimum guard band can be defined as $W_{GB} = \mathrm{BW}_{channel}/2 - \mathrm{SCS}(12N_{RB} + 1)/2$, where N_{RB} is the maximum number of resource blocks that can fit into the *BS channel bandwidth* and SCS denotes the subcarrier spacing. Note that an extra SCS/2 guard band is inserted on each side of the carrier due to the relation to the RF channel raster, which has a subcarrier-level granularity and is defined independent of the actual spectrum blocks. Therefore, it may not be possible to place an RF carrier exactly in the center of a spectrum block and an extra guard band would be required to ensure that the RF requirements can be met [10].

Some spectrum allocations may consist of fragmented blocks of spectrum. In the intra-band non-contiguous case, the BS transmits and receives over an RF bandwidth that is split in two (or more) separate subblocks with a subblock gap in between (see Fig. 5.3). A subblock is defined as one contiguous allocated block of spectrum for transmission and reception by the same BS. The lower subblock edge of the subblock bandwidth ($\mathrm{BW}_{channel\text{-}block}$) is defined as $f_{edge\text{-}block\text{-}low} = f_{C\text{-}block\text{-}low} - f_{offset\text{-}low}$. The upper subblock edge of the subblock bandwidth is defined as $f_{edge\text{-}block\text{-}high} = f_{C\text{-}block\text{-}high} + f_{offset\text{-}high}$. The subblock bandwidth $\mathrm{BW}_{channel\text{-}block}$ is defined as $\mathrm{BW}_{channel\text{-}block} = f_{edge\text{-}block\text{-}high} - f_{edge\text{-}block\text{-}low}$ (MHz). The lower and upper frequency offsets $f_{offset\text{-}block\text{-}low}$ and $f_{offset\text{-}block\text{-}high}$ depend on the transmission bandwidth configurations of the lowest and highest assigned edge component carriers within a subblock which are defined as $f_{offset\text{-}block\text{-}low} = (12N_{RB\text{-}low} + 1)\mathrm{SCS}_{low}/2 + W_{GB\text{-}low}$ (MHz) and $f_{offset\text{-}block\text{-}high} = (12N_{RB\text{-}high} - 1)\mathrm{SCS}_{high}/2 + W_{GB\text{-}high}$ (MHz), where $N_{RB\text{-}low}$ and $N_{RB\text{-}high}$ are the transmission bandwidth configurations for the lowest and highest assigned component carrier within a subblock, respectively. In the latter equations, SCS_{low} and SCS_{high} denote the subcarrier spacing for the lowest and highest assigned component carrier within a subblock, respectively; $W_{GB\text{-}low}$ and $W_{GB\text{-}high}$ are the minimum guard bands for the lowest and highest assigned component carriers, respectively. The subblock gap size between two consecutive subblocks W_{gap} is defined as $W_{gap} = f_{edge-block-low(n+1)} - f_{edge-blockn-high(n)}$ (MHz). Because the subblock gap starts from the inner edge of the channel bandwidth and not the center of the channel bandwidth, the subblock gap width is independent of the component carriers' channel bandwidth [5].

The frequency spacing between RF carriers depends on the deployment scenario, the size of the frequency block available and the *BS channel bandwidths*. The nominal channel spacing *DF* between two adjacent NR carriers is defined based on the frequency range and channel raster. For NR FR1 operating bands with 100 kHz channel raster, $DF = \left(BW_{channel(1)} + BW_{channel(2)}\right)/2$. For NR FR1 operating bands with 15 kHz channel raster, $DF = \left(BW_{channel(1)} + BW_{channel(2)}\right)/2 + \{-5, 0, 5\}$ kHz. For NR FR2 operating bands with 60 kHz channel raster, $DF = \left(BW_{channel(1)} + BW_{channel(2)}\right)/2 + \{-20, 0, 20\}$ kHz where $BW_{channel(1)}$ and $BW_{channel(2)}$ are the *BS channel bandwidths* of the two NR RF carriers. The channel spacing can be adjusted depending on the channel raster in order to optimize the system performance in a particular deployment scenario [5].

The channel spacing between adjacent component carriers for intra-band contiguously aggregated carriers is a multiple of least common multiple of channel raster and subcarrier spacing. The nominal channel spacing DF between two adjacent aggregated NR carriers is defined as follows [5]:

- For NR operating bands with 100 kHz channel raster

$$DF = \left\lfloor \frac{\left(BW_{channel(1)} + BW_{channel(2)} - 2\left|W_{GB_{channel(1)}} - W_{GB_{channel(2)}}\right|\right)}{0.6} \right\rfloor 0.3\ \text{(MHz)}$$

- For NR operating bands with 15 kHz channel raster

$$DF = \left\lfloor \frac{BW_{channel(1)} + BW_{channel(2)} - 2\left|W_{GB_{channel(1)}} - W_{GB_{channel(2)}}\right|}{0.015 \times 2^{n+1}} \right\rfloor 0.015 \times 2^{n}\ \text{(MHz)}$$

  and $n = \max(m_1, m_2)$

- For NR operating bands with 60 kHz channel raster

$$DF = \left\lfloor \frac{BW_{channel(1)} + BW_{channel(2)} - 2\left|W_{GB_{channel(1)}} - W_{GB_{channel(2)}}\right|}{0.06 \times 2^{n+1}} \right\rfloor 0.06 \times 2^{n}\ \text{(MHz)}$$

  and $n = \max\left(\mu_1, \mu_2\right) - 2$

In the above expressions, $BW_{channel(1)}$ and $BW_{channel(2)}$ represent the *BS channel bandwidths* of the two NR component carriers with values in MHz; $W_{GB_{channel(i)}}$ is the minimum guard band of the *i*th channel; μ_1 and μ_2 denote the subcarrier spacing configurations of the component carriers. The channel spacing for intra-band contiguous carrier aggregation can be adjusted to any multiple of least common multiple of the channel raster and SCS less than the nominal channel spacing to optimize performance in a particular deployment scenario. For intra-band non-contiguous carrier aggregation, the channel spacing between two NR component carriers in different subblocks must be larger than the nominal channel spacing [5].

5.2 Base Station Transceiver RF Characteristics and Requirements

5.2.1 General Base Station RF Requirements

The RF requirements in general are defined at the BS antenna connector. Those requirements are referred to as conducted requirements and are typically specified as absolute or relative power levels measured at the antenna connector. The OOB emission limits are often defined as conducted requirements. In active antenna systems, the RF requirements are defined as radiated requirement, which are measured over the air in the far-field of the antennas; thus they include the antenna patterns and directivity effects [1]. In OTA test procedures, the spatial characteristics of the BS including the antenna system are evaluated. In base stations equipped with AAS, where the active parts of the transceiver are integrated with the antenna system, it is not always possible to conduct the measurements at the antenna connector. For this reason, 3GPP Rel-13 specified the RF requirements for the AAS base stations in a set of separate RF specifications that are applicable to LTE and the previous generations. The radiated RF requirements and OTA testing are being specified for the new radio for the FR1 and FR2 bands, which have borrowed a large portion of the previously developed AAS specifications. Note that the term AAS is not used within the NR base station RF specifications, rather the requirements are specified for different BS types.

The AAS-type BS requirements are based on the generalized AAS BS radio architecture shown in Fig. 5.4 [5]. The architecture consists of a transceiver unit array that is connected to a composite antenna structure that contains a radio distribution network and an antenna array. The transceiver unit array comprises a number of transmitter and receiver units, which are connected to the composite antenna structure via a number of connectors on the transceiver array boundary (TAB). The TAB connectors correspond to the antenna connectors on a non-AAS BS and serve as a reference point for the conducted requirements. The radio distribution network is a passive unit which distributes or aggregates the transmitter outputs or the receiver inputs to the corresponding antenna elements, respectively. It must be noted that the actual implementation of an AAS BS may be different in terms of the physical location of different parts, array geometry, type of antenna elements, etc. Based on the architecture shown in Fig. 5.4, two types of requirements can be defined. Conducted requirements are defined for each RF characteristic at an individual or a group of TAB connectors. The conducted requirements are defined such that they are equivalent to the corresponding conducted requirement of a non-AAS BS, which implies that the performance of the system or the impact on other systems is expected to be the same. Radiated requirements, on the other hand, are defined based on OTA measurements conducted in the far-field of the antenna system. Since the spatial direction becomes relevant in this case, it is detailed for each requirement how it applies. The radiated requirements are defined with reference to a radiated interface boundary (RIB) in the far-field region of the antenna array [5,10].

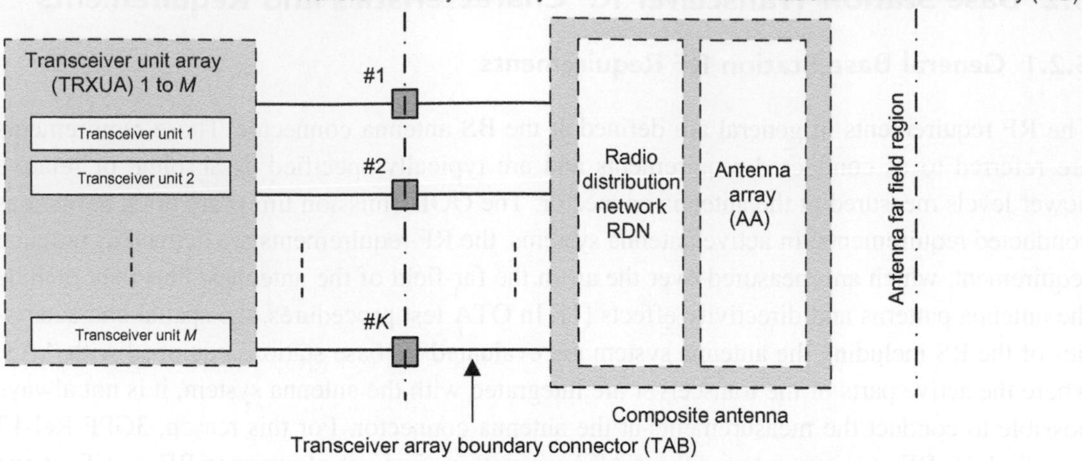

Figure 5.4
Radiated and conducted reference points for BS type 1-H [5]. *BS*, Base station.

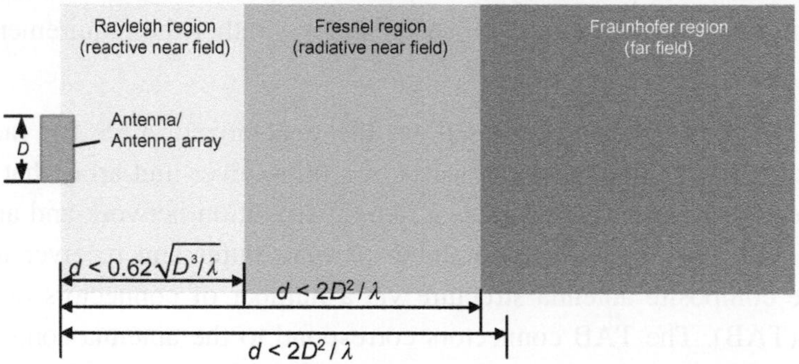

Figure 5.5
Illustration of the radiation regions of an antenna element/array.

In order to determine an appropriate point in the antenna array far-field region for conducting the OTA measurements, one must understand the radiation characteristics of the antenna systems and be able to define the minimum distance $d_{\min}$ in the far-field radiation region of an antenna array. In the theory of electromagnetics and antennas, the radiation regions surrounding an antenna element/array can be divided into three regions as follows (see Fig. 5.5):

- *Reactive near-field region*: In the immediate vicinity of the antenna, we have the reactive near-field. In this region, the fields are predominately reactive fields, which means the electric (E) and the magnetic (H) fields are out of phase by 90 degrees relative to each other. Note that for propagating or radiating fields, the fields are orthogonal and are in-phase. The boundary of this region is typically given as $d < 0.62\sqrt{D^3/\lambda}$, where

D is the maximum linear dimension of an antenna element/array, λ denotes the wavelength, and d is the distance from the antenna(s).

- *Radiating near-field (Fresnel) region*: The radiating near-field or Fresnel region is the region between the near- and far-fields. In this region, the reactive fields are not dominant, and the radiating fields begin to emerge. However, unlike the far-field region, the shape of the radiation pattern may vary noticeably with distance. This region is often identified by $0.62\sqrt{D^3/\lambda} < d < 2D^2/\lambda$. Note that depending on the values of d and the wavelength, this field may or may not exist.
- *Far-field (Fraunhofer) region*: The far-field is a region relatively far from the antenna element/array, where the shape of radiation pattern does not vary with distance, although the fields are attenuated proportional to d^{-1}, the power density degrades by d^{-2}. This region is dominated by radiated fields, with the E- and H-fields in a plane orthogonal to each other and to the direction of propagation which is the main characteristic of plane waves. If the maximum linear dimension of an antenna is D, the following three conditions must all be satisfied so that a point can be in the far-field region: $d > 2D^2/\lambda$, $d \gg D$ and $d \gg \lambda$. The first two inequalities ensure that the power radiated in a given direction from distinct parts of the antenna is approximately parallel. This helps ensure that the fields in the far-field region can be characterized as plane waves. Near a radiating antenna, there are reactive fields that typically have the E-fields and H-fields diminish with distance as d^{-2} or d^{-3}. The third inequality ensures that these near-fields are disappeared, and we are left with the radiating fields, which diminish with distance as d^{-1}.

The NR base stations can be categorized into three main configurations from RF requirements perspective as follows [5]:

- *BS type 1-C*: An NR BS operating in FR1 which is connected to the antennas via coaxial cables. The RF requirements for this BS type only consist of conducted requirements defined at individual antenna connectors. For BS type 1-C, the requirements are applied at the BS antenna connector (port A). If any external apparatus such as an amplifier, a filter, or the combination of such devices is used, the RF requirements apply at the far end antenna connector (port B).
- *BS type 1-H*: An NR BS operating in FR1 consisting of an integrated AAS with RF transceiver connected to antennas using a TAB connector. The requirement set for this BS type consists of conducted requirements defined at individual TAB connectors and OTA requirements defined at the RIB. For BS Type 1-H, the requirements are defined for two points of reference, signified by radiated requirements and conducted requirements.
- *BS type 1-O*: A connector-less AAS-type NR BS operating in FR1 whose RF requirement set only consist of OTA requirements defined at the RIB.
- *BS type 2-O*: A connector-less AAS-type NR BS operating in FR2 whose RF requirement set only consists of OTA requirements defined at the RIB.

For BS Type 1-O and BS Type 2-O, the radiated characteristics are defined over the air, where the operating band–specific radiated interface is referred to as the RIB. The radiated requirements are also referred to as OTA requirements. The (spatial) characteristics in which the OTA requirements apply are detailed for each requirement. Fig. 5.6 illustrates the NR BS configurations and the reference points for conducting connected or OTA measurements. Note that one of the RF configurations for FR1 does not require connectors between the RF transceivers and antennas according to BS Type 1-O; thus smaller equipment size and improved power efficiency can be expected relative to the LTE transceivers with AAS specified in 3GPP Rel-13. While the operation in FR2 has the advantage of wideband transmission in high-frequency bands in terms of RF configuration, the higher frequencies result in larger power losses at the connectors and cables; increased path loss and reduced coverage due to lower power density over wider channels. Therefore, higher antenna gains are necessary to compensate for the losses and to maintain a certain link budget and coverage. Since it would be more difficult to design and implement RF signal transceivers and antennas with high density in FR2, if conventional RF configuration with connectors is used, BS type 2-O RF configuration without connectors is defined for FR2 operation. The BS Type 2-O would allow implementation of beamforming over wide channel bandwidths to maintain coverage and to achieve high spectral efficiencies [29].

The RF performance requirements for BS types 1-C and 1-H are based on LTE-advanced specifications [11], when NR radio parameters are applied. However, BS types 1-O and 2-O have integrated radio transceivers and antennas with no connectors to conduct measurements; thus OTA specifications have been extended such that in the overall RF performance specifications, a reference point in the radiated space, referred to RIB, can be defined. The output power for various BS classes and types is shown in Table 5.3. It must be noted that the factor 10 log($N_{TXU-counted}$) is used to derive the rated carrier output power from output power per TAB connector for BS Type 1-H and no upper limits for output power has been specified for BS Type 2-O in 3GPP Rel-15.

In addition to the equivalent isotropic radiated power (EIRP) and the equivalent isotropic sensitivity (EIS) including the antenna characteristics in the beam direction, which were specified in 3GPP Rel-13 LTE, the total radiated power (TRP) is introduced as a new metric in NR RF specifications (Fig. 5.7). The TRP definition makes it possible to specify OTA requirements for power-related RF performance requirements such as the BS output power and spurious emissions. Fig. 5.7 illustrates the visualization of the EIRP, EIS, and TRP definitions.

The main BS RF performance specifications of the LTE and NR in FR1 and FR2 have been summarized and compared in Table 5.4. The RF specifications in FR1 are based on LTE specifications with the maximum channel bandwidth of 100 MHz. In FR2, the NR radio specifications support wider bands, lower latency, and faster response with maximum

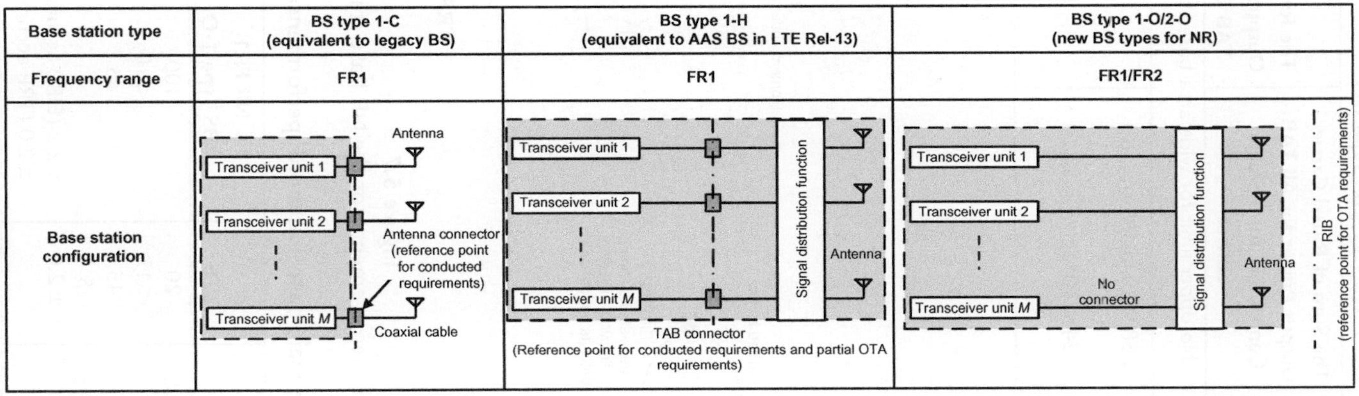

Figure 5.6
NR base station configurations [5,29].

Table 5.3: Output power per base station class and type (dBm) [5,14].

	BS Type 1-C	BS Type 1-H		BS Type 1-O
	The Rated Carrier Output Power per Antenna Connector	The Sum of Rated Carrier Output Power for All TAB Connectors for a Single Carrier	The Rated Carrier Output Power per TAB Connector	Rated Carrier TRP Output Power Declared per RIB
Wide area BS	No upper limit for wide area base station			
Medium range BS	≤ 38	$\leq 38 + 10\log(N_{TXU\text{-}counted})$	≤ 38	≤ 47
Local area BS	≤ 24	$\leq 24 + 10\log(N_{TXU\text{-}counted})$	≤ 24	≤ 33

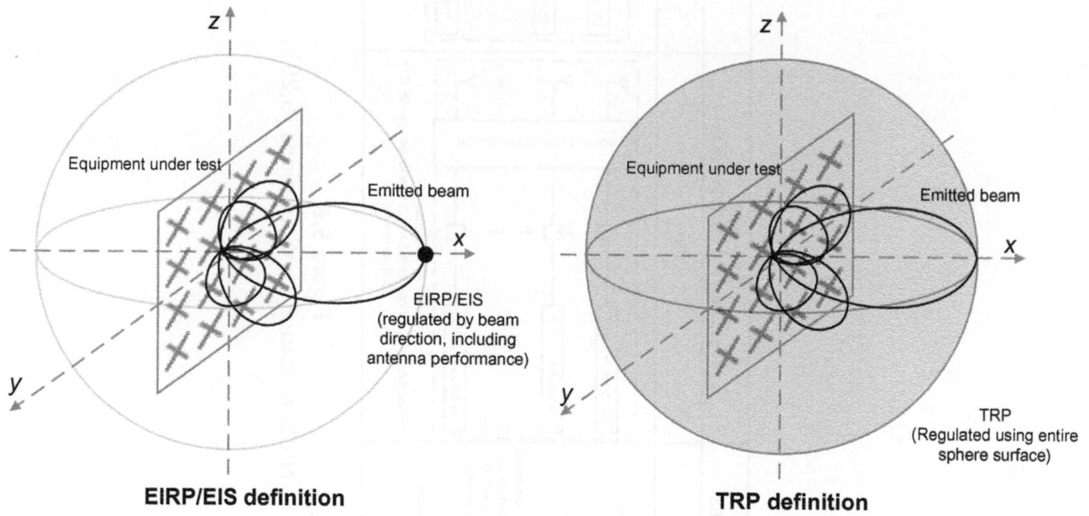

Figure 5.7
Illustration of the RF performance requirements for the NR base station and mobile station [29].

Table 5.4: Comparison of main base station RF performance specifications [5,29].

	LTE	NR FR1 BS Type 1-O	NR FR2 BS Type 2-O
Maximum channel bandwidth (MHz)	20	100	400
Transmitter transient period (μs)	<17	<10	<3
ACLR (dB)	45	45	28
NF (dB)	5	5	10
Transmit power deviation (dB)	± 2.0	± 2.2 (EIRP accuracy) ± 2.0 (TRP accuracy)	± 3.4 (EIRP accuracy) ± 3.0 (TRP accuracy)

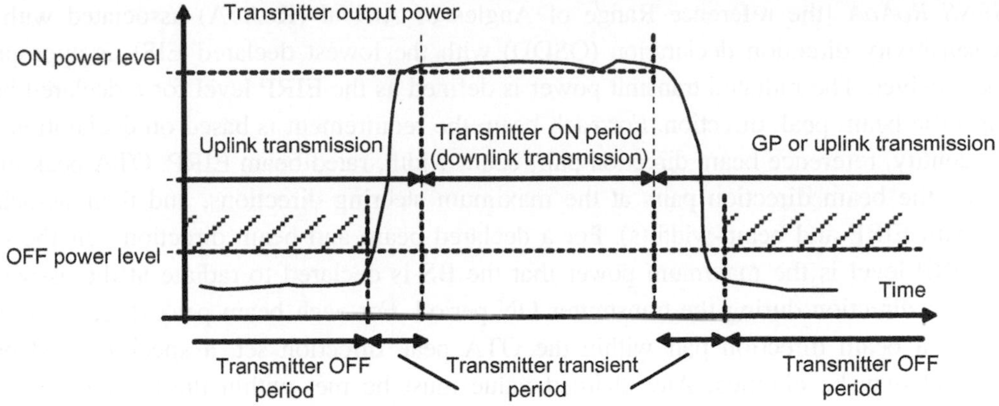

Figure 5.8
Illustration of transmitter transient period [5].

channel bandwidth of up to 400 MHz and transmitter transient period[3] of 3 μs, which reflects the ON/OFF switching time for a TDD system (see Fig. 5.8). The specifications further take into account the efficiency/performance degradation of gNB RF components in the mmWave bands, which are compensated by beamforming and high-gain antenna arrays, and by relaxing certain radio requirements such as adjacent channel leakage ratio (ACLR) in the transmitter side and noise figure (NF) in the receiver side. The transmit power fluctuation is controlled by TRP accuracy for per-carrier total power deviation and EIRP accuracy, accounting for the antenna performance.

5.2.2 Conducted and Radiated Base Station Transceiver Characteristics

The radiated RF requirements for the NR devices and base stations are primarily derived from the corresponding conducted RF requirements. Unlike the conducted requirements, the radiated requirements are inclusive of the antennas, thus the emission levels such as BS output power and unwanted emissions are defined either by incorporating the antenna gain as a directional requirement using an effective isotropic radiated power (EIRP) or by defining limits using TRP. The directional requirement is a requirement which is applied to a specific direction within the OTA coverage range for the transmitter and when the AoA of the incident wave of a received signal is within the *OTA reference sensitivity range of angles of arrival (REFSENS RoAoA)* (this is determined by the contour defined by the points at which the achieved EIS is 3 dB higher than the achieved EIS in the reference direction) or the

[3] The transmitter transient period is the time interval during which the transmitter is changing from the transmitter OFF period to the transmitter ON period or vice versa in a TDD base station [5].

minSENS RoAoA [the reference Range of Angles of Arrival (RoAoA) associated with the OTA sensitivity direction declaration (OSDD) with the lowest declared EIS] as appropriate for the receiver. The radiated transmit power is defined as the EIRP level for a declared beam at a specific beam peak direction. For each beam the requirement is based on declaration of a beam identity, reference beam direction pair, beam-width, rated beam EIRP, OTA peak directions set, the beam direction pairs at the maximum steering directions, and their associated rated beam EIRP and beam-width(s). For a declared beam and beam direction pair the rated beam EIRP level is the maximum power that the BS is declared to radiate at the associated beam peak direction during the transmitter ON period. For each beam peak direction associated with a beam direction pair within the OTA peak direction set, a specific rated beam EIRP level may be claimed. Any claimed value must be met within the accuracy requirements. The rated beam EIRP is only required to be declared for the beam direction pairs [5].

The TRP and EIRP metrics are related through the number of radiating antennas and are implementation specific, considering the geometry of the antenna array and the correlation between unwanted emission signals from different antenna ports. An EIRP limit results in different levels of total radiated unwanted emission power depending on the implementation. The EIRP limit per se cannot control the total amount of interference in the network, whereas a TRP requirement limits the total amount of interference in the network regardless of the specific BS implementation. In the case of passive systems, the antenna gain does not noticeably vary between the desired signal and the unwanted emissions. Thus EIRP is directly proportional to TRP and can be used interchangeably. For an active system such as NR, the EIRP may vary considerably between the desired signal and the unwanted emissions and also between implementations, thus EIRP (in this case) is not proportional to TRP and should not substitute TRP [10].

The RF requirements for BS type 1-O only consist of radiated (OTA) requirements, which are in general based on the corresponding conducted requirements with the exception of two additional radiated requirements; that is, *radiated transmit power* and *OTA sensitivity*. The BS type 1-H is defined with a set of requirements that are mainly related to conducted requirements and the additional radiated requirements, which are the same for BS type 1-O. The radiated transmit power is defined by considering the antenna array beamforming pattern in a specific direction as EIRP for each beam that the BS has declared to transmit, which in some sense relies on the accuracy of the declared EIRP level.

The OTA sensitivity requirement is based on declaration of one or more OSDDs,[4] related to an AAS BS receiver. The AAS BS receiver may optionally be capable of redirecting/changing the receiver target by means of adjusting BS settings, resulting in multiple

[4] OTA sensitivity directions declaration is a set of manufacturer declarations comprising one or more equivalent isotropic sensitivity (EIS) values (with related RAT and channel bandwidth), and the directions where they apply.

sensitivity RoAoA. The sensitivity RoAoA resulting from the current AAS BS settings is referred to as the active sensitivity RoAoA. This definition reflects the antenna array beamforming pattern in a specific direction as declared by EIS[5] level toward a receiver target. The EIS limit must be met not only in a single direction but also within a Range of Angle of Arrival (RoAoA) in the direction of the receiver target. Depending on the level of adaptivity for the AAS BS, the following two alternative declarations can be considered. If the receiver is direction adaptable, in the sense that the receiver target can be redirected, then the declaration contains a receiver target redirection range in a specified receiver target direction. The EIS limit should be met within the redirection range, which is evaluated at five declared sensitivity RoAoA within that range. If the receiver is not adaptive to direction and thus one cannot redirect the receiver target, the declaration consists of a single sensitivity RoAoA in a specified receiver target direction, in which the EIS limit should be met. Note that the OTA sensitivity is defined in addition to the reference sensitivity requirement, which exists both as conducted for BS type 1-H and radiated for BS type 1-O [1,10].

The RF requirements for BS type 2-O are radiated requirements for base stations in FR2. These requirements are identical to the radiated RF requirements defined for FR1 bands, except that the limits for many requirements are different. As for the device, the difference in coexistence at mmWave frequencies leads to more relaxed requirements on ACLR and adjacent channel selectivity (ACS). The implementation of mmWave technologies is more challenging compared to the more mature technologies in sub-6 GHz frequency bands. In the following, we provide an overview of the radiated RF requirements in FR2 bands [5,29]:

- *Output power level*: The maximum output power is the same for FR1 and FR2; however, it is scaled from the conducted requirement and expressed as TRP. In addition, there is a directional radiated transmit power requirement. The dynamic range requirement is defined similar to that of FR1.
- *Transmitted signal quality*: Frequency error, error vector magnitude (EVM), and time-alignment requirements are defined similar to those in FR1 and with the same limits.
- *Radiated unwanted emissions requirements*: Occupied bandwidth, spectrum mask, ACLR, and spurious emissions are defined similar to those for FR1. The latter features/metrics are based on TRP and have less stringent limits relative to those in FR1. The ACLR is on the order of 15 dB and relaxed compared to FR1 due to more favorable coexistence conditions.
- *Reference sensitivity and dynamic range*: These are defined similar to those in FR1; however, the levels are not comparable. There is in addition a directional OTA sensitivity requirement.

[5] Equivalent isotropic sensitivity is the sensitivity for an isotropic directivity device equivalent to the sensitivity of the discussed device exposed to an incoming wave from a defined AoA.

- *Receiver susceptibility to interfering signals*: ACS, in-band, and OOB blocking are defined in the same way as it was done for FR1; however, no narrowband blocking scenario is defined since there are only wideband systems in FR2. The ACS requirement is relaxed compared to FR1 due to more favorable coexistence condition.

The RF characteristics of the receiver and transmitter of a BS or a device are defined by the RF requirements that are partly specified by 3GPP. The BS is the RAN node that transmits and receives RF signals on one or more antenna connectors. An NR BS, defined in RF specifications [5], is not the same as a gNB, which is the corresponding logical node in the radio access network. However, the device in RF specifications is the same as the UE defined in RAN specifications [2,3,4]. The conducted RF requirements are defined for operating frequencies in FR1, whereas only radiated (OTA) requirements are defined for FR2 bands. The set of conducted RF requirements defined for NR are essentially the same as those defined for LTE. Some of the requirements related to the frequency bands and/or the system deployment are technology-agnostic and are derived from regional regulatory requirements. The NR-specific requirements related to flexible channel bandwidths and multiple OFDM numerologies have certain implications on the transmitter requirements for unwanted emissions, where the definition of the limits depends on the channel bandwidth. Such limits would be more difficult to define for a system where the BS may operate with multiple channel bandwidths and where the device may change its channel bandwidth during operation. The properties of the flexible OFDM-based physical layer have implications on the way that the transmitter modulation quality and the receiver selectivity and blocking requirements are defined. Note that the channel bandwidth in general is different for the BS and the UE. The type of transmitter requirements defined for the device is very similar to those defined for the BS. However, the output power levels are considerably lower for a device, while the restrictions on the device implementation are much higher due to considerations for complexity and cost of implementation.

The RF requirements are grouped into transmitter and receiver requirements. In addition, there are performance characteristics for base stations and devices that define the receiver baseband minimum performance for all physical channels under different propagation conditions. Each RF requirement has a corresponding test defined in the NR test specifications for the BS and the device [2–5]. These specifications define the test setup, procedure, signals, tolerances, etc., which are required to demonstrate the compliance with the RF and performance requirements.

The transmitter characteristics define the RF requirements for the desired device/BS transmit signal as well as the unwanted signal emissions outside the transmitted signal bandwidth. These requirements are specified in three categories: (1) output power level requirements set limits on the maximum permissible transmit power, the dynamic variation of the power level, and the transmitter ON/OFF state; (2) transmit signal quality requirements define the authenticity of the transmitted signal and the relation between multiple transmitter branches; and (3) unwanted emissions requirements set limits on the emissions outside the transmit signal bandwidth which are associated with the regulatory requirements and coexistence (spectrum sharing) with other systems. A summary of the device/BS transmitter and receiver characteristics is provided in Tables 5.5 and 5.6, respectively.

The set of receiver requirements for NR are similar to those defined for other systems such as LTE. The receiver characteristics can be categorized into three groups: (1) sensitivity and dynamic range requirements for receiving the desired signals; (2) receiver susceptibility to interfering signals defines receiver susceptibility to different types of interfering signals at different frequency offsets; and (3) unwanted emissions limits are also defined for the receiver.

Table 5.5: Summary of conducted NR transmitter characteristics [2–5,10].

Requirement Category	Base Station Requirements	UE Requirements
Output power level	Maximum output power Output power dynamics ON/OFF power (TDD)	Transmit power Output power dynamics Power control
Transmitted signal quality	Frequency error EVM Time alignment between transmitter branches	Frequency error Transmit modulation quality In-band emissions
Unwanted emissions	Operating band unwanted emissions ACLR and CACLR Spurious emissions Occupied bandwidth Transmitter intermodulation	Spectrum emission mask ACLR and CACLR Spurious emissions Occupied bandwidth Transmitter intermodulation

ACLR, Adjacent channel leakage ratio; EVM, error vector magnitude.

Table 5.6: Summary of conducted NR receiver characteristics [2–5,10].

Requirement Category	Base Station Requirements	UE Requirements
Sensitivity and dynamic range	Reference sensitivity Dynamic range In-channel selectivity	Reference sensitivity Power level Maximum input level
Receiver Susceptibility to Interfering Signals	Out-of-band blocking In-band blocking Narrowband blocking Adjacent channel selectivity Receiver intermodulation	Out-of-band blocking Spurious response In-band blocking Narrowband blocking Adjacent channel selectivity Receiver intermodulation
Unwanted emissions from the receiver	Receiver spurious emissions	Receiver spurious emissions

The channel bandwidths supported by a UE are determined by the NR operating bands as well as the transmitter and receiver RF requirements for those frequency bands. This is due to fact that the combination of maximum power and a large number of transmitted and/or received resource blocks can make it difficult to meet some of the RF requirements. The concept of network signaling (NS) of RF requirements has been used in LTE and continued to be used in NR, where a device can be informed during call setup of whether some specific RF requirements apply when the device is connected to a network. Additional RF requirements apply to the UE when a specific NS value (NS_*m*) is signaled to the device as part of the handover procedure or via a broadcast message. These requirements are associated with limitations and variations of RF parameters such as device output power, maximum channel bandwidth, and the number of transmitted resource blocks. The variations of the requirements are defined along with the NS_*m* in the UE RF specifications [2], where each value corresponds to a specific condition. The default value for all bands is NS_01. The NS_*m* values are related to the permissible power reduction referred to as additional maximum power reduction (A-MPR) and may further apply to transmission using the minimum number of resource blocks, depending on the channel bandwidth [10]. More specifically, each additional emission requirement is associated with a unique NS value indicated via RRC signaling by an NR frequency band number and an associated value in the field *additionalSpectrumEmission.* In NR radio specifications, the notion of indication or signaling of an NS value refers to the corresponding indication of an NR frequency band number (i.e., the IE field *freqBandIndicatorNR* and an associated value of *additionalSpectrumEmission* in the relevant RRC information elements) [2].

In order to accommodate different deployment scenarios for base stations, there are multiple sets of RF requirements for NR base stations, each applicable to a BS class. The NR BS classes include macro-cell, micro-cell, and pico-cell deployment scenarios; however, 3GPP does not identify the BS classes based on the latter terms, rather it refers to them as wide area, medium range, and local area base stations. The wide area BS class is a type of BS for macro-cell scenarios, with minimum distance of 35 m between the BS and the UE. This is the typical large cell deployment with high-tower or above-rooftop installations, providing wide area outdoor or indoor coverage. Medium range BS is a type of BS for micro-cell scenarios, with the minimum distance of 5 m between the BS and the UE. Typical deployments are outdoor below-rooftop installations, providing outdoor hotspot coverage and outdoor-to-indoor coverage through walls. Local area BS is a type of BS that is intended for pico-cell scenarios, with the minimum distance of 2 m between the BS and the UE. Typical deployments are indoor offices and indoor/outdoor hotspots, with the BS mounted on walls or ceilings. The associated deployment scenarios for each BS class are exactly the same for the BS with and without connectors. More specifically, for BS type 1-O and 2-O, BS classes are defined as *wide area* base stations characterized by requirements derived from macro-cell scenarios with a BS to UE minimum ground distance of 35 m; *medium range* base stations characterized by requirements derived from micro-cell scenarios with a BS to UE minimum ground distance of 5 m; and *local area* base stations are characterized by requirements derived from pico-cell scenarios with a BS to UE minimum ground distance of 2 m.

For BS types 1-C and 1-H, the BS classes are defined as *wide area* base stations characterized by requirements derived from macro-cell scenarios with a BS to UE minimum coupling loss of 70 dB; *medium range* base stations are characterized by requirements derived from micro-cell scenarios with a BS to UE minimum coupling loss of 53 dB; and *local area* base stations are characterized by requirements derived from pico-cell scenarios with a BS to minimum coupling loss of 45 dB [5].

The local area and medium range BS classes have different requirements compared to wide area base stations, mainly due to shorter minimum BS to device distance and lower minimum coupling loss. The maximum BS output power is limited to 38 dBm for medium range base stations and 24 dBm for local area base stations. This power is defined per antenna and per carrier. There is no maximum BS output power defined for wide area base stations. The spectrum mask, that is, the limits on the operating band unwanted emissions (OBUE), has lower limits for medium range and local area, consistent with the lower maximum output power levels. The receiver reference sensitivity limits are higher and more relaxed for the medium range and local area base stations. The receiver dynamic range and in-channel selectivity (ICS) are also adjusted accordingly for these BS classes. The limits on co-location for medium range and local area base stations are relatively more relaxed compared to the wide area base stations. The co-location refers to the additional blocking requirement which may be applied for the protection of NR BS receivers when another GSM, CDMA, UTRA, E-UTRA, or NR BS operating in a different frequency band is co-located with an NR BS. The requirement is applicable to all channel bandwidths supported by the NR BS. These requirements assume a 30 dB coupling loss between interfering transmitter and NR BS receiver and are based on co-location with base stations of the same class. All medium range and local area BS limits for receiver susceptibility to the interfering signals are adjusted considering the higher receiver sensitivity limit and the lower minimum coupling loss (BS to device) [5].

The requirements for base stations that are capable of multi-band operation are defined at the multi-band connector or multi-band RIB, where the RF requirements apply separately to each supported operating band. For BS types 1-C and 1-H that are capable of multi-band operation, various structures in the form of combinations of different transmitter and receiver implementations (multi-band or single-band) with mapping of transceivers to one or more antenna connectors for BS type 1-C or TAB connectors for BS type 1-H in different ways are possible [5].

The conducted transmitter characteristics are specified at the antenna connector for BS type 1-C and at the TAB connector for BS type 1-H. For BS type 1-H, the minimum number of supported geographical cells (i.e., geographical areas covered by beams) must be identified.

The minimum number of supported geographical cells (N_{cells}) corresponds to the BS setting with the minimum amount of cell splitting when transmitting on all TAB connectors for the operating band, or with the minimum number of transmitted beams. Each TAB connector of the BS Type 1-H supporting transmission in an operating band is mapped to one *TAB connector TX min cell group*. The *TAB connector TX min cell group* is an operating band–specific declared group of TAB connectors to which BS type 1-H conducted transit requirements are applied. In the latter definition, the group corresponds to the group of TAB connectors which are responsible for transmitting a cell when the BS type 1-H setting corresponding to the declared minimum number of cells with transmission on all TAB connectors supporting an operating band, but its existence is not limited to that condition. Note that the mapping of the TAB connectors to cells/beams is implementation dependent. The number of active transmitter units that are considered when calculating the conducted transmitter emissions limits ($N_{TXU\text{-}counted}$) for BS type 1-H is given as $N_{TXU\text{-}counted} = \min(N_{TXU\text{-}active}, 8N_{cells})$ where $N_{TXU\text{-}countedpercell}$ is used for scaling of basic limits and is calculated as $N_{TXU\text{-}countedpercell} = N_{TXU\text{-}counted}/N_{cells}$. The parameter $N_{TXU\text{-}active}$ depends on the actual number of active transmitter units and is independent of N_{cells} [5].

Some NR features require the BS to transmit from two or more antennas, for example, transmit diversity and multi-antenna transmission modes. For carrier aggregation, the carriers may also be transmitted from different antennas. In order for the device to properly receive the signals from multiple antennas, the timing relationship between any two transmitter branches is specified in terms of the maximum time-alignment error between transmitter branches. The maximum allowed error depends on the feature or combination of features in the transmitter branches.

In general, there is no maximum output power requirement for wide area base stations. As mentioned earlier, there is the maximum output power limit of 38 dBm for the medium range and 24 dBm for the local area base stations. In addition, 3GPP RF specifications specify a tolerance, which defines the extent to which the actual maximum power may deviate from the power level declared by the BS manufacturer. Power control is used to limit the interference level. A total power control dynamic range per resource element is specified for a BS, which is the difference between the power of a resource element and the average resource element power for a BS at maximum output power ($P_{max,c,TABC}$) for a specified reference condition (Table 5.7). There is also a dynamic range requirement for the total BS power. The BS total power dynamic range is the difference between the maximum and the minimum transmit power of an OFDM symbol for a specified reference condition (Table 5.8). For TDD operation, a power mask is defined for the BS output power, defining the OFF power level during the uplink subframes and the maximum time for the transmitter transition time between the transmitter ON and OFF states [5].

Transmit OFF power requirements only apply to TDD mode of operation of an NR BS. The transmitter OFF power is defined as the mean power measured over $70/N(\mu s)$ filtered with

Table 5.7: Resource element power control dynamic range [5].

Modulation Scheme Used on the Resource Element	Resource Element Power Control Dynamic Range (dB)	
	Down	Up
QPSK (PDCCH)	−6	+4
QPSK (PDSCH)	−6	+3
16QAM (PDSCH)	−3	+3
64QAM (PDSCH)	0	0
256QAM (PDSCH)	0	0

Table 5.8: Total power dynamic range [5].

BS Channel Bandwidth (MHz)	Total Power Dynamic Range (dB)		
	SCS 15 kHz	SCS 30 kHz	SCS 60 kHz
5	13.9	10.4	N/A
10	17.1	13.8	10.4
15	18.9	15.7	12.5
20	20.2	17	13.8
25	21.2	18.1	14.9
30	22	18.9	15.7
40	23.3	20.2	17
50	24.3	21.2	18.1
60	N/A	22	18.9
70	N/A	22.7	19.6
80	N/A	23.3	20.2
90	N/A	23.8	20.8
100	N/A	24.3	21.3

a square filter of bandwidth equal to the transmission bandwidth configuration of the BS (BW_{config}) centered around the assigned channel frequency during the *transmitter OFF period*. The parameter $N = SCS/15$, where SCS is the subcarrier spacing in kHz. For multi-band connectors and for single-band connectors supporting transmission in multiple operating bands, the requirement is only applicable during the transmitter OFF period in all supported operating bands. For BSs supporting intraband contiguous carrier aggregation, the transmitter OFF power is defined as the mean power measured over $70/N$ (μs) filtered with a square filter of bandwidth equal to the *aggregated BS channel bandwidth* $BW_{channel\_CA}$ centered around $\left(f_{edge\text{-}high} + f_{edge\text{-}low}\right)/2$ during the *transmitter OFF period*. In this case, parameter $N = SCS/15$, where SCS is the smallest supported subcarrier spacing in kHz in the *aggregated BS channel bandwidth* [5].

The requirements for transmitted signal quality specify the extent to which the transmitted signal from the BS or the device deviates from a reference signal. The transmitted signal impairments are caused by the transmitter RF components and particularly the nonlinear response of the power amplifier (PA). The signal quality of the BS or the device can be

Table 5.9: Error vector magnitude requirements for base station type 1-C and BS type 1-H [5].

Modulation Scheme for PDSCH	Required EVM (%)
QPSK	17.5
16QAM	12.5
64QAM	8
256QAM	3.5

assessed based on the requirements on EVM and frequency error as well as the device in-band emissions. The EVM is a measure of the error in the modulated signal constellation, calculated as the RMS of the error vectors over the active subcarriers, considering all symbols of the modulation scheme under consideration. It is expressed as a percentage value in relation to the power of the ideal signal (Table 5.9). In other words, EVM defines the maximum SINR that can be achieved at the receiver, if there are no additional impairments to the signal between the transmitter and the receiver. Since a receiver can remove some impairments of the transmitted signal such as time dispersion, the EVM is assessed after cyclic prefix removal and the equalization modules. In this way, the EVM evaluation includes a standardized model of the receiver. The frequency offset resulting from the EVM evaluation is averaged and used as a measure of the frequency error of the transmitted signal. Frequency error is the measure of the difference between the actual BS transmit frequency and the assigned frequency, wherein the same source is used for RF frequency and data clock generation.

The EVM is a measure of the difference between the ideal symbols and the measured symbols after the equalization. The EVM is averaged over all allocated downlink resource blocks with the modulation scheme under consideration in the frequency domain, and a minimum of 10 downlink subframes. Although the basic unit ofmeasurement is one subframe, the equalizer is calculated over 10 subframe measurement periods to reduce the impact of noise in the reference symbols. The boundaries of the 10 subframe measurement periods do not need to be aligned with radio frame boundaries. In FDD mode, the averaging in the time domain is performed over 10 subframes (the same frame) of the measurement period from the equalizer estimation step, whereas in TDD mode, the averaging in the time domain can be calculated from subframes of different frames and has a minimum of 10 subframes averaging length. The TDD frame special fields such as guard period are not included in the averaging window. This can be mathematically expressed as follows [5]:

$$\overline{\mathrm{EVM}_{frame}} = \left(\frac{\sum_{k=1}^{N_{dl}} \sum_{l=1}^{N_k} \mathrm{EVM}_{kl}^2}{\sum_{k=1}^{N_{dl}} N_k} \right)^{1/2}$$

where N_k is the number of resource blocks with the considered modulation scheme in the kth subframe and N_{dl} is the number of allocated downlink subframes in one frame. The EVM measurements start on the third symbol of a slot, when the first symbol of that slot is a downlink symbol.

The unwanted emissions consist of OOB emissions and spurious emissions according to ITU-R definitions. In ITU-R definition, OOB emissions are unwanted emissions immediately outside the *BS channel bandwidth* resulting from the modulation process and nonlinearity in the transmitter analog components but excluding spurious emissions. Spurious emissions are emissions which are caused by unwanted transmitter effects, such as harmonics, parasitic, intermodulation products, and frequency conversion products, but exclude OOB emissions [5].

The ITU-R defines the boundary between the OOB and spurious emission domains as a frequency separation relative to the center of the carrier by 2.5 × (or 250%) of the NR channel bandwidth. Any emission outside the channel bandwidth which occurs in the frequency range separated from the assigned frequency of the emission by less than 2.5 times the channel bandwidth of the emission will generally be considered an emission in the OOB domain. However, this frequency separation may be dependent on the type of modulation, the maximum symbol rate in the case of digital modulation, the type of transmitter, and frequency coordination factors. For example, in the case of some digital, broadband, or pulse-modulated systems, the frequency separation may need to differ from the 2.5 × factor. The transmitter nonlinearities may also spread in-band signal components into the frequency band of the OOB frequency range. Furthermore, the transmitter oscillator sideband noise also may extend into that frequency range. Since it may not be practical to isolate these emissions, their level will tend to be included during OOB power measurements.

All emissions, including intermodulation products, conversion products, and parasitic emissions, which fall within frequencies separated from the center frequency of the emission by 2.5 × or more of the necessary bandwidth of the emission will generally be considered as emissions in the spurious domain. However, this frequency separation may be dependent on the type of modulation, the maximum symbol rate in the case of digital modulation, the type of transmitter, and frequency coordination factors. For example, in the case of some digital, broadband, or pulse-modulated systems, the frequency separation may need to differ from the 2.5 × factor. For multi-channel or multi-carrier transmitters, where several carriers may be transmitted simultaneously from the PA or an AAS, the center frequency of the emission is taken to be the center of either the assigned bandwidth of the BS or of the −3 dB bandwidth of the transmitter, using the lesser of the two bandwidths.

The above separation of the requirements can be easily applied to systems with fixed channel bandwidth. It does; however, become more difficult for NR, which supports flexible bandwidths, implying that the frequency range where requirements apply would then vary

Table 5.10: Maximum offset of operating band unwanted emission outside the downlink operating bands [5].

BS Type	Operating Band (MHz)	Δf_{OBUE} (MHz)
BS Type 1-H	$f_{DL-high} - f_{DL-low} < 100$	10
	$100 \leq f_{DL-high} - f_{DL-low} \leq 900$	40
BS Type 1-C	$f_{DL-high} - f_{DL-low} \leq 200$	10
	$200 < f_{DL-high} - f_{DL-low} \leq 900$	40

with the channel bandwidth. The approach taken for defining the boundary in 3GPP is slightly different for BS and the device requirements. With the recommended boundary between OOB emissions and spurious emissions set at 2.5 × the channel bandwidth, third-order and fifth-order intermodulation products from the carrier will fall inside the OOB domain, which will cover a frequency range of twice the channel bandwidth on each side of the carrier. For the OOB domain, two overlapping requirements are defined for the BS and the device, namely, spectrum emissions mask (SEM) and ACLR.

The OOB emission requirement for the BS transmitter is specified both in terms of ACLR and OBUE. The maximum offset of the OBUEs mask from the operating band edge is denoted as Δf_{OBUE}. The OBUE defines all unwanted emissions in each supported downlink operating band plus the frequency ranges Δf_{OBUE} above and Δf_{OBUE} below each band. The unwanted emissions outside of this frequency range are limited by a spurious emissions requirement. The values of Δf_{OBUE} are defined in Table 5.10 for the NR operating bands [5].

The occupied bandwidth is the size of a frequency band below the lower and above the upper frequency edges, where the mean power emitted is equal to $\beta/2$ (in percent) of the total mean transmitted power. The value of $\beta/2$ is equal to 0.5%. The occupied bandwidth requirement is applicable during the transmitter ON period for a single carrier. There may also be regional requirements to declare the occupied bandwidth according to the above definition. The occupied bandwidth is a regulatory requirement that is specified for the equipment in some regions. It was originally defined by the ITU-R as the maximum bandwidth, outside of which emissions do not exceed a certain percentage of the total emissions. The occupied bandwidth for NR is equal to the channel bandwidth, outside of which a maximum of 1% of the emissions are allowed (0.5% on each side).

The adjacent channel leakage ratio is the ratio of the filtered mean power centered on the assigned channel frequency to the filtered mean power centered on an adjacent channel frequency. The requirements apply outside the BS RF bandwidth or radio bandwidth regardless of the type of transmitter considered (single-carrier or multi-carrier) and for all transmission modes supported by the manufacturer's specification. These requirements are applied during the transmitter ON period.

The spectrum of an OFDM signal decays slowly outside of the transmission bandwidth. Since the transmitted signal for NR occupies up to 98% of the channel bandwidth, it is not possible to meet the unwanted emission limits directly outside the channel bandwidth, if certain time-domain windowing or frequency-domain shaping are not used. Filtering is always used both in the form of time-domain digital filtering of the baseband signal or analog-domain filtering of the RF signal.

The nonlinear characteristics of the PA must be taken into account, since it is the cause of intermodulation products outside the channel bandwidth. A power backoff can be used to operate in linear region of the PA at the cost of lower power efficiency, thus the amount of backoff should be minimized in practice. For this reason, additional linearization schemes can be employed [e.g., digital predistortion (DPD) for the PA]. These are especially important for the BS, where there are fewer restrictions on implementation complexity and use of advanced linearization schemes is an essential part of controlling unwanted emissions.

The emission mask defines the permissible OOB spectrum emissions outside the transmission bandwidth. Due to support of flexible channel bandwidths by NR, the determination of the frequency boundary between OOB emissions and spurious emissions is more involved for the NR BS and the devices. For the NR BS, the problem of implicit variation of the boundary between OOB and spurious domains with the varying channel bandwidth is addressed by not defining an explicit boundary. The 3GPP solution for this problem is the concept of OBUE for the NR base stations instead of the spectrum mask that is usually defined for OOB emissions. The OBUE requirement applies over the entire BS transmitter operating band, plus an additional 10–40 MHz on each side, as shown in Fig. 5.9. All requirements outside of that range are set by the regulatory spurious emission limits, based on the ITU-R recommendations. As shown in Fig. 5.9, a large part of the OBUE is defined over a frequency range that for smaller channel bandwidths can be both in spurious and OOB domains. This means that the limits for the frequency ranges that may be in the spurious domain also have to align with the regulatory limits from ITU-R. The shape of the mask is generic for all channel bandwidths, which has to align with ITU-R limits starting 10–40 MHz from the channel edges. The OBUE is defined with a 100 kHz measurement bandwidth and is aligned to some extent with the corresponding emission masks for the LTE systems. In the case of carrier aggregation for a BS, the OBUE requirement is applied as for any multicarrier transmission, where the OBUE will be defined relative to the carriers on the edges of the RF bandwidth. In the case of non-contiguous carrier aggregation, the OBUE within a subblock gap is partly calculated as the cumulative sum of contributions from each subblock [10].

In addition to a SEM, the OOB emissions are defined by the ACLR requirement. The ACLR concept is very useful for analysis of coexistence between two systems that operate on adjacent frequencies. The ACLR defines the ratio of the power transmitted within the

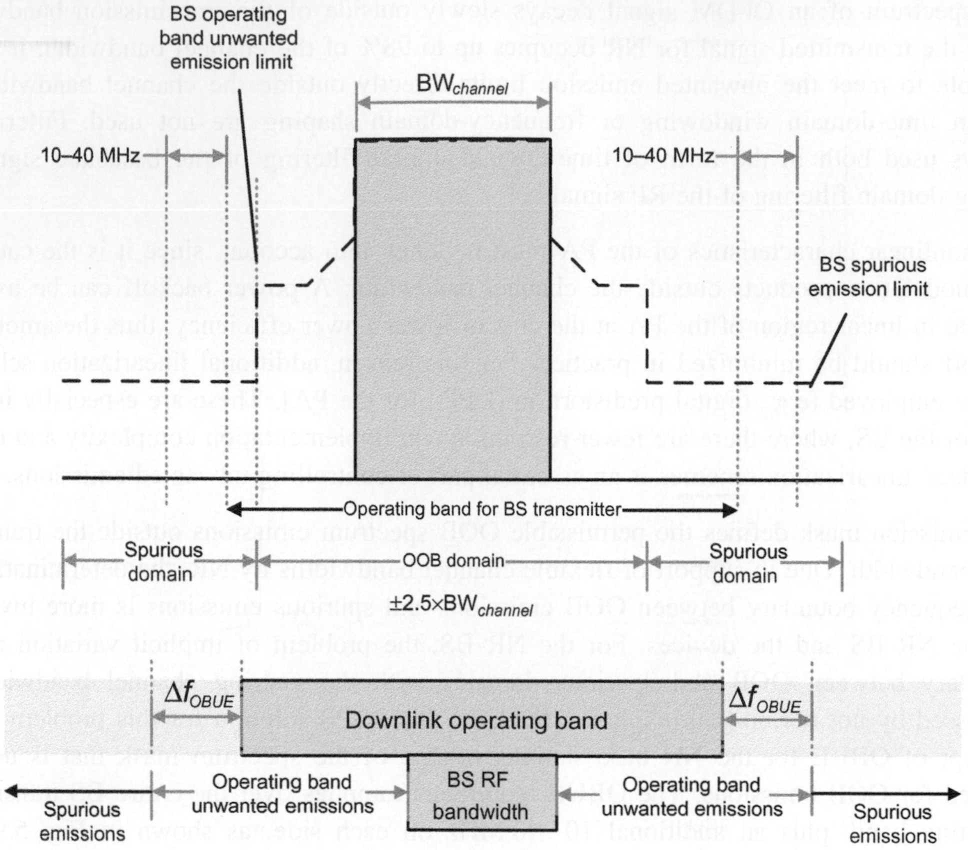

Figure 5.9
Illustration of the unwanted and spurious emissions applicable to NR base stations [5,8,9].

assigned channel bandwidth to the power of the unwanted emissions transmitted on an adjacent channel. There is a corresponding receiver requirement known as ACS, which defines the receiver's ability to suppress a signal on an adjacent channel. The definitions of ACLR and ACS are illustrated in Fig. 5.10 for a wanted and an interfering signal received in adjacent channels. The interfering signal's leakage of unwanted emissions at the wanted signal receiver is given by the ACLR and the ability of the receiver of the wanted signal to suppress the interfering signal in the adjacent channel is defined by the ACS. The two parameters when combined define the total leakage between two transmissions on adjacent channels. That ratio is called the adjacent channel interference ratio (ACIR) and is defined as the ratio of the power transmitted on one channel to the total interference received by a receiver on the adjacent channel, due to both the transmitter (ACLR) and the receiver (ACS). This relationship between the adjacent channel parameters is given as $\mathrm{ACIR} = 1/(1/\mathrm{ACLR} + 1/\mathrm{ACS})$. The ACLR and ACS can be defined with different

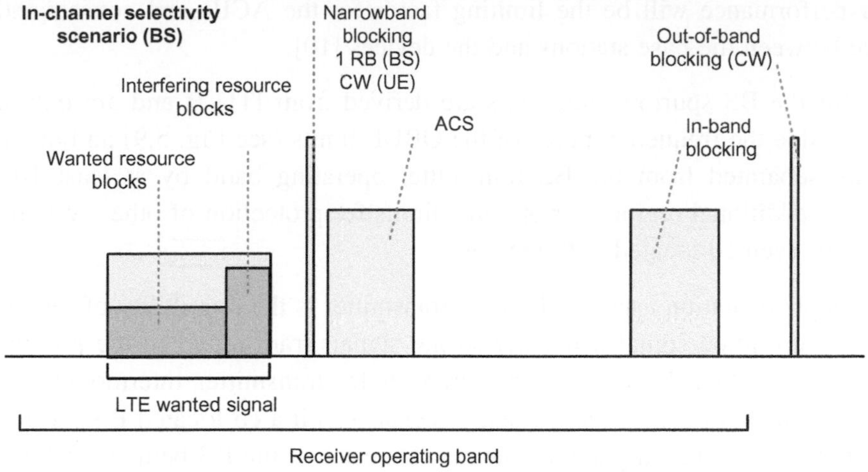

Figure 5.10
Illustration of the BS/UE requirements for receiver susceptibility to interference [5,10].

channel bandwidths for the two adjacent channels, which is the case for some requirements set for NR due to the bandwidth flexibility.

The ACLR limits for NR devices and base station have been derived based on extensive analysis of NR coexistence with NR or other systems on adjacent carriers. For an NR BS, there are ACLR requirements both for an adjacent channel with an NR receiver of the same channel bandwidth and for an adjacent LTE receiver. The ACLR requirement for NR BS is set to 45 dB. This is considerably more stringent than the ACS requirement for the device, which implies that in the downlink, the device receiver performance will be the limiting factor for ACIR and consequently for coexistence between the base stations and the devices. From a system perspective, this choice is cost-efficient since it moves implementation complexity to the BS, instead of requiring all devices to have high-performance RF components. In the case of carrier aggregation in a BS, the ACLR limits are applied as for any multi-carrier transmission, where the ACLR requirement will be defined for the carriers on the edges of the RF bandwidth. In the case of non-contiguous carrier aggregation where the subblock gap is very small that the ACLR requirements at the edges of the gap will overlap, a special cumulative ACLR (CACLR) requirement is defined for the gap. For CACLR, contributions from carriers on both sides of the subblock gap are taken into account in the calculation of CACLR limit. The CACLR limit is the same as the ACLR for the BS at 45 dB. The ACLR limits for the device are set based on the assumption that an NR or a UTRA receiver on the adjacent channel is used. In the case of carrier aggregation, the device ACLR requirement is applied to the aggregated channel bandwidth as opposed to per carrier. The ACLR limit for NR devices is set to 30 dB. This is considerably more relaxed compared to the ACS requirement for the BS, implying that in the uplink, the device transmitter performance will be the limiting factor for the ACIR and consequently for the coexistence between the base stations and the devices [10].

The limits for the BS spurious emissions are derived from ITU-R and are only defined in the region outside the frequency range of the OBUE limits (see Fig. 5.9) and in the frequencies that are separated from the BS transmitter operating band by at least 10–40 MHz. There are also additional regional or optional limits for protection of other systems that may coexist with or even co-located with the NR.

One of the implementation aspects of an RF transmitter is the possibility of intermodulation between the transmitted signal and other strong signals transmitted in the proximity of the BS or the device, which leads to the requirement for transmitter intermodulation. For the BS, the requirement is based on a stationary scenario with a co-located BS transmitter, with its transmitted signal appearing at the antenna connector of the BS being specified but attenuated by 30 dB. Since it is a stationary scenario, there are no additional unwanted emissions allowed, implying that all unwanted emission limits also have to be met in the presence of the interferer. For the device, there is a similar requirement based on a scenario with another device's transmitted signal appearing at the antenna connector of the device being specified but attenuated by 40 dB. The requirement specifies the minimum attenuation of the resulting intermodulation product below the transmitted signal [10].

The reference sensitivity requirement is meant to verify the receiver noise figure, which is a measure of how much the receiver's RF signal chain degrades the SNR of the received signal. For this reason a low-SNR transmission scheme using QPSK is chosen as a reference channel for the reference sensitivity test. The reference sensitivity is defined at a receiver input level where the throughput is 95% of the maximum throughput for the reference channel. For the device, reference sensitivity is defined in terms of the full channel bandwidth signals and with all resource blocks allocated to the wanted signal [5].

The reference sensitivity power level $P_{REFSENS}$ is the minimum mean power received at the antenna connector for BS Type 1-C or TAB connector for BS Type 1-H at which certain throughput requirement is met for a specified reference measurement channel. For NR, this throughput must be greater than 95% of the maximum throughput of the reference measurement channel. The reference sensitivity is a range of values that can be calculated as $P_{REFSENS} = 10\log(kT_0W_N) + \text{NF} + \text{SINR} + L_{IM}(\text{dBm})$ for the BS, where kT_0W_N is the thermal noise power in dBm, W_N is the noise equivalent bandwidth (approximately the system bandwidth), NF is the maximum receiver noise figure, SINR is the minimum SINR for the chosen MCS, and L_{IM} is the implementation margin. Note that the noise floor for any receiver is defined as $\text{Noise\_Floor} = 10\log(kT_0W_N) + \text{NF} = -174 + \text{NF} + 10\log W_N$ (dBm), where $\text{NF} = \text{SNR}_{in} - \text{SNR}_{out}$ (in dB) is a measure of SINR degradation by the components in the RF signal path including the RF filters and low-noise amplifier (LNA). The thermal noise power density $kT_0 = -174$ dBm/Hz is measured at typical room

Table 5.11: NR wide area/medium range/local area base station reference sensitivity levels [5].

BS Channel Bandwidth (MHz)	Subcarrier Spacing (kHz)	Reference Measurement Channel	NR Wide Area BS Reference Sensitivity Power Level (dBm) $P_{REFSENS}$	NR Medium Range BS Reference Sensitivity Power Level (dBm) $P_{REFSENS}$	NR Local Area BS Reference Sensitivity Power Level (dBm) $P_{REFSENS}$
5, 10, 15	15	G-FR1-A1-1	− 101.7	− 96.7	− 93.7
10, 15	30	G-FR1-A1-2	− 101.8	− 96.8	− 93.8
10, 15	60	G-FR1-A1-3	− 98.9	− 93.9	− 90.9
20, 25, 30, 40, 50	15	G-FR1-A1-4	− 95.3	− 90.3	− 87.3
20, 25, 30, 40, 50, 60, 70, 80, 90, 100	30	G-FR1-A1-5	− 95.6	− 90.6	− 87.6
20, 25, 30, 40, 50, 60, 70, 80, 90, 100	60	G-FR1-A1-6	− 95.7	− 90.7	− 87.7

temperature 290 K degree. The LTE specifications require NF $\leq$ 9 dB for the UE and NF $\leq$ 5 dB for the eNB, nevertheless, commercial RF components and UE receivers can achieve lower than these limits. In NR the implementation margin is 2 dB and the NF is 5 dB for wide area BS, 10 dB for medium range BS and 13 dB for local area BS [14].

In Table 5.11, $P_{REFSENS}$ is the power level of a single instance of the reference measurement channel. This requirement must be met for each consecutive application of a single instance of the reference measurement channel mapped to disjoint frequency ranges with a width corresponding to the number of resource blocks of each reference measurement channel, except for one instance that might overlap with one other instance to cover the full *BS channel bandwidth.*

The dynamic range is specified as a measure of the capability of the receiver to receive a wanted signal in the presence of an interfering signal at the antenna connector for BS Type 1-C or TAB connector for BS Type 1-H within the received *BS channel bandwidth.* Under this condition, certain throughput requirement must be met for a specified reference measurement channel. The interfering signal for the dynamic range requirement is an additive white Gaussian noise. For NR, this throughput must be greater than 95% of the maximum throughput of the reference measurement channel (Table 5.12). The dynamic range requirement for the device is specified as the maximum signal level at which the throughput requirement can be met. The wanted signal mean power is the power level of a single instance of the corresponding reference measurement channel. This requirement must be

Table 5.12: Wide area/medium range/local area base station dynamic ranges [5].

BS Channel Bandwidth (MHz)	Subcarrier Spacing (kHz)	Reference Measurement Channel	Wide Area BS Wanted Signal Mean Power (dBm)	Wide Area BS Interfering Signal Mean Power (dBm)/ BW_{config}	Medium Range BS Wanted Signal Mean Power (dBm)	Medium Range BS Interfering Signal Mean Power (dBm)/ BW_{config}	Local Area BS Wanted Signal Mean Power (dBm)	Local Area BS Interfering Signal Mean Power (dBm)/ BW_{config}	Type of Interfering Signal
5	15	G-FR1-A2-1	− 70.7	− 82.5	− 65.7	− 77.5	− 62.7	− 74.5	AWGN
	30	G-FR1-A2-2	− 71.4		− 66.4		− 63.4		
10	15	G-FR1-A2-1	− 70.7	− 79.3	− 65.7	− 74.3	− 62.7	− 71.3	AWGN
	30	G-FR1-A2-2	− 71.4		− 66.4		− 63.4		
	60	G-FR1-A2-3	− 68.4		− 63.4		− 60.4		
15	15	G-FR1-A2-1	− 70.7	− 77.5	− 65.7	− 72.5	− 62.7	− 69.5	AWGN
	30	G-FR1-A2-2	− 71.4		− 66.4		− 63.4		
	60	G-FR1-A2-3	− 68.4		− 63.4		− 60.4		
20	15	G-FR1-A2-4	− 64.5	− 76.2	− 59.5	− 71.2	− 56.5	− 68.2	AWGN
	30	G-FR1-A2-5	− 64.5		− 59.5		− 56.5		
	60	G-FR1-A2-6	− 64.8		− 59.8		− 56.8		
25	15	G-FR1-A2-4	− 64.5	− 75.2	− 59.5	− 70.2	− 56.5	− 67.2	AWGN
	30	G-FR1-A2-5	− 64.5		− 59.5		− 56.5		
	60	G-FR1-A2-6	− 64.8		− 59.8		− 56.8		
30	15	G-FR1-A2-4	− 64.5	− 74.4	− 59.5	− 69.4	− 56.5	− 66.4	AWGN
	30	G-FR1-A2-5	− 64.5		− 59.5		− 56.5		
	60	G-FR1-A2-6	− 64.8		− 59.8		− 56.8		
40	15	G-FR1-A2-4	− 64.5	− 73.1	− 59.5	− 68.1	− 56.5	− 65.1	AWGN
	30	G-FR1-A2-5	− 64.5		− 59.5		− 56.5		
	60	G-FR1-A2-6	− 64.8		− 59.8		− 56.8		
50	15	G-FR1-A2-4	− 64.5	− 72.2	− 59.5	− 67.2	− 56.5	− 64.2	AWGN
	30	G-FR1-A2-5	− 64.5		− 59.5		− 56.5		
	60	G-FR1-A2-6	− 64.8		− 59.8		− 56.8		
60	30	G-FR1-A2-5	− 64.5	− 71.4	− 59.5	− 66.4	− 56.5	− 63.4	AWGN
	60	G-FR1-A2-6	− 64.8		− 59.8		− 56.8		
70	30	G-FR1-A2-5	− 64.5	− 70.8	− 59.5	− 65.8	− 56.5	− 62.8	AWGN
	60	G-FR1-A2-6	− 64.8		− 59.8		− 56.8		
80	30	G-FR1-A2-5	− 64.5	− 70.1	− 59.5	− 65.1	− 56.5	− 62.1	AWGN
	60	G-FR1-A2-6	− 64.8		− 59.8		− 56.8		
90	30	G-FR1-A2-5	− 64.5	− 69.6	− 59.5	− 64.6	− 56.5	− 61.6	AWGN
	60	G-FR1-A2-6	− 64.8		− 59.8		− 56.8		
100	30	G-FR1-A2-5	− 64.5	− 69.1	− 59.5	− 64.1	− 56.5	− 61.1	AWGN
	60	G-FR1-A2-6	− 64.8		− 59.8		− 56.8		

met for each consecutive application of a single instance of the reference measurement channel mapped to disjoint frequency ranges with a width corresponding to the number of resource blocks of each reference measurement channel, except for one instance that might overlap another instance to cover the full *BS channel bandwidth* [5].

There are a set of requirements for the BS and the device, which define the receiver ability to receive a wanted signal in the presence of a stronger interfering signal. There are different interference scenarios depending on the frequency offset of the interferer from the wanted signal, where different types of receiver impairments will affect the performance.

A set of combinations of the interfering signals with different bandwidths can be defined which model the practical scenarios that are encountered within and outside of the BS/device receiver bandwidth. While the types of requirements are very similar for the BS and the device, the signal levels are different since the interference scenarios for the BS and device can vary. The following requirements are defined for NR base stations and the devices. In all cases where the interfering signal is an NR signal, it has the same or smaller bandwidth (less than 20 MHz) relative to the desired signal [5]:

- *In-band/Out-of-band Blocking*: Blocking refers to the scenario where strong interfering signals are present outside the operating band (OOB blocking) or inside the operating band (in-band blocking), but not adjacent to the desired signal. The in-band blocking characteristics is a measure of the receiver's ability to receive a desired signal in its assigned channel at the antenna connector for BS Type 1-C or TAB connector for BS Type 1-H in the presence of an unwanted interferer, which is an NR signal for general blocking or an NR signal with one resource block for narrowband blocking (see Fig. 5.10). In-band blocking includes interferers within 20−60 MHz range outside of the operating band for the base stations and within 15 MHz range of the operating band for the devices. These scenarios are modeled with a continuous wave (CW) signal for the out-of-band case and an NR signal for the in-band case. There are additional BS blocking requirements for the scenario in which the BS is co-located with another BS in a different operating band. For the UEs, a number of exceptions from the out-of-band blocking requirements are allowed, where the device is required to comply with more relaxed spurious response requirements.

  For a BS operating in non-contiguous spectrum in any operating bands, the in-band blocking requirements further apply in subblock gaps, if the subblock gap size is at least twice as wide as the interfering signal minimum offset. The interfering signal offset is defined relative to the subblock edges inside the subblock gap. For a multi-band connector, the blocking requirements apply in the in-band blocking frequency ranges for each supported operating band. For a BS operating in non-contiguous spectrum within any operating bands, the narrowband blocking requirements are further applied within the subblock gap, if the subblock gap size is at least as wide as the channel bandwidth of the NR interfering signal. The interfering signal offset is defined relative to the subblock edges inside the subblock gap. For a multi-band connector the narrowband blocking requirement is further applied inside any inter-RF bandwidth gap,[6] in case the inter-RF bandwidth gap size is at least as wide as the NR interfering signal. The interfering signal offset is defined relative to the base station RF bandwidth edges inside the inter-RF bandwidth gap.

- *Adjacent Channel Selectivity (ACS)*: This is a measure of the receiver's ability to receive a desired signal in its assigned channel frequency at the antenna connector for BS type 1-C or TAB connector for BS type 1-H in the presence of an adjacent channel signal with a specified center frequency offset from the interfering signal to the edge of the band of a victim system. For a BS operating in non-contiguous spectrum within any operating bands, the ACS requirement is further applied within the subblock gaps, in case the subblock gap size is at least as wide as the NR interfering signal. The interfering signal offset is defined relative to the subblock edges inside the subblock gap. For a multi-band connector, the ACS requirement is applied within the inter-RF bandwidth gap, in case the inter-RF bandwidth gap size is at least as wide as the NR interfering signal. The interfering signal offset is defined relative to the base station RF bandwidth edges inside the inter-RF bandwidth gap. This corresponds to a scenario where there is a strong signal in the channel adjacent to the desired signal and is closely related to the corresponding ACLR requirement. The adjacent interferer is an NR signal. For the device, the ACS is specified for two cases with a lower and a higher signal level.
- *In-Channel Selectivity (ICS)*: This is a measure of the receiver ability to receive a desired signal in its assigned resource block locations at the antenna connector for BS type 1-C or TAB connector for BS type 1-H in the presence of an interfering signal received with a larger power spectral density. Under this condition, certain throughput requirement (throughput must be greater than 95% of the maximum throughput of the reference measurement) must be met for a specified reference measurement channel. The interfering signal is an NR signal which is time-aligned with the desired signal. In other words, in this scenario, multiple signals with different power levels are received within the channel bandwidth, where the performance of the weaker desired signal is verified in the presence of the stronger interfering signal. The ICS requirement is only specified for the base stations (see Fig. 5.10).
- *Receiver Intermodulation*: The third and higher order harmonics resulted from mixing of the two interfering RF signals can produce an interfering signal in the band of the desired channel. Intermodulation response rejection is a measure of the capability of the receiver to receive a desired signal in its assigned channel bandwidth at the antenna connector for BS type 1-C or TAB connector for BS type 1-H in the presence of two

[6] Inter-RF bandwidth gap is the gap between the RF bandwidths in the two bands. Note that the inter-RF bandwidth gap may span a frequency range where other mobile operators can be deployed in bands *X* and *Y*, as well as the frequency range between the two bands that may be used for other services.

interfering signals which have a specific frequency relationship to the desired signal. In this scenario, there are two interfering signals in the proximity of the desired signal, where the interferers are a continuous wave signal and an NR signal. The purpose of this requirement is to test the receiver linearity. The interferers are placed at frequencies in such a way that the main intermodulation product falls within the desired signal channel bandwidth. There is also a narrowband intermodulation requirement for the BS where the CW signal is very close to the desired signal and the NR interferer is a single-resource-block signal (narrowband).

- *Receiver Spurious Emissions*: The receiver spurious emissions power is the power of emissions generated or amplified in a receiver unit that appear at the antenna connector (for BS Type 1-C) or at the TAB connector (for BS Type 1-H). The requirements are applicable to all base stations with separate RX and TX antenna connectors/TAB connectors. For systems operating in FDD mode, the test is performed when both TX and RX are ON, with the TX antenna connectors/TAB connectors terminated. For antenna connectors/TAB connectors supporting both RX and TX in TDD, the requirements are verified during the transmitter OFF period. For antenna connectors/TAB connectors supporting both RX and TX in FDD mode, the RX spurious emissions requirements are superseded by the TX spurious emissions requirements. For RX-only multi-band connectors, the spurious emissions requirements are subject to exclusion zones in each supported operating band. For multi-band connectors which transmit and receive in operating band supporting TDD mode, the RX spurious emissions requirements are verified during the TX OFF period and are subject to exclusion zones in each supported operating band. The number of active receiver units that are considered when calculating the conducted RX spurious emission limits ($N_{RXU-counted}$) for BS Type 1-H is given as $N_{RXU-counted} = \min(N_{RXU-active}, 8N_{cells})$ where $N_{RXU-counted-per-cell}$ is used for scaling of basic limits and is derived as $N_{RXU-counted-per-cell} = N_{RXU-counted}/N_{cells}$, where N_{cells} was defined earlier. The $N_{RXU-active}$ is the number of active receiver units and is independent of the declaration of N_{cells}. Since the receiver emissions are dominated by the transmitted signal, the receiver spurious emission limits are only applicable when the transmitter is not active, and also when the transmitter is active for an NR base station operating in FDD mode that has a separate receiver antenna connector [5].

5.3 UE Transceiver RF Characteristics and Requirements

5.3.1 General UE RF Requirements

An NR UE should support both standalone and non-standalone operational modes. For the non-standalone operation, the LTE carrier(s) is required to be the anchor carrier for the NR UEs. In addition to LTE and NR, there may be operators around the globe which rely on 3G and other 2G services. Those radio access technologies may utilize different frequency bands/channels for operation. The discrete (non-integrated) RF front-end architectures that support these RATs and their associated frequency bands may provide good RF performance; however, there would be implications related to cost, form factor, and hardware complexity. Multi-mode multi-band integrated RF architectures may overcome many of the challenges of the discrete design. Such architectures would combine several bands into a single transceiver chain, independent of radio access technology. The RF front-end subsystem comprises PAs, filters, duplexers, switches, oscillators, mixers, LNAs and other passive and active RF components that enable the device to conform to 3GPP and regional regulatory emission specifications. It is understood that various OEMs may combine RF front-end components in different groupings depending on component selection and functionality provided by those components for carrier aggregation and multi-connectivity purposes. Various architectures combine mid-frequency bands (1.7–2.1 GHz and 2.3–2.7 GHz) into a single module to simplify the implementation of carrier aggregation. Similarly, high-frequency bands (in sub-6 GHz) can be combined with Wi-Fi and share antennas.

Due to different transmit/receive configurations (e.g., antenna configurations), network deployments, and different downlink and uplink transmit power, the LTE coverage in general was limited in the uplink. According to the evaluation done for LTE Band 41 Power Class 2 (+26 dBm), the coverage asymmetry could be up to 5 dB based on the network deployment parameters. To improve the uplink coverage, an effective way is to increase the transmit power. Based on some initial analysis of the NR link budget, it appeares that it would suffer from the same issue. Therefore, operators proposed to specify Power Class 2 (+26 dBm) and Power Class 3 (+23 dBm) UE in 3GPP RAN Working Group 4 [2,25].

In contrast to the NR BS, the UE implementations utilize beamforming in FR2 and the RF front-end configurations in FR1 are not significantly different from LTE UEs except in the new 3.7 and 4.5 GHz band implementations. The early deployments of NR use non-standalone mode, thus the compliant UEs will have to implement both LTE and NR radio transceivers. In FR1, the RF performance is specified for the new radio parameters such as the maximum transmit power and receiver sensitivity. In this frequency range, the conducted requirements for antenna connectors are used similar to those of LTE. On the contrary, the FR2 implementations use integrated transceivers and antennas and the measurements cannot be conducted at the connectors, which was the case in the base stations. Therefore, the OTA requirements were introduced to allow UE RF performance measurements. The requirements are defined for EIRP and maximum transmit power in FR2 using the cumulative distribution function of the EIRP values obtained/measured, when performing beamforming, over a full sphere with the UE at the center of the sphere (see Fig. 5.11). These requirements were introduced to ensure (in statistical sense) that the beam can be aimed at the intended direction and toward the serving BS while fulfilling the uplink link budget requirement [29].

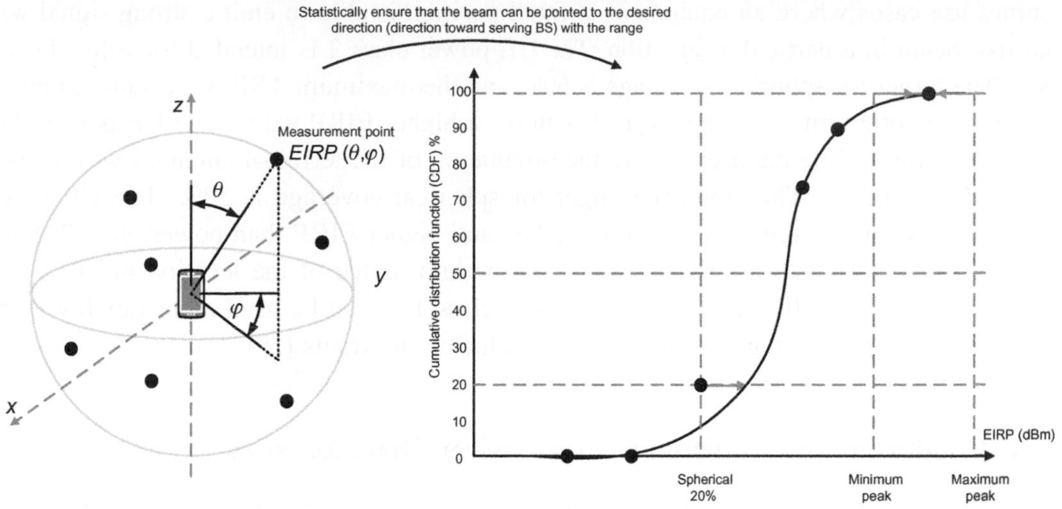

Figure 5.11
OTA EIRP evaluation using cumulative distribution function [29].

Table 5.13: UE power classes in FR2 [3,29].

FR2 Power Class	UE Type	EIRP (dBm)			TRP (dBm)
		Maximum	Minimum Peak	Spherical	Maximum
1	Fixed wireless terminal	55	40	32 (85%)	35
2	Vehicle-mounted terminal	43	29	18 (60%)	23
3	Mobile terminal	43	22.5	11.5 (50%)	23
4	Fixed wireless terminal	43	34	25 (20%)	23

In Fig. 5.11, the minimum peak value is defined as the value that at least one of the measured EIRP values would be exceeded. The spherical coverage is defined as $X\%$ CDF, where the value must be maintained over $(100-X)\%$ of the surface area of the sphere. The maximum value is defined as the value that the maximum measured EIRP must not exceed. The maximum EIRP is set based on regional regulatory requirements. The NR Rel-15 further considers methods to increase the effective range as well as increasing data rates for fixed wireless terminals. The NR UEs are categorized based on the power class in the standards specifications. There are four power classes for the NR devices operating in FR2, which are characterized by the transmit power and the spherical coverage as shown in Table 5.13.

The UE power class 1 is meant for transportable terminals and allows a maximum TRP of 35 dBm compared to other UE power classes with maximum permitted TRP of 23 dBm. It assumes use cases where an equipment at fixed position is able to emit a strong signal with a narrow beam in a particular direction. The UE power class 2 is intended for vehicular use cases. The target for spherical coverage is 60% and the maximum TRP is the same as power class 3 for mobile terminals; however, it requires a higher EIRP value. The UE power class 3 is defined for mobile handsets where the orientation of the terminal antennas with respect to the BS is random. Therefore, the target for spherical coverage is 50%. The UE power class 4 requires wider spherical coverage (20%) and higher EIRP than power class 2. Since this type of the terminal can be installed with the knowledge of the location of the BS, in contrast to the narrow-beam operation of power class 1, it can be used more flexibly in the scenarios involving moving platforms such as vehicles and trains [29].

5.3.2 Conducted and Radiated UE Transceiver Characteristics

The radiated RF requirements for the NR devices are primarily derived from the corresponding conducted RF requirements. Unlike the conducted requirements, the radiated requirements are inclusive of the antennas, thus the emission levels such as BS output power and unwanted emissions are defined either by incorporating the antenna gain as a directional requirement using an EIRP or by defining limits using TRP. The TRP and EIRP metrics are related through the number of radiating antennas and are implementation specific, considering the geometry of the antenna array and the correlation between unwanted emission signals from different antenna ports. An EIRP limit results in different levels of total radiated unwanted emission power depending on the implementation. The EIRP limit per se cannot control the total amount of interference in the network, whereas a TRP requirement limits the total amount of interference in the network regardless of the specific implementation. In the case of passive systems, the antenna gain does not noticeably vary between the desired signal and the unwanted emissions. Thus EIRP is directly proportional to TRP and can be used interchangeably. For an active system such as NR, the EIRP may vary considerably between the desired signal and the unwanted emissions and also between implementations, thus EIRP (in this case) is not proportional to TRP and should not substitute TRP [10].

The RF requirements in FR2 bands are specified separately for devices, because of the higher number of antenna elements for operation in FR2 and the high level of integration when mmWave technologies are used. In general, the RF requirements in FR2 are the same as the conducted RF requirements for FR1; however, the limits for certain requirements might be different. The difference in coexistence in mmWave bands leads to relatively relaxed requirements on ACLR and spectrum mask, which was demonstrated through coexistence studies performed in 3GPP. The mmWave implementations are more challenging than the more mature technologies in the frequency bands below 6 GHz. It should be noted that the channel bandwidths and numerologies supported in FR2 (see Table 5.1) are different from those used in FR1, making the comparison of the requirement levels, especially those related to the receiver somewhat difficult. The following is a list of the radiated RF requirements in FR2 [3,10]:

- *Output power level*: The maximum output power is expressed as TRP and/or EIRP. In FR2, the minimum output power and transmitter OFF power levels are higher than those in FR1. The radiated transmit power is an additional requirement, which unlike the maximum output power, is directional.
- *Transmitted signal quality*: Frequency error and EVM requirements are defined similar to those in FR1 and with the same limits.
- *Radiated unwanted emissions*: Occupied bandwidth, ACLR, spectrum mask, and spurious emissions are defined in the same way as for FR1. The latter two are based on TRP. The limits are relatively less stringent than those in FR1. The ACLR is approximately 10 dB more relaxed compared to FR1, due to more suitable coexistence conditions.
- *Reference sensitivity and dynamic range*: These are defined similar to the counterparts in FR1; however, the levels are not comparable.
- *Receiver susceptibility to interfering signals*: ACS, in-band, and OOB blocking are defined similar to FR1. There is no requirement for narrowband blocking, since there are only wideband systems in FR2. The ACS is approximately 10 dB more relaxed relative to its counterpart in FR1 due to better coexistence conditions (Fig. 5.12).

The device output power level is set by taking the following considerations into account [10]:

1. *UE Power Class* defines a nominal maximum output power for QPSK modulation. It may be different in different operating bands, but the main device power class is currently set to 23 dBm for all bands.
2. *Maximum Power Reduction (MPR)* defines the allowed reduction of maximum power level for certain combination of modulation scheme and resource block allocation. The

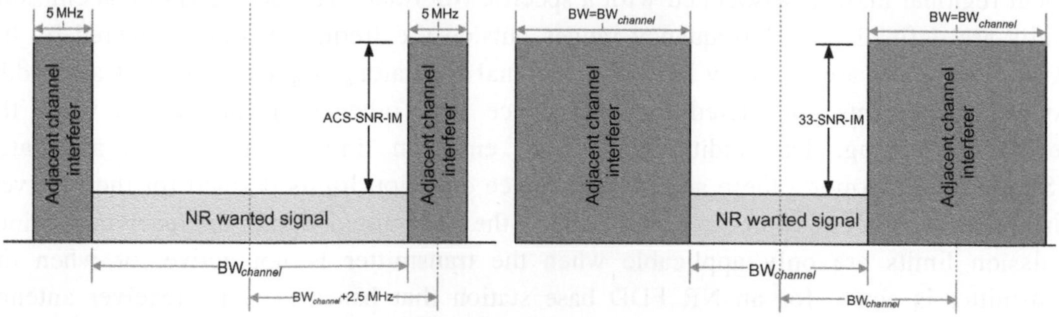

Figure 5.12
Illustration of UE ACS requirements [14].

UE is allowed to reduce the maximum output power due to higher order modulations and transmit bandwidth configurations. For UE power class 2 and 3 the allowed MPR are specified in [3] for channel bandwidths that meets both of the following criteria: Channel_Bandwidth ≤ 100 MHz and Relative Channel Bandwidth ≤ 4% (TDD bands) and ≤ 3% (FDD bands), where Relative Channel Bandwith = $2BW_{channel}/\left(f_{UL\_low} + f_{UL\_high}\right)$.

3. *Additional Maximum Power Reduction (A-MPR)* may be applied in some regions and is usually related to specific transmitter requirements such as regional emission limits and to certain RF carrier configurations. For those requirements, there is an associated NS value that identifies the permissible A-MPR and the associated conditions.

The minimum output power level configuration defines the device dynamic range. There is a definition of the transmitter OFF power level, which is applicable to conditions where the device is not allowed to transmit. There is also a general ON/OFF time mask specified, plus specific time masks for PRACH, PUCCH, SRS, and for PUCCH/PUSCH/SRS transitions. The device transmit power control is specified through requirements for the absolute power tolerance for the initial power setting, the relative power tolerance between two subframes, and the aggregated power tolerance for a sequence of power control commands. In-band emissions are emissions within the channel bandwidth. The requirement limits a device's transmission leakage into non-allocated resource blocks within the channel bandwidth. Unlike the OOB emissions, the in-band emissions are measured after cyclic prefix removal and FFT blocks, representing how a device transmitter affects a BS receiver in practice. For implementation purposes, it is not possible to define a generic device spectrum mask that does not vary with the channel bandwidth, thus the frequency ranges for OOB limits and spurious emissions limits do not follow the same principle as for the BS. The SEM extends to a separation Δf_{OOB} from the channel edges, as illustrated in Fig. 5.13. For 5 MHz channel bandwidth, this point corresponds to 250% of the operating bandwidth as recommended by ITU-R, but for wider channel bandwidths, it is set closer than 250%. The SEM is defined as a general mask and a set of additional masks that can be applied to reflect different regional requirements. Each additional regional mask is associated with a specific NS value. The device spurious emission limits are defined for all frequency ranges outside the frequency range covered by the SEM. The limits are not only based on regional regulatory requirements, but also additional requirements considered for coexistence with other frequency bands when the device is roaming. The additional spurious emission limits can have an associated NS value. Furthermore, there are BS and device emission limits defined for the receiver. Since the receiver emissions are dominated by the transmitted signal, the receiver spurious emission limits are only applicable when the transmitter is not active, or when the transmitter is active for an NR FDD base station that has a separate receiver antenna connector [10].

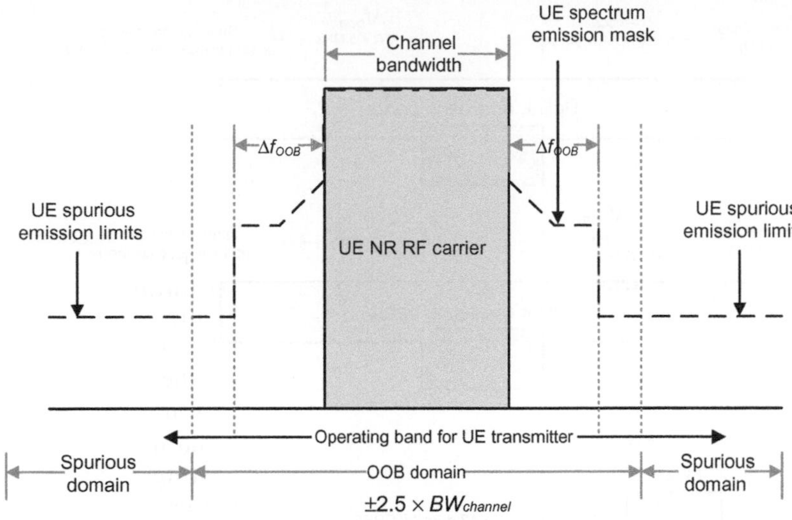

Figure 5.13
Illustration of the spectrum emission mask and spurious emissions requirements for the NR devices [3].

The UE output RF spectrum consists of the following regions that are regulated by 3GPP and regional regulatory requirements [14]:

- *Out-of-Band Emissions*: These are unwanted emissions occurring immediately outside of the assigned channel bandwidth, resulting from the modulation process and nonlinearity in the transmitter RF components. The OOB emissions are characterized in terms of the following metrics (see Fig. 5.14):
  - *Spectrum Emission Mask (SEM)*: The SEM is defined starting from each edge of the assigned NR channel bandwidth to $\pm(\mathrm{BW}_{channel} + 5\mathrm{MHz})$ for FR1 and starting from each edge of the assigned NR channel bandwidth to $\pm 2\mathrm{BW}_{channel}$ for FR2.
  - *Adjacent Channel Leakage Ratio (ACLR)*: The NR ACLR and UTRA ACLR requirements used for FR1 and the NR ACLR requirements are applied for FR2 bands.
- *Spurious Emissions*: These are caused by unwanted transmitter effects such as harmonics emission, parasitic emissions, intermodulation products, and frequency conversion products which cover a frequency range up to the fifth harmonic or 26 GHz for FR1 and a frequency range up to the second harmonic or the uplink operating band for FR2.

The UE reference sensitivity in FR1 is defined as a power level $P_{REFSENS} = 10\log(kT_0W_N) + \mathrm{NF} + \mathrm{SINR} + L_{IM} - G_{DIV}$ (dBm) where kT_0W_N is the thermal noise power in dBm, W_N is the noise equivalent bandwidth (approximately the UE bandwidth), NF is the maximum receiver noise figure, SINR is the minimum SINR for the chosen MCS, and L_{IM}

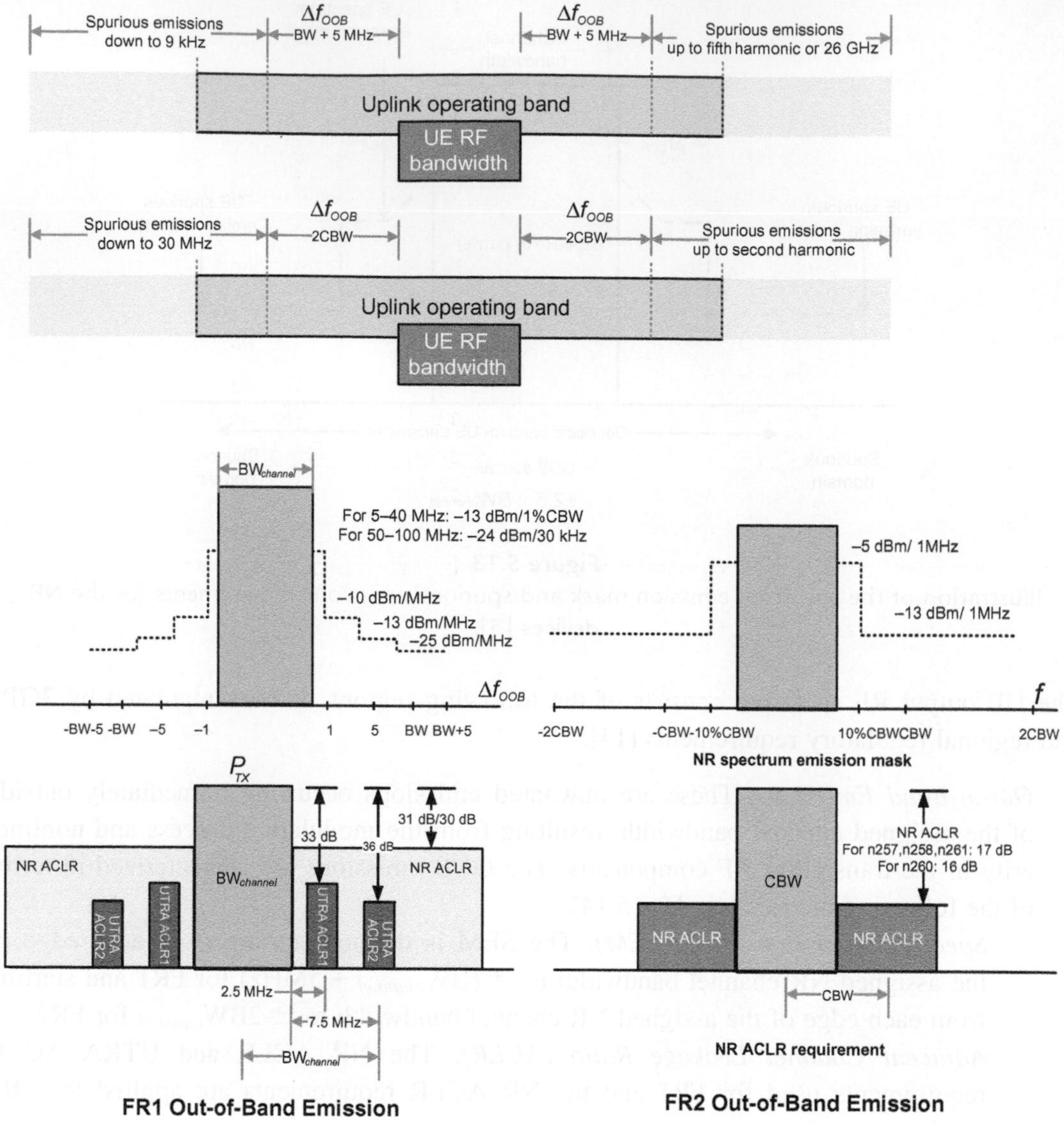

Figure 5.14
UE output RF spectrum and out-of-band emissions [2,3].

is the implementation margin. Note that the noise floor for any receiver is defined as Noise_Floor $= 10\log(kT_0W_N) + \text{NF} = -174 + \text{NF} + 10\log W_N$ (dBm), where $\text{NF} = \text{SNR}_{in} - \text{SNR}_{out}$ (in dB) is a measure of SINR degradation by the components in the RF signal path including the RF filters and LNA. The thermal noise power density $kT_0 = -174$ dBm/Hz is measured at typical room temperature 290 K degree. 3GPP assumes minimum SINR $= -1$ dB for the UE, the implementation margin of $L_{IM} = 2.5$ dB and a diversity gain of $G_{DIV} = 3$ dB for a UE for two RX antennas. The noise figure *NF* is 9 dB for LTE refarming bands, 10 dB for n77 (3.3–3.8 GHz), n78, and n79 bands, and 10.5 dB for n77 (3.8–4.2 GHz) band. In FR1, both 2RX REFSENS (for all bands) and 4RX REFSENS (for n7, n38, n41, n77, n78, n79) bands are defined. In FR2, REFSENS power level is the EIS level in the RX beam peak direction. In carrier aggregation and NR-LTE dual-connectivity scenarios, an additional relaxation is defined for the following cases: (1) band-combination-specific Δ_{RIB} for FR1 inter-band carrier aggregation; (2) aggregated *Channel BW* specific Δ_{RIB} for FR2 intra-band continuous carrier aggregation, that is, 0.5 dB for aggregated bandwidths greater than 800 MHz; and (3) band-combination-specific maximum sensitivity degradation (MSD)[7] for the band impacted by harmonic interference and intermodulation interference for the carrier aggregation and E-UTRA-NR dual-connectivity (EN-DC) scenarios [14]. Note that 3GPP allows certain amount of self-desensitization when TX emissions fall into the receiver band in non-contiguous carrier aggregation transmissions. This is done by relaxing the reference sensitivity requirements by an amount known as MSD when the UE transmits at the maximum power. However, these approaches (MPR and MSD) adversely impact the uplink link budget and throughput and are not the most appealing solutions. Alternatively, one could utilize higher quality RF components with good linear characteristics, which may considerably increase the overall radio implementation cost and size.

5.4 Radio Resource Management Specifications

Radio resource management (RRM) specifications are developed by 3GPP to ensure that the mobile terminals can maintain robust and reliable connection(s) to one or more base stations. The RRM specifications include mobility management, handover, cell quality measurements based on reference signals, or physical synchronization signals. For the purpose of handover to a neighbor cell or adding a new component carrier in carrier aggregation, the UE is required to conduct measurements on the neighboring cells signal quality using reference signal received power (RSRP) or reference signal received quality metrics (RSRQ). In LTE systems, all eNBs continuously transmit cell-specific reference signals (CRS), this would enable the terminals to measure the cell quality of the neighboring cells. However, NR eliminated always-on reference signals to reduce inter-cell interference and the overhead.

[7] The most basic receiver requirement is reference sensitivity. This is similar in concept to LTE/UMTS although there is a new dimension added with the introduction of the MSD parameter. This is a relaxation in the reference sensitivity level that applies when the UE is transmitting at maximum power (with MPR applied) using the maximum number of resource blocks allowed for the channel bandwidth. Under these conditions, it is expected that there will be a loss of sensitivity in the UE receiver.

In NR, the cell quality is measured by using SSBs. Each SSB comprises two synchronization signals and the physical broadcast channel which has a longer transmission periodicity compared to CRS. The SSB periodicity can be configured in the range of 5, 10, 20, 40, 80, and 160 ms for each cell; however, terminals do not need to measure cell quality with the same periodicity as the SSB and the appropriate measurement periodicity can be configured according to the channel conditions. This would avoid unnecessary measurements and conserve terminal power. The new SSB-based RRM measurement time configuration (SMTC)[8] window has been introduced to notify the terminals of the periodicity and the timing of the SSBs that the terminal must use for cell quality measurements. As shown in Fig. 5.15, the SMTC window periodicity can be set in the same range as the SSBs (i.e., 5, 10, 20, 40, 80, and 160 ms) and the duration of the window can be set to 1, 2, 3, 4, or 5 ms according to the number of SSBs transmitted on the cell that is being measured. When a UE is notified of the SMTC window by the gNB, it detects and measures the SSBs within that window and reports the measurement results back to the serving base station [6,29].

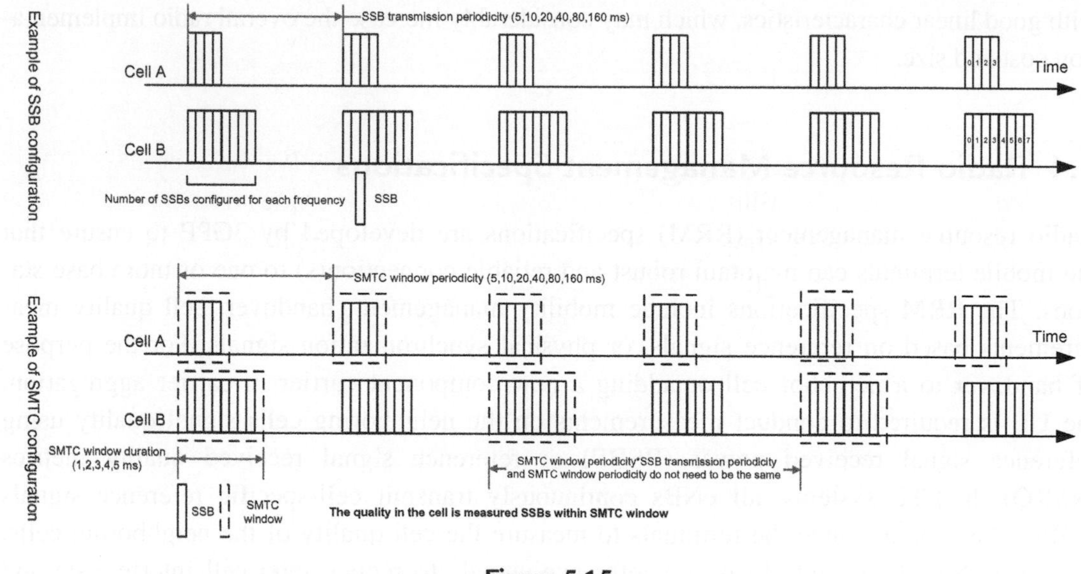

Figure 5.15
Example illustration of SSB and SMTC correspondence in NR [29].

[8] SS-block-based RRM measurement timing configuration or SMTC is the measurement window periodicity/duration/offset information for UE RRM measurement per-carrier frequency. For intra-frequency connected mode measurement, up to two measurement window periodicities, can be configured. For the idle mode measurements, a single SMTC is configured per carrier frequency. For interfrequency connected mode measurements, a single SMTC is configured per-carrier frequency.

Using the same RF transceiver for measuring neighbor-cell quality or other component carriers as well as transmitting/receiving data in the serving cell would make it possible to reduce implementation cost. Nevertheless, this would imply that data cannot be transmitted/received in the serving cell during measurement of other cells or component carriers at different frequencies. In LTE, the UE data transmission in the serving cell is suspended during a measurement gap, providing the UE the opportunity to tune its RF transceivers to conduct neighbor-cell quality measurements or measurement of other component carriers at different frequencies. The concept of measurement gaps is used in NR; however, the measurements are conducted on SSBs and the measurement gap configurations have been improved compared to LTE. In LTE, the measurement gap length (MGL) is fixed such that at least one primary/secondary synchronization signal occasion can be observed within the gap. In LTE, the primary/secondary synchronization signals are transmitted every 5 ms; thus the MGL of LTE is 6 ms, allowing 0.5 ms for RF tuning in the beginning and in the end of the measurement gap. The terminals detect synchronization signals within the MGL and identify the cell ID and the reception timing. The terminal later performs measurements on the CRS. In NR, the SMTC window duration can be set to match the SSB transmission. However, a fixed MGL could cause potential degradation of the serving cell throughput. As an example, if the SMTC window duration is 2 ms and the MGL is 6 ms, a 4 ms interval will not be available for data transmission and reception in the serving cell [6,29].

In order to minimize such performance degradation, the MGL in NR can be configured and can be set to 1.5, 3, 3.5, 4, 5.5, or 6 ms. Fig. 5.16 illustrates the concept of measurement gap configuration in NR, which shows different SMTC window and MGLs. The measurement gap repetition period (MGRP) in NR can be more flexibly configured compared to LTE (Table 5.14). The NR supports 20, 40, 80, and 160 ms MGRPs compared to LTE that only supports 40 and 80 ms. As we mentioned earlier, an RF transceiver tuning time is provisioned in the beginning and the end of the measurement gap, during which the terminal cannot conduct measurements or transmit/receive data. As shown in Fig. 5.17, the SMTC window and the measurement gap start at the same time. The start of the SMTC window may overlap with the RF retuning time and the UE cannot perform measurements during that time. As such, the new measurement gap timing advance function has been introduced, which enables all SSBs (to be measured) within the SMTC window to be used by advancing the start of the RF retuning gap.

The SMTC window and the measurement gaps are configured based on the SSB transmission timing for the candidate cell. However, in some cases when measuring other cells, the serving BS may not have such information. In such cases, it is not possible to properly set the SMTC window and measurement gap for the UE. Therefore, NR defines an SFN and frame timing difference (SFTD) measurement function, where the terminal measures the timing difference of the SFN and the frame boundary between the serving cell and the

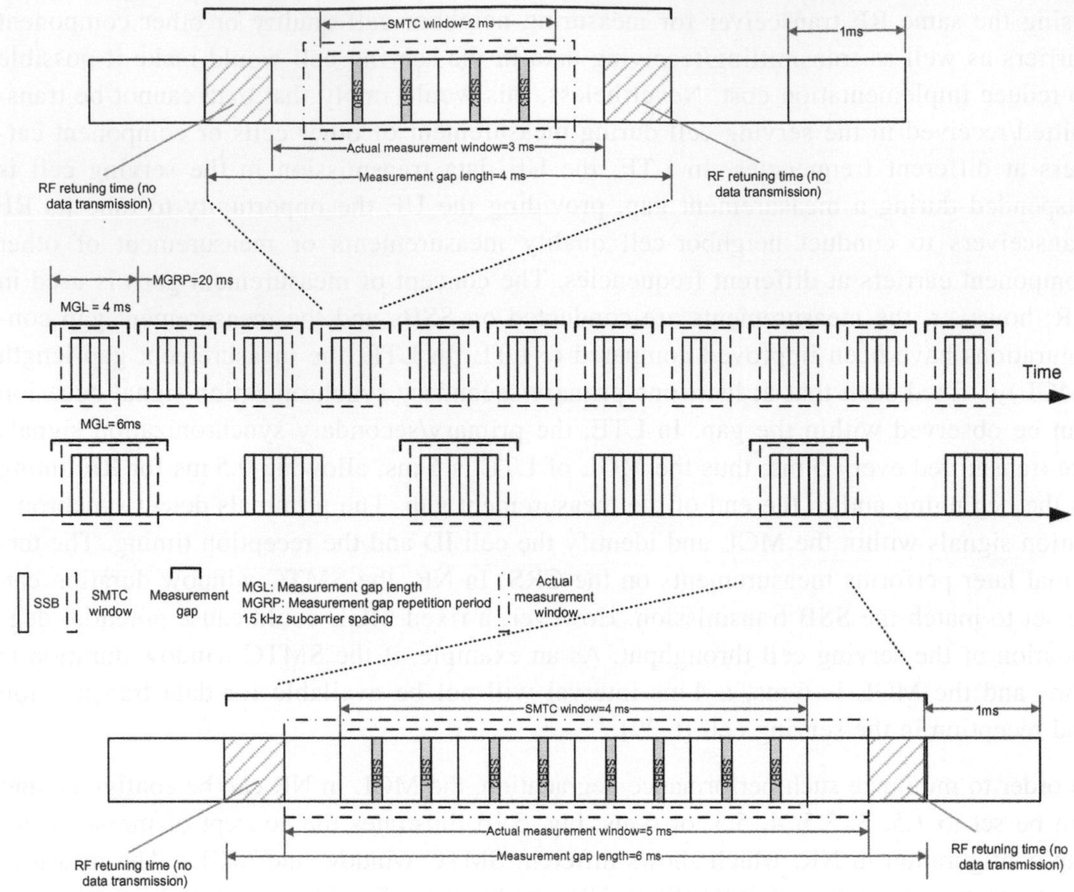

Figure 5.16
Illustration of the NR measurement gap configuration [29].

candidate cell and reports it to the serving gNB. The observed SFTD between an LTE PCell and an NR PSCell consists of two components as follows [7]:

- $SFN\_offset = (SFN_{PCell} - SFN_{PSCell}) \bmod 1024$, where SFN_{PCell} is the SFN of an LTE PCell and SFN_{PSCell} is the SFN of the NR PSCell from which the UE receives the closest starting time to the time when it receives the start of the PCell radio frame.
- $Frame\_boundary\_offset = \lfloor (T_{Frame\_Boundary\_PCell} - T_{Frame\_Boundary\_PSCell})/5 \rfloor$, where $T_{Frame\_Boundary\_PCell}$ is the time when the UE receives the start of a radio frame from the PCell, $T_{Frame\_Boundary\_PSCell}$ is the time when the UE receives the start of the radio frame, from the PSCell, that is closest in time to the radio frame received from the PCell. The unit of $(T_{Frame\_Boundary\_PCell} - T_{Frame\_Boundary\_PSCell})$ is the NR sampling time.

The BS configures SFTD measurements for an NR UE neighbor cells. For a terminal to perform SFTD measurements on neighboring cells to which it is not connected, measurement gaps are required. However, if the serving gNB does not know the SSB transmission timing

Table 5.14: NR measurement gap pattern configurations [6].

Gap Pattern ID	Measurement Gap Length (ms)	Measurement Gap Repetition Period (ms)
0	6	40
1	6	80
2	3	40
3	3	80
4	6	20
5	6	160
6	4	20
7	4	40
8	4	80
9	4	160
10	3	20
11	3	160
12	5.5	20
13	5.5	40
14	5.5	80
15	5.5	160
16	3.5	20
17	3.5	40
18	3.5	80
19	3.5	160
20	1.5	20
21	1.5	40
22	1.5	80
23	1.5	160

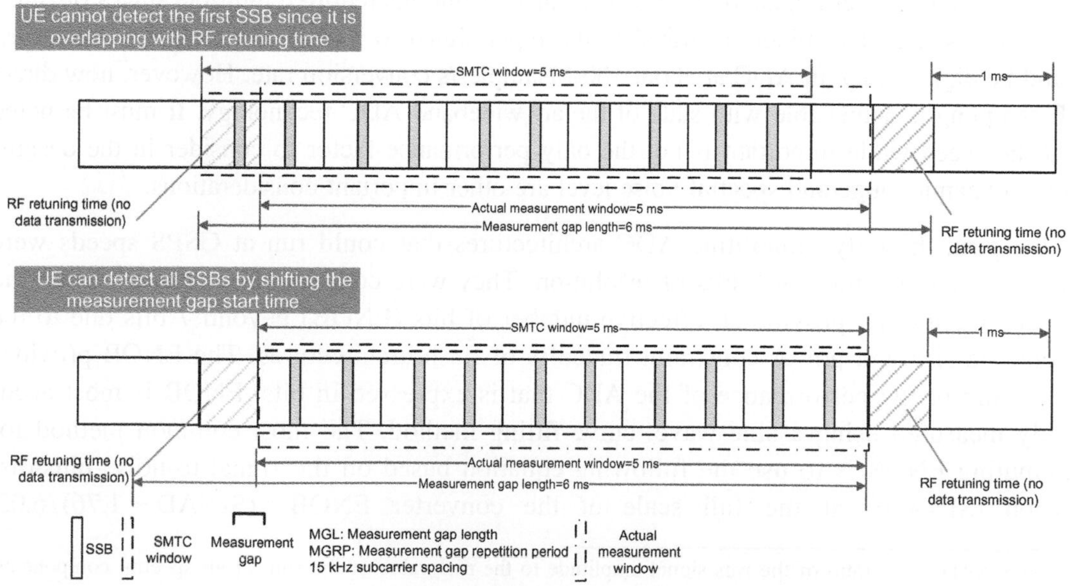

Figure 5.17

Illustration of the concept and usage of measurement gap timing advance [6,29].

of the candidate cell, it cannot configure such measurement gaps. In that case, the BS would need to configure the measurement gaps with different lengths until the terminal can detect the SSB of the candidate cell. This would increase the signaling overhead and the delay until the UE is able to find and connect to the target gNB. In order to avoid such inefficiency, a new procedure for performing SFTD measurement of candidate gNBs without using measurement gap is defined while maintaining UE's transmission/reception in the current cell. Fig. 5.18 shows an example of SFTD measurement which incorporates the aforementioned procedure. Instead of stopping transmission/reception of data by the RF device to perform measurements, the method uses separate RF device on the terminal, which is used if the UE supports carrier aggregation. When the RF transceiver that is used for measurement is turned on or off, the RF transceiver used for data transmission/reception is affected, thus the UE cannot transmit or receive in one subframe before and after the measurement window. However, the terminal can transmit/receive data in the current cell while SFTD measurements for neighboring gNBs are being performed.

5.5 Direct RF Sampling Transceivers

As analog-to-digital converter (ADC) and digital-to-analog converter (DAC) designs and architectures continue to advance using smaller geometry process nodes, a new class of giga-sample-per-second (GSPS) ADC and DAC products have begun to emerge. The ADCs/DACs that can directly sample the RF signal at gigahertz frequencies without the interleaving artifacts provide new solutions to systems for direct RF digitization of communications systems, instrumentation, radar applications, etc. The earlier solutions required multiple stages of filtering, synthesizers, and mixers to translate the input signal to a reference frequency that then could be digitized by an ADC at about 100-Msamples/s conversion rate. However, now direct RF sampling is achievable with state-of-the-art wideband ADC technology. It must be noted that the speed while important is not the only performance factor to consider in the designs as the dynamic range and spectral noise level are other important considerations.

In the past, the only monolithic ADC architectures that could run at GSPS speeds were flash converters with 6 or 8 bits of resolution. They were consuming excessive power and typically could not provide an effective number of bits (ENOB) beyond 7 bits due to the geometric size and power-constraint tradeoffs of flash architectures. The ENOB provides a measure of the performance of the ADC that is expressed in bits. ENOB is most accurately measured using a sine wave, curve-fitting method. The most common method for computing ENOB is to use the following equation based on the signal-to-noise-plus-distortion (SINAD)[9] at the full scale of the converter $\text{ENOB} = (\text{SINAD} - 1.76)/6.02$.

[9] The SINAD is the ratio of the rms signal amplitude to the rms value of the sum of all spectral components, including harmonics but excluding DC. The difference between SNR and SINAD is the energy contained in the first six harmonics.

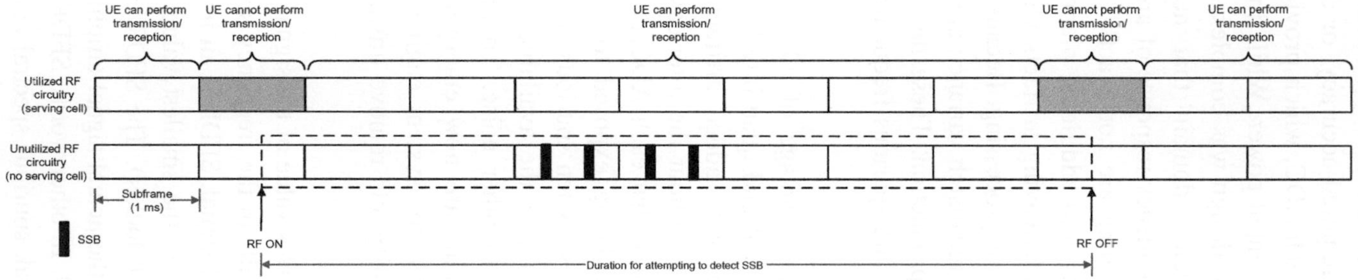

Figure 5.18
Example illustration of SFTD measurement without measurement gap [29]. *SFTD*, SFN and frame timing difference.

Therefore, the only way to sample higher dynamic range analog input signals above 1 GHz was to interleave multiple high-speed ADC cores with a sampling clock that had a staggered phase to each core with the required accuracy, or duty cycle. The analog input needed to be split and multiplexed to each ADC, which provided an opportunity for noise to enter the signal chain and reduce the input power. While this method may provide adequate results for some applications, the design was complex and yielded unwanted interleaving artifacts in the output frequency domain that needed digital filtering. The interleaving spurs can be seen in the frequency response of an FFT block, where the input offset, gain, bandwidth, and sample timing are not exactly matched across each of the internal interleaved ADC cores. This creates additional planning complexity for the system engineer to predetermine where interleaving artifacts will be seen in frequency and either avoid or remove them in digital postprocessing. Because each ADC core is discrete, the potential is high for manufacturing mismatch variance among these performance parameters during the life of a system in production. These mismatches cause imbalances in the periodicity of the incoming signal, and spurious frequencies are seen at the output of interleaved ADCs.

The new ADC technologies can take advantage of advanced architectures and algorithms that prevent the issues seen in dual and quad interleaved ADCs. Instead of using two interleaved ADCs at half speed, with added artifacts, the performance can be achieved in a single ADC at full speed without the interleaving spurs. Factory-trimmed algorithms and on-chip calibration ensure that each ADC operates to the expected high-performance standards, as opposed to being exposed to the mismatch variances seen from multiple discrete interleaved cores. When spurious frequencies are observed in an otherwise spectrally pure FFT, this reduces the available spurious free dynamic range (SFDR) of the carrier signal relative to other noise. To improve the SFDR of GSPS ADCs, new architectures and algorithms are now emerging beyond the use of interleaved cores. This removes the burden for system engineers to have dedicated ADC postprocessing routines that must identify and remove unwanted interleaving spurs (see Fig. 5.19).

The SFDR is defined as the ratio of the rms value of the signal to the rms value of the worst spurious signal regardless of where it falls in the frequency spectrum. The worst spur may or may not be a harmonic of the original signal. SFDR is an important specification in communications systems because it represents the smallest value of signal that can be distinguished from a large interfering signal (blocker). The SFDR can be specified with respect to full-scale (dBFS) or with respect to the actual signal amplitude (dBc). The definition of SFDR is shown graphically in Fig. 5.19. In other words, SFDR is the ratio of the rms value of the signal to the rms value of the peak spurious spectral component for the analog input that produces the worst result.

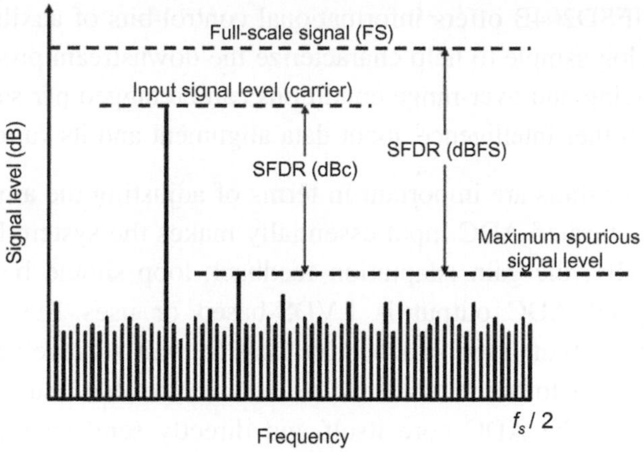

Figure 5.19
Illustration of SFDR concept.

Multi-GSPS converters with 10, 12, or 14-bit resolution generate multiple output data lanes. The use of low-voltage differential signaling (LVDS)[10] data could require 30 parallel lanes of 1-Gbps data for a 2.5 GSPS, 12-bit ADC, in which managing 30 differential LVDS pairs per ADC can be very challenging in terms of routing and maintaining matched lengths on a system layout. Equivalent data can be sent with only six or eight differential lanes using JESD204B/C,[11] a high-speed serializer/de-serializer (SERDES) standard designed specifically for converter interfaces. JESD204B/C provides a means to output data at high speeds on fewer data lines without the matched timing board complexities of many high-speed LVDS lanes. Since the data sent over JESD204B/C is framed, based on an embedded clock and control characters, the routing of the lower count serial lanes is much more tolerant of timing skew than LVDS. This eliminates the need to adjust output timing on every I/O of the system

[10] Low-voltage differential signaling (LVDS) is a high-speed digital interface that has become the solution for many applications that demand low-power consumption and high noise immunity for high data rates. Since its standardization under ANSI/TIA/EIA-644, LVDS has been implemented in a diverse set of applications and industries. The LVDS standard provides guidelines that define the electrical characteristics for the driver output and receiver input of an LVDS interface but does not define a specific communication protocol, required process technology, media, or voltage supply. The general, non-application-specific nature of the standard has been conducive to the adoption of LVDS across a wide variety of commercial and military applications.

[11] A standardized serial interface between data converters (ADCs and DACs) and logic devices (FPGAs or ASICs). JESD204B supports serial data rates up to 12.5 Gbps, a mechanism to achieve deterministic latency across the serial link, and uses 8B/10B encoding for SerDes synchronization, clock recovery, and DC balance. In Revision A, the main goal was to support both single and multiple lanes per convertor device. In Revision B, the added features were programmable deterministic latency, usage of device clock as main clock source, and data rate up to 12.5 Gbps. In the latest Revision C, the data rate is increased up to 32 Gbps, and three link layers have been introduced as 64B/66B, 64B/80B, and 8B/10B where 8B/10B link layer is same as JESD204B link layer. JESD204C defines layer-wise relationship with the IEEE Ethernet model.

board. Additionally, JESD204B offers informational control-bits of auxiliary data that can be appended to each analog sample to help characterize the downstream processing. In this manner, trigger time-stamping and over-range conditions can be tagged per sample so that a back-end FPGA can have further intelligence about data alignment and its validity.

The adaptive gain algorithms are important in terms of adjusting the amplitude of an analog input signal, since a saturated ADC input essentially makes the system blind in its ability to decipher signals. Ideally, the gain adaptation feedback loop should be as fast as possible. Whether the high-speed ADC output is LVDS-based or uses JESD204B/C, the added latency of this digital output often can be too long to wait to receive the saturated data, detect the issue, and react to the condition. One solution to this issue is to use a variable-level comparison within the ADC core itself and directly send an immediate output flag upon occurrence of an over-range condition. This technique bypasses the latency of the longer back-end output stage, which shortens the feedback time to the amplifier, allowing for a faster adaptive gain cycle. In addition to this fast, over-range detection output the over-range samples can be appended with alert bits, using the JESD204B interface, to let downstream system processing make appropriate decisions about the data.

A wideband ADC not only offers the benefits of broadband sampling, but also may provide more data than needed in some applications. For high-sample-rate systems that do not need to observe a large frequency spectrum, digital down-conversion (DDC) allows a subsampling and filtering strategy for decimating the amount of data output from the GSPS ADC. Downstream processing then observes a smaller portion of the frequency spectrum. The DDCs are often implemented after the ADC in the signal chain. This not only consumes more resources in an FPGA but also requires the full bandwidth to be transmitted between the ADC and FPGA. Instead of transmitting and processing the sampled data in an FPGA, the DDC filtering can be done within the ADC to see just one-eighth or one-sixteenth of the total bandwidth. When used in conjunction with a synthesized numerically controlled oscillator (NCO), the precise placement of the converter's DDC filter in the band can be tuned with accurate resolution. This permits a lower output rate and eliminates the need to move and process large amounts of unwanted data in an FPGA. When two DDCs are available, each with a unique NCO, they can alternately be stepped across the spectrum to sweep for expected signals, without loss of visibility. This is often typical in some radar applications.

5.5.1 Nyquist Zones and Sampling of Wideband Signals

In the theory of sampling, a signal must be sampled at more than twice the maximum frequency $f_{\max}$ in the spectrum of the signal, in order to be able to regenerate the signal from its sampled values. This sampling frequency is known as the Nyquist rate $f_s \geq 2f_{\max}$. If the signal is sampled below the Nyquist rate, aliasing occurs. As a general rule, for a sample rate of f_s, one would not be able to differentiate between any sine waves with frequencies

$Nf_s/2 \pm \Delta f$, where N is an integer. In this case, the signal power for these frequencies will concentrate at $f_s/2 - \Delta f$. As long as, for each Δf, there is only a signal at one of the aliasing frequencies, the signal has not been irreversibly corrupted. That means the signal can still be reconstructed, if one knows which of the frequencies that alias occurs $f_s/2 \pm \Delta f$ were occupied by the desired signal. In this case, the aliasing decisively can be useful, which leads to the concept of the Nyquist zones. However, if the desired signal has significant power at both $f_s/2 - \Delta f$ and $f_s/2 + \Delta f$ frequencies, then aliasing occurs and you no longer have the information to distinguish these two components since these formerly distinct signal frequencies now appear as a single frequency, thus it is not possible to determine the power of each component. For a sampling frequency f_s, various zones exist where, if a signal is band-limited to that zone, the original spectrum can be recovered. As shown in Fig. 5.20, at the sampling frequency f_s, one can define zones where a band-limited signal in a higher Nyquist zone will alias down into the first Nyquist zone. For even-numbered Nyquist zones, the spectrum will appear in reversed order, and for odd-numbered zones, it appears in the original order. This means, for example, a 100 MHz band ranging from 700 to 800 MHz need only be sampled at 100 MHz to be characterized. As long as you know which Nyquist zone is occupied, you have all information to reconstruct the original signal. One issue is that the boundaries of the Nyquist zones are fundamentally tied to the sampling rate, which implies that the bandwidth of the desired signal must fit within a Nyquist zone.

According to Poisson summation formula,[12] the samples of the Fourier transform of function $v(t)$ are sufficient to create a periodic extension of $V(f)$ as $V(f) = \sum_{k=-\infty}^{\infty} V(f - kf_s)$. The latter equation can be interpreted as the sum of the replicas of $V(f)$ shifted to integer multiples of f_s (see Fig. 5.20). Fig. 5.20 also illustrates an example of the Nyquist frequency $f_s/2$. Since these signals are repeated at multiples of f_s, then for bandwidths greater than $f_s/2$ aliasing and information loss will occur. The integer multiples of the Nyquist frequency determines the Nyquist Zones. In the literature, the Nyquist Zones are only defined for the positive frequency spectrum.

In other words, the FFT of a discrete-time signal can be divided into an infinite number of $f_s/2$ frequency bands, also known as Nyquist zones. The frequency spectrum between DC (zero frequency) and $f_s/2$ is known as the first Nyquist zone. The frequency spectrum repeats itself over different Nyquist zones. Note that the even-numbered Nyquist zones appear as mirror images of the odd-numbered Nyquist zones. Aliasing in ADCs is a ramification of the sample-and-hold processing of the analog signal at the input stage. In the

[12] In mathematics, the Poisson summation formula is an equation that relates the Fourier series coefficients of the periodic summation of a function to values of the function's continuous Fourier transform. Consequently, the periodic summation of a function is completely defined by discrete samples of the original function's Fourier transform. Conversely, the periodic summation of a function's Fourier transform is completely defined by discrete samples of the original function.

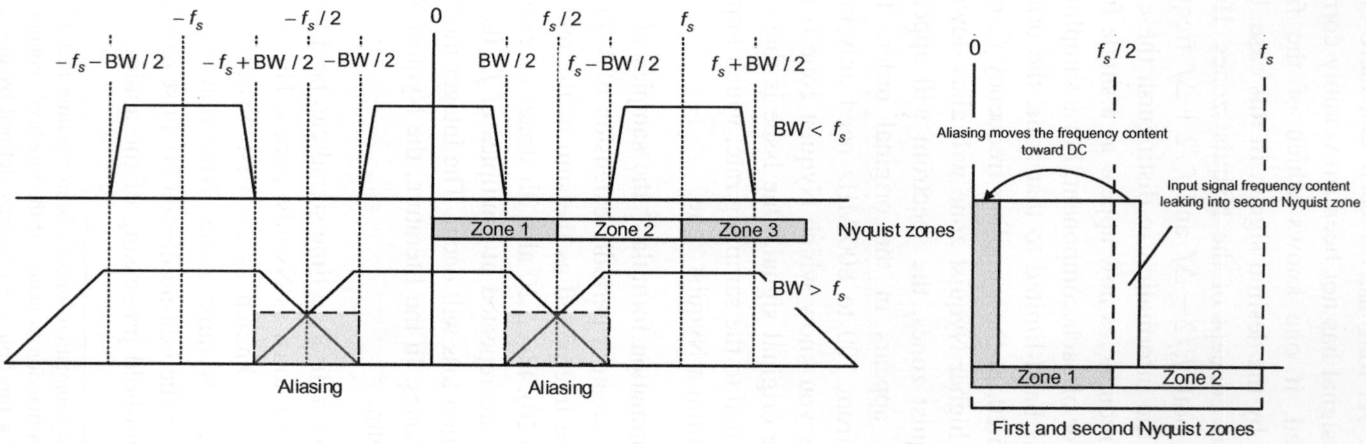

Figure 5.20
Illustration of the Nyquist zone concept [15].

digital signal processing domain, sample-and-hold processing is equivalent to convolution of the frequency spectrum of the impulse train (due to the sampling clock) with the frequency spectrum of the analog input. This convolution results in periodicity of the frequency spectrum that is observed over various Nyquist zones, as previously explained. When the input signal contains frequency components above the Nyquist frequency ($f_s/2$), the adjacent Nyquist zones start overlapping and result in aliasing. Aliasing in DACs is a consequence of the zero-order-hold processing (used to avoid code-dependent output glitches) of the discrete-time samples at the output stage. The zero-order-hold processing in the DSP domain is equivalent to convolution of the $\sin(x)/x$ type of frequency spectrum (of the rectangular function appearing due to holding discrete-time samples) with the DAC core's output-impulse train frequency spectrum (of varying amplitude, in general). As in the case of ADCs, the periodicity of the output spectrum over different Nyquist zones can be attributed to this convolution (see Fig. 5.21).

Data converters are divided into two main categories, namely, Nyquist rate and oversampling. In the Nyquist rate type, the input occupies a large fraction of the available bandwidth (Nyquist zone), whereas in the oversampling type, the input occupies only a small fraction of the Nyquist zone, simplifying the antialiasing filter (AAF) design and lowering the quantization noise at the expense of higher sampling rate (see Fig. 5.22).

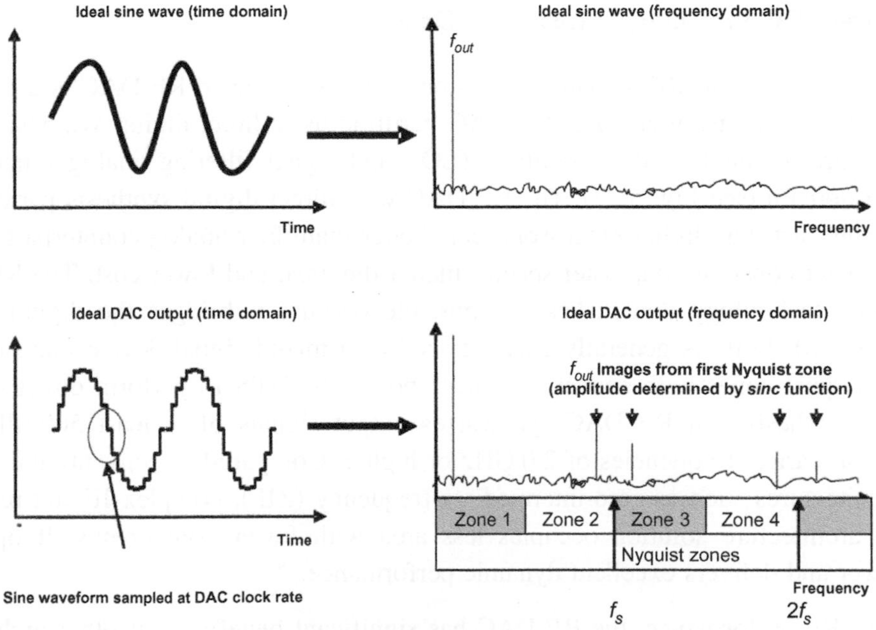

Figure 5.21
Example illustration of Nyquist zones [15].

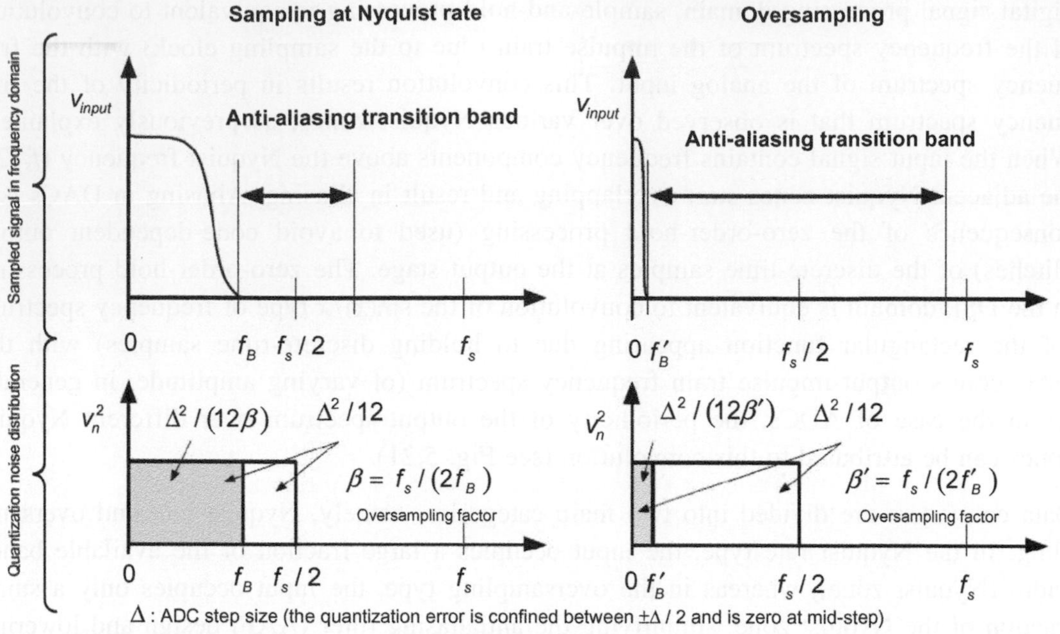

Figure 5.22
Illustration of Nyquist rate and oversampling data conversion [15].

5.5.2 Direct RF Sampling Transmitter Design

Direct digital domain to RF domain conversion with a high-speed RF DAC is a technology disruption for BS transmitters. The RF DAC partition uses direct digital synthesis to move the quadrature modulator, local oscillator (LO), and signal filtering analog functions into the digital domain (see Fig. 5.23). The RF DAC with direct digital synthesis partition capitalizes on the fact that digital processors scale better than their analog counterparts in terms of lower power consumption, faster speed, smaller die area, and lower cost. The RF DAC is the enabling technology that makes this possible because it bridges the digital-to-analog domains. An RF DAC is generally characterized as a mixed signal device that operates in multiple Nyquist zones with conversion rates above 1.5 GSPS to perform direct digital-to-RF signal synthesis. An RF DAC synthesizes output signals of at least 500 MHz signal bandwidth at carrier frequencies of 2.0 GHz or higher. Compared to conventional RF transmitter architectures such as zero-intermediate-frequency (ZIF), complex-IF and real-IF, the RF DAC architecture solution occupies less area with fewer components. It operates at lower power and delivers excellent dynamic performance.

In terms of RF performance, the RF DAC has significant benefits over other architectures. The digital up-conversion with digital filtering implemented in direct digital synthesis eliminates gain-phase errors and achieves perfect carrier suppression with no LO leakage. The result is excellent EVM performance when transmitting high-order modulations like

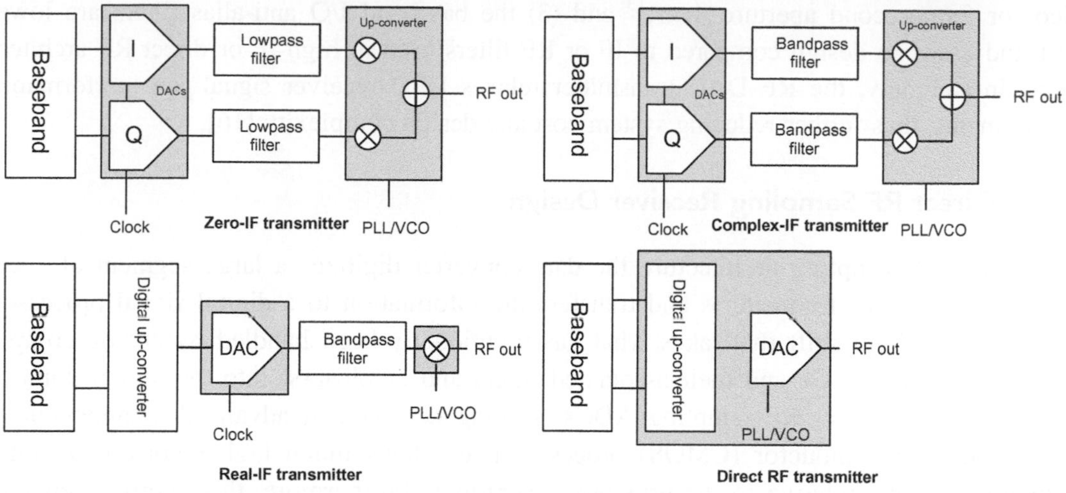

Figure 5.23
Illustration of RF transmitter architecture options [16].

64QAM. The quadrature NCO makes the RF DAC an agile transmitter capable of tuning across the entire spectrum of LTE/NR sub-6 GHz bands (and mmWave bands using an external mixer and filter). Since the RF DAC is broadband and frequency agile with high dynamic range, a single device can synthesize multi-carrier, multi-band, and multi-standard signals including GSM, WDMA, LTE and NR. As a result, a fully digital software-defined radio can be realized using common hardware across multiple BS hardware platforms [16].

Another advantage of direct-conversion RF DAC technology is enabling lower cost digital predistortion (DPD) observation receiver. Macro-cell base stations typically use DPD techniques for RF PA linearization. This requires a PA observation receiver channel to monitor the PA output. The observation receiver detects PA distortion products and works with a predistortion block to compensate for intermodulation and adjacent channel leakage power. The DPD bandwidth expansion requires the DPD observation receiver bandwidth to be five times the data bandwidth. As an example, in 100 MHz carrier aggregation scenarios, this means that the DPD bandwidth must be at least 500 MHz. Also, the observation receiver cannot add impairments to the observed signal because they cannot be discerned from the main transmitter (TX) path impairments. Consequently, the DPD observation path must have excellent linearity which adds cost and complexity. Conversely, if the main TX path has negligible impairments, then the DPD path impairments can be corrected. Since the RF DAC does not introduce gain or phase errors, there are negligible TX path impairments. Therefore, a low-cost and lower performance DPD receiver such as a ZIF receiver can be employed. There are three reasons that the ZIF architecture has lower cost compared to high-IF or direct RF sampling: (1) quadrature demodulation enables a lower conversion rate, baseband sampling, dual-channel high-speed ADC because it only needs to quantize one-half of the DPD expansion bandwidth; (2) the ADC samples baseband signals, which means that the ADC does not need pico- or femtosecond aperture jitter;[13] and (3) the baseband I/Q anti-alias filters are lower cost and easier to design compared to IF or RF filters used in high-IF or direct RF architectures. In summary, the RF DAC transmitter relaxes DPD receiver signal-path performance requirements, thus further reducing system cost and design complexity [16].

5.5.3 Direct RF Sampling Receiver Design

In a direct RF-sampling architecture the data converter digitizes a large segment of spectrum directly at RF frequencies and transfers the information to a digital signal processor. This is a paradigm shift that takes what has traditionally been handled by analog components (i.e., mixers, LOs and their associated filters and amplifiers) into the digital domain. A new class of direct RF-sampling ADCs is being designed in advanced complementary metal-oxide semiconductor (CMOS) processes that allows much higher conversion rates with lower power relative to the previous generations. Furthermore, this approach enables higher integration, which is used for a low-power, multi-giga-bit serial interface, and on-chip DDC, leading to very size- and power-efficient digital interconnect between the data converter and digital processor. As shown in Fig. 5.24, in a super-heterodyne receiver, the input signal, which resides at an RF frequency, is down-converted to a lower IF frequency. It is then digitized prior to digital filtering and demodulation. Depending on the application, the input signal can range from 700 MHz to several GHz, while IF typically ranges from zero to 500 MHz. The classic super-heterodyne receiver consists of bandpass filters (BPFs), a LNA, mixer and LO, an IF amplifier, and ADC AAF [15,17].

In a direct-conversion receiver, the RF-sampling ADC replaces the signal chain from the mixer, which greatly simplifies the overall receiver design. The first RF BPF is typically a

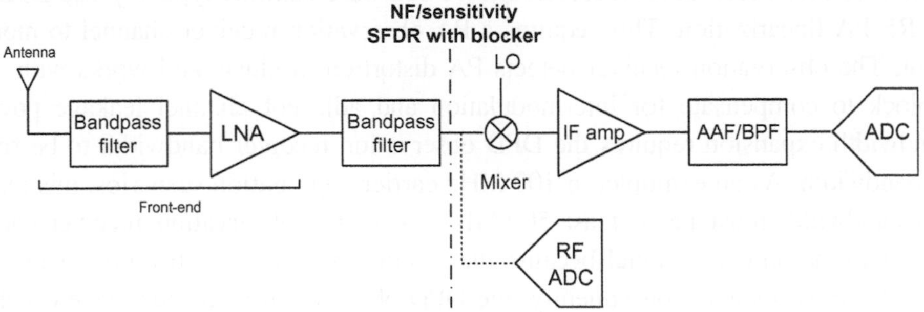

Figure 5.24
Performance comparison between super-heterodyne and direct RF-sampling receivers [15].

[13] Aperture jitter or aperture uncertainty is the sample-to-sample variation in aperture delay that can be manifested as frequency-dependent noise on the ADC input. Aperture delay is a measure of the difference in the delay between the analog path and the encode path. It is measured by observing the time from the 50% point of the rising edge of the sample clock to the time at which the input signal is actually sampled.

wideband, preselect filter, which provides most of the OOB rejection. It prevents signals that are far from the actual passband from saturating the analog front-end (AFE). The LNA increases the amplitude of weak signals. The lower its NF, the less it degrades the overall NF of the receiver based on the cascaded NF concept. The second RF filter is a narrowband filter prior to the mixer typically a surface acoustic wave (SAW) type. It suppresses nearby OOB interferers at the mixer image locations ($mf_{LO} \pm nf_{RF}$) that would fall into the IF passband. The mixer translates the RF signal to IF frequencies. The RF and LO signals mix to produce a difference frequency known as IF frequency ($f_{IF} = |f_{RF} - f_{LO}|$). The undesired spurs and images (e.g., a half-IF) need to be filtered out either before or after the mixer stage. It also can be used to convert the single-ended input signal to a differential signal for the ADC. The LO is tuned to the desired frequency spacing above or below the RF signal and is injected into the mixer.

In a direct RF-sampling receiver, the LO essentially turns into the ADC's sampling clock. However, similar to the IF-sampling converter, the RF ADC clock also requires very good phase noise properties. The IF amplifier adds gain to the input signal by reducing the ADC's impact to the receiver NF. Moreover, it compensates for the BPF attenuation. It can also be used as a buffer to drive the ADC's capacitive load. In many cases, this stage has multiple gain steps that are digitally controlled to provide an automated gain control loop, which enhances the system's dynamic range. The anti-aliasing filter limits noise and distortion contribution from the IF amplifier. More importantly, it filters the ADC's alias bands. This filter may need a sharp roll-off. The implementation is more feasible at IF than RF frequencies because the ratio of alias frequency to the desired signal is larger. Thus aliasing frequencies are further away from the filter cutoff frequency. The ADC digitizes the input signal. It needs a fast sampling rate to accommodate the filter roll-offs, typically at least three to five times the signal bandwidth. Its SNR needs to be good to ensure minimal impact on the receiver noise figure. The SFDR must be sufficient so that spurs caused by in-band and out-of-band interferers do not dominate the overall noise level.

When evaluating a transition from an IF-sampling-based to a direct RF-sampling-based receiver design, one must examine the overall impact on the receiver sensitivity, as well as performance in a blocking condition. Since the RF-sampling ADC replaces the signal chain from the mixer onward, the comparisons provided are focused at the mixer input, assuming the same AFE for both designs (see Fig. 5.24). The parameters also can be calculated at the antenna input for a given LNA in the same manner using the respective gains and losses of the amplifier and filters. The receiver sensitivity is a measure of how it can recover and process very weak input signals. Weak input signals cannot be demodulated, if the receiver noise within the demodulated bandwidth is larger than the received signal itself. The transmitter and receiver are often completely independent, thus raising the desired signal amplitude above the noise floor is not always possible. The only option to improve receiver sensitivity is to reduce its noise floor or, in other words, to improve its noise figure. In the sensitivity condition, the receiver operates at the maximum gain. In the presence of a strong blocker/interferer the gain needs to be reduced in order to avoid saturation of the input (desensitization). Therefore, the receiver NF will be much stronger, which impacts the minimum signal that can be recovered in a blocking condition.

The images from the mixer at $mf_{LO} \pm nf_{RF}$ can fall into the band of interest that requires adequate filtering at RF level. This ensures that an OOB blocker at the image locations does not generate a mixing product within the desired band. Furthermore, the phase noise of the LO frequency is also mixed with the RF signal (interferer). This increases the noise power with in-band blockers, increasing the amount of noise leaking from the blocker into the wanted carrier bandwidth. The IF gain amplifier generates second- and third-order harmonic distortion (HD2 and HD3) products from the strong interferer, which can fall in the signal band of interest. These low-order harmonics are typically either frequency controlled by choosing an appropriate IF frequency or attenuated by the following BPF.

Several impairments originate from the data converter. Similar to the IF amplifier, the ADC generates strong low-order harmonic distortions (HD2 and HD3) from the blocker. The IF frequency can be chosen such that these harmonics are located out-of-band for an in-band blocker. For an OOB blocker, adequate external filtering should be provided. High-order harmonics set a spur floor that cannot be controlled and needs to be taken into consideration. The phase noise of the sampling clock (jitter) also mixes with the interferer. It scales with $20 \log(f_S/f_{IN})$, so the higher the input frequency range (IF) with a fixed ADC sampling rate, the larger its impact on the overall system noise. The RF-sampling ADC may achieve its fast clock rate using interleaving techniques. In this case, the interleaving spurs need to be considered not only for the noise analysis with in-band blockers but also for the filter design for out-of-band interferers.

5.6 Considerations for Implementation and Operation in mmWave Bands

5.6.1 Semiconductor Technologies for mmWave Operation

The mmWave RF analog implementations have been historically dominated by III–V[14] monolithic microwave integrated circuit (MMIC)[15] semiconductor technologies which primarily use

[14] A substance that can act as an electrical conductor or insulator depending on chemical alterations or external conditions. Examples include silicon, germanium, and gallium arsenide. It is called III–V materials since semiconductor elements are in groups III and V of the periodic table of chemical elements.

[15] An MMIC is a type of integrated circuit device that operates in microwave frequencies. These devices typically perform functions such as mixing, power amplification, low-noise amplification, and high-frequency switching. Inputs and outputs on MMIC devices are matched to a characteristic impedance of 50 ohms.

gallium arsenide (GaAs).[16] These are ideally suited for the RF front-ends of mmWave systems such as PAs and LNAs, as well as enabling oscillators with excellent phase noise characteristics. More recently, SiGe-heterojunction bipolar transistor (HBT)-based technologies have attracted increasing interest for emerging mmWave markets, as f_T or $f_{\max}$[17] of the HBT devices has exceeded 200 GHz. The performance of SiGe-HBT[18] is no longer the limiting factor for a mmWave transceiver front-end integration for small-cell scenarios with limited output power, rather the quality factors of the on-chip passive devices, such as inductor, capacitor, and transmission lines for matching and tuning and their accurate characterization in the mmWave frequency domain, are the limiting factors. However, the latest products in the market demonstrate sufficient quality, which is a trade-off between performance and cost. The E-band backhaul for macro-cells usually require support of high-order QAM modulation, which can be achieved with GaAs components or a combination of SiGe/BiCMOS[19] transmitter/receiver and GaAs PAs, LNAs, and voltage-controlled oscillators (VCOs) [23].

The CMOS[20] technology promises higher levels of integration at reduced cost, if volumes scale to multi-million parts per year. Several recent developments, particularly the need for the availability of chipsets operating in mmWave frequencies, have increased efforts to enable CMOS circuit blocks to operate in mmWave bands, as the CMOS transistor f_T approaches to 400 GHz. However, the performance for point-to-point links is worse than that achieved by SiGe[21] or GaAs components in terms of phase noise and NF for the same distances typically greater than 100 m [23].

[16] Gallium arsenide (GaAs) is a compound of the elements gallium and arsenic. It is a III−V direct bandgap semiconductor with a zinc blende crystal structure. GaAs is used in the manufacturing of devices such as microwave frequency integrated circuits, monolithic microwave integrated circuits, and infrared light-emitting diodes. GaAs is often used as a substrate material for the epitaxial growth of other III−V semiconductors including indium GaAs, aluminum GaAs, and others.

[17] The transit frequency is a measure of the intrinsic speed of a transistor and is defined as the frequency where the current gain reduces to one.

[18] HBT, a type of bipolar junction transistor (BJT) which uses differing semiconductor materials for the emitter and base regions, creating a heterojunction.

[19] BiCMOS is an evolved semiconductor technology that integrates two formerly separate semiconductor technologies, those of the BJT and the CMOS transistor, in a single integrated circuit device.

[20] CMOS is a technology for manufacturing integrated circuits. CMOS technology is used in microprocessors, microcontrollers, etc., as well as analog circuits such as image sensors and data converters. Two important characteristics of CMOS devices are high noise immunity and low static power consumption. Since one transistor of the pair is always off, the combination draws significant power only momentarily during switching between ON and OFF states. Consequently, CMOS devices do not dissipate heat as other forms of logic. CMOS also allows a high density of logic functions on a chip. It was primarily for this reason that CMOS became the most used technology to implement very-large-scale integration chips.

[21] Silicon germanium (SiGe) is an alloy with any molar ratio of silicon and germanium, which is commonly used as a semiconductor material in integrated circuits for HBTs or as a strain-inducing layer for CMOS transistors. This relatively new technology offers opportunities in mixed-signal and analog integrated circuit design and manufacturing. SiGe is also used as a thermoelectric material for high temperature applications process.

Fully depleted/partially depleted silicon on insulator (RF-SOI/FD-SOI) is a semiconductor technology which realizes fully/partially depleted CMOS devices on SOI and combines the flat structure of 2D CMOS transistors with the fully depleted mode of SOI. The SOI has a good power-performance-area-cost (PPAC) value. It uses a unique type of substrate material whose thickness is controlled to atomic scale and could provide excellent transistor performance. The FD-SOI technology can operate in mmWave frequencies, thus it is gaining more attention in practical 5G mmWave applications.

Packaging processes which are available and in production for RF analog components are Quad Flat No-leads (QFN) Package, embedded Wafer-Level-Ball (eWLB) Grid Array Package,[22] and Flip Chip.[23] However, the existing mmWave RF components are mostly bare die or modules. The eWLB examples with System in Package (SiP) demonstrate that assembly and packaging using the eWLB technology offer outstanding system integration capabilities. This includes the integration of different chips and the design of integrated passive components such as resistors, inductors, and transformers either in the redistribution layer (RDL)[24] or using through encapsulate via (TEV).[25] Antennas can also be integrated into the package. Other technologies where silicon wafer level technology and back-end merge are through silicon via (TSV)[26] and die embedding in laminate technologies. TSV technologies are typically combined with RDLs, for example, for silicon interposer. However, a major obstacle against their broad adoption is cost [23].

The evolution of RF analog components integration for mmWave applications depends on several factors including the following:

- Permissible output power and EIRP of the system including the antennas
- Phase noise required for standard modulation schemes (BPSK, QPSK, 16QAM, ..., 256QAM, etc.)
- Noise figure
- Power consumption
- Form factor and cost

[22] eWLB grid array is a packaging technology for integrated circuits. The package interconnects are applied on an artificial wafer made of silicon chips and a casting compound.

[23] Flip chip is a method for interconnecting semiconductor devices to external circuitry with solder bumps that have been deposited onto the chip pads. The solder bumps are deposited on the chip pads on the top side of the wafer during the final wafer processing step. In order to mount the chip to external circuitry, it is flipped over so that its top side faces down and aligned so that its pads align with matching pads on the external circuit, and then the solder is reflowed to complete the interconnect. This is in contrast to wire bonding, in which the chip is mounted upright and wires are used to interconnect the chip pads to external circuitry.

[24] An RDL is an extra metal layer on a chip that makes the I/O pads of an integrated circuit available in other locations. When an integrated circuit is manufactured, it usually has a set of I/O pads that are wire-bonded to the pins of the package.

[25] TEV technology is an interconnect technology process.

[26] TSV technology is interconnect technology process.

Silicon transistors cannot compete with III–V compounds (GaAs, InP, GaN) for low-noise performance, linearity, and output power at frequencies above 20 GHz. A GaAs mmWave LNA yields an average NF of 2.5 dB, which is much lower than state-of-the art SiGe LNA of 5 dB. The output power levels (P_{sat}) of over 30 dBm can be achieved with GaAs in E-Band, while SiGe-HBTs can reach 19 dBm (P_{sat}) [23].

Silicon RFICs allow the integration of multiple application-specific functionalities on a single silicon chip (RF application-specific integrated circuit or ASIC) with excellent yield and uniformity plus the possibility to integrate different calibration schemes to mitigate RF impairments, which are not possible or more complex to implement in GaAs. The level of integration is a factor to be considered. A high level of integration makes the chip very specific and could increase development time in a first design but reduces production test and simplifies module assembly. A good compromise for high-end applications (e.g., E-band high power, 256QAM) is to use compound semiconductors for the front-ends (LNA of the receiver input and PA of the transmitter output) and silicon semiconductors for the lower frequency mixed signal functions and control/digital elements. The III–V compound devices can realize systems to expand the use of the electromagnetic spectrum above 90 GHz. However, improvements in the high-frequency capability of CMOS/BiCMOS technology have made it possible to consider it as a low-cost, lower performance alternative to III–V compound devices.

5.6.2 Analog Front-End Design Considerations

The baseband modems utilize digital signal processing to perform channel coding and complex-valued modulation and demodulation in order to transmit extremely high data rate streams over RF carriers in mmWave bands. For backhaul applications with data rates ranging from 1 to 10 Gbps, the systems typically utilize channel bandwidths of 250 and 500 MHz which have recently increased to 1 and 2 GHz to support integrated access and backhaul applications. The baseband modulation schemes, generating complex-valued symbols, range from BPSK, QPSK, 16QAM and up to 256QAM (or higher) for high-performance systems. Therefore, it is necessary to perform real-time I/Q ADC (receive) and I/Q DAC (transmit) at the interface between the analog and digital domains. Moreover, such data conversions require low jitter baseband clocks in order to minimize inter-symbol interference. The combination of I/Q ADC, I/Q DAC, and baseband phase-locked loop (PLL)[27] is known as AFE as shown in Fig. 5.25. The performance of the AFE has a significant impact on the overall performance of the modem in terms of the receiver dynamic range, transmitter spurious emissions, and end-to-end bit error rate/packet error rate (BER/PER).

[27] A PLL is a closed-loop frequency-control system based on the phase difference between the input clock signal and the feedback clock signal of a controlled oscillator. The main blocks of the PLL are the phase-frequency detector, loop filter, voltage-controlled oscillator, and counters, such as a feedback counter, a prescale counter, and postscale counters.

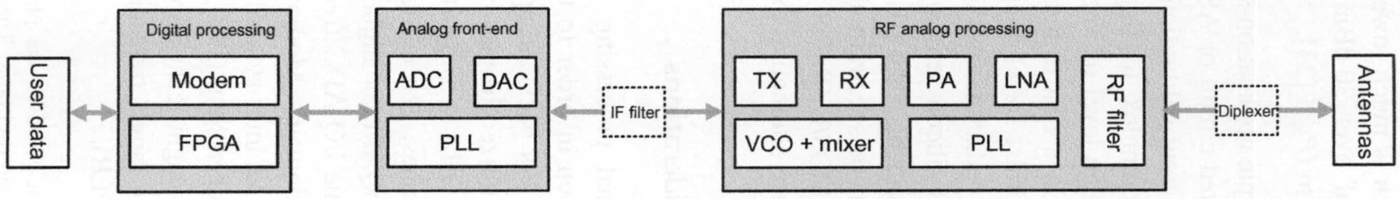

Figure 5.25
High-level system architecture of point-to-point mmWave radio [23].

A quadrature I/Q ADC is required in order to sample the IF (or baseband in case of zero-IF receiver) I/Q analog signals down-converted from the mmWave receiver. Typical channel bandwidths range from 250/500 MHz for traditional backhaul systems to 1 GHz or higher for the 5G systems, thus the sample rates would be in the excess of 500 MHz depending on the required over-sample ratio for a given modem implementation. The required resolution or ENOB is primarily a function of modulation scheme which ranges from 4 bits for QPSK to 5 bits for 16QAM, and 6 bits for 64QAM. There is a need for additional dynamic range to support OFDM waveform compared to single-carrier waveform due to power summation over multiple subcarriers (crest factor). This requires an additional 5–6 dB dynamic range for OFDM waveform compared to single-carrier waveform for a given modulation index. In order to determine the dynamic range of the ADC for a given modulation scheme and receiver, the noise floor of ADC and the receiver noise must be carefully modeled. The noise floor of the ADC is defined by the power level of the (quantization) noise produced by the ADC itself when referred back to the receiver input, that is, the actual ADC noise power divided by the gain of the front-end. Conventionally, this is set to a level approximately 10 dB below the total effective noise power at the input to the receiver so that the effect of the ADC is to increase the front-end NF by 0.5 dB. The receiver noise power is just thermal noise power (KTB) raised by the receiver NF in dB [23].

In addition to the receiver noise, a margin must be added for the minimum SNR requirement at the input to the demodulator for a given modulation and coding scheme. This must be followed by an additional margin related to the crest factor of the signal. A receiver extra power margin is defined, which is the amount by which the average signal power is allowed to increase before any gain reduction needs to be applied by the receiver automatic gain control (AGC). The receiver extra power margin determines the minimum AGC step requirement for any given mode. The larger the power margin, the larger the value that can be assigned to the AGC step size. Therefore, the ADC dynamic range is given by $\mathrm{DR}_{ADC} = 9.636 + \mathrm{SNR} + \mathrm{CF} + \mathrm{PH}\ (\mathrm{dB})$, where SNR is the minimum signal-to-noise ratio required at the demodulator, *CF* is the crest factor, and *PH* is the ADC power headroom. A typical dynamic range based on an example QPSK modulation would be approximately 36 dB, if the ADC power headroom is restricted to 14 dB. This analysis may be extended to higher order modulations in order to determine the optimum ADC dynamic range and resolution to support 16QAM, 64QAM, or 256QAM. The above ADC dynamic range is related to ENOB based on the following equation $\mathrm{ENOB} = \left[\mathrm{DR}_{ADC} - 10\log_{10}\left(1.5f_s/\mathrm{BW}_{channel}\right)\right]/6\ (\mathrm{Bits})$, where f_s and $\mathrm{BW}_{channel}$ are the sampling frequency and channel bandwidth, respectively. In general, we can observe that doubling of I or Q resolution (i.e., 4× in modulation domain) requires an additional 6 dB or one additional bit of ENOB.

A quadrature I/Q DAC is required to generate the baseband I/Q analog signals for up-conversion, filtering, and amplification in the mmWave transmitter. The typical bandwidths and sample rates are the same as those needed for the I/Q ADC. However, the key technical parameter for the transmitter is related to the spurious emission requirement. It can be shown that a DAC resolution of 7 bits or higher are typically required. The dynamic range requirement of the DACs is largely determined by the spurious emission requirement of the transmitter. The dynamic range requirement is also dependent on the power margin necessary to cope with both gain variations in the mmWave transmitter chain as well as the digital power control. The DAC dynamic range requirement can be calculated as $DR_{DAC} = -S_p - L_{IF} + CF + PH$ (dB), where S_p is the spurious emissions requirement, L_{IF} is the IF filter attenuation, CF is the crest factor of the wanted signal, and PH is the required DAC power headroom. This equation applies to all modulation and coding schemes, thus if the power headroom is fixed, the dynamic range becomes dependent on only the crest factor.

A baseband PLL is required to generate the local I/Q sampling clocks required for the I/Q ADC and DAC. Depending on the channel bandwidth, the sample clock frequency can range from 500 MHz to 3 GHz or higher. The phase noise of this baseband clock may have an impact on the overall system performance. To evaluate this effect, consider the single-sideband phase noise characteristics of an example commercial ADC clock PLL with a 40 MHz crystal reference, which achieves an rms phase noise of 1.57 degrees in the bandwidth ranging from 100 Hz to 4.5 GHz. This phase noise can be translated into an equivalent SNR value of 31.24 dB in the Nyquist band between $\pm f_s/2$ using the following equation $SNR_{\theta dB} = 10\log_{10}(180/\pi\theta)$ where f_s is the sampling frequency. The required phase noise performance of the ADC clock PLL can be further estimated from an analysis of the receiver SNR at reference sensitivity for all modulation schemes and later defining the acceptable additional reduction in SNR that could be tolerated as a result of the additional phase noise contribution of the sampling clock. If the additional reduction of SNR is set to 0.1 dB, it can be shown that the corresponding rms phase noise of the PLL would be 2.08 degrees and the rms time jitter for a sample clock of 2.64 GHz would be 2.2 ps. Therefore, a reasonable jitter requirement for the baseband PLL should be less than 2 ps [23].

In order to evaluate the performance impact of high-frequency operation on data converters, a figure of merit (FoM) is defined, which combines several performance metrics into a single value. For ADCs the construction of a fair FoM is non-trivial, not only because an ADC is rather a complex device with many attributes but also because the exact trade-off between the relevant metrics is often non-obvious. Most of the commonly used FoMs for ADCs share two basic attributes: (1) they are purposefully kept simple and mainly focus on power dissipation, conversion rate, and resolution and (2) they are constructed using first-order statistics, limited tradeoffs and avoid empirically derived parameters as much as possible. More complex FoMs (considering area, supply voltage, process technology, etc.) have been proposed in the literature [22]; nevertheless, their empirical and data-driven nature makes them difficult to use. Ultimately, it is important to realize that an FoM is not meant to capture every aspect of a specific data converter. A direct comparison between two converters should always be made using a table, where the FoM is just one of several entries. One of the widely used FoMs for quantifying the trade-off between ADC speed, resolution, and power dissipation is Walden FoM (FoM = $P2^{-ENOB}/f_s$). This FoM asserts that power P increases linearly with conversion rate f_s. This is justified from first-order physics (power = energy/time) but it tends to fall apart as the sampling speed approaches the technology limits. In terms of resolution, this FoM assumes that each added effective bit (ENOB) doubles the power dissipation. This trade-off was chosen empirically and tended to match experimental outcomes reasonably well for a number of years. In recent years, however, it has been observed that moderate-to-high-resolution converters tend to follow a 4× power per effective bit trend, a trade-off that is consistent with noise-limited analog circuits. To capture the 4× per bit trade-off, one can simply modify the denominator of the latter FoM expression and replace 2^{-ENOB} with 2^{-2ENOB}. However, a different way of formulating the same trade-off in logarithmic scale results in Schreier FoM (FoM = SNDR(dB) + $10\log(f_s/2P)$) where in its original form takes the dynamic range DR of the data converter in dB and adds 10 times the logarithm of the bandwidth BW to power ratio. Thus a 6 dB increase in dynamic range is assumed to reflect the 4× increase in power at the same efficiency. Also note that the proper unit of this FoM is dB/Joule. The dynamic range is replaced with SNDR (to capture distortion) and BW is replaced with $f_s/2$. While it is certainly true that the power dissipation of a data converter depends on its actual acquisition bandwidth (which may be different from $f_s/2$), it is difficult to argue that there should be a strict proportionality [22].

5.6.3 Power Consumption Considerations

Reducing power consumption is an important objective in the design of the network entities and the devices, but it is more critical for the devices due to their limited power supply. The new NR features affect the power consumption of the UE's major hardware components such as antenna tuning unit, RF front-end, RFIC, baseband modem, and application processor. In this section, we identify the main factors affecting the power consumption of UE's major hardware components and discuss methods to make their operation power-efficient. As we discussed earlier, the integration of RF front-end components can significantly reduce the [insertion] loss and improve the power efficiency of the AFE.

Table 5.15 shows the impact of NR features on UE key components' power consumption. The PA is one of the key components of the AFE. For PAs, the efficiency (usually noted as power-added efficiency or PAE[28]) is used to represent the power consumption

[28] PAE is a metric for rating the efficiency of a power amplifier that takes into account the effect of the gain of the amplifier. PAE will be very similar to efficiency when the gain of the amplifier is sufficiently high. But if the amplifier gain is relatively low, the amount of power that is needed to drive the input of the amplifier should be considered in a metric that measures the efficiency of the said amplifier.

Table 5.15: Impact of NR features on UE key components' power consumption [25].

5G Features	Antenna Tuning Unit	RF Front-End	RF Integrated Circuit	Baseband Processing	Application Processor
[Scalable] Bandwidth	x	x	x	x	x
UL-MIMO	x	x	x	x	x
High-power UE	x	x	x	x	x
Bandwidth part	x	x	x	x	
Discontinuous reception				x	
Cross-slot scheduling				x	
Multiple PDCCH monitoring occasions				x	
Measurement				x	
System information acquisition				x	
Paging				x	
RAN-based notification area update				x	

characteristic. The relationship between the efficiency and power consumption can be given as $\text{PAE} = (P_{OUT} - P_{IN})/P_{dc} = (P_{OUT} - P_{IN})/(V_{dc}I_{dc})$. There are many parameters affecting the efficiency of the PA including backoff power, load line loss, topology, etc. For a typical load line, the loss is between 0.3 and 0.7 dB, which means it will degrade the PAE by 7%–15%. Another factor affecting the efficiency of the PA is the backoff power. For the topology of traditional linear PAs, for example, Class-A/Class-AB/Class-F, the PAE decease when the backoff from the peak power increases. There are several methods for improving the efficiency of the PAs including the following [25]:

- *Average power tracking (APT)*: This is a technique that can be utilized to adapt the supply voltage to a PA on a timeslot basis in order to reduce power consumption of the PA. In APT method, the supplied voltage of the PA is adjusted according to output power level so that the linearity of the PA is maintained while the efficiency is improved. Although limited by the speed of the DC–DC converter, the APT voltage must be kept constant in each time slot, which means the efficiency cannot be improved, if the PA constantly operates at the highest power.
- *Envelope tracking (ET)*: The ET is a power supply adaptation technique that maximizes the energy efficiency by keeping the PA in compression mode over the entire transmission cycle, as opposed to the peaks, by dynamically adjusting the supply voltage to the PA. The effect of OFDM-based signals on PAPR motivated the use of ET technique to achieve considerable energy saving in high-PAPR transmitters. More commercial ET PAs have been developed for 4G/5G UE transmitters.
- *High voltage supply*: With higher supply voltage, the portion of knee voltage of the transistors can be minimized, which means the PA will achieve higher efficiency. Another advantage of high supply voltage is that the load line can be increased.

With increasing of the load line, the associated loss will be lower, which further improves the efficiency. However, because the supply voltage of the handset is 3.8 V at nominal condition, another DC–DC converter is needed to achieve a higher voltage. It should be noted that the DC–DC's efficiency is always less than 100%.

- *Power combination techniques*: The PAs often need to support very large output power; therefore power combination techniques are used to combine the outputs of small-PAs. Since the combiners are always at the last stage of the power amplification, the efficiency of the combiners will affect the overall PA efficiency. There are many power combination techniques including voltage or current combination. However, regardless of the combination technique used, low-loss, high-Q components must be utilized. Among combination techniques the push–pull combiner has the advantage of doubling the impedance which as to be matched, lowering the loss of the impedance transformation.
- *Doherty amplifier*: The Doherty amplifier uses a configuration known as active load-pull technique, which can modify the RF load of the PA in different RF power levels. Thus when the power backs off from the peak power, the efficiency will not drop significantly comparing to the traditional PA topologies. Each Doherty amplifier includes two devices, which are referred to as the main and the auxiliary devices, where the RF output power is the combined power of both devices. As the input drive level is reduced, both devices contribute to the output power until a certain point is reached, typically 6 dB below the maximum composite power, where the auxiliary amplifier shuts down and generates no RF power, reducing the DC power consumption. The Doherty amplifier has some disadvantages. The AM–AM and AM–PM distortion of the amplifier cannot be maintained constant as the auxiliary amplifier turning ON and OFF with the changes of the input power. Therefore DPD method is always used in the Doherty PAs. Furthermore, the performance of the Doherty amplifier is very sensitive to the load. As a result, the Doherty amplifier topology is always adopted in the BS PA design.
- *Reconfigurable technique*: The reconfigurable technique allows the PA to adapt to different configuration in different bands and different modes. With this technique, the load line, bias point, and other configurations can be adjusted. Comparing with wideband design, this technique will improve the efficiency at specified modes.
- *Digital pre-distortion*: This technique can further reduce power consumption. The DPD technology improves both power consumption and linearity.

The baseband processing in an NR UE modem has to handle a significantly increased volume of data compared to its LTE counterpart, and this is reflected in increased power consumption at the highest data rates. Some early studies have estimated the baseband contribution to UE power consumption at peak throughput, as follows [25]:

- For UE configuration (2 × 100 MHz, 4 × 4 DL, 2 × 2 UL), the power at peak throughput for FDD and TDD modes are 4500 and 2970 mW, respectively.
- For UE configuration (1 × 100 MHz, 4 × 4 DL, 2 × 2 UL) the power at peak throughput for FDD and TDD modes are 2250 and 1485 mW, respectively.

That study was based on a very simple set of assumptions. In practice, the baseband processing of two aggregated RF carriers is not exactly twice the power of a single-carrier baseband. More recent studies suggest that these estimates were reasonably accurate for the single-carrier UE. However, power consumption at peak throughput is only one aspect of UE power consumption. In an active connection, a UE will spend a large percentage of time monitoring the downlink control channel in slots which may not contain any data for the UE. Therefore, it is important to minimize the power consumption during PDCCH monitoring occasions. This can be achieved by a combination of cross-slot scheduling, bandwidth part adaptation, and MIMO restriction. The UE can further reduce its average power consumption in the connected mode (at the expense of higher latency) by entering a DRX cycle to reduce the time spent on monitoring PDCCH.

The applications processor (AP) in a smartphone comprises a number of processing cores and graphical processors which cooperate to support the computational requirements of the active applications. Multiple cores support a mixture of clock rates and processing capabilities; thus power consumption can vary significantly depending on the number of cores that are active and the set of applications that are running. The highest processing loads are usually associated with display-intensive applications, and the combined power consumption of the applications processor and display can exceed 1000 mW in some scenarios. The AP power consumption can in many cases be considered independently from the UE modem power. Present-day application data requirements are relatively modest in NR, even streaming of high-definition video to the UE uses a comparatively small proportion of the data bandwidth that is available. However, the applications themselves can have a significant impact on modem power consumption. Interactive gaming applications can involve frequent transfer of small packets of data to give a real-time response, and if the update frequency falls within the inactivity timer period of the DRX cycle, the modem will be awake most of the time with increased power consumption. Updates from different applications, if their timing is not coordinated, would increase the proportion of DRX cycles in which the UE is active and thereby increase power consumption.

5.6.4 Antenna Design Considerations

Large antenna arrays at the BS are necessary for 5G systems and in particular for operation in mmWave bands. Large number of antennas at the UE side are also expected to guarantee acceptable received signal levels for 5G services. As the number of antennas increases, it is advantageous to use 2D and 3D structures for the arrays. This reduces the required space and also enables spatial separation and beamforming in two or three dimensions. For example, a 64-element uniform linear array with an inter-element spacing of $\lambda/2$ could occupy a horizontal span of 3 m at 2 GHz, which reduces to 1.5 m if dual-polarized antenna elements are utilized. In contrast an 8×4 dual-polarized array can be accommodated in a 0.6 m $\times$ 0.4 m space and can spatially resolve users and form beams in 3D space. Antenna arrays may be arranged in a number of different ways, the most common architectures being uniform linear arrays, uniform rectangular arrays, uniform circular arrays, and stacked uniform circular arrays. The system performance of these architectures is often measured in terms of beam gain and half-power beam-width both in azimuth and in elevation planes. For massive MIMO systems, it is important to capture as many degrees of freedom of the channel as possible with a high number of effective antennas. For compact antenna configurations, this often leads to antenna structures that have shown to be effective for multi-dimensional channel characterization and parameter estimation. The number of antennas on the UE will not be as large as the corresponding number at the BS; however, the number of UE antennas can be much larger than what is used today in mmWave frequencies. The studies suggest that the performance improves as the array aperture increases, but the impact of the aperture is mainly visible when the users are closely grouped (i.e., they have high correlation). Furthermore, there is in general a good channel resolvability in the sense that the larger the aperture, the greater the resolvability. One important aspect to remember is that, for physically large arrays, there can be large differences in received power levels over the array, which affects user resolvability. For beamforming solutions, the antenna elements in an array must be placed close together. All the analog components (phase shifters, LNAs, PAs, etc.) should be tightly packed behind the antenna elements (active antenna systems). The dense packing of antenna elements creates two main effects: (1) spatial correlation and (2) mutual coupling [21].

The NR operation in mmWave bands necessitates the use of active antenna systems and large antenna arrays in the base stations and the devices. While this goal is enabled by relatively small-sized antenna arrays in mmWave frequencies, it significantly increases the implementation complexity and pushes the limits of antenna design and semiconductor technologies. The fully integrated mmWave transceivers and antennas require sophisticated design and consideration of power efficiency, component placement and routing, mutual couplings, and thermal constraints both at the component level and system level within a small area or volume. These considerations directly affect the achievable performance and the RF requirements and are applicable to the NR base stations and the devices, given that in mmWave frequencies, the BS/UE transceiver design and development will have less differences compared to frequency bands below 6 GHz. The extremely wide bandwidths available at mmWave frequencies set forth serious challenges for the data converters and the data conversion interfaces between the analog and digital domains in both receivers and transmitters.

5.7 Sub-6 GHz Transceiver and Antenna Design Considerations

The evolution of 4G to 5G and the reliance on the existing technology allows several LTE solutions to be integrated into the initial rollout of the NR, providing immediate benefits without waiting for future releases. The early rollouts are also supported by the use of

EN-DC functionality where an NR gNB is always associated with an LTE eNB. In Rel-15 NR, 4 × 4 downlink MIMO, particularly at frequencies above 2.5 GHz, which include n77/78/79 and B41/7/38, will be mandatory. The presence of four MIMO layers not only enables extended downlink data rates but also means there will be four separate antennas in the UE. An additional feature that is driven by mobile operators, however not mandatory, is the deployment of 2 × 2 uplink MIMO. Having 2 × 2 uplink MIMO in UE requires two 5G NR transmit PAs to transmit from separate antennas. This is particularly beneficial in cases where higher frequency TDD spectrum (in sub-6 GHz) is used as is the case with n41, n77, n78, and n79 as well as other TDD bands. The effective doubling of the uplink data rate enables shorter uplink bursts and flexible use of the 5G frame timing to increase the number of downlink slots, potentially increasing downlink data rates. However, when the downlink data rate increases, the uplink is challenged by the requirement of fast feedback from the UE [28].

A further use of the available second transmit path is a new transmission mode known as 2TX coherent transmission. This effectively uses the principles of diversity, which are strongly leveraged on the downlink side of the network and enable up to 1.5–2 dB of additional transmit diversity gain, which is critical to address the fundamental uplink-limited network performance. Such improvements in uplink translate into an approximately 20% increase in the range at the cell edge. In addition to improving cell edge performance, 2 × 2 uplink MIMO improves spectrum efficiency. Since 5G NR is mostly a TDD technology above 2 GHz, and TDD cells are likely to be configured in a highly asymmetrical configuration with priority given to downlink, improving spectrum efficiency is the key to delivering high cell capacity.

In the initial-phase of Rel-15 deployment, mobile operators emphasized the need to establish a framework for the dual-connectivity non-standalone mode of operation. In principle, network deployment with dual-connectivity NSA means that the 5G systems are overlaid by an existing LTE network. Dual-connectivity implies that the control and synchronization between the BS and the UE are provided by a 4G network, while the 5G network is a complementary radio access network tied to the 4G anchor. In this model, the 4G anchor establishes the connection using the existing 4G network which later can be complemented with a 5G NR user-plane connection (see Fig. 5.26). The addition of the new radio, together with the existing LTE multi-band carrier aggregation mode, strains the system performance, size, and interference mechanisms, posing additional challenges to be resolved when designing the NR RF front-end. Depending on 1TX or 2TX carrier frequencies and their relative spacing, intermodulation distortion products may fall into the LTE anchor receiver frequency band and cause LTE receiver desensitization.

A simplified view of NSA option-3a network topology is shown in Fig. 5.26, which is expected to be used in early rollouts of 5G networks, where the mobility is handled by LTE

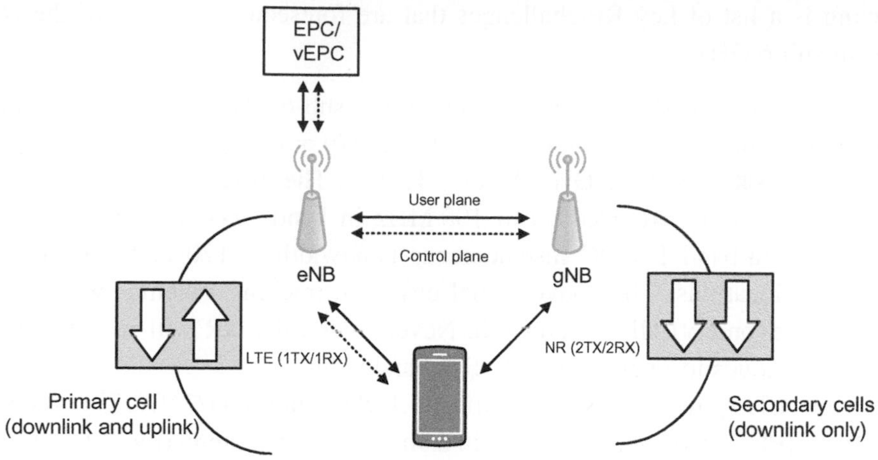

Figure 5.26
Non-standalone option-3a dual-connectivity network architecture [28].

anchors (control and user planes). This architecture leverages the LTE legacy coverage to ensure continuity of service delivery and the progressive rollout of 5G cells. It certainly seems the most plausible method of implementing 5G, while at the same time ensuring that the integrity of data connections is maintained in areas where the backhaul and network infrastructure have not yet upgraded to 5G. However, this requires UE by default to support simultaneous dual uplink transmissions of LTE (1TX/1RX) and NR (2TX/2RX) carriers in all possible combinations of standardized bands and radio access technologies (FDD, TDD, SUL, SDL), which would increase the implementation complexity where multiple separate radios and bands functioning in a small device form factor. When combined with an LTE TDD anchor point, the network operation may be synchronous, in which case the operating modes will be constrained to 1TX/2TX and 1RX/2RX, or asynchronous which will require 1TX/2TX, 1TX/2RX, 1RX/2TX, 1RX/2RX. When the LTE anchor is an FDD carrier, the TDD/FDD inter-band operation will require simultaneous 1TX/1RX, 1TX/2RX, and 2TX/1RX/2RX. In all cases, since control-plane information is transported over LTE radio bearers, it is critical to ensure that LTE anchor point uplink traffic is reliable [28].

The RF front-end developers try to mitigate the interference as much as possible to allow optimal signal transmission/reception in the UE. The complex nature of dual transmit LTE/NR concurrency and 5G-capable UE constitutes an even greater challenge for the NR RF front-end. The second phase of NR Rel-15 deployments will include standalone operation, which uses a 5G core network that will not require backward compatibility to LTE. However, the assumption is that the initial implementation of 5G NR starts with non-standalone, which is the main deployment strategy for refarmed bands, since the standalone systems are anticipated to be deployed the spectrum above 3 GHz.

The following is a list of key RF challenges that are foreseen as a result of the NR implementations in sub-6 GHz:

- *Wider channel bandwidth*: The new bands in the sub-6 GHz range feature much larger percent bandwidth[29] (n77 = 24%, n78 = 14%, n79 = 12.8%) than the existing bands (B41 = 7.5%, B40 = 4.2%, 5 GHz Wi-Fi = 12.7%). The instantaneous signal modulation bandwidth for NR is extended up to 100 MHz in bands n41, n77, n78, and n79. The contiguous intra-band EN-DC instantaneous bandwidth is 120 and 196 MHz for non-contiguous allocations. The conventional envelope tracking schemes would have difficulty going beyond 60 MHz bandwidth. Nevertheless, the NR will require new envelop tracking techniques to operate at 100 MHz bandwidth.
- High-power UE (power class 2 specific to TDD bands n41/77/78/79): As mentioned earlier, the HPUE or power class 2 (+26 dBm at a single antenna) will increase radiated output power by +3 dB relative to power class 3 operation. The PAs will need to be designed to meet higher output power with more complex waveforms. Optimized system design will be critical to achieving minimal post-PA loss in order to achieve HPUE benefits.
- *High-power transmission of NR inner-band allocations away from channel edge*: The NR would require less MPR or power backoff, when reduced waveform allocations are at a specified offset away from the channel edge. This enables much higher power across uplink modulation orders and addresses a fundamental coverage issue in LTE networks, which was related to the uplink power-limited transmission and SNR for reduced resource block allocations at the cell edge.
- *NR waveforms and 256QAM uplink*: The new NR waveforms, especially CP-OFDM, have a higher peak-to-average power ratio and will require more power backoff relative to the LTE waveforms. The 256QAM modulation is going to be used in the uplink to increase the data rates. This will challenge the RF front-end to maintain the total EVM below 3% including the PA and transceiver effects. Other issues such as in-band distortion, frame rate, and clipping must be managed to achieve maximal efficiency.
- *Cost-effective support for* 4 × 4 *downlink MIMO,* 2 × 2 *uplink MIMO and coherent 2TX transmission modes*: The 4 × 4 downlink MIMO is required in 3GPP for n7, n38, n41, n77, n78, n79 bands either operating as a standalone band or as part of a band combination. This feature has been prioritized due to the significant benefits of doubling the downlink data rate and spectral efficiency, as well as the up to 3 dB receive diversity gain versus 2 × 2 downlink modes.

[29] Percent bandwidth provides a normalized measure of how much frequency variation a system or component can tolerate. As we go higher in frequency, the absolute bandwidth will naturally increase, while its percent bandwidth will decrease. The percent bandwidth can be expressed as the ratio of the bandwidth over the carrier frequency, thus a filter with 1 GHz passband centered at 10 GHz will have 10% bandwidth, while a filter with 10 GHz bandwidth at 100 GHz will have the same percent bandwidth.

- *New 5G NR spectrum*: The new bands for sub-6 GHz will extend the frequency from 3 up to 6 GHz in the device. The increase in frequency will require improvements in the entire radio front-end as the industry tries to maintain current performance while operating at a higher frequency. There will also be new antenna multiplexing and tuning challenges, as well as in-device coexistence with 5 GHz Wi-Fi.

In the following, we focus on a practical example of a sub-6 GHz radio front-end supporting n77, n78, and n79 frequency bands. This example is intended to illustrate the important criteria for designing a 5G NR radio in sub-6 GHz (see Fig. 5.27). We will also discuss the impact of 5G spectrum, waveform, and modulation on the constituent components of the radio front-end module. The 3GPP requirements for n77 and n79 spectrum indicate 100 MHz instantaneous bandwidth for the component carrier in the uplink. This is much more rigorous than current LTE standards that use carrier aggregation of 20 MHz base channels to extend support for 40–60 MHz [28].

It is anticipated that early 5G systems will require the PA to operate with APT mode to accommodate the wider bandwidth signals. Therefore, users can expect a 100 MHz channel when the PA is operating under APT conditions. Conversely, conventional ET is challenged to perform in 40–60 MHz bandwidth. In order to extend the ET modulator bandwidth to reach 100 MHz, additional power consumption would be required, in addition to addressing amplitude/phase delay mismatch sensitivity, management of memory effects, limitations in

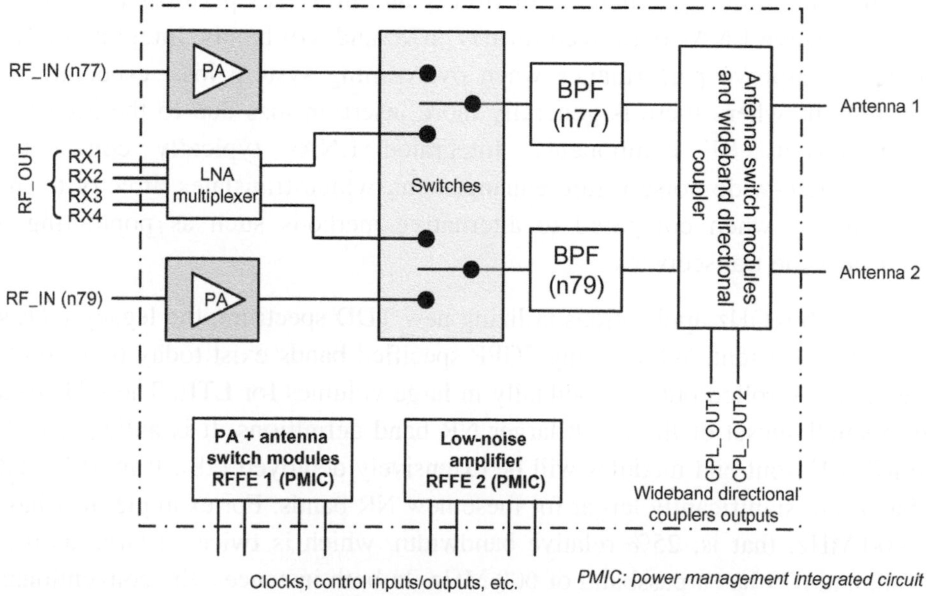

Figure 5.27
Example of sub-6 GHz 5G NR transmit/receive front-end architecture [28].

capacitive supply loading, OOB [transmitter] emissions, and intermodulation leakage into the LTE anchor band. Although, there are several promising techniques under development to extend ET to wider operating bandwidth, it will take several years before they can be commercialized. Designers are thus left with the challenge of delivering a better performing PA at two to three times the present state-of-the-art instantaneous bandwidth, while operating with a higher peak-to-average power ratio of CP-OFDM and over much larger passbands than present filters for LTE sub-3 GHz.

Aside from the wider channel bandwidth, operators have shown significant interest in high-power UE capabilities, especially as it pertains to TDD bands in the sub-6 GHz range. At present, there is some uncertainty regarding whether the bands will require 2×2 uplink (two transmitters operating at the same time) or a single transmitter would suffice, which means that the PAs will not only need to deliver industry-grade output powers compared to their LTE counterparts, but also they will have to achieve that over a wider bandwidth and at higher frequencies. Meeting higher output power at higher frequencies without ET scheme has created some serious design issues. In order to meet the new, challenging performance requirements of wider channel bandwidth and HPUE, new PA topologies have been developed that deliver linear PA performance at higher frequencies and over much wider channel bandwidths. These new architectures must be capable of significantly outperforming their LTE counterparts under more rigorous operating conditions.

When implementing the sub-6 GHz modules, integrating the receiver LNA functionality inside the module allows considerable flexibility and improves performance. In Fig. 5.27, there are two receive LNAs optimized for n77, n78, and n79 bands. Integrated LNAs have been shown to improve performance when overcoming system loss, especially in high-frequency regions where there is generally more insertion loss due to the high-frequency roll-off of various RF components. Integrated LNAs typically contribute about 1.5–2.0 dB system-wide noise figure enhancement, which translates directly to improved receive sensitivity when compared to alternative methods such as populating discrete LNAs at or near the transceiver.

In the case of sub-6 GHz applications utilizing new TDD spectrum, the legacy LTE systems are virtually non-existent. While many 3GPP specified bands exist today (e.g., B42/43/48), they have yet to be rolled out commercially in large volumes for LTE. The LTE bands only represent a small subset of the much larger NR band definitions. It is anticipated that n77, n78, and n79 RF front-end modules will be extensively deployed [28]. It must be noted that the passbands are significantly larger for these new NR bands. For example, n77 has a passband of 900 MHz; that is, 25% relative bandwidth, which is twice as large as the 5 GHz Wi-Fi band, and n79 has a passband of 600 MHz. In both instances, the conventional acoustic filters are not well suited for these extremely wide passbands. There are additional complexities that will determine the extent of the NR wideband filter requirements. For

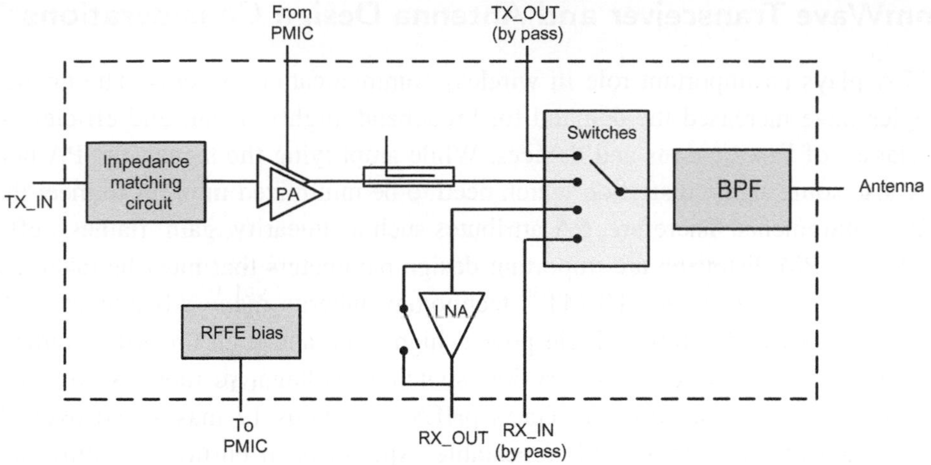

Figure 5.28
Example of 5G NR n77/n78 RF front-end block diagram [26].

example, one can derive a simple filter response, if an ideal environment with a separate high-band antenna and no coexistence requirements are assumed. On the other hand, if we consider a more complex radio environment, such as a multi-radio platform with simultaneous Wi-Fi transmission, the filter requirements become more stringent. Thus it is important to take note of the radio environment, antenna topology, and in-device coexistence requirements in order to specify the optimum filter. In other words, the filter design and antenna topology into which the front-end module will be subsequently integrated must be customized to the specific use case or application (see Fig. 5.28).

Antenna configurations play a significant role in the development and deployment of 5G products. While market requirements are slowly becoming clearer, there is already some uncertainty as to which of the optional features will be supported. One feature is the fast hopping sounding reference signal, which uses the transmitter to send a series of known symbols across all of the downlink receive antennas in the UE in order to better calibrate the MIMO channel and improve the downlink SNR. This process is key to enhanced MIMO and beamforming operations. The SRS carrier switching was recently introduced in LTE Rel-14 to assist the eNB in obtaining the channel state information of secondary TDD cells in LTE carrier aggregation scenarios. The SRS transmit switching allows the UE to route its SRS transmissions to all other available antenna ports. Assuming the channel reciprocity assumption holds, which should be the case for TDD operation, this feature enables gNB to estimate CSI on secondary downlink-only cells. By applying this concept to multi-user MIMO, the network performance is further enhanced, which in turn will improve the user experience.

5.8 mmWave Transceiver and Antenna Design Considerations

The RF PA plays an important role in wireless communication systems. The emerging 5G technologies have increased the demand for broadband, highly linear, and efficient PAs for various classes of base stations and devices. While amplifying the signal, the PA nonlinearity may cause some signal distortion which need to be minimized in order to meet the transmitter RF requirements. Therefore, PA attributes such as linearity, gain, flatness, efficiency, AM–AM/AM–PM distortion are important design parameters that must be taken into consideration. The GaAs E-mode pHEMT[30] technology inherits many advantages of D-mode pHEMT technology and exhibits single power, high span, and high linearity characteristics. The E-mode pHEMT device has a very low switching voltage, so there is no limit on the offset margin. An (E-mode) 0.15 μm GaAs pHEMT contains 17 masks and over 200 process steps. Each of these steps needs repeatable experiment to ensure the entire process is reliable and commercial grade [26].

The GaAs-based devices have dominated most of the market due to the maturity of the technology in the microwave frequency bands. However, an InP-device has a higher breakdown voltage, higher average electron velocity, larger discontinuity in the bandgap at the interface of InAlAs/InGaAs heterojunction, higher two-dimensional electron gas (2DEG) concentration,[31] and higher electron mobility in the channel, making it more suitable for

---

[30] A high-electron-mobility transistor (HEMT) is a field-effect transistor incorporating a junction between two materials with different band gaps (i.e., a heterojunction) as the channel instead of a doped region (as is generally the case for MOSFET). A commonly used material combination is GaAs with AlGaAs, depending on the application of the device. Devices incorporating more indium generally show better high-frequency performance, while in recent years, gallium nitride HEMTs have attracted attention due to their high-power performance. The HEMTs are used in integrated circuits as digital ON/OFF switches. HEMT transistors are able to operate at higher frequencies than ordinary transistors, up to mmWave frequencies. Ideally, the two different materials used for a heterojunction would have the same lattice constant (spacing between the atoms). In practice, the lattice constants are typically slightly different (e.g., AlGaAs on GaAs), resulting in crystal defects. A HEMT where this rule is violated is called a pHEMT or pseudo-morphic HEMT. This is achieved by using an extremely thin layer of one of the materials so thin that the crystal lattice simply stretches to fit the other material. This technique allows the construction of transistors with larger bandgap differences than otherwise possible, giving them better performance. Another way to use materials of different lattice constants is to place a buffer layer between them. This is done in the mHEMT or metamorphic HEMT, an advancement of the pHEMT. The buffer layer is made of AlInAs, with the indium concentration graded so that it can match the lattice constant of both the GaAs substrate and the GaInAs channel.

[31] A 2DEG is a scientific model in solid-state physics. It is an electron gas that is free to move in two dimensions, but tightly confined in the third dimension. This tight confinement leads to quantized energy levels for motion in the third direction, which can then be ignored for most problems. Thus, the electrons appear to be a 2D sheet embedded in a 3D coordinate system. The analogous construct of holes is called a two-dimensional hole gas, and such systems have many useful and interesting properties.

high-frequency applications. For low-voltage applications, InP HBT is expected to replace GaAs HBT. InP HBT is made of narrow bandgap material InGaAs. The bias voltage is as low as 0.5–0.6 V. In addition, InP HBT also exhibits better heat conduction characteristics, better heat dissipation capability (approximately 1.5 times of GaAs substrate), and higher peak electron velocity, allowing higher performance at lower voltages. Although InP/InGaAs HBT with lattice matching on the InP substrate has superior high-frequency and high-speed characteristics, and lower turn-on voltage than GaAs HBT, it suffers from small substrate size, high cost, and fragility of the InP substrate, limiting its large-scale and low-lost production. The capability to obtain high-quality and large-size semi-insulating InP substrate is the key to lower cost. Therefore, to meet the increasing demand for InP devices, the fundamental task is to grow high-quality and large-diameter InP single crystal substrate [26].

There is a growing interest in GaAs mHEMT technology due to the cost, fragility, and process compatibility issues of InP substrates. The GaAs mHEMT technology grows a relatively thick InAlAs layer between the channel and the GaAs substrate. The composition of indium changes from a positive value to zero, thus the mismatch of the lattice is alleviated. After adding a buffer layer with graded component, the component percentage of indium in the channel layer can be chosen almost arbitrarily between 30% and 60%, and the device performance is optimized with a great degree of freedom. The mHEMT can be considered as InP HEMT technology on GaAs substrate. It shows superior performance relative to InP in terms of low noise, allowing GaAs to solidify its position at the low-end of mmWave frequencies, and further its way into the high-end range of mmWave spectrum [26].

A 220–320 GHz mHEMT MMIC amplifier was reported in the literature using high indium composition in InGaAs as channel to achieve high-carrier density and mobility. Double InAlAs barrier layers and bilateral delta doping are designed to increase channel electron confinement and to reduce contact resistance. The buffer layer with linear component change is used to achieve lattice matching. The f_T and f_{max} are 515 and 700 GHz, respectively, for the device with 35 nm gate length and 20 μm gate width. The linear gain of the two-stage amplifier is more than 10.5 dB in the 220–320 GHz frequency band, and the small signal gain is 13.5 dB at 330 GHz. In another study, the device with 20 nm gate length was reported, and the f_T was 660 GHz for the mHEMT with 2×10 μm gate width. GaAs mHEMT has been able to work into the H-band and enter the low-end of the THz region. It has become an important component of the sub-mmWave applications with a cost advantage [26].

The mmWave RF components are expected to be deployed in large scale and high density and to meet the multi-band and high-frequency requirements. This would underline the importance of mmWave filter technologies. The acoustic filter technologies continue to evolve to meet the challenges of the global transition to 5G networks. SAW[32] and bulk acoustic wave (BAW)[33] filter technologies are widely used in today's mobile device filtering functions. Under the existing framework, SAW filters are widely used in 2G, 3G, and 4G wireless technologies, which has the advantages of design flexibility, good frequency selection characteristics, and small size; and can utilize the same production process as the integrated circuits. The frequency characteristic of the SAW filter is closely related to the pitch of the inter-digital transducers electrode, thus the higher the frequency, the smaller the distance required between the electrodes. The SAW filter quality drops rapidly when the frequency exceeds 2.5 GHz. Therefore, the SAW filter is mostly used in below 2.5 GHz applications. The BAW filters are gaining traction beyond 2.5 GHz. In contrast to SAW filters, they are insensitive to temperature changes, with low insertion loss and high OOB rejection, making them a good candidate for 4G/5G wireless systems' RF filtering applications. However, most BAW filters are used below 6 GHz, although they can reach to a maximum frequency of 20 GHz. The mmWave micro-electro-mechanical systems (MEMS) filter[34] on semiconductor substrate, uses the semiconductor manufacturing process, and constitutes a filter structure with high-Q factor and low-loss characteristics, as well as good performance in mmWave bands. However, regardless of whether it is using GaAs substrate or silicon substrate, the cost is relatively high. Under the conditions of ensuring a high degree of consistency, mass production is still more difficult [26].

As for the mmWave LNAs, GaAs pHEMT has excellent noise and linearity performance. Due to its high insulation rate, GaAs substrate improves chip inductor quality factor (Q), which helps reduce the overall NF of the terminal's receiver. While GaAs pHEMT has excellent noise and gain performance, it has not been widely used in the early years due to

[32] A SAW filter is a filter whereby the electrical input signal is converted to an acoustic wave by inter-digital transducers (IDTs) on a piezoelectric substrate such as quartz. The IDTs consist of interleaved metal electrodes which are used to launch and receive the waves, so that an electrical signal is converted to an acoustic wave and then back to an electrical signal. The most common group of SAW filters are bandpass filters, which are in very widespread use in radio systems. There are many types with differing advantages, such as low shape factor, low insertion loss, small size, or high-frequency operation. The wide variety of types is possible because almost arbitrary shapes can be defined on the surface with very high precision. SAW filters are limited to frequencies from about 50 MHz up to 3 GHz.

[33] Unlike SAW filters, the acoustic wave in a BAW filter propagates vertically. In a BAW resonator using a quartz crystal as the substrate, metal patches on the top and bottom sides of the quartz excite the acoustic waves, which bounce from the top to the bottom surface to form a standing acoustic wave. The frequency at which resonance occurs is determined by the thickness of the slab and the mass of the electrodes. At the high frequencies in which BAW filters are effective, the piezo layer must be only micro-meters thick, requiring the resonator structure to be made using thin-film deposition and micro-machining on a carrier substrate. The BAW filter size also decreases with higher frequencies, which makes ideal for most wireless applications. In addition, BAW design is far less sensitive to temperature variation even at broad bandwidths.

[34] MEMS are millimeter/sub-millimeter devices, realizing a certain transduction function between two (or more) distinct physical domains, among which the mechanical is always involved. More specifically, regardless of the specific function it is conceived for, a MEMS device always features tiny structural parts that move, bend, stretch, deform, and/or contact together. These characteristics make micro-system devices particularly suitable for the realization of a wide variety of micro-sized sensors and actuators.

its cost of production. However, with the recent improvement of the process, GaAs pHEMT has become the mainstream for the development of LNAs. For cost and performance considerations, the mainstream process is the gate length of 0.25 μm. GaAs pHEMT technology allows implementation of gate lengths in the order of 0.1 μm. As an example, the NF and gain of a prototype LNA based on the latter technology is 1.3 and 7 dB, respectively, at 40 GHz. Another prototype LNA developed by one of the leading RF companies has 0.13 μm gate length and a cutoff frequency of 110 GHz with an NF of less than 0.5 dB in 8–12 GHz band [26].

Switches are the smallest active electronic components and play a critical role in the overall operation and performance of wireless communication systems. Switches route the transmit and receive signals at different frequencies, enable monitoring and calibration, and are the baseline component for functions such as phase shifting and step attenuation. The 5G wireless systems will continue reliance on solid-state switching. The switch performance is evaluated by a number of different specifications. In general, a switch must have a low insertion loss, high linearity, high port-to-port isolation, high-power handling, and a fast switching time. In mmWave systems, broadband frequency operation becomes increasingly important. To facilitate the transition to mmWave, switches must be able to support multiple bands, such as dual bands (28 and 39 GHz) or even triple bands (24, 28, and 39 GHz). In addition to multiple bands, each band has rigorous requirements on frequency-response variation. For example, a 28 GHz mmWave application will have a wide bandwidth and the system frequency response (gain) may not meet the flatness requirement. Furthermore, a true wideband support implies minimum variation across the supported frequency bands [26].

5.9 Large Antenna Array Design and Implementation in Sub-6 GHz and mmWave Bands

It is desirable for an antenna system to be able to focus its radiated energy in a particular direction in order to maximize the signal power toward a particular user device. Using a simple $\lambda/2$ dipole as an example, the gain can be increased using two methods [18,19]:

- Antenna aperture: By increasing the aperture (or size of the antenna), the (larger) antenna becomes more directive due to the periodic current distribution across the antenna. Although this method does not require external circuitry for control, the direction of the beam is fixed and the number of sidelobes increases.
- Antenna array: If the single dipole element is repeated according to the periodicity of the current distribution, an antenna array is created. The amplitude and phase of the signals to individual elements can be adjusted to control both the beam direction and sidelobe levels, creating a phased-array. This results in a significantly more complex feeding network with relatively higher losses.

The beam steering capabilities of an antenna array can create both a high-gain beam toward a specific direction as well as creating a null in a specific direction in order to mitigate interference in a MU-MIMO system. Therefore, in addition to phase shifting, weighting of the signal amplitude is applied to reduce the side lobes. For example, a symmetric linear tapering of the signal amplitudes in a linear antenna array results in sidelobes that are 10–15 dB lower, but it increases the beam-width of the main lobe by approximately 5 degrees.

As we discussed in Chapter 4, there are three types of beamforming architectures used for antenna arrays (see Fig. 5.29):

- *Analog beamforming (ABF)*: The traditional way to form beams is to use attenuators and phase shifters as part of the analog RF circuitry where a single data stream is divided into separate paths. The advantage of this method is that only one RF chain (PA, LNA, filters, switch/circulator) is required. The disadvantage is the loss from the cascaded phase shifters at high power.
- *Digital beamforming (DBF)*: DBF assumes that there is a separate RF chain for each antenna element. The beam is formed by matrix-based operations in the baseband where

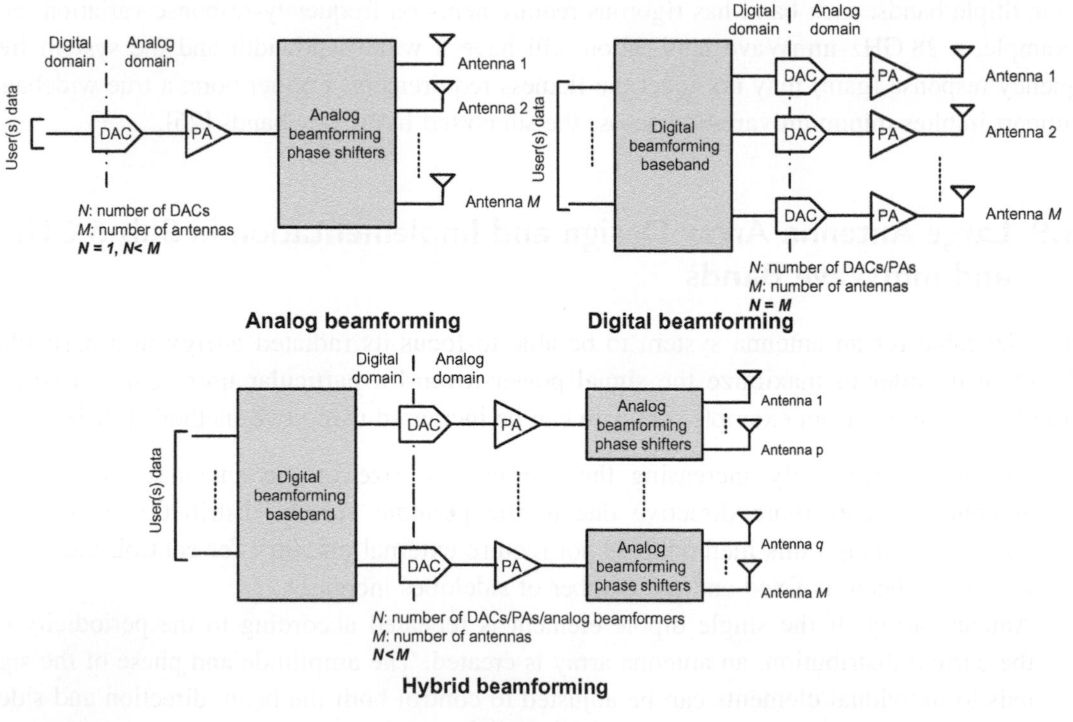

Figure 5.29
Beamforming architectures for active antenna systems [18].

digital amplitude and phase weighting is performed. For frequencies lower than 6 GHz, this is the preferred method since the RF chain components are comparatively inexpensive and can combine MIMO and beamforming into a single array. For frequencies of 28 GHz and above, the PAs and ADCs are very lossy for standard CMOS components. If alternative semiconductor technologies, such as GaAs and gallium nitrate, are used, the losses decrease at the expense of higher cost.

- *Hybrid beamforming (HBF)*: HBF combines DBF with ABF in order to allow the flexibility of MIMO plus beamforming while reducing the cost and losses of the beamforming unit (BFU). Each data stream has its own [separate] analog BFU with a set of M antennas. If there are N data streams, then there are $N \times M$ antennas. The analog BFU loss due to phase shifters can be mitigated by replacing the adaptive phase shifters with a selective beamformer such as a Butler matrix. One proposed architecture uses the digital BFU to steer the direction of the main beam while the analog BFU steers the beam within the digital envelop.

A 32-element linear antenna array (with $\lambda/2$ antenna spacing) is simulated and reported in [19] and the array factor is illustrated in Fig. 5.30, where the transceiver number is 4, and the antenna number per transceiver is 8. The maximum array gain is achieved at azimuth 90 degrees. The first curve in Fig. 5.30 depicts the array factor of ABF consisting of eight

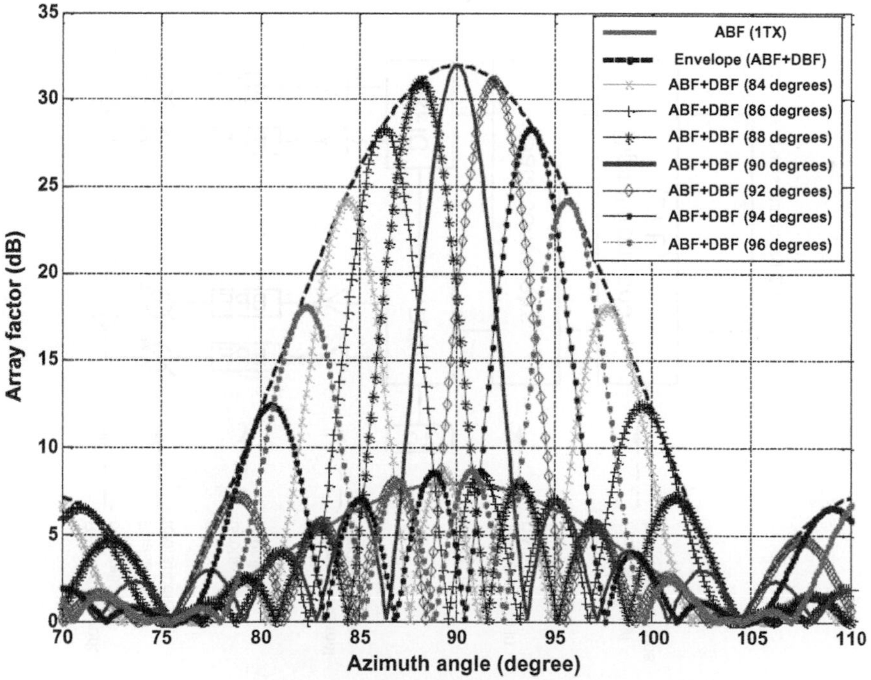

Figure 5.30
Comparison between analog, digital, and hybrid beamforming [18,19].

antenna elements with the main beam direction at azimuth 90 degrees. The second curve in the figure is the array factor of the 32-element hybrid analog and digital beamforming. It can be seen that the amplitude of the second curve is exactly four times of that of the first curve. With same ABF per transceiver, seven DBF designs are shown, with the main beam direction of azimuth 84, 86, 88, 90, 92, 94, and 96 degrees, respectively. As shown in Fig. 5.30, the main beam direction can be controlled with HBF design. Some observations are summarized as follows [19]:

1. The further the main beam direction of the HBF is away from the ABF main beam direction, the smaller the gain. Therefore, the coverage of the beams is limited.
2. The HBF designed for 90 degrees is actually the ABF with all antennas.
3. Users located at certain AoD φ within the range of 84–96 degrees will observe that the HBF designed for direction φ has the largest signal power.

Due to the sensitivity of the antenna array beam steering to the phase differences between the antenna elements, each array must be calibrated for the following tolerances (see Fig. 5.31) [18]:

- Phase: Phase error can have a significant effect on the antenna beam depending on its statistical properties. If the phase error is uniformly distributed across the array, then the main beam direction does not change. Instead, the nulls that are often used to block

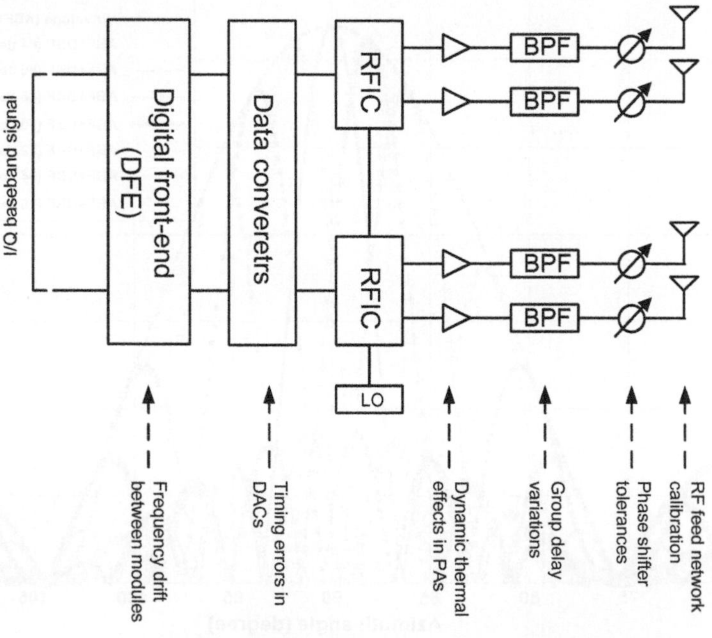

Figure 5.31
Array calibration: static and dynamic tolerances in active antenna systems [18].

interference are severely affected, resulting in 10–20 dB loss. If there is more deterministic phase error distribution, this will steer the beam in a different direction. Phase error can be caused by manufacturing tolerances in the RF feeding network, thermal effects in the PAs and LNAs, and group delay variations in the filters. It is recommended to keep the phase error between antenna elements below ± 5 degrees (commercial specification for AAS).

- Amplitude: Amplitude error does not affect the direction of the beam, but rather the peak gain and the sidelobe levels and is generally due to the thermal effects on the active components (PA and LNA). Recommended error should be below ± 0.5 dB (commercial specification for AAS).
- Timing/frequency: Depending on the circuit architecture, if a common LO network is not used between modules, there will be frequency drift in addition to the timing errors in the ADCs. Recommended level of frequency drift is 0.5 ppm (commercial specification for AAS).

The different phases in active antenna array development require different measurement and verification methods, thereby using various approaches to measure a massive MIMO system will require different test interfaces to both the complete antenna array and individual antenna elements. Antenna arrays with 64 or more elements (corresponding to an 8×8 cross-polarized antenna array) may not provide any individual antenna connectors in the final assembly. In earlier phases of the product design, however, antenna elements are typically accessible with connectors to verify the S-parameters of individual antennas. Verification and qualification of antenna arrays is required in all product development phases from initial R&D design to final production test. Mutual coupling between antenna elements has an adverse effect on network capacity. Therefore, simultaneous multi-port passive (conducted) measurements for accurate characterization are required [18].

In the design phase, when antenna connectors are still accessible, a vector network analyzer can be used to measure the mutual coupling between the antenna array elements. For antenna arrays, the most common measurement with a vector network analyzer is the S-parameter measurements (both transmission and reflection coefficients). The S-parameters include magnitude and phase information, which can be used to measure both near-field and far-field quantities. For a two-port system, S_{11} represents the reflection coefficient (reflected power at antenna 1 divided by injected power at antenna 1) and S_{21} denotes transmission coefficient (transmitted power at antenna 2 divided by injected power at antenna 1). An example of an antenna array is shown in Fig. 5.32 with 64 dual-polarized antennas and 128 antenna ports. Due to the large number of antenna ports, connecting cables and subsequent calibration of the vector network analyzer is difficult and time-consuming [18].

Fig. 5.33 illustrates the effect of antenna mutual coupling on the capacity of the cell. We compare a uniform linear array with a 2D planner array. It is shown that the element

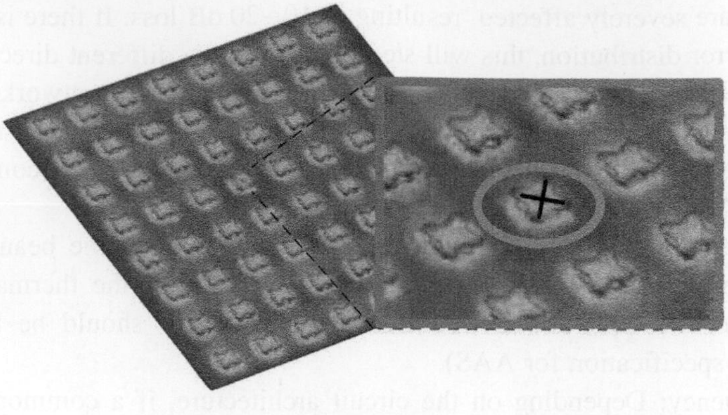

Figure 5.32
64-Element planar antenna array and mutual coupling between array elements [18].

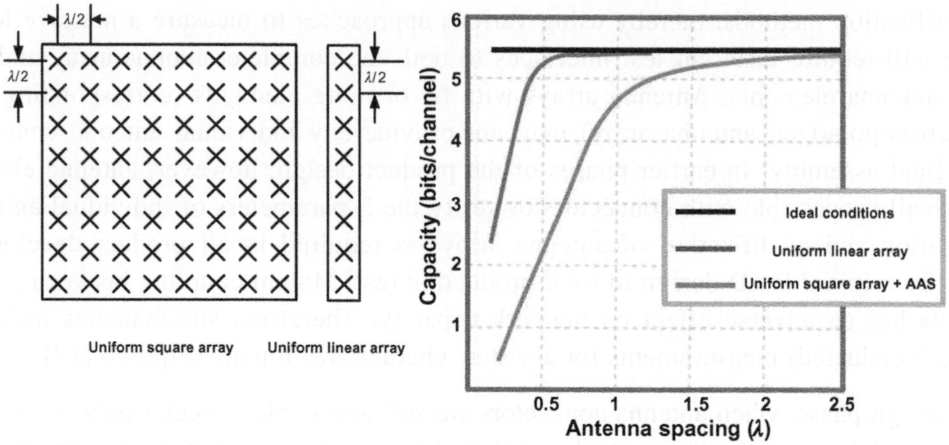

Figure 5.33
Network capacity as a function of mutual coupling in two antenna arrays [20].

spacing between elements on a 2D planner array needs to be three times greater than that between elements on a uniform linear array to maintain the same network capacity. A large antenna array to implement massive MIMO in both sub-6 GHz and mmWave frequency ranges does not provide antenna connectors due to added complexity and cost, physical size limitations, and resulting insertion loss. Consequently, the OTA tests are required. The OTA tests measuring the 3D antenna pattern can be performed either in the near-field or in the far-field. The measurements in the antenna near-field allow smaller anechoic chambers for the measurement; however, it requires an additional near-field to far-field transformation for antenna gain patterns.

As we mentioned in the previous sections, the OTA measurements can be used in pre-production and post-production phases. In the pre-production phase, the following OTA measurements need to be conducted [18]:

- Gain patterns: Gain patterns are either 2D from one of the three principal planes (E1, E2, or H-plane) or a complete 3D pattern. For antenna arrays with one or more beams, the 3D gain pattern is more useful.
- Radiated power: The EIRP is used to measure an AAS either as a UE or a BS. For UE testing, TRP is used instead where TRP is the weighted integral of the effective radiated power values over a sphere.
- Receiver sensitivity: Receiver sensitivity is characterized by the equivalent isotropic sensitivity (EIS) and measures the block error rate as a function of the receive power equal to the specified receiver sensitivity.
- Transceiver and receiver characterization: Each individual transceiver in the AAS needs to be verified through an OTA interface. This includes a range of measurements for both the transmitter (maximum output power, EVM, ACLR, spurious emissions, intermodulation) and the receiver (sensitivity, dynamic range, band selection, in-band/out-of-band/narrowband blocking, ACS). It is assumed that each transceiver will turn on for individual verification.
- Beam steering and beam tracking: Due to the high path loss and limited range of a mmWave wireless system, precise beam tracking and fast beam acquisition is required for mobile users. While in the antenna implementations of the existing cellular technologies static beam pattern characterization was sufficient, mmWave systems will further require dynamic beam measurement systems.

In the post-production phase the following tests need to be conducted [18]:

- Antenna calibration: In order to accurately form beams, the phase misalignment between RF signal paths needs to be less than ± 5 degrees. This measurement can be performed for both passive and active antenna systems using either a phase-coherent receiver to measure the relative difference between all antenna elements. This is then compiled into a lookup table for the AAS to use as a reference for beam generation or to calibrate the internal self-calibration circuits inside the AAS unit.
- Transceiver calibration: Due to lack of RF ports on some massive MIMO systems, the individual transceivers will need to be calibrated using OTA techniques. This includes both transmitters and receivers.
- Five-point beam test: The AAS manufacturer specifies a reference beam direction, maximum EIRP, and accordingly EIRP values for each declared beam. For conformance, the maximum EIRP point, and four additional points corresponding to the most extreme steering positions are measured.

- Functional tests: This is the final test performed on the completely assembled unit in production. It can consist of a simple radiated test, a five-point beam test, and aggregate transceiver functionality, such as an EVM measurement of all transceivers.

References

3GPP Specifications[35]

[1] 3GPP TS 37.105, Active Antenna System (AAS) Base Station (BS) Transmission and Reception (Release 15), December 2018.

[2] 3GPP TS 38.101-1, NR User Equipment (UE) Radio Transmission and Reception; Part 1: Frequency Range 1 Standalone (Release 15), December 2018.

[3] 3GPP TS 38.101-2, NR User Equipment (UE) Radio Transmission and Reception; Part 2: Frequency Range 2 Standalone (Release 15), December 2018.

[4] 3GPP TS 38.101-3, NR User Equipment (UE) Radio Transmission and Reception; Part 3: Frequency Range 1 and Frequency Range 2 Interworking Operation With Other Radios (Release 15), December 2018.

[5] 3GPP TS 38.104, NR Base Station (BS) Radio Transmission and Reception (Release 15), December 2018.

[6] 3GPP TS 38.133, NR Requirements for Support of Radio Resource Management (Release 15), January 2019.

[7] 3GPP TS 38.215, NR, Physical Layer Measurements (Release 15), December 2018.

ITU-R Recommendations[36]

[8] ITU-R Recommendation SM.328, Spectra and Bandwidth of Emissions, May 2006.

[9] ITU-R Recommendation SM.329, Unwanted Emissions in the Spurious Domain, September 2012.

Articles, Books, White Papers, and Application Notes

[10] E. Dahlman, S. Parkvall, 5G NR: The Next Generation Wireless Access Technology, Academic Press, August 2018.

[11] S. Ahmadi, LTE-Advanced: A Practical Systems Approach to Understanding 3GPP LTE Releases 10 and 11 Radio Access Technologies, November 2013, Academic Press.

[12] Global Mobile Suppliers Association (GSA) Report, Spectrum for Terrestrial 5G Networks: Licensing Developments Worldwide, January 2019.

[13] Ian Beavers, Giga-Sample ADCs Promise, Direct RF Conversion, Electronic Design, September 2014.

[14] 3GPP RWS-180011, NR Radio Frequency and Co-existence, Workshop on 3GPP Submission Toward IMT-2020, October 2018.

[15] T. Neu, Direct RF Conversion: From Vision to Reality, Texas Instruments Incorporated Application Note, May 2015.

[16] D. Anzaldo, Application Note, LTE-Advanced Release-12 Shapes New eNB Transmitter Architecture: Part 2, Analog Integration Challenge, Maxim Integrated Products, Inc., March 2015.

[17] C. Pearson, High Speed, Digital to Analog Converters Basics, Texas Instruments Incorporated, Application Note, October 2012.

[18] M. Kottkamp, C. Rowell, Antenna Array Testing—Conducted and Over the Air: The Way to 5G, Rohde & Schwarz White Paper, November 2016.

[35] 3GPP specifications can be accessed at the following URL: http://www.3gpp.org/ftp/Specs/archive/.

[36] ITU-R recommendations can be access at the following URL: https://www.itu.int/rec/.

[19] S. Han, et al., Large-scale antenna systems with hybrid analog and digital beamforming for millimeter wave 5G, IEEE Commun. Mag. 53 (1) (January 2015).

[20] F. Rusek, et al., Scaling up MIMO: opportunities and challenges with very large arrays, IEEE Signal Proc. Mag. 30 (1) (January 2013).

[21] M. Shafi, et al., 5G: a tutorial overview of standards, trials, challenges, deployment, and practice, IEEE J. Sel. Areas Commun. 35 (6) (June 2017).

[22] B. Murmann, The race for the extra decibel, a brief review of current ADC performance trajectories, IEEE Solid-State Circ. Mag. 7 (3) (2015).

[23] U. Rüddenklau, mmWave Semiconductor Industry Technologies: Status and Evolution, ETSI White Paper, (July 2016).

[24] B. Peterson, D. Schnaufer, 5G Fixed Wireless Access Array and RF Front-End Tradeoffs, *Microwave Journal*, February, (2018).

[25] Global TD-LTE Initiative, GTI 5G Device Power Consumption White Paper, (February 2019).

[26] Global TD-LTE Initiative, GTI 5G Device RF Component Research Report, (January 2019).

[27] G. Lloyd, Linearization of RF Front-ends, Rohde & Schwarz White Paper, (September 2016).

[28] K. Walsh, White Paper 5G New Radio Solutions: Revolutionary Applications Here Sooner Than You Think, Skyworks Solutions, Inc., (September 2018).

[29] Y. Sano, et al., 5G radio performance and radio resource management specifications, NTT DoCoMo Tech. J. 20 (3) (January 2019).

CHAPTER 6

Internet of Things (NB-IoT and Massive MTC)

Internet of Things (IoT) is an interconnected network of physical objects (devices) that interact with humans or other physical objects and systems, where communications among machines can be performed without human interaction or supervision. The IoT is considered a driving force behind recent improvements in wireless communications technologies such as 3GPP LTE and NR in order to satisfy the stringent requirements of massive machine-type communication (mMTC) diverse applications. The IoT use cases have been divided into two main categories: massive IoT and mission-critical IoT. The massive IoT aims to connect a very large number of devices, for example, remote indoor or outdoor sensors, to the cloud-based control and monitoring systems, where the main requirements are low cost, low power consumption, good coverage, and architectural scalability. In contrast, mission-critical IoT applications, for example, remote healthcare, remote traffic monitoring/control, and industrial control, require very high availability and reliability as well as very low latency.

Narrowband IoT (NB-IoT) is an LTE-based cellular radio access technology that provides low-power wide-area (LPWA) connectivity in licensed spectrum. There are several short-range technologies in unlicensed spectrum, including Bluetooth,[1] ZigBee,[2] etc., and other LPWA technologies, including SigFox,[3] LoRaWAN,[4] etc., which have been developed to address vertically different use cases of industrial IoT. The 3GPP design targets for Rel-13 were long device battery life, low device complexity to ensure low cost, support for massive number of devices, and coverage enhancements to be able to reach devices in basements and other inaccessible locations. The NB-IoT is optimized for machine-type traffic. It was designed to be as simple as possible in order to reduce device complexity and to minimize power consumption. Indoor penetration and rural coverage are among the key requirements for the IoT applications, and support of operation in sub-1 GHz bands is considered as crucial for IoT deployments. The NB-IoT further supports improved battery life through

[1] Bluetooth Special Interest Group, https://www.bluetooth.com/.

[2] Zigbee Alliance, http://www.zigbee.org/.

[3] Sigfox, https://www.sigfox.com/.

[4] LoRa Alliance, https://www.lora-alliance.org/.

5G NR. DOI: https://doi.org/10.1016/B978-0-08-102267-2.00006-3

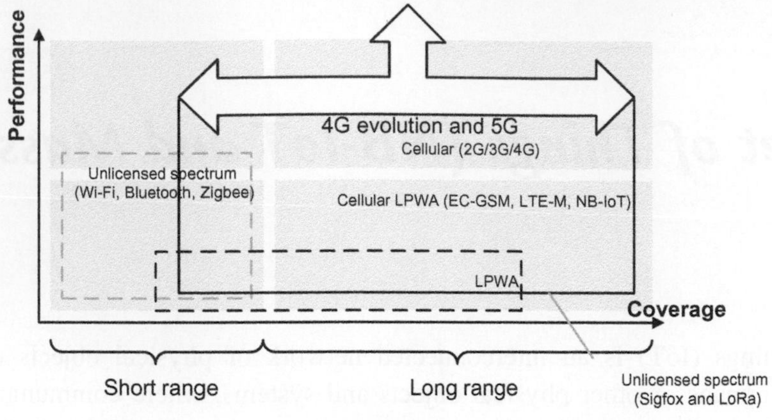

Figure 6.1
Scope and use cases of various wireless access technologies in the IoT domain [10,16].

extended discontinuous reception (eDRX) and power-saving mode, and improved capacity via multi-carrier operation [22].

The NB-IoT specifications have continued to evolve beyond Rel-13, with support for multicasting and positioning, and are considered the main technology platform to satisfy 5G mMTC applications in Rel-15 and beyond. In 3GPP LTE Rel-13, NB-IoT was standardized for providing LPWA connectivity for mMTC applications. In LTE Rel-14, NB-IoT was further enhanced to provide additional features such as increased positioning accuracy, higher peak data rates, a lower device power class, improved non-anchor carrier operation, multicast, and authorization of the coverage enhancements. To achieve enhanced coverage up to 164 dB maximum coupling loss (MCL),[5] a significant amount of network resources is required due to excessive code repetitions. With authorization of the coverage enhancements feature, the network can restrict the use of coverage enhancements to a UE based on its subscription information. Fig. 6.1 shows the scope and applicability of different wireless technologies when considering the coverage and throughput requirements. While 3GPP and non-3GPP wireless access technologies may equally find application in certain deployment scenarios, 3GPP wireless technologies can provide better integration with the existing and forthcoming cellular access and core networks.

In March 2018, 3GPP decided that no NR-based solution will be studied or specified for the LPWA use cases and those use cases would be exclusively addressed by the evolving 3GPP

[5] Maximum coupling loss is a measure of the attenuation of the radio signal between the transmitter and receiver. Maximum coupling loss is the largest attenuation that the system can support with a defined level of service. This can also be used to define the coverage of the service.

LTE-M and NB-IoT standards. This important decision was motivated by operators' and major vendors' implementation and deployment plans for LTE-M and NB-IoT in 2018 + . In essence, the operators and vendors did not want an NR-based solution for mMTC use cases to distract the market and jeopardize their ongoing developments plans around the world. It was further decided that the enhancements (if any) to coexistence between NR and LTE-M/NB-IoT beyond what is supported in 3GPP Rel-15 can be studied as part of 3GPP Rel-16 [9]. Considering the 3GPP decision, LTE-based NB-IoT solution will be the 3GPP technology to address LPWA applications, and thus we will study the theoretical and practical aspects of this technology in this chapter.

6.1 General Aspects and Use Cases

The IoT market is expected to cover several industry segments and a wide range of applications with different quality of service (QoS) requirements. Fig. 6.2 shows the partitioning of IoT applications into two broad segments namely massive IoT (mMTC) and mission-critical

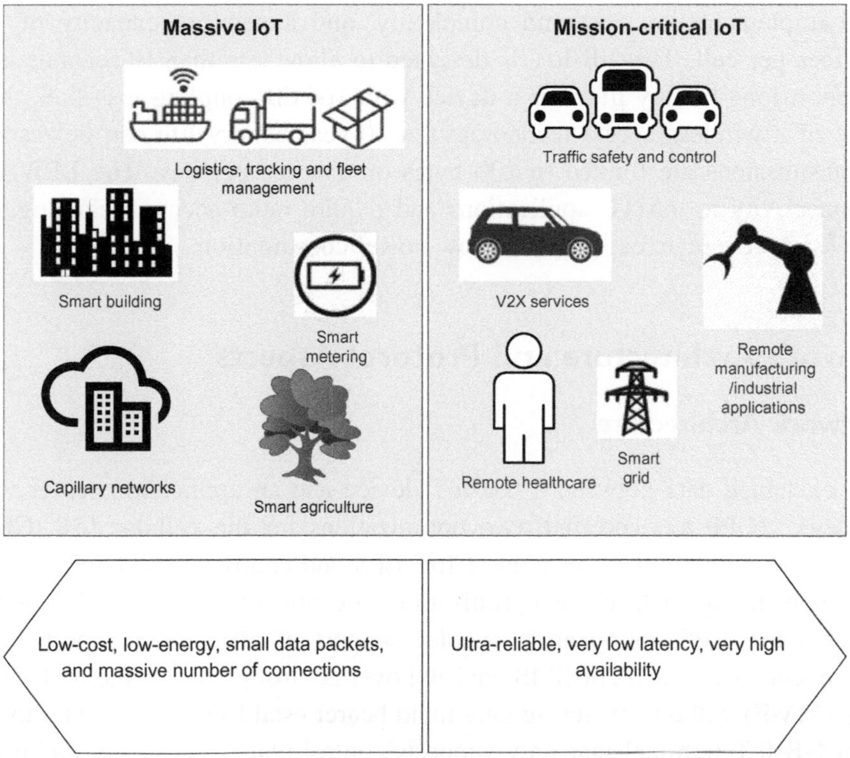

Figure 6.2
Differing requirements for massive and mission-critical IoT applications [10].

IoT (URLLC), in order to underline their distinct requirements. Massive IoT comprises applications involving exchange of small and infrequent data packets to/from numerous standalone devices. Hence, for economic viability, such devices must be implemented with low-cost and low-power consumption without stringent latency requirement. Consequently, the network technology providing connectivity to such devices must satisfy these requirements and support high capacity. On the other hand, mission-critical IoT category consists of time-sensitive applications, with latency requirements in the order of fraction of a millisecond, such as remote surgery and vehicle-to-vehicle communications. These applications demand high reliability, availability, and very low latency.

Server/cloud-based big data analytics and machine learning are central to the majority of IoT business models. IoT uses MTC to harvest data and route control messages between widely distributed things (e.g., sensors or actuators) and cloud-based intelligence. Many topologies include gateway nodes as aggregation points between IoT devices and the cloud. NB-IoT uses a narrow channel bandwidth which can be deployed either in-band with LTE, in the LTE guard-bands, or in reused GSM spectrum. The enhancements to LTE baseline have been selected such that NB-IoT can be deployed via software upgrades to the existing LTE or GSM infrastructure. The NB-IoT requirements include 10-year battery life, very low device implementation cost and complexity, and a network capacity of more than 50,000 devices per cell. The NB-IoT is designed to allow less than 10 seconds latency and aims to support long battery life. For a device with 164 dB coupling loss (i.e., the measure of coverage of a wireless access technology), a 10-year battery life can be reached, if the UE data transmissions are limited to 200 bytes on average per day. The LPWA networks provide connectivity to mMTC applications and require radio access technologies that can deliver extensive coverage, capacity, and low power consumption.

6.2 Network Architecture and Protocol Aspects

6.2.1 Network Architecture

In order to exchange data between a NB-IoT device and an application server via NB-IoT access network, 3GPP has specified two optimizations for the cellular IoT (CIoT) in the evolved packet system (EPS): user-plane CIoT EPS and control-plane CIoT EPS optimizations, as shown in Fig. 6.3. These optimizations are not limited to NB-IoT devices. The control-plane CIoT EPS optimization enables support of efficient transport of user data [Internet protocol (IP), non-IP, or SMS packets] over control plane via the mobility management entity (MME) without triggering data radio bearer establishment in order to reduce the latency. An NB-IoT terminal that only supports control-plane CIoT EPS optimization is a UE that does not support user-plane CIoT EPS optimization and S1-U data transfer, but may support other CIoT EPS optimizations. The user-plane CIoT EPS optimization enables

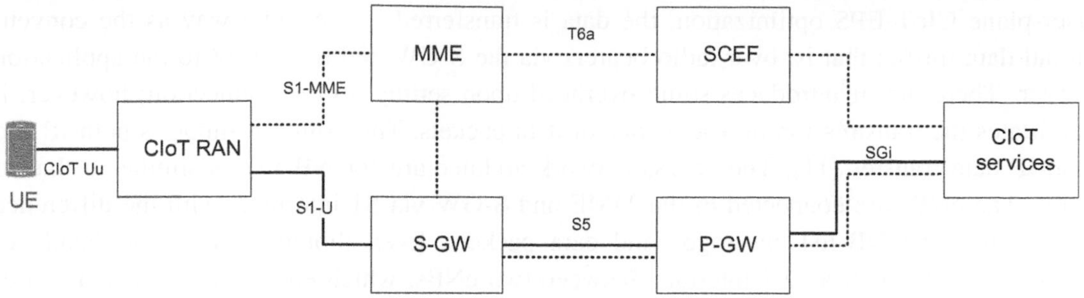

Figure 6.3
NB-IoT network architecture [11,19].

support for transition from EMM-IDLE mode to EMM-CONNECTED mode without invoking the service request procedure [6].

In other words, user-plane CIoT EPS optimization allows radio resource control (RRC) connections to be suspended and resumed. In LTE, a new RRC context must be established when the UE moves from idle mode to connected mode. Reestablishing the context only takes about a 100 ms, but for IoT devices that only send a few bytes, it is a very significant overhead. Therefore, a method has been specified to preserve the context of an RRC connection and suspend it on the mobile device and network side rather than releasing it. In LTE, the control plane is used for management tasks such as authentication and connection establishment, while data is transmitted over the user plane. For high-data-rate transmissions, the overhead of setting up a connection on the control plane may be negligible. But, when dealing with extremely small data packets, setting up the connection for a short transmission has a significant impact on the overall connection time and power consumption. To reduce this overhead, control-plane CIoT EPS optimization specifies a mechanism to include user data packets within signaling messages. By including the payload in the signaling message, the overhead is significantly reduced [11].

As shown in Fig. 6.3, in the control-plane CIoT EPS optimization, the uplink data is transferred from the eNB (CIoT RAN) to the MME. At that point, the data may either be transferred via the S-GW to the P-GW, or to the service capability exposure function (SCEF). The latter is only possible for non-IP data packets. From these nodes, the data is ultimately forwarded to the application server (CIoT services). The downlink data is transmitted over the same route in the reverse direction. However, there is no data radio bearer setup and data packets are instead sent on the signaling radio bearer (SRB). This solution is suitable for transmission of infrequent and small data packets. The SCEF is a new network entity introduced to enable machine-type communications. It is used for delivery of non-IP data over control plane and provides an abstract interface for the network services; that is, authentication and authorization, discovery and access network capabilities. With the user-plane CIoT EPS optimization, the data is transferred in the same way as the conventional data traffic; that is, over radio bearers via the S-GW and the P-GW to the application server. Therefore, it introduces some overhead upon setting up the connection; however, it facilitates the transmission of a sequence of data packets. This solution supports both IP and non-IP data transfer [11]. The access network architecture for NB-IoT is similar to that of LTE. The eNBs are connected to the MME and S-GW via S1 interface, with the difference of carrying the NB-IoT messages and data packets. Even though there is no handover defined, there is still an X2 interface between two eNBs, which enables a fast resume after the UE transitions to the idle state.

In the control-plane CIoT EPS optimization, the data exchange between the NB-IoT terminal and the eNB is performed at RRC level. In the downlink and uplink, the data packets may be piggybacked in the *RRCConnectionSetup* and *RRCConnectionSetupComplete* messages, respectively. For larger payloads, data transfer may continue using *DLInformationTransfer* and *ULInformationTransfer* messages. In all of these messages, there is a byte array containing non-access stratum (NAS) information, which in this case corresponds to the NB-IoT data packets. As a result, this procedure is transparent to the eNB, and the UE's RRC sublayer forwards the content of the received *DLInformationTransfer* directly to its upper layer. The *dedicatedInfoNAS* message is exchanged between the eNB and the MME via the S1-MME interface. For this data transfer mechanism, security on access stratum (AS) level is not applied. Since there is no RRC connection reconfiguration, it may immediately start after or during the RRC connection setup or resume procedure. The RRC connection has to be terminated afterwards with the RRC connection release. In the user-plane CIoT EPS optimization, data is transferred over the conventional user plane through the network; that is, the eNB forwards the data to the S-GW or receives it from this node. In order to minimize the UE complexity, only one or two dedicated radio bearers may be simultaneously configured. One needs to distinguish between two cases:

1. If the previous RRC connection was released with a possible resume operation indicated, the connection may be requested as a resume procedure. If the resume procedure is successful, then the security is established with updated keys and the radio bearer is set up similar to the previous connection.
2. If there was no previous RRC connection release with a resume indication, or if the resume request was not accepted by the eNB, the security and radio bearer have to be reestablished.

6.2.2 Modes of Operation

NB-IoT may be deployed as a standalone system using any available spectrum exceeding 180 kHz. It may also be deployed within an LTE spectrum allocation, either inside an LTE channel or in the guard-band. These different deployment scenarios are illustrated in Fig. 6.4. The deployment scenario, standalone, in-band, or guard-band should be transparent to a UE when it is first turned on and scans for an NB-IoT carrier. Similar to existing LTE UEs, an NB-IoT UE is only required to search for a carrier on a 100 kHz raster. An NB-IoT carrier that is intended for facilitating UE initial access and synchronization is referred to as an anchor carrier. The 100 kHz UE search raster implies that for in-band deployments, an anchor carrier can only be placed in certain physical resource blocks (PRBs). For example, in a 10 MHz LTE carrier, the indices of the PRBs that are best aligned with the 100 kHz grid and can be used as an NB-IoT anchor carrier are 4, 9, 14, 19, 30, 35, 40, and 45. The PRB indexing starts from index 0 for the PRB occupying the lowest frequency within the LTE channel bandwidth. In the in-band operation, the assignment of resources between LTE and NB-IoT is not fixed. However, not all frequencies; that is, resource blocks (RBs) within the LTE carrier are allowed to be used for cell connection, rather they are restricted to the values shown in Table 6.1.

Fig. 6.5 illustrates the deployment options of NB-IoT in conjunction with a 10 MHz LTE system. The PRB immediately after the DC subcarrier, that is, PRB_25, is centered at 97.5 kHz; that is, at a distance of 6.5 subcarriers from the center of the band. Since the LTE DC subcarrier is placed on the 100 kHz raster, the center of PRB_25 is 2.5 kHz from the nearest 100 kHz grid. The spacing between the centers of two neighboring PRBs above the

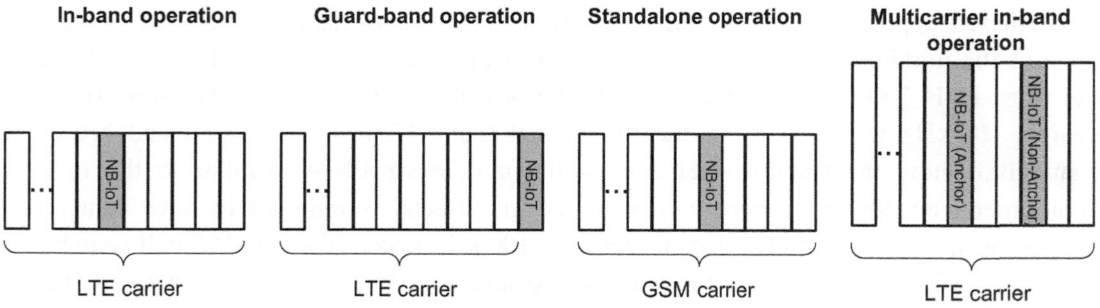

Figure 6.4
Operation modes for NB-IoT [11–15].

Table 6.1: Permissible LTE PRB indices for NB-IoT in-band operation [1].

LTE System Bandwidth (MHz)	3	5	10	15	20
Permissible LTE PRB indices for NB-IoT in-band operation	2, 12	2, 7, 17, 22	4, 9, 14, 19, 30, 35, 40, 45	2, 7, 12, 17, 22, 27, 32, 42, 47, 52, 57, 62, 67, 72	4, 9, 14, 19, 24, 29, 34, 39, 44, 55, 60, 65, 70, 75, 80, 85, 90, 95

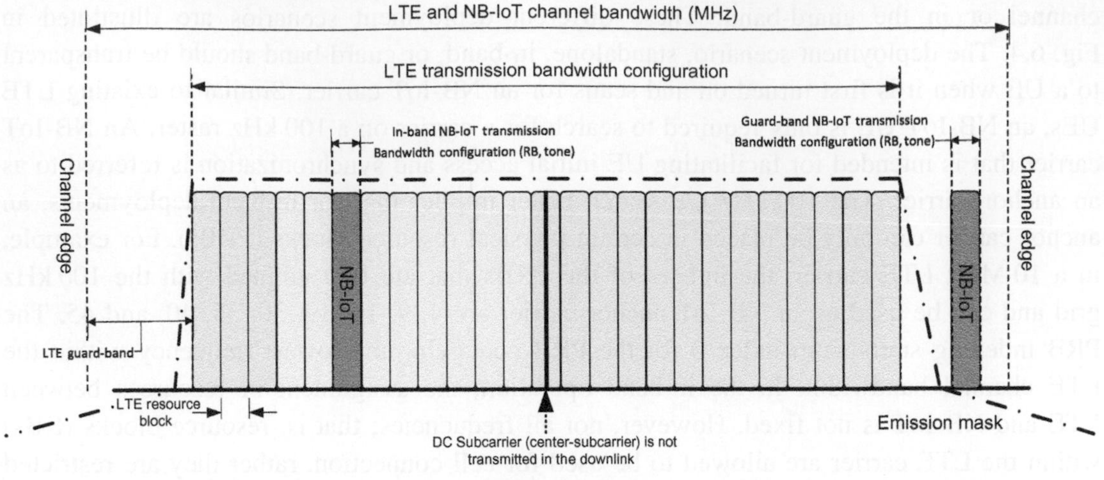

Figure 6.5

Definition of channel bandwidth and transmission bandwidth configuration for in-band/guard-band NB-IoT operation [1].

DC subcarrier is 180 kHz. As a result, the PRB indices 30, 35, 40, and 45 are all centered at 2.5 kHz from the nearest 100 kHz grid. It can be shown that for LTE carriers of 10 and 20 MHz, there exists a set of PRB indices that are all centered at 2.5 kHz from the nearest 100 kHz grid, whereas for LTE carriers of 3, 5, and 15 MHz bandwidth, the PRB indices are centered approximately 7.5 kHz away from the 100 kHz raster. It must be noted that none of the middle six PRBs of the LTE carrier can be assigned as an NB-IoT anchor carrier (e.g., PRB_25 of 10 MHz LTE, despite the fact that its center is 2.5 kHz away from the nearest 100 kHz raster). This is due to the fact that LTE synchronization and broadcast channels occupy the resource elements in the middle six PRBs. Similar to the in-band deployment, an NB-IoT anchor carrier in the guard-band is required to have center frequency no more than 7.5 kHz distant from the 100 kHz raster. The NB-IoT cell search and initial acquisition are designed for a UE to be able to synchronize to the network in the presence of a raster offset up to 7.5 kHz. Multi-carrier operation of NB-IoT is also supported. Since one NB-IoT anchor carrier is necessary to enable UE initial access/synchronization, the additional carriers do not need to be near the 100 kHz raster grid. These additional carriers are referred to as secondary carriers. As we mentioned earlier, NB-IoT carrier occupies 180 kHz of bandwidth, which corresponds to one PRB in an LTE system [2]. With this selection, the following operation modes are possible (see Fig. 6.4):

- *Standalone*: A scenario where currently used GSM frequencies with 200 kHz bandwidth can be utilized for NB-IoT deployment.
- *Guard-band*: A deployment scenario which utilizes the unused RBs within an existing LTE carrier's guard-band.
- *In-band*: A scenario where PRBs within an existing LTE channel bandwidth are utilized for NB-IoT deployment.

In the in-band operation, the assignment of resources between LTE and NB-IoT is not fixed. However, as we mentioned earlier, some of the physical LTE RBs cannot be used for NB-IoT deployment.

The NB-IoT system is designed to operate in the following LTE operating bands: 1, 2, 3, 4, 5, 8, 11, 12, 13, 14, 17, 18, 19, 20, 21, 25, 26, 28, 31, 66, 70, 71, 72, 73, and 74 [1]. 3GPP specifications do not specify how to allocate the RBs between LTE and NB-IoT. However, downlink synchronization and paging can only be established on certain RBs. The RBs located at the center of the band cannot be used due to LTE transmission of the downlink synchronization signals and broadcast channel. Owing to capacity limitations, NB-IoT is not designed for 1.4 MHz channel bandwidth. The RBs allocated for a cell connection are referred to as anchor carriers. For the actual exchange of data (in the connected state), other RBs (non-anchor carriers) can be assigned.

For an LTE service provider, the in-band option provides the most efficient NB-IoT deployment scenario because if there is no IoT traffic, the PRB(s), available for an NB-IoT carrier, may be allocated to LTE services. Note that NB-IoT can be fully integrated with the existing LTE infrastructure. This allows the base station scheduler to multiplex LTE and NB-IoT traffic in the same spectrum. Fig. 6.6 shows the results of a coexistence study where the impact of NB-IoT uplink transmission in in-band and guard-band modes on an LTE [victim] system throughput is given in terms of cumulative distribution function (CDF) of throughput. The results suggest that there is a negligible impact on LTE operation as a result of NB-IoT deployment [17].

6.2.3 Protocol Structure

The NB-IoT protocol stack is a functionally reduced version of the LTE protocols. The NB-IoT further uses a different bearer structure. SRBs are partly reused from LTE; that is, the SRB0 is used for RRC messages transmitted over the common control channel, and the SRB1 is used for transport of RRC and NAS messages using the dedicated control channel. However, there is no SRB2 defined for NB-IoT. In addition, a new SRB, the SRB1bis is defined, which is implicitly configured with SRB1 using the same configuration, rather without the packet data convergence protocol entity. This bearer type plays the role of the SRB1 until the security architecture is activated, after that SRB1bis is not used anymore. This also implies that for the control-plane CIoT EPS optimization, only SRB1bis is used, because there is no security activation in this mode. The NB-IoT user-plane and control-plane protocol stack are the same as those of LTE with functionalities optimized for NB-IoT sustainable operation (see Fig. 6.7).

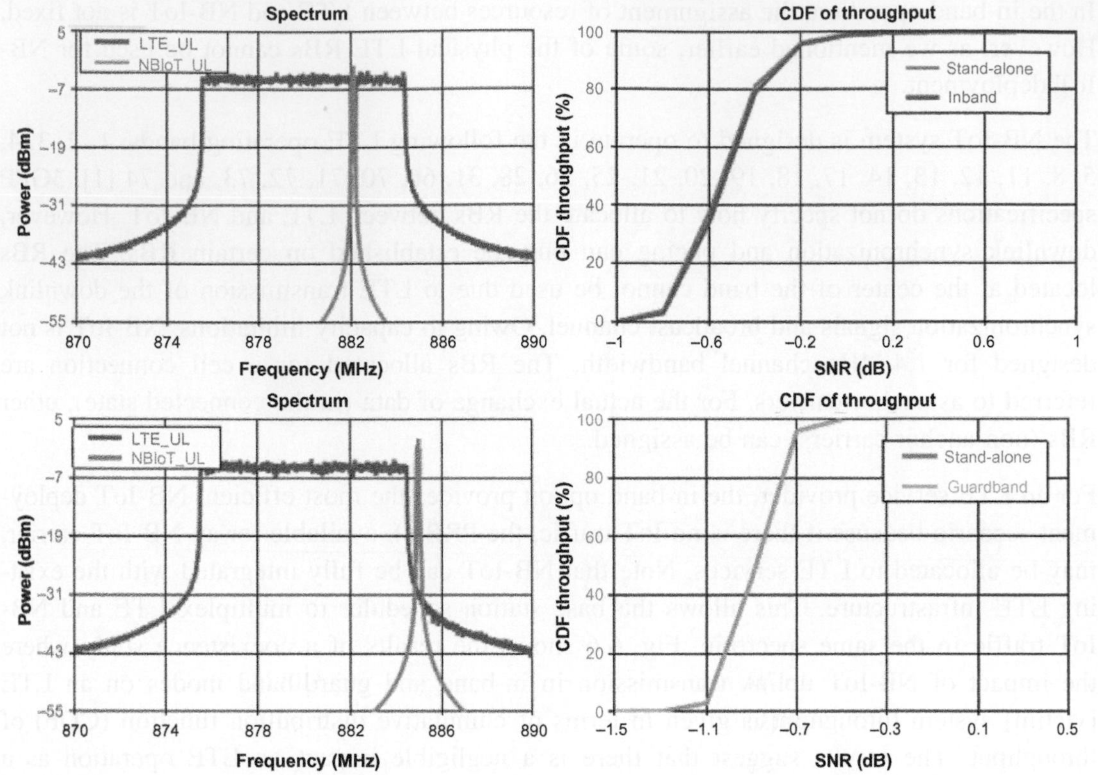

Figure 6.6

Example of in-band/guard-band NB-IoT operation and uplink coexistence analysis (10 MHz victim LTE system and aggressor NB-IoT device). A total of 1000 LTE subframes were used to derive CDF of the throughput [16,17].

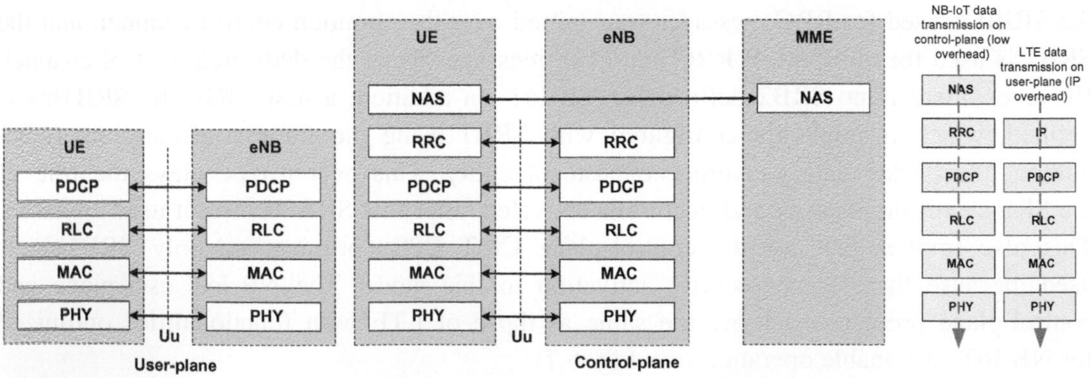

Figure 6.7

NB-IoT user-plane and control-plane protocol structure [6].

6.3 Physical Layer Aspects

6.3.1 Frame Structure

The downlink multiple access scheme of NB-IoT is based on orthogonal frequency division multiple access with the same 15 kHz subcarrier spacing as LTE with the slot, subframe, and frame durations configured as 0.5, 1, and 10 ms, respectively. Furthermore, the slot format in terms of cyclic prefix length and number of OFDM symbols per slot are also identical to those in LTE. In principle, an NB-IoT carrier can occupy one LTE PRB in the frequency domain. Reusing the same OFDM numerology as LTE ensures coexistence with LTE in the downlink. The uplink of NB-IoT supports both multi-tone and single-tone transmissions. Multi-tone transmission is based on single-carrier frequency division multiple access (SC-FDMA) with the same 15 kHz subcarrier spacing, slot, and subframe length as LTE. Single-tone transmission supports two numerologies: 15 and 3.75 kHz. The 15 kHz subcarrier spacing is identical to that of LTE, ensuring coexistence with LTE in the uplink. The 3.75 kHz single-tone numerology uses 2 ms slot duration. Similar to the downlink, an uplink NB-IoT carrier uses a total system bandwidth of 180 kHz [2,11]. In addition to the system frames, the concept of hyper frames is further defined, which counts the number of system frame periods; that is, it is incremented each time the system frame number (SFN) wraps. Similar to the SFN, the hyper frame number (HFN) is a 10-bit counter; thus the hyper frame period spans 1024 system frame periods, corresponding to a time interval of approximately 3 hours. Because NB-IoT only uses a single PRB in the downlink, the physical channels are only multiplexed in the time domain, as shown in Fig. 6.8.

6.3.2 Narrowband Primary Synchronization Signal

Unlike LTE, the NB-IoT physical channels and signals are primarily multiplexed across time. Fig. 6.8 illustrates how NB-IoT subframes are allocated to different physical channels and signals. The first three OFDM symbols are not used for NB-IoT, because they carry the

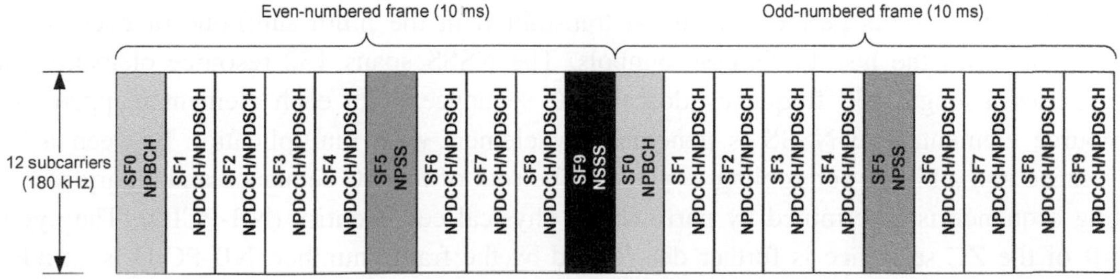

Figure 6.8
NB-IoT frame structure [14,25].

physical control format indicator channel (PCFICH) and physical downlink control channel (PDCCH) in LTE when NB-IoT is operated in the in-band mode. Note that during the time when the UE synchronizes to the narrowband primary synchronization signal (NPSS) and narrowband secondary synchronization signal (NSSS), it may not know the operation mode; consequently, this guard time applies to all modes. In addition, both synchronization signals are punctured by the LTE cell-specific reference signals. It is not specified which of the antenna ports is used for the synchronization signals; this may even change between any two subframes. Each NB-IoT subframe spans over one PRB (i.e., 12 subcarriers) in the frequency domain and 1 ms in the time domain. The NPSS and NSSS are used by an NB-IoT UE to perform cell search, which includes time and frequency synchronization, and cell identity detection. Since the legacy LTE synchronization sequences occupy six PRBs, they could not be reused for NB-IoT, thus a new design was introduced. The NPSS is transmitted in the fifth subframe of every 10 ms frame, using the last 11 OFDM symbols in the subframe. The NPSS detection is one of the most computationally intensive operations from a UE perspective. To allow efficient implementation of NPSS detection, NB-IoT uses a hierarchical sequence. The signal itself consists of a single length-11 Zadoff–Chu (ZC) sequence that is either directly mapped to the 11 lowest subcarriers (the 12th subcarrier is null in the NPSS) or is inverted before the mapping process. After successful detection of the NPSS, an NB-IoT UE is able to determine the frame boundaries of a downlink transmission. For each of the 11 NPSS OFDM symbols in a subframe, either $\mathbf{p}$ or $-\mathbf{p}$ is transmitted, where $\mathbf{p}$ is the base sequence generated based on a length-11 ZC sequence with root index 5. Each of the length-11 ZC sequence is mapped to the lowest 11 subcarriers within the NB-IoT PRB. The sequence $d_l(n)$ used for the NB primary synchronization signal is generated from a frequency domain ZC sequence according to $d_l(n) = S(l)e^{-j\pi un(n+1)/11}; \quad n = 0, 1, \ldots, 10$, where the ZC root sequence index $u = 5$ and $S(l)$ for different symbol indices l is given as $[S(3)\ S(4) \ldots S(13)] = [1 \quad 1 \quad 1 \quad 1 \quad -1 \quad -1 \quad 1 \quad 1 \quad 1 \quad -1 \quad 1]$ for normal cyclic prefix [2]. The structure of NPSS is illustrated in Fig. 6.9.

6.3.3 Narrowband Secondary Synchronization Signal

The NSSS has 20 ms periodicity and is transmitted in the ninth subframe of every other frame and uses the last 11 OFDM symbols. The NSSS spans 132 resource elements and comprises a length-132 frequency domain ZC sequence, with each element mapped to a resource element. The NSSS is generated by element-wise multiplication between a ZC sequence and a binary scrambling sequence. The root of the ZC sequence and binary scrambling sequence is determined by narrowband physical cell identity (NB-PCID). The cyclic shift of the ZC sequence is further determined by the frame number. NB-PCID is an additional input parameter so that it can be derived from the sequence. There are a total of 504 distinct NB-PCID values. The NSSS is transmitted in the last subframe of each

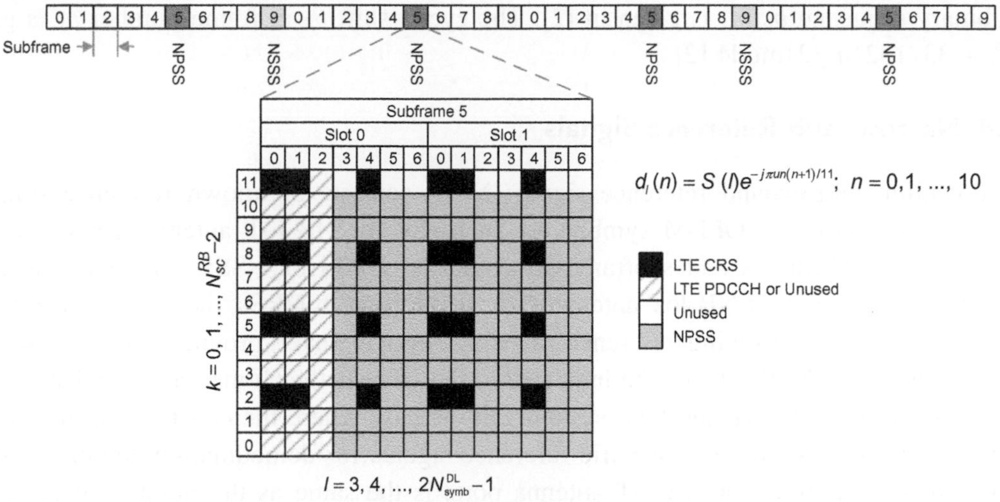

Figure 6.9
Structure of NPSS in time and frequency [2,25].

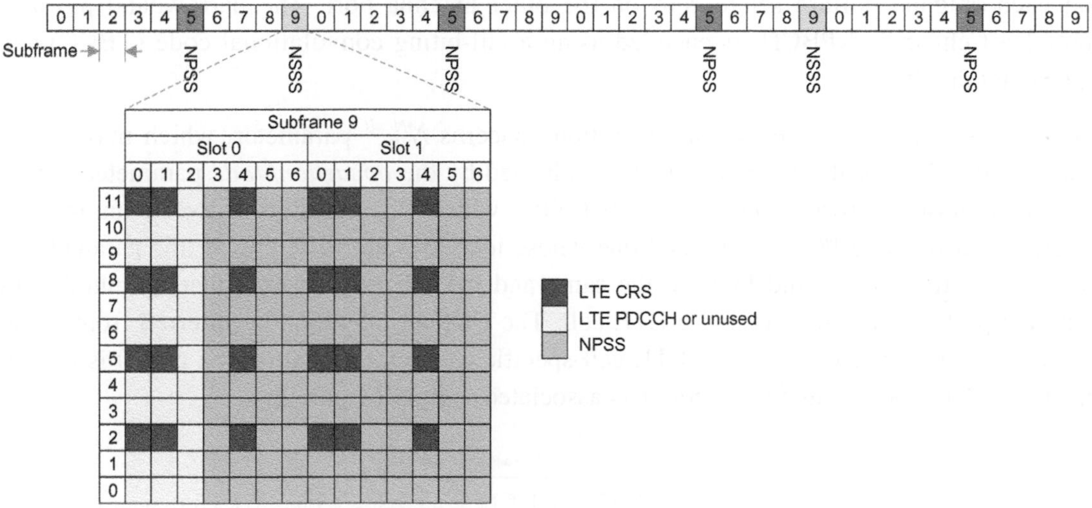

Figure 6.10
Structure of NSSS in time and frequency domains [2,25].

even-numbered radio frame, as shown in Fig. 6.10. The sequence $d(n)$ used for the NSSS is generated from a frequency domain ZC sequence according to the following:

$$d(n) = b_q(m)e^{-j2\pi\theta_f n}e^{-j\pi un'(n'+1)/131} \quad n = 0, 1, \ldots, 131$$

$$n' = n \bmod 131; \quad m = n \bmod 128; \quad u = N_{ID}^{Ncell} \bmod 126 + 3; \quad q = \left\lfloor \frac{N_{ID}^{Ncell}}{126} \right\rfloor$$

The binary sequence $b_q(m)$ is given in [2]. The cyclic shift θ_f in frame number n_f is given by $\theta_f = 33/132(n_f/2)\text{mod}4$ [2].

6.3.4 Narrowband Reference Signals

The downlink narrowband reference signal (NRS) consists of known reference symbols inserted in the last two OFDM symbols of each slot for NB-IoT antenna ports 0 and 1, except invalid subframes and subframes transmitting NPSS or NSSS. There is one NRS transmitted per downlink NB-IoT antenna port. In addition to NRSs, the physical layer supports narrowband positioning reference signals. Physical layer provides 504 unique cell identities using the NSSS. It will be indicated to the UE whether it may assume that the cell ID is identical for NB-IoT and LTE systems. In the case where the cell IDs are identical, a UE may use the downlink cell-specific reference signals for demodulation and/or measurements when the number of NB-IoT antenna ports is the same as the number of downlink cell-specific reference signals antenna ports [5]. The NRS time–frequency mapping, shown in Fig. 6.11, is additionally cyclically shifted by $N_{ID}^{Ncell}\text{mod}6$ in the frequency domain. When NRS is transmitted on two antenna ports, then the resource elements that are used for NRS on each antenna portare set to zero on the other antenna port. The narrowband physical broadcast channel (NPBCH) is encoded using a tail-biting convolutional code (TBCC) and QPSK modulated.

An important point on the in-band operation concerns N_{ID}^{Ncell} parameter, which may be the same as the PCI for the LTE cell. This is indicated by the *opeartionMode* parameter in narrowband master information block (MIB-NB), which distinguishes between an in-band operation with same PCI as LTE and one whose identities are different. If this parameter is set to true, then N_{ID}^{Ncell} and PCI are the same and the UE may assume that the number of antenna ports is the same as in the LTE cell. The channel may then be inferred from either reference signal set. In that case, LTE cell-specific reference signal (CRS) port 0 is associated with NRS port 0, and CRS port 1 is associated with NRS port 1.

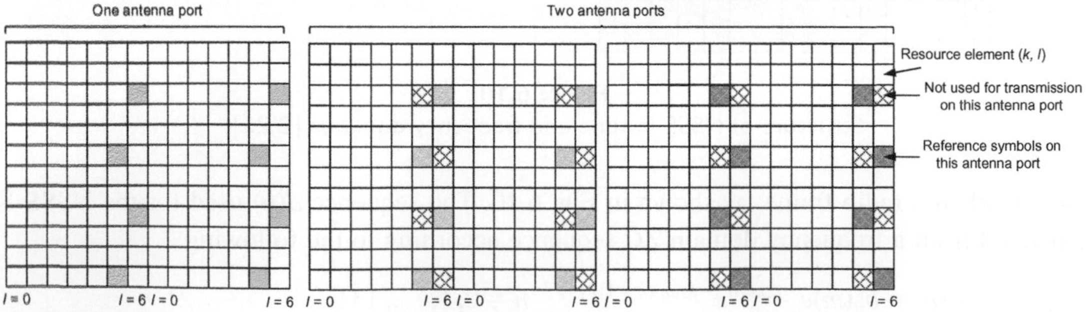

Figure 6.11
NRS subcarrier mapping [2,11,19].

6.3.5 Narrowband Physical Broadcast Channel

The NPBCH carries the MIB-NB. The MIB-NB contains 34 bits and is transmitted over a time period of 640 ms; that is, 64 radio frames, and includes the following information:

- 4 bits indicating the most significant bits of the SFN, the remaining least significant bits (LSBs) are implicitly derived from the MIB-NB starting point.
- 2 bits representing the two LSBs of the HFN.
- 4 bits for the narrowband system information block type 1 (SIB1-NB) scheduling and size.
- 5 bits indicating the system information value tag.
- 1 bit representing whether access class barring is applied.
- 7 bits indicating the operation mode with the mode-specific values.
- 11 reserved bits for future extensions.

Fig. 6.12 shows the mapping of NPBCH to physical resources. After physical layer baseband processing, the resulting MIB-NB is split into eight blocks. The first block is transmitted on the first subframe (SF0) and repeated in SF0 of the next seven consecutive radio frames. In SF0 of the following radio frame, the same procedure is performed for the second block. This process is continued until the entire MIB-NB is transmitted. The use of SF0 for all transmissions ensures collision avoidance of NPBCH with MBSFN transmissions in LTE, if NB-IoT is deployed in in-band operation mode.

The MIB-NB and *SystemInformationBlockType1-NB* use fixed scheduling. The periodicity of MIB-NB is 640 ms in comparison to the periodicity of MIB in LTE, which is 40 ms. The periodicity of SIB1-NB is 2560 ms relative to the periodicity of LTE SIB1 that is 80 ms. The MIB-NB contains the information required to acquire SIB1. SIB1-NB contains the information to acquire other SIBs (see Table 6.2 for a list of NB-IoT SIBs). The broadcast control channel and other logical channels cannot be transmitted in the same subframe. The NB-IoT UE is not required to detect SIB changes in RRC_CONNECTED state. The NPBCH consists of eight independent 80 ms blocks. A block is always transmitted in SF0 of a radio frame and then repeated eight times once per radio frame. The NPBCH is not transmitted in the first three symbols to avoid collision with the LTE control channels. The NPBCH symbols are mapped around the NRS and the LTE CRS resources, where it is always assumed that two antenna ports are defined for NRS and four antenna ports for CRS. This assumption is necessary, because the UE obtains the actual antenna port information only after detecting the MIB-NB. The reference signal location in the frequency domain is given by N_{ID}^{Ncell}, provided by the NSSS. Although the N_{ID}^{Ncell} may be different than PCI in the in-band operation, its range is restricted so that it points to the same frequency locations, thus the CRS cyclic shift in the frequency domain is known to the UE.

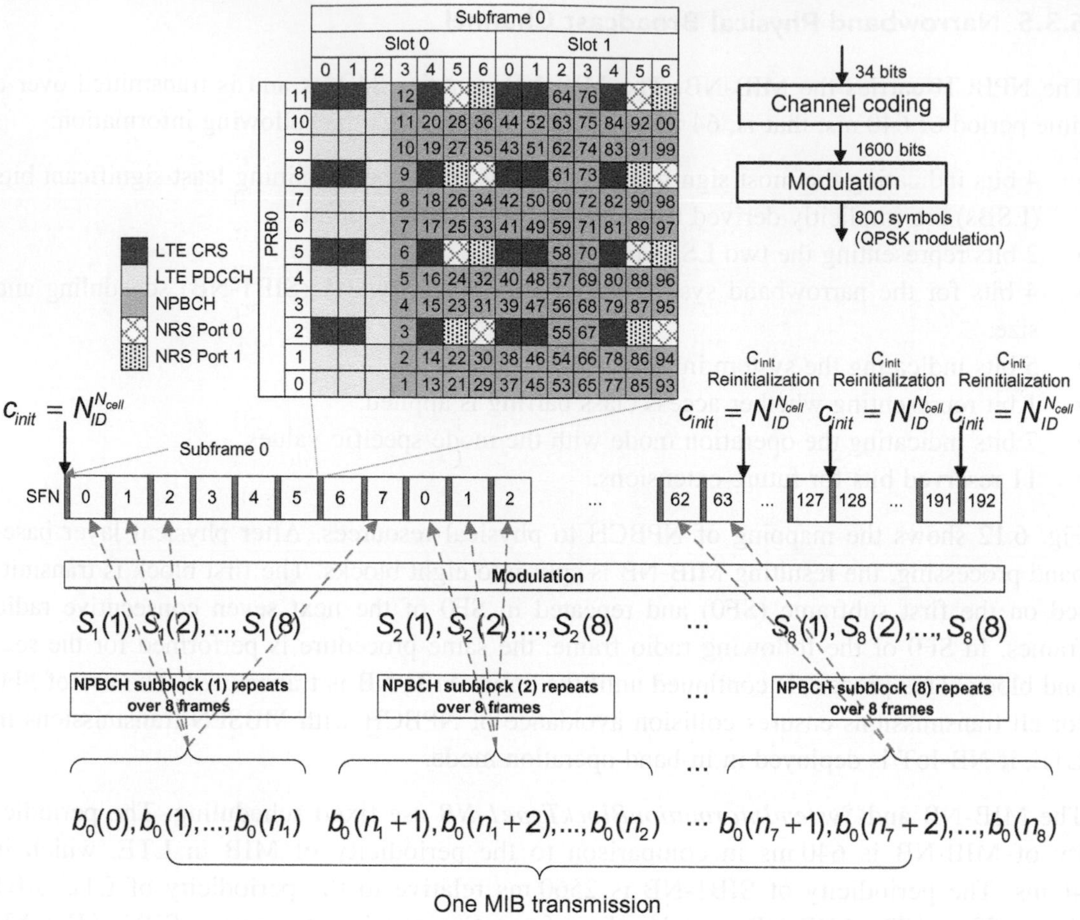

Figure 6.12
Processing and resource mapping of NPBCH [2,3,25].

Table 6.2: MIB and SIB specified for NB-IoT [8].

Message	Content
MIB-NB	Essential information required to receive further system information
SIB1-NB	Cell access and selection, other SIB scheduling
SIB2-NB	Radio resource configuration information
SIB3-NB	Cell reselection information for intra-frequency, inter-frequency
SIB4-NB	Neighbor cell-related information relevant for intra-frequency cell reselection
SIB5-NB	Neighbor cell-related information relevant for inter-frequency cell reselection
SIB14-NB	Access barring parameters
SIB16-NB	Information related to GPS time and coordinated universal time (UTC)

6.3.6 Narrowband Physical Downlink Control Channel

In NB-IoT, the downlink control information (DCI) is carried by the narrowband physical downlink control channel (NPDCCH). The basic resource allocation unit for NPDCCH is defined as narrowband control channel element (NCCE), which is the basic resource unit (RU) allocated for DCI transport. NPDCCH carries scheduling information for downlink and uplink data channels and further carries HARQ ACK/NACK information for the uplink data channel as well as paging indication and random-access response (RAR) scheduling information. There are two different formats for NPDCCH; that is, Format 0 and Format 1. NPDCCH Format 0 occupies only one NCCE, whereas NPDCCH Format 1 occupies two NCCEs. As shown in Fig. 6.13, there are only a few subframes that can be allocated to carry NPDCCH or narrowband physical downlink shared channel (NPDSCH). To reduce UE complexity, all downlink channels use LTE TBCC. Furthermore, the maximum transport block size (TBS) of NPDSCH is limited to 680 bits. The NRS is used to provide phase reference for coherent demodulation of the downlink channels. The NRS resources are time and frequency multiplexed with information bearing symbols in subframes carrying NPBCH, NPDCCH, and NPDSCH, using eight resource elements per subframe per antenna port. The NPDCCH structure is depicted in Fig. 6.13. In this example, we show the mapping for an in-band NB-IoT operation assuming a single antenna port in the LTE cell and two antenna ports in NB-IoT. The parameter $l_{NPDCCHstart}$, the LTE control region size signaled by NB-SIB1 [4], indicates the OFDM start symbol, which helps avoid conflict with the LTE control channel in the in-band operation. For the guard-band and standalone operation modes, the control region size is by default zero, which provides more resource elements for the NPDCCH. In order for the UE to find the control information with reasonable amount of decoding complexity, NPDCCH is grouped into the following search spaces:

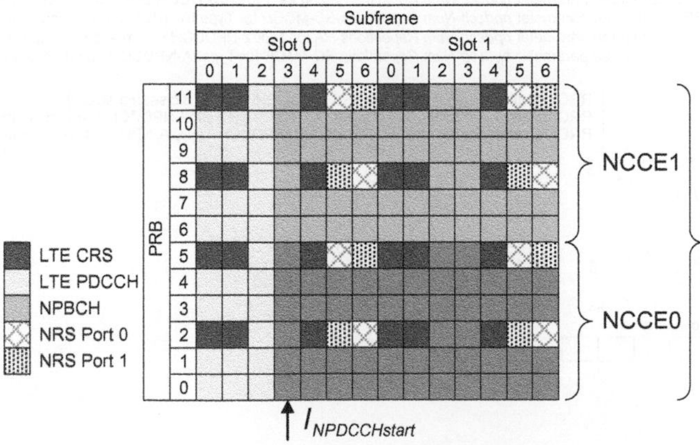

NPDCCH format	Number of NCCEs
0	1
1	2

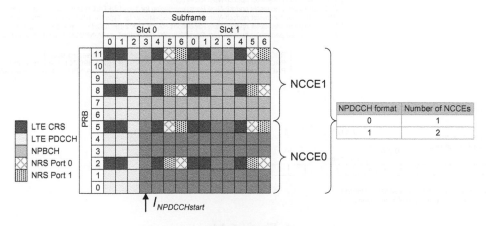

Figure 6.13
NPDCCH structure in time and frequency [2,25].

- Type-1/1A common search spaces used for paging
- Type-2/2A common search spaces used for random access
- UE-specific search space

A UE is not required to simultaneously monitor an NPDCCH UE-specific search space and a Type-1 or Type-2 NPDCCH common search space. An NPDCCH search space $NS_k^{(L',R)}$ at aggregation level (AL) $L' \in \{1, 2\}$ and repetition level $R \in \{1, 2, 4, 8, 16, 32, 64, 128, 256, 512, 1024, 2048\}$ is defined by a set of NPDCCH candidates where each candidate is repeated in a set of R consecutive NB-IoT downlink subframes excluding subframes used for transmission of system information messages starting with subframe k [4]. Thus each NPDCCH may be repeated several times with an upper limit configured by the RRC signaling. In addition, Type-2 common search space and UE-specific search space are configured by RRC, whereas the Type-1 common search space is identified by the paging opportunity subframes. The location of NPDCCH (i.e., the subframes used for transmitting NPDCCH) is determined via a process as illustrated in Fig. 6.14. In case of Type-1 NPDCCH common search space, $k = k_0$, which is determined from locations of NB-IoT paging opportunity subframes.

Different radio network temporary identifiers (RNTIs) are assigned to each UE, one for random access (RA-RNTI), one for paging (P-RNTI), and a UE-specific temporary identifier (C-RNTI) provided during the random-access procedure. These identifiers are implicitly indicated (scrambled) in the NPDCCH CRC. Therefore, the UE has to search for a specific

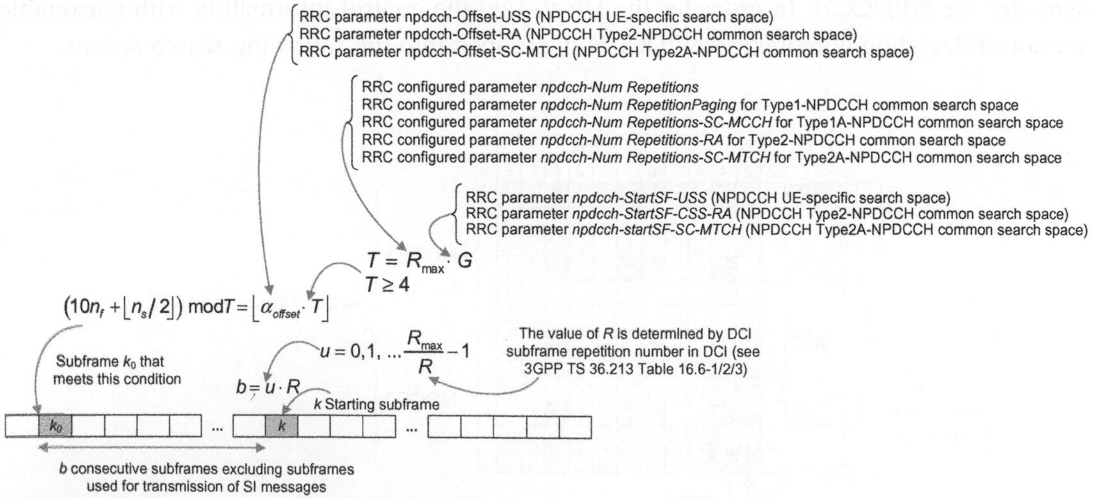

Figure 6.14
Search spaces and determination of NPDCCH location [4,25].

RNTI and, if found, decode the NPDCCH. Three DCI formats have been specified for NB-IoT, namely DCI format N0, N1, and N2.

When an NB-IoT UE receives a NPDCCH, it can distinguish different formats in the following manner. DCI format N2 is implicitly indicated, wherein the CRC is scrambled with the P-RNTI. If the CRC is scrambled with the C-RNTI, then the first bit in the message indicates whether it contains DCI format N0 or N1. For the case that the CRC is scrambled with the RA-RNTI, the content is a restricted DCI format N1 including only those fields required for the random-access response. The scheduling delay is included in DCI formats N0 and N1; that is, the time interval between the end of NPDCCH and the start of NPDSCH or the start of narrowband physical uplink shared channel (NPUSCH). This delay is at least five subframes for the NPDSCH and eight subframes for the NPUSCH. For downlink transmission via DCI format N2, the scheduling delay is fixed and equal to 10 subframes [4] (see Table 6.3).

In order to determine whether there is any data sent through NPDSCH to an NB-IoT UE or to detect any uplink grant for NPUSCH, the UE should monitor and try to decode various regions within downlink subframes. There is no information explicitly sent by the network regarding which regions the UE needs to monitor. The UE should monitor all possible regions that are allowed for NPDCCH and attempt to decode the information; that is, blind decoding. However, the UE does not need to decode every possible combination of resource elements within a subframe. There are a certain set of predefined regions in which a PDCCH could be allocated. The UE monitors only those predefined regions which are called NPDCCH search spaces.

Since an NB-IoT UE does not know in advance when a DCI will be sent to it, it must monitor NPDCCH subframes and attempt to decode them, in order to detect a pending downlink transmission. The NPDCCH configuration for a UE is primarily defined by how often the UE should start to monitor NPDCCH subframes and the maximum number of subframes that it should monitor. These two parameters are denoted by NPDCCH period T and maximum NPDCCH repetitions R_{max} (Fig. 6.15). In the example shown in Fig. 6.16, the NPDCCH period is assumed to start from a subframe at time $t = 0$. Therefore, the UE starts to monitor

Table 6.3: Various NB-IoT DCI formats [3].

DCI Format	Size (bits)	Content
N0	23	Uplink grant
N1	23	NPDSCH scheduling RACH procedure initiated by NPDCCH order
N2	15	Paging and direct indication

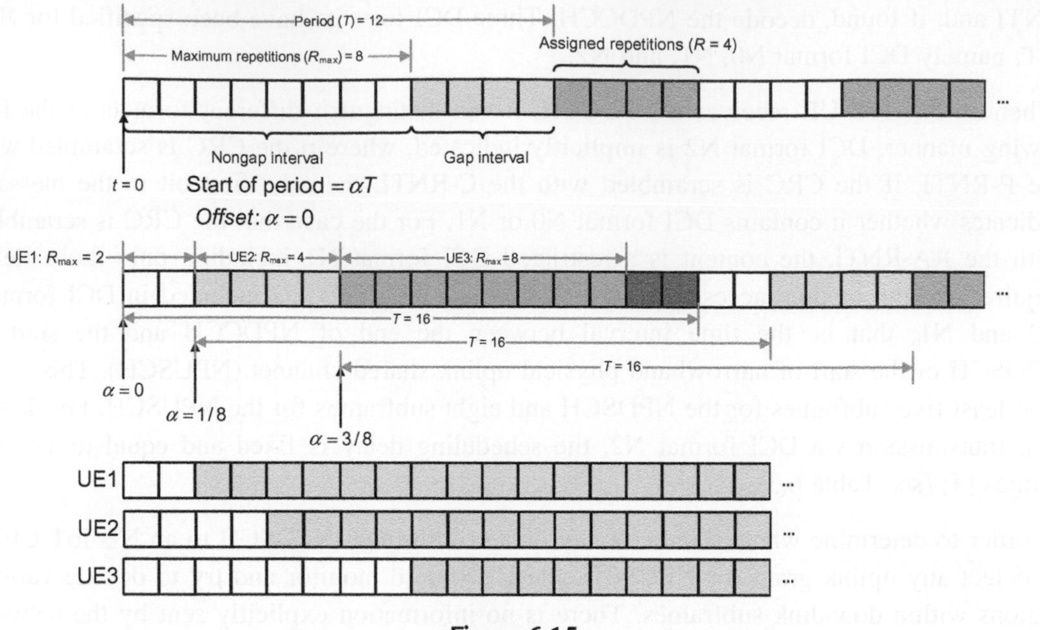

Figure 6.15
Example NPDCCH configuration for an NB-IoT UE and the use of an offset to create non-overlapping NPDCCH configurations [18,19].

NPDCCH subframes at the following time instances $t = 0, T, 2T, \ldots$. While there are T subframes in one period, the UE monitors NPDCCH subframes for a maximum duration equal to R_{max} subframes, alternatively referred to as the non-gap interval. The duration of the remaining $T - R_{max}$ subframes are known as gap interval. Fig. 6.16 further shows that at the start of the second period, the UE has received four repetitions of the NPDCCH, which it would then start decoding to check for any relevant control information. The choice of R_{max}, for a given period T, has the following implications. A large R_{max} implies higher downlink signal-to-interference plus noise ratio (SINR), resulting in better downlink coverage. A large R_{max} further indicates more opportunities to schedule the UE within period T. In heavy network loading scenarios, this may decrease the waiting time for the UE to be scheduled. The maximum number of scheduling opportunities per second can be represented by R_{max}/T. A trade-off must be made between increasing R_{max} value and the increased UE energy consumption since the UE receiver needs to be active and monitor NPDCCH subframes for a longer time duration. The beginning of the period may be shifted with respect to $t = 0$ using the offset parameter α. The number of subframes by which the period is shifted is denoted by αT. Fig. 6.15 shows how different α values can be used to create non-overlapping NPDCCH configurations for three UEs having the same value of period $T = 16$. Note that the term non-overlapping means that the non-gap intervals of the UEs do not overlap with each other. As shown in the figure, the NPDCCH periods of UE-2 and UE-3 start two and six subframes

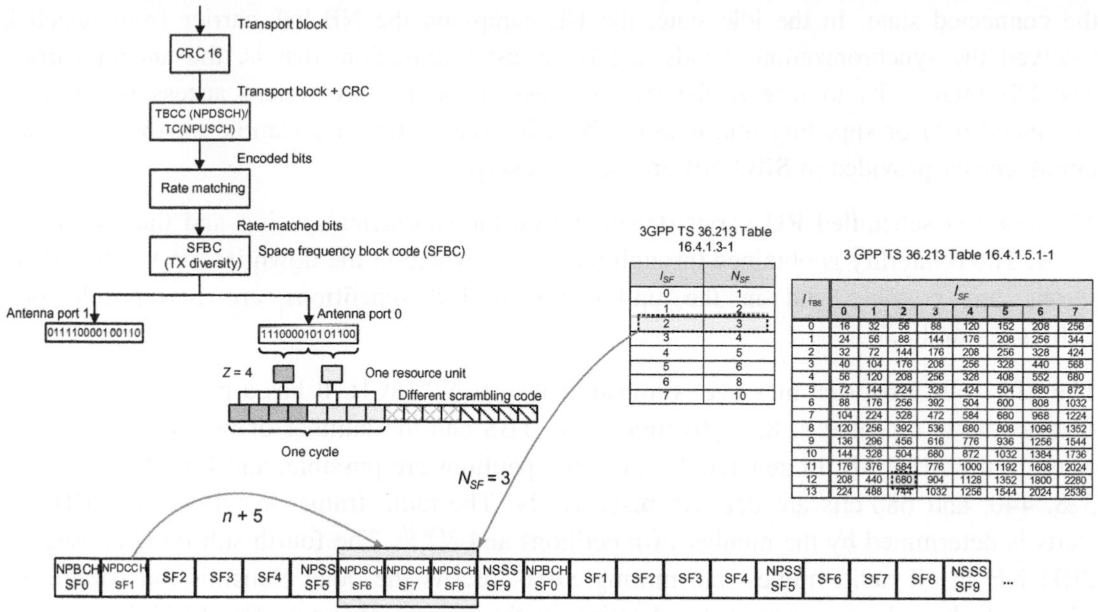

Figure 6.16
NPDSCH processing and mapping procedure [25].

later, respectively, compared to that of UE-1. It must be noted that, in an overlapping configuration, if any of the UEs receive one or more NPDCCH subframes, it may partially or fully block the non-gap interval of the remaining UEs, thus blocking the scheduling opportunity of those UEs. Furthermore, the likelihood of this blocking increases with larger R_{max} values [4].

6.3.7 Narrowband Physical Downlink Shared Channel

The NPDSCH is the traffic channel and carries SIBs, upper layer data, and RAR messages. The NPDSCH subframe has the same structure as the NPDCCH shown in Fig. 6.16. It starts at a configurable OFDM symbol $l_{NPDCCHstart}$ and is mapped around the NRS and LTE CRS during in-band operation, where the value of parameter$l_{NPDCCHstart}$ is provided by RRC signaling for the in-band operation, and is zero otherwise. A maximum TBS of 680 bits is specified for NB-IoT applications. The mapping of a transport block spans N_{SF} subframes. The transport block is repeated providing N_{RP} identical copies using a subframe interleaving mechanism for an optimized reception at the NB-IoT UE, as shown in Fig. 6.16. Both values, N_{SF} and N_{RP}, are signaled via DCI. The resulting subframe sequence is mapped to $N_{SF}N_{RP}$ consecutive subframes defined for NPDSCH (see Fig. 6.16). For the downlink, there is no automatic acknowledgment to a transmission and the eNB indicates this information in DCI. In that case, the UE transmits the acknowledgment using NPUSCH format 2. The associated timing and subcarrier mapping is indicated in this DCI, as well. There is a multi-carrier support for all operation modes, which means that another carrier may be used when the UE is in the connected state. In the idle state, the UE camps on the NB-IoT carrier from which it received the synchronization signals and broadcast information, that is, the anchor carrier. The UE then waits to receive the paging message or to start system access for mobile-originated data or signaling transmission, both by transmitting a preamble on the associated uplink carrier provided in SIB2-NB broadcast message.

The smallest scheduled RU varies depending on the subcarrier spacing and the number of tones. The reliability is obtained through repetitions where in the downlink up to 2048 repetitions are possible and in the uplink up to 128 repetitions are permissible (see Tables 6.4 and 6.5).

The SIB1-NB broadcast message is transmitted over NPDSCH. It has a period of 256 radio frames and is repeated 4, 8, or 16 times. The TBS and the number of repetitions are indicated in the MIB-NB. In general, 4, 8, or 16 repetitions are possible, and four TBSs of 208, 328, 440, and 680 bits are defined, respectively. The radio frame on which the SIB1-NB starts is determined by the number of repetitions and N_{ID}^{Ncell}. The fourth subframe is used for SIB1-NB in all radio frames transmitting SIB1-NB. As the other transmission parameters are fixed, there is no associated indication in the control channel. The SIB1-NB content may only be changed on each modification period, which has a length of 4096 radio frames; that is, every 40.96 seconds. This corresponds to four SFN periods, which is the reason for indication of two LSBs of the HFN in the MIB-NB. If such a modification occurs, it is

Table 6.4: Resource allocation in NB-IoT [2].

Subcarrier Spacing (kHz)	Number of Tones	Link Direction	Resource Unit Size (ms)
15	12	Uplink/Downlink	1
	6	Uplink	2
	3	Uplink	4
	1	Uplink	8
3.75	1	Uplink	32

Table 6.5: Uplink resource allocation in NB-IoT [2,4].

Uplink Data Type	Subcarrier Spacing (kHz)	Subcarriers per RU	Slots per RU	Duration of RU (ms)	Number of Resource Elements per RU
Uplink data	3.75	1	16	32	96
	15	1	16	8	96
		3	8	4	144
		6	4	2	144
		12	2	1	144
Uplink control information (ACK/NACK)	3.75	1	4	8	16
	15	1	4	2	16

indicated in the NPDCCH using DCI format N2. Although sent over the NPDSCH, the SIB1-NB resources are mapped similar to the MIB-NB shown in Fig. 6.12; that is, excluding the first three OFDM symbols, which is necessary because the UE obtains the start of the resource mapping from SIB1-NB.

One of the key features of NB-IoT is coverage enhancement through signal repetition. For typical NB-IoT use cases, the signal-to-noise ratio (SNR) operation point is below 0 dB. Under this operating condition, the channel estimation error is the dominating factor affecting the receiver performance. It can be shown that the effective SNR after combining N_{RP} repetitions is given as follows [18]:

$$SNR_{effective} = \frac{N_{RP}(\sigma^2 + \gamma)}{(\sigma^2 + \gamma + \sigma^2/\gamma)(1 + \sigma^2/2\gamma)}$$

where γ denotes SNR per transmission, σ^2 represents the variance of channel estimation error, and N_{RP} is the number of repetitions.

6.3.8 Narrowband Physical Random-Access Channel

Similar to LTE, the narrowband physical random-access channel (NPRACH) is used by an unsynchronized UE to inform the eNB of its desire to establish a connection. The random-access procedure is a contention-based and collision-prone mechanism in which the UE selects and transmits a random-access preamble to the base station. The NPRACH preamble consists of four symbol groups, with each symbol group comprising one cyclic prefix and five symbols, as shown in Fig. 6.17. The length of the cyclic prefix is 66.67 μs (Format 0) for cell radius up to 10 km and 266.7 μs (Format 1) for cell radius up to 40 km. Each symbol, with fixed symbol value 1, is modulated on a 3.75 kHz tone with symbol duration of 266.67 μs. However, the tone frequency index changes from one symbol group to another. The waveform of NPRACH preamble is referred to as single-tone frequency hopping. An example of NPRACH frequency hopping is illustrated in Fig. 6.18. To support coverage

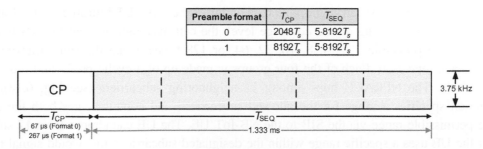

Preamble format	T_{CP}	T_{SEQ}
0	$2048T_s$	$5 \cdot 8192T_s$
1	$8192T_s$	$5 \cdot 8192T_s$

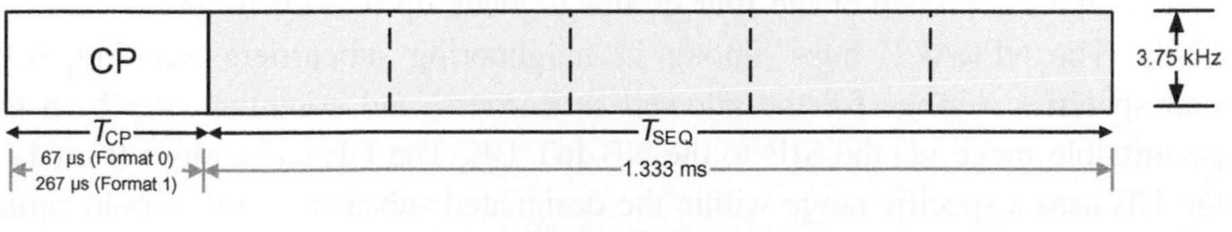

Figure 6.17
Time domain structure of NPRACH [2].

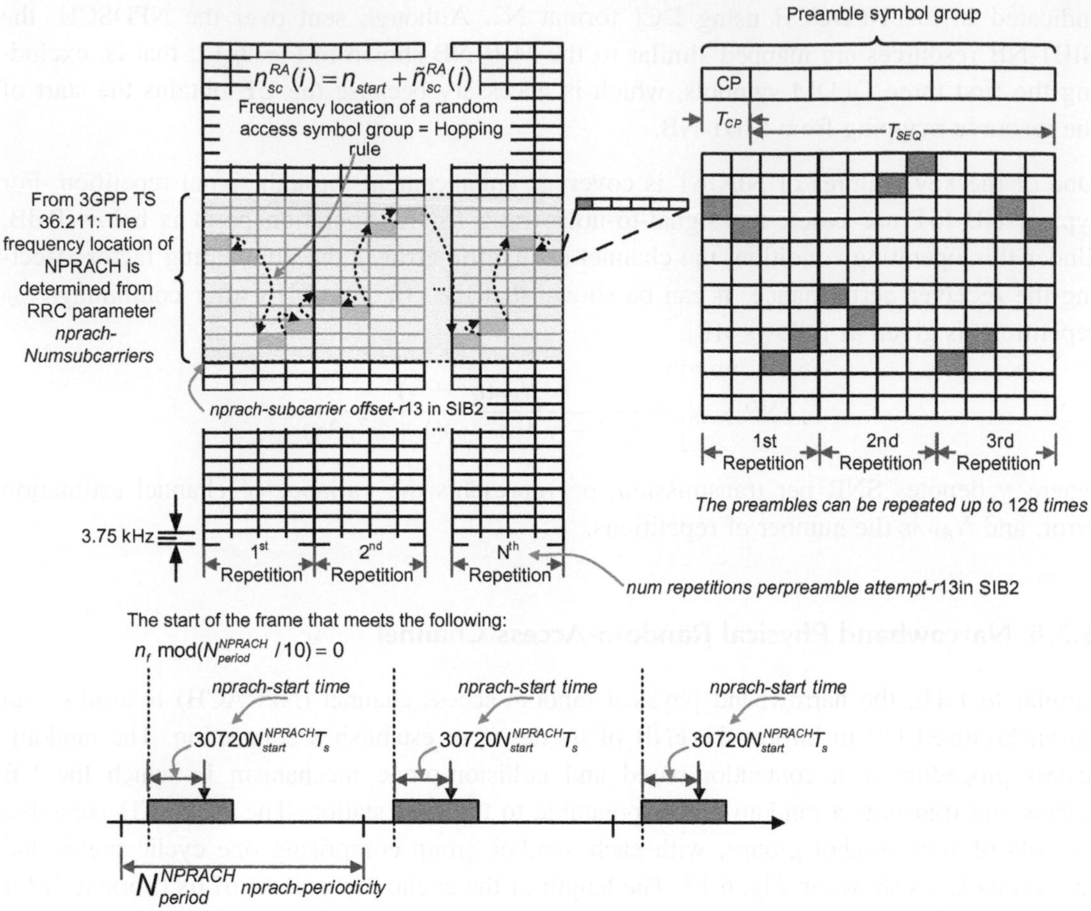

Figure 6.18
Illustration of NPRACH time and frequency domain resource allocation and transmission timing [2,25].

extension, a NPRACH preamble can be repeated up to 128 times. The NPUSCH has two formats. Format 1 is used for carrying uplink data and uses the same LTE turbo code for forward error correction. Depending on the coverage level, the cell may indicate that the NB-IoT UE must repeat the preamble 1, 2, 4, 8, 16, 32, 64, or 128 times, using the same transmission power in each repetition. Each of the four groups is made up of a cyclic prefix and four identical symbols. The NPRACH hops among 12 neighboring subcarriers (see Fig. 6.18). The base station specifies a range for the allowed subcarriers, and communicates both the delay and the permissible range via the SIB to the NB-IoT UE. The UE can choose from 12 subcarriers. If the UE uses a specific range within the designated subcarriers, this would signal to the base station that the UE supports multi-tone transmission format.

The NPRACH resources are separately provided for each coverage enhancement group. They consist of the assignment of time and frequency resources and occur periodically, where an NPRACH periodicity between 40 ms and 2.56 seconds may be configured. Their start time within a period is provided in the system information, whereas the number of repetitions and the preamble format determine their end. In the frequency domain, subcarrier spacing of 3.75 kHz is utilized. The NPRACH resources occupy a contiguous set of 12, 24, 36, or 48 subcarriers and are located on a discrete set of subcarrier ranges. Depending on the cell configuration, the resources may be further partitioned into resources used by UEs supporting multi-tone transmission for msg3 and UEs that do not support it.

The physical random-access preamble is based on single-subcarrier frequency-hopping symbol groups. A symbol group is illustrated in Fig. 6.18, which consists of a cyclic prefix of length T_{CP} and a sequence of five identical symbols with total length T_{SEQ}. The preamble consists of four symbol groups transmitted without gaps and is transmitted N_{RP}^{NPRACH} times. The NPRACH transmission can only start $30720\, N_{start}^{NPRACH} T_s$ time units after the start of a radio frame whose frame number satisfies $n_f \bmod (N_{period}^{NPRACH}/10) = 0$. Following the transmission of $256(T_{CP} + T_{SEQ})$ time units, a gap is inserted [2]. The NPRACH preamble formats, that is, format 0 and format 1, differ in the cyclic prefix length. The five symbols have a duration of $T_{SEQ} = 1.333\text{ms}$, appended with a cyclic prefix of $T_{CP} = 67\ \mu s$ for format 0 and $T_{CP} = 267\ \mu s$ for format 1, making a total length of 1.4 and 1.6 ms, respectively. The preamble format to be used is broadcast in the system information.

The NPRACH starting subcarriers which are allocated to UE-initiated random access are divided into two sets of subcarriers, $\{0, 1, \ldots, N_{sc\_cont}^{NPRACH} N_{MSG3}^{NPRACH} - 1\}$ and $\{N_{sc\_cont}^{NPRACH} N_{MSG3}^{NPRACH}, \ldots, N_{sc\_cont}^{NPRACH} - 1\}$, where the second set would indicate the UE support for multi-tone msg3 transmission. The frequency location of the NPRACH transmission is constrained within $N_{sc}^{RA} = 12$ subcarriers. Frequency hopping is used within the 12 subcarriers, where the frequency location of the ith symbol group is given by $n_{sc}^{RA}(i) = n_{start} + \tilde{n}_{sc}^{RA}(i)$ where $n_{start} = N_{sc-offset}^{NPRACH} + \lfloor n_{init}/N_{sc}^{RA} \rfloor N_{sc}^{RA}$ and

$$\tilde{n}_{sc}^{RA}(i) = \begin{cases} \left[\tilde{n}_{sc}^{RA}(0) + f(i/4)\right] \bmod N_{sc}^{RA} & i \bmod 4 = 0 \text{ and } i > 0 \\ \tilde{n}_{sc}^{RA}(i-1) + 1 & i \bmod 4 = 1,3 \text{ and } \tilde{n}_{sc}^{RA}(i-1) \bmod 2 = 0 \\ \tilde{n}_{sc}^{RA}(i-1) - 1 & i \bmod 4 = 1,3 \text{ and } \tilde{n}_{sc}^{RA}(i-1) \bmod 2 = 1 \\ \tilde{n}_{sc}^{RA}(i-1) + 6 & i \bmod 4 = 2 \text{ and } \tilde{n}_{sc}^{RA}(i-1) < 6 \\ \tilde{n}_{sc}^{RA}(i-1) - 6 & i \bmod 4 = 2 \text{ and } \tilde{n}_{sc}^{RA}(i-1) \geq 6 \end{cases}$$

$$f(t) = \left(f(t-1) + \left[\sum_{n=10t+1}^{10t+9} c(n) 2^{n-(10t+1)} \right] \bmod \left(N_{sc}^{RA} - 1\right) + 1 \right) \bmod N_{sc}^{RA}$$

$$f(-1) = 0$$

where $\tilde{n}_{sc}^{RA}(0) = n_{init} \mathrm{mod} N_{sc}^{RA}$ and n_{init} denotes the subcarrier selected by the MAC sublayer from $\{0, 1, \ldots, N_{sc}^{NPRACH} - 1\}$, and the pseudo-random sequence $c(n)$ is specified in [2]. Note that the pseudo-random sequence generator is initialized with $c_{init} = N_{ID}^{Ncell}$ [2]. The NPRACH can be transmitted only with a specific timing within a NPRACH period as illustrated in Fig. 6.18, where *nprach-StartTime* and *nprach-Periodicity* parameters are configured by higher layers via SIB2-NB. The RACH procedure for NB-IoT will be described later in layer 2 aspects.

6.3.9 Narrowband Physical Uplink Shared Channel

The NPUSCH transports two types of information: uplink data via NPUSCH format 1 and uplink control information (UCI) via NPUSCH format 2. The latter always uses one subcarrier and is always BPSK modulated. It carries HARQ ACK/NACK corresponding to NPDSCH, whereas NPUSCH format 1 can use one or more subcarriers. For single-tone $\pi/2$-BPSK or $\pi/4$-QPSK modulation is used, while for multi-tone QPSK modulation is used. The NPUSCH can repeat data up to 128 times to improve the link budget. NPUSCH format 1 uses the same LTE turbo code (minimum code rate 1/3) for forward error correction. The maximum TBS of NPUSCH format 1 is 1000 bits, which is much lower than that in LTE. NPUSCH format 2 uses a repetition code for error control (minimum code rate of 1/16). NPUSCH format 1 supports multi-tone transmission based on the same legacy LTE numerology. In this case, the UE can be allocated with 12, 6, or 3 tones. While only the 12-tone format is supported by legacy LTE UEs, the six-tone and three-tone formats are introduced for NB-IoT UEs, which due to coverage limitation cannot benefit from higher UE bandwidth allocation. Moreover, NPUSCH supports single-tone transmission based on either 15 or 3.75 kHz numerology. To reduce the peak-to-average power ratio (PAPR), single-tone transmission uses $\pi/2$-BPSK or $\pi/4$-QPSK with phase continuity between symbols. As shown in Fig. 6.19, NPUSCH format 1 uses the same slot structure as LTE PUSCH with seven OFDM symbols per slot and the middle symbol as the demodulation reference signal (DMRS), whereas NPUSCH format 2 consists of seven OFDM symbols per slot, but uses the middle three symbols as DMRS. The DMRS is used for channel estimation and coherent detection. The smallest unit to map a transport block is the resource unit whose definition depends on the NPUSCH format and subcarrier spacing. For NPUSCH format 1 and 3.75 kHz subcarrier spacing, an RU consists of one subcarrier in the frequency domain, and 16 slots in the time domain; that is, an RU has a length of 32 ms. For the 15 kHz subcarrier spacing, there are four options: 1, 3, 6, and 12 subcarriers over 16, 8, 4, and 2 slots, respectively (see Fig. 6.19). For NPUSCH format 2, the RU is always composed of one subcarrier with a length of four slots. As a result, for 3.75 kHz subcarrier spacing, the RU has 8 ms duration and for 15 kHz subcarrier spacing, the RU spans over 2 ms.

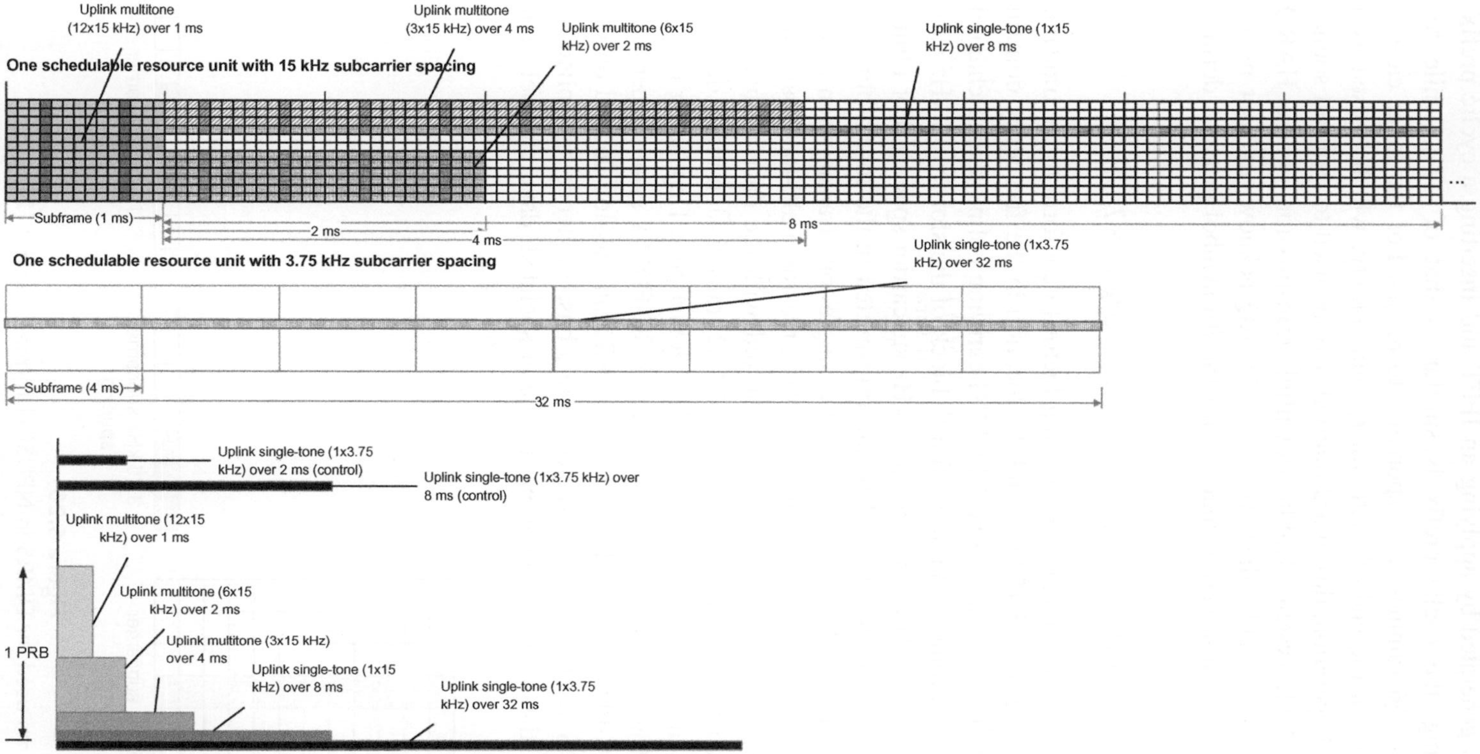

Figure 6.19
NPUSCH time–frequency mapping and uplink slot structure [11,19].

The signal in time domain is generated by applying an IFFT and inserting a cyclic prefix. For 15 kHz subcarrier spacing, the cyclic prefix is similar to that of LTE, while for 3.75 kHz, the cyclic prefix is 256 samples, corresponding to 8.3 μs. For the latter case, a period of 2304 samples (75 μs) at the end of each slot remains empty, which is used as a guard interval. For the in-band operation, this guard interval may be used to transmit sounding reference signals in the LTE system. Unlike downlink transmission, where HARQ ACK/NACK transmission is configurable, there is always HARQ acknowledgment in the associated downlink slot. The random-access signal $s(t)$ for the ith symbol group is defined as follows [2]:

$$s_i(t) = \beta_{NPRACH} \exp\left[j2\pi\left(n_{sc}^{RA}(i) + k_0\Delta f/\Delta f_{RA} + 1/2\right)\Delta f_{RA}(t - T_{CP})\right]$$

where $0 \leq t < T_{SEQ} + T_{CP}$. β_{NPRACH} is an amplitude scaling factor to conform to the transmit power P_{NPRACH}, $k_0 = -N_{sc}^{UL}/2$, $\Delta f_{\mathrm{RA}} = 3.75$ kHz, and the location in the frequency domain controlled by the parameter $n_{sc}^{RA}(i)$. For single-tone NPUSCH carrying uplink shared channel (UL-SCH), the uplink DMRSs are transmitted in the fourth block of the slot for 15 kHz subcarrier spacing, and in the fifth block of the slot for 3.75 kHz subcarrier spacing. For multi-tone NPUSCH carrying UL-SCH, the uplink DMRSs are transmitted in the fourth block of the slot. The length of the uplink DMRS sequence is 16 for single-tone transmission and is equal to the size (number of subcarriers) of the assigned resource for multi-tone transmission. In the uplink, the DMRS is defined and it is multiplexed with the data so that it is only transmitted in the RUs containing data. There is no MIMO transmission mode defined for the uplink; consequently, all transmissions use a single antenna port. Depending on the NPUSCH format, DMRS is transmitted in either one or three SC-FDMA symbols per slot. As shown in Fig. 6.20, the SC-FDMA symbols used for DMRS transmission depend on the subcarrier spacing. The DMRS symbols are constructed from a base sequence multiplied by a phase factor. They have the same modulation as the associated data channel. For

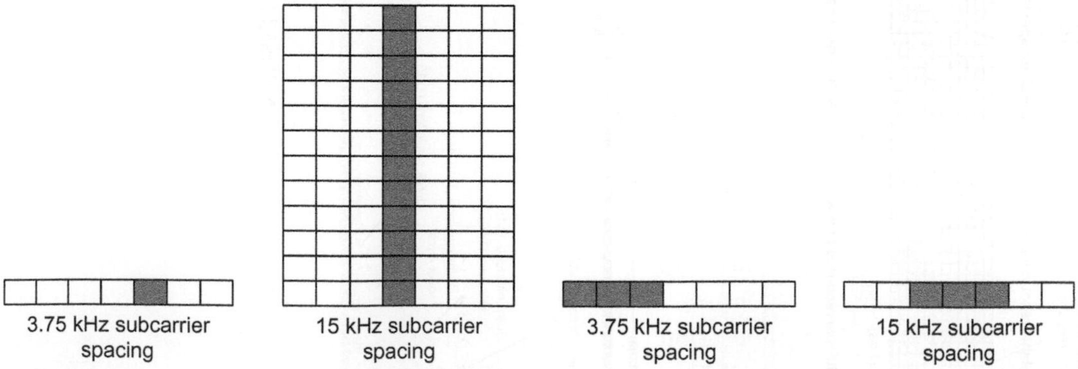

Figure 6.20
Resource elements used for DMRS in NPUSCH format 1 and 2 [2].

NPUSCH format 2, DMRS symbols are spread with the same orthogonal sequence as defined for the LTE PUCCH formats 1, 1a, and 1b.

6.4 Layer 2/3 Aspects

As in physical layer, there has been a number of changes in LTE L2/L3 operation to enable NB-IoT including introduction of new RRC messages and information elements. Some of the baseline LTE functions are not supported in NB-IoT including public safety notifications, inter-RAT mobility, and security activation for transfer of RRC context information, measurement configuration and reporting, and self-configuration and self-optimization as well as measurement logging and reporting for network performance optimization. The UE in RRC_CONNECTED can be configured, via UE-specific RRC signaling, with a non-anchor carrier for all unicast transmissions. The UE in RRC_IDLE, based on broadcast/multicast signaling, can use a non-anchor carrier for single-cell point-to-multipoint reception. The UE in RRC_IDLE can, based on broadcast signaling, use a non-anchor carrier for paging reception. The UE in RRC_IDLE or RRC_CONNECTED, based on broadcast signaling, can use a non-anchor carrier for random access. If the non-anchor carrier is not configured for the UE, all transmissions occur on the anchor carrier.

6.4.1 HARQ Protocol and Scheduling

To enable low-complexity UE implementation, NB-IoT allows only one HARQ process in both downlink and uplink, and allows longer UE decoding time for both NPDCCH and NPDSCH. Asynchronous, adaptive HARQ procedure is adopted to support scheduling flexibility. An example is illustrated in Fig. 6.21. The scheduling commands are conveyed through DCI, which is carried by NPDCCH using aggregation-level (AL)-1 or AL-2 for transmission. With AL-1, two DCIs are multiplexed in one subframe, otherwise each subframe carries only one DCI (i.e., AL-2), resulting in a lower coding rate and improved coverage. Further coverage enhancement can be achieved through repetition. Each repetition occupies one subframe. DCI can be used for scheduling downlink data or uplink data. In the case of downlink data, the exact time offset between NPDCCH and the associated NPDSCH is indicated in the DCI. Since NB-IoT devices are expected to have reduced computing capability, the time offset between the end of NPDCCH and the beginning of the associated NPDSCH is at least 4 ms. In comparison, LTE PDCCH schedules PDSCH in the same subframe. After receiving NPDSCH, the UE needs to send HARQ ACK/NACK using NPUSCH format 2. The resources of NPUSCH carrying HARQ ACK/NACK are also signaled in DCI. Considering the limited computing resources in an NB-IoT device, the time offset between the end of NPDSCH and the start of the associated HARQ ACK/NACK is at least 12 ms. This offset is longer than that between NPDCCH and NPDSCH because the size of the transport block carried in NPDSCH can be as large as 680 bits, which is much

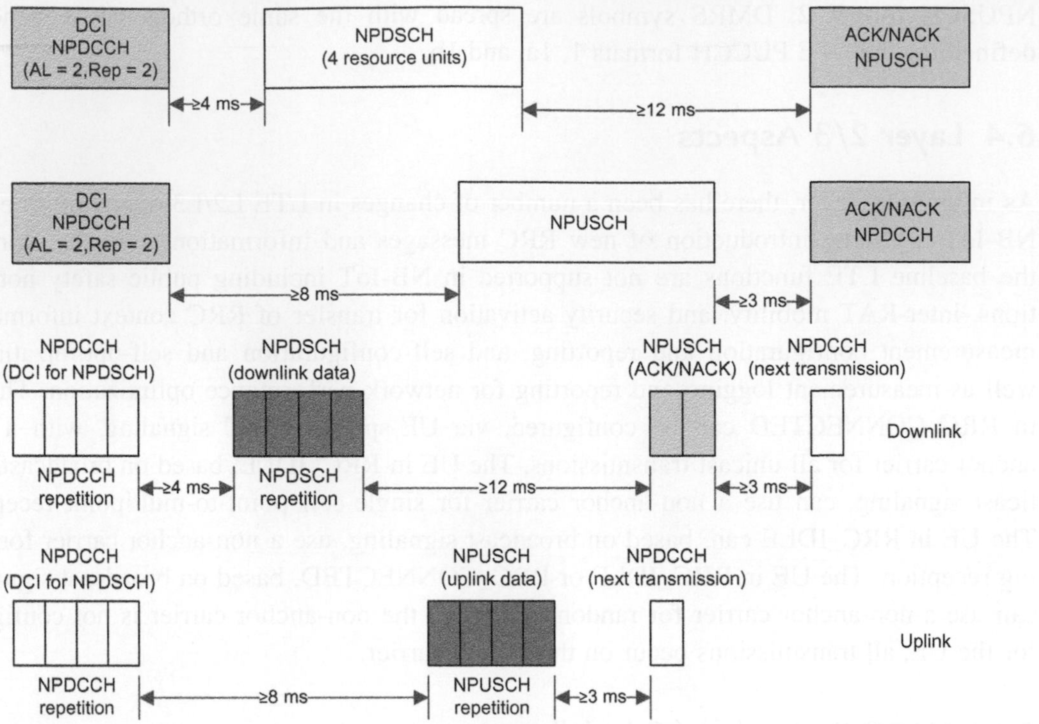

Figure 6.21
Illustration of the NB-IoT HARQ operation [14].

larger than DCI that is only 23 bits long. The DCI for uplink scheduling grant needs to specify which subcarriers are allocated for the UE. The time offset between the end of NPDCCH and the beginning of the associated NPUSCH is at least 8 ms. Upon completion of the transmission of NPUSCH, the UE monitors NPDCCH to check whether NPUSCH was received correctly by the base station, or a retransmission is required [4,14].

6.4.2 Physical, Logical, and Transport Channels

NB-IoT physical, transport, and logical channel structure are similar to LTE, with the exception that there is no physical uplink control channel in NB-IoT and the UCI content is carried in NPUSCH. In the downlink, NB-IoT provides the following physical signals and channels. Unlike LTE, these NB-IoT physical channels and signals are primarily multiplexed in time [6].

- NPSS
- NSSS
- NPBCH

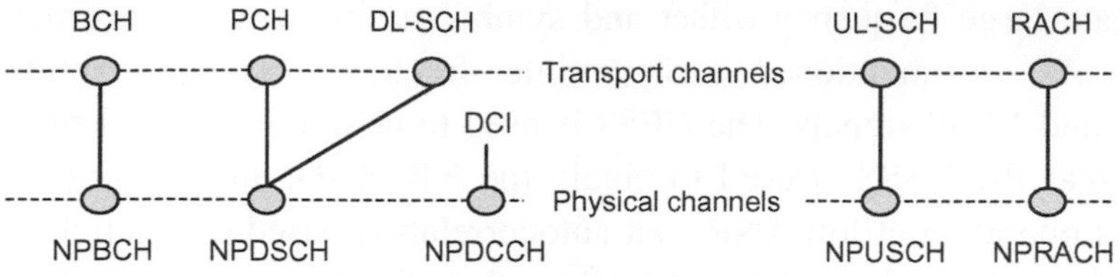

Figure 6.22
NB-IoT channel structure [6].

- NRS
- NPDCCH
- NPDSCH

The NB-IoT includes the following channels in the uplink:

- NPRACH
- NPUSCH

NPRACH is a newly designed channel since LTE PRACH uses a bandwidth of 1.08 MHz, which is more than NB-IoT uplink bandwidth (Fig. 6.22).

6.4.3 Cell Search and Random-Access Procedure

Synchronization is an important aspect in cellular network operation. When a UE is powered on, it needs to find and camp on a suitable cell. For this purpose, the UE must synchronize to downlink symbol, subframe, and frame timing, as well as to the carrier frequency. In order to synchronize with the carrier frequency, the UE needs to correct any frequency offset that is present due to local oscillator inaccuracy, and to perform symbol timing alignment with the frame structure from the base station. In addition, due to the presence of multiple cells, the UE needs to detect a particular cell on the basis of an NB-PCID. As a result, a typical synchronization procedure consists of determining the timing alignment, correcting the frequency offset, obtaining the correct cell identity, and the absolute subframe and frame number reference. The NB-IoT technology is intended to be used for low-cost UEs and at the same time, to provide extended coverage for UEs deployed in environments with high penetration loss. The low-cost UEs inevitably utilize less expensive crystal oscillators that can have carrier frequency offsets (CFOs) as large as 20 ppm. Deployment of NB-IoT in the in-band or guard-band mode introduces an additional raster offset as large as 2.5 or 7.5 kHz, resulting in even higher CFO values. Despite of the large frequency offset, an NB-IoT UE must be able to perform accurate synchronization at low SNR conditions.

Time–frequency synchronization in NB-IoT follows the same principles as in LTE; nevertheless, it incorporates improvements in the design of the synchronization sequences in order to mitigate large frequency offset and symbol timing estimation error issues in low SNR regions. As we mentioned earlier, time–frequency synchronization is achieved through NPSS and NSSS signals. The NPSS is used to obtain the symbol timing and correct the CFO, whereas the NSSS is used to obtain the NB-PCID, and the timing within 80 ms block. For UEs operating at low SNRs, an autocorrelation based on a single 10 ms received segment would not be sufficient for detection, thus for more accurate detection, the UE must coherently accumulate several sequences over multiple 10 ms segments. Due to large initial CFO value, the sampling time at the UE is different from the actual sampling time, the difference being proportional to the CFO. For UEs with limited coverage, more number of accumulations might be required to successfully achieve downlink synchronization.

Following the synchronization procedure, the UE proceeds to the acquisition of the MIB. The NPBCH consists of eight self-decodable subblocks, and each subblock is repeated eight times. The design is intended to provide successful acquisition for coverage-limited UEs. Subsequent to symbol timing detection and CFO compensation, in the in-band and guard-band deployments, there is still an additional raster offset, as high as 7.5 kHz, which needs to be compensated. The presence of raster offset results in either overcompensation or undercompensation of the carrier frequency. As a result, the symbol timing drifts in either the forward or backward direction depending on whether the carrier frequency was over-compensated or undercompensated. This may cause a severe degradation in the performance of NPBCH detection.

During cell selection, the UE measures the received power and quality of the NRS. These values are then compared to cell-specific thresholds provided by the SIB1-NB. If the UE is in coverage of a cell, it will try to camp on it. Depending on the received NRS power, the UE may have to start a cell reselection. The UE compares this power to a reselection threshold, which may be different for the intra-frequency and the inter-frequency scenarios. All required parameters are received from the serving cell; thus there is no need to acquire system information from other cells. Among all cells fulfilling the cell-selection criteria, the UE ranks the cells with respect to the excess power over a threshold. A hysteresis is added in this process in order to prevent frequent cell reselection, and also a cell-specific offset may be applied for the intra-frequency scenarios. In contrast to LTE, there are no priorities for different frequencies. The UE selects the highest ranked cell which is deemed suitable.

In NB-IoT, random-access procedure serves multiple purposes such as initial access when establishing a radio link and scheduling request. Among others, the main objective of random access is to achieve uplink synchronization, which is important for maintaining uplink orthogonality in NB-IoT. As shown in Fig. 6.23, the contention-based random-access procedure in NB-IoT consists of four steps: (1) UE transmits a random-access preamble; (2) the network transmits a RAR message that contains the timing advance command and scheduling of uplink resources for the UE to use in the third step; (3) the UE transmits its identity

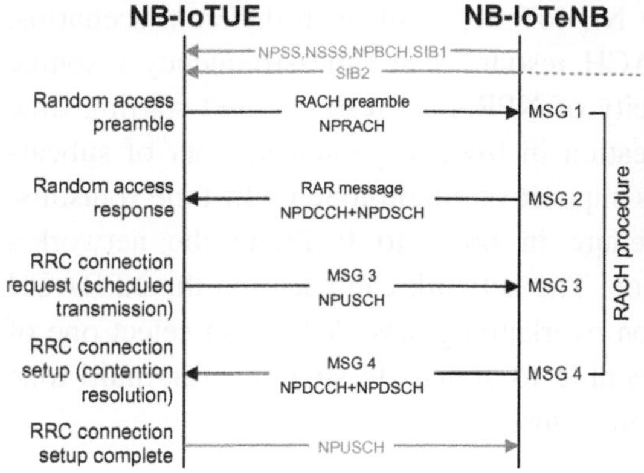

SIB2 parameters	
nprach-Periodicity	NPRACH resource periodicity
nprach-SubcarrierOffset	Frequency location of the first subcarrier allocated to NPRACH
nprach-NumSubcarriers	Number of subcarriers allocated to NPRACH
nprach-NumCBRA-StartSubcarriers	Number of starting sub-carriers allocated to contention based NPRACH random access
numRepetitionsPerPreambleAttempt	Number of NPRACH repetitions per attempt
nprach-StartTime	NPRACH starting time
nprach-SubcarrierMSG3-RangeStart	Fraction for calculating starting subcarrier index for the range of NPRACH subcarriers reserved for indication of UE support for multi-tone msg3 transmission

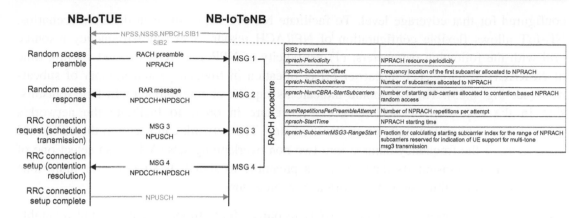

Figure 6.23
NB-IoT RACH procedure (these messages are repeated according to the UE coverage enhancement level) [6].

to the network using the scheduled resources; and (4) the network transmits a contention resolution message to resolve any contention due to multiple UEs transmitting the same random-access preamble in the first step. More specifically, upon transmission of the preamble, the UE first calculates its RA-RNTI from the transmission time. It then looks for a PDCCH with the DCI format N1 scrambled with the expected RA-RNTI, in which the RAR message is signaled. The UE expects this message within the response window, which starts three subframes after the last preamble subframe and has a coverage enhancement dependent length given in SIB2-NB. If the preamble transmission was not successful, that is, the associated RAR message was not received, the UE transmits another RACH preamble. This process is repeated up to a maximum number, which depends on the coverage enhancement level. For the case that this maximum number is reached without success, the UE proceeds to the next coverage enhancement level, if this level is configured. If the total number of access attempts is reached, an associated failure is reported to the RRC sublayer. With the RAR message, the UE receives a temporary C-RNTI and the timing advance command. Consequently, the forthcoming msg3 is already time aligned, which is necessary for transmission over the NPUSCH. Further, the RAR message provides the uplink grant for msg3, containing all relevant data for msg3 transmission. The remaining procedure is performed similar to LTE; that is, the UE sends an identification and upon reception of the contention resolution message, indicating successful completion of the random-access procedure.

The network can configure up to three NPRACH resource configurations in a cell in order to serve UEs in different coverage scenarios (based on the measured path loss). In each configuration, a repetition value is specified to increase the reliability of random-access preamble transmission. The UE measures the downlink received signal strength to estimate its coverage level, and transmits a random-access preamble in the NPRACH resources configured for that coverage level. To facilitate NB-IoT deployment in different scenarios, NB-IoT allows flexible configuration of NPRACH resources in time–frequency resource grid with the following parameters: (1) periodicity of NPRACH resource and starting time of NPRACH resource; (2) subcarrier offset (location in frequency) and number of subcarriers (see Fig. 6.23). The UE should indicate its support of single-tone/multi-tone transmission in the first step of random-access procedure in order to facilitate the network's scheduling of uplink transmission in the third step. The network can partition the NPRACH subcarriers in the frequency domain into two non-overlapping sets. A UE can select one of the two sets to transmit its random-access preamble to signal its supports for multi-tone transmission in the third step of random-access procedure.

As we mentioned earlier, a set of PRACH resources (e.g., time, frequency, and preamble sequences) is provided for each coverage level. The PRACH resources per coverage level are configurable by the system information. The UE selects PRACH resources based on the coverage level estimated using downlink signal measurements, for example, RSRP [5]. The UE MAC will reattempt the process at a higher coverage level RACH preamble set, if it does not receive the RAR message after the anticipated number of attempts at a certain level. If the contention resolution is not successful, the UE will continue to use the same coverage level RACH preamble set [2,21].

6.4.4 Power-Saving Modes

Power-save mode (PSM) is a power conserving mechanism in NB-IoT that allows the devices to skip the periodic paging channel monitoring cycles between active data transmissions, letting the device enter a sleep state. However, the device becomes unreachable when PSM is active; therefore, it is best utilized by device-originated or scheduled applications, where the device initiates communication with the network. Assuming there is no device-terminated data, an NB-IoT device can remain in PSM state for a long time, with the upper limit determined by the maximum value of the tracking area update (TAU) timer. During the PSM active state, the access stratum at the device is turned off, and the device would not monitor paging messages or perform any radio resource management measurements. In addition, the PSM enables more efficient low-power mode entry/exit, as the device remains registered with the network and its NAS state is maintained during the PSM without the need to spend additional cycles to setup registration/connection after each PSM exit event. Examples of applications that can take advantage of PSM include smart meters, sensors, and any IoT devices that periodically send data to the network. When a device initiates PSM with the network, it provides two preferred internal timers (T3324 and T3412), where the PSM time is the difference between these timers. The network may accept these values or set different ones. The network then retains state information and the device remains registered with the network. If a device wakes up and sends data before the expiration of the time interval it

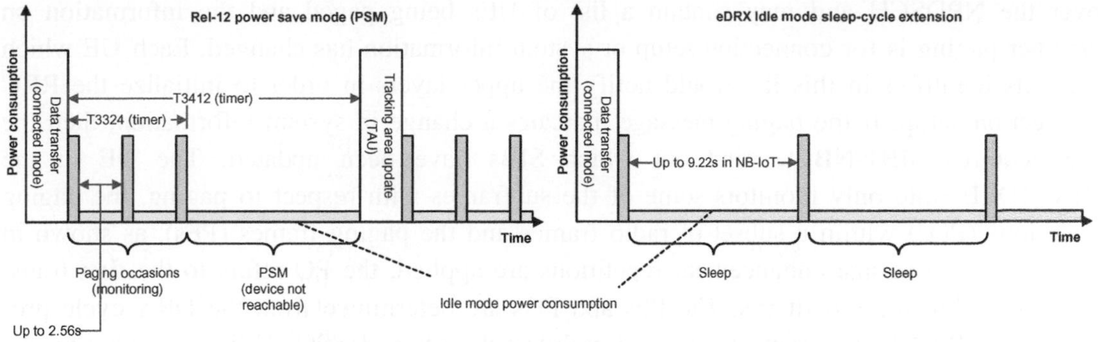

Figure 6.24
Illustration of PSM and eDRX operation [21,24].

agreed with the network, a reattach procedure is not required. However, in a similar manner to a radio module that has been powered off, a radio module in PSM cannot be contacted by the network while it is in sleep mode [7,8,21].

One problem with PSM mechanism is the ability to support device-terminated traffic, as the UE is unreachable when it is in PSM state. The device would become reachable by the network as the TAU timer expires (Fig. 6.24), which can introduce significant latency for device-terminated traffic. While periodic TAU can be configured to occur more frequently to match the UE's delay requirement, such configurations would result in additional signaling overhead from unnecessary periodic TAU procedures and increased device power consumption. To address this shortcoming of PSM, the extended DRX (eDRX) was introduced in 3GPP Rel-13. The eDRX is an extension of LTE DRX feature which can be used with NB-IoT devices to reduce power consumption. The eDRX can be used without PSM or in conjunction with it to obtain additional power savings. It allows the time interval during which a device is not listening to the network to be greatly extended. For an NB-IoT application, it might be acceptable not to be reachable for a few seconds or longer. Although it does not provide the same level of power saving as PSM, for some applications, the eDRX may provide a good compromise between device reachability and power consumption reduction. Fig. 6.24 illustrates the concept and the operation of PSM and eDRX mechanisms. In eDRX, the DRX cycle is extended up to and beyond 10.24 seconds in idle mode, with a maximum value of 2621.44 seconds. For NB-IoT, the maximum value of the DRX cycle is 10,485.76 seconds [7,21].

6.4.5 Paging and Mobility

Paging is used to notify an idle-mode UE of a pending downlink traffic, to establish an RRC connection and to indicate a change in system information. A paging message is sent over the NPDSCH and may contain a list of UEs being paged and the information on whether paging is for connection setup or system information has changed. Each UE which finds its identifier in this list would notify the upper layers in order to initialize the RRC connection setup. If the paging message indicates a change in system information, then the UE acquires SIB1-NB to find out which SIBs have been updated. The UE in the RRC_IDLE state only monitors some of the subframes with respect to paging, the paging occasions (POs) within a subset of radio frames and the paging frames (PFs), as shown in Fig. 6.25. If coverage enhancement repetitions are applied, the PO refers to the first transmission within the repetitions. The PFs and POs are determined from the DRX cycle provided in SIB2-NB, and the international mobile subscriber identity (IMSI) provided by the universal subscriber identity module (USIM) card. The DRX is the discontinuous reception of downlink control channel that is used to save the UE battery life. DRX cycles of 128, 256, 512, and 1024 radio frames are supported, corresponding to a time interval between 1.28 and 10.24 seconds. Since the algorithm for determining the PFs and POs also depends on the IMSI, different UEs have different POs, which are uniformly distributed across time. It is sufficient for the UE to monitor one PO within a DRX cycle, if there are several POs therein, the paging is repeated in every one of them. As we stated earlier, the concept of extended DRX may be applied for NB-IoT, as well. If eDRX is supported, then the time interval in which the UE does not monitor the paging messages may be considerably extended up to almost 3 hours. Correspondingly, the UE must know on which HFN and on which time interval within this HFN, that is, the paging transmission window (PTW), it has to monitor the paging. The PTW is defined by a start and stop SFN. Within a PTW, the determination of the PFs and POs is done in the same way as for the non-eDRX.

The NB-IoT supports stationary and low-mobility UEs as most of the NB-IoT nodes are sensors and devices that hardly move. Handover is not supported in NB-IoT, thus when an

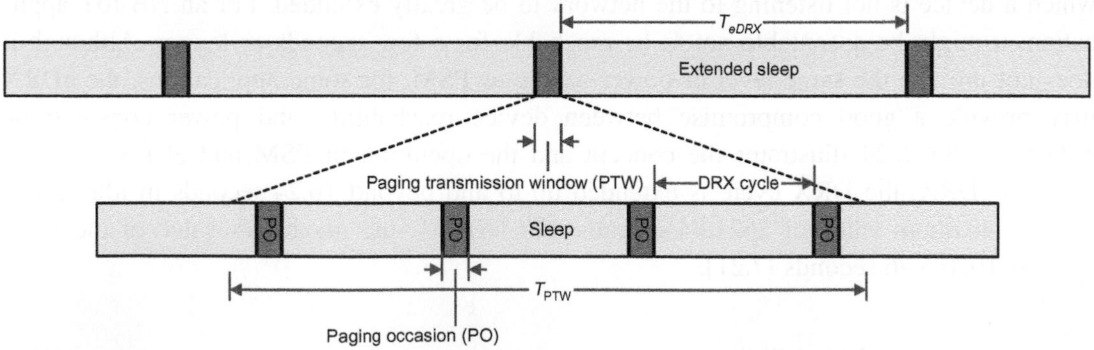

Figure 6.25
Illustration of the correspondence of PO and PTW.

NB-IoT UE moves out of the coverage area of the serving cell, it will experience a radio link failure (RLF). As we mentioned earlier, an NB-IoT UE may support data transport via control plane. The UE may also support data transfer via the user plane, and when it does, the RRC connection reestablishment procedure is supported, which means that after an RLF is detected, the UE attempts to find a suitable cell through cell selection. If the UE finds a suitable cell, it will try to reestablish the connection on that cell and resume the data transfer. The RRC reestablishment intends to hide the temporary loss of the radio interface to the upper layers. The RRC reestablishment for a UE only supporting data transfer via the control plane was added in 3GPP Rel-14 [6,8].

6.5 Implementation and Deployment Considerations

The NB-IoT specifications meet a number of challenging requirements including a greater coverage area, longer device battery life, and lower device cost resulting from the small and intermittent data transmissions. The reduced peak data rate requirements make it possible to employ a simple radio and baseband processing in the receiver chain. With half-duplex operation of the NB-IoT, the duplex filter in a typical LTE device can be replaced by a simple switch in addition to fewer oscillators for frequency synthesis. The use of simplified downlink convolutional coding instead of the LTE turbo code would allow a low-complexity baseband decoding process. The main candidate architectures are the zero-IF and low-IF receivers, which combine analog frontends and digital baseband signal processing on a single chip. However, each of these architectures has some structural issues that must be resolved. In the zero-IF receiver, the desired signal is degraded by time variant DC offset caused by local oscillator leakage and self-mixing. In the low-IF receiver, nonideal hardware results in amplitude and phase mismatches between the I and Q signal paths, which results in degradation of the desired signal with leakage from the interference signal. In the generic low-IF receiver architecture, the incoming RF signal in the antenna is filtered by the band selection filter and amplified by a low-noise amplifier. The quadrature demodulator down-converts the RF signal to the complex low-IF signal, which is represented by in-phase and quadrature components. The IF signals pass through low-pass filters and then sampled by the analog to digital converters (ADCs). After the ADC sampling and conversion, the digitized IF signal is down-converted to the baseband, yielding digital complex signals. Using a moderately low-IF frequency, this architecture can avoid DC offset and $1/f$ noise issues that frequently arise from zero-IF receivers. However, it also reintroduces image issues. Image cancellation can be achieved after the low-noise amplifier, but requires narrowband filtering, thus significantly increasing the complexity and cost of the device. The latter issue can be addressed by complex mixing and subsequently and filtering techniques in the low-IF receiver.

In order to meet the stringent link budget requirements of NB-IoT, low-cost low-complexity single-chip devices have been developed by various vendors [20]. The integration of the power amplifier (PA) and antenna switch simplifies routing by reducing the number of RF components in the frontend. The PA design with a low PAPR is possible with the use of the single-tone transmission technique. This enables the implementation of an RF chip that includes an efficient on-chip PA that may be operated near its saturation point for maximum output power. While there is a trade-off between an integrated on-chip PA and an external PA, one can analyze the effect of PA nonlinearity on error vector magnitude (EVM), and thus on the NB-IoT uplink coverage, using an RF and baseband cross-domain simulation technique. In that case, the baseband LTE signal is generated, which supports both single-tone and multi-tone transmission. The baseband signal is filtered by two digital filters and fed into the modulator to generate a spectrum centered at the carrier frequency. The signal is then amplified using an amplifier with certain characteristics. The linearity of the PA can be modeled by setting the 1 dB compression point to an appropriate value. After the signal is amplified by the PA, it is demodulated by the receiver to determine the EVM. For single-tone transmission, the EVM values are very small, less than 0.08% for a 3.75 kHz subcarrier spacing and less than 0.9% for 15 kHz subcarrier spacing [16,17]. Therefore, we can conclude that the nonlinearity of PA has slight effect on the EVM for single-tone transmission mode. The studies suggest that PAPR is 4.8, 5.7, and 5.6 dB for a signal with 3, 6, and 12 tones, respectively; thus we can conclude that the nonlinearity of PA has unfavorable effects on EVM for multi-tone transmissions. From this study one can conclude that, in case of single-tone transmission, some of the PAPR reduction circuit inside the chip can be removed which significantly reduces chip design complexity. Considering the key aspects of NB-IoT applications, devices that only support single-tone transmission combined with an on-chip nonlinear PA are much more advantageous in ultra-low-power and low-cost applications.

The NPUSCH provides two subcarrier spacing options: 15 and 3.75 kHz. The additional option of using 3.75 kHz provides deeper coverage to reach challenging locations such as inside buildings, where there is limited signal strength. The data subcarriers are modulated using BPSK and QPSK with a phase rotation of $\pi/2$ and $\pi/4$, respectively. Selection of the number of subcarriers for a resource unit can be 1, 3, 6, or 12 to support both single-tone and multitone transmission. The narrowband downlink physical resource block has 12 subcarriers with 15 kHz spacing, providing180 kHz transmission bandwidth (see Fig. 6.26). It only supports a QPSK modulation scheme. To facilitate low-complexity decoding for downlink transmission in devices, the turbo coding was traded off with TBCC .

NB-IoT achieves an MCL 20 dB higher than LTE. Coverage extension is achieved by increasing the reliability of data transmission through increasing the number of repetitions.

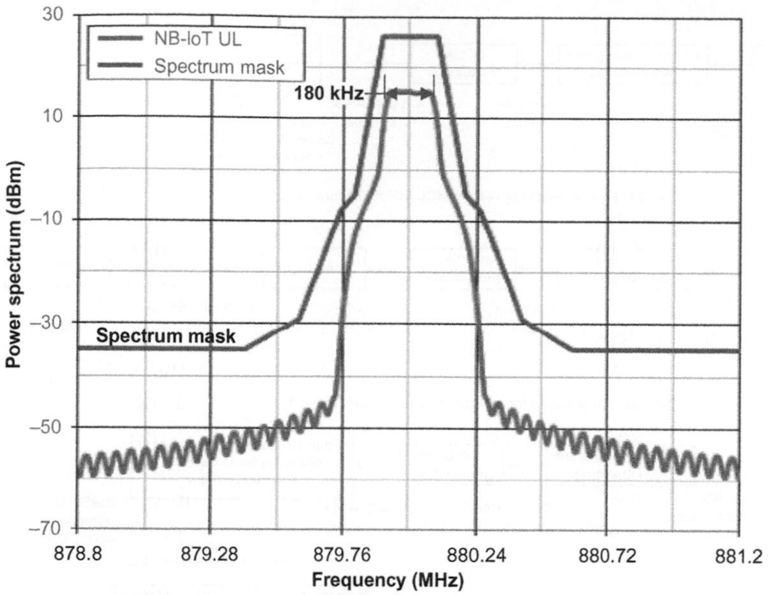

Figure 6.26
Example of NB-IoT NPUSCH transmission spectrum with 15 kHz subcarrier spacing [16,17].

Coverage enhancement is ensured also by introducing single-tone NPUSCH transmission and $\pi/2-$ BPSK modulation to maintain close to 0 dB PAPR, thereby reducing the unrealized coverage potential due to PA backoff. The single-tone NPUSCH transmission with 15 kHz subcarrier spacing provides a layer-1 data rate of approximately 20 bps when configured with the highest repetition factor, that is, 128, and the lowest modulation and coding scheme. The NPDSCH can provide a layer-1 data rate of 35 bps when configured with repetition factor 512 and the lowest modulation and coding scheme. These configurations support close to 170 dB coupling loss. In comparison, the LTE network is designed for approximately 142 dB coupling loss.

Fig. 6.27 shows an example of NPUSCH format 1 and 2 signal processing stages in order to achieve 144, 154, and 164 dB MCL in the uplink. The information bits can be mapped to NPUSCH format 1 based on LTE uplink SC-FDMA waveform. Furthermore, it depicts the processing stages for achieving 164 dB MCL based on NPUSCH format 2. In the latter case, following the use of $\pi/2-$ BPSK modulation, there are 3840 symbols which are repeated six times, appended with cyclic prefix and guard time to occupy one slot, requiring 3840 slots in total. For every six slots with data symbols, one slot of DMRS is added. As a result, 640 DMRS slots are added. Overall, it requires 4480 slots, that is, 2240 subframes [23].

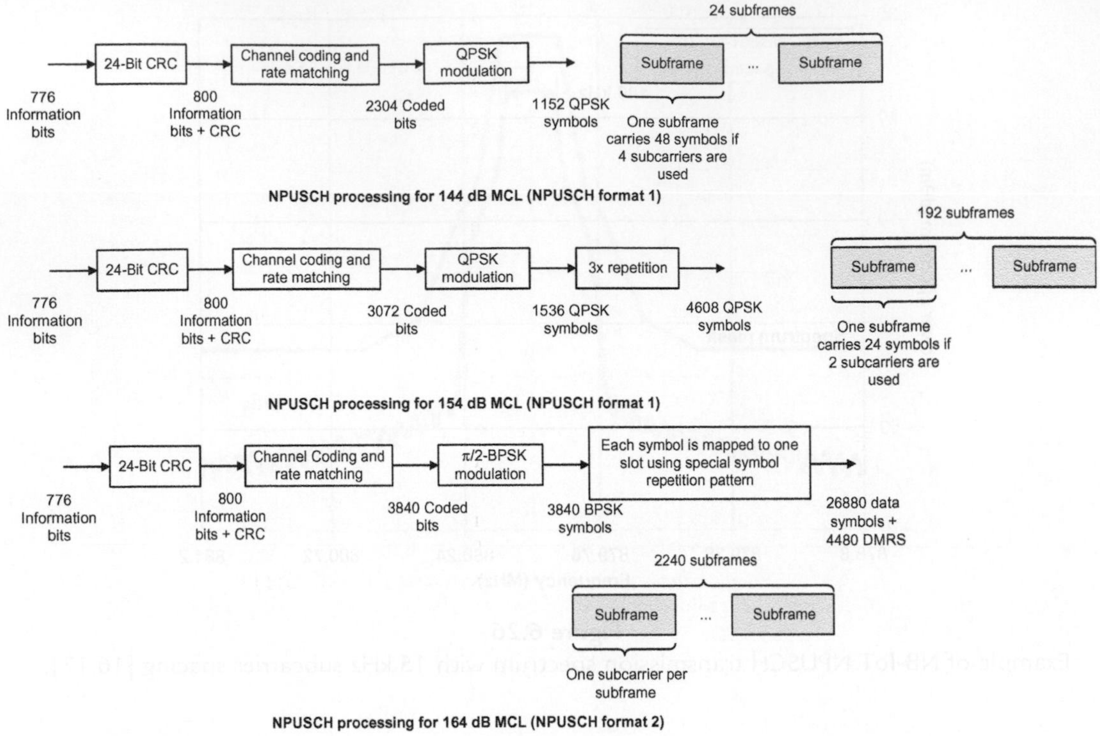

Figure 6.27
Example NB-IoT uplink processing to achieve various MCL values [23].

References

3GPP Specifications[6]

[1] 3GPP TS 36.104, Evolved universal terrestrial radio access (E-UTRA), Base station (BS) radio transmission and reception(Release 15), June 2018.

[2] 3GPP TS 36.211, Evolved universal terrestrial radio access (E-UTRA), Physical channels and modulation (Release 15), June 2018.

[3] 3GPP TS 36.212, Evolved universal terrestrial radio access (E-UTRA), Multiplexing and channel coding (Release 15), June 2018.

[4] 3GPP TS 36.213, Evolved universal terrestrial radio access (E-UTRA), Physical layer procedures (Release 15), June 2018.

[5] 3GPP TS 36.214, Evolved universal terrestrial radio access (E-UTRA), Physical layer measurements (Release 15), June 2018.

[6] 3GPP TS 36.300, Evolved universal terrestrial radio access (E-UTRA) and evolved universal terrestrial radio access network (E-UTRAN). Overall description; Stage 2 (Release 15), June 2018.

[7] 3GPP TS 36.321, Evolved universal terrestrial radio access (E-UTRA), Medium access control (MAC) protocol specification (Release 15), June 2018.

[6] 3GPP specifications can be accessed at the following URL: http://www.3gpp.org/ftp/Specs/archive/.

[8] 3GPP TS 36.331, Evolved universal terrestrial radio access (E-UTRA), Radio resource control (RRC); Protocol specification (Release 15), June 2018.
[9] 3GPP TR 38.825, Study on NR industrial Internet of Things (IoT) (Release 16), September 2018.

Articles, Books, White Papers, and Application Notes

[10] Cellular Networks for Massive IoT, Ericsson White Paper, January 2016.
[11] J. Schlienz, D. Raddino, Narrowband Internet of Things, Rohde & Schwarz Whitepaper, August 2016.
[12] A. Höglund, et al., Overview of 3GPP Release 14 enhanced NB-IoT, IEEE Network, November/December 2017.
[13] M. Chen, et al., Narrow band Internet of Things, IEEE Access, vol. 5, 2017.
[14] Y. Wang, et al., A primer on 3GPP narrowband Internet of Things. IEEE Commun. Mag., 55 (3) (2017).
[15] J.H. Wu, CAT-M & NB-IoT design and conformance test, Keysight Technologies White Paper, June 2017.
[16] Keysight Technologies, the Internet of Things: Enabling technologies and solutions for design and test. Application Note, February 2016.
[17] S. Shin, NB-IoT System modeling: simple doesn't mean easy, Keysight Technologies White Paper, August 2016.
[18] A. Puschmann, et al., Implementing NB-IoT in software, experiences using the srsLTE library, in: Proceedings of the Wireless Innovation Forum Europe, 2017.
[19] B. Schulz, Narrowband Internet of Things measurements, Rohde & Schwarz Application Note, June 2017.
[20] Evolution to NB-IoT and LTE-M, Global Mobile Suppliers Association (GSA) Report, August 2017.
[21] 5G Americas White Paper, LTE and 5G Technologies Enabling the Internet of Things, December 2016.
[22] NB-IoT, A sustainable technology for connecting billions of devices, Ericsson Technology Review, April 2016.
[23] 3GPP R1-160085, NB-IoT NB-PUSCH design, Ericsson, January 2016.
[24] 5G Americas White Paper, Wireless Technology Evolution Towards 5G: 3GPP Release 13 to Release 15 and Beyond, February 2017.
[25] NB-IoT, ShareTechnote. <http://www.sharetechnote.com/>.

CHAPTER 7

Vehicle-to-Everything (V2X) Communications

The concept of connected car has recently emerged, which provides new services to drivers via wireless communications that is considered as one of the most distinctive features of next generation vehicles. Vehicles wirelessly connected to other vehicles and pedestrians within proximity can identify the possibility of collisions by exchanging information such as speed, direction and their location. The vehicles can also communicate with a network entity in charge of traffic control so that they can be informed of weather, and road hazards or receive guidance on the speed and route for traffic flow optimization. The automotive industry is evolving toward connected and autonomous vehicles that have the potential to improve safety and to reduce traffic congestion while reducing the environmental impacts of the vehicles. A key enabler of this evolution is vehicle-to-everything (V2X) communications, which allows a vehicle to communicate with its surroundings, other vehicles, pedestrians, road-side equipment, and the Internet. Using V2X, critical information can be exchanged among vehicles to improve situation awareness and to avoid accidents. Furthermore, V2X provides reliable access to the vast information available in the cloud. For example, real-time traffic reports, sensor data, and high-definition mapping data can be shared and accessed, which are useful not only for improving today's driving experience, but also will be essential for navigating self-driving vehicles in the future.

V2X communications enables the exchange of information between vehicles and between vehicles and other nodes (infrastructure and pedestrians). The 3GPP Rel-12 device-to-device communication served as the basis for the V2X work in Rel-14. In Rel-14, 3GPP specified cellular V2X (C-V2X) communications with two complementary transmission modes: direct communications between vehicles and network communications. 3GPP Rel-16 focuses on continuation of LTE-based cellular V2X and tries to address advanced use cases. These include vehicle platooning, enhanced vehicle to infrastructure features, extended sensors, advanced driving (to enable semi-automated or fully-automated driving), and remote driving. C-V2X users can benefit from the existing widely-deployed cellular infrastructure. However, since the availability of cellular infrastructure cannot always be guaranteed, C-V2X defines transmission modes that enable direct communication using the sidelink channel over the PC5 interface.

5G NR. DOI: https://doi.org/10.1016/B978-0-08-102267-2.00007-5

This chapter provides a technical overview of LTE-based V2X standards as of 3GPP Rel-15 and further provides insights into the emerging NR-based V2X technologies in 3GPP Rel-16 that are expected to accelerate the realization of advanced V2X communications and to improve transportation and commute experience. The NR-based V2X is required to enable very high throughput, high reliability, low latency, and accurate positioning use cases beyond the basic safety features provided by LTE-based V2X services. Some of the use cases will involve 5G working in tandem with other technologies, including cameras, radar, and light detection and ranging (LIDAR)[1] The Rel-16 C-V2X includes use cases such as advanced driving categories identified in 3GPP, ranging/positioning, extended sensors, platooning, and remote driving.

7.1 General Aspects and Use Cases

The V2X communications enable the vehicles to exchange data with each other and the infrastructure, with the goal of improving road safety, traffic efficiency, and the availability of infotainment services. V2X communications, as defined in 3GPP, consist of four use cases; namely, vehicle-to-vehicle (V2V), vehicle-to-infrastructure (V2I), vehicle-to-network (V2N), and vehicle-to-pedestrian (V2P), as shown graphically in Fig. 7.1. It is implied that

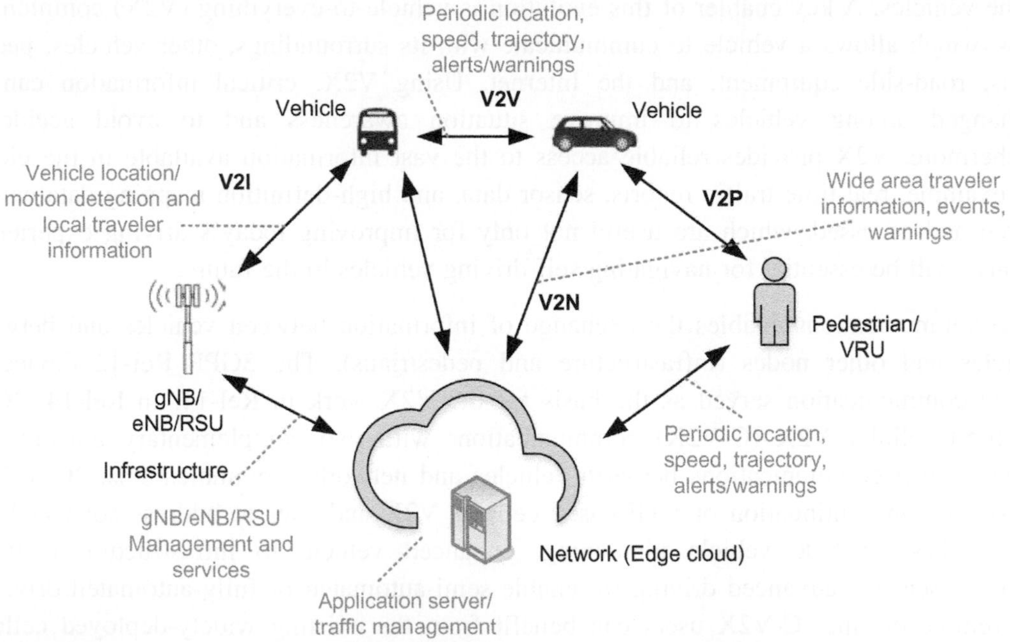

Figure 7.1
Illustration of various V2X use cases [1,9,13,14].

[1] Light detection and ranging is a remote sensing method used to examine the surface of the earth.

these wireless connections are generally bidirectional; that is, the V2I and V2N also allow the infrastructure to send messages to the vehicles. The V2V and V2P transmissions are typically based on broadcast capability between vehicles or between vehicles and vulnerable road users, for example, pedestrians and cyclists, for providing information about location, velocity, and direction to avoid accidents. The V2I transmission is between a vehicle and a road-side unit (RSU). The V2N transmission is between a vehicle and a V2X application server which is in the cloud. An RSU may be used to extend the range of a V2X message received from a vehicle by acting as a forwarding node. The V2I may include communication between vehicles and traffic control devices near road work sites. V2N may include communication between vehicle and the server via 4G/5G network, such as for traffic operations.

There are two prominent V2X technologies available today, which have been designed to operate in 5.9 GHz intelligent transportation systems (ITS) spectrum. The IEEE 802.11p standard, which is commonly referred to as dedicated short-range communication (DSRC) and ITS-G5 in the United States and Europe, respectively, was developed as an extension of IEEE 802.11a standard. IEEE 802.11p specification was completed in 2012 and uses a half-clocked version of IEEE 802.11a, leveraging a radio technology that is over two decades old and was originally developed for wireless Ethernet cable replacement and not for high-speed mobile applications. IEEE 802.11p performance results indicate its limited range and unreliable large-scale field performance, derived from susceptibility to congestion and lack of minimum performance guarantees, which would limit its usefulness. Most of the present-day cars are equipped with several active sensors, including camera, radar, and LIDAR, which compel the V2X wireless sensor to provide longer range and reliability, especially in non-line-of-sight (NLoS) scenarios where other vehicles and buildings obstruct the vehicle's vision systems. In the meantime, 3GPP has continued to evolve LTE device-to-device (D2D) technology, specified as part of Rel-12, and to optimize it for automotive applications in Rel-14, also referred to as C-V2X communications, over proximate radio interface (PC5) or sidelink and/or the LTE Uu interface. The C-V2X incorporates LTE's mobility support and further extends the baseline standard for automotive applications, while learning from IEEE 802.11p issues and shortcomings. The C-V2X includes both direct communications and network-based communications. Fig. 7.2 shows various scenarios for sidelink and direct link communications where UEs are located in-coverage and out-of-coverage of a cell.

The future transportation technologies will include connected and intelligent vehicles, which can cooperate with each other and a transportation infrastructure that can provide safer and more convenient commute/travel experience. A wide range of use cases that require longer range or higher throughput can be supported with LTE-based/NR-based V2X communications. C-V2X is the technology developed in 3GPP and is designed to operate in the following two modes:

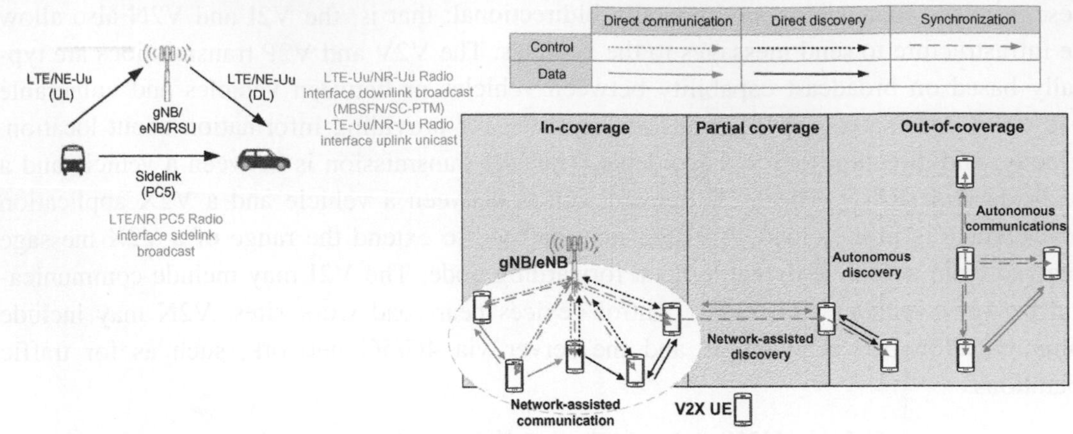

Figure 7.2
Various V2X scenarios [12–14].

- *Device-to-device:* This mode is sidelink communication between two or more devices without network involvement.
- *Device-to-network:* This mode uses the traditional cellular links to enable cloud services as part of the end-to-end solution by means of network slicing architecture for vertical industries.

Some key performance metrics for low-latency local communications that are addressed by LTE-based V2X as part of Rel-14 include support of vehicle speeds up to 160 km/h and relative speeds of 280 km/h; extended range to provide the driver(s) with sufficient response time (e.g., 4 seconds); message sizes for periodic broadcast messages between two vehicles/UEs with payloads of 50–300 bytes and for event-triggered messages up to 1200 bytes; and maximum latency of message transfer of 100 ms between two UEs and 1000 ms for messages sent via a network server [27]. In subsequent releases, more stringent requirements for latency, range, speed, reliability, location accuracy, and message payloads have been specified in order to support more advanced use cases. For example, NR-based V2X systems are expected to provide end-to-end latency between vehicles of less than 5 ms and guarantee more than 99% reliable packet delivery within a short-to-medium range (80–200 m).

RSUs are the new network nodes that are part of the LTE-based V2X communications system. These entities can be co-located with eNBs or operate in standalone mode. The V2I communication mode allows RSUs to monitor traffic-related conditions, such as traffic signals and tolls and subsequently notify the surrounding vehicles. The evolution path for LTE-based V2X is going to exploit the NR features and to provide ubiquitous coverage as the 5G technologies are designed and deployed. 3GPP evolved V2X (eV2X) has identified several new use cases/applications as follows [1,21,25,26]:

- *Platooning:* In this scenario, vehicles (dynamically) form a platoon while moving together in the same direction. Vehicles in the platoon obtain information from the leading vehicle for managing the platoon. Platooning allows the vehicles to form a tightly coordinated group with significantly reduced inter-vehicle distance, thus increasing road capacity and efficiency. It also improves fuel efficiency, reduces accident rate, and enhances productivity by allowing the drivers to perform other tasks. Vehicles within a platoon must be able to frequently exchange information (e.g., to share information such as vehicle's speed and direction) and to send event notifications such as the intent for braking or acceleration. There are several aspects of platooning that must be supported through reliable V2V communications such as joining and leaving a platoon; that is, to allow a vehicle to signal its intention to join or to leave a platoon at any time while the platoon is active, and to support additional signaling in order to complete join/leave operations; announcement and warning to indicate formation and existence of the platoon so that nearby vehicles can select to join the platoon or to avoid disruptions to the platoon; steady-state operation group communication to support the exchange of platoon management messages and further to indicate braking, acceleration, road selection, change of platoon leader, etc. Given the small target inter-vehicle distance while the vehicles are traveling at relatively high speed, V2V communication must be able to support reliable and secure message exchange to ensure effective and safe platooning operation. The following are some key V2V communication requirements in order to support platooning [13,14]:
  - 25 ms end-to-end communication latency among a group of vehicles (10 ms for the highest degree of automation[2]).
  - 90% message reliability, and 99.99% for the highest degree of automation relative longitudinal position accuracy of less than 0.5 m is required.
  - 10–30 messages per second during broadcasting.
  - Dynamic communication range control to improve resource efficiency given the varying platoon size, and to limit message distribution for privacy reasons.
- *Advanced driving:* Vehicle/RSU shares data that is collected through local sensors with nearby vehicles, allowing the vehicles to coordinate their paths and to avoid accidents. Advanced driving enables semi-automated or fully automated driving. Each vehicle shares its intention to change path and speed with vehicles in proximity, thus inter-vehicle distance adjustments are required. The key requirements for communication between two vehicles employing advanced driving mechanisms include:
  - Large system bandwidth to support burst transmission of large data packets
  - 10 ms latency for highest degree of automation
  - 99.99% message reliability for highest degree of automation

[2] In an autonomous vehicle scenario, the vehicle's on-board computers are fully capable of performing all driving operations on their own, with no human monitoring or intervention.

- *Extended sensors:* Extended sensors enable the exchange of raw or processed data (e.-g., cameras, radar, LIDAR) gathered from vehicle sensors or live video images among vehicles, road site units, devices of pedestrian, and V2X application servers. The vehicles can increase the knowledge of the environment/road conditions beyond of what their own sensors can detect and have a broader and holistic view of the driving conditions. The sensor data that a vehicle can share ranges from photo of a road hazard to real-time video stream. The availability of sensor data from multiple separate sources enhances situation awareness of the vehicles and pedestrians, and thus improves road safety. Extended sensors further enable new features such as cooperative driving and precise positioning, which are necessary for autonomous driving.
- *Remote driving:* Enables a remote driver or a V2X application to operate a remote vehicle for those passengers who cannot drive by themselves or remote vehicles located in dangerous environments. For a case where route variation is limited and routes are predictable, such as public transportation, driving based on cloud computing can be used. High reliability and low latency are the main requirements of this use case. Remote driving enables the remote control of a vehicle by a human operator or by a cloud-based application, via V2N communication. There are several scenarios that can leverage remote driving, including:
  - Provide a fallback solution for autonomous vehicles. An example is during the initial autonomous vehicle deployment when a vehicle is in an unfamiliar environment and has difficulty navigating.
  - Provide remote driver services to the youths, elderly, and others who are not licensed or able to drive.
  - Enable fleet owners to remotely control their vehicles. Examples including moving trucks from one location to another, delivering rental cars to customers, and providing remotely driven taxi services.
  - Enable cloud-driven public transportation and private shuttles, all of which are particularly suitable for services with predefined stops and routes. Remote driving can reduce the cost of fully autonomous driving for certain use cases because of the less stringent technical requirements (e.g., smaller number of in-vehicle sensors and less computation requirements for sophisticated algorithms). The following are V2X requirements for supporting remote driving:
    - Data rate up to 1 Mbps downlink and 25 Mbps in the uplink.
    - Ultra-high reliability at 99.999% or higher [similar to ultra-reliable and low-latency communication (URLLC) use case].
    - End-to-end latency of 5 ms between the V2X application server and the vehicle.
    - Support vehicular speeds of up to 250 km/h.

Remote control will be required when an obstacle blocks an autonomous driving vehicle, rendering it unable to decide about a pathway or approach to safely navigate around it. Examples of obstacles include lanes that are blocked due to a recent accident, double-parked cars not allowing the vehicle to pass without crossing the ingress/egress yellow lines, or unexpected situations where the vehicle is unable to determine a safe action or a way forward. When the vehicle encounters such conditions, it will stop or find a minimum risk position and then will request assistance from a remote control operator to take control and navigate around the obstacle. The remote controller would need to understand the obstacle and determine the path that the vehicle must take. In this case, the controller will utilize the photos or streaming sensor information (e.-g., video, LIDAR, radar) that has been made available by the vehicle. Once cleared of the obstacle, the video stream to the controller will stop and the vehicle reasserts full control toward destination.

NR V2X is not intended to replace the services offered by LTE V2X, rather it is expected to complement LTE V2X for advanced V2X services and to support interworking with LTE V2X. From 3GPP RAN technology development point of view, the focus and scope of NR V2X is to target advanced V2X applications. However, this does not imply that NR V2X capability is necessarily restricted to advanced services. It is up to the regional regulators, car manufacturers, equipment vendors, and automotive industry, in general, to deploy the technology suited for their intended services and use cases. NR V2X is planned as 3GPP V2X phase 3 and would support advanced services beyond those supported in Rel-15 LTE V2X. The advanced V2X services would require enhancing the NR baseline system and developing a new NR sidelink to meet the stringent requirements of the new use cases. NR V2X system is expected to have a flexible design to support services with low latency and high reliability requirements, considering the higher system capacity and extended coverage enabled by the baseline NR system. The flexibility of NR sidelink framework would allow easy extension of NR system to support the future development of further advanced V2X services. More specifically, the NR V2X enhancements include the following areas [11]:

- *Sidelink design:* Identify technical solutions for NR sidelink to satisfy the requirements of advanced V2X services, including support of sidelink unicast, sidelink groupcast, and sidelink broadcast; study NR sidelink physical layer structure and procedure(s); study sidelink synchronization mechanism; study sidelink resource allocation mechanism; and study sidelink L2/L3 protocols.
- *Uu enhancements for advanced V2X use cases:* Evaluate whether Rel-15 NR Uu and/or LTE Uu interfaces could support advanced V2X use cases and identify enhancements, if any, that are needed to meet advanced V2X use cases.
- *Uu-based sidelink resource allocation/configuration (LTE V2X mode 3 and mode 4):* Identify necessary enhancements of LTE Uu and/or NR Uu to control NR sidelink through the cellular network and further identify the necessary enhancements of NR Uu to control LTE sidelink from the cellular network.

- *RAT/interface selection for operation:* Study if additional mechanisms are required to decide whether LTE PC5, NR PC5, LTE Uu, or NR Uu should be utilized for operation.
- *Quality of service (QoS) management:* Study technical solutions for QoS management of the radio interface (including Uu and sidelink) used for V2X operations.
- *In-device coexistence:* Study the feasibility of the coexistence mechanisms when NR sidelink and LTE sidelink technologies are implemented in the same vehicle for the non-co-channel scenarios such as advanced V2X services provided by NR sidelink while coexisting with V2X service provided by LTE sidelink in different channels.
- *Sidelink operating bands:* Sidelink frequency bands include both unlicensed ITS bands and licensed bands in FR1 and/or FR2. The target is to have a common sidelink design for both FR1 and FR2.

For groupcast V2X communication, the following radio-layer enhancements are considered to improve sidelink communication performance in distributed resource allocation mode:

- Group radio-layer feedback
  - When transmitter node sends a sidelink transmission/message to a group of UEs, the UE that has not successfully received physical sidelink shared channel (PSSCH) sends a NACK on physical sidelink control channel (PSCCH) resource reserved by the transmitter for acknowledgment.
- Group radio-layer (re-)transmission
  - If the UEs in a group detect a NACK from at least one of the group members, the UEs can retransmit successfully the received packet on a resource that can be either reserved by the UE which is failed to receive or by the original source of transmission. Note that for groupcast communication, the same principles of sensing and resource selection as well as resource reservation can be reused. The group radio-layer feedback and (re-)transmissions can be developed using the same sidelink channel access and resource selection mechanisms.

The latency of V2X systems is categorized into transmission time interval (TTI) dependent and TTI-independent, depending on whether the latency is proportional to or independent of the transmission time interval of the radio air-interface [15]. Each data transmission, that is, control signaling, scheduling, HARQ (re-)transmission, and so on, consumes at least one TTI. The V2X services can be categorized into three groups: (1) safety-related services, (2) non-safety-related services, and (3) automated driving-related services. The safety-related services handle real-time safety messages, such as warning messages (e.g., abrupt brake warning message) to reduce the risk of car accidents. In this type of services, timeliness and reliability are the key requirements. On the other hand, non-safety-related services are intended to optimize the traffic flow on the road so that travel time is reduced. Therefore,

Table 7.1: Performance requirements of different V2X use cases [1].

Use Case	V2X Mode	End-to-End Latency	Reliability	Data Rate per Vehicle (kbps)	Communication Range
Cooperative awareness	V2V/V2I	100 ms to 1 second	90–95%	5–96	Short-to-medium
Cooperative sensing	V2V/V2I	3 ms to 1 second	>95%	5–25,000	Short
Cooperative maneuver	V2V/V2I	<3–100 ms	>99%	10–5000	Short-to-medium
Vulnerable road user	V2P	100 ms to 1 second	95%	5–10	Short
Traffic efficiency	V2N/V2I	>1 second	<90%	10–2000	Long
Tele-operated driving	V2N	5–20 ms	>99%	>25,000	Long

these services enable more efficient driving experience with no stringent requirements in terms of latency and reliability. For the safety-related services, if we consider the frequency of periodic messages (e.g., from 1 to 10 messages/second) and the reaction time of most drivers (e.g., from 0.6 to 1.4 seconds), then the maximum allowable end-to-end latency must not exceed 100 ms. In fact, depending on the service type, the latency requirement may even be less than 100 ms (e.g., 20 ms for a pre-crash sensing warning). In addition to these types of services, automated driving-related services are being developed as a key transformation in the automotive industry. These services require more rigorous latency limits, data rates, and positioning accuracy. Therefore, the latency requirements for automated driving-related services are more stringent than those required for safety-related services. For example, automated overtaking or high-density platooning services have a 10 ms latency requirement. Table 7.1 lists the V2X use cases and the corresponding latency and data rate requirements [15].

V2V communication is conceptually based on D2D communications that was specified as part of proximity-based services (ProSe) in 3GPP Rel-12/Rel-13. The D2D feature provided public safety UEs the option to communicate directly. As part of ProSe, a new D2D interface (designated as PC5 interface or sidelink) was defined, which has been subsequently enhanced for vehicular use cases, specifically addressing high-speed (up to 250 km/h) and high node-density scenarios. Therefore, some fundamental modifications to PC5 have been made, including additional reference symbols to support high Doppler frequencies associated with relative speeds of up to 500 km/h and high frequency bands (e.g., 5.9 GHz ITS band being the main target) [1,24].

In order to support distributed scheduling, a sensing mechanism with semi-persistent transmission has been introduced. The C-V2V traffic from a device is mostly periodic in nature. This was utilized to sense congestion on a resource and estimate future congestion on that resource. Based on this estimation, the resources are reserved in advance. This technique optimizes the use of the channel by enhancing resource separation between transmitters that are using overlapping resources. The design is scalable for different bandwidths. There are

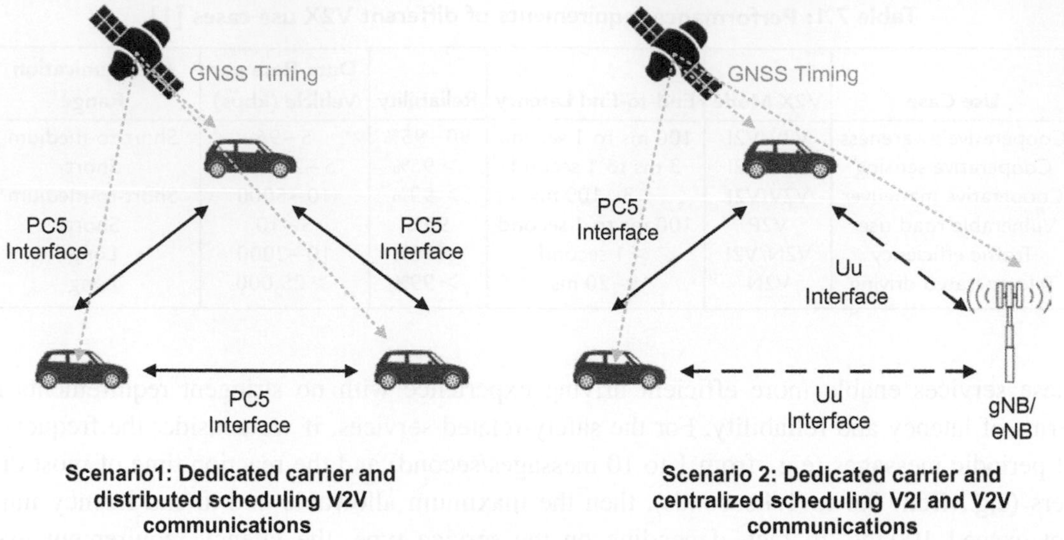

Figure 7.3
LTE V2X use cases and scheduling options [1].

two high-level deployment scenarios that are currently defined, as illustrated in Fig. 7.3. Both scenarios use a dedicated carrier for V2V communications, meaning that the target band is only used for PC5-based V2V communications. In both use cases, global navigation satellite system (GNSS) is used for time synchronization. In the first scenario, the scheduling and interference management of V2V traffic is supported based on distributed algorithms (referred to as mode 4) implemented between the vehicles. As we mentioned earlier, the distributed algorithm is based on sensing with semi-persistent transmission. Furthermore, a new mechanism where resource allocation is dependent on geographical information is introduced. Such a mechanism counters near-far effect arising due to in-band emissions. In the second scenario, the scheduling and interference management of V2V traffic is assisted by eNBs (referred to as mode 3) via control signaling over the Uu interface. The eNB assigns the resources that are used for V2V signaling in a dynamic manner.

The goal of the next-generation V2X communication is to enable accident-free cooperative automated driving by efficiently using the available roadways. To achieve this goal, the communication system will need to enable a diverse set of use cases, each with a specific set of requirements. The V2X feature was initially introduced with IEEE 802.11p and supported a limited set of basic safety services. IEEE 802.11 group has initiated a new project called IEEE 802.11bd which is tasked to enhance IEEE 802.11p and develop similar features as NR V2X, while backward compatible to IEEE 802.11p [28]. 3GPP Rel-14 V2X supported a wider range of applications and services, including low-bandwidth safety

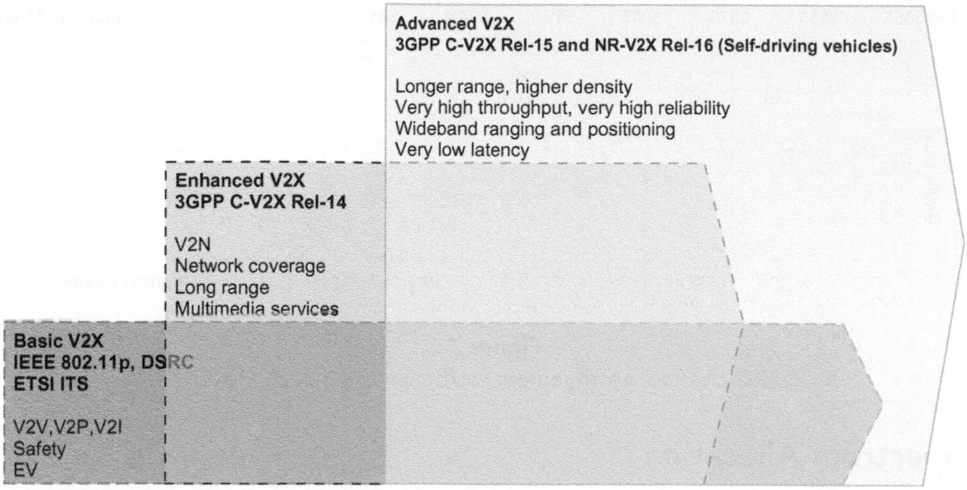

Figure 7.4
V2X technology evolution [13,14,25].

Table 7.2: High-level comparison of attributes for DSRC, LTE V2X, and NR V2X [13,14].

Features	DSRC/IEEE 802.11p	Rel-14 LTE C-V2X	(5G) (Rel-15/16) C-V2X
Out-of-network operation	Yes	Yes	Yes
Support for V2V	Yes	Yes	Yes
Support for safety-critical uses	Yes	Yes	No
Support for V2P	Yes	Yes	Yes
Support for V2I	Limited	Yes	Yes
Support for multimedia services	No	Yes	Yes
Network coverage support	Limited	Yes	Yes
Global economies of scale	No	Yes	Yes
Regulatory/testing efforts	Yes	Limited	No
Very high throughput	No	No	Yes
Very high reliability	No	No	Yes

applications and high-bandwidth applications such as infotainment. 3GPP Rel-15 and 16 enable even more V2X services by providing longer range, higher density, very high throughput and reliability, accurate positioning, and ultra-low latency. Fig. 7.4 summarizes these features and shows the evolution of V2X technologies and how IEEE 802.11p, LTE-based, and NR-based V2X may coexist over time. Table 7.2 compares DSRC, Rel-14 C-V2X, and 5G C-V2X key features at a high level. The analysis of radio link performance of LTE V2X and IEEE 802.11p indicates that the likelihood of successful delivery of warning messages between two vehicles both equipped with LTE V2X (PC5) is notably greater than utilizing IEEE 802.11p technology under the same test conditions.

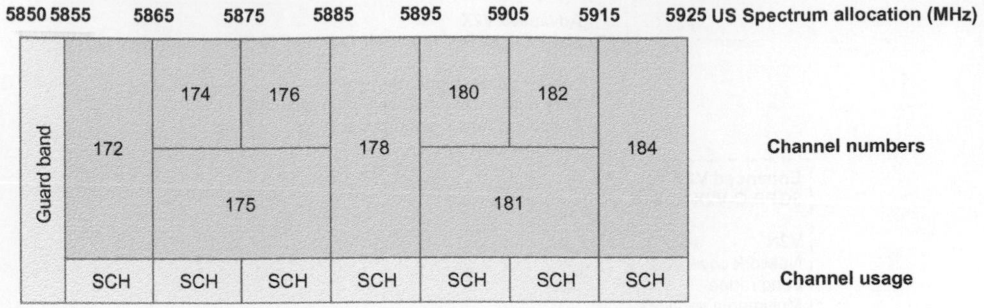

Figure 7.5
DSRC channel arrangement in the United States [13,14].

7.2 Spectrum Allocation

Some DSRC-based systems have been deployed or are expected to be deployed in 5.9 GHz band or in 700 MHz band. As shown in Fig. 7.5, the United States allocation divides the entire 75 MHz ITS band into seven non-overlapping 10 MHz channels, with 5 MHz reserved as the guard band. There is one control channel (CCH), corresponding to channel number 178 and six service channels (SCHs) for the DSRC in this band. The pair of channels (174 and 176, and 180 and 182) can be combined to form a single 20 MHz channel in either channel 175 or channel 181, respectively. Channel 172 is reserved exclusively for critical public safety communications, while channel 184 is reserved for high-power public safety use cases. The remaining channels can be used for non-safety applications. The V2X spectrum demands highly depend on environment (vehicle density, applications, traffic model, etc.), radio access technology, and target communication range and performance metrics.

System-level analysis shows that approximately 20–30 MHz of bandwidth is needed in typical highway scenarios for target range of 300–320 m based on LTE V2X sidelink communication [10]. Discussions on ITS spectrum allocation adjustments in low and high bands are in progress to allocate new spectrum for V2X as follows (see Fig. 7.6):

- *Low-band:* Potential use of 5.925–6.425 GHz in addition to 5.855–5.925 GHz.
- *High band:* Frequency allocation above 6 GHz in Europe may be updated—center frequency of ITS band may be shifted to 64.80 GHz and extended to 2.16 GHz of bandwidth (to avoid overlapping with two IEEE 802.11ad channels and to increase ITS bandwidth).

In the quest for increased network capacity, reduced cell sizes and the use of wider system bandwidths (whether contiguous or via carrier aggregation) have been actively investigated in the past decade. The use of mmWave spectrum for sidelink communications is believed to have several advantages for V2I communication, wherein transitory, high data rate

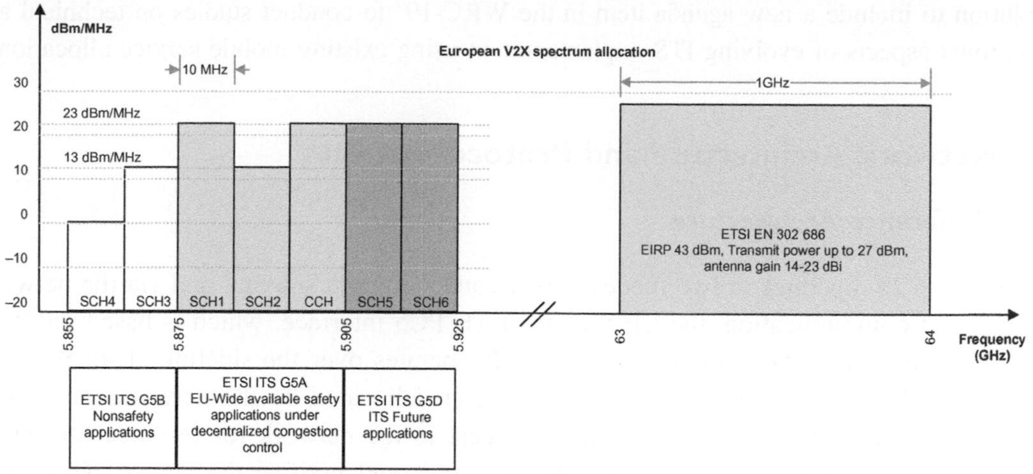

Figure 7.6
Region-specific ITS spectrum allocation in Europe [24,25].

connection can be established between vehicle and nearby base stations (small cells) to exchange delay-insensitive data (e.g., map updates and infotainment data in the downlink, and collected traffic and sensor information for large-scale traffic monitoring in the uplink); and for directional V2V communication for supporting specific use cases, such as communication between adjacent vehicles in a platoon. Although mmWave communication is very attractive from the data throughput perspective, it creates challenges for the physical layer. Due to high propagation loss (path loss) and its susceptibility to shadowing, the use of mmWave bands is deemed suitable for mainly short range (a few hundred meters) and point-to-point LoS communications. Furthermore, since the Doppler spread is linear function of the carrier frequency, the amount of Doppler spread in the 60 GHz band will be 10–30 × of that in 2–6 GHz band. Radio signals used for communication with the cellular network and between vehicles can also be used for position estimation. The achievable positioning accuracy can be significantly higher in 5G networks relative to legacy LTE networks due to higher signal bandwidths and dense network deployments, which enables LoS communication with a high probability. Estimation of the relative position between vehicles is directly useful for certain use cases such as platooning, but can also serve as input to the cellular network in order to improve the estimation of vehicles' absolute position [16].

Global ITS spectrum is currently under further study within ITU-R Working Party 5A.[3] This study is considered essential in improving the safety and efficiency of roads and highways. Specifically, the ITU World Radio communication Conference 2015 (WRC-15) adopted a

[3] ITU-R Study Group 5 (SG 5) Working Party 5A (WP 5A): Land mobile service excluding IMT; amateur and amateur-satellite service https://www.itu.int/en/ITU-R/study-groups/rsg5/rwp5a

resolution to include a new agenda item in the WRC-19[4] to conduct studies on technical and operational aspects of evolving ITS implementation using existing mobile service allocations.

7.3 Network Architecture and Protocol Aspects

7.3.1 Reference Architecture

LTE-based V2X operates in two modes: direct connection via sidelink and via the network. In the direct communication, the UE uses the LTE PC5 interface, which is based on 3GPP Rel-12 ProSe. The D2D communication in LTE operates over the sidelink channel in two different modes. In the first mode, the resources are allocated by the network (i.e., eNB). Devices that intend to transmit will send a request to the network for resource allocation, and the network subsequently allocates the resources and notifies the device. This mode requires additional signaling for every transmission, thus increasing transmission latency. In the second operation mode, devices select the resource autonomously. The autonomous mode reduces latency, but some issues related to possible collisions and interference may arise. Optimization of resource allocation procedures has been considered with emphasis on the autonomous mode due to its lower latency. Further enhancements to accommodate high-speed/high-Doppler, high-vehicle density, improved synchronization, and decreased message transfer latency have been studied. This mode is suitable for proximal direct communications over short distances and for V2V safety applications that require low latency; for example, advanced driver-assistance systems or situational awareness. This mode can work in and out-of-network coverage (see Fig. 7.2).

Network-based communication uses the LTE Uu interface between the vehicle (i.e., the UE) and the eNBs. The UEs send unicast messages via the eNB to an application server, which in turn re-broadcasts them via evolved multimedia broadcast multicast service (eMBMS) for all UEs in the relevant geographical area. This mode uses the existing LTE radio access and is suitable for latency-tolerant use cases (e.g., situational awareness, mobility services). It is possible to deploy the network-based mode in an operator licensed spectrum, while the direct PC5-based mode can be deployed in unlicensed spectrum.

LTE uses unicast and broadcast bearers for data transmission. Broadcast is applicable, if the mobile operator has deployed eMBMS. LTE can complement the short-range communication path for V2X provided by other technologies (e.g., DSRC/IEEE 802.11p). This type of broadcast transmission can potentially cover more vehicles in network coverage because the network can control the broadcast range, which is more suitable for the V2I/V2N type of services. Mobile operators can provide additional value-added services to subscribed drivers including traffic jams/blocked roads further ahead, real-time map/3D building/landmarks,

[4] World Radiocommunication Conference 2019 (WRC-19) https://www.itu.int/en/ITU-R/conferences/wrc/2019

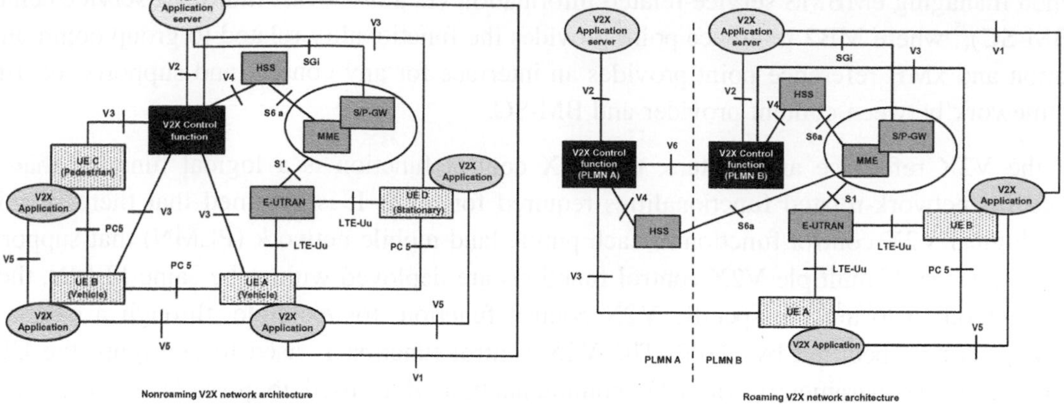

Figure 7.7
Roaming/non-roaming network architecture for LTE-based V2X communication [2].

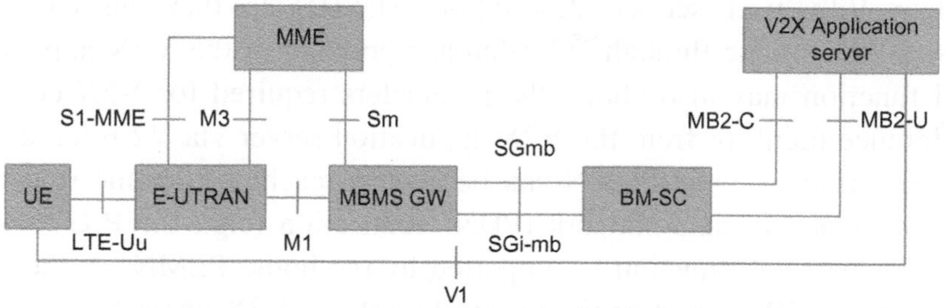

Figure 7.8
Reference architecture for eMBMS for LTE-Uu-based V2X communication via MB2 [2].

updates for the area, or suggested speed. These two operation modes may be used by a UE independently for transmission and reception; for example, a UE can use eMBMS for reception without using LTE-Uu for transmission. A UE may also receive V2X messages via LTE-Uu unicast downlink.

V2X application servers (e.g., for traffic management services) in different domains can communicate with each other to exchange V2X messages. The interface between V2X application servers and the methods of the exchange of messages between V2X application servers is out of scope of 3GPP specifications. ProSe discovery feature can be used by a V2X-enabled UE and it is up to UE implementation. The RSU is not an architectural entity, but an implementation option. The high-level roaming and non-roaming reference architectures for PC5 and LTE-Uu-based V2X communications are shown in Fig. 7.7.

Fig. 7.8 shows the high-level reference architecture with eMBMS for LTE-based V2X communication. The V2X application server may apply either MB2 or xMB reference points when managing eMBMS service-related information via a broadcast multicast service center (BM-SC),[5] where MB2 reference point provides the functionality related to group communication and xMB reference point provides an interface for any content and supports security framework between content provider and BM-SC.

In the V2X reference architecture, the V2X control function is a logical function that is used for network-related functionalities required for V2X. It is assumed that there is only one logical V2X control function in each public land mobile network (PLMN) that supports V2X services. If multiple V2X control functions are deployed within the same PLMN, then the method to locate the specific V2X control function, for example, through a database lookup, is not specified by 3GPP. The V2X control function is used to configure the UEs with necessary parameters for V2X communication. It is used to provide the UEs with network-specific parameters that allow the UEs to use V2X service in that PLMN. The V2X control function is also used to configure the UEs with parameters that are needed when the UEs are not served by an LTE network. The V2X control function may also be used to obtain V2X user service descriptions (USDs) so that the UEs can receive eMBMS-based V2X traffic through V2 reference point from the V2X application server. V2X control function may also obtain the parameters required for V2X communications over PC5 reference point, or from the V2X application server via V2 reference point. The V2X control function in home PLMN can be always reached, if home-routed configuration is applied for packet data network (PDN) connection (e.g., the P-GW is part of the home PLMN), when such function is supported by the home PLMN. In the case of local breakout (e.g., the P-GW is part of the visited PLMN), a V2X control function proxy can be deployed by the visited PLMN to support UE to home V2X control function logical connection, if inter-PLMN signaling is required. The UE is not aware of these transactions and it will not know which access point name (APN) can be used for communication with the V2X control function unless the specific APN information is configured in the UE indicating that this APN provides signaling connectivity between the UE and the home V2X control function. The V2X control function of the home PLMN is discovered through interaction with the domain name service (DNS) function [2].

The V2X application server supports receiving uplink data from the UE over a unicast connection and delivering data to the UE(s) in a target area using a unicast and/or eMBMS delivery mechanism as well as mapping of geographic location information to the appropriate target eMBMS service area identities for the broadcast. It is further responsible for mapping from geographic location information to the appropriate [target] E-UTRAN cell global identifier (ECGI), which is used to globally identify the cells. The

[5] Broadcast multicast service center is an evolved multimedia broadcast multicast service network entity located in the core network and is responsible for authorization and authentication of content providers, charging, and overall configuration of the data flow through the core network.

V2X application server handles the mapping from UE provided ECGI to appropriate target eMBMS service area identities for the broadcast; provides the appropriate ECGI(s) and/or eMBMS service area identities to BM-SC, which is preconfigured with the local eMBMS information (e.g., Internet protocol (IP) multicast address, multicast source, common tunnel endpoint identifier); and is preconfigured with the local eMBMS IP address and port number for the user plane. The V2X application server is also in charge of sending local eMBMS information to the BM-SC; requesting BM-SC for allocation/de-allocation of a set of temporary mobile group identities; requesting BM-SC for activating/deactivating/modifying the eMBMS bearer; providing the V2X USDs so that the UE can receive eMBMS-based V2X traffic from V2X control function; providing the parameters for V2X communications over PC5 reference point to V2X control function; or providing the parameters for V2X communications over PC5 reference point to UE [2].

The reference points or network interfaces shown in Fig. 7.7 can be further described as follows [2]:

- *V1:* This reference point is between the V2X application in the UE and in the V2X application server, and it is not defined in 3GPP specifications.
- *V2:* This reference point is between the V2X application server and the V2X control function in the operator's network. The V2X application server may connect to V2X control functions belonging to multiple PLMNs.
- *V3:* This reference point is between the UE and the V2X control function in UE's home PLMN. It is based on the service authorization and provisioning part of the PC3 reference point. It is applicable to both PC5 and LTE-Uu-based V2X communication and optionally eMBMS and LTE-Uu-based V2X communication.
- *V4:* This reference point is between the home subscriber server (HSS) and the V2X control function in the operator's network.
- *V5:* This reference point is between the V2X applications in the UEs, and is not specified in 3GPP specifications.
- *V6:* This reference point is between the V2X control function in the home PLMN and the V2X control function in the visited PLMN.
- *PC5:* This reference point is between the UEs for direct communication over user-plane for V2X service.
- *S6a:* In case of V2X service, S6a is used to download V2X service-related subscription information to mobility management entity (MME) during E-UTRAN attach procedure or to inform MME subscription information in the HSS has changed.
- *S1-MME:* In case of V2X service, it is used to convey the V2X service authorization from the MME to eNB.
- *xMB:* This reference point is between the V2X application server (e.g., content provider) and the BM-SC.

- *MB2:* This reference point is between the V2X application server and the BM-SC.
- *SGmb/SGi-mb/M1/M3:* The SGmb/SGi-mb/M1/M3 reference points are internal to the eMBMS system.
- *LTE-Uu:* This reference point is between the UE and the E-UTRAN.

The V2N and communication via the network represent suitable applications for the network slicing. For instance, autonomous driving or safety/emergency services would require a URLLC network slice. Meanwhile, some infotainment services or personal mobility services would require either a best-effort slice or an eMBB. A vehicle can access different slices at the same time, with passengers watching a high-definition movie while a background application detects a road hazard and triggers an emergency message for the cars behind or nearby to slow down or stop to prevent an accident. Fig. 7.9 illustrates this scenario. The slices could come from one device or multiple devices. In the case of one device, 3GPP has defined that a given device can support up to eight different slices with a common access and mobility management function (AMF) for all slices and a session management function (SMF) per slice. In the example shown in Fig. 7.9, there are three network slices attached to the same device sharing the same AMF instance. The first slice is massive machine-type communication, which sends data to the core and the PDN. The second slice offers caching at the edge, while the third slice provides access to an edge V2X application.

The C-V2X can operate outside of network coverage using direct communication without requiring provisioning of a universal subscriber identity module (USIM). To enable USIM-less communication, automobile manufacturers will preconfigure the vehicle device with parameters necessary for out-of-network operation, including authorization to use V2X; a list of authorized application classes and the associated frequencies to use; radio parameters for use over the direct link; and configuration for receiving V2X messages via cellular

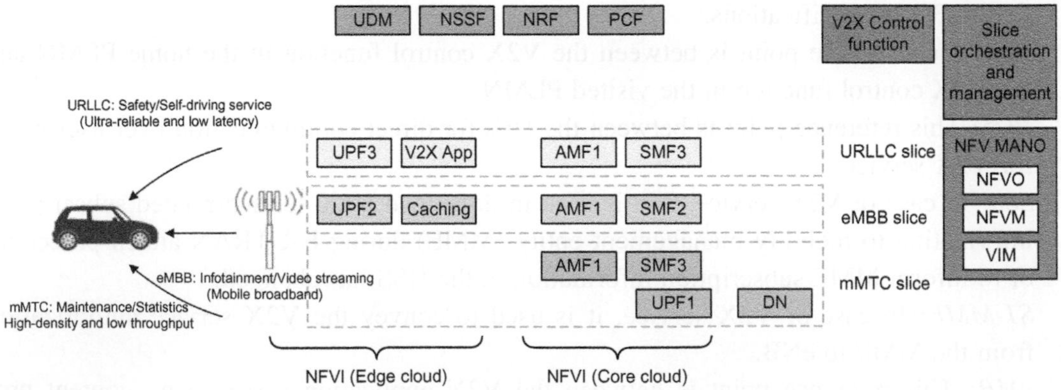

Figure 7.9
Example of NR V2X network slicing [13,14].

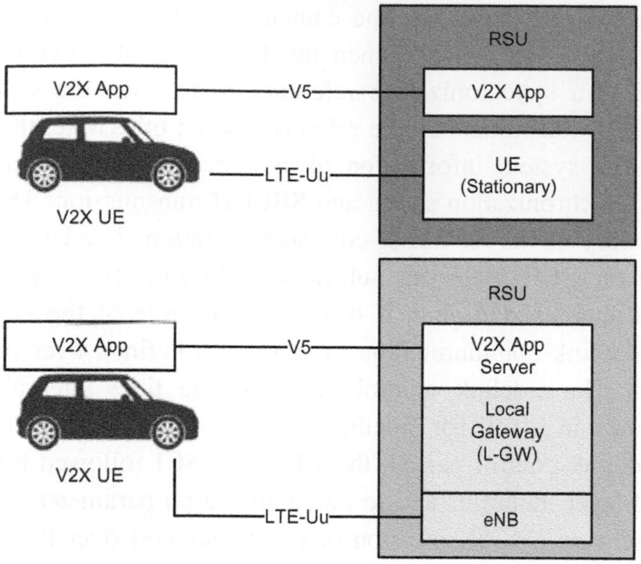

Figure 7.10
RSU implementation options [20].

broadcast (i.e., eMBMS). Direct USIM-less communication allows C-V2X to support critical safety services when network coverage is unavailable or if the vehicle does not have an active cellular subscription. These parameters can also be securely updated by the vehicle manufacturers or the mobile operators.

For V2I applications, the infrastructure that includes an RSU can be implemented in an eNB or a stationary UE. Fig. 7.10 shows two RSU implementation options and the functional entities in each case.

7.3.2 Sidelink and Radio Access Protocols

Sidelink communication is a mode of operation whereby UEs can communicate with each other directly over the PC5 interface. This operation mode is supported when the UE is served by an eNB/gNB or outside of the network coverage. The use of sidelink communication was originally limited to those UEs that were authorized for public safety operation; however, the application of sidelink communication was further extended to V2X services. In order to perform synchronization for out-of-coverage devices, the originating device may act as a synchronization source by transmitting sidelink broadcast control channel (SBCCH) and synchronization signals. The SBCCH carries the most essential system information needed to receive other sidelink channels and signals. In LTE V2X, the SBCCH is transmitted along with a synchronization signal with a fixed periodicity of 40 ms [6].

When the UE is in network coverage, the content of SBCCH are derived from the parameters signaled by the serving eNB. When the UE is out-of-network coverage and if it selects another UE as a synchronization reference node, then the system information is obtained from SBCCH transmitted by the reference node; otherwise, the UE uses preconfigured parameters. The system information block type 18 (SIB18) provides the resource information for the synchronization signal and SBCCH transmission. There are two preconfigured subframes every 40 ms for out-of-coverage operation. The UE receives the synchronization signal and the SBCCH in one subframe and transmits the synchronization signal and the SBCCH in another subframe, if it assumes the role of the synchronization node. The UE performs sidelink communication in subframes defined over the duration of sidelink control period. The sidelink control period is the time interval during which the resources are allocated in a cell for sidelink control information (SCI) and data transmission. Within the sidelink control period, the UE sends SCI followed by sidelink data. SCI indicates a physical layer identifier and resource allocation parameters; for example, modulation and coding scheme (MCS), location of the resource(s) over the duration of sidelink control period, timing alignment, etc. The UE performs transmission and reception over LTE-Uu and PC5 (without sidelink discovery gap) where Uu transmission/reception is the highest priority, followed by PC5 sidelink communication transmission/reception and then by PC5 sidelink discovery announcement/monitoring, which is considered the lowest priority. The UE further performs transmission/reception over Uu and PC5 (with sidelink discovery gap) starting with Uu transmission/reception for RACH, followed by PC5 sidelink discovery announcement during a sidelink discovery gap for transmission; non-RACH Uu transmission; PC5 sidelink discovery monitoring during a sidelink discovery gap for reception; non-RACH Uu reception; and PC5 sidelink communication transmission/reception [6].

The UE radio protocol structure for sidelink communication consists of user-plane and control-plane protocols. Fig. 7.11 shows the protocol stack for the user plane (the access

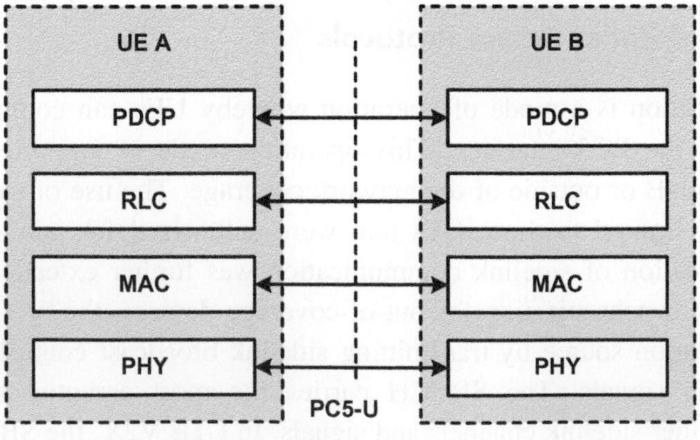

Figure 7.11
User-plane protocol stack for sidelink communication [6].

stratum protocol stack over PC5 interface), where packet data convergence protocol (PDCP), radio link control (RLC), and medium access control (MAC) sublayers, which are terminated at the other UE, perform similar functions defined for LTE protocols with some exceptions. The user plane corresponding to sidelink communication does not support HARQ feedback; uses radio link control (RLC) unacknowledged mode (UM) and the receiving UE must maintain at least one RLC UM entity per transmitting UE; a receiving RLC UM entity used for sidelink communication does not need to be configured before reception of the first RLC UM protocol data unit (PDU); and robust header compression (RoHC) unidirectional mode is used for header compression in PDCP for sidelink communication. A UE may establish multiple logical channels. In that case, the logical channel ID (LCID) included in the MAC subheader uniquely identifies a logical channel with one source layer-2 ID and destination layer-2 ID combination [7]. The parameters for logical channel prioritization are not configured. The access stratum is provided by ProSe per-packet priority (PPPP)[6] of the PDU transmitted over PC5 interface by higher layers. Note that there is one PPPP associated with each logical channel [6]. The user-plane protocol stack and functions are further used for V2X sidelink communication. In addition, for V2X sidelink communication, the sidelink traffic channel (STCH) is used. Non-V2X data (e.g., public safety) is not multiplexed with V2X data transmitted over the resources configured for V2X sidelink communication. The access stratum is provided by the PPPP of a PDU transmitted over PC5 interface by upper layers. The packet delay budget (PDB), which refers to the permissible latency of data packets transported between UE and P-GW, of the PDU can be determined from the PPPP. The low PDB is mapped to the high priority PPPP value. The existing logical channel prioritization based on PPPP is used for V2X sidelink communication [6].

The control-plane protocol stack for the sidelink broadcast channel (SL-BCH) is also used for V2X sidelink communication (see Fig. 7.12). A UE that supports V2X sidelink communication can operate in two modes for resource allocation: (1) scheduled resource allocation,

[6] In 3GPP Rel-13, ProSe per-packet priority (PPPP) was introduced to enable QoS differentiation across different traffic streams corresponding to different sidelink logical channels. PPPP has eight values ranging from one to eight, and each PPPP value represents the priority at which the associated traffic should be treated over the sidelink. Each data packet to be transmitted is assigned a PPPP value selected by the application layer. The UE then performs logical channel prioritization such that the transmission of the data associated with higher PPPP is prioritized. PPPP is also applied to transmission pool selection in the case of UE autonomous resource selection. The network can configure one or multiple PPPP for each transmission pool in the list of pools. Then, for each MAC PDU to transmit on the sidelink, the UE selects a transmission pool associated with the PPPP [2]. For downlink to sidelink mapping, which occurs when the relay UE receives traffic from the eNB, it identifies whether the packet should be relayed, by referring to the destination IP address of the packet. The relay UE then assigns a priority value called PPPP to the received packet to be relayed. The priority assignment is based on the mapping information representing the association between the QoS class identifier values of downlink bearers and the priority values. The QoS class identifier-to-priority mapping information is provisioned to the relay UE by the network.

which is characterized by the UE's need for radio resource control (RRC) connection establishment to transmit data and the UE's request for transmission resources from the eNB. The eNB schedules transmission resources for transmission of SCI and data. Sidelink semi-persistent scheduling (SPS) is supported for scheduled resource allocation; and (2) UE autonomous resource selection, which is characterized by the UE's self-selection of resources from resource pools and transport format selection to transmit SCI and data. In the latter case, if mapping between the zones and V2X sidelink transmission resource pools is configured, the UE selects V2X sidelink resource pool based on the zone in which the UE is located. The UE further performs sensing for (re)selection of sidelink resources. Based on sensing results, the UE (re)selects some specific sidelink resources and reserves multiple sidelink resources. The UE can perform up to two parallel and independent resource reservation processes. It is also allowed to perform a single resource selection for its V2X sidelink transmission [6].

A UE does not establish and maintain a logical connection to receiving UEs before one-to-many sidelink communication. Higher layer protocols establish and maintain a logical connection for one-to-one sidelink communication, including ProSe UE-to-network relay operation. The control plane functions for establishing, maintaining, and releasing the logical connections for unicast sidelink communication are shown in Fig. 7.12.

Sidelink discovery is defined as the procedure used by the UE supporting sidelink discovery to find other UE(s) in its proximity, using LTE direct radio signals via PC5 (see Fig. 7.13). Sidelink discovery is supported regardless of whether the UE is within network coverage or out-of-coverage. The service was originally limited to ProSe-enabled public safety UEs to perform sidelink discovery when they were out-of-network coverage where the allowed operating frequency was preconfigured in the UE, and is used even when UE is out-of-network coverage in that frequency. The preconfigured frequency is the same frequency as

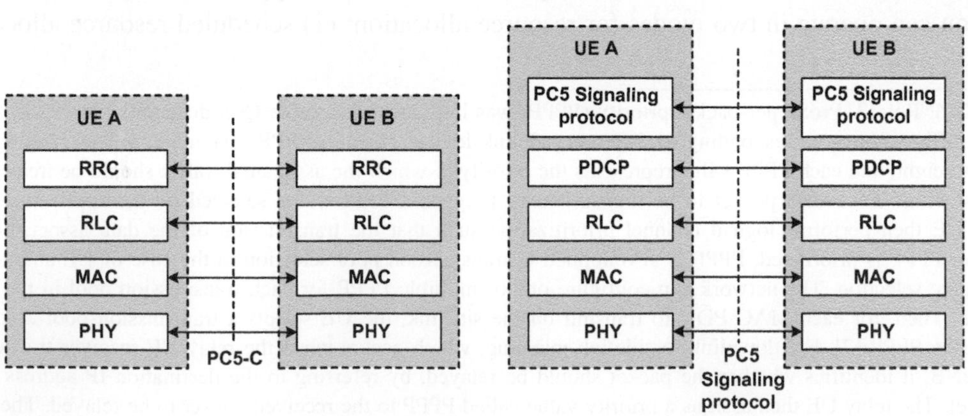

Figure 7.12
Control-plane protocol stack for sidelink broadcast and unicast transmissions [6].

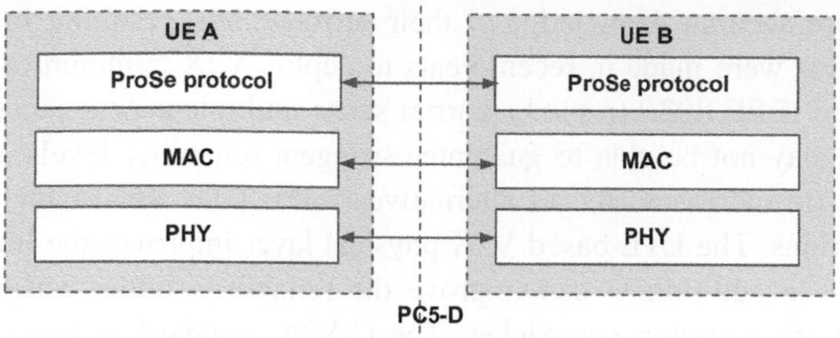

Figure 7.13
PC5 interface for sidelink discovery [6].

the public safety ProSe carrier[7] [6]. The AS protocol stack for sidelink discovery consists of only MAC and PHY. The AS layer interfaces with upper layer (ProSe protocol) where the MAC sublayer receives the discovery message from the upper layer (ProSe protocol). The IP layer is not used for transmitting the discovery message. The MAC layer determines the radio resource to be used for announcing the reception of the discovery message from upper layer and further generating the MAC PDU carrying the discovery message and sending the MAC PDU with no MAC header to the physical layer for transmission in the predetermined radio resources [7]. The content of discovery message is transparent to the access stratum, over which no distinction is made for sidelink discovery models and types; however, higher layer protocols detect whether the sidelink discovery notification is related to public safety. A UE can participate in announcing and monitoring of discovery message in both RRC_IDLE and RRC_CONNECTED states following eNB configuration, where the UE announces and monitors its discovery message subject to the half-duplex constraint. In order to perform synchronization, the UE(s) participating in announcing of discovery messages may act as a synchronization source by transmitting SBCCH and a synchronization signal based on the resource information allocated for synchronization signals provided in SIB19. There are three range classes. The upper layer authorization provides applicable range class of the UE. The maximum allowed transmission power for each range class is signaled in SIB19. The UE uses the applicable maximum allowed transmission power corresponding to its authorized range class. This sets an upper limit on the configured transmit power based on open-loop power control parameters.

7.4 Physical Layer Aspects

V2X communications enable information exchange between vehicles and between vehicles and the infrastructures and/or pedestrians. The information exchange will provide the

[7] Public safety ProSe carrier is the carrier frequency used for public safety sidelink communication and public safety sidelink discovery.

vehicles with more accurate knowledge of their surroundings, resulting in improved traffic safety. Some efforts were made in recent years to deploy V2X communications using IEEE 802.11p. However, IEEE 802.11p uses a carrier sense multiple access scheme with collision avoidance which may not be able to guarantee stringent reliability levels and network scalability as the traffic increases. As an alternative, 3GPP LTE Rel-14 included support for V2X communications. The LTE-based V2X physical layer improves the link budget relative to IEEE 802.11p. In addition, it can improve the reliability, under certain conditions, by adding redundant transmission per packet. The C-V2X standard includes two radio interfaces. The LTE-based Uu radio interface supports vehicle-to-infrastructure communications, while the PC5 interface supports V2V communications based on LTE-based sidelink communications. LTE sidelink (or D2D communication) was originally introduced in Rel-12 for public safety, and included two modes of operation: mode 1 and mode 2 (see Fig. 7.14). Both modes were designed with the objective of prolonging the battery life of mobile devices at the cost of increased latency. Connected vehicles require highly reliable and low-latency V2X communications; therefore, modes 1 and 2 are not suitable for vehicular applications. 3GPP Rel-14 introduced two new communication modes (modes 3 and 4) specifically designed for V2V communications (see Fig. 7.15). In mode 3, the cellular network selects and manages the radio resources used by vehicles for their direct V2V communications. In mode 4, vehicles autonomously select the radio resources for their direct V2V communications. In mode 4, the UE (the vehicle) can operate without cellular coverage; therefore, this is considered the baseline V2V mode since safety applications cannot depend

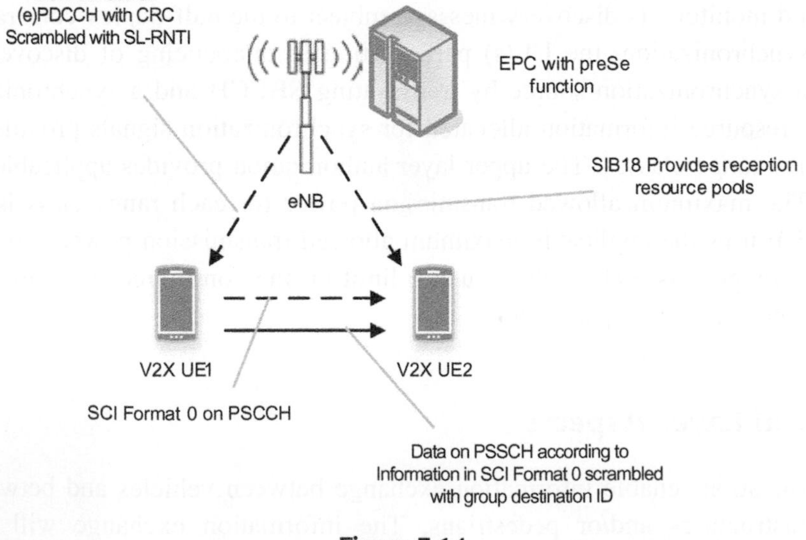

Figure 7.14
Scheduling transmission resources for direct communication, mode 1 [18,19].

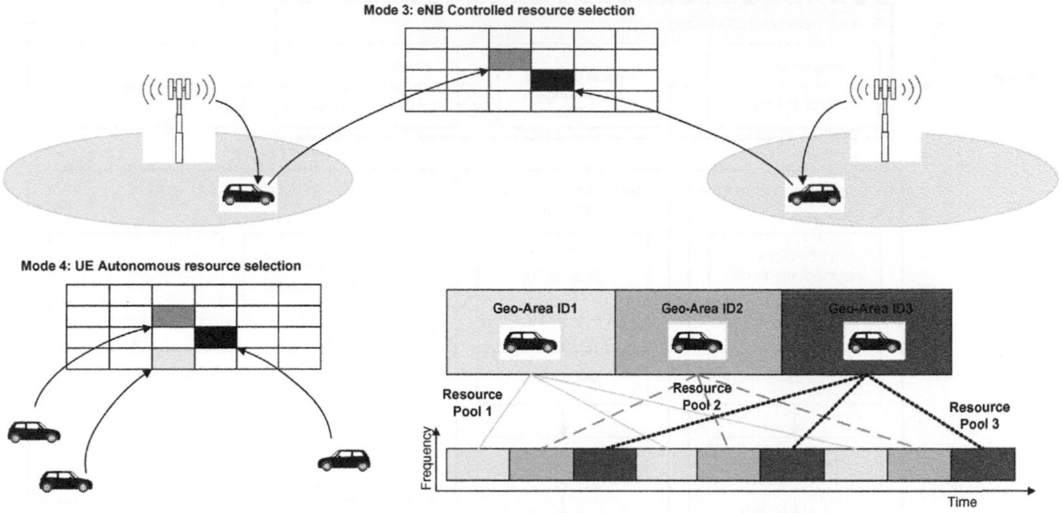

Figure 7.15
Illustration of LTE V2X modes 3 and 4 operation [25].

on the availability of cellular coverage. Mode 4 includes a distributed scheduling scheme for vehicles to select their radio resources and includes the support for distributed congestion control. In mode 3, the vehicle reports its location and coordinates to assist the eNB in scheduling, whereas in mode 4, the vehicle location information can be used to restrict sidelink resource selection (to enable spatial reuse in a distributed system).

The C-V2X utilizes LTE uplink multiple access scheme (SC-FDMA), and supports 10 and 20 MHz channels. Each channel is divided into subframes, resource blocks (RBs), and subchannels. Subframes are 1 ms long. The resource block is the smallest unit of frequency resources that can be allocated to a user. It is 180 kHz wide in frequency (12 subcarriers with 15 kHz subcarrier spacing). The C-V2X defines subchannels as a group of RBs in the same subframe, and the number of RBs per subchannel can vary. Subchannels are used to transmit data and control information. The data is transmitted in the units of transport blocks (TBs) over PSSCH, and the SCI messages are transmitted over PSCCH. A TB contains a full packet to be transmitted; for example, a beacon or cooperative awareness message. A node that wants to transmit a TB must also transmit its associated SCI, which is also referred to as a scheduling assignment. The SCI includes information such as the MCS that is used to transmit the TB, the number of RBs, and the resource reservation interval for SPS. This information is critical for other nodes to be able to receive and decode the TB; thus the SCI must be correctly received. A TB and its associated SCI must always be transmitted in the same subframe. The overall LTE-based V2X sidelink physical layer processing is shown in Fig. 7.16.

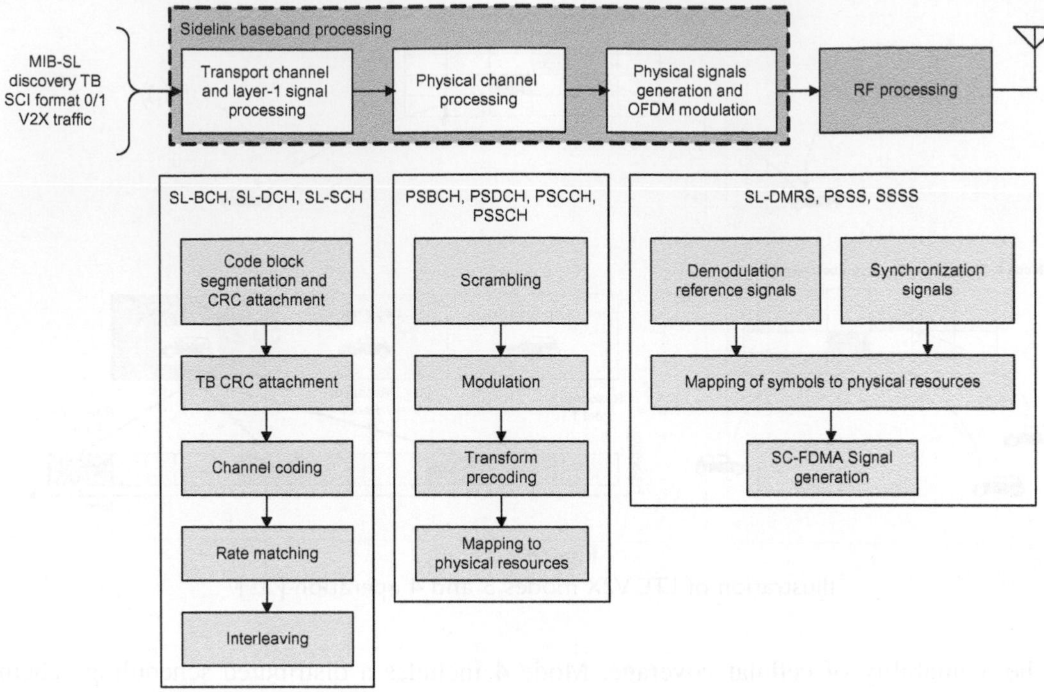

Figure 7.16
LTE V2X sidelink physical layer processing [17,22].

The C-V2X specifies two subchannelization schemes, as shown in Fig. 7.17: (1) adjacent PSCCH + PSSCH, where the SCI and TB are transmitted in adjacent RBs. For each SCI + TB transmission, the SCI occupies the first two RBs of the first subchannel utilized for the transmission. The TB is transmitted in the RBs following the SCI, and can occupy several subchannels depending on its size. It will also occupy the first two RBs of the following subchannels; and (2) non-adjacent PSCCH + PSSCH, where the RBs are divided into pools. One pool is dedicated to transmit only SCIs, and the SCIs occupy two RBs. The second pool is reserved to transmit only TBs and is divided into subchannels. The TBs can be transmitted using QPSK or 16QAM modulation, whereas the SCIs are always transmitted using QPSK. The C-V2X uses turbo coding with normal cyclic prefix. There are 14 OFDM symbols per subframe, and four of these symbols are dedicated to the transmission of demodulation reference signals (DM-RS) in order to improve robustness against the Doppler effect at high speeds. The reference signals are transmitted in the third, sixth, ninth, and twelfth symbol of the subframe. The maximum transmit power is 23 dBm, and the standard specifies a sensitivity-power-level requirement at the receiver of −90.4 dBm and a maximum input level of −22 dBm [17].

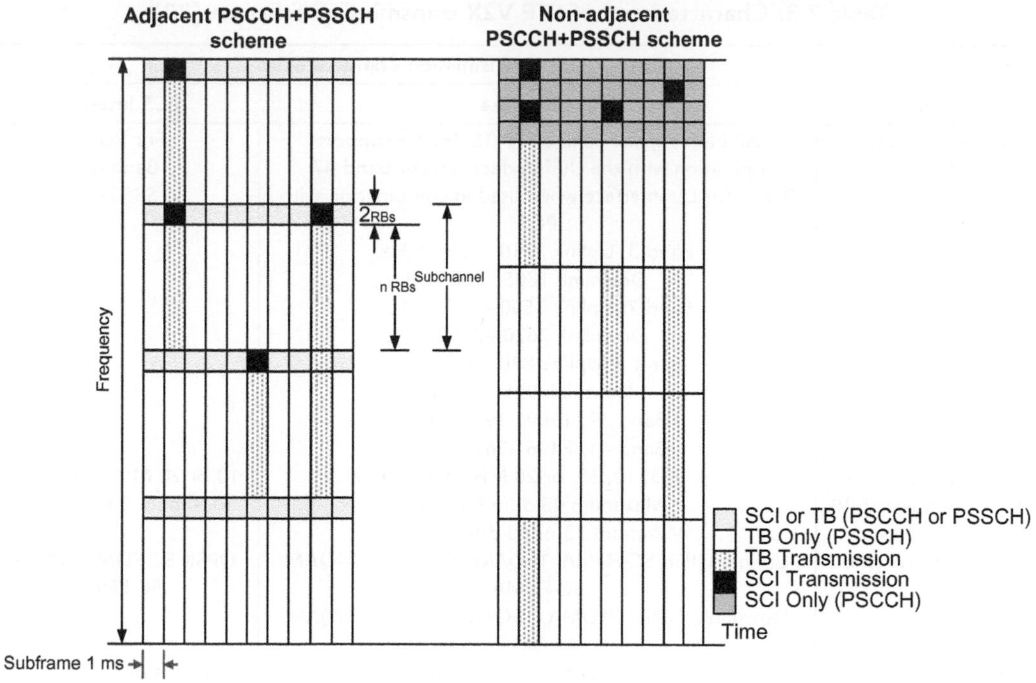

Figure 7.17
C-V2X frame structure and resource allocation schemes [17].

Vehicles communicate using sidelink or V2V communications in mode 4 and autonomously select their radio resources, independent of cellular coverage. When the vehicles are under cellular coverage, the network decides how to configure the V2X channel and informs the vehicles through the sidelink V2X configurable parameters. The message includes the carrier frequency of the V2X channel, the V2X resource pool, synchronization references, the subchannelization scheme, the number of subchannels per subframe, and the number of RBs per subchannel. When the vehicles are not within the cellular coverage, they utilize a preconfigured set of parameters to replace the sidelink V2X configurable parameters. However, the standard does not specify a specific value for each parameter. The V2X resource pool indicates which subframes of a channel are utilized for V2X. The rest of the subframes can be utilized by other services. The standard includes the option to divide the V2X resource pool based on geographical areas. In this case, vehicles in an area can only utilize the pool of resources that have been assigned to those areas (see Table 7.3).

Sidelink transmissions use the same basic transmission scheme as the uplink; however, sidelink is limited to single-cluster transmission for all sidelink physical channels. Furthermore, sidelink uses one symbol gap at the end of each sidelink subframe. For V2X sidelink communication, PSCCH and PSSCH are transmitted in the same subframe (see Fig. 7.18). The

Table 7.3: Characteristics of LTE V2X transmission schemes [20].

Item	Transmission Characteristics	
	Uu Interface	PC5 Interface
Operating frequency range	All bands specified in 3GPP TS 36.101 support operation with the Uu interface, except band 47 Bands for Uu interface when used in combination with PC5 Band 3: Uplink 1710–1785 MHz Downlink 1805–1880 MHz Band 7: Uplink 2500–2570 MHz Downlink 2620–2690 MHz Band 8: Uplink 880–915 MHz Downlink 925–960 MHz Band 39: 1880–1920 MHz Band 41: 2496–2690 MHz	For Rel-14 Band 47: 5855–5925 MHz
RF channel bandwidth	1.4, 3, 5, 10, 15, or 20 MHz per channel	10 or 20 MHz per channel
RF transmit power/EIRP	Maximum 43 dBm for eNB Maximum 23 or 33 dBm for UE	Maximum 23 or 33 dBm
Modulation scheme	Uplink: QPSK SC-FDMA, 16QAM SC-FDMA, 64QAM SC-FDMA Downlink: QPSK OFDMA, 16QAM OFDMA, 64QAM OFDMA	QPSK SC-FDMA, 16QAM SC-FDMA
Forward error correction scheme	Convolutional coding and turbo coding	Convolutional coding and turbo coding
Data transmission rate	Uplink: From 1.4 to 36.7 Mbps for 10 MHz channel Downlink: From 1.4 to 75.4 Mbps for 10 MHz channel	From 1.3 to 15.8 Mbps for 10 MHz channel
Scheduling	Centralized scheduling by eNB	Centralized scheduling or distributed scheduling
Duplex method	FDD or TDD	TDD

sidelink physical layer processing of transport channels differs from uplink transmission in two aspects: (1) scrambling in physical sidelink discovery channel (PSDCH) and PSCCH processing is not UE-specific and (2) 64QAM and 256QAM modulation schemes are not supported for the sidelink. The PSCCH is mapped to the sidelink control resources and indicates resource and other transmission parameters used by a UE for PSSCH. For PSDCH, PSCCH, and PSSCH demodulation, reference signals similar to uplink DM-RS are transmitted in the fourth symbol of the slot in normal cyclic prefix and in the third symbol of the slot for extended cyclic prefix. The sidelink DM-RS sequence length equals the size (number of subcarriers) of the assigned resources in the frequency domain. For V2X sidelink communication, reference signals are transmitted in the third and the sixth symbol of the first slot and the second and the fifth symbol of the second slot in normal cyclic prefix. For PSDCH and PSCCH, the reference signals are generated based on a fixed base sequence, cyclic shift, and orthogonal cover code. For V2X sidelink communication, cyclic shift for PSCCH is randomly selected in each transmission [6].

For in-coverage operation, the power spectral density of the sidelink transmissions is determined by the eNB. For measurement on the sidelink, the following basic UE measurement quantities are defined [18,19]:

- *Sidelink reference signal received power (S-RSRP)* is defined as the linear average over the power contributions [in (Watts)] of the resource elements that carry DM-RSs associated with physical sidelink broadcast channel (PSBCH), within the six middle-band PRBs of the relevant subframes. The reference point for the S-RSRP is the antenna connector of the UE. The reported value must not be lower than thc corresponding S-RSRP of any of the individual diversity branches, if receive diversity is utilized by the UE.
- *Sidelink discovery reference signal received power (SD-RSRP)* is defined as the linear average over the power contributions (in [Watts]) of the resource elements that carry DM-RSs associated with PSDCH for which CRC has been validated. The reference point for the SD-RSRP is the antenna connector of the UE. If receive diversity is used by the UE, the reported value will be lower than the corresponding SD-RSRP of any of the individual diversity branches.
- *PSSCH reference signal received power (PSSCH-RSRP)* is defined as the linear average over the power contributions [in (Watts)] of the resource elements that carry DM-RSs associated with PSSCH, within the PRBs indicated by the associated PSCCH. The reference point for the PSSCH-RSRP is the antenna connector of the UE. If receive diversity is used by the UE, the reported value must not be lower than the corresponding PSSCH-RSRP of any of the individual diversity branches.
- *Sidelink reference signal strength indicator (S-RSSI)* is defined as the linear average of the total received power [in (Watts)] per SC-FDMA symbol observed by the UE only in the configured subchannels over SC-FDMA symbols $(1, 2, \ldots, 6)$ of the first slot and SC-FDMA symbols $(0, 1, \ldots, 5)$ of the second slot of a subframe. The reference point for the S-RSSI is the antenna connector of the UE. If receive diversity is used by the UE, the reported value should not be lower than the corresponding S-RSSI of any of the individual diversity branches.

In the sidelink, there is no HARQ feedback and retransmissions are always performed in a predefined/configured manner. Measurement gaps and sidelink discovery transmission during a sidelink discovery gap for transmission are of higher priority than HARQ retransmissions, and whenever a HARQ retransmission collides with a measurement gap or sidelink discovery transmission during a sidelink discovery gap for transmission, the HARQ retransmission does not take place.

Vehicles (or the UEs) select their subchannels in mode 4 using the sensing-based SPS scheme specified in Rel-14 [6]. A vehicle reserves the selected subchannel(s) for several consecutive reselection counter packet transmissions. This counter is randomly set between

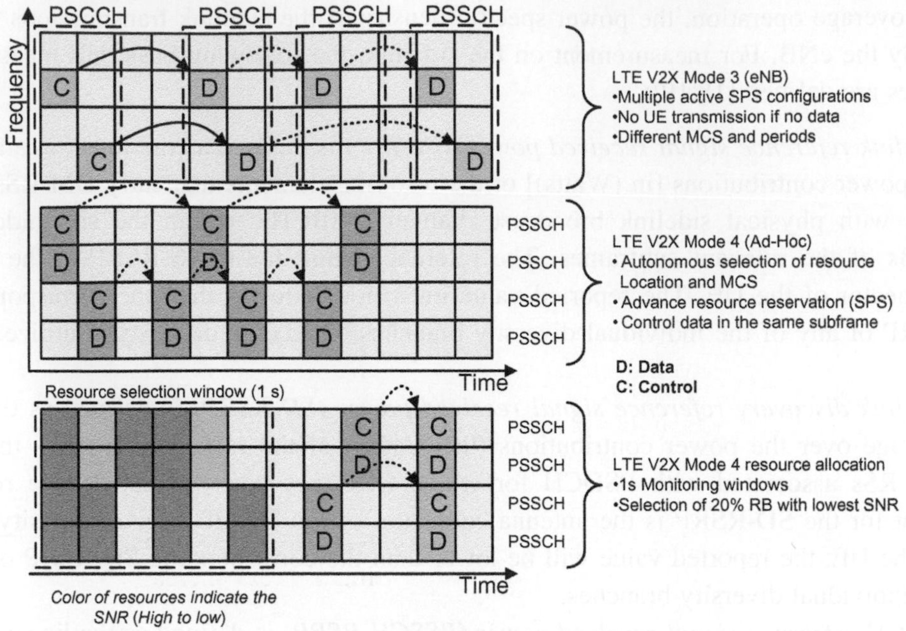

Figure 7.18
Comparison of LTE V2X modes 3 and 4 [25].

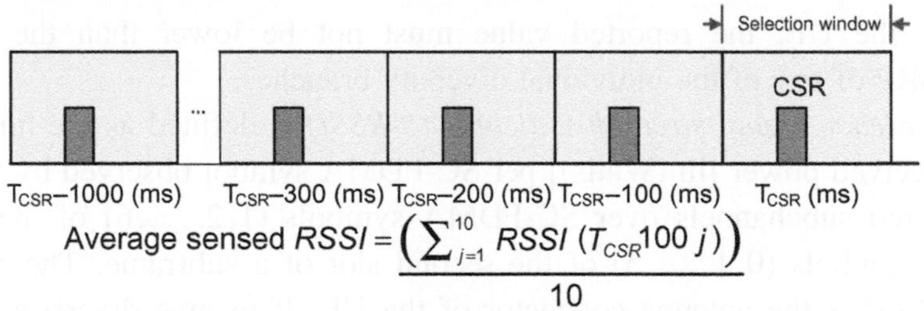

Figure 7.19
Calculation of the average RSSI of a candidate resource [17].

5 and 15, and the vehicle includes its value in the SCI. After each transmission, the reselection counter is decremented by one. When it is equal to zero, new resources must be selected and reserved with probability $(1-p)$ where the UE/vehicle can set the value of p between 0 and 0.8. New resources also need to be reserved, if the size of the packet to be transmitted does not fit in the subchannel(s) previously reserved. The reselection counter is randomly chosen every time that new resources are reserved. Packets can be transmitted every 100 subframes or in multiples of 100 subframes with a minimum of one packet per subframe. Each UE/vehicle includes its packet transmission interval in the resource reservation field of its SCI. The semi-persistent reservation of resources and the inclusion of the

reselection counter and packet transmission interval in the SCI would help other vehicles to estimate the [potentially] unused subchannels when making their own reservation, resulting in reduced risk of packet collision.

Let us assume that a vehicle needs to reserve new subchannels at time $T_{reservation}$. It can reserve subchannels between $T_{reservation}$ and the maximum latency of 100 ms. This interval is referred to as the selection window (see Fig. 7.19). Within the selection window, the vehicle identifies candidate single-subframe resources (CSR) to be reserved by all groups of adjacent subchannels within the same subframe where the SCI + TB information to be transmitted would fit. The vehicle analyzes the information it has received in the previous 1000 subframes before time instant $T_{reservation}$ and creates a list of permissible CSRs.

LTE V2X mode 4 provides an option for each packet to be transmitted twice to increase the reliability. The sensing-based SPS scheme randomly selects a CSR from a candidate list for the redundant transmission of the SCI + TB. 3GPP Rel-14 includes a variant of the sensing-based SPS scheme for pedestrian-to-vehicle communications, where pedestrians broadcast their presence using mobile devices. Since the sensing process increases the battery consumption, the standard provides an option to only sense a fraction of the 1000 subframes (1000 ms) before $T_{reservation}$. The mobile devices can only select CSRs in the sensed subframes using the sensing-based SPS scheme. LTE V2X supports two types of sidelink transmission: (1) single shot without resource reservation and (2) multi-shot with resource reservation. Both types of transmission follow sensing and resource selection procedure. The timing diagram for LTE V2X mode 4, dedicated sensing and resource selection, is illustrated in Fig. 7.20 with sensing (1000 ms) and resource (re)-selection (up to 100 ms) windows. The concept of sensing and resource reservation is further depicted in Fig. 7.21.

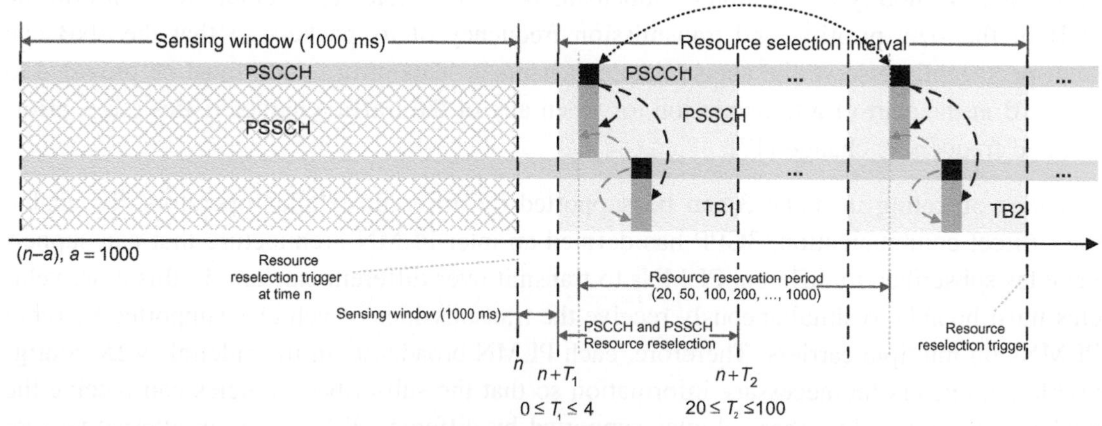

Figure 7.20
Mode 4 UE autonomous sensing and resource selection [25].

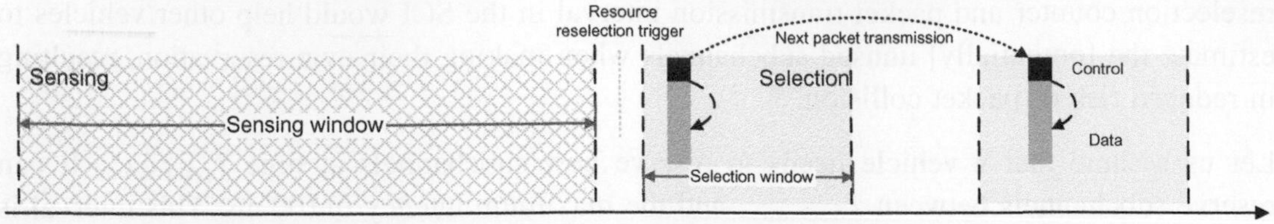

Figure 7.21
Sensing and resource reservation for collision avoidance in LTE V2X [6].

Vehicles also communicate using sidelink or V2V communications in mode 3. However, the selection of subchannels is managed by the base station and not by each vehicle as opposed to mode 4 (see Fig. 7.18). Mode 3 is only available when vehicles are within the network coverage. 3GPP has defined network architecture enhancements to support V2X. One of these enhancements is the V2X control function that is used by the network in mode 3 to manage radio resources and to provide vehicles with the sidelink V2X configurable parameters. Mode 3 utilizes the same subchannel arrangements as defined for mode 4. Vehicles using mode 3 must also transmit an associated SCI/TB, and the transmission of the SCI/TB must take place in the same subframe. In contrast to mode 4, the standard does not specify a resource management algorithm for mode 3. Each operator can implement its own algorithm that should fall under one of two categories: (1) Dynamic scheduling where the vehicles request subchannels from the eNB for each packet transmission, which increases the signaling overhead, and delays the packet transmission until vehicles are notified of their assigned subchannels; and (2) SPS-based scheduling where the base station reserves subchannels for the periodic transmissions of a vehicle. However, in contrast to mode 4, the eNB decides how long the reservation should be maintained and it can activate, deactivate, or modify reservation of subchannels for a vehicle. The vehicle must inform the eNB of the size, priority, and transmission frequency of its packets so that the eNB can semi-persistently reserve the appropriate subchannels. This information must be provided to the eNB at the start of a transmission, or when any of the traffic characteristics (size, priority, and frequency) change [17].

Vehicles operating in mode 3 can be supported by different cellular operators. To enable their direct communication, 3GPP has defined an inter-PLMN architecture that can support vehicles subscribed to different PLMNs to transmit over different carriers. In this case, vehicles must be able to simultaneously receive the transmissions of vehicles supported by other PLMNs on multiple carriers. Therefore, each PLMN broadcasts in the sidelink V2X configurable parameters the necessary information so that the subscribed vehicles can receive the packets transmitted by other vehicles supported by different PLMNs. In an alternative scenario, the vehicles may be supported by different PLMNs sharing the same carrier, where each PLMN is assigned a fraction of the resources of the carrier. The standard does not

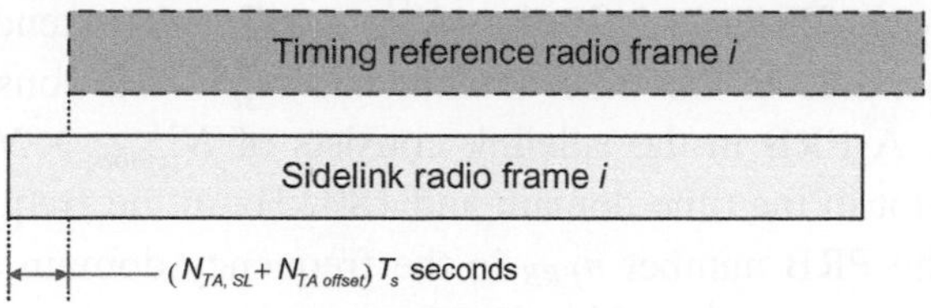

Figure 7.22
Sidelink timing alignment [3].

specify how the resources should be allocated among the PLMNs, but introduces a coordination mechanism (through the V2X control function) between PLMNs to avoid packet collisions [17].

The frame timing synchronization is an important consideration. Transmission of the ith sidelink radio frame from the UE starts at $(N_{TA,SL} + N_{TA\ offset})T_s$ seconds before the start of the corresponding timing reference frame at the UE (see Fig. 7.22). The UE is not required to receive sidelink or downlink transmissions earlier than $624T_s$ following the end of a sidelink transmission. The parameter $N_{TA,SL}$ differs between channels and signals where $N_{TA,SL} = N_{TA}$ for PSSCH in sidelink transmission mode 1 and 0 otherwise [3].

7.4.1 Sidelink Physical Resources and Resource Pool

In LTE-based sidelink communication, physical layer transmissions are organized in the form of radio frames with duration $T_{frame} = 10$ ms, each consisting of 20 slots of duration $T_{slot} = 0.5$ ms. A sidelink subframe consists of two consecutive slots. A physical channel or signal is transmitted in a slot and is described by a resource grid of $N_{RB}^{SL}N_{sc}^{RB}$ subcarriers and N_{symbol}^{SL} SC-FDMA symbols. The sidelink bandwidth is equal to the uplink bandwidth $N_{RB}^{SL} = N_{RB}^{UL}$ if the cell-selection criterion is satisfied for a serving cell having the same uplink carrier frequency as the sidelink; otherwise, a preconfigured value is used. The sidelink cyclic prefix is configured independently for type 1 discovery, type 2B discovery, sidelink transmission mode 1, sidelink transmission mode 2, control signaling, and PSBCH as well as synchronization signals. The configuration is done per resource pool for discovery, sidelink transmission mode 2, and control signaling. The PSBCH and synchronization signals always use the same cyclic prefix value. Normal cyclic prefix is only supported for PSSCH, PSCCH, PSBCH, and synchronization signals for a sidelink configured with transmission mode 3 or 4 [3].

Each resource element in the resource grid is uniquely defined by the index pair $\left\{(k,l)|k=0,\ldots,N_{RB}^{SL}N_{sc}^{RB}-1;\ l=0,\ldots,N_{symbol}^{SL}-1\right\}$ in a slot where the first and the second indices represent the frequency and time, respectively. The resource elements that are not used for transmission of a physical channel or a physical signal in a slot are set to zero.

A physical resource block (PRB) is defined as $N_{symbol}^{SL} = 7$ or 6 (extended cyclic prefix) consecutive SC-FDMA symbols in the time domain and $N_{sc}^{RB} = 12$ consecutive subcarriers in the frequency domain. A PRB in the sidelink consists of $N_{symbol}^{SL} \times N_{sc}^{RB}$ resource elements, corresponding to one slot in the time domain and 180 kHz in the frequency domain [3]. The relationship between the PRB number n_{PRB} in the frequency domain and resource elements (k, l) in a slot is given by $n_{\text{PRB}} = \lfloor k/12 \rfloor$.

A key concept in LTE-based sidelink communication is the resource pool, which defines a subset of available subframes and RBs for either sidelink transmission or reception. Sidelink communication is a half-duplex scheme and a UE can be configured with multiple transmit resource pools and multiple receive resource pools. The resource pools are configured semi-statically by RRC signaling. When data is sent using a resource pool, the actual transmission resources are selected dynamically from within the pool using one of the two following modes:

- Transmission mode 1, where the serving eNB identifies the resources via downlink control information (DCI) format 5 that has to be sent to the transmitting UE. This mode requires the UE to be fully connected to the network; that is, in RRC_CONNECTED state.
- Transmission mode 2, where the transmitting UE self-selects the resources according to certain rules aimed at minimizing the risk of collision. This mode can be used when the UE is in connected state, idle state, or out of network coverage.

There are two types of resource pools: (1) reception resource pools and (2) transmission resource pools. These are either signaled by the eNB for the in-coverage cases, or preconfigured for the out-of-coverage scenarios. Each transmission resource pool has an associated reception resource pool in order to enable bidirectional communication. However, within a cell, there may be more reception resource pools than transmission resource pools, allowing reception from the UEs in the neighboring cells or from the UEs that are out-of-coverage. Fig. 7.23 illustrates the LTE V2X resource pool structure.

A sidelink direct communication resource pool is configured semi-statically using layer-3 signaling. The physical resources (subframes and RBs) associated with the pool are partitioned into a sequence of repeating hyper frames known as PSCCH periods or alternatively referred to as the scheduling assignment period or sidelink control period. Within a PSCCH period, there are separate subframe pools and RB pools for control (PSCCH) and data (PSSCH). The PSCCH subframes always precede those for PSSCH transmission. This is analogous to the symbol layout of the physical downlink control channel and physical downlink shared channel OFDM symbols within a single downlink subframe, where the control region precedes the data subchannel. The PSCCH carries SCI messages, which describe the dynamic transmission properties of the PSSCH that follows. The receiving UE

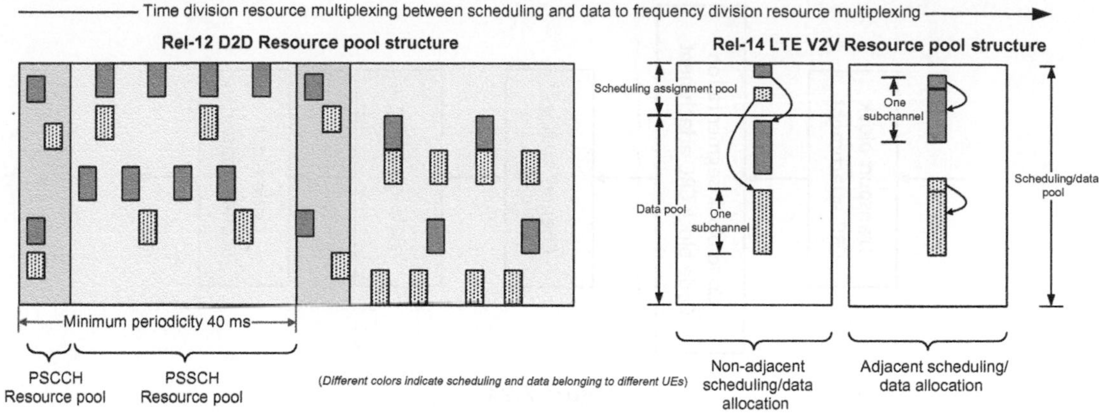

Figure 7.23
LTE V2X resource pool structure [13,14].

searches all configured PSCCH resource pools for SCI transmissions of interest. A UE can be a member of more than one sidelink communications group.

For PSSCH, the number of current slot in the subframe pool $n_{ss}^{PSSCH} = 2n_{ssf}^{PSSCH} + i$, where $i \in \{0, 1\}$ is the number of current slot within the current sidelink subframe $n_{ssf}^{PSSCH} = j\text{mod}10$, where j is equal to the subscript of l_j^{PSSCH} for sidelink transmission modes 1 and 2, respectively; and $i \in \{0, 1\}$ is the number of the current slot within the current sidelink subframe $n_{ssf}^{PSSCH} = k\text{mod}10$ in which k is equal to the subscript of t_k^{SL} for sidelink transmission modes 3 and 4. The last SC-FDMA symbol in a sidelink subframe is used as a guard period and is not used for sidelink transmission [3,5].

7.4.2 Sidelink Physical Channels

7.4.2.1 Physical Sidelink Shared Channel

In LTE-based sidelink communication, the processing of the sidelink shared channel (SL-SCH) follows the procedures for LTE downlink shared channel processing with some differences as follows: (1) data arrives at the channel coding unit in the form of a maximum of one TB in every transmission time interval; (2) in the step of code block concatenation, the sequence of coded bits corresponding to one TB after code block concatenation is referred to as one codeword; and (3) physical uplink shared channel (PUSCH) interleaving is applied without any control information in order to apply a time-first rather than frequency-first mapping, where $C_{mux} = 2\left(N_{symb}^{SL} - 1\right)$. For SL-SCH configured by higher layers for V2X sidelink, $C_{mux} = 2\left(N_{symb}^{SL} - 2\right)$ is used [3,5] (Fig. 7.24).

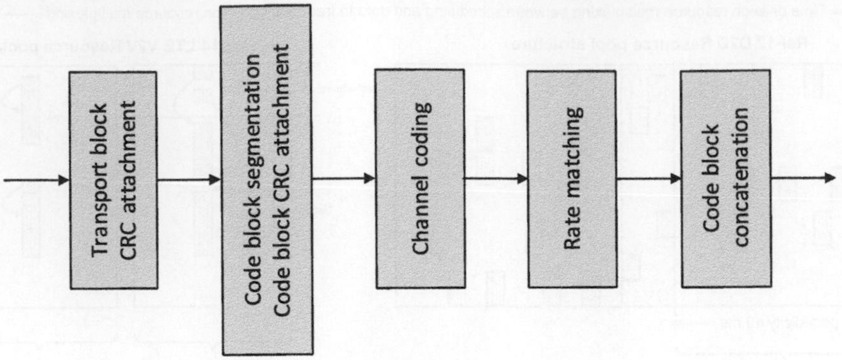

Figure 7.24
Transport block processing for PSSCH [4].

The PSSCH processing begins with the scrambling of the block of bits $b(0),\ldots,b(N_{bit}-1)$, where N_{bit} is the number of bits transmitted on the PSSCH in one subframe. The scrambling sequence generator is initialized with $c_{init}=n_{ID}^{X}2^{14}+n_{ssf}^{PSSCH}2^{9}+510$ at the start of each PSSCH subframe. For sidelink transmission modes 1 and 2, $n_{ID}^{X}=n_{ID}^{SA}$ is the destination identity obtained from the sidelink control channel, and for sidelink transmission modes 3 and 4 $n_{ID}^{X}=\sum_{i=0}^{L-1}p_i 2^{L-1-i}$ with p and L equal to the decimal representation of CRC on the PSCCH transmitted in the same subframe as the PSSCH [3]. The output of the scrambling function is modulated using QPSK or 16QAM modulation and layer-mapped assuming a single antenna port [3]. Transform precoding is then performed with M_{RB}^{PSSCH} and M_{sc}^{PSSCH} parameters followed by precoding for single antenna port transmission.

The block of complex-valued modulated symbols $z(0),\ldots,z(M_{symb}^{p}-1)$ are power-adjusted using the scaling factor β_{PSSCH} and sequentially mapped to PRBs on antenna port p starting with $z(0)$ allocated for PSSCH transmission. The resource elements (k,l) used for the latter frequency-first mapping should not be designated for transmission of reference signals, starting with the first slot in the subframe. Resource elements in the last SC-FDMA symbol within a subframe are counted in the mapping process but not used for transmission. If sidelink frequency hopping is disabled, the set of PRBs used for transmission is given by $n_{PRB}=n'_{VRB}$, where n'_{VRB} is given in [3]. If sidelink frequency hopping with predefined hopping pattern is enabled, the set of PRBs used for transmission is given by the SCI associated with a predefined pattern such that only inter-subframe hopping is used. The number of subbands $N_{sb}\in\{1,2,4\}$ is configured by RRC signaling. The parameters $N_{RB}^{HO}\in\{0,\ldots,110\}$ is given by higher layers; $n_s=n_{ss}^{PSSCH}$ where n_{ss}^{PSSCH} is given in [3]; and $CURRENT\_TX\_NB=n_{ssf}^{PSSCH}$. The pseudo-random sequence generator is initialized at the start of each slot with $n_{ss}^{PSSCH}=0$ and the initialization value $c_{init}\in\{0,1,\ldots,503,510\}$ is determined by *hoppingParameter-r12* RRC parameter; the quantity n_{VRB} is replaced by n'_{VRB}; and for sidelink transmission mode 1, $N_{RB}^{UL}=N_{RB}^{SL}$, for sidelink transmission mode 2, $N_{RB}^{UL}=M_{RB}^{PSSCH\_RP}$ where $M_{RB}^{PSSCH\_RP}$ in given in [5]; the quantity n_{PRB} is replaced by n'_{PRB}; and the physical RB to use for transmission $n_{PRB} = m^{PSSCH}_{n'_{PRB}}$ with m^{PSSCH}_{j} given in [3].

7.4.2.2 Physical Sidelink Control Channel

As we mentioned earlier, the SCI messages are transmitted over PSCCH in LTE-based sidelink communication. The fields defined in the SCI formats below are mapped to the information bits a_0 to a_{A-1} such that each field is mapped in the order in which it appears in the description, with the first field mapped to the lowest order information bit a_0 and each successive field mapped to higher order information bits.

There are two SCI formats for transmission of the SCI content as follows [4]:

- *SCI format 0* carries 1-bit frequency hopping flag; RB assignment and hopping resource allocation $\lceil \log_2 [N^{SL}_{RB}(N^{SL}_{RB}+1)/2] \rceil$ bits where for PSSCH hopping N_{SL_hop} most significant bits (MSBs) are used to obtain the value of $\tilde{n}_{PRB}(i)$ and $(\lceil \log_2(N^{SL}_{RB}(N^{SL}_{RB}+1)/2) \rceil - N_{SL\_hop})$ bits provide the resource allocation in the subframe. For non-hopping PSSCH, $(\lceil \log_2(N^{SL}_{RB}(N^{SL}_{RB}+1)/2) \rceil)$ bits provide the resource allocation in the subframe. The SCI format 0 further includes 7-bit time resource pattern; 5-bit modulation and coding; 11-bit timing advance indication; and 8-bit group destination ID.
- *SCI format 1* is used for the scheduling of PSSCH and carries 3-bit priority; 4-bit resource reservation; frequency resource location of initial transmission and retransmission $\lceil \log_2(N^{SL}_{subchannel}(N^{SL}_{subchannel}+1)/2) \rceil$ bits; 4-bit time gap between initial transmission and retransmission; 5-bit modulation and coding; 1-bit retransmission index; and a number of reserved bits to adjust the size of SCI format 1 to 32 bits. The reserved bits are set to zero.

The block of bits $b(0), \ldots, b(N_{bit}-1)$, where N_{bit} is the number of bits transmitted on the physical sidelink control channel in one subframe are scrambled with the scrambling sequence generator initialized to $c_{init} = 510$ at the start of each PSCCH subframe and then QPSK-modulated. Layer mapping with single antenna port and transform precoding are performed on complex-valued modulated symbols. Transform precoding is performed similar to LTE uplink with M^{PUSCH}_{RB} and M^{PUSCH}_{sc} parameters replaced by M^{PSCCH}_{RB} and M^{PSCCH}_{sc} values, respectively. For transmission on a single antenna port, precoding is defined by $z^{(0)}(i) = y^{(0)}(i)$, where $i = 0, 1, \ldots, M^{layer}_{symb} - 1$ and $y^{(0)}(i)$ denotes the transform-precoded complex-valued symbols.

The block of complex-valued symbols $z(0), \ldots, z(M^{p}_{symb}-1)$ are multiplied by β_{PSCCH} in order to adjust the transmit power and are sequentially mapped, starting with $z(0)$, to the PRBs on antenna port p that have been assigned for transmission of PSCCH. The latter frequency-first resource mapping must avoid resources that are designated to transmission of reference signals, starting with the first slot in the subframe. Resource elements in the last SC-FDMA symbol within a subframe are considered in the mapping process but are not used for transmission.

The radio resources for direct communication can be selected by the device autonomously or will be scheduled by the network. In case the device has acquired SIB18 and has further a passive connection with the network in RRC_IDLE, the device would select radio resource from the broadcast resource pool in SIB18. Similar to direct discovery case, a UE would have to transit to the RRC_CONNECTED state when no valid (transmit) resource pool are provided by SIB18. In this case, a ProSe UE information indication is sent by the terminal to the network, indicating the intent to use the direct communication capability. In response, the network will assign a sidelink radio network temporary identifier (SL-RNTI) to the device. The network then uses the SL-RNTI and the downlink control channel to assign a transmission grant to the device with the new defined DCI format 5. The DCI format 5 is used for scheduling of PSCCH and contains some SCI format 0 fields that are used for scheduling of PSSCH. The DCI format 5 information fields include 6-bit resource indication for PSCCH; and 1-bit transmit power control command for PSCCH and PSSCH, as well as SCI format 0 fields including frequency hopping flag; RB assignment and hopping resource allocation; and time resource pattern [5]. Similarly, the DCI format 5A is used for scheduling of PSCCH, containing some SCI format 1 fields to schedule PSSCH. The DCI format 5A information fields include 3-bit carrier indicator; the lowest index of the subchannel allocation to the initial transmission $\lceil \log_2(N^{SL}_{subchannel}) \rceil$ bits as well as SCI format 1 information fields including frequency resource location of initial transmission and retransmission; time gap between initial transmission and retransmission; and 2-bit sidelink index [this field is present only for TDD uplink/downlink configuration 0–6]. When DCI format 5A CRC is scrambled with sidelink semi-persistent scheduling V-RNTI (SL-SPS-V-RNTI[8]), it would further include 3-bit sidelink SPS configuration index and 1-bit activation/release indication [4].

For sidelink transmission mode 1, if a UE is configured via RRC signaling to receive DCI format 5 with the CRC scrambled by the SL-RNTI, the UE is required to decode PDCCH/ePDCCH according to the combination defined in Table 7.4. For sidelink transmission mode 3, if a UE is configured by higher layers to receive DCI format 5A with the CRC scrambled by the SL-V-RNTI or SL-SPS-V-RNTI, the UE must decode the PDCCH/ePDCCH according to the combination defined in Table 7.4, and it is not expected to receive DCI format 5A with size larger than DCI format 0 in the same search space that DCI format 0 is defined.

The PSCCH carrying SCI format 0 is transmitted in two subframes within the configured resource pool occupying only one RB pair. The 7-bit time resource pattern determines

[8] Semi-persistently scheduled sidelink transmission for V2X sidelink communication, which is used for activation, reactivation, deactivation, and retransmission.

Table 7.4: PDCCH/ePDCCH configured by various RNTIs [5].

DCI Format	Search Space
DCI format 5	For PDCCH: Common and UE-specific by C-RNTI For ePDCCH: UE-specific by C-RNTI
DCI format 5A	For PDCCH: Common and UE-specific by C-RNTI For ePDCCH: UE-specific by C-RNTI

which subframes are used for transmission of PSSCH. A subframe indicator bitmap of variable length is defined, where the length of this bitmap depends on the duplex mode; i.e., FDD or TDD, and in case of TDD which UL/DL configuration is used. In case of FDD, the bitmap is 8 bits long. Up to 128 different time resource patterns define how these 8 bits are used. The RB allocation for PSSCH follows the same principles defined for LTE Rel-8 while interpreting the RB assignment and hopping allocation information provided by SCI format 0. The information is transmitted four times. For mode 2, the device would autonomously select resources from the transmission resource pool provided in SIB18. If a device is out-of-coverage, it can only autonomously select resources from a preconfigured resource pool.

Transmissions in PSSCH follow a time resource pattern, which is a subframe indication bitmap with fixed length N_{TRP} (e.g., eight subframes) repeated over the length of PSSCH, to identify which subframes are used by a transmitting UE. Each time resource pattern is identified by an index I_{TRP} corresponding to the predefined subframe indication bitmap. In order to mitigate the throughput degradation due to inter-cell/inter-user interference, each transmission on PSSCH is performed with four HARQ processes without feedback. Thus, each TB transmission on PSSCH requires four subframes to be carried. In other words, in the case of PSSCH, different parameters are used to specify the time and frequency resources. This differs from PSCCH, which signals the subframes and PRB to be used by a single value.

The subframes associated with PSSCH transmission are indicated by the time resource pattern index I_{TRP}. This index is used to look up a bitmap from a set of tables, with the choice of table depending on the duplexing mode. The selected bitmap is denoted by $\left(b'_0, b'_1, \ldots, b'_{N_{TRP}-1}\right)$ where N_{TRP} is 6, 7, or 8 depending on the table. This bitmap is repeated to form an extended bitmap $\left(b_0, b_1, \ldots, b_{L_{PSSCH}-1}\right)$ where $b_j = b'_{j \bmod N_{TRP}}$ covers the entire PSSCH subframe pool. The subframes used for PSSCH transmission are selected by l_j^{PSSCH} values in this extended bitmap to obtain the final subframe set denoted by $\left(n_0^{PSSCH}, n_1^{PSSCH}, \ldots, n_{N_{PSSCH}-1}^{PSSCH}\right)$, where N_{PSSCH} value is a multiple of 4, and denotes the number of subframes that can be used for PSSCH transmission in the PSCCH period. This is consistent with the fact that each TB transmitted within this interval will be sent four times using the fixed HARQ redundancy version sequence (0,2,3,1) [5].

7.4.2.3 Physical Sidelink Discovery Channel

In LTE-based sidelink communication, the processing of the SL-DCH follows the downlink shared channel with the following differences: (1) data arrives at the channel coding unit in the form one TB per each transmission time interval; (2) in the step of code block concatenation, the sequence of coded bits corresponding to one TB after code block concatenation is referred to as one codeword; and (3) PUSCH interleaving is applied without any control information in order to apply a time-first rather than frequency-first mapping such that $C_{mux} = 2\left(N_{symb}^{SL} - 1\right)$. The block of bits $b(0), \ldots, b(N_{bit} - 1)$, where N_{bit} is the number of bits transmitted on the PSDCH in one subframe, are scrambled with the scrambling sequence generator initialized with $c_{init} = 510$ at the start of each PSDCH subframe and then QPSK-modulated. The layer mapping, transform precoding, precoding, and the mapping to the physical resources are similar to PSCCH processing described earlier.

7.4.2.4 Physical Sidelink Broadcast Channel

Fig. 7.25 shows the processing stages of SL-BCH transport channel in LTE-based sidelink communication. The broadcast channel data in a TB is processed through channel coding module, which includes CRC attachment to the TB, channel coding, and rate matching. Since the latter processing is in the uplink direction of LTE, following the rate matching, LTE PUSCH interleaving is used without multiplexing with control information. A time-first rather than frequency-first mapping is applied, where $C_{mux} = 2\left(N_{symbol}^{SL} - 3\right)$. For SL-BCH configured by higher layers for V2X sidelink, $C_{mux} = 2\left(N_{symbol}^{SL} - 2\right) - 3$ is used. The entire TB containing the sidelink broadcast channel is used to calculate the 16-bit CRC. Information bits inclusive of the attached 16-bit CRC are encoded using tail biting convolutional code (i.e., a TBCC with constraint length 7 and coding rate 1/3) and rate matched.

The block of bits $b(0), \ldots, b(N_{bit} - 1)$, where N_{bit} is the number of bits transmitted on the PSBCH in one subframe, are scrambled so that the scrambling sequence generator is initialized at the start of every PSBCH subframe with $c_{init} = N_{ID}^{SL}$ and then QPSK-modulated. Layer mapping with single antenna port and transform precoding are performed on

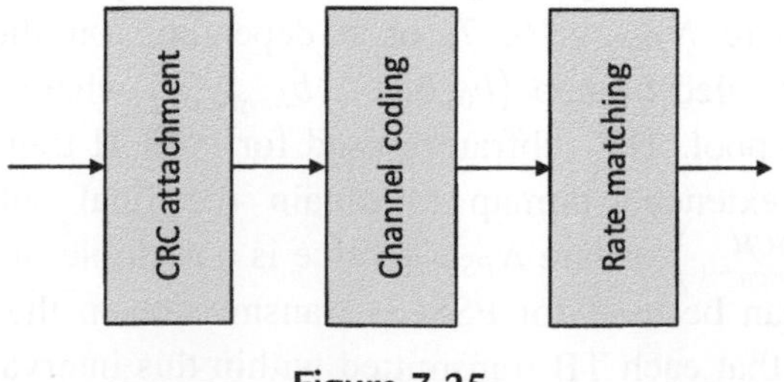

Figure 7.25
Physical sidelink broadcast channel processing [4].

complex-valued modulated symbols. Transform precoding is performed similar to LTE uplink with parameters M_{RB}^{PSBCH} and M_{sc}^{PSBCH}. For transmission on a single antenna port, precoding is defined by $z^{(0)}(i) = y^{(0)}(i)$, where $i = 0, 1, \ldots, M_{symb}^{layer} - 1$ and $y^{(0)}(i)$ denotes the transform-precoded complex-valued symbols. The block of complex-valued symbols $z(0), \ldots, z(M_{symb}^{p} - 1)$ are multiplied by an amplitude scaling factor β_{PSBCH} in order to adjust the transmit power, and sequentially mapped to PRBs on antenna port p. The PSBCH utilizes the same set of RBs as the synchronization signal. The frequency-first mapping to the PRBs would avoid resources designated for transmission of reference signals or synchronization signals, starting with the first slot in the subframe. The resource element index k is determined by $k = k' - 36 + N_{RB}^{SL} N_{sc}^{RB}/2, \quad k' = 0, 1, \ldots, 71$ since the last symbol of the subframe is used as a gap and not used for transmission.

7.4.3 Sidelink Physical Signals

7.4.3.1 Demodulation Reference Signals

In LTE V2X, the DM-RSs associated with PSSCH, PSCCH, PSDCH, and PSBCH are transmitted similar to LTE PUSCH with different parameters and antenna ports. The set of physical RBs used in the mapping process are identical to the corresponding PSSCH/PSCCH/PSDCH/PSBCH transmission. As shown in Fig. 7.26, for sidelink transmission modes 3 and 4 on PSSCH and PSCCH, the mapping uses symbols $l = 2, 5$ for the first slot in the

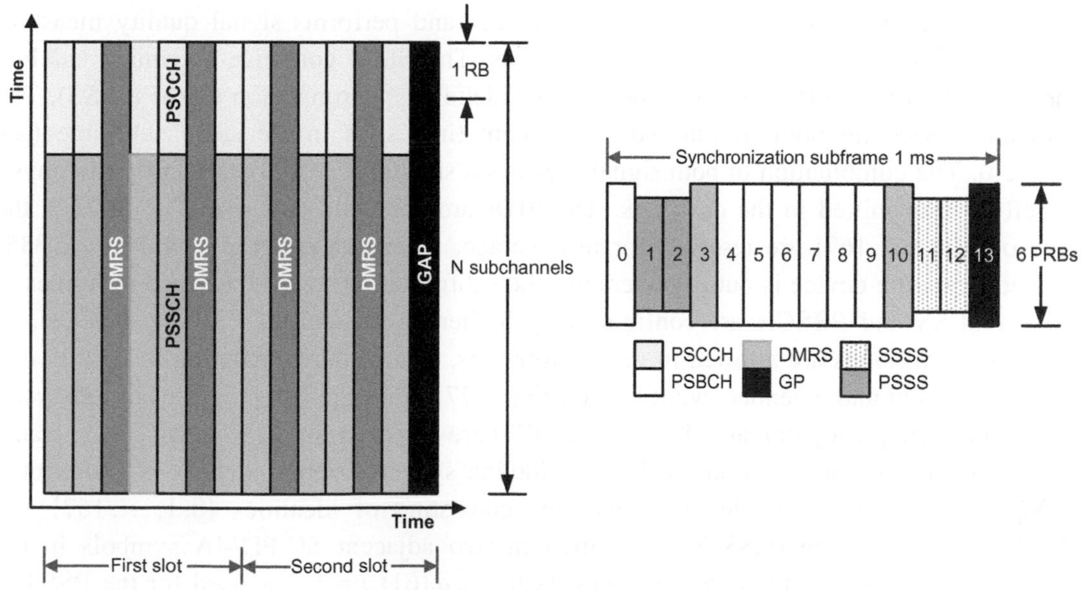

Figure 7.26
C-V2X frame structure and location of various physical channels and signals [3].

subframe and symbols $l = 1, 4$ for the second slot in the subframe. For sidelink transmission modes 3 and 4 on PSBCH, the mapping uses symbols $l = 4, 6$ for the first slot in the subframe and symbol $l = 2$ for the second slot in the subframe. For sidelink transmission modes 1 and 2, the pseudo-random sequence generator used for scrambling is initialized at the start of each slot where $n_{ss}^{PSSCH} = 0$, whereas for sidelink transmission modes 3 and 4, the pseudo-random sequence generator is initialized at the start of each slot where $n_{ss}^{PSSCH} \mathrm{mod} 2 = 0$. For sidelink transmission modes 3 and 4, the quantity n_{ID}^{X} is equal to the decimal representation of the CRC on PSCCH transmitted in the same subframe as PSSCH according to $n_{ID}^{X} = \sum_{i=0}^{L-1} p_i 2^{L-1-i}$ with parameters p and L defined in [3].

7.4.3.2 Synchronization Signals

Time and frequency synchronization are important aspects of cellular communication for interference management and mitigation, which are further extended to V2X communication. A V2X-enabled device first needs to determine if it is in coverage of the network. A device is defined to be in-coverage based on signal quality measurement using the RSRP measurement performed on the downlink synchronization signals. When the measured RSRP values are above a specific threshold, the device considers itself in coverage and uses the base station downlink synchronization signals for timing and frequency alignment. This threshold is defined as part of broadcast system information. If the received signal quality measurement falls below the threshold, the device would start transmission of sidelink synchronization signals (SLSS) and the PSBCH. These signals have a periodicity of 40 ms. Assuming the device cannot detect an eNB, due to possibly being out-of-coverage, the device starts looking for SLSS from other devices and performs signal quality measurements (S-RSRP) on those synchronization signals. The SLSS comprise a primary sidelink synchronization signal (PSSS) and a secondary sidelink synchronization signal (SSSS). The PSSS and SSSS are both transmitted in adjacent time slots in the same subframe (see Fig. 7.26). The combination of both signals defines a sidelink ID (SID), similar to the physical cell ID transmitted in the downlink. The SIDs are split into two sets. The SIDs in the range of $\{0, 1, \ldots, 167\}$ are reserved for in-coverage, whereas the SIDs $\{168, 169, \ldots, 335\}$ are used when the device is out-of-coverage. The subframes to be used as radio resources to transmit SLSS and PBSCH are configured by higher layers and no PSDCH, PSCCH, or PSSCH transmissions are allowed in these subframes. The resource mapping is slightly different for normal and extended cyclic prefix. Fig. 7.27 shows the mapping for normal cyclic prefix. In the frequency domain, the inner six RBs are reserved for SLSS and PSBCH transmission. More specifically, a physical layer sidelink synchronization identity is represented by $N_{ID}^{SL} \in \{0, 1, \ldots, 335\}$, divided into two sets consisting of identities $\{0, 1, \ldots, 167\}$ and $\{168, 169, \ldots, 335\}$. The PSSS is transmitted in two adjacent SC-FDMA symbols in the same subframe. Each of the two sequences $d_i(0), \ldots, d_i(61), i = 1, 2$ is used for the PSSS in the two SC-FDMA symbols with root index $u = 26$ if $N_{ID}^{SL} \leq 167$ and $u = 37$, otherwise. The sequence $d(n)$ used for the primary synchronization signal is derived from a frequency

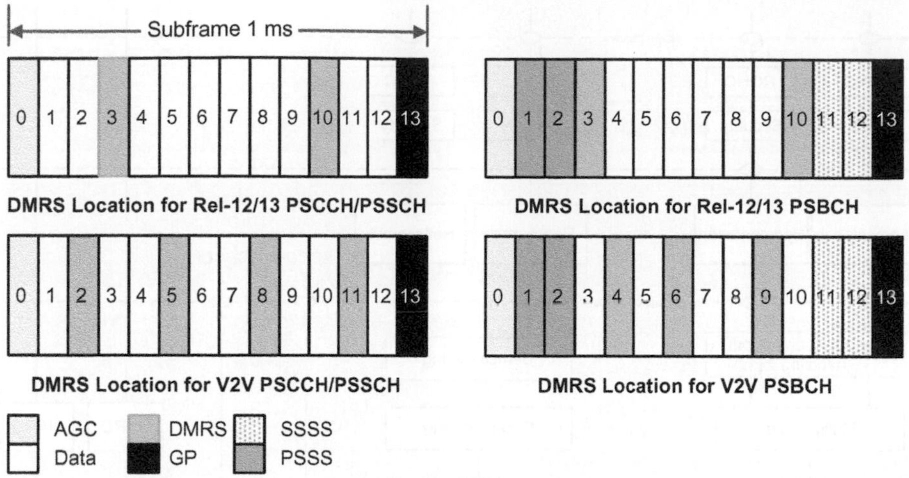

Figure 7.27
LTE V2X sidelink frame structure.

domain Zadoff-Chu sequence according to $d_u(n) = \exp\left[-j\pi un(n+1)/63\right] \forall n = 0, 1, \ldots, 30$ and $d_u(n) = \exp\left[-j\pi u(n+1)(n+2)/63\right] \forall n = 31, 32, \ldots, 61$, where u denotes the Zadoff-Chu root sequence index [3]. The sequence $d_i(n)$ is multiplied with an amplitude scaling factor of $\sqrt{72/62}\beta_{PSBCH}$ and mapped to resource elements on a single antenna port in the first slot of the subframe according to $a_{k,l} = d_i(n) \forall n = 0, \ldots, 61; k = n - 31 + N_{RB}^{SL} N_{sc}^{RB}/2$ with $l = 1, 2$ (normal cyclic prefix) and $l = 0, 1$ (extended cyclic prefix).

The SSSS is transmitted in two adjacent SC-FDMA symbols in the same subframe. Each of the two sequences $d_i(0), \ldots, d_i(61) \forall i = 1, 2$ is used for the SSSS. The sequence $d(0), \ldots, d(61)$ used for the second synchronization signal is an interleaved concatenation of two length-31 binary sequences. The concatenated sequence is scrambled with a scrambling sequence given by the primary synchronization signal assuming subframe 0 with $N_{ID}^{(1)} = N_{ID}^{SL} \bmod 168$ and $N_{ID}^{(2)} = \lfloor N_{ID}^{SL}/168 \rfloor$ for transmission modes 1 and 2, and subframe 5 for transmission modes 3 and 4 [3]. The sequence $d_i(n)$ is multiplied with the amplitude scaling factor β_{SSSS} in order to adjust the transmit power and mapped to resource elements on a single antenna port in the second slot in the subframe according to $a_{k,l} = d_i(n) \forall n = 0, \ldots, 61; k = n - 31 + N_{RB}^{SL} N_{sc}^{RB}/2$ with $l = 4, 5$ (normal cyclic prefix) and $l = 3, 4$ (extended cyclic prefix) [3].

7.5 Layer 2/3 Aspects

The LTE layer-2 functions are divided into three sublayers: MAC, RLC, and PDCP. Fig. 7.28 depicts the layer-2 structure for the LTE-based sidelink. In this figure, the service access points (SAP) for peer-to-peer communication are marked with circles at the interface

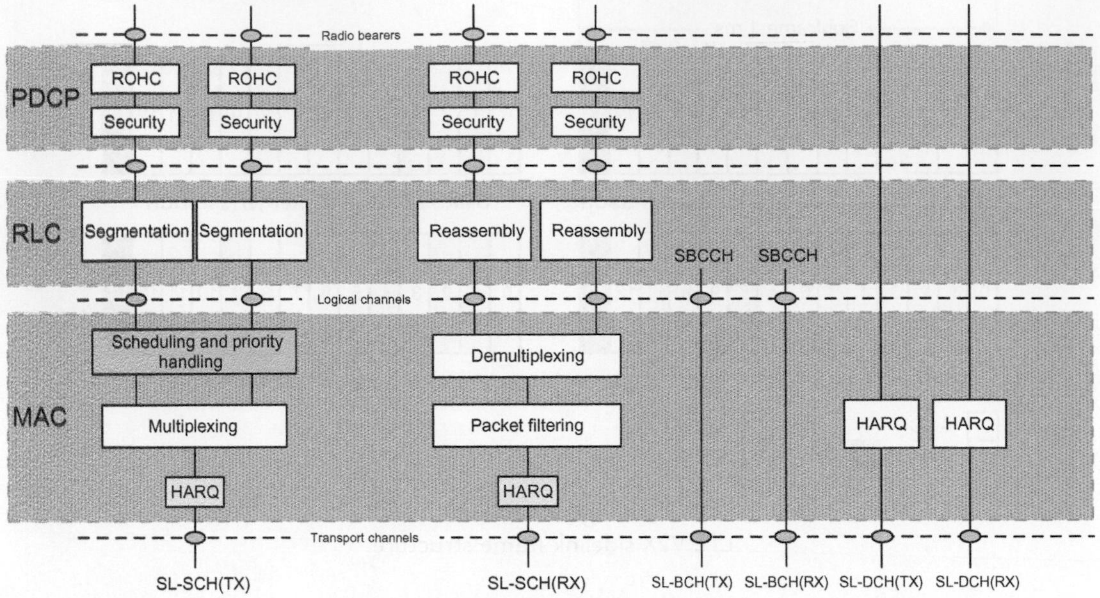

Figure 7.28
Layer-2 structure for sidelink [6].

between sublayers. The SAP between the physical layer and the MAC sublayer provides the transport channels. The SAPs between the MAC sublayer and the RLC sublayer provide the logical channels. The multiplexing of several logical channels (i.e., radio bearers) on the same transport channel is performed by the MAC sublayer. In sidelink communications, only one TB is generated per TTI. The sidelink specific services and functions of the MAC sublayer include radio resource selection and packet filtering for sidelink communication and V2X sidelink communication [6].

MAC sublayer provides different types of services that are represented by logical channels, where each logical channel is defined by the type of information it conveys. The logical channels are generally classified into two groups: (1) control channels (for the transfer of control-plane information) and (2) traffic channels (for the transfer of user-plane information), as shown in Fig. 7.29. The SBCCH is a sidelink logical channel for broadcasting sidelink system information from one UE to another UE(s). STCH is a point-to-multipoint channel, for transfer of user information from one UE to another UE(s). This channel is used only by sidelink communication-capable UEs and V2X sidelink communication-capable UEs. Point-to-point communication between two sidelink communication-capable UEs is also realized with an STCH.

As shown in Fig. 7.29, STCH logical channel can be mapped to SL-SCH transport channel and SBCCH logical channel can be mapped to SL-BCH transport channel. Sidelink

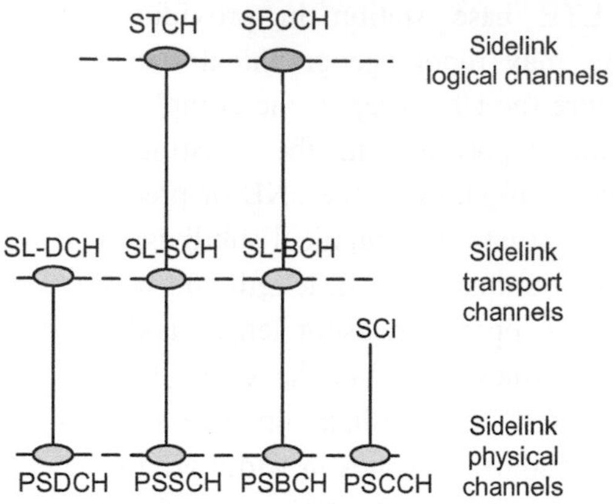

Figure 7.29
Mapping between sidelink logical, transport, and sidelink physical channels [6].

transport channels include SL-BCH, which is characterized by a predefined transport format; SL-DCH that is characterized by a fixed size, predefined format, and periodic broadcast transmission, as well as support for both UE autonomous resource selection and scheduled resource allocation by the eNB. It is subject to collision risk due to support of UE autonomous resource selection; however, no collision is expected when UE is allocated dedicated resources by the eNB. It further supports HARQ combining, but there is no support for HARQ feedback. The sidelink transport channels further include SL-SCH, which has similar characteristics as the SL-DCH, as well as support for dynamic link adaptation by varying the transmit power, modulation, and coding [6]. In the sidelink, no HARQ feedback is used and the retransmissions are always performed in a predefined/configured number. Furthermore, the measurement gaps and sidelink discovery transmission during a sidelink discovery gap for transmission are of higher priority than HARQ retransmissions; that is, whenever a HARQ retransmission collides with a measurement gap or sidelink discovery transmission during a sidelink discovery gap for transmission, the HARQ retransmission does not happen.

One of the functions of the RRC sublayer is broadcasting the system information. Both LTE RRC states, RRC_IDLE and RRC_CONNECTED, support sidelink transmission and reception; sidelink discovery announcement and monitoring; and V2X sidelink transmission and reception. The *SystemInformationBlockType18* contains information related to sidelink communication; *SystemInformationBlockType19* contains information related to sidelink discovery; and *SystemInformationBlockType21* contains information related to V2X sidelink communication. The LTE RAN uses SL-RNTI and SL-V-RNTI to identify sidelink communication scheduling and V2X sidelink communication scheduling, respectively [6].

In order to assist an LTE base station in providing sidelink resources, a UE in RRC_CONNECTED state may report geographical location information to the serving eNB. The eNB can configure the UE to report the complete UE geographical location information based on periodic reporting via the existing measurement report signaling. Geographical zones can be configured by the eNB or preconfigured. When zones are (pre-)configured, the area is divided into geographical subdivisions using a single fixed reference point; that is, geographical coordinates (0,0), length and width. The UE determines the zone identity by means of modulo operation using length and width of each zone, number of zones in length, number of zones in width, the single fixed reference point, and the geographical coordinates of the UE's current location. The length and width of each zone, number of zones in length, and number of zones in width are provided by the eNB when the UE is in-coverage, and preconfigured when the UE is out-of-coverage. The zone is configurable for both in-coverage and out-of-coverage. In an in-coverage scenario, when the UE uses autonomous resource selection, the eNB can provide the mapping between zone(s) and V2X sidelink transmission resource pools via RRC signaling. For out-of-coverage UEs, the mapping between the zone(s) and V2X sidelink transmission resource pools can be preconfigured. If the mapping between zone(s) and V2X sidelink transmission resource pool is (pre-)configured, then the UE selects transmission sidelink resources from the resource pool corresponding to the zone where it is presently located. The zone concept is not applied to exceptional V2X sidelink transmission pools as well as reception pools. Resource pools for V2X sidelink communication are not configured based on priority.

For V2X sidelink transmission, during handover, transmission resource pool configurations including exceptional transmission resource pool for the target cell can be signaled in the handover command to minimize the transmission interruption. In this way, the UE may use the V2X sidelink transmission resource pools of the target cell before the handover is completed, if either synchronization is performed with the target cell in case eNB is configured as synchronization source or synchronization is performed with GNSS in case GNSS is configured as synchronization source. If the exceptional transmission resource pool is included in the handover command, the UE uses randomly selected resources from the exceptional transmission resource pool, starting from the reception of handover command. If the UE is configured with scheduled resource allocation in the handover command, then it continues to use the exceptional transmission resource pool while the timer associated with the handover is running. If the UE is configured with autonomous resource selection in the target cell, it continues to use the exceptional transmission resource pool until the sensing results on the transmission resource pools for autonomous resource selection are available. For exceptional cases (e.g., during radio link failure, during transition from RRC_IDLE to RRC_CONNECTED, or during change of dedicated V2X sidelink resource pools within a cell), the UE may select and temporarily use resources in the exceptional pool provided in serving cell's SIB21 message or in dedicated signaling based on random selection. During cell reselection, the RRC_IDLE UE may use the randomly selected resources from the exceptional transmission resource pool of the reselected cell until the sensing results on the transmission resource pools for autonomous resource selection are available [6].

To avoid interruption in receiving V2X messages due to delay in acquiring reception resource pools broadcast from the target cell, synchronization configuration and reception resource pool configuration for the target cell can be signaled to RRC_CONNECTED UEs in the handover command. For RRC_IDLE UE, it is up to UE implementation to minimize V2X sidelink transmission/reception interruption time associated with acquisition of SIB21 message of the target cell.

A UE is considered in-coverage on the carrier used for V2X sidelink communication, whenever it detects a cell on that carrier. If the UE that is authorized for V2X sidelink communication is in-coverage on the frequency used for V2X sidelink communication or if the eNB provides V2X sidelink configuration for that frequency (including the case where the UE is out-of-coverage on that frequency), the UE uses the scheduled resource allocation or UE autonomous resource selection according to eNB configuration. When the UE is out-of-coverage on the frequency used for V2X sidelink communication and if the eNB does not provide V2X sidelink configuration for that frequency, the UE may use a set of transmission and reception resource pools preconfigured in the UE. The V2X sidelink communication resources are not shared with other non-V2X data transmitted over sidelink. An RRC_CONNECTED UE may send a *sidelink UE information* message to the serving cell, when it wishes to establish V2X sidelink communication in order to request sidelink resources [6].

If the UE is configured by upper layers to receive V2X sidelink communication and V2X sidelink reception resource pools are provided, then the UE will receive sidelink communication on the allocated resources. The reception of V2X sidelink communication on different carriers or from different PLMNs can be supported by incorporating multiple receivers in the UE. For sidelink SPS, up to eight SPS configurations with different parameters can be configured by the eNB and all SPS configurations can be simultaneously active. The activation/deactivation of SPS configuration is signaled via PDCCH by the eNB. The existing logical channel prioritization based on PPPP is used for sidelink SPS. The UE can provide supplementary information to the eNB which configures the reporting of such information for V2X sidelink communication. The latter information includes traffic characteristic parameters (e.g., a set of preferred SPS interval, timing offset with respect to subframe 0 of the SFN 0, PPPP and maximum TB size based on the observed traffic pattern) related to the SPS configuration. The UE supplementary information can be reported regardless of whether SPS is configured. Triggering of UE supplementary information transmission is implementation specific. For instance, the UE can report its supplementary information when a change in estimated periodicity and/or timing offset of packet arrival occurs. The scheduling request mask per traffic type is not supported for V2X sidelink communication. The serving cell can provide synchronization configuration for the V2X sidelink carrier. In this case, the UE follows the synchronization configuration received from serving cell. If no cell is detected on the carrier used for V2X sidelink communication and the UE does not receive synchronization configuration from serving cell, the UE follows preconfigured synchronization procedure. There are three possible synchronization nodes; that is, eNB, UE, and GNSS. In case GNSS is configured as synchronization source, the UE utilizes the universal time and the (pre)configured direct frame number (DFN) offset to calculate DFN and subframe number. If the eNB timing is configured as synchronization reference for the UE in order to perform synchronization and conduct downlink measurements, the UE follows the cell associated with the acquired frequency when in-coverage. The UE can indicate the current synchronization reference type to the eNB. One transmission pool for scheduled resource allocation is configured, considering the synchronization reference of the UE.

For controlling channel utilization, the network can indicate how the UE adapts its transmission parameters for each transmission pool depending on the channel busy ratio (CBR). The UE measures all configured transmission resource pools including the exceptional resource pool. If a resource pool is (pre)configured such that a UE always transmits PSCCH and PSSCH in adjacent RBs, then the UE measures PSCCH and PSSCH resources together. If a resource pool is (pre)configured such that a UE may transmit PSCCH and the corresponding PSSCH in non-adjacent RBs in a subframe, then PSSCH resource pool and PSCCH resource pool are measured separately. A UE in RRC_CONNECTED state can be configured to report CBR measurement results. For CBR reporting, periodic reporting and event-triggered reporting are supported. Two reporting events are introduced for event-triggered CBR reporting. In case PSSCH and PSCCH resources are placed non-adjacently, only PSSCH resource pool measurement is used for event-triggered CBR reporting. In case PSSCH and PSCCH resources are placed adjacently, CBR measurements of both PSSCH and PSCCH resources are used for event-triggered CBR reporting. Event-triggered CBR reporting is triggered by overloaded threshold and/or less-loaded threshold. The network can configure which of the transmission pools the UE needs to report [6,8].

A UE (regardless of its RRC state) performs transmission parameter adaptation based on the CBR. If PSSCH and PSCCH resources are placed non-adjacently, only PSSCH pool measurement is used for transmission parameter adaptation. However, if PSSCH and PSCCH resources are placed adjacently, CBR measurements of both PSSCH and PSCCH resources are used for transmission parameter adaptation. When CBR measurements are not available, the default transmission parameters are utilized. The exemplary adapted transmission parameters include maximum transmission power, range of the number of retransmissions per TB, range of PSSCH RB number, range of MCS, and the maximum limit on channel occupancy ratio. The transmission parameter adaption applies to all transmission resource pools including exceptional resource pools [5,6].

In V2X sidelink communication, sidelink transmission and/or reception resources including exceptional resource pool are provided via dedicated signaling, SIB21 message, and/or pre-configuration for different frequencies in both scheduled resource allocation and UE autonomous resource selection scenarios. The serving cell may signal only the frequency on which the UE should acquire the resource configuration for V2X sidelink communication. If multiple frequencies and associated resource information are provided, then it is up to UE to select a frequency among the permissible frequencies. The UE does not use preconfigured transmission resource, if it detects a cell providing resource configuration or inter-carrier resource configuration for V2X sidelink communication. An RRC_IDLE UE may prioritize the frequency that provides cross-carrier resource configuration for V2X sidelink communication over other choices during cell reselection. If the UE supports multiple transmission chains, it may simultaneously transmit on multiple carriers via PC5. In that case, a mapping between V2X service types and the suitable V2X frequencies is configured by upper layers. For scheduled resource allocation, the eNB can schedule a V2X transmission on a frequency based on the sidelink buffer status report, in which the UE includes the destination index uniquely associated with a frequency reported by the UE to the eNB in *sidelink UE information* message [6,8].

The UE may receive the V2X sidelink communication of other PLMNs. The serving cell can directly signal the resource configuration for V2X sidelink communication in an inter-PLMN operation or indirectly via the frequency on which the UE may acquire the inter-PLMN resource configuration. Note that the V2X sidelink communication transmission in other PLMNs is not permissible. When uplink transmission overlaps in time domain with V2X sidelink transmission on the same frequency, the UE prioritizes the latter over the former, if the PPPP of sidelink MAC PDU is lower than a (pre-)configured PPPP threshold; otherwise, the UE prioritizes the uplink transmission over the V2X sidelink transmission. When uplink transmission overlaps in time domain with V2X sidelink transmission in different frequency, the UE may prioritize the V2X sidelink transmission over the uplink transmission or may reduce uplink transmission power, if the PPPP of sidelink MAC PDU is lower than a (pre-)configured PPPP threshold; otherwise, the UE prioritizes the uplink transmission over the V2X sidelink transmission or reduces V2X sidelink transmission power. However, if uplink transmission is prioritized by upper layer or random-access procedure is performed, the UE prioritizes uplink transmission over any V2X sidelink transmission irrespective of the sidelink MAC PDU's PPPP [6].

Resource pool for transmission of a pedestrian UE (P-UE) may be overlapped with resources dedicated for V2X sidelink communication. For each transmission pool, the resource selection mechanism (i.e., random selection or partial sensing-based selection) is configured. If the P-UE is configured to choose either random selection or partial sensing-based selection for a transmission pool, then it is up to the UE to select a specific resource selection mechanism. If the eNB does not specify a random selection pool, the P-UEs that only support random selection cannot perform sidelink transmission. In exceptional resource pool, the P-UE uses random selection. The P-UE can send *sidelink UE information* message to indicate that it requests resource pools for pedestrian-related V2X sidelink transmission. It is not mandatory for P-UE to support zone-based resource selection. The P-UE reports whether it supports zone-based resource selection as part of UE capability signaling. The P-UEs do not perform CBR measurements; however, they may adjust the transmission parameters based on the default transmission parameter configuration, which can be provided to the P-UE via RRC signaling [6].

The LTE V2X messages can be delivered in unicast mode via non-guaranteed bit rate (non-GBR) or GBR bearers. To meet the QoS message delivery requirements for V2X services, a non-GBR QoS class identifier (QCI) value and a GBR QCI value are used. For broadcast V2X messages, single-cell point-to-multipoint or multimedia broadcast single frequency network transmission can be used. The reception of downlink broadcast V2X messages on different carriers/PLMNs can be supported by having multiple receiver chains in the UE. A GBR QCI value is used for the delivery of V2X messages over eMBMS bearers [6].

7.6 LTE/NR V2X Security

3GPP LTE has one of the most advanced security mechanisms among all wireless technologies and NR V2X will continue to maintain very high standards for V2X security. 3GPP has been developing capabilities that will benefit V2X, starting with enhancements in LTE to support V2X use cases. 5G will evidently support more advanced use cases; therefore, two requirements are particularly important: (1) the need for direct, ad hoc, broadcast, secure communication without any a priori configuration of security by the network and (2) management of identities for user privacy from the network or other third parties. There are two types of LTE transport-level security mechanisms: (1) LTE security protecting the UE signaling and communications with the LTE network, and (2) LTE D2D or ProSe communications security. LTE security uses a symmetric keying scheme for data protection between the UE and the network. For user-plane data, only confidentiality (encryption) is applied; there is no integrity protection on application-layer messages exchanged with the network. For ProSe/D2D communications, each group of devices conduct one-to-many, confidentiality protected data transfers. This is done using a group symmetric key, which is provisioned by the network to all member devices. There is no integrity protection on the user data and, because the key is shared by all, there is no way to positively identify which of the group members has sent the data.

V2X sidelink communication uses the following identities [6]:

- *Source layer-2 ID:* Identifies the sender of the data in the sidelink and V2X sidelink communications. The source layer-2 ID is 24 bits long and is used together with destination layer-2 ID and LCID for identification of the RLC UM entity and PDCP entity in the receiver.

- *Destination layer-2 ID:* Identifies the target of the data in the sidelink and V2X sidelink communications. For sidelink communication, the destination layer-2 ID is 24 bits long which is split in the MAC sublayer into two bit strings. The first bit string is the least significant bit part (8 bits) of destination layer-2 ID and is forwarded to physical layer as group destination ID. This identifies the target of the intended data in SCI and is used for filtering of packets at the physical layer. The second bit string is the MSB part (16 bits) of the destination layer-2 ID which is carried within the MAC header. This is used for filtering of packets at the MAC sublayer. In the case of V2X sidelink communication, destination layer-2 ID is not split and is carried within the MAC header.

No access stratum signaling is required for group formation and to configure source layer-2 ID, destination layer-2 ID and group destination ID in the UE. These identities are either provided by higher layer or derived from identities provided by higher layer. In case of groupcast and broadcast, the ProSe UE ID provided by higher layer is used directly as the source layer-2 ID and the ProSe layer-2 group ID provided by higher layer is used directly as the destination layer-2 ID in the MAC sublayer. In the case of one-to-one communications, the ProSe UE ID provided by higher layer is used directly as the source layer-2 ID or the destination layer-2 ID in the MAC sublayer. In the case of V2X sidelink communication, higher layer provides source layer-2 ID and destination layer-2 ID.

5G network security is still under development by 3GPP for device-to-network communications. The C-V2X security was not updated with Rel-15. However, the enhancements developed for network access and associated security aspects will apply to the network-enabled mode of communication. As for the direct mode of operation, the security design for the Rel-14 C-V2X is expected to remain unchanged, namely specifying the reuse of the application-level security already defined by IEEE for DSRC systems.

In particular, LTE Rel-14 does not support vehicle identity privacy when it sends V2X traffic via the network. Architectural changes would be required to support user/vehicle anonymity from the operator. It is expected that 5G core network will enable network mode V2X operation privacy. 5G security is going to define device-to-network authentication methods and transport; specify provisioning and storage of 3GPP credentials for devices; and specify network functions and protocols necessary for secure device operation within the operator network. The V2X architecture relies on the security relationship between device and its home operator network. In addition, secondary authentication schemes can support industrial and virtual private networks that wish to deploy their own authentication methods and credentials, thus it is possible to deploy public key infrastructure (PKI)[9]

[9] A public key infrastructure (PKI) is a set of rules, policies, and procedures needed to create, manage, distribute, use, store, and revoke digital certificates and manage public key encryption. In cryptography, a PKI is an arrangement that binds public keys with respective identities of entities. The binding is established through a process of registration and issuance of certificates at and by a certificate authority.

security, where devices use a digital certificate to authenticate themselves to the network and vice versa. 5G security design is being developed to achieve the following goals [23]:

- Enhanced subscriber/device privacy: This is an improvement over LTE in that the new subscriber permanent identifier (SUPI) is never allowed to be transmitted over the air. Instead, the sensitive part of the SUPI is sent over the radio link protected against spoofing and tracking, which was not the case with the 3G/4G international mobile subscriber identity.
- Enhanced security support at the network level, along with capability for flexible authentication and authorization schemes: A new network function is specified, known as security anchor function (SEAF), which maintains the security anchor deep in a network (in a physically secure location). The SEAF also provides flexibility in deploying other network entities such as AMF and SMF. In Rel-15, the SEAF is co-located with the AMF.
- Support various types of devices that have different security capabilities and requirements: Secondary authentication enables support for non-3GPP access links and sessions authorized by third-party servers, such as for industrial IoT, V2X, and automation. This functionality is related to the access control for network slices. For example, various IoT devices may require different credential provisioning and authentication methods. The details for this type of industrial scenarios are being investigated.
- Support user data and signaling encryption and integrity protection: These features are essential for a secure system. A new feature is data path integrity protection, which may be especially important for certain new services such as industrial IoT. There is also the potential for the user-plane security to be terminated in the network instead of the base station.
- Separate device credential management and access authentication from data session setup and management: This split results in a separate security context between device and the AMF which is used for mobility management. A different security context is established for session management, between device and the SMF, used to authorize access of the device to specific services; for example, network slices or specific data networks such as enterprise networks. This new type of access control employs a separate and flexible authentication and authorization procedure; for example, extensible authentication protocol (EAP).[10]
- Support secure slicing: There could be several services instantiated as a network slice, each with different security requirements. The access to a slice is granted based on the primary authentication and subscription information, but this authorization is carried out

[10] Extensible authentication protocol (EAP), defined in IETF RFC 5247, is an authentication framework frequently used in wireless networks and point-to-point connections which provides an authentication framework for transport and usage of keying material and parameters generated by EAP methods. Each protocol that uses EAP defines a method to encapsulate EAP messages within that protocol's messages.

by the respective SMF. Therefore, the access security is contained within that network slice and does not rely on the AMF, which may serve multiple slices that may have different security requirements. Moreover, an attack mounted on one slice does not result in an increased risk for an attack on a different slice of the same network.

In a nutshell, what 5G security is trying to achieve is increased user privacy, robustness to cyber-attacks on the network, and better device hardware security. These goals can be achieved with stronger authentication/authorization schemes between device and network, both radio access and core network functions, secure credential provisioning and storage on device, and new network functions that support device-to-network communications security.

Cellular-based V2X systems treat latency as the most important performance metric because the level of protection decreases as the delay in receiving safety information increases. While the delay in non-mission-critical applications may be tolerated to some extent, delayed information in V2X communications could result in serious automobile accidents and injuries. The volume of traffic in V2X communication is much smaller than other applications in cellular systems. Sensing information or safety notifications transmitted via a V2X link can be carried in small packets, thus high-speed data transmissions are less important in V2X systems. The device-to-core network (AMF) signaling is integrity- and confidentiality-protected. The device link to the radio access network is also protected for both signaling and data traffic. The V2X system can leverage the 5G system security for vehicle device authorization, authentication, and access to the network.

7.7 Implementation and Deployment Considerations

The DSRC would require the deployment of tens of thousands of RSUs embedded or attached to roadway infrastructure to enable an effective network along the country roads. This would be a challenge in rural areas considering the vast distances involved. State highway administrations and other highway authorities would be responsible for deploying, managing, and operating the RSUs and the associated infrastructure network and interconnections such as fiber or copper backhaul. After considering how to map each V2V service to different 3GPP technologies, the conclusion is that Rel-14 is only used for basic safety, while NR-based V2X is used for advanced services to avoid duplication or replacement of LTE-based Rel-14 functionalities. Therefore, it is expected that in the beginning of the V2X service deployment, there will be LTE-based Rel-14 V2X devices which later evolve into dual-mode UEs that support both LTE and NR V2X services.

Several V2X use cases require vehicles to communicate with an infrastructure. The DSRC security relies on a public key infrastructure that distributes and manages digital certificates for vehicles. This means that vehicles need to have access to this infrastructure, which in the case of DSRC is provided via the RSUs. The RSU may also be used by vehicles to communicate with the V2X application server. Owing to various factors, the deployment of RSUs might be limited. It is therefore unrealistic to expect the provision of ubiquitous coverage of roadways via DSRC equipment in near future. One challenge with the deployment of V2X technologies is that there is uncertain business incentive for network providers. In the United States, where V2V deployments may be mandated, there is still lack of clarity regarding the plan to implement the infrastructure that would utilize the balance of the channels, and which entities should manage the security network functions. Since government funding has been a driving force for many V2X pilot programs, it is unclear if commercial business models can be applied to accelerate infrastructure deployment or if deployments will be managed by the governments. In some regions, the government agencies are mainly promoting pilot projects in order to benefit the economy of cities and regions by improving traffic efficiency, reducing emissions, and minimizing the risk of crashes. With C-V2X, mobile operator involvement can make additional, commercially motivated services available. The beneficiaries of these can be both subscribers and road operators.

References

3GPP Specifications[11]

[1] 3GPP TS 22.185. Service requirements for V2X services; Stage 1 (Release 15); June 2018.

[2] 3GPP TS 23.285. Architecture enhancements for V2X services (Release 15); June 2018.

[3] 3GPP TS 36.211. Evolved universal terrestrial radio access (E-UTRA). Physical channels and modulation (Release 15); June 2018.

[4] 3GPP TS 36.212. Evolved universal terrestrial radio access (E-UTRA). Multiplexing and channel coding (Release 15); June 2018.

[5] 3GPP TS 36.213. Evolved universal terrestrial radio access (E-UTRA). Physical layer procedures (Release 15); June 2018.

[6] 3GPP TS 36.300. Evolved universal terrestrial radio access (E-UTRA) and evolved universal terrestrial radio access network (E-UTRAN); Overall description; Stage 2 (Release 15); March 2019.

[7] 3GPP TS 36.321. Evolved universal terrestrial radio access (E-UTRA). Medium access control (MAC) protocol specification (Release 15); June 2018.

[8] 3GPP TS 36.331. Evolved universal terrestrial radio access (E-UTRA). Radio Resource Control (RRC); Protocol Specification (Release 15); June 2018.

[9] 3GPP TR 36.885. Study on LTE-based V2X services (Release 14); July 2016.

[10] 3GPP TR 37.885. Study on evaluation methodology of new vehicle-to-everything V2X use cases for LTE and NR (Release 15); June 2018.

[11] 3GPP TR 38.885. Study on NR vehicle-to-everything (V2X) (Release 16); March 2019.

Articles, Books, White Papers, and Application Notes

[12] Recommendation ITU-R M.2084-0. Radio interface standards of vehicle-to-vehicle and vehicle-to-infrastructure communications for Intelligent Transport System applications; September 2015.

[11] 3GPP specifications can be accessed at the following URL: http://www.3gpp.org/ftp/Specs/archive/

[13] 5G Americas White Paper. V2X cellular solutions; October 2016.
[14] 5G Americas White Paper. Cellular V2X communications towards 5G; March 2018.
[15] K. Lee, et al., Latency of cellular-based V2X: Perspectives on TTI-proportional latency and TTI-independent latency, IEEE Access 5 (2017).
[16] M. Boban, et al., Use cases, requirements, and design considerations for 5G V2X, IEEE Vehicular Technol Mag, December 2017.
[17] R. Molina-Masegosa, J. Gozalvez, LTE-V for sidelink 5G V2X vehicular communications, IEEE Vehicular Technol Mag, December 2017.
[18] Rohde & Schwarz White Paper. LTE-advanced (Release 12) technology introduction; September 2014.
[19] Rohde & Schwarz White Paper. Device to device communication in LTE; September 2015.
[20] Husain S, et al. An overview of standardization efforts for enabling vehicular-to-everything services. IEEE Conference on Standards for Communications and Networking (CSCN); September 2017.
[21] S. Chen, et al., Vehicle-to-everything (V2X) services supported by LTE-based systems and 5G, IEEE Commun Standards Mag, June 2017.
[22] H. Seo, et al., LTE evolution for vehicle-to-everything services, IEEE Commun Mag, June 2016.
[23] Kousaridas A, et al. Recent advances in 3GPP networks for vehicular communications. IEEE Conference on Standards for Communications and Networking (CSCN); September 2017.
[24] ETSI TS 102 687 V1.2.1. Intelligent transport systems (ITS); Decentralized congestion control mechanisms for intelligent transport systems operating in the 5 GHz range; Access Layer Part; April 2018.
[25] Khoryaev A. Evolution of cellular-V2X (C-V2X) technology use cases, technical challenges, and radio-layer solutions for connected cars. IEEE ComSoc Webinar; March 2018.
[26] 5GAA Automotive Association. An assessment of LTE-V2X (PC5) and 802.11p direct communications technologies for improved road safety in the EU; December 2017.
[27] Qualcomm Technologies Inc. Accelerating C-V2X commercialization; 2017.
[28] Gaurang Naik et al., IEEE 802.11bd & 5G NR V2X: Evolution of Radio Access Technologies for V2X Communications, Cornell University Online Library, March 2019.

CHAPTER 8

Operation in Unlicensed and Shared Spectrum

The RF spectrum is divided into licensed and license-exempt/unlicensed bands. The traditional cellular communications systems used to exclusively operate in licensed spectrum. However, due to the insufficiency and cost of licensed spectrum below 6 GHz, the telecommunication industry has shifted attention to the unlicensed bands to deploy supplementary uplink/downlink carriers or standalone systems. The use of unlicensed spectrum has been increasingly considered by cellular operators as a complementary mechanism to offload traffic from licensed-band RF carriers to increase the overall system throughput. Various approaches to cellular operation in unlicensed spectrum have been investigated and trialed in the past few years. Two practical cellular technologies for communication in unlicensed spectrum, the LTE-U and the LTE-based license-assisted access (LAA), have been investigated by LTE-U Forum and 3GPP. The use of Wi-Fi as a complementary carrier to offload cellular networks was studied earlier and specified by 3GPP under LTE-Wi-Fi link aggregation (LWA) work item [12].

Operation in unlicensed spectrum is subject to various limitations and restrictions which are regional and band specific. The typical limits are in terms of total transmit power, power spectral density (PSD), carrier bandwidth, and duty cycle that each device can use. In addition, sharing protocols may also be specified in some bands to protect other systems in the band or to allow efficient sharing. As an example, dynamic frequency selection (DFS) aims to protect radar systems and listen-before-talk (LBT) allows efficient spectrum sharing by minimizing inter-user interference in unlicensed spectrum. 3GPP LTE began to explore 5 GHz unlicensed band for offloading cellular traffic in Rel-12. The non-standard LTE-U and standard LAA both use LTE technology as the baseline in unlicensed spectrum anchored on a licensed carrier to increase user throughput and system capacity. It must be noted that LTE-U can only be deployed in selected regions due to lack of LBT support whereas LTE-based LAA uses a similar LBT procedure as IEEE 802.11 systems. In addition to LTE-U and LAA, the standalone operation of LTE in the unlicensed spectrum has been considered by some operators.[1]

[1] MulteFire Alliance: https://www.multefire.org

5G NR. DOI: https://doi.org/10.1016/B978-0-08-102267-2.00008-7

The 60 GHz band has yet to attain significant large-scale industry traction although the specification for IEEE 802.11ad[2] system was finalized in 2012. The 60 GHz frequency bands have been used for wireless backhaul and indoor and outdoor point-to-point communication. The unlicensed 5 GHz spectrum provides about 500 MHz of usable bandwidth. A large number of Wi-Fi deployments based on the IEEE 802.11 are in 5 GHz band and are mainly used for broadband indoor applications. Cellular operators actively use Wi-Fi to offload traffic in dense hotspots using unlicensed spectrum; however, due to a high level of inter-cell interference, Wi-Fi-based offloading scenarios are not attractive for outdoor use cases.

Multi-antenna transmission and beamforming with large antenna arrays are some of the key features of the new radio systems, which can benefit the NR-unlicensed operation in the sense that transmission power can be reduced due to highly directional antennas. In addition, directional transmission leads to lower inter-user/inter-cell interference, which in turn improves the coexistence conditions. For typical network deployment where the elevation angle of the antenna is manually tilted to one fixed direction, the maximum antenna gain is only achieved in the tilted elevation direction while lower antenna gain is seen in other directions. When both elevation and horizontal beamforming are utilized (FD-MIMO), the maximum beamforming gain can be obtained in each direction via beam tracking procedures. Beamforming can improve the link budget and can increase signal-to-interference-plus-noise ratio (SINR) performance. Moreover, the interference between different systems in the unlicensed band can be minimized using highly directional antennas due to the significantly lower collision probability compared to the omnidirectional transmission cases. As a result, higher spatial reuse would be possible, increasing the system throughput as well as improving spectrum efficiency.

Some licensed frequency bands are designated for commercial use while others are designated for public safety. The assignment of a frequency band to a user allows the user to utilize that frequency and bandwidth for stated purposes using predefined emission parameters. Licensed spectrum allows exclusive use of certain frequencies in specific geographic locations, meaning that when someone is granted the right by a regulatory body to communicate at certain frequencies and in certain locations, everyone else is prohibited from using that frequency in that location. In contrast, in the spectrum that is designated as license-exempt or unlicensed, users can operate without a license but must comply with the regulatory constraints. For instance, the regulations limit the transmit power and the effective isotropic radiated power (EIRP) to minimize interference to other cochannel systems. The unlicensed frequency bands were originally allocated for industrial, scientific, and medical (ISM bands) applications.

[2] IEEE Standards Association: https://standards.ieee.org/findstds/standard/802.11ad-2012.html

To coexist with Wi-Fi in the unlicensed spectrum, some enhancements in the LTE and NR systems were required including a mechanism for channel sensing based on LBT, discontinuous transmission on a carrier with limited maximum transmission duration, DFS for radar avoidance in certain bands, and multi-carrier transmission across multiple unlicensed channels. The DTX and LBT have had major impacts on various aspects of LTE functionalities including downlink/uplink physical channel design, channel state information (CSI) estimation and reporting, HARQ operation, and radio resource management. In 5 GHz band, about 500 MHz of unlicensed spectrum is available for LAA use. This unlicensed band can be divided into multiple channels of 20 MHz bandwidth. The selection of LAA carrier(s) with minimal interference is the first step for an LAA node to coexist with Wi-Fi systems in the unlicensed spectrum. However, when a large number of nodes are present, interference avoidance cannot be guaranteed through channel selection, thereby, sharing carriers between different technologies is required. The carrier selection can be further performed periodically by adding or removing unlicensed carriers as required. These carriers are then configured and activated as secondary cells (SCells) for use by LAA-enabled UEs.

The 3GPP NR has initiated a study item in Rel-16 to investigate the required amendments in the Rel-15 NR to extend the operation to the unlicensed bands in sub-6 GHz and above 6 GHz. Meanwhile, the LTE track has continued to enhance the LAA and non-homogeneous LTE/Wi-Fi carrier aggregation solutions, which global operators have already started to deploy. In this chapter, we will review the general aspects and use cases of the operation in unlicensed spectrum including identifying the unlicensed frequency bands, overview of IEEE 802.11 operation, which is important for understanding the coexistence studies, as well as the necessary NR and LTE enhancements to address the unlicensed band operation from various aspects, including network architecture, radio access protocol structure, and L1/L2 processing, as well as deployment and implementation considerations.

8.1 General Aspects and Use Cases

8.1.1 Unlicensed and Shared Spectrum

Unlicensed/license-exempt bands are a spectrum that has been defined for use collectively by an undetermined number of independent users without registration or individual permission. For unlicensed bands, the regulatory bodies establish rules for how applications, technologies, and industries must use the spectrum that allows applications and users to coexist with limited interference to each other. The rules are defined openly with no limitation on technologies and applications other than requirements to avoid/reduce destructive interference. With unlicensed spectrum, there is no process for establishing the right of use, and therefore the band may be utilized by any device that is compliant with usage rules such as maximum power levels, bandwidth limitations, and duty cycles. The use of unlicensed spectrum is an important complement for all 5G systems and deployments, particularly in small cell deployments. Table 8.1 provides a summary of such rules for 5 GHz band utilization across the world [9,13,15,21].

In recent years, short-range communication technologies such as Bluetooth,[3] ZigBee,[4] and Wi-Fi[5] have utilized unlicensed bands for short-range communication services. The frequency mapping of unlicensed bands is country dependent, which determines the type of technology used within designated parts of the spectrum. Despite the original intent, radio communications in the ISM bands are possible as long as the communication systems are designed to tolerate the inter-user interference as well as the cochannel interference potentially from other communication systems operating in the same band. Since the advent of mobile devices, more and more short-range, low-power, low-cost wireless communication systems, such as cordless phones, Wi-Fi, Bluetooth, and ZigBee, have utilized some of the unlicensed bands in 902–928 MHz, 2.40–2.4835 GHz, and 5.725–5.875 GHz. The growing number of wireless applications in the ISM bands has motivated the wireless industry to increase the amount of spectrum available for unlicensed use. In the United States, the Federal Communications Commission (FCC) made 300 MHz of spectrum available in 5.15–5.25 GHz (UNII-1), 5.25–5.35 GHz (UNII-2A), including 5.725–5.825 GHz (UNII-3), for use by a new category of unlicensed equipment. The latter band is partially overlapped with the ISM band (5.725–5.875 GHz), hence, is sometimes referred to as Unlicensed National Information Infrastructure (U-NII)/ISM. The FCC further made available additional 255 MHz spectrum in 5.47–5.725 MHz (UNII-2C) band. This aligns the U-NII frequency band in the United States with other parts of the world, thereby allowing the same product to be used in most parts of the world. The frequency ranges 5.250–5.350 and 5.470–5.725 GHz in U-NII-2 are used by radar systems worldwide; thus the use of these bands requires DFS techniques to avoid adverse effects on radar operation [20]. In certain geographical regions, such as the European Union and Japan, support of LBT rule is mandatory to reduce the interference to other users operating in the same band. The LBT medium access rule prevents a transmitter from continuous transmission and dominating the communication channel, rather it requires the transmitter to check for other radio activities in the channel prior to transmission.

In the United States, the use of 5 GHz unlicensed spectrum is subject to FCC regulations. At present, unlicensed wireless systems can access bands 5.15–5.25 GHz (UNII-1), 5.25–5.35 GHz (UNII-2A), 5.47–5.725 GHz (UNII-2C), and 5.725–5.85 GHz (UNII-3). In addition, bands 5.35–5.47 GHz (UNII-2B) and 5.85–5.925 GHz (UNII-4) are also being

---

[3] Bluetooth Special Interest Group: https://www.bluetooth.com/
[4] ZigBee Alliance: https://www.zigbee.org/
[5] Wi-Fi Alliance: https://www.wi-fi.org/

Table 8.1: 5 GHz unlicensed band designations worldwide [18].

Region (Total Bandwidth)	5 GHz								
	UNII-1 100 MHz	UNII-2A 100 MHz	UNII-2B 120 MHz	UNII-2C 255 MHz			UNII-3 125 MHz		UNII-4 75 MHz
	5.15–5.25 GHz	5.25–5.35 GHz	5.35–5.47 GHz	5.47–5.59 GHz	5.59–5.65 GHz	5.65–5.725 GHz	5.725–5.825 GHz	5.82–5.85 GHz	5.85–5.925 GHz
China (325 MHz)	Indoor	Indoor DFS/TPC					Indoor/outdoor DFS/TPC		
Europe (455 MHz)	Indoor LBT	Indoor/outdoor DFS/TPC LBT		Indoor/outdoor DFS/TPC LBT					
Japan (455 MHz)	Indoor LBT	Indoor/outdoor DFS/TPC LBT		Indoor/outdoor DFS/TPC LBT					
Korea (480 MHz)	Indoor	Indoor/outdoor DFS/TPC		Indoor/outdoor DFS/TPC			Indoor/outdoor		
The United States (580 MHz)	Indoor/outdoor	Indoor/outdoor DFS/TPC		Indoor/outdoor DFS/TPC			Indoor/outdoor		

considered for unlicensed use. The FCC has some regulations regarding transmission bandwidth, maximum transmit power, out-of-band emission, power spectrum density, transmit power control (TPC), and DFS for each unlicensed band. For example, the maximum transmit power is 24 dBm in the UNII-1 and UNII-2A bands, and 30 dBm in the UNII-2C and UNII-3 bands. In addition to the maximum transmit power, TPC may further limit the output power of a transmitter to minimize interference to users of other wireless technologies. In fact, TPC is required for both the UNII-2A and UNII-2C bands. The DFS is used for unlicensed devices to detect radar signals and to change their operating channels whenever radar activity is detected. The DFS should be adopted in the UNII-2A and UNII-2C bands to protect radar signals.

Unlike LTE-based LAA/eLAA that only provide support for the 5 GHz unlicensed band, the NR is required to cover a wide range of unlicensed and shared licensed frequency bands, where regulatory and inter-RAT coexistence requirements may differ for each band. This includes 3.5 GHz band that has been designated as a shared band in the United States as well as the 5 and 60 GHz bands that are unlicensed bands. Moreover, the system design needs to consider the vastly different wireless channel characteristics for the lower carrier frequency bands such as sub-6 GHz as well as higher carrier frequency bands such as 60 GHz [26].

5G networks are expected to operate over a wide range of licensed and unlicensed frequencies in low, medium, and high spectrum bands, but some of those frequencies are yet to be specifically defined by 3GPP and ITU-R. It can be assumed that most of the bands currently being used for 4G networks will be ultimately re-allocated to 5G technologies. Meanwhile, worldwide activities have already begun to explore a number of bands between 24.25 and 86 GHz that are being studied for the 2019 World Radiocommunication Conference (WRC), as well as on bands not included in the WRC agenda item. In the United States, the FCC is planning to make 64–71 GHz band available for unlicensed use based on the same rules applicable to the unlicensed 57–64 GHz band. In addition, FCC is studying several other bands in 24 GHz and above. It further plans to make some bands available for unlicensed applications using the same rules applicable to the unlicensed 57–64 GHz band. However, it has been asked to reconsider allocating the entire 64–71 GHz band to unlicensed operations. 3GPP also has a study item on 5G in unlicensed bands below and above 6 GHz. Unlicensed bands above 52.6 GHz covering the FCC 64–71 GHz frequency range will be considered to the extent that waveform design principles remain unchanged relative to that in below 52.6 GHz bands. The FCC has allowed access to spectrum for next-generation wireless broadband in the 28 GHz (27.5–28.35 GHz), 37 GHz (37–38.6 GHz), and 39 GHz (38.6–40 GHz) bands, as well as an unlicensed band at 64–71 GHz. The new rules make available almost 11 GHz of spectrum consisting of 3.85 GHz of the licensed spectrum and 7 GHz of the unlicensed spectrum [31]. As shown in Fig. 8.1, 60 GHz unlicensed spectrum comprises up to four non-overlapping channels, where each channel has a

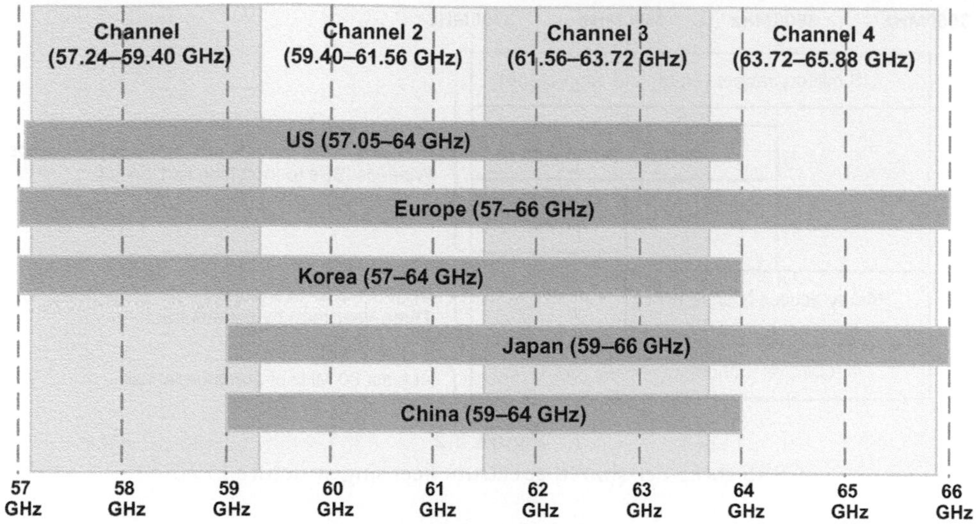

Figure 8.1
Worldwide 60 GHz unlicensed spectrum allocation [26].

bandwidth of 2.16 GHz and enables four channels in Europe; three channels in the United States, Canada, South Korea, and Japan, and two channels in China. The requirements and regulations concerning the EIRP for 60 GHz vary by geographic areas and are different from the ones in 5 GHz. For example, the requirement for peak EIRP is 43 dBm in the United States and 40 dBm in Europe where the maximum PSD is also required to be 13 dBm/MHz. The main challenge of operation in 60 GHz band is excessive path loss and blockage relative to operation in 5 GHz band. However, the large propagation loss at 60 GHz can be mitigated by increasing the antenna array gain obtained by large array sizes. Fig. 8.1 summarizes the worldwide availability and channelization of 60 GHz spectrum.

The US FCC established Citizen Broadband Radio Service (CBRS) for shared commercial use of the 3.5 GHz (3550–3700 MHz) band with the incumbent military radars and fixed satellite stations in 2015. Dynamic spectrum sharing rules have been defined to make additional spectrum available for flexible wireless broadband use while ensuring interference protection and uninterrupted use by the incumbent users. Under the plan, a three-tier sharing framework coordinates spectrum access among the incumbent military radars and satellite ground stations and new commercial users. The three tiers are as follows: incumbent, priority access license (PAL), and general authorized access (GAA) users, as shown in Fig. 8.2.

The incumbent military radar systems, satellite ground stations, and wireless ISPs are always protected from possible interference from the lower-tier PAL and GAA users. Tier-2 PAL users have the next highest priority access and are protected from GAA users.

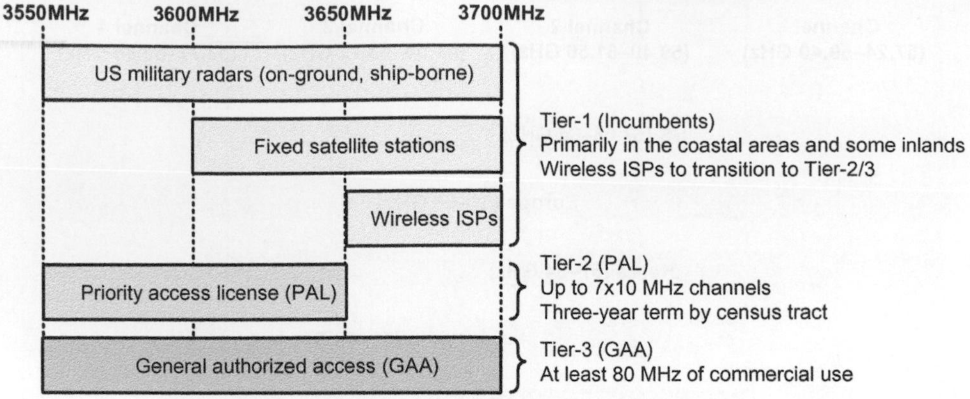

Figure 8.2
CBRS three-tier shared spectrum licensing structure [19].

PAL licenses within the 3550–3650 MHz portion of the band are assigned based on spectrum auctions. Each PAL license covers a 10 MHz channel for a single census tract (i.e., a small, relatively permanent statistical subdivision of a county) for a 3-year term. For any given census tract, up to seven total PALs may be assigned. With over 50,000 census tracts in the United States, each PAL spectrum license is expected to cost much less and encourage participation from a variety of participants. It should be noted that a PAL frequency range may change over time based on incumbent activity. The lowest-tier GAA users are permitted to use any portion of the 3.5 GHz band not assigned to higher-tier users. With an open-access rule, GAA provides free access to the spectrum similar to unlicensed spectrum. Because PAL licenses are limited to a maximum of 70 MHz in any given census tract, a minimum of 80 MHz bandwidth is available for GAA use when not in use by the incumbent users. While GAA operation does not require a costly license, GAA operators must coordinate their use of the spectrum through the dynamic spectrum sharing system [19].

As shown in Fig. 8.3, a key element of the CBRS spectrum sharing architecture is the spectrum access system (SAS). A SAS maintains a database of all CBRS base stations, also known as CBRS devices (CBSDs), including their tier status, geographical location, and other pertinent information to coordinate channel assignments and manage potential interferences. To mitigate possible interference to tier-1 military radar systems, environmental sensors known as environmental sensing capability (ESC) will be deployed in strategic locations near naval stations, mostly along coastal regions, to detect incumbent activities. When incumbent use is detected, the ESC alerts the SAS, which then directs CBSDs utilizing impacted CBRS channels in that area to move to other channels. The cloud-based SAS enforces the three-tier spectrum sharing mechanism based on FCC rules via centralized, dynamic coordination of spectrum channel assignments across all CBRS base stations in a region. The CBRS rule-making defines two classes of base stations: class-A and class-B. A

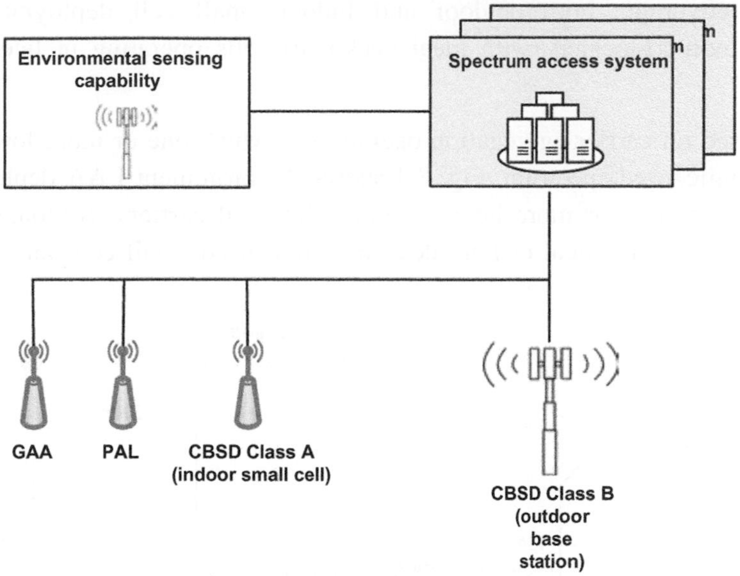

Figure 8.3
CBRS network architecture [19].

class-A base station can be an indoor or low-power outdoor small cell with a maximum conducted power of 24 dBm (per 10 MHz) and a maximum EIRP of 30 dBm. This type of small cell is similar to commercial enterprise-class small cells with 250 mW transmit power and a typical 2 dBi omnidirectional antenna or up to 6 dBi directional antenna. The class-B base station is meant for outdoor use with a maximum EIRP of 47 dBm. The highly directional antenna makes the outdoor CBRS base stations more suitable for fixed wireless use cases. While indoor and outdoor base stations can be assigned to either GAA or PAL, more indoor GAA deployments are expected until ESC certification and PAL auctions are concluded [19].

8.1.2 Use Cases and Deployment Models

The introduction of carrier aggregation in LTE-advanced required the distinction between a primary cell (PCell) and SCell. The PCell is the main cell with which a UE communicates and maintains its connection with the network. One or more SCells can be allocated to and activated for the UEs supporting carrier aggregation for bandwidth extension. Because the unlicensed carrier is shared by multiple systems, it can never match the licensed carrier in terms of mobility, reliability, and quality of service (QoS). Hence, in LAA, the unlicensed carrier is considered only as a supplemental downlink or uplink SCell assisted by a licensed PCell via carrier aggregation. LAA deployment scenarios encompass scenarios with and without macro-coverage, both outdoor and indoor small cell deployments, and both co-located and non-co-located (with ideal backhaul) cells operating in licensed and unlicensed carriers.

The LAA is based on carrier aggregation operation in which one or more low-power SCells operate in the unlicensed spectrum. Fig. 8.4 shows the prominent LAA deployment scenarios, where there are one or more licensed and unlicensed carriers. Although the backhaul for the small cells can be ideal or non-ideal, the unlicensed small cell only operates in the

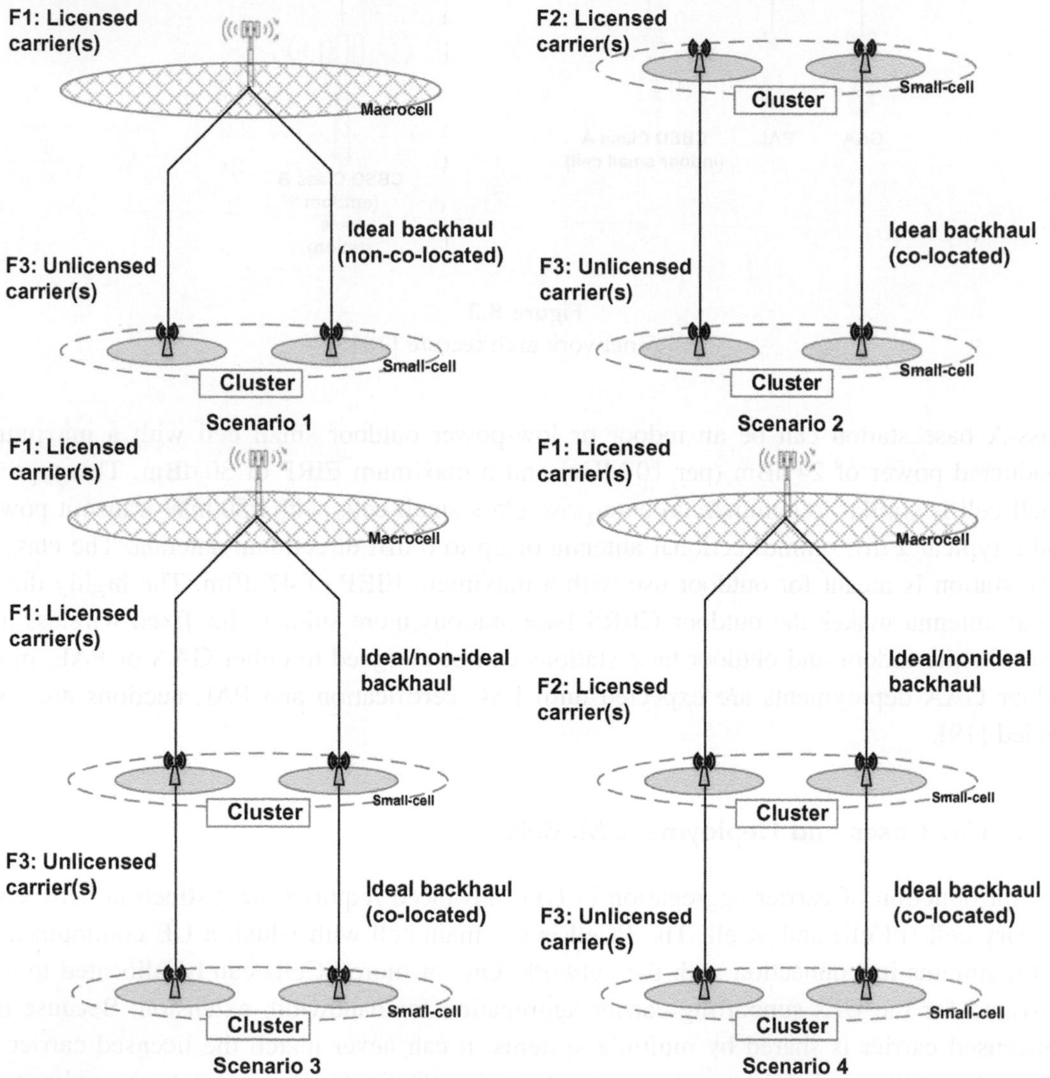

Figure 8.4
LAA deployment scenarios [12].

context of the carrier aggregation through ideal backhaul with a licensed cell. In scenarios where carrier aggregation is used within the small cell with carriers in both licensed and unlicensed bands, the backhaul between macrocell and small cell can be ideal or non-ideal.

The deployment scenarios depicted in Fig. 8.4 can be described as follows. In scenario 1, carrier aggregation between licensed macrocell (F1) and unlicensed small cell (F3) is used, whereas in scenario 2, carrier aggregation between licensed small cell (F2) and unlicensed small cell (F3) without macrocell coverage is intended. In scenario 3, licensed macrocell and small cell (F1) with carrier aggregation between licensed small cell (F1) and unlicensed small cell (F3) is utilized, and in scenario 4, licensed macrocell (F1), licensed small cell (F2), and unlicensed small cell (F3) and further carrier aggregation between licensed small cell (F2) and unlicensed small cell (F3) is used. In the latter scenario, if there is ideal backhaul between macrocell and small cell, there can be carrier aggregation between macrocell (F1), licensed small cell (F2), and unlicensed small cell (F3). If dual connectivity is enabled, there can be dual connectivity between macrocell and small cell. In the study to support deployment in the unlicensed spectrum for the above scenarios, carrier aggregation functionalities are used as a baseline to aggregate PCell/PSCell[6] on a licensed carrier and SCell on an unlicensed carrier. When non-ideal backhaul is applied between a macrocell and a small cell cluster in scenarios 3 and 4, a small cell on an unlicensed carrier must be aggregated with a small cell on a licensed carrier in the small cell cluster through ideal backhaul.

The NR-unlicensed (NR-U) study item has considered different unlicensed bands or shared bands such as 2.4, 3.5, 5, 6, 37, and 60 GHz band. Some bands are available globally whereas others are only regionally available. The NR-U study item does not target sub-gigahertz unlicensed bands due to the lack of sufficient spectrum to support efficient NR-U operation. Five deployment scenarios have been identified for NR-U as follows [12]:

- Scenario A: Carrier aggregation between licensed-band NR (PCell) and NR-U (SCell)
  - NR-U SCell may have both downlink and uplink or only downlink coverage
- Scenario B: Dual connectivity between licensed-band LTE (PCell) and NR-U (PSCell)
- Scenario C: Standalone NR-U
- Scenario D: An NR cell with a downlink in an unlicensed band and uplink in a licensed band

[6] In LTE Rel-10 and Rel-11 carrier aggregation, PUCCH was configured only on the PCell. Therefore, uplink control information, that is, HARQ ACK/NACK, CSI of SCells were transmitted on PUCCH of the PCell, if not multiplexed with SCell's own PUSCH. In LTE dual connectivity, it was determined not suitable to carry UCIs of SeNB in PUCCH of PCell in MeNB due to backhaul latency. Therefore, PUCCH is configured on a special SCell of SeNB called Primary SCell or PSCell. The PSCell is never deactivated and RACH procedure needs to be initiated upon its initial configuration.

- Scenario E: Dual connectivity between licensed-band NR (PCell) and NR-U (PSCell)

The references to sub-7 GHz bands are intended to include unlicensed bands in 6 GHz region that are under consideration by some regulatory bodies. NR-U is trying to identify additional functionalities that may need a new physical layer design (except channel access procedures) for operation in the unlicensed spectrum that are applicable to a particular frequency range (e.g., sub-7 GHz, 7–52.6 GHz, above 52.6 GHz). Optimizations for some frequency bands may be necessary due to different requirements for each band such as PSD limitation or occupied channel bandwidth (OCB), defined as the bandwidth containing 99% of the signal power, which falls between 80% and 100% of the nominal channel bandwidth. Channel bandwidths below 5 MHz are not included in this study. The study further targets the design of channel access procedures for frequency bands based on coexistence and regulatory considerations applicable to the band. The study includes identification of procedures for technology-neutral channel access for frequency bands that may become available subject to new regulations. If the absence of a Wi-Fi system cannot be guaranteed in sub-7 GHz band where NR-U is operating, the baseline assumption is that NR-U operating bandwidth is an integer multiple of 20 MHz. The following physical layer procedures may be impacted as a result of the study for NR-U: HARQ operation; configured grant support in NR-U; and channel access mechanisms. For the latter, LTE-based LAA LBT mechanism is used as the baseline. The study may further encompass enhancement of the baseline LBT mechanism; techniques to cope with directional antennas/transmissions; receiver-assisted LBT; on-demand receiver-assisted LBT, for example, receiver-assisted LBT enabled only when needed; and techniques to enhance spatial reuse, preamble detection, and enhancements to baseline LBT mechanism above 7 GHz.

The citizens broadband radio service can potentially lower the barrier to entry for non-traditional wireless carriers. The limited propagation characteristics of the 3.5 GHz spectrum facilitate indoor and floor-by-floor deployment options that could compete with the existing Wi-Fi networks. Due to the significantly lower cost of PALs compared to licensed spectrum costs, private operators now have access to 150 MHz of spectrum on every floor. This may allow enterprise applications, industrial Internet of things, and densely populated venues. Trials are already happening in the industrial IoT and smart home areas. PALs fit the need for local connectivity in remote or temporary locations for industrial complexes such as mines, power plants, oil platforms, factories, and warehouses. Private and localized LTE deployments combine the QoS of LTE and the low cost of unlicensed spectrum. Aside from industrial IoT players and general network operators looking for more available bandwidth, the SAS model for CBRS encourages more players to participate in the eventual deployment of 5G technology. Low license fees and neutral hosts allow non-traditional cellular carriers to build private networks independent of exclusively licensed frequencies or heavily congested unlicensed spectrum in 5 GHz band.

8.1.3 Principles of IEEE 802.11 Operation

IEEE 802.11 family of standards is a group of wireless local area network radio access technologies developed by IEEE 802 LAN/MAN Standards Committee. These standards define non-synchronous contention-based multiple access schemes based on carrier sense multiple access with collision avoidance (CSMA/CA). Unlike LTE, Wi-Fi takes a decentralized approach to initiate transmissions from different devices. The rule is to listen before you talk, and if your transmission collides with another transmission, wait a random period before you try again. Therefore, when a Wi-Fi device wants to begin a transmission, it senses the medium and performs a clear channel assessment (CCA). If the channel is detected to be free for a time duration referred to as distributed interframe space, or DIFS, the transmission proceeds. Otherwise, the Wi-Fi device draws a random number, between 0 and 16 (or between 0 and 32 for IEEE 802.11b/g), starts a counter and backs-off from transmission while the channel is busy. When the counter reaches zero, the device that was attempting to transmit starts transmission over the channel. However, if other devices were also sensing the carrier at the same time and tried to transmit, a collision may occur. When a transmission fails (which is detected by the absence of an ACK from the receiver), a random backoff number is drawn and the process repeats. With every backoff, the random counter value is doubled, that is, increasing as 16, 32, 64, etc. This random-access process, referred to as CSMA/CA, is illustrated in Fig. 8.5.

Most wireless systems operating in unlicensed spectrum employ CSMA as the basis for channel access. In this case, the most commonly used medium access mechanism is distributed coordination function (DCF), which is based on CSMA/CA. The DCF concept is illustrated in Fig. 8.5. A transmitting node senses the channel, and if the channel is idle for a certain time duration, that is, the DCF interframe space (DIFS), the node starts to transmit; otherwise, it continues to monitor the channel until it becomes idle for a time duration equal to DIFS. At this time, the node generates a random backoff timer, uniformly distributed within a contention window (CW). The random backoff helps avoid potential collisions, which may happen when two or more nodes are simultaneously waiting for the channel to be cleared. The backoff timer is decremented as long as the channel is idle but remains frozen when a transmission is detected and is reactivated after the DIFS period of time as soon as the channel is free. A node refrains from transmission until its backoff timer expires. Note that the DCF mechanism tries to ensure that only one transmission is present in a channel at any time, and each node has a fair share of the channel. The channel use for each node at a particular time is not guaranteed. Thus there is no deterministic schedule for transmission, reflecting the random and contentious nature of communication in the unlicensed spectrum. Reliable services and efficient resource usage are typically hard to achieve in unlicensed operation. This property is very different from that of operation in a licensed spectrum.

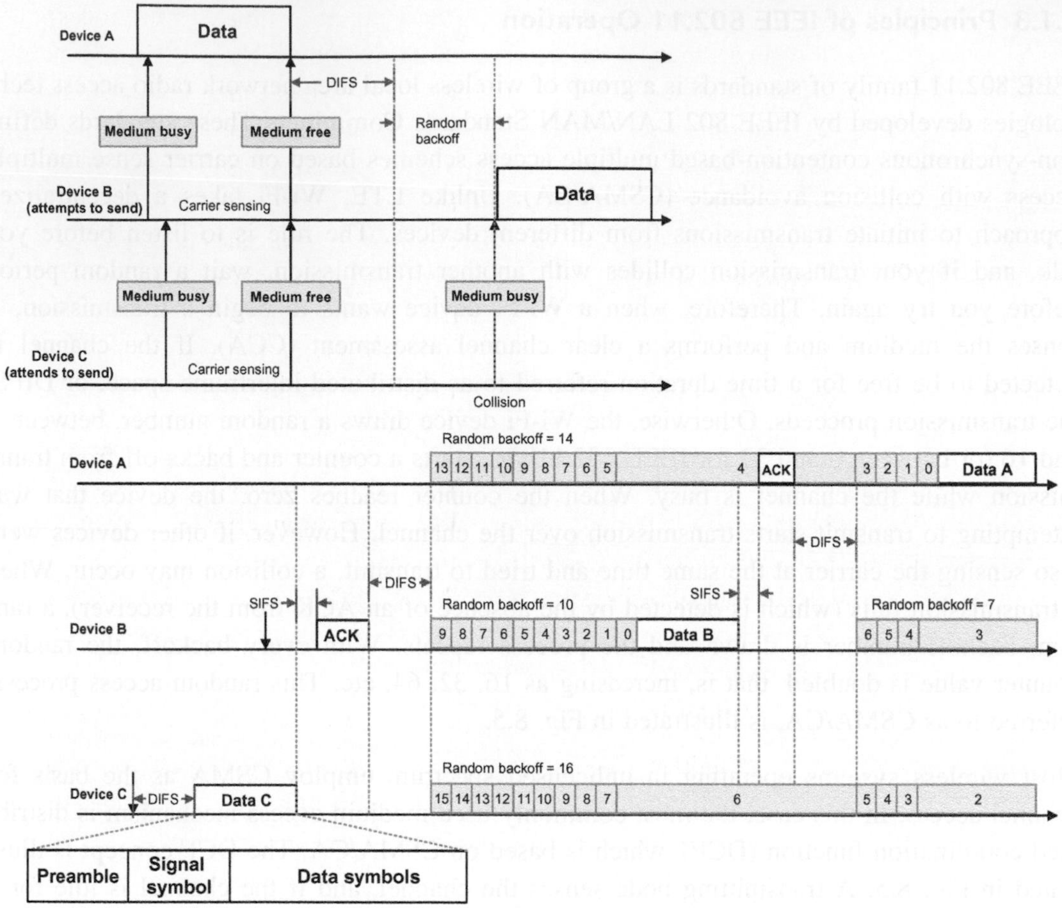

Figure 8.5
Illustration of the asynchronous Wi-Fi channel access concept [30].

A generic IEEE 802.11 frame structure consists of a set of preambles (new and legacy preambles), a signal field, and multiple data symbols. The preamble is a special waveform designed for signal identification, AGC adjustment, timing and frequency synchronization, and channel estimation. The preamble is particularly suited for activity detection in a channel because waveform detection is 10–20 dB more sensitive than the energy-detection-based CCA. Furthermore, because the IEEE 802.11 frame structure does not have a fixed timing, the preamble detection is crucial for a receiver to synchronize to the incoming frame. The signal symbol following the preamble contains information that includes the modulation and coding scheme and the total number of octets for the following data symbols that carry a MAC PDU. Depending on the payload, the frame can be very short (e.g., ACK frame) or long in the case of data frame for user traffic. Each IEEE 802.11 MAC PDU contains a transmission duration field for informing the neighboring nodes of the medium occupancy time of the current burst. This is the amount of time that all nodes must wait, if they receive it. A local timer, called a network allocation vector (NAV), of a neighboring node is updated after the node reads the duration value from the ongoing transmission. This node avoids medium access until the NAV counter expires. Using this virtual medium sensing mechanism, Wi-Fi utilizes a special clear-to-send (CTS)-to-self message to deal with the newer versions of Wi-Fi frames coexisting with a legacy node. CTS-to-self is a standard Wi-Fi CTS message except that it is addressed to the transmitting node itself, as the name implies. Nevertheless, it is meant for the neighboring nodes, if it is detected. A new generation Wi-Fi node transmits a CTS-to-self frame immediately before transmitting. The duration field of the CTS-to-self packet contains the time of the following traffic frame, thereby providing more effective protection of the subsequent frame than relying on physical medium sensing.

IEEE 802.11 supports two network architecture types namely infrastructure and ad-hoc modes. The basic service set (BSS) is the basic building block of an IEEE 802.11 network. Direct association of stations in ad-hoc network forms an independent BSS or IBSS. The interconnection of a number of BSS through a distributed system creates an extended service set. IEEE 802.11 specifications define multiple physical layers and a common MAC layer for wireless local area networks (as shown in Fig. 8.6). Another important aspect of this family is that they use unlicensed spectrum in 2.4, 5, and 60 GHz for operation.

The IEEE 802.11 family comprises many technologies that have evolved from direct sequence spread spectrum (DSSS) and complementary code keying (CCK) in the first generation to OFDM waveforms combined with advanced coding and modulation techniques and spatial division multiplexing (SDM) multi-antenna schemes in the latest generations that, depending on channel conditions, can provide data rates in the excess of a few gigabits per second within short distances. Table 8.2 summarizes the key physical layer characteristics of IEEE 802.11 air-interface technologies. There are other IEEE 802.11 family members that each provides an extension to the baseline standard by adding new features such as handover and roaming, mesh networking, QoS, security, regulatory, and measurement for various regions of the world [30].

In IEEE 802.11, the stations and the access point are not synchronized except when they exchange data or control information in the downlink or uplink. An LBT method combined with CSMA/CA is used to gain access to the medium and to ensure collision avoidance with other contenders wishing to access the shared medium. The stations either passively scan the beacons transmitted by nearby access points or actively scan neighboring access points by transmitting a probe signal.

In IEEE 802.11, there are two options for medium access. The first is a centralized control scheme that is referred to as point coordination function (PCF), and the second is a contention-based approach known as DCF. The PCF mode supports time-sensitive traffic

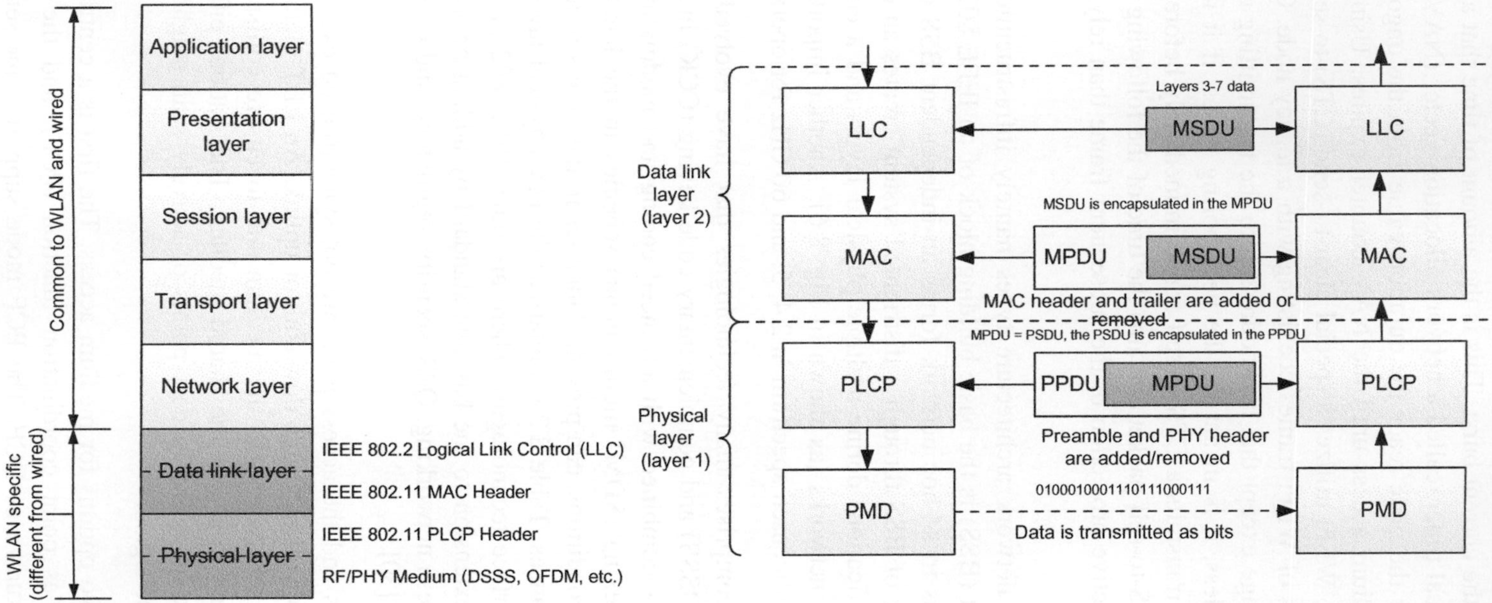

Figure 8.6
OSI and IEEE 802.11 protocol relationships [30].

Table 8.2: Evolution of IEEE 802.11 air-interface technologies [30].

Standard	Frequency (GHz)	Bandwidth (MHz)	Transmission Scheme	Highest Order Modulation	Coding Rate	Spatial Streams	Peak Data Rate (Mbps)
802.11	2.4	25	DSSS	–	Convolutional coding with coding rates 1/2, 2/3, 3/4, and 5/6	1	1, 2
802.11b	2.4	25	DSSS/CCK	–		1	1–11
802.11a	5	20	OFDM	64QAM		1	6–54
802.11g	2.4	25	OFDM, DSSS/CCK	64QAM		1	1–54
802.11n	2.4 and 5	20, 40	OFDM/SDM	64QAM		1–4	6.5–600
802.11ac	5	20, 40, 80, 160	OFDM/SDM/ MU-MIMO	256QAM		1–8	433 (80 MHz and one spatial stream) 6933 (160 MHz and 8 spatial streams)
802.11ad	60	2160	OFDM	256QAM		1	Up to 6912
802.11ax	2.4 and 5	20, 40, 80, 160	OFDMA/ SDM/MU-MIMO	1024QAM	Convolutional coding and LDPC with coding rates 1/2, 2/3, 3/4, and 5/6	1–8	600.4 (80 MHz and one spatial stream) 9607.8 (160 MHz and 8 spatial streams)

flows where the access points periodically send beacon frames to communicate network management and identification, which is specific to that Wi-Fi node. Between the sending of these frames, PCF splits the timeframe into a contention-free period and a contention period. If PCF is enabled on the remote station, it can transmit data during the contention-free polling periods. However, the main reason why this approach has not been widely adopted is because the transmission times are not predictable. The other approach, DCF, relies on the CSMA/CA scheme to send/receive data. Within this scheme, the MAC layer sends instructions for the receiver to look for other stations' transmissions. If it sees none, then it sends its packet after a given interval and waits for an acknowledgment. If one is not received, then it knows its packet was not successfully delivered. The station then waits for a given time interval and checks the channel before retrying to send its data packet. This can be achieved because every packet that is transmitted includes a value indicating the length of time that transmitting station expects to occupy the channel. This is noted by any station that receives the signal, and only when this time has expired other stations consider transmitting. Once the channel appears to be idle, the prospective transmitting station must wait for a period equal to the DCF interframe spacing. If the channel has been active, it must first wait for a time consisting of the DIFS plus a random number of backoff slot times. This is to ensure that, if two stations are waiting to transmit, then they do not transmit together, and then repeatedly transmit together. A time known as contention window is used for this purpose. This is a random number of backoff slots. If a transmitter intending to transmit senses that the channel becomes active, it must wait until the channel becomes free. If the channel is still busy, the transmitter continues waiting a random period for the channel to become free, but this time allowing a longer CW (see Fig. 8.7). While the system works well in preventing stations to transmit together, the result of using this access system is that, if the network usage level is high, then the time that it takes for data to be successfully transferred increases. This results in the system appearing to become slower for the users. In view of this, Wi-Fi may not provide a suitable QoS in its current form for systems where real-time data transfer is required.

To introduce QoS, a new MAC layer was developed as part IEEE 802.11e. The user traffic is assigned a priority level prior to transmission. There are eight user priority levels. The transmitter prioritizes the data by assigning one of the four access categories. The QoS-enabled MAC layer has combined features from the DCF and PCF schemes into hybrid coordination function (HCF). In this approach, the modified elements of the DCF are termed the enhanced distributed channel access (EDCA), whereas the elements of the PCF are termed the HCF controlled channel access. A new class of interframe space called an arbitration interframe space (AIFS) has been introduced for EDCA. This is chosen such that the higher the priority the message, the shorter the AIFS and associated with this there is also a shorter CW. The transmitter then gains access to the channel in the normal manner, but in view of the shorter AIFS and shorter CW, the higher the chance of it gaining access

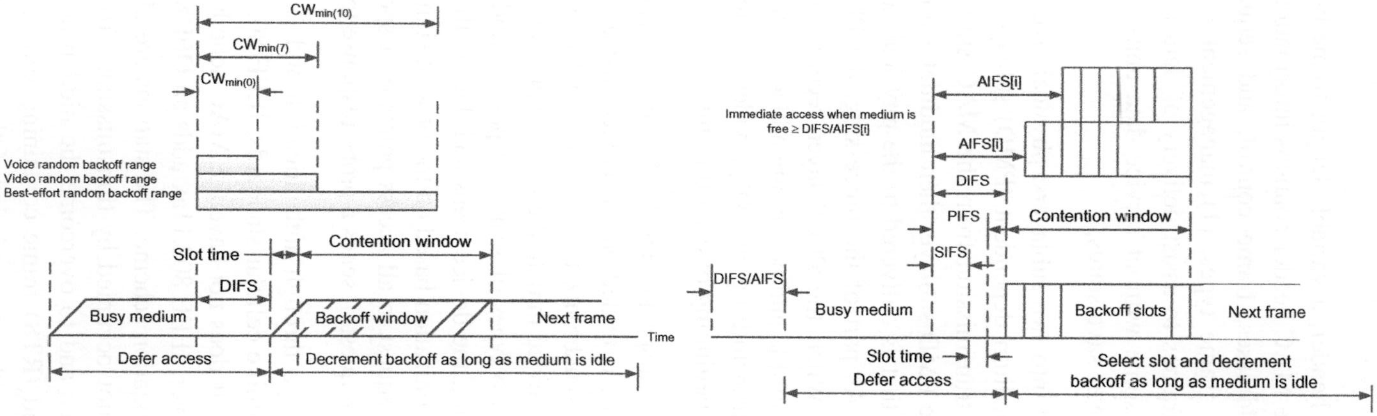

Figure 8.7
Medium access methods in IEEE 802.11 [30].

to the channel. Although statistically a higher priority message will usually gain the channel, this will not always be the case.

IEEE 802.11 frame contains a MAC header, a variable length frame body (0–2304 bytes) and 32-bit frame check sequence. The MAC header contains information related to the type of frame, source and destination addresses, frame control, and sequence control. IEEE 802.11 standard defines three major frame types: (1) management frames which do not carry service data units, (2) control frames to assist delivery of data frames, and (3) some data frames with service data units, some without service data units, and a null frame to inform the access point of client's power save status.

IEEE 802.11 physical layer is divided into two sublayers: physical layer convergence protocol (PLCP) sublayer and physical medium dependent (PMD) sublayer (see Fig. 8.6). The PLCP sublayer receives a frame for transmission from the MAC sublayer and creates the PLCP protocol data unit (PPDU). The PMD sublayer then modulates and transmits the data as bits. The MAC protocol data unit that is delivered to the physical layer is referred to as the PLCP service data unit (PSDU). As part of the processing, the PLCP sublayer adds a preamble and header to the PSDU. When the PLCP layer receives the PSDU from the MAC layer, the appropriate PLCP preamble and header are added to the PSDU to create the PPDU. When transmitting data, the transmitting station provides the receiving station special synchronization sequences at the beginning of each transmission.

The access point periodically broadcasts a special signal called beacon (once every 102.4 ms). When the Wi-Fi communication module is turned on, the device first detects and decodes the beacon signal and establishes physical synchronization with the sender. After establishing synchronization, the access point and the device initiate the authentication procedure followed by the association procedure. There are two types of scanning: passive and active scanning. As shown Fig. 8.8, during passive scanning, the device scans and detects the beacon signal from the access point and establishes synchronization based on the beacon signal. In the active scan mode, the device broadcasts a probe request to all access points or a specific one. If there is any access point that detects the probe request, it sends a probe response to the device.

Once the client and the access point go through authentication and association procedure, the client can send or receive data. Unlike cellular standards, IEEE 802.11 does not support dedicated control/traffic channels and it does not have the MAC scheduling functionality in previous generations of Wi-Fi. However, IEEE 802.11ax adds an OFDMA scheduling capability and more granular resource allocation scheme. The stations are allowed to transmit at any time as long as the medium is not occupied by transmissions from other stations. To determine whether the medium is free and to overcome the hidden-node problem, the station transmits a short request to send (RTS) frame containing source address, destination address, and the duration of upcoming data transmission. Other stations located around the transmitting station may receive the RTS burst and check if the RTS is meant for them. If the RTS is meant for a station and the medium is free, the receiving device would transmit

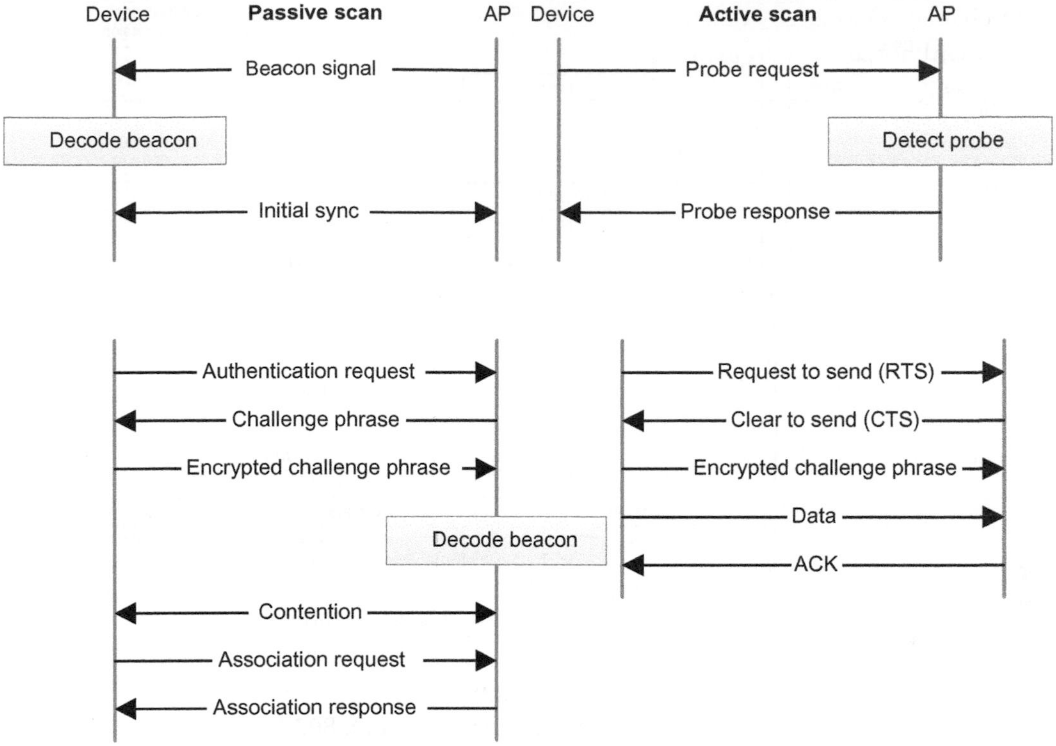

Figure 8.8
Scanning and association procedures [30].

a CTS frame containing the duration of the transaction. At this time, other devices in the vicinity of the communicating stations know that the medium will be occupied for certain time duration and set their NAV counters, which is a MAC-level contention control timer accordingly so that it would not try sensing the medium and transmitting during that period (see Fig. 8.9).

When a packet arrives at the MAC layer from higher layers, a sequence number is assigned to it, and if the packet length is larger than a single MAC frame, it is segmented into multiple fragments. In this case, a fragment number is assigned to each segment. When a packet is segmented into multiple MAC frames, those fragmented frames are assigned the same sequence number and different values for the fragment number. IEEE 802.11 can transmit a maximum of 2304 bytes of higher layer data.

8.1.4 LTE and NR Solutions for Operation in Unlicensed Spectrum

The traditional methods of interworking between cellular networks and Wi-Fi have proved to be cumbersome, with practical limitations in handling the mobility of the IP flows

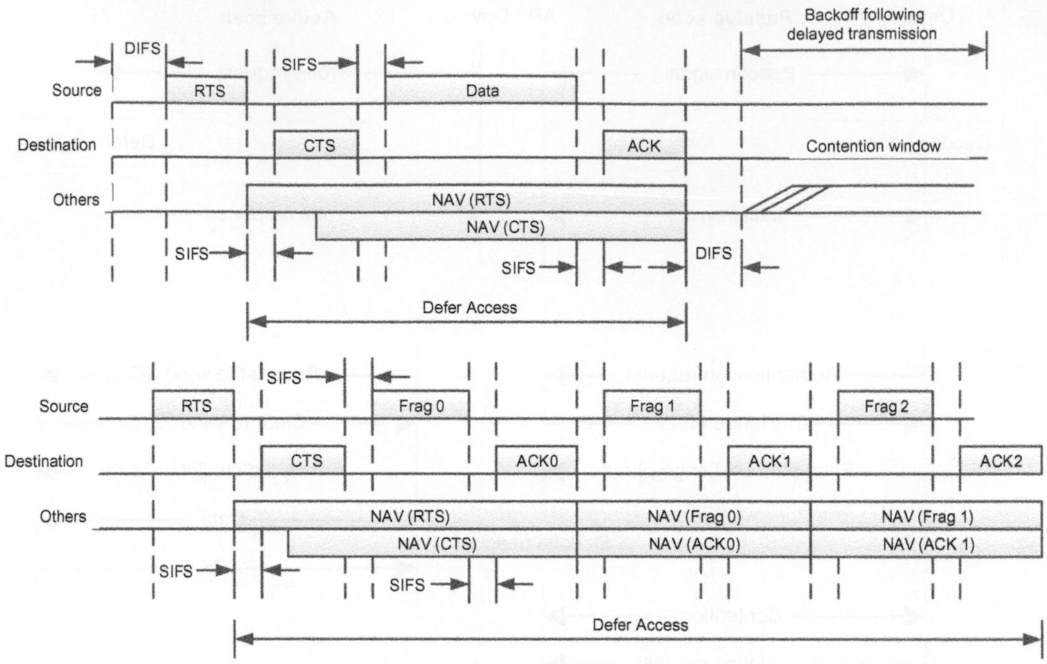

Figure 8.9
Medium access and contention control in IEEE 802.11 [30].

between cellular and Wi-Fi networks. All these schemes have suffered from complexity to define which Wi-Fi networks can be selected for traffic offload, and in which condition a Wi-Fi network should be selected so that it provides the best performance. For these needs, 3GPP has specified access network discovery and selection function (ANDSF), where an ANDSF server in the operator network controls the connection manager at the device. The ANDSF may have location-dependent network selection policies, defining how and in which priority order to select Wi-Fi networks. It is noted that the UE local operating environment and user preferences are important. The ANDSF has recently included comparative signal and load thresholds to instruct the UE when it should use the cellular base station and when to use the selected Wi-Fi access points for any given flow. Due to the lack of commercial interest in ANDSF, 3GPP has decided to integrate Wi-Fi into cellular networks by RAN-level features, which can provide scalability as a Wi-Fi access node acts in a similar manner as an eNB does, rather in the packet data network. A Wi-Fi access point may be integrated into an eNB. This type of interworking enables timely accounting of radio aspects in the offloading decisions and transparent integration of the Wi-Fi access with a single node connection to the cellular core network (in this case, the eNB), which not only reduces control signaling but simplifies network management operations. In this section, we focus on 3GPP RAN-level features, for example, LTE-Wi-Fi aggregation (LWA) and LTE-Wi-Fi radio-level integration with IP security tunnel (LWIP) introduced by 3GPP in Rel-13 and extended in Rel-14 [16,22]. In both features, the aggregated LTE link (licensed

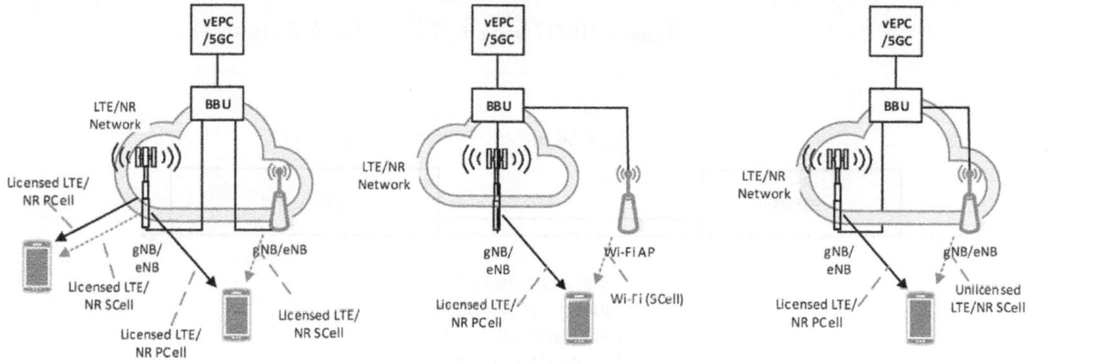

LTE/NR Carrier Aggregation using Licensed Bands (Link Level at MAC Sub-layer) | Licensed LTE/NR + Unlicensed Wi-Fi Carrier Aggregation (Link Level at PDCP Sub-layer) | Licensed LTE/NR + Unlicensed LTE/NR Carrier Aggregation (Link Level at MAC Sub-layer)

Figure 8.10
Comparison of various solutions for cellular operation in unlicensed spectrum [18].

operation) provides robust mobility and coverage, whereas the aggregated Wi-Fi link (unlicensed operation) allows routing of high data rate traffic, providing higher user throughput and more efficient use of both types of radio access. Despite their commonalities, LWA and LWIP have significant differences, mainly in the protocol layer at which the aggregation occurs and in the adopted user-plane security mechanisms. Fig. 8.10 summarizes various unlicensed access mechanisms that have been studied in 3GPP. In the following sections, we discuss in more details various schemes that have been used to enable LTE and NR access and operation in unlicensed spectrum.

8.1.4.1 License-Assisted Access

Licensed assisted access was introduced in LTE Rel-13. It uses carrier aggregation in the downlink to combine an LTE carrier in the unlicensed spectrum with LTE carrier(s) in the licensed spectrum (see Fig. 8.12). This aggregation of spectrum provides higher data rates, improved indoor connectivity, and network capacity. This is done by maintaining a persistent anchor in the licensed spectrum that carries the control and signaling information and combining it with one or more carriers from the unlicensed spectrum. The LAA has been designed as a single global solution that can adapt to unique regional regulatory requirements while coexisting with Wi-Fi and other unlicensed-band radios. For regulatory compliance and coexistence with other devices operating in the unlicensed band, devices supporting LAA must utilize LBT, which is mandated in Europe and Japan. The LBT used in LAA is similar to the method used by Wi-Fi nodes. A new frame structure (type 3) was defined exclusively for LAA cells, supporting discontinuous time-limited transmissions to ensure that the LTE access nodes do not dominate the channel. It also enables the use of incomplete subframes below 1 ms for more flexible adaptation to transmission opportunities after a successful LBT. The enhanced LAA (eLAA) in Rel-14 extended the downlink-only LAA to the uplink. For LAA operation in some geographical regions (e.g., Japan),

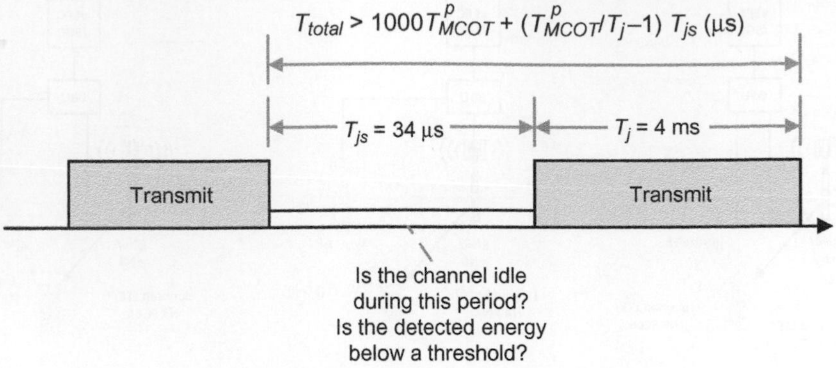

Figure 8.11
Discontinuous transmission in LAA [22].

discontinuous transmission is part of the specification. An eNB can transmit again on the LAA SCell for a maximum duration of $T_j = 4$ ms immediately after sensing the channel to be idle for a sensing interval of $T_{js} = 34\ \mu s$ as shown in Fig. 8.11.

The Rel-13 LAA functionality relies on LBT, frame structure type 3 along with discontinuous transmission with limited use of channel, use of incomplete subframes, downlink only, no PBCH (PBCH is not transmitted within frame structure type 3), use of Band 46 (5.150–5.925 GHz), and discovery reference signals (DRS) to enable radio resource management. In discontinuous transmission, the first subframe in a burst can last 1 ms (starts at symbol 0) or 0.5 ms (starts at symbol 7), which is signaled to the UE via RRC message *subframeStartPosition* in the dedicated physical configuration. The last subframe in a burst can last 1 ms or the duration of a special subframe with normal cyclic prefix (3, 6, 9, 10, 11, 12) OFDM symbols, which is signaled to the UE via DCI format 1C scrambled with CC-RNTI [8].

In LAA, the configured set of serving cells for a UE always include at least one SCell operating in the unlicensed spectrum. The LAA SCells act as regular SCells where the eNB and the UE use LBT before performing transmission in the SCell. The combined time of transmissions by an eNB may not exceed 50 ms in any contiguous 1 second period on an LAA SCell. The LBT type (i.e., type 1 or type 2 uplink channel access) which the UE applies is signaled via uplink grant for physical uplink shared channel (PUSCH) transmission on LAA SCells. For type 1 uplink channel access on autonomous uplink (AUL)[7], the eNB signals the *channel access priority class* for each logical channel and the UE must select the lowest *channel access priority class* of the logical channel(s) with MAC SDU multiplexed into the MAC PDU. The MAC control elements, except padding buffer status report (BSR), use the highest *channel access priority class*. For type 2 uplink channel access on AUL, the UE

[7] In autonomous uplink, uplink transmissions are allowed without requiring a prior scheduling request or an explicit scheduling grant from the eNB.

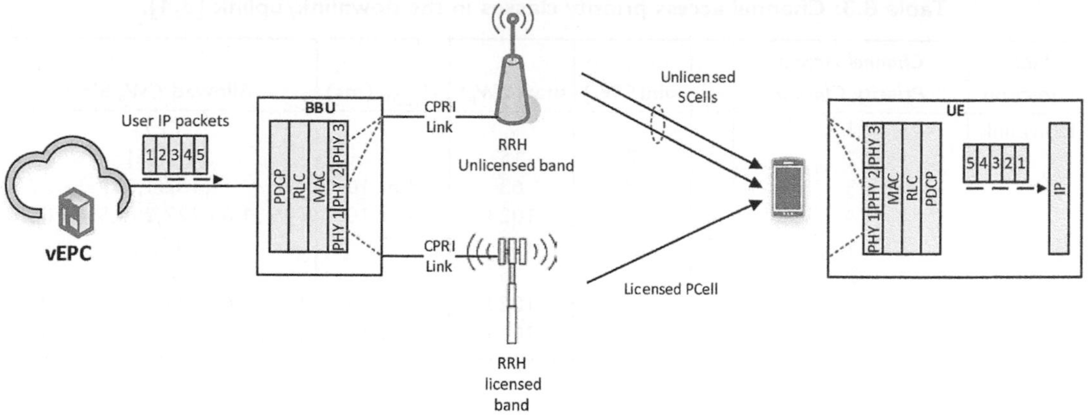

Figure 8.12
Illustration of LAA operation concept [18].

may select logical channels corresponding to any *channel access priority class* for uplink transmission in the subframes signaled by eNB in common downlink control signaling. For uplink LAA operation, the eNB does not schedule the UE more than the minimum number of subframes necessary to transmit the traffic corresponding to the selected *channel access priority class* or lower than the following: (1) The *channel access priority class* signaled in uplink grant based on the latest buffer status report and received uplink traffic from the UE if type 1 uplink channel access procedure is signaled to the UE. (2) The *channel access priority class* used by the eNB based on the downlink traffic, the latest BSR and received uplink traffic from the UE, if type 2 uplink channel access procedure is signaled to the UE [5].

Four *channel access priority classes* are defined as shown in Table 8.3, which can be used when performing uplink and downlink transmissions on unlicensed carriers. In LTE-based LAA, each *channel access priority class*, used by user traffic, is associated with the standardized LTE QoS class identifiers (QCIs); i.e., channel access priorities 1,2,3,4 are associated with QCI values (1,3,5,65,66,69,70),(2,7),(4,6,8,9),(-), respectively [5]. A non-standardized QCI (operator defined QCI) should use a suitable *channel access priority class* associated with the standardized QCIs, which best matches the traffic class. If a downlink transmission on physical downlink shared channel (PDSCH) is intended for which channel access has been obtained using *channel access priority class* $p \in \{1, 2, 3, 4\}$, the LTE system must ensure that the transmission duration does not exceed the minimum duration needed to transmit all available buffered traffic; and further the transmission duration does not exceed the maximum channel occupancy time (MCOT)[8] as defined in Table 8.3 for each *channel*

[8] Maximum MCOT is the maximum continuous transmission time after channel sensing, while Wi-Fi may transmit for much shorter time duration. The maximum continuous transmission time in license-assisted access is limited to 8 or 10 ms regardless of frame size and data rate.

Table 8.3: Channel access priority classes in the downlink/uplink [3,4].

Link Direction	Channel Access Priority Class p	m_p	$\min(CW_p)$	$\max(CW_p)$	T^p_{MCOT} (ms)	Allowed CW_p Sizes
Downlink	1	1	3	7	2	{3,7}
	2	1	7	15	3	{7,15}
	3	3	15	63	8 or 10	{15,31,63}
	4	7	15	1023	8 or 10	{15,31,63,127,255,511,1023}
Uplink	1	2	3	7	2	{3,7}
	2	2	7	15	4	{7,15}
	3	3	15	1023	6 or 10	{15,31,63,127,255,511,1023}
	4	7	15	1023	6 or 10	{15,31,63,127,255,511,1023}

access priority class. The downlink burst in the above refers to the continuous transmission by LTE after a successful LBT. For uplink PUSCH transmission, there is no additional restriction on the UE regarding the type of traffic that can be carried in the scheduled subframes [10].

LTE Rel-15 further enhanced the uplink transmission in an LAA SCell in the unlicensed spectrum by specifying support for multiple uplink starting and ending point in a subframe, and support for AUL transmission, including channel access mechanisms, core and RF requirements for base stations and UEs, and RRM requirements. The key functionalities introduced in LTE Rel-15 include additional starting and ending points for PUSCH transmissions on an LAA SCell (starting PUSCH transmission at the slot boundary; ending PUSCH transmission after the third symbol or at the slot boundary; and UE-based selection of the starting point for PUSCH transmission at the subframe or slot boundary depending on successful channel access); AUL access where a UE can be RRC-configured with a set of subframes and HARQ processes that it may use for autonomous PUSCH transmissions [8,11]. The AUL operation is activated and released with DCI format 0A (transmission mode [TM1]) or DCI format 4A (TM2). The UE skips an AUL allocation if there is no data in the uplink buffers. The PRB allocation and MCS as well as DMRS cyclic shift and orthogonal cover code are indicated to the UE via AUL activation DCI. The UE informs the eNB along with each AUL transmission of the selected HARQ-process ID, new data indicator, redundancy version, UE ID, PUSCH starting and ending points, as well as whether the UE-acquired channel occupancy time can be shared with the eNB. The eNB may provide the UE with HARQ feedback for AUL-enabled HARQ processes, transmit power command, and transmit PMI [5,6].

LTE Rel-13 introduced UE RSSI measurements with configurable measurement granularity and time instances of the reports which can be used for the assessment of hidden nodes by an eNB which is located near specific UEs. For example, if UE measurement shows a high RSSI value when the serving cell is inactive due to LBT, it can imply the presence of hidden nodes and can be considered for channel (re)selection. The DFS is a regulatory requirement for certain frequency bands in some regions, for example, to detect interference from radar systems and to avoid cochannel operation with these systems by selecting a different carrier on a relatively slow time scale. The corresponding time scales for DFS are in the order of seconds and can, therefore, be at an even slower time scale than carrier selection. This functionality is an implementation issue and does not impact the specifications. As we mentioned earlier, LBT procedure is defined as an algorithm by which a UE performs one or more CCAs prior to transmitting on the channel. It is the LAA equivalent of the DCF and EDCA MAC protocols in Wi-Fi (see Fig. 8.13).

3GPP has specified four categories of LBT as follows:

- Cat-1: no LBT
- Cat-2: LBT without random backoff
- Cat-3: LBT with random backoff and fixed CW
- Cat-4: LBT with random backoff and variable CW

A simple approach to ensure coexistence with Wi-Fi is to make the LAA LBT procedure for both data and discovery reference signal[9] as similar as possible to IEEE 802.11 DCF and EDCA protocols. This was the underlying design principle in Rel-13 LAA LBT mechanism for detecting the presence of Wi-Fi which can be summarized as follows:

- An LAA-enabled node must sense the carrier to be idle for a random number of 9 μs CCA slots prior to data transmission.
- If the energy in a CCA slot is sensed to be above an energy detection (ED) threshold, then the process is suspended and the counter is stopped. The backoff process is resumed and the counter is decremented once the carrier has been idle for the duration of the deferred period. The energy detection threshold is both channel type and output power related, that is, -72 dBm for 23 dBm PUSCH and -62 dBm for DRS. An important component of LBT design is the choice of ED threshold, which determines the level of sensitivity to declare the existence of ongoing transmissions. 3GPP has studied mechanisms to adapt the ED threshold. An eNB accessing a carrier on which LAA SCell(s) transmission(s) are performed, sets the ED threshold $ED_{threshold}$ to less than or equal to the maximum ED threshold $\max(ED_{threshold}) = \min(T_{\max} + 10\text{ dB}, \eta_r)$, if the absence of any other technology sharing the carrier can be guaranteed on a long-term

[9] A downlink transmission by the eNB may include a discovery signal and without physical downlink shared channel on a carrier on which license-assisted access SCell(s) transmission(s) are performed, immediately after sensing the idle channel for a sensing interval $T_{drs} = 25\ \mu s$ and if the duration of the transmission is less than 1 ms. The parameter T_{drs} consists of a duration $T_f = 16\ \mu s$ immediately followed by one slot duration $T_{slot} = 9\ \mu s$ and T_f includes an idle slot duration T_{slot} at start of T_f. The channel is considered idle for T_{drs}, if it is sensed to be idle during the slot durations of T_{drs}.

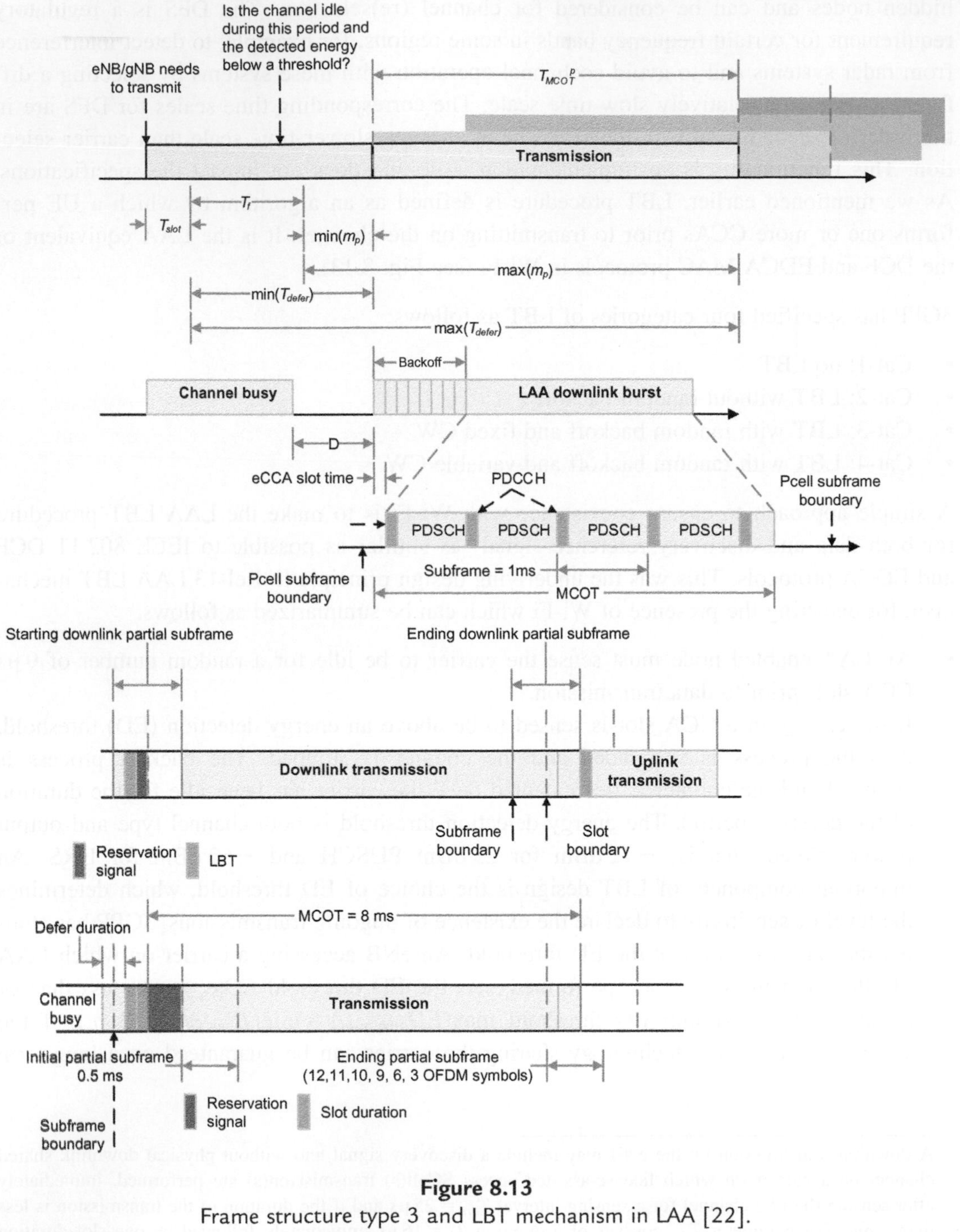

Figure 8.13
Frame structure type 3 and LBT mechanism in LAA [22].

basis, where η_r is maximum ED threshold defined by regulatory requirements in dBm when such requirements are defined; otherwise [4]:

$$ED_{threshold} = \max\left\{\begin{array}{l} -72 + 10\log_{10}(BW\ \mathrm{MHz}/20\ \mathrm{MHz})\ \mathrm{dBm}, \\ \min\left\{\begin{array}{l} T_{\max}, \\ T_{\max} - T_A + \left(P_H + 10\log_{10}(BW\ \mathrm{MHz}/20\mathrm{MHz}) - P_{TX}\right) \end{array}\right\} \end{array}\right\}$$

where P_H is a reference power equal to 23 dBm, $T_A = 10$ dB for transmission(s) including PDSCH or $T_A = 5$ dB for transmissions including discovery reference signal transmission(s) and excluding PDSCH, P_{TX} is the configured maximum transmit power for the carrier in dBm, and it is given by $T_{\max} = -75\ \mathrm{dBm/MHz} + 10\log_{10}(BW)$, where BW is the channel bandwidth in MHz. In a nutshell, the energy detection threshold can be increased if the bandwidth increases and/or the transmit power decreases. The detected energy level needs to be below this threshold for a certain period of time with a slot duration $T_{slot} = 9\ \mu s$ and defer time $T_{defer} = T_f + m_p$, where $T_f = 16(\mu s) m_p$ is based on the *channel access priority class* and is, therefore, traffic-type dependent and has a duration of at least one slot. The defer time is required to specifically protect Wi-Fi ACK/NACK transmission between the access points and the clients. As a result, the channel needs to be idle for an initial CCA period of 34 μs and a maximum wait time of 88 μs before an LAA-capable eNB can start its transmission. If the channel is sensed to be clear, the transmitter can only transmit for a limited amount of time defined as the maximum channel occupancy time T^p_{MCOT}. If the channel is sensed to be occupied during that time or after a successful transmission, the enhanced CCA period is started by generating a random number that is within the contention window. Because there are different traffic types (VoIP, video, background traffic, etc.), different *channel priority access classes* have been defined with different CW sizes and T^p_{MCOT} (see Table 8.3).

- If the most recent downlink transmission burst showed 80% or more decoding errors, as reported via HARQ feedback (NACKs) from UEs, then the CW is doubled for the next LBT (see Fig. 8.14).
- Once downlink transmission is complete, a new random backoff is chosen and used with the next transmission.
- A single, short CCA period of 25 μs can be used to transmit control information without accompanying data such as DRS. This is aligned with the CCA duration used for Wi-Fi beacon frames.
- Rel-13 LTE defined an LAA equivalent to the four Wi-Fi priority classes in the form of four sets of minimum and maximum CW sizes, maximum channel occupancy times, and deferred period CCA slots.

An LAA/Wi-Fi coexistence study suggests that the coexistence performance is more sensitive to factors that affect the channel occupancy (e.g., control signals) rather than to the

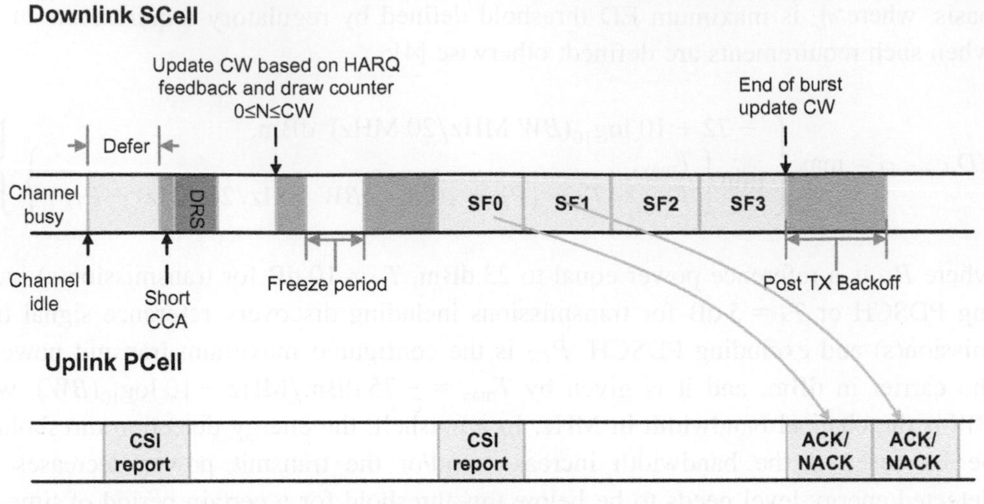

Figure 8.14

An example of LAA downlink transmission with LBT CW updates based on HARQ ACK/NACK feedback. The UE provides HARQ feedback and CSI reports on the licensed carrier [17,23].

choice of parameters in the LBT CCA and backoff algorithms. Consequently, the coexistence is highly affected by the behavior of the upper-layer protocols; bursty traffic pattern; HARQ-based CW slot update; and either CTS-to-self or support for lower Wi-Fi energy detection thresholds, which seems to be a fundamental feature to be supported by LAA to allow coexistence with Wi-Fi and protect the LAA performance in the presence of hidden nodes [12].

To minimize interference to Wi-Fi transmissions, an LTE system can mute its operation in almost blank subframes (ABS). These subframes are called almost blank because LTE can still transmit some broadcast signals, control signals, and synchronization signals over these subframes. However, these signals only use a small fraction of system resources in the time/frequency domain and reduced power. Therefore, cochannel interference is much less during the ABS transmission periods.

The LTE-based LAA defines methods for UEs in the LAA SCell coverage to provide radio resource management measurements and reporting. The Rel-12 discovery reference signals, originally designed for small cells, are being used with certain modifications in LTE-based LAA SCells. As shown in Fig. 8.15, the LTE-based LAA DRS can be transmitted within a periodically occurring time window called the DRS measurement timing configuration occasion that has a duration of 6 ms and a configurable period of 40, 80, or 160 ms. The radio frame and subframe in which DRS can be transmitted depends on the RRC parameters (*dmtcOffset, dmtcPeriodicity*) that are signaled to the UE. The LTE-based LAA DRS can be transmitted following a single idle observation interval of at least 25 μs. To compensate for

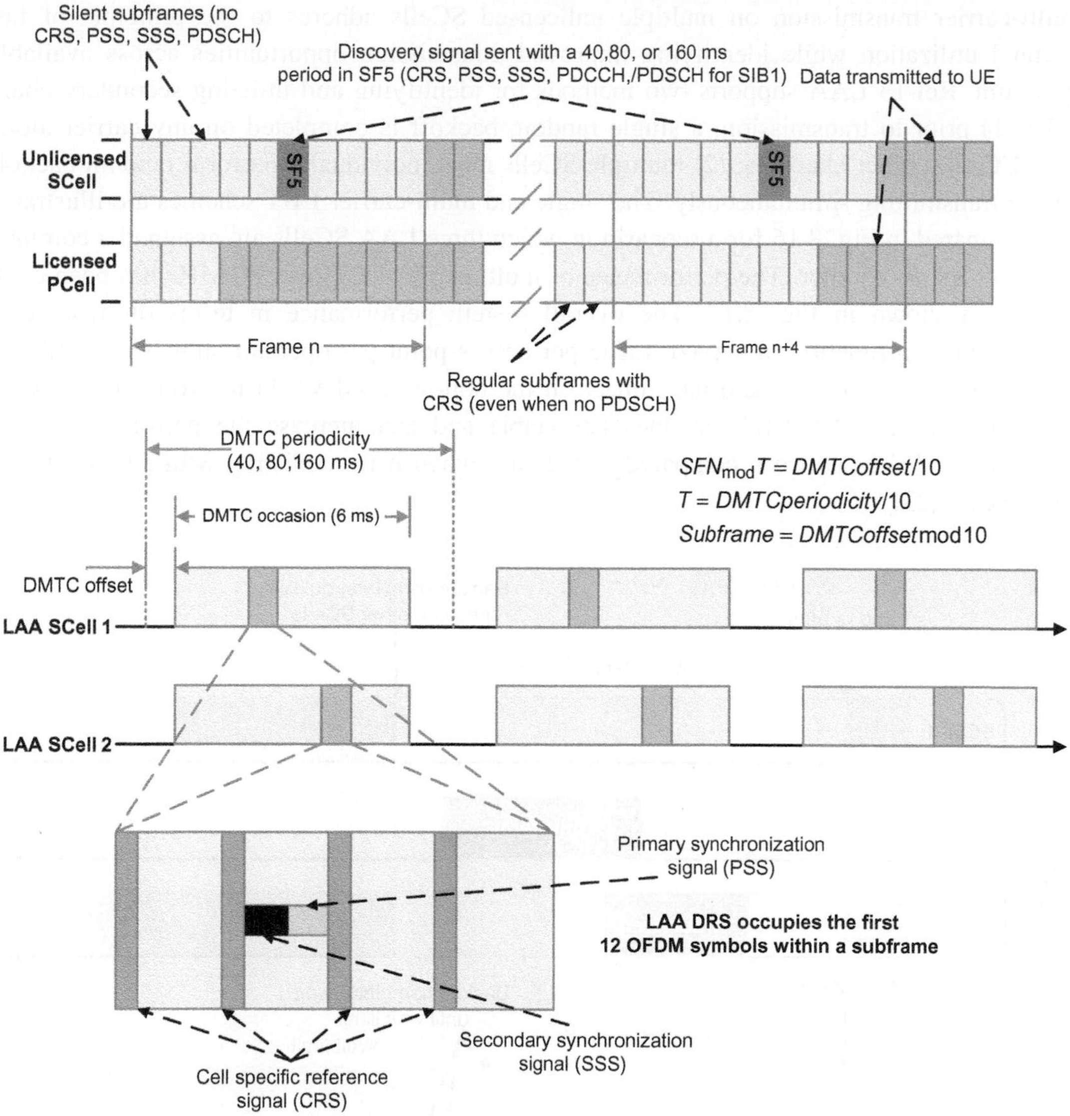

Figure 8.15
Structure and timing of discovery signals in LAA SCell [14,20].

potential DRS transmission blocking due to LBT and increase the probability of successful of DRS transmission, the network can attempt DRS transmission in any subframe within the DMTC occasion [26].

LAA operation on multiple unlicensed carriers is a key requirement for maximizing throughput. IEEE 802.11ac/ax supports transmission bandwidths of up to 160 MHz, which would span eight contiguous 20 MHz unlicensed channels in the 5 GHz band. The LAA multi-carrier transmission on multiple unlicensed SCells adheres to the principle of fair channel utilization while identifying improved transmission opportunities across available spectrum. Rel-13 LAA supports two methods for identifying and utilizing secondary channels: (1) prior to transmission, a single random backoff is completed on any carrier along with CCA on other channels; (2) multiple SCells must individually perform random backoff before transmitting simultaneously. The single and multi-carrier LBT schemes are illustrated and compared in Fig. 8.16 for a scenario in which three LAA SCells are assigned a common random backoff counter. The performance of multi-carrier LBT over 80 MHz has been evaluated and shown in Fig. 8.17. The overall system performance in terms of mean user throughput as a function of served traffic per access point per operator show that, from the coexistence point of view and the impact on the non-replaced Wi-Fi network, both classes of multi-channel LAA LBT schemes are viable and can increase the performance of a multi-carrier Wi-Fi network compared to the case when it is coexisting with another Wi-Fi network [17,23].

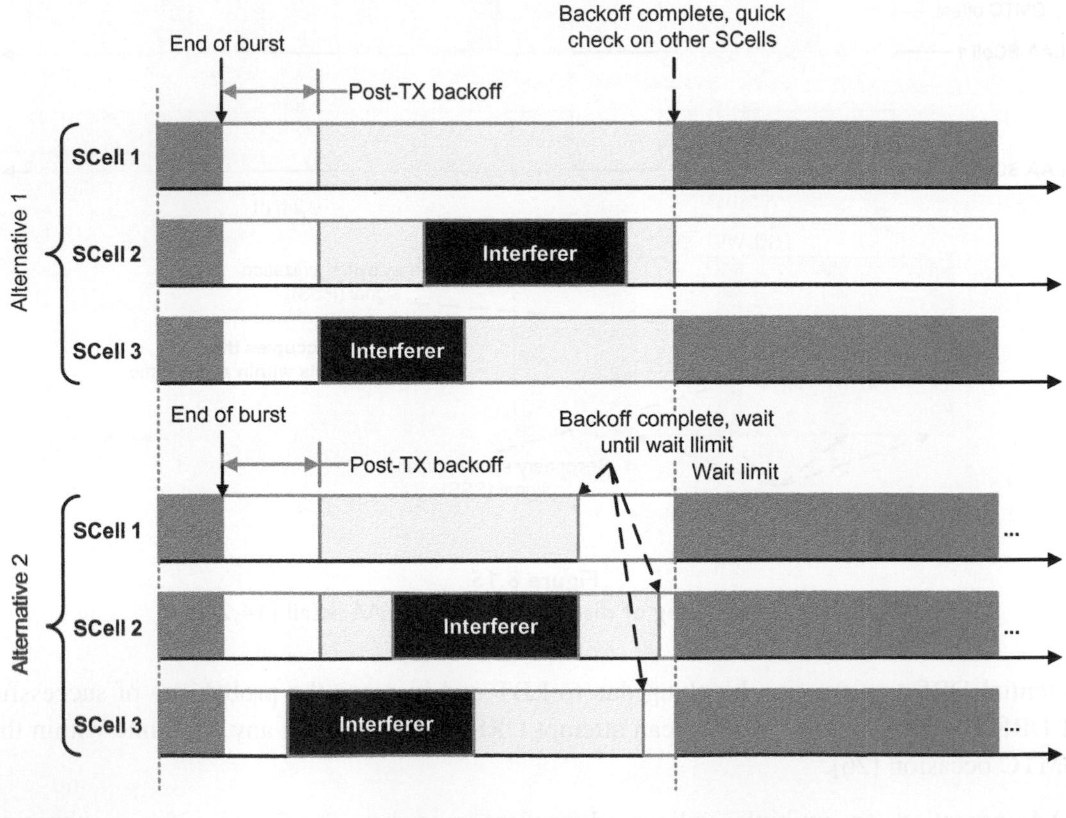

Figure 8.16
Comparison of LAA multi-carrier LBT access schemes with single and multiple random backoff channels [17,23,31,32].

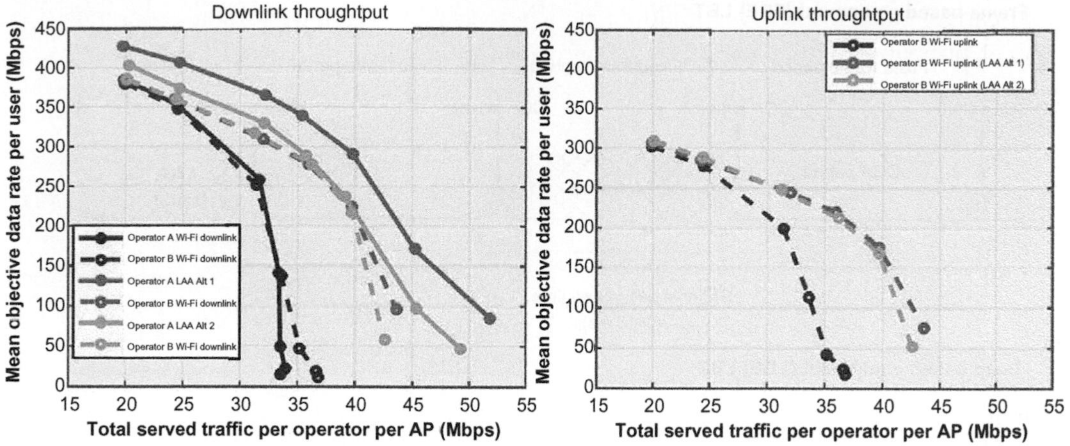

Figure 8.17

Mean user throughput versus served traffic per access point per operator for the indoor multi-carrier deployment scenario with FTP traffic using up to 80 MHz transmission bandwidth [17,23].

As mentioned earlier, LBT is a mechanism used by devices operating in the unlicensed bands to determine the presence of other signals in the channel prior to transmission and to avoid collisions with other transmissions. This protocol allows several users and different technologies to use the same channel without pre-coordination. In LAA, the LBT procedure is initialized when the SCell has data to transmit. Prior to that, the cell is completely turned off to all transmissions, including cell reference signals. Then, the LAA cell will initiate a CCA to determine if the channel is idle. If there are no signals detected in the channel, then the transmission can proceed. This procedure is illustrated in Fig. 8.18. If the channel is not idle, the device performs a slotted random backoff procedure, in which a random number of slots are withdrawn from the contention window. The contention window is increased exponentially with the occurrence of collisions and is reset to the minimum value once the transmission succeeds. Given the random nature of the backoff procedure, different devices will have different backoff intervals, improving channel adaptation [14,20].

8.1.4.2 LTE-Wi-Fi Aggregation

An alternative approach to deploying LTE in the unlicensed spectrum, which was more acceptable to the Wi-Fi industry, was through the aggregation of LTE and Wi-Fi carriers. This solution was meant to enhance LTE performance by partly offloading traffic to an available Wi-Fi link. In this scheme, an LTE payload is split at the PDCP level and some traffic is tunneled over the Wi-Fi link while the remaining traffic is sent over the native LTE connection. The LWA approach uses Wi-Fi access points to augment LTE RAN by encapsulating LTE data in IEEE 802.11 MAC frames, such that it appears as a Wi-Fi frame

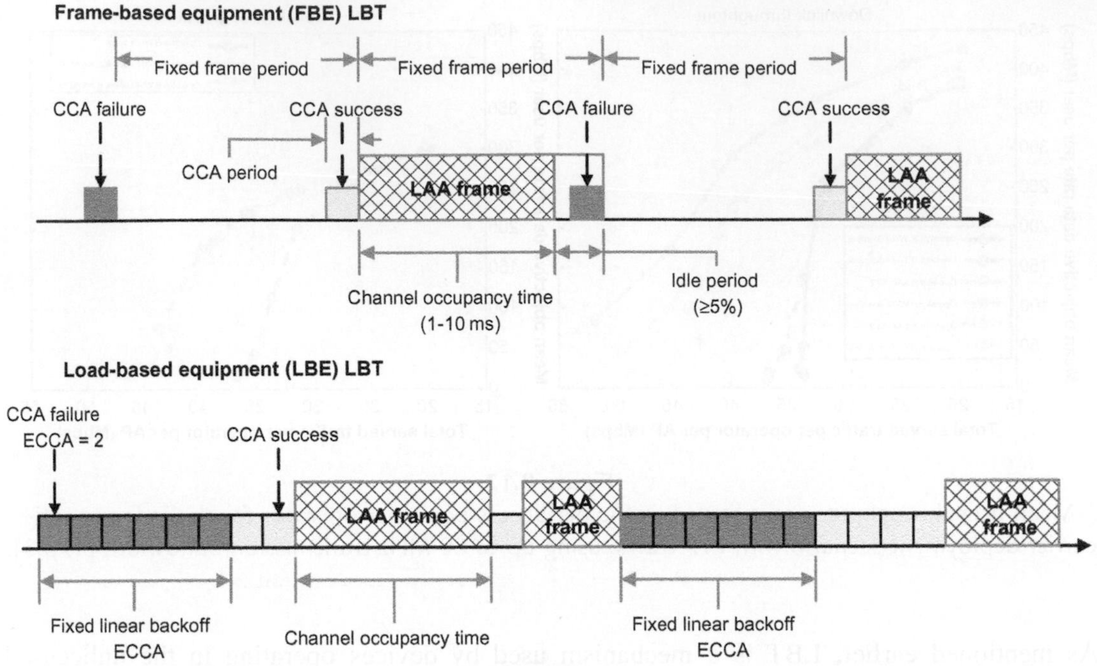

Figure 8.18
Illustration of channel sensing before transmission until the channel is unoccupied [14,20].

to another network. This method would allow both technologies to operate in their respective spectrum, that is, Wi-Fi operating in the unlicensed band and LTE continues to operate in a licensed band, while there are no changes in their respective access mechanisms. This is the significant difference compared to LTE-unlicensed and LAA schemes. In the case of LWA solution, the aggregation takes place at the radio link level. The Wi-Fi access points can use the LTE core network functions (e.g., authentication, security, etc.). It will be discussed in Section 8.2 that the LWA architecture consists of LWA eNB, LWA-aware Wi-Fi access points, and LWA UEs. The LWA eNB and Wi-Fi AP can be co-located or non-co-located. If non-co-located, data is delivered through IP tunnel between two systems. The LWA eNB performs scheduling of PDCP packets and transmits some of the IP packets over the LTE air-interface and others over the Wi-Fi after encapsulating them in Wi-Fi MAC frames. All packets, received over either LTE or Wi-Fi, are then aggregated at the PDCP sublayer of the LWA UE. The Wi-Fi APs are connected to LWA eNBs and report information on channel conditions to the LWA eNB. The LWA eNB then determines whether to use the Wi-Fi link or not. The LWA eNB can improve LTE performance by managing radio resources in real-time according to the RF and load conditions of both LTE and Wi-Fi. A Wi-Fi AP can work as a native Wi-Fi AP while not serving the LWA purpose. This

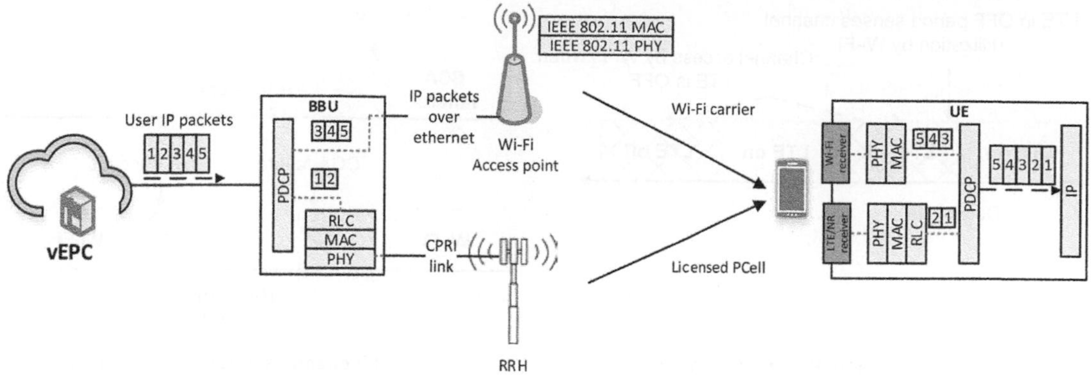

Figure 8.19
Illustration of LWA principle of operation [18].

eliminates potential coexistence issues in the unlicensed bands. However, as LTE data must be split at eNB and then aggregated at UE, the involved nodes such as eNB, Wi-Fi AP, and UE must be LWA-enabled. The LWA architecture, protocol, and operations will have to be defined as well. Fig. 8.19 shows the high-level architecture and operation of LWA scheme [18].

8.1.4.3 LTE Unlicensed and MulteFire

In 2014, a group of companies formed the LTE-U Forum to develop technical specifications based on LTE Rel-12 in the 5 GHz band. LTE-U Forum initially focused on non-LBT markets without modifying the LTE physical layer and MAC functionalities. The deployment scenarios included supplemental LTE downlink operating in an unlicensed spectrum comprising multiple contiguous 20 MHz bands. The LTE network with an uplink and downlink anchored on the licensed spectrum (up to 20 MHz) for delivering essential control signaling and synchronization related to radio resource management and connection/mobility provides reliability, mobility, and coverage, whereas the unlicensed spectrum was exclusively used for increasing the downlink data rates.

LTE-U was designed to comply with the US regulations and targets other regions of the world with no LBT requirements. However, it does not necessarily meet the needs of other regions which do require LBT feature such as Europe and Japan. While LTE-U and LAA are both LTE-based solutions for operation in the unlicensed band, they differ in the way they meet the regulatory requirements. LTE-U is well suited for regions that do not require LBT and address coexistence issues with Wi-Fi using an algorithm called carrier sensing adaptive transmission (CSAT) in the LTE-U eNB, which is based on the use of duty cycle that adapts the duration of the ON period by sensing the average power in the channel as shown in Fig. 8.20. LTE-U also limits the duration of the ON period to 20 ms so that low

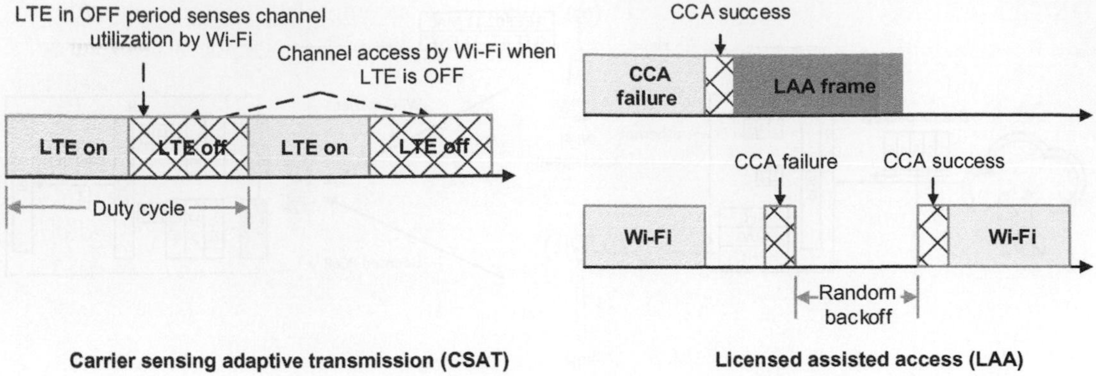

Figure 8.20
Comparison of CSAT and LBT mechanisms [17,23].

latency applications can coexist. On the other hand, LAA is a global solution and includes support for LBT and a more flexible frame structure to better adapt to channel availability. LAA is generally preferred by operators as it is the only method that provides full cellular capabilities, whereas LWA or LWIP are less frequently used because they require either a Wi-Fi network or partnership with a Wi-Fi operator.

In an ultra-dense deployment of Wi-Fi and LTE-U small cells, there is a possibility that no clean channel can be found. In such cases, LTE-U can share the channel with the neighboring Wi-Fi or another LTE-U system by using the CSAT algorithm. Typical cochannel coexistence techniques in unlicensed bands such as LBT and CSMA, used by Wi-Fi systems, are based on the contention-based access. In these techniques, transmitters are expected to sense the medium to ensure that there is no activity prior to transmission. The goal of these algorithms is to provide coexistence across different technologies in a time-division multiplexing (TDM) manner. LTE-U in unlicensed spectrum uses CSAT scheme that relies on the same concept of TDM coexistence and medium sensing. In CSAT, the small cell senses the medium for longer (than LBT and CSMA) duration (up to 200 ms), and according to the observed medium activities, the algorithm proportionally switches off LTE transmission. In particular, CSAT defines a time cycle where the small cell transmits in a fraction of the cycle and remains off in the remaining duration. The duty cycle of transmission relative to OFF interval is determined by the sensed medium activity of other technologies. The CSAT is conceptually similar to CSMA except that it has longer latency, an impact that is mitigated by avoiding channels where Wi-Fi APs use for discovery signals and QoS-enforced traffic. The LTE-U, which is deployed on the SCell, is periodically activated and deactivated using LTE MAC control elements. The procedures and timeline are chosen to ensure compatibility with legacy LTE UE behavior. During the LTE-U OFF period, the channel is released to neighboring Wi-Fi APs, which can resume normal transmissions. The small cell will measure Wi-Fi medium utilization during the LTE-U OFF period and adaptively adjust

ON/OFF duty cycle. The adaptive cycle can effectively accommodate the activation/de-activation procedures while controlling the data transmission delay. Because the anchor carrier in license band is always available, the supplemental downlink carrier in the unlicensed band can be used opportunistically. When the downlink traffic of the small cell exceeds a certain threshold and there are active users within the unlicensed band coverage area, the supplementary downlink (SDL) carrier can be turned on for offloading the traffic. When the traffic load can be managed by the primary carrier or there is no user within the unlicensed band coverage area, the SDL carrier is turned off. Opportunistic SDL mitigates the interference from LTE continuous cell-specific reference signal transmission in an unlicensed channel.

MulteFire was proposed by the MulteFire Alliance in late 2015 as a standalone version of LTE for small cells, which operates in the unlicensed spectrum and can provide service to users with or without a universal subscriber identity module card (i.e., an operator subscription). It combines the advantages of LTE technology and the simplicity of Wi-Fi deployments and can be deployed either by traditional mobile operators or neutral hosts. It specifies two different architectures: (1) a PLMN access mode, which allows mobile network operators to extend their coverage into the unlicensed band, especially in the case where a licensed spectrum is not available at certain locations; and (2) a neutral host network access mode, which is similar to Wi-Fi, a self-contained network deployment that provides access to the Internet. Because of the nature of transmission in the unlicensed band and the need to adhere to the LBT requirements, MulteFire has introduced some modifications to the radio interface of LTE [16].

8.1.4.4 NR-Unlicensed

The NR-U study item in Rel-16 is identifying and evaluating solutions for the NR when operating in the unlicensed spectrum. 3GPP will further investigate the coexistence with NR-based systems in the licensed and unlicensed bands as well as with LTE-based LAA and other incumbent RATs to ensure compliance with regulatory requirements. 3GPP Rel-15 NR introduced a flexible frame structure with various subcarrier spacing options and slot formats. It further introduced mini-slot configuration, whereby PDSCH resource allocation can start from almost any symbol in a slot. The NR-U is expected to support both baseline NR Type A (slot-based) and Type B (non-slot-based) resource allocation schemes. Even with restricted mini-slot length of 2, 4, and 7 symbols for downlink, the PDSCH transmission (without gap) can still start at almost any symbol position, if we allow multiple mini-slots to be allocated within a slot. In the uplink, however, the mini-slot length and starting symbol are not restricted, thus uplink transmission in NR-U (without gap) can start at any symbol position.

For sub-7 GHz unlicensed bands, 15 kHz SCS with normal cyclic prefix that was supported in LTE-based LAA and the same SCS must be supported in NR-U to achieve similar coverage when the component carrier bandwidth is 20 MHz. On the other hand, if a larger SCS is used, the OFDM symbol duration will be shorter, which increases the channel access opportunities, reduces latency, and enhances resource utilization. A larger SCS can facilitate single wideband-carrier operation. In carrier aggregation scenarios (similar to LTE-based LAA), the device performs LBT per component carrier and then transmits on any available component carrier. Operating a single carrier allows reducing the control overhead and avoiding the use of inter-carrier guard bands. For the same system bandwidth, a larger SCS requires a smaller FFT size, hence reducing the implementation complexity. The studies suggest that an NR system with SCS of 60 kHz operating as a wideband carrier obtains about 60% average downlink user throughput gain and 80% average uplink user throughput gain over LTE-based LAA system for the same system bandwidth [14,20]. The performance gain comes from the finer granularity of channel occupation in the time domain. In addition, the processing delay for the short symbol demodulation is reduced, so retransmission for short symbols in the same MCOT becomes possible and HARQ combining gain can be achieved with low latency. Furthermore, the overhead of guard band with 60 kHz SCS wideband carrier is also smaller than that of LTE-based LAA. However, in carrier aggregation scenarios, the performance gain of 60 kHz SCS relative to LTE-based LAA diminishes because of the spectrum efficiency loss from guard bands in 60 kHz SCS and 20 MHz bandwidth. The occupied channel bandwidth of an SS/PBCH block with subcarrier spacing 15 and 30 kHz is 3.6 and 7.2 MHz, respectively. An SS/PBCH block with subcarrier spacing 15 kHz in 5 GHz unlicensed band cannot meet the OCB requirements because system/nominal bandwidth in most cases would be greater than 5 MHz. The NR-U should support similar SCS in 6–7 GHz unlicensed bands. To support 60 kHz SCS for all NR-U signaling in operation below 6 GHz, some modifications in Rel-15 NR are required because NR does not support SS/PBCH block transmission using 60 kHz. In addition, the candidate SS/PBCH positions for a half frame are not defined for 60 kHz SCS. Note that the use of 60 kHz SCS for PRACH transmission in NR is also restricted to frequencies above 6 GHz. For outdoor deployments, even if the output power is limited, considerable delay spread can still be observed. For 60 kHz SCS with normal CP, the cyclic prefix is 1.17 μs corresponding to a range of 175.5 m, which is very typical in outdoor environments. The cyclic prefix of 60 kHz SCS with extended CP is 4.17 μs corresponding to a range of 625.5 m, which would be sufficient for most outdoor small cell deployments.

In NR, the SCS for each channel or signal is configured through the BWP information element (IE) configuration, and the subcarrier spacing in the BWP IE is used for all channels and reference signals unless explicitly configured. The reference SCS of the slot structure configuration is included in SIB1. Furthermore, the SCS for SIB1, Msg.2/Msg.4 for initial access, and broadcast SI-messages is included in the MIB. For NR-U, these SCS configuration schemes can be reused directly. In licensed bands, NR supports dynamic and semi-statically configured slot configuration. The semi-static DL/UL assignment is done via cell-specific and UE-specific RRC configuration. With the cell-specific configuration, the UE is

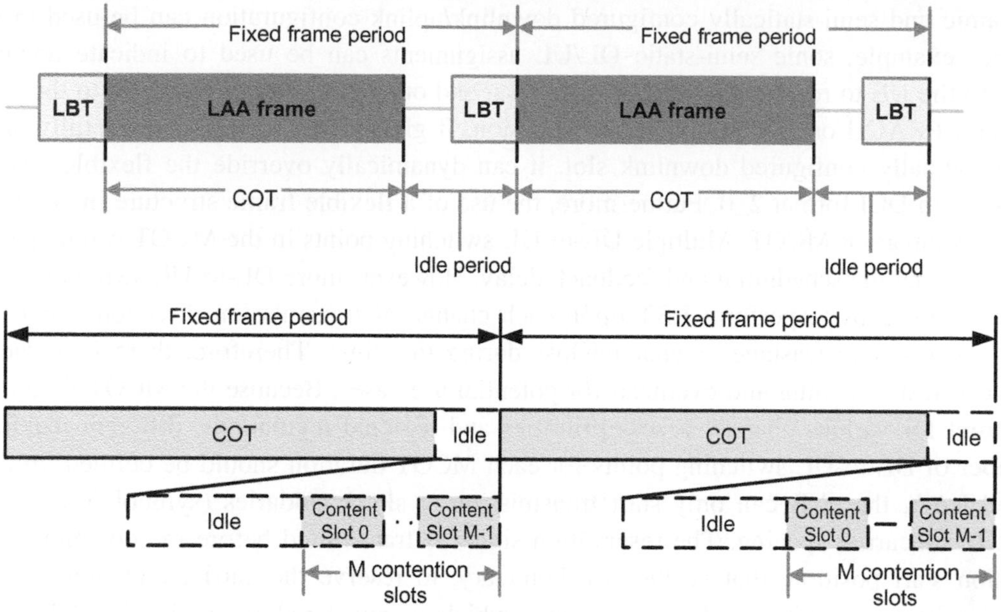

Figure 8.21
Illustration of timing relation for FBE [24,25].

allocated flexible resources that can be assigned to downlink or uplink using UE-specific configuration or dynamic signaling. The UE-specific RRC signaling or dynamic signaling includes per-slot-basis indication that can override the unknown allocation in the cell-specific configuration. Resources in the slot configuration that are without downlink/uplink indication are considered flexible resources.

Frame-based equipment (FBE)[10] is a channel access mechanism wherein the transmit/receive frame structure has a periodic timing with a periodicity known as the fixed frame period (FFP), which is between 1 and 10 ms. The transmitting device performs a single-shot LBT before starting transmission in a channel at the beginning of an FFP, and the channel occupancy time (COT) associated with a successful LBT (see the example shown in Fig. 8.21). In addition to the load-based equipment (LBE) operation mode, NR-U can also support the FBE operation mode for use cases in which other LBE-based networks (e.g., Wi-Fi, LTE-based LAA) are excluded such as in a factory or private NR-U network. In comparison to the LBE mode, the FBE operation mode can achieve higher spectrum utilization in such scenarios considering much simpler LBT process. The necessity of signaling to indicate FFP and channel occupancy time to the UE is being studied.

[10] FBE is a scheme where the transmit/receive frame structure is not directly demand-driven but has fixed timing as opposed to LBE where the transmit/receive frame structure is not fixed in time but demand-driven.

Dynamic and semi-statically configured downlink/uplink configuration can be used in NR-U. For example, some semi-static DL/UL assignments can be used to indicate downlink slots to the UE to receive the SS/PBCH blocks, and other slots can be assigned to the UE to transmit PRACH or granted uplink transmission. If gNB performs LBT successfully on the semi-statically configured downlink slot, it can dynamically override the flexible resource (s) through DCI format 2_0. Furthermore, the use of a flexible frame structure in NR-U can take advantage of MCOT. Multiple DL-to-UL switching points in the MCOT can be permitted to reduce the scheduling and feedback delay. However, more DL-to-UL switching points create more gaps to perform LBT upon each change of transmission direction which may result in resource wastage or channel loss during this time. Therefore, there is a tradeoff between fast switching and overhead for potential use cases. Because the MCOT duration is different for various channel access priorities and regional regulations, different maximum number of DL-to-UL switching points for each MCOT duration should be defined. In LTE-based LAA, the eNB can only start transmission at slot boundaries (symbol 0 or 7) with 15 kHz subcarrier spacing. The reservation signal is transmitted before the downlink transmission start position, that is, the slot boundary, to reserve the carrier and prevent other devices from occupying the frequency band, which is considered an overhead in LTE-based LAA system. Therefore, the overhead of reservation signal should be minimized. In NR-U, different slot types including slot and mini-slot (non-slot-based scheduling) can be utilized to minimize the reservation signal length. The gNB or UE should be able to send DL/UL transmissions as soon as possible whenever LBT/CCA procedure is completed, resulting in different slot duration. As illustrated in Fig. 8.22, the gNB can choose the type of the first slot, that is, mini-slot based on LBT/CCA success location and slot boundary, and then select the type of the following slots with a larger size. The type of the last slot(s) is chosen according to the remaining time within the MCOT. In the example shown in Fig. 8.22, the remainder of the MCOT is not enough for a normal slot, thus two mini-slots are allocated [17,28].

In the downlink, because a CORESET can be configured to start at any symbol, gNB can start downlink transmission from the nearest symbol after a successful LBT such that the reservation signal overhead can be minimized. However, this will cause increased blind detection complexity for the UE. Therefore, NR-U would rely on an efficient mechanism for the UE to detect the beginning of a downlink transmission. In LTE-based LAA, the initial signal was discussed to facilitate the detection of downlink burst. The NR supports non-slot-based PUSCH scheduling in the uplink and the start symbol can be dynamically indicated by DCI. Therefore, the UE can transmit PUSCH starting from the indicated symbol. However, this position may not align with the nearest symbol where LBT succeeds. Consequently, more reservation signals may be needed for the uplink. Because NR already supports multiple start positions and durations, it can provide sufficient transmission flexibility for NR-U [10,33].

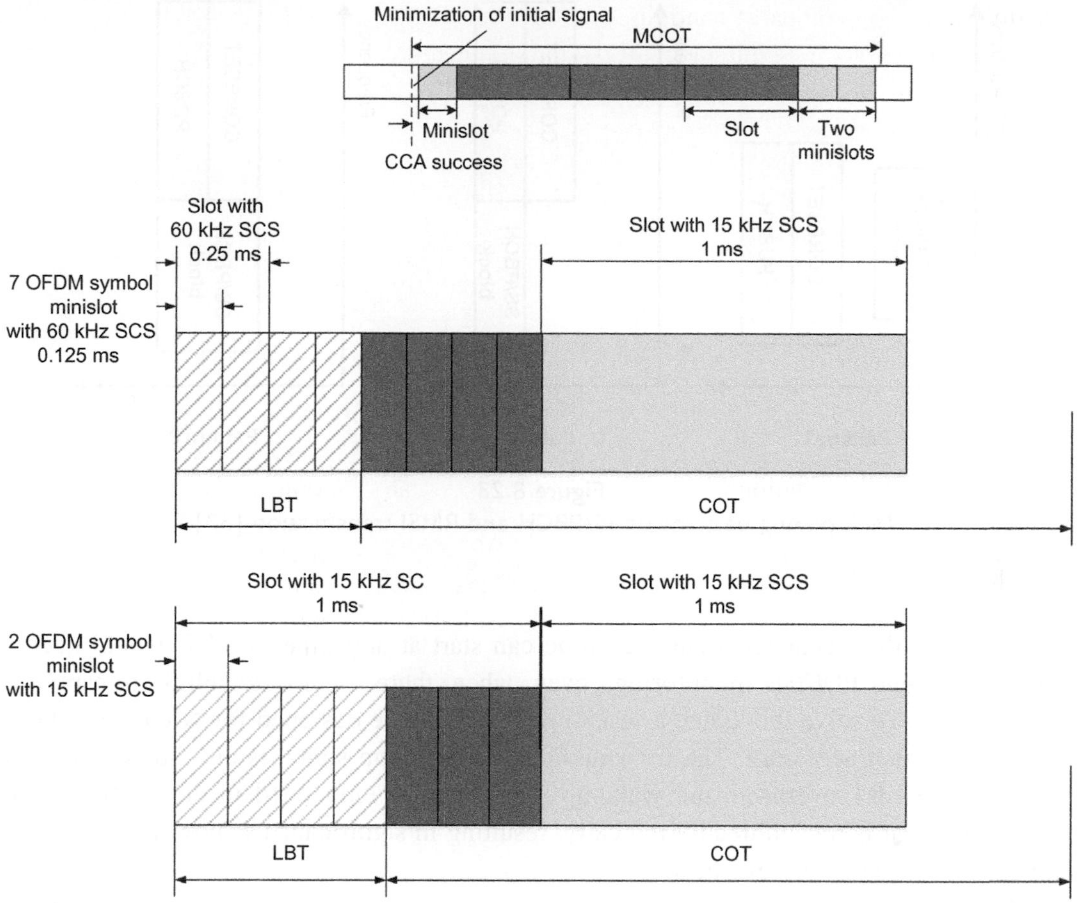

Figure 8.22
Multiple slot durations in an MCOT with different SCS [17,23].

In NR licensed operation, there are three patterns for multiplexing the SS/PBCH block and the RMSI transmissions as shown in Fig. 8.23. The number of PRBs for SS/PBCH blocks is 20 whereas the minimum number of PRBs for RMSI is 24. Using a 20 MHz bandwidth, the OCB requirement can be met if the numerology of SS/PBCH block and RMSI is set to 60 kHz. In that case, Pattern 1 can be considered for NR-U in the sub-7 GHz bands. The TDM method in Pattern 1 allows the RMSI to provide consecutive transmissions between SS/PBCH blocks over the time span of DRS. In contrast, if 30 kHz SCS is selected, the bandwidth of SS/PBCH blocks would only be about 40% of a 20 MHz. In this case, a frequency division multiplex pattern such as Pattern 2 or 3 is preferred in terms of OCB requirement. When RMSI is multiplexed with SS/PBCH blocks using Pattern 2 or 3, the RMSI may not completely fill the gaps between the SS/PBCH block transmissions.

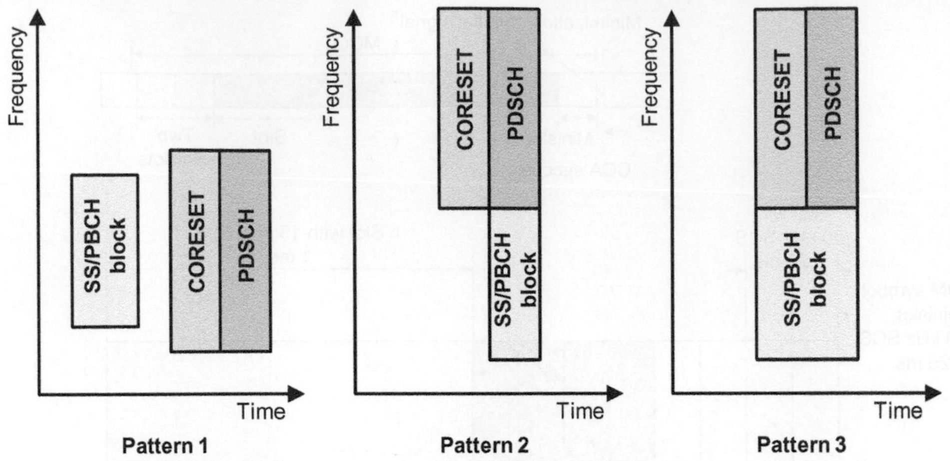

Figure 8.23
Multiplexing patterns of SS/PBCH and RMSI transmission [33].

In NR-U, because a channel occupancy time can start at any time, the UE would need to perform frequent PDCCH monitoring, even when there is no downlink transmission intended for it. To solve this issue, a wake-up signal may be designed for NR-U to wake up the UE only when necessary. The transmission of a wake-up signal is also subject to LBT. To reduce the LBT overhead, the wake-up signal can be embedded in an existing signal such as DRS that is transmitted periodically, resulting in significant UE power saving while maintaining low signaling overhead.

Similar to LTE, NR supports SRS-based frequency-selective scheduling and periodic/aperiodic SRS transmission. NR also supports SRS-based downlink beamforming and semi-persistent SRS transmission, which are not supported in LTE. In NR-U, the transmission time of the SRS resource as well as the configuration of SRS resource and resource mapping adapt to LBT and channel availability. Due to the uncertainty of channel availability, periodic SRS would not be feasible. The aperiodic SRS which is triggered by DCI has more flexibility for NR-U operation. The SRS transmission should also meet the OCB requirements; thus NR-U can only support wideband and BWP/subband SRS transmission.

In LTE-based LAA revisions, downlink and uplink transmissions start and end at certain points. More specifically, using frame structure type 3, a downlink or uplink transmission can start at slot boundaries. Because LBT may end at any time which is not necessarily aligned with the symbol boundaries, the interval between the end of LBT and the start of a PDCCH/PUSCH transmission may be wasted. The partial ending subframe reduces overhead by defining more ending points, that is, OFDM symbol positions 2, 5, 8, 9, 10, 11, and 13 for the downlink and 6, 12, and 13 for the uplink can be used as partial subframe ending

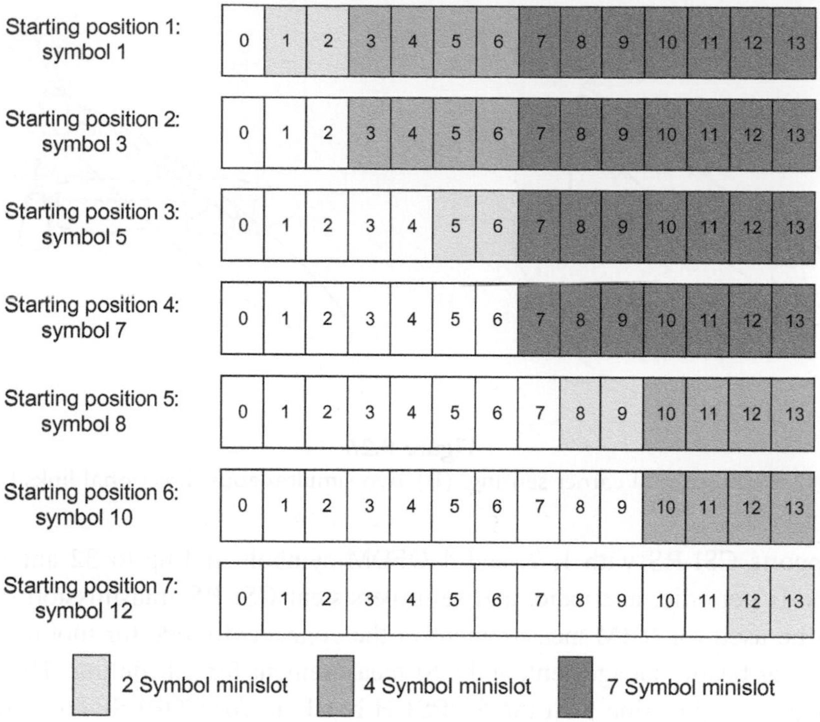

Figure 8.24
Flexible starting point with nonuniform minislot patterns [33].

points. The NR-U provides more frequent start/end points than LTE-based LAA allowing more flexible transmission as shown in Fig. 8.24.

As we mentioned earlier, LBT is a spectrum sharing mechanism that can operate across different RATs. However, it suffers from the hidden node and the exposed node problems. These issues arise when the coverage of the nodes in a network are different. The problem worsens when the sensing and transmit coverages are different. This may occur when an omnidirectional antenna pattern is used for carrier sensing whereas a directional antenna (or array) pattern is used for transmission, resulting in a higher node exposure likelihood. In the example depicted in Fig. 8.25, device A senses the carrier using omnidirectional antenna pattern and overhears the transmission from device C to D. Device A subsequently refrains from transmitting to its targeted receiving device B, which is unnecessary because the transmission from A to B using highly directional beams does not interfere with the transmission from C to D. If the direction of the communication is known, directional carrier sensing may be useful; however, this would create a different problem as shown in Fig. 8.25. The effects of directionality on carrier sensing and transmission are studied in NR-U to ensure optimal system operation and performance [34].

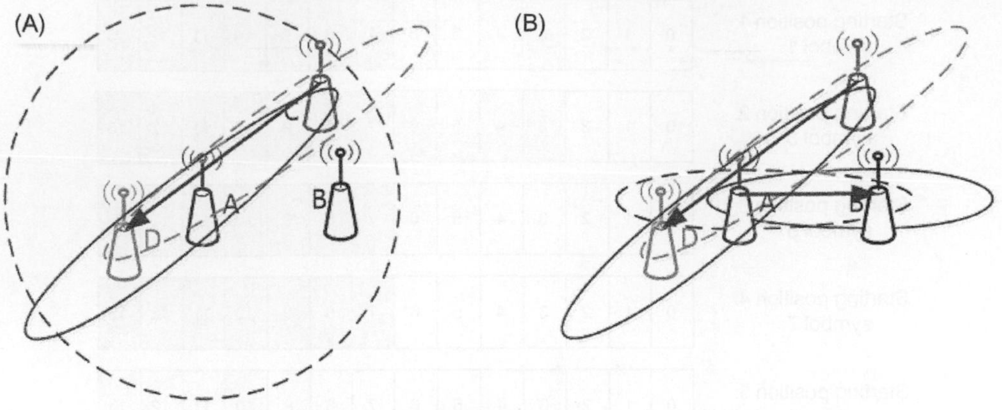

Figure 8.25
(A) Omnidirectional carrier sensing. (B) Two simultaneous directional links [34].

The NR supports CSI-RS with 1, 2, and 4 OFDM symbols and up to 32 antenna ports. It further supports periodic, aperiodic, and semi-persistent CSI-RS transmission patterns. The CSI-RS can be used for RRM measurement in the connected mode for mobility, the RSRP measurement for beam management, and CSI measurement for scheduling. The CSI-RS can be transmitted in conjunction with the SS/PBCH block and/or CORESET/RMSI using time or frequency multiplexing following an LBT process. If frequency-domain multiplexing is used, a CSI-RS resource is configured on the resources that are not occupied by SS/PBCH block and CORESET. Due to the uncertainty of channel availability for transmission of CSI-RS and the corresponding measurement report, the periodic CSI measurement and report is not feasible. Furthermore, the interference condition in the unlicensed spectrum may fluctuate severely and is unpredictable due to the larger delays between the measurement and the report as well as beamforming-based transmission.

In LTE-based eLAA, block interlaced frequency division multiple access (B-IFDMA) was introduced in the uplink transmission to ensure compliance with OCB and maximum PSD requirements while maintaining a transmit power level that can support the desired cell coverage. In NR-U, given similar regulatory requirements, the B-IFDMA serves as the baseline design for the uplink transmission. NR supports a large number of channel bandwidth combinations and subcarrier spacings, which makes the realization of a unified B-IFDMA design very challenging. A typical B-IFDMA design can be characterized by three parameters: the number of subcarriers per block N_{sc}, number of blocks per interlaces L, and number of interlaces per symbol N as illustrated in Fig. 8.26. In LTE-based eLAA, $N_{RB} = 100$ and with RB-based interlaced design, $N_{sc} = 12, L = 10$, and $N = 10$. The set of subcarriers S_n allocated for a specific interlace n can be represented as $S_n = \{N_{sc}(Nl + n) + m | 0 \leq m < N_{sc}, 0 \leq l < \lfloor N_{RB}/N \rfloor\}$. Note that it is not always possible to

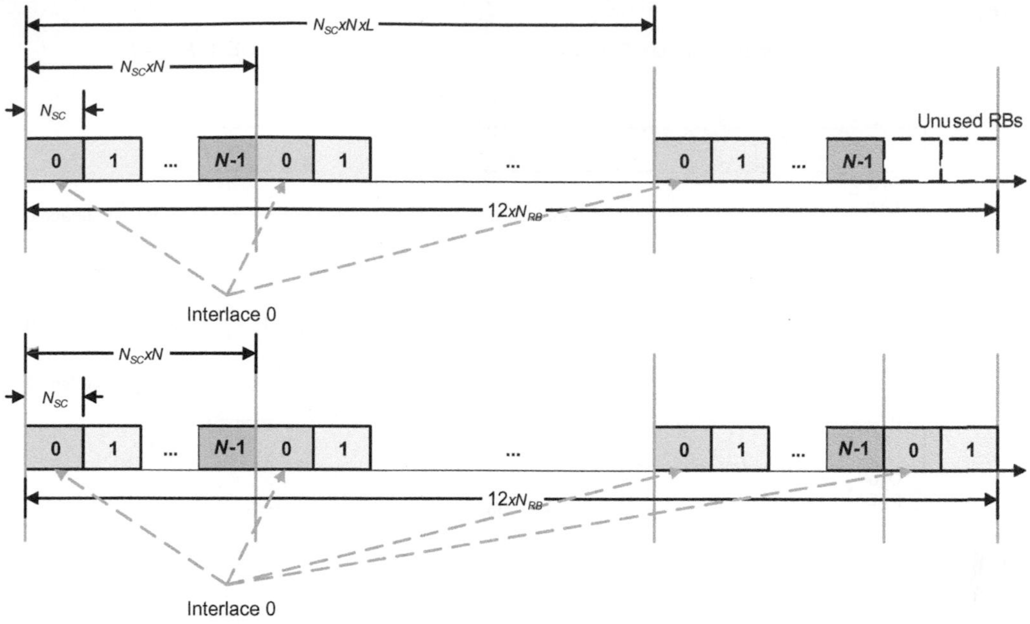

Figure 8.26
B-IFDMA design parameters and various design options [35].

divide N_{RB} by N. As an example, in NR, for a channel bandwidth of 20 MHz and subcarrier spacing of 15 kHz, $N_{RB} = 106$, and if the LTE-based eLAA parameters are maintained, then six RBs will not be used by any interlace. To avoid resource wastage, one can assign the remaining blocks to some of the interlaces. For example, if we assign R remaining blocks to the first R interlaces, then the set of subcarriers S'_n allocated for a specific interlace n can be represented as $S'_n = \{N_{sc}(Nl + n) + m | 0 \leq m < N_{sc}, 0 \leq l < (N_{RB} - n - 1)/N\}$. In this case, the number of blocks per interlace is a function of n as shown in Fig. 8.26. To meet the OCB requirement, one can design interlaces such that all interlaces occupy channel bandwidths larger than what the minimum OCB requires. In that case, the interlace parameters N_{sc}, L, and N need to be carefully chosen. In particular, assuming nominal channel bandwidth B and subcarrier spacing Δf, the minimum normalized OCB among all interlaces is given by $B_{\min} = N_{sc}\left(\left[\lfloor 12N_{RB}/(N_{sc}N)\rfloor - 1\right]N + 1\right)\Delta f/B$. As an example, in LTE-based eLAA, $\Delta f = 15$ kHz and $B = 20$ MHz, thus $B_{\min} = 0.819$, which indicates that eLAA interlace scheme can only satisfy 80% of the minimum OCB requirements. If we consider an NR numerology with $\Delta f = 60$ kHz, $B = 20$ MHz, and $N_{RB} = 24$ and if we adopt RB-based interlace $N_{sc} = 12$ to satisfy 80% of the minimum OCB requirements, the maximum number of permissible interlaces would be $N = 2$. For some NR bandwidth and SCS combinations, due to regulatory requirements and design constraints, the number of interlaces available

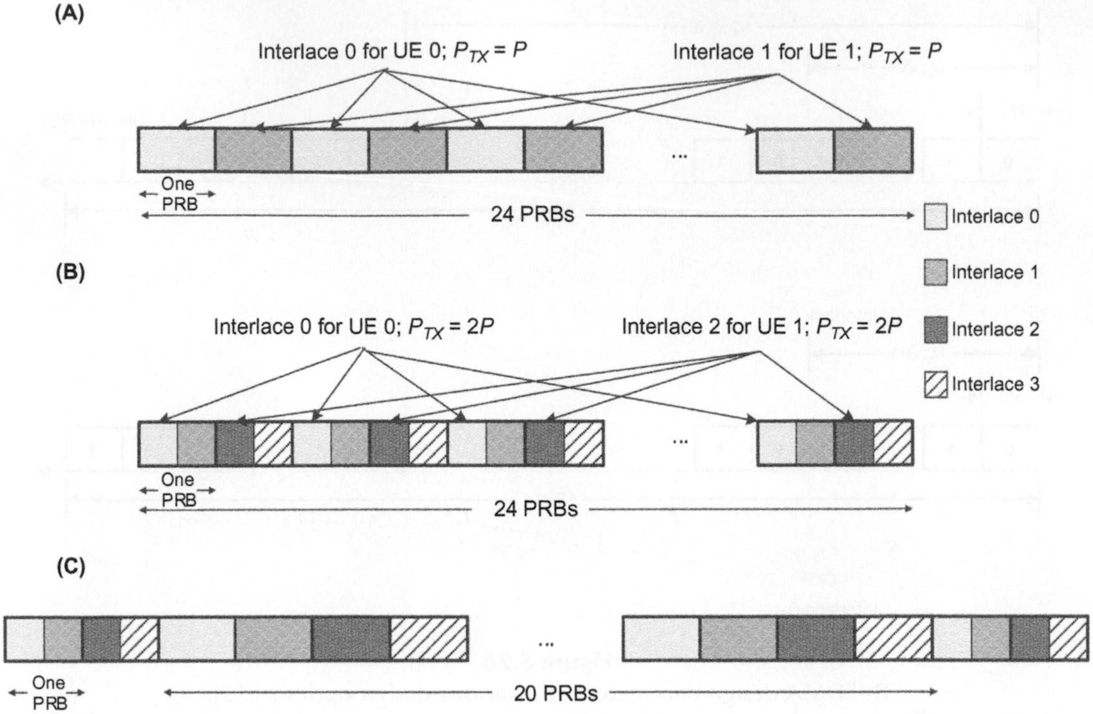

Figure 8.27
RE-group-level interlace design [35]. (A) Uniform RB level interlace. (B) Uniform RE group level interlace. (C) Nonuniform RE-group level interlace.

per symbol can be very limited; in such cases, RE-group-level interlace schemes may be utilized (see Fig. 8.27).

In Rel-15 NR, five PUCCH formats are defined, where PUCCH formats 0 and 2 are short PUCCH formats which occupy up to two OFDM symbols; PUCCH formats 1, 3, and 4 are long PUCCH formats which occupy between 4 and 14 OFDM symbols. For PUCCH formats 0 and 1, the number of UCI bits is 1 or 2, whereas, for PUCCH formats 2, 3, and 4, the UCI bits can be very large. For PUCCH formats 2 and 3, the maximum number of occupied PRBs is 16, whereas PUCCH formats 0, 1, and 4 use only one PRB. Considering the use of LBT and the regulatory requirements on OCB, NR-U makes special considerations about UCI payload size and transmission efficiency as well as UE multiplexing capacity. To overcome the uncertainty of LBT and to improve the spectral efficiency in the unlicensed band, the network may schedule uplink transmission of multiple UEs within the same channel occupancy time. However, due to the OCB requirement, the number of interlaces per symbol may be limited. As described earlier, a larger SCS value leads to a more restrictive interlace design.

8.2 Network Architecture and Protocol Aspects

In the previous section, we described the prominent methods that cellular industry has considered for access to the unlicensed spectrum, among which LWA is the only scheme that has network architecture implications that will be discussed in this section. The LWA has been supported in 3GPP specs since Rel-13, wherein a UE in RRC_CONNECTED state is configured by the eNB to utilize radio resources of LTE and Wi-Fi. Two scenarios are supported depending on the backhaul connection between LTE and Wi-Fi namely non-co-located LWA scenario for a non-ideal backhaul and co-located LWA scenario for an ideal/internal backhaul. The overall architecture for the non-co-located LWA scenario is illustrated in Fig. 8.28, where the Wi-Fi Termination (WT) terminates the Xw interface for Wi-Fi.

In LWA, the radio protocol architecture that a particular bearer uses depends on the LWA backhaul scenario and the way the bearer is set up. The LWA supports two bearer types: (1)

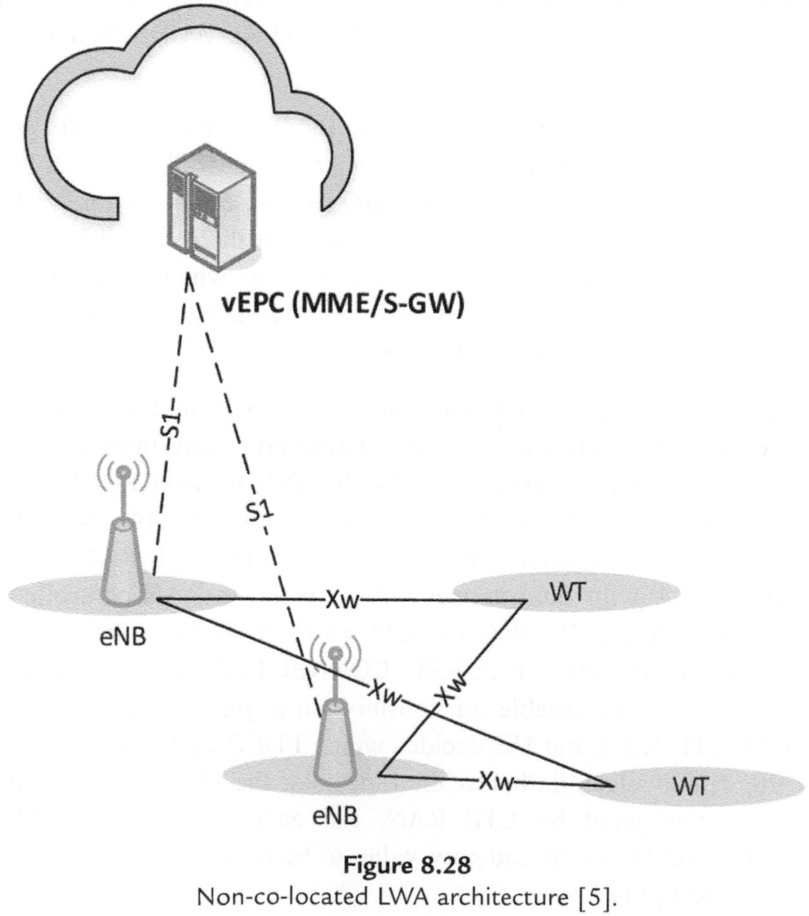

Figure 8.28
Non-co-located LWA architecture [5].

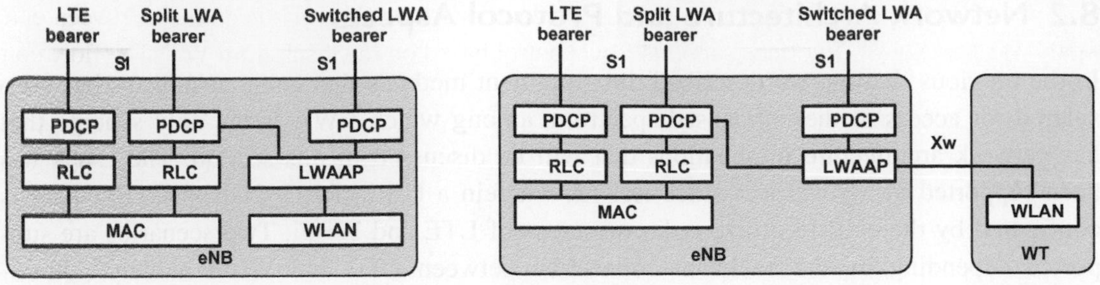

Figure 8.29
LWA radio protocol architecture for the colocated and non-co-located scenarios [5,7].

split LWA bearer and (2) switched LWA bearer. These two bearer types are depicted in Fig. 8.29 for the co-located and the non-co-located scenarios. In the non-co-located LWA scenario, the eNB is connected to one or more WTs via Xw interface. In the co-located LWA scenario, the interface between LTE and Wi-Fi is implementation specific. For LWA, the only required interfaces to the core network are S1-U and S1-MME which are terminated at the eNB. The Wi-Fi has no core network interface.

In the downlink and for the PDUs that are sent over the Wi-Fi link, the LTE-Wi-Fi aggregation adaptation protocol (LWAAP) entity generates an LWAAP PDU (i.e., a PDU with data radio bearer (DRB) ID generated by LWAAP entity for transmission over Wi-Fi) containing a DRB identity and the WT. An LWA UE determines that the received PDU belongs to an LWA bearer and uses the DRB identity to determine to which LWA bearer the PDU belongs to. In the uplink and for the PDUs sent over Wi-Fi, the LWAAP entity in the UE generates LWAAP PDU containing a DRB identity.

The LWA supports split bearer operation where the PDCP sublayer handles sequential delivery of upper-layer PDUs based on the reordering procedure introduced for dual connectivity. An LWA UE may be configured by the eNB to send PDCP status report or LWA status report in cases where feedback from WT is not available. Note that only RLC AM and RLC UM can be configured for an LWA bearer. The LTE RAN does not simultaneously configure LWA with dual connectivity, LWIP, or RAN-controlled LTE-Wi-Fi interworking for the same UE. If LWA and RAN-assisted Wi-Fi interworking are simultaneously configured for the same UE in RRC_CONNECTED, the UE only utilizes LWA. For LWA bearer, if the data available for transmission is greater than or equal to a threshold configured by LTE RAN, the UE decides which PDCP PDUs are sent over Wi-Fi or LTE links. If the data available is below the threshold, the UE transmits PDCP PDUs on LTE or Wi-Fi as configured by LTE RAN. For each LWA DRB, LTE RAN may configure the IEEE 802.11 access category value to be used for the PDCP PDUs that are sent over Wi-Fi in the uplink.

In the non-co-located LWA scenario, Xw user-plane interface (Xw−U) is defined between eNB and WT. The Xw−U interface supports flow control based on feedback from WT. The flow control function is applied in the downlink when an E-UTRAN radio access bearer (E-RAB) is mapped to an LWA bearer, that is, the flow control information is provided by the WT to the eNB to control the downlink user data flow to the WT for the LWA bearer. The operation, administration and maintenance (OAM) configures the eNB with the information of whether the Xw downlink delivery status provided by a connected WT concerns LWAAP PDUs successfully delivered to the UE or successfully transferred toward the UE. The Xw−U interface is used to deliver LWAAP PDUs between eNB and WT. In an LWA architecture, the S1-U terminates at the eNB and, if Xw−U user data bearers are associated with E-RABs for which the LWA bearer option is configured, the user-plane data is transferred from the eNB to the WT using the Xw−U interface. Fig. 8.30 shows the user-plane connectivity of eNB and WT in LWA, where the S1-U terminates at the eNB and the eNB and the WT are interconnected via Xw−U.

In the non-co-located LWA scenario, Xw control-plane interface (Xw−C) is defined between eNB and WT. The application layer signaling protocol is referred to as Xw application protocol (Xw−AP). The Xw−AP protocol supports the transfer of Wi-Fi metrics from WT to eNB; LWA for a UE in ECM-CONNECTED state; establishment, modification, and release of a UE context at the WT; control of user-plane tunnels between eNB and WT for a specific UE for LWA bearers; general Xw management and error handling functions including error indication, setting up Xw, resetting Xw, and updating the WT configuration data. The eNB-WT control-plane signaling for LWA is performed through Xw−C interface. There is only one S1-MME connection per LWA UE between the eNB and the MME. The coordination between eNB and WT is performed via Xw interface signaling.

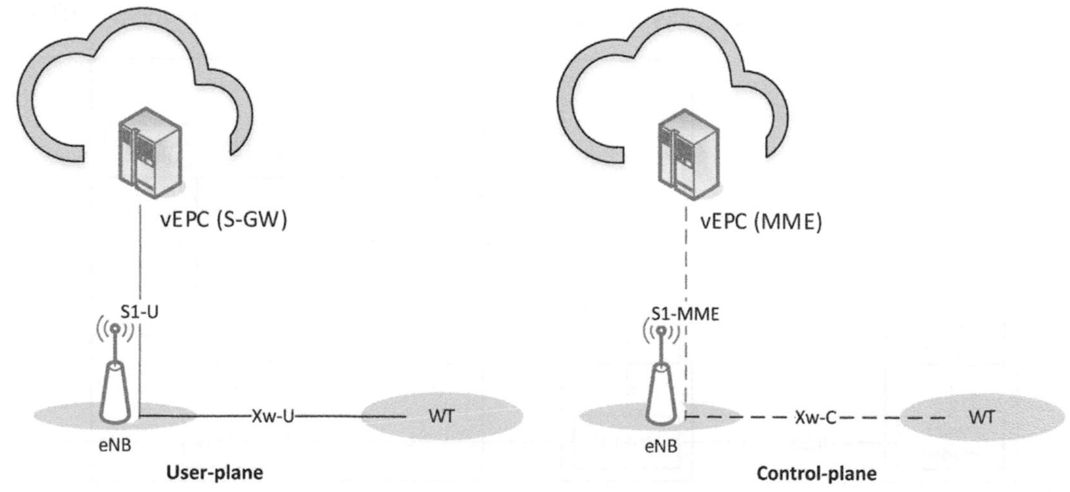

Figure 8.30
User-plane and control-plane connectivity of eNB and WT for LWA [10].

Fig. 8.30 shows control-plane connectivity of eNB and WT, where the S1-MME is terminated at the eNB and eNB and the WT are interconnected via Xw−C.

The Wi-Fi mobility set is a set of one or more Wi-Fi access points identified by their BSSID/ESSID/SSIDs (these identifiers are used to describe different sections of a Wi-Fi network), in which Wi-Fi mobility mechanisms apply while the UE is configured with LWA bearer(s), that is, the UE may perform mobility between Wi-Fi APs belonging to the mobility set without informing the eNB. The eNB provides the UE with a Wi-Fi mobility set. When the UE is configured with a Wi-Fi mobility set, it will attempt to connect to a Wi-Fi AP whose identifiers match the ones in the configured mobility set. The UE mobility relative to Wi-Fi APs not belonging to the UE mobility set is controlled by the eNB, for example, updating the Wi-Fi mobility set based on measurement reports provided by the UE. A UE is connected to only one mobility set at a time. All Wi-Fi APs belonging to a mobility set share a common WT which terminates Xw−C and Xw−U. The termination points for Xw−C and Xw−U may differ. The Wi-Fi identifiers belonging to a mobility set may be a subset of all Wi-Fi identifiers associated with the WT [5].

The UE supporting LWA may be configured by the LTE RAN to perform Wi-Fi measurements. Wi-Fi measurement object can be configured using Wi-Fi identifiers (BSSID, ESSID, or SSID), Wi-Fi carrier information and Wi-Fi band (2.4, 5, and 60 GHz). The Wi-Fi measurement reporting is triggered using RSSI measurement. A Wi-Fi measurement report contains RSSI and Wi-Fi identifier and may contain Wi-Fi carrier information, Wi-Fi band, channel utilization, station count, admission capacity, backhaul rate, and an indication whether the UE is connected to the Wi-Fi [5].

The LTE and Wi-Fi integration architecture requires Wi-Fi-specific core network nodes and interfaces (as shown by dotted lines in Fig. 8.31). However, the LWA scheme is different

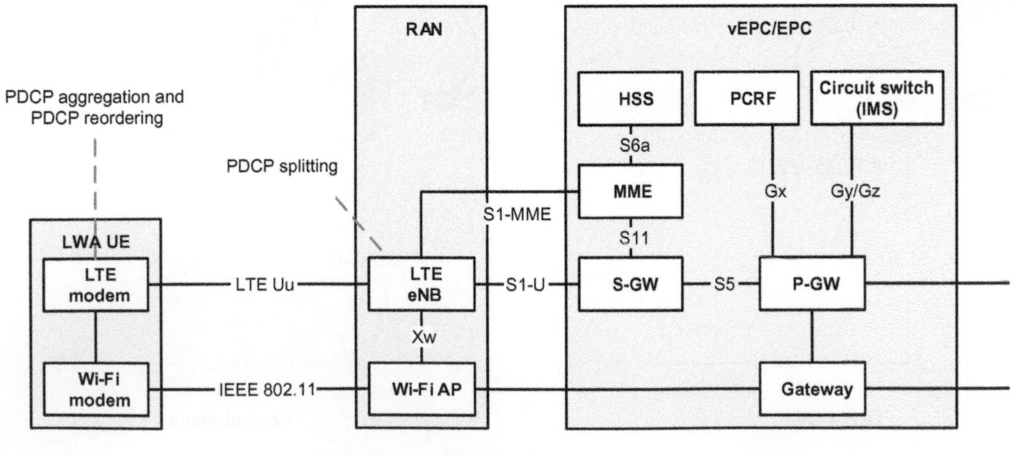

Figure 8.31
LWA network architecture (non-co-located case) [5,18].

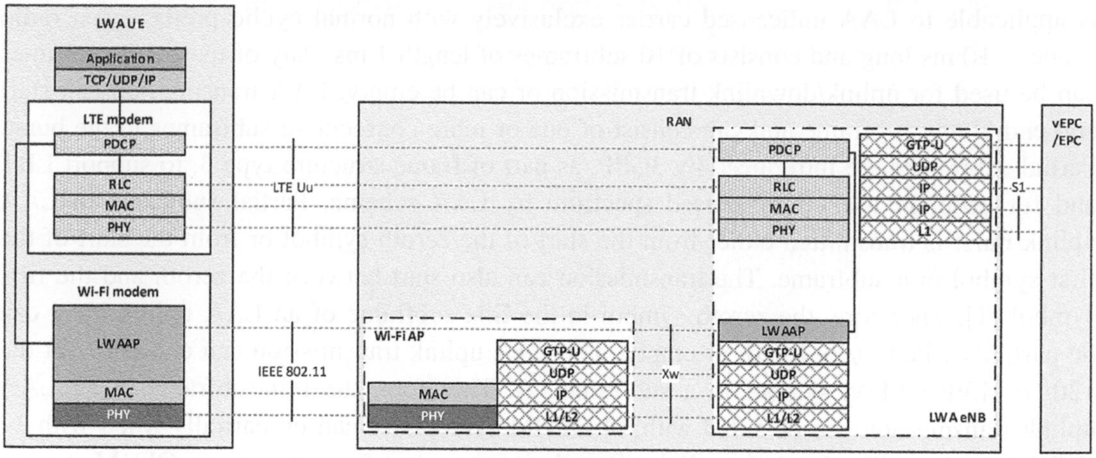

Figure 8.32
LWA protocol architecture (non-co-located case, user-plane) [5,18].

because LTE and Wi-Fi are aggregated at the radio link level. The LWA protocol architecture, data aggregation at the PDCP sublayer, signaling and interfaces between eNB and Wi-Fi AP, etc., have been specified by 3GPP and illustrated in Figs. 8.31 and 8.32.

The LWA architecture introduced a new interface Xw which was defined for communication between eNB and Wi-Fi AP, which is similar to X2, where user data is delivered through GTP tunnel, while control messages are delivered as Xw−AP messages over SCTP. Upon arriving at eNB, downlink user traffic is split in PDCP sublayer and is forwarded over LTE and Wi-Fi. Some PDCP packets are delivered via data radio bearer over the LTE radio link and other packets are delivered to Wi-Fi AP by eNB, which adds a DRB ID to the packets to indicate which DRB they belong to and delivers the LWA PDUs to Wi-Fi AP through the IP tunnel established over Xw [5]. Wi-Fi AP then sets the packet type to PDCP and forwards the LWA PDU to LWA UE over IEEE 802.11 interface. Upon receiving IEEE 802.11 frames, the LWA UE forwards the frames to LTE PDCP sublayer, if the packet type is set to PDCP. The PDCP sublayer then collects PDCP packets received from LTE and Wi-Fi that belong to the same LWA bearer by checking their DRB IDs and aggregates them through reordering. The LWA adaptation protocol supports LWA operation as shown in Fig. 8.32 [5].

8.3 Physical Layer Aspects

To support a flexible LAA operation, a new type 3 frame structure was introduced in LTE Rel-13, in which UE considers each subframe as empty unless downlink transmission is detected in that subframe. For LAA, the LBT procedure can be completed at any time and a downlink transmission may not start/end at the subframe boundary. Frame structure type 3 is applicable to LAA unlicensed carrier exclusively with normal cyclic prefix. Each radio frame is 10 ms long and consists of 10 subframes of length 1 ms. Any of these 10 subframes can be used for uplink/downlink transmission or can be empty. LAA transmission can start and end at any subframe and can consist of one or more consecutive subframes in the burst. Partial subframes are introduced by 3GPP, as part of frame structure type 3, to support LBT and for efficient use of unlicensed spectrum by LAA scheme. Partial subframe in LAA uplink burst is transmitted either from the start of the zeroth symbol or from the start of the first symbol in a subframe. The transmission can also start between the zeroth and the first symbol [1]. Therefore, the zeroth symbol in the first subframe of an LAA uplink burst can be partially filled, fully filled, or empty. The LAA uplink transmission can end either at the 12th or 13th OFDM symbol in a subframe [2]. Therefore, the last subframe in an LAA uplink transmission can be filled with 14 OFDM symbols or can be partially filled with 13 OFDM symbols. The LAA downlink transmission can start from the zeroth OFDM symbol (subframe boundary) or from the seventh OFDM symbol (second slot starting position) of a subframe [3,4]. The LAA downlink transmission can either end at the subframe boundary or at any existing downlink pilot time slot (DwPTS) symbol (in frame structure type 2). Therefore, the last subframe can be completely occupied with 14 OFDM symbols or consist of any of DwPTS symbols, that is, 3, 6, 9, 10, 11, or 12 OFDM symbols [1]. Example configurations of frame structure type 3 are shown in Fig. 8.33 [1].

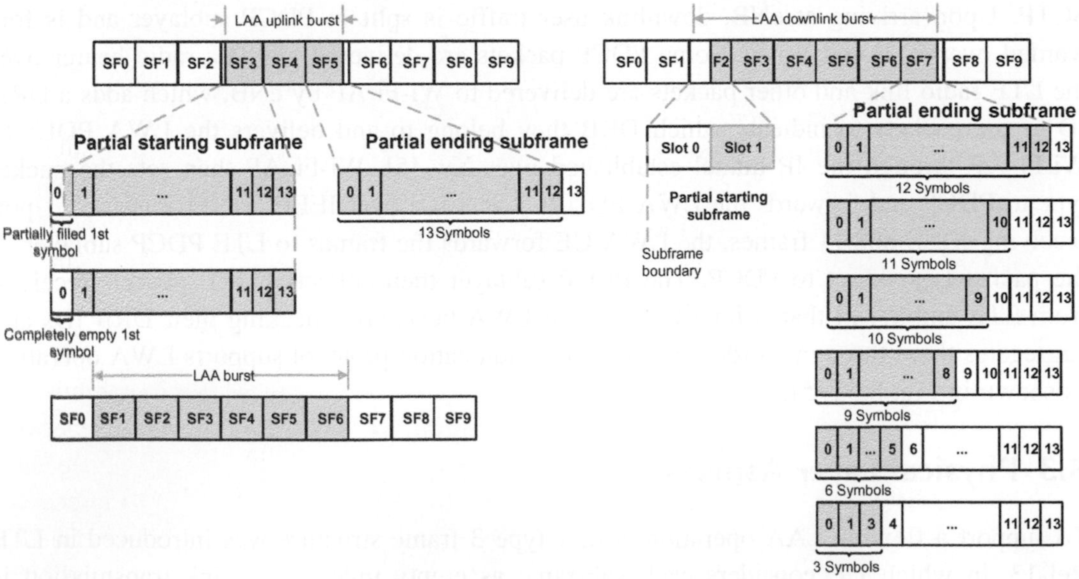

Figure 8.33
Example configurations of frame structure type 3 [1].

Cell selection and synchronization rely on the reception of the synchronization signals, that is, primary and secondary synchronization signal (PSS/SSS) as well as cell-specific reference signals (CRS). In particular, PSS/SSS can be used for physical cell identity detection, and the CRS can be used to further improve the performance of cell ID detection. These synchronization signals are further used to acquire coarse and fine time/frequency synchronization, although large time/frequency offset between two successive downlink bursts is unlikely due to multiple DRS transmission opportunities within a DRS measurement timing configuration (DMTC) occasion which has a duration of 6 ms. Therefore, the synchronization based on DRS in LAA systems can achieve reliable performance. The presence of downlink subframe must be detected by the LAA UE because the eNB does not always transmit in LAA scenarios. The exact detection method depends on UE implementation.

LTE-based LAA supports transmission modes (TMs) with CRS-based CSI feedback, including TM1, TM2, TM3, TM4, and TM8, and those with CSI-RS-based CSI feedback, including TM9 and TM10 [3]. The CSI-RS/CSI-IM for CSI measurement is present in the configured periodic CSI-RS/CSI-IM subframes within downlink transmission bursts. Similar to the legacy LTE systems, both periodic and aperiodic CSI reports are supported in LAA scenarios. Unlike the legacy LTE system where the CRS/CSI-RS transmission power, or energy per resource element, is fixed, CRS/CSI-RS transmission power on LAA SCell is only fixed within a downlink transmission burst and can vary across downlink transmission bursts. Thus, the UE should not average CRS/CSI-RS measurements across different transmission bursts. The UE could either rely on CRS detection or common control signaling to differentiate the downlink bursts. LTE supports two different scheduling approaches, namely cross-carrier scheduling and self-scheduling. With cross-carrier scheduling, the control information including scheduling indication on PDCCH and the actual data transmission on PDSCH take place on different carriers, whereas they are transmitted on the same carrier in the case of self-scheduling. Due to the uncertainty of channel access opportunities on unlicensed carriers, the synchronous uplink HARQ protocol with fixed timing relation between retransmissions was difficult to use in LAA. Thus, asynchronous HARQ protocol is used for LAA downlink and uplink. For LAA uplink, in particular, UEs would need to rely on the uplink grant from eNB for (re)transmissions.

Enhanced licensed-assisted access was introduced in 3GPP Rel-14. It defines how a UE can access 5 GHz band to transmit data in the uplink direction. Because all uplink transmissions in LTE are scheduled and controlled by the serving eNB, the channel contention between devices and the LBT scheme, which was defined in LAA for downlink operation, needed to be adapted to the uplink direction. Furthermore, the regulatory requirements for access to 5 GHz band vary in different regions. For example, ETSI requires that the OCB, defined by 3GPP as the bandwidth containing 99% of the signal power, to be between 80% and 100% of the declared nominal channel bandwidth [29]. As an initial approach, multi-cluster PUSCH transmission, which was specified in 3GPP Rel-10, was considered to fulfill this

ETSI requirement [29]. Multi-cluster PUSCH allows two clusters of resource blocks to be scheduled far enough from each other to fulfill, for example, the 80% bandwidth requirement. However, further studies in 3GPP suggested that the latter was not an efficient solution. Furthermore, the PSD limits defined for the 5G GHz band was another limitation of the multi-cluster PUSCH approach as shown in Fig. 8.34. The LTE-based eLAA supports multi-subframe scheduling, where an uplink grant can schedule PUSCH transmissions in a number of consecutive subframes ranging from two to four subframes. It further supports DCI formats 0A and 4A for single subframe scheduling and DCI formats 0B and 4B for multi-subframe scheduling for single-layer and multi-layer PUSCH transmissions, respectively (see Fig. 8.35).

As an example, the 10 dBm/MHz defined by the ETSI over 5150–5350 MHz frequency range is shown in Fig. 8.34, where ETSI allows a PSD of 17 dBm/MHz (over 5470–5725 MHz) if TPC can be applied [29]. The US FCC defines 11 dBm/MHz PSD for 5150–5350 MHz frequency range. As a result, 3GPP adopted the principle of B-IFDMA

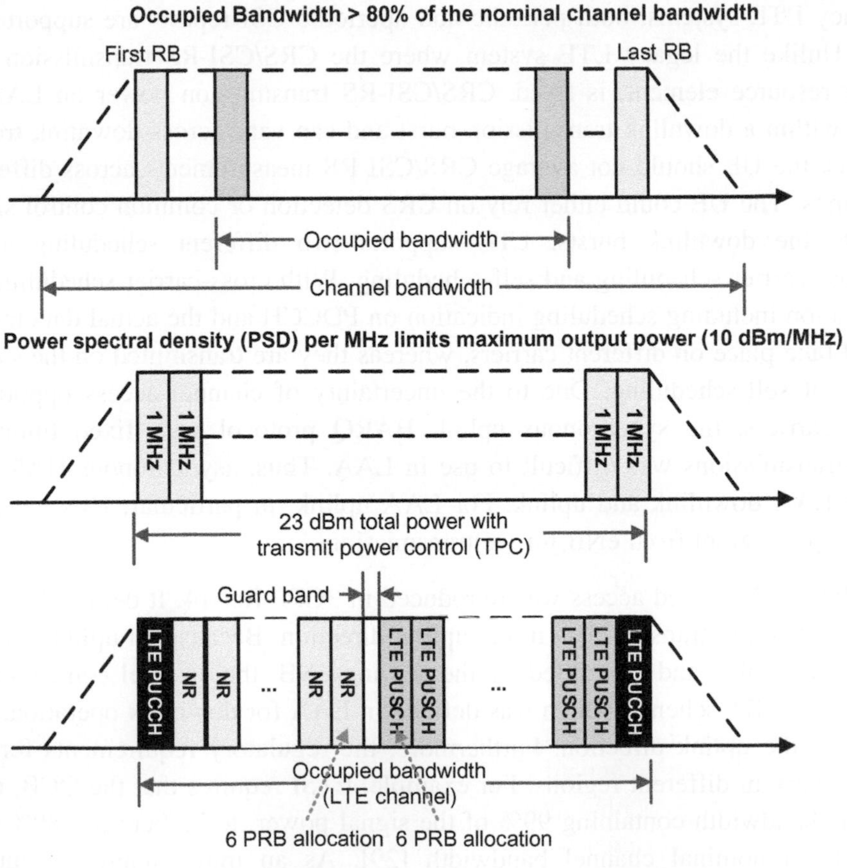

Figure 8.34
ETSI requirements for transmission 5 GHz band [22,27,29].

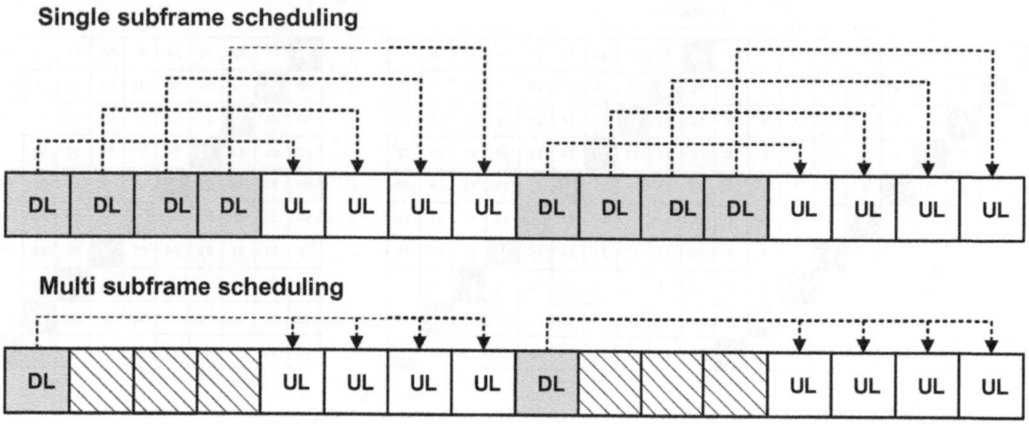

Figure 8.35
Multi-subframe scheduling versus single subframe scheduling [23].

for eLAA. The available number of resource blocks is organized in interlaces that are equally spaced in the frequency domain. A UE can now transmit on one or multiple interlaces. Similar to LTE uplink transmissions, the total number of allocated resource blocks must be multiples of 2, 3, and 5 to minimize the complexity of the DFT precoding. To incorporate this specific access mode into the standard, 3GPP Rel-14 specified a new uplink resource allocation type 3 that is only applicable for LAA. As a result, two new scheduling grants (DCI format 0A/0B and 4A/4B) were introduced. A DCI transports downlink, uplink or sidelink scheduling information, requests for aperiodic CQI reports, LAA common information, or uplink power control commands for one cell and one RNTI. The RNTI is implicitly encoded in the CRC. DCI formats 0A and 4A schedule single subframes for single antenna (0A) and multi-antenna (4A) transmission, whereas DCI formats 0B and 4B allow scheduling of up to four consecutive subframes for SISO and MIMO, respectively. These new uplink grants provide the UE with a resource indication value *RIV* that is either 5 or 6 bits (5 bits for 10 MHz channel and 6 bits for 20 MHz channel). The *RIV* represents a start value RB_{start} and the actual number of allocated resource blocks L. For a 20 MHz LTE channel, corresponding to 100 resource blocks, there are 10 interlaces with 10 RBs/interlace. Interlace 0 contains resource blocks $(0, 10, 20, 30, 40, 50, 60, 70, 80, 90)$ as shown in Fig. 8.36. If the UE transmits on 4 out of the 10 available RBs per interlace $L = 4$, then the *RIV* is calculated as $RIV = N(L-1) + RB_{start}$. With $N = 10$, RB_{start} can have a value between 0 and 6 and *RIV* can be between 30 and 36. The *RIV* is signaled via DCI to the UE. In addition to this and other information, the DCI format also includes the method on how to access the channel, that is, how to perform LBT in the uplink. As shown in Table 8.3, there are two channel access types defined for eLAA, whose usage is signaled with the uplink scheduling grant (DCI formats 0A, 0B, 4A, 4B). The type 1 channel access procedure is identical to the procedure for LAA. The difference is that there are separate

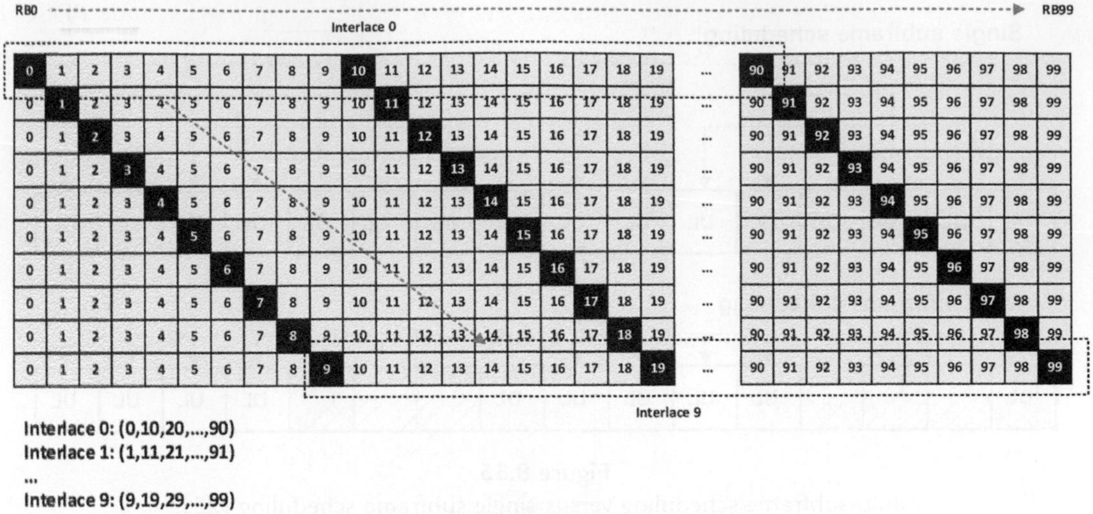

Figure 8.36
eLAA interlaces for 20 MHz LTE channel [22].

channel access priority classes defined for the uplink. Type 2 is a procedure similar to transmitting DRS in the downlink. After sensing that the channel is idle for 25 μs, the device can start its PUSCH transmission [22].

It must be noted that DCI format 0A is used for the scheduling of PUSCH in an LAA SCell and DCI format 0B is used for the scheduling of PUSCH in each of the multiple subframes in an LAA SCell. Furthermore, DCI format 1C is used for very compact scheduling of one PDSCH codeword, notifying multicast control channel change, notifying single-cell multicast control channel change and direct indication, reconfiguring TDD, and LAA common information, whereas DCI format 4 is used for the scheduling of PUSCH in an LAA SCell with multi-antenna port transmission mode. DCI format 4B is used for the scheduling of PUSCH with multi-antenna port transmission mode in each of the multiple subframes in an LAA SCell.

The LTE-based LAA supports two alternative approaches to multi-carrier LBT. In the first approach, the eNB is required to designate a carrier requiring LBT with random backoff and the eNB can sense other configured carriers with single interval LBT, only if the eNB completes the LBT with random backoff on the designated carrier. In the second approach, the eNB performs LBT with random backoff on more than one unlicensed carriers and can transmit on the carriers that have completed the LBT with potential self-deferral to align transmissions over multiple carriers. An LTE eNB can access multiple carriers on which LAA SCell(s) transmission(s) are performed using either Type A or Type B procedures as follows [4]:

- *Type A multi-carrier access procedure*: The eNB performs channel access on each carrier $c_i \in C$, where C is a set of carriers on which the eNB intends to transmit and

$i = 0, 1, \ldots, N_{carrier} - 1$ is the carrier index. The eNB transmission may include PDSCH/PDCCH/ePDCCH on an LAA SCell(s) carrier(s), after sensing the channel to be idle during the slot durations of defer period T_{defer} and after the counter N is initialized and adjusted by sensing the channel for additional slot duration(s). The counter N is determined for each carrier c_i and is denoted as N_{c_i}. In Type A1, the counter N is independently determined for each carrier. If the absence of another technology sharing the carrier cannot be guaranteed over a long period of time, and when the eNB stops transmission on any carrier $c_j \in C \forall c_i \neq c_j$, it can continue to decrement N_{c_i} when idle slots are detected either after waiting for a duration of $4T_{slot}$ or after reinitializing N_{c_i}. In Type A2, the counter N_{c_j} is determined for the carrier $c_j \in C$, where c_j is the carrier that has the largest contention window CW_p value. For each carrier c_i, $N_{c_i} = N_{c_j}$. When the eNB stops transmission on any carrier for which N_{c_i} is determined, it reinitializes N_{c_i} for all carriers.

- *Type B multi-carrier access procedure*: A carrier $c_j \in C$ is selected by the eNB by randomly choosing c_j from C before each transmission on multiple carriers $c_i \in C \forall i = 0, 1, \ldots, N_{carrier} - 1$, or the eNB selects c_j less frequently, where C is a set of carriers on which the eNB may transmit. To transmit on a carrier $c_i \neq c_j \forall c_i \in C$, the eNB performs carrier sensing for a minimum sensing interval $T_{mc} = 25\ \mu s$ prior to transmitting on carrier c_j, and it may transmit on carrier c_i immediately after sensing the carrier c_i to be idle for at least the sensing interval T_{mc}. The carrier c_i is considered to be idle for T_{mc} if the channel is sensed to be idle during all time intervals in which such idle sensing is performed on carrier c_j in a given time interval T_{mc}. The eNB will not continuously transmit on a carrier $c_i \neq c_j \forall c_i \in C$ for a period exceeding T^p_{MCOT}, where the value of the latter parameter is determined using the channel access class used for carrier c_j. In Type B1, a single CW_p value is maintained for the set of carriers C. For determining CW_p for performing channel access on carrier c_j, if at least 80% of HARQ feedback corresponding to PDSCH transmission(s) in reference subframe k of all carriers $c_i \in C$ are determined as NACK, then CW_p is increased for each priority class $p \in \{1, 2, 3, 4\}$ to the next higher permissible value. In Type B2, a CW_p value is maintained independently for each carrier $c_i \in C$. For determining N_{init} for the carrier c_j, CW_p value of carrier $c_{j1} \in C$ is used, where c_{j1} is the carrier with largest CW_p among all carriers in set C.

Discontinuous transmission on an unlicensed carrier with limited maximum transmission duration has an impact on some LTE/NR essential functionalities that support AGC setting, time and frequency synchronization of the UEs, and channel reservation. Channel reservation refers to the transmission of signals by an LAA node after gaining channel access via a successful LBT operation, so that other nodes that receive the transmitted signal with energy above a certain threshold sense the channel to be occupied.

8.4 Layer 2/3 Aspects

In this section, we discuss the L2/L3 aspects and impacts of unlicensed band operation in LTE and NR-based systems. The uncertainty of the cell availability for HARQ retransmissions on unlicensed carriers would create some limitations for normal HARQ operation. The uncertainty may arise due to the LBT operation needed to acquire the channel or because the maximum transmission duration for LAA is exceeded. Fig. 8.37 illustrates an example in which the maximum transmission duration of an LAA cell has been reached before the third retransmission of a HARQ process can be performed. There are two ways to address this issue. One approach is to keep the HARQ retransmissions on the same LAA cell or to ensure that the HARQ process is completed within the maximum transmission duration of an LAA cell as shown in Fig. 8.37. A new HARQ process is started when the LAA cell reacquires the channel after an LBT operation. Alternatively, a retransmission can be delayed until the LAA cell acquires the channel again. The RLC retransmission may be invoked if HARQ transmission is not successfully completed. Another alternative would be to move HARQ retransmissions to another cell (note that each carrier in the carrier aggregation framework is referred to as a cell). The retransmission may also be performed via another cell which would be either the PCell or another SCell. This would change the baseline LTE HARQ protocol as it may be linked to two or more cells.

In synchronous HARQ, the UE identifies the HARQ process that is associated with the current transmission time interval (TTI). Depending on whether spatial multiplexing is used, each TTI has one or two associated HARQ processes, and for the identified HARQ

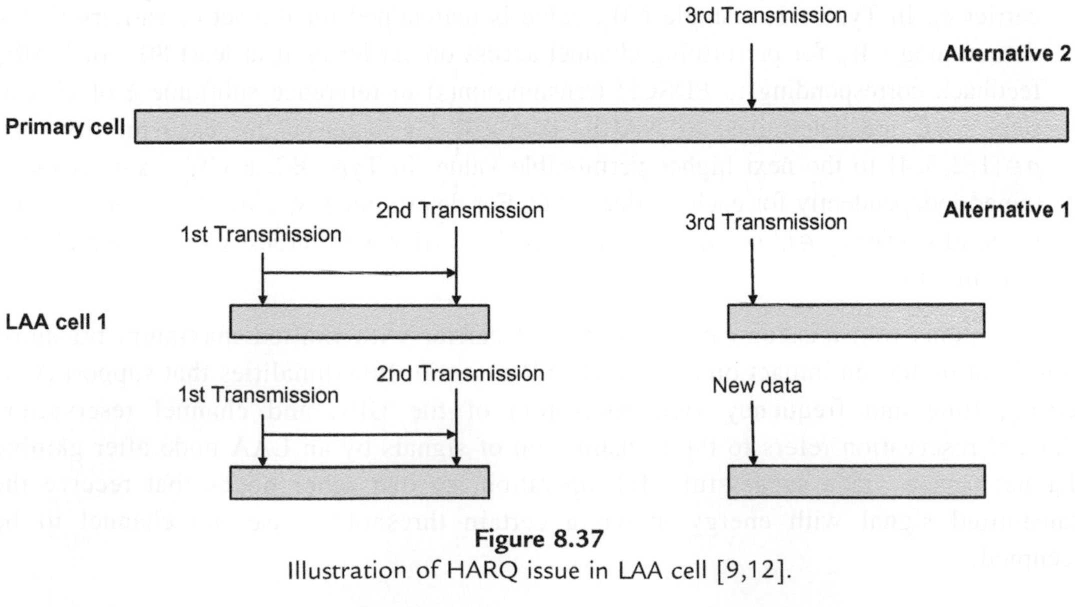

Figure 8.37
Illustration of HARQ issue in LAA cell [9,12].

processes, the UE will perform a (re)transmission. The association between TTIs and HARQ processes relies on uplink HARQ being synchronous and is derived from the timing relation. If there has been an initial transmission in a certain subframe for a certain HARQ process, then $T_{round\text{-}trip}$ later, the same HARQ process is considered for retransmission. In LTE, the UE receives uplink HARQ feedback from the eNB on PHICH. PHICH received in subframe k relates to the transmission in the subframe $k-4$. If the UE receives NACK on PHICH (but does not receive a grant), a non-adaptive retransmission is automatically triggered. LTE-based LAA uses asynchronous HARQ in the uplink. To enable asynchronous HARQ, the eNB needs to know the HARQ process that the UE is using when performing a transmission/retransmission to correctly choose the soft-buffer where the received transmission should be combined. Hence, the eNB needs to indicate to which HARQ process a grant is related and the redundancy version that should be used so that the UE uses the correct HARQ process with the correct redundancy version when performing a transmission or retransmission. Therefore, with the uplink asynchronous HARQ protocol, all transmissions or retransmissions are scheduled via (e)PDCCH. The process index is indicated in the HARQ process index field in the uplink grant. For synchronous uplink HARQ, the number of HARQ processes is not explicitly specified. Instead, the supported number is derived from the HARQ timing. For asynchronous uplink HARQ, a maximum number of HARQ processes may need to be specified. The exact number of uplink HARQ processes in the eNB may be left to implementation [12].

In synchronous uplink HARQ, the UE can expect to receive uplink grants at a specified time as the HARQ process follows a fixed pattern. It should be noted that the UE monitors (e)PDCCH once in every HARQ RTT even if the UE has received ACK in PHICH. With asynchronous uplink HARQ, the UE does not know when to expect grants as the eNB may send them at any time. In addition, if HARQ buffer is not flushed, the UE would never stop monitoring (e)PDCCH according to the current mechanism. Therefore, with asynchronous uplink HARQ, the DRX behavior needs to be modified. In carrier aggregation, the same DRX operation applies to all configured and activated serving cells (i.e., identical active time for PDCCH monitoring). In other words, a common DRX is applied to all the serving cells. The difference in the case of LAA is that due to LBT there is no guarantee that the channel is obtained for scheduling the UE at the exact moment desired by the eNB. In addition, even if CCA succeeds, the eNB transmitter can only occupy the channel for a limited duration due to limited maximum transmission duration requirement. This means that the DRX timers (on-duration, inactivity timer) should be long enough or DRX cycles should be short enough to allow time for obtaining access to the channel.

In LAA, there are scenarios where multiple operators may operate in the same frequency channel. If these operators do not coordinate the allocation of physical cell identifier (PCI) values across their cells, it may lead to PCI confusion or PCI collision cases. PCI confusion refers to the case where a UE discovers and configures the LAA cell of another operator with the same PCI value as its own operator. PCI collision concerns a UE that is in the coverage area of two (or more) cells which have the same PCI value where the cells belong to different operators. If the same PCI value is used for cells on the same carrier frequency in the same area, PCI confusion or collision may occur. In LTE and NR, the number of PCIs is limited to 504 and 1008, respectively. If operators do not coordinate PCI assignment in their cells, the probability of PCI confusion or collision depends on the number of LAA cells of other operators the UE can find in the PCell coverage area, increasing the risk of collision when there is a dense deployment of LAA cells in a large PCell coverage area. PCI collision will happen with a lower probability than PCI confusion because PCI collision happens when the coverage areas of two cells with the same PCI partially overlap. Considering the LTE requirement for carrier aggregation where the UE is only required to handle a maximum timing difference of approximately 30 μs between the PCell and an SCell, the UE may not be able to receive anything from the other operator's LAA cell, if the downlink signal of that cell arrives later than $\pm$30 μs of the UE's PCell. The probability that the other operator's LAA cell overlaps with the UE's PCell is 6% assuming random timing of cells. Furthermore, for self-scheduling, the UE will not decode downlink assignments of the other operator's cell unless the downlink transmissions are scrambled with a C-RNTI value matching the C-RNTI assigned to the UE. Considering that there are 65536 C-RNTIs (16-bit values), if the other operator serves 20 UEs in a cell, the probability that the same C-RNTI is used is roughly 0.03%. Therefore, the probability that the UE can decode a downlink message from another operator's LAA cell is 0.0018% [14,20].

In the event of a PCI confusion when self-scheduling from the LAA cell, the probability that the UE could decode downlink is negligible (0.0018% in the scenario described earlier). However, in case of cross-carrier scheduling from the PCell, the UE would never be able to decode the downlink data for which the UE has received a downlink assignment as the data is sent in another cell. In the uplink, the UE will not be able to acquire uplink synchronization and would therefore not be able to transmit uplink traffic in the other operator's cell. However, it may send unnecessary random-access preambles to another operator's LAA cell (e.g., if it receives a PDCCH order from the PCell). It is expected that the eNB can detect PCI confusion by observing that the UE is reporting good quality for a cell, but no data communication is succeeded for this UE. The eNB can then resolve the confusion by changing the PCI for the problematic cell(s).

In the event of a PCI collision, there is a non-negligible likelihood that both LAA cells become unusable to the UEs that fall in their common coverage area. This would result in the UE not being able to receive the downlink and/or transmit in uplink while in those cells. It is also expected that in some cases, that is, not in hidden-node cases where the two cells with the same PCI are hidden from each other but are heard by the UEs, the network can detect PCI collision by listening to carriers and avoid the collision by changing the PCI values of their cells. PCI confusion and PCI collision can be avoided completely if operators coordinate the PCI values for their LAA cells. Otherwise, the probability of occurrence would be scenario dependent.

In LWA, the UE mobility is transparent to the eNB while in a Wi-Fi mobility set, that is, a group of Wi-Fi access points that are controlled by one WT logical entity. As long as the UE moves between APs of the same mobility set, it does not need to inform the eNB about its movement. When the UE leaves the mobility set, the eNB may change the access point based on the Wi-Fi measurements, thus it informs the UE of its decision.

Coexistence with other technologies is very important for LAA, and therefore accessing and using already congested frequencies/channels that are used by Wi-Fi access points and clients should be avoided. Because efficient radio resource management is critical for the overall performance of LAA scheme, LTE defines signal quality criteria such as reference signal received power [RSRP (dBm)] and reference signal received quality [RSRQ (dB)] metrics to effectively quantify the [shared] channel conditions. The RSSI serves as the key performance indicator for interference on a given carrier. To measure RSSI, the DRS needs to be present. However, because DRS is subject to LBT, any RSSI measurement report of an LAA-capable UE needs to include a time stamp indicating when the measurements were conducted. Therefore, higher layers configure an RSSI measurement time configuration (RMTC) with a measurement period [40, 80, 160, 320, or 640 ms], a subframe offset [0,. . .,639] and a measurement duration [1, 14, 28, 42, or 70 OFDM symbols]. The device averages the RSSIs over the measurement duration and conducts measurements according to the signaled periodicity. The device reports the average RSSI as well as the channel occupancy (CO), which is defined as the percentage of measured RSSI samples above a predefined threshold that is also signaled by higher layers. Both metrics provide an indication of the load and interference condition on the given LAA SCell. An example RMTC configuration with an averaging granularity of one OFDM symbol, measurement duration of 70 symbols or 5 ms with a periodicity of 40 ms, and a total measurement period comprising three measurement durations or 120 ms is illustrated in Fig. 8.38.

The Rel-16 NR-U supports 4-step and 2-step RACH procedures, where the 2-step RACH refers to the procedure that can complete contention-based RACH (CBRA) in two steps. The use of 2-step RACH may be beneficial due to less LBT impact with the reduced number of messages. The NR-U further supports contention-free RACH (CFRA) and CBRA for both 2-step and 4-step RACH. On SCells, CFRA is supported as a baseline, while both CBRA and CFRA are supported on SPCells. In the 4-step RACH procedure, the messages in time order are named as msg1, msg2, msg3, msg4, and in 2-step RACH, the messages are identified as msgA and msgB. A single RACH procedure will be used and thus multiple RACH procedures in parallel will not be supported for NR-U. As a baseline, the random-access response to msg1 will be on SPCell and msg3 is assumed to use a predetermined HARQ ID. In legacy RACH, the counters for preamble transmission and power ramping are increased with every

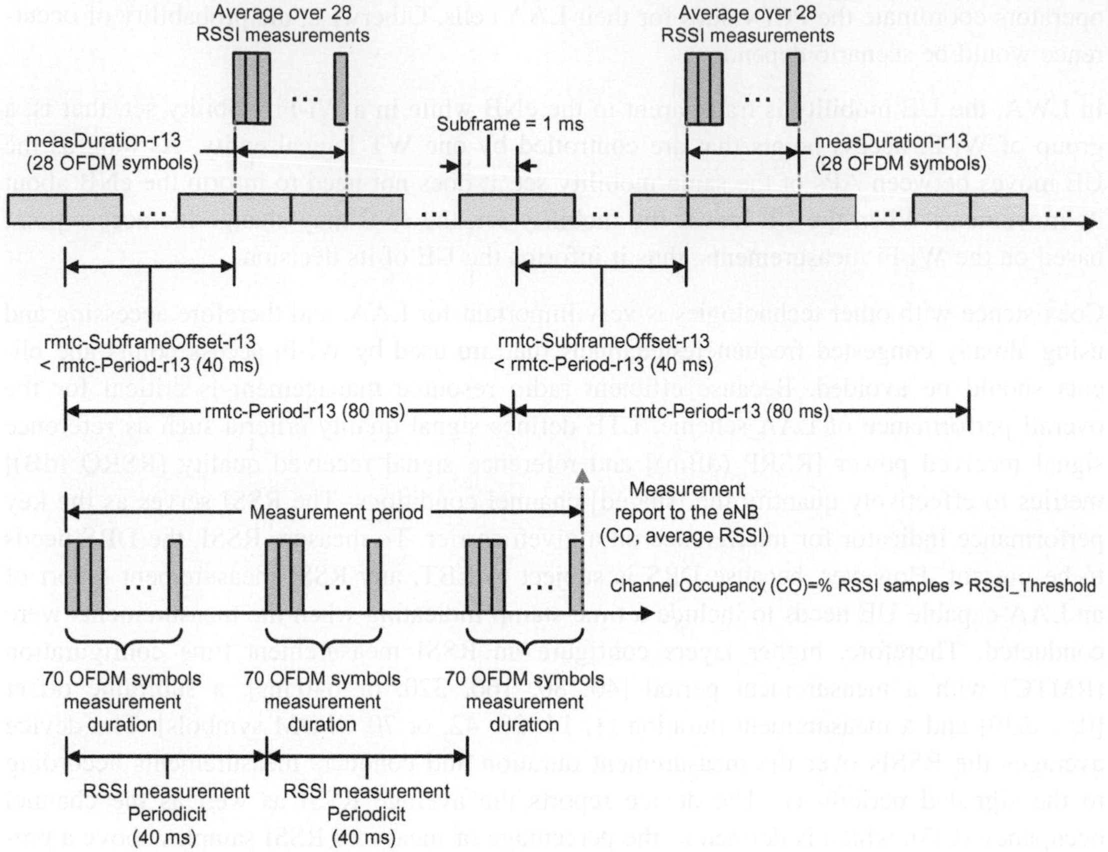

Figure 8.38
RMTC configuration for channel occupancy measurements based on R

attempt. In NR-U, power ramping is not applied when preamble is not transmitted due to LBT failure. This will require an indication from the physical layer to the MAC sublayer. In addition, ra-ResponseWindow is not started when the preamble is not transmitted due to LBT failure. It is assumed that ra-ContentionResolutionTimer may need to be extended with larger values to overcome the LBT impact. For 2-step RACH, the msgA is a signal to detect the UE and a payload while the second message is for contention resolution for CBRA with a possible payload. The msgA will include the equivalent information, which is transmitted in msg3 for 4-step RACH [12].

8.5 Implementation and Deployment Considerations

The main challenge in the implementation and use of LTE/NR in the unlicensed bands, particularly in 5 GHz band, is the coexistence with already deployed Wi-Fi networks, where LTE/

NR operation would adversely impact the performance of Wi-Fi systems, while the performance of LTE/NR would remain unchanged due to reliance of Wi-Fi systems on CSMA/CA mechanism. The issue is caused because of different channel usage and access procedures of these technologies. LTE/NR is designed to operate in the licensed bands based on the assumption that one operator has exclusive control of a given spectrum. They will continuously transmit with minimum time gap even in the absence of user traffic. LTE/NR also has an almost continuously transmitting protocol, as well as a periodically transmitting protocol to transmit a variety of control and reference signals. Wi-Fi, on the contrary, is designed to coexist with other technologies through random backoff and channel sensing. As a result, Wi-Fi users would have a slight chance to sense a clear channel and to transmit.

To ensure fair spectrum sharing and [practically] minimum inter-system interference among different wireless technologies operating in the unlicensed spectrum, 3GPP has specified a number of Wi-Fi coexistence mechanisms. These mechanisms operate in time, frequency, or power domains. In the frequency and time-domain schemes, the goal is to separate transmissions of LTE and Wi-Fi in frequency and time, respectively, while in the power domain, the goal is to adjust the output power of LTE nodes to a tradeoff between LTE throughput and opportunistic Wi-Fi transmission. Prior to any specification work on LAA, 3GPP conducted studies to investigate the feasibility of LTE operating in unlicensed bands [9]. The focus of those studies was fair sharing and coexistence with Wi-Fi systems where the criterion used to ensure coexistence was that an LAA network does not impact existing Wi-Fi neighbors more than another Wi-Fi network.

We discussed earlier that there are two design options for LTE-based LAA LBT, that is, asynchronous and synchronous LBT. The main difference between them lies in the fact that the asynchronous LBT is based on the current DCF protocol. In this case, the LBT scheme may use IEEE 802.11 RTS/CTS signals to ensure that the channel is idle just at that moment. However, synchronous LBT is considered as a special version of asynchronous LBT, wherein, data subframes are synchronized with the licensed LTE carrier. This LBT approach required minimal changes to the LTE specifications and could use inter-cell interference coordination (ICIC) mechanism already defined in earlier releases of LTE to manage the interference among LTE base stations. The ICIC mechanism is illustrated in Fig. 8.39. In this figure, different shades represent different frequencies in the outer sectors where the same frequency is used in the inner sectors.

Deterministic channel sharing relies on LTE centralized scheduling to periodically turn off its transmission so that Wi-Fi users can access the shared channel. Among time-domain coexistence mechanisms, we have discussed CSAT and blank-subframe allocation. A blank-subframe is an LTE subframe where transmission is muted so that Wi-Fi users can access the channel. Similar to CSAT, a blank-subframe allows time-domain sharing between LTE unlicensed and Wi-Fi networks. In each radio frame, the eNB can configure a certain

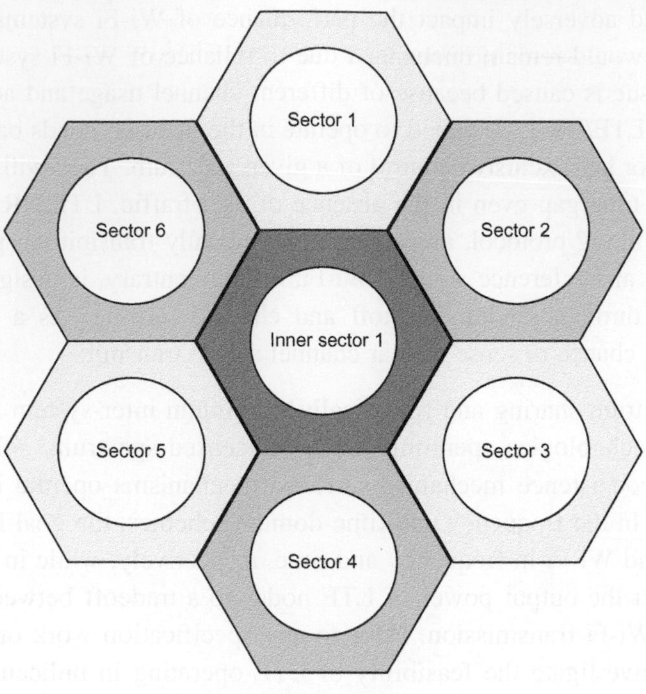

Figure 8.39
Illustration of LTE ICIC scheme where cell-center and cell-edge users have frequency reuse factor of 1 and 3, respectively.

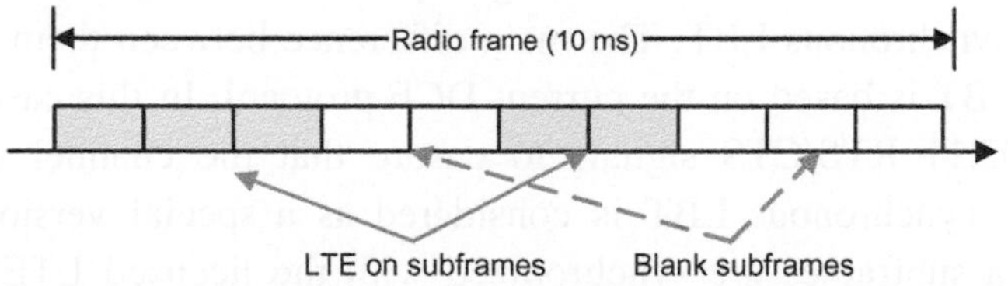

Figure 8.40
Example of blank-subframe allocation in an LTE radio frame [25].

number of blanked subframes based on the measurement of Wi-Fi's traffic load (see the example in Fig. 8.40). Fairness can be achieved by adjusting the number of blank subframes in each radio frame. Blank subframe offers more flexibility than CSAT as the ratio between the non-blank and blank subframes can be dynamically adjusted at the frame level, which is shorter than a CSAT cycle. Moreover, the positions of these blank subframes in each frame do not need to be consecutive. A blank-subframe is similar to the ABS used for enhanced ICIC (eICIC) mechanism in LTE-based heterogeneous networks. An ABS is a subframe

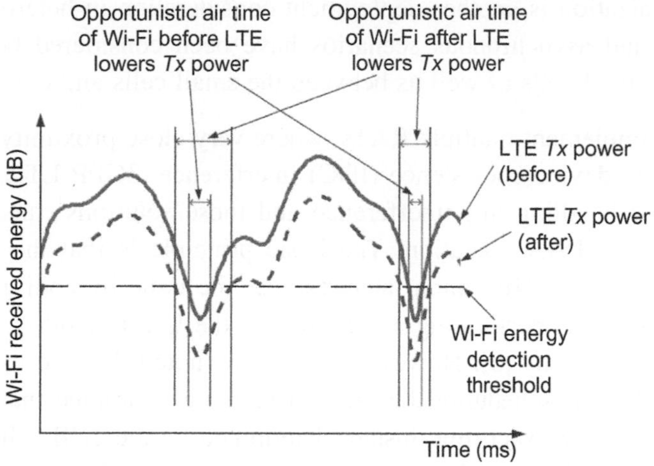

Figure 8.41
An example of Wi-Fi transmission opportunities before and after LTE power reduction [25].

during which only control and reference signals are transmitted with reduced transmit power. In contrast to ABS, a blank subframe does not include transmission of control and reference signals.

The coexistence between LTE and Wi-Fi networks by adjusting the output power of LTE nodes is an alternative method to minimize the interference. The Wi-Fi nodes typically employ energy detection to determine the activities of other users. More specifically, if the aggregate received energy is above a threshold, a Wi-Fi node would consider the channel to be busy and would postpone its transmission. For LTE/Wi-Fi coexistence, one may increase the transmission opportunity of Wi-Fi nodes by reducing the output power of the LTE nodes. As illustrated in Fig. 8.41, when the LTE transmit power is reduced, the transmission window for a Wi-Fi node becomes larger and the Wi-Fi node can more opportunistically transmit. On the other hand, the reduction in LTE transmit power will also result in lower LTE throughput due to the decrease of the SINR as a result of increased Wi-Fi transmissions.

Considering the potentially large number of deployed Wi-Fi APs and/or LTE-based LAA nodes, the backhaul and inter-node connectivity is another key challenge in such heterogeneous networks. An ideal backhaul (a dedicated point-to-point connection) is considered a link which provides (one-way) transport latency less than 2.5 ms and a throughput of up to 10 Gbps. Other types of backhaul links are considered non-ideal. The unlicensed and licensed carriers in ideal backhaul deployments can be co-located or inter-connected.

Inter-node synchronization is another deployment consideration in heterogeneous networks. Both synchronous and asynchronous scenarios have been considered between LTE-based LAA and/or Wi-Fi small cells as well as between the small cells and the macrocell(s).

Modern UEs often implement multiple RATs, where very close proximity of the RF components would cause in-device coexistence (IDC) interference. 3GPP LTE Rel-11 introduced several solutions for handling this interference and those solutions can be used to protect Wi-Fi networks during LAA operation. The basic principle is that the UE indicates IDC interference to the serving eNB which later resolves the issue by configuring the UE with an appropriate DRX cycle, performing a handover of the UE to another cell, or completely releasing one or more SCells [5]. However, it must be noted that the use of LBT in LAA would complicate the UE scheduling because there is no guarantee that a channel can be obtained for the UE at the exact time instant determined by the eNB. The LBT mechanism also limits the duration when the channel can be occupied; therefore the DRX timers should be adjusted to be long enough or the DRX cycles should be short enough to allow time for obtaining the channel access.

In LTE, when the UE detects an IDC condition, it would initially try to solve the problem internally. If this fails, the UE can inform the eNB that it is experiencing an IDC condition. Note that the detection of IDC condition in a UE is implementation specific. The UE identifies the frequencies that are suffering from IDC interference. If the UE determines that the IDC problem can be solved in a TDM-manner (i.e., by multiplexing the use of the interfering transceivers across time) the UE can indicate a TTI bitmap or DRX cycles that are affected by IDC interference to the eNB. When the eNB receives the indication, it can solve the problems by performing a handover of the UE to other frequencies, removing the problematic cell or configuring the UE with a new DRX configuration which would solve the problem. The existing IDC solutions can be used to support Wi-Fi background scanning during LAA operation. The existing IDC solutions can also be used to indicate interference problems for cases where the UE intends to use Wi-Fi on the same or adjacent carrier to the unlicensed carrier. If the eNB does not support IDC, the only way for the UE to enable Wi-Fi transmission would be to perform detach and attach procedures, and changing its capabilities to indicate that LAA is not supported. However, from a system operation viewpoint, having the UE perform detach and attach procedures is considered undesirable. Hence, eNB should enable IDC indications and respond to the IDC requests from the multi-radio UEs.

The QoS of some radio bearers might suffer when LAA is used due to support of LBT as there can be various interference sources in the unlicensed spectrum such as other RATs and LAA nodes of other operators. To improve the QoS, the characteristics of an LAA cell should be considered when mapping traffic from radio bearers to carrier(s). For example, it is better not to send critical control information, delay-sensitive data or guaranteed bit rate bearers through LAA cells, if the LBT operation is required.

References

3GPP Specifications[11]

[1] 3GPP TS 36.211, Evolved universal terrestrial radio access (E-UTRA), Physical Channels and Modulation (Release 15), June 2018.

[2] 3GPP TS 36.212, Evolved universal terrestrial radio access (E-UTRA), Multiplexing and channel coding (Release 15), June 2018.

[3] 3GPP TS 36.213, Evolved universal terrestrial radio access (E-UTRA), Physical layer procedures (Release 15), June 2018.

[4] 3GPP TS 37.213, Physical layer procedures for shared spectrum channel access (Release 15), June 2018.

[5] 3GPP TS 36.300, Evolved universal terrestrial radio access (E-UTRA) and evolved universal terrestrial radio access network (E-UTRAN); Overall Description; Stage 2 (Release 15), June 2018.

[6] 3GPP TS 36.321, Evolved universal terrestrial radio access (E-UTRA), Medium Access Control (MAC) Protocol Specification (Release 15), June 2018.

[7] 3GPP TS 36.323, Evolved universal terrestrial radio access (E-UTRA), Packet Data Convergence Protocol (PDCP) specification (Release 15), June 2018.

[8] 3GPP TS 36.331, Evolved universal terrestrial radio access (E-UTRA), Radio Resource Control (RRC); Protocol Specification (Release 15), June 2018.

[9] 3GPP TR 36.889, Study on licensed-assisted access to unlicensed spectrum (Release 13), June 2015.

[10] 3GPP TS 38.300, NR; NR and NG-RAN overall description, Stage 2 (Release 15), June 2018.

[11] 3GPP TS 38.323, NR; Packet data convergence protocol (PDCP) specification (Release 15), June 2018.

[12] 3GPP TR 38.889, Study on NR-based access to unlicensed spectrum (Release 15), December 2018.

Articles, Books, White Papers, and Application Notes

[13] B. Ren, et al., Cellular communications on license-exempt spectrum, IEEE Communications Magazine, May 2016.

[14] Keysight Technologies Application Note, 4G LTE-A in Unlicensed Band – Use Cases and Test Implications, December 2017.

[15] J. Zhang, et al., *LTE on license-exempt spectrum*, IEEE Commun. Surveys Tutor. 20 (1) (First Quarter 2018).

[16] M. Labib, et al., Extending LTE into the unlicensed spectrum: technical analysis of the proposed variants, IEEE Communications Standards Magazine, December 2017.

[17] A. Mukherjee, et al., Licensed-assisted access LTE: co-existence with IEEE 802.11 and the evolution toward 5G, IEEE Communications Magazine, June 2016.

[18] Netmanias Report, Analysis of LTE – Wi-Fi Aggregation Solutions, NMC Consulting Group, March 2016.

[19] K. Mun, CBRS: new shared spectrum enables flexible indoor and outdoor mobile solutions and new business models, Mobile Experts CBRS White Paper, March 2017.

[20] A. Savoia, LTE in the unlicensed spectrum, Keysight Technologies, March 2016.

[21] H.-J. Kwon, et al., Licensed-assisted access to unlicensed spectrum in LTE Release 13. IEEE Communications Magazine, February 2017.

[22] Rohde & Schwarz White Paper, LTE-Advanced Pro Introduction, eMBB Technology Components in 3GPP Release 13/14, May 2018.

[23] R. Karaki, et al., Uplink performance of enhanced licensed assisted access (eLAA) in unlicensed spectrum, in: IEEE Wireless Communications and Networking Conference (WCNC), 2017.

[24] B. Chen, et al., Co-existence of LTE-based LAA and Wi-Fi on 5GHz with corresponding deployment scenarios: a survey, IEEE Commun. Surveys Tutor. 19 (1) (First Quarter 2017).

[11] 3GPP specifications can be accessed at the following URL: http://www.3gpp.org/ftp/Specs/archive/

[25] Y. Huang, et al., Recent advances of LTE/Wi-Fi co-existence in unlicensed spectrum, IEEE Netw. (March/April 2018).
[26] B.L. Ng, et al., Unified access in licensed and unlicensed bands in LTE-A Pro and 5G, APSIPA Transactions on Signal and Information Processing (SIP), vol. 6, Cambridge University Press, July 2017.
[27] T. Levanen, et al., 5G new radio and LTE uplink co-existence. IEEE Wireless Communications and Networking Conference (WCNC), April 2018.
[28] A. Mukherjee, et al., System architecture and co-existence evaluation of licensed-assisted access LTE with IEEE 802.11, in: 2015 IEEE International Conference on Communication Workshop (ICCW), June 2015.
[29] ETSI EN 301.893 v1.7.1, Broadband radio access networks (BRAN); 5 GHz high performance RLAN, 2012.
[30] Part 11: Wireless LAN medium access control (MAC) and physical layer (PHY) specifications, IEEE Std 802.11-2018, May 2018.
[31] 4G Americas, LTE aggregation & unlicensed spectrum, November 2015.
[32] 5G Americas, Spectrum landscape for mobile services, November 2017.
[33] 3GPP R1-1808318, Discussion on frame structure for NR-U, ZTE, August 2018.
[34] 3GPP R1-1808058, NR numerology and frame structure for unlicensed bands, Huawei, HiSilicon, August 2018.
[35] 3GPP R1-1808274, On channel access procedure in NR-U, MediaTek, August 2018.

Index

Note: Page numbers followed by "*f*" and "*t*" refer to figures and tables, respectively.

E

I

N

T

U

V

Printed and bound by CPI Group (UK) Ltd, Croydon, CR0 4YY
03/10/2024
01040313-0007